[redacted]
36156000545341

W9-BBS-560

For Reference

Not to be taken from this room

A Manual of New Mineral Names 1892–1978

A Manual of New Mineral Names 1892–1978

Edited by
Peter G. Embrey and John P. Fuller
British Museum (Natural History)

88582

British Museum (Natural History)
Oxford University Press

ST. THOMAS AQUINAS
COLLEGE LIBRARY
SPARKILL, N.Y. 10976

© Trustees of the British Museum (Natural History) 1980
All rights reserved
Except for brief quotations in a review, this book, or part thereof, must not be reproduced in any form without permission in writing from the publisher

British Library Cataloguing in Publication Data

A manual of new mineral names, 1892–1978.
1. Mineralogy – Terminology
I. Embrey, Peter Godwin
II. Fuller, John P. III. British Museum – (Natural History)
549′.03 QE357
ISBN0–19–858501–2

Library of Congress Catalog Card Number 80–40204

First published 1980 by

British Museum (Natural History)
Cromwell Road, London SW7 5BD
and

Oxford University Press, Walton Street, Oxford OX2 6DP

OXFORD LONDON GLASGOW
NEW YORK TORONTO MELBOURNE WELLINGTON
KUALA LUMPUR SINGAPORE JAKARTA HONG KONG TOKYO
DELHI BOMBAY CALCUTTA MADRAS KARACHI
IBADAN NAIROBI DAR ES SALAAM CAPE TOWN

Published in the USA by
Oxford University Press, New York

Filmset in 'Monophoto' Times 10 on 11 pt by
Richard Clay (The Chaucer Press) Ltd, Bungay, Suffolk
and printed in Great Britain by Fletcher & Son Ltd, Norwich

A Manual of New Mineral Names 1892–1978

Editors' Foreword

This work is a collected edition of the thirty **Lists of New Mineral Names** that have been published in the *Mineralogical Magazine*.

The first List appeared in 1897, compiled (as were the following twenty Lists) by Dr L J Spencer, with the object of providing a convenient reference to names not contained in the sixth edition (1892) of Dana's *System of Mineralogy*. Of this first compilation, Spencer wrote: 'Although it can scarcely be hoped to make such a list complete, it may still be useful for reference. Probably not more than a third of [these names] are new minerals, which could, when completely determined, stand as distinct species.' The usefulness proved itself at once, and the Lists became an established feature of the last number of each volume of the Magazine. Containing critical insight, they are more than abstracts and have proved their worth as a valuable research tool to generations of mineralogists.

For the past twenty years, the Commission on New Minerals and Mineral Names of the International Mineralogical Association has been having an increasingly beneficial effect on mineralogical nomenclature, and most editors of earth-science journnals now require the description of a new mineral species to have received approval by the Commission before publication. The data required in a submission to the Commission have been specified (*Amer. Mineral.* 1970, **55**, 1016), and its findings are published. (References are given in the 2nd Appendix to *Chemical Index of Minerals*, 1974, p. vii.) The Commission adopted Levinson's proposals (*Amer. Mineral*, 1966, **51**, 152) for the naming of rare-earth minerals, and has also appointed subcommittees to recommend revision of the nomenclature of certain groups of minerals; its approved rulings have been published for the pyrochlore (*Amer. Mineral.* 1977, **62**, 403) and amphibole (*Mineral. Mag.* 1978, **42**, 533 and elsewhere) groups. Despite this work of good housekeeping, however, there remain large numbers of imperfectly described minerals or varieties which only investigation in the laboratory can authenticate or relegate to the synonymy.

Most mineralogical reference works have omitted these unsatisfactory minerals, or have contained guesses at synonymy rarely based on experiment. There must, however, be many minerals that have been assigned to the synonymy, or to varietal status, but which investigation by modern methods may prove to be distinct species. It is a prime purpose of a work such as the present one to keep these 'twilight' minerals alive until such time as they be fully reinstated or decently buried. Some entries are of interest away from mineralogy, and we may instance the mention of nearly-forgotten putative chemical elements such as canadium, masrium (masrite), and odenium (odenite). It has often been claimed, following a misreading of J D Dana (*Syst. Min.*, 5th edn, p.xxxiii), that imperfectly characterized minerals not in currency for fifty years or more should have their names lapse; rather than inflict injustice on a worthy author who did the best he could at the time, it seems preferable to treat each case of rediscovery on its merits.

In addition to names applied to mineral species or varieties in the strict sense, Spencer's Lists contained many other types of entry including mis-spellings, variant transliterations, and some of the names given to industrial products and artificial minerals (often gem simulants) that might be mistaken for those of minerals. Rock names, on the other hand, which with their -ite termination might even more readily be confused, were almost entirely omitted. These rather loose guide lines have been followed by Hey in the 22nd to 30th Lists (see *Mineral. Mag.* 1974, **39**, 903). The names given by Gagarin and Cuomo to some previously unnamed minerals, without further investigation by the authors, were included in the 19th List; but attempts to introduce extensive new systems of nomenclature have been largely ignored (e.g. Povarennykh, 1966, reviewed in *Amer. Mineral.* 1967, **52**, 931; Aymé, 1901, see entry 'Carbite').

Our involvement with the present compilation arose from the collaboration of one of us (PGE) with Dr Hey in the preparation of recent Lists, and a recognition of the need for a cumulative index. It rapidly became clear that an alphabetized reprinting of the full entries would be preferable, with editorial comments and references to significant later investigation, and a separate index to the authors of the names. In some respects the result is a companion volume to Chester's *A Dictionary of the Names of Minerals* (New York, 1896), particularly since the dates run on with minimal overlap, and will thus prove invaluable to the authors of the new mineral names which are currently being coined at the rate of about fifty each year.

In the course of collation we have inevitably found a few errors, additional to those recorded in the *Mineralogical Magazine*, and all variations from the original text are indicated by square brackets [ ]. Inevitably some new errors will have crept in that we have failed to correct in proof and for these we offer apologies. We also regret errors that we may have made in the index to authors; some spellings of names and completeness of initials have differed from one List to another, only becoming evident when brought together, and we have undoubtedly failed to recognize some changes of name on marriage. We have incorporated some of the footnotes in the original text, but have omitted both the prefatory comments and the chemical classifications that accompanied many of the Lists. The reader will find much of interest in the former, and may use the *Chemical Index of Minerals* and its two Appendices in place of the latter. To have included this material would have greatly increased the size of the book.

R
549.03
M

Acknowledgements

Permission to publish the collected Lists in this form has been given by the Council of the Mineralogical Society of Great Britain and by their present compiler, Dr M H Hey. It should be noted in this connection that there is no infringement of the copyright claimed by the McGraw-Hill Book Company, Inc. for G E English's work *Descriptive List of the New Minerals 1892–1938* (New York and London, 1939), which attracted the comment in a review by Dr L J Spencer: 'Much of the information has been copied, often word for word, from the Mineralogical Magazine and Abstracts' (M.A. **7**, 313). We also thank Dr Hey for his encouragement and many helpful comments.

A

Aanerödite. Variant of Ånnerödite; *A.M.*, **9**, 62. 10th List

Abelsonite. A. Pabst, P. A. Estep, E. J. Dwornik, R. B. Finkelman, and C. Milton, 1975. *Geol. Soc. Amer. Abstr. Program*, **7**, 1221. Tiny purple flakes from a drill core from the Green River Formation, Uintah County, Utah, proved to be a nickel porphyrin, $C_{32}H_{36}N_4Ni$. Anorthic, *a* 8·51, *b* 11·18, *c* 7·29 Å, α 90° 53′, β 114° 8′, γ 79° 59′. Named for Dr. P. H. Abelson. [*A.M.*, **61**, 502.] 29th List

Abernathyite. M. E. Thompson, B. Ingrim, and E. B. Gross, 1956. *Amer. Min.*, vol. 41, p. 82. $K(UO_2)AsO_4.4H_2O$, tetragonal, meta-autunite group. Small yellow crystals from Temple Mts., Utah. Named after Mr. Jess Abernathy, who found the material. [*M.A.*, **13**, 86.] 21st List

Abkhazite. N. E. Efremov, 1940. *Trans. Inst. Geol. Sci., Acad. Sci. USSR*, 1938 [i.e. 1940], no. 11 (*Min. Geochem. Ser.*, no. 3), p. 37 (абхазит). A monoclinic amphibole-asbestos in which the ratio (Mg,Fe):Ca is 4:3 (instead of 5:2 as in tremolite and actinolite). Named from the province Abkhaziya (Абхазия),Georgia, Transcaucasia. [*M.A.*, **8**, 2.] 18th List [Is Tremolite; *Amph.* (1978)]

Ablykite. I. D. Sedletzky and S. M. Yusupova, 1940. *Compt. Rend.* (*Doklady*) *Acad. Sci. URSS*, vol. 26, p. 244 (Ablick clay, ablikite), p. 946 (ablykite). A clay mineral close to halloysite containing some K, Mg, and Ca, with distinctive thermal and X-ray characters. $RO.2R_2O_3.5SiO_2.6H_2O$. Named from the locality, Ablyk, Uzbekistan. [*M.A.*, **8**, 147, 342.] 16th List

Absite. A. W. G. Whittle, 1954. *Mining Rev. South Australia Dept. Mines*, no. 97 (for 1952), p. 99 (absite), p. 142 (absite, thorium brannerite). Variety of brannerite (9th List) containing $ThO_2$12·81%.$2UO_2.ThO_2 7TiO_2.5H_2O$. From Crockers Well, South Australia. Deposit located with air-borne scintillometer, hence the name. [*M.A.*, **13**, 87.] 21st List

Abukumalite. S. Hata, 1938. *Sci. Papers Inst. Phys. Chem. Research*, Tokyo, 1938, vol. 34, p. 1018. Phospho-orthosilicate of yttrium and calcium, $CaYt_2(Si,P)_2O_8$, hexagonal and isomorphous with britholite. Named from the locality, Abukuma range, Fukushima, Japan. F. Machatschki (*Zentr. Min., Abt. A*, 1939, p. 161) regards it as a yttrium silicate apatite. [*M.A.*, **7**, 225, 395.] 15th List [Is Britholite-(Y); Fleischer's Gloss.]

Acarbodavyne. G. Cesaro, 1911. *See* Akalidavyne. 10th List

Acetamide. B. I. Srebrodolskii, 1975. *Zap.* **103**, 328 (Ацетамид). Natural acetamide, $CH_3.CONH_2$, occurs in dry weather on the waste piles of a coal mine in the Lvov–Volynsk basin, identical with the synthetic compound. [*M.A.*, 76-876; *A.M.*, **61**, 338.] 29th List

Achavalite. J. Olsacher, 1939. *Bol. Fac. Cienc. Univ. Córdoba*, vol. 2, no. 3–4, p. 73 (achavalita). R. Herzenberg, ibid., 1944, vol. 7, no. 4, p. 3 of reprint. Iron selenide FeSe, occurring with other selenides at Cacheuta, Argentina. Named after Prof. Luis Achaval (1870–1938) of Córdoba. [*M.A.*, **10**, 8.] 18th List [Orthorhombic; *M.A.*, **3**, 19]

Achiardite. (R. Koechlin, *Mineralogisches Taschenbuch der Wiener Mineralogischen Gesellschaft*, 1911, pp. 12, 62; *Min. Petr. Mitt.*, 1912, vol. xxxi, p. 91 (Achiardit).) Synonym of Dachiardite (G. D'Achiardi, 1906; 4th list). 6th List

Achlusite. W. F. Petterd, 1910. *Catalogue of the Minerals of Tasmania*, 3rd edit., Hobart, 1910, p. 191; *Papers Roy. Soc. Tasmania*, 1910, p. 191. A green alteration-product of topaz resembling steatite in appearance, but near soda-mica in composition. Derivation not stated, but no doubt from ἀχλύς, mist, alluding to the cloudy alteration of the clear topaz 6th List. [A mixture: *M.M.*, **38**, 902.]

Achrite. Variant of or error for achirite, *syn.* of dioptase. R. Webster, *Gems*, 1962. 28th List

Achromaite. F. Kretschmer, 1918. *Jahrb. Geol. Reichsanstalt, Wien*, vol. 67 (for 1917), p. 115 (Achromait). A variety of hornblende, perfectly colourless in micro-sections, forming a constituent of weigelite (hornblende-peridotite) from Weigelsberg, Moravia. Named from ἀχρώματος, colourless. 12th List

Acmite-augite. F. Zambonini, 1910. *Mineralogia Vesuviana. Mem. Accad. Sci. Fis. Mat. Napoli*, vol. xiv. pp. 153, 155 (acmiteaugite). The same as aegirine-augite (H. Rosenbusch, 1902; 2nd List), but brown in colour. 6th List

Acrochordite. *See* Akrochordite. 10th List

Adamite. (*Mineral Resources, US Geol. Survey*, for 1903, 1904, p. 1015; Thorpe's *Dictionary of Applied Chemistry*, 2nd edit., 1912, vol. i, p. 47; *The Mineral Industry*, New York, 1917, vol. xxv for 1916, p. 31). Trade-name for artificial corundum manufactured for abrasive purposes. Compare Alundum (5th List) and Aloxite (7th List). Not the Adamite of C. Friedel, 1866. 8th List

Adandit, error for Anandit. (C. Hintze, *Handb. Min., Erg. III*, p. 517). 25th List

Adigeite. N. E. Efremov, 1939. *Compt. Rend.* (*Doklady*) *Acad. Sci. URSS*, vol. 22, p. 433 (Adigeite). A mineral of the serpentine group with the composition $5MgO.3SiO_2.3\frac{1}{2}–4H_2O$, from Mt. Tkhach, northern Caucasus. This is perhaps the 'new variety of serpentine' described, without name, by the same author in *Bull. Acad. Sci. URSS, Sér. Géol.*, 1938, p. 107. [*M.A.*, **7**, 221, 370.] 15th List

Aegerite. (*Mineral Resources, US Geol. Survey*, for 1910, 1911, part ii, p. 836.) Trade-name for a bitumen allied to elaterite. 6th List

Ægirine-augite. H. Rosenbusch, *Mikrosk. Physiogr.* 3rd edit., 1892, I, 537; Rosenbusch-Iddings, 4th edit., 1898, p. 257. Intermediate between augite and ægirite. 2nd List

Ægirine-diopside. W. C. Brøgger, *Skrifter Vid.-Selsk. Christiana*, I, *Math.-natur. Kl.* 1898, no. 6, p. 169. The same as ægirine-augite. 2nd List

Aegirine-hedenbergite. F. von Wolff, 1904. *Centralblatt Min.*, 1904, p. 214 (Aegirin-Hedenbergit). A monoclinic pyroxene intermediate in composition between ægirine and hedenbergite. H. Rosenbusch (*Mikroskopische Physiographie d. Mineralien*, 4th edit., 1905, vol. i, part 2, p. 218) prefers the form Hedenbergit-Ägirin. 4th List

Aeonite. (*Mineral Resources, US Geol. Survey*, for 1909, 1911, part ii, p. 733.) Trade-name for a bitumen very similar to elaterite. 6th List

Afghanite. P. Bariand, F. Cesbron, and R. Giraud, 1968. *Bull. Soc. franç. Min. Crist.*, vol. 91, p. 34. A hexagonal aluminosilicate of calcium and alkalis, near $(Na,Ca,K)_{12}(Si,Al)_{16}O_{34}(Cl,SO_4,CO_3)_4.0{\cdot}6H_2O$, from a lapis-lazuli mine, Sar-e-Sang, Badakshan province, Afghanistan. The powder diagram is near that of cancrinite and the unit cells appear to be related. Named from the locality. 25th List [Transbaikalia; *M.A.*, 76-1946]

Afwillite. J. Parry & F. E. Wright, 1925. *Min. Mag.*, vol. 20, p. 277. Hydrous calcium silicate, $3CaO.2SiO_2.3H_2O$, as colourless, monoclinic crystals from Kimberley, South Africa. Named after Mr. Alpheus Fuller Williams, of Kimberley, by whom the mineral was found. 10th List

Agardite. J. E. Dietrich, M. Orliac, and F. Permingeat, 1969. *Bull. Soc. franç. Min. Crist.* **92**, 420. Greenish-blue hexagonal crystals from the Bou-Skour copper mine, Jebel Sarhro,* Ouazazate,* Morocco, have the composition $(Yt,CaH)Cu_6(AsO_4)_3(OH)_6.3H_2O$, and are the rare-earth analogue of mixite. Named for J. Agard. [*M.A.*, 70-1649.] The proposal by K. Walenta (*Chemie der Erde*, **29**, 36, 1970) to transfer the name chlorotile (Frenzel, 1875), of uncertain connotation, to this mineral was rejected before publication by the Commission on New Minerals and Mineral Names of the International Mineralogical Association. 26th List [*corr. *M.M.*, **38**; *A.M.*, **55**, 1447]

Agathocopalite. J. Paclt, 1953. *Tschermaks Min. Petr. Mitt.*, ser. 3, vol. 3, p. 342. A recent resin (copalite) containing agathic acid ($C_{20}H_{30}O_4$), from the conifer *Agathis*. [*M.A.*, **12**, 306.] 21st List

Aglaurite. R. Handmann, 1907. *Zeits. Min. Geol. Stuttgart*, vol. i, p. 78. Orthoclase-felspar with a fine blue reflection forming a constituent of quartz-porphyry (Aglauritporphyr) from Teplitz, Bohemia. Named from ἄγλαυρος = ἀγλαός, bright. 5th List

Agrellite. J. Gittins, M. G. Bown, and D. Sturman, 1976. *C.M.*, **14**, 120. White pseudomonoclinic crystals, *a* 7·773, *b* 18·942, *c* 6·984 Å, α 90·148°, β 116·84°, γ 94·145°, occur in a metamorphosed alkaline rock in Villedieu Township, Temiskaming, Quebec. Sp. gr 2·90. $4[NaCa_2Si_4O_{10}F]$. Named for Dr. S. O. Agrell. 29th List

Agrinierite. F. Cesbron, W. L. Brown, P. Bariand, and J. Geffroy, 1972. *Min. Mag.* **38**, 781. Orange orthorhombic crystals, *a* 14·04, *b* 24·07, *c* 14·13 Å, *Cmmm*, D 5·7, occur with uranophane in cavities in gummite at Margnac, France. Composition $8[2(K_2,Ca,Sr)O.6UO_3.8H_2O]$. α ∥ [001], γ ∥ [010]. Named for H. Agrinier. [*M.A.*, 72-3346.] 27th List [*A.M.*, **58**, 805]

Ahlfeldite. R. Herzenberg, 1935. *Zentr. Min., Abt. A*, 1935, pp. 189, 279 (Ahlfeldit). Nickel selenate (or selenite) occurring as a reddish alteration product of blockite (q.v.), from Colquechaca, Bolivia. Perhaps identical with cobaltomenite. Named after Dr. Friedrich Ahlfeld, of Marburg, Germany. [*M.A.*, **6**, 147.] 14th List [$NiSeO_3.2H_2O$; *Canad. Min.* **12**, 304]

Aidyrlite. M. N. Godlevsky, 1934. *Mém. Soc. Russ. Min.*, ser. 2, vol. 63, p. 338 (Айдырлит), p. 334 (Aidyrlit). Hydrated silicate of aluminium and nickel, $2NiO.2Al_2O_3.3SiO_2.7\frac{1}{2}H_2O$, as compact turquoise-blue masses, from the river Aidyrly (p. Айдырлиы), Urals. Named from locality. [*M.A.*, **6**, 150.] 14th List

Aidyrlyit. Mistransliteration of Айдырлит, aidyrlit [14th List], appearing consistently in Strunz *Min. Tab.* (all 5 editions). [A mixture, *M.A.*, **11**, 177.] 28th List

Aikinite. R. P. Greg and W. G. Lettsom, *Mineralogy of Great Britain and Ireland*, 1858, p. 354. Pseudomorph of wolframite after scheelite from Cornwall, named by A. Lévy [but not found in any of his publications]. Named after Arthur Aikin (1773–1854). Not the aikinite of E. J. Chapman, 1843. 14th List

Ajatit, German transliteration of Аятит, Ayatite (24th List) (C. Hintze, *Handb. Min. Erg.* III, p. 517). 25th List

Ajkaite. (L. Zechmeister, *Math. Természettud. Értesitö*, Budapest, 1926, vol. 43, p. 332 (ajkait); L. Zechmeister and V. Vrabély, *Ber. Deutsch. Chem. Gesell.*, 1926, vol. 59, Abt. B, p. 1426). The same as ajkite (*Bull. Soc. Min. France*, 1878, vol. 1, p. 126; abstract from . . .?). A fossil resin containing 1·5% sulphur and no succinic acid, from Ajka, com. Veszprém, Hungary. [*M.A.*, **3**, 362.] 11th List

Ajoite. W. T. Schaller and A. C. Vlisidis, 1958. *Amer. Min.*, vol. 43, p. 1107. Bluish-green laths, plates, or massive, from Ajo, Pima Co., Arizona, near $Cu_6Al_2Si_{10}O_{29}.5\frac{1}{2}H_2O$. Named from the locality. [*M.A.*, **14**, 198.] 22nd List

Akaganéite. M. Nambu, 1961. *Min. Mag.*, vol. 33, p. 270. Natural β-FeOOH, occurring at the Akagané mine, Iwate prefecture, Japan, and named from the locality. Synthetic material described by A. L. Mackay, *Min. Mag.*, 1960, vol. 32, p. 545; natural β-FeOOH first described from Belgium by R. Van Tassel, *Bull. Soc. belge Géol.*, 1959, vol. 68, p. 360 [*M.A.*, **15**, 133], and from India by K. C. Chandy, *Indian Minerals*, 1961, vol. 40, p. 197. 23rd List [*A.M.*, **48**, 711]

Akalidavyne. G. Cesàro, 1911. *Mém.* (*in 8°*) *Acad. R. Belgique, Cl. des Sci.*, ser. 2, vol. 3, fasc. 2, p. 6. Criticizing F. Zambonini's Natrodavyne (1910; 6th List), it is pointed out that in all analyses of davyne there is an excess of sodium over potassium; if, therefore, 'natrodavyne' is to imply the absence of potassium, then *Akalidavyne* would be a more suitable name. Further, as carbon dioxide may be present or absent in davyne, it is necessary to distinguish between *carbodavyne* and *acarbodavyne*. 10th List

Akatoreïte. P. B. Read and A. Reay, 1971. *Amer. Min.* **56**, 416 (Akatoreite). Orange-brown sheaves of prisms from Akatore Creek, Dunedin, New Zealand, are anorthic, *a* 8·344, *b* 10·358, *c* 7·627 Å, α 104° 29′, β 93° 38′, γ 103° 57′, D 3·48, and optics α 1·698, colourless, β 1·704, pale yellow, γ 1·720, canary yellow. Ideal composition, assuming Mn divalent, $[Mn_9Si_8Al_2O_{24}(OH)_8]$ with some deficiency of Si and excess (OH). Named for the locality. [*M.A.*, 71-3103.] 27th List

Akdalaite. E. P. Shpanov, G. A. Sidorenko, and T. I. Stolyarova, 1970. *Zap.* 99, 333 (Акдалаит). Aggregates of tabular hexagonal crystals in skarns in the Solvech fluorite deposit. Karagandin region. Kazakhstan, have *a* 12·87, *c* 14·97 Å, and a composition near $4Al_2O_3.H_2O$. ω 1·747, ε 1·741. The X-ray data are near those of artificial $5Al_2O_3.H_2O$ (tohdite, (q.v.); Yamaguchi *et al.*, 1964), but show some extra weak lines. 27th List [*A.M.*, **56**, 635]

Akrochordite. G. Flink, 1922. *Geol. För. Förh.* Stockholm, vol. 44, p. 773 (Akrochordit), p. 776 (Akrochordite). Abstr. in *Amer. Min.*, 1923, vol. 8, p. 167, and *Bull. Soc. Franç. Min.*, 1923, vol. 46, p. 74, give the form Acrochordite. Hydrated basic arsenate of manganese and magnesium, $Mn_3As_2O_8.MnOH.Mg(OH)_2.5H_2O$, occurring as small, reddish-brown, spherical aggregates at Långban, Sweden. The optical characters indicate monoclinic symmetry. Named from ἀκροχορδών, a wart. [*M.A.*, **2**, 51.] 10th List [*Ark. Min. Geol.*, **4**, 425]

Aksaite. L. N. Blazko, V. V. Kondrateva, and Ya. Ya. Yarzhemski, 1962. Зап. Всесоюз. Мин. Общ. (*Mem. All-Union Min. Soc.*), vol. 91, p. 447 (Аксаит); *M.A.*, **16**, 65. Orthorhombic blades in impure halite from Ak-sui, Kazakhstan. Formula given as $5[Mg_2B_{10}O_{17}.8H_2O]$ or $4[Mg_3B_{14}O_{24}.10H_2O]$, but M. Mrose and M. Fleischer (*Amer. Min.*, 1963, vol. 48, p. 210) point out that the X-ray powder data agree reasonably well with synthetic $MgB_6O_{10}.5H_2O$ (Lehmann and Papenfuss, 1959), and this is confirmed by J. R. Clark and R. C. Erd (*Amer. Min.*, 1963, vol. 48, p. 930); the unit cell contains 8 of these formulae, leading to a calculated sp. gr. of 1·97 (obs., 2·07, 2·37). Named from the locality. [*M.A.*, **16**, 65.] 23rd List

Aktaschit, Germ. trans. of Акташит, aktashite (26th List). Hintze, 2. 28th List

Aktashite. V. I. Vasilev, 1968. [Вопрос. металлогении ртути. Изд. 'Наука']; abstr. *Zap.* **99**, 64 (1970). (Акташит). Sulpharsenite of Cu and Hg, with Cu around 24%, Hg 33%, As (+ Sb) 20%, and S 23%, occurring at Aktash in the high Altai. The description is incomplete, and the name was published against the recommendation of the New Minerals Commission of the All-Union Mineralogical Society. 26th List [Trigonal $Cu_6Hg_3As_5S_{12}$; *A.M.*, **58**, 562]

Al-antigorite. H. Strunz, 1970. *Min. Tab.* 5th edit., 458 (Al-Antigorit). The naturally occurring aluminian serpentine ('Al-serpentine'; cf. aluminous-serpentine, 20th List) from the Lake Superior region [*M.A.*, **15**, 411; **19**, 268]. An unnecessary name for a variety. 28th List

Al-chamosite. [L. M. Miropolsky, 1936. *Uchen. Zapis. Kazan. Univ.*, vol. 96, no. 3, p. 70.] O. M. Shubnikova, *Trans. Inst. Geol. Sci. USSR*, 1938, no. 11 (*Min. Geochem. Ser.*, no. 3), (Алшамозит, алюминиевый шамозит, Al-chamosite). An oolitic chamosite rich in alumina (Al_2O_3 37·14%), from Bashkir republic, Russia. [*M.A.*, **9**, 185.] 17th List

Al-lizardite. S. W. Bailey and S. A. Tyler, 1960. *Econ. Geol.* **55**, 150 [*M.A.*, **15**, 411]. An unnecessary name for aluminian lizardite, from the Lake Superior region. Cf. Al-antigorite. 28th List

Al-nontronite. H. Strunz, 1957. *Min. Tabellen*, 3rd edit., p. 311 (Al-Nontronit). Unnecessary name for aluminian nontronite. 22nd List

Al-saponite, variant of Aluminium-saponite (22nd List) (I. Kostov, *Mineralogy*, 1968, p. 373). 25th List

Alaite. K. A. Nenadkevich, 1909. *Bull. Acad. Sci. Saint-Pétersbourg*, ser. 6, vol. iii, p. 185 (Алаить). Hydrated vanadic oxide, $V_2O_5.H_2O$, forming blood-red, mossy growths with silky lustre. Found with turanite (q.v.) in the neighbourhood of the Alai Mountains, Russian Central Asia. 5th List

Alamosite. C. Palache and H. E. Merwin, 1909. *Amer. Journ. Sci.*, ser. 4, vol. xxvii, p. 399; *Zeits. Kryst. Min.*, vol. xlvi, p. 513. Lead meta-silicate, $PbSiO_3$, occurring as snow-white, radially fibrous masses. Crystals are monoclinic, though apparently not isomorphous with wollastonite. From Alamos, Sonora, Mexico. Prepared artificially by S. Hilpert and P. Weiller, *Ber. Deutsch. Chem. Ges.*, 1909, vol. xlii, p. 2969. 5th List [Tsumeb, *M.A.*, 75-3479.]

Alasanite. *Zap.*, **104**, 606. Erroneous transliteration of Алазанит, alazanite. 30th List

Alazanite. R. V. Ivanitskii, R. A. Akhvlediani, E. I. Kakhadze, and A. I. Tsepin, 1973. *Dokl. acad. nauk. SSSR*, **213**, 688 (Алазанит). An abundant mineral in ores of the Kakhetin deposit, USSR, is intermediate in composition between pyrrhotine and FeS_2, but the analyses (by microprobe) give very low totals. X-ray powder data are indexed on a cell with *a* 4·506, *b* 5·511, *c* 3·406 Å, very near that of marcasite. Named for the Alazani river, Georgia. [The name should not have been given.—*M.F.*] [*A.M.*, **60**, 161; *Zap.*, **104**, 606.] 29th List

Albrittonite. W. W. Crook III and L.–A. Marcotty, 1978. *A.M.*, **63**, 410. A supergene alteration halo above a veinlet of Co and Ni minerals in a serpentine quarry near Oxford, Llano Co., Texas, contains red-violet monoclinic crystals, *a* 8·899, *b* 7·065, *c* 6·644 Å, *β* 97·25°, space group $C2/m$. Composition $2[CoCl_2.6H_2O]$, with small amounts of Ni and Cu. Sp. gr. 1·897, RIs α 1·525, β 1·550, pale red, ∥ [010], γ 1·576, reddish-violet, γ:[100] − 3°. Named for C. C. Albritton. 30th List

Aldanite. M. M. Bespalov, 1941. *Soviet Geology*, 1941, no. 6, p. 107 (алданит). $k\mathrm{ThO_2}.l\mathrm{UO_2}.m\mathrm{UO_3}.n\mathrm{PbO}$, cubic, thorianite group. Named from the locality, Aldan region, eastern Siberia (18th List). [*M.A.*, **12**, 460.] 20th List [Also 18th List (later ref.)]

Aldshanit, Germ. trans. of Алджанит, aldzhanite (27th List). Hintze, 2. 28th List

Aldzhanite. N. P. Avrova, V. M. Bocharov, I. I. Khalturina, and Z. R. Ynosova, 1968. [Геол. развед. месторожд. тверд. полез. Ископ. Каз., 1969, 169], abstr. Реф. журн., геол., 1969, abstr. no. 6V233 (Альджанит). An incompletely analysed orthorhombic mineral, a 12·76, b 14·59, c 8·19 (Å or kX?), D 2·21, in insoluble residues from a carnallite–bischofite rock; no locality or etymology given. B_2O_3 23·6%, Ca 15 to 20%, Cl 10%. [*A.M.*, **56**, 1122; *Zap.*, **100**, 86.] 27th List

Alexandrolite. S. M. Losanitsch, *Chem. News*, LXIX. 243, 1894; and *Ber. deutsch. chem. Ges.* XXVIII (3), 2631, 1895. A green, hydrated silicate of Al and Cr, resulting, with blue miloschine, from the decomposition of avalite. Servia. 1st List

Alexjejewite. A. Karnojitzky, *Zeits. Kryst. Min.* XXIV, 504, 1895 (V. Aleksyeev, *Verh. russ. min. Ges.* XXIX, 201, 1892). A wax-like hydrocarbon. [Kaluga Gov.] Russia. [*Min. Mag.*, **11**, 236.] 1st List

Aliettite. F. Veniale and H. W. van der Marel, 1969. [*Proc. Int. Clay Conf.*, Tokyo, **1**, 233], abstr. *A.M.*, **57**, 598. A regularly interstratified talc–saponite mineral from Monte Chiaro and other localities in the Taro valley region, Italy (*M.A.*, **8**, 517; **15**, 332) is now named for A. Alietti. 27th List

Alite, A. E. Törnebohm [*Ueber die Petrographie des Portland-Cements*. Stockholm, 1897, 34 pp.] *Neues Jahrb. Min.* 1899, I, Ref. 485; *Zeits. Kryst. Min.* 1900, XXXII, 610. Silicate and aluminate of calcium, $x(3CaO.SiO_2) + (9CaO.2Al_2O_3)$. Probably rhombic with hexagonal habit. An important constituent of Portland-cement clinkers; with it are smaller amounts of other crystalline substances of similar composition, namely, *belite*, *felite*, and *celite*. 2nd List [See Hatrurite.]

Alkali-apatite. D. McConnell, 1938. *Amer. Min.*, vol. 23, pp. 10, 17 (alkali-apatites). To include the apatite-like minerals dehrnite and lewistonite (12th List). [*M.A.*, **7**, 88.] 15th List

Alkali-beryl. S. L. Penfield, 1884. *Amer. Journ. Sci.*, ser 3, vol. 28, p. 25 ('alkalies in beryl'). A. Lacroix, *Bull. Soc. Franç. Min.*, 1908, vol. 31, p. 236 ('Béryl riche en alcalis'). Certain beryls contain appreciable amounts of alkalis: Li_2O up to 2·00%, Na_2O 4·22%, K_2O 2·25%, Cs_2O 4·56%, Rb_2O 1·34%. *See* Caesium-beryl. [*M.A.*, **1**, 76, 446; **3**, 448; **4**, 138, 315.] 12th List [*M.A.*, **16**, 644; 73-2872]

Alkali-chlorapatite. H. Wondratschek, 1963. *Neues Jahrb. Min., Abh.*, vol. 99, p. 155 (Alkali-Chlor-Apatit). A name for an artificial member of the apatite group prepared by F. Zambonini (1923). 23rd List

Alkali-garnet. W. C. Brøgger and H. Bäckström, *Zeits. Kryst. Min.*, 1890, vol. xviii, p. 215 (Alkaligranate, *pl.*). A general term for members of the sodalite group; these being closely related crystallographically and chemically to the true garnets. (Cf. the artificial lagoriolite = natrongranat of J. Morozewicz, 1898.) 3rd List

Alkali-hastingsite, Alkali-ferrohastingsite, Alkali-femaghastingsite. M. Billings. 1928. *See* Ferrohastingsite. 12th List

Alkali-montmorillonite. W. Noll, 1936. *See* Calcium-montmorillonite. 16th List

Alkali-oxyapatite. V. I. Vlodovets, 1933. *Trans. Arctic Inst.*, vol. 12, p. 84 (щелочной окси-апатит), p. 99 (alkaline oxyapatite). O. M. Shubnikova, *Trans. Lomonosov Inst.*, 1936, no. 7, p. 328 (щелочной оксиапатит, alkali-oxyapatite). Apatite containing Na_2O 1·37% replacing CaO, and O partly replacing (F,OH). 20th List

Alkali-pyromorphite. H. Wondratschek, 1963. *Neues Jahrb. Min., Abh.*, vol. 99, p. 114 (Alkali-Pyromorphit). A group name for the artificial compounds $Pb_4A(XO_4)_3$, where A is an alkali metal or Tl. 23rd List

Alkali-spinel. H. von Eckermann, 1922. *Geol. För. Förh., Stockholm*, vol. 44, p. 757. A dark-green spinel from Mansjö Mt., Sweden, containing small amounts of alkalis (Na_2O 1·38, K_2O 1·31%). [*M.A.*, **2**, 185.] 10th List

Alkanasul. J. Westman, 1931. *Bol. Minero Soc. Nac. Mineria*, Santiago de Chile, vol. 43 (año 47), p. 433; *Zeits. Prakt. Geol.*, 1932, vol. 40, p. 110. Hydrous basic sulphate of aluminium, potassium, and sodium, $K_2SO_4.Na_2SO_4.2Al_2(SO_4)_3.6Al(OH)_3.6H_2O$, yellowish-white to bluish-grey, massive, occurring in large amount near Salamanca, Chile. Named from the chemical composition. [Evidently identical with natroalunite.] [*M.A.*, **5**, 200.] 13th List

Allargentum. P. Ramdohr, 1950. *Die Erzmineralien und ihre Verwachsungen*, Berlin, pp. 205, 261, 264; *Fortschr. Min.*, 1950, vol. 28 (for 1949), p. 69 („Alargentum"). Hexagonal ε-modification of silver containing antimony (Ag,Sb), Sb 8–15%, as an exsolution product in dyscrasite from Cobalt, Ontario. Named from ἄλλος, another, and argentum, silver. [*M.A.*, **11**, 312.] 19th List [$Ag_{1-x}Sb_x$; *A.M.*, **56**, 638]

Allcharite. B. Ježek, 1912. *Zeits. Kryst. Min.*, vol. li, p. 275 (Allcharit). Small, acicular, orthorhombic crystals, resembling stibnite in appearance, found with vrbaite (q.v.) on specimens of realgar and orpiment from Allchar, Macedonia. Chemical composition not determined. Named after the locality. 6th List [Is Goethite; *A.M.*, **54**, 1498]

Alleghanyite. C. S. Ross and P. F. Kerr, 1932. *Amer. Min.*, vol. 17, p. 7. Manganese silicate, $5MnO.2SiO_2$, as pink orthorhombic grains, considered to be distinct from tephroite [$2MnO.SiO_2$]. Named from the locality, Alleghany County, North Carolina. [*M.A.*, **5**, 50.] 13th List [Wales; *M.M.*, **27**, 40]

Allenite. G. Gagarin and J. R. Cuomo, 1949, loc. cit. p. 13 (allenita). $MgSO_4.5H_2O$ in the chalcanthite group; dehydration product of epsomite ($MgSO_4.7H_2O$) as distinct from hexahydrite ($MgSO_4.6H_2O$, 6th List). Named after Dr. Eugene Thomas Allen (1864–), formerly chemist in the Geophysical Laboratory, Washington, who analysed in 1927 material from 'The Geysers', Sonoma Co., California. Not to be confused with allanite (T. Thomson, 1810). Synonyms, magnesium-chalcanthite, pentahydrite (qq.v.). Allenite is a trade-name for tungsten carbide tools. [*M.A.*, **11**, 517.] 19th List

Allevardite. S. Caillère and S. Hénin, 1950. *Compt. Rend. Acad. Sci., Paris*, vol. 230, p. 668; S. Caillère, A. Mathieu-Sicaud, and S. Hénin, *Bull. Soc. Franç. Min. Crist.*, 1950, vol. 73, pp. 141, 193. A paper-like mineral from La Table in Savoie, near Allevard in Isère, previously referred to kaolinite, mica, and palygorskite. Named from the locality. Alternative names suggested are caillérite, déribérite, tablite, and tabulite (qq.v.). [*M.A.*, **5**, 521, **8**, 225, **10**, 43, 371, **11**, 127, 190.] 19th List [Is Rectorite; *A.M.*, **49**, 446]

Allingite. A. Tschirch and E. Aweng [*Archiv f. Pharmacie*, 1894, CCXXXII, 660–688] *Journ. Chem. Soc. Abstracts*, 1895, LXVIII (i), 385. A variety of amber (succinite) from Switzerland. 2nd List

Allite. H. Harrassowitz, 1926. *Laterit, Material und Versuch erdgeschichtlicher Auswertung*, Berlin 1926, p. 255 (Allit, *plur.* Allite). A rock-name to include both bauxite and laterite. Later (*Metall und Erz*, Halle, 1927, vol. 24, p. 589) bauxite with $Al_2O_3.H_2O$ is distinguished as monohydrallite (Monohydrallit) and laterite with $Al_2O_3.3H_2O$ as trihydrallite (Trihydrallit). These, although suggestive of mineral-names (and given so in error in *Chem. Zentr.*, 1926, vol. 1, p. 671), are proposed as rock-names; from aluminium and λίθος. Similarly, siallites (1926, p. 252, Siallit, from Si, Al, λίθος), to include kaolinite and allophanite, are rocks composed of the aluminium silicates kaolin and allophane. 11th List

Allodelphite. P. Quensel and H. von Eckermann, 1931. *Geol. För. Förh. Stockholm*, vol. 52 (for 1930), p. 639 (Allodelphite). Hydrous silico-arsenite of manganese, $5MnO.2Mn_2O_3.As_2O_3.SiO_2.5H_2O$ or $MnO.(Mn_2,As_2,MnSi)O_3.H_2O$, as red-brown crystals, probably orthorhombic, from Långban, Sweden. Named from ἄλλος, other, and ἀδελφός, brother, on account of its relationship to synadelphite. [*M.A.*, **4**, 496.] 12th List [Is Synadelphite; *A.M.*, **22**, 526]

Allokite. B. J. S. Collins, 1955, *Diss. Univ. Illinois*, vol. 15, no. 11, p. 2163. Composite name for a clay mineral with layer structure intermediate between allophane and kaolinite. 21st List

Allophane-chrysocolla, *see* Allophane-evansite.

Allophane-evansite, Allophane-chrysocolla. C. S. Ross and P. F. Kerr, 1934. *Prof. Paper U.S. Geol. Surv.*, no. 185–G, p. 147 (allophane-evansite). Mixtures of amorphous minerals. [*M.A.*, **6**, 136.] Compare phosphate-allophane, 17th List. 19th List

Allophanoids. S. J. Thugutt, 1911. *Spraw. Tow. Nauk. Warszawa*, 1911, vol. iv, p. 222 (alofanoid); *Centralblatt Min.*, 1912, 35 (Allophanoide). Clays of the allophane, halloysite, and montmorillonite groups. 6th List

Almashite. G. Murgoci, 1924. *Les ambres roumains, Correspondance économique roumaine, Bulletin officiel, Ministère de l'Industrie et du Commerce*, Bucarest, 1924, vol. 6, no. 6, pp. 10, 16. A. P. Iancoulesco. *Les richesses minières de la nouvelle Roumanie*, Paris, 1928, p. 229. C. Doelter, *Handbuch d. Mineralchemie*, 1931, vol. 4, pt. 3, p. 935 (Almaschit). Green (almashite I) and black (almashite II) varieties of amber, differing from romanite, from the Almas (Almash) valley near Piatra in Moldavia. The green variety contains C 82·15, H 10·94, O 2·57, S 0·33, ash 3·51, and the black C 79·45, H 10·23, O 3·00, S 1·40, ash 5·52. Named from the locality. [*M.A.*, **4**, 297.] 12th List

Almeraite. L. Tomás & J. Folch, 1914. *Butll. Inst. Catalana Hist. Nat.*, vol. 11, p. 11; L. Tomás, *Els minerals de Catalunya, Treballs Inst. Catalana Hist. Nat., Barcelona*, vol. for 1919–1920, p. 221 (Almeraita, Almeraïta). The formula $KCl.NaCl.MgCl_2.H_2O$ is deduced from an analysis of a reddish, semitransparent, crystalline, granular aggregate from the salt deposits at Suria, prov. Barcelona. The crystalline form could not be determined, and the mineral is given doubtfully as a new species allied to carnallite. Named after Dr. Jaume Almera, of Barcelona. Not to be confused with the Almeriite of S. Calderón, 1910 (6th List). [*M.A.*, **2**, 116.] 10th List [Remains unconfirmed]

Almeriite. S. Calderón, 1910. *Los Minerales de España*, vol. ii, p. 206 (Almeriita). A hydrated basic sulphate of aluminium and sodium, $Al_2(SO_4)_3.Na_2SO_4.5Al(OH)_3.H_2O$, the formula being the same as that of calafatite (q.v.) with sodium in place of potassium. A compact, white material, resembling halloysite in appearance, from Almeria. 6th List [Is Natroalunite; *M.M.*, **33**, 353]

Aloisiite. L. Colomba, 1908. *Rend. R. Accad. Lincei, Roma*, ser. 5, vol. xvii, sem. 2, p. 233. A hydrated sub-silicate of calcium, ferrous iron, magnesium, sodium, and hydrogen, $(R'', R'_2)_4 SiO_6$, occurring in an amorphous condition, intimately mixed with calcium carbonate, in a palagonite-tuff at Fort Portal, Uganda. Named in honour of H.R.H. Prince Luigi Amedeo of Savoy, Duke of Abruzzi. Aloisius or Aloysius is a Latin form of Luigi or Lewis. 5th List [Status doubtful]

Alomite. Trade-name for the fine blue sodalite quarried at Bancroft, Ontario, Canada, for use as an ornamental stone. Named alomite [not allomite] after Charles Allom, of White, Allom & Co., marble merchants, London. Also called 'Princess Blue'. (*The Quarry*, London, 1908, vol. xiii, p. 257. W. G. Renwick, *Marble and marble working*, London, 1909, p. 115.) 5th List

Alouchtite. *Bull. Soc. Franç, Min. Crist.*, 1956, vol. 79, p. 339. French spelling of alushtite from Russian алуштитъ (10th List). A mixed clay mineral. 21st List

Aloxite. (*The Mineral Industry*, New York, vol. xxii for 1913, 1914, p. 1.) Trade-name for a form of fused crystalline alumina, or artificial corundum, manufactured by the Carborundum Company, and used for abrasive purposes. 7th List

Alpha-cerolite. I. I. Ginzburg and I. A. Rukavinshnikova, 1950. *Mém. Soc. Russe Min.*, vol. 79, p. 33 (Альфа-керолит). *See* Beta-cerolite. [*M.A.*, **11**, 405.] 19th List

Alpha-vredenburgite. B. Mason, 1943. *Geol. För. Förh. Stockholm*, vol. 65, p. 263 (Alpha-vredenburgite), pp. 157, 268 (α-vredenburgite). Homogeneous tetragonal $(Mn,Fe)_3O_4$, as distinct from the usual mixture of cubic and tetragonal $(Mn,Fe)_3O_4$ distinguished as β-vredenburgite. [*M.A.*, **9**, 37.] 17th List

Alterite. Tj. H. van Andel, 1950. *Rhine sediments, Diss. Groningen*, p. 45 (alterites), p. 46 (epidote-alterite. &c.). D. Carroll, *Amer. Min.*, 1957, vol. 42, p. 110. A loose term for weathered (altered) grains of heavy minerals, usually not identifiable from their optical characters. 21st List

Althausite. G. Raade and M. Tysseland, 1975. *Lithos*, **8**, 215. Cleavable masses in serpentine–magnesite deposits at Modum, Norway, composition $Mg_2PO_4(OH,F,O)$, have *a* 8·258, *b* 14·383, *c* 6·054 Å, space group *Pna* 2_1. α 1·588 ∥ [010], β 1·592, γ 1·598 ∥ [001]. Named for Professor E. Althaus. 29th List

Alumianite. Variant of Alumian; *A.M.*, **8**, 51. 10th List

Aluminatchromit. K. Spangenberg, 1943. *Zeits. prakt. Geol.* **51**, 22. A chrome ore from Tampadel, Zobten, Silesia, with Cr only slightly >Al and Mg:Fe about 3·5. Synonym of alumo-chrompicotite (15th List). 27th List

Aluminatspinelle. H. Strunz, 1957. *Min. Tabellen*, 3rd edit., p. 352. Group name for those spinels in which Al is the principal trivalent metal. 24th List

Aluminiocopiapit, variant of or error for Aluminocopiapit (H. Strunz, *Min. Tabellen*, 3rd edit., 1957, p. 211). 22nd List

Aluminium-autunite. C. Frondel, 1951. *Amer. Min.*, vol. 36, p. 671 (aluminum-autunite). Synonym of sabugalite (q.v.). [*M.A.*, **11**, 412.] 19th List

Aluminium-epidote. M. Goldschlag, 1916. Doelter's *Handbuch d. Mineralchemie*, vol. 2 (part 2), p. 821. The isomorphous molecules $HCa_2Al_3Si_3O_{13}$ and $HCa_2Fe_3Si_3O_{13}$ which enter into the composition of epidote are distinguished as Aluminiumepidot and Eisenepidot respectively. F. Zambonini (*Boll. Com. Geol. Italia*, 1920, vol. 47, p. 81) has independently suggested Aluminioepidoto (= clinozoisite) and Ferriepidoto. Eisenepidot has been translated Iron-epidote (*M.A.*, 1922, **1**, 346). 10th List

Aluminium-Ferroanthophyllit, error for aluminian Ferroanthophyllite (C. Hintze, *Handb. Min., Erg.-Bd. II*, p. 654). 24th List

Aluminiumglauconite. H. Borchert and H. Braun, 1963. *Chem. Erde*, vol. 23, p. 82 (Aluminiumglaukonit). An unnecessary name for highly aluminian glauconite. 25th List

Aluminium phosphocristobalite. R. J. Manly, *Jr.*, 1950. *A.M.*, **35**, 108 (Aluminum phosphocristobalite). *Syn.* of phosphocristobalite. 28th List

Aluminium phosphotridymite. R. J. Manly, *Jr.*, 1950. *A.M.*, **35**, 108 (Aluminum phosphotridymite). *Syn.* of phosphotridymite. 28th List

Aluminium-saponite. H. Strunz, 1957. *Min. Tabellen*, 3rd edit., p. 313 (Aluminium-Saponit). Unnecessary name for aluminian saponite. 22nd List

Aluminium-sepiolite. L. E. R. Rogers, J. P. Quirke, and K. Norrish, 1956. *Journ. Soil Sci.*, vol. 7, p. 177. Unnecessary name for aluminian sepiolite. 22nd List

Aluminiumskorodit, Alumoskorodit, Alumskorodit. K. F. Chudoba, 1954. Hintze, *Min., Ergbd. 2*, pp. 8, 10, 11. Synonyms of Aluminoscorodite (18th List). Unnecessary complications. 20th List [Also 23rd List]

Aluminium spinel. N. E. Filonenko, I. V. Lavrov, O. V. Andreeva, and R. L. Pevzner, 1957. Доклады Акад. Наук СССР (*Compt. Rend. Acad. Sci. URSS*), vol. 115, p. 583 (Глиноземистой шпинел). Octahedral crystals obtained in the manufacture of synthetic corundum are stated to give chemical analyses corresponding to Al_3O_4. [*M.A.*, **13**, 648.] 22nd List

Alumino-barroisite. *Amph.* (1978), **14**, 5. 30th List

Aluminobetafite. T. Kawai, 1960. [*Journ. Chem. Soc. Japan, Pure Chem. Sect.*, vol. 81, p. 1219], abstr. in *M.A.*, **16**, 62; *Amer. Min.*, 1963, vol. 48, p. 1183. Dark grey cubic crystals from Kaijo, Manchuria, are regarded as an aluminian betafite. 'Data inadequate' (M. Fleischer). Named from the composition. 23rd List [Remains doubtful; *A.M.*, **62**, 406]

Alumino-chrysotile. D. P. Serdyuchenko, 1945. *Compt. Rend.* (*Doklady*) *Acad. Sci. URSS*, vol. 46, p. 117. Chrysotile containing Al_2O_3 5·68%, in the series serpentine-parakaolinite, $H_4(Mg_3,Al_2)Si_2O_9$; from Caucasus. [*M.A.*, **9**, 185.] 17th List

Aluminocopiapite. L. G. Berry, 1947. *Univ. Toronto Studies, Geol. Ser.*, no. 51 (for 1946), p. 29. A variety of copiapite in which X in the formula $X(OH)_2Fe_4'''(SO_4)_6.nH_2O$ is mainly Al (Al_2O_3 1·72–4·45%). Compare Ferricopiapite, 15th List. [*M.A.*, **10**, 99.] 18th List

Alumino-ferro-hornblende. *Amph.* (1978), syn. of ferro-hornblende. 30th List

Aluminohydrocalcite, error for alumohydrocalcite (*C.M.*, **10**, 88). 26th List [Also 28th List]

Alumino-katophorite. *Amph.* (1978), **14**, 13. 30th List

Alumino-magnesio-hornblende. *Amph.* (1978), syn. of magnesio-hornblende. 30th List

Aluminoscorodite. T. Ito and K. Sakurai, 1947. *Wada's minerals of Japan*, 3rd edit., Tokyo, p. 309 (Aluminoscorodite, Alumskorodit, and Japanese script). A variety of scorodite containing Al_2O_3 5·11%, $(Fe,Al)AsO_4.2H_2O$. The same as aluminum-bearing scorodite [*M.A.*, **9**, 142] = aluminous-scorodite [*M.A.*, **9**, 262] = aluminian scorodite [*M.A.*, **10**, 353] with Al_2O_3 5·76% from Oregon. [*M.A.*, **10**, 351.] 18th List

Alumino-taramite. *Amph.* (1978), **14**, 17. 30th List

Alumino-tschermakite. *Amph.* (1978), **10**, 9. 30th List

Aluminous-serpentine. H. S. Yoder, 1952. *Amer. Journ. Sci.*, Bowen volume, p. 579 (Aluminous-Serpentine, aluminous serpentine, Al-serpentine). Artificially produced, platy habit, related to antigorite or clinochlore. Compare Alumino-chrysotile, 17th List. [*M.A.*, **12**, 83.] 20th List

Alumino-winchite. *Amph.* (1978), **14**, 1. 30th List

Alumo-aeschynite. E. M. Eskova, A. G. Zhabin, and G. N. Mukhitdinov, 1964. [Мин. геохнм. редк. элем. Вишнев. Горы (*Min. geochem. rare elements in the Vishnevaya Mts., Urals*), Moscow (Издат. „Наука“)]: abstr. *Amer. Min.*, 1965, vol. 50, p. 2101; Зап. Всесоюз. Мин. Общ. (*Mem. All-Union Min. Soc.*), 1965, vol. 94, p. 672 (Алюмоэшинит, alumoeschynite). An unnecessary name for aluminian Aeschynite. 24th List

Alumoantigorite, Alumochrysotile, Alumodeweylite. N. Efremov. 1951. *Fortschr. Min.*, vol. 29–30 (for 1950–1951), p. 85 (Alumoantigorit, Alumochrysotil, Alumodeweylith). Serpentine minerals containing Al_2O_3 2·55–6·27%. Compare Alumino-chrysotile (17th List). 19th List

Alumoberesowskit. H. Strunz, 1957. *Min. Tabellen*, 3rd edit., p. 352. Synonym of Alumoberezovite. Aluminian berezovskite, a variety of chromite. Berezovskite (15th List) was originally named Beresofite (9th List), a name that had already been used twice; the aluminian variety was named Алюмо-Березовит (Alumoberezovite). 24th List

Alumo-berezovite, Alumo-chrompicotite. S. A. Vakhromeev *et alii*, 1936. *Trans. All-Union Sci. Research Inst. Econ. Min. USSR*, no. 85, p. 225 (алюмо-березовит, алюмо-хромпикотит). Members of the spinel group with the composition $(Fe,Mg)O.(Cr,Al)_2O_3$ (alumoberezovite) and $(Mg,Fe)O.(Cr,Al)_2O_3$ (alumo-chrompicotite). *See* Berezovskite. [*M.A.*, **7**, 151.] [Also *see* Alumochromite.] 15th List

Alumobritholite. M. A. Kudrina, V. S. Kudrin, and G. A. Sidorenko, 1961. [Геол. Мест. Редк. Элем. (*Geol. Depos. Rare Elem.*), vol. 9, p. 108], abstr. in Зап. Всесоюз. Мин. Общ. (*Mem. All-Union Min. Soc.*, 1962, vol. 91, p. 200 (Алюмобритолит, alumobritolite), and in *Amer. Min.*, 1961, vol. 46, p. 1514. $(Ca,Ce,Y)_3(Al,Fe)_2\{(Si,Al,P)O_3\}_3(F,O)$, from an unspecified Siberian locality; an aluminian variety of britholite. *See also* Pravdite. [*M.A.*, **16**, 553.] 23rd List

Alumo-chalcosiderite. A. Jahn and E. Gruner, 1933. *Mitt. Vogtländ. Gesell. Naturfor.*, vol. 1, no. 8, p. 19 (Alumo-Chalkosiderit). A variety of chalcosiderite containing some alumina (Al_2O_3 10·45%) replacing ferric oxide. [*M.A.*, **5**, 391.] 13th List

Alumochromite. A. G. Betekhtin, 1934. *Ann. Inst. Mines*, Leningrade, vol. 8, p. 38 (алюмохромит), p. 63 (Alumchromit). A member of the spinel group with the composition $Fe(Cr,Al)_2O_4$. Other hypothetical members of the group are named Magnoferrichromite, $(Fe,Mg)(Cr,Fe)_2O_4$ (магноферрихромит, Magnoferrichromit); Ferrichromspinel, $Mg(Cr,Al,Fe)_2O_4$ (феррихромшпинель, Ferrichromspinel); and Ferrichrompicotite, $(Fe,Mg)(Cr,Al,Fe)_2O_4$ (феррихромпикотит, Ferrichrompikotit). Several other names of this type for the spinel group are given by A. K. Boldyrev, *Kurs opisatelnoi mineralogii*, Leningrad, 1935, part 3, p. 115 (*see* Cobaltochrompicotite). 14th List

Alumo-chrompicotite, *see* Alumo-berezovite.

Alumochrysocolla. F. V. Chukhrov, B. B. Zvyagin, A. I. Gorshakov, L. P. Ermilova, and E. S. Rudnitskaya, 1968. Изв. Акад. наук СССР (*Bull. Acad. Sci. URSS*), no. 6, 29 (Алюмохризоколла). Superfluous name for an aluminian chrysocolla. 28th List

Alumochrysotile, *see* Alumoantigorite.

Alumocobaltomelane. I. I. Ginzburg and I. A. Rukavishnikova, 1951, Мин. древн. выветривания Урала (*Minerals of the ancient zone of weathering of the Urals*), Moscow, 1951, p. 128 (алюмокобальтомелан). The names alumocobaltomelane, cobaltomelane, cobaltonickelemelane, cryptonickelmelane, nickel-cobaltomelane, and nickelemelane are given to a variety of admitted mixtures of manganese and other oxides. 'The names are . . . merely mineralogical waste baskets' (M. Fleischer, *Amer. Min.*, 1961, vol. 46, p. 767). 22nd List

Alumodeweylite. N. E. Efremov, 1939. *Compt. Rend. (Doklady) Acad. Sci. URSS*, vol. 24,

p. 287 (alumodeveillites). Hydrated silicates of aluminium and magnesium intermediate in composition between montmorillonite and sepiolite, or mixtures of these. [*M.A.*, **7**, 419.] 15th & 19th Lists. Also *see* Alumochrysotile.

Alumoeschynite, undesirable spelling variant of Alumo-aeschynite (Зап. Всесоюз. Мин. Общ. (Mem. All-Union Min. Soc.), 1965, vol. 94, p. 672). 24th List

Alumoferrichrysocolla. F. V. Chukhrov, B. B. Zvyagin, A. I. Gorshakov, L. P. Ermilova, and E. S. Rudnitskaya, 1968. Изв. Акад. наук СССР (*Bull. Acad. Sci. URSS*), no. 6, 29 (Алюмоферрихризоколла). A superfluous name for an aluminian ferrian chrysocolla. 28th List

Alumoferroascharite. D. P. Serdyuchenko, 1956. *Mém. Soc. Russ. Min.*, vol. 85, p. 292 (алюмоферроашарит). Variety of ascharite containing Al_2O_3 6·47, FeO 8·79, Fe_2O_3 4·30%. Formula $(Mg,Fe)(B,Al)O_2.OH$. [*M.A.*, **13**, 522 (alumoferroasharite).] 21st List [A mixture; *A.M.*, **49**, 1501]

Alumogel. O. Pauls, 1913: *Zeits. prakt. Geol.*, Jahrg. xxi, p. 545. Amorphous aluminium hydroxide of indefinite composition forming the main constituent of bauxite. Compare Diasporogelite. 7th List. H. Strunz, *Min. Tab.*, 1941, p. 111. 'AlOOH + aq'. Synonym of kliachite (5th List) and sporogelite (6th List). Compare Siderogel (19th List). 20th List

Alumogoethite. S. I. Beneslavsky, 1957. Доклады Акад. Наук СССР (*Compt. Rend. Acad. Sci. URSS*), vol. 113, p. 1130 (Алюмогетит). Aluminian goethite, occurring in bauxites. An unnecessary name. 22nd List

Alumohematite. S. I. Beneslavsky, 1957. Доклады Акад. Наук СССР (*Compt. Rend. Acad. Sci. URSS*), vol. 113, p. 1130 (Алюмогематит). K. F. Chudoba, Hintze, *Handb. Min.*, 1959, Erg.-Bd. II, p. 655 (Alumohaematit). An aluminian hematite in bauxites. An unnecessary name. 22nd List

Alumohydrocalcite. G. A. Bilibin, 1926. *Zap. Ross. Min. Obshch.* (*Mém. Soc. Russe Min.*), ser. 2, vol. 55, p. 243 (Алюмогидрокальцит, Alumohydrocalcite). Hydrated carbonate of calcium and aluminium $CaO.Al_2O_3.2CO_2.5H_2O$ or $CaH_2(CO_3)_2.2Al(OH)_3.H_2O$, as white chalky masses consisting of radially fibrous spherulites, with monoclinic symmetry, from Siberia. Named from the composition. [*M.A.*, **3**, 472.] 11th List [*A.M.*, **13**, 569]

β-Alumohydrocalcite. A. Morawiecki, 1961. [*Przegląd geol.*, vol. 9, p. 382], abstr. in *Amer. Min.*, 1963, vol. 48, p. 212, and in *Bull. Soc. franç. Min. Crist.*, 1964, vol. 87, p. 111. A mineral from shales at Nowa Ruda, Dolny Śląsk, Poland, has the composition of alumohydrocalcite, $(CaAl_2(CO_3)_2(OH)_4.3H_2O)$, but the fibres give straight extinction. The distinction from alumohydrocalcite is doubtful. 23rd List [Remains doubtful; *M.A.*, 78-868]

Alumolimonite. G. A. Bilibin, 1928. A. K. Boldyrev, *Kurs Opisat. min.*, Leningrad, 1928, no. 2, p. 97 (алюмолимонит). S. I. Beneslavsky, *Doklady Acad. Sci. USSR*, 1957, vol. 113, p. 1130. Mixture of aluminium and iron hydroxides. [*M.A.*, **13**, 520.] 21st List

Alumoludwigite. N. N. Pertsev and S. M. Aleksandrov, 1964. Зап. всесоюз. мин. общ. (*Mem. All-Union Min. Soc.*), vol. 93, p. 13 (Алюмолюдвигит). An unnecessary name for aluminian ludwigite. 25th List

Alumolyndochite. S. A. Gorzhevskaya, G. A. Sidorenko, and A. I. Ginzburg, 1974. *Abstr. Zap.* **105**, 76 (Алюмолиндокит). Unnecessary name for aluminian lyndochite. 30th List

Alumomaghaemite, alumomaghemite. S. I. Beneslavskii, 1957. Доклады акад. наук СССР (*Compt. Rend. Acad. Sci. URSS*), vol. 113, p. 1130 (Алюмомаггемит). An unnecessary name for aluminian maghemite. 25th List

Alumomaghemite, *see* Alumomaghaemite.

Alumomelanocerite. I. I. Kupriyanova and G. A. Sidorenko, 1963. Докл. Акад. Наук СССР (*Compt. Rend. Acad. Sci. URSS*), vol. 148, p. 212 (Алюмомеланоцерит). Aluminian melanocerite. 'A superfluous name' (Зап. Всесоюз. Мин. Общ. (*Mem. All-Union Min. Soc.*), 1964, vol. 93, p. 456). 23rd List

Alumopharmacosiderite. G. Hägele and F. Machatschki, 1937. *Fortschr. Min. Krist. Petr.*, vol. 21, p. 77 (Alumopharmakosiderit). Artificially prepared cubic crystals $Al_3(AsO_4)_2(OH)_3.5H_2O$, or perhaps $Al_5As_3O_{12}(OH)_6.6H_2O$, analogous to pharmacosiderite with aluminium in place of iron. 14th List

Alumoskorodit, *see* Aluminiumskorodit.

Alumospencite. I. I. Kupriyanova and G. A. Sidorenko, 1963. Докл. Акад Наук СССР (*Compt. Rend. Acad. Sci. URSS*), vol. 148, p. 212 (Алюмоспенсит). Aluminian spencite. 'A superfluous name' (Зап. Всесоюз. Мин. Общ. (*Mem. All-Union Min. Soc.*), 1964, vol. 93, p. 456) (Алюмоспенсит, alumospensite). 23rd List

Alumotrichite. (P. Groth, *Min.-Samml. Strassburg*, 1878, p. 257). White, fibrous. Chili. [Probably the same as kalinite]. 2nd List

Alumotungstite. Th. G. Sahama, 1971, in M. Fleischer, *Glossary of Mineral Species*, 3. $(W,Al)_{16}(O,OH)_{46}.xH_2O$, the aluminium analogue of yttrotungstite (19th List). 28th List

Alumskorodit, *see* Aluminiumskorodit.

Alumyte. G. H. Kinahan, *Journ. R. Geol. Soc. Ireland*, 1889, VIII, 66; *Trans. Manchester Geol. Soc.* 1895, XXIII, 165. Alum-clay ('bauxite') of Co. Antrim. 2nd List

Alundum. Trade-name for artificial corundum manufactured in the electric furnace from bauxite by the Norton Emery Wheel Company, of Worcester, Mass., since 1904. The material is used as an abrasive, and the name is no doubt a contraction of aluminium and carborundum (4th List). (*Mineral Industry*, New York, for 1906, 1907, vol. xv, p. 28.) 5th List

Alushtite. A. E. Fersman, 1914. *In* P. A. Dvoichenko, *Minerals of the Crimea* (Russ.), Zap. Krym. Obshch. Est. i Lyub. Prir. (*Mem. Crimean Soc. Sci. & Nat.*), vol. 4, p. 105 (Алуштить); also published separately as a book, Petrograd, 1914; abstract in *Zeits. Krist.*, 1923, vol. 57, p. 591 (Aluschtit). A hydrated aluminium silicate near to kaolinite, containing H_2O 13·7% and a little magnesia. It occurs as bluish or greenish crusts, nests, and veins in quartz veins in black clay-slates near Alushta and elsewhere in the Crimea. Named from the locality. 10th List [Contains essential Ca and Mg; *M.A.*, **16**, 279; **19**, 268]

Alvanite. E. A. Ankinovich, 1959. Зап. Всесоюз. Мин. Общ. (*Mem. All-Union Min. Soc.*), vol. 88, p. 157 (Альванит). Light blue-green rosettes, monoclinic, near $Al_3VO_4(OH)_6.2\frac{1}{2}H_2O$, in the argillaceous anthraxolitic vanadiferous deposits of Kurumsak and Balasanskandyk, Karatau, Kazakhstan. Named from the composition, *al*uminium *van*adate. [*M.A.*, **14**, 280.] An anion has probably been overlooked, as the mineral is stated to give off acid water when heated. 22nd List [*A.M.*, **44**, 1325]

Alvarolite. W. Florencio, 1952. *Anais Acad. Brasil. Cienc.*, vol. 24, p. 261 (Alvarolita). Perhaps monoclinic $MnTa_2O_6$ dimorphous with manganotantalite [cf. *Min. Mag.* **27**, 162]. Named after Admiral Alvaro Alberto da Motta e Silva, director of the national research council, Brazil. [*M.A.*, **12**, 305.] 20th List. Subsequently shown to be manganotantalite ([*Anais Acad. Brasil. Cienc.*, 1955, vol. 27, p. 7], abstr. in *Amer. Min.*, 1956, vol. 41, p. 168). 23rd List

Amakinite. I. T. Kozlov and P. P. Levshov, 1962. Зап. Всесоюз. Мин. Общ. (*Mem. All-Union Min. Soc.*), vol. 91, p. 72 (Амакинит). Occurs as thin veins in kimberlite from the 'Lucky Eastern' pipe, presumably in Yakutia, and has the composition $(Fe,Mg)(OH)_2$; rhombohedral: oxidizes rapidly in air. Named from the Amakin expedition (*Amer. Min.*, 1962, vol. 47, p. 1218; *M.A.*, **15**, 541). 23rd List

Amargosite. J. Melhase, 1926. *Engin. Mining Journ.-Press*, New York, vol. 121, p. 841. Trade-name for a bentonite clay from Amargosa river, Inyo Co., California. *See* Otaylite. [*M.A.*, **3**, 143.] 11th List

Amarillite. H. Ungemach, 1933. *Compt. Rend. Acad. Sci. Paris*, vol. 197, p. 1133; *Bull. Soc. Franç. Min.*, [1934], vol. 56 (for 1933), p. 303. Hydrated sulphate of sodium and ferric iron $NaFe(SO_4)_2.6H_2O$, as yellow monoclinic crystals from Tierra Amarilla, Chile. Named from the locality. [*M.A.*, **5**, 390.] 13th List [*A.M.*, **21**, 270]

Amatrice. Trade-name for a green gem-stone from Utah, consisting of variscite, utahlite, or wardite, or a mixture of these, in a matrix of quartz, chalcedony, &c. So called (A-matrix) because it is an American matrix gem-stone. (D. B. Sterrett, *Mineral Resources of the United States*, for 1907, 1908, part ii, p. 832. J. Wodiska, *A book on precious stones*, New York, 1909, p. 185.) 5th List

Ambatoarinite. A. Lacroix, 1916. *Bull. Soc. franç. Min.*, vol. xxxviii (for 1915), p. 265. An orthorhombic carbonate of cerium metals and strontium, $5SrCO_3.4(Ce,La,Di)_2(CO_3)_3.(Ce,La,Di)_2O_3$, occurring with celestite, monazite, felspar, etc., as a constituent of a crystalline limestone at Ambatoarina, Madagascar. Compare Ancylite (G. Flink, 1900; 2nd List.) 8th List

Amberine. (D. B. Sterrett, Gems and precious stones in 1913, *Mineral Resources of the United States*, for 1913, 1914, part ii, p. 652.) Local trade-name, used by Mr. Joseph Ward, of Barstow, California, for a yellowish-green chalcedony from the Death Valley region, California. 7th List

Ameghinite. L. F. Aristarain and C. S. Hurlbut, Jr., 1967. *Amer. Min.*, 1967, vol. 52, p. 935. Colourless monoclinic crystals from the Tincalayu borax deposit, Salta, Argentina, have the composition $Na_2B_6O_{10}.4H_2O$. Named for F. and C. Ameghino. [*Bull.* **91**, 97.] 25th List

Ameletite. P. Marshall, 1929. *Min. Mag.*, vol. 22, p. 174. Silicate with chloride of aluminium and sodium, approximate formula $9Na_2O.6Al_2O_3.12SiP_2.\frac{1}{2}NaCl$, as minute hexagonal crystals in phonolite from Dunedin, New Zealand. Named from ἀμελής, neglected, because the mineral had long been unrecognized. 12th List [A mixture; *M.A.*, **18**, 279]

Amensite. I. C. Jahanbagloo and T. Zoltai, 1968, *A.M.*, **53**, 23. Error for amesite; not to be confused with amansite (an error for amausite, *see Chem. Index*, 1955, 2nd edit., 326). 28th List

Amianthinite. R. Kirwan, *Elements of Mineralogy*, 2nd edit., 1794, vol. i, p. 164. Asbestiform actinolite. 3rd List

Aminoffite. C. S. Hurlbut. 1937. *Geol. För. Förh. Stockholm*, vol. 59, p. 290 (Aminoffite). Hydrous silicate of calcium, beryllium, and aluminium, $Ca_8Be_3AlSi_8O_{28}(OH).4H_2O$, as colourless tetragonal crystals, related to meliphane, from Sweden. Named after Dr. Gregori Aminoff (1883–) of Stockholm. [*M.A.*, **7**, 119.] 15th List [*A.M.*, **23**, 293]

Ammersooite. H. W. van der Marel, 1954. *Soil Science*, vol. 78, pp. 172, 176 (ammersooite). A variety of illite capable of fixing potassium in Dutch soils. Named from the potash experimental field near Ammerzoden (locally called Ammersooien), Gelderland, Holland. 21st List

Ammonia-nitre. Dana's *Mineralogy*, 7th edit., 1951, vol. 2, p. 305 (ammonia niter), Ammonium nitrate $(NH_4)NO_3$, nitrammite of C. U. Shepard, 1857, of doubtful occurrence in a Tennessee cave. 19th List

Ammonioborite. W. T. Schaller, 1931. *Amer. Min.*, vol. 16, p. 114. Hydrous borate of ammonium, $(NH_4)_2O.5B_2O_3.5H_2O$, differing optically from larderellite, with which it occurs at Laderello [Lardarello; *M.M.*, **40**], Toscana. Named from the composition. 12th List

Ammoniojarosite. E. V. Shannon, 1927. *Amer. Min.*, vol. 12, p. 424. Hydrated sulphate of ammonium and ferric iron, $(NH_4)_2O.3Fe_2O_3.4SO_3.6H_2O$, as ochre-yellow nodules consisting of minute flattened grains, probably rhombohedral, from Utah. A member of the jarosite isomorphous group, with ammonium oxide in place of potash, etc. [*M.A.*, **3**, 470.] 11th List

Ammonium-analcite, Ammonium-leucite, Ammonium-natrolite, Ammonium-stilbite, &c. F. W. Clarke and G. Steiger, *Amer. Journ. Sci.*, 1900, ser. 4, vol. ix, pp. 117, 345; ibid., 1902, vol. xiii, p. 27; *Bull. U.S. Geol. Survey*, 1902, No. 207. Artificial derivatives of analcite, etc., with ammonium in place of sodium, potassium, or calcium. 3rd List

Ammonium-aphthitalite. K. F. Chudoba, 1959. Hintze, *Handb. Min.*, Erg.-Bd. II, p. 658 (Ammonium-Aphthitalit). An unnecessary name for the ammonian aphthitalite of Frondel. [*M.A.*, **11**, 302.] 22nd List [Also 20th List]

Ammonium boltwoodite, sodium boltwoodite. R. M. Honea, 1961. *Amer. Min.*, vol. 46, p. 12. The ammonium and sodium analogues of boltwoodite, $K_2(UO_2)_2(SiO_3)_2(OH)_2.5H_2O$; the term potassium boltwoodite is applied to boltwoodite itself. 22nd List

Ammonium-cryolite. C. Hintze, *Handb. Min.*, 1913, vol. 1, pt. 2, p. 2524 (Ammoniumkryolith). C. Brosset, *Arkiv Kemi, Min. Geol.*, 1946, vol. 21 A, no. 9, p. 8 (ammonium cryolite). Artificial ammonium fluo-aluminate near $(NH_4)_3AlF_6$, analogous to cryolite, but cubic. *See* Potassium-cryolite. [*M.A.*, **4**, 362, **10**, 16, 208.] 18th List

Ammonium gastunite, *see* Gastunite (of Honea). 22nd List

Ammonium-glaserite. H. Strunz, 1957. *Min. Tabellen*, 3rd edit., p. 195 (Ammonium-Glaserit). Syn. of Ammonium-aphthitalite (q.v.). 22nd List

Ammonium-leucite, *see* Ammonium-analcite.

Ammonium-mica. J. W. Gruner, 1939. *Amer. Min.*, vol. 24, p. 428 (Ammonium mica). An artificial product containing up to 2·4% NH_4 obtained by the action of H_2O_2 and NH_4OH on vermiculite. [*M.A.*, **7**, 479.] 18th List [Also 20th List]

Ammonium-natrolite, -stilbite, *see* Ammonium-analcite.

Ammonium-syngenite. J. D'Ans, 1906. *Ber. Deutsch. Chem. Gesell.*, vol. 39 (part 3), p. 3326 (Ammoniumsyngenit). A. E. Hill and N. S. Yanick, *Journ. Amer. Chem. Soc.*, 1935, vol. 57, p. 650 (Ammonium Syngenite). An artificial compound, $(NH_4)_2SO_4.CaSO_4.2H_2O$, isomorphous with syngenite. *See* Koktaite. [*M.A.*, **10**, 352.] 18th List

Ammonium-uranospinite. *See* hydrogen-uranospinite. 19th List

Amosite. A. L. Hall, 1918. *Geol. Survey S. Africa*, Mem. no. 12, p. 20 et seq., *Trans. Geol. Soc. S. Africa*, 1919, vol. xxi, p. 8. A monoclinic amphibole-asbestos rich in iron (FeO 32–44%) sometimes containing sodium, and near to cummingtonite and grünerite in composition. Material of commercial quality occurs in large amount over a wide area in the Lydenburg and Pietersburg districts, Transvaal. Named from the Amosa asbestos mine, this word being formed of the initial letters of the company 'Asbestos Mines of South Africa'. 8th List [Original material is Actinolite and Cummingtonite; *A.M.*, **33**, 308. Cf. Montasite]

Ampangabeite. A. Lacroix, 1912. *Compt. Rend. Acad. Sci. Paris*, cliv, p. 1044; *Bull. Soc. franç. Min.*, 1912, vol. xxxv, p. 194 (ampangabéite). A tantalo-niobate (containing but little titanium) of uranium, iron, yttrium, etc., occurring as large rectangular (tetragonal or orthorhombic) prisms in pegmatite at Ampangabe, Madagascar. It approximates in composition to the heterogeneous ånnerödite. 6th List [Is Samarskite, *A.M.*, **46**, 770; *M.A.*, **15**, 289]

Analbite. A. N. Winchell, 1925. *Journ. Geol. Chicago*, vol. 33, p. 717; Elements of optical mineralogy, 2nd edit., 1927, part 2, pp. 315, 325. Suggested as a more suitable name than anorthoclase, but restricted to those anorthoclases that contain less than 10 mol. % $KAlSi_3O_8$. 11th List

—— H. L. Alling, 1936. *Interpretative petrology of igneous rocks*, p. 59 (low-temperature triclinic β-$NaAlSi_3O_8$). 20th List

—— F. Laves, 1952. *Journ. Geol. Chicago*, vol. 60, p. 570 (low-temperature albite; p. 571, anperthite containing exsolved analbite; p. 572, anplagioclases). Not the analbite of A. N. Winchell (11th List) or of H. L. Alling (above). [*M.A.*, **12**, 136.] 20th List

Analcidite. C. Hintze, 1897. *Handbuch d. Mineralogie*, vol. ii, p. 1714 (Analcidit). The more correct derivation of analcite, from ἄναλκις, ἀνάλκιδος, weak. 4th List

Anandite. D. B. Pattiaratchi, E. Saari, and Th. G. Sahama, 1967. *Min. Mag.*, vol. 36, p. 1; J. F. Lovering and J. R. Widdowson, ibid., 1968, vol. 36, p. 871. A black pseudohexagonal monoclinic mineral from the Wilagedera iron ore body, North-Western Province, Ceylon is related to the brittle micas, but contains S^{2-}, having a composition near $Ba_2(Fe^{2+},Mg)_6(Si,Fe^{3+})_8(O,OH,S)_{24}$. Named for Ananda Coomaraswamy. [*A.M.*, **52**, 1586; Зап., **97**, 77; *Bull.* **90**, 603.] 25th List

Anapaite. A. Sachs, *Sitz.-ber. Akad. Wiss. Berlin*, 1902, p. 18; *Zeits. angew. Chem.*, 1902, vol. xv, p. 111; *80er Jahres-Ber. Schlesisch. Ges.*, for 1902, 1903, Abt. II, Sect. *a*, p. 3 (Anapait). J. Loczka, *Zeits. Kryst. Min.*, 1903, vol. xxxvii, p. 438. Pale green, glassy crystals on limonite, from Anapa, Black Sea. Anorthic. $FeCa_2(PO_4)_2.4H_2O$. Differs from messelite only in containing a little more water. *See* tamanite. 3rd List

Anarakite. D. Adib and J. Ottemann, 1972. *Neues Jahrb. Min.*, Monatsh. 335. Green crystals from the Kali-Kafi ore deposit, Anarak, Iran, with the composition $(Cu,Zn)_2(OH)_3Cl$, are regarded as a zincian variety of a fourth polymorph, additional to atacamite, paratacamite, and botallackite, and indexed on a monoclinic cell. Named for the locality. [The powder data are identical with those of paratacamite and can be indexed on the rhombohedral cell of the latter: the mineral is possibly a zincian paratacamite—*M.H.H.*] [*M.A.*, 73-1934; *Bull.*, **96**, 245; *A.M.*, **58**, 560; *Zap.*, **102**, 445.] 28th List

Anchi-zeolite. W. E. Richmond, 1937. *Amer. Min.*, vol. 22, p. 291 (anchi-zeolite). A term to include such minerals as prehnite, datolite, babingtonite, etc., formed during a late hydrothermal phase of igneous activity, before the zeolite phase. Named from ἄγχι, near, and zeolite. [*M.A.*, **7**, 118.] 15th List

Ancylite. G. Flink, *Medd. om Grönland*, 1900, XXIV, 49. Small yellow crystals of octahedral habit with curved faces. Rhombic. $4Ce(OH)CO_3.3SrCO_3.3H_2O$. [Narsarsuk] South Greenland. 2nd List [Contains Ca, *Min. Record* **2**, 18]

Andeclase. A. N. Winchell, 1925. *Journ. Geol. Chicago*, vol. 33, p. 726. P. Niggli, *Lehrbuch Min.*, 1926, vol. 2, p. 536 (Andeklas). A contraction of andesine-oligoclase for felspars of the plagioclase series ranging in composition from $Ab_{70}An_{30}$ to $Ab_{60}An_{40}$. 11th List

Andersonite. J. Axelrod, F. Grimaldi, C. Milton, and K. J. Murata, 1948. Program and Abstracts, *Min. Soc. Amer.*, 29th Annual Meeting, p. 4; *Bull. Geol. Soc. Amer.*, vol. 59, p. 1310; *Amer. Min.*, 1949, vol. 34, p. 274. Hydrous sodium-calcium-uranyl carbonate, $Na_2CaUO_2(CO_3)_3.nH_2O$, from Arizona. Named after Charles Alfred Anderson (1902–) of the United States Geological Survey. [*M.A.*, **10**, 452.] 18th List [*A.M.*, **36**, 1]

Andorite. J. A. Krenner, *Math. és term.-tud. Ertesito*, XI. 119, 1892. *Min. Mag.*, **11**, 286. $2PbS.Ag_2S.3Sb_2S_3$. Rhombic. [Felsöbánya], Hungary. 1st List [*See* Sundtite]

Andreattite. R. C. Mackenzie, 1954. *Anal. Edafol. y Fisiol. Veget.*, Madrid, vol. 13, p. 122 (andreattita); *Potassium-Symposium*, Berne, 1955, p. 130 (andreattite). The clay minerals illidromica, illite-hydromica of C. Andreatta (19th List) are divided into: hydromica-Al, hydromica-Mg, a layered series between illite and hydromuscovite (or hydrobiotite); and andreattite-Al, andreattite-Mg, a non-interstratified series between illite (or mica) and smectite. Named after Prof. Ciro Andreatta (1906–) of Bologna. 21st List

Andremeyerite. Th. G. Sahama, J. Siivola, and P. Rehtijärvi, 1973. *Bull. Geol. Soc. Finland*, **45**, 1. Pale emerald-green monoclinic crystals (*a* 7·464, *b* 13·794, *c* 7·093 Å, β 118° 15′) in vesicles of lava from Mt. Nyiragongo, Zaïre; $4[BaFe_2Si_2O_7]$. Named for André Meyer. [*A.M.*, **59**, 381.] 28th List

Anemolite. J. Barnes and W. F. Holroyd, *Trans. Manchester Geol. Soc.*, 1896, vol. xxiv, p. 232 (anemolites). A curved and upturned form of stalactite of calcium carbonate from the Derbyshire limestone caves. Named from ἄνεμος, wind, and λίθος, stone, because of the influence of varying currents of air in determining the direction which the stalactites take. 3rd List

Anemousite. H. S. Washington and F. E. Wright, 1910. *Amer. Journ. Sci.*, ser. 4, vol. xxix, p. 61. A plagioclase-felspar found, together with linosite (q.v.), as loose crystals in a volcanic tuff on the Island of Linosa, off the coast of Tunis. The silica percentage is rather lower than that required by the albite-anorthite series, and the composition is explained as a mixture of albite and anorthite together with soda-anorthite (carnegieite, q.v.) in the ratio 8:10:1. Named from the ancient Greek name of the island. 5th List

Angaralite. A. Meister, 1910. *Explorations géologiques dans les régions aurifères de la Sibérie, Région aurifère d'Iénisséi*, Livraison IX, 1910, p. 506 (ангаралитъ), p. 667 (Angaralite); abstract in *Zeits. Kryst. Min.*, 1914, vol. liii, p. 597 (*Angaralith*). Silicate of aluminium, iron, and magnesium, $2MgO.5(Al,Fe)_2O_3.6SiO_2$, occurring as thin, black plates (optically uniaxial and positive, and perhaps hexagonal) in metamorphosed limestone near the Angara river, Siberia. (*See* Tatarkaite.) 7th List [A chlorite near ripidolite; *A.M.*, **50**, 2111]

Angelardite. A. Lacroix, *Minéralogie de la France*, 1910, vol. iv, p. 524. The correct form of anglarite (of F. von Kobell, 1831); so named after the locality, Angelard (not Anglar), in the commune of Compreignac, Haute-Vienne, France. A blue, hydrated phosphate of iron usually identified with vivianite, but regarded by Lacroix as representing a distinct species, near to ludlamite. 6th List

Angelellite. P. Ramdohr, F. Ahlfeld, and F. Berndt, 1959. *Neues Jahrb. Min.*, Monatsh., p. 145 (Angelellit); *see also* K. Weber, ibid., p. 152. Dark brown anorthic crystals in the fissures of a fumarole on Cerro Pululus, north-west Argentina, appear to be $Fe_4As_2O_{11}$. Named after Dr. V. Angelelli, Director of the Geol. Survey of Argentina. [*M.A.*, **14**, 343.] 22nd List [*A.M.*, **44**, 1322]

Angleso-barite. *See* Hokutolite. 6th List

Angolite. E. Breusing, *Neues Jahrb. Min.* 1900, *Beil.-Bd.* XIII, 265. Breithaupt's manganocalcite from Schemnitz was shown by Des Cloizeaux to be a mixture of carbonates with an anorthic hydrated silicate of manganese. To the latter the name *Angolith* is now given, and the composition assigned to it is $H_2Mn_3(SiO_3)_4 + H_2O$. It appears to be a zeolite. 2nd List

Anhydrobiotite, Anhydromuscovite. F. Rinne, 1925. *Zeits. Krist.*, vol. 61, p. 122 (Anhydrobiotit, Anhydromuscovit). Artificially dehydrated biotite and muscovite. Cf. Meta-. [*M.A.*, vol. 2, p. 505.] 10th List

Anhydrokainite. [E. Jänecke, 1913. Kali, *Zeits. Kalisalze*, Halle a. S., vol. 7, p. 140.] M. Rózsa, *Centralblatt Min.*, 1916, p. 510 (Anhydrokainit). Anhydrous chloride and sulphate of potassium and magnesium, $KMgClSO_4$, produced by the dehydration of kainite by the intrusion of basalt into the Prussian salt-deposits. M. Rózsa terms it also Basaltkainit. 9th List

Anhydrokaolin. R. Schwarz and G. Trageser, 1932. *Chemie der Erde*, vol. 7, p. 583 (Anhydrokaolin). Artificially dehydrated kaolin which passes with change in volume to metakaolin (10th List). [*M.A.*, **5**, 361.] 13th List

Anhydromuscovite, *see* Anhydrobiotite.

Anhydrosaponite. A. Weiss, G. Koch, and U. Hofmann, 1955. *Ber. Deutsch. Keram. Gesell.*, vol. 32, pp. 16–17 (Anhydrosaponit). Artificially dehydrated saponite (from Grosschlattengrün, Bavaria). An endothermal peak at 600°C. is attributed to loss of water from the OH group, with further change at 800° to enstatite + Al_2O_3. This is criticized by R. C. Mackenzie, *Min. Mag.*, 1957, vol. 31, pp. 676–678. 21st List

Anilite. N. Morimoto, K. Koto, and Y. Shimazaki, 1969. *Amer. Min.* **54**, 1256. Orthorhombic Cu_7S_4, occurring as an intergrowth with djurleïte at the Ani mine, Akita, Japan, in prismatic or platy crystals closely resembling chalcosine; has also been obtained synthetically. Inverts to a digenite-type structure on grinding. Named for the locality. [*M.A.*, 70-1640.] 26th List

Ankoleite, syn. of Meta-ankoleïte (24th List) (I. Kostov., *Mineralogy*, 1968, p. 476). 25th List

Ankylite. G. Flink, *Meddelelser om Grönland*, 1901, Hefte xxiv, plate III. [The reprint of Flink's paper is dated 1899, but was not issued till 1900; Hefte xxiv is dated 1901.] The same as ancylite. 3rd List

Anophorite. W. Freudenberg, 1908. *Mitt. Badisch, geol. Landesanst.*, 1908, vol. vi, p. 45 (Anophorit). A variety of alkali-hornblende from the shonkinite of the Katzenbuckel, Baden. It resembles catophorite in its pleochroism, but differs in chemical composition (containing more MgO and less FeO) and optical orientation. Named from ἀνώφορος, ascending, because the angle of optical extinction is less than that of catophorite. 5th List [Titanian calcian magnesio-arfvedsonite, *Amph.* (1978)]

Anorthite-haüyne. L. H. Borgström, 1930. *Zeits. Krist.*, vol. 74, p. 119 (Anorthithauyn). A hypothetical molecule, $3CaAl_2Si_2O_8.2CaSO_4$, corresponding with haüyne ($3Na_2Al_2Si_2O_8.2CaSO_4$), to explain the composition of the sodalite group. Part of this molecule has the composition of anorthite, hence the name. [*M.A.*, **4**, 355.] 12th List

Anorthoclase-sanidine. *See* Sanidine-anorthoclase. 12th List

Anosovite. K. Kh. Tagirov, 1951. A. A. Rusakov and G. S. Zhdanov, *Doklady Acad. Sci. USSR*, 1951, vol. 77, p. 411 (Аносовит); D. S. Belyankin and V. V. Lapin, ibid., 1951, vol. 80, p. 421. Titanium oxide Ti_3O_5 as black orthorhombic crystals and in blast-furnace slag. Named after the Russian metallurgist P. P. Anosov. [*M.A.*, **11**, 415, 536.] 19th List

Anperthite, anplagioclases, *see* Analbite (of Laves).

Ansilit. *Zeits. Kryst. Min.*, 1906, vol. xli, p. 184. Error for ancylite, owing to the name having been transliterated into Russian and back again. 4th List

Antamokite. A. D. Alvir, 1928. *Engin. & Mining Journ. New York*, 1928, vol. 125, p. 616; *Amer. Min.*, 1928, vol. 13, p. 491; *Philippine Journ. Sci.*, 1930, vol. 41, p. 137. Telluride of gold with trace of silver, differing from calaverite in its micro-chemical reactions. Named from the locality, [Benguet mine], Antamok, Mountain Province, Philippine Islands. [*M.A.*, **4**, 250.] 12th List [A mixture; *M.A.*, **10**, 183]

Antarcticite. T. Torii and J. Ossaka, [1965]. *Science*, vol. 149, p. 975. Acicular aggregates crystallizing from the brine of Don Juan Pond, Victoria Land, Antarctica, prove to be the well-known hexagonal salt $CaCl_2.6H_2O$. Named for the locality. [*A.M.*, **50**, 2098.] 24th List

Antarkticite, variant of Antarcticite (24th List) (C. Hintze, *Handb. Min.*, Erg. III, p. 520). 25th List

Anthodite. N. Kashima, 1965. [*Mem. Ehime Univ.*, Sect. 2, Ser. D, **5**, 79]; abstr. *M.A.*, **18**, 282. 'Flower-like dripstone'; an alternation of calcite and aragonite. 28th List

Anthoinite. N. Varlamoff, 1947. *Ann. (Bull.) Soc. Géol. Belgique*, vol. 70, p. B 153. Hydrous aluminium tungstate, $Al_2O_3.2WO_3.3H_2O$, as white chalky material, from Belgian Congo. Named after Raymond Anthoine, mining geologist of Bruxelles. [*M.A.*, **10**, 145, 495.] 18th List [*A.M.*, **43**, 384]

Anthonyite. S. A. Williams, 1963. *Amer. Min.*, vol. 48, p. 614. Monoclinic, lavender-coloured pleochroic crystals from the Centennial mine, Calumet, Michigan, have the composition $Cu(OH,Cl)_2.3H_2O$ with $OH \gg Cl$. The mineral is unstable in dry air. Named for Prof. John W. Anthony. [*M.A.*, **16**, 54.] (Cf. Calumetite.) 23rd List

Anthracene. R. Rost, 1935. *Věda Přirodní*, Praha, vol. 16, p. 204 (antracen). An organic compound ($C_{14}H_{10}$) found by the burning of pyritous shale in Bohemia. [*M.A.*, **6**, 357.] 14th List

Antiglaucophane. W. N. Lodochnikov, 1933. *Problems of Soviet Geology*, vol. 2, p. 132 (антиглаукофан), p. 148 (antiglaucophane). Like glaucophane but differing (ἀντι-) somewhat in its optical characters. Compare Pseudoglaucophane (11th List). 13th List

Antimon-luzonite. S. Stevanović, *Zeits. Kryst. Min.*, 1903, vol. xxxvii, p. 235 (Antimon-Luzonit, Stibio-Luzonit). A massive mineral, with a reddish colour and absence of cleavage, from Peru. In composition, $Cu_3(As,Sb)S_4$, it is intermediate between luzonite and famatinite. It is identical with the 'famatinite' from Chile analysed by A. Frenzel in 1875. 3rd List

Antimonpearceite. C. Frondel, 1963. *Amer. Min.*, vol. 48, p. 565. $(Ag,Cu)_{16}(Sb,As)_2S_{11}$, the antimony end-member corresponding to polybasite, and members of this series with $Sb > As$. Dimorphous with polybasite, which has a unit-cell eight times as large. [*M.A.*, **16**, 546.] 23rd List [But *see A.M.*, **50**, 1507]

Antimonpyrochlore. F. Machatschki, 1932. *Chemie der Erde*, vol. 7, pp. 67, 75 (Antimonpyrochlor). The pyrochlore-roméite group of minerals with the same type of cubic structure and the general formula $X_2Z_2(O,OH,F)_7$, where X = Na, Ca, Ce, etc., and Z = Nb, Ta, Ti, Sb, includes the chemical varieties niobpyrochlore (pyrochlore proper), tantalpyrochlore (microlite), niobtantalpyrochlore (neotantalite), antimonpyrochlore (roméite, atopite, schneebergite, weslienite), Bleiantimonpyrochlor (monimolite?), titanantimonpyrochlore (mauzeliite, lewisite), and a hypothetical titanpyrochlore. [*M.A.*, **5**, 185.] 13th List [I.M.A. review of pyrochlore group; *A.M.*, **62**, 403–410]

Antimonwesterveldite. *Zap.*, **105**, 5, contents. Erroneous transliteration of сурьмянистий вестервелдит, antimonian westerveldite. 30th List

Antiperthite. F. E. Suess, 1905. *Jahrb. geol. Reichsanst. Wien*, vol, liv, pp. 419, 425 (Antiperthite, *pl.*). Regular intergrowths of two felspars in which orthoclase is the enclosed mineral and plagioclase the host; being the reverse of the case which holds in microperthite. 4th List

Antofagastite. C. Palache and W. F. Foshag, 1938. *Amer. Min.*, vol. 23, p. 85; M. C. Bandy, ibid., p. 705. Hydrated copper chloride, $CuCl_2.2H_2O$, as bluish-green orthorhombic crystals from province Antofagasta, Chile. Named from the locality. [*M.A.*, **7**, 59, 223.] 15th List [Is Eriochalcite; *M.M.*, **29**, 44]

Antunesit, *see* Antunit.

Antunit, Antunesit, spelling variants of Antunezite (H. Strunz, *Min. Tabellen*, 3rd edit., 1957, p. 354). 24th List

Aplowite. J. L. Jambor and R. W. Boyle, 1965. *Canad. Min.*, vol. 8, p. 166. Bright pink, fine grained, with moorhouseite (*q.v.*) as an efflorescence on sulphides at the Magnet Cove Barium Corporation mine, Walton, Nova Scotia, have the composition $(Co, Mn, Ni)SO_4.4H_2O$ with Co:Mn:Ni 2:1:1, monoclinic and isomorphous with ilesite and rozenite. The name is given for A. P. Low, and defined to include all members of the series with Co as principal cation. [*A.M.*, **50**, 809.] 24th List

Apoanalcite. C. Oftedahl, 1947. *Norsk Geol. Tidsskrift*, vol. 26, p. 215. H. Neumann, ibid., 1949, vol. 27, p. 171. Hydrated sodium aluminium silicate $NaAl(Al,Si)SiO_6.1\frac{1}{2}H_2O$, as red masses in syenite-pegmatite from Nordmarka, Norway; possibly an alteration product of analcime. [*M.A.*, **10**, 145, 510.] 18th List [Is largely Natrolite; *A.M.*, **39**, 406]

Apricotine. (D. B. Sterrett, *Mineral Resources, US Geol. Survey*, for 1909, 1911, part ii, p. 802; W. T. Schaller, ibid., for 1917, 1918, part ii, p. 148.) Trade-name for yellowish-red, apricot-coloured quartz pebbles from near Cape May, New Jersey, used as gem-stones. 8th List

Aquacrepit, *error for* Aquacreptit (Hintze, *Handb. Min.*, Erg.-Bd. II, p. 659). 22nd List

Arakawaite. Y. Wakabayashi and K. Komada, 1921. *Journ. Geol. Soc. Tōkyō*, vol. 28, p. 211. Hydrated copper-zinc phosphate, $4CuO.2ZnO.P_2O_5.6\frac{1}{2}H_2O$, occurring as bluish-green, monoclinic crystals in the Arakawa mine, Japan. Named from the locality. Differs from Veszelyite in containing no arsenic. [*M.A.*, **1**, 250.] 9th List [Cf. Kipushite]

Aramayoite. L. J. Spencer, 1926. *Min. Mag.*, vol. 21, p. 156. E. Kittl, *Revista Minera de Bolivia*, 1927, vol. 2, p. 53 (Aramayoita). Sulphantimonite and sulphobismuthite of silver, $Ag_2S.(Sb,Bi^*)_2S_3 = Ag(Sb,Bi)S_2$, as black pseudo-tetragonal plates with perfect basal cleavage and steep pyramidal cleavages. X-ray examination (K. Yardley, *Min. Mag.*, vol. 21, p. 163) proves it to be triclinic. From the Animas mine of the Compagnie Aramayo de Mines en Bolivie. Named after Señor Don Felix Avelino Aramayo, formerly Managing Director of the Company. [*M.A.*, **3**, 269.] 11th List [*A.M.*, **12**, 265] [*Corr. in *M.M.*, **22**]

Arandisite. F. C. Partridge, 1930. *Trans. Geol. Soc. S. Africa*, vol. 32 (for 1929), p. 171. T. W. Gevers, ibid., p. 169. Hydrous basic silicate of tin, $3SnSiO_4.2SnO_2.4H_2O$, as apple-green masses with cassiterite and quartz from near Arandis, Swakopmund district, South-West Africa. Named from the locality. [*M.A.*, **4**, 248.] 12th List [? A mixture; *A.M.*, **15**, 274]

Archerite. P. J. Bridge, 1977. *M.M.*, **41**, 33. Tetragonal crystals of $(K,NH_4)H_2PO_4$ occur in the Petrogale Cave, near Madura Motel (31°54′S, 127°0′E), Western Australia. Formed from bat guano. Sp. gr. 2·23, ω 1·513, ε 1·470. Named for M. Archer. [*M.A.*, 77-2182; *A.M.*, **63**, 593.] 30th List

Arcubisite. S. Karup-Møller, 1976. *Lithos*, **9**, 253. Rare grains with galena in the Ivigtut cryolite deposit, Greenland, have composition $BiCuAg_6S_4$. Presumably named for the composition, *Ar*gentum, *Cu*prum, *Bis*mutum. [*M.A.*, 78-2116; *A.M.*, **63**, 424.] 30th List

Ardealite. J. Schadler, 1931. *In* F. Halla, *Zeits. Krist.*, vol. 80, p. 349; J. Schadler, *Centr. Min.*, Abt. A, 1932, p. 40 (Ardealit). Hydrated double salt of calcium sulphate and acid phosphate, $CaHPO_4.CaSO_4.4H_2O$, isomorphous with brushite ($CaHPO_4.2H_2O$) and gypsum ($CaSO_4.2H_2O$), but not a mechanical mixture of these minerals; occurring in phosphate deposits in a cave in Transylvania, the old Romanian name for which is Ardeal. Named from the locality. [*M.A.*, **5**, 49.] 13th List [*A.M.*, **17**, 251]

Ardmorite. (H. S. Spence, *Mines Branch*, Canada, 1924, no. 626, p. 14.) Trade-name for a bentonite clay from Ardmore, South Dakota. Other names of a similar character are 'Refinite', 'Wilkinite', etc. *See* Amargosite, Otaylite. 11th List

Arduinite. E. Billows, 1912. Two pamphlets both dated Padova, 1912, and entitled 'Analisi di alcuni minerali del Veneto, Nota I, Arduinite, un nuovo minerale', but one of 11 pp. and the other of 14 pp. One of them is stated to be an extract from *Riv. Min. Crist. Ital.*, vol. xli, but the paper does not appear in that or in earlier volumes of that periodical. A red, radially fibrous zeolite, $H_{16}Na_4CaAl_2Si_8O_{30}$, from Val dei Zuccanti, Venetia. Named after the Venetian geologist Giovanni Arduino (1714–1795). 6th List [Is mordenite; *M.A.*, **11**, 293]

Argental. D. R. Hudson, 1943. *Metallurgia*, Manchester, vol. 29, p. 56. Syn. of moschellandsbergite and landsbergite (17th List), erroneously taking the adjective from 'mercure argental' (L. Cordier, 1802). [*M.A.*, **9**, 55.] 18th List

Argentoaikinite, Argentocosalite, Argentocuprocosalite, Argentogoongarrite, Argentolillianite. A. A. Godovikov, 1972, abstr. *Zap.*, **103**, 617 (1974) (Аргентоайкинит, Аргентокозалит, Аргентокупрокоапит, Аргентогунгаррит, Аргентолиллианит). Unnecessary, invalid names based solely on literature analyses. 29th List

Argentoalgodonite. G. A. Koenig, 1903. *Proc. Amer. Phil. Soc.*, vol. xlii, p. 229; *Zeits. Kryst. Min.*, 1904, vol. xxxviii, p. 537. An artificially prepared copper arsenide containing some silver $(Cu,Ag)_6As$. 4th List

Argentocuproauride. L. V. Razin and K. V. Yurkina, 1971. [Геол. рудн. мин. no. 1, 93 (Аргентокупроаурид)], abstr. *Zap.*, **102**, 437. A superfluous name for an argentian variety of cuproauride (15th List). 28th List

Argentocuproaurite. L. V. Razin, 1975. *Trudy Mineral. Muz. akad. nauk SSSR*, **24**, 93 (Аргентокупроаурит). Two grains, one from Noril'sk and one from Talnakh, have compositions $Au_{50}Cu_{21}Ag_{17}Rh_6Pd_6$ and $Au_{47}Ag_{26}Cu_{24}Pd_3$. Indexed as primitive cubic (eight lines), *a* 4·073 Å. [Can be indexed as face-centred cubic . . . except for the weakest reflection at 1·673 Å. Confirmation by single-crystal methods . . . is necessary to justify characterization as a new mineral species—*L.J.C.* Only two grains, of appreciably differing composition, but essentially cuprian electrum—*M.H.H.*] [*A.M.*, **62**, 593.] 30th List

Argentodomeykite. G. A. Koenig, 1903. *Proc. Amer. Phil. Soc.*, vol. xlii, p. 229; *Zeits. Kryst. Min.*, 1904, vol. xxxviii, p. 537. An artificially prepared copper arsenide containing some silver $(Cu,Ag)_3As$. 4th List

Argentojarosite. W. T. Schaller, 1923. *Journ. Washington Acad. Sci.*, 1923 (June), vol. 13, p. 233. C. A. Schempp, *Amer. Journ. Sci.*, 1923 (July), ser. 5, vol. 6, p. 73 (Argento-jarosite). A mineral from Utah resembling jarosite, but with silver in place of potassium, $Ag_2O.3Fe_2O_3.4SO_3.6$*H_2O. [*M.A.*, **2**, 148.] [*Corr. in *M.M.*, **23**.] 10th List [*A.M.*, **8**, 230]

Argentopercylite. [H. O. Schulze.] (*Chem.-Zeit.* 1892, XVI, 1952.) The same as boleite. 2nd List

Argento-Perrylit, original erroneous spelling of Argentopercylite (2nd List). *Chemiker Zeitung*, 1893, vol. 16, p. 1952. 23rd List

Argyrojodit. T. Barth and G. Lunde, 1926. *Zeits. Physik. Chem.*, vol. 122, p. 308. Error for Iodargyrite = Iodyrite. 11th List

Aristarainite. C. S. Hurlbut and R. C. Erd, 1974. *A.M.*, **59**, 647. Small monoclinic crystals, *a* 18·869, *b* 7·531, *c* 7·810, *β* 97° 43′, space group $P2_1/a$, in a matrix of borax and kernite from the Tincalayu borax deposit, Salta, Argentina; composition $2[Na_2MgB_{12}O_{20}.10H_2O]$. Named for Dr. L. F. Aristarian. [*M.A.*, 75-547; *Bull.*, **97**, 497; *Zap.*, **104**, 609.] 29th List

Arizonite. C. Palmer, 1909. *Amer. Journ. Sci.*, ser. 4, vol. xxviii, p. 353. Ferric meta-titanate, $Fe_2O_3.3TiO_2$, found as irregular masses, with sub-conchoidal fracture, dark steel-grey colour, and metallic to sub-metallic lustre, in a pegmatite-vein near Hackberry in Arizona. It is probably monoclinic, and is distinct from ilmenite (ferrous titanate, $FeO.TiO_2$). 5th List [A mixture; *A.M.*, **35**, 117]

—— H. G. Hanks, before 1878. [Hinton, *Handbook to Arizona*, 1878], quoted in A. L. Flagg, *Mineralogical journeys in Arizona* (Scottsdale, Arizona), 1958, p. 23. Name for 'a type of ore, discovered in Yavapai county . . . on a claim known at the time as the Sumner'. A mixture: 'The principal vein matter is micaceous iron, iodide of silver, gold, sulphurets of iron and antimony.' Not the Arizonite of C. Palmer, 1909 (5th List). 22nd List

Arkelite. W. B. Blumenthal, 1958. *Chem. Behavior of Zirconium.* Princeton, 1962. The cubic phase of ZrO_2 (cf. *M.A.*, **7**, 131). 23rd List

Armalcolite. A. T. Anderson, T. E. Bunch, E. N. Cameron, S. E. Haggerty, F. R. Boyd, O. B. James, K. Keil, M. Prinz, P. Ramdohr, and A. El Goresy, 1970. *Proc. Apollo XI Lunar Sci. Conf.* **1**, 55. A member of the pseudobrookite family, $4[(Mg,Fe)Ti_2O_5]$, with $Mg \approx Fe^{2+}$, intermediate between $Fe^{2+}Ti_2O_5$ and karrooite, occurring in lunar rocks from Tranquillity Base. *a* 9·743, *b* 10·02, *c* 3·74 Å. Named for N. A. Armstrong, E. E. Aldrin, and M. Collins. [*A.M.*, **55**, 2136; *M.A.*, 71-1383; *Zap.*, **100**, 618.] 27th List

Armangite. G. Aminoff and R. Mauzelius, 1920. *Geol. För. Förh. Stockholm*, vol. 42, p. 301. Ortho-arsenite of manganese, $Mn_3(AsO_3)_2$, occurring as black, rhombohedral crystals at Långban, Sweden. Named from the chemical composition. [*M.A.*, **1**, 124.] 9th List [*A.M.*, **6**, 64]

Armenite. H. Neumann, 1939. *Norsk Geol. Tidsskrift*, vol. 19, p. 312 (Armenite). Hydrated alumino-silicate of calcium and barium, $BaCa_2Al_6Si_8O_{28}.2H_2O$, as colourless pseudo-hexagonal (orthorhombic?) crystals from Armen mine, Kongsberg, Norway. Named from the locality. [*M.A.*, **7**, 468.] 15th List [*A.M.*, **26**, 235]

Armstrongite. N. V. Vladykin, V. I. Kovalenko, A. A. Kashaev, A. N. Sapozhnikov, and V. A. Pisarskaya, 1973. Докл. **209**, 1185 (Армстронгит). A brown monoclinic mineral (*a* 14·04, *b* 14·16, *c* 7·81 Å, *β* 109° 33′) from a pegmatite in the Khan–Bogdinskii massif is near $CaZr(Si_6O_{15}).2{\cdot}5H_2O$. Named for N. Armstrong. [*M.A.*, 74-504; *A.M.*, **59**, 208; *Zap.*, **103**, 364.] 28th List

Aromite. G. Mueller, 1964. *Rept. 22nd Internat. Geol. Congr.*, India, part 1, p. 46. An insoluble, infusible bitumen yielding aromatic hydrocarbons on pyrolysis. Not to be confused with aromite of Darapsky, 1900, a doubtful sulphate of Mg and Al. 26th List [Darapsky, 1890; *M.M.*, **40**]

Arrojadite. D. Guimarães, 1925. *Publicação da Inspectoria de Obras Contra as Seccas*, Rio de

Janeiro, 1925, no. 58, p. 1 (Arrojadita). J. B. de A. Ferraz, *Bull. Soc. Franç. Min.*, 1927, vol. 50, p. 16 (arrojadite). Phosphate of sodium, iron, manganese, etc., $4R'_3PO_4.9R''_3P_2O_8$, as dark green masses (monoclinic) from Brazil. It agrees with a mineral from Black Hills, South Dakota, regarded by W. P. Headden in 1891 as a new mineral near triphylite. Named after the Brazilian geologist, Dr. Miguel Arrojado Lisbôa. [*M.A.*, **3**, 113.] 11th List [$(Na,K,Ca)_2(Fe,Mn)_5(PO_4)_4$; *A.M.*, **35**, 70]

Arschinowit, *see* Arshinovite. 22nd List

Arsenatapatit. Synonym of svabite. *See* Bleiapatit. 20th List

Arsenatbelowit, German transliteration of Арсенат-беловит (arsenate-belovite, 21st List) (Hintze, *Handb. Min.*, 1959, Erg.-Bd. II, p. 660). 22nd List

Arsenate-belovite. L. K. Yakhontova* and G. A. Siderenko, 1956. *Mém. Soc Russ. Min.*, vol. 85, p. 297 (арсенат-беловит). The two minerals, belovite (E. I. Nefedov, 1953) and belovite (L. S. Borodin and M. E. Kazakova, 1954) (20th List), are distinguished as arsenate-belovite and phosphate-belovite (фосфат-беловит) respectively. [*M.A.*, **13**, 523.] [*Corr. in *M.M.*, **31**.] 21st List [*A.M.*, **42**, 583; **50**, 813. Is Talmessite (q.v.)]

Arsenbrackebuschite. W. Hofmeister and E. Tilmanns, 1976. *Fortschr. Mineral.* **54**, *Beib.* **1**, 38. K. Schmetzer and W. Berdesinski, *Neues Jahrb. Min.*, Monatsh. 193 (1978). Small platelets from Tsumeb, SW Africa, and earthy aggregates from the Clara mine, Oberwolfach, Schwarzwald, are monoclinic, *a* 7·764, *b* 6·045, *c* 9·022 Å, *β* 112·5°, space group $P2_1/m$. Composition [$Pb_2(Fe^3,Zn)(AsO_4)_2(OH,H_2O)$] RIs $\alpha < 2{\cdot}04 < \gamma$. The arsenate analogue of brackebuschite, named accordingly. 30th List

Arsendestinezite. J. Kratochvíl, 1958. *Topograf. Min. Čzech.* **2**, 136. $Fe_2^{3+}AsO_4SO_4(OH).nH_2O$, from Kaňk, Kutná Hora, Czechoslovakia, regarded as the arsenic analogue of destinezite; now renamed bukovskýite (q.v.). [*A.M.*, **54**, 994.] 26th List

α-Arsenic sulphide. A. H. Clark, 1970. *Amer. Min.* **55**, 1338 (Alpha-arsenic sulfide). The high-temperature modification of AsS occurs at the Alacrán mine, Pampa Larga, Chile. It is very similar to realgar (*β*-AsS), but gives a different X-ray powder pattern. [*M.A.*, 71-548.] 27th List

Arsenioardennite. F. Zambonini, 1922. *Rend. R. Accad. Lincei, Cl. Sci. Fis.*, Roma, ser. 5, vol. 31, sem. 1, p. 151. The end-members of the ardennite series are distinguished as arsenioardennite and vanadioardennite according to the predominance of arsenic or vanadium. [*M.A.*, **2**, 44.] 10th List

Arsenobismite. A. H. Means, 1916. *Amer. Journ. Sci.*, ser. 4, vol. xli, p. 127 (Arseno-Bismite). A hydrated bismuth arsenate, $2Bi_2O_3.As_2O_5.2H_2O$, forming yellowish-green crystalline aggregates in a mixed friable ore from the Mammoth mine, Tintic district, Utah. This formula is deduced from a very unsatisfactory analysis, and the characters as given are not sufficient to distinguish the supposed new mineral from atelestite and rhagite. The name, which is suggested by the chemical composition, is not intended to imply an arsenical variety of bismite. 8th List [A valid species; *A.M.*, **28**, 536]

Arsenoestibio. F. Pardillo, 1947. *Tratado de mineralogía*, translation of 12th edit. of F. Klockmann and P. Ramdohr, Barcelona, p. 315; G. Gagarin and J. R. Cuomo, 1949, loc. cit., p. 5. Spanish form of stibarsen (16th List). *See* Wretbladite. 19th List

Arsenoferrite. H. Baumhauer, 1912. *Zeits. Kryst. Min.*, vol. li, p. 143 (Arsenoferrit). Cubic-dyakisdodecahedral iron arsenide, $FeAs_2$, isomorphous with iron-pyrites; but represented only by pseudomorphous crystals in the gneiss of the Binnenthal, Switzerland. 6th List [Almost certainly Cafarsite (q.v.) *Schweiz. Min. Pet. Mitt.*, **46**, 373]

Arsenoklasite. G. Aminoff, 1931. *Kungl. Svenska Vetenskapsakad. Handl.*, ser. 3, vol. 9, no. 5, p. 52 (Arsenoklasite). Basic arsenate of manganese, $Mn_3(AsO_4)_2.2Mn(OH)_2$, as red orthorhombic cleavages from Långban, Sweden. Named from ἀρσενικόν, orpiment, and κλάσις, breaking. [*M.A.*, **4**, 496.] 12th List [*A.M.*, **17**, 25. Arsenoclasite in Dana's Syst. Min., 7th edit., and in Fleischer's *Gloss.*]

Arsenomarcasite. W. T. Schaller, 1930. *Amer. Min.*, vol. 15, p. 567. Suggested as a more suitable name for arsenopyrite, the mineral being orthorhombic and not cubic. Synonym of mispickel. 12th List

Arsenomiargyrite. F. M. Jaeger and H. S. van Klooster, 1912. *Zeits. Anorg. Chem.*, vol. lxxviii, p. 265 (Arsenomiargyrit). The artificially produced sulpharsenite of silver, $AgAsS_2$, corresponding with the sulphantimonite miargyrite. No doubt identical with the mineral smithite (R. H. Solly, 1905; 4th List). 6th List

Arsenopalladinite. F. A. Bannister, G. F. Claringbull, and M. H. Hey, 1955. M. H. Hey. *Index of Minerals* (*Brit. Mus.*), 1955, pp. 23, 339. G. F. Claringbull and M. H. Hey, *Min. Soc. Notice*, 1956, no. 94; *M.A.*, **13**, 237. Palladium arsenide, Pd_3As, hexagonal. With gold from Itabira, Brazil. Named from the composition. 21st List [$Pd_5(As,Sb)_2$, *M.M.*, **39**, 529]

Arsenostibite. P. Quensel, 1937. *Geol. För. Förh. Stockholm*, vol. 59, p. 148 (arsenostibite). Variant of arsenstibite (G. J. Adam, 1869) for an amorphous alteration product of allemontite consisting of hydrated oxides of arsenic and antimony. [*M.A.*, **6**, 487.] 14th List [Is arsenian Stibiconite; *A.M.*, **37**, 982]

Arsenosulvanite. A. G. Betekhtin, 1941. *Mém. Soc. Russ. Min.*, vol. 70, p. 161 (арсеносульванит). V. I. Mikheev*, ibid., p. 165. Isomorphous with sulvanite (2nd List) with vanadium largely replaced by arsenic, $Cu_3(As,V)S_4$, cubic; from Mongolia. [*M.A.*, **12**, 460.] 20th List [*A.M.*, **40**, 368. Cf. Lazarevićite] [*Corr. in *M.M.*, **40**]

Arsenothorite. E. M. Bonshtedt-Kupletskaya, 1961. Зап. Всесоюз. Мин. Общ. (*Mem. All-Union Min. Soc.*), vol. 90, p. 108 (Арсеноторит, arsenothorite). Translation of the Chinese name Shen-t'u-shih (Shentulite, q.v.). 22nd List

Arsenpolybasit. S. L. Penfield, *Zeits. Kryst. Min.* 1896, XXVII, 66. Arsenical polybasite = pearceite. 2nd List

Arsenpolybasite. C. Frondel, 1963. *Amer. Min.*, vol. 48, p. 565. $(Ag,Cu)_{16}(As,Sb)_2S_{11}$, the arsenic end-member corresponding to polybasite, and members of this series with As > Sb. Dimorphous with pearceite, with a unit-cell eight times as large. Not to be confused with Arsenpolybasit of Penfield (1896; 2nd List), a synonym of Pearceite. [*M.A.*, **16**, 546.] 23rd List [*A.M.*, **50**, 1507]

Arsen-rösslerite. O. M. Friedrich and J. Robitsch, 1939. *Zentr. Min.*, Abt. A, 1939, p. 143 (Arsen-Rößlerit). Synonym of rösslerite ($MgHAsO_4.7H_2O$), which is used as a group name to include also phosphor-rösslerite (q.v.). [*M.A.*, **7**, 317.] 15th List

Arsen-stibiconite. H. Strunz, 1957. *Min. Tabellen*, 3rd edit., p. 146. (Arsen-Stibiconit). Syn. of Arsenstibite. 22nd List

Arsenstruvite. (Author?). H. Strunz, *Min. Tab.*, 1949, p. 172 (Arsenstruvit). $NH_4MgAsO_4.6H_2O$, like struvite with As in place of P. Presumably an artifical product. 20th List [Also Arsen-Struvit, *Min. Tab.*, 3rd edit., 22nd List]

Arsensulfurite. F. Rinne, Centralblatt Min., 1902, p. 499 (Arsensulfurit). Amorphous sulphur containing much arsenic (29·22%). A volcanic sublimation from Java. 3rd List

Arsensulvanit, *error for* Arsenosulvanit (20th List) (Hintze, *Handb. Min.*, Erg.-Bd. II, p. 660). 22nd List

Arsentsumebite. *Bull. Soc. franç. Min.*, 1935, vol. 58, p. 4 (arsentsumébite). A variety of tsumebite (6th List) containing some arsenate in place of phosphate of lead and copper, from Tsumeb, South-West Africa. 14th List

—— L. Vésignié, 1935. *Bull.*, **58**, 4 (14th List). This material, supposed to be the arsenate analogue of tsumebite (6th List) was shown by C. Guillemin (*Bull.*, **79**, 15 and 71 (1956)) to be duftite. The arsenate analogue of tsumebite has since been found by R. A. Bideaux, M. C. Nichols, and S. A. Williams (*A.M.*, **51**, 258 (1966)), and the name is revived (Arsentsumebit) for this new material by K. F. Chudoba (Hintze, *Erg.*, **4**, 5 (1974)). 28th List

Arsenuranocircite. H. Strunz, 1957. *Min. Tabellen*, 3rd edit., p. 253 (Arsen-Uranocircit). Name proposed by Strunz for the arsenic analogue of uranocircite, then unknown in nature. L. N. Belova, *2nd UN Internat. Conf. Peaceful Uses Atomic Energy*, 1958, vol. 2, p. 294, proposed the same name for a natural barium uranyl arsenate from a pitchblende-molybdenum deposit (locality not stated), but this mineral belongs to the meta- series: 'it should have been called meta-arsenuranocircite' (M. Fleischer, *Amer. Min.*, 1959, vol. 44, p. 466). [*M.A.*, **14**, 199, 344.] Synonym of Heinrichite [Metaheinrichite] (q.v.); it is not clear which name has priority as applied to a natural mineral. 22nd List [Heinrichite seems to be tacitly adopted]

Arsenuranospathite. K. Walenta, 1963. *Jb. geol. Landesamt Baden-Württemberg*, **6**, 113 (Arsen-Uranospathit). The arsenic analogue of uranospathite occurs in small amounts at

Menzenschwand and Wittichen, Schwarzwald, Germany. Tetragonal, *a* 7·16, *c* 30·37 Å, space group $P4_2n$, 2·54 g.cm^{-3}, *β* 1·538, *γ* 1·542, $2V_\alpha$ 52°. Composition $[HAl(UO_2)^4$-$(AsO_4)_4.40H_2O]$ *M.M.*, **42**, 117. Named from the composition. [*M.A.*, 78-2117] 30th List

Arsenuranylite. L. N. Belova, 1958. Зап. Всесоюз. Мин. Общ. (*Trans. All-Union Min. Soc.*), vol. 87, p. 598 (Арсенуранилит). The arsenate analogue of phosphuranylite, which it closely resembles except for a deeper orange colour. Formula $Ca(UO_2)_4(AsO_4)_2(OH)_4.6H_2O$. Locality not stated. Named from the composition and in analogy with phosphuranylite. [*M.A.*, **14**, 282, 344.] 22nd List [*A.M.*, **44**, 208]

Arsenvanadinite. H. Strunz, *Min. Tab.*, 1941, p. 156 (Arsenvanadinit). Syn. of endlichite. 18th List

Arshinovite. E. G. Razumnaya, G. A. Smelyanskaya, K. G. Korolev, and G. V. Pokulnis, 1957. [Мет. иссл. мин. сырья (*Methods of study of raw materials*),Gosgeoltekhizdat, p. 45]; abstr. in Зап. Всесоюз. Мин. Общ. (*Mem. All-Union Min. Soc.*), 1958, vol. 87, p. 486 (аршиновит, arshinovite); Hintze, *Handb. Min.*, Erg.-Bd. II, 1959, p. 660 (Arschinowit). Yet another name for a partially metamict zircon. [*M.A.*, **14**, 277, 345; *Amer. Min.*, **44**, 210.] 22nd List

Arthurite. R. J. Davis and M. H. Hey, 1964. *Min. Mag.*, vol. 33, p. 937. Thin apple-green crusts on quartz from Hingston Down Consols mine, Calstock, Cornwall, consist of an intimate mixture of pharmacosiderite and $Cu_2Fe_4(AsO_4)_3(OH)_7.6H_2O$. Named for Sir Arthur Russell and Mr. Arthur W. G. Kingsbury. 23rd List [$Cu_2Fe_4\{(As,P,S)O_4\}_4(O,OH)_4.8H_2O$; *M.M.*, **37**, 520]

Artinite. L. Brugnatelli, *Rend. Ist. Lombardo*, 1902, ser. 2, vol. xxxv, p. 874; *Centralblatt Min.*, 1903, p. 144. A loose aggregate of snow-white tufts* which are composed of minute prismatic crystals [*see* L. Brugnatelli, *Centr. Min.*, 1903, p. 663]*. The optical characters agree with orthorhombic symmetry. In composition, $MgCO_3.Mg(OH)_2.3H_2O$, near to hydrogiobertite, which, however, is shown to be a mixture of at least two minerals. In peridotite from the Val Lanterna, Val Tellina. Named after Prof. Ettore Artini, of Milan. [*Corr. in *M.M.*, **14**] 3rd List [Monoclinic]

Arzrunite. A. Dannenberg, *Zeits. Kryst. Min.* 1899, XXXI, 230; Abstr. *Min. Mag.* XII, 308. A bright blue-green crystalline crust. $PbSO_4.PbO + 3(CuCl_2.H_2O) + Cu(OH)_2$. Rhombic [Buena Esperanza mine, Challacollo, Tarapaca] Chili. 2nd List [Named for A. Arzruni (1847–1898)]

Asbecasite. S. Graeser, 1966. *Schweiz. Min. Petr. Mitt.*, vol. 46, p. 367; *see also Urner Mineralien Freund*, 1966, Jahrb. 4, heft 4. Yellow rhombohedral crystals in clefts of orthogneisses of Monte Leone, Binnental, Switzerland, have a composition near $Ca_3Be(Ti,Sn)Si_2As_6O_{19}$ [the formula given in the publications cited does not agree well with the cell-dimensions and density—*M.H.H.*]. [*A.M.*, **52**, 1584; Зап., **97**, 75.] 25th List [Loc. Mt Cherbadung. $Ca_3(Ti,Sn)As_6Si_2Be_2O_{20}$, *Lincei-Rend. Sci. fis. mat. e nat.*, **46**, 457]

Asbophite. N. D. Sobolev, 1930. [*Min. Syre*, Moscow, vol. 5, no. 9, p. 1181.] V. N. Lodochnikov, *Trans. Centr. Geol. Prosp. Inst. USSR*, 1936, no. 38, pp. 34, 517, 604 (асбофит), p. 729 (Asbophite). O. M. Shubnikova, *Trans. Lomonosov Inst. Acad. Sci. USSR, Ser. Min.*, 1937, no. 10, p. 195. A variety of chrysotile-asbestos, $42H_4(Mg,Fe)Mg_2Si_2O_9 + (Al,Fe)_2O_3 + 8SiO_2$; sp. gr. 2·5847, *α* 1·557, *γ* 1·569. A contraction of asbestos + ophite. 15th List

Ascharite. W. Feit, *Chemiker-Zeitung*, XV. 327, 1891. $3Mg_2B_2O_5.2H_2O$. Aschersleben, Prussia. [*latin* Ascharia] 1st List [Identical with szájbelyite; *M.A.*, **9**, 123; 74–480; *A.M.*, **49**, 224]

Ashcroftine. M. H. Hey and F. A. Bannister, 1932. *Nature*, London, vol. 130, p. 858. *Min. Mag.*, 1933, vol. 23, p. 305. A zeolite of the composition $NaKCaAl_4Si_5O_{18}.8H_2O$, previously described as kalithomsonite (10th List), but shown to be tetragonal and to have no relation to thomsonite. From Greenland. Named after Mr. Fredrick Noel Ashcroft, of London. 13th List [Not a zeolite; $KNaCaYt_2Si_6O_{12}(OH)_{18}.4H_2O$; *M.M.*, **37**, 515]

Ashtonite. E. Poitevin, 1932. *Amer. Min.*, vol. 17, p. 120. A zeolite related to ptilolite, but with 2RO instead of 1RO, the formula being $2(Ca,Na_2)O.Al_2O_3.9SiO_2.5H_2O$. As radiating masses from British Columbia. Named after the Honourable Wesley Ashton Gordon, Minister of Mines for Canada. [*M.A.*, **5**, 50.] 13th List [Is a strontian mordenite; *M.M.*, **38**, 383]

Askanite. D. S. Belyankin, 1934. *Trav. Inst. Pétrogr. Acad. Sci. URSS*, no. 6, p. 110 (Асканит), p. 114 (askanite). O. M. Shubnikova, *Trans. Lomonosov Inst. Acad. Sci. USSR*,

1937, no. 10, 201 (Асканит, Ascanite). D. S. Belyankin and V. P. Ivanova, *Compt. Rend. (Doklady) Acad. Sci. URSS*, 1938, vol. 18, p. 279 (Askanite). A montmorillonite-like clay occurring as a decomposition product of volcanic ash. Named from the locality, Askana (Аскана), Ozurgety district, Georgia, Transcaucasia. [*M.A.*, **7**, 426.] 15th List

Asoproit. Germ. trans. of Азопроит azoproite (26th List). Hintze, 6. 28th List

Asphaltite. W. P. Blake, 1890. *Trans. Amer. Inst. Mining Engin.*, vol. xviii, p. 576. G. H. Eldridge, *22nd Ann. Rep. United States Geol. Survey, for 1900–1*, 1901, part i, p. 220. The same as asphalt or asphaltum with the mineralogical termination *ite*. Includes the solid bituminous hydrocarbons known as albertite, impsonite (q.v.), gilsonite, grahamite, nigrite, and uintahite. The word asphaltite appears in Greek and in English dictionaries as an adjective, meaning bituminous or asphaltic. 5th List

Astrapia. John Walker, 1781. *Schediasma Fossilium, in usus Academicos Edinburgi*, p. 13. Radiating baryte from the Isle of Sheppey, Kent. 23rd List

Astridite. H. W. V. Willems, 1934. *De Ingenieur in Nederlandsch-Indië*, vol. 1, part 4 (Mijnbouw en Geologie), p. 120 (Astridiet). An ornamental stone consisting mainly of chromojadeite from New Guinea. Named after Astrid (1905–1935), Queen of Belgium. [*M.A.*, **6**, 502.] 14th List

Astrolite. R. Reinisch, 1904. *Centralblatt Min.*, 1904, p. 108 (Astrolith). Small, greenish-yellow spheres with radially fibrous structure, which occur embedded in fragments of carbonaceous quartz-schist, limestone, and shale in a basalt-tuff at Neumark, Saxon Vogtland. $(Al,Fe)_2Fe''(Na,K)_2(SiO_3)_5.H_2O$. Named from ἄστρον, stars, and λίθος, stone; known locally as 'Sternle'. 4th List [Is muscovite; *A.M.*, **57**, 993]

As-tsumebite. L. Fanfani and P. F. Zanazzi, 1967. *M.M.*, **36**, 522. A provisional name for the arsenate analogue of tsumebite described by R. Bideaux *et al.* (*A.M.*, **51**, 258 (1966)). *Syn.* of arsentsumebite (q.v.). 28th List

Athabascaite. D. C. Harris, L. J. Cabri, and S. Kaiman, 1970. *Canad. Min.* **10**, 207. Lath-shaped grains included in or replacing umangite, and stringers in carbonate vein material, at the Martin Lake mine, Uranium City, Lake Athabasca, Saskatchewan, are orthorhombic, *a* 8·227, *b* 11·982, *c* 6·441 Å. Composition $4[Cu_5Se_4]$. Named for the locality. [*A.M.*, **55**, 1444; **56**, 632; *M.A.*, 71–2328; *Zap.*, **100**, 617.] 27th List

Atheneïte. A. M. Clark, A. J. Criddle, and E. E. Fejer, 1974. *M.M.*, **39**, 528. A few grains in the arsenopalladinite concentrates from Itabira, Minas Gerais, Brazil, are hexagonal, *a* 6·798, *c* 3·483 Å, with composition $2[(Pd,Hg)_3As]$. This material was supposed to be arsenopalladinite (which is in fact anorthic), and the cell-dimensions given by Claringbull and Hey (*M.A.*, **13**, 237) refer to atheneïte. Named, in oblique reference to its palladium content, from Pallas Athēnē. [*M.A.*, 74-1444.] 28th List

Atokite. P. Mihalik, S. A. Hiemstra, and J. P. R. de Villiers, 1975. *C.M.*, **13**, 146. Pd_3Sn_{1-x}; *see* Rustenbergite for details. 29th List

Attapulgite. J. de Lapparent, 1935. *Compt. Rend. Acad. Sci. Paris*, 1935, vol. 201, p. 483; 1936, vol. 202, p. 1728; 1936, vol. 203, pp. 482, 596. A variety of fuller's earth containing alumina and magnesia, $x[9SiO_2.6(Mg,Fe'',Ca)O.12H_2O] + y[9SiO_2.2(Al,Fe''')_2O_3.12H_2O]$, from Attapulgus, Georgia, USA. P. F. Kerr (*Amer. Min.*, 1937, vol. 22, p. 534) identifies it with montmorillonite. [*M.A.*, **6**, 150, 346, 413.] 14th List [Shown to be palygorskite; *X-ray identification and structures of clay minerals*, London, 1951, p. 234]

Attapulgite-palygorskite. A. Haji-Vassiliou and J. H. Puffer, 1975. *A.M.*, **60**, 328. Syn. of Palygorskite. [*A.M.*, **60**, 1132.] 29th List

Auricupride. P. Ramdohr, 1950. *Fortschr. Min.*, vol. 28 (for 1949), p. 69 (Auricuprid). Synonym of cuproauride (15th List). [*M.A.*, **11**, 312.] 19th List [Auricupride is the accepted name, Fleischer's *Gloss.*]

Aurobismuthinite. G. A. Koenig, 1912. *Journ. Acad. Nat. Sci. Philadelphia*, ser. 2, vol. xv, p. 423. A lead-grey, massive, cleavable mineral from unknown locality, consisting mainly of bismuth sulphide with Au 12·27% and Ag 2·32%; formula, $(Bi,Au,Ag_2)_5S_6$. This may represent a mixture of $(Bi,Au,Ag_2)S$, or possibly of a gold–silver alloy, and bismuthinite (Bi_2S_3). *See* Stibiobismuthinite. 7th List

Aurocuproite. L. V. Razin, 1975. *Trudy Mineral. Muz. akad. nauk SSSR*, **24**, 93 (Аурокупроит). Two grains from Noril'sk are formulated $Cu_{54}Au_{38}Pd_8$ and $Cu_{53}Au_{39}Pd_8$.

Indexed as primitive cubic (eight lines), *a* 3·862 Å. [. . . may be indexed as face-centered cubic . . . except for the weakest reflection at 1·070 Å. Confirmation by single-crystal methods . . . is necessary to justify characterization as a new species—*L.J.C.* Essentially auroan copper—*M.H.H.*] [*A.M.*, **62**, 593] 30th List

Aurorite. A. S. Radtke, C. M. Taylor, and D. F. Hewett, 1967. *Econ. Geol.*, vol. 62, p. 186. Minute grains in calcite from the Aurora mine, Hamilton, Nevada, give X-ray powder data near that of chalcophanite. Their composition is close to $(Mn,Ag,Ca)Mn_3O_7.3H_2O$; the name aurorite, from the locality, is proposed for the Mn^{2+} analogue of chalcophanite, the natural material being an argentian aurorite. [*A.M.*, **52**, 1581; Зап., **97**, 70; *Bull.*, **91**, 97.] 25th List

Aurosirita. G. Gagarin and J. R. Cuomo, 1949, loc. cit., p. 4. Contraction in Spanish form of aurosmirid, aurosmiridium (14th and 17th Lists). 19th List

Aurosmirid. O. E. Zvyagintsev, 1934. *Doklady Akad. Nauk CCCP* (*Compt. Rend. Acad. Sci. URSS*), vol. 4, p. 178 (ауросмирид), p. 179 (Aurosmirid). Solid solution of gold and osmium in cubic iridium (as distinct from a solid solution of iridium, etc., in hexagonal osmium). Found as silver-white grains in residues of Uralian platinum insoluble in aqua regia. A contraction of aurum-osmiumiridium. [*M.A.*, **6**, 51.] 14th List

Aurosmiridium. Dana's *System of mineralogy*, 1944, 7th edit., vol. 1, p. 111. Another spelling of aurosmirid (14th List). 17th List

Aurostibite. A. R. Graham and S. Kaiman, 1951. *Program & Abstracts, Min. Soc. Amer.*, Nov. 1951, p. 15; *Bull. Geol. Soc. Amer.*, 1951, vol. 62, p. 1444; *Amer. Min.*, 1952, vol. 37, pp. 292, 461. $AuSb_2$, cubic with pyrite structure, as minute grains in gold ores from Canada. [*M.A.*, **11**, 414.] 19th List

Austinite. L. W. Staples, 1935. *Amer. Min.*, vol. 20, pp. 112, 199. Basic arsenate of calcium and zinc, $CaZn(OH)AsO_4$, as colourless bisphenoidal orthorhombic crystals, from Utah. Identical with brickerite (13th List), F. Ahlfeld and R. Mosebach, *Zentr. Min.*, Abt. A, 1936, p. 289. Not to be confused with austenite (F. Osmond, 1895), a constituent of manufactured steel. Named after Professor Austin Flint Rogers (1877–), of Stanford University, California. [*M.A.*, **6**, 53, 345, 384.] 14th List

Auxite. E. W. Hilgard, 1916. *Proc. National Acad. Sci. U.S.A.*, vol. ii, p. 11. A name suggested, but discarded in favour of Lucianite (q.v.). From αὔξω, to increase. 8th List

Avelinoite. M. L. Lindberg and W. T. Pecora, 1954. *Science* (*Amer. Assoc. Adv. Sci.*), vol. 120, p. 1074. Hydrous sodium ferric phosphate, $NaFe_3'''(PO_4)_2(OH)_4.2H_2O$, tetragonal, from Brazil. Named after Avelino Ignacio de Oliveira, Brazilian geologist. [*M.A.*, **12**, 512.] 20th List [Is Cyrilovite; *A.M.*, **42**, 586]

Avicennite. Kh. N. Karpova, E. A. Konkova, E. D. Larkin, and V. F. Saveliev, 1958. [Доклады Акад. Наук Узб. ССР (*Compt. Rend. Acad. Sci. Uzbek SSR*), no. 2, p. 23]; abstr. in Зап. Всесоюз. Мин. Общ. (*Mem. All-Union Min. Soc.*), 1959, vol. 88, p. 319 (Авиценнит, Avicennite). Minute black crystals; cubic, essentially Tl_2O_3; provisional formula $7Tl_2O_3.Fe_2O_3$, with *a* 9·12 Å and $z = 4$ [but artificial Tl_2O_3 has only 16 Tl_2O_3 per unit cell]. From Dzhuzumli, Mt. Zirabulaksk region, Bukhara. Named after the alchemist Avicenna (Abu Ali Ibn Sina), who lived in Bukhara, Tadzhikistan. [*M.A.*, **14**, 278.] 22nd List [*A.M.*, **44**, 1324]

Avogadrite. F. Zambonini, 1926. *Atti* (*Rend.*) *R. Accad. Lincei, Roma, Cl. Sci. fis. mat. nat.*, ser. 6, vol. 3, p. 644. G. Carobbi, ibid., 1926, vol. 4, p. 382. Potassium fluoborate, KBF_4, containing up to 9·5% $CsBF_4$, occurring as minute orthorhombic crystals of the baryte type in a mixed saline sublimation on Vesuvian lava. Named after the Italian physicist, Amedeo Avogadro (1776–1856). [*M.A.*, **3**, 238.] 11th List

Ayasite. J. D. Buddhue, 1939. *Popular Astronomy*, Northfield, Minnesota, vol. 47, p. 97. The names rustite and ayasite are suggested for the rust or iron-shale formed by the oxidation of meteoritic irons. Named from the Sanskrit *ayas*, iron. [*M.A.*, **7**, 266.] 15th List

Ayatite. A. K. Gladskovskii and I. N. Ushatinskii, 1961. [Труды Горно-геол. инст. Уральск. фил. Акад. наук СССР (*Proc. Mining-geol. Inst. Ural Div. Acad. Sci. USSR*), vol. 56, p. 114], abstr. Зап. Всесоюз. Мин. Общ. (*Mem. All-Union Min. Soc.*), 1966, vol. 95, p. 311 (Аятит). Merely finely dispersed Corundum. 24th List

Azoproite. A. A. Konev, V. S. Lebedeva, A. A. Kashaev, and Z. F. Ushchapovskaya, 1970.

Зап. Всесоюз. Мин. Общ. (*Mem. All-Union Min. Soc.*), **99**, 225. (Азоироит). ($Mg_{7\cdot27}Fe^{2+}_{0\cdot53}Mn_{0\cdot01})(Fe^{3+}_{1\cdot49}Ti^{4+}_{1\cdot43}Mg^{2+}_{1\cdot00})B_{4\cdot06}O_{19\cdot99}$* orthorhombic, from Tazheran; a highly titanian member of the ludwigite group. The ideal formula is written $4MgO.(TiO_2,MgO).B_2O_3$. [*M.A.*, 70-3432.] [*Corr. in *M.M.*, **39**.] 26th List

Azovskite. N. E. Efremov, 1937. *Trans. Lomonosov Inst. Acad. Sci. USSR*, no. 10, p. 151 (Азовскит), p. 154 (Asovskite), p. 155 (azovskite). Hydrated ferric phosphate, $FePO_4.2Fe(OH)_3.3H_2O$, as dark brown masses with pitchy lustre in the iron ore on the Taman shore of the Sea of Azov (Азовское Море). Named from the locality. [*M.A.*, **7**, 59.] 15th List

Azurchalcedony. G. F. Kunz, 1907. *See* Azurlite. 5th List

Azurite. An objectionable trade name for a sky-blue smithsonite. R. Webster, *Gems*, 1962, p. 753. (Cf. L. J. Spencer, *A.M.*, **22**, 683.) 28th List

Azurlite or Azurchalcedony. G. F. Kunz, 1907. [*New York Acad. Sci.*, April, 1907] quoted in *Mineral Industry*, New York, for 1907, 1908, vol. xvi, p. 792. Chalcedony coloured blue by chrysocolla, from Arizona; used as a gem-stone. 5th List

Azurmalachite. G. F. Kunz, 1907. *Engin. and Mining Journ. New York*, vol. lxxxiv, p. 296; *Mineral Industry*, New York, for 1907, 1908, vol. xvi, p. 792. D. B. Sterrett, *Mineral Resources of the United States*, for 1907, 1908, part ii, p. 797; for 1908, 1909, part ii, p. 809. A popular term for a mixture of azurite and malachite in concentric bands; used as a gem-stone. From Arizona. 5th List

B

Ba-priderite, *see* Barium priderite.

Bababudanite. W. F. Smeeth, [1911?]. *Rec. Mysore Geol. Dept.*, for 1907–1908, vol. ix, p. 85. A soda-amphibole, $2NaFe'''(SiO_3)_2.Fe''Mg_3(SiO_3)_4$, closely allied to riebeckite, but differing from this in the slightly greater angle of optical extinction (a:c′ = 7–9°) and in the character of the pleochroism (*a* Prussian blue, *b* purple tending to violet, *c* yellow with tinge of green). Occurs with cummingtonite in quartz-magnetite-schists in the Bababudan Hills, Kadur district, Mysore. 6th List [Syn. of Magnesio-riebeckite, *Amph.* 1978.]

Babeffit, a misleading though exact transliteration of Бабеффит, Babefphite (= Ba-Be-F-Phosphate) (C. Hintze, *Handb. Min.*, Erg. III, p. 524). 25th List

Babefphite. A. S. Nazarova, N. N. Kuznetsova, and D. P. Shaskin, 1966. Доклады акад. наук СССР (*Compt. Rend. Acad. Sci. URSS*), vol. 167, p. 895 (Бабеффит). Tetragonal $Ba_4Be_5(PO_4)_4OF_4.0{\cdot}3$ to $0{\cdot}4$ H_2O, occurring in an unnamed rare-metal-fluorite deposit in Siberia. Named from the composition (BaBeFPh[osphorus]). [*M.A.*, **18**, 48; *A.M.*, **51**, 1547; Зап , **97**, 72; Bull., **90**, 115.] 25th List

Babepfite. Variant transliteration of Бабеффит, Babefphite (25th List), (the two *φ*–s are for fluorine and phosphorus, F and P. *Dokl. Acad. Sci. USSR* (*Earth Sci. Sect.*), 1967, **167**, 152. 28th List

Babylonian quartz. (P. Groth, *Mineraliensammlung*, Strassburg, 1878, p. 100 (Babylonquarz). *Century Dictionary*, New York, 1889.) Synonym of babel-quartz. 4th List

Bacalite. J. D. Buddhue, 1935. *Rocks and Minerals*, Peekskill, NY, vol. 10, p. 171. A variety of amber from Baja (Lower) California, Mexico. Named from the locality in a contracted form. (Not to be confused with the artificial product bakelite.) [*M.A.*, **6**, 503.] 14th List

Bäckströmite. G. Aminoff, 1919. *Geol. För. Förh. Stockholm*, vol. 41, p. 473 (Bäckströmit). Abstract in *Amer. Min.*, 1920, vol. 5, p. 88 gives the form Baeckstroemite. An orthorhombic modification of manganous hydroxide, $Mn(OH)_2$, dimorphous with the rhombohedral pyrochroite. The two modifications occur together, often in regular intergrowth, at Långban, Sweden. The material examined had, however, not only been transformed into pyrochroite, but had also suffered oxidation by exposure to the air, the manganese being mainly in the form of sesquioxide. The orthorhombic crystals therefore represent double pseudomorphs. Named after Professor Helge Bäckström, of Stockholm. Earlier referred to as pseudopyrochroite (q.v.). 9th List [*M.A.*, **1**, 3] [Is Hydrohausmannite (q.v.), *A.M.*, **38**, 762]

Bacillarite. H. Strunz, *Min. Tab.*, 1941, p. 227 (Bacillarit), given as a synonym of leverrierite, A. Orlov, *Tsch. Min. Petr. Mitt.*, 1942, vol. 54, p. 225 (Bacillarites). An ill-defined clay mineral with bacillary structure, originally described by C. Feistmantel in 1869 as a fossil 'Bacillarities problematicus'. [*M.A.*, **10**, 24, 29.] 18th List

Baddeckite. G. C. Hoffmann, *Amer. J. Sci.* 1898, VI, 274; *Rept. Geol. Survey, Canada, for 1896*, 1898, IX, R 11. A copper-red, highly ferruginous variety of muscovite occurring as scales in clay near Baddeck, Nova Scotia. 2nd List [A mixture; *A.M.*, **11**, 5]

Baddeleyite. W. B. Clarke, 1882, in A. Liversidge, *Description of the minerals of New South Wales* (Sydney) p. 116. Syn. of ilvaite. Not the well-known baddeleyite of Fletcher (1892; 1st List). 30th List

Baddeleyite. L. Fletcher, *Nature*, XLVI. 620, 1892; XLVII. 283, 1892; *Min. Mag.* X. 148, 1893. E. Hussak, *Tsch. Min. Mitth.* XIV. 395, 1895. [*Min. Mag.*, **11**, 110.] ZrO_2. Monosymmetric. Ceylon and Brazil. 1st List

Badenite. P. Poni, *Ann. Sci. Univ. Jassy*, 1900, vol. i, p. 29; *Anal. Acad. Române*, Bukarest, 1900, vol. xxii; (abstract, this vol. [*Min. Mag.*, **13**], p. 207). A massive steel-grey mineral which appears to be smaltite (or safflorite) with part of the arsenic replaced by bismuth (4·76%); $(Co,Ni,Fe)_2(As,Bi)_3$. From Badeni, Roumania. 3rd List [Doubtful validity]

Baeckstroemite. *See* Bäckströmite. 9th List

Baeumlerite. O. Renner, 1912. *Centralblatt Min.*, 1912, p. 106 (Baeumlerit). A colourless, highly deliquescent, and optically biaxial salt, $KCl.CaCl_2$, occurring in the Prussian salt-deposits. Named after — Baeumler, General-Director of the Heldburg Salt Company. F. Zambonini (*Centralblatt Min.*, 1912, 270) suggests that this is identical with the cubic chlorocalcite (of A. Scacchi, 1872) from Vesuvius. 6th List

Bafertisite. Peng Ch'i-Jui, 1959. [*Ti-chih K'o-hseuh*, vol. 10, p. 289]; abstr. in *Amer. Min.*, 1960, vol. 45, p. 754. Also in [*Sci. Record* (*Pekin*), vol. 3, p. 652]; abstr. in *Amer. Min.*, 1960, vol. 45, p. 1317, and in Зап. Всесоюз. Мин. Общ. (*Mem. All-Union Min. Soc.*), vol. 90, p. 105 (Бафертисит). $BaFe_2TiSi_2O_9$, in orthorhombic crystals distinct from taramellite. Named from the composition, Ba–Fe(r)–Ti–Si. 22nd List [$Ba(Fe,Mn)_2TiSi_2O_7(O,OH)_2$; *A.M.*, **57**, 1005]

Bagotite. [T. Egleston, *Catalogue of Minerals*, (1887), 1889, p. 192; A. H. Chester, *Dictionary of the Names of Minerals*, 1896, p. 25; *Student's Index to the British Museum Collection of Minerals*, since 1885]. Green pebbles from Bagot, Ontario, have been in the British Museum since 1882; these are labelled bagotite, and are identified with the lintonite variety of thomsonite. Molybdenite comes from the same locality. 1st List

Bahianite. P. B. Moore and T. Araki, 1976. *Neues Jahrb. Min. Abh.* **126**, 113. Tan to cream-coloured favas from Paramirim, Agua Quente, Bahia, Brazil, are often vuggy. Monoclinic, *a* 9·406, *b* 11·54, *c* 4·410 Å, β 90·94°, space group *C2/m*. Composition $2[Sb_3Al_5O_{14}(OH)_2]$, with some Be. α 1·81, β 1·87, γ 1·92. Named from the locality. 30th List

Baikovite. A. V. Rudneva, 1958. [Акад. Наук СССР, Инст. Геол. Руд. Месторожд. Петрог., Мин., Геохим., Инст. Хим. Силикат, 1958, p. 285]; abstr. in *Amer. Min.*, 1959, vol. 44, p. 907; Hintze, *Handb. Min.*, 1960, Erg.-Bd. II, p. 925 (Baikowit). The artificial spinel $MgTi_2O_4$. Named after A. A. Baikov. [*M.A.*, **15**, 45.] 22nd List

Bakerite. W. B. Giles, *Min. Mag.*, 1903, vol. xiii, p. 353. A massive borosilicate of calcium, $8CaO.5B_2O_3.6SiO_2.6H_2O$, from the borax mines of California. Named after Mr. R. C. Baker, of Nutfield (Surrey), by whom the mineral was found. 3rd List [Monoclinic; *Amer. Min.*, **41**, 689; *M.A.*, 76–3619]

Balavinskite. Ya. Ya. Yarzhemski, 1966. [Акад. наук СССР, Сибир. отдел. Всес. науч.-иссл. инст. Галургии. Изд. 'Наука']; abstr. *Amer. Min.* **54**, 575 (1969). Only a formula, $Sr_2B_6O_{11}.4H_2O$, and refractive indices are cited. [*Bull.*, **92**, 517.] 26th List

Balawinskit, Germ. trans. of Балавинскит, balavinskite (26th List). Hintze, 6. 28th List

Balchaschit, Balkaschit, German transliterations of Балхашит, Balkhashite. 24th List

Baldaufite. F. Müllbauer, 1925. *Zeits. Krist.*, vol. 61, p. 334 (Baldaufit). Hydrated pohosphate of ferrous iron, etc. $(Fe,Mn,Ca,Mg_3(PO_4)_2.3H_2O$, as flesh-red, monoclinic crystals resembling wenzelite (q.v.), from Hagendorf, Bavaria. Named after Dr. Richard Baldauf, of Dresden, who possesses the only specimen of the mineral yet found. [*M.A.*, **2**, 418.] 10th List [Is Hureaulite; *M.M.*, **20**, 447]

Balipholite. Anon., 1975. [*Sci. Geol. Sinica*, 100], abstr. *M.A.*, 75-3590. $BaMg_2LiAl_3Si_4O_{12}(OH)_8$, space group *Ccca*, *a* 13·00, *b* 20·24, *c* 5·16 Å, from an unnamed locality in China. Named from the composition. [*A.M.*, **61**, 338.] 29th List

Balkanite. V. A. Atanassov and G. N. Kirov, 1973. *A.M.*, **58**, 11. A steel-grey mineral from the Sedmochislenitsi mine, Bulgaria, with tennantite, bornite, chalcopyrite, etc., is orthorhombic, *a* 10·62, *b* 9·42, *c* 3·92 Å. Composition $[Cu_9Ag_5HgS_8]$, checked by synthesis. Named for the locality. [*M.A.*, 73-288; *Zap.*, **103**, 355.] 28th List

Balkhashite. S. V. Kumpan, *Bull. Geol. Prosp. Service USSR*, 1931, vol. 50, no. 7, p. 95 (балхашит), p. 100 (balkhashite). G. L. Stadnikov and Z. Vosdshinskaya, *Brennstoff-Chemie*, 1930, vol. 11, p. 414; G. L. Stadnikov, ibid., 1933, vol. 14, p. 227 (Balchaschit). A rubbery bitumen (elaterite) similar to coorongite formed by algae in lake Balkhash (Балхаш), Siberia. 13th List [Var. of Coorongite; *M.A.*, 69-2188]

Bambollaite. D. C. Harris and E. W. Nuffield, 1972. *C.M.*, **11**, 738. Microcrystalline aggregates with klockmannite and tellurites at the Bambolla mine, Moctezuma, Sonora, Mexico, are tetragonal, *a* 3·865, *c* 5·632 Å; composition $Cu(SeTe)_2$. Named from the locality. [*Bull.*, **96**, 234; *A.M.*, **58**, 805; *Zap.*, **102**, 443; *M.A.*, 74-1445.] Name first used by R. V. Gaines, *Min. Record*, 1970, **1**, 41, and the erroneous form bombollalite by I. Kostov, *Mineralogy*, 1968, p. 150. 28th List

Banalsite. W. C. Smith, F. A. Bannister, and M. H. Hey, 1944, *Min. Mag.*, vol. 27, p. 33; *Nature, London*, vol. 154, p. 336. W. C. Smith, *Min. Mag.*, 1945, vol. 27, p. 63. A barium-felspar with sodium, $BaNa_2Al_4Si_4O_{16}$, as orthorhombic crystals from Benallt mine, Wales. Named from the chemical formula. 17th List

Bandylite. C. Palache and W. F. Foshag, 1938. *Amer. Min.*, vol. 23, p. 87; M. C. Bandy, ibid., p. 704. Hydrated borate-chloride of copper, $CuB_2O_4.CuCl_2.4H_2O$, as dark blue tetragonal crystals from Chile. Named after Mark Chance Bandy, mining engineer, who collected the mineral. [*M.A.*, **7**, 59, 223.] 15th List

Bannisterite. M. L. Smith and C. Frondel, 1968. *Min. Mag.*, vol. 36, p. 893. The 'ganophyllite' from Benallt, Carnarvonshire (Campbell Smith, 1948) and from Franklin, New Jersey, includes two distinct minerals: ganophyllite, identical with the original material from the Harstig mine, Pajsberg, Sweden, and a very similar mineral with a smaller unit cell, different optical orientation, and composition near $(Na,K,Ca)(Mn,Fe^{2+},Zn,Mg)_8(Si,Al)_{14}O_{28}(OH)_{16}$. Named for F. A. Bannister. [*M.A.*, **19**, 314.] 25th List

Baotite. Peng Ch'i-Jui, 1959. [*Ti-chih K'o-hsueh*, vol. 10, p. 289]; abstr. in *Amer. Min.*, 1960, vol. 45, p. 750 (Pao-t'ou-k'uang). V. I. Simonov, Кристаллография, 1960, vol. 5, p. 542 (Баотит), with reference to E. I. Semenov and Huan Wen-Sin, *Ti-chih K'o-hsueh*, vol. 10. E. I. Semenov, Huan Wen-Sin, and T. A. Kapitinova, Доклады Акад. Наук СССР, 1961, vol. 136, p. 915 (Баотит, p. 758, baotite). Silicate of Ba, Ti, and Nb, with chloride; tetragonal. Found at Paotow (Баотоу), Inner Mongolia; named from the locality. 22nd List [$Ba_4(Ti,Nb)_8ClO_{16}Si_4O_{12}$; *A.M.*, **46**, 466]

Bararite. Dana's *Mineralogy*, 7th edit., 1951, vol. 2, p. 106. Hexagonal fluosilicate of ammonium $(NH_4)_2SiF_6$, occurring, with cryptohalite (cubic $(NH_4)_2SiF_6$), over a burning coal seam at Barari, Jharia coalfield, India. [*M.A.*, **3**, 308; **6**, 42.] 19th List

Baratovite. V. D. Dusmatov, E. I. Semenov, A. P. Khomyakov, A. V. Bykova, and N. Kh. Dzhafarov, 1975. *Zap.*, **104**, 580 (Баратовит). Plates in quartz–albite–aegirine veins in the Dara-Pioz massif, Tadzhikistan, are monoclinic, *a* 16·90, *b* 9·73, *c* 20·91 Å, β 112° 30′; cleavage {001}; $\alpha \approx \beta$ 1·672, γ 1·673, 2Vγ 60°; α 55° to the normal to (001). Composition $4[(K,Na)Ca_8Li_2(Ti,Zr)_2Si_{12}O_{37}F]$. Named for R. B. Baratov. 29th List

Barbertonite. C. Frondel, 1940. *Amer. Min., Program and Abstracts*, December 1940, p. 6; *Amer. Min.*, 1941, vol. 26, pp. 196, 311. Hydrous basic carbonate of magnesium and chromium, $Cr_2Mg_6(OH)_{16}CO_3.4H_2O$, as hexagonal scales dimorphous with the rhombohedral stichtite. Named from the locality, Barberton, Transvaal. [*M.A.*, **8**, 51, 100.] 16th List

Barbierite. W. T. Schaller, 1910. *Amer. Journ. Sci.*, ser. 4, vol. xxx, p. 358; *Bull. Soc. franç. Min.*, 1910, xxxiii, p. 320; *Journ. Washington Acad. Sci.*, 1911, vol. i, p. 114; *Zeits. Kryst. Min.*, 1912, vol. 1, p. 347; *Bull. U.S. Geol. Survey*, 1912, No. 509, p. 40. Monoclinic soda-felspar ($NaAlSi_3O_8$) isomorphous with orthoclase and dimorphous with albite. The existence of this has been established by the analyses of soda-rich orthoclases (one from Kragerö, Norway, containing only 1·15% K_2O) made by Philippe Barbier, Professor of Chemistry in the University of Lyons. Compare Cryptoclase. 6th List [Is Microcline; *Zeits. Krist.*, **109**, 241. Cf. Monalbite]

Barbosalite. M. L. Lindberg and W. T. Pecora, 1954. *Science* (*Amer. Assoc. Adv. Sci.*), vol. 119, p. 739. Hydrous ferrous ferric phosphate, $Fe''Fe'''(PO_4)_2(OH)_2$, as black grains from Brazil. Named after Prof. A. L. de M. Barbosa, School of Mines, Minas Geraes. *See* Ferroferri-lazulite. [*M.A.*, **12**, 408.] 20th List [$Fe^{2+}Fe^{3+}_2(PO_4)_2(OH)_2$, monoclinic; *A.M.*, **40**, 952]

Bardiglione. Comte de Bournon, *Trans, Geol. Soc.* 1811, I, 77. The same as anhydrite. 2nd List

Bardolite. J. Morozewicz, 1924. *Bull. Soc. Franç. Min.*, vol. 47, p. 49; *Spraw. Polsk. Inst. Geol.* (= *Bull. Serv. Géol. Pologne*), 1924, vol. 2, p. 217 (bardolit). A dark-green, chlorite-like mineral occurring as a primary constituent in diabase at Bardo, central Poland. It resembles biotite in the high potash (K_2O 4·67%) but contains H_2O 20%. Empirical formula $K_2O.5MgO.FeO.2Fe_2O_3.Al_2O_3.12SiO_2.21H_2O$. Named from the locality. [*M.A.*, **2**, 343, 417, 433.] 10th List [May by Stilpnomelane; *A.M.* **10**, 134]

Bariandite. F. Cesbron and H. Vachey, 1971. *Bull.*, **94**, 49. A black fibrous mineral in the uranium–vanadium deposit of Mounana, Gabon, is monoclinic, *a* 11·70, *b* 3·63, *c* 20·06 Å, β 101° 30′, D 2·7. Composition $2[V_2O_4.4V_2O_5.12H_2O]$. The finest fibres are pleochroic, greenish brown ∥ [010], bottle green ∥ [010]. Named for P. Bariand. [*M.A.*, 71-2329; *Bull.*, **94**, 571.] 27th List [*A.M.*, **57**, 1555]

Barićite. B. D. Sturman and J. A. Mandarino, 1976. *Progr. Abstr. Mineral. Assoc. Canada*, 65. A ferroan variety of the magnesium analogue of vivianite occurs in fractures in an ironstone formation in the region of Big Fish River and Blow River, Yukon Territory; *a* 10·075, *b* 13·416, *c* 4·670 Å, *β* 104° 52′. 2[(Mg,Fe)PO_4.4H_2O]. Named for Dr. L. Barić of Zagreb. 29th List

Bariohitchcockite. E. T. Wherry, 1916. *Proc. US Nat. Museum*, vol. li, p. 83. Taking the name hitchcockite (C. U. Shepard, 1856) [= plumbogummite] for the group-name of the isomorphous series of minerals with the general formula $R'_2O.3Al_2O_3.2P_2O_5.8H_2O = R'_2H_4[Al(OH)_2]_6(PO_4)_4$, the several end-members become: when R′ = ½Ba, bariohitchcockite (= gorceixite); when R′ = ½Sr, strontiohitchcockite (= hamlinite = goyazite); when R′ = K, kaliohitchcockite; and when R′ = Na, natrohitchcockite. The last two of these salts have been met with in only small proportions in isomorphous mixtures. 8th List

Bariomicrolite. I.M.A. Subcomm. pyrochlore group, 1978. *A.M.*, **62**, 404. A systematic name to replace rijkeboerite. 30th List

Bariopyrochlore. I.M.A. Subcomm. pyrochlore group, 1978. *A.M.*, **62**, 404. A systematic name to replace pandaite. 30th List

Barium-adularia. T. Yoshimura, H. Shirozu, and M. Kimura, 1954. *Mem. Fac. Sci. Kyushu Univ.*, ser. D, vol. 4, p. 163. Adularia from Japan containing BaO 3·36%. Compare barium-orthoclase (4th List) and barium-sanidine (16th List). 21st List

Barium-albite. T. Yoshimura, 1939. *Journ. Fac. Sci. Hokkaido Univ.*, Ser. 4, vol. 4, p. 385 (Bariumalbite). A soda-potash-felspar containing $BaAl_2Si_2O_8$ 14%. [*M.A.*, **7**, 413.] 15th List

Barium-alumopharmacosiderite. K. Walenta, 1966. *Tschermaks Min. Petr. Mitt.*, ser. 3, vol. 11, p. 121. Pale yellow cubes from Neubulach, Schwarzwald, Germany, gave Al, Fe, Ba, and As on qualitative analysis, and an X-ray powder pattern near that of synthetic $BaAl_4(AsO_4)_3(OH)_5.5H_2O$. The mineral is assumed to be $Ba(Al,Fe)_4(AsO_4)_3(OH)_5.5H_2O$ and an analogue of pharmacosiderite (which, however, has 6 or 7 H_2O). [*M.A.*, **18**, 285; *A.M.*, **52**, 1584; Зап., **97**, 74.] 25th List

Barium anorthite. *See* celsian. 1st List

—— S. R. Nockolds and E. G. Zies, 1933. *Min. Mag.*, vol. 23, p. 454 (barium anorthite). A plagioclase felspar containing BaO 5·5%, from Broken Hill, New South Wales. 13th List

Barium-aragonite. H. Strunz, *Tab. Min.*, 1941, p. 118 (Barium-aragonit). Syn. of alstonite. 18th List

Barium-autunite. J. Beintema, 1938. *Rec. Trav. Chim. Pays-Bas*, vol. 57, p. 171 (barium autunite). Artificial base-exchange product of autunite. *See* Calcium-autunite. Synonym of Bariumphosphoruranit, barium-uranite, uranocircite. [*M.A.*, **4**, 307, **7**, 237, **11**, 108.] 19th List

Barium carbonate-apatite. Samad Mohseni-Koutchesfehani and Gerard Montel, 1961. *Compt. Rend. Acad. Sci. Paris*, vol. 252, p. 1161 (carbonate-apatite barytique). Synthetic $Ba_{10}(PO_4)_6CO_3$, obtained by the action of CO_2 on barium hydroxy-apatite. 22nd List

Barium-francevillite. V. P. Rogova, G. A. Sidorenko, and N. N. Kuznetsova, 1966. *Zap.*, **95**, 448. Material that does in fact fall within the definition of Francevillite (21st List), but is somewhat richer in Ba than the original, was described as a new mineral. *Syn.* of Francevillite. [*M.A.*, **19**, 55.] 28th List

Barium-Hamlinite (*Chem. Zentralblatt*, 1918, pt. I, p. 858; *Fortsch. Min. Krist. Petr.*, 1920, vol. 6, p. 68). Analyses of three phosphate pebbles ('favas') found with diamond in Brazil were given by O. C. Farrington, *Amer. Journ. Sci.*, 1916, ser. 4, vol. 41, p. 356 [*M.A.*, **1**, 256]. One of these he identifies with gorceixite, another is clearly a mixture, whilst the third with the composition $2BaO.4Al_2O_3.3P_2O_5.11H_2O$ is regarded as doubtful. This 'composition suggests that of . . . hamlinite with barium replacing strontium'. In the German abstract of Farrington's paper it appears as Barium-Hamlinit. 9th List.

Barium heulandite (heulandite baritica). D. Lovisato, *Rend. R. Accad. Lincei*, VI (1), 260, 1897. Heulandite with 2·55% BaO. Sardinia. 1st List

Barium-lamprophyllite. O. B. Dudkin, 1959. Зап. Всесоюз. Мин. Общ. (*Mem. All-Union Min. Soc.*), vol. 88, p. 713 (Бариевый лампрофиллит); ibid., 1961, vol. 90, p. 111 (barium-lamprophyllite). An unnecessary name for barian lamprophyllite. 22nd List

Barium-muscovite. L. H. Bauer and H. Berman, 1933. *Amer. Min.*, vol. 18, p. 30. A variety of muscovite containing barium (BaO 9·89%). Synonym of oellacherite (J. D. Dana, 1867). [*M.A.*, **5**, 284.] 13th List

Barium-nepheline. A. S. Ginzberg, 1915. *Ann. Inst. Polytechn. Pétrograde*, vol. 23. Artificially prepared hexagonal $BaAl_2Si_2O_8$, a high-temperature analogue of celsian. [*M.A.*, **2**, 153; *Min. Mag.*, **26**, 243.] 16th List

Barium-orthoclase, *see* Baryta-orthoclase.

Barium-oxyapatite. H. Wondratschek, 1963. *Neues Jahrb. Min.*, Abh., vol. 99, p. 159 (Bariumoxyapatit). The compound $Ba_{10}(PO_4)_6O$, synthesized by H. Bauer, 1959. 23rd List

Barium-parisite. *See* cordylite. 2nd List

Barium-pharmacosiderite. K. Walenta, 1966. *Tschermaks Min. Petr. Mitt.*, ser. 3, vol. 11, p. 121. Yellow-brown cubic crystals with limonite and baryte from the Clara mine, Schwarzwald, Germany, contain Fe, As, and Ba, and give an X-ray pattern near that of pharmacosiderite, but with lower symmetry, probably tetragonal. The material is assumed to be $BaFe_4(AsO_4)_3(OH)_5.5H_2O$ and an analogue of pharmacosiderite, which is $R.Fe_4(AsO_4)_3(OH)_4.6$ or 7 H_2O. [*M.A.*, **18**, 285; *A.M.*, **52**, 1585; Зап., **97**, 74.] 25th List

Barium-phlogopite. H. von Eckermann, 1925. *Tschermaks Min. Petr. Mitt.*, vol. 38, p. 282 (Barium phlogopites), p. 286 (Baryumphlogopite). A phlogopite containing BaO 1·28%, with golden sub-metallic lustre, from Mansjö Mtn., Sweden. 11th List

Barium-phosphuranylite. V. Ross, 1956. *Amer. Min.*, vol. 41, p. 818. Artificial $Ba(UO_2)_4(PO_4)_2(OH)_4.8H_2O$, the barium analogue of phosphuranylite. Named from the composition. *See also* Bergenite. 22nd List [Is Bergenite, *A.M.*, **45**, 909; Cf. Phurcalite]

Barium-plagioclase. S. R. Nockolds and E. G. Zies, 1933. *Min. Mag.*, vol. 23, p. 448 (barium plagioclase), p. 454 (barium anorthite, 13th List; not the barium anorthite = celsian of H. Sjögren, 1895, 1st List). E. R. Segnit, *Min. Mag.*, 1946, vol. 27, p. 171 (barium-plagioclase). Plagioclase containing BaO 3–5% from Broken Hill, New South Wales. 17th List

Barium-priderite. K. Norrish, 1951. *Min. Mag.*, vol. 29, p. 500 (Ba-priderite). Hintze, *Min.*, 1954, Ergbd. 2, p. 31 (Barium-Priderit). The natural mineral priderite (19th List) contains BaO 6·7, K_2O 5·6%. Ba-priderite and K-priderite (q.v.) were prepared artificially. 20th List

Barium-sanidine. E. S. Larsen *et alii*, 1941. *Bull. Geol. Soc. Amer.*, vol. 52, p. 1849 (barium sanidine). A variety of sanidine containing BaO 5% from the Highwood Mountains, Montana. [*M.A.*, **8**, 243.] 16th List

Barium uranophane. L. N. Belova, 1958. *Proc. 2nd UN Conf. Peaceful Uses Atomic Energy*, vol. 2, p. 295. A silicate of Ba and U; X-ray data resemble those of cuprosklodowskite [*M.A.*, **14**, 344.] Relation to uranophane uncertain and description incomplete; 'needs further study' (M. Fleischer, *Amer. Min.*, 1959, vol. 44, p. 466). Cf. Зап. Всесоюз. Мин. Общ. (*Mem. All-Union Min. Soc.*), 1961, vol. 90, p. 109 (Бариевый уранофан, barium uranophane). 22nd List

Barium-vanadium muscovite. K. G. Snetsinger, 1966. *Amer. Min.*, vol. 51, p. 1623. A barian muscovite, the state of oxidation of which suggests the presence of both V^{4+} and V^{3+} in addition to Fe^{2+} and Ti^{4+}, occurs in a quartz–graphite schist at Silver Knob, near Yosemite Valley, California. 25th List

Barnesite. Malcolm Ross, 1959. *Amer. Min.*, vol. 44, p. 322. The unnamed sodium analogue of hewettite described by A. D. Weeks and M. E. Thompson (*USGS Bull. 1009–B*, 1954, p. 57); named for W. H. Barnes. Much 'metahewettite' is barnesite. 22nd List [$Na_2V_6O_{16}.3H_2O$; *A.M.*, **48**, 1187]

—— A proprietary trade-name for rare-earth oxides used for polishing. J. Sinkankas, *Gemstone and Mineral Data Book*, 1972, p. 67. Not the barnesite (of Ross, 1959). [22nd List.] 28th List

Barracanite. R. Schneider, *Journ. prakt. Chem.* 1895, LII, 557. Those specimens of cubanite, from Barracanao, Cuba, which give the formula $CuFe_2S_4$, instead of $CuFe_2S_3$ as in the original cubanite. *Cupropyrite* is given as an alternative name. 2nd List

Barrerite. E. Passaglia and D. Pongiluppi, 1975. *M.M.*, **40**, 208; *see also Lithos*, **7**, 69. An orthorhombic zeolite from Capa Pula, Sardinia, originally described as a sodian stellerite, has Na > Ca and space group *Amma* (stellerite has *Fmmm*). Named for Dr. R. M. Barrer. [*M.A.*, 75-3591.] 29th List

Barringerite. P. R. Buseck, 1969. *Science*, **165**, 169. Hexagonal $(Fe,Ni)_2P$ occurs along the contacts of schreibersite and troilite in the Ollague meteorite. [*A.M.*, **55**, 317; *M.A.*, 70-1647.] 26th List

Barringtonite. B. Nashar, 1965. *Min. Mag.*, vol. 34, p. 370. Nodular incrustations of $MgCO_3.2H_2O$ occur on olivine basalt under Rainbow Falls, Semphill Creek, Barrington Tops, New South Wales. The mineral is formed at about 5°C; it is anorthic. Named from the locality [*A.M.*, **50**, 2103; J. A. Mandarino notes that the optics are very near those of lansfordite, and the calculated density unexpectedly high]. 24th List

Barroisite. G. Murgoci, 1922. *Compt. Rend. Acad. Sci. Paris*, vol. 175, pp. 373, 426. A dark-green amphibole intermediate between glaucophane and hornblende. [*M.A.*, **2**, 221.] 10th List [*Amph.* (1978)]

Barsanovite. M. D. Dorfman, V. V. Ilyukhin, and T. A. Burova, 1964. Докл. Акад. Наук СССР (*Compt. Rend. Acad. Sci. URSS*), vol. 155, p. 1164 (Ба рсановит). Near $(Ca,Na)_9(Fe,Mn)_2(Zr,Nb)_2Si_2(O,Cl)_{37}$; monoclinic; in a pegmatite from Petrelius, Khibina massif, Kola peninsula. Named for G. P. Barsanov. [*M.A.*, **16**, 549.] 23rd List [Is Eucolite; *A.M.*, **54**, 1499]

Barthite. M. Henglein and W. Meigen, 1914. *Centralblatt Min.*, 1914, p. 353 (Barthit). A hydrated arsenate of zinc and copper, $3ZnO.CuO.3As_2O_5.2H_2O$, occurring as grass-green, crystalline (monoclinic?) crusts in dolomite at Guchab, Otavi, South-West Africa. Named after Mr. — Barth, mining engineer at Guchab, who collected the material. 7th List [Is cuprian Austinite; *A.M.*, **30**, 550]

Baryta-orthoclase. J. E. Strandmark, 1904. *Geol. Fören. Förh. Stockholm*, 1903, vol. xxv, p. 289; 1904, vol. xxvi, p. 97. Celsian, $BaAl_2Si_2O_8$, is shown to be monoclinic and to form a series of mixed crystals with orthoclase. Celsian ranges in composition from Ce to Ce_1Or_2, the other members of the series being classed as baryta-orthoclases (baryt-kalifältspater); those having the composition Ce_1Or_2—Ce_1Or_6 are referred to hyalophane, whilst those with still less barium (Ce_1Or_6—Or) are called barium-bearing orthoclases (barythaltiga kalifältspater). J. P. Iddings (Rock Minerals, New York, 1906, p. 233) refers to minerals of the last of these divisions as barium-orthoclase. 4th List

Baryt-Flussspath. *See* Fluobaryt. 7th List

Baryt-hedyphane. G. Lindström, *Geol. För. Förh. Stockholm*, 1879, vol. 4, p. 266 (Barythaltig hedyfan från Långban); British Museum, *Index of minerals*, 9th edit., 1881 (Barythedyphane), from a dealer's label; ibid., 27th edit., 1936 (Baryt-hedyphane). Variety of hedyphane containing BaO 8·03%. Later named calcium-barium-mimetite (q.v.). 18th List

Barytheulandit. *Neues Jahrb. Min.* 1898, I, Ref. 447. The same as barium heulandite. 2nd List

Barytoanglesite. P. Ramdohr, 1947. *Abhandl. Deutsch. Akad. Wiss. Berlin*, for 1945–1946, no. 4, p. 22 (Barytoanglesit). Variety of anglesite containing some barium. [*M.A.*, **10**, 253.] 18th List

Barytolamprophyllite. Tze-chung Peng and Chien-hung Chang, 1965. [*Scientia Sinica*, vol. 14, p. 1827]; abstr. in *Amer. Min.*, 1966, vol. 51, p. 1549. Monoclinic $(Na,K)_6(Ba,Ca,Sr,Mn)_3(Ti,Fe,Mg)_7(Si,Al)_8O_{32}(F,OH,O,Cl)_4$, the barium analogue of lamprophyllite, occurring in ijolite in the Lovozero intrusive, Kola Peninsula, USSR. [*Bull.*, **90**, 116.] 25th List

Basaltkainit. *See* Anhydrokainite. 9th List

Basaluminite. F. A. Bannister and S. E. Hollingworth, 1948. *Nature*, London, vol. 162, p. 565. Hydrous basic aluminium sulphate, $2Al_2O_3.SO_3.10H_2O$, as white compact material lining crevices in ironstone from Irchester, Northamptonshire. Felsőbányite with the same formula has a different X-ray pattern. So named because more basic than aluminite. *See* Hydro-basaluminite. [*M.A.*, **10**, 452.] 18th List [Cf. Meta-aluminite]

Basiliite. L. J. Igelström, *Geol. För. Förh.* XIV. 307, 1892; and *Zeits. Kryst. Min.* XXII. 470, 1894. Hydrated antimonate of manganese. Sjö mine, [Örebro] Sweden. 1st List [A mixture; *A.M.*, **58**, 562]

Basitom-Glanz. A. Breithaupt, 1832. *Vollständige Charakteristik des Mineral-Systems*, 3rd edit., p. 267 (Staurotyper Basitom-Glanz). Apparently so named because of the supposed existence of a basal cleavage. An obsolete synonym of freieslebenite. 7th List

Basobismutite. K. A. Nenadkevich, 1917. *Bull. Acad. Sci. Pétrograd*, ser. 6, vol. xi (1), p. 454 (базобисмутить). A basic bismuth carbonate, $2Bi_2O_3.CO_2.H_2O$, from Transbaikal, Siberia. 8th List [Is Bismutite; *M.A.*, **9**, 6]

Bassanite. F. Zambonini, 1910. Mineralogia Vesuviana. *Mem. R. Accad. Sci. Fis. Mat. Napoli*, ser. 2, vol. xiv, no. 7, p. 327. White, opaque crystals with the form of gypsum, found in blocks ejected in 1906 from Vesuvius. Anhydrous calcium sulphate ($CaSO_4$), differing from anhydrite, and possibly identical with the hexagonal calcium sulphate produced by the artificial dehydration of gypsum. Named after Francesco Bassani, Professor of Geology in the University of Naples. 6th List [$2CaSO_4.H_2O$; *M.M.*, **35**, 354. Cf. Miltonite, Mirupolskite, Vibertite]

β-Bassanite. H. Strunz, 1966. *Min. Tabellen*, 4th edit., p. 260. A name for the hexagonal, high-temperature polymorph of $CaSO_4.\frac{1}{2}H_2O$ reported by O. W. Florke (1952). P. Gay (*Min. Mag.*, 1965, vol. 35, p. 345) concludes that the existence of this polymorph is very improbable. 24th List

Bassetite. A. F. Hallimond, 1915. *Mineralogical Magazine*, vol. xvii, p. 221. A yellow 'uranium-mica' from the Basset mines, Redruth, Cornwall, hitherto referred to autunite, but now shown to be distinct from the original autunite from Autun in France. It is monoclinic with the probable composition $Ca(UO_2)_2(PO_4)_2.8H_2O$. Named after the locality. 7th List [$Fe^{2+}(UO_2)_2(PO_4)_2.8(?)H_2O$; *M.A.*, **7**, 529; *M.M.*, **30**, 344]

Bastinite. D. J. Fisher, 1945. *Program and Abstracts, Min. Soc. Amer.*, 26th Annual Meeting, p. 9; *Rep. Invest. Geol. Surv. South Dakota*, 1945, no. 50, p. iv; *Amer. Min.*, 1946, vol. 31, p. 192. Lithium phosphate with some iron and manganese, as colourless triclinic crystals on fractured surfaces of lithiophilite, from Custer, South Dakota. Named after Prof. Edson Sunderland Bastin (1878–) of the University of Chicago. [*M.A.*, **9**, 262.] 17th List [Is lithian Hureaulite; *A.M.*, **49**, 398]

Bastnaesite-(Yt). D. A. Mineev, T. I. Lavrishcheva, and A. V. Bykova, 1970. *Zap.*, **99**, 328 (Бастнезит(Y)). Fine-grained brick-red pseudomorphs up to 8 cm long after gagarinite (23rd List). Yt. 40·1% of total rare-earth metals. [*A.M.*, **57**, 594.] 27th List

Batavite. E. Weinschenk, *Zeits. Kryst. Min.* XXVIII. 157–160, 1897. A scaly decomposition product, perhaps related to the micas or chlorites. [Passau dist.] Bavaria. 1st List

Batchelorite. W. F. Petterd, 1910. *Catalogue of the Minerals of Tasmania*, 3rd edit., Hobart, 1910, p. 22; *Papers Roy. Soc. Tasmania*, 1910, p. 22. A greenish, foliated mineral, from Mt. Lyell mine, Tasmania; it had previously been described as pyrophyllite, which it resembles in appearance, but it differs from this in containing more silica, the formula being approximately $Al_2O_3.2SiO_2.H_2O$. Named in memory of the late Mr. W. T. Batchelor, formerly manager of the Mt. Lyell mine. 6th List [Essentially muscovite; *M.M.*, **31**, 700]

Batisite. S. M. Kravchenko and E. V. Vlasova, 1959; Доклады Акад. Наук СССР (*Compt. Rend. Acad. Sci. URSS*), vol. 128, p. 1046 (Батисит); ibid., 1960, vol. 133, p. 657. $Na_2BaTi_2(Si_2O_7)_2$, dark brown orthorhombic crystals from the Inaglina pegmatite, Central Aldan. Isostructural with shcherbakovite, but contains no Nb. Named from the composition, Ba–Ti–Si. 22nd List [*A.M.*, **45**, 908, 1317]

Bauerite. F. Rinne, 1911. *Ber. k. sächs. Ges. Wiss.*, Leipzig, Math.-Physis. Kl., vol. lxiii, p. 445 (Bauerit). The crystalline end-product, consisting mainly of hydrated silica, which results from the artificial or natural bleaching (baueritization) of biotite. It retains the form and optically uniaxial character of the original biotite. Named after Professor Max Bauer, of Marburg. 6th List

Baumhauerite. R. H. Solly, *Min. Mag.*, 1902, vol. xiii, p. 151; 1903, vol. xiii, p. 339; *Zeits. Kryst. Min.*, 1903, vol. xxxvii, p. 321. Crystals resembling in general appearance the several other sulpharsenites of lead from the Binnenthal dolomite. Monoclinic. $4PbS.3As_2S_3$. Named after Professor Heinrich Baumhauer, of Freiburg, Switzerland. 3rd List [Anorthic, perhaps two modifications; *see* refs. in *2nd App. to Chem. Index of Minerals* (1974), p. 18]

Baumhauerite-I. W. Nowacki, 1966. *Jahrb. Naturhist. Mus. Bern*, 1966–1968, 71. *Syn.* of Baumhauerite. 28th List

Baumhauerite-II. H. Rösch and E. Hellner, 1959. *Naturwiss.* **46**, 72. An *artificial* product in the system PbS–As_2S_3, with twice the *a*-axis of Baumhauerite. Also occurs naturally. 28th List

Baumite. C. Frondel and J. Ito, 1975. *Neues Jahrb. Min., Abh.* **123**, 111. A septechlorite, $(Mg_{9\cdot0}Al_{1\cdot3}Mn_{0\cdot9}Zn_{0\cdot7}Fe_{0\cdot1})(Si_{6\cdot3}Al_{1\cdot7})O_{20}(OH)_{16}$, from Franklin, New Jersey. [*M.A.*, 75-3592; *A.M.*, **61**, 174.] 29th List

Bauranoite. V. P. Rogova, L. N. Belova, G. P. Kiziyarov, and N. N. Kusnetsova, 1973. *Zap.*, **102**, 75 (Бауранoит). Veinlets of minute brown grains from an unnamed Russian locality; $BaO.2UO_3.4–5H_2O$. Named from its composition. [*M.A.*, 74-505; *Bull.*, **96**, 234; *A.M.*, **58**, 1111; *Zap.*, **103**, 358.] 28th List

Bavenite. E. Artini; *Rend. Accad. Lincei*, Roma, 1901, ser. 5, vol. x, sem, 2, p. 139. Radiating tufts of white needles in the drusy cavities of the Baveno granite. Monoclinic, but twinned to simulate orthorhombic symmetry. $3CaO.Al_2O_3.6SiO_2.H_2O$. Approaches pilinite in composition. 3rd List [$Ca_4BeAl_2Si_9O_{24}(OH)_2$; *A.M.*, **17**, 409; **38**, 988]

Bayate. (E. F. Burchard, *Trans. Amer. Inst. Mining Metall. Engin.*, 1920, vol. 63, pp., 54, 58; D. F. Hewett and E. V. Shannon, *Amer. Journ. Sci.*, 1921, ser. 5, vol. 1, pp. 492, 497.) A local name for a brown ferruginous jasper occurring with the manganese-ores of Cuba. 9th List

Bayerite. R. Fricke, 1928. *Zeits. Anorg. Chem.*, 1928, vol. 175, p. 252; 1929, vol. 179, p. 287 (Bayerit). G. F. Hüttig and O. Kostelitz, ibid., 1930, vol. 187, p. 4. An artificial form of aluminium hydroxide, $Al(OH)_3$, metastable with respect to gibbsite, obtained in the K. J. Bayer process for the purification of bauxite. [*M.A.*, **4**, 30, 169, 305.] 12th List [Israel; *M.M.*, **33**, 723]

Bayleyite. J. Axelrod, F. Grimaldi, C. Milton, and K. J. Murata, 1948. *Program and Abstracts, Min. Soc. Amer.*, 29th Annual Meeting, p. 4; *Bull. Geol. Soc. Amer.*, vol. 59, p. 1310; *Amer. Min.*, 1949, vol. 34, p. 274. Hydrous magnesium-uranyl carbonate, $Mg_2UO_2(CO_3)_3.nH_2O$, from Arizona. Named after William Shirley Bayley (1861–1943) of the University of Illinois. [*M.A.*, **10**, 452.] 18th List

Baylissite. K. Walenta, 1976. *Schweiz. Min. Petrogr. Mitt.* **56**, 187. Colourless crusts of $K_2Mg(CO_3)_2.4H_2O$ occur in the cable tunnel of Gerstenegg/Sommerloch, in the Grimsel area, Switzerland. Monoclinic, *a* 12·37, *b* 6.24, *c* 6·86 Å, α 1·462, γ 1·531. [*M.A.*, 77-2183] 30th List

Bazirite. J. R. Hawkes, R. J. Merriman, R. R. Harding and D. P. F. Darbyshire, 1975. *In* R. K. Harrison, *Expeditions to Rockall*, 1971–1972, *Rept. Inst. Geol. Sci. no.* 75/1. A colourless mineral forming about 0·1% of the Rockall aegirine-granite gives data matching artificial $BaZrSi_3O_9$. Named for the composition. [Isostructural with benitoite and pabstite.—*M.F.*, *M.A.*, 75-2520; *A.M.*, **61**, 175.] 29th List

Bazzite. E. Artini, 1915. *Atti (Rend.) R. Accad. Lincei, Roma*, ser. 5, vol. xxiv, sem. 1, p. 313. Minute, sky-blue, hexagonal prisms and barrel-shaped crystals occurring sparingly in the drusy cavities of the granite of Baveno, Piedmont. A preliminary chemical analysis shows them to consist of silicate of scandium with other rare-earth metals, iron, and a little sodium. Named after the engineer E. Bazzi, who collected the material. 7th List [$Be_3(Sc,Fe)_2Si_6O_{18}$, the Sc analogue of beryl; *M.A.*, **13**, 484; **18**, 115]

Be-vesuvianite, *see* Beryllium-vesuvianite.

Beaconite. M. E. Wadsworth, *Rept. State Board Geol. Survey, Michigan, for 1891–2*, 1893, p. 171. A fibrous variety of talc resembling asbestos, from Beacon, Michigan. 2nd List

Bearsite. E. V. Kopchenova and G. A. Sidorenko, 1962. Зап. Всесоюз. Мин. Общ. (*Mem. All-Union Min. Soc.*), vol. 91, p. 442 (Беарсит). Monoclinic Be_2AsO_4OH, the arsenic analogue of moraesite, in the oxidation zone of an ore deposit in Kazakhstan. Named from the composition (*Amer. Min.*, 1963, vol. 48, p. 210; *M.A.*, **16**, 68). 23rd List

Beaverite. B. S. Butler and W. T. Schaller, 1911. *Journ. Washington Acad. Sci.*, vol. i, p. 26; *Amer. Journ. Sci.*, 1911, ser. iv, vol. xxxii, p. 418; *Zeits. Kryst. Min.*, 1912, vol. 1, p. 114; *Bull. US Geol. Survey*, 1912, no. 509, p. 77. Hydrous sulphate of copper, lead, ferric iron (and aluminium), $CuO.PbO.Fe_2O_3.2SO_3.4H_2O$, occurring as canary-yellow, earthy and friable masses. Under the microscope it shows minute, six-sided plates. Named from the locality, Beaver County, Utah. 6th List

Beckelite. J. Morozewicz, 1904. *Rozpr. Akad. Kraków*, Ser. A, vol. xliv, p. 216; *Bull. Intern. Acad. Sci. Cracovie*, 1905, année 1904, p. 485 (beckelicie, Beckelith, béckélite); *Min. Petr. Mitt.* (*Tschermak*), 1905, vol. xxiv, p. 120. Wax-yellow, octahedral or rhombic-dodecahedral crystals resembling pyrochlore in appearance and physical characters. Occurs in a dyke modification of elaeolite-syenite near Mariupol on the Sea of Azov, Russia. Calcium cero-lanthano-didymo-silicate, $Ca_3(Ce,La,Di,Y)_4(Si,Zr)_3O_{15}$. Named after Professor Friedrich Becke, of Vienna. 4th List [Is probably Britholite; *M.M.*, **31**, 455]

Beckerite. E. Pieszczek [*Archiv f. Pharmacie*, [iii], XIV, 433–436] *Journ. Chem. Soc. Abstracts*, 1881, XL, 687. A brown resin occurring with Prussian amber. 2nd List

Becquerelite. A. Schoep, 1922. *Compt. Rend. Acad. Sci. Paris*, vol. 174, p. 1240 (becquerélite). Uranium hydroxide, $UO_3.2H_2O$, as minute, yellow, orthorhombic crystals and crusts on pitchblende from Katanga. Named after Antoine Henri Becquerel (1852–1908). [*M.A.*, **1**, 377.] 9th List [$CaU_6O_{19}.11H_2O$; *A.M.*, **45**, 1026]

Bedenite. N. E. Efremov, 1935. [*Min. Syre, Moscow*, 1935, no. 9, p. 15]; *Mém. Soc. Russe Min.*, 1937, ser. 2, vol. 66, p. 479 (Беденит), p. 485 (Bedenite). O. M. Shubnikova, *Trans. Lomonosov Inst. Acad. Sci. USSR*, 1937, no. 10, p. 183. An asbestiform mineral of the anthophyllite group, $H_2Ca_2Mg_4AlFe'''Si_8O_{26}$, from Beden Mtn., northern Caucasus. Named from the locality. [*M.A.*, **7**, 170.] 15th List [Ferrian actinolitic hornblende, *Amph. Sub-Comm.*]

Befanamite. A. Lacroix, 1923. *Minéralogie de Madagascar*, vol. 3, p. 311. The scandium end-member of the thortveitite group (containing also zirconium) from Befanamo, Madagascar, as distinct from the Norwegian thortveitite which contains much yttrium-earths. Named from the locality. [*M.A.*, **2**, 146.] 10th List

Behierite. J. Behier, 1960. [*Rep. Malgache, Rapport Annual Serv. Géol.*, p. 181]; abstr. in *Amer. Min.*, 1961, vol. 46, p. 767. Two small crystals from Manjaka, Madagascar, are thought, on X-ray data, to be $TaBO_4$. Named after M. Jean Behier of the Serv. géol., Madagascar. A premature name. 22nd List [A valid species; *A.M.*, **47**, 414]

Behoite. A. J. Ehlmann and R. S. Mitchell, 1970. *Amer. Min.* **55**, 1. β-$Be(OH)_2$ occurs as colourless orthorhombic crystals in alteration zones round gadolinite in the Rode Ranch pegmatite, Llano County, Texas. Named from the composition. 26th List

Beidellite. E. S. Larsen and E. T. Wherry, 1925. *Journ. Washington Acad. Sci.*, vol. 15, p. 465; C. S. Ross and E. V. Shannon, ibid., p. 467; *Journ. Amer. Ceramic Soc.*, 1926, vol. 9, p. 93. Hydrous metasilicate of aluminium, $Al_2O_3.3SiO_2.4H_2O$, probably orthorhombic, occurring as a gouge-clay (previously referred to leverrierite) at Beidell, Saguache Co., Colorado. Named from the locality. *See* Iron-beidellite. [*M.A.*, **3**, 8; **3**, 73; **3**, 314; **3**, 413.] 11th List

Beiyinite. T. L. Ho, 1935. *Bull. Geol. Soc. China*, vol. 14, p. 279 (Beiyinite). An undetermined mineral presumed to contain rare-earths (La, Ce, Yt, Er) as minute grains (tetragonal?) in fluorite from Beiyin Obo, Inner Mongolia. With oborite (q.v.), named from the locality. [*M.A.*, **6**, 151.] 14th List [Remains doubtful]

Belbaite. V. I. Vernadsky, 1913. *See* Elbaite. 7th List

Beldongrite. L. L. Fermor, 1909. *Mem. Geol. Survey India*, vol. xxxvii, pp. lxix, 115. A black, pitch-like mineral closely allied to psilomelane. Formula, perhaps $6Mn_3O_5.Fe_2O_3.8H_2O$. Probably an alteration product of spessartite. From Beldongri, Nágpur district, Central Provinces, India. 5th List

Belgite. R. Panebianco, 1916. *Riv. Min. Crist. Italiana*, vol. xlvii, p. 13 (Belgito). Synonym of willemite (Zn_2SiO_4), which was named by A. Lévy in 1830 after William I (1772–1844), king of the Netherlands. Writing in Esperanto, the author objects to naming minerals after kings, preferring a name derived from the locality. He, however, overlooks the fact that this mineral is not from Belgium, but from the neutral state of Moresnet. 8th List

Belite. *See* under alite. 2nd List.

Bellidoite. L. A. de Montreuil, 1975. *Econ. Geol.* **70**, 384. Anhedral grains in calcite at Habri, Moravia, are tetragonal, *a* 11·52, *c* 11·74 Å, composition $Cu_{2\cdot01}Se$; identical with synthetic β-Cu_2Se. Named for E. Bellido Bravo. [*A.M.*, **60**, 736; *Bull.*, **98**, 318.] 29th List

Bellingerite. H. Berman and C. W. Wolfe, 1940. *Amer. Min.*, vol. 25, p. 505. Hydrous copper iodate, $3Cu(IO_3)_2.2H_2O$, as green triclinic crystals from Chuquicamata, Chile. Named after Mr. Herman Carl Bellinger, vice-president of the Chile Exploration Company. [*M.A.*, **8**, 3.] 16th List

Bellite. W. F. Petterd, 1905. *Rep. Secr. Mines, Tasmania*, for 1904, p. 83. 'Chromo-arsenate of lead' occurring as velvety tufts of bright red or yellow, hexagonal needles or as powdery encrustations at Magnet, Tasmania. [The analysis suggests a mixture of crocoite, mimetite, and quartz.] Named after W. R. Bell, a prospector in Tasmania. The name bellite has long been in use for an explosive. 4th List [A mixture; *M.M.*, **38**, 902; possibly a chromatian mimetite; *Bull. Br. Mus. nat. Hist.* (*Miner.*), **2**, 8]

Belmontite. G. Küstel. R. Koechlin, *Mineralogisches Taschenbuch der Wiener Mineralogischen Gesellschaft*, 1911, p. 16 (Belmontit). A yellow mineral, said to be a silicate of lead, occurring with stetefeldtite at Belmont, Nevada. Named from the locality. A specimen so labelled was presented by Mr. Küstel in 1873 to the Royal Natural History Museum of Vienna. 6th List

Belomorite, Беломорит. Trade-name for moonstone from the White Sea (Белое Море). A. E. Fersman, *Precious and coloured stones of USSR*, Leningrad, 1925, vol. 2, p. 30. [*M.A.*, **9**, 146.] 17th List

Belovite. E. I. Nefedov, 1953. *Mém. Soc. Russ. Min.*, ser. 2, vol. 82, p. 317 (Беловит). '$Ca_3(Ca,Mg)(AsO_4)_2.2H_2O$', near roselite. [*M.A.*, **12**, 352.] 20th List [Syn. of Talmessite; *A.M.*, **50**, 813; *see* Arsenate-belovite]

—— L. S. Borodin and M. E. Kazakova, 1954. *Doklady Acad. Sci. USSR*, vol. 96, p. 613 (Беловит). An apatite-like mineral from pegmatite, hexagonal, $(Sr,Ce,Na,Ca)_{10}(PO_4)_6(O,OH)_2$. Named after academician N. V. Belov, Н. В. Белов, of Moscow. [*M.A.*, **12**, 461.] 20th List [*A.M.*, **40**, 367. The accepted name for this species, Fleischer's *Gloss.*]

Belyankinite. V. I. Gerasimovsky and M. E. Kazakova, 1950. *Doklady Acad. Sci. USSR*, vol. 71, p. 925 (Белянкинит). $2CaO.12TiO_2.\frac{1}{2}Nb_2O_5.ZrO_2.SiO_2.28H_2O$. Platy yellowish-brown masses, optically biaxial, in nepheline-syenite-pegmatite from Kola peninsula, Russia. Named after Prof. Dmitry Stepanovich Belyankin Дмитрий Степанович Белянкин (1876–). [*M.A.*, **11**, 123.] 19th List [*A.M.*, **37**, 882. Cf. Mangan-belyankinite, Gerasimovskite]

Belyankite. M. D. Dorfman, 1950. *Doklady Akad. Sci. USSR*, vol. 75, p. 852 (белянкит). $Ca_3Al_2F_{12}.4H_2O$ as white monoclinic crystals in kaolinized granite from Kazakhstan. Named after Prof. Dmitry Stepanovich Belyankin Дмитрий Степанович Белянкин (1876–). [*M.A.*, **11**, 311.] 19th List [Is Creedite; *A.M.*, **37**, 785; **39**, 405]

Bemagalite, a name considered, but rejected, for Taaffeite (*see* 19th List). 24th List

Bemmelenite. F. V. Chukhrov, 1936. [Colloids in the earth's crust (Russ.), *Acad. Sci. USSR*, 1936, p. 103.] O. M. Shubnikova, *Trans. Inst. Geol. Sci., Acad. Sci. USSR*, 1938 [i.e. 1940], no. 11 (*Min. Geochem. Ser.*, no. 3), p. 7 (Беммеленит, Bemmelenite). Colloidal ferrous carbonate, the amorphous analogue of chalybite. Named after Prof. Jakob Maarten van Bemmelen (1830–1911) of Leiden, who described the material occurring in peat deposits (*Zeits. Anorg. Chem.*, 1900, vol. 22, p. 316). 16th List

Benitoide. (J. Escard, *Les pierres précieuses*, Paris, 1914, p. 210 (Bénitoïde).) Variant of benitoite (G. D. Louderback, 1907). 9th List

Benitoite. G. D. Louderback, 1907. *Bull. Dep. Geol. Univ. California*, vol. v, p. 149; ibid., 1909, vol. v, p. 331. Also papers and notes by R. Arnold, H. Baumhauer, W. C. Blasdale, C. Hlawatsch, E. H. Kraus, C. Palache, and A. F. Rogers, references to which are given in Louderback's later paper. B. Ježek, *Bull. Intern. Acad. Sci. Bohême*, 1909, année xiv, p. 213. An acid titano-silicate of barium $BaTiSi_3O_9$, forming beautiful, sapphire-blue, transparent crystals, suitable for cutting as gems, and the only representative (except Ag_2HPO_4, H. Dufet, 1886) of the ditrigonal-bipyramidal class of the hexagonal system. Occurs with neptunite in natrolite veins traversing schistose rocks near the source of the San Benito River, in San Benito Co., California. 5th List

Benjaminite. E. V. Shannon, 1924. *Proc. United States National Museum*, vol. 65, art. 24, p. 1. A sulphobismuthite of lead, silver, and copper, $Pb_2(Ag,Cu)_2Bi_4S_9$, belonging to the klaprotholite group, and occurring as grey masses in white quartz from Nevada. Named after Dr. Marcus Benjamin, of the United States National Museum. [*M.A.*, **2**, 337.] 10th List [Redefined; monoclinic $2[(Ag,Cu)_3(Bi,Pb)_7S_{12}]$; *C.M.*, **13**, 394, 402]

Benstonite. F. Lippmann, 1961. *Naturwiss.*, vol. 48, p. 550 (Benstonit); preliminary note in *Fortschr. Min.*, 1961, vol. 39, p. 81. A rhombohedral carbonate from a baryte mine in Hot

Spring Co., Arkansas, USA, has unit-cell contents $3[Ca_7(Ba,Sr)_6(CO_3)_{13}]$. Named after Mr. O. J. Benston. The name is unfortunately near bentonite. 22nd List [*A.M.*, **47**, 585. Perhaps $MgCa_6Ba_6(CO_3)_{13}$, *Min. Record* **1**, 140]

Bentonite. W. C. Knight, *Engineering and Mining Journ.* New York, 1898, LXVI, 491, 638. An impure clay occurring in the Fort Benton shales, Wyoming. Previously called *taylorite* (q.v.). 2nd List

Berborite. E. I. Nefedov, 1967. Доклады акад. наук СССР (*Compt. Rend. Acad. Sci. URSS*), vol. 174, p. 189 (Берборит). Small trigonal crystals from an unnamed skarn deposit have the composition $Be_2BO_3(OH,F).H_2O$. Named from the composition. [*A.M.*, **53**, 348; Зап., **97**, 71.] 25th List

Beresofskite, *see* Berezovskite.

Beresowite. J. Samoiloff, *Bull. Soc. Nat. Moscou*, 1897, p. 290. *Beresovite*, Dana's *Appendix*, 1899. Indistinct dark red crystals resembling melanochroite. $2PbO.3PbCrO_4.PbCO_3$. Beresovsk, Urals. 2nd List [A pseudomorph, *Bull. Br. Mus. nat. Hist.* (*Miner.*), **2**, 8]

Berezovskite. E. S. Simpson, 1932. A key to mineral groups, species and varieties, London, p. 8 (Beresofskite). To replace the earlier name, beresofite (E. S. Simpson, *Min. Mag.*, 1920, vol. 19, pp. 101–105), for a magnesium variety of chromite. Beresofite of C. U. Shepard (1844) is an obsolete synonym of crocoite, $PbCrO_4$. Beresovite (Beresowit) of J. (= Y. V.) Samoilov (1897; 2nd List) is $2PbO.3PbCrO_4.PbCO_3$. Beresite of G. Rose (1837) is a micro-granite (quartz-aplite). All are named from the locality, Berezovsk (Березовск), a mining village 13 km. NE of Ekaterinburg (now Sverdlovsk), Urals. *See* Alumo-berezovite. 15th List [*See* previous entry]

Bergalith, a rock-name (Johannsen, *Descr. Petr. Ign. Rocks*, vol. 4, p. 379), is incorrectly given as a synonym of Deeckëite (7th List) (H. Strunz, *Min. Tabellen*, 2nd edn, p. 244). 24th List

Bergenite. H. W. Bültemann and G. H. Moh, 1959. *Neues Jahrb. Min.*, Monatsh., p. 232 (Bergenit). Synonym of barium-phosphuranylite (q.v.); occurs naturally at Bergen an der Trieb, Saxony, with other uranium minerals; named from the locality, the older name being rejected as implying a barian phosphuranylite rather than the barium analogue. 22nd List [*A.M.*, **45**, 909. The accepted name, Fleischer's *Gloss.*; cf. Phurcalite]

Berillite. M. Fleischer, *Amer. Min.*, vol. 40, p. 787. Direct transliteration of бериллит. (In Russian beryl is берилл.) Synonym of beryllite (20th List). A. H. Chester (*Dict. 1896*, p. 30) gives berylite as a synonym of beryl. 21st List

Berillosodalite, *see* Beryllosodalite.

Berinel, a name considered, but rejected, for Taaffeite (*see* 19th List). 24th List

Berkeyite. P. F. Kerr, 1926. *Jewelers' Circular*, New York, vol. 92, p. 67. A blue gem-stone from Brazil, afterwards (G. F. Kunz, ibid., p. 86) identified as lazulite. Named after Prof. Charles Peter Berkey (1865–), of Columbia University, New York. 11th List

Bermanite. C. S. Hurlbut, 1936. *Amer. Min.*, vol. 21, p. 657. Hydrated basic phosphate of manganese, etc. $(Mn,Mg)_5''(Mn,Fe)_8'''(PO_4)_8(OH)_{10}.15H_2O$, as minute reddish-brown orthorhombic crystals in triplite from Arizona. Named after Dr. Harry Berman of Harvard University, Cambridge, Massachusetts. [*M.A.*, **6**, 442.] 14th List [$Mn^{2+}Mn_2^{3+}(PO_4)_2(OH)_2.4H_2O$, monoclinic; *A.M.*, **53**, 416]

Bernalite. G. Mueller, 1964. *Rept. 22nd Internat. Geol. Congr., India*, part 1, 47. An olefinic bitumen, named for Professor J. D. Bernal. 26th List

Berndtite. G. H. Moh, 1966. *Fortschr. Min.*, vol. 42, p. 211. β-SnS_2, hexagonal, described from Serro de Potosi, Bolivia (G. H. Moh and F. Berndt, *Neues Jahrb. Min.*, Monatsh., 1964, p. 94) is named for F. Berndt. [*A.M.*, **50**, 2107.] 24th List [*A.M.*, **51**, 1551; **58**, 347; **60**, 739]

Berryite. E. W. Nuffield and D. C. Harris, 1965. *Canad. Min.*, vol. 8, p. 400 (abstr.). Lath-like crystals on the type specimen of Cuprobismutite (Hillebrand, 1884) from Park County, Colorado, and on specimens from Nordmark, Sweden, are monoclinic, probable composition $Pb_2(Cu,Ag)_3Bi_5S_{11}$. Named for L. G. Berry. [*A.M.*, **51**, 532.] 24th List [*A.M.*, **52**, 928; USSR, *M.A.*, 78-851]

Berthonite. H. Buttgenbach, 1923. *Ann. Soc. Géol. Belgique*, vol. 46, *Bull.*, p. 212. Sulphantimonite of lead and copper $5PbS.9Cu_2S.7Sb_2S_3$ or $2(Pb,Cu_2)S.Sb_2S_3$, occurring as finely granular masses filling fissures in iron-ore at Slata, Tunisia. Named after Mr. Louis Berthon,

Chief Engineer of the Department of Mines, Tunisia. [*M.A.*, **2**, 149.] 10th List [Is Bournonite; *A.M.*, **27**, 109; **32**, 485]

Bertossaite. O. von Knorring and M. E. Mrose, 1966. *Canad. Min.*, vol. 8, p. 668. A pale pink massive mineral from the Buranga lithium pegmatite, Rwanda, gives the formula $(Li,Na)_2(Ca,Fe,Mn)Al_4(PO_4)_4(OH,F)_4$; orthorhombic. It is the calcium analogue of palermoite. [*A.M.*, **52**, 1583; Зап., **97**, 73; *Bull.*, **91**, 48.] 25th List

Beryllite. M. V. Kuzmenko, 1954. *Doklady Acad. Sci. USSR*, vol. 99, p. 451 (бериллит). Hydrous silicate of beryllium, $Be_3SiO_4(OH)_2.H_2O$, as an alteration product of epididymite. [*M.A.*, **12**, 569.] 20th List [*A.M.*, **40**, 787]

Beryllium-felspar. M. E. Kazakova, 1946. *Compt. Rend.* (*Doklady*) *Acad. Sci. URSS*, vol. 54, p. 623 (beryllium feldspar). E. I. Kutukova, ibid., p. 724 (beryllium microcline). Felspar containing BeO 1·2%, with bavenite from the decomposition of beryl and plagioclase in the emerald mines, Urals. [*M.A.*, **10**, 247.] 18th List

Beryllium-humite. H. Rosenbusch, 1905. *Mikroskopische Physiographie d. Mineralien*, 4th edit., vol. i, part 2, p. 193 (Berylliumhumite, *pl.*). Humite containing a small amount of beryllium (BeO, 1%) and no fluorine, $Mg_5(MgOH)_2(SiO_4)_3$. Analysed by P. Jannasch and J. Locke (1894), and described by R. W. Schäfer (1895) from the serpentine of the Allalin district, Wallis, Switzerland. 4th List

Beryllium microcline, *see* Beryllium-felspar.

Beryllium-orthite. P. Quensel, 1945. Arkiv Kemi, *Min. Geol.*, vol. 18A, no. 22 (Berylliumorthit). Beryllium-bearing (BeO 3·83%) orthite, regarded as a synonym of muromontite. [*M.A.*, **9**, 257.] 17th List

Beryllium-Sodalith. H. Strunz, 1966. *Min. Tabellen*, 4th edit., p. 455. Variant of Beryllosodalite (22nd List). 24th List

Beryllium-vesuvianite, Be-vesuvianite. C. Palache and L. H. Bauer, 1930. *Amer. Min.*, vol. 15, p. 31; L. H. Bauer, ibid., p. 199. Idocrase from Franklin Furnace, New Jersey, containing BeO 9·20%. [*M.A.*, **4**, 329.] 12th List

Beryllosodalite. E. I. Semenov and A. V. Bykova, 1960. Доклады Акад. Наук СССР (*Compt. Rend. Acad. Sci. URSS*), vol. 133, p. 1191 (Бериллосодалит). *Abstr. Bull. Soc. franç. Min. Crist.*, 1961, vol. 84, p. 205 (Berillosodalite). Sodalite with partial replacement of Al_2 by BeSi, near $Na_4BeAlSi_4O_{12}Cl$, from the Lovozero massif, Kola peninsula. What is evidently the same mineral is also described, unnamed, from Ilimaussaq, Greenland, by H. Sørensen, *Rept. 21st session Internat. Geol. Congr.*, Norden, 1960, part 17, p. 31; tetragonal. 22nd List [Is Tugtupite; *A.M.*, **46**, 241]

Berzelite. Variant of Berzeliite (of O. B. Kühn, 1840); not the berzelite of E. D. Clarke, 1818, nor of A. Lévy, 1837. *A.M.*, **9**, 62. 10th List

Beta-cerolite. I. I. Ginzburg and I. A. Rukavishnikova, 1950. *Mém. Soc. Russe Min.*, vol. 79, p. 33 (Бэта-керолит). $MgSiO_3.H_2O$, differing from alpha-cerolite (q.v.) in optical, thermal, and X-ray data. [*M.A.*, **11**, 405]. 19th List

Betafite. A. Lacroix, 1912. *Compt. Rend. Acad. Sci. Paris*, vol. cliv, p. 1042; *Bull. Soc. franç. Min.*, 1912, vol. xxxv, pp. 88, 234. A member of a group of cubic minerals, niobo-tantalo-titanates of uranium, etc., including also blomstrandite (of G. Lindström, 1874) and samiresite (q.v.); they are closely allied to pyrochlore and hatchettolite, but differ from the former in containing only little lime and rare earths, and from the latter in containing titanium. Betafite is a hydrated niobate and titanate of uranium, and occurs in pegmatite near Betafo in Madagascar. Named after the locality. 6th List [*A.M.*, **46**, 1519; **62**, 403]

Beta-roselite. C. Frondel, 1955. *Amer. Min.*, vol. 40, p. 828. $Ca_2Co(AsO_4)_2.2H_2O$, triclinic, previously labelled as 'roselite' from Schneeberg, Saxony: Dimorphous with monoclinic roselite. [*M.A.*, **13**, 8.] 21st List

Beta-uranophane. C. Frondel, 1956. *Amer. Min.*, vol. 41, p. 551 (beta-uranophane). The monoclinic polymorph of orthorhombic uranophane, $Ca(UO_2)_2(SiO_3)_2(OH)_2.5H_2O$. [*M.A.*, **13**, 204.] Synonym of β-uranotile [*M.A.*, **6**, 149; **7**, 409]. 21st List [Replaces β-uranotile as the accepted name]

Beta-vredenburgite. M. Fleischer, 1944. *Amer. Min.*, vol. 29, p. 247. β-Vredenburgite of B. Mason, *Geol. För. Förh. Stockholm*, 1943, vol. 65, pp. 157, 268. *See* Alpha-vredenburgite. 17th List

Betechtinite, German transliteration of Бетехтинит (Betekhtinite, 21st List). H.-J. Bautsch, *Neues Jahrb. Min.*, Monatschr., 1962, p. 21. 23rd List

Betecktinite, error for Betechtinite. (*Min. Depos.* **5**, 33 (1970).) 26th List

Betekhtinite. A. Schüller and E. Wohlmann, 1955. *Geologie Zeitschr. East Berlin*, vol. 4, p. 535 (Betechtinit). E. Hörne, *Fortschr. Min.*, 1957, vol. 35, p. 50. $Cu_{10}PbS_6$, orthorhombic needles in ores from Mansfeld, Germany. Named after academician A. G. Betekhtin (A. Г. Бетехтин), mineralogist and geochemist, of Moscow. [*M.A.*, **13**, 85.] 21st List [*A.M.*, **41**, 371. $Cu_{10}(Fe,Pb)S_6$]

Betpakdalite. L. P. Ermilova and V. M. Senderova, 1961. Зап. Всесоюз. Мин. Общ. (*Mem. All-Union Min. Soc.*), vol. 90, p. 425 (Бетпакдалит). Near $CaO.Fe_2O_3.As_2O_5.5MoO_3.14H_2O$; perhaps essentially an arsenomolybdate of Ca and Fe, as minute lemon-yellow crystals in a muscovite-quartz greisen from the oxidation zone of the Karaoba tungsten deposit (Central Kazakhstan). 22nd List [*A.M.*, **47**, 172]

Beusite. C. S. Hurlbut Jr., and L. F. Aristarain, 1968. *Amer. Min.* **53**, 1799. $(Mn,Fe)_3(PO_4)_2$, the manganese analogue of graftonite, as red-brown crystals from San Luis Province, Argentina, at Los Aleros, Amanda, and San Salvador. Named for Professor A. Beus. [*M.A.*, 69-2395; *Bull.*, **92**, 511; *Zap.*, **99**, 79.] 26th List [Cf. Magniophilite]

Beyerite. C. Frondel, 1943. *Amer. Min.*, vol. 28, p. 532. E. W. Heinrich, ibid., 1946, vol. 31, p. 198. Carbonate of bismuth (and calcium) as minute tetragonal crystals and earthy masses from Schneeberg, Saxony, and Pala, California. Named after Adolph Beyer (1743–1805), mining engineer of Schneeberg, Saxony, who in 1805 recognized 'luftsaures Wismuth'. Not to be confused with bayerite (12th List). [*M.A.*, **9**, 6.] 17th List [$(Ca,Pb)Bi_2(CO_3)_2O_2$, Colorado; *A.M.*, **32**, 660]

Bialite. H. Buttgenbach. 1929. *Ann. Soc. Géol. Belgique*, Publ. Congo Belge, vol. 51 (for 1927–8), p. c117. Hydrous phosphate of aluminium, magnesium, and calcium, as white orthorhombic needles, from Belgian Congo. Perhaps a magnesian variety of tavistockite. Named after Captain Lucien Bia (1852–1892), a pioneer explorer in Belgian Congo. [*M.A.*, **4**, 148.] 12th List [Is Wavellite; *M.M.*, **37**, 123]

Bianchite. C. Andreatta, 1930. *Atti* (*Rend.*) *R. Accad. Lincei, Cl. Sci. fis. mat. nat. Roma*, ser. 6, vol. 11, p. 760. Abstract in *Chem. Zentr.*, 1930, vol. 2, p. 1355, gives the name incorrectly as Biankit. Hydrous double sulphate of zinc and iron, $FeZn_2(SO_4)_3.18H_2O$, as a white crystalline crust (probably monoclinic) from Raibl, Italy (formerly in Carniola). Named after Prof. Angelo Bianchi, of Padova. [*M.A.*, **4**, 341.] 12th List [*A.M.*, **15**, 538]

Bicchulite. C. Hemi, I. Kusachi, K. Henni, P. A. Sabine, and B. R. Young, 1973. *Min. Journ.* [*Japan*], **7**, 243. An alteration product of gehlenite at Fuka, Okayama Prefecture, Japan, and an isotropic mineral in a coating on a wollastonite rock from Carneal, Northern Ireland, are identical with the artificial gehlenite hydrate of Carlson (1964). Cubic, *a* 8·837 Å; $4[Ca_2Al_2SiO_7.H_2O]$. Named for the Bicchu (= Bitchu) township. 28th List [*See also A.M.*, **63**, 58]

Bidalotite. B. Rama Rao and L. Rama Rao, 1937. *Proc. Indian Acad. Sci.*, Sect. B, vol. 5, p. 290. An orthorhombic pyroxene differing from hypersthene in containing Al_2O_3 (5–10%) and in its optical characters. Named from the locality, Bidaloti, Mysore. [*M.A.*, **7**, 11.] 15th List [Is Anthophyllite; *A.M.*, **33**, 310]

Bideauxite. S. A. Williams, 1970. *Min. Mag.* **37**, 637. Well-formed cubic crystals of $Pb_2AgCl_3(F,OH)_2$ occur on and replacing boléite on a few specimens from the Mammoth-St. Anthony mine, Tiger, Pinal County, Arizona. Named for its discoverer, Richard A. Bideaux. [*M.A.*, 70-2610.] 26th List [*A.M.*, **57**, 1003]

Bikitaite. C. S. Hurlbut, 1957. *Amer. Min.*, vol. 42, p. 792. Hydrous silicate, $LiAlSi_2O_6.H_2O$, monoclinic, as white granular aggregates with eucryptite in lithia-pegmatite from Bikita, Southern Rhodesia. Named from the locality. 21st List [*A.M.*, **43**, 768]

Bilibinite. E. Z. Buryanova, 1958. Зап. Всесоюз. Мин. Общ. (*Mem. All-Union Min. Soc.*), vol. 87, p. 667 (Билибинит). A black amorphous hydrated silicate of uranium, lead, lanthanons, etc., near nenadkevite. Locality not given. Named after Ya. A. Bilibin. [*M.A.*, **14**, 280.] There are too many names for ill-defined amorphous uranium silicates already. 22nd List [Perhaps Coffinite; *A.M.*, **44**, 692]

Bilinite. J. Šebor, 1913. [*Sborník Klubu přírodovědeckého*, Prag, 1913, no. II.] Abstract in

Neues Jahrb. Min., 1914, vol. i, ref. 395 (Bílinit). An iron-alum, $Fe''Fe'''_2S_4O_{16}.24H_2O$, the iron analogue of halotrichite, occurring as white to yellowish, radially-fibrous masses in lignite at Schwaz, near Bilin, Bohemia. 7th List

Billietite. J. F. Vaes, 1947. *Ann. (Bull.) Soc. Géol. Belgique*, vol. 70, p. B 214. Hydrous barium uranate as amber-yellow orthorhombic plates resembling becquerelite, from Katanga. Named in memory of Valère Louis Billiet (1903–1945) of Ghent. [*M.A.*, **10**, 146, 255.] 18th List [$BaU_6O_{19}.11H_2O$; *A.M.*, **45**, 1026]

Billingsleyite. C. Frondel and R. M. Honea, 1968. *Amer. Min.* **53**, 1791. $Ag_7(As,Sb)S_6$, as fine-grained lead-grey aggregates from the North Lily mine, East Tintic district, Utah; orthorhombic. Named for P. Billingsley. [*M.A.*, 69-2383; *Bull.*, **92**, 511; *Zap.*, **99**, 73.] 26th List

Binarite. An obsolete synonym of marcasite. Named, no doubt, from the Latin binarius, since the crystals are usually twinned. The German form Binarkies, attributed to Weiss, appears in the textbooks of Glocker (1831, p. 452), Hartmann, and Quenstedt. The form Binarit is quoted by A. Frenzel, *Mineralogisches Lexicon für das Königreich Sachsen*, 1874, p. 197. 6th List

Binghamite. Author and date? 'A beautiful gemstone unique to Minnesota is the crystalline quartz replacement of fibrous goethite locally named binghamite in honor of William J. Bingham of St. Paul, who, with his son, discovered this attractive material in 1936' (J. Sinkankas, *Gemstones of North America*, 1959, 346). 26th List

Biolite. E. I. Parfenova and E. A. Yarilova, 1958. [Почвоведение (Pedology), no. 12, p. 28], abstr. in *M.A.*, **14**, 98. A group name for minerals formed by biological action. 23rd List

Biopyribole. A. Johannsen, 1911. *Journ. Geol. Chicago*, vol. xix, p. 319. A contraction of the words biotite, pyroxene, and amphibole, suggested for use as a group name. Similarly, Pyribole (q.v.). 6th List

—— D. R. Veblen, P. R. Buseck, and C. W. Burnham, 1977. *Science*, **198**, 359. A term for structures intermediate between the single and double chains of the pyroxenes and amphiboles and the sheets of the micas. [Not the biopyribole of Johannsen (6th List).] [*M.A.*, 78-2118.] 30th List

Biringuccite. C. Cipriani, 1961. *Atti (Rend.) Accad. Naz. Lincei, Cl. Sci. fis. mat. nat.*, ser. 8, vol. 30, pp. 74 and 235; vol. 31, p. 141. $Na_4B_{10}O_{17}.4H_2O$, in recent incrustations at Larderello, Tuscany; monoclinic. Named for the alchemist V. Biringucci (1480–1539) (*Amer. Min.*, 1963, vol. 48, p. 709; *M.A.*, **16**, 373). 23rd List

Birmite. *Dana's Appendix*, 1899. The same as burmite. 2nd List

Birnessite. L. H. P. Jones and A. A. Milne, 1956. *Min. Mag.*, vol. 31, p. 283. A manganese oxide, near $(Na_{0.7},Ca_{0.3})\ Mn_7O_{14}.2{\cdot}8H_2O$, optically uniaxial negative, with X-ray pattern similar to δ-MnO_2 and manganous manganite [*M.A.*, **9**, 227; **10**, 105]. From manganese pan in gravel at Birness, Aberdeenshire. Named from the locality. [*M.A.*, **13**, 237.] 21st List

Birunite. S. T. Badalov and I. M. Golovanov, 1957. [Доклады Акад. Наук Узбек ССР (*Compt. Rend. Acad. Sci. Uzbek SSR*), no. 12, p. 17]; abstr. Зап. Всесоюз. Мин. Общ. (*Mem. All-Union Min. Soc.*), 1959, vol. 88, p. 320 (Бирунит, Birunite). Silicate, carbonate, and sulphate of calcium, occurring with but distinct from thaumasite, from the Almylyk orefield, Uzbekistan. Named after the medieval Uzbek alchemist Abu-r-Raikhana Al-Biruni. [*M.A.*, **14**, 279.] 22nd List

Bisbeeite. W. T. Schaller, 1915. *Journ. Washington Acad. Sci.*, vol. v, p. 7; Third Appendix to 6th edit. of Dana's *System of Mineralogy*, p. 14. A fibrous, orthorhombic, pale-blue hydrated silicate of copper, $CuSiO_3.H_2O$, resulting from the hydration of shattuckite (q.v.). Named from the locality, Bisbee, Arizona. 7th List [Status doubtful]

Bismoclite. E. D. Mountain, 1935. *Min. Mag.*, vol. 24, p. 59. Bismuth oxychloride, pale grey, compact (tetragonal), from South Africa. Named from the chemical composition BiOCl. 14th List

Bismostibnit. H. Strunz, 1966. *Min. Tabellen*, 4th edit., p. 455. Synonym of Horobetsuite (22nd List). 24th List

Bismuth-jamesonite. M. S. Sakharova, 1955, *Trudy Min. Mus. Acad. Sci. USSR*, no. 7, p. 122 (бисмутовый джемсонит). Sulphosalt, $PbS.(Bi,Sb)_2S_3$, as lead-grey acicular crystals, with ustarasite (q.v.) from Siberia. Variety of jamesonite containing some bismuth and very little iron. [*M.A.*, **13**, 164.] 21st List [Is Sakharovaite; *A.M.*, **45**, 1134]

Bismuthmicrolite. N. E. Zalashkova, 1957. Труды Инст. Мин. Геохим. Крист. Редк. Элем. (*Trans. Inst. Min. Geochem. Cryst. Rare Elements*), no. 1, p. 77 (Висмутомикролит). Abstr. Зап. Всесоюз. Мин. Общ. (*Mem. All-Union Min. Soc.*), 1958, vol. 87, p. 479 (Висмутомикролит, bismuthmicrolite); *Amer. Min.*, 1958, vol. 43, p. 1223; *Bull. Soc. franç. Min. Crist.*, 1959, vol. 82, p. 90 (bismuthomicrolite); Hintze, *Handb. Min.*, Erg.-Bd. II, 1960, p. 926 (Bismutomikrolith). A bismuthian microlite (3% Bi_2O_3) from the Altai Mts. [*M.A.*, **14**, 276.] An unnecessary name. 22nd List.

Bismuth-parkerite. J. W. du Preez, 1944. *Ann. Univ. Stellenbosch*, vol. 22, sect. A, p. 101 (Bismuth-Parkerite). Parkerite (14th List), originally described as a nickel sulphide, has been found to contain also bismuth and lead. The end-members bismuth-parkerite and lead-parkerite form mixed crystals in the system $Ni_3Bi_2S_2$–$Ni_3Pb_2S_2$. [*M.A.*, **9**, 5, 126.] 17th List

Bismutodiaphorite. A. A. Godovikov, 1972. Abstr. in *Zap.*, **103**, 617 (1974) (Бисмутодиафорит). Unnecessary, invalid name based on an old analysis near $(Ag,Cu)_6Pb_4Bi_6S_{16}$. 29th List

Bismutoniobite. G. Frenzel, 1955. *Neues Jahrb. Min.*, Monatshefte, p. 243 (Bismutoniobit). End-member of the series $Bi(Ta,Nb)O_4$. Compare bismutotantalite, 12th List. 21st List

Bismutoplagionite. E. V. Shannon, 1920. *Amer. Journ. Sci.*, ser. 4, vol. 49, p. 166; *Chem. News*, 1920, vol. 120, p. 234; *Proc. US Nat. Museum*, 1920, vol. 58, p. 598. A lead-grey, indistinctly fibrous mineral, perhaps orthorhombic, with the composition $5PbS.4Bi_2S_3$. This is the same ratio as in plagionite, but with bismuth in place of antimony; hence the name. Plagionite has, however, a granular, rather than a fibrous, structure. [*M.A.*, **1**, 75, 151.] 9th List [Is Cannizzarite; *A.M.*, **23**, 790]

Bismutosmaltine. A. Frenzel, *Tsch. Min. Mitth.* XVI. 524, 1897. $Co(As,Bi)_3$. Cubic [Zschorlau, near] Schneeberg, Saxony. 1st List

Bismutosmaltite. E. T. Wherry, *Journ. Washington Acad. Sci.*, 1920, vol. 10, p. 495. Variant of Bismutosmaltine (A. Frenzel, 1897). 9th List

Bismutospherite. Variant of Bismutosphaerite. *A.M.*, **9**, 62. 10th List

Bismutotantalite. E. J. Wayland and L. J. Spencer, 1929. *Min. Mag.*, vol. 22, p. 185. Tantalate (and niobate) of bismuth, $Bi_2O_3.Ta_2O_5$, as black orthorhombic crystals from Uganda. Named from the chemical composition on analogy with stibiotantalite. *See* Ugandite. 12th List [*A.M.*, **14**, 312; **15**, 201]

Biteplatinite and Bitepalladite. Huang Van-Kang, Yeh Hsien-Hsien, Chang Yuen-Ming, Chuang Tsan-Fu, and Fang Chun-Ming, 1974. [*Geochimica*, **4**, 258], abstr. *A.M.*, **61**, 174. Unnecessary names for Pt-rich moncheïte and Pd-rich merenskyite respectively, from an unnamed locality. 29th List

Bitumenite. T. S. Traill, 1853, *Trans. R. Soc. Edinb.* 1857, XXI, 7. The same as torbanite. Groth (*Tab. Uebers. Min.* 2nd–4th edit.) gives *bituminite*. 2nd List

Bityite. A Lacroix, 1908. *Compt. Rend. Acad. Sci. Paris*, vol. cxlvi, p. 1371; *Bull. Soc. franç. Min.*, vol. xxxi, p. 241. Small, yellowish-white, pseudo-hexagonal prisms occurring with tourmaline in pegmatite near Mt. Bity [or Ibity] in Madagascar. A basic orthosilicate, $10SiO_2.8Al_2O_3.5\frac{1}{2}(Ca,Be,Mg)O.1\frac{1}{2}(Li,Na,K)_2O.7H_2O$, allied to staurolite. 5th List [Cf. Bowleyite. Margarite group; *M.A.*, **14**, 136; *A.M.*, **35**, 1091]

Bixbite. (A. Eppler, *Die Schmuck- und Edelsteine*, Stuttgart, 1912, p. 253 (Bixbit).) A gooseberry-red beryl found to the SW of Simpson Spring, Utah. 7th List

Bixbyite. S. L. Penfield and H. W. Foote, *Amer. Journ. Sci.* IV. (ser. 4), 105, 1897. Ferrous manganite. $FeO.MnO_2$. Cubic. [35 miles SW of Simpson] Utah. 1st List [Cf. Kurnakite, Partridgeite, Sitaparite]

Bjarebyite. P. B. Moore, D. H. Lund, and K. L. Keester, 1973. *Min. Record*, **4**, 282. Emerald green monoclinic crystals (*a* 8·930, *b* 12·073, *c* 4·917 Å, *β* 100·15°) from the Palermo pegmatite, North Groton, New Hampshire, gave an electron-probe analysis agreeing with a formula $BaFeMnAl_2(PO_4)_3(OH)_3$, with some Sr and Mg. 28th List [*A.M.*, **59**, 873]

Blackeit, error for Blakeit (of Dana, 1850) (H. Strunz, *Min. Tabellen*, 1st edit., 1941, p. 229, and later edits.). 24th List

Blakeite. C. Frondel and F. H. Pough, 1944. *Amer. Min.*, vol. 29, p. 211. Anhydrous ferric tellurite as reddish-brown microcrystalline (cubic?) crusts from Goldfield, Nevada. Named after Professor William Phipps Blake (1826–1910), who was the first to recognize tellurium minerals in California. [*M.A.*, **9**, 62.] 17th List

—— G. Gagarin and J. R. Cuomo, 1949, loc. cit., p. 10 (blakeita). Titano-zirconate of Th, U, Ca, Fe, etc., described as zirkelite from Ceylon by G. S. Blake and G. F. H. Smith (*Min. Mag.*, **16**, 309), but differing in chemical composition and also apparently in crystalline form from the original zirkelite [q.v.] from Brazil (*Min. Mag.*, **11**, 86, 180). Named after George Stanfield Blake (1876–1940), formerly government geologist in Palestine. Not the blakeite of J. D. Dana, 1850, nor of C. Frondel and F. H. Pough, 1944 (17th List). 19th List

Blanchardite. M. F. Strong, 1964. *The Mineralogist*, vol. 32, no. 3, p. 5. In a description of the Old Hansonburg or Blanchard lead mine, Bingham, New Mexico, it is stated that 'A (?) new mineral from the claim, *blanchardite* is being studied by Dr. Clifford Frondel'. [*M.A.*, **17**, 233.] 24th List [Is Brochantite, *A.M.*, **58**, 562]

Blanfordite. L. L. Fermor, 1906. *Trans. Mining and Geol. Inst. India*, vol. i, p. 78. A monoclinic pyroxene containing some sodium, manganese, and iron, occurring with manganese ores in the Central Provinces, India. The strong pleochroism (rose-pink to sky-blue) is a prominent character. Named after the late Dr. William Thomas Blanford (1832–1905). 4th List [*M.A.*, **16**, 645; **17**, 392; 71-1309]

Blastonite. R. E. van Alstine, 1944. *Econ. Geol.*, vol. 39, p. 117. Local name for brecciated fluorite in a matrix of cryptocrystalline quartz from Newfoundland. [*M.A.*, **9**, 252.] 17th List

Blätterserpentin. P. Groth, *Tab. Uebers. Min.* 4th edit. 1898, p. 135. Lamellar serpentine = antigorite. 2nd List

Bleiantimonspießglanze. H. Strunz. *Min. Tabellen*, 3rd edit., pp. 86, 109. A group name, including fülöppite, zinckenite, robinsonite, plagionite, heteromorphite, jamesonite, parajamesonite, semseyite, boulangerite, meneghinite, and sakharovaite (bismuth-jamesonite). 22nd List

Bleiapatit, etc. F. Machatschki, 1953. *Spezielle Mineralogie*, Wien, p. 330. In the apatite family, pyromorphite is named Bleiapatit. Similarly, Arsenatapatit = svabite, Bleiarsenatapatit = mimetite, Bleivanadatapatit = vanadinite, Silikatsulfatapatit = wilkeite, Cererdensilikatapatit = britholite, Yttererdensilikatapatit = abukumalite. [*M.A.*, **12**, 232.] 20th List [Also 22nd List]

Bleiarsenapatit, Syn. of mimetite. *See* Bleiapatit.

Bleiarsenspießglanze. H. Strunz, 1957. *Min. Tabellen*, 3rd edit., pp. 86, 108. A group name, including sartorite, baumhauerite, liveingite, rathite, dufrenoysite, lengenbachite, jordanite, geocronite, and gratonite. 22nd List

Blei-Barysilit, original form of Lead Barysilite (q.v.). 28th List

Blei-Brom-Apatit. H. Wondratschek, 1959. Original form of Brompyromorphite (q.v.). 23rd List

Blei-Jod-Apatit. H. Wondratschek, 1959. Original form of Iodopyromorphite (q.v.). 23rd List

Bleikupferarsen, cited by H. Strunz (*Min. Tabellen*, 4th edit., 1966, p. 456) as a synonym of Duftite (9th List) was stated by C. Guillemin (*Bull. Soc. franç. Min. Crist.*, 1956, vol. 79, p. 75) to be merely a name on a dealer's label on a specimen bearing a mixture of Malachite and α-Duftite (22nd List). 24th List

Bleikupferspießglanze. H. Strunz, 1957. *Min. Tabellen*, 3rd edit., pp. 86, 106. A group name, including seligmannite, bournonite, aikinite, hammarite, lindströmite, gladite, rézbányite, and wittite. 22nd List

Bleimalachit. S. F. Glinka, *Centralblatt Min.*, 1901, p. 281. Acicular crystals of orthorhombic habit, but monoclinic and twinned. $2CuCO_3.Cu(OH)_2.PbCO_3$. Altai Mountains. *See* plumbomalachite. 3rd List

Bleiromeit. F. Machatschki, 1953. *Spez. Min.*, Wien, p. 328. Syn. of Monimolite. 22nd List

Bleisilberspießglanze. H. Strunz, 1957. *Min. Tabellen*, 3rd edit., pp. 86, 107. A group name, including hutchinsonite, fiźlyite, ramdohrite, andorite, freieslebenite, diaphorite, owyheeite, schirmerite, and benjaminite. 22nd List

Bleispießglanze. H. Strunz, 1957. *Min. Tabellen*, 3rd edit., pp. 86, 108. A group name, subdivided into Bleiarsenspießglanze, Bleiantimonspießglanze, and Bleiwismutspießglanze (qq.v.). 22nd List

Bleivanadatapatit. Syn. of Vanadinite. *See* Bleiapatit. 22nd List

Bleiwismutspießglanze. H. Strunz, 1957. *Min. Tabellen*, 3rd edn, pp. 86, 111. A group name, including ustarasite, galenobismutite, cannizzarite, cosalite, kobellite, weibullite, bursaite, [and bonchevite]. 22nd List

Bleizinkchrysolith. P. P. Heberdey, *Zeits. Kryst. Min.*, 1892, vol. xxi, p. 56. An artificial product (from a crystallized slag) with the composition $PbZnSiO_4$ and isomorphous with chrysolite. 3rd List

Bleizinkolivenit. Author ?; quoted by C. Guillemin, *Bull. Soc. franç. Min. Crist.*, 1956, vol. 79, p. 71. Synonym of α-Duftite. 23rd List

Bliabergite. L. J. Igelström, *Zeits. Kryst. Min.* XXVII. 603, 1897; spelt bliabergsite in *Geol. För. Förh.* XVIII. 41, 1896. M. Weibull, *Geol. För. Förh.* XVIII. 515, 1896. Shown by Weibull to be a brittle mica near ottrelite. Bliaberg, Sweden. 1st List

Blixite. O. Gabrielson, A. Parwell, and F. E. Wickman, 1958. *Arkiv. Min. Geol.*, vol. 2, p. 411 (Blixit). A basic lead chloride, $Pb_4Cl_2O_3$ or $Pb_{16}Cl_8(O,OH)_{16-x}$ with x about 2·6, occurring as a fissure mineral at Långban, Sweden. Named after Dr. Ragnar Blix. [*M.A.*, **14**, 416.] 22nd List [*A.M.*, **45**, 908]

Blockite. R. Herzenberg, 1935. *Zentr. Min.*, Abt. A, 1935, p. 189; R. Herzenberg and F. Ahlfeld, ibid., p. 277 (Blockit). Nickel and copper selenide $(Ni,Cu)Se_2$, as shelly masses, from Colquechaca, Bolivia. F. A. Bannister and M. H. Hey (*Amer. Min.*, 1937, vol. 22, p. 319) identify it with penroseite (11th List). Named after the mining engineer, Hans Block, who collected the material. [*M.A.*, **6**, 147, 490.] 14th List

Blomstrandine. W. C. Brøgger, 1906. *Videnskabs-Selsk. Skrifter*, Kristiania, 1906, no. 6, p. 98 (Blomstrandin). A titano-niobate of yttrium-earths, thorium, uranium, etc., occurring at Hitterö, Arendal, and other localities in Norway, as large orthorhombic crystals, which had previously (W. C. Brøgger, 1879) been referred to aeschynite. Named after the late Professor Christian Wilhelm Blomstrand (1826–1897), of Lund, Sweden, whose analyses showed it to be distinct from aeschynite. It is dimorphous with polycrase and isomorphous with priorite (q.v.). Not to be confused with the blomstrandite of G. Lindström (1874), which is a hydrated titano-niobate and -tantalate of uranium, etc. 4th List [Aeschynite-(Y), Fleischer's *Gloss.*]

Blomstrandinite. A. N. Winchell, 1927. *Elements of optical mineralogy*, 2nd edit., part 2, p. 161. The same as Blomstrandine (of W. C. Brøgger, 1906; 4th List) with a double termination. Not the Blomstrandite of G. Lindström, 1874, also named after C. W. Blomstrand (1826–1897). 12th List

Blueite. S. H. Emmens, *Journ. Amer. Chem. Soc.* XIV. 207, 1892; S. L. Penfield, *Amer. Journ. Sci.* XLV. 496, 1893. Shown by Penfield to be nickeliferous pyrites. Sudbury, Canada. 1st List

Blythite. L. L. Fermor, 1926. *Rec. Geol. Survey India*, vol. 59, p. 204. E. V. Shannon, *Journ. Washington Acad. Sci.*, 1927, vol. 17, p. 444. A hypothetical garnet molecule $3MnO.Mn_2O_3.3SiO_2$. Named after Mr. T. R. Blyth (d. 1911), assistant curator of the Geological Survey of India. Other garnet molecules assumed are $3FeO.Fe_2O_3.3SiO_2$ (named skiagite, q.v.), and $3MnO.Fe_2O_3.3SiO_2$ (to which the name calderite of H. Piddington, 1850, is restricted). [*M.A.*, **3**, 308; **3**, 458.] 11th List

Boakite. A trade-name for a brecciated green and red jasper. R. Webster, *Gems*, 1962, p. 753. 28th List

Bobkovite. Y. V. Kazitzyn, 1955. [*Kristallografiya, Acad. Sci. USSR*, no. 4, p. 116.] Abstracts in *Mém. Soc. Russ. Min.*, 1956, vol. 85, p. 376 (Бобковит); *Amer. Min.*, 1957, vol. 42, p. 440. A variety of opal, SiO_2 89·20, H_2O 3·24%, with small amounts of Al, Fe, Ca, Mg, K. Named after N. A. Bobkov, crystallographer, of Leningrad. [*M.A.*, **13**, 521.] 21st List

Bobkowit. German transliteration of Бобковит, bobkovite (21st List). (Hintze, *Handb. Min.*, 1959, Erg.-Bd. II, p. 670.) 22nd List

Bobrovkite. N. K. Vuisotzkiĭ (= N. Wyssotzky), 1913. *Mem. Comité Géol. St.-Pétersbourg*, nouv. sér., livr. 62, p. 106 (бобровкить). p. 668 (Bobrowkit). L. Duparc and M. N. Tikonowitch*, Le platine, Genève, 1920, p. 193 (Bobrowkite). An alloy of nickel and iron, Ni_5Fe_2, found as fine scales in the platiniferous sands of the Bobrovka river, Nijni-Tagil, Urals. Not proved to differ from awaruite, etc. [*Corr. in *M.M.*, **19**.] 9th List

Bodenbenderite. E. Rimann, 1928. *Sitzungsber. Abhandl. Naturwiss. Gesell. Isis*, Dresden, Festschrift Richard Baldauf, p. 42 (Bodenbenderit). Silicate and titanate of aluminium, yttrium, manganese, etc., $(Mn,Ca)_4Al[(Al,Yt)O][Si,Ti)O_4]_3$, as flesh-red cubic crystals resembling garnet, from the Sierra de Córdoba, Argentina. Named after Prof. Wilhelm (Guillermo) Bodenbender (1857–), of Córdoba, Argentina, [*M.A.*, **3**, 472.] 11th List [A mixture; *A.M.*, **34**, 608]

Bodenzeolith. *See* geolyte. 3rd List

Bodhanowiczyte. Error for Bohdanowiczyte (q.v.). 25th List

Boehmite. J. de Lapparent, 1927. *Comp. Rend. Acad. Sci. Paris*, vol. 184, p. 1662. Aluminium hydroxide, assumed to be $Al_2O_3.H_2O$, as microscopic orthorhombic crystals (distinct from diaspore and isomorphous with lepidocrocite) forming a constituent of some bauxites. It is assumed to be identical with the 'bauxite' examined by J. Böhm of Berlin in 1925; hence the name. [*M.A.*, **3**, 369; **3**, 430.] 11th List [*A.M.*, **13**, 72. *See* Böhmite]

Böggildit. Variant of Bøggildite (20th List). (H. Strunz, *Min. Tabellen*, 3rd edit., 1957, p. 234.) 22nd List

Bøggildite. R. Bøgvad, 1952 (MS.). A. H. Nielsen, *Acta Chem. Scand.*, 1954, vol. 8, p. 136. A fluoride $Na_2Sr_2Al_2(PO_4)F_9$ from the Greenland cryolite deposit. Named after Ove Balthasar Bøggild (1872–), emeritus professor of mineralogy in the university of Copenhagen. [*M.A.*, **12**, 14, 353.] 20th List [*A.M.*, **39**, 848; **41**, 959]

Bohdanowiczyte*. M. Bonas and J. Ottemann, 1967. [*Przegl. geol. Polska*, vol. 5, p. 340], abstr. *Bull. Soc. franç. Min. Crist.*, 1968, vol. 91, p. 101. An inadequately described bismutoselenide of silver from Kletna, Poland. [*Corr. in *M.M.*, **37**.] 25th List [$AgBiSe_2$; *A.M.*, **55**, 2135.]

Böhmite. A more correct form of Boehmite (11th List), named by J. de Lapparent (1927) in honour of Dr. Johann Böhm of Berlin: corrected in *Neues Jahrb. Min.*, 1928, Abt. A, Ref. I, p. 84 (Böhmit). The correct form has been fairly widely used in Germany (Hintze. *Handb. Min.*, Erg.-Bd., p. 86; H. Strunz, *Min. Tabellen*, 2nd and 3rd edits.; F. Machatschki, *Spez. Min.*, Wien; etc.), but hardly ever in Britain or America; it is perhaps not too late to correct this spelling. 22nd List

Bokite. E. A. Ankinovich, 1963. Зап. Всесоюз. Мин. Общ. (*Mem. All-Union Min. Soc.*), vol. 92, p. 51 (Бокит). Black massive material from shales in the Balasauskandyk area. Kara-Tau, Kazakhstan, is near the ill-defined Corvusite (13th List; Henderson and Hass, 1933), but has a different V^{4+}/V^{5+} ratio and more Fe and Al. Its composition is near $KAl_3Fe_6V_6^{4+}V_{20}^{5+}O_{76}.30H_2O$. Named for I. I. Boky. [*M.A.*, **16**, 285.] 23rd List [*A.M.*, **48**, 1180]

Boksputite. E. D. Mountain, 1935. *Min. Mag.*, vol. 24, p. 62. Carbonate of lead and bismuth, $6PbO.Bi_2O_3.3CO_2$, pale yellow, compact. Named from the locality, Boksput farm, Gordonia, South Africa. 14th List [A mixture; *A.M.*, **32**, 365]

Boldyrevite. G. Gagarin and J. R. Cuomo, 1949, loc. cit., p. 8 (boldyrevita). $NaCaMgAl_3F_{14}.4H_2O$ as yellow crusts, optically isotropic, in fumaroles on lava from the Klyuchevsky volcano in Kamchatka, described by S. I. Naboko in 1941 [*M.A.*, **8**, 342]. Named after Anatolii Kapitonovich Boldyrev Анатолий Капитонович Болдырев (1883–1946), professor of crystallography and mineralogy in the Mining Institute of Leningrad. 19th List

Boleslavite. C. Haranczyk, 1961. *Bull. Acad. Polon. Sci., sér. sci. géol. géogr.*, vol. 9, p. 85. A superfluous name for finely divided (colloidal) galena. [*M.A.*, **16**, 546.] 23rd List

Bolidenite. P. Ramdohr and O. Ödman, *Neues Jahrb. Min.*, Abt. A. 1940, Beil.-Bd. 75, p. 317 (Bolidenit). Local name for falkmanite (q.v.), and also other ores, at the Boliden mines, north Sweden. 15th List

Bolivarite. L. Fernández Navarro and P. Castro Barea, 1921. *Bol. Soc. Española Hist. Nat.*, vol. 21, p. 326 (bolivarita). A hydrous aluminium phosphate, $AlPO_4.Al(OH)_3.H_2O$, occurring as greenish-yellow, cryptocrystalline crusts on granite near Pontevedra, Spain. Named after the Spanish entomologist Ignacio Bolívar. [*M.A.*, **1**, 378.] 9th List

Bolivianite. A. Pauly, 1923. *Anal. Soc. Cient. Argentina*, vol. 96, p. 273 (bolivianita); *Centr. Min.*, Abt. A, 1926, p. 43 (Bolivianit). Incompletely described as a sulphide of tin, copper, and iron forming black trigonal crystals [evidently identical with stannite]. Named after the locality, Bolivia, where the name has been in use for some years (*see Mining Journ.* London, 1914, vol. 104, p. 147). Not the Bolivian of A. Breithaupt, 1866, = Bolivianite, J. D. Dana, 1868. [*M.A.*, **3**, 112; **3**, 370.] 11th List

Boltwoodite. C. Frondel and J. Ito, 1956. *Science* (*Amer. Assoc. Adv. Sci.*), vol. 124, p. 931. Hydrous potassium uranyl silicate, $K_2(UO_2)_2(SiO_3)_2(OH)_2.5H_2O$, analogous to sklodowskite with K in place of Mg, orthorhombic or monoclinic, as yellow fibres from Utah. Named after Bertram Borden Boltwood (1870–1927), professor of radiochemistry at Yale University; discoverer of ionium. [*M.A.*, **13**, 380.] 21st List [*A.M.*, **46**, 12]

Bombollalite. Error for bambollaite, q.v. 28th List

Bonaccordite. S. A. Dewaal, E. A. Viljoen, and L. C. Calk, 1974. *Trans. Geol. Soc. S. Africa*, **77**, 373. The nickel analogue of ludwigite occurs as clusters of reddish-brown tiny prisms with other nickel minerals in the Bon Accord area, Barberton Mountain Land, Transvaal. *a* 9·213, *b* 12·229, *c* 3·001 Å, space group assumed to be *Pbam*. Composition Ni_2FeBO_5. Named for the locality. [*A.M.*, **61**, 502.] 29th List

Bonamite. A jeweller's trade-name for an apple-green calamine ($ZnCO_3$), resembling chrysoprase in colour, from Kelly, New Mexico. Called 'bonamite' by Goodfriend Brothers, of New York [no doubt in playful allusion to their own name]. (D. B. Sterrett, *Mineral Resources of the United States*, for 1908, 1909, part ii, p. 839.) 5th List

Bonattite. C. L. Garavelli, 1957. *Rend. Soc. Min. Ital.*, vol. 13, p. 269. $CuSO_4.3H_2O$, monoclinic. Partly dehydrated chalcanthite ($CuSO_4.5H_2O$) from Elba. Named after Prof. Stephano Bonatti of Pisa. 21st List

Bonchevite. I. Kostov, 1958. *Min. Soc. Notice*, 1958, no. 101; *Min. Mag.*, vol. 31, p. 821. Sulphosalt, $PbBi_4S_7$, orthorhombic. From Rhodope Mts., Bulgaria, in quartz-scheelite veins. Named after George Bonchev, Георгий Бончев (1866–1955), formerly professor of mineralogy and petrology, University of Sofia. 21st List [Perhaps a mixture; *A.M.*, **55**, 1449; *M.A.*, **19**, 222; 71-523]

Boodtite. L. De Leenheer, 1936. *Natuurwet. Tijdschr. Gent*, vol. 18, p. 77 (boodtiet). Hydrated oxide of cobalt, copper, and iron, $5Co_2O_3.CuO.Fe_2O_3.11H_2O$, as friable grey-black masses, from Katanga. Named after Anselm Boëthius de Boodt (1550–1634) of Bruges, author of 'Gemmarum et lapidum historia' (1609). [*M.A.*, **6**, 343.] 14th List [Impure Heterogenite; *M.M.*, **33**, 253]

Boothite. W. T. Schaller, *Bull. Dep. Geol. Univ. California*, 1903, vol. iii, p. 207. A copper sulphate with $7H_2O$, instead of $5H_2O$ as in chalcanthite; $CuSO_4.7H_2O$. Monoclinic, and isomorphous with melanterite. Occurs with other sulphates as an alteration product of chalcopyrite in the Alma mine, Leona Heights, Alameda Co., California. Named after Mr. Edward Booth, of the University of California. 3rd List

—— J. Fröbel, 1843. *Grundzüge eines Systemes der Krystallologie*. A nickel arsenide from Richelsdorf described by Booth in 1836, later included in chloanthite. Not to be confused with the boothite of Schaller (1903; 3rd List). 27th List

Borax, octahedral. *See* 'Octahedral borax'. 4th List

Borcarite. N. N. Pertsev, I. V. Ostrovskaya, and I. B. Nikitina, 1965. Зап. Всесоюз. Мин. Общ. (*Mem. All-Union Min. Soc.*), vol. 94, p. 180 (Боркарит, borcarite). Dense blue-green masses in Kotoite marbles from an unnamed locality in Siberia are monoclinic, with composition $Ca_4Mg(HBO_3)_4(HCO_3)_2$. Named for the composition. [*M.A.*, **17**, 398; *A.M.*, **50**, 2097.] 24th List

Borgehlenit. Original form of Boron-gehlenite (q.v.). 28th List

Borgniezite. P. de Béthune, 1956. *Ann.* (*Bull.*) *Soc. Géol. Belgique*, vol. 80, p. в 63 (borgniézite); *Compt. Rend. Acad. Sci. Paris*, 1956, vol. 243, p. 1133 (borgniezite). A soda-amphibole with special optical characters (pleochroism, high extinction), occurring with aegirine in carbonatite and surrounding schists at Lueshe, Kivu, Belgian Congo. Previously described, but without name, by the author in *Mém. Inst. Géol. Univ. Louvain*, 1952, vol.

16, p. 278. Named after Georges Borgniez of Auderghem, Belgium. 21st List [Soda amphibole; *Amph.* (1978)]

Borgströmite. M. Saxén, 1921, *Meddel. Geol. Fören. Helsingfors*, for 1919–1920, p. 20 (borgströmit). The spelling Borgstroemite is given in *Amer. Min.*, 1923, vol. 8, p. 187. A basic ferric sulphate $Fe_2O_3.SO_3.3H_2O$, occurring as a yellow, earthy weathering product of pyrites deposits at Otravaara, Finland. Named after Prof. Johan Leonard Henrik Borgström of Helsingfors. E. Posnjak and H. E. Merwin, *Journ. Amer. Chem. Soc.*, 1922, vol. 44, p. 1977, suggest that this is identical with their artificial salt $3Fe_2O_3.4SO_3.9H_2O$, slightly contaminated with limonite. [*M.A.*, **2**, 10.] 10th List [Is jarosite; *M.M.*, **31**, 408]

Borishanskite. L. V. Razin, L. S. Dubakina, V. I. Meshchankina, and V. D. Begizov, 1975. *Zap.*, **104**, 57 (Боришанскит). Small (40 to 150 μm) grains in the Talnakh ore deposit, USSR, are orthorhombic, *Ccm*2₁, *a* 7·18, *b* 8·62, *c* 10·66 Å; 16 [PdPbAs]. Named for S. I. Borishanskii. [*A.M.*, **61**, 502.] 29th List

Borkarit, variant of Borcarite (24th List) (C. Hintze, *Handb. Min.*, Erg. III, p. 531). 25th List

Bornemanite. Yu. P. Menshikov, I. V. Bussen, E. A. Goiko, N. I. Zabavnikova, A. N. Merkov, and A. P. Khomyakov, 1975. *Zap.*, **104**, 322 (Борнеманит). Yellowish platy coatings on lomonosovite, or rarely in natrolite, occur in the Jubilee (Юбилей) pegmatite of the Lovozero massif. Orthorhombic, *a* 5·48, *b* 7·10, *c* 48·2 Å, *Ibmm* or *Ibm*2; composition near $4[Na_7BaTi_2NbSi_4O_{17}FPO_4]$. Named for I. D. Borneman-Starynkevich. [*M.A.*, 76-878; *A.M.*, **61**, 338.] 29th List

Bornhardtite. P. Ramdohr and M. Schmitt, 1955. *Neues Jahrb. Min.*, Monatshefte, no. 6, p. 141 (Bornhardtit). Cobalt selenide, Co_3Se_4, cubic (linnaeite group) from Trogtal, Harz. Named after Dr. W. Bornhardt, mine manager. [*M.A.*, **13**, 5.] 21st List [*A.M.*, **41**, 164]

Bornite, Orange. Murdoch, 1916. This ore-microscopic term has been used for material later identified as renierite, mawsonite, or stannite (*A.M.*, **50**, 900; *M.A.*, **16**, 540). 28th List

Boron-edenite. J. A. Kohn and J. E. Comeforo, 1955. *Amer. Min.*, vol. 40, p. 410 (fluor-boron edenite), p. 411, (boron edenite), p. 413 (boron-edenite). Artificial $NaCa_2Mg_5(Si_{3·5}B_{0·5}O_{11})_2F_2$, containing B_2O_3 3·91%. [*M.A.*, **13**, 486.] 21st List

Boron-gehlenite. H. Bauer, 1962. *Neues Jahrb. Min., Monatsh.* 127 (Borgehlenit). *Artificial* $Ca_2B_2SiO_7$, isomorphous with Gehlenite and giving mix-crystals with the latter. Name modified by P. Černý, C.M. 1970, **10**, 636. 28th List

Boron-melilite. J. Tarney, A. W. Nicol, and G. F. Marriner, 1973. *M.M.*, **39**, 158. An *artificial* product, $Ca_2SiB_2O_7$, obtained by heating datolite to 700°C. Tetragonal, *a* 7·14, *c* 4·82 Å; isostructural with melilite and named accordingly. [*M.A.*, 73-3730.] 28th List

Boron-phlogopite. R. A. Hatch, R. A. Humphrey, and E. C. Worden, 1956. *US Bureau of Mines*, report 5283, p. 28 (boron-phlogopite). Artificial $KMg_3BSi_3O_{10}F_2$. [*M.A.*, **13**, 448.] 21st List

Borovskite. A. A. Yalovai, A. F. Sidorov, N. S. Rydashevskii, and I. A. Budko, 1973. *Zap.*, **102**, 427 (Боровскит) Pd_3SbTe_4, cubic (*a*5·794 Å), in pentlandite–chalcopyrite–pyrrhotine ores of the Khautovaar deposit, Karelia. Named for I. B. Borovsky. [*Zap.*, **102**, 425; **103**, 356.] 28th List [*A.M.*, **59**, 873]

Bortz. (E. H. Kraus and W. F. Hunt, *Mineralogy*, 1920, pp. 188, 192.) A corruption of the plural (as often used in the trade) of bort. 9th List

Boryslawite. *British Museum Index of minerals*, 18th edit., 1895 (Boryslavite), from a dealer's label (1890). C. Hintze, *Handb. Min.*, 1933, vol. 1, Abt. 4, pt. 2, p. 1362 (Boryslawit). A hard brittle variety of ozocerite from Boryslaw, Galicia. 18th List

Börzsönyite. F. Papp, 1933. *Földtani Közlöny*, Budapest, vol. 62 (for 1932), p. 61 (Börzsönyit). Suggested as an alternative for the names wehrlite (J. J. N. Huot, 1841) and pilsenite (A. Kenngott, 1853) for an incompletely determined bismuth telluride [probably identical with tetradymite] from Börzsöny = Deutsch-Pilsen, comitat Hont, Hungary; because the name wehrlite (F. Kobell, 1838) is also applied to a peridotite rock [earlier thought to be the mineral ilvaite], and the German name Deutsch-Pilsen for the locality is now not recognized in Hungary. 13th List

Bosphorite. P. A. Dvoichenko, 1914. ['Minerals of the Crimea' (Russ.), *Zap. Krym. Obshch. Est. i Lyub. Prir.* (*Mem. Crimean Soc. Sci. & Nat.*), vol. 4, pp. 113–114; also published separately as a book, Petrograd, 1914]; abstract in *Zeits. Krist.*, 1925, vol. 61, p. 586 (Bosphorit). Hydrated ferric phosphate $3Fe_2O_3.2P_2O_5.17H_2O$, as a compact yellow encrustation on limonite from Yanysh-Takil, Kerch Peninsula. As an alteration product of vivianite, it had been described by S. P. Popov (*Trav. Mus. Géol. Acad. Sci. St.-Pétersbourg*, 1911, vol. 4, p. 173; abstract in *Zeits. Kryst. Min.*, 1913, vol. 52, p. 611). 10th List

Bostrichites. R. Jameson, 1800. *Min. Scot. Isles*, vol. 1, p. 11. An early synonym of Prehnite. 23rd List

Botesite. K. Vrba, 1897. *Ottův Slovník Naučný* [*Otto's Encyclopaedia*], Praha, 1897, vol. 11, p. 230. F. Slavík in C. Doelter's *Handbuch d. Mineralchemie*, 1926, vol. 4, pt. 1, p. 868 (Botesit). Synonym of Hessite. The name was used by K. Vrba on MS. labels in the University mineral collection at Praha in 1882: it is unknown in Hungarian literature. Named from the locality, Mt. Botes, Transylvania. 12th List

Botryogenite. Variant of Botryogen. *A.M.*, **8**, 51. 10th List

Bouazzerite. Name found on a dealer's label by J. Paclt, *Neues Jahrb. Min.*, Monatshefte, 1953, p. 188 (Bouazzerit). A ferriferous variety of stichtite from Bou Azzer, Morocco. [*M.A.*, **9**, 121; **12**, 239.] 20th List

Bouglisite. E. Cumenge [A. Lacroix, *Bull. Mus. d'Hist. Nat. Paris*, 1895, p. 42]. F. A. Genth, *Amer. Journ. Sci.* XLV. 32, 1893. Described by Genth as a mixture of anglesite and gypsum. $2PbSO_4 + CaSO_4 + 2H_2O$. Boleo, Lower California. 1st List

Bourgeoisite. R. Breñosa, *Anal. Soc. Española Hist. Nat.*, 1885, vol. xiv, p. 129 (Bourgeoisita). Described as a tetragonal modification of calcium silicate dimorphous with wollastonite. In a devitrified glass of doubtful origin. Named after Dr. L. Bourgeois, of Paris. 3rd List

Boussingaultin. J. Fröbel, 1843. *See* chrysargyrite (this List). 27th List [Not to be confused with Boussaingaultite]

Bowleyite. H. P. Rowledge and J. D. Hayton, 1948. *Journ. Roy. Soc. W. Australia*, vol. 33 (for 1946–1947), p. 45. Hydrous silicate of Al, Ca, Be (BeO 7·30%), $3(Ca,Be)O.2Al_2O_3.3SiO_2.2H_2O + n(Li,Na)_2O$, as white compact material with beryl in pegmatite, from Londonderry, Western Australia. Named after H. Bowley, late government Chemist and Mineralogist of Western Australia. [*M.A.*, **10**, 508.] 18th List [Is Bityite (q.v.); *A.M.*, **35**, 1091; *M.A.*, **14**, 136]

Bowmanite. R. H. Solly, 1904. *Nature*, vol. lxxi, p. 118; *Min. Mag.*, 1905, vol. xiv, p. 80. Small, honey-yellow, rhombohedral crystals from the white, crystalline dolomite of the Binnenthal in Switzerland. Named after Herbert Lister Bowman, of Oxford, and proved by him (*Min. Mag.*, 1907, vol. xiv, p. 389) to be identical with hamlinite. 4th List

Boydite. (W. F. Foshag, *Amer. Min.*, 1931, vol. 16, p. 338.) Local name for a borate mineral in California, since identified as probertite (12th List). [*M.A.*, **5**, 52.] 13th List

Bracewellite. C. Milton, D. [E.] Appleman, E. C. T. Chao, F. Cuttitta, J. L. Dinnin, E. J. Duvornik, M. Hall, B. L. Ingram, and H. J. Rose, Jr., 1967. *Geol. Soc. Amer., Progr. Ann. Meeting*, p. 151. CrOOH, isostructural with goethite, a major constituent of the mixture of chromium oxides occurring in alluvial gravels of the Merume river, Mazaruni district, Guayana (Merumite, 18th List). Named for S. Bracewell. 25th List [Cf. Grimaldiite, Guyanaite]

Bradleyite. J. J. Fahey, 1941. *Amer. Min.*, vol. 26, p. 646. A double salt of sodium phosphate and magnesium carbonate, $Na_3PO_4.MgCO_3$, as very fine-grained material in saline oil-shale from Wyoming. Named after Dr. Wilmot Hyde Bradley (1899–) of the United States Geological Survey. [*M.A.*, **8**, 229.] 16th List

Braggite. F. A. Bannister, 1932. *Min. Mag.*, vol. 23, p. 198. Sulphide of platinum, palladium, and nickel, (Pt,Pd,Ni)S, tetragonal, as minute grains in the concentrates of the Bushveld norite, Transvaal. Named after Sir William Henry Bragg and Professor William Lawrence Bragg, as being the first new mineral to be discovered by X-ray methods. 13th List [*A.M.*, **17**, 455; **18**, 79]

Braitschite. O. B. Raup, A. J. Gude 3rd, E. J. Dwornik, F. Cuttitta, and H. J. Rose Jr., 1968. *Amer. Min.* **53**, 1081. White, microcrystalline, hexagonal, in nodules in anhydrite rock in the Cane Creek potash mine, Moab, Grand County, Utah; formula given as $(Ca,Na_2)_7Ln_2B_{22}O_{43}.7H_2O$, but the observed density does not fit. [Empirical cell contents

$Ca_{6\cdot4}Na_{0\cdot9}Ln_{2\cdot1}B_{22\cdot9}O_{44\cdot3}.7\cdot1H_2O$ *M.H.H.*] Named for Professor Otto Braitsch. [*M.A.*, 69-615; *Bull.*, **92**, 511; *Zap.*, **98**, 325.] 26th List

Brammallite. F. A. Bannister, 1943. *Min. Mag.*, vol. 26, p. 304. A micaceous mineral differing from illite (15th List) in containing soda in excess of potash, also called sodium-illite; from crevices in coal-measure shales from Llandebie, South Wales. Named after Dr. Alfred Brammall (1879–) of the Imperial College of Science and Technology, London. 16th List

Brandãosite. A. Mário de Jesus, [1936]. *Com. Serv. Geol. Portugal*, vol. 19 (for 1933), p. 132 (Brandãosite). An almandine-spessartine with a formula, $4RO.R_2O_3.4SiO_2$, different from that of garnet. From Mangualde, Portugal. Named after the Portuguese crystallographer, Vicente de Souza Brandão (1863–1916). [*M.A.*, **6**, 441.] 14th List

Brannerite. F. L. Hess and R. C. Wells, 1920. *Journ. Franklin Inst.*, vol. 189, pp. 225, 779; *Chem. News*, vol. 120, p. 253, vol. 121, p. 22. A complex titanate of uranium with small amounts of rare-earths; written as a metatitanate, the formula is $6(Ca,Fe,UO,TiO)TiO_3 + 8(Th,Zr,UO)Ti_2O_6 + Yt_2Ti_3O_9 + 3H_2O$. Found as grains and rough prisms (tetragonal or orthorhombic?) in gold placers in Stanley Basin, central Idaho. Named after Dr. John Casper Branner (1850–1922), formerly President of Leland Stanford University, California. [*M.A.*, **1**, 22, 122.] 9th List

Brannockite. J. S. White Jr., J. E. Arem, J. A. Nelen, P. B. Leavens, and R. W. Thomssen, 1973. *Min. Record*, **4**, 73. Thin hexagonal plates from the Foote Company's spodumene mine at King's Mountain, Cleveland County, North Carolina, have *a* 10·0167, *c* 14·2452 Å; composition $2[(K,Na)Li_3Sn_2Si_{12}O_{30}]$, a lithium–tin member of the osumilite group. Named for K. C. Brannock. [*M.A.*, 73-4078; *A.M.*, **58**, 1111.] 28th List

Brasilianite. F. H. Pough and E. P. Henderson, 1945. *Mineração e Metalurgia*, Rio de Janeiro, vol. 8, p. 334 (Brasilianita); *Anais Acad. Brasileira Cienc.*, vol. 17, p. 15 (Brasilianite); *Amer. Min.*, vol. 30, p. 572 (Brazilianite). Hydrous phosphate of aluminium and sodium, $Al_3Na(PO_4)_2(OH)_4$, as yellow-green monoclinic crystals of gem quality, from Brazil. Named from the locality. [Not the brazilianite of J. Mawe, 1818 (= wavellite), A. H. Chester, 1896.] [*M.A.*, **9**, 186.] 17th List

Brassite. F. Fontan, M. Orliac, F. Permingeat, R. Pierrot, and R. Stahl, 1974. *Bull.*, **96**, 365. White cryptocrystalline crusts and powdery coatings on specimens from Jachymov, Czechoslovakia, and other localities are identical with synthetic $MgHAsO_4.4H_2O$. Orthorhombic, *a* 7·472, *b* 10·891, *c* 16·585 Å. Named for R. Brasse. [*M.A.*, 74-3462.] 28th List [*A.M.*, **60**, 145]

Braunite-II. P. R. de Villiers and F. H. Herbstein, 1967. *A.M.*, **52**, 20. Material from South Africa with the composition of Braunite but with a *c*-axis double that of Braunite is probably an ordered phase, and is named provisionally. [*M.A.*, **18**, 281.] 28th List

Bravoite. W. F. Hillebrand, 1907. *Amer. Journ. Sci.*, ser. 4, vol. xxiv, p. 151; *Journ. Amer. Chem. Soc.*, vol. xxix, p. 1028. D. F. Hewett, *Trans. Amer. Min. Engin.*, 1910, vol. xl, p. 286. A highly nickeliferous (Ni, 18%) pyrites, $(Fe,Ni)S_2$, disseminated as grains in vanadium-ore (patronite) from Minasragra, near Cerro de Pasco, Peru. Named after José J. Bravo, of Lima, Peru. 5th List

Brazilianite. Now the accepted spelling of Brasilianite (q.v.).

Brazilite. E. Hussak, *Neues Jahrb. Min.* II. 141, 1892; *Min. Mag.* X. 158; XI. 110. Synonym of baddeleyite. 1st List [Jacupiranga, São Paulo, Brazil]

—— (1) Used commercially since about 1884 for an oil-bearing rock from Bahia (L. Fletcher, *Mineralogical Magazine*, 1893, vol. x, p. 160).
(2) E. Hussak, 1892 (first list), synonym of baddeleyite, monoclinic zirconia, ZrO_2.
(3) Used commercially since about 1916 for the fibrous, mamillated form of zirconia which is perhaps distinct from baddeleyite (H. C. Meyer, *Mineral Foote-Notes*, Philadelphia, March 1917, p. 2; W. T. Schaller, ibid., March 1918, p. 2; E. H. Rodd, *Journ. Soc. Chem. Industry*, 1918, vol. xxxvii, p. 213 R). *See* Caldasite and Zirkite. 8th List

Breadalbanite. (*Catalogue of the Mineral Collections in the Museum of Practical Geology*. By W. W. Smith and others. London, 1864, p. 180.) A variety of hornblende from Perthshire, Scotland. T. Egleston (*Catalogue of Minerals and Synonyms*, Washington, 1887, *Bull. US National Mus.*, 1889, No. 33) gives breadalbaneite (p. 13) and breadalbanite (p. 33). 3rd List [Hornblende, *Amph.* (1978)]

Bredigite. C. E. Tilley and H. C. G. Vincent, 1948. *Min. Mag.*, vol. 28, p. 255. Calcium orthosilicate, Ca_2SiO_4, orthorhombic high-temperature α'-form, in the dolerite-chalk contact-zone at Scawt Hill, Co. Antrim, and in the gabbro-limestone contact-zone on the island of Muck, Inverness-shire. Named after Max Albert Bredig (1902–), of New York, formerly of Berlin. 18th List [*A.M.*, **33**, 786. Now γ'-Ca_2SiO_4. Cf. Larnite]

Brezianit, error for Brezinait (Hintze, 13). 28th List

Brezinaite. T. E. Bunch and L. H. Fuchs, 1969. *Amer. Min.* **54**, 1509. Cr_3S_4, monoclinic, occurs as dull grey grains in the metal of the Tucson meteorite, adjacent to the silicate inclusions. Named for Aristides Brezina. [*M.A.*, 70-2612.] 26th List

Brianite. L. H. Fuchs, E. Olsen, and E. P. Henderson, 1966. *Abstr. 29th Ann. Meeting Meteoritical Soc.*, p. 12. *Geochimica Acta*, 1967, vol. 31, p. 1711. Tiny grains with panethite, whitlockite, etc. in pockets in the Dayton meteorite (a siderite) prove to be orthorhombic $Na_2CaMg(PO_4)_2$; also obtained synthetically. Named for Brian Mason. [*A.M.*, **52**, 309; **53**, 508; *Bull.*, **91**, 300.] 25th List [*A.M.*, **60**, 717]

Briartite. J. Francotte, J. Moreau, R. Ottenburgs, and C. Lévy, 1965. *Bull. Soc. franç. Min. Crist.*, vol. 88, p. 432. Small grains with chalcopyrite, renierite, tennantite, and blende in the Prince Leopold mine, Kipushi, Katanga, are tetragonal, composition $Cu_2(Fe,Zn)GeS_4$ or $Cu(Fe,Zn,Ge)S_2$, probably isostructural with chalcopyrite or stannite. [*M.A.*, **17**, 499.] 24th List [*A.M.*, **51**, 1816]

Brickerite. Barrande-Hesse, 1932. *In* F. Ahlfeld, *Neues Jahrb. Min.*, Abt. A, 1932, Beil.-Bd. 66, pp. 42, 44 (Brickerit). Arsenate of zinc and calcium as white radially fibrous crusts from Bolivia. [*M.A.*, **5**, 200.] 13th List [Is Austinite; *A.M.*, **22**, 71; **23**, 347]

Brindleyite. Z. Maksimović and D. L. Bish, 1978. *A.M.*, **63**, 484. A name to replace nimesite (28th List), which was not approved by the New Minerals Commission of the I.M.A. New analyses and X-ray data are provided. The mineral is related to berthierine rather than to amesite. Named for G. W. Brindley. 30th List

Britholite. C. Winther, *Medd. om Grönland*, 1900, XXIV, 190. Rhombic; pseudohexagonal by twinning. $3[4SiO_2.2(Ce,La,Di,Fe)_2O_3.3(Ca,Mg)O.H_2O.NaF].2[P_2O_5.Ce_2O_3]$. [Naujakasik] S. Greenland. 2nd List [Apatite group, *M.A.*, **7**, 395. Cf. Abukumalite]

Brocchite. A. Scacchi, 1840. *Annali Civili del Regno delle Due Sicilie*, Napoli, vol. xxiii, p. 15; *Mem. R. Accad. Sci. Napoli, Classe delle Sci. Nat.*, 1852, vol. vi, p. 268; *Ann. Chem. Phys. (Poggendorff)*, 1853, Erg.-Band iii, p. 183. A synonym of chondrodite. It was applied by Scacchi to his 'Type II' of humite. Named after Giovanni Battista Brocchi (1772–1826). 6th List

Brocenite and β-Brocenite. Kuo Chi-Ti, Wang I-Hsien, Wang Hsien-Chueh, Wang Chung-Kang, and Hou Hung-Chuan, 1973. [*Geochimica*, **2**, 86], abstr. in *A.M.*, **60**, 485 (1975). Unnecessary names for the cerium analogue of fergusonite, from an unnamed locality in northern China. The names should be Fergusonite-(Ce) and β-Fergusonite-(Ce). [*See also A.M.*, **62**, 397.] 29th List

Brockite. F. G. Fisher and R. Meyrowitz, 1962. *Amer. Min.*, vol. 47, p. 1346. $(Ca,Th,Ln)\{(PO_4),(CO_3)\}.H_2O$, from Wet Mountains, Colorado; the carbonate is probably not essential, and the mineral appears to be the hexagonal polymorph of Grayite (which is orthorhombic), and essentially $CaTh(PO_4)_2.2H_2O$. Rhabdophane group. Named for M. Brock of the US Geol. Survey. [*M.A.*, **16**, 286.] 23rd List

Brodrickite. H. C. Dake, 1941. *The Mineralogist*, Portland, Oregon, vol. 9, p. 443. A micaceous mineral, apparently an alteration product of phlogopite, from Bolton, Massachusetts. Named after Mr. John H. Brodrick, of Clinton, Massachusetts, who collected the material. [*M.A.*, **8**, 230.] 16th List

Broggite. G. A. Fester and J. Cruellas, 1935. *Bol. Soc. Geol. Peru*, vol. 7, p. 14 (Broggita). A variety of asphaltum from Peru. Named after Jorge Alberto Broggi, Inspector-General of Mines in Peru. [*M.A.*, **6**, 443.] 14th List

Bromatacamite. P. Chirvinsky, 1906. *Bull. Univ. Kiev*, p. 1 (Бромистый атакамитъ); *Zeits. Kryst. Min.*, 1909, vol. 46, p. 293. The artificial compound $Cu_2Br(OH)_3$, subsequently shown to be an analogue of Botallackite, not of Atacamite, and renamed accordingly (*see below*). 23rd List

Brombotallackite. H. R. Oswald, 1961. *Helvet. Chim. Acta*, vol. 44, p. 2103. The artificial compound $Cu_2Br(OH)_3$ formerly termed Bromatacamite (q.v.) is really the bromine analogue of Botallackite, and is renamed accordingly. 23rd List

Bromcarnallite. A. de Schulten, 1897. *Bull. Soc. Chim. Paris*, ser. 3, vol. 17, p. 166 (Carnallites bromées). K. R. Andress and O. Saffe, *Zeits. Krist.*, 1939, vol. 101, p. 451 (Bromkarnallit). Artificial mixed crystals $KMg(Cl,Br)_3.6H_2O$. Carnallite from New Mexico, Utah, and Spain contains Br 0·1–0·29%. [*M.A.*, **7**, 490; **10**, 304, 344.] 18th List

Bromchlorargyrite. H. Strunz, *Min. Tab.*, 1941, p. 85 (Bromchlorargyrit). Syn. of embolite. 18th List

Bromellite. G. Aminoff, 1925. *Zeits. Krist.*, vol. 62, p. 122 (Bromellit). Beryllium oxide, BeO, as white hexagonal crystals from Långban, Sweden. Named after the Swedish mineralogist, Magnus von Bromell (1679–1731). [*M.A.*, **3**, 5.] 11th List [USSR, *M.A.*, 75–1351]

Bromkarnallit, *see* Bromcarnallite.

Brompyromorphite. H. Wondratschek, 1959. *Zeits. anorg. Chem.*, vol. 300, p. 41 (Blei-Brom-Apatit). The artificial compound $Pb_5(PO_4)_3Br$. 23rd List

η'-bronze. A. H. Clark, 1972. *Neues Jahrb. Min.*, Monatsh, 108. (η'–Cu_6Sn_5). A natural occurrence of this alloy in oxidized tin ores at Panasqueira, Beira Beixa, Portugal, is recorded [*M.A.*, 73-811; *Bull.*, **96**, 244; *Zap.*, **102**, 437.] 28th List

Brostenite. P. Poni, *Ann. Sci. Univ. Jassy*, 1900, vol. i, p. 53; *Anal. Acad. Române*, Bukarest, 1900, vol. xxii; (abstract, this vol., p. 207). Black friable masses occurring as an alteration product of rhodochrosite, near Brosteni, Roumania. The results of three analyses show it to be a hydrated manganite of iron and manganese near chalcophanite, but of variable composition. 3rd List [A mixture; *A.M.*, **60**, 489]

Brownmillerite. E. Spohn, 1932. [*Dissertation*, Berlin, 1932]; *Zement*, Charlottenburg, 1932, vol. 21, p. 702 (Brownmiller'sche Verbindung), p. 732 (Brownmillerit). S. Solacolu, *Zement*, 1932, vol. 21, p. 301 (Brownmillerit). Tetracalcium aluminoferrite, $4CaO.Al_2O_3.Fe_2O_3$, first prepared by W. C. Hansen, L. T. Brownmiller, and R. H. Bogue, *Journ. Amer. Chem. Soc.*, 1928, vol. 50, p. 396, and afterwards detected in Portland cement, and later in dolomite-silica fire-bricks (J. R. Rait, *Second Rep. Refractory materials, Iron and Steel Inst.*, 1942, *Special Rep. no. 28*, p. 66; *Nature*, London, 1942, vol. 150, p. 134). Named after Dr. Lorrin Thomas Brownmiller (1902–) of the Alpha Portland Cement Company, Easton, Pennsylvania. 16th List [*A.M.*, **50**, 2106]

Brüggenite. M. E. Mrose, G. E. Ericksen, and J. W. Marinenko, 1971. *Progr. Abstr. Geol. Soc. Amer. Ann. Meet.*, 653. $Ca(IO_3)_2.H_2O$ occurs with lautarite, to which it dehydrates, in the Chilean nitrate deposits. Monoclinic, *a* 8·51, *b* 10·00, *c* 7·50 Å, *β* 95° 20′, $P2_1/c$. [*A.M.*, **57**, 597, 1911] 27th List

Brugnatellite. E. Artini, 1909. *Rend. R. Accad. Lincei*, Roma, ser. 5, vol. xviii, sem. 1, p. 3; *Riv. Min. Crist. Italiana*, xxxvii, p. 119. A hydrated ultra-basic carbonate.

$$Mg_6FeCO_{20}H_{21} = MgCO_3.5Mg(OH)_2.Fe(OH)_3.4H_2O.$$

Flesh-red, lamellar masses with perfect micaceous cleavage. Optically uniaxial; rhombohedral or hexagonal. Found in an asbestos quarry in Val Malenco, Lombardy. Named after Dr. Luigi Brugnatelli, professor of mineralogy in the University of Pavia. 5th List [Cf. Manasseite, Sjögrenite]

Brunckite. R. Herzenberg, 1938. *Zentr. Min.*, Abt. A, 1938, p. 373. Zinc sulphide as white amorphous (gel) material from Peru. Named after Bergrat Otto Brunck (1866–), of Freiberg, Saxony. [*M.A.*, **7**, 264.] 15th List [Is Sphalerite; *A.M.*, **36**, 383]

Brünnichite. C. L. Giesecke, MS. catalogue. C. G. Gmelin, *Vet. Akad. Handl. Stockholm*, 1816, p. 171 (Brünnikit). O. B. Bøggild, *Mineralogia Groenlandica*, Meddel, om Grønland, 1905, vol. 32, p. 554 (Brünnichit). A zeolite from Greenland shown by Gmelin's analysis to be apophyllite. Named after the Danish mineralogist, Morten Thrane Brünnich (1737–1827). 11th List

Brunogeierite. J. Ottemann and B. Nuber, 1972. *Neues Jahrb. Min., Monatsh*, 263. Grey encrustations on tennantite in the Tsumeb ores, SW Africa, have the composition $(Ge,Fe)Fe_2O_4$, with Ge $\gg$ Fe. Cubic, *a* 8·409, spinel structure. Named for Dr. Bruno Geier. [*M.A.*, 73-805; *Bull.*, **96**, 234; *A.M.*, **58**, 348; *Zap.*, **102**, 445.] 28th List

Brunsvigite. J. Fromme, *Min. petr. Mitt.* (*Tschermak*), 1902, vol. xxi, p. 171 (Brunsvigit). A chloritic mineral, near metachlorite, occurring as fine scaly masses in the gabbro of Radauthal, Harz (Brunswick). 3rd List

Bruyerite. G. Tacnet, 1956. [*Soc. Hist. Nat. Creusot*, vol. 14, no. 4]: abstr. *Amer. Min.*, 1958, vol. 43, p. 624. Also [*Feder. franç. Soc. Sci. Nat.*, no. 5, p. 121], abstr. *Bull. Soc. franç. Min. Crist.*, 1958, vol. 81. p. 154. Black concretionary material, mainly calcite, from Queue de Bruyère, Breuil reservoir, Le Creusot, France. Named from the locality. An unnecessary name. 22nd List

Buchwaldite. E. Olsen, J. Erlichman, T. E. Bunch, and P. B. Moore, 1977. *A.M.*, **62**, 362. Minute inclusions in troilite nodules in the Cape York meteorite (a siderite) are orthorhombic, *a* 5·167, *b* 9·259, *c* 6·737 Å, space group $Pmn2_1$. Composition 4[$NaCaPO_4$]. α 1·607, β 1·610, γ 1·616, elongation of the fibres +. An artificial product, a principal component of Rhenaniaphosphat fertilizer, was formulated $NaCaPO_4$ and named rhenanite (15th List). Named for V. Buchwald. [*M.A.*, 78-880.] 30th List

Buddingtonite. R. C. Erd, D. E. White, J. J. Fahey, and D. E. Lee, 1964. *Amer. Min.*, vol. 49, p. 811. The monoclinic ammonium feldspar, $NH_4AlSi_3O_8$, occurs in andesite and hydrothermally altered rocks at Sulphur Bank mercury mine, Lake County, California, as compact masses pseudomorphous after plagioclase, and as small crystals in cavities. Below about 370°C the mineral carries appreciable zeolitic water, up to about $NH_4AlSi_3O_8.0{\cdot}5H_2O$. 24th List [Idaho; *M.A.*, 75-2481]

Buergerite. G. Gagarin and J. R. Cuomo, 1949, loc. cit., p. 7 (buergerita). The 15*R* polymorph $Zn_{15}S_{15}$ of wurtzite. Named after Dr. Martin Julian Buerger (1903–) of the Massachusetts Institute of Technology, Cambridge, Mass. [*M.A.*, **10**, 532; **11**, 126, 128.] 19th List [This usage is defunct]

—— G. Donnay, C. O. Ingamells, and B. Mason, 1966. *Amer. Min.*, vol. 51, p. 198. Tourmaline in which the end-member $NaFe_3^{3+}Al_6Si_6B_3O_{30}F$ is predominant. A tourmaline from Mezquitic, San Luis Potosi, Mexico, approaches this composition. Named for M. J. Buerger. [*M.A.*, **17**, 767; Зап., **96**, 78.] 25th List [*Amer. Min.*, **61**, 1029]

Buetschliite. C. Milton and J. M. Axelrod, 1946. *Bull. Geol. Soc. Amer.*, vol. 57, p. 1218; *Amer. Min.*, 1947, vol. 32, pp. 204, 607. Hydrous potassium and calcium carbonate, $3K_2CO_3.2CaCO_3.6H_2O$, probably hexagonal, formed by the hydration of fairchildite (q.v.) in the fused wood-ash of burnt trees. Named after Johann Adam Otto Bütschli (1848–1920), formerly professor of zoology at Heidelberg, who prepared the compound artificially. [Not the bütschliite of R. Lang, 1914; 7th List.] [*M.A.*, **10**, 101, 252.] 18th List

Bukovite. Z. Johan and M. Kravček, 1971. *Bull.*, **94**, 529. Rare dark brown to black grains, up to 2 mm, in the ore deposits of Bukov and Petrovice, Moravia, are tetragonal, *a* 3·976, *c* 13·70 Å; composition [$Cu_{3+x}Tl_2FeSe_{4-x}$], with *x* up to 0·28. Named for the locality. [*M.A.*, 72-3334.] 27th List [*A.M.*, **57**, 1910. $(Cu,Fe)Tl_2Se_2$]

Bukovskýite. F. Novak, P. Povondra, and J. Vtělenský, 1967. *Acta Univ. Carolinae*, Geol. no. 4, 297. The mineral from Kaňk, Kutná Hora, formerly known as arsendestinezite (this List) is distinct from both destinezite and sarmientite, and is accordingly renamed in honour of Professor Antonin Bukovsky. Composition $Fe_2^{3+}AsO_4SO_4(OH).7H_2O$. [*A.M.*, **54**, 576, 991; *Zap.*, **97**, 617; *Bull.*, **92**, 512.] 26th List

Buldymite. A. S. Amelandov and K. N. Ozerov, 1934. [*Min. Syre*, Moscow, 1934, no. 2, p. 24.] O. M. Shubnikova, *Trans. Lomonosov Inst. Acad. Sci. USSR*, 1936, no. 7, p. 320 (Булдымит, Buldymite). A variety of vermiculite, $K_2O.7MgO.2FeO.3(Al,Fe)_2O_3.9SiO_2.6H_2O$. A brown scaly mineral intermediate between biotite and vermiculite; swells up on heating. From corundum-plagioclase veins at Buldymsk, Sverdlovsk district, Ural. Named from the locality. 15th List

Bultfonteinite. J. Parry, A. F. Williams, and F. E. Wright, 1932. *Min. Mag.*, vol. 23, p. 145. Hydrous calcium silicate and fluoride, $2Ca(OH,F)_2.SiO_2$ or $2Ca(OH)_2.2SiO_2.Ca(OH)_2.CaF_2$. Pink spherules of radiating triclinic needles from the Bultfontein diamond mine, Kimberley, South Africa. Named from the locality. *See* Dutoitspanite. 13th List [*A.M.*, **17**, 455; **18**, 32; *M.M.*, **30**, 569; *M.A.*, **16**, 612]

Bungonite. I. Iwasa, 1877. [*Gakugéisirin*, no. 57.] Z. Harada, *Journ. Fac. Sci. Hokkaido Univ.*, Sapporo, ser. 4, 1936, vol. 3, p. 324 (Bungonit). An incorrectly determined mineral from Japan, afterwards (Z. Sasamoto, 1895) identified with kämmererite. Presumably named from the locality, Bungo, Japan. 14th List

Bunkolite. K. Kinoshita, 1927. [*Journ. Geol. Soc. Japan*, **34**, 52], quoted in *Intro. Jap. Min.* 109 (1970), *Geol. Surv. Japan*. A massive hydrated silicate of Mn^{2+} and Mn^{3+} from the Takayama mine, a branch of the Bunko mine, Hiroshima Prefecture, Japan: 'believed to be a variety of penwithite.' [*Zap.*, **101**, 286.] 28th List

Bunsite, error for Bunsenite. (*Contr. Min. Petr.* 1976, **56**, 1). 29th List

Buonnemite, error for Vuonnemite (*Zap.*, **102**, 423). 28th List

Burangaite. O. von Knorring, M. Lehtinen, and Th. G. Sahama, 1977. *Bull. Geol. Soc. Finland*, **49**, 33. Blue monoclinic prisms from Buranga, Rwanda, *a* 25·09, *b* 5·048, *c* 13·45 Å, *β* 110·91°, space group *C*2/c, are related to dufrenite. Composition $2[(Na,Ca)_2(Fe^2,Mg)_2Al_{10}(OH,O)_{12}(PO_4)_8.4H_2O]$. *α* 1·611, *β* 1·635, *γ* 1·643 ‖ [010], *α*:[001] 11°, $2V_\alpha$ 58°, strongly pleochroic. Named from the locality. [*M.A.*, 78-881] 30th List

Burbankite. W. T. Pecora and J. H. Kerr, 1953. *Amer. Min.*, vol. 38, p. 1169. $(Ca,Sr,Ba,Ce,Na)_6(CO_3)_5$ as pale yellow hexagonal crystals with other rare-earth carbonates from Montana. Named after Wilbur Swett Burbank (1898–) of the United States Geological Survey. [*M.A.*, **12**, 301.] 20th List [Quebec, *C.M.*, **12**, 342. Cf. Carbocernaite]

Burkeite. J. E. Teeple, 1921. *Journ. Indust. Engin. Chem.*, Easton, Pa., vol. 13, p. 251. W. C. Blasdale, *Journ. Amer. Chem. Soc.*, 1923, vol. 45, p. 2942. A. F. Rogers, *Amer. Journ. Sci.*, 1926, ser. 5, vol. 11, p. 473. Sodium sulphate and carbonate, $2Na_2SO_4.Na_2CO_3$, as orthorhombic crystals, obtained artificially by heating above 25°C the brine of Searles Lake, San Bernardino Co., California, or a solution containing Na_2SO_4, Na_2CO_3, and NaCl. Named after Mr. W. E. Burke, chemist of the American Trona Corporation. [*M.A.*, **3**, 162.] 11th List [*A.M.*, **20**, 50]

Burmite. F. Noetling, *Records Geol. Survey India*, XXVI. 31, 1893. O. Helm, ibid. XXV. 180, 1892; XXVI, 61, 1893; and *Schriften Ges. Danzig*, VIII. 63, 1894. An amber-like resin from Upper Burma. 1st List

Bursaite. R. Tolun, 1955. *Bull. Min. Res. Inst. Turkey*, no. 46–47, p. 124. Sulphosalt, $Pb_5Bi_4S_{11}$, as small grey prisms, monoclinic(?). Named after the locality, Bursa (Brusa), NW Turkey. Evidently a synonym of cosalite ($Pb_2Bi_2S_5$). [*M.A.*, **13**, 380.] 21st List [Status uncertain. Cf. *A.M.*, **57**, 328; *M.A.*, **15**, 435; 75-2506]

Buryktalskite. I. I. Ginzburg, 1960. Кора выветривания (Crust of weathering), vol. 3, p. 56 (Бурыктальскит). The X-ray patterns of the mixtures called nickelemelane, cobaltomelane, etc. (qq.v.), include lines of goethite, cryptomelane, and elizavetinskite (q.v.); Ginzburg subtracts these and defines buryktalskite by the pattern with lines at 4·88 (10), 4·66 (10), 4·61 (10), 1·482 (10), 9·17 (7), 3·09 (7), 1·834 (7), 1·689 (7). Named from its occurrence in the ores of Buryktal (Бурыктал). 'Many of the lines attributed to "buryktalskite" can be assigned to strong lines of pyrolusite, lithiophorite, or cryptomelane; others cannot be assigned with any confidence' (M. Fleischer, *Amer. Min.*, 1961, vol. 46, p. 767). Inadequately defined and very doubtful; the name ought not to have been given. 22nd List

Buserite. R. Giovanoli, W. Feitknecht, and F. Fischer, 1971. *Helv. Chim. Acta*, **54**, 1112. The '10 Å Manganite' from deep-sea nodules described by W. Buser (in *Oceanography*, Amer. Ass. Adv. Sci., 1959, 962) is named. 28th List

Buszite. E. Steinwachs, 1929. *Centr. Min.*, Abt. A, 1929, p. 202 (Buszit). From qualitative tests on a small amount of material, apparently a silicate of the rare-earths neodymium, praseodymium, erbium, and europium. A single small ditrigonal bipyramidal crystal, has been found in SW Africa. Named after Prof. Karl Heinrich Emil Georg Busz [1863–1930], of Münster, Westphalia. [*M.A.*, **4**, 149.] 12th List [Is. Basknäsite; *M.A.*, **12**, 229]

Butlerite. C. Lausen, 1928. *Amer. Min.*, vol. 13, p. 211. Hydrous ferric sulphate, $Fe_2O_3.2SO_3.5H_2O$, as minute orange-yellow orthorhombic crystals formed by the burning of pyritic ore in a mine in Arizona. Named after Prof. Gurdon Montague Butler (1881–1961), of the University of Arizona. 11th List [Cf. Parabutlerite]

Bütschliite. R. Lang, 1914. *Neues Jahrb. Min.*, Beil.-Bd. xxxviii, p. 150 (Bütschliit). Amorphous calcium carbonate represented by the freshly precipitated material and also present in the hard parts of certain organisms. Named after Otto Bütschli, Professor of Zoology in the University of Heidelberg. 7th List

Buttgenbachite. A. Schoep, 1925. *Compt. Rend. Acad. Sci. Paris*, vol. 181, p. 421; *Bull. Soc. Chim. Belgique*, vol. 34, p. 313; *Ann. Soc. Géol. Belgique*, 1927, vol. 49, *Bull.* p. в 308, vol.

50, p. B 215. Hydrous chloride and nitrate of copper, '$18CuO.3Cl.N_2O_5.19H_2O$', forming a felt of sky-blue needles resembling connellite; from Belgian Congo. Named after Prof. Henri Jean François Buttgenbach (1874–), of Bruxelles. H. Buttgenbach (*Ann. Géol. Soc. Belgique*, 1926, vol. 50, p. B 35) shows the crystals to be hexagonal and isomorphous with connellite; he writes the formula $2CuCl_2.Cu(NO_3)_2.15Cu(OH)_2.4H_2O$. [*M.A.*, **3**, 6; **3**, 270; **3**, 372.] 11th List [*A.M.*, **11**, 216; **12**, 381. $Cu_{19}Cl_4(NO_3)_2(OH)_{32}.2H_2O$; *M.M.*, **29**, 280]

Byströmite. G. Gagarin and J. R. Cuomo, 1949, loc. cit., p. 6 (byströmita). Monoclinic magnetic pyrites of Anders Byström, 1945, as distinct from hexagonal pyrrhotine. [*M.A.*, **9**, 224.] 19th List [This usage is defunct]

—— B. Mason and C. J. Vitaliano, 1950. *Progr. & Abstr. Min. Soc. Amer.*, p. 16; *Amer. Min.*, 1951, vol. 36, p. 320; 1952, vol. 37, p. 53 (bystromite). Magnesium antimonate, $MgSb_2(O,OH)_6$, tetragonal, massive, pale blue-grey, from El Antimonio, Sonora, Mexico. Named after Anders Byström, Swedish mineral chemist, who determined the crystal structure of the artificial compound. [*M.A.*, **11**, 188, 516.] 19th List

Bytownorthite. A. N. Winchell, 1925. *Journ. Geol. Chicago*, vol. 33, p. 726; Elements of optical mineralogy, 2nd edit., 1927, part 2, p. 319. A contraction of bytownite-anorthite for felspars of the plagioclase series ranging in composition from $Ab_{20}An_{80}$ to $Ab_{10}An_{90}$. 11th List

C

Ca-gümbelite, variant of Calcium-gümbelite (q.v.), 28th List

Ca-hureaulite. A Volborth, 1954. *Geologi*, Helsinki, vol. 2, no. 2, p. 5 (Ca-hureauliitti). $CaMn_5(PO_4)_4.4H_2O$, from Eräjärvi, Finland. Hureaulite is $H_2Mn_5(PO_4)_4.4H_2O$, and no evidence is adduced that the material is really a variety of or related to hureaulite. 22nd List

Ca-Illite. H. Strunz, 1957. *Min. Tabellen*, 3rd edit., p. 308 (Ca-Illit). 22nd List

Ca-Langbeinit, unnecessary variant of calcium langbeinite (q.v.). Hintze, 15. 28th List

Ca-ursilite, variant of Calcium-ursilite (22nd List) (I. Kostov, *Mineralogy*, p. 327). 25th List

Cadmium-dolomite. J. R. Goldsmith, 1958. *Bull. Geol. Soc. Amer.*, vol. 69, p. 1570. A name for $CdMg(CO_3)_2$, the cadmium analogue of dolomite; obtained synthetically. 22nd List

Cadmium olivine. H. Hayashi, N. Nakayama, M. Yoshida, T. Kozuka, M. Mizuno, K. Yamamoto, T. Yamamoto, and T. Noguchi, 1964. [*Rept. Govt. Indust. Res. Inst.*, Nagoya, vol. 13, p. 285]; abstr. *Min. Journal.* [Japan], 1965, vol. 4, p. 322. Artificial Cd_2SiO_4. 24th List

Cadmiumspat. H. Strunz, *Min. Tab.*, 1941, p. 117. Syn. of otavite (4th List; *M.A.*, **8**, 366.) 18th List

Cadmoselite. E. Z. Buryanova, G. A. Kovalev, and A. I. Komkov, 1957. *Mém. Soc. Russ. Min.*, vol. 86, p. 626 (кадмоселит). Cadmium selenide, CdSe, hexagonal, minute black grains. Named from the composition. (*Min. Mag.* **31**, 963, correction **31**, viiib; *M.A.*, **14**, 59; *Bull. Soc. franç. Min. Crist.*, 1958, vol. 81, p. 238; *Amer. Min.*, 1958, vol. 43, p. 623.) 21st and 22nd Lists

Cadwaladerite. S. G. Gordon, 1941. *Notulae Naturae, Acad. Nat. Sci. Philadelphia*, no. 80. Hydrous basic aluminium chloride, $Al(OH)_2Cl.4H_2O$, as amorphous grains in halite from Cerro Pintados, Chile. Named after Charles M. B. Cadwalader, President of the Academy of Natural Sciences of Philadelphia. [*M.A.*, **8**, 187.] 16th List [*A.M.*, **27**, 144]

Caeruleofibrite. *See* Ceruleofibrite. 9th List

Caesium astrophyllite. A. F. Efimov, V. D. Dusmatov, A. A. Ganzeev, and Z. [T.] Kataeva, 1969. *In* A. A. Ganzeev, A. F. Efimov, and N. G. Semenova, Геохимия (*Geokhimiya*) 1969, 340 (Цезийастрофиллит). A provisional name for the material subsequently described as Caesium kupletskite (27th List). 28th List

Caesium-beryl. S. L. Penfield, 1888. *Amer. Journ. Sci.*, ser. 3, vol. 36, p. 317 ('the cæsium beryl of Norway, Maine'). A variety of alkali-beryl (q.v.) containing Cs_2O up to 4·56%, usually as pink crystals of tabular habit. Afterwards named vorobyevite (V. I. Vernadsky, 1908; 5th List) and morganite (G. F. Kunz, 1911; 6th List). Compare also rosterite (F. Zambonini and V. Caglioti, 1928; *M.A.*, **4**, 138). [*M.A.*, **1**, 76; **3**, 448; **4**, 96, 315.] 12th List

Caesium-biotite. F. L. Hess and J. J. Fahey, 1932. *Amer. Min.*, vol. 17, p. 173 (Caesium biotite). A variety of biotite from South Dakota containing Cs_2O 3·14%. [*M.A.*, **5**, 192.] 13th List

Caesium kupletskite. A. F. Efimov, V. D. Dusmatov, A. A. Ganzeev, and Z. T. Kataeva, 1971. Докл. Акад. наук СССР (*Compt. Rend. Acad. Sci. URSS*), **197**, 1394 (Цезийкуплетскит). Rosettes of platy anorthic crystals from the Alai alkalic province have a 5·41, b 11·74, c 21·16 Å, α 89°, β 90°, γ 102° 23′. Composition $(Cs,K,Na)_3(Mn,Fe,Li)_7(Ti,Nb)_2Si_8O_{24}$-$(O,OH,F)_7$, the caesium analogue of kupletskite (21st List). α yellow-green, β 1·726, yellow-brown, γ 1·758 ∥ [100], brown. [*A.M.*, **57**, 328.] 27th List

Caesium-spodumene. P. Quensel, 1939. *Geol. För. Förh. Stockholm*, vol. 60 (for 1938), p. 625 (caesium spodumene), p. 626 (caesium-spodumene). *See* Diaspodumene. [*M.A.*, **7**, 335.] 15th List

Cafarsite. S. Graeser, 1966. *Schweiz. Min. Petr. Mitt.*, vol. 46, p. 367; *see also Urner*

Mineralien Freund, Jahrg. 4, heft 4. Dark brown cubic crystals, from the Monte Leone, Binnatal, Switzerland, have a composition near $(Ca,Mn)_5Fe_2Ti_2(AsO_4)_8.2H_2O$. [The formula given in the publications cited does not agree well with the cell-dimensions and density—*M.H.H.*] [*M.A.*, **18**, 207; *A.M.*, **52**, 1584, Зап. **97**, 75; *Bull.*, **90**, 604.] 25th List [Cf. Arsenoferrite] [Perhaps $4[Ca_{5\cdot9}Mn_{1\cdot7}Fe_3Ti_3(AsO_3)_{12}.4\text{–}5H_2O]$; *Schweiz. Min. Petr. Mitt.*, **57**, 1]

Cafetite. A. A. Kukharenko, V. V. Kondrateva, and V. M. Kovyazina, 1959. Зап. Всесоюз. Мин. Общ. (*Mem. All-Union Min. Soc.*), vol. 88, p. 444 (Кафетит). Orthorhombic radiating crystals in a pyroxenite from Africanda, Kola peninsula; approximately $(Ca,Mg)(Fe,Al)_2Ti_4O_{12}.4H_2O$. Named from the composition, Ca–Fe–Ti. [*M.A.*, **14**, 501.] 22nd List [*A.M.*, **45**, 476]

Cahnite. C. Palache, 1921. 'Holdenite and cahnite, two new minerals from Franklin Furnace, N.J.' (title only), *Amer. Min.*, 1921, vol. 6, p. 39. Named after Mr Lazard Cahn, of Colorado Springs, Colorado (10th List). C. Palache and L. H. Bauer, *Amer. Min.*, 1927, vol. 12, pp. 77, 149. Hydrous boro-arsenate of calcium, $4CaO.B_2O_3.As_2O_5.4H_2O$, as white tetragonal sphenoids from Franklin Furnace, New Jersey. Named after M. Lazard Cahn, of Colorado Springs, who first recognized the minute crystals. [*M.A.*, **3**, 365.] 11th List [$Ca_2(AsO_4)B(OH)_4$; *M.M.*, **32**, 666]

Caillerite. *Bull. Soc. Franç. Min. Crist.*, 1950, vol. 73, p. 147 (caillérite). G. T. Faust and M. Fleischer, *Amer. Min.*, 1952, vol. 37, p. 135 (caillèrite). Alternative name suggested for allevardite (q.v.) [now Rectorite]. Named after Mlle Simonne Caillère, of the Nat. Hist. Mus., Paris. 19th List

Calafatite. S. Calderón, 1910. *Los Minerales de España*, vol. ii, p. 205 (Calafatita). A hydrated basic sulphate of aluminium and potassium, containing rather more water than alunite: formula $Al_2(SO_4)_3.K_2SO_4.5Al(OH)_3.H_2O$. As white compact masses it occurs abundantly in Almeria. Named after Mr. Juan Calafat León, of the Museum of Natural Sciences, Madrid. 6th List [Is Alunite; *Inst. Invest. Geol. Lucas Mallada*, Estud. Geol., **18**, no. 1–2, 111]

Calbenite. Syn. of myrickite (6th List), a *var.* of chalcedony. R. Webster, *Gems*, 1962, p. 755. 28th List

Calcantite. Italian (and Spanish, calcantita) spelling of chalcanthite. 19th List

Calc-clinoenstatite, Calc-clinobronzite, Calc-clinohypersthene. G. T. Prior, 1920. *Min. Mag.*, vol. 19, pp. 57, 62, 63. To replace the terms 'enstatite-augite', etc., of W. Wahl (1906; 5th List) for those varieties of clinoenstatite, etc., (W. Wahl, 1906; 5th List) that contain appreciable (though small as compared with diopside, etc.) amounts of lime. They are constituents of certain meteoric stones. 9th List

Calciborite. E. S. Petrova, 1955. [*Min. Syre* (*Min. Resources*), no. 2, p. 218.] Abstracts in *Mém. Soc. Russ. Min.*, 1956, vol. 85, p. 76 (кальциборит, calciborite); *Amer. Min.*, 1956, vol. 41, p. 815. Calcium borate, $Ca_5B_8O_{17}$, monoclinic? White radial aggregates in drill-cores from limestone skarn, Urals. Named from the composition. [*M.A.*, **13**, 208.] *See* Frolovite. 21st List

Calciclase. A. Johannsen, 1926. *Journ. Geol. Chicago*, vol. 34, p. 841. Members of the plagioclase series between pure anorthite and $Ab_{10}An_{90}$. *See* Sodaclase. 11th List

Calcioaegirine. D. P. Serdyuchenko, A. V. Glebov, and V. A. Pavlov, 1961. [Изв. Акад. наук СССР, сер. геол. (*Bull. Acad. Sci. USSR*, geol. ser.) no. 2, p. 87]; abstr. Зап. Всесоюз. Мин. Общ. (*Mem. All-Union Min. Soc.*), 1965, p. 680. (Кальциоэгирин, calcioaegirine). The *hypothetical* end-member $CaFe_2^{3+}(SiO_3)_4$. 24th List

Calcio-åkermanite. A. N. Winchell, Optical mineralogy, 2nd edit., New York, 1927, p. 267 (Calcium-akermanite). O. M. Shubnikova and D. V. Yuferov, *Spravochnik po novym mineralam*, Leningrad, 1934, p. 71 (кальцио-акерманит, calcio-åkermanite): The hypothetical molecule $Ca_3Si_2O_7$, i.e. åkermanite with Mg replaced by Ca. [Cf. *M.A.*, **2**, 427.] 14th List

Calcioancylite. A. E. Fersman, 1922. *Compt. Rend. Acad. Sci. Russie*, p. 60 (кальциоанцилит); *Trans. Northern Sci. Econ. Exped.*, no. 16 (Sci. Techn. Dept. Supreme Council of National Economy, no. 8), Moscow & Petrograd, 1923, pp. 16, 41, 68, 72 (кальциоанцилит). G. P. Chernik, *Bull. Acad. Sci. Russie*, 1923, ser. 6, vol. 17, p. 83 кальциевый анцилит, Calcioancylite). A variety of ancylite [i.e. calcian Ancylite] with strontium partly replaced by calcium (CaO 4·36%). [*M.A.*, **2**, 262–263, 407.] 10th List

Calciobaryt. H. Strunz, *Min. Tab.*, 1941, p. 130. Syn. of calcareobarite (T. Thomson, 1836). 18th List

Calciobiotite. F. Zambonini, 1919. *Mem. Descr. Carta Geol. Italia*, 1919, vol. 7, pt. 2, p. 124. A pale-coloured variety of biotite rich in calcium (CaO 14·33%) occurring in blocks of metamorphosed limestone in the pipernoid tuff of Campania, Italy. [*M.A.*, **1**, 107.] 9th List

Calciocancrinite. F. Zambonini, 1910. Mineralogia Vesuviana. *Mem. R. Accad. Sci. Fis. Mat. Napoli*, ser. 2, vol. xiv, No. 7, p. 202. Synonym of Kalkcancrinit (J. Lemberg, 1876) and lime-cancrinite (Dana, System, 1892, p. 428). 6th List

Calcio-carnotite. E. T. Wherry, 1914. *Science*, New York, new ser., vol. xxxix, p. 576; *Bull. United States Geol. Survey*, no. 580, p. 149. W. F. Hillebrand, H. E. Merwin, and F. E. Wright, *Proc. Amer. Phil. Soc.*, 1914, vol. liii, p. 38. A synonym of tyuyamunite (6th List). *See* Kalio-carnotite. 7th List

Calciocatapleiite. A. M. Portnov, V. T. Dybinchik, and L. S. Solntseva, 1972. Докл. **202**, 430 (Кальциокатаплеит, Calciokatapleite). $CaZrSi_3O_9.H_2O$, the calcium analogue of catapleiite, described but not named in 1964 (*M.A.*, **16**, 648; *A.M.*, **49**, 1153) is now named. (Cf. *M.M.*, **35**, 1129.) [*Zap.*, **102**, 453.] 28th List

Calciocelsian. E. R. Segnit, 1946. *Min. Mag.*, vol. 27, p. 169. Celsian, containing CaO 4%, from Broken Hill, New South Wales. 17th List

Calcio-chondrodite. E. R. Buckle and H. F. W. Taylor, 1958. *Amer. Min.*, vol. 43, p. 818. $Ca_5(SiO_4)_2(OH)_2$, the calcium analogue of chondrodite; synthetic. 22nd List

Calciocopiapite. M. A. Kashkai and R. M. Aliev, 1960. [Труды Азербайдж. Геогр. Общ. (*Trans. Azerbaidzhan Geogr. Soc.*) 1960, p. 49]; abstr. in Зап. Всесоюз. Мин. Общ. (*Mem. All-Union Min. Soc.*), 1962, vol. 91, p. 196 (Кальциокопиапит, calciocopiapite), and in *Amer. Min.*, 1962, vol. 47, p. 807. The calcium member of the copiapite family, $CaFe_4(SO_4)_6(OH)_2.10H_2O$, occurring at Dashkesan, Middle Caucasus. Named from the composition. *See also* Tusiite. [*M.A.*, **16**, 553.] 23rd List [*A.M.*, **47**, 807]

Calciodialogite. T. Nicolau, 1910. Synonym of Calciorhodochrosite (q.v.). 6th List

Calcio-gadolinite. T. Nakai, 1938. *Bull. Chem. Soc. Japan*, vol. 13, p. 591. A variety of gadolinite [i.e. calcian Gadolinite] with rare-earths partly replaced by calcium (CaO 11·91%), from Japan. [*M.A.*, **7**, 264.] 15th List

Calcio-jarosite. H. Strunz, 1957. *Min. Tabellen*, 3rd edit., p. 197 (Calcio-Jarosit). Syn. of Calcium jarosite (19th List). 22nd List

Calciolyndochite. S. A. Gorzhevskaya, G. A. Sidorenko, and A. I. Ginzburg, 1974. Abstr. *Zap.*, **105**, 76 (Кальциолиндокит). Unnecessary name for calcian lyndochite. 30th List

Calcio-olivine. O. M. Shubnikova and D. V. Yuferov, 1934. *Spravochnik po novym mineralam*, Leningrad, 1934, p. 67 (Кальциооливин), p. 163 (Calcio-olivin). Translation of Lime-olivine and Kalk-Olivin (11th List). 14th List

Calciopalygorskite. A. Fersmann, 1908. *Bull. Acad. Sci. Saint-Pétersbourg*, ser. 6, vol. ii, p. 274 (Calciopalygorskit). A 'mountain-leather' from Strontian, Argyllshire, containing much calcium (CaO, 10%, according to T. Thomson's analysis, 1836). 5th List

Calciorhodochrosite. T. Nicolau, 1910. *Anuarul Inst. Geol. României*, vol. iii, p. 39 (Calciorhodochrosită, Calciodialogită), p. 41 (Germ., Calziorhodochrosit, Calziodiallogit). Mixed carbonates of manganese and calcium, occurring intimately intermixed with rhodonite in manganese ore from Roumania. 6th List

Calciorinkite. H. Strunz, 1970. *Min. Tab.* 5th edit., 513. Variant of calciumrinkite (14th List), *syn.* of götzenite (21st List). 28th List

Calciosamarskite. H. V. Ellsworth, 1928. *Amer. Min.*, vol. 13, pp. 65, 66. A variety of samarskite containing 4·76 to 7·56% CaO, from Ontario, Canada. [*M.A.*, **3**, 471.] 11th List [status uncertain; *A.M.*, **62**, 406]

Calcioscheelite. E. T. Wherry, 1914. *Proc. United States National Museum*, vol. xlvii, p. 504. Synonym of scheelite. 7th List

Calcio-spessartine. O. M. Shubnikova and D. V. Yuferov, *Spravochnik po novym mineralam*, Leningrad, 1934, p. 66 (Кальциоспессартит, Calc-spessartite), p. 163 index (Calcio-spessartite). Variant of calc-spessartite (11th List). 14th List

Calciotalc. D. P. Serdyuchenko, 1959. Зап. Всесоюз. Мин. Общ. (*Mem. All-Union Min. Soc.*), vol. 88, p. 298 (Кальциотальк). Abstr. Hintze, *Handb. Min.*, Erg.-Bd. II, 1960, p. 926 (Calciotalk, Kalciotalk, Kalziotalk). A variety containing 13% CaO, occurring with normal talc replacing actinolite in the phlogopite deposit of the Medviezhy river, Aldan region, Yakutia. Named from the composition. [*M.A.*, **14**, 280.] N. V. Belov considers that the mineral is really a brittle mica, $CaMg_2Si_4O_{10}(OH)_2$, and would be better named Magnesium margarite (q.v.). [*M.A.*, **14**, 280.] 22nd List

Calciotantalite. E. S. Simpson, 1907. *Rep. Australian Assoc. Adv. Sci.*, vol. 11, p. 452; *Amer. Min.*, 1928, vol. 13, p. 465. A calciferous (CaO 7·78%) tantalite from Western Australia. [*M.A.*, **4**, 184.] 12th List [A mixture; *M.A.*, 72-2276]

Calciothomsonite. S. G. Gordon, 1923. *Proc. Acad. Nat. Sci. Philadelphia*, vol. 75, p. 273; *Amer. Min.*, 1923, vol. 8, p. 126. A variety of thomsonite from Franklin, New Jersey, with $CaO:Na_2O = 5:1$. Also applied (S. G. Gordon, *Proc. Acad. Nat. Sci. Philadelphia*, 1924, vol. 76, p. 107) to the hypothetical end-member $CaO.Al_2O_3.2SiO_2.3H_2O$ of the thomsonite series. The terms Kalkthomsonit and Natronthomsonit had previously been applied by C. F. Rammelsberg (*Handbuch d. Mineralchemie*, 2nd suppl., 1895, p. 389) to the end-members $CaAl_2(SiO_4)_2.2\frac{1}{2}H_2O$ and $Na_2Al_2(SiO_4)_2.2\frac{1}{2}H_2O$ of the thomsonite group. [*M.A.*, **2**, 361, 528.] 10th List

Calciouraconite. A. K. Boldyrev, 1935. *Kurs opisatelnoi mineralogii*, Leningrad, part 3, p. 83 (Кальцио-ураконит). Variety of uraconite containing 3% CaO. 20th List

Calciouranoite. V. P. Rogova, L. N. Belova, G. P. Kiziyarov, and N. N. Kusnetsova, 1974. *Zap.*, **103**, 108 (Кальцураноит). Material with the composition $CaO.2UO_3.5H_2O$ gives X-ray powder data very near those of metacalciouranoite (q.v.), which has only 2 H_2O. Named from the composition. 28th List

Calciriebeckite. *Dokl. Acad. Sci. USSR* (*Earth Sci. Sect.*), 1966, **169**, 195. Erroneous translation of Кальциевый рибекит, calcian riebeckite (Докл. 1966, **169**, 1162). [*M.A.*, 69-3260.] 28th List

Calcirtite, erroneous transliteration of Кальциртит (Calzirtite, 22nd List). Crystallography (translation of Кристаллография), 1961, vol. 6, p. 155. 23rd List

Calcistrontite. Von der Marck, *Verh. Ver. Rheinl.* Bonn, 1882, XXXIX, Corr.-bl. 84. Supposed to be $3CaCO_3.2SrCO_3$, but shown by H. Laspeyres (*Zeits. Kryst. Min.* 1896, XXVII, 41) to be a mechanical mixture of calcite and strontianite. Westphalia. 2nd List.

Calcium-. *See also* Ca-, Calcio-.

Calcium-analcime. A. Steiner, 1955. *Min. Mag.*, vol. 30, p. 695. Artificial $CaAl_2Si_4O_{12}.2H_2O$, analogous to analcime with Ca in place of Na. Synonym of wairakite (q.v.). 20th List

Calcium-autunite. J. G. Fairchild, 1929. *Amer. Min.*, vol. 14, p. 265 (calcium autunite), p. 269 (calcium-autunite). Artificially prepared autunite in which calcium can be replaced by Na, K, Ba, Mn, Cu, Ni, Co, Mg. Synonym of autunite. *See* Barium-autunite, Lead-autunite, Sodium-autunite. [*M.A.*, **4**, 307.] 19th List

Calcium-barium-mimetite. H. Strunz, *Min. Tab.*, 1941, p. 156 (Calciumbariummimetesit). Syn. of baryt-hedyphane (q.v.). 18th List

Calcium catapleiite. A. M. Portnov, 1964. Докл. акад. наук СССР (*Compt. Rend. Acad. Sci. URSS*), vol. 154, p. 607 (Кальциевый катаплеит. The calcium end-member of the catapleiite series. [*M.A.*, **16**, 648; *A.M.*, **49**, 1153 (Calcium catapleite).] 24th List

Calcium chondrodite. R. M. Gan'ev, Yu. A. Kharitonov, V. V. Ilyukhin, and N. V. Belov, 1969. *Soviet Physics—Doklady*, **14**, 946. This is a mistranslation of Кальциевый хондродит (Докл. **188**, 1821) correctly translated calcian chondrodite, but the latter is in fact a misnomer, the material described being Calcio-chondrodite (22nd List). 28th List

Calciumcylit. (*Chem. Zentr.*, 1926, vol. 1, p. 1787). Error for Calcioancylite (A. E. Fersman, 1922; 10th List). 11th List

Calcium-gümbelite. G. Frenzel, 1971. *Neues Jahrb. Min., Abh.* **115**, 164. (Calcium-Gümbelit, Ca-Gümbelit). A calcian variety of gümbelite (= hydromuscovite). [*M.A.*, 73-677.] 28th List

Calciumhilgardite-3Tc, Calcium hilgardite-2M(Cc). O. Braitsch, 1959. *Beitr. Min. Petr.*, vol. 6, p. 233 (3Tc-Calciumhilgardit; 2M(Cc)-Calciumhilgardit). Systematic names for the anorthic polymorph parahilgardite and the monoclinic polymorph hilgardite. M. Fleischer

(*Amer. Min.*, 1959, vol. 44, p. 1102) points out that the letters and numbers serving to distinguish polymorphic structures should always be written as suffixes, not as prefixes (cf. wurtzite, the micas, and SiC). 22nd List

Calcium-jarosite. D. P. Serdyuchenko, 1951. *Doklady Acad. Sci. USSR*, vol. 78, p. 347 (кальциевый ярозит), p. 348 (Ca-ярозит). An impure jarosite from Caucasus containing CaO 1·58%, corresponding to 27·4% Ca-jarosite $Ca[Fe_3(SO_4)_2(OH)_6]_2$, with K-jarosite, Na-jarosite, and 50·1% carphosiderite. [*M.A.*, **11**, 366.] 19th List

Calciumkatapleit. H. Strunz, 1966. *Min. Tabellen*, 4th edit., p. 459. Variant of Calcium catapleiite (q.v.). 24th List

Calcium-langbeinite. G. W. Morey, J. J. Rowe, and R. O. Fournier, 1964. *Journ. Inorg. Nucl. Chem.* **26**, 53. *Artificial* $K_2Ca_2(SO_4)_3$ is cubic and isomorphous with langbeinite above 200°C., anisotropic and biaxial below 200°C. 28th List

Calcium-larsenite. C. Palache, L. H. Bauer, and H. Berman, 1928. *Amer. Min.*, vol. 13, p. 142. Zinc, lead, and calcium silicate, $(Pb,Ca)ZnSiO_4$, white and massive from Franklin Furnace, New Jersey. Like larsenite (q.v.) with some calcium replacing lead. [*M.A.*, **3**, 469.] 11th List [Now named Esperite, q.v.]

Calcium lipscombite. D. McConnell, 1963. *Amer. Min.*, vol. 48, p. 300. Tetragonal $(Ca,Fe^{\cdot\cdot})(Fe^{\cdot\cdot\cdot},Al)_2(PO_4)_2(OH,F)_2$, formed by heating Richellite at 500°C. 23rd List

Calcium lazulite. T. L. Watson, 1921. *Journ. Washington Acad. Sci.*, vol. 11, p. 389. The variety of lazulite from Graves Mountain, Georgia, and Keewatin, Canada, containing about 3% of lime. [*M.A.*, **1**, 377.] 9th List [Calcian lazulite, probably a mixture; *A.M.*, **35**, 8]

Calcium-melilite. A. N. Winchell, 1933. *Optical mineralogy*, 3rd edit., New York, 1933, p. 208 (Calcium-melilite). A hypothetical molecule, $Ca_3Al_2Si_4O_{14}$, to interpret the composition of mixed crystals of the melilite group. 14th List

Calcium-montmorillonite. W. Noll, 1936. *Chemie der Erde*, vol. 10, p. 137 (Ca-Montmorillonit). An artificially prepared clay mineral with calcium in place of magnesium. I. D. Sedletzky, *Compt. Rend.* (*Doklady*) *Acad. Sci. URSS*, 1940, vol. 26, p. 154 (calcium montmorillonite), records its presence in Russian saline soils and gives a formula $(OH)_8(Al_2Ca_3)Si.CaSi_4O_{10}$, as distinct from Mg-montmorillonite, $(OH)_8(Al_2Mg_3)Si.MgSi_4O_{10}$. W. Noll (pp. 135, 137) also mentions Mg-, Na-, K-, and alkali-montmorillonites. U. Hofmann and W. Bilke, *Kolloid-Zeits.*, 1936, vol. 77, pp. 243, 244, obtained Ca-, H-, and Na-montmorillonites as base-exchange products of bentonitic montmorillonite. [*M.A.*, **6**, 353; **7**, 97; **8**, 146.] 16th List

Calcium-pectolite. E. S. Larsen, 1917. *Amer. Journ. Sci.*, ser. 4, vol. 43, p. 465 (calcium pectolite). R. Koechlin, *Min Taschenb. Wien. Min. Gesell.*, 2nd edit., 1928, p. 15 (Calciumpektolith). Syn. of eakleite = xonotlite. [*M.A.*, **1**, 206; **2**, 253.] 18th List

Calcium-Pharmakosiderit. K. Walenta, 1966. *Tschermaks Min. Petr. Mitt.*, ser. 3, **11**, 154, fn. The calcium analogue of pharmacosiderite occurs at Krunkelbachtal, Menzenschwand, Südschwarzwald. Cf. alumopharmacosiderite, 14th List. 28th List

Calcium-pyromorphite. H. Strunz, *Min. Tab.*, 1941, p. 156 (Calciumpyromorphit). Syn. of polysphaerite. 18th List

Calcium-rinkite. I. D. Borneman-Starynkevich, 1935. *Materials Geochem. Khibina tundra, Acad. Sci. USSR*, pp. 48, 57, 65 (кальциевый ринкит); P. N. Chirvinsky, ibid., pp. 82, 88 (Kalzium-Rinkit). Fluo-titano-silicate of calcium and sodium, $3CaTiO_3.10(Ca,Na_2,H_2)SiO_3.3CaF_2$, isomorphous with rinkite with calcium in place of cerium earths; from the Kola peninsula, Russia. Perhaps identical with hainite (J. Blumrich, 1893; 1st List; incorrectly spelt as Gainit, гаинит). [*M.A.*, **6**, 343.] 14th List [Is Götzenite; *M.A.*, **15**, 132; *A.M.*, **45**, 221]

Calcium seidozerite. Erroneous translation of Кальциевый сейдозерит, calcian seidozerite [*M.A.*, **17**, 78] (*Soviet Physics—Crystallography*, 1966, **10**, 565) [*M.A.*, **17**, 736]. Cf. calcium catapleiite (24th List), of which it is a replacement product. 28th List

Calcium-strontianite. H. Strunz, *Min. Tab.*, 1941, p. 118 (Calcium-strontanit). The same as calciostrontianite (A. Cathrein, 1888), syn. of emmonite (T. Thomson, 1836). 18th List

Calciumuranoite and Caltsuranoite, *Zap.*, **103**, 103 (1974), variant transliterations of Кальцуранoит, Calciouranoite (28th List). 29th List

Calcium-uranospinite. M. E. Mrose, 1953. *Amer. Min.*, vol. 38, p. 1159 (calcium-uranospinite). Synonym of uranospinite, as applied to the artificial product. Compare hydrogen-uranospinite, etc. (19th List). [*M.A.*, **12**, 445.] 20th List [also 22nd List]

Calcium-urcilite (error), Calcium-ursilite, *see* Ursilite. 22nd List

Calcjarlite. A. S. Povarennykh, 1973. Конституция и свойства минералов, сб. 7, 131–135. $Na(Ca,Sr)_3Al_3(OH)_2F_{14}$, with Sr < Ca; the calcium analogue of jarlite. Named from the composition. 28th List [*A.M.*, **59**, 873]

Calclacite. R. Van Tassel, 1945. *Bull. Musée Roy. Hist. Nat. Belgique*, vol. 21, no. 26, p. 2. Hydrous chloride and acetate of calcium, $CaCl_2.Ca(C_2H_3O_2)_2.10H_2O$, formed as a fibrous efflorescence on certain limestones stored in wooden drawers. Named from the composition Ca,Cl,Ac. [*M.A.*, **10**, 101.] 18th List [*A.M.*, **32**, 254]

Calcomenita. G. Gagarin and J. R. Cuomo, 1949, loc. cit., p. 15. Spanish form of chalcomenite. [*M.A.*, **11**, 129.] 19th List

Calcotephroite. C. Palache, 1935. *Prof. Paper U.S. Geol. Survey*, no. 180, p. 80. Local name for what appears to be an impure variety of glaucochroite ($CaMnSiO_4$). 14th List

Calcowulfenite. V. Zepharovich, 1884. *Zeits. Kryst. Min.*, vol. 8, p. 583 (Kalkhaltige Wulfenit). A. Krantz's label, 1884 (Kalkwulfenit). Student's index to the collection of minerals, British Museum, 13th edit. 1886 (Calco-wulfenite); ibid., 27th edit., 1936 (Calcowulfenite). A variety of wulfenite containing CaO 1·24%, from Carinthia. 18th List

Calc-pyralmandite. L. L. Fermor, 1938. *Rec. Geol. Surv. India*, vol. 73, pp. 154, 156; *Indian Assoc. Cultiv. Sci.*, 1938, Spec. Publ. no. 6, p. 18. *See* Gralmandite. 15th List

Calc-spessartite. L. L. Fermor, 1926. *Rec. Geol. Survey India*, vol. 59, pp. 203, 205. The names calc-spessartite, ferro-calderite, ferro-spessartite, magnesia-blythite, mangan-almandite, mangan-grandite, pyralmandite, and spalmandite (qq.v.) are suggested for garnets of intermediate composition. *See* Blythite. 11th List

Calcurmolite. A. S. Povarennykh, 1962, in Мин. Таблицы, the Russian translation of H. Strunz's *Min. Tabellen*, at pp. 210 and 394. The unnamed mineral $Ca(UO_2)_3(MoO_4)_3(OH)_2.8H_2O$ described by L. S. Rudnitskaya (Зап. Всесоюз. Мин. Общ. (*Mem. All-Union Min. Soc.*), 1961, vol. 90, p. 101), is named from its composition. *See also* Зап. Всесоюз. Мин. Общ. (*Mem. All-Union Min. Soc.*), 1963, vol. 92, p. 464; *M.A.*, **16**, 458. 23rd List [*A.M.*, **49**, 1152]

Calcybeborosilite. A. S. Povarennykh and V. D. Dusmatov, 1970. Инфракрасные спектры поглощения новых минералов из щелочных пегматитов Средней Азий. Сь.: Конституция и свойства минералов, Киев, no. 4, стр. 7. (Калькибеборосилит). The unnamed Mineral A of Semenov, near $(Ln,Ca)_2(B,Be)_2Si_2O_8(OH)_2$ from the Alaisk ridge, Tadzhikistan (*A.M.*, **49**, 443) is named according to Povarennykh's chemical system (*A.M.*, **58**, 968). 28th List

Caldasite. O. A. Derby in T. H. Lee, 1917. *Revista Soc. Brasileira Sci.*, no. 1, p. 31; *Amer. Journ. Sci.*, 1919, vol. xlvii, p. 126. Zirconia-ore or rock consisting mainly of baddeleyite or of a mixture of zircon and orvillite (q.v.), from the Caldas district, Minas Geraes, Brazil. *See* Zirkite. 8th List

Californite. G. F. Kunz, 1903. *Amer. Journ. Sci.*, ser. 4, vol. xvi, p. 397. F. W. Clarke and G. Steiger, *Bull. US Geol. Survey*, 1905, no. 262, p. 72. Massive, green idocrase occurring as large masses in California. It resembles jade in appearance, and is used as an ornamental stone. 4th List

Calingastite. V. Angelelli and R. A. Trelles, 1938. [*Bol. Obras Sanitarias de la Nación*, Buenos Aires, nos. 8–10, p. 40.] S. G. Gordon, *Notulae Naturae, Acad. Nat. Sci. Philadelphia*, 1941, nos. 89 and 92. A zinciferous variety of melanterite $(Fe,Zn,Cu)SO_4.7H_2O$, containing FeO 16·67, ZnO 8·42, CuO 1·29%, from sulphate deposits between San Juan and Calingasta, Argentina. Named from the locality. Cf. zinc-copper-melanterite (9th List). [*M.A.*, **8**, 187.] 16th List

Calkinsite. W. T. Pecora and J. H. Kerr, 1953. *Amer. Min.*, vol. 38, p. 1169. Hydrous carbonate of rare-earths, $(La,Ce,Nd,Pr)_2(CO_3)_2.4H_2O$, as minute pale-yellow orthorhombic plates from Montana. An alteration product of burbankite (q.v.). Named after Frank Cathcart Calkins (1878–) of the United States Geological Survey. [*M.A.*, **12**, 301.] 20th List

Callaghanite. C. W. Beck and J. H. Burns, 1953. *Progr. & Abstr. Min. Soc. Amer.*, p. 10;

Amer. Min., 1954, vol. 39, p. 316, 630. $Cu_4Mg_4Ca(OH)_{14}(CO_3)_2.2H_2O$, blue monoclinic crystals in dolomite-rock from Gabbs, Nevada. Named after Dr. Eugene Callaghan, director of the New Mexico Bureau of Mines. [*M.A.*, **12**, 304, 410.] 20th List

Calogerasite. C. P. Guimarães, 1944. *Mineração e Metalurgia*, Rio de Janeiro, vol. 8, p. 135; *Anais Acad. Brasileira Cienc.*, vol. 16, p. 255 (Calogerasita). Synonym of simpsonite (15th List). Named after João Pandiá Calogeras (1870–1935). [*M.A.*, **9**, 127, 186.] 17th List

Calomelite. G. Gagarin and J. R. Cuomo, 1949, loc. cit., p. 8 (calomelita). Synonym of calomel. 19th List

Caltsuranoite, *see* Calciouranoite.

Calumetite. S. A. Williams, 1963. *Amer. Min.*, vol. 48, p. 614. Azure-blue spherules and sheaves of orthorhombic scales with good basal cleavage from the Centennial mine, Calumet, Michigan, have the composition $Cu(OH,Cl)_2.2H_2O$, with OH $\gg$ Cl. Named from the locality. [*M.A.*, **16**, 547.] (Cf. Anthonyite.) 23rd List

Calziodiallogit, Calziorhodochrosit, *see* Calciorhodochrosite.

Calzirtite. T. B. Zdorik, G. A. Sidorenko, and A. V. Bykova, 1961. Доклады Акад. Наук СССР (*Compt. Rend. Acad. Sci. URSS*), vol. 137, p. 681 (Кальциртит, calzirtite). $CaZr_3TiO_9$, tetragonal; from the East Siberian massif. Named from the composition, *cal*cium-*zir*conium-*ti*tanium. 22nd List [*A.M.*, **46**, 1515]

Camermanite. M. E. Denaeyer and D. Ledent, 1950. *Bull. Soc. Franç. Min. Crist.*, vol. 73, p. 483; 1952, vol. 75, p. 231. Hexagonal form of K_2SiF_6, dimorphous with cubic hieratite, in encrustations of potassium salts at chemical works. 19th List

Camsellite. H. V. Ellsworth and E. Poitevin, 1921. *Trans. R. Soc. Canada*, ser. 3, vol. 15, sect. 4, p. 1. Hydrated borate of magnesium, $2MgO.B_2O_3.H_2O$, forming white, fibrous (orthorhombic?) masses in serpentine from British Columbia. Named after Mr. Charles Camsell, of the Geological Survey of Canada. [*M.A.*, **1**, 375.] 9th List [Is Szájbelyite; *M.A.*, **9**, 123]

Canadium. A. G. French, 1911. *Chem. News*, vol. civ, p. 283. An alleged new element of the platinum group, said to have been found as white crystalline grains in the Nelson district, British Columbia, Canada. Named after the country of origin. This reported discovery has not been confirmed (*Ann. Rep. Minister of Mines, British Columbia, for 1911, 1912*, pp. 157, 165; *Mining Journ. London*, 1913, vol. ci, p. 344). 6th List

Canasite. M. D. Dorfman, D. L. Rogachev, Z. I. Goroshchenko, and E. I. Uspenskaya, 1959. Труды Мин. Муз. Акад Наук СССР (*Trans. Min. Mus. Acad. Sci. USSR*), vol. 9, p. 158 (Канасит). Silicate and fluoride of calcium and sodium, monoclinic; occurs associated with fenaksite (q.v.) at Khibina, Kola peninsula. Named from the composition, Ca–Na–Si. [*M.A.*, **14**, 414; *Amer. Min.*, **45**, 253.] May be the Mineral 6 of Dorfman (*Amer. Min.*, **44**, 910). 22nd List

Canavesite. G. Ferraris, M. Franchini-Angela, and P. Orlandini, 1978. *Can. Mineral.*, **16**, 69. Rosettes of fibres ‖ [010] on fractures of ludwigite and magnetite skarns from the Vola Gera tunnel, Brosso, Torino, Italy, are monoclinic, *a* 23·49, *b* 6·164, *c* 21·91 Å, β 114·91°. Composition $12[Mg_2CO_3HBO_3.5H_2O]$. α 1·485, β 1·494, γ 1·505 ‖ [010]. Named from the district, Canavese. 30th List

Canbyite. A. C. Hawkins and E. V. Shannon, 1924. *Amer. Min.*, vol. 9, p. 1. Hydrated ferric silicate, $H_4Fe'''_2Si_2O_9.2H_2O$; the crystalline equivalent of the amorphous hisingerite. Named after William Marriott* Canby (1831–1904)*, of Wilmington, Delaware. [*M.A.*, **2**, 253.] [*Corr. in *M.M.*, **21**.] 10th List [*A.M.*, **41**, 816]

Canfieldite. S. L. Penfield, *Amer. Journ. Sci.* XLVII. 451, 1894; XLVI. 107, 1893. [*Min. Mag.* X. 336; XI. 40.] $4Ag_2S.(Sn,Ge)S_2$. Cubic. [La Paz] Bolivia. 1st List

Cannizzarite. F. Zambonini, O. De Fiore, and G. Carobbi, 1925. *Rend. Accad. Sci. Fis. Mat. Napoli*, ser. 3, vol. 31, p. 28; *Annali R. Osservatorio Vesuviano*, 1925, ser. 3, vol. 1 (for 1924), p. 34. Sulpho-bismuthite of lead. $PbS.2Bi_2S_3$, as small needles, probably orthorhombic from fumaroles on Vulcano, Lipari Islands. Named after the celebrated Italian chemist, Stanislao Cannizzaro (1826–1910). An abstract in *Chem. Zentr.*, 1925, vol. 2, p. 2048, gives the spelling Cannizzarith. [*M.A.*, **3**, 10.] 11th List

Capillitite. V. Angelelli, 1948. F. Ahlfeld and V. Angelelli, *Las especies minerales de la República Argentina*, Jujuy, 1948, p. 143 (capillitita; misprint, capillita). E. E. Galloni, *Amer. Min.*, 1950, vol. 35, p. 562 (ferroan zincian rhodochrosite). M. M. Radice, *Notas*

Museo La Plata, 1949, vol. 56, Geol. no. 56, p. 231 (carbonato blanco). $(Mn,Zn,Fe)CO_3$, containing MnO 29·80, ZnO 14·88, FeO 13·93%. A variety of rhodochrosite from Capillitas, Catamarca. [*M.A.*, **11**, 119, 188, 403.] 19th List

Capreite. R. Bellini, 1922. *Boll. Soc. Geol. Ital.*, vol. 40 (for 1921), p. 228 (Capreite). A black encrustation on the walls of a limestone cave on the sea-shore of the island of Capri, Italy. A fetid calcite similar to pelagosite. Named from the locality. 14th List

Carbapatite. P. N. Tschirwinsky, 1906. *Annu. Géol. Min. Russie*, vol. viii, p. 251 (карбапатитъ), p. 256 (Carbapatit). A name later withdrawn (*Centralblatt Min.*, 1907, p. 283) in favour of podolite (q.v.). 4th List

Carbin. S. A. Vishnevskii and N. A. Palchik, 1975. [*Geol. Geofiz.* 65], abstr. *M.A.*, **75**, 3450. 'The third allotrope of carbon', occurring in rocks of the Popigai impact structure, USSR: distinct from diamond, graphite, and lonsdaleite. 29th List

Carbite. V. Aymé, 1901. *Essai de nomenclature minéralogique*, Hanoi, p. 34. A. D. Gonsalves, *Carbite o diamante*; *estudo geologico das zonas diamantiferas da Bahia*, Bahia, 1911. A contraction of the word carbon with the mineral termination *ite*, applied to both diamond and graphite. Also used as a trade-name for an explosive. Aymé works through the whole mineral kingdom on these lines, producing such names as 'plochlocarboxite' [i.e. Pb,Cl,C,O] for phosgenite, 'hyalcalmanfersiloxite' for piedmontite. Unfortunately, some of his compound abbreviations have the appearance of real names, e.g. aloxite (for corundum; cf. 7th List), mersulite (HgS), merselite (HgSe), mertelite (HgTe), cusulite (Cu_2S or CuS), etc. 9th List

Carboborite. Hsien-Te Hsieh, Tze-Chiang Chien, and Lai-Pau Liu, 1964. [*Scientia Sinica*, vol. 13, p. 813]; abstr. *Amer. Min.*, 1965, vol. 50, p. 262, and in *M.A.*, **17**, 75. $Ca_2MgCO_3B_2O_3(OH)_4.8H_2O$, monoclinic, from an unspecified lacustrine borate deposit in China. Named from the composition. 24th List

Carbocer. I. D. Borneman-Starynkevich, 1933. *Khibina Apatite*, Leningrad, 1933, vol. 6, p. 272 (Карбоцер, Carbocer). O. M. Shubnikova and D. V. Yuferov, *Spravochnik po novym mineralam*, Leningrad, 1934, p. 157. A carbonaceous mineral containing 8·2% of rare-earths; burns with loss on ignition 74·36%. Occurs in kondrikovite from the Kola peninsula, Russia. Named from carbon and cerium. [*M.A.*, **6**, 342.] 14th List

Carbocernaite. A. G. Bulakh, V. V. Kondrateva, and E. N. Baranova, 1961. Зап. Всесоюз. Мин. Общ. (*Mem. All-Union Min. Soc.*), vol. 90, p. 42 (Карбоцернаит). A carbonate of Ca, Sr, Na, and lanthanons (mainly La and Ce), $(Na,Ca,Sr,Ln)CO_3$; orthorhombic. From Vuorijärvi*, Kola peninsula. Named from the composition, *carbon*-*cer*ium-*na*trium. The name is uncomfortably near carbocer. [* Corr. in *M.M.*, **33**.] 22nd List [Ontario, major fluorine; *C.M.*, **11**, 812]

Carbodavyne. G. Cesàro, 1911. *See* Akalidavyne. 10th List

Carbonate-apatite. R. Brauns, 1916. *Neues Jahrb. Min.*, Beilage-Band xli, pp. 60, 73 (Carbonatapatit). A member of the apatite group with the composition $3Ca_3P_2O_8.CaCO_3$. The same as carbapatite (P. N. Chirvinsky, 1906; 4th List: = podolite = dahllite). Crystals from the Laacher See district, Rhine, are, however, said to differ from these in having an optically *positive* biaxial shell surrounding an optically negative uniaxial nucleus. *See* Sulphate-apatite. 8th List

Carbonate-cyanotrichite. E. A. Ankinovich, I. I. Gekht, and R. I. Zaitseva, 1963. Зап. Всесоюз. Мин. Общ. (*Mem. All-Union Min. Soc.*), vol. 92, p. 458 (Карбонат-цианотрихит). Pale blue fibrous aggregates from NW Kara-Tau give X-ray powder data corresponding to cyanotrichite, but contain carbonate replacing a large proportion of the sulphate, and approximate to $(Cu,Zn)_{3\cdot7}Al_{2\cdot3}(C_{0\cdot67}S_{0\cdot33})O_{2\cdot98}(OH)_{1\cdot02}(OH)_{12}.2H_2O$ (*Amer. Min.*, 1964, vol. 49, p. 441; *M.A.*, **16**, 548). 23rd List

Carbonate-fluorapatite. D. McConnell, 1973. *Apatite* (Springer), p. 88. Syn. of francolite. 30th List

Carbonate-fluor-chlor-hydroxyapatite. P. G. Cooray, 1970. *Amer. Min.*, **55**, 2040. A carbonatian apatite (1·2% CO_2) with F ≈ Cl ≈ OH, from Metale, Ceylon. [*M.A.*, 71-2322.] 27th List

Carbonate-hydrotalcite. D. M. Roy, R. Roy, and E. F. Osborn, 1953. *Amer. Journ. Sci.*, vol. 251, p. 355. Synonym of hydrotalcite. *See* Nitrate-hydrotalcite. [*M.A.*, **12**, 196.] 20th List

Carbonate-marialite and Carbonate-meionite. L. H. Borgström, 1914. *Zeits. Kryst. Min.*, vol. liv, p. 252. The chemical composition of the minerals of the scapolite group is explained by the isomorphous mixing of the following hypothetical molecules:

Chloride-marialite (Chloridmarialit)	$NaCl.3NaAlSi_3O_8$
Sulphate-marialite (Sulfatmarialit)	$Na_2SO_4.3NaAlSi_3O_8$
Carbonate-marialite (Karbonatmarialit)	$Na_2CO_3.3NaAlSi_3O_8$
Carbonate-meionite (Karbonatmejonit)	$CaCO_3.3CaAl_2Si_2O_8$
Sulphate-meionite (Sulfatmejonit)	$CaSO_4.3CaAl_2Si_2O_8$.

The existence of an *oxide-meionite* (oxydmejonit) molecule, $CaO.3CaAl_2Si_2O_8$, assumed by G. Tschermak (1883), is doubted. R. Brauns (*Neues Jahrb. Min.*, 1915, vol. ii, ref. p. 141), however, suggests that there may be *oxydhydratmejonit* and *oxydhydratmarialith*. Compare Silvialite. 7th List

Carbonate-meionite, *see* Carbonate-marialite.

Carbonate-sodalite. R. Brauns, 1916. *Neues Jahrb. Min.*, Beilage-Band xli, p. 73 (Carbonatsodalith). A hypothetical molecule assumed to explain the presence of carbonic acid (CO_2 1·27%, together with Cl 1·08 and SO_3 7·97) in noselite from the Laacher See, Rhine. 8th List

Carbonate-whitlockite. C. Frondel, 1943. *Amer. Min.*, vol. 28, p. 230; Dana's *Mineralogy*, 7th edit., 1951, vol. 2, p. 686. Whitlockite ($Ca_3P_2O_8$) containing some CO_2, in phosphate-rock, West Indies. 19th List

Carbonite. G. Mueller, 1964. *Rept. 22nd Internat. geol. Congr., India*, part 1, 46. An insoluble, infusible bitumen. Not to be confused with carbonite of Heinrich, 1875, a natural coke. 26th List

Carbonyl. G. Vavrinecz, 1939. *Földtani Közlöny*, Budapest, vol. 69, pp. 82, 98 (Carbonyl). Carbon monoxide, CO, as a natural gas. [*M.A.*, **7**, 471.] 15th List

Carborundum. E. G. Acheson, 1893. *Journ. Franklin Inst. Philadelphia*, vol. cxxxvi, p. 194. Silicide of carbon, CSi, formed artificially as brilliant, rhombohedral crystals and extensively used as an abrasive agent. Named from carbon and corundum, because, before it had been analysed, it was thought to be a compound of carbon and alumina. More recently (1904) the same substance has been isolated from a meteoric iron and named moissanite (q.v.). 4th List

Carburan. A. N. Labuntzov, 1934. O. M. Shubnikova and D. V. Yuferov, *Spravochnik po novym mineralam*, Leningrad, 1934, p. 157 (Карбуран, Carburan). P. K. Grigoriev, *Trans. Geol. Prosp. Inst. USSR*, 1935, no. 37, p. 29 (карбуран). A carbonaceous mineral containing C 60·96, H_2O 28·93, ash 9·51% (the ash contains UO_3 54·20, PbO 17·01, Fe_2O_3 6·01%), occurring with uraninite in pegmatite in Karelia, Russia. Related to thucholite (11th List). Named from carbon and uranium. [*M.A.*, **6**, 437.] 14th List

Cardenite. D. M. C. MacEwan, 1954. *Clay Min. Bull.*, vol. 2, p. 120. 'A trioctahedral montmorillonoid derived from biotite' in soil-clay at Carden Wood and other places in Aberdeenshire. A mixed alteration product previously described, without name, by G. F. Walker in *Min. Mag.*, **29**, 72. 20th List [Perhaps var. of Saponite; *A.M.*, **40**, 137]

Cardosonite. I. Asensio Amor, 1955. *Estudios Geológicos*, Madrid, no. 25, p. 37 (cardosonita). A mineral (Fe_2O_3 54·07, FeO nil, H_2O 9·21%) of the dufrenite series, from Coruña, Spain, with X-ray pattern distinct from frondelite and rockbridgeite. Named after Gabriel Martín Cardoso (–1954), professor of crystallography in the University of Madrid. [*Amer. Min.*, **41**, 165; *M.A.*, **13**, 487.] 21st List

Caringbullite. *M.A.*, 78-884, error for claringbullite. 30th List

Carletonite. G. Y. Chao, 1971. *Amer. Min.*, **56**, 354 (abstr.) and 1855; **57**, 765. A tetragonal mineral from Mt. St. Hilaire, Quebec, *a* 13·178, *c* 16·695 Å, D 2·45, *P4/mbm*, has the ideal composition $8[KNa_4Ca_4Si_8O_{18}(CO_3)_4(OH,F).H_2O]$, with considerable vacancies in the K, Na, Ca, CO_3, and (OH,F) sites. ω 1·521, ε 1·517. Named for Carleton University, Ottawa. Pink to pale blue. Perfect {001} cleavage; no distinct crystals. [*M.A.*, 72-2329, 3335.] 27th List

Carlfriesite. S. A. Williams and R. V. Gaines, 1975. *M.M.*, **40**, 127. Primrose yellow crusts lining cavities at the Bambollita (= La Oriental) mine, Sierra La Huerta, Moctezuma,

Sonora, Mexico, have space group *Cc* or *C2/c*, *a* 12·585, *b* 5·658, *c* 9·985 Å, *β* 115° 35′; composition 4[$H_4Ca(TeO_3)_3$]. Named for Carl Fries, *Jr.* [*M.A.*, 75-3593.] 29th List

Carlinite. A. S. Radtke and F. W. Dickson, 1975. *A.M.*, **60**, 559. Small grains with quartz and hydrocarbons in limestone at the Carlin mine, Eureka County, Nevada, have space group *R*3, *a* 12·12, *c* 18·175 Å, composition 27[Tl_2S]. Named for the locality. [*M.A.*, 76-879.] 29th List

Carlosite. G. D. Louderback, 1907. *Bull. Dep. Geol. Univ. California*, vol. v, p. 153; ibid., 1909, vol. v, p. 354. A supposed new mineral, found with benitoite (q.v.) in San Benito Co., California, which was very soon afterwards identified with the neptunite of Greenland. Named from the San Carlos peak, one of the highest points in the locality. 5th List

Carlsbergite. V. F. Buchwald and E. R. D. Scott, 1971. *Nature* (*Phys. Sci.*), **233**, 113. Oriented platelets and grains in kamacite in Descubridora and other siderites consist of CrN, cubic, *a* 4·16 Å, *Fm*3*m*. Named for the Carlsberg Foundation. [*M.A.*, 72-2330; *A.M.*, **57**, 1311.] 27th List

Carlsfriesite. *Bull.*, **99**, 340, error for carlfriesite (29th List). 30th List

Carnegieite. H. S. Washington and F. E. Wright, 1910. *Amer. Journ. Sci.*, ser. 4, vol. xxix, p. 52. A triclinic felspar closely allied to the plagioclases, which from its composition, $Na_2Al_2Si_2O_8$, may be described as soda-anorthite (S. J. Thugutt, 1895; 2nd List), sodium taking the place of calcium in the anorthite formula. It has been prepared artificially; and, in isomorphous intermixture with albite and anorthite, its presence is assumed to explain the composition of anemousite (q.v.). Named after Andrew Carnegie, founder of the Carnegie Institution of Washington, DC, where the mineral was determined. The compound $Na_2Al_2Si_2O_8$ is dimorphous, being also prepared artificially as hexagonal crystals closely allied to nepheline. 5th List

Carnevallite. B. H. Geier and J. Ottemann, 1970. *Min. Depos.*, **5**, 29. Sulphide of Cu and Ga, near Cu_3GaS_4, and some Fe and Zn, in small (40 μm) grains in the Tsumeb ores. Only optical data and an electron-probe analysis given. 26th List

Carnotite. C. Friedel and E. Cumenge, *Compt. Rend. Acad. Sci. Paris*, 1899, CXXVIII, 532; *Bull. Soc. Chim.*, 1899, XXI, 328; *Bull. Soc. franç. Min.*, 1899, XXII, pp. 26 and 26 *bis*; *Chem. News*, 1899, LXXX, 16. W. F. Hillebrand and F. L. Ransome, *Amer. J. Sci.*, 1900, X, 120. A bright canary-yellow substance impregnating sandstone in [Montrose Co.] Western Colorado. Friedel gives the composition as $2U_2O_3$*.V_2O_5.K_2O.$3H_2O$, but Hillebrand shows it to be a mixture of minerals. [*Corr. in *M.M.*, **15**.] 2nd List

—— *See* Silico-Carnotite. 9th List

Carobbiite. H. Strunz, 1956. *Rend. Soc. Min. Ital.*, vol. 12, p. 212 (Carobbiit). Potassium fluoride, KF, with some NaCl, etc., NaCl structure. Named after Prof. Greido Carobbi of Firenze (Florence) who had earlier [*M.A.*, **6**, 444] described the material in fumarole deposits from Vesuvius. [*M.A.*, **13**, 382.] 21st List [*A.M.*, **42**, 117]

Carpathite. G. L. Piotrovsky, 1955. *Min. Sbornik Lvov Geol. Soc.*, no. 9, p. 120 (карпатит). Hydrocarbon $C_{33}H_{17}O$, yellow monoclinic crystals in contact-zone between diorite-porphyrite and slate from Trans-Carpathians. Named from the locality. Distinct from curtisite [*M.A.*, **12**, 173, 418] from the same locality. (*See* Curtisitoids.) [*M.A.*, **13**, 208.] 21st List [Is nearly pure coronene; *A.M.*, **61**, 1055. Syn. Pendletonite, q.v. Also spelled Karpatite]

Carrboydite. W. W. Barker, M. Bussell, A. B. Fletcher, R. E. T. Hill, D. R. Hudson, E. H. Nickel, A. R. Ramsden, and M. R. Thornber, 1975. [*Ann. Rept. C.S.I.R.O. Min. Res. Lab., Australia*, 1974–1975, 12], abstr. *A.M.*, **61**, 366. Felted aggregates about 0·1 mm across from the Carr Boyd mine, Western Australia, have composition near $Ni_4Al_2SO_4(OH)_{12}.3H_2O$ and are related to woodwardite. Named from the locality. [*M.A.*, 76-1985.] 29th List

Cäsiumbiotit, variant of Caesium biotite (13th List). (H. Strunz, *Min. Tabellen*, 3rd edit., 1957, p. 305.) 22nd List

Cassidyite. J. S. White, E. P. Henderson, and B. Mason, 1967. *Amer. Min.*, vol. 52, p. 1190. $Ca_2(Ni,Mg)(PO_4)_2.2H_2O$, the nickel analogue of collinsite, occurring among the weathering products of the Wolf Creek sideritic meteorite. [*Bull.*, **91**, 98.] 25th List

Castaingite. A. Schüller and J. Ottermann, 1963. *Neues Jahrb. Min.*, Abh., vol. 100, p. 317 (Castaingit). $CuMo_2S_5$, intergrown with molybdenite; hexagonal, with molybdenite-like crystal structure. From Mansfeld, Anhalt, Germany. Named for R. Castaing. 23rd List [Perhaps Molybdenite; *A.M.*, **50**, 264]

Caswellite. A. H. Chester, *Trans. NY Acad. Sci.* XIII. 181, 1894. [*M.M.*, **11**, 243.] An altered biotite, allied to clintonite. [Trotter mine, Franklin Furnace.] New Jersey. 1st List [Largely manganoan andradite; *A.M.*, **51**, 1119]

Catalinaite. Variant of catalinite (6th List), a locality variety of jasper. R. Webster, *Gems*, 1962, p. 754. 28th List

Catalinite. (D. B. Sterrett, *Mineral Resources US Geol. Survey for 1910, 1911*, part ii, p. 898.) Local trade-name for beach pebbles, used as gems, from Santa Catalina Island, California. 6th List

Cataphorite, *see* Catophorite.

Catapleite, variant of or error for Catapleiite (*Amer. Min.*, 1964, vol. 49, p. 1153). 24th List [Proposed variant; *A.M.*, **9**, 62. 10th List]

Catophorite. W. C. Brøgger, *Die Eruptivgest. d. Kristianiagebietes, Skrifter Vid.-Selsk.* I, *Math.-natur. Kl.* 1894, No. 4, pp. 27–39, 73; 1898, No. 6, 169 (*Katoforit*); Abstr. *Min. Mag.* XI. 115. Dana's *Appendix*, 1899, *Cataphorite*. A soda-iron amphibole between barkevikite and arfvedsonite in its optical characters. Occurs in the grorudite of the Christiania district. A. Osann, *Tsch. Min. Mitth.* 1896, XV, 450; Abstr. *Min. Mag.* XI, 250. Texas. 2nd List [Katophorite is now the accepted spelling]

Catoptrite. *See* Katoptrite. 8th List

Cattierite. P. F. Kerr, 1945. *Amer. Min.*, vol. 30, pp. 483, 498. Cobalt disulphide, CoS_2, cubic with pyrite structure, from Shinkolobwe, Belgian Congo. Named after Felicien Cattier, chairman of the Union Minière du Haut Katanga. [*M.A.*, **9**, 188, 224.] 17th List

Caustobiolites, Caustolites. Group-names for combustible organic substances. H. Potonié, *Die Entstehung der Steinkohle und Kaustobiolithe*, Berlin, 1910, p. 1. A. W. Grabau, *Textbook of geology*, New York, 1920, part 1, p. 270 (Caustolith). J. Paclt, *Tschermaks Min. Petr. Mitt.*, 1953, ser. 3, vol. 3, p. 332 (Caustolites). [*M.A.*, **12**, 306.] 20th List

Caustolites, *see* Caustobiolites.

Cavansite. L. W. Staples, H. T. Evans, Jr., and J. R. Lindsay, 1967. *Progr. Ann. Meeting Geol. Soc. Amer.*, p. 211. Radiating greenish-blue needles from near Owyhee Dam, Malheur County, Oregon, and Goble, Columbia County, Oregon, are orthorhombic; composition $Ca(V^{4+}O)Si_4O_{10}.6H_2O$. Named from the composition (Ca,van[adium],Si). [*A.M.*, **53**, 510; *Bull.*, **91**, 300.] 25th List [*A.M.*, **58**, 405. $Ca(VO)Si_4O_{10}.4H_2O$]

Cayeuxite. Z. Sujkowski, 1936. *Arch. Min. Tow. Nauk. Warszaw.* (*Arch. Min. Soc. Sci. Varsovie*), vol. 12, p. 122 (cayeuxyt), p. 138 (cayeuxite). Pyritic nodules rich in As, Sb, Ge, Mo, Ni, etc., from Lower Cretaceous shales in the Carpathians. Named after Prof. Lucien Cayeux (1864–) of Paris. [*M.A.*, **6**, 344.] 14th List

Caysichite. D. D. Hogarth, G. Y. Chao, A. G. Plant, and H. R. Steacy, 1974. *C.M.*, **12**, 293. Crusts up to 2 mm thick on fractures in a granite pegmatite at the Evans–Lou feldspar mine, 22 miles N of Ottawa, are orthorhombic, *a* 13·282, *b* 13·925, *c* 9·727 Å; $4[(Yt,Ca)_4Si_4O_{10}(CO_3)_3.4H_2O]$. Named for its composition (Ca,Y,Si,C,H). 28th List

Ce-britholite. A. V. Rudneva, A. V. Nikitin, and N. V. Belov, 1962. Докл. **146**, 1182 (Се-бритолит). *Syn.* of Cefluosil, which is shown to be closely related to Britholite. [*M.A.*, **17**, 395.] 28th List

Ce-Vesuvian. H. Strunz, 1970. *Min. Tab.*, 5th edit., 399. An unnecessary name for cerian idocrase (vesuvianite). 28th List

Cebollite. E. S. Larsen and W. T. Schaller, 1914. *Journ. Washington Acad. Sci.*, vol. iv, p. 480. A hydrated silicate of calcium, aluminium, etc., $H_4Al_2Ca_5Si_3O_{16}$, forming a dull, compact, white to greenish, fibrous (orthorhombic?) aggregate. It occurs as an alteration product of melilite at Cebolla Creek, Gunnison Co., Colorado. Named after the locality. Cf. Deeckeite. 7th List

Cedarite. R. Klebs, *Jahrb. k. Preuss. geol. Landesanst. for 1896*, 1897, XVII (ii), 199. An amber-like resin from Cedar Lake, Canada, previously described under the name chemawinite. 2nd List

Cefluorosil, variant of Cefluosil (Цефтосил, 22nd List), a cerian variety of Britholite. Докл. Акад. Наук СССР (*Compt. Rend. Acad. Sci. URSS*), 1962, vol. 146, p. 1182. 23rd List

Cefluosil. A. V. Rudneva and T. Ya. Malysheva, 1961. Доклады Акад. Наук СССР (*Compt. Rend. Acad. Sci. URSS*), vol. 136, p. 191 (Цефтосил; p. 7, cephtosyl). Silicate and fluoride of cerium earths and Ca, from a slag. Named from the composition церий-фтор-силикат (*ce*rium-*fluo*r-*sil*icate). 22nd List [A cerian phosphate-free britholite; *M.A.*, **17**, 395]

Celanite. A. V. Rudneva and T. Ya. Malysheva, 1961. Доклады Акад. Наук СССР (*Compt. Rend. Acad. Sci. URSS*), vol. 136, p. 191 (Целанит; p. 7, coelanite). Oxide of Ti, Al, and lanthanons, cubic, isolated from a slag. Named from the principal rare earths present, церия (ceria), лантана (lanthana), and неодима (neodymia). Composition, X-ray data, and origin agree exactly with ceraltite (q.v.), which has priority. The name is easily confused with ceylonite (Цейлонит). 22nd List

Celedonite. M. F. Heddle, *The Mineralogy of Scotland*, Edinburgh, 1901, vol. i, p. 60; vol. ii, p. 145. An incorrect spelling of Celadonite. 8th List

Celite. *See* under alite. 2nd List

Celsian. H. Sjögren, *Geol. För. Förh.* XVII. 578, 1895. Barium anorthite. $BaO.Al_2O_3.2SiO_2$. Anorthic. [Jakobsberg] Sweden. 1st List [Cf. Paracelsian]

Cementite. [H. M. Howe, 1890 (*The Metallurgy of Steel*).] F. Osmond [*Bull. Soc. Encour. Indust. Paris*, 1895, X, 480] *Nature*, 1895, LII, 367; *Zeits. Kryst. Min.* 1896, XXVII, 538. A carbide of iron occurring in steel. Probably Fe_3C, and possibly identical with cohenite (Groth, *Tab. Uebers. Min.* 4th edit. 1898, p. 16). To various other microscopic constituents of steel have been given the names, *austenite*, *ferrite*, *martensite*, *perlite* (*pearlyte*), *sorbite* and *troostite*. 2nd List

Cephtosyl. Erroneous transliteration of Цефтосил, cefluosil (q.v.). 22nd List

Cerafolite. H. H. Lohse, 1958. *Diss. Kiel*, referred to in *Neues Jahrb. Min.*, Abh., 1961, vol. 97, p. 113 (footnote). The residue when Koenenite is leached with water; near $Mg_5Al_4(OH)_{22}.4H_2O$. 23rd List

Ceraltite. V. V. Lapin, H. H. Kyrtseva, and D. N. Knyazeva, 1960. Доклады Акад. Наук СССР (*Compt. Rend. Acad. Sci. URSS*), vol. 134, p. 1195 (Цералтит). Oxide of Ti, Al, and lanthanons, with a perovskite structure, isolated from slags. Named from the composition, Ce(r)–Al–Ti. 22nd List

Cerapatite. A. E. Fersman, 1926. *Neues Jahrb. Min.*, Abt. A, Beil-Bd. 55, pp. 40, 45 (Cerapatit); *Amer. Min.*, 1926, vol. 11, p. 293 (Cerium-apatite). O. M. Shubnikova and D. V. Yuferov, *Spravochnik po novym mineralam*, Leningrad, 1934, p. 109 (церапатит, cerapatite). A variety of apatite containing 3·18% of rare-earths (Ce_2O_3 1·33%) from the Kola peninsula, Russia. [Cf. *M.A.*, **2**, 408, 409; **3**, 235; **6**, 233, 312.] 14th List [Cf. Britholite]

Cerargerite, error for Cerargyrite (A. S. Eakle, *Bull. Dept. Geol. Univ. California*, 1912, vol. 7, p. 1). 24th List [Now Chlorargyrite; *A.M.*, **49**, 224]

Cerepidote. H. Rosenbusch, 1905. *Mikroskopische Physiographie d. Mineralien*, 4th edit., vol. i, part 2, p. 286 (Cerepidot). Synonym of allanite. So named because it is a member of the epidote group containing cerium. 4th List

Cererdensilikatapatit. F. Machatschki, 1953. *Spez. Min.*, Wien, p. 330. Syn. of Britholite (2nd List). 22nd List

Cererdenthoriumeuxenit. F. Machatschki, 1953. *Spez. Min.*, Wien, p. 319. Syn. of Aeschynite. 22nd List

Cerfluorite. T. Vogt, 1914. *Neues Jahrb. Min.*, vol. ii, p. 15 (Cerfluorit). Artificial, cubic mixed crystals of calcium and cerium fluorides, $(Ca_3,Ce_2)F_6$, analogous to the mineral yttrofluorite (T. Vogt, 1911; 6th List), and mixed with this forming the mineral yttrocerite. 7th List

Cergadolinite. W. C. Brøgger, 1922. *Vid.-Selsk. Skrifter*, Kristiania, I. *Mat.-Nat. Kl.*, 1922, no. 1, p. iii (Cergadolinit). A gadolinite from Fyrrisdal, Norway, rich in cerium oxides (Ce_2O_3 23·40%). [i.e. cerian Gadolinite.] [*M.A.*, **2**, 25.] 10th List

Cerianite. A. R. Graham, 1955. *Amer. Min.*, vol. 40, p. 560; *Contr. Canadian Min.*, vol. 5, part 7. Minute greenish-yellow octahedra in carbonate rock from Lackner*, Sudbury, Ontario. Cubic CeO_2 with some ThO_2. Named from its relation to thorianite and uraninite. [*M.A.*, **13**, 6.] [*Corr. in *M.M.*, **31**.] 21st List

Ceriopyrochlore. *I.M.A. Subcomm.* pyrochlore group, 1978. *A.M.*, **62**, 404. A systematic name to replace marignacite. 30th List

Cerium-ankerite. H. Strunz, *Min. Tab.*, 1941, p. 118 (Ceriumankerit, Coddazit [sic]). Syn. of codazzite (11th List). 18th List

Čermíkite. B. Ježek, *Mineralogie*, Praha, 1932, (*Nat. Hist.*, vol. 6), p. 1260 (Čermíkit); R. Rost, *Bull. Internat. Acad. Sci. Bohême*, 1937, vol. —, preprint p. 2 (Čermíkite). The Czech spelling of tschermigite (F. Kobell, 1853), from the locality, Čermíky = Tschermig, Bohemia. [*M.A.*, **7**, 11.] 15th List

Černýite. S. A. Kissin, D. R. Owens, and W. L. Roberts, 1978. *Can. Mineral.*, **16**, 139. Intergrowths with kësterite from the Tanco mine, Bernic Lake, Manitoba, and the Hugo mine, Keystone, Pennington Co., S. Dakota, are tetragonal, *a* 5·5330, *c* 10·8266 Å, space group $I\bar{4}2m$. Composition $2[Cu_2(Cd,Zn,Fe)SnS_4]$, with Cd:Zn:Fe from 0·37:0·33:0·29 to 0·77:0·14:0·10. The cadmium analogue of stannite, but not miscible with it. Named for P. Černý. 30th List

Cerorthite. A variety of orthite rich in cerium (Ce_2O_3 20%) from Finland. *Bull. Soc. Franç. Min.*, 1933, vol. 56, p. 188. Student's index to the collection of minerals, British Museum, 27th edit., 1936 (Cerorthite), from a dealer's label (Cer-Orthit). 18th List

Cerotungstite. Th. G. Sahama, O. von Knorring, and M. Lehtinen, 1970. *Bull. Geol. Soc. Finland*, no. 42, 223. The cerium analogue (Ce = 10%, Nd = 6%) of yttrotungstite (19th List), and named accordingly. A secondary mineral in tungsten ore from Nyamulilo mine, Kigezi, Uganda. [*M.A.*, 72-3336; *Zap.*, **100**, 623.] 27th List [*A.M.*, **57**, 1558]

Cerphosphorhuttonite. A. S. Pavlenko, L. P. Orlova, and M. V. Akhmanova, 1965. [Труды Мин. Муз. Акад. Наук СССР (*Proc. Min. Mus. Acad. Sci. USSR*), vol. 16, p. 166]; abstr. *Amer. Min.*, 1965, vol. 50, p. 2099; *M.A.*, **17**, 503; Зап. Всесоюз. Мин. Общ. (*Mem. All-Union Min. Soc.*), 1966, vol. 95, p. 320 (церфосфорхаттонит). A mineral almost exactly midway between Huttonite (monoclinic $ThSiO_4$) and Monazite (monoclinic $CePO_4$) occurs in amazonite pegmatite at an unspecified locality in S-E Siberia. 24th List

Ceruleite. H. Dufet, *Bull. Soc. franç. Min.*, 1900, vol. xxiii, p. 147 (céruléite). Turquoise-blue, clayey masses. Under the microscope it is seen to be minutely crystalline. $CuO.2Al_2O_3.As_2O_5.8H_2O$. Huanaco, Chile. Named from the Latin, *caeruleus*, sky-blue. 3rd List [$Cu_2Al_7(AsO_4)_4(OH)_{13}.11{\cdot}5H_2O$; *M.A.*, 77-2184]

Cerulene. (*Review of Mining Operations, South Australia, for 1910, 1911*, no. 13, p. 9.) Trade-name for a form of calcium carbonate coloured green and blue by malachite and chessylite; found near Bimbowrie, South Australia, and used as a gem-stone. Named from the Latin, *caeruleus*, sky-blue. 8th List

Ceruleofibrite. E. F. Holden, 1922. *Amer. Min.*, vol. 7, p. 80 (on p. 47 of the same volume, the alternative spelling Caeruleofibrite is given). A basic chloro-arsenate of copper, $CuCl_2.\frac{1}{3}Cu_3As_2O_8.6Cu(OH)_2$, forming blue, fibrous tufts of orthorhombic needles in cuprite from Bisbee, Arizona. Named from the Latin *caeruleus*, sky-blue, and *fibra*, a fibre. 9th List [Is Connellite; *A.M.*, **9**, 55]

Cesarolite. H. Buttgenbach and C. Gillet, 1920. *Ann. Soc. Géol. Belgique*, vol. 43, *Bull.*, p. 239 (Cesàrolite). Acid lead manganate, $H_2PbMn_3O_8$, occurring as steel-grey, spongy masses in galena from Tunisia. Named after Prof. Giuseppe Cesàro of Liége. [*M.A.*, **1**, 201.] 9th List [*A.M.*, **5**, 211]

Cesbronite. S. A. Williams, 1974. *M.M.*, **39**, 744. Bright green crusts of steep bipyramidal crystals from the Bambollita mine, Moctezuma, Sonora, Mexico, are orthorhombic, *a* 8·624, *b* 11·878, *c* 5·872 Å, cell-contents $2[Cu_5(TeO_3)_2(OH)_6.2H_2O]$. Named for F. Cesbron. 28th List

Cesium kupletskite, *see* Caesium kupletskite. 27th List

Chacaltaite. M. Kołaczkowska, 1936. *Spraw. Tow. Nauk. Warsaw.* (*Compt. Rend. Soc. Sci. Varsovie*), Cl. II, vol. 29, p. 71 (Czakaltait), p. 72 (Chacaltaïte). A green chlorite-like mineral previously described as pinite [*M.A.*, **6**, 473], from which it differs in its X-ray pattern. Named from the locality, Chacaltaya, Bolivia. [*M.A.*, **7**, 226.] 15th List [Is muscovite; *A.M.*, **55**, 1437]

Chakassit, German transliteration of Хакассит, Khakassite (12th List; = Alumohydrocalcite) (C. Hintze, *Handb. Min.*, Erg. III, p. 537). 25th List

Chalcoalumite. E. S. Larsen and H. E. Vassar, 1925. *Amer. Min.*, vol. 10, p. 79. Hydrous aluminium copper sulphate, $CuSO_4.4Al(OH)_3.3H_2O$, forming turquoise-green, botryoidal crusts on limonite stalactities from Bisbee, Arizona. Perhaps anorthic. Named from the chemical composition [*M.A.*, **2**, 520.] 10th List [*M.A.*, 72-534]

Chalcocite, Tetragonal. A. H. Clark and R. H. Sillitoe, 1971. *Neues Jahrb. Min., Monatsh.* 418. A tetragonal phase of composition $Cu_{1\cdot 96}S$, occurring at Mina María, Quebrada Puquios, Atacama, Chile; breaks down to djurleïte. [*M.A.*, 72-1369.] 28th List [*Sulphide Mineralogy* (M.S.A. 1974), CS-60]

Chalcocyanite. Dana's *Mineralogy*, 7th edit., 1951, vol. 2, p. 578. Anhydrous cupric sulphate $CuSO_4$. To replace the name hydrocyanite (A. Scacchi, 1873), because the pure mineral contains no water (but is then white, not blue). Named from *χαλκός*, copper, and *κύανος*, dark blue. 19th List

Chalcokyanite. H. Strunz, 1961. *Amer. Min.*, vol. 46, p. 758. Variant of chalcocyanite; synonym of hydrocyanite. The substitution of kyan- for cyan- in those mineral names where it represents Greek, *κυανός*, to avoid the mispronunciation sian-, is too late by a century or more, in view of the innumerable chemical names involving the radical CN. Moreover, chalcokyanite suggests a variety of kyanite. 22nd List

Chalcolamprite. G. Flink, *Medd. om Grönland*, 1900, XXIV, 160. A member of the pyrochlore group occurring as small regular octahedra with a metallic coppery lustre on the surface. $R''Nb_2O_6F_2 + R''SiO_3$; (R = Ca,Zr,Na,Ce,etc.). [Narsarsuk] S Greenland. Cf. endeiolite below. 2nd List [Impure pyrochlore; *A.M.*, **62**, 407]

Chalcomalachite. Author and date? A mixture of calcite and malachite. R. Webster, *Gems*, 1962, p. 754. 28th List

Chalconatronite. C. Frondel and R. J. Gettens, 1955. *Science* (*Amer. Assoc. Adv. Sci.*), vol. 122, p. 75. R. J. Gettens and C. Frondel, *Studies in Conservation* (*London*), 1955, vol. 2, no. 2, p. 64. $Na_2Cu(CO_3)_2.3H_2O$, probably monoclinic, as a greenish-blue incrustation on ancient bronze objects from Egypt. Named from *χαλκος*, copper, and natron, soda. [*M.A.*, **13**, 6, 379.] 21st List [*A.M.*, **40**, 943]

Chalcopentlandite. H. Pauly, 1958. *Medd. Grønland*, vol. 157, no. 2, p. 32. An assumed high-temperature phase, now represented by aggregates of pentlandite and chalcopyrite (about 10% of the latter), which are believed to be the product of exsolution. From a nickeliferous pyrrhotine deposit at Igdlukúnguaq, Greenland. Named from the composition. 22nd List

Chalcothallite. E. I. Semenov, H. Sørensen, M. S. Bessmertnaya, and L. E. Novorossova, 1967. *Medd. Grønland*, **181**, no. 5, 13. Metallic grey lamellar aggregates from Nakalaq, Ilimaussaq, Greenland; Cu_3TlS_2. Named for the composition. [*A.M.*, **53**, 1775; *M.A.*, 69-1528; *Bull.*, **92**, 316; *Zap.*, **97**, 612.] 26th List

Chalkocyanit. Germanized version of Chalcocyanite (19th List). (H. Strunz, *Min. Tabellen*, 3rd edit., 1957, p. 193.) 22nd List

Chalkonatrit, Chalkonatronit. Erroneous and corrected Germanized variants of Chalconatronite (21st List). (Hintze, *Handb. Min.*, Erg.-Bd. II, p. 682; H. Strunz, *Min. Tabellen*, 3rd edit., 1957, p. 179.) 22nd List

Chalkopissite. (R. Koechlin, *Min. Taschenbuch*, Wien, 1911, p. 20; P. Niggli, *Lehrbuch Min.*, Berlin, 1920, p. 648 (Chalkopissit).) The same as Pitchy-copper-ore = Copper-pitchblende = Kupferpecherz. A mixture of chrysocolla and limonite. Evidently from *χαλκός*, copper, and *πίσσα*, pitch. 10th List

Challantite. C. C. Ramusino and G. Giuseppetti, 1973. [*Soc. Ital. Sci. Nat. Mus. Civito Milano*, **64**, 451], abstr. *A.M.*, **60**, 736. A yellow powder in samples of quartzite from an abandoned gold mine at Challant St. Anselme, Valle d'Ayas, Aosta, Italy, matches synthetic $6Fe_2(SO_4)_3.Fe_2O_3.63H_2O$. Named for the locality. 29th List

Chalmeleonite. Presumably an error for chameleonite, author and date? 'Change colour tourmaline.' R. Webster, *Gems*, 1962, p. 754. 28th List

Chalmersite. E. Hussak, *Centralblatt Min.*, 1902, p. 69 (Chalmersit). Brilliant, bronze-yellow, acicular crystals very similar to copper-glance in habit, twinning, and angles; they are magnetic like pyrrhotite. Orthorhombic. $CuFe_3S_4 = Cu_2S.Fe_6S_7$ (but only 0·016 gram of material analysed). Associated with pyrrhotite, etc. at the Morro Velho goldmine, Minas Geraes, Brazil. Named after Mr. G. Chalmers, the superintendent of the mine. 3rd List [Is Cubanite; *M.A.*, **2**, 235, 236; **3**, 468, 469]

Chambersite. R. M. Honea and F. R. Beck, 1962. *Amer. Min.*, vol. 47, p. 665. The manganese analogue of boracite, occurring in tetrahedra in brines from Barber's Hill salt dome, Chambers County, Texas; $Mn_3B_7O_{13}Cl$. Named from the locality. [*M.A.*, **16**, 66.] 23rd List

Changbaiite. *Acta Geol. Sinica*, 1978, 63. Pale brown tabular crystals in kaolin veinlets in granite near Changbai Mtn., E Kirin, China, are rhombohedral, *a* 10·499, *c* 11·553, space group *R3m*. Composition 9[$PbNbO_6$]. Sp. gr. 6·47, ω 2·476, ε 2·485. Named for the locality. 30th List

Chantalite. H. Sarp, J. Deferne, and B. W. Liebich, 1977. *Schweiz. Mineral. Petrogr. Mitt.*, **57**, 149. Grains associated with vuagnatite (30th List), prehnite, etc., from a rodingite dyke in SW Turkey, are tetragonal, *a* 4·945, *c* 23·27 Å, space group $I4_1/a$. Composition 4[$CaAl_2\,SiO_4(OH)_4$]. ω 1·653, ε 1·642. Named for the discoverer's wife. [*M.A.*, 78-3469] 30th List

Chaoite. A. El Goresy, 1970. *Naturwiss.*, **56**, 493. The hexagonal polymorph of carbon, *a* 8·948, *c* 14·078 Å, reported by A. El Goresy and G. Donnay (*Science*, **161**, 363 (1968)) from graphite gneiss from the Ries Crater, Bavaria, and by G. P. Vdovykin (Геохимия, 1969, 1145) is named for E. C. T. Chao. [*A.M.*, **54**, 326; **55**, 1067; *Zap.*, **100**, 614.] 27th List [Cf. Lonsdaleite, Carbin]

Chapapote. *See* Malthite. 5th List

Chapmanite. T. L. Walker, 1924. *Univ. Toronto Studies, Geol. Ser.*, no. 17, p. 5; preliminary abstract in *Amer. Min.*, vol. 9, p. 66. Hydrous ferrous silico-antimonate $5FeO.5SiO_2.Sb_2O_5.2H_2O$, as an olive-green, pulverulent material (probably orthorhombic) with native silver from South Lorrain, Ontario. Named after Edward John Chapman (1821–1940), formerly Professor of Mineralogy at Toronto, Canada. [*M.A.*, **2**, 336.] 10th List [$SbFe_2(SiO_4)_2(OH)$; *A.M.*, **49**, 1499]

Charoite. L. V. Nikol'skaya, A. I. Novozhilov, and M. I. Samioilovich, 1976. *Izvest. akad. nauk SSSR*, ser. geol. no. 10, 116. A complex alkali silicate occurring with tinaksite (24th List) in Transbaikal. Deep violet, fibrous. Named for the Charo River. [*M.A.*, 78-4923.] 30th List

Chasovrite. S. V. Potapenko, 1952. [*Bull. Acad. Architecture*, Ukraine, p. 43.] Criticized by I. D. Sedletzky, *Mém. Soc. Russ. Min.*, 1954, ser. 2, vol. 83, p. 292 (часоврит). A variety of clay mineral (glinite, q.v.) from the Chasovyar deposit in Ukraine. Named from the locality. [*M.A.*, **13**, 5.] 21st List

Chavesite. J. Murdoch, 1958. *Amer. Min.*, vol. 43, p. 1148. An anorthic hydrated calcium manganese phosphate (not analysed quantitatively for lack of sufficient material) occurring in the Boqueirao pegmatite, Borborema, Paraíba, Brazil; characterized by optical and X-ray data; perhaps isostructural with monetite. [*M.A.*, **14**, 199.] Named after Dr. Onofre Chaves of Brazil. 22nd List

Chelkarite. N. P. Avrova, V. M. Bocharov, I. I. Khalturina, and Z. R. Yunosova, 1968. [Геол. развед. месторожд. тверд. полез. ископ. Каз. 1969, 169], abstr. Реф. журн., геол. 1969, abstr. no. 6V 233 (челкарит). Prismatic crystals in insoluble residues from a carnallite–bischofite rock are orthorhombic, *a* 13·69, *b* 20·84, *c* 8·26 (Å or kX?), *Pbca*. Two analyses differ considerably, but are near $(Ca,Mg)_4B_3O_7Cl_3.11H_2O$. No locality or etymology is given. [*A.M.*, **56**, 1122; *Zap.*, **100**, 86.] 27th List

Chengbolite. W.-C. Sun, C.-L. Li, Y.-W. Jen, Y. T. Chang, W.-L. Cheng, S.-F. Yuan, and C.-L. Change, 1973. [*Acta Geol. Sinica*, **1**, 89]; abstr. *M.A.*, 74-2472. $PtTe_{2-x}$, with x up to about $\frac{1}{3}$ and a little Pd; distinct from moncheïte (23rd List); *a* 4·041, *c* 5·220 Å [tetragonal or hexagonal?]. 28th List [Var. of Moncheïte, Fleischer's *Gloss.*]

Cheralite. S. H. U. Bowie and J. E. T. Horne, 1953. *Min. Mag.*, vol. 30, p. 93. A member of the monazite group rich in thorium (ThO_2 31·50%), $(Ce,La,Th,U,Ca)(P,Si)O_4$, as green monoclinic crystals in pegmatite from Travancore. Named from Chera (Kerala), an ancient kingdom in SW India. 20th List [*A.M.*, **38**, 734; **39**, 403]

Cherlbutite. Error for Hurlbutite, resulting from back-transliteration through Херлбутит (*Dokl. Acad. Sci. USSR*, 1967, **176**, 152). 28th List

Chernikite. G. Gagarin and J. R. Cuomo, 1949, loc. cit., p. 10 (chernikita). Tantalo-tungstate of Ti, Ca, Fe, etc. from Kola, analysed in 1927 by G. P. Chernik Г. П. Черник. [Dana, 7th edit., **1**, 741.] 19th List

Chernovite. B. A. Goldin, N. P. Yuskin, and M. V. Fishman, 1967. Зап. всесоюз. мин. общ. (*Mem. All-Union Min. Soc.*), vol. 96, p. 699 (Черновит, chernovite). $YAsO_4$, the arsenate analogue of xenotime, in tetragonal crystals from the Urals. Named for A. A. Chernov. 25th List [*A.M.*, **53**, 1777]

Chernykhite. G. Ankinovich, E. A. Ankinovich, I. V. Rozhdestvenskaya, and V. A. Frank-Kamenetskii, 1972. *Zap.*, **101**, 451 (Черныхит). Veinlets and dark-green leaflets up to 5 mm across in shales in Karatau, Kazakhstan, are monoclinic, *a* 5·29, *b* 9·182, *c* 20·023 Å, *β* 95° 41′; composition $(Ba,K)_{1-x}(V^{3+},Al,V^{4+},Mg)(Si,Al)_4O_{10}(OH)_2$, with $x \approx \frac{1}{2}$, the barium analogue of roscoelite. Named for V. V. Chernykh. [*A.M.*, **58**, 966; *M.A.*, 74-506; *Zap.*, **102**, 455.] Nothing to do with Chernikite (of Gagarin and Cuomo), 19th List. 28th List

Chernyshevite. English equivalent of tschernichewite (4th List). Named after T. H. Чернышевъ (1856–1914). 18th List

Cherskite. B. A. Gavrusevich, 1935. *Trans. Lomonosov Inst. Acad. Sci. USSR*, Ser. Min. no. 5, p. 100 (черскит); O. M. Shubnikova, ibid., 1937, no. 10, p. 223 (Черскит, Cherskite). An undescribed manganese mineral with phlogopite, diopside, etc., from Slyudyanka, Transbaikalia. 15th List

Chervetite. P. Bariand, F. Chantret, R. Pouget, and A. Rimsky, 1963. *Bull. Soc. franç. Min. Crist.*, vol. 86, p. 117. $Pb_2V_2O_7$, in small monoclinic crystals at the Mounana uranium mine, Republic of Gabon. Named for J. Chervet. [*M.A.*, **16**, 373.] 23rd List [*A.M.*, **48**, 1416; *M.A.*, 74-3452]

Chesterite. D. R. Veblen, P. R. Buseck, and C. W. Burnham, 1977. *Science*, **198**, 359. An orthorhombic mineral from Chester, Vermont, *a* 18·61, *b* 45·31, *c* 5·30 Å, space group $A2_1ma$, belonging to the biopyribole group (30th List). Composition $(Mg,Fe)_{17}Si_{20}O_{54}(OH)_6$. Named for the locality. [*M.A.*, 78-3473.] 30th List

Chevkinite. Committee on Nomenclature, *Amer. Min.*, 1924, vol. 9, pp. 61, 62; L. J. Spencer, *Min. Mag.*, 1925, vol. 20, p. 357. Another spelling for Tschewkinit of G. Rose, 1839. The form Tscheffkinite in common use is a mixed German and French transliteration; but, curiously, German authors (W. Haidinger, 1845; C. F. Naumann, 1846; P. Groth, 1898) have generally used the French *ff*, whilst French authors (A. Dufrénoy, 1845; A. Des Cloizeaux, 1862; A. Lacroix, 1913) have taken the German *w*. The Russian form is Чевкинитъ, after the Russian General, Konstantin Vladimirovich Chevkin (Константинъ Владимiровичъ Чевкинъ, 1802–1875). Chief of the Department of Mines of Russia. A Spanish form is Cherfquinita. 10th List [Cf. Perrierite]

Chiklite. S. A. Bilgrami, 1955. *Min. Mag.*, vol. 30, p. 634. A manganese amphibole in manganese ore from Chikla, Bhandara, India. Named from the locality. 20th List [Manganoan ferri-ferro-richterite, *Amph. Sub-Comm.*]

Childro-eosphorite. H. Strunz and M. Fischer, 1957. *Neues Jahrb. Min.*, Monatshefte, p. 78 (Childro-Eosphorit), p. 79 (Childrenit-Eosphorit). Midway between childrenite and eosphorite in the isomorphous series $AlPO_4.(Fe,Mn)(OH)_2.H_2O$. From Hagendorf, Bavaria. 21st List

Chile-loeweite. W. Wetzel, 1928. *Chemie der Erde*, vol. 3, p. 388 (Chile-Loeweït). Minute trigonal crystals found in Chile saltpetre ('caliche'), differing optically and in composition ($K_2Na_4Mg_2(SO_4)_5.5H_2O$) from loeweite. [*M.A.*, **3**, 554.] 11th List [Is Humberstonite (q.v.); *A.M.*, **55**, 1518]

Chilkinite, Schilkinit. French (*Bull. Soc. franç. Min. Crist.*, 1951, vol. 74, p. 182) and German (Hintze, *Min.*, 1954, Ergbd. 2, p. 349) transliterations of шилкинит, shilkinite (16th List). 20th List [also 19th List]

Chillagite. A. T. Ullmann, 1913. *Journ. R. Soc. New South Wales*, vol. xlvi (for 1912), p. 186; C. D. Smith and L. A. Cotton, ibid., p. 207. Tungstate and molybdate of lead, $PbWO_4.PbMoO_4$, intermediate between stolzite and wulfenite, forming yellow, platy, tetragonal crystals. Named after the locality, Chillagoe, Queensland. In a mineral-dealer's circular (F. Krantz of Bonn, November 1912; advertisement in *Nature*, London, September 12, 1912) the name *Lionit* or *Lyonite* has been applied to this mineral, after Mr. D. Lyon, the manager of the Christmas Gift mine in which it was found. 7th List

Chinglusuite. V. I. Gerasimovsky, 1938. *Bull. Acad. Sci. URSS, Cl. Sci. Math. Nat., Sér. Géol.*, 1938, p. 153 (Чинглусуит), p. 156 (Tchinglusuite). Hydrated titano-silicate of manganese, sodium, etc., $2(Na,K)_2O.5(Mn,Ca)O.3(Ti,Zr)O_2.14SiO_2.9H_2O$, as black amorphous

ST. THOMAS AQUINAS
COLLEGE LIBRARY
SPARKILL, N.Y. 10976

(metamict) grains. Named from the locality, Chinglusuai (Чинглусуай) river, Kola peninsular, Russia. [*M.A.*, **7**, 222.] 15th List

Chinkolobwite. A. Schoep, 1923. *Bull. Soc. Belge Géol.*, vol. 33, p. 87; *Bull. Soc. Chim. Belgique*, vol. 32, p. 345; *Bull. Soc. Belge Géol. Bruxelles*, 1924, vol. 33 (for 1923), p. 186. Hydrated silicate of uranium, perhaps dimorphous with soddite (9th List) from which it differs in its optical characters. Named from the locality Chinkolobwe, Katanga, Belgian Congo. [*M.A.*, **2**, 250, 342.] 10th List [Is Sklodowskite; *A.M.*, **9**, 156]

Chinoite. C. W. Beck and D. B. Givens, 1953. *Amer. Min.*, vol. 38, p. 191. A basic copper phosphate from Chino mine, New Mexico. Later proved to be identical with libethenite. [*M.A.*, **12**, 132, 462.] 20th List

Chiropterite. O. Abel, 1922. [*Höhlenkundliche Vorträge*, Heft 7, Wien.] O. Abel and G. Kryle, *Die Drachenhöhle bei Mixnitz*, Spelaeol. Monogr. Wien, 1931, vol. 7–8, p. 182 (Chiropterit). G. E. Hutchinson, *Bull. Amer. Mus. Nat. Hist.*, 1950, vol. 96, p. 381. Bat guano in caves; named from chiroptera, order of bats. The Times, February 1, 1932 ('Battite'). 19th List

Chirvinskite. N. K. Platonov, 1941. *Compt. Rend.* (*Doklady*) *Acad. Sci. URSS*, vol. 33, p. 361 (chirvinskite). A metamorphosed bitumen near shungite and anthraxolite, representing a stage in the alteration to graphite from Caucasus. Named after Prof. Petr Nikolaevich Chirvinsky, Петр Николаевич Чирвинский (1880–). [*M.A.*, **9**, 125.] 17th List

Chizeuilite. A. Lacroix, 1910. *Minéralogie de la France*, vol. iv, pp. 594, 905. A supposed new mineral afterwards identified with andalusite. Occurs as colourless prisms at Chizeuil, near Chalmoux, Saône-et-Loire. 6th List

Chkalovite. V. I. Gerasimovsky, 1939. *Compt. Rend.* (*Doklady*) *Acad. Sci. URSS.*, vol. 22, p. 259 (chkalovite). Sodium and beryllium silicate, $Na_2Be(SiO_3)_2$, orthorhombic, from the Kola peninsula, Russia. Named after the polar aviator, Valery Pavlovich Chkalov. [*M.A.*, **7**, 314.] 15th List [*A.M.*, **25**, 380]

Chlopinite. *See* Khlopinite. 14th List

Chlor-amphibole. G. A. Krutov, 1936. *See* Dashkesanite. 14th List

Chlorarsenian. P. Groth, *Tab. Uebers. Min.* 4th edit. 1898, p. 171. The same as chloroarsenian. 2nd List

Chlorhastingsite. G. A. Krutov and R. A. Vinogradova, 1966. Доклады акад. наук СССР (*Compt. Rend. Acad. Sci. URSS*), vol. 169, p. 204 (Хлоргастингсит). Hastingsite from E Sayan, containing 1–2% Cl. [Зап., **97**, 77] 25th List

Chloride-marialite. L. H. Borgström, 1914. *See* Carbonate-marialite. 7th List

Chloritite. J. Samojlov, 1906. *Materialui Geol. Rossii* (*Imperial Mineralogical Society, Petrograd*), vol. xxiii, p. 204 (Хлорититъ); abstract in *Neues Jahrb. Min.*, 1907, vol. ii, ref. p. 196 (Chloritit). A name proposed for V. I. Vernadsky's hypothetical chlorite acid, $H_2Al_2SiO_6$ (*Min. Mag.*, 1902, vol. xiii, p. 128). The name *α-chloritite* is applied to a scaly chloritic mineral, from the Donetz Basin in SE Russia, approximating to the above in composition, $4Al_2O_3.5SiO_2.7H_2O$. 7th List

Chloritoserpentine. *See* Serpochlorite. 17th List

Chlormanganokalite. H. J. Johnston-Lavis, 1906. *Nature*, vol. lxxiv, p. 104. A. Lacroix, *Compt. Rend. Acad. Sci. Paris*, 1906, cxlii, p. 1251; H. J. Johnston-Lavis and L. J. Spencer, *Nature*, 1907, vol. lxxvi, p. 215. Chloride of manganese and potassium, $MnCl_2.4KCl$, found as yellow, deliquescent, rhombohedral crystals in blocks ejected from Mt. Vesuvius. 4th List

Chlormankalite. A. N. Winchell, 1927. *Elements of optical mineralogy*, 2nd edit., part 2, p. 33. A suggested abbreviation of Chlormanganokalite (4th List). 11th List

Chlormimetesit. H. Wondratschek, 1963. *Neues Jahrb. Min.*, Abh., vol. 99, p. 113. Synonym of Mimetite. 23rd List

Chlornatrokalite. H. J. Johnston-Lavis, 1906. *Nature*, vol. lxxiv, p. 174. Potassium and sodium chloride, 6KCl.NaCl, found as cubic crystals with chlormanganokalite in blocks ejected from Mt. Vesuvius. [Specimens of 'chlornatrokalite' sent by Dr. Johnston-Lavis to the British Museum consist of cubes of sylvite and of halite in association—*L.J.S.*] 4th List

Chloroarsenian. L. J. Igelström, *Geol. För. Föhr.* XV. 471, 1893; and *Zeits. Kryst. Min.* XXII.

468, 1894. Arsenate of manganese? Sjö mine, Sweden. 1st List [Is Allactite; *Geol. Fören. Förh.*, **94**, 426]

Chlorohastingsite. *Dokl. Acad. Sci. USSR* (*Earth Sci. Sect.*), 1966, **169**, 116. Erroneous transliteration of Хлоргастингсит, Chlorhastingsite (25th List). 28th List

Chloropal. P. Schneiderhohn, 1965. *Tschermaks Min. Petr. Mitt.*, ser. 3, **10**, 385. A mixture of $\frac{1}{4}$ nontronite and $\frac{3}{4}$ opal-cristobalite. Not to be confused with the Chloropal of Berhardi and Brandes, 1822, for which the current name is nontronite. [*M.A.*, **17**, 494]. 28th List

Chlorophoenicite. W. F. Foshag and R. B. Gage, 1924. *Journ. Washington Acad. Sci.*, vol. 14, p. 362. Basic arsenate of manganese, zinc, etc., $R_3As_2O_8.7R(OH)_2$, as monoclinic crystals from Franklin Furnace, New Jersey. The crystals are pale-green by daylight and pale purplish-red in artificial light; hence the name, from *χλωρός*, green, and *φοῖνιξ*, *φοίνικος*, purple-red. [*M.A.*, **2**, 337.] 10th List [$(Mn,Zn)_5AsO_4(OH)_7$]

Chloroxiphite. L. J. Spencer, 1923. *Min. Mag.*, vol. 20, p. 75. Oxychloride of lead and copper, $2PbO.Pb(OH)_2.CuCl_2$, occurring as green, blade-like, monoclinic crystals embedded in mendipite from the Mendip Hills, Somerset. Named from *χλωρός*, green and *ξίφος*, a blade or straight sword. An abstract in *Amer. Min.*, 1924, vol. 9, p. 96, gives, somewhat unnecessarily, the spelling chloro-ziphite. 10th List

Chloroxyapatite. R. D. Morton, 1962. *Norsk geol. Tidsskr.*, vol. 41, p. 223. Artificial $Ca_{10}(PO_4)_6(O,Cl_2)$. 23rd List

Chloro-ziphite. *See* Chloroxiphite. 10th List

Chlorpyromorphit. H. Wondratschek, 1963. *Neues Jahrb. Min.*, Abh., vol. 99, p. 113. Synonym of Pyromorphite. 23rd List

Chlor-spodiosite. F. K. Cameron and W. J. McCaughey, 1911. *Journ. Physical Chem.*, Ithaca, NY, vol. xv, p. 464/8. The chlorine analogue, $Ca_3(PO_4)_2.CaCl_2$, of spodiosite, prepared artificially as orthorhombic crystals. The natural mineral is distinguished as fluor-spodiosite. 6th List [also 20th List. *Min. Mag.*, **30**, 166]

Chlor-utahlite. (D. B. Sterrett. *Mineral Resources U.S. Geol. Survey, for 1908*, 1909, part ii, p. 853.) The same as utahlite (G. F. Kunz, 1895; 1st List), the prefix being, no doubt, added because of the characteristic green colour of the stone. A synonym of variscite. 6th List

Chlor-utalite. Variant of or error for chlor-utahlite (6th List). R. Webster, *Gems*, 1962, p. 754. 28th List

Chlorvanadinit. H. Wondratschek, 1963. *Neues Jahrb. Min.*, Abh., vol. 99, p. 113. Synonym of Vanadinite. 23rd List

Chlorvoelckerite, synonym of Chloroxyapatite. [*M.A.*, **15**, 528.] 23rd List

Chocolate-stone. *See* lacroisite. 3rd List

Chocolite. J. Garland, 1894. *Trans. Inst. Mining and Metallurgy*, 1893–1894, vol. ii, pp. 128, 224. A chocolate-coloured nickel-ore from New Caledonia; a hydrated silicate of iron, nickel, and magnesium, related to garnierite. 4th List

Chondrostibian. L. J. Igelström, *Geol. För. Föhr.* XV. 343, 1893; and *Zeits. Kryst. Min.* XXII. 43, 1893. Hydrated antimonate of manganese and iron. Sjö mine, Sweden. 1st List

Choschiit (Germanized back-transliteration of Хошиит (= Hoshiite). Зап. всесоюз. мин. общ. (*Mem. All-Union Min. Soc.*), 1965, vol. 94, p. 672; C. Hintze, *Handb. Min.*, Erg. III, p. 537. 25th List

Choubnikovite. *Bull. Soc. Franç. Min. Crist.*, 1955, vol. 78, p. 216. French transliteration of шубниковит, shubnikovite (20th List). 21st List

Chowachsit, German transliteration of Ховахсит, khovakhsite (22nd List). Hintze, *Handb. Min.*, Erg.-bd. II, p. 873. 23rd List

Christensenite. T. F. W. Barth and A. Kvalheim, 1944. *Sci. Results Norwegian Antarctic Expeditions 1927–28, Norske Vidensk.–Akad. Oslo*, no. 22. A solid solution of nepheline (5%) in tridymite, from Deception Island, Antarctic. Named after Lars Christensen (1884–), consul and shipowner of Sandefjord, Norway, who instituted and financed the expeditions. [*M.A.*, **9**, 261.] 17th List [Var. of Tridymite; *A.M.*, **38**, 866]

Christite. A. S. Radtke, F. W. Dickson, J. F. Slack, and K. L. Brown, 1977. *A.M.*, **62**, 421.

Orange to crimson plates and crystals occur with other Tl, As, and Sb minerals in the Carlin gold deposit, Elko, Nevada. Monoclinic, *a* 6·113, *b* 16·188, *c* 6·111 Å, *β* 96·71°, space group $P2_1/n$. Composition 4[$TlHgAsS_3$], synthesized. Named for C. L. Christ. [*M.A.*, 78-883.] 30th List

Chromamesite. [I. A. Zimin, *Mineralnoe Syre*, Moscow, 1935, no. 10, p. 61.] O. M. Shubnikova, *Trans. Inst. Geol. Sci. USSR*, 1938, no. 11 (*Min. Geochem. Ser.*, no. 3), p. 12 (Хромистый амезит, Chrom-amesite), p. 13 (хромамезит), p. 35 (Chromamesite). Analysis of amesite from the Urals containing Cr_2O_3 0·77% is interpreted as a mixture of the molecules $H_4Mg_2Al_2SiO_9$ (amesite) 95·6%, $H_4Fe_2Al_2SiO_9$ (daphnite) 3%, and $H_4Mg_2Cr_2SiO_9$ (chromamesite) 1·4%. 17th List

Chromatite. F. J. Eckhardt and W. Heimbach, 1963. *Naturwiss.*, vol. 50, p. 612. Finely crystalline citron-yellow crusts from clefts in limestones on the Jerusalem–Jericho highway are identified by X-ray data as $CaCrO_4$. Named from the composition (*Amer. Min.*, 1964, vol. 49, p. 439). 23rd List

Chrombiotit. H. Strunz, 1957. *Min. Tabellen*, 3rd edit., p. 304 (Chrombiotit). Hypothetical chromian biotite; an unnecessary name. 22nd List

Chrom-brugnatellite. L. Hezner, 1912. *Centralblatt Min.*, 1912, p. 570 (Chrom-Brugnatellit). A scaly, violet mineral occuring in serpentine at Dundas, Tasmania. In composition, $2MgCO_3.5Mg(OH)_2.2Cr(OH)_3.4H_2O$, it is analogous to brugnatellite with chromium in place of iron. Possibly identical with stichtite (q.v.). 6th List

Chromdisthene. V. S. Sobolev and N. V. Sobolev, 1967. [Геол. рудн. месторожд. (Geol. ore-deposits), no. 2, p. 10]; abstr. *Amer. Min.*, 1968, vol. 52, p. 349. Synonym of Chrome-kyanite (15th List). 25th List

Chrome-acmite. A. Holmes, 1937. *Trans. Geol. Soc. South Africa*, vol. 39 (for 1936), p. 405. Suggested as a constituent molecule $NaCrSi_2O_6$ in chrome-diopside (containing Cr_2O_3 2·03, Na_2O 1·37%) from the kimberlite at Jagersfontein, Orange Free State. [*M.A.*, **7**, 188.] 15th List

Chrome-antigorite. H. Strunz, *Min. Tab.*, 1941, p. 204 (Chrom-Antigorit). Antigorite containing some chromium. 18th List

Chrome-beidellite. D. P. Serdyuchenko, 1933. *Zap. Ross. Min. Obshch.* (*Mém. Soc. Russe Min.*), ser. 2, vol. 62, p. 380 (хромовый байделит), p. 391 (chrome-beidellite). A variety of beidelite containing chromium (Cr_2O_3 5·02%) and grading to wolchonskoite (volkonskoite), $(Cr,Fe^{III},Al)_2O_3.3SiO_2.nH_2O$; from northern Caucasus. [*M.A.*, **5**, 486.] 13th List

Chrome-cerussite. *Catalogue of the minerals of Tasmania*, revised edit., 1969, 28. Unnecessary new name for the variety chromiferous cerussite of the 1910 edit. [*Min. Mag.*, **38**, 902.] 27th List

Chrome-epidote. P. Eskola, 1933. *Compt. Rend. Soc. Géol. Finlande*, 1933, no. 7, pp. 35, 38 (chrome epidote). Variant of chromepidote (10th List) for tawmawite (5th List) or the compound $H_2Ca_4Cr_6Si_6O_{26}$. Varieties richer in aluminium are 'chrome clinozoisite', and those richer in iron 'chrome pistazite'. [*M.A.*, **6**, 47.] 14th List

Chrome-ferrimontmorillonite. G. S. Gritsaenko, 1946. [Зап. Всесоюз. Мин. Общ. (*Mem. All-Union Min. Soc.*), vol. 75, p. 150], abstr. in C. Hintze, *Handb. Min.*, Erg.-Bd. II, p. 79 (Chrom-Ferrimontmorillonit). Synonym of Chrome-nontronite (13th List). 24th List

Chrome fluorite. R. T. Liddicoat, 1970. *Gems and Gemmology*, **13**, 231. An undesirable term for a green fluorite from South America. 28th List

Chrome-halloysite. G. S. Gritsaenko and S. V. Grum-Grzhimailo, 1949. *Mém. Soc. Russe Min.*, vol. 78, p. 61 (хромовый галлуазит). A pale blue variety of halloysite containing Cr_2O_3 0·59%. [*M.A.*, **11**, 346, chromium halloysite.] 19th List

Chrome-idocrase. British Museum (Natural History), *The Student's Index to the Collection of Minerals, 1914*, 25th edit., p. 9. An emerald-green variety of idocrase containing chromium, from the Montreal Chromite pit, Black Lake, Megantic Co., Quebec, and from the Monetnaya estate, Ekaterinburg, Urals. 7th List

Chrome-jadeite. E. Gübelin, 1965. *Journ. Gemmology*, vol. 9, p. 372. Syn. of Chromojadeite (13th List). [*M.A.*, **17**, 377.] 24th List

Chrome-kaolinite. T. Sudo and T. Anzai, 1942. *Proc. Imp. Acad. Tokyo*, vol. 18, p. 403

(Chrome-kaolinte [*sic*]). Z. Harada, *Journ. Fac. Sci. Hokkaido Univ.*, Ser. 4, 1948, vol. 7, p. 195 (Chrome Koalin [*sic*]), p. 200 (Chrome Kaolin). A green variety of kaolinite containing Cr_2O_3 0·41–1·12%, occurring with nickel ores, and previously thought to be garnierite, from Urakawa, Sizuoka prefecture, Japan. 19th List

Chrome-kyanite. K. N. Ozerov and N. A. Bykhover, 1936. *Trans. Centr. Geol. Prosp. Inst. USSR*, no. 82, p. 72 (хром-кианит), p. 100 (chrome cyanite). A green variety of kyanite containing chromium (Cr_2O_3 1·81%) from Yakutia, Siberia. [*M.A.*, **7**, 49.] 15th List

Chrome-magnetite. J. J. Frankel, 1942. *South African Journ. Sci.*, vol. 38, p. 153. A magnetic mineral containing Cr_2O_3 12·34%, associated with non-magnetic chromite and titaniferous magnetite at Lydenburg, Transvaal. Compare ishkulite* (16th List). [*M.A.*, **10**, 493.] 18th List [*Corr. *M.M.*, **40**]

Chrome-nontronite. D. P. Serdyuchenko, 1933. *Zap. Ross. Min. Obshch.* (*Mém. Soc. Russe Min.*), ser. 2, vol. 62, p. 376 (хромовый нонтронит), p. 390 (chrome-nontronite). A variety of nontronite (chloropal) containing chromium and grading to wolchonskoite (volkonskoite), $(Cr,Fe^{III},Al)_2O_3.3SiO_2.nH_2O$. [*M.A.*, **5**, 486.] 13th List

Chrome-phengite. P. de Wijkerslooth, 1943. *Maden Tetkik ve Arama Enstitüsü Mecmuası*, Ankara, vol. 8, p. 256 (kromfengit, Turkish), p. 261 (Chromphengit, German). A chromium-bearing mica enclosed in chromite ore. [*M.A.*, **9**, 244.] 18th List

Chrome phlogopite. Chang Pao-Kwei and Lin Kuo-Cheng, 1974. [*Geochimica*, **1**, 71], abstr. *A.M.*, **60**, 161 (1975). An unnecessary name for a chromian phlogopite from Honan, China. 29th List

Chromepidote. F. Zambonini, 1920. *Boll. Com. Geol. Italia*, vol. 47, p. 80 (cromepidoto). An alternative name for tawmawite (5th List). 10th List

Chrome-pyrophyllite. H. Meixner, 1961. *Chemie der Erde*, vol. 21, p. 1 (Chrom-Pyrophyllit). Unnecessary name for a green chromian pyrophyllite (3% Cr_2O_3) from Mühlbach, Salzburg. 22nd List

Chrome-tourmaline. A. Cossa and A. Arzruni, 1883, *Zeits. Kryst. Min.*, vol. 7, p. 1 (Chromturmalin). A. G. Gill, *Johns Hopkins Circulars*, 1889, vol. 8, p. 101 (chrome-tourmaline). Variety of tourmaline from Urals (Cr_2O_3 10·86%) and Maryland (Cr_2O_3 4·32%). 18th List

Chrome-tremolite. P. Eskola, 1933. *Compt. Rend. Soc. Géol. Finlande*, 1933, no. 7, p. 31 (chrome-tremolite), p. 39 (chrome tremolite). *Bull. Soc. Franç. Min.*, 1933, vol. 56, p. 188 (chromtrémolite). Tremolite containing Cr_2O_3 1·61%, from Outokumpu, Finland. [*M.A.*, **6**, 47.] 14th List

Chrome-vesuvian. S. M. Kurbatov, 1922. *Bull. Acad. Sci. Russie*, 1922, ser. 6, vol. 16, p. 414; 1923, vol. 17, p. 115; 1925, vol. 19, p. 482 (хромовый везувиан). C. Hintze, *Handbuch der Mineralogie*, Ergänzungsband, 1936, p. 116 (Chromvesuvian). Synonym of chrome-idocrase (7th List). [*M.A.*, **3**, 354.] 14th List

Cromferrit, variant of Chromoferrite (C. Hintze, *Handb. Min.*, Erg.-Bd. II, p. 684). 24th List

Chromidokras, variant of Chrome-idocrase (C. Hintze, *Handb. Min.*, Erg.-Bd. II, p. 115). 24th List

Chrominium. D. Adib and J. Ottemann, 1970. *Min. Depos.*, **5**, 86. Deep red monoclinic crystals from the T. Khuni mine, Anarak, Iran, have the composition Pb_2CrO_5. The description is seriously inadequate: no X-ray powder data are given and the material has not been compared with synthetic Pb_2CrO_5. The name 'refers to the chemical composition' [it may also allude to minium, but is objectionable, as it suggests a new element. It also appears as 'chromium'. Almost certainly identical with phoenicochroite—*M.H.H.*] 26th List

Chromitite. M. Z. Jovitschitsch, 1908. *Sitzungsber. Akad. Wiss. Wien, Math.-naturwiss. Klasse*, vol. cxvii, Abt. II b, p. 818; *Monatshefte für Chemie, etc.*, Wien, 1909, vol. xxx, p. 44 (Chromitit). Oxide of chromium and iron, $FeCrO_3 = Fe_2O_3.Cr_2O_3$, found as shining octahedra in the sands of streams from the Kopaonik Mountains, Servia. Differs from chromite in containing ferric oxide in place of ferrous oxide [and therefore possibly an altered chromite]. 5th List

Chromitspinelle. H. Strunz, 1957. *Min. Tabellen*, 3rd edit., p. 364. Group-name for those spinels in which Cr is the principal trivalent metal (cf. Chromspinellids). 24th List

Chrom-Klinochlor, German transliteration of Хромклинохлор, Chrome-clinochlore (H. Strunz, *Min. Tabellen*, 3rd edit., 1957, p. 318). 24th List

Chrom-Lanarkit. H. Strunz, 1966. *Min. Tabellen*, 4th edit., p. 247. An unnecessary and possibly incorrect name for the unnamed mineral, possibly a chromatian Lanarkite, from Leadhills, Scotland, described by A. K. Temple. (*Trans. Roy. Soc. Edin.*, 1956, vol. 63, p. 85; *A.M.*, **45**, 909.) 24th List

Chromloeweite. W. Wetzel, 1928. *Chemie der Erde*, vol. 3, p. 390 (Chromloeweït). Minute trigonal crystals found in Chile saltpetre ('caliche') where referred by the author in 1923 to dietzeite, in 1924 to an iron sulphate, and he now suggests that they may be a 'chromloeweite'. [*M.A.*, **3**, 554.] 11th List

Chrommuscovit. H. Strunz, 1957. *Min. Taballen*, 3rd edit., p. 304 (Chrommuskovit). Syn. of Fuchsite. 22nd List

Chromocyclite. C. Klein, *Jahrb. Min.*, 1892, vol. ii, pp. 176, 220 (Chromocyclit); C. Hintze, *Handb. d. Min.*, 1897, vol. ii, p. 1733 (Chromocyklit). Apophyllite which shows coloured interference rings in convergent polarized light, as distinct from the black and white rings of the leucocyclite variety. 3rd List

Chromohercynite. A. Lacroix, 1920. *Bull. Soc. franç. Min.*, vol. 43, p. 69. A mineral of the spinel group with the composition $FeCr_2O_4.(Fe,Mg,Mn)Al_2O_4$, i.e. an isomorphous mixture in equal molecular proportions of chromite and hercynite, hence the name. Occurs as black, granular masses in Madagascar. [*M.A.*, **1**, 123.] 9th List

Chromojadeite. A. Lacroix, 1930. *Bull. Soc. Franç. Min.*, vol. 53, p. 226 (Chromojadéite). Synonym of tawmawite (A. W. G. Bleeck, 1907; 5th List) for a chromiferous jadeite from Tawmaw, Burma. [*M.A.*, **5**, 71.] 13th List

Chromopicotite. A. Lacroix, 1910. *Minéralogie de la France*, vol. iv, p. 311. The same as Chrompicotite (T. Petersen, 1869). A member of the spinel group intermediate between chromite and picotite, the formula being $(Fe,Mg)(Cr,Al)_2O_4$. 6th List

Chromphengit, *see* Chrome-phengite.

Chromrutile. S. G. Gordon and E. V. Shannon, 1928. *Amer. Min.*, vol. 13, p. 69. A variety of rutile containing chromium (Cr_20_3 16·61%). 11th List [Now Redledgeite, q.v.]

Chromspinell. P. de Wijkerslooth, 1943. [*Maden Tektik ve Arama Enstitüsü Mecmuası*, Ankara, vol. 8, p. 254], abstr. in *M.A.*, **9**, 244. Synonym of Chromite. Not to be confused with the old term Chromespinel, a synonym of Picotite. 23rd List

Chromspinellide. J. Bauer, J. Fiala, and R. Hřichová, 1963. *Amer. Min.*, vol. 48, p. 540 (the chromspinelides). A group name or general term for the chromium spinels. (*M.A.*, **16**, 540, chromspinellid). Formed on the basis of Spinellide (9th List), a general term for all spinels. 23rd List

Chromspinellids. V. D. Ladieva, 1964. [*Chem. comp. and intern. struct. min.*, Kiev, p. 192]; abstr. *M.A.*, **17**, 389. A group-name for chromian Spinels (cf. Spinellids, 9th List, and Chromitspinelle, 24th List). 24th List

Chromsteigerite. E. A. Ankinovich, 1963. [Труды Инст. геол. Наук, Акад. наук Казах ССР (*Proc. Inst. Geol. Kazakh SSR*), vol. 7, p. 207]; abstr. *Amer. Min.*, 1964, vol. 49, p. 1774; Зап. Всесоюз. Мин. Общ. (*Mem. All-Union Min. Soc.*), 1965, vol. 94, p. 680 (Хромштейгерит). An unnecessary name for a chromian Steigerite from Kurumsak. 24th List

Chromturmalin, *see* Chrome-tourmaline.

Chrysargyrite. J. Fröbel, 1843. *Grundzüge eines Systemes der Krystallologie*. Synonym of electrum. Subdivided into boussingaultin, trinitatin, hyperythrin, pyrrhochrysit, gironit, and michelottin, with different Au:Ag ratios. 27th List

Chrysophrase. Green artificially dyed chalcedony. R. Webster, *Gems*, 1962, p. 754. Either an error for chrysoprase, or a coinage to suggest it. 28th List

Chubutite. H. Corti, 1918. *Anal. Soc. Quim. Argentina*, vol. vi, p. 65; E. Rimann, ibid., p. 323 (Chubutita). A reddish-yellow, tetragonal (?) oxychloride of lead, $7PbO.PbCl_2$, from Chubut, Argentina. Evidently identical with Lorettoite (q.v.). 8th List

Chuchrovit, an inconsistent transliteration of Чухровит, Chukhrovite (22nd List) (H. Strunz, *Min. Tabellen*, 4th edit., 1966, p. 148). 24th List

Chudobaite. H. Strunz, 1960. *Neues Jahrb. Min.*, Monatsh., p. 1 (Chudobait). $(Na,K)(Mg,Zn)_2H(AsO_4)_2.4H_2O$, in anorthic crystals from the second oxidation zone at Tsumeb, SW Africa. Named after Prof. K. F. Chudoba. [*M.A.*, **14**, 500.] 22nd List [*A.M.*, **45**, 1130]

Chukhrovite. L. P. Ermilova, V. A. Moleva, and R. F. Klevtsova, 1960. Зап. Всесоюз. Мин. Общ. (*Mem. All-Union Min. Soc.*), vol. 89, p. 15 (Чухровит). A cubic mineral occurring in the Kara-Oba molybdenum deposit, Central Kazakhstan, approximates to $Ca_6Al_3(Y,Ln)_2(SO_4)_2F_{23}.20H_2O$. Named after Dr. F. Kh. Chukhrov. Perhaps the same as an unnamed cubic mineral from Greenland described by O. B. Bøggild (*Zeits. Kryst. Min.*, 1913, vol. 51, p. 608). 22nd List [*A.M.*, **45**, 1132]

Churchillite. A. Dufrénoy, *Min.*, 2nd edit., 1856, III, 280. The same as mendipite. From Churchill, Mendip Hills. 2nd List

Ciempozuelite. A. de Areitio y Larrinaga, *Anal. Soc. Española Hist. Nat.*, 1873, vol. ii, p. 393 (Ciempozuelita). $3Na_2SO_4.CaSO_4$. From the salt mines at Ciempozuelos, Madrid. [Possibly a mixture of glauberite ($Na_2SO_4.CaSO_4$) and thenardite (Na_2SO_4), both of which occur at this locality.] 3rd List

Cirolite. A trade name for *artificial* yttrium aluminate, $Yt_3Al_5O_{12}$. R. Webster, *priv. comm.* Other trade-names include: yttrogarnet (19th List); triamond, diamonite (*et sim.*) (q.v.); also astrilite, YAG (yttrium aluminium garnet), Di'Yag, geminair, and Linde simulated diamond. (M. J. O'Donoghue, *priv. comm.*). 28th List

Clarain. M. C. Stopes, 1919. *See* Fusain. 8th List

Claringbullite. E. E. Fejer, A. M. Clark, A. G. Couper, and C. J. Elliott, 1977. *M.M.*, **41**, 433. Blue platy crystals from N'changa, Zambia, from Bisbee, Arizona, and from M'sesa mine, Kambove, Katanga, are hexagonal, *a* 6·671, *c* 9·183 Å. Composition $2[Cu_4Cl(OH)_7.nH_2O]$, with *n* about 0·5, ω 1·782, ε 1·780. Named for G. F. Claringbull. [*M.A.*, 78-884.] 30th List

Clarite. R. Potonié, 1924. *Kohlenpetrographie*, Berlin, p. 33 (Clarit). Variant of clarain (8th List). Not the clarite of F. Sandberger, 1874. The names Durain, Fusain, and Vitrain (8th List) are similarly altered to Durit, Fusit, and Vitrit. Several other variations of these names are given by E. Stach, *Lehrbuch der Kohlenpetrographie*, Berlin, 1935. 14th List

Clarkeite. C. S. Ross, E. P. Henderson, and E. Posnjak, 1931. *Amer. Min.*, vol. 16, pp. 114, 213. Hydrous uranate of sodium and lead, $RO.3UO_3.3H_2O$, dark brown with waxy lustre, forming with gummite a zone of alteration around uraninite from North Carolina. Named after Prof. Frank Wigglesworth Clarke (1847–1931), of Washington, DC. [*M.A.*, **4**, 498.] 12th List [$(Na,Ca,Pb)_2U_2(O,OH)_7$; *A.M.*, **41**, 131]

Clayite. J. W. Mellor, 1909. *Trans. English Ceramic Soc.*, Longton, vol. viii, p. 28. The non-crystalline variety of kaolinite, $Al_2O_3.2SiO_2.2H_2O$, of which china-clay and most other clays are largely composed. Not the clayite of W. J. Taylor, 1859. 5th List

Cleïte. S. J. Thugutt, 1945. *Arch. Min. Tow. Nauk. Warszaw.* (*Arch. Min. Soc. Sci. Varsovie*), vol. 15 (for 1939–1945), p. 233 (cleïte, French), p. 236 (kleit, Polish). Variants of clayite (5th List). [*M.A.*, **10**, 23.] 18th List

Cliachite. J. F. A. Breithaupt, 1847. *Handbuch d. Mineralogie*, vol. iii, p. 896 (Picites Cliachites; Cliachit). A ferruginous bauxite from Cliache (= Kljake) in Dalmatia. F. Cornu in 1909 (5th list) used this name (Kliachite) for the colloidal aluminium hydroxides occurring in bauxite. *See* Kljakite and Sporogelite. 6th List

Cliffordite. R. V. Gaines, 1969. *Amer. Min.*, **54**, 697. Bright yellow octahedra from the San Miguel mine, Moctezuma, Sonora, Mexico, agree in all respects with synthetic $U^{iv}Te_3O_8$. Named for Dr. Clifford Frondel. [*M.A.*, 70-750.] 26th List [UTe_3O_9; *A.M.*, **57**, 597]

Clino-amphibole. P. Eskola, 1922. *Journ. Geol. Chicago*, vol. 30, p. 293. A collective name for the monoclinic amphiboles, analogous to Clinopyroxene (3rd List). 10th List

Clino-anthophyllite. W. Layton and R. Phillips, 1960. *Min. Mag.*, vol. 32, p. 659. The doubtful, synthetic, monoclinic polymorph of anthophyllite; distinct from cummingtonite, to which calcium is essential. 22nd List [Magnesio-cummingtonite, *Amph.* (1978)]

Clino-antigorite. H. Strunz, 1957. *Min. Tabellen*, 3rd edit., p. 322 (Klinoantigorit). Synonym of Antigorite. 24th List

Clinoaugite. *See* orthoaugite. 3rd List

Clinobarrandite. D. McConnell, 1940, *Amer. Min.*, vol. 25, p. 719. Hydrous aluminium ferric phosphate, $(Al,Fe)PO_4.2H_2O$, shown by X-rays to be monoclinic, and so dimorphous with the orthorhombic barrandite with which it is intimately intermixed, from Manhattan, Nevada. [*M.A.*, **8**, 50.] 16th List [Perhaps aluminian Phosphosiderite]

Clinoberthierine. H. Strunz, 1966. *Min. Tabellen*, 4th edit., p. 403 (Klinoberthierin). Synonym of Berthierine. 24th List

Clinobisvanite. P. J. Bridge and M. W. Pryce, 1974. *M.M.*, **39**, 847. Monoclinic (*a* 5·186, *b* 11·708, *c* 5·100 Å, *β* 90° 26′) $BiVO_4$, a dimorph of pucherite, occurs at Yinnietharra and other localities in Western Australia as yellow powder and orange aggregates. Named for the composition and symmetry. 28th List

Clinobronzite. W. Wahl, 1906. *See* Clinoenstatite. W. Wahl, loc. cit. (Klinobronzit). A. Lacroix, loc. cit. (Clinobronzite). 5th List

Clino-chevkinite. S. Bonatti and G. Gottardi, 1953. *Rend. Soc. Min. Ital.*, vol. 9, p. 242 (clino-chevkinite); ibid., 1954, vol. 10, p. 224 (clinochevkinite). Monoclinic chevkinite from Urals (A. K. Boldyrev, 1924, *M.A.*, **3**, 405), as distinct from orthorhombic chevkinite from Madagascar (A. Lacroix, 1915). *See* Ortho-chevkinite. [*M.A.*, **12**, 240, 498.] 20th List

Clinochlorite. Variant of clinochlore; *A.M.*, **8**, 51. 10th List

Clino-chrysotile, ortho-chrysotile. E. J. W. Whittaker, 1951. *Acta Cryst.*, vol. 4, p. 187; 1952, vol. 5, p. 143. Monoclinic and orthorhombic forms of chrysotile, as determined by X-rays. [*M.A.*, **11**, 430, 539.] 20th List [Cf. Pecoraite]

Clinoenstatite. W. Wahl, 1906. *Die Enstatitaugite*, Helsingfors, May 1906, p. 141; *Min. Petr. Mitt.* (*Tschermak*), 1907, vol. xxvi, p. 121; *Öfvers. Finska Vet. Soc. Förh.*, 1908, vol. 1, no. 2, p. 1 (Klinoenstatit). A. Lacroix, *Bull. Soc. Sci. Nat. Ouest France*, September 1906, ser. 2, vol. vi, p. 94 (Clinoenstatite). E. T. Allen and others, *Amer. Journ. Sci.*, 1909, ser. 4, vol. xxvii, pp. 30, 45 (clino-enstatite). F. Zambonini, *Zeits. Kryst. Min.*, 1909, vol. xlvi, pp. 12, 601. F. E. Wright, *Zeits. Kryst. Min.*, 1909, vol. xlvi, p. 599.

A monoclinic pyroxene with the chemical composition of enstatite; that is, dimorphous with enstatite, and differing from diopside in containing no calcium. It occurs in meteoric stones, and has been prepared artificially (β-$MgSiO_3$). The magnesium-diopside of H. Rosenbusch (1905; 4th List) approximates to this.

Similarly, clinobronzite and clinohypersthene are dimorphous forms of the magnesium-iron meta-silicates bronzite and hypersthene respectively. 5th List [Cf. Clinoferrosilite]

Clinoenstenite. A. N. Winchell, 1923, *Amer. Journ. Sci.*, ser 5, vol. 6, p. 512. A species name for the isomorphous series $MgSiO_3$—$FeSiO_3$ of monoclinic pyroxenes, comprising clinoenstatite and clinohypersthene. *See* Enstenite. [*M.A.*, **2**, 220.] 10th List

Clinoeulite. W. Schreyer, D. Stepto, K. Abraham, and W. F. Müller, 1978. *Contrib. Mineral. Petrol.*, **65**, 351. A metamorphosed iron formation in the Vredefort structures, S Africa, contains a clinopyroxene of composition $(Fe_{1\cdot48}Mg_{0\cdot37}Mn_{0\cdot08}Ca_{0\cdot05}Al_{0\cdot05})_{\Sigma 1\cdot99}Si_{2\cdot01}O_6$, space group $P2_1/x$. The monoclinic dimorph of eulite (18th List, a var. of orthoferrosilite). [*M.A.*, 78-3470.] 30th List

Clinoferrosilite. N. L. Bowen, 1935. *Amer. Journ. Sci.*, ser. 5, vol. 30, pp. 481, 492. N. F. M. Henry, *Min. Mag.*, 1935–1937, vol. 24, pp. 225, 528. A monoclinic pyroxene with the composition $FeSiO_3$, ferrous metasilicate, found as minute crystals in the lithophysae of obsidians. [*M.A.*, **6**, 261.] 14th List [Cf. Orthoferrosilite, Ferrosilite, Clinoenstatite]

Clinoguarinite. G. Cesàro, 1932. *Mém. (8°) Acad. Roy. Belgique, Cl. Sci.*, vol. 12, fasc. 3, p. 18. The optical orientation of guarinite points to two varieties, orthorhombic and monoclinic, which are termed orthoguarinite and clinoguarinite respectively. They occur intergrown in the same crystal and the former is no doubt a minutely twinned form of the latter, as in orthoclase and microcline. [*M.A.*, **5**, 426.] 13th List [A doubtful silicate and fluoride of Na, Ca, and Zr]

Clinohedrite. (1). A. H. Chester, *Dictionary of the Names of Minerals*, 1896. Breithaupt's Clinoëdrit (= tetrahedrite).

(2). S. L. Penfield and H. W. Foote, *Amer. J. Sci.*, 1898, V, 289; *Zeits. Kryst. Min.*, 1899, XXX, 587 (*Klinoëdrit*); Abstr. *Min. Mag.*, XII, 133. Monoclinic, domatic class. $Zn(OH)$-$Ca(OH)SiO_3$. [Trotter mine] Franklin Furnace, New Jersey. 2nd List

Clinoholmquistite. I. V. Ginzburg, 1965. [Труды Мин. Муз. Акад. Наук СССР (*Proc. Min.*

Mus. Acad. Sci. USSR), vol. 16, p. 73], abstr. Зап. Всесоюз. Мин. Общ. (*Mem. All-Union Min. Soc.*), 1966, vol. 95, p. 326 (Клиногольмквистит, Clinoholmquistite). The monoclinic polymorph of Holmquistite (7th List), from an unnamed Siberian pegmatite. 24th List

Clinohypersthene. W. Wahl, 1906. *See* Clinoenstatite. W. Wahl, loc. cit. (Klinohypersthen). A. Lacroix, loc. cit., p. 93 (Clinohypersthène). 5th List

Clinojimthompsonite. D. R. Veblen, P. R. Buseck, and C. W. Burnham, 1977. *Science*, **198**, 359. A monoclinic mineral from Chester, Vermont, *a* 9·87, *b* 27·24, *c* 5·32 Å, *β* 109·5°, space group *C*2/*c*, belonging to the biopyribole group and dimorphous with jimthompsonite (both 30th List). Composition $(Mg,Fe)_{10}Si_{12}O_{32}(OH)_4$. Named for J. Thompson. [*M.A.*, 78-3473] 30th List

Clinoptilolite. W. T. Schaller, 1923. Report of meeting in *Amer. Min.*, 1923, vol. 8, p. 94. Further mention in *Amer. Min.*, 1923, vol. 8, p. 169; *Proc. US National Museum*, 1924, vol. 64, art. 19, p. 8. A monoclinic zeolite dimorphous with the orthorhombic ptilolite, $(Ca,Na_2)O.Al_2O_3.9SiO_2.6H_2O$, and distinct from mordenite. Crystals from Wyoming, described by L. V. Pirsson (1890) as mordenite, are referred to this species. So named because the optical extinction is inclined (*κλίνω*). 10th List [Distinct species, related to heulandite; *A.M.*, **45**, 341, 351]

Clinopyroxene. *See* orthoaugite. 3rd List

Clinosafflorite. D. Radcliffe and L. G. Berry, 1971. *Canad. Min.*, **10**, 877. A monoclinic polymorph of safflorite; occurs intergrown with skutterudite at Cobalt, Ontario. A sub-cell has *a* 5·040, *b* 5·862, *c* 3·139 Å, *β* 90° 13′, $P2_1/n$. The natural material has Co:Fe:Ni::0·76:0·14:0·10. [*M.A.*, 72-2331.] 27th List [*A.M.*, **57**, 1552]

Clinoscorodite. H. Strunz and K. Sztrókay, 1939. *Zentr. Min.*, Abt. A, 1939, p. 277 (Klinoskorodit). A supposed monoclinic form dimorphous with orthorhombic scorodite. [*M.A.*, **7**, 509.] 15th List

Clino-sklodowskite. Author? H. Strunz, *Min. Tab.*, 3rd edit., 1957, p. 276 (Klino-Sklodowskit). Monoclinic, $Mg(H_3O)_2[UO_2 \mid SiO_4]_2.3H_2O$, as distinct from sklodowskite, orthorhombic (?), $Mg[UO_2 \mid SiO_3OH]_2.5H_2O$. 21st List

Clinostrengite. G. Gagarin and J. R. Cuomo, 1949, loc. cit., p. 14 (clinostrengita). Monoclinic $Fe'''PO_4.2H_2O$, dimorphous with orthorhombic strengite. Syn. of phosphosiderite and metastrengite (q.v.). 19th List

Clino-triphylite. P. Quensel, 1937. *Geol. För. Förh. Stockholm*, vol. 59, p. 81 (clino-triphylite). A form of triphylite, $Li(Fe,Mn)PO_4$, with polysynthetic twinning and optical extinction suggesting divergence from orthorhombic symmetry. [*M.A.*, **6**, 485.] 14th List

Clino-ungemachite. M. A. Peacock and M. C. Bandy, 1936. *Amer. Min.*, 1936, vol. 21, no. 12, part 2, p. [2]; 1937, vol. 22, p. 207. Monoclinic (pseudo-rhombohedral) crystals closely related to ungemachite (q.v.) [*M.A.*, **6**, 443.] 14th List [*A.M.*, **23**, 314]

Clinovariscite. G. Gagarin and J. R. Cuomo, 1949, loc. cit., p. 14 (clinovariscita). Monoclinic $AlPO_4.2H_2O$, dimorphous with orthorhombic variscite. Synonym of metavariscite (10th List) [*M.A.*, **8**, 51.] 19th List

Clinozoisite. See klinozoisite. 1st List [Now the accepted spelling]

Cl-Tyretskite. R. von Hodenberg and R. Kühn, 1977. *Kali und Steinsalz*, **7**, 165. Rosette-like aggregates in the sylvine and halite of the Boulby mine, Co. Durham, are anorthic, *a* 6·297, *b* 6·464, *c* 6·56 Å, *α* 74·14°, *β* 61·58°, *γ* 61·26°. Composition near $[Ca_2B_5O_8Cl(OH)_2]$, sp. gr. 2·69. Differs from tyretskite (24th List) in having $Cl(OH)_2$ instead of $(OH)_3$. [Without evidence of discontinuity, or that the Cl is ordered, this is an unnecessary name for a chlorian tyretskite—*M.H.H.*] [*A.M.*, **63**, 598.] 30th List

Co-Ludwigite. W. Götz and V. Herrmann, 1966. *Naturwiss.*, **53**, 475. An *artificial* analogue of Ludwigite, with the composition $Co_2^{2+}Co^{3+}O_2BO_3$. [*M.A.*, **18**, 86.] 28th List

Coalingite. F. A. Mumpton, H. W. Jaffe, and C. S. Thompson, 1965. *Amer. Min.*, vol. 50, p. 1893. Soft reddish-brown platelets, optically uniaxial and probably hexagonal, occur in the surface weathering zone of the New Idria serpentinite. Fresno and San Benito Counties, California. The composition is near $Mg_{10}Fe_2{}^{3+}CO_3(OH)_{24}.2H_2O$. Named from the nearby town of Coalinga. Much 'Ferrobrucite' is probably coalingite. [*M.A.*, **17**, 605.] 24th List

Cobalt-cabrerite. H. Meixner, 1951. *Neues Jahrb. Min.*, Monatshefte, 1951, p. 17 (Kobaltcabrerit). Mixed crystals, $(Co,Mg)_3(AsO_4)_2.8H_2O$, [magnesian Erythrite], with optical data intermediate between those of erythrite and hörnesite. *See* Nickel-cabrerite. [*M.A.*, **11**, 312.] 19th List

Cobalt-chalcanthite. E. S. Larsen and M. L. Glenn, 1920. *See* Zinc-copper-chalcanthite. 9th List

Cabalt chrysotile. W. Noll, H. Kircher, and W. Sybertz, 1958. *Naturwiss.*, vol. 45, p. 489 (Kobaltchrysotil). *Beitr. Min. Petr.*, 1960, vol. 7, p. 232. An artificial product, $Co_3Si_2O_5(OH)_4$, isostructural with chrysotile. 22nd List

Cobalt-löllingite. R. J. Holmes, 1942. *Science*, New York, vol. 96, p. 90 (cobaltiferous löllingite), p. 92 (cobalt-löllingite). Synonym of safflorite, which always contains iron, $(Co,Fe)As_2$, and gives the same X-ray pattern as löllingite. [*M.A.*, **8**, 380.] 16th List

Cobalt-manganese-spar. C. Bergemann, 1857. *Verh. Naturhist. Ver. Rheinl. Westph.*, vol. 14, p. 111 (Kobalt-Manganspath). *British Museum Catalogue of minerals*, 1st edit., 1863 (Cobalt manganese spar). Rhodochrosite from Rheinbreitbach containing $CoCO_3$ 3·71%. 18th List

Cobalt-melanterite. E. S. Larsen and M. L. Glenn, 1920. *Amer. Journ. Sci.*, ser. 4, vol. 50, p. 230. Synonym of bieberite. *See* Zinc-copper-melanterite. 9th List

Cobaltnickelpyrite. M. Henglein, 1914. *Centralblatt Min.*, 1914, p. 129 (*Kobaltnickelpyrit*). A member of the pyrites group containing considerable amounts of nickel (11·7–17·5%) and cobalt (6·6–10·6%), $(Fe,Ni,Co)S_2$; occurring as steel-grey, pentagonal-dodecahedral crystals at Müsen, Westphalia.

V. I. Vernadsky (*Centralblatt Min.*, 1914, p. 494) points out that he had previously, in his Russian textbook on mineralogy (1910), applied this name to cobaltiferous (2·0–3·5%) and nickeliferous (2·2–5·8%) iron-pyrites. If these are not identical, he suggests that Henglein's mineral is perhaps $(Co,Ni,Fe)S_2$ intermixed with some iron-pyrites. The same name (Cobalt-nickel pyrites) has appeared, as a synonym of linnaeite, in the *British Museum Index of Minerals* since 1863. 7th List

Cobaltoadamite. A. Lacroix, 1910. *Minéralogie de la France*, vol. iv, p. 424. A variety of adamite from Cape Garonne, Var, of a pale rose-red to carmine colour, and containing some cobalt isomorphously replacing zinc (analysed by A. Damour, 1868). 6th List

Cobaltocalcite. F. Millosevich, 1910. *Rend. R. Accad. Lincei*, Roma, ser. 5, vol. xix, sem. 1, p. 92. A bright red variety of calcite containing cobalt (CoO, 1·27%), occurring as crystalline masses at Capo Calamita, Elba. 5th List

——. Dana's *Mineralogy*, 7th edit., 1951, vol. 2, p. 175. To replace the generally accepted name sphaerocobaltite (A. Weisbach, 1877) for rhombohedral $CoCO_3$. Not the cobaltocalcite of F. Millosevich, 1910 (5th List), a red cobaltiferous variety of calcite. 19th List

Cobaltochrompicotite. A. K. Boldryev, 1935. *Kurs opisatelnoi mineralogii*, Leningrad, part 3, p. 115 (Кобальтохромпикотит). A member of the spinel group containing cobalt, $(Mg,Fe,Co)(Cr,Al)_2O_4$(?). Many other compound names are here given for the spinel group. *See* Alumochromite. 14th List

Cobalt-olivine. C. W. F. T. Pistorius, 1963. *Neues Jahrb. Min.*, Monatsh., p. 30 (Cobalt-olivine; p. 31, Kobaltolivin). Artificial Co_2SiO_4, isostructural with Olivine. 24th List

Cobaltomelane, Cobalto-nickelemelane. I. I. Ginzburg and I. A. Rukavishnikova, 1951 (кобальтомелан, кобальто-никелемелан). *See* Alumocobaltomelane. 22nd List

Cobalto-nickelemelane, *see* Cobaltomelane.

Cobaltorhodochrosite. *Bull. Soc. Franç. Min.*, 1936, vol. 59, p. 385. Cobaltiferous rhodochrosite from Schneeberg, Saxony. The same as cobalt-manganese-spar (q.v.). 18th List

Cobalto-sphaerosiderite. R. Reissner, 1935. *Zentr. Min.*, Abt. A, 1935, p. 173 (Cobalto-Sphärosiderit, Kobalt-Oligonspat). A peach-blossom-red rhombohedral carbonate containing $FeCO_3$ 40·48, $MnCO_3$ 19·11, $MgCO_3$ 21·06, $CoCO_3$ 14·44, $CaCO_3$ 4·34, $ZnCO_3$ 0·61%. [*M.A.*, **6**, 151.] 14th List

Cobalt pentlandite. O. Kuovo, M. Huhma, and Y. Vuorelainen, 1959. *Amer. Min.*, vol. 44 p. 897. The cobalt analogue of pentlandite, Co_9S_8, from Northern Karelia. 22nd List [*A.M.*, **50**, 2107]

Cobalt-pimelite. C. W. F. T. Pistorius, 1963. *Neues Jahrb. Min.*, Monatsh., p. 30. The cobalt analogue of Alipite (Pimelite of Schmidt); artificial. 24th List

Cobalt-pyrite. K. Johansson, 1924. *Arkiv Kemi, Min. Geol.*, vol. 9, no. 8, p. 2 (Koboltpyrit). Octahedra of pyrite containing 13·90% Co from Gladhammer, Sweden. Not cobalt-pyrites, a synonym of linnæite. Compare Cobaltnickelpyrite of M. Henglein, 1914 (7th List). [*M.A.*, **2**, 339.] 10th List

Cobalt skutterudite. E. H. Roseboom, 1962. *Amer. Min.*, vol. 47, p. 310. The pure end-member, $CoAs_3$, of the Skutterudite series. 23rd List

Cobaltsmithsonite. G. A. Bilibin, 1927. *Mém. Soc. Russ. Min.*, ser. 2, vol. 56, p. 34 (Кобальтсмитсонит), p. 36 (Cobaltsmithsonite). The cobaltiferous smithsonite (CoO 10·25%) from Boleo, Lower California, Mexico, analysed by C. H. Warren in 1898. 12th List

Cobalt-talc. C. W. F. T. Pistorius, 1963. *Neues Jahrb. Min.*, Monatsh., p. 30 (Cobalt-talc; p. 31, Kobalttalkum). Artificial $Co_3Si_4O_{10}(OH)_2$, isostructural with Talc. 24th List

Cochranite. J. E. Stead, 1918. *Journ. Iron and Steel Inst.*, vol. xcvii, p. 171. Artificially-produced titanium dicyanide, $Ti(CN)_2$, found as minute, dark-blue cubes in blast-furnace 'bears'. It was first found in quantity by Mr. Alfred O. Cochrane at the iron-works of Messrs. Cochrane & Co., at Ormesby near Middlesborough.

It is formed under the same conditions, and sometimes together with, the copper-red cubes of titanium cyano-nitride, $Ti(CN)_2.3Ti_3N_2$. This [now included with osbornite] was named *sorbite* (after Henry Clifton Sorby, 1826–1908) by H. M. Howe (1890), a term afterwards withdrawn, as the same name was given by F. Osmond (1895) for one of the transition conditions in carbon-steel. A sugar also bears the name sorbite (from Lat. *Sorbus*, the service-tree). 8th List [Correct formula Ti(C,N), with $C \approx N$; *M.M.*, **26**, 36]

Cocinerite. G. J. Hough, 1919. *Amer. Journ. Sci.*, ser. 4, vol. 48, p. 206. A massive, silver-grey mineral with the composition Cu_4AgS, from the Cocinera mine, Ramos, San Luis Potosi, Mexico. [*M.A.*, **1**, 18.] 9th List [A mixture; *A.M.*, **52**, 1214; *M.A.*, **19**, 140]

Coconinoite. E. J. Young, A. D. Weeks, and R. Meyrowitz, 1966. *Amer. Min.*, vol. 51, p. 651. Pale yellow microcrystalline $Fe_2Al_2(UO_2)_2(PO_4)_4SO_4(OH)_2.20H_2O$, probably monoclinic, from various localities in Utah and from Arizona, Wyoming, and New Hampshire. Named for Coconino County, Arizona. 24th List

Codazzite. R. L. Codazzi, 1925. *Notas mineralógicas y pertrográficas*, Bogotá, 1925, p. 10, plate 4; *Los minerales de Colombia*, Bogotá, 1927, p. 94 (codazzita). A rhombohedral carbonate near to ankerite but containing up to 7% $(Ce,La,Di)CO_3$; from the emerald mine of Muzo, Colombia. Named after the Italian geographer Agostino Codazzi (1793–1859), who made the first map of Colombia. 11th List

Coelanite. Erroneous transliteration of Целанит, celanite (q.v.). 22nd List

Coelestobaryt. H. Puchelt and G. Müller, 1964. *Sedimentology and Ore Genesis. Developments in Sedimentology*, **2**, 144. *Syn.* of strontiobarite. Not to be confused with celestobarite, a *syn.* of barytocelestine. 28th List

Coesite. R. B. Sosman, 1954. *Science* (*Amer. Assoc. Adv. Sci.*), vol. 119, p. 738 (coesite), p. 739 (silica C). A high-pressure form of dense (sp. gr. 3·01) silica, probably triclinic, prepared under dry conditions by L. Coes, after whom it is named. [*M.A.*, **12**, 410.] 20th List [Yakutia, *M.A.*, 78-818; South Africa, *M.A.*, 78-819; *see also* under Stishovite]

Coffinite. A. D. Weeks and M. E. Thompson, 1954. [*Bull. US Geol. Surv.*, no. 1009–B, p. 31] abstract in *Amer. Min.*, 1954, vol. 39, p. 1037. L. R. Steiff, J. W. Stern, and A. M. Sherwood, *Science* (*Amer. Assoc. Adv. Sci.*), 1955, vol. 121, p. 608. Uranium silicate, $U(SiO_4)_{1-x}(OH)_{4x}$, tetragonal, as a black impregnation in sandstone on the Colorado plateau. Named after Reuben Clare Coffin of the Colorado Geological Survey. [*M.A.*, **12**, 566–567, 586–587.] 20th List

Cokeite. A. Lacroix, 1910. *Minéralogie de la France*, vol. iv, p. 648 (cokéite). The same as carbonite. A native coke produced by the action of igneous intrusions or of earth movements on coal-seams, or by the spontaneous combustion of coal. 6th List

Colerainite. E. Poitevin and R. P. D. Graham, 1918. *Canada, Geol. Survey, Museum Bulletin, No. 27*, p. 66; E. Poitevin, *Trans. Roy. Soc. Canada*, 1918, ser. 3, vol. xii, sect. iv and v, p. 37. Hydrated silicate of magnesium and aluminium, $4MgO.Al_2O_3.2SiO_2.5H_2O$, forming colourless, thin, hexagonal (optically uniaxial) crystals which are usually aggregated in white rosettes or botryoidal forms; occurs as veins in serpentine. Named from the locality, Coleraine township, Megantic Co., Quebec. 8th List [Probably a chlorite]

Collbranite. D. F. Higgins, 1918. *Economic Geology*, vol. xiii, p. 19. A black acicular mineral forming stellar aggregates in crystalline limestone in the Suan mining district, central Korea. It was identified by B. Kotô (1910) as ilvaite; the microscopical characters, however, suggest a highly ferriferous pyroxene of the hedenbergite type, possibly near $FeSiO_3$, but no analysis is given. Named after Mr. H. Collbran and his son A. H. Collbran, of the Suan mine. 8th List [Is Ludwigite; *A.M.*, **6**, 86]

Collieite. R. Brown, 1927. *Trans. Dumfriesshire & Galloway Nat. Hist. Antiq. Soc.*, ser. 3, vol. 13 (for 1925–1926), p. 72. The calcium vanado-pyromorphite from Leadhills, Scotland, analysed in 1889 by Prof. John Norman Collie, FRS, of London, after whom it is named. [*M.A.*, **4**, 468.] 12th List

Collinsite. E. Poitevin, 1927. *Bull. Geol. Survey Canada*, 1927, no. 46, p. 5. The name first appeared, in a title only, in *Trans. Roy. Soc. Canada*, 1924, vol. 18, Proc. p. xlii. Hydrated phosphate of calcium, magnesium, and iron, $Ca_2(Mg,Fe)P_2O_8.2\frac{1}{2}H_2O$, forming concentric layers with 'quercyite' in phosphorite nodules from British Columbia. The cleavages and optical orientation point to triclinic symmetry; isomorphous with messelite. Named after Dr. William Henry Collins (1878–), Director of the Geological Survey of Canada [*M.A.*, **3**, 470.] 11th List

Colloid-calcite, etc. *See* Kolloid-calcite. 14th List

Collusite. K. Kinoshita, 1944. [*Journ. Japanese Assoc. Min. Petr. Econ. Geol.*, vol. 31, p. 11]; Z. Harada, *Journ. Fac. Sci. Hokkaido Univ.*, ser. 4, 1948, vol. 7, p. 151. A variety of tetrahedrite containing Sn 3·21% from Japan. Error for colusite (13th List). 19th List

Colombianite. G. Gagarin and J. R. Cuomo, 1949, loc. cit., p. 4 (colombianita). Gold-amalgam (Hg 57·4%) from Colombia, South America, analysed by H. Schneider in 1848. 19th List

Colomite. J. Blake, 1876. *Proc. California Acad. Sci.*, vol. vi (for 1875), p. 150. A supposed chromium mica, from Coloma, California, which was re-described by the author in the same year under the name of roscoelite or vanadium mica. 5th List

Columbomicrolite. J. E. de Villiers, 1941. *Amer. Min.*, vol. 26, p. 501. Microlite containing niobium in place of tantalum, from Eshowe, Natal. [*M.A.*, **8**, 188.] 16th List [Unnecessary syn. of pyrochlore; *A.M.*, **62**, 406]

Columbotantalite. L. Van Wambeke, 1958. *Bull. Soc. belge Géol.*, vol. 67, p. 383 (colombotantalite). A non-committal term for members of the columbite–tantalite series. 22nd List

Colusite. A name used locally for tin-bearing ores at Butte, Montana. At first thought to be a stanniferous tennantite (H. Schneiderhöhn and P. Ramdohr, *Lehrbuch der Erzmikroskopie*, 1931, vol. 2, p. 433. R. E. Landon, *Amer. Min.*, 1932, vol. 17, p. 575; 1933, vol. 18, p. 114), but later described (R. E. Landon and A. H. Mogilnor, *Amer. Min.*, 1933, vol. 18, p. 528. W. H. Zachariasen, ibid., p. 534) as a mineral of the zinc-blende group with the composition $(Cu,Fe,Mo,Sn,Zn)_4(S,As,Te,Sb)_{3-4}$. Named from the Colusa claim, which was one of the earliest in the Butte district. [*M.A.*, **5**, 388.] 13th List [*A.M.*, **24**, 369]

Combeite. T. G. Sahama and K. Hytönen, 1957. *Min. Soc. Notice*, no. 98; *Min. Mag.*, 1957, vol. 31, p. 503; *M.A.*, **13**, 555. $Na_4Ca_3Si_6O_{16}(OH,F)_2$, rhombohedral. In nephelinite from Kivu, Belgian Congo. Named after Arthur Delmar Combe (1893–1949), Geological Survey of Uganda. 21st List [*A.M.*, **43**, 791]

Compreignacite. J. Protas, 1964. *Bull. Soc. franç. Min. Crist.*, vol. 87, p. 365; H. Brasseur *ibid.*, p. 629; M. M. Granger and J. Protas, *ibid.*, 1965, vol. 88, p. 251. Tiny yellow orthorhombic crystals with other oxidation products of pitchblende ore of the Margnac deposit, Compreignac, France, are identical with synthetic $K_2O.6UO_3.11H_2O$ (or possibly $10\,H_2O$) and isostructural with Billietite. Named from the locality (the second c should presumably be hard). [*M.A.*, **17**, 182; *A.M.*, **50**, 807.] 24th List

Comstockite. G. Gagarin and J. R. Cuomo, 1949, loc. cit., p. 13 (comstockita). $(Mg,Cu,Zn)SO_4.5H_2O$, containing ZnO 5·60, MgO 9·40, CuO 9·00, H_2O 39·07%, from the Comstock lode, Nevada. Syn. of zinc-magnesia-chalcanthite (C. Milton and W. D. Johnston, *Econ. Geol.*, 1938, vol. 33, p. 761). (15th List) 19th List [Cuprian zincian Pentahydrite]

Comuccite. C. Doelter, 1925. *Handbuch d. Mineralchemie*, vol. 4, part 1, p. 481 (Comuccit). A sulphantimonite of lead with some iron, $18PbS.7FeS.15Sb_2S_3$, from Sardinia. Named after Dr. Probo Comucci, of Firenze, who analysed the mineral in 1916. [No doubt identical with jamesonite.] Incorrectly spelt Cornuccit [*M.A.*, **3**, 469]. 11th List [Is Jamesonite; *A.M.*, **43**, 1225]

Conchilites. T. L. Tanton, 1944. *Trans. Roy. Soc. Canada*, ser. 3, vol. 38, sect. 4, p. 99. Shell-shaped concretions of limonite. Named from *κόγχη* shell, and *λίθος* stone. [*M.A.*, **9**, 127.] 17th List

Conchite. Agnes Kelly, *Nature*, 1900, LXII, 239; *Min. Mag.*, 1900, XII, 363; *Sitz-Ber. Akad. München*, 1900, 187. A form of calcium carbonate; optically uniaxial and negative, but with higher refractive indices than calcite, and no cleavage or twinning. Occurs as various animal secretions (molluscs, etc.); the fur of kettles, and in the hot springs of Carlsbad, Bohemia. 2nd List

Congolite. E. Wendling, R. von Hodenberg, and R. Kühn, 1972. [*Kali und Steinsalz*, **6**, 1]. Small (0·2 mm) red crystals of trigonal iron boracite, from water-insoluble fraction of Cretaceous salt drill core from Brazzaville, Congo. $6[(Fe_{2\cdot68}Mg_{0\cdot24}Mn_{0\cdot08})B_7O_{13}Cl]$; $R3c$ or $R\bar{3}c$, a_{rh} 8·6042 Å, α 60° 10′; ω 1·731, ε 1·755. Artificial compound known previously. A third polymorph complicating the α- and β-ericaite picture (21st List; H. Strunz, *Min. Tab.*, 4th edit., 1966, 236 and 5th edit., 1970, 266). Named for the locality. [*M.A.*, 72-3337; *A.M.*, **57**, 1315.] 27th List

Coolgardite. A. Carnot, *Compt. rend. Acad. Sci.*, Paris, 1901, vol. cxxxii, p. 1300; *Bull. Soc. franç. Min.*, 1901, vol. xxiv, p. 360; *Ann. des Mines*, sér. 9, vol. xix, p. 533. Described as a sesquitelluride of gold, silver, and mercury from Kalgoorlie, East Coolgardie gold-field, Western Australia. Proved to be a mixture of coloradoite (HgTe), and the gold and silver tellurides, calaverite, sylvanite, and petzite (L. J. Spencer, *Min. Mag.*, 1903, vol. xiii, p. 268). 3rd List

Cooperite. F. Wartenweiler, 1928. In R. A. Cooper, *Journ. Chem. Metall. Mining Soc. South Africa*, 1928, vol. 28, p. 283. H. Schneiderhöhn, *Centr. Min.*, Abt. A, 1929, p. 193; *Chem. Erde*, 1929, vol. 4, pp. 268, 275. P. A. Wagner, *The platinum deposits and mines of South Africa*, 1929, pp. 12, 18, 226, 237. H. R. Adam, *Trans. Geol. Soc. South Africa*, 1931, vol. 33 (for 1930), p. 104. Platinum sulphide PtS_2, as minute pyritohedral-cubic crystals isomorphous with sperrylite, occurring in the platiniferous norite of the Bushveld, Transvaal. Named after R. A. Cooper, of Johannesburg, by whom the mineral was first described (loc. cit., 1928, p. 281). Not the cooperite of [G.J.] Adam, 1869. Also trade-name for an alloy of nickel, zirconium, tungsten, etc., used for cutting-tools. [*M.A.*, **4**, 10, 145, 149, 500.] 12th List [Correct formula PtS; *M.M.*, **23**, 188]

Copper-chalcanthite. A. N. Winchell, 1942. *Elements of mineralogy*, p. 292; *Amer. Min.*, 1949, vol. 34, p. 223 (Copper chalcanthite). Syn. of chalcanthite ($CuSO_4.5H_20$), which with siderotil ($FeSO_4.5H_2O$) and cobalt-chalcanthite = bieberite ($CoSO_4.5H_2O$) is classed as sub-species under a species 'chalcanthite'. [*M.A.*, **10**, 568.] 18th List

Copper-melanterite; Copper-zinc-melanterite. E. S. Larsen and M. L. Glenn, 1920. *See* Zinc-copper-melanterite. 9th List

Copper vermiculite. W. A. Bassett, 1958. *Amer. Min.*, vol. 43, p. 1112. Vermiculite from the Roan Antelope mine, Northern Rhodesia, contains up to 7% Cu replacing Mg; similar material with up to 10% Cu has been obtained artificially. An unnecessary name for cuprian vermiculite. 22nd List

Copper-zinc-epsomite. C. Milton and W. D. Johnston, 1937. *Amer. Min.*, vol. 22, no. 12, part 2, p. 10; 1938, vol. 23, p. 175. 15th List

Corderoite. E. E. Foord, P. Berendsen, and L. O. Storey, 1974. *A.M.*, **59**, 652. A massive orange-pink mineral, darkening in light, replacing cinnabar, occurs in playa sediments near the Cordero mine, Humboldt County, Nevada, and is identical with synthetic $4[Hg_3S_2Cl_2]$; space group $I2_13$, a 8·94 Å. Named for the locality. [*M.A.*, 75-551; *Bull.*, **97**, 501; *Zap.*, **104**, 608.] 29th List

Cordierite-pinite. H. Gemböck, *Zeits. Kryst. Min.*, 1898, XXIX, 305; 1899, XXXI, 248. Pinite when derived from cordierite. 2nd List

Cordobaite, cited by H. Strunz, *Min. Tabellen*, 3rd edit., 1957, p. 365, without reference, as a synonym of Brannerite (9th List). 24th List

Cordylite. G. Flink, *Medd. om Grönland*, 1900, XXIV, 42. Wax-yellow, club-shaped hexagonal crystals, isomorphous with parisite. $(CeF)_2Ba(CO_3)_3$. Also called *barium-parisite*. [Narsarsuk] S Greenland. 2nd List

Corencite. C. Mira, 1939. [*Rev. Matér. Const., Trav. Publ.*, no. 359, p. 87.] P. F. Kerr and P. K. Hamilton, *Amer. Petrol. Inst.*, Clay mineral standards, 1949, rep. no. 1, p. 16. Synonym of nontronite. 21st List

Corindite. (A. Bigot, *Trans. Ceramic Soc. Stoke-on-Trent*, 1918, vol. 17 (for 1917–18), p. 267). Trade-name for an artificial product consisting mainly of corundum ($Al_2O_3$69% with SiO_2,Fe_2O_3,TiO_2, etc.) prepared by a process (patented in France in 1914 by N. Lecesne) of fusing bauxite, and used as a refractory material and as an abrasive. (Compare Alundum, Aloxite, Adamite, etc.) 9th List

Cornetite. H. Buttgenbach, 1917*. *Les Minéraux et les Roches*, Liège, p. 452 (Cornètite), p. 521 (cornétite). Described, but without name, by G. Cesàro, *Annales Soc. Géol. Belgique*, 1912, vol. xxxix, Bull. p. 241; Annexe to vol. xxxix (Publ. relatives au Congo Belge), p. 41. Phosphate of copper and cobalt occurring as small, blue, orthorhombic crystals in l'Étoile du Congo cupper mine, Katanga, Belgian Congo. Named after the Belgian geologist Jules Cornet. [*Corr. in *M.M.*, **19**.] 8th List [$Cu_3PO_4(OH)_3$]

Cornubite. G. F. Claringbull, M. H. Hey, and A. Russell, 1958. *Min. Mag.*, vol. 31, p. 792 [*M.A.*, **13**, 558]. Basic copper arsenate, $Cu_5(AsO_4)_2(OH)_4$, dimorphous and associated with cornwallite. The name was provisionally given as cornubianite (*Min. Soc. Notice*, 1957, no. 99), earlier applied to a metamorphic rock, hornfels (H. S. Boase, *Trans. Roy. Geol. Soc. Cornwall*, 1832, vol. 4, pp. 390, 394). Named from the locality, *Cornubia, Cornubian*, medieval Latin, Cornwall, Cornish. 21st List

Cornuccit, *see* Comuccite.

Cornuite. A. F. Rogers, 1917. *Journ. Geol. Chicago*, vol. xxv, p. 537. A glassy, green or bluish-green copper silicate, mCuO.$n$$SiO_2$.$x$$H_2O$, the amorphous equivalent of chrysocolla (of which microcrystalline and crystallized material is known). Named in memory of Dr. Felix Cornu (1882–1909), who wrote on colloidal minerals. 8th List

—— F. V. Hahn, 1925. *Centr. Min.*, Abt. A, 1925, p. 353; 1926, p. 199; *Kolloid-Zeits.*, 1925, vol. 37, p. 303 (Cornuit). A yellow gelatinous substance, apparently an albumen with 97% water, found in fissures in the diatomite deposit of the Lüneburger Heide, Hanover. Named in memory of Dr. Felix Cornu (1882–1909). [The substance is perhaps a fungoidal growth.] Not the Cornuite of A. F. Rogers, 1917 (8th List). [*M.A.*, **3**, 114.] 11th List

Coronadite. W. Lindgren and W. F. Hillebrand, 1904. *Amer. Journ. Sci.*, ser. 4, vol. xviii, p. 448; *Bull. US Geol. Survey*, 1905, no. 262, p. 42; *Profess. Paper US Geol. Survey*, 1905, no. 43, p. 103. A black, metallic mineral with finely fibrous structure; not unlike psilomelane in appearance. Manganite of lead and manganese, (Pb,Mn)O.3MnO_2. Occurs in fairly large amount in the Coronado vein, Clifton-Morenci district, Arizona. Named after the Spanish explorer, Francisco Vasquez de Coronado, who visited the region in 1540. 4th List

Corrensite. F. Lippmann, 1954. *Heidelberg. Beitr. Min. Petr.*, vol. 4, p. 134 (Corrensit). A 'swelling' chloritic mineral determined by X-rays in the finest fraction of red Keuper clay. Named after Prof. Carl Wilhelm Correns (1893–) of Göttingen. 20th List [*A.M.*, **40**, 137; **46**, 769]

Corundolite. A trade name for (colourless) *synthetic* spinel, MgO.2Al_2O_3 [presumably in allusion to the excess of alumina compared with natural spinel]. R. Webster, *priv. comm.* Cf. erinide, (dialite), rozircon, strongite (this List); other trade names include alumag and magalux (M. J. O'Donoghue, *priv. comm.*). 28th List

Corundophyllite. A. Des Cloizeaux, *Manuel Min.*, 1862–1893; A. de Lapparent, *Cours Min.*, 1st–4th edit., 1884–1908; variable entries in text and indexes, corundophilite, corundophyllite. J. Orcel, *Compt. Rend. Acad. Sci.*, Paris, 1924, vol. 178, p. 1730 (corundophyllite). Error for corundophilite, named from φίλος, friend of corundum, of which it is an alteration product. [*M.A.*, **3**, 56, 372.] 20th List

Corunuvite, Corvunuvite, errors for Cornubite (*Intro. Jap. Min.*, 32 (1970), Geol. Surv. Japan). [*Zap.*, **101**, 284.] 28th List

Corvusite. E. P. Henderson and F. L. Hess, 1933. *Amer. Min.*, vol. 18, p. 199. Hydrated vanadic pentoxide and dioxide, $V_2O_4.6V_2O_5.xH_2O$, as bluish-black compact material in sandstone from Utah. Named on account of the colour from the Latin corvus, a raven. [*M.A.*, **5**, 293.] 13th List

Cosmochlore. See kosmochlor. 1st List [The accepted spelling. Cf. Ureyite]

Cosmolite. Synonym of meteorite; probably much earlier than M. Selga, *Publ. Manila Observatory*, 1930, vol. 1, no. 9, p. 25 (kosmolite), p. 51 (cosmolita, cosmolito), applied to meteorites and tektites. F. C. Leonard, *Popular Astronomy*, Northfield, Minnesota, 1944, vol. 52, p. 352, reprinted in *Contributions to the Society for Research on Meteorites*, 1945, vol. 3 (for 1944), p. 161, 'in view of the cosmic origin of meteorites, it is perhaps unfortunate that they were not termed *cosmolites*'. [*M.A.*, **4**, 442; **9**, 289.] 17th List

Costibite. L. J. Cabri, D. C. Harris, and J. M. Stewart, 1970. *Amer. Min.*, **55**, 10. Type willyamite (cubic (Co,Ni)SbS, with Co > Ni) from the Consols Lode, Broken Hill, New South Wales, also carries lamellae of CoSbS, orthorhombic with space group $P2_1mn$ and distinct from paracostibite. [*M.A.*, 70-2607.] 26th List

Coulsonite. J. A. Dunn, 1937. *Mem. Geol. Surv. India*, vol. 49, p. 21. J. A. Dunn and A. K. Dey, *Trans. Mining Geol. Inst. India*, 1937, vol. 31, p. 131. A vanadiferous iron ore assumed to have the composition $FeO.(Fe,V)_2O_3$. It occurs in magnetite from Bihar, India, and was first mentioned under the name vanado-magnetite (q.v.). Named after Dr. Arthur Lennox Coulson, of the Geological Survey of India. [*M.A.*, **6**, 489.] 14th List [*A.M.*, **47**, 1284]

Courtzilite. (*17th Ann. Rept. US Geol. Survey, 1895–6*, Part III, 752). A kind of asphaltum. 2nd List

Courzite. S. J. Thugutt, 1945. *See* Kurtzite. 18th List

Cousinite. J. F. Vaes, 1958. [*Geol. en Mijnbouw*, vol. 20, p. 449 (Cousiniet)]; abstr. *Amer. Min.*, 1959, vol. 44, p. 910. An inadequately described molybdate of uranium or of uranium and magnesium. Compare Moluranite (21st List), Umohoite (20th List). 22nd List

Covellin, Blaubleibender. G. Frenzel, 1959. *Neues Jahrb. Min., Abh.*, **93**, 115. A copper sulphide from Messina, Transvaal, approximating to $Cu_{1\cdot40}S$. [*M.A.*, **15**, 352.] 28th List

Cowlesite. W. S. Wise and R. W. Tschernich, 1975. *A.M.*, **60**, 951. Clusters of thin colourless blades in basalt from Goble, Columbia County, Oregon, are orthorhombic, *a* 11·27, *b* 15·25, *c* 12·61 Å, α 1·512 || [010], β 1·515, γ 1·517 || [001], cleavage {010}, elongation [001]. Subsequently identified from several other localities. Zeolite group. $6[CaAl_2Si_3O_{10}.6H_2O]$; ρ 2·14 g. cm^{-3}. Named from John Cowles. 29th List

Cr-Zr-armalcolite. I. M. Steele and J. V. Smith, 1972. *Nature* (*Phys. Sci.*), **237**, 105. An oxide of Ti, Fe, Cr, and Zr, near Phase X of Peckett *et al.* [*M.A.*, 73-2950], from lunar fines 14166, 6 (Apollo 14 mission). [*M.A.*, 73-2756.] 28th List

Craigite. S. L. Miller, 1970. *Science*, **165**, 489. Because air bubbles are not present in the Antarctic ice at depths greater than 1200 m, although gas is still evolved on melting, it is assumed that the known cubic compounds $4O_2.23H_2O$ and $4N_2.23H_2O$ are present. Named for H. Craig. [*A.M.*, **55**, 1071.] 27th List

Crandallite. G. F. Loughlin and W. T. Schaller, 1917. *Amer. Journ. Sci.*, ser. 4, vol. xliii, p. 69. A hydrated phosphate of calcium and aluminium, $CaO.2Al_2O_3.P_2O_5.5H_2O$, forming compact, greyish masses with fibrous structure and probably resulting from the alteration of goyazite. From the Tintic district, Utah. Named after Mr. M. L. Crandall, mining engineer of Provo, Utah. 8th List [Syn. Pseudowavellite, q.v.]

Creaseyite. S. A. Williams and R. A. Bideaux, 1975. *M.M.*, **40**, 227. Small green spherules from Tiger, Arizona, from Wickenburg, Arizona, and from Caborca, Sonora, Mexico, are orthorhombic, *a* 12·483, *b* 21·395, *c* 7·283 Å; composition $4[Cu_2Pb_2(Fe,Al)_2Si_5O_{17}.6H_2O]$. Named for S. C. Creasey. [*M.A.*, 75-3595; *A.M.*, **61**, 503.] 29th List

Creedite. E. S. Larsen and R. C. Wells, 1916. *Proc. Nat. Acad. Sci. USA*, vol. ii, p. 362. Hydrated fluoride and sulphate of calcium and aluminium, $2CaF_2.2Al(F,OH)_3.CaSO_4.2H_2O$, the formula being the same as that of gearksutite (with which the new mineral occurs) with the addition of $CaSO_4$. The cleavages and optical characters point to monoclinic symmetry. Occurs as white to colourless grains embedded in kaolinite near Wagon Wheel Gap, Colorado. This locality lies near the centre of the Creede quadrangle of the United States Geological Survey; hence the name. 8th List [*M.A.*, **18**, 87]

Creniadite, cited by H. Strunz, *Min. Tabellen*, 3rd edit., 1957, p. 366, without reference, as a synonym of Kaolinite from Colorado. 24th List

Crenite. D. A. Wells, 1852. *Proc. Amer. Assoc. Adv. Sci.*, vol. vi, p. 230. Stalactitic calcite coloured yellow by organic matter, which was identified with crenic acid or crenate of calcium. [J. Fromme (*Jahresber. Ver. Naturwiss. Braunschweig*, 1897, vol. x, p. 104) found 0·231% of apocrenic acid in chestnut-brown, translucent crystals of calcite from the Harz.] 4th List

Creolite. A superfluous name for a red and white banded jasper from California. R. Webster, *Gems*, 1962, p. 755. 28th List

Crestmoreite. A. S. Eakle, 1917*. *Bull. Dept. Geol. Univ. California*, vol. x, p. 344. Hydrated calcium silicate, perhaps $CaSiO_3.H_2O$, occurring as compact, show-white masses as an alteration product of wilkeite (7th List) in blue calcite at Crestmore, Riverside Co., California. Named after the locality. [*Corr. in *M.M.*, **19**.] 8th List [A mixture; *A.M.*, **39**, 405]

Críptosa. (S. Calderón, *Los Minerales de España*, 1910, vol. ii, p. 474.) Spanish form of Cryptoclase (q.v.). 6th List

Cromepidoto, *see* Chromepidote.

Crossite. C. Palache, *Bull. Dept. Geol. Univ. Calif.*, I. 181, 1894. [*M.M.*, **11**, 35.] A soda amphibole between riebeckite and glaucophane. California. 1st List

Crusite. Error for or variant of crucite (of Delamétherie), a *syn.* of chiastolite. R. Webster, *Gems*, 1962, p. 755. 28th List

Cryohalite. (Author?). A. E. Fersman and O. M. Shubnikova, *Geochem. Min. Companion*, Moscow, 1937, p. 160 (криогалит, $NaCl.10H_2O$). Eutectic mixture of ice and hydrohalite ($NaCl.2H_2O$). Named from *κρύος*, frost, and halite. Compare cryohydrate (F. Guthrie, 1874). *See* Maakite. [*M.A.*, **12**, 240.] 20th List

Cryolithionite. N. V. Ussing, 1904. *Overs. K. Danske Videnskab. Selsk. Forhandl.*, 1904, p. 3. Large, colourless rhombic-dodecahedra found in the cryolite of Greenland. $Li_3Na_3Al_2F_{12}$. So named because of its relation to cryolite and the large amount of lithium (5·35%) it contains. 4th List

Cryophillite, error for Cryophyllite. *Amer. Min.*, 1963, vol. 48, p. 435. 23rd List

Cryptoclase. V. Souza-Brandão, 1909. *Communic. Commissão do Serviço Geologico de Portugal*, vol. vii, p. 137 (Fr. cryptose; Germ. Kryptoklas). A variety of albite which by repeated twinning according to the albite-law simulates monoclinic symmetry. It bears the same relation to albite that orthoclase does to microcline. Anorthoclase belongs to the isomorphous series of which orthoclase and cryptoclase are the end-members. 6th List

Cryptomelane. W. E. Richmond and M. Fleischer, 1942. *Amer. Min.*, vol. 27, p. 607. L. S. Ramsdell, ibid., p. 611. Potassium-manganese manganate giving an X-ray pattern distinct from psilomelane (Ba–Mn manganate). Named from *κρυπτός*, hidden, and *μέλᾱς*, *-ᾰνος*, black. [*M.A.*, **8**, 310.] 16th List [Cf. Ebelmenite]

Cryptonickelemelane. K. K. Nikitin, 1960. Кора выветривания (Crust of weathering), vol. 3, p. 42 (криптоникелемелан). *See* Alumocobaltomelane. [A mixture.] 22nd List

Cryptotilite. E. T. Wherry, 1925. *Amer. Min.*, vol. 10, p. 143. Variant of Kryptotile (A. H. Chester, 1896; Kryptotil, A. Sauer, 1886). 11th List

Crystolon. (*The Mineral Industry*, New York, 1913, vol. xxi, for 1912, p. 774. *Mineral Resources, US Geol. Survey, for 1914*, 1915, part ii, p. 568.) Trade-name for an artificially-produced crystalline carbide of silicon, CSi, used for abrasive purposes. *See* Carborundum (4th List). 8th List

Cs-Beryll. H. Strunz, 1966. *Min. Tabellen*, 4th edit., p. 361. Synonym of Caesium beryl (Penfield, 1888), Vorobievite (5th List). 24th List

Csiklovaite. S Koch, 1948. *Acta Min. Petr., Publ. Min. Petr. Inst. Univ. Szeged*, vol. 2, p. 11 (csiklovait), p. 19 (csiklovaite). Sulphide and telluride of bismuth Bi_2TeS_2, in the series $Bi_2Te_3 \rightarrow Bi_2Te_2S \rightarrow Bi_2TeS_2 \rightarrow Bi_2S_3$; isomorphous with the associated tetradymite (Bi_2Te_2S), but differing in colour and etching reactions. Named from the locality Csiklova (Ciclova), Banat, Romania. [*M.A.*, **10**, 446, 509.] 18th List [*A.M.*, **35**, 333]

Ctypeite. *Bull. Soc. franç. Min.*, 1901, vol. xxiv, p. 456, abstracts (ctypéite). The same as ktypeite (A. Lacroix, 1898). 3rd List

Cubaite. F. Vidal y Careta, *Crónica Cientifica, Barcelona*, XIII. 497, 1890; L. F. Navarro, *Anal. Soc. Españ. Hist. Nat.* XXI. *Actas*, p. 120, 1893. A supposed cubic form of silica; shown by Navarro to be rhombohedra of quartz. [Guanabacoa, Cuba.] *See* guanabaquite. 1st List

Cubeite. Dana's *Appendix*, 1899. *See* Kubeit. 2nd List

Cubosilicite. L. Bombicci, *Mem. Accad. Sci. Bologna*, 1899 [v], VIII, 67. The smalt-blue cubes from Tresztya, Transylvania, usually supposed to be pseudomorphs of chalcedony after fluor, are considered to be a definite form of pseudo-cubic silica related to melanophlogite and cristobalite. [Cf. guanabaquite.] 2nd List

Cumengeite. E. Mallard, *Bull. Soc. fran. Min.* XVI. 184, 1893; E. Cumenge, *Compt. Rend.* CXVI. 898, 1893; A. Lacroix, *Bull. Mus. d'Hist. Nat. Paris*, 1895, p. 39. [*M.M.*, **11**, 164.] $PbCl_2.CuO.H_2O$. Tetragonal. Boleo, Lower California. 1st List

Cumengite. P. Groth, *Tab. Uebers. d. Min.*, 1898, 4th edit., p. 55 (Cumengit); E. S. Dana, *1st appendix to the 6th edit. of Dana's System of Mineralogy*, 1899, p. 21. The same as the cumengeite of Mallard (1893, named after Mr. E. Cumenge); not the cumengite of Kenngott (1853). 3rd List [This variant spelling is becoming common]

Cuproadamite. A. Lacroix, 1910. *Minéralogie de la France*, vol. iv, p. 424. A variety of adamite from Cape Garonne, Var, of a sea-green colour and containing much copper (analysed by F. Pisani, 1870), thus forming a passage to the isomorphous mineral olivenite. 6th List

Cuproarquerite. R. Koechlin, *Min. Taschenb. Wien. Min. Gesell.*, 2nd edit., 1928, p. 19 (Cuproarquerit). Silver-amalgam containing some copper, from Chile. 18th List

Cupro-asbolane. L. De Leenheer, 1938. *Ann. Service Mines*, Katanga, vol. 8 (for 1937), p. 35 (cupro-asbolane). A cobaltiferous wad (asbolane) containing also copper, from Katanga, Belgian Congo. [*M.A.*, **7**, 419.] 15th List

Cuproauride. M. P. Lozhechkin, 1939. *Compt. Rend.* (*Doklady*) *Acad. Sci. URSS*, vol. 24, p. 454 (cuproauride). The 'gold cupride' of Karabash, Ural, is a mixture of Cu_3Au_2 (63%) and $AgAu_4$. The former, as a new mineral, is called cuproauride. [*M.A.*, **7**, 515.] 15th List [Accepted name now Auricupride, q.v.]

Cuprobinnite. A Weisbach, *Char. Min.*, 1880, p. 42. The same as binnite (= tennantite). 2nd List

Cuprobismuthit, variant of Cuprobismutite (H. Strunz, *Min. Tabellen*, 3rd edit., 1957, p. 104). 22nd List

Cuproboulangerite. S. S. Smirnov, 1933. *Trans. United Geol. Prospecting Service USSR*, no. 327, pp. 140, 284, 338 (купро-буланжерит). O. M. Shubnikova, *Trans. Lomonosov Inst. Acad. Sci. USSR*, 1937, no. 10, p. 175 (Купробуланжерит, Cuproboulangerite). A cupriferous variety of boulangerite from Transbaikalia. 15th List

Cuprocannizzite, *see* Cuprocosalite.

Cuprocassiterite. T. Ulke, *Trans. Amer. Inst. Mining Engineers*, XXI. 240, 1892; W. P. Headden, *Amer. Journ. Sci.*, XLV. 108, 1893.

'$4SnO_2 + Cu_2Sn(OH)_6$.' Shown by Headden to be a decomposition product of stannite. [Etta,] S Dakota. 1st List [A mixture, mainly pseudomalachite. *Mineralogy of the Black Hills* (1965), p. 158]

Cuprocopiapite. M. C. Bandy, 1938. *Amer. Min.*, vol. 23, p. 738; L. G. Berry, ibid., 1939, vol. 24, p. 182. A variety of copiapite containing 6% CuO, from Chile. [*M.A.*, **7**, 223.] 15th List [Now regarded as a species]

Cuprocosalite, Cuprolillianite, Cuprocannizzite, Cuproselencannizzarite. A. A. Godovikov, 1972. Abstr. in *Zap.*, **103**, 617 (1974) (Купрокозалит, Купролиллианит, Купроканниццарит, Купроселенканниццарит). Unnecessary, invalid names based solely on literature analyses. 29th List

Cuprocuprite. V. I. Vernadsky, 1910. *Opuit opisatelnoĭ Mineralogii*, St. Petersburg, vol. i, part 3, p. 416 (купрокупритъ); *Centralblatt Min.*, 1912, p. 760 (Cuprocuprit). Native copper containing admixed or dissolved cuprous oxide ($Cu + Cu_2O$). 6th List

Cuprogoslarite. A. F. Rogers, *Kansas Univ. Quart.*, 1899, VIII, A 105 (*Cupro-Goslarite*). Cupriferous goslarite. 2nd List

Cuprohalloysite. L. K. Yakhontova, 1961. Труды Мин. Муз. Акад. Наук СССР (*Proc. Min. Mus. Acad. Sci. USSR*), no. 11, p. 123 (Купрогаллуазит). A cuprian halloysite from the Dashkesan ores, named from the composition; it has not been shown that the material is monomineralic or the copper an essential constituent (Зап. Всесоюз. Мин. Общ. (*Mem. All-Union Min. Soc.*), 1962, vol. 91, p. 903). 23rd List

Cuproiodargyrite (Cupro-Jodargyrit). H. Schulze, *Chemiker-Zeitung*, XVI. 1952, 1892. CuI.AgI. [San Agustin mine, Huantajaya] Chili. 1st List

Cuprojarošite. J. Kokta, 1937. *Sborník Klubu Přírodovědeckého v Brně*, vol. 19 (for 1936). p. 76 (Kuprojarošit). A variety of melanterite containing copper (CuO 4·40%) and magnesium (MgO 4·29%). *See* Jarošite. [*M.A.*, **7**, 316.] 15th List

Cuprokirovite. G. N. Vertushkov, 1939. *Bull. Acad. Sci. URSS, Sér. Géol.*, 1939, no. 1, p. 109 (купрокировит), p. 114 (cuprokirovite). A variety of melanterite containing MgO 3·36% and CuO 3·18%, $(Fe,Mg,Cu)SO_4.7H_2O$, resulting from underground fires in the Kalata mine, Kirovgrad, Ural. Named after Mr. S. M. Kirov (С. М. Киров). *See* Kirovite. [*M.A.*, **7**, 418.] 15th List

Cuprolillianite, *see* Cuprocosalite.

Cuprolovchorrite or Cuprovudyavrite. P. N. Chirvinsky, 1935. *Materials Geochem. Khibina tundra, Acad. Sci. USSR*, p. 87 (медистый ловчоррит, медистый вудъяврит), p. 89 (Kupfferlovčorrit, Kupfferwudjavrit). An emerald-green amorphous mineral occurring with lovchorrite (11th List) and vudyavrite (q.v.) in the Kola peninsula, Russia. [*M.A.*, **6**, 343.] 14th List

Cupro-mangano-aphthitalite. G. Bianchini, 1937. *Rend. Accad. Sci. Fis. Mat. Napoli*, ser. 4, vol. 7, p. 43 (aftitalite cupro-manganesifera). O. M. Shubnikova, *Trans. Inst. Geol. Sci. USSR*, 1940, no. 31, p. 53 (купро-мангано-афтиталит, cupro-mangano-aphthitalite). Aphthitalite from Vesuvius containing Cu 1·13, Mn 0·61%. [*M.A.*, **9**, 144, 125.] 19th List

Cupromontmorillonite. M. Fleischer, 1951. *Amer. Min.*, vol. 36, p. 793, abstract of the original paper, interpreting the Russian name medmontite (q.v.). 19th List. F. V. Chukhrov and F. Y. Anosov, 1950. *Mém. Soc. Russ. Min.*, ser. 2, vol. 79, p. 26 (медмонтит, купромонтмориллонит). I. D. Sedletzky, Priroda, *Acad. Sci. USSR*, 1950, vol. 39, no. 10, p. 48 (купромонтмориллонит). Correction to 19th List. [*M.A.*, **12**, 239.] 20th List

Cuproplatinum. P. A. Wagner, 1929. *The platinum deposits and mines of South Africa*, p. 11. A variety of platinum containing 8–13% of copper, occurring as thin shells around grains of ferro-platinum from the Urals. It was first described as 'copper-bearing platinum' by A. N. Zavaritzky, *Matér. Géol. gén. appl., Comité Géol.*, Leningrad, 1928, no. 108, p. 55 [*M.A.*, **6**, 365]. 14th List

Cuproplumbite. F. K. Biehl, 1919. *Inaug.-Diss. Münster* (*Westf.*), pp. 50, 57 (Cuproplumbit). Basic copper-lead arsenate, $2R_3As_2O_8.3R(OH)_2.xH_2O$, where $x = 0$, 1, or 2, occurring as green pseudomorphous crusts at Tsumeb, SW Africa. Not the Cuproplumbite of A. Breithaupt, 1944. Compare Duftite and Parabayldonite. [*M.A.*, **1**, 203.] 9th List

Cupropyrite. R. Schneider, *Journ. prakt. Chem.*, 1895, LII, 557. An alternative name for barracanite (q.v.). 2nd List [Unrelated to CuS_2; cf. Villamaninite]

——. (E. T. Wherry, *Journ. Washington Acad. Sci.*, 1920, vol. 10, p. 494.) An impure form of chalcopyrite. 9th List

Cuprorivaite. C. Minguzzi, 1938. *Periodico Min. Roma*, vol. 9, p. 333 (cuprorivaite); *Atti X Congr. Internaz. Chim. Roma*, 1938, vol. 2, p. 725; *Chimica e Industria*, Milano, 1938, vol. 20, p. 278. Hydrous silicate of copper, calcium, aluminium, and sodium, as small blue grains from Vesuvius. Named from a supposed relation to rivaite (6th List). [*M.A.*, **7**, 225, 470.] 15th List [Tetragonal $CaCuSi_4O_{10}$; *A.M.*, **47**, 409]

Cuproselencannizzarite, *see* Cuprocosalite.

Cuprosklodowskite. H. Buttgenbach, 1933. *Ann. Soc. Géol. Belgique*, vol. 56, Bull. p. в 331 (cuprosklodovskite). Hydrous silicate and uranate of copper as yellowish-green orthorhombic needles with the characters of sklodowskite [$MgO.2UO_3.2SiO_2.7H_2O$; 10th List] but with CuO in place of MgO; from Katanga, Belgian Congo. [*M.A.*, **5**, 389.] 13th List [*A.M.*, **19**, 235]

Cuprospinel. E. H. Nickel, 1973. *C.M.*, **11**, 1003. Irregular grains and lamellar intergrowths with hematite on an oxidized ore dump at Baie Verte, Newfoundland, are cubic, *a* 8·369, with spinel structure; composition $CuFe_2O_4$; also with some Mg and Al. Named from the composition and structure. [*M.A.*, 73–2941; *Bull.*, **96**, 236; *A.M.*, **59**, 381; *Zap.*, **103**, 357.] 28th List

Cuprostibite. C. Sørensen, E. I. Semenov, M. S. Bessmertnaya, and E. B. Khalezova, 1969. Зап. Всесоюз. мин. общ. (*Mem. All-Union Min. Soc.*), **98**, 716 (Купростибит). The intermetallic compound Cu_3Sb_2 occurs at Ilimaussaq, West Greenland. Named from the composition. [*M.A.*, 70-3427.] 26th List [Cu_2(Sb,Tl); *A.M.*, **55**, 1810]

Cuprovanadinite. E. M. Yanishevsky, 1931. *Trans. Geol. Prospecting Service USSR*, fasc. 109, p. 19 (купрованадинит). A variety of vanadinite containing copper (CuO 1·55%) from Kazakhstan. Not the cuprovanadite (= chileite) of [G. J.] Adam, 1869. 14th List

Cuprovudyavrite, *see* Cuprolovchorrite.

Cuprozincite. F. K. Biehl, 1919. *Inaug.-Diss. Münster, Westf.*, p. 30 (Cuprozinkit), p. 37 (Cupro-Zinkit). Basic carbonate of copper and zinc. $(Cu,Zn)CO_3.(Cu,Zn)(OH)_2$. Named from the composition, though not related to zincite. Compare Paraurichalcite. [*M.A.*, **1**, 203] 9th List

Cuprozippeite. A. K. Boldyrev, 1935. *Kurs opisatelnoi mineralogii*, Leningrad, part 3, p. 83 (Купроциппеит). Variety of zippeite containing 5% CuO. 20th List

Curiénite. F. Cesbron and N. Morin, 1968. *Bull. Soc. franç. Min. Crist.*, **91**, 453. The lead analogue of francevillite, $Pb(UO_2)_2(VO_4)_2.5H_2O$, occurs as a microcrystalline powder on crystals of francevillite at the Mounana mine, Gabon, and has been obtained synthetically. Named for Professor Hubert Curien [*A.M.*, **54**, 1220; *M.A.*, 69-1537; *Bull.*, **92**, 316; *Zap.*, **99**, 80.] 26th List

Curite. A. Schoep, 1921. *Compt. Rend. Acad. Sci.*, Paris, vol. 173, p. 1186. Hydrated uranate of lead and uranyl, $2PbO.5UO_3.4H_2O$, occurring as orange-yellow, acicular crystals in Katanga, Belgian Congo. Named after Pierre Curie (1859–1906). [*M.A.*, **1**, 249.] 9th List

Curtisite. F. E. Wright and E. T. Allen, 1926. *Amer. Min.*, vol. 11, p. 67. A hydrocarbon, $C_{60}H_{40}O$, as a greenish deposit, orthorhombic (or monoclinic), from hot springs at Skaggs Springs, Sonoma Co., California. Named after Mr. P. L. Curtis of Skaggs Springs, who collected the material. [*M.A.*, **3**, 239.] 11th List [A mixture of hydrocarbons, *A.M.*, **61**, 1055]

Curtisitoids. I. V. Grinberg and V. M. Shimansky, 1954. *Min. Sbornik Lvov Geol. Soc.*, no. 8, p. 107 (кертиситоиди). Group name for hydrocarbons allied to curtisite (11th List) and carpathite (q.v.) differing in colour, composition, and melting-point. [*M.A.*, **13**, 120.] 21st List [*See A.M.*, **61**, 1055]

Curzite, S. J. Thugutt, 1949. *Rocznik Polsk. Tow. Geol.* (*Ann. Soc. Géol. Pologne*), vol. 18 (for 1948), p. 5 (curzite), p. 14 (courzite), p. 35 (kurcyt). Another spelling of kurtzite (18th List). 19th List

Cuspidite. Variant of Cuspidine; *A.M.*, **8**, 51. 10th List

Custerite. J. B. Umpleby, W. T. Schaller, and E. S. Larsen, 1913, *Amer. Journ. Sci.*, ser. 4, vol. xxxvi, p. 385; *Zeits. Kryst. Min.*, 1914, vol. liii, p. 321. A monoclinic, hydrous fluosilicate of calcium, $Ca_2(OH,F)_2SiO_3$, occurring as finely granular masses, resembling greenish marble in appearance, at a limestone-granite contact in Custer County, Idaho. Named from the locality. 7th List [Is Cuspidine; *A.M.*, **33**, 100]

—— Зап. Всесоюз. Мин. Общ. (*Mem. All-Union Min. Soc.*), 1958, vol. 87, p. 76; *Bull. Soc. franç. Min. Crist.*, 1958, vol. 81, p. 154. Improper transliteration of Кёстерит (Kësterite or Kyosterite). Not to be confused with Custerite of Umpleby, Schaller, and Larsen (7th List). 22nd List

Cyclite. G. Mueller, 1964. *Rept. 22nd Internat. Geol. Congr., India*, part 1, 46. A fusible, partly soluble bitumen, believed to have a purely polycyclic structure. Sub-divided into bernalite, elaterite, and mutabilite. 26th List

Cyclowollastonit. H. Strunz, 1965. *Min. Tabellen*, 4th edit., p. 358. Synonym of Pseudowollastonite (4th and 7th Lists). 24th List

Cylindrite. A. H. Chester, *Dictionary of the Names of Minerals*, 1896. The same as kylindrite. 2nd List [Now the accepted spelling]

Cymrite. W. C. Smith, F. A. Bannister, and M. H. Hey, 1949. *Min. Mag.*, vol. 28, p. 676. Barium aluminium silicate, $BaAlSi_3O_8(OH)$, as hexagonal crystals from the Benallt manganese mine, Wales. Named from Cymru, the Welsh name for Wales. 18th List [Alaska; *A.M.*, **49**, 158; USSR. $(Ba,H_3O)Al_2Si_2(O,OH)_8.H_2O$; *M.A.*, 69-2896]

Cyrilovite. M. Novotný and J. Staněk, 1953. *Acta Acad. Sci. Nat. Moravo-Silesicae*, vol. 25, p. 325 (cyrilovit). Brown tetragonal crystals, $4Fe_2O_3.3P_2O_5.5\frac{1}{3}H_2O$; in pegmatite from Cyrilov, Moravia. Named from the locality. [*M.A.*, **12**, 512.] 20th List [$NaFe_3(PO_4)_2(OH)_4.2H_2O$; *A.M.*, **42**, 204, 586; cf. Wardite]

Czakaltait. Polish form of Chacaltaite (q.v.). 15th List

Cziklovaite, variant of Csiklovaite (18th List) (I. Kostov, *Mineralogy*, 1968, p. 163). 25th List

D

D'Achiardit, variant of Dachiardite (4th List) (H. Strunz, *Min. Tabellen*, 1st edit., 1941, p. 217). 24th List

Dachiardite. G. D'Achiardi, 1906. *See* Zeolite mimetica. 4th List [Now the accepted name]

Dadsonite. J. L. Jambor, 1969. *Min. Mag.*, **37**, 437. The mineral formerly described as Mineral Q, from Yellowknife, North West Territories, and that described as Mineral QM, from Madoc, Ontario, and from Pershing County, Nevada, are identical; monoclinic, composition $Pb_{11}Sb_{12}S_{22}$. Named for A. S. Dadson. [*M.A.*, 70-752.] 26th List [*A.M.*, **55**, 1445]

Daiton-sulphur. T. Wada, 1916. *Minerals of Japan* (in Japanese), 2nd edit., p. 19 (Daiton-sulphur). A monoclinic sulphur of prismatic habit, differing from β-sulphur and γ-sulphur, described by M. Suzuki (*Journ. Geol. Soc. Tokyo*, 1915, vol. 22, p. 343) from the volcano Daiton, Taiwan (= Formosa). [*M.A.*, **1**, 63.] 9th List [α-sulphur pseudomorphous after β-sulphur; *M.A.*, **11**, 511]

Dakeite. E. S. Larsen, 1937. *Mineralogist*, Portland, Oregon, vol. 5, no. 2, p. 7. E. S. Larsen and F. A. Gonyer, *Amer. Min.*, 1937, vol. 22, p. 561. A complex mineral $3CaCO_3.Na_2SO_4.UO_3.10H_2O$, occurring as green-yellow nodules in surface efflorescences of gypsum in Wyoming. Named after Dr. Henry C. Dake, of Portland, Oregon. [*M.A.*, **6**, 488.] 14th List [Is Schröckingerite; *A.M.*, **24**, 317]

Dallasite. A superfluous name for an unspecified green and white rock from Vancouver Island, British Columbia. R. Webster, *Gems*, 1962, p. 755. 28th List

Dalyite. R. Van Tassel, 1952. *Min. Mag.*, vol. 29, p. 850. Potassium zirconium silicate, $K_2ZrSi_6O_{15}$, triclinic, from Ascension Island, Atlantic. Named after Reginald Aldworth Daly (1871–), emeritus professor of geology, Harvard University. 19th List [*A.M.*, **37**, 1071]

Damburite. An objectionable trade-name for artificial light-red corundum. R. Webster, *Gems*, 1962, p. 755. Not to be confused with danburite. 28th List

D'Ansite. H. Autenreith and G. Braune, 1958. *Naturwiss.*, vol. 45, p. 362 (D'Ansit). See also H. Strunz, *Neues Jahrb. Min.*, Monatsh., 1958, p. 152. A name for the cubic artificial product $MgSO_4.3NaCl.9Na_2SO_4$, near to but distinct from hanksite. Named after Prof. J. D'Ans. [*M.A.*, **14**, 283.] [*A.M.*, **43**, 1221.] 22nd List [China; *M.A.*, 76-2835]

Daomanite. Yu Tsu-Hsiang, Lin Shu-Jen, Chao Pao, Fang Ching-Sung, and Huang Chi-Shun, 1974. [*Acta Geol. Sin.*, **2**, 202], abstr. *M.A.*, 75-2522. $CuPtAsS_2$, orthorhombic, *a* 8·085, *b* 5·905, *c* 7·314 Å, from an undisclosed locality in China. [*A.M.*, **61**, 184.] 29th List

Darapiozite. E. I. Semenov, V. D. Dusmatov, A. P. Khomyakov, A. A. Boronkov, and M. E. Kazakova, 1975. *Zap.*, **104**, 583 (Дарапиозит). Grains up to 5 mm in the pegmatites of the Dara-Pioz massif, Tadzhikistan, are hexagonal, *a* 10·32, *c* 14·39 Å; ω 1·585, ε 1·575. Composition $2[KNa_2LiMnZnZrSi_{12}O_{30}]$. Named for the locality. 29th List

Darapiosite. Erroneous transliteration of Дарапиозит, darapiozite. *Zap.*, **104**, 583. [*A.M.*, **61**, 1053.] 30th List

Darlingite. (*Trans. R. Soc. Victoria*, 1866, VII, 80; *Proc. R. Soc. Victoria*, 1897, N.S. IX, 86). A kind of lydian stone from Victoria. 2nd List

Dashkesanite. G. A. Krutov, 1936. *Bull. Acad. Sci. URSS, Cl. Sci. Mat. Nat., Sér. Géol.*, p. 341 (дашкесанит, хлорсдержащий амфибол), p. 371 (dashkessanite, chlorine amphibole). A chlorine amphibole of the hastingsite group, containing Cl 7·24%. Named from the locality, Dashkesan, Transcaucasia. [*M.A.*, **6**, 438.] 14th List [Chlor potassian hastingsite, *Amph.* (1978)]

Darlingite. (*Trans. R. Soc. Victoria*, 1866, VII, 80; *Proc. R. Soc. Victoria*, 1897, N.S. IX, 86). 1943, no. 4, reprint p. 22. A siliceous montmorillonitic clay of sedimentary origin, containing 25% of organic silica, as distinct from bentonite of volcanic origin. Named from the locality Castelnovo della Daunia, Puglia, Italy. [*M.A.*, **9**, 263.] 17th List

Davidite. D. Mawson, 1906. *Trans. R. Soc. South Australia*, vol. xxx, p. 191. An incompletely described mineral [possibly identical with ilmenite] from South Australia. Named after Professor T. W. Edgeworth David, of Sydney University. 4th List [$(Fe,La,U,Ca)_6(Ti,Fe)_{15}(O,OH)_{36}$, trigonal; *A.M.*, **46**, 700. Cf. Ferutite, Mavudzite]

Davisonite. Dana's *Mineralogy*, 7th edit., 1951, vol. 2, p. 939. Correction of the name 'dennisonite' as given by E. S. Larsen and E. V. Shannon, 1930 (12th List), after John Mason Davison (1840–1915) of the University of Rochester, New York. [*Min. Mag.*, **11**, 226; *M.A.*, **4**, 343.] 19th List [Description under Dennisonite; *A.M.*, **37**, 362]

Davyno-cavolinite. G. Cesàro, 1914. *Bull. Acad. R. Belgique, Cl. des Sci.*, 1914, p. 268; *Mém. (in-8°) Acad. R. Belgique, Cl. des Sci.*, 1920, ser. 2, vol. 4, pp. 20, 57. Hexagonal crystals from Vesuvius with positive birefringence (0·0053) intermediate between the values for davyne and cavolinite. 10th List

Dayingite. Yu Tsu-Hsiang, Lin Shu-Jen, Chao Pao, Fang Ching-Sung, and Huang Chi-Shun, 1974. [*Acta Geol. Sin.* **2**, 202], abstr. *M.A.*, 75-2522. Dodecahedral crystals from an undisclosed locality in China, *a* 9·697 Å, space group *Fm3m*; Composition $CuCoPtS_4$. [*A.M.*, **61**, 184.] 29th List

Dedolomite. M. Warrak, 1974. *Journ. Geol. Soc.*, **130**, 229. Calcite formed by replacement of dolomote. A superfluous term [*M.A.*, 74-3600.] 29th List

Deeckeite. J. Soellner, 1913. *Mitt. Badischen Geol. Landesanstalt*, vol. vii, p. 436 (Deeckëit). A hydrated silicate, $(H,K,Na)_2(Mg,Ca)(Al,Fe)_2(Si_2O_5)_5.9H_2O$, allied to ptilolite and mordenite, occurring as a pseudomorph after melilite in a dyke rock (bergalith) in the Kaiserstuhl, Baden. The characteristic 'peg-structure' of melilite is a result of the partial alteration of this mineral to deeckeite. Named after Professor Wilhelm Deecke, Director of the Geological Survey of Baden. 7th List

Deerite. S. O. Agrell, M. G. Bown, and D. McKie, 1965. *Amer. Min.*, 1965, vol. 50, p. 278 (abstr.). Black monoclinic crystals in metamorphic rocks of the Franciscan formation, Laytonville district, Mendocino County, California, are near $Fe_{13}^{2+}Fe_7^{3+}Si_{13}O_{44}(OH)_{11}$. Named for Prof. W. A. Deer. 24th List [Near $(Fe,Mn)_{12}Si_8(O,OH)_{32}$; *A.M.*, **62**, 1262]

Dehrnite. E. S. Larsen and E. V. Shannon, 1930. *Amer. Min.*, vol. 15, pp. 303, 324. Hydrous phosphate of calcium and sodium, perhaps $7CaO.Na_2O.2P_2O_5.H_2O$ for soda-dehrnite from Dehrn, Nassau (p. 305), or $14CaO.K_2O.Na_2O.4P_2O_5.2H_2O.CO_2$ from Utah (p. 325), as white crystalline (hexagonal) crusts. Named from the locality. [*M.A.*, **4**, 342, 344.] 12th List

Dekalbite. F. R. Van Horn, 1926. *Amer. Min.*, vol. 11, p. 54 (De Kalbite). A name suggested for colourless diopside free from iron $CaMg(SiO_3)_2$, diopside being limited to $Ca(Mg,Fe)(SiO_3)_2$. Dekalbite and diopside in the pyroxene group then correspond with tremolite and actinolite in the amphibole group. Named from DeKalb, St. Lawrence Co., New York. [V. I. Vernadsky, 1913 (7th List), gave the name Kalbaite for tourmaline from this locality.] 11th List

Delatorreite. F. S. Simons and J. A. Straczek, 1958. *US Geol. Surv. Bull.*, no. 1057, p. 1. Synonym of Todorokite (14th List). (*Amer. Min.*, 1960, vol. 45, p. 1175.) 22nd List

Delatynite. J. Niedźwiedzki, 1908. *Kosmos*, Lemberg, vol. xxxiii, p. 531 (delatynit), p. 535 (Delatinit). A variety of amber from Delatyn in the Galician Carpathians, differing from succinite in containing rather more carbon (79·93%), less succinic acid (0·74–1·67%), and no sulphur. 6th List

Delhayelite. Th. G. Sahama and Kai Hytönen, 1959. *Min. Mag.*, vol. 32, p. 6. Orthorhombic laths in a melilite-nephelinite lava from Mt. Shaheru, Kivu, Congo; near $(Na,K)_4Ca_5Al_6Si_{32}O_{80}.18H_2O.3(Na_2,K_2)(Cl_2,F_2,SO_4)$. Named after F. Delhaye, a Belgian geologist and a pioneer in the exploration of North Kivu. 22nd List [*A.M.*, **44**, 1321]

Dellaite. S. O. Agrell, 1965. *Min. Mag.*, vol. 34, p. 1. Small grains in metamorphosed limestone at Kilchoan, Ardnamurchan, Scotland; a single crystal gave an X-ray oscillation photo identical to that obtained from a synthetic calcium silicate, phase γ of D. M. Roy (1958), to which the formula $Ca_6Si_3O_{11}(OH)_2$ was assigned by Glasser and Roy (1959). Named for Della M. Roy. [*A.M.*, **50**, 2104; J. A. Mandarino comments: 'There is little doubt in my mind that this substance is a new mineral. The data presented, however, are of such a preliminary nature that I question the advisability of naming the mineral. Certainly more information is needed on the crystallography and, if an analysis is not possible, a comparison of the X-ray powder data of the natural and analyzed synthetic material would be desirable.'] 24th List

Delorenzite. F. Zambonini, 1908. *Rend. R. Accad. Sci. Fis. Mat. Napoli*, ser. 3, vol. xiv, p. 113; *Riv. Min. Crist. Italiana*, vol. xxxiv, p. 74; *Zeits. Kryst. Min.*, vol. xlv, p. 76. Black, orthorhombic crystals resembling polycrase (but differing from this in containing no niobium), from pegmatite at Craveggia, Piedmont. $2FeO.UO_2.2Y_2O_3.24TiO_2$, or expressed as a meta-titanate $2FeTiO_3.U(TiO_3)_2.2Y_2(TiO_3)_3.7(TiO)TiO_3$. Named after Giuseppe De Lorenzo, professor of geology in the University of Naples. 5th List [Is Tanteuxenite; *M.M.*, **32**, 308]

Delrioite. M. E. Thompson and A. M. Sherwood, 1959. *Amer. Min.*, vol. 44, p. 261. $CaSrV_2O_7.3H_2O$, as a microcrystalline efflorescence on sandstone on a dump of the Jo Dandy mine, Montrose Co., Colorado. Named after A. M. del Rio, who first discovered vanadium (erythronium) in North America. [*M.A.*, **14**, 282.] 22nd List [$CaSrV_2O_6(OH)_2.3H_2O$, monoclinic; *A.M.*, **55**, 185; cf. Metadelrioite]

Deltaite. E. S. Larsen and E. V. Shannon, 1930. *Amer. Min.*, vol. 15, p. 321. Hydrous phosphate of calcium and aluminium, $8CaO.5Al_2O_3.4P_2O_5.14H_2O$, as a constituent of variscite nodules from Utah. Named from the Δ-like (trigonal) form of the mineral in intimate intergrowth with pseudowavellite [*M.A.*, **4**, 344.] 12th List [A mixture; *A.M.*, **46**, 467]

Demesmaekerite. F. Cesbron, B. Backet, and R. Oosterbosch, 1965. *Bull. Soc. franç. Min. Crist.*, vol. 88, p. 422. Green anorthic crystals in the Cu–Co deposit of Musonoi, Kolwezi, Katanga, have the composition $Pb_2Cu_5(UO_2)_2(SeO_3)_6(OH)_62H_2O$. Named for M. G. Demesmaeker. [*M.A.*, **17**, 505.] 24th List

Denhardtite. H. Potonié, 1905. *Monatsber. Deutsch. geol. Ges.*, 1905, p. 259 (Denhardtit). A pale yellow, waxy hydrocarbon similar to pyropissite. It is derived from plants and forms a stratum in loam in British East Africa. Named after the brothers Clemens and Gustav Denhardt, by whom it was collected in 1878. 4th List

Denningite. J. A. Mandarino, S. J. Williams, and R. S. Mitchell, 1961. *Canad. Min.*, vol. 7, p. 340; *Amer. Min.*, 1961, vol. 45, p. 1201; 1962, vol. 47, p. 1484. Colourless to pale green tetragonal plates and platy masses from Moctezuma, Sonora, Mexico, have the composition $(Mn,Ca,Zn)Te_2O_5$. Named for Prof. R. M. Denning. [*M.A.*, **16**, 551.] 23rd List [*A.M.*, **48**, 1419]

Dennisonite. E. S. Larsen and E. V. Shannon, 1930. *Amer. Min.*, vol. 15, p. 322. Hydrous phosphate of calcium and aluminium, $6CaO.Al_2O_3.2P_2O_5.5H_2O$, as white crusts (hexagonal?) in variscite nodules from Utah. Named after 'J. M. Dennison who first described wardite', i.e. John Mason Davison (1840–1915), of the University of Rochester, New York. [*M.A.*, **4**, 344.] 12th List [Name later corrected to Davisonite (q.v.)]

Deodatite. K. W. Nose, 1790. *Orographische Briefe über das Siebengebirge* (*Niederrheinische Reise*), Frankfurt am Mayn, 1790, part 2, p. 198; *Peschreibung einer Sammlung von meist vulkanischen Fossilien die Deodat-Dolomieu im Jahre 1791 von Maltha aus . . . versandt*, Frankfurt a. M., 1797, pp. 27, 43, 80; *Revision des Beschlusses der Kritik über die Theorie der Geologie*, Bonn, 1835, p. 7. A bluish mineral found in the trass of the Lower Rhenish district, at first regarded as perhaps the same as prehnite, and by some authors referred to pleonaste or haüyne (J. R. Zappe, *Mineralogisches Hand-Lexicon*, Wien, 1817, vol. i, p. 257). In 1797 Nose, apparently extending the use of the name to other substances, considered deodatite to be identical with pitchstone, at the same time changing the name to *dolomian*. Both names, now obsolete, are after Déodat G. S. T. G. de Dolomieu (1750–1801). 8th List

Derbylite. E. Hussak and G. T. Prior, *Min. Mag.*, XI. 176, 1897; XI. 85, 1895. $FeO.Sb_2O_5 + 5(FeO.TiO_2)$. Rhombic. [Tripuhy, near Ouro Preto, Minas Gerais] Brazil. 1st List [Monoclinic, $2[Fe_4^{3+}Ti_3^{4+}Sb^{3+}O_{13}(OH)]$; *A.M.*, **62**, 396]

Déribérite. E. Lemoine, 1950, *Bull. Soc. Franç. Min. Crist.*, vol. 73, p. 146. Alternative name suggested for allevardite [Now Rectorite] (q.v.) Named after Maurice Déribéré of Paris, who gave a description of the mineral [*M.A.*, **10**, 43]. 19th List

Derriksite. F. Cesbron, R. Pierrot, and T. Verbeek, 1971. *Bull.*, **94**, 534. Rare malachite-green crystals on selenian digenite at Musonoi, Katanga, are *Pnmm* or *Pnm*2, *a* 5·57, *b* 19·07, *c* 5·96 Å. $2[Cu_4UO_2(SeO_3)_2(OH)_6.H_2O]$; α 1·77 ∥ [100], β ∥ [010]. Named for J. Derriks. [*M.A.*, 72-3338.] 27th List

Dervillite. R. Weil, 1941. *Rev. Sci. Nat. Auvergne*, Clermont-Ferrand, vol. 7, p. 110. An incompletely determined black metallic mineral containing Sb, Pb, (Bi?), S, as minute monoclinic crystals from Markirch, Alsace. Named after Dr. Henri Derville of Strasbourg university. [*M.A.*, **10**, 353.] 18th List

Descloizeauxita. G. Gagarin and J. R. Cuomo, 1949, loc. cit., p. 15. Variant of descloizite (A. Damour, 1854), named after Alfred Louis Olivier Legrand Des Cloizeaux (1817–1897). 19th List

Despujolsite. C. Gaudefroy, M.-M. Granger, F. Permingeat, and J. Protas, 1968. *Bull. Soc. franç. Min. Crist.*, vol. 91, p. 43. Lemon-yellow hexagonal crystals from Tachgagalt, Morocco, have the composition $Ca_3Mn^{4+}(SO_4)_2(OH)_6.3H_2O$, and are isomorphous with schaurteïte (24th List). Named for P. Despujols. 25th List [*A.M.*, **54**, 326]

Devadite. L. L. Fermor, 1938. *Proc. Nat. Inst. Sci. India*, vol. 4, p. 275. A hypothetical mineral, $5Mn_3O_4.5Mn_2O_3.8Fe_2O_3$, which on breaking down gave rise to a mineral with the vredenburgite structure, but containing more iron than the vredenburgite originally described. Named from the locality, Devada, Vizagapatam, Madras. *See* Garividite. [*M.A.*, **7**, 169.] 15th List [Is Vredenburgite; *A.M.*, **29**, 74]

Devillite. Dana, *Syst. Min.*, 7th edit, 1951, p. 590. Synonym of devilline (F. Pisani, 1864). Liable to be confused with similar names with ending *-ite*. 21st List

Devitrite. G. W. Morey and N. L. Bowen, 1931. *The Glass Industry*, New York, vol. 12, p. 133; *Journ. Soc. Glass Technology*, Sheffield, 1931, vol. 15, abstr. p. A 133. Orthorhombic $Na_2O.3CaO.6SiO_2$, common in devitrified commercial glasses. Perhaps the same as reaumurite (6th List). [*M.A.*, **3**, 167.] 13th List

Dewindtite. A Schoep, 1922. *Compt. Rend. Acad. Sci.*, Paris, vol. 174, p. 623. Hydrated phosphate of uranium and lead, $8UO_3.4PbO.3P_2O_5.12H_2O$, occurring as a canary-yellow powder at Kasolo, Katanga. Named after a Belgian geologist, the late Dr. Jean Dewindt. [*M.A.*, **1**, 377.] 9th List [Perhaps a mixture; *A.M.*, **41**, 915. Cf. Stasite]

Dhanrasite. S. R. N. Murthy, 1967. *Current Science*, vol. 36, p. 295. A superfluous name for stannian garnets, and in particular for a stannian pyrope-almandine from the Dhanras hills, Gaya district, Bihar. Named from the locality. [*A.M.*, **53**, 509.] 25th List

Diaboleite. L. J. Spencer, 1923. *Min. Mag.*, vol. 20, p. 78. Oxychloride of lead and copper, $2Pb(OH)_2.CuCl_2$, occurring as bright-blue, tetragonal crystals with chloroxiphite (q.v.). Named from διά, apart or distinct from boleite. 10th List

Dialite. Trade-name for a doublet cut stone, with a crown of colourless *artificial* spinel and a base of strontium titanate. R. Webster, *priv. comm.* Also known as 'Carnegiegem'; cf. 'Laser gem', a similar doublet but with an *artificial* corundum crown. M. J. O'Donoghue, *priv. comm.* 28th List

Diaman-, Diamon-. Many artificial substances have received trade-names starting thus, implying relationship to diamond because of their hardness or lustre. Diamonite, diamonair, diamone, diamonique, diamonte, diamanite: *artificial* yttrium aluminate, $Yt_3Al_5O_{12}$ (cf. yttrogarnet, 19th List; cirolite and triamond, 28th List). Diamontina: *artificial* strontium titanate, $SrTiO_3$ (cf. fabulite, marvelite, and symant, 28th List). Diamondite, diamantin(e), diamontine: *artificial* Al_2O_3 (*see* under zircolite, 28th List, for other names). Diamantine: 'a crystallized boron abrasive powder'. Ref.: M. J. O'Donoghue, *priv. comm.*; J. Sinkankas, *Gemstones and Mineral Data Book*, 1972, p. 70; R. Webster, *Gems*, 1962, p. 755, also *priv. comm.* Also, diamantane $C_{10}H_{16}$, a 3-dimensional hydrocarbon. 28th List

Diamonesque. *See* under phianite (30th List). 30th List

Diaspodumene. P. Quensel, 1939. *Geol. För. Förh. Stockholm*, vol. 60 (for 1938), p. 626 (diaspodumene, caesium diaspodumene, caesium-diaspodumene). A symplektic intergrowth of normal spodumene and a hypothetical caesium-spodumene. [*M.A.*, **7**, 335.] 15th List

Diasporogelite. F. Tućan, 1913. *Centralblatt Min.*, 1913, p. 768 (Diasporogelit). A more correct form of the name Sporogelite (6th List). Similarly, *Pyritogelit*, *Limonitogelit*, *Hämatitogelit*, and *Gibbsitogelit*, for the colloidal forms corresponding to the minerals, pyrites, etc. E. T. Wherry (*Centralblatt Min.*, 1913, p. 519) had previously proposed to distinguish these colloidal forms as *κ-Diaspor, κ-Pyrit*, etc., from the initial letter of κόλλα. 7th List

Dickite. C. S. Ross and P. F. Kerr, 1930. *Amer. Min.*, vol. 15, p. 34; *Journ. Amer. Ceramic Soc.*, 1930, vol. 13, p. 151; *Prof. Paper US Geol. Survey*, 1931, no. 165–E, p. 157. The crystallized kaolin mineral, $Al_2O_3.2SiO_2.2H_2O$ from Anglesey, distinguished by its optical characters from both kaolinite and nacrite. Named after Allan Brugh Dick (1833–1926), of London, who described the mineral as kaolinite in 1888. [*M.A.*, **4**, 247.] 12th List

Dicksbergite. L. J. Igelström, *Geol. För. Förh.*, XVIII. 231, 1896. Shown by M. Weibull and A. Upmark (Ibid., XVIII. 523, 1896) to be rutile. Dicksberg [Ransät parish, Wermland] Sweden. 1st List

Diderichite. J. F. Vaes, 1947. *Ann.* (*Bull.*) *Soc. Géol. Belgique*, vol. 70, p. B 224. Slightly hydrated uranium carbonate as yellow-green fibrous crusts, orthorhombic, from Katanga. Named after Nobert Diderich, who was the first to study the copper and iron ores of Katanga. [*M.A.*, **10**, 146.] 18th List [Is rutherfordine; *M.A.*, **13**, 205]

Didjumolite. *See* Didymolite. 5th List

Didymolite. A. Meister, 1908. *Verh. Russ. Mineral. Ges.*, ser. 2, vol. xlvi, p. 151 (Дидюмолитъ). Monoclinic, twinned crystals in *crystalline limestone at contact with* nepheline-syenite from Siberia. $2CaO.3Al_2O_3.9SiO_2$. Named from δίδυμος, a twin, and λίθος, a stone. The German translation of the title on the wrapper of the journal gives the form Didjumolit. [*⁻*Corr. in *M.M.*, **16**.] 5th List [A plagioclase; *A.M.*, **50**, 2111]

Dienerite. C. Doelter, 1926. *Handbuch d. Mineralchemie*, vol. 4, part 1, p. 718 (Dienerit). Nickel arsenide, Ni_3As, found as greyish-white cubes. Named after Prof. Carl Diener (1862–1928) of Vienna, by whom the mineral was found at Radstadt, Salzburg. Described by O. Hackl, 1921 [*M.A.*, **2**, 54]. [*A.M.*, **12**, 46.] 11th List

Diestit. *Zeits. Kryst. Min.*, 1901, vol. xxxiv, p. 720; *Chem. Central-Blatt*, 1901, Jahrg. 72, vol. ii, p. 828. The name as vandiestite (q.v.). 3rd List

Dietzeite. A Osann, *Zeits. Kryst. Min.*, XXIII. 588, 1894. To replace Dietze's name jodchromate (*ibid.*, XIX. 449, 1891). $7Ca(IO_3)_2.8CaCrO_4$. Monosymmetric. [Esp. caliche deposit 'Pampa del Pique III', Oficina Lautaro, Atacama dist.] Chili. 1st List [$Ca_2CrO_4(IO_3)_2$]

Diocroma. A. Scacchi (unpublished note read at Accad. Sci. Naples, April 27, 1841). F. Zambonini, *Mineralogia Vesuviana. Mem. R. Accad. Sci. Fis. Mat. Napoli*, 1910, ser. 2, vol. xiv, no. 7, pp. 270, 271 (diocroma). A name applied to the small crystals of zircon from Monte Somma which change in colour, when heated, from sky-blue to orange-yellow. 6th List

Diopside-jadeite. H. S. Washington, 1922. *Proc. US Nat. Museum*, vol. 60, art. 14, pp. 6, 9. A pyroxene (jade from Mexico) intermediate between jadeite and diopside in composition. The jadeite from Burma, being jadeite proper, is distinguished as soda-jadeite (q.v.). [*M.A.*, **1**, 382.] 9th List [Renamed Tuxtlite (q.v.)]

Dioptasite. Variant of Dioptase; *A.M.*, **8**, 51. 10th List

Dipingite, error for dypingite (q.v.). [*Bull.*, **94**, 571.] 27th List

Dipyrite. A. N. Winchell, 1924. *Amer. Min.*, vol. 9, p. 110. The well-known mineral name Dipyre disguised and confused with pyrite by adding *-ite*. Not the Dipyrite of T. A. Readwin, 1867 (a synonym of pyrrhotine). 10th List

Discachatae. M. F. Heddle, *Trans. Geol. Soc. Glasgow*, 1900 [1901], vol. xi, pp. 167–169; Heddle, *Mineralogy of Scotland*, 1901, vol. i, pp. 70–72 (Discachatæ). This and other names (haemachatae, haema-ovoid-agates, oonachatae) are proposed for structural varieties of agate, defined by the enclosure of discoidal or ovoidal patches of cacholong, etc., sometimes of a red colour. 3rd List

Disth-sillimanite. A. Leyreloup, 1974. *Contr. Min. Petr.*, **46**, 17. A supposed polymorph of Al_2SiO_5 intermediate between kyanite and sillimanite. [*M.A.*, 75-659.] 29th List

Distrene. A trade-name for a variety of polystyrene. R. Webster, *Gems*, 1962, p. 755. Not to be confused with disthene (syn. of kyanite). 28th List

Ditroite. A variant of ditröyte, a syn. of sodalite (and of nepheline). R. Webster, *Gems*, 1962, p. 755. 28th List

Dixenite. G. Flink, 1920. *Geol. För. Förh. Stockholm*, vol. 42, p. 436 (Dixenit). Arsenite and silicate of manganese, $(MnOH)_2Mn_3SiO_3(AsO_3)_2$, occurring as rhombohedral scales in haematite at Långban, Sweden. Named from δίς, twice, and ξένος, a stranger, in allusion to the unusual association of silica and arsenic trioxide. [*M.A.*, **1**, 149.] 9th List

Dixeyite. V. Marmo, 1959. *Schweiz. Min. Petr. Mitt.*, vol. 39, p. 125 (Dixeyit). Isotropic grains in an amphibolite from Belihun, Kangeri hills, Sierra Leone, are believed to be a cubic hydrated aluminium silicate, near $Al_2O_3.4–5SiO_2.3–4H_2O$. Named after Dr. F. Dixey. 22nd List

Djalindit, erratic German transliteration of Джалиндит, Dzhalindite (23rd List) (H. Strunz, *Min. Tabellen*, 4th edit., 1966, p. 200). 24th List

Djalmaite. C. P. Guimarães, 1939. *Ann. Acad. Brasil. Sci.*, vol. 11, p. 347 (DJALMAITA), p. 350 (DJALMAÍTA). *Amer. Min.*, 1941, vol. 26, p. 343 (Djalmaite). Hydrous tantalate of uranium, as yellow octahedra from Brazil. The tantalum analogue of betafite. Named after Dr. Djalma Guimarães, of the Geological and Mineralogical Survey of Brazil. [*M.A.*, **8** , 2, 100.] 16th List [Is Microlite; *A.M.*, **25**, 440; **26**, 343. Re-named uranmicrolite; *A.M.*, **62**, 406]

Djerfisherite. L. H. Fuchs, 1966. *Science*, vol. 153, p. 166. $K_3(Cu,Na)(Fe,Ni)_{12}S_{14}$, cubic, occurring in the Kota-Kota and St. Mark's meteorites; identified with the 'Mineral C' of Ramdohr (1963; *M.A.*, **17**, 57). Named for D. J. Fisher. [Зап., **97**, 67.] 25th List [*A.M.*, **51**, 1815]

Djeskasganite, aberrant transliteration of Джезказганит (Dzhezkazganite). Зап. Всесоюз. Мин. Общ. (*Mem. All-Union Min. Soc.*), 1963, vol. 92, p. 566. 23rd List

Djevalite. Trade name for cubic ZrO_2, stabilized by CaO(?), produced by the Djevahirdjian Company, Switzerland, as a gem diamond substitute. Cf. phianite (30th List). (K. Nassau, *Lapidary J.*, 1977, p. 904). 30th List

Djouloukoulite, French transliteration of Джулукулит, dzhulukulite (q.v.), (*Bull. Soc. franç. Min. Crist.*, 1958, vol. 81, p. 334). 22nd List

Djurleïte. E. H. Roseboom Jr. and N. Morimoto, 1962. *Amer. Min.*, vol. 47, p. 1181, and *Min. Journ.* (*Japan*), vol. 3, p. 338. The synthetic polymorph of $Cu_{2-x}S$ described by S. Djurle (1958) and named $Cu_{1.96}S$–III has been found occurring naturally at a number of localities, and is named for Djurle. [*M.A.*, **16**, 180, 282.] 23rd List [*A.M.*, **48**, 215]

Dnieprovskite. P. I. Skornyakov, 1944. Reported by M. N. Ionov, [Труды Всесоюз. Магадан. Н.-И. Инст. (*Trans. All-Union Magadan Sci. Res. Inst.*), 1957, no. 19, p. 9]; abstr. Зап. Всесоюз. Мин. Общ. (*Mem. All-Union Min. Soc.*), 1959, vol. 88, p. 311 (Днепровскит, dneprovskite). A wholly unnecessary name given to wood-tin of a radial structure. Named from the Dniepropetrovsk deposit. [*M.A.*, **14**, 278.] 22nd List

Doelterite. A. Lacroix, 1913. *Nouv. Arch. Muséum Hist. Nat. Paris*, ser. 5, vol. v, p. 334 (dœltérites). The titanium dioxide commonly shown in analyses of laterites is presumably present in a hydrated colloidal form: when soluble in hydrochloric acid it is perhaps the ortho-titanic acid H_4TiO_4, or, when soluble only in sulphuric acid, the meta-titanic acid H_2TiO_3. This constituent of laterite is not visibly recognizable even under the microscope. Named after Professor Cornelius August Doelter y Cisterich, of Vienna. Cf. Paredrite. 8th List

Doloresite. T. W. Stern, L. R. Stieff, H. T. Evans, and A. M. Sherwood, 1957. *Amer. Min.*, vol. 42, p. 587. H. T. Evans and M. E. Mrose, *Acta Cryst.*, 1958, vol. II, p. 57. Hydrous vanadium oxide, $3V_2O_4.4H_2O$, monoclinic (pseudo-orthorhombic). Dark-brown alteration product of montroseite in sandstone from several mines on the Colorado Plateau. Named from the Dolores river, Colorado. 21st List [*A.M.*, **45**, 1144]

Donathite. E. Seeliger and A. Mücke, 1969. *Neues Jahrb. Min.*, *Monatsh.*, 49. A 'chromite' from Hestmandö Island, Norway, is distinctly anisotropic and splitting of X-ray diffractions shows that it is tetragonal; composition near $Fe^{2+}_{0.8}Mg_{0.2}Cr_{1.3}Fe^{3+}_{0.7}O_4$, with a little Zn. Named for M. Donath, who described the material in 1930. [*A.M.*, **54**, 1218; *M.A.*, 70-2615; *Zap.*, **99**, 75.] 26th List [*M.M.*, **41**, 351]

Donbassite. E. K. Lazarenko, 1940. *Compt. Rend.* (*Doklady*) *Acad. Sci. URSS*, vol. 28, p. 519 (donbassites). A group of hydrous alumino-silicates, $H_{14}Al_8Si_5O_{29}$, etc., with small amounts of Fe, Mg, Ca, Na, closely resembling pyrophyllite. Previously described as α-chloritite (7th List). Named from the locality, Donetz basin, Ukraine. [*M.A.*, **8**, 53.] 16th List

Dosulite. T. Yoshimura, 1967. [*Sci. Rept. Fac. Sci. Kyushu Univ.*, ser. D, vol. 9, special issue no.1], abstr. *Min. Abstr.*, 69-1096. An undesirable name for chocolate-coloured isotropic manganese oxide ore. 25th List

Doughtyite. W. P. Headden, 1905. *Proc. Colorado Sci. Soc.*, vol. viii, p. 66. Hydrated basic aluminium sulphate, $Al_2(SO_4)_3.5Al_2(OH)_6.21H_2O$, formed abundantly as a white precipitate by the interaction of the waters (an alkaline water and one containing aluminium sulphate) of the Doughty Springs, in Delta Co., Colorado. Named after Mr. Doughty, the owner of the springs. 4th List

Doverite. W. L. Smith, J. Stone, D. D. Riska, and H. Levine, 1955. *Science* (*Amer. Assoc. Adv. Sci.*), vol. 122, p. 31. Fluo-carbonate of yttrium and calcium, $YtFCO_3.CaCO_3$, as fine-grained aggregates giving an X-ray pattern similar to that of synchysite. From an iron mine at Dover, Morris Co., New Jersey. Named from the locality. [*M.A.*, **13**, 7.] 21st List [Synchysite-(Y); *A.M.*, **51**, 154]

Downeyite. R. B. Finkelman and M. E. Mrose, 1977. *A.M.*, **62**, 316. Gases escaping through vents in burning culm banks near Forestville, Pennsylvania, deposit hygroscopic crystals of SeO_2. Tetragonal, *a* 8·36, *c* 5·07 Å, uniaxial +. Previous supposed natural occurrences ('selenolite') cannot have been SeO_2, which deliquesces in a few minutes when exposed to air. Named for W. F. Downey. [*M.A.*, 78-885.] 30th List

Dresserite. J. L. Jambor, D. G. Fong, and Ann P. Sabina, 1970. *Canad. Min.*, **10**, 84. $Ba_2Al_4(CO_3)_4(OH)_8.3H_2O$, the barium analogue of dundasite, occurs as spherical aggregates of orthorhombic fibres at St.-Michel, Montreal Island, Quebec, in cavities in an alkalic sill. Named for J. A. Dresser. 26th List [*A.M.*, **55**, 1447. Cf. Hydrodresserite, Strontiodresserite]

Drewite. (R. M. Field, *Carnegie Inst. Washington, 1920, Year book no. 18* (*for 1919*), p. 197; E. M. Kindle, *Pan-Amer. Geol.*, 1923, vol. 39, pp. 368, 369). A form of calcium carbonate, evidently identical with calcite, formed by bacterial precipitation from sea-water, as described by George Harold Drew (1881–1913) in *Journ. Marine Biol. Assoc. Plymouth*, 1911, n. ser., vol. 9, p. 142. The name dates back to 1911, but I have been unable to find the earliest bibliographical reference. 11th List

Droogsmansite. H. Buttgenbach, 1925. *Ann. Soc. Géol. Belgique*, vol. 48, Bull. p. B 219; *Mém. Soc. Roy. Sci. Liège*, 1925, ser. 3, vol. 13, no. 5, p. 72. An undetermined uranium mineral forming small orange-yellow globules, from Kasolo, Katanga. Named after Mr. Hubert Droogmans, President of the Comité Spécial of Katanga. [*M.A.*, **3**, 6.] 11th List

Drysdallite. F. Čech, M. Rieder, and S. Vrána, 1973. *Neues Jahrb. Min.*, Monatsh. 433. Soft black polycrystalline masses from the uranium deposit of Kapijimpanga, Solwezi, Zambia, are identical with synthetic $MoSe_2$. Hexagonal, *a* 3·287, *c* 12·929 Å. Named for A. R. Drysdall. 28th List [*A.M.*, **59**, 1139]

Dschalindit, standard German transliteration of Джалиндит, Dzhalindite (23rd List) (H. Strunz, *Min. Tabellen*, 4th edit., 1966, p. 466). 24th List

Dscheskasganit, German transliteration of Джезказганит, Dzhezkazganite (23rd List) (H. Strunz, *Min. Tabellen*, 4th edit., 1966, p. 466). 24th List

Dschulukulit, German transliteration of Джулукулит, Dzhulukulite (22nd List) (H. Strunz, *Min. Tabellen*, 4th edit., 1966, p. 466). 24th List

Dubuissonite. C. Baret, 1904. *Bull. Soc. Sci. Nat. de l'ouest de la France*, Nantes, ser. 2, vol. iv, p. 141. A pink clay from near Nantes, analysed by A. Damour in 1885 and placed by A. Lacroix (*Minéralogie de la France*, 1895, vol. i, p. 483) under montmorillonite; it differs from this, however, in its resistance to acids and in its fusibility. Named after Mr. Dubuisson, mineralogist and founder of the museum at Nantes. 4th List

Dufreniberaunite. E. T. Wherry, 1914. *Proc. United States National Museum*, vol. xlvii, p. 509. A hydrated ferric (and manganic) phosphate from Hellertown, Pennsylvania, intermediate in composition between dufrenite and beraunite, and possibly a mixture of these species, but referred by the author to the species beraunite. 7th List

Duftite. O. Pufahl, 1920. *Centralblatt Min.*, p. 289 (Duftit). Basic copper-lead arsenate, $2Pb_3(AsO_4)_2.Cu_3(AsO_4)_2.4Cu(OH)_2$, occurring as a pale olive-green, crystalline encrustation on chessylite from Tsumeb, SW Africa. Named after G. Duft, director of the Otavi mines. [Near bayldonite.] Compare Parabayldonite. [*M.A.*, **1**, 150.] 9th List [$PbCuAsO_4(OH)$; the preferred name for α-Duftite (next entry); *A.M.*, **42**, 123]

α-Duftite, β-Duftite. C. Guillemin, 1956. *Bull. Soc. franç. Min. Crist.*, vol. 79, p. 70. Duftite is divided into two species, α-Duftite, space-group *Pnma* (D_{2h}^{16}), isomorphous with mottramite, and β-Duftite, space-group $P2_12_12_1(D_2^4)$, isomorphous with conichalcite. The material from Cap Garonne, Var, is α-Duftite, while that from Mapimi, Mexico, from Brandy Gill,

Cumberland, and from St. Nicholas, Ter-de-Belfort, is β-Duftite. The β-Duftite from Mapimi, studied in detail, contains 3% CaO and has lower refractive indices and density and slightly different cell-dimensions from lime-free α-Duftite from Tsumeb. 22nd List

Dumontite. A. Schoep, 1924. *Compt. Rend. Acad. Sci. Paris*, vol. 179, p. 693. Hydrated phosphate of uranium and lead, $2PbO.3UO_3.P_2O_5.5H_2O$, as ochre-yellow, orthorhombic crystals from Katanga. Named after the Belgian geologist André Hubert Dumont (1809–1857). [*M.A.*, **2**, 383.] 10th List [*See A.M.*, **41**, 915]

Dundasite. W. F. Petterd, *Catalogue of Minerals of Tasmania*, 1893, p. 26; and *Papers and Proc. Roy. Soc. Tasmania*, for 1893, 26, 1894. 'Hydrous carbono-phosphate of lead and alumina.' [Adelaide Proprietory mine, Dundas, Tasmania]. 1st List [Orthorhombic $PbAl_2(CO_3)_2(OH)_4.H_2O$; *M.M.*, **38**, 564]

Dunhamite. E. E. Fairbanks, 1946. *Econ. Geol.*, vol. 41, p. 767. A pale brown alteration product, perhaps $PbO.TeO_2$, of altaite from Organ Mts., New Mexico. Named after Dr. Kingsley Charles Dunham (1910–) of the Geological Survey of Great Britain, who first noticed it (*Bull. New Mexico School of Mines*, 1935, no. 11, p. 158). *See* microdunhamite. [*M.A.*, **10**, 100.] 18th List

Duparcite. S. E. Nicolet, 1932. *Schweiz. Min. Petr. Mitt.*, vol. 12, p. 543 (Duparcite). E. Brandenberger, ibid., p. 545 (Duparcit). Radiating groups of long prisms in hornstone from Morocco with characters very near those of idocrase and genevite (11th List). Named after Professor Louis Duparc (1866–1932), of Genève. [*M.A.*, **5**, 292.] 13th List [Is Genevite; *M.A.*, **5**, 292.] [var. of Idocrase; *M.A.*, **8**, 251]

Duplexite. H. P. Rowledge and J. D. Hayton, 1948. *Journ. Roy. Soc. W. Australia*, vol. 33 (for 1946–1947), p. 49. Hydrous silicate, $6CaO.4BeO.Al_2O_3.14SiO_2.2H_2O$, as white rosettes with beryl in pegmatite from Londonderry, Western Australia. Differing from bavenite in containing rather more BeO (7%). Named after S. Duplex, manager of the felspar quarry, who found the material. [*M.A.*, **10**, 508.] 18th List [Is Bavenite; *A.M.*, **38**, 988]

Durain. M. C. Stopes, 1919. *See* Fusain. 8th List

Duranusite. Z. Johan, C. Laforêt, P. Picot, and J. Feraud, 1973. *Bull.*, **96**, 131. Minute grains with arsenic and realgar at Duranus, Alpes Maritimes, France, are orthorhombic, *a* 3·576, *b* 6·759, *c* 10·074 Å; composition $2[As_4S]$. Named from the locality. 28th List [*A.M.*, **60**, 945]

Dussertite. J. Barthoux, 1925. *Compt. Rend. Acad. Sci. Paris*, vol. 180, p. 299. Hydrated arsenate of ferric iron and calcium $Ca_3Fe_3(OH)_9(AsO_4)_2$, as minute hexagonal crystals and green crystalline crusts from Algeria. [Not proved to differ from arseniosiderite.] Named after Mr. — Dussert. [*M.A.*, **2**, 419.] 10th List [$BaFe_3(AsO_4)_2(OH)_5$; *A.M.*, **28**, 63]

Dutoitspanite. A. F. Williams, 1932. *The genesis of the diamond*, London, vol. 1, pp. 171, 172. Synonym of bultfonteinite (q.v.). Named from the Dutoitspan diamond mine, Kimberley, South Africa, one of the localities of the mineral. [*M.A.*, **5**, 97.] 13th List

Duttonite. M. E. Thompson, C. H. Roach, and R. Meyrowitz, 1956. *Science* (*Amer. Assoc. Adv. Sci.*), vol. 123, p. 990; *Amer. Min.*, 1957, vol. 42, p. 455. H. T. Evans and M. E. Mrose, *Bull. Geol. Soc. Amer.*, 1956, vol. 67, p. 1693 (abstract); *Acta Cryst.*, 1958, vol. 11, p. 58. Vanadium hydroxide, $VO(OH)_2$ or $V_2O_4.2H_2O$, monoclinic (pseudo-orthorhombic), as minute pale-brown scales; an alteration product of montroseite in sandstone from Colorado. Named after Clarence Edward Dutton (1841–1912), an early worker on the Geological Survey in Colorado. [*M.A.*, **13**, 378, 379, 524.] 21st List

Dypingite. G. Raade, 1970. *Amer. Min.*, **55**, 1457. White globular aggregates from the Dypingdal serpentine–magnesite deposit, Snarum, Norway, have α 1·508, β 1·510, γ 1·516. Composition $Mg_5(CO_3)_4(OH)_2.5H_2O$. Named for the locality. [*M.A.*, 71-1384; *Bull.*, **94**, 571; *Zap.*, **100**, 88.] 27th List

Dysanalite. Variant of Dysanalyte; *A.M.*, **8**, 51. 10th List

Dysklaukite*. J. Fröbel, 1843. *Grundzüge eines Systemes der Krystallologie*. A cobaltian arsenian ullmannite from Eisern and Frensburg, Westphalia. [*Corr. in *M.M.*, **39**.] 27th List

Dyslaukite. Error for dysklaukite. 27th List

Dzhalindite. A. D. Genkin and I. V. Muraveva, 1963. Зап. Всесоюз. Мин. Общ. (*Mem. All-Union Min. Soc.*), vol. 92, p. 445. A yellow-brown alteration product of Indite (q.v.) at the Dzhalind ore deposit, Little Khingan ridge, Far Eastern Siberia, is identified as $In(OH)_3$, agreeing closely with the synthetic product. Named from the locality (*Amer. Min.*, 1963, vol. 49, p. 439; *M.A.*, **16**, 457). 23rd List

Dzhezkazganite. E. M. Poplavko, I. D. Merchukova, and C. S. Zak, 1962. Докл. Акад. Наук СССР (*compt. Rend. Acad. Sci. URSS*), vol. 146, p. 433 (Джезказганит). An incompletely described mineral from the Dzhezkazgan copper ores, Kazakhstan, containing 40 to 50% Re and 15 to 20% Cu, probably an alloy or a sulphide; appears to be amorphous to X-rays. Named from the locality. (*Amer. Min.*, 1963, vol. 48, p. 209; *M.A.*, **16**, 180; **16**, 618). 23rd List [Composition and status uncertain; *see* refs. in *M.M.*, **35**, 871]

Dzhulukulite. N. N. Shishkin, 1958. Доклады Акад. Наук СССР (*Compt. Rend. Acad. Sci. URSS*), vol. 121, p. 724 (Джулукулит). Superfluous name for a nickelian cobaltite from Lake Dzhulu-kul, SW Tuva; named from the locality. [*M.A.*, **14**, 140.] 22nd List

E

Eakerite. P. B. Leavens, J. S. White, Jr., and M. H. Hey, 1970. *Min. Record*, **1**, 92. Prismatic crystals in spodumene pegmatite at King's Mt., North Carolina, are monoclinic, *a* 15·829, *b* 7·721, *c* 7·438 Å, *β* 101° 9′, $P2_1/a$; D 2·93; α 1·584 ∥ [010], *β* 1·586, *γ* 1·600; $2[Ca_2SnAl_2Si_6O_{16}(OH)_6]$. Named for J. Eaker. [*A.M.*, **56**, 637; *M.A.*, 71-2330.] 27th List

Eakleite. E. S. Larsen, 1917. *Amer. Journ. Sci.*, ser. 4, vol. xliii, p. 464. A pale pink, fibrous mineral from California resembling pectolite in appearance and perhaps a calcium pectolite, $5CaO.5SiO_2.H_2O$. Named after Professor Arthur Starr Eakle, of the University of California. 8th List [Is xonotlite; *M.A.*, **2**, 253]

Eardleyite*. Author? *Filer's C 15 Mineral Catalog*, Redlands, California, 1961. A dull blue-green massive mineral from Utah is believed to be $(Ni,Zn)_6(Al,Fe)_2CO_3(OH)_{16}.4H_2O$, the nickel-zinc analogue of hydrotalcite. [*Corr. in *M.M.*, **33**.] 22nd List [*A.M.*, **47**, 807]

Eardlyite. Error for eardleyite. 22nd List

Earlandite. F. A. Bannister, 1936. *Discovery Reports*, Cambridge, vol. 13, p. 67. Hydrated calcium citrate, $Ca_3(C_6H_5O_7)_2.4H_2O$, as small warty nodules in deep-sea deposits from the Weddell Sea, Antarctic. Named after Mr. Arthur Earland, of London, who isolated the material. [*M.A.*, **6**, 341.] 14th List [*A.M.*, **21**, 71]

Eastonite. (*The Mineral Collector*, New York, 1899, VI, 118.) A silver-white mica similar in composition to a vermiculite. Easton, Pennsylvania. 2nd List

—— A. N. Winchell, 1925. *Amer. Journ. Sci.*, ser. 5, vol. 9, p. 323. The theoretical 'magnesium di-alumina mica' with the composition $H_4K_2Mg_5Al_4Si_5O_{24}$. A biotite from Easton, Pennsylvania, approximates to this formula (analysis by J. Eyerman, 1904). This name has been earlier applied (1899; 2nd List. S. G. Gordon, *Mineralogy of Pennsylvania*, 1922, pp. 109, 119) to a silver-white vermiculite occurring as an alteration-product of biotite at this locality. 10th List

Ebelmenite. (*Collection de Minéralogie du Muséum d'Histoire Naturelle, Paris, Guide du Visiteur*, 2nd edit., 1900, p. 29.) A variety of psilomelane containing potassium. Named after Jacques Joseph Ebelmen (1814–1852), who analysed a specimen in 1841. 4th List [Cf. Cryptomelane]

Echellite. N. L. Bowen, 1920. *American Mineralogist*, vol. 5, p. 1. A zeolitic mineral occurring as small, white, spheroidal masses of radiating fibres in a basic igneous rock from northern Ontario. The optical characters point to orthorhombic symmetry. $(Ca,Na_2)O.2Al_2O_3.3SiO_2.4H_2O$. Named from the French échelle, ladder, in allusion to the stepped ratios, 1, 2, 3, 4, in the chemical formula. [*M.A.*, **1**, 25.] 9th List [Is Thomsonite; *A.M.*, **18**, 31]

Eckermannite. O. J. Adamson, 1942. *Geol. För. Förh. Stockholm*, vol. 64, p. 329; ibid., 1944, vol. 66, p. 194. An alkali-amphibole containing Na_2O 11·30, K_2O 2·41%, $Na_4Mg_2AlFe'''(Si_4O_{11})_2(O,OH,F)_2$, from Norra Kärr, Sweden. Named after Professor Claes Walther Harry von Eckermann (1886–) of Stockholm. [*M.A.*, **9**, 87.] 17th List

Eckrite. J. Ravier, 1951. *Bull. Soc. Franç. Min. Crist.*, vol. 74, p. 10. A variety of soda-amphibole near arfvedsonite in its optical characters but near glaucophane in chemical composition. Named from the locality, Eqe (pronounced éckré), west Greenland. [*M.A.*, **11**, 365.] 19th List

Ectropite. *See* Ektropite. 8th List

Edgarite. R. C. Morris, 1962. *Amer. Min.*, vol. 47, p. 1079. A provisional name for material subsequently shown to be Osarizawaite (q.v.). 23rd List

Edinite. An objectionable *syn.* (author and date?) of prase. R. Webster, *Gems*, 1962, p. 755. Not to be confused with edenite. 28th List

Efremovite. G. Gagarin and J. R. Cuomo, 1949, loc. cit., p. 14 (efremovita). Calcium ferri-phosphate кальциевый ферри-фосфат, $2CaO.3Fe_2O_3.2P_2O_5.10H_2O$ + 16aq., of N. E. Efremov H. E. Ефремов, 1936, from Taman peninsula, Kuban. [*M.A.*, 6, 439.] 19th List

Eglestonite. A. J. Moses, *Amer. Journ. Sci.*, 1903, ser. 4, vol. xvi, p. 253. Cubic oxychloride of mercury, $Hg_6Cl_3O_2$. The brownish-yellow colour of the crystals quickly changes to black on exposure to light. Named after the late Professor Thomas Egleston, of Columbia University, New York. *See* terlinguaite. 3rd List [16[$Hg_6Cl_3O_2H$]; *A.M.*, **62**, 396]

Egueiite. A. Lacroix, 1910. *Minéralogie de la France*, vol. iv, p. 536 (eguéïite). G. Garde, *Description géologique des régions situées entre le Niger et le Tchad*, Paris, 1911, p. 263 (Egueïite), p. 264 (égueïite). Described without name by G. Garde, *Compt. Rend. Acad. Sci. Paris*, 1909, vol. cxlviii, p. 1618. A hydrated basic phosphate of ferric iron with a little calcium and aluminium, occurring as small nodules, with a fibro-lamellar internal structure, embedded in clay in Eguei, Sudan. It is allied to borickite and foucherite (q.v.), and is probably an altered vivianite. 6th List

Egyrinaugite. A. N. Winchell, 1923. *Amer. Journ. Sci.*, ser. 5, vol. 6, pp. 518, 519. An incorrect form of Aegirine-augite [from Ægir]. 10th List

Ehrenwerthite. F. Cornu, 1909. *Zeits. prakt. Geol.*, Jahrg. xvii, p. 82 (Ehrenwerthit). Colloidal hydroxide of iron with the same composition ($Fe_2O_3.H_2O$) as the crystalline mineral goethite; occurring as pseudomorphs after iron-pyrites. Named after Professor Josef von Ehrenwerth, of Leoben, Styria. 5th List

Eichbergite. O. Grosspietsch, 1911. *Centralblatt Min.*, 1911, p. 433 (Eichbergit). A massive iron-grey mineral, $(Cu,Fe)_2S.3(Bi,Sb)_2S_3$, from the magnesite deposits of Eichberg, Semmering Pass, Austrian Alps. 6th List

Eicosyl alcohol. C. J. Kelly, 1970. *A.M.*, **55**, 2118. The fluorescent substance present in the red calcite of Deutsch-Altenburg, Austria, is eicosyl alcohol, $C_{20}H_{41}OH$. [*M.A.*, 71-2301.] 28th List

Eicotourmaline. W. N. Lodochnikov, 1933. *Problems of Soviet Geology*, vol. 2, p. 132 (эйкотурмалин), p. 148 (eicotourmaline). Like (εἰκός) tourmaline, but optically biaxial and containing no boron. 13th List

Eisen-Åkermanit. K. Hofmann-Degen, *Sitzungsber. Heidelberg. Akad. Wiss. Math.-naturw. Kl.*, 1919, Abt. A, Abh. 14, p. 39; abstr. in *Zeits. Krist.*, 1922, vol. 57, p. 105. A crystallized slag approximating to the composition $2CaO.FeO.2SiO_2$, corresponding with åkermanite ($2CaO.MgO.2SiO_2$). 9th List [*See* Iron-åkermanite]

Eisenalabandin. P. Ramdohr. 1957. *Neues Jahrb. Min.*, vol. 91, p. 89 (Eisenalabandin), p. 90 (Fe-alabandin). Alabandine containing much FeS in solid solution, (Mn,Fe)S. As minute grains with native iron and troilite in basalt from Bühl, Kassel. Abstr. in *Amer. Min.*, **43**, 378 (Iron-alabandite). Ferroalabandine more suitable for international use. 21st List

Eisenandradit. *See* Iron-andradite. 11th List

Eisenanthophyllit. *See* Iron-anthophyllite. 8th List

Eisenbeidellit. *See* Iron-beidellite. 11th List

Eisen-Berlinit. H. Strunz, 1966. *Min. Tabellen*, 4th edit., p. 467. Translation of Iron berlinite (20th List). 24th List

Eisenchrysotil. F. Machatschki, 1953. *Spezielle Mineralogie*, Wien, p. 350. Given as a synonym of 'Greenolith' (error for greenalite, 4th List). Not the same as ferro-chrysotile (14th List) and iron-serpentine (15th List). 20th List [Also 22nd List]

Eisencordierit. H. Bücking, *Ber. Senckenberg. Naturf. Ges. Frankfurt-am-Main*, 1900, p. 15. Cordierite with magnesia largely replaced by ferrous iron, as is the case in the cordierite of the 'vitrified' sandstones altered by contact with basalt in central Germany. 3rd List

Eisenenstatit. F. Machatschki, 1953. *Spez. Min.*, Wien, p. 313. Syn. of Hypersthene. 22nd List

Eisenepidot. M. Goldschlag, 1915. *Anzeiger Akad. Wiss. Wien*, Jahrg. 52, p. 270. *See* Aluminium-epidote. [*M.A.*, 1, 346.] 10th List

Eisengedrite. C. Doelter, 1913. *See* Iron-gedrite. 9th List

Eisenglaukonit. H. Borchert and H. Braun, 1963. *Chem. Erde*, vol. 23, p. 82. An unnecessary name for highly ferrian glauconite. 25th List

Eisenhypersthen. M. Saxén, 1925. *See* Iron-hypersthene. 14th List

Eisen-magnesium-retgersite. K. F. Chudoba, 1960. Hintze, *Handb. Min.*, Erg.-Bd. II, p. 928 (Eisen-magnesium-retgersit). The ferroan magnesian retgersite of E. N. Eliseev and S. I. Smirnova (Зап. Всесоюз. Мин. Общ. [*Mem. All-Union Min. Soc.*], 1958, vol. 87, p. 3). 22nd List

Eisenmonticellit. *See* Kalkeisenolivin. 11th List

Eisenpickeringit. H. Meixner and W. Pillewizer, 1937. *Zentr. Min.*, Abt. A, 1937, p. 266 (Eisenpickingerit [*sic*]). Pickeringite containing some iron replacing magnesium. Synonym of Ferropickeringite (q.v.). *See* Mangan-pickeringite. [*M.A.*, **7**, 12.] 15th List

Eisenpyrochroit. G. Flink, 1919. *Geol. För. Förh. Stockholm*, vol. 41, p. 436. An iron-bearing variety of pyrochroite from Långban, Sweden, differing from the ordinary type in the acicular habit of its crystals and in the mode of alteration. [*M.A.*, **1**, 124.] 9th List

Eisenrichterit. M. Belowsky, 1905. *Zeits. Deutsch. Geol. Gesell.*, vol. 57, p. 33. A fibrous blue amphibole from Greenland, allied to richterite ('astochite') with iron in place of manganese. (The composition was calculated from a rock analysis after deducting 76% of felspar.) *See* Iron-richterite, Ferririchterite. 20th List

Eisenrömerit. *See* ferrorömerite. 3rd List

Eisen (III)-Spinelle. H. Strunz, 1970. *Min. Tab.* 5th edit., 177. A group name for the ferric iron spinels. 28th List

Eisenstrigovit. *Neues Jahrb. Min.*, Ref. I, 1936, p. 467. German translation of Iron-strigovite (q.v.). 14th List

Eisenwagnerite, translation of Iron-wagnerite (23rd List) (C. Hintze, *Handb. Min.*, Erg. III, p. 545). 25th List

Eisenwolframit. F. Machatschki, 1953. *Spez. Min.*, Wien, p. 314. Syn. of Ferberite. 22nd List

Eitelite. C. Milton, J. M. Axelrod, and F. S. Grimaldi, 1954. *Bull. Geol. Soc. Amer.* [*1955*], vol. 65, no. 12 (for 1954), p. 1286 (abstr.). $Na_2O.MgO.2CO_3$ hexagonal, from an oil well in Utah. Named after Wilhelm (Hermann Julius) Eitel (1891–), director of the Institute for Silicate Research in the University of Toledo, Ohio. [*M.A.*, **12**, 511.] 20th List [*A.M.*, **40**, 326]

Ekanite. B. W. Anderson, G. F. Claringbull, R. J. Davis, and D. K. Hill, 1961. *Nature*, vol. 190, p. 997. $(Th,U)(Ca,Fe,Pb)_2Si_8O_{20}$; metamict, recrystallizing on heating to a tetragonal phase. From Eheliyagoda, Ratnapura district, Ceylon. Named after the discover, Mr. F. L. D. Ekanayake. 22nd List [Formula and status in doubt; *M.A.*, **19**, 177; 69-1044; 73-1298; 73-2802. Cf. Iraqite]

Ektropite. G. Flink, 1917. *Geol. För. Förh.*, vol. xxxix, p. 426 (Ektropit). Abstr. in *Amer. Journ. Sci.*, 1917, vol. xliv, p. 484, and *Amer. Min.*, 1917, vol. ii, p. 128 give the spelling Ectropite. Hydrated silicate of manganese, with some magnesium, ferrous iron, and calcium, $12RO.8SiO_2.7H_2O$, forming small, brown, monoclinic crystals of tabular rectangular habit. Occurs with garnet in magnetite in the Norrbotten iron mine at Långban, Sweden. Named from ἐκτροπή, a turning aside, an evasion, on account of the difficulty experienced in determining the characters. 8th List [Is Caryopilite; *A.M.*, **49**, 446]

Elatolite. A. E. Fersman, 1922. *Compt. Rend. Acad. Sci. Russie*, p. 59; *Bull. Acad. Sci. Russie*, 1923, ser. 6, vol. 17, p. 251; *Trans. Northern Sci. Econ. Exped.*, no. 16 (*Sci. Techn. Dept. Supreme Council of National Economy, no. 8*), Moscow and Petrograd, 1923, pp. 16, 32 (элатолит). Primary (magmatic) calcium carbonate corresponding to the α-calcium carbonate (stable above 970°C.) of H. E. Boeke [This has, however, not been confirmed, *Min. Abstr.*, vol. 2, p. 218]. Named from ἐλάτη, fir, and λίθος, stone, on account of the dendritic form of the skeletal crystals. [*M.A.*, **2**, 262–264, 381.] 10th List

Elbaite. V. I. Vernadsky, 1913. *Zeits. Kryst. Min.*, vol. liii, p. 283 (Elbait). The chemical composition of the tourmalines is expressed by the mixing of three hypothetical molecules (additive derivatives of the kaolin ring), namely $A = M'_8Al_4B_2Si_4O_{21}$, $B = M'_{14}Al_2B_2Si_4O_{21}$, and $C = M'_2Al_6B_2Si_4O_{21}$. These hypothetical components of tourmaline are named *kalbaite*, *belbaite*, and *elbaite* respectively. The pale red tourmaline from the Island of Elba consists of nearly pure elbaite, and that from DeKalb in New York of nearly pure kalbaite; hence these names. Belbaite is apparently from elbaite with *B* prefixed. The name elbaite has previously been applied (C. L. Giesecke, 1832) to ilvaite from Elba. 7th List [Now one of the accepted tourmaline species, $Na(Li,Na)_3Al_6(BO_3)_3Si_6O_{13}(OH)_4$, with Buergerite (q.v.), Dravite, Liddicoatite (q.v.), and Schorl]

Elbrussite. I. J. Mickey, 1930. *Centr. Min.*, Abt. A, 1930, p. 300 (Elbrussit). Hydrous silicate of aluminium, ferric and ferrous iron, with some magnesia and alkalis, belonging to the nontronite-beidellite group. Named from the locality, Mt. Elbrus (Эльбрус), Caucasus. [*M.A.*, **4**, 342.] 12th List

El Doradoite. S. L. Watkins, 1912. [*Los Angeles Mining Review*, January 13, 1912]; *American Mineralogist*, 1917, vol. ii, p. 26. Trade-name for a blue variety of quartz used as a gemstone; from El Dorado Co., California. W. T. Schaller (*Mineral Resources, US Geol. Survey, for 1917, 1918*, part ii, p. 153) spells the name Eldoradoite, and states it to be iridescent quartz. 8th List

Eldoradoite. An objectionable and superfluous name (author and date?) for a blue chalcedony. R. Webster, *Gems*, 1962, p. 755. Not the (equally objectionable) El Doradoite (Watkins, 1912; 8th List), a blue (yellowish, acc. to Webster) quartz. 28th List

Elfestorpite. P. Groth, *Tab. Uebers. Min.* 4th edit., 1898, p. 172. The same as elfstorpite. 2nd List

Elfstorpite. L. J. Igelström, *Geol. För. Förh.* XV. 472, 1893, and *Zeits. Kryst. Min.* XXII. 468, 1894. Hydrated arsenate of manganese? 'Rhombic.' Sjö mine, Sweden. 1st List [Status remains doubtful]

Elizavetinskite. V. I. Mikheev, 1957. [Рентгометрический оредрелитель минералов (X-ray determinative tables for minerals), Госгеолтехиздат, Moscow, p. 409 (елизаветинскит)]; abstr. *Amer. Min.*, 1961, vol. 46, p. 767. A black powdery material in clay from the Elizavetinsk deposit, Sverdlovsk; the X-ray pattern is near that of lithiophorite, but the composition is assumed to be (Mn,Co)OOH. 'The name has no standing. Every strong line of the X-ray pattern corresponds closely to the published data for lithiophorite, a mineral known to contain appreciable amounts of cobalt and nickel' (M. Fleischer, *Amer. Min.*, 1961, vol. 46, p. 767). 22nd List

Elkonite. M. L. Tainter, G. Kulchar, and A. B. Stockton, 1940. *Journ. Amer. Pharmac. Assoc.*, vol. 29, p. 306. A colloidal clay from Elko, Nevada. [*M.A.*, **8**, 4.] 16th List

Ellestadite. D. McConnell, 1937. *Amer. Min.*, vol. 22, p. 977. An end-member of the apatite group resembling wilkeite in which P_2O_5 is replaced by SO_3 (20·69) and SiO_2 (17·31%). From Crestmore, California. Named after Dr. R. B. Ellestad, of the University of Minnesota. [*M.A.*, **7**, 14, 88, 475.] 15th List

Ellsworthite. T. L. Walker & A. L. Parsons, 1923. *Univ. Toronto Studies, Geol. Ser.*, no. 16, p. 13. Hydrated metatitanoniobate of uranium, calcium, and iron, $RO.Nb_2O_5.2H_2O$, occurring as yellow or brown, optically isotropic masses in pegmatite at Hybla, Ontario. Named after Dr. Hardy Vincent Ellsworth, of the Geological Survey of Canada. [*M.A.*, **2**, 248, 280.] 10th List [Is Betafite; *A.M.*, **46**, 1519; syn. of uranpyrochlore; *A.M.*, **62**, 406]

Ellweilerite. H. W. Bültemann, 1960. [*Der Aufschluss*, vol. 11, no. 11, p. 281]; abstr. *Amer. Min.*, 1961, vol. 46, p. 465. Thin tabular crystals with zeunerite from Bühlskopf, Ellweiler, Birkenfeld, Germany, contain Na, U, and As, and appear to be identical with Sodium uranospinite (19th List); named from the locality. The name Sodium uranospinite has priority. 22nd List

Elpidite. G. Lindström, *Geol. För. Förh.* XVI. 330, 1894; G. Nordenskiöld, *ibid.* p. 343. $Na_2O.ZrO_2.6SiO_2.3H_2O$. Rhombic. [Narsarsuk] Greenland. 1st List

Elyite. S. A. Williams, 1972. *Amer. Min.* **57**, 364. Tiny prismatic to fibrous violet monoclinic crystals, with langite, serpierite, and galena in the Caroline tunnel of the Silver King mine, Ward, Nevada. $2[Pb_4CuSO_4(OH)_8]$; *a* 14·248, *b* 5·768, *c* 7·309 Å, β 100° 26′, $P2_1/a$, D about 6; α 1·990, β 1·993 ∥ [010], γ 1·994. Named for J. Ely. [*M.A.*, 72-3339.] 27th List

Emaldine, *see* Emildine.

Embreyite. S. A. Williams, 1972. *Min. Mag.* **38**, 790. Orange to brown drusy crusts on a number of old specimens from Berezov, Siberia, carrying crocoite, phoenicochroite, and vauquelinite, prove to be a chromate and phosphate of lead with some Cu. Monoclinic, *a* 9·755, *b* 5·636, *c* 7·135 Å, β 103° 5′, D 6·45; $[Pb_5(CrO_4)_2(PO_4)_2.H_2O]$; α 2·20, β and γ 2·36, β ∥ [010]. Named for P. G. Embrey. [*M.A.*, 72-3340.] 27th List [*A.M.*, **58**, 806]

Emeleusite. B. G. J. Upton, P. G. Hill, O. Johnsen, and O. V. Petersen, 1978. *M.M.*, **42**, 31. Colourless glassy crystals from Igdlutalik, Julianehåb, Greenland, are orthorhombic, *a* 10·073, *b* 17·350, *c* 14·010 Å, space group *Acam* or *Aba*2. Trillings are common. Composition $4[Li_2Na_4Fe_2{}^3Si_{12}O_{30}]$, ρ 2·775 g.cm^{-3}. α 1·596 ∥ [010], β 1·597 ∥ [100], γ 1·597 ∥ [001]. Named for C. H. Emeleus. [*M.A.*, 78-2119] 30th List

Emeraldine. A trade-name for green dyed chalcedony. R. Webster, *Gems*, 1962, p. 755. Not to be confused with emeraldite (28th List), nor with emeraudine, a syn. of dioptase. 28th List

Emeraldite, error for Smaragdite ((Доклады Акад. Наук СССР [*Compt. Rend. Acad. Sci. URSS*], 1960, vol. 130, p. 485). 22nd List

——, Emeralite. A trade-name for a pale green tourmaline. R. Webster, *Gems*, 1962, p. 755. Not to be confused with emeraldite (of?) (22nd List), an error for smaragdite, nor with emeraldine (28th List), nor yet with emerilite (variant of emerylite, a *var.* of margarite). 28th List

Emildine. J. S. van der Lingen, 1928. *South African Journ. Sci.*, vol. 25, pp. 12, 15 (Emildine), p. 13 (Emaldine), p. 14 (Emilite). A spessartine garnet from the granites of SW Africa, found on spectroscopic analysis to contain yttrium, but no chromium and little or no magnesium (cf. erinadine). Named after the author's son Emil, the *d* in the termination being from false analogy with almandine. [*M.A.*, **4**, 83.] 12th List

Empressite. R. D. George in MS., quoted by W. M. Bradley, 1914. *Amer. Journ. Sci.*, ser. 4, vol. xxxviii, p. 163; ibid., 1915, vol. xxxix, p. 223. A telluride of silver, AgTe, found as pale bronze-coloured, finely granular masses in the Empress-Josephine mine, Kerber Creek district, Colorado.

W. T. Schaller (*Journ. Washington Acad. Sci.*, 1914, vol. iv, p. 497) suggests that this is identical with the muthmannite of F. Zambonini (1911, 6th List), being a pure muthmannite, whilst the original muthmannite should be called an auric muthmannite, (Ag,Au)Te. 7th List [*A.M.*, **49**, 325; *M.A.*, **17**, 492; 69-1497]

Enalite. K. Kimura and Y. Miyake, 1932. *Journ. Chem. Soc. Japan*, vol. 53, p. 93 (Japanese; English title on wrapper); abstr. in *Amer. Min.*, 1933, vol. 18, p. 223. Hydrous silicate of thorium and uranium, $(Th,U)O_2.nSiO_2.2H_2O$, tetragonal, a variety of uranothorite, as orange-yellow grains with monazite in sands in the Ena district, Gifu prefecture, Japan. Named from the locality. [*M.A.*, **5**, 293.] 13th List

Endeiolite. G. Flink, *Medd. om Grönland*, 1900, XXIV, 166. A member of the pyrochlore group occurring in small regular octahedra. $R''Nb_2O_6(OH)_2.R''SiO_3$; ($R''$ = Ca,Na,Zr,Ce, etc.). [Narsarsuk] S. Greenland. Cf. chalcolamprite. 2nd List [Probably impure pyrochlore; *A.M.*, **62**, 406]

Endellite. L. T. Alexander, G. T. Faust, S. B. Hendricks, H. Insley, and H. F. McMurdie, 1943. *Amer. Min.*, vol. 28, p. 1. The 'hydrated halloysite' of S. B. Hendricks [*M.A.*, **7**, 422], $H_4Al_2Si_2O_9.2H_2O$; the name halloysite being incorrectly limited to $H_4Al_2Si_2O_9$, which has been called metahalloysite [14th List, *M.A.*, **6**, 181; **7**, 96]. Named after Prof. Kurd Endell (1887–) of the Technical High School, Berlin, incorrectly stated to be one of the discoverers of the material. Compare hydrohalloysite and hydrokaolin (15th List). [*M.A.*, **8**, 342.] 16th List [*A.M.*, **40**, 1110]

Endiopside. H. H. Hess, 1941. *Amer. Min.*, vol. 26, p. 519. A contraction of enstatite-diopside (W. Wahl, 1906; 5th List) for a clinopyroxene intermediate in composition between enstatite and diopside. [*M.A.*, **8**, 234.] 16th List

Endothermite. D. S. Belyankin, 1938. *See* Monothermite. 15th List

Enelectrite. T. L. Walker, 1934. *Univ. Toronto Studies, Geol. Ser.*, no. 36, p. 11; *Amer. Min.*, 1935, vol. 20, p. 195. Minute colourless monoclinic crystals, presumably a hydrocarbon, embedded in amber (*ἤλεκτρον*) from Manitoba. [*M.A.*, **6**, 52.] 14th List

Englishite. E. S. Larsen and E. V. Shannon, 1930. *Amer. Min.*, vol. 15, p. 328. Hydrous phosphate of calcium, aluminium, and potassium, $4CaO.K_2O.4Al_2O_3.4P_2O_5.14H_2O$, as thin colourless layers in variscite nodules from Utah; probably orthorhombic, with micaceous cleavage. Named after Mr. George Letchworth English, of Rochester, New York. [*M.A.*, **4**, 344.] 12th List

Enigmatite. Variant of Aenigmatite; *A.M.*, **9**, 62. 10th List

Enstatite-augite. W. Wahl, 1906. *Die Enstatitaugite*, Helsingfors, 1906; *Min. Petr. Mitt.*

(*Tschermak*), 1907, vol. xxvi, p. 1; *Öfvers. Finska Vet. Soc. Förh.*, 1908, vol. 1, no. 2, p. 11. A group of monoclinic pyroxenes intermediate in optical properties and chemical composition between the enstatite group and the calcium-bearing monoclinic pyroxenes (diopside, augite, etc.), and yet distinct from clinoenstatite (q.v.). For the various members of the group several compound names are given, e.g. enstatite-diopside, bronzite-augite, augite-bronzite, hypersthene-hedenburgite, etc. 5th List

Enstenite. A. N. Winchell, 1923. *Amer. Journ. Sci.*, ser. 5, vol. 6, p. 506. A species name for the isomorphous series $MgSiO_3$—$FeSiO_3$ of orthorhombic pyroxenes. A contraction of enstatite-hypersthene. [*M.A.*, **2**, 220.] 10th List

Eosite. Author and date? A trade-name for a white and red quartzite rock, containing pyrite crystals. R. Webster, *Gems*, 1962, p. 173. Not the eosite (of Schrauf, 1871). Cf. leonite. 28th List

Epidesmine. V. Rosický and S. J. Thugutt, 1913. *Spraw. Tow. Nauk. Warsawa*, vol. vi, pp. 225, 231; *Centralblatt Min.*, 1913, p. 422 (Epidesmin). An orthorhombic zeolite, $3(Ca,Na_2,K_2)Al_2Si_6O_{16}.20H_2O$, dimorphous with stilbite (Germ. Desmin), occurring as minute crystals at Schwarzenberg, Saxony. The name is suggested from the similar dimorphous pair epistilbite and heulandite (Germ. Stilbit). 7th List [Is Stilbite; *A.M.*, **53**, 1066]

Epididymite. G. Flink, *Geol. För. Förh.* XV. 201, 1893; and *Zeits. Kryst. Min.* XXIII. 353, 1894. [*M.M.*, **11**, 100] $BeNaHSi_3O_8$. Rhombic. [Narsarsuk] Greenland. 1st List

Epidote-orthite. V. M. Goldschmidt, 1911. *Centralblatt Min.*, 1911, p. 5 (Epidot-Orthite); *Vid.-Selsk. Skrifter*, Kristiania, *I. Mat.-Nat. Kl., 1912, for 1911*, no. 1, p. 416 (Epidot-Orthit). T. Vogt, *Vid.-Selsk. Skrifter*, Kristiania, *I. Mat.-Nat. Kl.*, 1922, no. 1, p. 24; J. Schetelig, ibid., p. 138 (Epidotorthit). Orthite-like minerals from Norway which in their optical characters are intermediate between orthite and epidote. [*M.A.*, **2**, 25.] 10th List

Epiianthinite. A. Schoep and S. Stradiot, 1947. *Amer. Min.*, vol. 32, p. 344. Uranic hydroxide, $yUO_3.xH_2O$, as yellow orthorhombic crystals, an alteration product of ianthinite (11th List); from Belgian Congo. [*M.A.*, **10**, 145.] 18th List [Is Schoepite; *A.M.*, **44**, 1104]

Epileucite. A. N. Zavaritzky, 1934. *Compt. Rend. Acad. Sci. URSS*, 1934, vol. 3, p. 645 (эпилейцит), p. 650 (epileucite). Pseudomorphs of orthoclase and muscovite after leucite, in distinction from pseudoleucite (pseudomorphs of orthoclase and nepheline after leucite). [*M.A.*, **6**, 418.] 14th List

Epinatrolite. S. J. Thugutt, 1911. *Centralblatt Min.*, 1911, p. 408 (Epinatrolith). A less stable form of natrolite which loses its water at a lower temperature, but is otherwise identical in morphological and optical characters and chemical composition with normal natrolite. The distinction is based on the micro-chemical colouring reactions with the partially dehydrated material.

In the slightly earlier Polish publications (*Spraw. Tow. Nauk. Warszawa*, 1910, vol. iii, p. 414; 1911, vol. iv, p. 77) the author used the term *metanatrolite* for this modification; but as this name had been previously used by F. Rinne in 1890 for partially dehydrated natrolite, he renamed it epinatrolite. 6th List

Epiramsayite. V. I. Gerasimovsky, 1956, Геохимия, no. 5 (Эпирамсаит); translation in *Geochemistry* (Геохимия), 1956, no. 5, p. 494. $Na_2Ti_2Si_3O_{11}.H_2O$; no details. 23rd List

Episcolecite. S. J. Thugutt, 1949. *Rocznik Polsk. Tow. Geol.* (*Ann. Soc. Géol. Pologne*), vol. 18 (for 1948), p. 15 (episcolecite), p. 35 (metaskolecyt). A suggested dimorphous form of scolecite as a derivative of sodalite. Cf. Metascolecite (1st List; *for* F. Rinne, 1894 *read* F. Rinne, 1890), epinatrolite and metanatrolite (6th List). 19th List

Epi-sericite. J. Jakob, 1933. *Schweiz. Min. Petr. Mitt.*, vol. 13, p. 82 (Epi-Sericit). Sericite formed in the epi- (upper) zone of metamorphism of U. Grubenmann (1907). 14th List

Epistolite. O. B. Boeggild, *Medd. om Grönland*, 1900, XXIV, 183. Large thin rectangular plates with a perfect cleavage parallel to the large face which is silver-white with a strong pearly lustre. Monoclinic. $19SiO_2.4TiO_2.5Nb_2O_5.(Ca,Mg,Fe,Mn)O.10Na_2O.21H_2O.4NaF$. [Narsarsuk] South Greenland. 2nd List [Isomorphous with Murmanite (q.v.); *M.A.*, **14**, 370; near $Na_2(Nb,Ti)_2Si_2O_9.nH_2O$]

Epithomsonite. S. J. Thugutt, 1949, *Rocznik Polsk. Tow. Geol.* (*Ann. Soc. Géol. Pologne*), vol. 18 (for 1948), p. 5 (epithomsonite), p. 17 (metameric thomsonite), p. 35 (metatomsonit). A suggested dimorphous form of thomsonite as a derivative of sodalite, identified with the metathomsonite (q.v.) of M. H. Hey (1932). Cf. epinatrolite and metanatrolite (6th List). [*M.A.*, **11**, 291.] 19th List

Eremeevite, standard English transliteration of Еремеевит, replacing Eremeyevite. 23rd List

Eremeyevite. Committee on Nomenclature, *Amer. Min.*, 1924, vol. 9, pp. 61, 62. Another spelling for Jeremejevite (Jérémélewite of A. Damour, 1883), named after the Russian mineralogist Pavel Vladimirovich Eremyeev (Павелъ Владиміровичъ Еремѣевъ, 1830–1899). The Russian form is Еремѣевитъ (Eremyeevite, according to the system of transliteration adopted in this Magazine). 10th List

Ericaite. Author? H. Strunz, *Min. Tab.*, 2nd edit., 1949, p. 135; 3rd edit., 1957, p. 185 (Ericait, Manganboracit, α-Ericait). Abstr. in *Amer. Min.*, 1956, vol. 41, p. 372. Few analyses of boracite show the presence of MnO (up to 2·32%); iron-boracite = huyssenite contains FeO (up to 35·26%). Etymology? 21st List

β-Ericaite. *See* entry for congolite. 27th List

Ericssonite. P. B. Moore, 1967. *Canad. Min.* **9**, 301. Orthorhombic '$BaMn_3Fe^{3+}Si_2O_7OH$', perhaps in error for $Ba\dot{M}n_2Fe^{3+}(Si_2O_7)O.OH$, related to lamprophyllite, from Långban, Sweden. [*A.M.*, **53**, 1426; *Bull.*, **91**, 305; *Zap.*, **98**, 330.] 26th List [Cf. Orthoericssonite]

Erikite. O. B. Boeggild, 1903. *Contributions to Mineralogy, Min. Geol. Mus. Univ. Copenhagen*, no. 2, p. 93; *Meddelelser om Grønland*, 1904, vol. xxvi, p. 93. Brown and opaque, orthorhombic crystals from the nephelite-syenite near Julianehaab in Greenland, when examined in thin section under the microscope, were seen to consist of an intergrowth of two substances, of which the predominant one (erikite) is yellow, strongly refracting and birefringent, whilst the other is probably hydronephelite. An analysis of this mixture corresponds with the formula $8SiO_2.4P_2O_4.4(Ce,La,Di)_2O_3.3Al_2O_3.CaO.3Na_2O.11H_2O$. Named after Erik the Red, who discovered Greenland in AD 986. 4th List

—— E. I. Semenov, 1959. (Мат. Мин. Колск. Полуост., Акад. Наук СССР (*Mater. Min. Kola Peninsula*), vol. 1, p. 91); abstr. in *Amer. Min.*, 1962, vol. 47, p. 419. A name for a mineral from the Kola peninsula, considered to be related to Rhabdophane on the grounds of a certain resemblance in their powder patterns; but the data fit Monazite as well as they do Rhabdophane. Not to be confused with the Erikite of O. B. Bøggild (4th List), which has been shown to be Monazite (*Amer. Min.*, 1959, vol. 44, p. 1329; *M.A.*, **14**, 370). [*A.M.*, **45**, 1135; **47**, 419.] An inappropriate name for a doubtful species. 23rd List

Erinadine. J. S. van der Lingen, 1928. *South African Journ. Sci.*, vol. 25, p. 13. A metamorphic garnet from the granite-slate contact in the Cape Peninsula, found on spectroscopic analysis to contain yttrium together with chromium and magnesium (cf. emildine). Named after the author's daughter Erina, the *d* in the termination being from false analogy with almandine. The same author had earlier used the preoccupied name erinite (Erinit: *Centr. Min.*, Abt. A, 1926, p. 182) in the belief that the mineral contained the hypothetical radioactive element 'hibernium'. [*M.A.*, **4**, 83; **2**, 162.] 12th List

Erinide. A trade name for a yellowish-green *artificial* spinel. R. Webster, *Gems*, 1962, p. 756. Not to be confused with erinadine, with the erinites of van der Lingen, of Haidinger, of Thomson, or of Beudant, nor with erinoid, a casein resin. *See* under corundolite, 28th List

Erionite. A. S. Eakle, *Amer. J. Sci.*, 1898, VI, 66; *Zeits. Kryst. Min.*, 1898, XXX, 176. A zeolite occurring as fine woolly hairs. Rhombic. Chemical formula the same as that of stilbite, but with more alkalies and less lime. Oregon. 2nd List [Relation with Offretite and Levyne; *A.M.*, **61**, 853]

Erlan. A. Breithaupt, *Vollst. Char. Min.-Syste.*, 1823, pp. 64, 208. Massive; green; a mixture of silicates. Erla, Saxony. 2nd List

Erlichmanite. K. G. Snetsinger, 1971. *Amer. Min.* **56**, 1501. A few grains of a cubic mineral, *a* 5·60 Å, with ferroplatinum in placers at the MacIntosh mine, Trinity River, Humboldt Co., California, prove to be OsS_2, matching synthetic material. Also found in a platinum sample from Ethiopia. Named for J. Erlichman. [*M.A.*, 72-1398.] 27th List

Ermakite. P. L. Dravert, 1926. *Bull. West Siberian Branch Russ. Geogr. Soc.* (Изв. Зап.-Сиб. отд Русс. Географ. Общ.), vol. 5, p. 137 (Ермакит, Ermakite). A brown waxy clay with the approximate composition $(Al,Fe)_2O_3.3SiO_2.2H_2O$, from the banks of the Irtysh river near Omsk, Siberia. Named after Ermaka Timofeevicha (Ермака Тимофеевича), who met her death in this river in the year 1584. 14th List

Ernite. — Franck, 1911. [*Rev. Asoc. Rural Uruguay*, Montevideo, 1911, p. 88 (Ernita).] H. Himmel, *Zentr. Min.*, Abt. A, 1938, p. 243 (Ernita). A supposed new mineral from Uruguay, later identified with grossular. Named after the author's wife. [*M.A.*, **7**, 317.] 15th List

—— *see* Oehrnite.

Ernstite. E. Seeliger and A. Mücke, 1970. *Neues Jahrb. Min., Monatsh.* 289. A yellow-brown oxidation product of eosphorite, from Karibib, SW. Africa, has *a* 13·32, *b* 10·497, *c* 6·969 Å, *β* 90° 22′, very near the eosphorite cell, but with the *a* and *b* axes interchanged. *α* 1·678, yellow-brown, *β* 1·706, red-brown, *γ* 1·721, pale yellow, ‖ [010]; D 3·07; $8[Mn^{2+},Fe^{3+})AlPO_4(O,OH)_2.?H_2O]$. Named for Th. Ernst. [*A.M.*, **56**, 637; *Zap.*, **100**, 623.] 27th List

Errite. J. Jakob, 1923. *Schweiz. Min. Petr. Mitt.*, vol. 3, p. 231 (Errit). A variety of parsettensite (q.v.) containing an extra $Mn(OH)_2$ group, the formula being $7MnO.8SiO_2.9H_2O$. Named from the locality, Val d'Err, Grisons. [*M.A.*, **2**, 251.] 10th List

Erzbergite. E. Hatle, *Mitth. naturwiss. Ver. Steiermark*, 1892, vol. xxviii (Jahrg. 1891), p. 295 (Aragonit-Calcit-Sinterbildungen (Erzbergit)). A calcareous deposit from Erzberg, Eisenerz, Styria, consisting of alternate layers of calcite and aragonite. 3rd List

Esaidrite. (C. Andreatta, *Atti* (*Rend.*) *R. Accad. Sci. Lincei*, Roma, 1932, ser. 6, vol. 16, p. 62.) Italian form of Hexahydrite ($MgSO_4.6H_2O$, of R. A. A. Johnston, 1911; 6th List). 13th List

Eschinite. Variant of Aeschynite; *A.M.*, **9**, 62.

Eschwegeite. D. Guimarães, 1926. *Bol. Inst. Brasileiro Sci.*, vol. 2, p. 1 (Echwegeita), p. 2 (Eschwegeita). Hydrated tantalo-niobotitanate of yttrium (and erbium), $2Ta_2O_5.4Nb_2O_5.10TiO_2.5Yt_2O_3.7H_2O$, as dark-red, optically-isotropic material found as pebbles in the upper Rio Doce, Brazil. Named after Baron Wilhelm Ludwig von Eschwege (1777–1855), a pioneer in Brazilian geology and mineralogy. [*M.A.*, **3**, 113.] 11th List [Tantalian Polycrase; *A.M.*, **36**, 927]

Eskebornite. P. Ramdohr, 1948 (MS.). H. Strunz, *Min. Tab.*, 2nd edit., 1949, p. 78 (Eskebornit). P. Ramdohr, *Fortschr. Min.*, 1950, vol. 28 (for 1949), p. 70; *Die Erzmineralien und ihre Verwachsungen*, Berlin, 1950, p. 418. At first described as hexagonal iron selenide FeSe [cf. achavalite, 18th List]. Later as cubic and containing also copper. Named from the locality, Eskeborn level, Tilkerode, Harz. [No relation to bornite.] [*M.A.*, **11**, 118, 312.] 19th List [Canada, $CuFeSe_2$, *C.M.*, **10**, 786]

Eskerite, error for eakerite (*Min. Record*, **1**, 94). 27th List

Eskimoite. S. Karup-Møller, 1977. *Bull. Geol. Soc. Denmark*, **26**, 41. Monoclinic, *a* 13·459, *b* 30·194, *c* 4·100 Å, *β* 93·35°, composition $[Bi_{15}S_{36}]$. From Ivigtut, Greenland, and named for the original inhabitants of Greenland. [*M.A.*, 78-899.] 30th List

Eskolaite. O. Kuovo and Y. Vuorelainen, 1958. *Amer. Min.*, vol. 43, p. 1098. Chromium oxide, Cr_2O_3, isomorphous with hematite, from Outokumpu, Finland. Named after Prof. Pentti Eskola. [*M.A.*, **14**, 198.] Essentially identical with Merumite (18th List), which has priority, but merumite was imperfectly described and was believed to be hydrated; the name eskolaite is to be preferred. 22nd List

Esmeraldaite. A. S. Eakle, *Bull. Dep. Geol. Univ. California*, 1901, vol. ii, p. 320. Pod-shaped masses of a coal-black colour and bright vitreous lustre; it has a glassy fracture, and in thin splinters transmits yellowish-red light; streak, yellowish-brown. The mean of several analyses gives, after deducting 17% of impurities, the formula $Fe_2O_3.4H_2O$. Found in earthy limonite in Esmeralda Co., Nevada. 3rd List

Esperite. P. B. Moore and P. H. Ribbe, 1965. *Amer. Min.*, vol. 50, p. 1170. The mineral named Calcium-larsenite by Palache, Bauer, and Berman (1918) is shown to be monoclinic and not related structurally to Larsenite, and is therefore re-named Esperite for Esper S. Larsen, Jr. [*M.A.*, **17**, 606.] 24th List

Esporogelita. G. Gagarin and J. R. Cuomo, 1949, loc. cit., p. 12. Spanish form of sporogelite (6th List). 19th List

Estibioluzonita. G. Gagarin and J. R. Cuomo, 1949, loc. cit., p. 16. Spanish form of stibioluzonite. 19th List

Estramadurite. (H. E. Roscoe and C. Schorlemmer, *Treatise on Chemistry*, 1877, vol. i, p. 459 (estramadourite); 1878, vol. ii, part 1, p. 205 (estramadurite). T. E. Thorpe, *Dictionary of Applied Chemistry*, 1890, vol. i, p. 402; vol. iii, p. 184.) 'Phosphorite and estramadurite are massive varieties of apatite which occur in Estramadura [i.e. Estremadura or Extremadura], in Spain' (Roscoe and Schorlemmer, 1878). 6th List

Euban. A. Breithaupt (P. Groth, *Min.-Samml. Strassburg*, 1878, 258). A variety of quartz from Euba, Saxony. 2nd List

Euchlorine. Used by A. Scacchi at the Naples Museum for an emerald-green incrustation found in 1869 on Vesuvian lava. The text-books of F. Pisani (1875, p. 338) and other French authors state that it is a chloride and sulphate of copper. The first complete description (E. Scacchi, *Rend. Accad. Sci. Napoli*, 1884, XXIII, 159) gives it as rhombic with the composition $(K,Na)_2SO_4.2CuSO_4.CuO$. An analysis by Rammelsberg (*Min.-Chem. Erg.-Heft*, 1886, p. 87) gives the formula $4(K,Na)_2SO_4.6CuSO_4.3Cu(OH)_2$. 2nd List

Euclasite. Variant of Euclase; *A.M.*, **8**, 51. 10th List

Eudialite. Variant of Eudialyte; *A.M.*, **8**, 51. 10th List

β-Eukryptit. *See* Pseudo-eucryptite. 7th List

Eulebrite, error for Culebrite. Dana, *Syst. Min.*, 3rd edit., 1850, 481. 28th List

Eulite. A. Poldervaart, 1947. *Min. Mag.*, vol. 28, p. 165. F. Walker and A. Poldervaart, *Bull. Geol. Soc. Amer.*, 1949, vol. 80, p. 632. Orthopyroxenes containing 70–90 mol.% $FeSiO_3$. A contraction of eulysite (εὔλυτος, easy to dissolve or break), a rock in which the mineral occurs. 18th List

Eunicite. J. E. de Paiva Netto*, 1955. [*Engenharia, mineria e metalurgia*, vol. 22, no. 128, p. 99 (eunicita).] Abstr. in *Mém. Soc. Russ. Min.*, 1956, vol. 85, p. 382 (эвнисит) and *Amer. Min.*, 1957, vol. 42, p. 441. A variety of montmorillonite from the decomposition of melaphyre at Serra de Botucatu, São Paulo, Brazil*. Etymology? [*Corr. in *M.M.*, **31**] 21st List

Eutectoperthite, Eutecto-oranite. H. L. Alling, 1921. *See* Hypoperthite. 14th List

Evaporate, Evaporite. A. W. Grabau, 1920. *Principles of salt deposition*, New York and London, pp. 14, 79 (evaporates), p. 15 (evaporate). C. P. Berkey, *Bull. Geol. Soc. China*, 1922, vol. 1, p. 24 (evaporate changed to evaporite); reprinted in *New York State Mus. Bull.*, 1924, no. 251, p. 116. Salts deposited from solution by evaporation, as distinct from precipitates deposited by reaction. [Cf. distillate, crystallate, precipitate.] [*M.A.*, **1**, 322; **2**, 307; **10**, 184, 567; *Min. Mag.*, **28**, 621.] 19th List

Eveite. P. B. Moore, 1967. *Canad. Min.*, **9**, 301; *Arkiv Min. Geol.*, **4**, 473 (1969); *Amer. Min.*, **53**, 1841 (1968). Green orthorhombic crystals of Mn_2AsO_4OH, from Långban, Sweden, are isostructural with adamite. Named in allusion to this relation. [*A.M.*, **53**, 1426; **55**, 319; *M.A.*, 69-2397, 2398; *Bull.*, **91**, 305; *Zap.*, **98**, 328; **99**, 79.] 26th List

Evenkite. A. V. Skropyshev, 1953. *Doklady Acad. Sci. USSR*, vol. 88, p. 719 (эвенкит). Paraffin wax, $C_{21}H_{44}$*, as white, optically biaxial scales in a vein of sulphide ores from the district of the Evenki people, Lower Tunguska river, Siberia. [*M.A.*, **12**, 305.] [*Corr. in *M.M.*, **30**. Misprint in original paper] 20th List [Monoclinic $C_{24}H_{50}$, *n*-tetracosane; *A.M.*, **50**, 2109]

Ewaldite. G. Donnay, J. D. H. Donnay, and M. H. Hey, 1971. *Tschermaks min. petr. Mitt.*, **15**, 185; G. Donnay and H. Preston, ibid., 201 (struct.). Two-phase polycrystals from the Green River formation, Wyoming, consist of mckelveyite (24th List) and a bluish-green mineral with *a* 5·284, *c* 12·78 Å, $P6_3mc$, D 3·25; $2[Ba(Ca,Ln,Na,K)(CO_3)_2]$; ω 1·646, dark blue-green, ε 1·572, pale yellow-green. Named for P. P. Ewald. [*A.M.*, **56**, 2156; *M.A.*, 71-3105 and 3106.] 27th List

Expandite. Also hemiexpandite, oligoexpandite, plioexpandite, and pleistoexpandite. F. Scheffer, H. Fölster, and B. Meyer, 1961. *Chemie der Erde*, **21**, 232. Group names for three-layer clay minerals with differing swelling characteristics. 28th List

Eylettersite. L. Van Wambeke, 1972. *Bull.*, **95**, 98. White nodules in the Kobokobo pegmatites, Kivu, Congo, are a thorian member of the crandallite family. *a* 6·98, *c* 16·7 Å, $R\bar{3}m$. Composition variable, near $[Th_{0\cdot3-0\cdot4}Pb_{0\cdot1}(U,Ca,Sr)_{0\cdot1}H_{0\cdot2}Al_{3\cdot5}(PO_4)_{1\cdot3-1\cdot4}$-

$(SiO_4)_{0 \cdot 1}(OH)_{8 \cdot 1 - 8 \cdot 5}$]. The name is for Mme Van Wambeke. [*M.A.*, 72-3341.] 27th List [*A.M.*, **56**, 1366; **59**, 208]

Ezcurrite. S. Muessig and R. D. Allen, 1957. *Econ. Geol.*, vol. 52, p. 426. Hydrous sodium borate, $2Na_2O.5B_2O_3.7H_2O$, with characters very similar to those of kernite ($Na_2B_4O_7.4H_2O$). Associated with kernite, borax, tincalconite in the Tincalayu mine, prov. Salta, Argentina. Named after Juan Manuel de Ezcurra, manager of the mine. *See* Metakernite (14th List). [*M.A.*, **13**, 623.] 21st List [*A.M.*, **42**, 919]

F

F-micas, F-muscovite, F-phlogopite, F-pyrophyllite, F-talc, *see* Fluor-mica.

Fabianite. H. Gaertner, K.-L. Roese, and R. K. Kühn, 1962. *Naturwiss.*, vol. 49, p. 230; *Kali und Steinsalz*, 1962, vol. 3, p. 285. Monoclinic crystals from halite at Rehden, Diepholz, Germany, have the composition CaB_3O_5OH, but are distinct from the synthetic compound. Named for the geologist. H.-J. Fabian. [*M.A.*, **16**, 181.] 23rd List [*A.M.*, **48**, 212]

Fabulit, cited without reference by H. Strunz, *Min. Tabellen*, 4th edit., 1966, p. 470; synthetic $SrTiO_3$; presumably a trade-name. 24th List

Fabulite. A trade-name for *artificial* $SrTiO_3$. H. Strunz, *Min. Tab.*, 1970, 5th edit., 525. Other trade-names include diamontina, (dialite), marvelite, symant, and zeathite (all 28th List); also diagem, Bal de Feu, dynagem, Kenneth Lane jewel, and lustigem. M. J. O'Donoghue and R. Webster, both *priv. comm.* 28th List

Faheyite. M. L. Lindberg and K. J. Murata, 1952. *Bull. Geol. Soc. Amer.* [*1953*], vol. 63, no. 12 (for 1952), p. 1275 (abstr.); *Amer. Min.*, 1953, vol. 38, pp. 263, 349. $(Mn,Mg,Na)Be_2Fe_2'''(PO_4)_4.6H_2O$, hexagonal, as white fibres in pegmatite from Brazil. Named after Dr. Joseph John Fahey (1901–), geochemist, United States Geological Survey. A preliminary notice of the title of the paper, *Amer. Min.*, Nov–Dec., 1952, vol. 37, p. x, gives the form faheylite. [*M.A.*, **12**, 131.] 20th List [*A.M.*, **49**, 395]

Fairbanksite. F. Morgan, 1965. *Rocks and Minerals*, vol. 40, p. 586. This name (for E. E. Fairbanks) is improperly given to unidentified microscopic crystals in shrinkage cracks of concretions at Greenbelt, Maryland. No data are given. 24th List

Fairburnite. Yet another locality variety of agate, from Fairburn, South Dakota. R. Webster, *Gems*, 1962, p. 756; J. Sinkankas, *Gemstones of North America*, 1959, p. 349 (Fairburn agate). 28th List

Fairchildite. C. Milton and J. M. Axelrod, 1946. *Bull. Geol. Soc. Amer.*, vol. 57, p. 1218; *Amer. Min.*, 1947, vol. 32, pp. 204, 607. Potassium and calcium carbonate, $K_2CO_3.CaCO_3$, hexagonal, formed by the fusion of wood-ash in burnt trees. On hydration it yields buetschliite (q.v.) and calcite. Named after John Gifford Fairchild (1882–) of the United States Geological Survey. [*M.A.*, **10**, 101, 252.] 18th List

Falcondoite. G. Springer, 1976. *Can. Mineral.*, **14**, 407. The nickel analogue of sepiolite occurs near Bonao, Dominican Republic. Orthorhombic, *a* 13·5, *b* 26·9, *c* 5·24 Å. Composition $2[(Ni,Mg)Si_{12}O_{30}(OH)_4.12H_2O]$. Named from the Falcondo mining company. [*M.A.*, 78-886.] 30th List

Falkenstenite. T. F. W. Barth, 1945. *Skrifter Norske Videnskaps-Akad. Oslo, I. Mat.-Naturv. Kl.*, 1945, no. 8, p. 13. An incompletely determined tetragonal zeolite, related to ashcroftine and gonnardite, forming the groundmass (40%) of variolite at Falkensten, Oslo fjord. Named from the locality. [*M.A.*, **10**, 5.] 18th List

Falkmanite. P. Ramdohr and O. Ödman, 1940. *Neues Jahrb. Min.*, Abt. A., Beil.-Bd. 75, p. 315 (Falkmanit). J. E. Hiller, *Zeits. Krist.*, 1939, vol. 102, p. 138. Sulphantimonite of lead, $3PbS.Sb_2S_3$, as acicular monoclinic crystals. Named after Oscar Carl August Falkman (1877–), director of the Boliden mines in north Sweden, one of the localities where the mineral was found. It had previously been known locally as bolidenite. [*M.A.*, **7**, 468, 513.] 15th List [Is Boulangerite; *A.M.*, **40**, 1155]

Farallonite. N. N. Kohanowski, 1953. *Mines Mag.*, Denver, Colorado, vol. 43, no. 2, p. 19. Sky-blue chalky material, monoclinic?, '$2MgO.W_2O_5.SiO_2.nH_2O$?'; an alteration product of wolframite in the Farallon mine, Tasna, Bolivia. Named from the locality. [*M.A.*, **12**, 306.] 20th List

Faratsihite. A. Lacroix, 1915. *Bull. Soc. franç. Min.*, vol. xxxvii (for 1914), p. 231. A canary-yellow, compact mineral resembling nontronite in colour and kaolinite in structure, and intermediate in composition, $(Al,Fe)_2O_3.2SiO_2.2H_2O$, between these two species. Occurs with opal filling veins in decomposed phonolitic trachyte near Faratsiho, Madagascar. Named from the locality. 7th List [A mixture; *A.M.*, **20**, 475; **24**, 529]

Farölith, error for Faröelite (H. Strunz, *Min. Tabellen*, 1st edit., 1941, p. 239). 24th List

Farringtonite. E. R. DuFresne and S. K. Roy, 1961. *Geochimica Acta*, vol. 24, p. 198. $Mg_3(PO_4)_2$, monoclinic, occurring in the Springwater pallasite (*Amer. Min.*, 1961, vol. 46, p. 1513; *M.A.*, **15**, 212). 23rd List

Faseriges SiO_2. A. and A. Weiss, 1954. *Naturwiss.*, **41**, 12. An *artificial* fibrous orthorhombic modification of SiO_2 formed by heating SiO in air or *in vacuo*: *Icma*, *a* 4·72 Å, *b* 5·16, *c* 8·36; D 1·95–1·98. Not to be confused with faserquarz, a fibrous chatoyant variety of quartz. [*M.A.*, **12**, 515.] 28th List

Faserkohle, *see* Fusain.

Fasernephrit. J. Uhlig, 1910. *Neues Jahrb. Min.*, 1910, vol. ii, p. 91. The nephrite of this newly discovered occurrence in the Radauthal, Harz, shows in part a fibrous structure owing to the parallel, rather than matted, aggregation of the actinolite fibres. This material has already been described under the name nephritoid (J. Fromme, 1909; 5th List). 6th List

Faserserpentin. P. Groth, *Tab. Uebers. Min.*, 4th edit., 1898, p. 135. The same as chrysotile. 2nd List

Fauserite. H. Strunz, 1957. *Min. Tabellen*, 3rd edit., p. 202. An hypothetical orthorhombic polymorph of $MnSO_4.7H_2O$; not to be confused with Fauserite of Breithaupt (1865), which appears to have been a manganoan epsomite (*Min. Mag.*, **22**, 511). 22nd List

Faustite. R. C. Erd, M. D. Foster, and P. D. Proctor, 1953. *Amer. Min.*, vol. 38, p. 964. The zinc analogue of turquoise; $(Zu,Cu)Al_6(PO_4)_4(OH)_8.5H_2O$, containing ZnO 7·74, CuO 1·61%. Fine-grained, apple-green masses from Nevada. Named after Dr. George Tobias Faust (1908–) of the United States Geological Survey. [*M.A.*, **12**, 302.] 20th List

Fauyasit. Error for Faujasit (F. Machatschki, *Spez. Min.*, Wien, pp. 345, 362). 22nd List

Fe-Åkermannit, *see* Iron-åkermanite.

Fe-alluaudite, *see* Ferrialluaudite.

Fe-pennantite. R. A. H[owie], 1970. *M.A.*, **70**, 2356. An analogue of pennantite (17th List) with Fe $\approx$ Mn from Ushatan, Atasui, Kazakhstan, described by M. M. Kayupova (*Zap.*, 1967, **96**, 155) under the name Железистый пеннантит (ferroan pennantite). 28th List

Fe-spodumene. S. Šćavničar and G. Sabatier, 1957. *Bull. Soc. franç. Min. Crist.*, vol. 80, p. 308 (spodumène-Fe); *M.A.*, **13**, 636 (Fe-spodumene). The synthetic compound $LiFeSi_2O_6$, the iron analogue of spodumene. 22nd List

Fedorite. A. A. Kukharenko *et al.*, 1968. [The Caledonian ultrabasic rocks and carbonatites of the Kola Peninsula and northern Karelia. *Izd. 'Neda', Moscow.*, pp. 479–481]; abstr. *Amer. Min.*, 1967, vol. 52, p. 561. Pseudo-hexagonal tablets from the Turii peninsula, Kola, near $Ca(Na,K)_{1\frac{1}{4}}(Si,Al)_4O_9OH.1\frac{1}{2}H_2O$. Named for E. S.* Fedorov. [Зап., **96**, 77; *Bull.*, **90**, 610.] [*Corr. in *M.M.*, **40**.] 25th List [*A.M.*, **52**, 561]

Fedorovskite. S. V. Malinko, D. P. Shashkin, and K. V. Yurkina, 1976. *Zap.*, **105**, 71 (Федоровскит). Brown grains or layers of fibres in kurchatovite–sakhaite masses in the Solongo ore deposit, Buryat S.S.R., have *a* 8·96, *b* 13·15, *c* 8·15 Å, space group *Pbam*, ρ 2·60 g. cm^{-3}. Composition $4[Ca_2(Mg,Mn)_2B_4O_7(OH)_6]$. 29th List

Fedorowite. C. Viola, *Neues Jahrb. Min.*, 1899, I. 121. C. Viola and E. H. Kraus, *Zeits. Kryst. Min.*, 1900, XXXIII, 36. '*Fedorovite*', Dana's *Appendix*, 1899. A monoclinic pyroxene near diopside in chemical composition, but near ægirite in optical characters. In all kinds of igneous rocks in the province Rome. 2nd List

Feitknechtite. O. Bricker, 1965. *Amer. Min.*, vol. 50, p. 1296. Natural and synthetic Hydrohausmannite (q.v.) always consists of a mixture of Hausmannite and β-MnOOH; for the latter the name Feitknechtite is proposed after W. Feitknecht. 24th List

Feldspathides. A. Michel-Lévy, *Structures et classification des roches éruptives*, Paris, 1889, p. 38. A collective term to include leucite, nepheline, melilite, sodalite, haüyne, nosean, present in alkalic rocks. Later altered to felspathoids. Cf. Lenad (5th List), Feloid (10th List), Foid (q.v.). 21st List

Felite. *See* under alite. 2nd List

Feloid. E. T. Hodge, 1924. *Univ. Oregon Publ.*, vol. 2, no. 7, p. 35. 'A group of minerals comprising the feldspars and felspathoids.' Cf. Quarfeloids. [*M.A.*, **2**, 439.] 10th List

Femaghastingsite. M. Billings, 1928. *See* Ferrohastingsite. 12th List

Femolite. K. V. Skvortsova, G. A. Sidorenko, A. D. Dara, N. I. Silanteva, and M. M. Medoeva, 1964. Зап. Всесоюз. Мин. Общ. (*Mem. All-Union Min. Soc.*), vol. 93, p. 436 (Фемолит). Colloform material with composition near $(Mo,Fe)S_2$ (Fe near 6·5%) gives a weak X-ray pattern near that of Molybdenite, from which it shows minor differences. There are no valid grounds for regarding this as homogeneous; a superfluous name for an inadequately characterized substance. [*A.M.*, **50**, 261.] 23rd List

Fenaksite. M. D. Dorfman, D. L. Rogachev, Z. I. Goroshchenko, and A. V. Mokretsova, 1959. Труды Мин. Муз. Акад. Наук СССР (*Trans. Min. Mus. Acad. Sci. USSR*), vol. 9, p. 152 (Фенаксит). A pale rose monoclinic silicate of ferrous iron and alkalis from a pegmatite associated with an ijolite–urtite intrusion at Khibina, Kola peninsula. Named from the composition, Fe–Na–K–Si. [*M.A.*, **14**, 414.] May be identical with the Mineral 5 of Dorfman (*Amer. Min.*, **44**, 910). The name is easily confused with phenakite. 22nd List [*A.M.*, **45**, 252; $(K,Na)_4(Fe,Mn)_2(Si_4O_{10})_2(OH,F)$]

Fenghuangite. Chi-Jui Peng and Yuan-Lung Liu, 1962. [*Scientia Sinica*, vol. 11, p. 577], abstr. in *Amer. Min.*, 1963, vol. 48, p. 210, and in Зап. Всесоюз. Мин. Общ. (*Mem. All-Union Min. Soc.*), 1963, vol. 92, p. 574 (Фенгуангит, fenghuangite). Variant spelling of Fenghuanglite (22nd List). [*M.A.*, **16**, 183.] 23rd List [Also Strunz, *Min. Tab.*, 4th edit., 1966, p. 470 (fenghuangit). 24th List]

Fenghuanglite. Ch'i-Jui Peng, 1959. [*Ti-chih K'o-hsuẹh*, vol. 10, p. 289]; abstr. *Amer. Min.*, 1960, vol. 45, p. 754 (feng-huang-shih); Зап. Всесоюз. Мин. Общ. (*Mem. All-Union Min. Soc.*), 1961, vol. 90, p. 108 (Фынченит [fȳnchenit], feng-huang-shih). Unnecessary name for a thorian britholite. 22nd List [Cf. Fynchenite, another variant]

Feng-huang-shih, *see* Fenghuanglite.

Fengluanite. Yu Tsu-Hsiang, Lin Shu-Jen, Chao Pao, Fang Ching-Sung, and Huang Chi-Shun, 1974. [*Acta Geol. Sinica*, **2**, 202], abstr. *M.A.*, 75-2522. Orthorhombic, *a* 11·03, *b* 3·37, *c* 6·13 Å, composition $Pd_3(As,Sb)$, from an unnamed locality in China. Apparently an antimonian variety of guanglinite (q.v.). [*A.M.*, **61**, 184.] 29th List

Feranthophyllite. A. N. Winchell, 1933. *Optical mineralogy*, part 2, 3rd edit., p. 241. A contraction of Ferroanthophyllite (9th List). 13th List

Ferantigorite. E. S. Simpson, 1937. *Journ. Roy. Soc. W. Australia*, vol. 23, p. 22. A contraction of ferro-antigorite (11th List). 15th List

Feraxinite. A. N. Winchell, 1927. *Elements of optical mineralogy*, 2nd edit., New York, p. 254. A rather unnecessary contraction of ferroaxinite (5th List). 14th List

Ferchevkinite. Z. Rubai and F. Liangming, 1976. *Geochimica*, **4**, 244. Unnecessary name for a ferroan chevkinite. [*A.M.*, **63**, 426.] 30th List

Ferdisilicite, *see* Fersilicite.

Ferganite. I. A. Antipov, 1908. *Gornyi Zhurnal*, St. Petersburg, year lxxxiv, vol. iv, p. 259 (Ферганитъ); abstr. in *Neues Jahrb. Min.*, 1909, vol. ii, ref. p. 38 (Ferghanit). A hydrated uranium vanadate, $U_3(VO_4)_2.6H_2O$, containing also a small amount of lithium. Found as sulphur-yellow scales, together with other uranium minerals, in province Fergana, Russian Turkistan. Related to carnotite. 5th List

α-Fergusonite, synonym of Fergusonite as distinct from β-Fergusonite (q.v.). 23rd List

β-Fergusonite. S. A. Gorzhevskaya, G. A. Sidorenko, and I. E. Smorchkov, 1961. [Геол. Мест. Редк. Элем. (*Geol. Depos. Rare Elem.*), vol. 9, p. 28], abstr. in Зап. Всесоюз. Мин. Общ. (*Mem. All-Union Min. Soc.*), 1962, vol. 91, p. 190 β-фергусонит, β-fergusonite), and in *Amer. Min.*, 1961, vol. 46, p. 1516. An orthorhombic or monoclinic polymorph of Fergusonite. [*M.A.*, **16**, 552.] 23rd List [*A.M.*, **60**, 485. Monoclinic]

Fermorite. G. T. Prior and G. F. H. Smith, 1910. *Nature*, London, vol. lxxxiii, p. 513. Arsenate, phosphate, and fluoride of calcium and strontium, isomorphous with apatite. Occurs as white, crystalline masses in the Indian manganese-ore deposits. Named after Dr. Lewis Leigh Fermor, of the Geological Survey of India. 5th List

Fernandinite. W. T. Schaller, 1915. *Journ. Washington Acad. Sci.*, vol. v, p. 7; 3rd Appendix to 6th edit. of Dana's *System of Mineralogy*, p. 29. A massive, dull-green, hydrated vanadate of calcium and vanadyl, $CaO.V_2O_4.5V_2O_5.14H_2O$. From Minasragra, Peru. Named after Mr. Eulagio E. Fernandini, the former owner of the vanadium deposit at this locality. 7th List

Feroxyhyte. F. V. Chukhrov, B. B. Zvyagin, A. I. Gorshkov, L. P. Ermilova, V. V. Korovashkin, E. S. Rudnitskaya, and N. Yu. Yakubovskaya, 1976. *Izvest. akad. nauk SSSR*, ser. geol., no. 5, 5 (Фероксигит, feroxygite). Artificial δ-FeOOH consists of a magnetic, ordered phase and a non-magnetic disordered phase, both hexagonal. The latter, disordered, phase occurs in Fe-Mn concretions from the Pacific Ocean, and from the Baltic, White, and Kara Seas, also in some soils. The unit cell, *a* 2·93, *c* 4·60 Å, contains a single FeOOH. The material readily transforms to goethite on exposure. Named for *Fe*rrum, *Oxy*gen, *Hy*drogen. [Possible relationship to ferrihydrite (28th List), which has *a* $5{\cdot}08 = 2{\cdot}93\sqrt{3}$, and *c* $9{\cdot}4 = 2 \times 4{\cdot}7$ Å, is not discussed—*M.H.H.*] [*A.M.*, **62**, 1057; *M.A.*, 77-3144; 78-2120.] 30th List

Ferrazite. T. H. Lee and L. F. de Moraes, 1919. *Amer. Journ. Sci.*, ser. 4, vol. 48, p. 353; *Chem. News*, London, 1921, vol. 122, p. 54. Hydrated phosphate of lead and barium, $3(Pb,Ba)O.P_2O_5.8H_2O$, occurring as compact, dark yellowish-white pebbles ('favas') in the Brazilian diamond deposits. Named after Dr. Jorge Belmiro de Araujo Ferraz, of the Geological Survey of Brazil. [*M.A.*, **1**, 18.] 9th List

Ferriallophane. F. A. Nikolaevski, 1914. *Bull. Acad. Sci.*, St. Pétersbourg, ser. 6, vol. viii (1), p. 147 (Ферри-аллофанъ), p. 150 (Ферриаллофанъ). A variety of allophane containing much iron (Fe_2O_3 21–25%), occurring as dark-brown, colloidal masses in dolomite near Moscow. This and the ochreous clays, sinopite, melinite, ochran, and plinthite, are included in a group of *ferriallophanoids* (cf. allophanoids, S. J. Thugutt, 1911; 6th List). 7th List

Ferri-alluaudite. Dana's *Mineralogy*, 7th edit., 1951, vol. 2, p. 674. Fe-alluaudite of P. Quensel, 1937 [*M.A.*, **6**, 485]. 19th List

Ferrialunogen. H. Strunz, *Min. Tab.*, 1941, p. 135 (Ferrialunogen), p. 239 (Ferroalunogen). Alunogen with some alumina replaced by ferric oxide. Identified with tecticite (Tekticit). 18th List

Ferri-annite. D. R. Wones, 1963. *Amer. Journ. Sci.*, **261**, 581 (Ferriannite); G. Donnay *et al.*, *Acta Cryst.*, **17**, 1369 (1964), Ferri-annite. *Artificial* $KFe_3^{2+}Fe^{3+}Si_3O_{12}(OH)_2$, the ferric analogue of annite. 26th List

Ferri-barroisite. *Amph.* (1978), **14**, 7. 30th List

Ferri-beidellite. O. M. Shubnikova and D. V. Yuferov, *Spravochnik po novym mineralam*, Leningrad, 1934 (ферри-бейделлит, ferri-, iron-, Eisen-beidellite). Translation of Iron-beidellite (11th List). 14th List

Ferri-berthierine. H. Strunz, 1957. *Min. Tabellen*, 3rd edit., p. 323 (Ferri-Berthierin). An oxidized product obtained by heating berthierine (= ferro-berthierine) to 400°C. in air. The ferric chamosite of G. W. Brindley and R. F. Youell (*Min. Mag.*, **30**, 57). 22nd List

Ferribiotite. H. Meixner, 1939. *Fortschr. Min. Krist. Petr.*, vol. 23, p. xliv (Ferribiotite, *pl.*). F. Angel and A. Marchet, ibid., p. xxxviii (Ferri-Biotite, *pl.*). Biotite rich in ferric iron. 15th List

Ferribraunite. B. Wasserstein, 1943. *Econ. Geol.*, vol. 38, p. 393. Intermediate between braunite and sitaparite. [*M.A.*, **9**, 38.] 18th List

Ferric chamosite. G. W. Brindley and R. F. Youell, 1953. *Min. Mag.*, vol. 30, p. 57 (ferric chamosite). Oxidized product of ferro-chamosite (q.v.) when heated in air or weathered. May be a distinct species (? berthierine, *Min. Mag.*, **30**, 645). 20th List

Ferri-chamosit, variant of Ferric chamosite (20th List) (H. Strunz, *Min. Tabellen*, 3rd edit., 1957, p. 372). 24th List (Also 20th List)

Ferrichinglusuite. A. I. Soklakov and M. D. Dorfman, 1964. [Минералы СССР, **15**, 167]; abstr. *M.A.*, **18**, 160. A metamict mineral from Khibina with Fe:Mn about 2:1, compared with about 1:4 in chinglusuite (15th List). The evidence that the mineral is a variety of chinglusuite is not convincing. German variant ferritschinglusuit, Strunz, *Min. Tab.*, 1970, 5th edit., p. 526 (cf. tschinglusuite, 20th List). 28th List

Ferrichlorite. L. O. Stankevich, 1957. Минер. Сборник Львовск. Геол. Общ. (*Min. Mag.*

Lvov Geol. Soc.), no. 11, p. 159 (Феррихлорит). A group name. [*M.A.*, **14**, 141.] Internationalized form of Eisenchlorit (of Holzner, 1938). 22nd List

Ferrichrompicotite, Ferrichromspinel. A. G. Betekhtin, 1934. *See* Alumochromite. 14th List

Ferrichromspinel, *see* Ferrichrompicotite.

Ferrichrysocolla. F. V. Chukhrov, B. B. Zvyagin, A. I. Gorshakov, L. P. Ermilova, and E. S. Rudnitskaya, 1968. Изв. Акад. наук СССР (*Bull. Acad. Sci. URSS*), no. 6, 29 (Феррихризоколла). Superfluous name for a ferrian chrysocolla. 28th List

Ferricopiapite. L. G. Berry, 1938. *Amer. Min.*, vol. 23, no. 12, part 2, p. 3; 1939, vol. 24, p. 182. A variety of copiapite in which X in the formula $X(OH)_2.Fe'''_4(SO_4)_6.nH_2O$ is mainly ferric iron. Similarly, ferrocopiapite and magnesiocopiapite when X is mainly ferrous iron or magnesium. 15th List

Ferri-diopside. H. G. Huckenholz, J. F. Schairer, and H. S. Yoder, Jr., 1969. *Min. Soc. Amer. Spec. Paper*, no. 2, 163. In *synthetic* preparations, diopside can contain in solid solution up to 33 wt% of $CaFe_2{}^{3+}SiO_6$. The exact connotation intended for the name ferri-diopside, whether for this end-member or for the stable ferrian diopsides, is not clear. 26th List

Ferriedenite. A. N. Winchell, 1949. *Amer. Min.*, vol. 34, p. 225. Hypothetical molecule $NaCa_2Fe_3Fe_2O_2Si_7AlO_{22}$ in hornblendes, differing from ferroedenite (q.v.) in containing some ferric iron. [*M.A.*, **10**, 568.] 18th List

Ferri-Eisen-Turmalin. K. F. Chudoba, 1974. Hintze, 28. An unnecessary name for the ferric iron tourmaline of Frondel *et al.* (*A.M.*, **51**, 1501). Not buergerite, 25th List. 28th List

Ferriepidote. F. Zambonini, 1920. *See* Aluminium-epidote. 10th List

Ferrierite. R. P. D. Graham, 1918. *Trans. Roy. Soc. Canada*, ser. 3, vol. xii, sect. iv and v, p. 185. A zeolite mineral forming spherical aggregates of white, pearly, orthorhombic blades, occurring with chalcedony and calcite in basalt at Kamloops Lake, British Columbia. Chemically, it is related to ptilolite and mordenite, but with magnesium in place of calcium, the formula being $(Mg,Na_2,H_2)_4Al_2(Si_2O_5)_5.6H_2O$. Named after Dr. Walter F. Ferrier, of Ottawa, who discovered the mineral. 8th List [*M.A.*, 76-807; *A.M.*, **61**, 60. $(Na,K)_2MgAl_3Si_{15}O_{36}(OH).9H_2O$]

Ferrifayalite. I. V. Ginzburg, G. A. Lisitsina, A. T. Sadikova, and G. A. Sidorenko, 1962*. Труды Мин. Муз. Акад. Наук СССР (*Proc. Min. Mus. Acad. Sci. USSR*), vol. 13, p. 16 (Феррифаялит). $(Fe,Fe,Mn)_{2-x}SiO_4$, with 32 to 47% Fe_2O_3 and 27 to 12% FeO, from the Cherkassk massif, Kuraminsk, Siberia. The homogeneity of the material is doubtful, and the name premature (Зап. Всесоюз. Мин. Общ. (*Mem. All-Union Min. Soc.*), 1964, vol. 93, p. 455). [*Corr. in *M.M.*, **39**.] 23rd List [Cf. Laihunite]

Ferrigarnierite. E. F. Alekseeva and M. N. Godlevsky, 1937. *Mém. Soc. Russe Min.*, ser. 2, vol. 66, p. 99 (ферригарниерит). Garnierite containing 5·34% Fe_2O_3, from Novo-Cheremshansky, Ural. [*M.A.*, 7, 214.] 15th List

Ferrigedrite. D. P. Serdyuchenko, 1936. *Bull. Acad. Sci. URSS, Cl. Sci. Math. Nat., Sér. Géol.*, 1936, p. 693 (ферригедрит), p. 696 (ferrihedrite [*sic*]). An orthorhombic amphibole analogous to gedrite, with Fe_2O_3 in place of Al_2O_3. [*M.A.*, **7**, 10.] 15th List [**Ferri-gedrite**, *Amph.* (1978)]

Ferri-gehlenite. P. Niggli, 1922. Abstr. in *Zeits. Krist.*, vol. 57, p. 105 (Ferri-Gehlenit). A hypothetical end-member, $2CaO.Fe_2O_3.SiO_2$, of the melilite group, corresponding to gehlenite ($2CaO.Al_2O_3.SiO_2$). 9th List

Ferriglaucophane. M. B. Ramachandra Rao, 1939. *Rec. Mysore Geol. Dept.*, vol. 37, p. 68 (ferriglaucophane), p. 77 (ferri glaucophane). Intermediate in composition between glaucophane and riebeckite. [*M.A.*, 7, 470.] 15th List [Magnesio-riebeckite, *Amph.* Sub-Comm.]

Ferri-halloysite. N. E. Efremov, 1936. *Mém. Soc. Russ. Min.*, ser. 2, vol. 65, p. 224 (ферри-
p. 232 (ferri-halloysite). A variety of halloysite containing iron. 14th List

Ferrihastingsite. A. N. Winchell, 1949. *Amer. Min.*, vol. 34, p. 225. Hypothetical molecule $NaCa_2Fe_2(Al,Fe)_3O_2Si_6Al_2O_{22}$, differing from ferrohastingsite (12th List) in containing some ferric iron. [*M.A.*, **10**, 568.] 18th List

Ferrihedrite, error for Ferrigedrite (q.v.).

Ferri-hidalgoite. K. F. Chudoba, 1959. Hintze, *Handb. Min.*, Erg.-Bd. II, p. 702 (Ferri-

Hidalgoit). An unnecessary name for the ferrian hidalgoite of R. L. Smith, F. S. Simons, and A. C. Vlisidis (1953). 22nd List

Ferrihydrite. F. V. Chukhrov, B. B. Zvyagin, A. I. Gorshkov, L. P. Ermilova, and E. S. Rudnitskaya, Изв. Акад. наук СССР, сер. геол. 1971, no. 1, 3 (Ферригидрит). The phase $HFe_5O_8.4H_2O$ described by W. E. Towe and R. M. Bradley, *Journ. Colloid Sci.*, 1967, **24**, 386, occurs naturally in two deposits in the Altai, the Belousovsk and the Leninogrosk; *a* 5·08, *c* 9·4 Å. Named from the composition. [*Zap.*, **101**, 281.] 28th List [*A.M.*, **60**, 945]

Ferri-hydroxykeramohalite. A. Dubanský, 1956. *Chem. Listy*, Praha, vol. 50, p. 1350 (ferri-hydroxykeramohalit). $(Al,Fe)_4(SO_4)_5(OH)_2.21H_2O$, with Al_2O_3 12·54, Fe_2O_3 11·58%. (*See* Hydroxykeramohalite.) 21st List

Ferri-ilmenite. R. Chevallier, J. Bolfa, and S. Mathieu, 1955. *Bull. Soc. Franç. Min. Crist.*, vol. 78, p. 310 (ferri-ilménites). Ilmenite containing up to 33% Fe_2O_3 in solid solution, as distinct from titano-hématites (titanhaematite, 17th List) containing up to 33% TiO_2, in the isomorphous rhombohedral series Fe_2O_3–$FeTiO_3$. 21st List

Ferrikalite. R. Scharizer, 1927. *Zeits. Krist.*, vol. 65, p. 15 (Ferrikalit). An artificially prepared potassium ferric sulphate, $K_6Fe_2(SO_4)_6 + x$ aq., analogous to ferrinatrite (= ferronatrite). 11th List

Ferrikaolinite. D. P. Serdyuchenko, 1945. *Compt. Rend.* (*Doklady*) *Acad. Sci. URSS*, vol. 46, p. 118. Hypothetical molecule $H_4Fe_2Si_2O_8$, corresponding to kaolinite. 17th List

Ferri-katophorite. *Amph.* (1978), **14**, 14. 30th List

Ferrikerolite. I. A. Rukavishnikova, 1956. [*Kora vyvetrivaniya, Acad. Sci USSR*, no. 2, p. 141.] Abstr. in *Mém. Soc. Russ. Min.*, 1957, vol. 86, p. 125 (феррикеролит, ferrikerolite). Compact, green, variety of kerolite (cerolite), with karpinskite (q.v.) in serpentine. Urals. 21st List

Ferri-metahalloysite. Yu. F. Pekun, 1956, according to H. Strunz (*Min. Tabellen*, 3rd edit., 1957, p. 325). An unnecessary name for ferrian metahalloysite. 22nd List

Ferrimolybdite. P. Pilipenko, 1914. [Festschrift V. I. Vernadsky, *Beilage z. d. Mater. z. Kenntnis d. geol. Baues d. Russ. Reichs*, Moskau.] Abstr. in *Neues Jahrb. Min.*, 1915, vol. ii, ref. 191 (Ferrimolybdit). The mineral long known as molybdic ochre or molybdite was shown by W. T. Schaller in 1907 to be a hydrated ferric molybdate ($Fe_2O_3.3MoO_3.7\frac{1}{2}H_2O$) rather than molybdenum trioxide, and he suggested that if the latter compound be ever found as a mineral the former should receive a distinctive name: this is now supplied by P. Pilipenko, whose analysis of material from Siberia leads to the formula $2Fe_2O_3.7MoO_3.19H_2O$. The analogous term ferritungstite has been introduced by Schaller (1911; 6th list) under similar circumstances. 7th List [*A.M.*, **48**, 14; $Fe_2(MoO_4)_3.nH_2O$]

Ferrimontmorillonite. I. Z. Korin, 1939. *Bull. Acad. Sci. URSS, Sér. Géol.*, 1939, no. 6, p. 149 (Ферримонтмориллонит), p. 158 (ferrimontmorillonite). V. P. Ivanova, D. S. Belyankin Jubilee vol., *Acad. Sci. USSR*, 1946, p. 102. The correct form of ferromontmorillonite (15th List). $Fe_2O_3.4SiO_2.2H_2O$. Syn. of nontronite. [*M.A.*, **9**, 240; **10**, 554.] 18th List

Ferri-muscovite. W. Wahl, 1925. Fennia (*Bull. Soc. Géogr. Finlande*), Helsingfors, vol. 45, no. 20, p. 85 (Ferri-Muskovit). The ferric molecule $H_2KFe_3'''Si_3O_{12}$, corresponding to muscovite ($H_2KAl_3Si_3O_{12}$). The term 'Ferro-muscovite' is applied (erroneously) to lepidomelane, and 'Ferro-ferri-muscovite' to monrepite (q.v.). 11th List

Ferrinatrite. R. Scharizer, 1905. *Zeits. Kryst. Min.*, vol. xli, p. 209 (Ferrinatrit). To replace the name ferronatrite (J. B. Mackintosh, 1889), since the mineral to which it is applied is a hydrated sodium ferric (not ferrous) sulphate. 4th List [Trigonal $Na_3Fe(SO_4)_3.3H_2O$]

Ferri-Orthochamosit. F. Novák and Z. Valcha, 1964. [*Sborn. geol. věd. Techn.-geochem.*, vol. 3, p. 7], abstr. *M.A.*, **17**, 183. Synonym. of Ferri-berthierine (22nd List). 25th List

Ferri-orthoclase. W. Wahl, 1925. Fennia (*Bull. Soc. Géogr. Finlande*), Helsingfors, vol. 45, no. 20, p. 48 (Ferri-Orthoklas). The ferric molecule $KFe'''Si_3O_8$ corresponding to orthoclase ($KAlSi_3O_8$). It was prepared artificially by P. G. Hautefeuille (1888) and is present (Fe_2O_3 up to 2·88%) in the golden-yellow orthoclase 'orthose ferrifère' (A. Lacroix, 1922) of Madagascar. Iron-orthoclase [*M.A.*, **3**, 480]. 11th List

Ferripalygorskite. H. Strunz, 1957. *Min. Tabellen*, 3rd edit., p. 328 (Ferripalygorskit). Internationalized form of Eisenpalygorskit (Koechlin). 22nd List

Ferri-paraluminite. P. P. Pilipenko, 1927. [Ученые Записки Саратов. Университета, *Mem. Univ. Saratov*, 1927, vol. 6, p. 171.] O. M. Shubnikova and D. V. Yuferov, *Spravochnik po novym mineralam*, Leningrad, 1934, p. 149 (Железистый паралюминит, Ferri-paraluminite, Eisenparaluminit), p. 164 ('Ferro-paraluminite'). Abstr. in *Neues Jahrb. Min.*, ref. I, 1928, p. 298 (Eisenparaluminit). A variety of paraluminite containing iron (Fe_2O_3 13·39%), $2(Al,Fe)_2O_3.SO_3.15H_2O$; deposited as a greenish-grey slime (earthy when air-dried), together with paraluminite and aluminite, by springs from Cretaceous rocks near Saratov. 14th List

Ferriphengite. K. Kanehira and S. Banno, 1960. *Journ. Min. Soc. Japan*, vol. 66, p. 654. A variety of muscovite from the Iimori district, Japan, near $K_2(Mg,Fe)Fe^{\cdots}Al_3Si_7O_{20}(OH)_4$. 22nd List

Ferri-phlogopite. M. Sambonsugi, 1958. *Journ. Min. Soc. Japan*, vol. 3, pp. 634 (Japanese), 801 (English summary). A brown mica from Teshirogi district, Fukushima prefecture, having 15% Fe_2O_3 and very little FeO. [*M.A.*, **14**, 343.] 22nd List

Ferripléonaste. J. Babkine, F. Conquéré, and J.-C. Vilminot, 1968. *Compt. Rend. Acad. Sci. Paris*, **266**D, 1455. An unnecessary name for a ferrian ceylonite (pleonaste) with 5–9% Fe_2O_3. [*M.A.*, 69-3306.] 26th List

Ferripumpellyite. P. B. Moore, 1971. *Lithos*, **4**, 98. A name for the Fe^{3+} analogue of pumpellyite, $Ca_2MgFe^{3+}Si_3O_{11}(OH)_2.H_2O$. *Hypothetical* end-member. [*A.M.*, **56**, 2158.] 27th List

Ferripurpurite. W. T. Schaller, 1907. *Amer. Journ. Sci.*, ser. 4, vol. xxiv, p. 154. Purpurite (L. C. Graton and W. T. Schaller, 1905; 4th List, 1907) is a hydrated manganic ferric phosphate, $2(Mn,Fe)PO_4.H_2O$, in which either manganese or iron may predominate. For the end members of this isomorphous series the names ferripurpurite and manganipurpurite are suggested. 5th List

Ferripyroaurite. H. Meixner, 1937. *See* Ferropyroaurite. 15th List

Ferri-reddingite. P. B. Moore, 1964. *Amer. Min.*, vol. 49, p. 1122. Synonym of Landesite (12th List). 24th List

Ferririchterite. S. A. Bilgrami, 1955. *Min. Mag.*, vol. 30, p. 641. Suggested as a more appropriate name for the manganese amphibole juddite. Cf. ferrorichterite (18th List). *See* Eisenrichterit, iron-richterite. 20th List [Manganoan magnesio-arfvedsonite, *Amph.* (1978)]

Ferrisalites. T. E. Khmaruk and I. B. Shcherbakov, 1963. [Матер. петрограф. мінер. Украiн. крист. щита, Акад. наук УССР]; abstr. in Зап. Всесоюз. Мин. Общ. (*Mem. All-Union Min. Soc.*), 1965, vol. 94, p. 199 (Феррисалиты, ferrisalites). A group-name for certain clinopyroxenes high in Fe_2O_3 and Al_2O_3 and low in SiO_2. Not to be confused with Ferrosalite (16th List), a magnesian Hedenbergite [in *M.A.*, **17**, 408, the name is misprinted ferrosalite]. 24th List

Ferrisaponite. H. Strunz, 1957. *Min. Tabellen*, 3rd edit., p. 313 (Ferri-Saponit). A group name to include griffithite and lembergite (of Sudo). 22nd List

Ferri-sarcolite. P. Niggli, 1922. Abstr. in *Zeits. Krist.*, vol. 57, p. 105 (Ferri-Sarkolithe). Hypothetical end-members, $3CaO.Fe_2O_3.3SiO_2$ and $3Na_2O.Fe_2O_3.3SiO_2$, of the melilite group, corresponding to sarcolite ($3CaO.Al_2O_3.3SiO_2$) and soda-sarcolite ($3Na_2O.Al_2O_3.3SiO_2$). The possibility of a 'ferric iron sarcolite' ($3CaO.Fe_2O_3.3SiO_2$) was suggested by W. T. Schaller, *Bull. US Geol. Survey*, 1916, no. 610, p. 119. 9th List

Ferrisepiolite. H. Strunz, 1956. *Fortschr. Min.*, vol. 34, p. 48 (Ferrisepiolith). Group name to include gunnbjarnite and xylotile. 22nd List

Ferri-sericite. H. Strunz, 1957. *Min. Tabellen*, 3rd edit., p. 372 (Ferri-Sericit). Internationalized form of Iron-sericite (20th List). 22nd List [Also 24th List]

Ferriserpentines, mistaken translation of Железистые серпентины, ferruginous serpentines (Доклады Акад. Наук СССР [*Compt. Rend. Acad. Sci. URSS*], 1960, vol. 130, p. 1325). 22nd List

Ferri-sicklerite. P. Quensel, 1937. *Geol. För. Förh. Stockholm*, vol. 59, p. 85 (ferri-sicklerite). An intermediate member of the series triphylite–ferri-sicklerite–heterosite in which iron predominates over manganese, the parallel series in which manganese predominates being lithiophilite–sicklerite (Mn-sicklerite)–purpurite. [*M.A.*, **6**, 485.] 14th List [Cf. Sicklerite. $Li(Fe,Mn)PO_4$; *A.M.*, **22**, 875; **26**, 681. Pure crystals, Morocco; *M.A.*, 77-2171]

Ferristilpnomelane. A. N. Winchell, 1951. *Optical mineralogy*, part 2, 4th edit., p. 390. Oxidized ferrostilpnomelane (15th List); synonym of chalcodite. 19th List

Ferrisymplesite. T. L. Walker and A. L. Parsons, 1924. *Univ. Toronto Studies, Geol. Ser.*, no. 17, p. 17. Hydrous ferric arsenate, $3Fe_2O_3.2As_2O_5.16H_2O$, occurring as amber-brown, resinous particles in erythrite from Cobalt, Ontario. Named on account of its relation to symplesite ($3FeO.As_2O_5.8H_2O$), of which it is perhaps an oxidation product. [*M.A.*, **2**, 382.] 10th List [*A.M.*, **10**, 134]

Ferritchromit. K. Spangenberg, 1943. *Zeits. prakt. Geol.*, **51**, 23. The grey, magnetic outer zone on many chromite grains from Tampadel, Zobten, Silesia, is enriched in Fe^{2+} and Fe^{3+} and depleted in Mg^{2+} and Al^{3+}, and is termed Ferritchromit as distinct from the inner, fresh Aluminatchromit (q.v.). No analysis is given; probably the same as the ferrichrom-picotite of Betekhtin (14th List). Cf. Ferritspinelle (24th List). 27th List

Ferri-taramite. *Amph.* (1978), **14**, 15. 30th List

Ferrite. H. M. Howe, 1890. *The Metallurgy of Steel*, New York, vol. i, pp. 164, 165. Pure or nearly pure metallic iron detected under the microscope as a crystalline constituent of manufactured iron and steel. The same term has been applied by V. I. Vernadsky (*Opuit opisatelnoĭ Mineralogii*, St. Petersburg, 1908, vol. i, pp. 156, 162 (Феррит); *Centralblatt Min.*, 1912, 759 (Ferrit)) to native iron; for example, the terrestrial iron from Uifak, Disko Island, Greenland.

Unfortunately, however, the name ferrite had been earlier used in different senses. Chemically, it signifies a salt in which ferric oxide plays the part of an acid; for example, barium ferrite, $BaFe_2O_4$. Mineralogically, it was used by H. Vogelsang in 1872 for an amorphous iron hydroxide of unknown composition occurring as red and yellow patches in decomposed igneous rocks; and by M. F. Heddle in 1882 for ferruginous pseudomorphs after olivine. 6th List

—— K. Kristoffersen, 1950. *K. Norske Vid. Selsk. Skrifter*, for 1947, no. 4, p. 14 (grønnbrunt nåleformig mineral (ferritt), dikalsiumferritt $2CaO.Fe_2O_3$, tetrakalsiumaluminatferritt $4CaO.Al_2O_3.Fe_2O_3$), p. 308 (Ferrite, dicalcium alumoferrite). Used alternatively as a mineral or chemical name for green to brown needles, Ca_2AlFeO_5 to $Ca_2Fe_2O_5$, present in basic slags. Those of the composition Ca_2AlFeO_5 are identical with brownmillerite (16th List). [*M.A.*, **11**, 191.] 19th List

Ferrite-spinels. P. Ramdohr, 1950. *Erzmineralien*, Berlin, p. 657 (Ferritspinelle). Collective name for minerals of the magnetite series, ferrite compounds RFe_2O_4 (R = Mg,Fe,Zn,Mn), in the spinel group. 20th List [Also 24th List]

Ferrithorite. I. E. Starik, L. L. Kravchenko, and O. S. Melikova, 1941. *Compt. Rend. (Doklady) Acad. Sci. URSS*, vol. 32, p. 254. A ferriferous (Fe_2O_3 12·02, FeO 3·55%) variety of thorite from Kirghizia. The same as ferrothorite (11th List). [*M.A.*, **8**, 302.] 16th List

Ferrititanbiotit, *see* Ferriwotanite.

Ferritremolite. A. N. Winchell, 1949. *Amer. Min.*, vol. 34, p. 224; *Optical mineralogy*, part 2, 4th edit., 1951, p. 437. Hypothetical molecule $Ca_2Fe_3''Fe_2'''O_2Si_8O_{22}$, in oxyhornblende. Compare ferrotremolite (13th List). 19th List [Ferri-ferro-actinolite, *Amph.* (1978)]

Ferritschermakite. A. N. Winchell, 1949. *Amer. Min.*, vol. 34, p. 224; *Optical mineralogy*, part 2, 4th edit., 1951, p. 437. Hypothetical molecule $Ca_2Fe''(Al,Fe''')_4O_2Si_6Al_2O_{22}$, in oxyhornblende. Cf. ferrotschermakite (17th List). 19th List

Ferritspinelle, *see* Ferrite-spinels.

Ferritungstite. W. T. Schaller, 1911. *Journ. Washington Acad. Sci.*, vol. i, p. 24; *Amer. Journ. Sci.*, 1911, ser. 4, vol. xxxii, p. 161; *Zeits. Kryst. Min.*, 1912, vol. 1, p. 112; *Bull. US Geol. Survey*, 1912, no. 509, p. 83. A yellow ochre closely resembling tungstite (tungstic-ochre), but consisting of hydrated ferric tungstate, $Fe_2O_3.WO_3.6H_2O$. Under the microscope it shows minute hexagonal plates which are optically isotropic on the base. Occurs as an alteration-product of wolframite in the State of Washington. 6th List [Cubic $Ca_2Fe^{2+}_2Fe^{3+}_2(WO_4)_7.9H_2O$; *A.M.*, **42**, 83]

Ferri-turquoise. E. E. Fairbanks, 1942. *The Mineralogist*, Portland, Oregon, vol. 10, p. 44. A variety of crystallized turquoise containing Fe_2O_3 5%, from Lynch, Virginia. [*M.A.*, **8**, 270.] 16th List

Ferri-winchite. *Amph.* (1978), **14**, 3. 30th List

Ferriwotanite. H. Meixner, 1939. *Fortschr. Min. Krist. Petr.*, vol. 23, p. xliv (Ferriwotanite = Ferrititanbiotite, *pl.*). F. Angel and A. Marchet, ibid., p. xxxvii (Ferri-Wotanit). Biotite rich in ferric iron and titanium. (Cf. wodanite. 9th List; wotanite, 14th List.) 15th List

Ferroactinolite. N. Sundius, 1946. *Årsbok Sveriges Geol. Undersök.*, vol. 40, p. 7. A hypothetical molecule $Ca_2Fe_5''Si_8O_{22}(OH)_2$ to explain the composition of the amphibole group. [*M.A.*, **10**, 70.] 18th List [*Amph.* Sub-Comm.]

Ferro-åkermanite. O. M. Shubnikova and D. V. Yuferov, *Spravochnik po novym mineralam*, Leningrad, 1934, p. 71 (ферроакерманит, ferro-, iron-, Eisen-åkermanite). Translation of Eisen-Åkermanit (9th List) and Iron-åkermanite (10th List). 14th List

Ferroalabandine. *See* Eisenalabandin. 21st List

Ferro-alluaudite. D. J. Fisher, 1957. *Amer. Min.*, vol. 42, p. 661. Synonym of Varulite. 23rd List

Ferro-alumino-barroisite. *Amph.* (1978), **14**, 6. 30th List

Ferro-alumino-tschermakite. *Amph.* (1978), **10**, 10. 30th List

Ferro-alumino-winchite. *Amph.* (1978), **14**, 2. 30th List

Ferroalunite. G. A. Gvakhariya and Yu. I. Nazarov, 1963. [Сообщ. Акад. наук Груз. ССР (*Comm. acad. sci. Georgian SSR*), vol. 32, p. 381 (Ферроалунит)]; abstr. *M.A.*, **17**, 402. An unnecessary and inaccurate name for ferrian (not ferroan) Alunite from the Madneul copper and baryte–lead–zinc deposits. 24th List

Ferroalunogen, *see* Ferrialunogen.

Ferroamesite. G. L. Dschang, 1931. *Chem. Erde*, vol. 6, p. 427 (Ferroamesit). A hypothetical molecule $H_4Fe_2Al_2SiO_9$ (corresponding with amesite $H_4Mg_2Al_2SiO_9$) to explain the composition of the chlorites. [*M.A.*, **5**, 39.] 13th List

Ferroankerite. I. Mincheva-Stefanova and M. Gorova, 1967. [*Bull. Geol. Inst., Ser. Geochem. Min. Petr.* (Bulgaria), **16**, 95]; abstr. *M.A.*, **18**, 282. An unnecessary name for a member of the dolomite family, $Ca_2FeMg(CO_3)_4$. 28th List

Ferroanthophyllite. E. V. Shannon, 1921. *Proc. US Nat. Museum*, vol. 59, p. 397. The iron end-member of the anthophyllite series, occurring as greyish-green, asbestiform fibres in Idaho. Similarly, magnesioanthophyllite (p. 401) for the magnesium end-member. The same as iron-anthophyllite (C. H. Warren, 1903) and Eisenanthophyllite (J. Palmgren, 1917; 8th List). [*M.A.*, **1**, 253.] 9th List [**Ferro-anthophyllite**, *Amph.* (1978)]

Ferro-antigorite. A. N. Winchell, 1926. *Amer. Journ. Sci.*, ser. 5, vol. 11, p. 284 (ferro-antigorite); *Amer. Min.*, 1928, vol. 13, p. 166 (ferroantigorite). A hypothetical molecule $H_4Fe''_3Si_2O_9$ (corresponding with antigorite $H_4Mg_3Si_2O_9$) to explain the composition of the chlorites. *See* Iron-antigorite. [*M.A.*, **3**, 373.] 11th List

Ferroaugite. H. H. Hess, 1941. *Amer. Min.*, vol. 26, pp. 517, 577, *passim*. A variety of augite rich in iron and distinct from pigeonite. [*M.A.*, **8**, 234.] 16th List

Ferroaxinite. W. T. Schaller, 1909. Second Appendix to 6th edit. of Dana's *System of Mineralogy*, p. 11. Axinite consists of isomorphous mixtures of ferroaxinite, $8SiO_2.2Al_2O_3.2FeO.H_2O.4CaO.B_2O_3$, and manganoaxinite, $8SiO_2.2Al_2O_3.2MnO.H_2O.4CaO.B_2O_3$. 5th List [Cf. Manganaxinite, Tinzenite; *M.A.*, 71-2989]

Ferrobabingtonite. R. A. Vinogradova and I. N. Plyusinna, 1967. [Вест. Москов. унив., сер. 4, Геол. (*Rept. Moscow Univ.*, ser. 4, geol), 54]; abstr. *Amer. Min.*, **53**, 1064 (1968). Superfluous synonym of babingtonite. 26th List

Ferro-berthierine. H. Strunz, 1957. *Min. Tabellen*, 3rd edit., p. 323 (Ferro-Berthierin). Syn. of Berthierine. The ferrous chamosite of G. W. Brindley and R. F. Youell (*Min. Mag.*, **30**, 57). 22nd List

Ferrobrucite. A. Lacroix, 1909. *Minéralogie de la France*, vol. iii, p. 402. Brucite (MgO_2H_2) containing a small amount of ferrous oxide. The same as Eisenbrucit of F. von Sandberger, 1880. 6th List

Ferrobustamite. P. A. Rapoport and C. W. Burnham, 1973. *Zeits. Krist.*, **138**, 419. 'Iron rhodonite' from Skye described by C. E. Tilley (*A.M.*, **33**, 736; *M.A.*, **11**, 16; but Tilley actually used the name Iron-wollastonite, 14th List; cf. ferrowollastonite, 19th List) is shown to have the bustamite structure and is renamed. Data of Rutstein (*A.M.*, **56**, 2040) indicate a gap in the series ferrowollastonite–ferrobustamite. [*A.M.*, **59**, 632.] 28th List

Ferro-calderite. L. L. Fermor, 1926. *See* Calc-spessartite. 11th List

Ferrocarpholite. W. P. de Roever, 1951. *Amer. Min.*, vol. 36, p. 736. $H_4FeAl_2Si_2O_{10}$, orthorhombic, analogous to carpholite with Fe in place of Mn. [*M.A.*, **11**, 413.] 19th List

Ferrochalcanthite. G. A. Yurgenson, 1971. *Zap.*, **100**, 359 (Феррохалькантит). $(Fe,Cu)SO_4.5H_2O$, with Fe ≈ Cu; syn. of Ferrocuprochalcanthite, internationalized form of iron-copper-chalcanthite, 9th List. [*M.A.*, 72-3309.] 28th List

Ferro-chamosite. G. W. Brindley and R. F. Youell, 1953. *Min. Mag.*, vol. 30, p. 57 (ferrous chamosite). Synonym of chamosite. When heated in air or weathered it changes to ferri-chamosite (q.v.). 20th List

Ferrochromite. J. Beckenkamp, 1921. *See* Talc-spinel. 9th List

Ferro-chrysotile. F. V. Syromyatnikov, 1934. *Bull. Soc. Nat. Moscou*, ser. 2, vol. 42 (*Sect. Geol.*, vol. 12), p. 568 (ферро-хризотил), p. 574 (ferro-chrysotile). The molecule $H_4Fe_3Si_2O_9$ present in ferruginous chrysotile. [*M.A.*, **6**, 259.] 14th List

Ferro-clinoholmquistite. Amph. (1978). 6.4. 30th List

Ferrocopiapite. L. G. Berry, 1938. *See* Ferricopiapite. 15th List [Syn. of Copiapite]

Ferrocordierite. W. Schreyer, 1966. *Fortschr. Min.*, vol. 42 (for 1964), p. 213. (Ferrocordierit). Internationalized form of Eisencordierit (3rd List), Iron-cordierite (10th List). 24th List [Renamed Sekaninaite (q.v.)]

Ferrodickinsonite. D. J. Fisher, 1954. *Amer. Min.*, vol. 39, p. 676 (ferroan dickinsonite), p. 840 (ferro dickinsonite); ibid., 1955, vol. 40, p. 1107 (ferrodickinsonite); *Science* (*Amer. Assoc. Adv. Sci.*), 1955, vol. 121, p. 312. Alternative name for arrojadite (11th List). A dickinsonite richer in iron. [*M.A.*, **12**, 561.] *See* Manganodickinsonite. 21st List

Ferrodolomite. A. N. Winchell, 1927. *See* Magnesiodolomite. 14th List

Ferro-eckermannite. R. Phillips and W. Layton, 1964. *Min. Mag.*, vol. 33, p. 1097. The hypothetical amphibole end-member $Na_3Fe_4^{2+}AlSi_8O_{22}(OH)_2$. 23rd List

Ferroedenite. N. Sundius, 1946. *Årsbok Sveriges Geol. Undersök.*, vol. 40, p. 6. A hypothetical molecule $NaCa_2Fe_5''AlSi_7O_{22}(OH)_2$ to explain the composition of the amphibole group. [*M.A.*, **10**, 70.] 18th List

Ferroepsomite. G. N. Vertushkov, 1939. *Bull. Acad. Sci. URSS, Sér. Géol.*, 1939, no. 1, p. 110 (ферроэпсомит), p. 115 (ferroepsomite). A variety of epsomite containing up to 30% of the tauriscite [orthorhombic $FeSO_4.7H_2O$] molecule. [*M.A.*, **7**, 418.] 15th List

Ferro-ferri-barroisite. *Amph.* (1978), **14**, 8. 30th List

Ferroferrichromite. V. D. Ladieva, 1964. [*Chem. comp. and intern. struct. min.*, Kiev, p. 192]; abstr. *M.A.*, **17**, 389. An unnecessary name for a ferrian Chromite or chromian Magnetite. 24th List

Ferro-ferri-lazulite. M. A. Gheith, 1953. *Amer. Min.*, vol. 38, p. 612 (ferrous ferric lazulite). Artificially produced monoclinic iron phosphate approximating to $Fe''Fe_2'''(PO_4)_2(OH)_2$, isomorphous with lazulite with Fe in place of Mg and Al. Later named barbosalite (q.v.). At high temperature it changes over to a tetragonal modification, lipscombite (q.v.). [*M.A.*, **12**, 238.] 20th List

Ferroferrimargarite. A. I. Ginzburg, 1955. *Trudy Min. Mus. Acad. Sci. USSR*, no. 7, p. 75 (ферроферримаргарит). A brittle mica, chemical variety of margarite. [*M.A.*, **13**, 209, 521.] 21st List

Ferroferrite. J. Beckenkamp, 1921. *See* Talc-spinel. 9th List

Ferro-ferri-tschermakite. *Amph.* (1978), **10**, 12. 30th List

Ferro-ferri-winchite. *Amph.* (1978), **14**, 4. 30th List

Ferrofillowite. D. J. Fisher, 1955. *Bull. Geol. Soc. Amer.*, vol. 66, p. 1558. Na-Fe-Mn phosphate. 'The Fe″ compound ferrodickinsonite (q.v.), heated in vacuum, inverts irreversibly to ferrofillowite.' Synonym of fillowite (Brush and Dana, 1879). 21st List

Ferro-friedelite. H. Strunz, 1957. *Min. Tabellen*, 3rd edit., p. 327 (Ferro-Friedelit). To replace the name Ferroschallerite (12th List), as this mineral is regarded as a variety of friedelite and not of schallerite (cf. *M.A.*, **13**, 125). 22nd List

Ferrogedrite. C. E. Tilley, 1939. *Geol. Mag. London*, vol. 76, p. 329. An aluminous orthorhombic amphibole (gedrite) rich in ferrous iron. Compare ferroanthophyllite (9th List). 15th List [**Ferro-gedrite**, *Amph.* (1978)]

Ferroglaucophane. A. Miyachiro, 1957. *Journ. Fac. Sci. Univ. Tokyo*, sect. 2, vol. 11, p. 57. A name for the clino-amphibole end-member $Na_2Fe_3^{\cdot\cdot}Al_2Si_8O_{22}(OH)_2$. 22nd List [**Ferro-glaucophane**, *Amph.* (1978)]

Ferrohalotrichite. E. Z. Vieira de Mello, 1969. [*Brasil, Min. Inter. Supt. Desenvolvimento Nordeste, Dept. Recurrsos Naturais, Div. Geol. Ser. Especial*, no. 10, 1 (Ferrohalotriquita)], abstr. *A.M.*, **56**, 1122 (1971). A superfluous name for halotrichite. 27th List

Ferrohastingsite. M. Billings, 1928. *Amer. Min.*, vol. 13, p. 287. The original hastingsite, from Hastings Co., Ontario, in which ferrous oxide predominates largely over magnesia. In Magnesiohastingsite magnesia predominates, and intermediate members of the group are distinguished as Femaghastingsite (p. 292). Other varieties containing more alkalis are distinguished as Alkali-hastingsite, including Alkali-ferrohastingsite and Alkali-femaghastingsite (p. 294). [*M.A.*, **4**, 39.] 12th List [Syn. of Hastingsite, *Amph.* (1978)]

Ferrohedenbergite. A. Poldervaart, 1947. *Min. Mag.*, vol. 28, pp. 159, 161. F. Walker and A. Poldervaart, *Bull. Geol. Soc. Amer.*, 1949, vol. 80, p. 632. A variety of hedenbergite richer in iron. 18th List

Ferrohexahydrite. J. Kubisz, 1958. *Bull. Acad. Polon. Sci., Sér. sci. chim., géol., géogr.*, vol. 6, p. 459. A name for the hypothetical monoclinic end-member $FeSO_4.6H_2O$. 22nd List [Cf. Hexahydrite; *A.M.*, **48**, 433]

Ferro-holmquistite. *Amph.* (1978), **3**, 8. 30th List

Ferro-hornblende. *Amph.* (1978), **10**, 14. 30th List

Ferrohortonolite. W. A. Deer and L. R. Wager, 1939. *Amer. Min.*, vol. 24, p. 25. Members of the olivine group containing 70–90 mol.% of Fe_2SiO_4. [*M.A.*, **7**, 447.] 15th List

Ferrohumite. N. K. Skakovsky, 1929. [*Min. Syre*, Moscow, no. 8, p. 913.] O. M. Shubnikova, *Trans. Lomonosov Inst. Acad. Sci. USSR*, 1937, no. 10, p. 138 (Феррогумит, Ferrohumite). Humite containing FeO 44·47% from Sadon mine, northern Caucasus. This mineral was later referred to knebelite by L. A. Vardanyantz, 1934 [*M.A.*, **6**, 426]. 15th List

Ferrohydrite. F. V. Chukhrov, 1936. [*Colloids in the earth's crust* (*Russ.*), *Acad. Sci. USSR*, 1936, p. 97.] O. M. Shubnikova, *Trans. Inst. Geol. Sci., Acad. Sci. USSR*, 1938 [i.e. 1940], no. 11 (*Min. Geochem. Ser.*, no. 3), p. 6 (Феррогидрит, Ferrohydrite). Colloidal hydrous iron oxide occurring in the mud of salt lakes. 16th List

Ferrohypersthene. H. H. Hess and A. H. Phillips, 1940. *Amer. Min.*, vol. 25, p. 285. Members of the enstatite-orthoferrosilite series between hypersthene (80–50 mol.% enstatite) and orthoferrosilite (12–0% En). Compare iron-hypersthene (14th List). [*M.A.*, **8**, 18.] 16th List

Ferro-johannsenite. K. F. Chudoba, 1959. Hintze, *Handb. Min.*, Erg.-Bd. II, p. 705 (Ferro-Johannsenit). An unnecessary name for the ferroan johannsenite of V. T. Allen and J. J. Fahey (1953). 22nd List

Ferro-kaersutite. *Amph.* (1978), **10**, 16. 30th List

Ferrolazulite. W. E. Tröger, 1952. *Tabellen opt. Bestim. gesteinsbild. Minerale*, p. 145 (Ferrolazulith). Variety of lazulite containing some FeO replacing MgO. Distinct from scorzalite (18th List) and iron-lazulite (19th List). 20th List

Ferrolite. Author and date? A name proposed for a black iron slag to be used as gemstones. R. Webster, *Gems*, 1962, p. 756. 28th List

Ferrolizardite. Ping-Wen Chia and Che Cheng, 1964. [*Ti chih Hseuh Pao*, vol. 44, p. 86]; abstr. *Amer. Min.*, 1965, vol. 50, p. 2102; *M.A.*, **17**, 504. An unnecessary name for a ferroan Lizardite from the neighbourhood of Peiping, China. 24th List

Ferroludwigite. B. S. Butler and W. T. Schaller, 1917. *See* Magnesioludwigite. 8th List [Is Vonsenite; *A.M.*, **14**, 102]

Ferromagnesite. H. Strunz, *Min. Tab.*, 1941, p. 117 (Ferromagnesit). Syn. of mesitite. Similarly, for other mixed crystals of the calcite group: Zinkocalcit, Kobaltcalcit, Mangancalcit, Strontiumcalcit, Bariumcalcit, Zinkrhodochrosit, Ferrorhodochrosit, Kalkrhodochrosit, Zinksiderit, Mangansiderit, Calciumsiderit, Ferrosmithsonit, Mangansmithsonit. 18th List

Ferromangandolomite. J. A. Smythe and K. C. Dunham, 1947. *Min. Mag.*, vol. 28, p. 65. Hypothetical molecule $(Fe,Mn)CO_3.CaCO_3$ as an end-member of the ankerite series. 18th List

Ferro-miyashiroite. R. Phillips and W. Layton, 1964. *Min. Mag.*, vol. 33, p. 1097. The hypothetical amphibole end-member $Na_3Fe_3^{2+}Al_3Si_7O_{22}(OH)_2$. 23rd List

Ferromontmorillonite. I. I. Ginzburg and A. I. Ponomarev*, 1939. *Bull. Acad. Sci. URSS, Sér. Géol.*, 1939, no. 1, p. 85 (ферромонтмориллонит), p. 94 (ferromontmorillonite). Synonym of nontronite. [*M.A.*, **8**, 40.] [*Corr. in *M.M.*, **28**.] 15th List

Ferronemalite. A. Fersmann, 1911. *Bull. Acad. Sci. St.-Pétersbourg*, ser. 6, vol. v, p. 551 (Ферронемалить). Abstr. in *Chem. Zentralblatt*, 1911, vol. i, p. 1879 (Ferronemalith). A variety of nemalite from Caucasus containing some ferrous oxide (FeO 5%). *See* Ferrobrucite. 6th List

Ferroniobite. K. F. Chudoba, 1959. Hintze, *Handb. Min.*, Erg.-Bd. II, p. 705 (Ferroniobit). Syn. of Ferrocolumbite (of Simpson). 22nd List

Ferro-ortho-titanate. F. Mogensen, 1946. *See* Ulvöspinel. 18th List

Ferropallidite. R. Scharizer, *Zeits. Kryst. Min.*, 1903, vol. xxxvii, p. 547 (Ferropallidit). A white, granular ferrous sulphate, $FeSO_4.H_2O$, occurring with römerite in Chile. Named from *ferrum*, iron, and *pallidus*, pale. 3rd List

Ferro-paraluminite, *see* Ferri-paraluminite.

Ferropargasite. J. F. G. Wilkinson, 1961. *Amer. Min.*, vol. 46, p. 341. This name is preferred to hastingsite for the clino-amphibole end-member $NaCa_2(Fe_4Al)(Si_6Al_2)O_{22}(OH)_2$, 'inasmuch as the term hastingsite for many years has referred to Fe-rich amphiboles with small amounts of MgO, or else to various calciferous amphiboles with variable FeO/MgO ratios' [hastingsite has also been used for the end-member $NaCa_2Fe_4^{\cdot\cdot}Fe^{\cdot\cdot\cdot}Al_2Si_6O_{22}(OH)_2$ (*M.A.*, **10**, 70)]. There are already too many names for particular 'end-members' of the clino-amphibole series. 22nd List [**Ferro-pargasite**, redefined (not hastingsite), *Amph.* (1978), 10, 6, 30th List]

Ferropericlase. C. E. Tilley, 1951. *Min. Mag.*, vol. 29, p. 629. Iron-bearing variety of periclase, (Mg,Fe)O. Compare magnesiowüstite (14th List). 19th List

Ferrophengite. A. N. Winchell, 1949. *Amer. Min.*, vol. 34, p. 223. A hypothetical molecule $K_2FeAl_3(OH)_4Si_7AlO_{20}$, classed as a sub-species of muscovite. Compare picrophengite (q.v.). [*M.A.*, **10**, 568.] 18th List

Ferrophlogopite. V. A. Frank-Kamenetsky, 1958. Referred to by R. C. Mackenzie, *Clay Min. Bull.*, 1959, vol. 4, p. 61. Undefined; presumably a variety of biotite. 22nd List

Ferropickeringite. G. N. Vertushkov, 1939. *Bull. Acad. Sci. URSS, Sér. Géol.*, 1939, no. 1, p. 110 (ферропиккерингит), p. 114 (ferropickeringite). A variety of pickeringite containing up to 30% of the halotrichite molecule. Synonym of Eisenpickeringit (q.v.). [*M.A.*, **7**, 418.] 15th List

Ferropicotite. A. Lacroix, 1910. *Minéralogie de la France*, vol. iv, p. 306. A variety of spinel differing from picotite in containing ferric oxide in place of chromic oxide, the formula being $(Fe,Mg)(Al,Fe)_2O_4$. 6th List

Ferropigeonite. W. N. Benson, 1944. *Trans. Roy. Soc. New Zealand*, vol. 74, p. 115 (ferropigeonite), p. 117 (ferro-pigeonite). Pigeonite (restricted to material with optic axial angle 2V 0–30°) is subdivided into magnesian pigeonite, pigeonite, and ferropigeonite. Pigeonite of other authors with 2V 30–45° is called subcalcic augite. [*M.A.*, **9**, 151.] 17th List

Ferroplatinum. V. I. Vernadsky, *Opuit opisatelnoĭ Mineralogii*, St. Petersburg, 1908, vol. i, p. 156 (Ферроплатина), *Centralblatt Min.*, 1912, p. 759 (Ferroplatin). Synonym of Eisenplatin of Breithaupt (1832) = Iron-platinum (Dana, 1854). 6th List

Ferroplumbite. *Chem. Abstr.* (*Amer. Chem. Soc.*), 1925, vol. 19, p. 2621. Error for Plumboferrite (L. J. Igelström, 1881; A. Aminoff, 1925 [*M.A.*, **4**, 89]). 15th List

Ferroprehnite. R. A. A. Johnston, 1913. *Victoria Memorial Museum, Geol. Survey Canada*, Bull. no. 1, p. 98; *A List of Canadian mineral occurrences, Canada, Dept. of Mines* (*Geol. Survey*), 1915, *Memoir 74* (*Geol. Ser. 61*), p. 183. A variety of prehnite rich in iron (Fe_2O_3 6·58%) from Baffin Island, Arctic Canada. 7th List

Ferropseudobrookite. Author? *Artificial* $Fe^{2+}Ti_2O_5$, a member of the pseudobrookite family. Name also used in preliminary reports for armalcolite (q.v.); cf. karrooite (22nd List). [*A.M.*, **55**, 2136.] 27th List

Ferropumpellyite. P. B. Moore, 1971. *Lithos*, **4**, 98. The Fe^{2+} analogue of pumpellyite, $Ca_2Fe^{2+}AlSi_3O_{11}(OH)_2.H_2O$. [*A.M.*, **56**, 2158.] 27th List

Ferropyroaurite, ferripyroaurite. H. Meixner, 1937. *Zentr. Min.*, Abt. A, 1937, p. 370 (Ferropyroaurit, Ferripyroaurit). The white 'Eisenbrucite' of F. Sandberger, 1880, from Siebenlehn, Saxony, belongs to the hydrotalcite group and is called ferropyroaurite, $MgCO_3.2Fe(OH)_2.5Mg(OH)_2.4H_2O$. On oxidation this passes to gold-yellow or brown ferripyroaurite, $MgCO_3.2Fe(OH)_3.5Mg(OH)_2.4H_2O$. [*M.A.*, **7**, 218.] 15th List

Ferrorhabdite. (A. Lacroix, *Minéralogie de la France*, 1909, vol. iii, p. 618.) Synonym of rhabdite. 7th List

Ferrorhodochrosite. (Author?). A. E. Fersman and O. M. Shubnikova, *Geochem. Min. Companion*, 1937, p. 213 (ферроpодохрозит). Ferriferous variety of rhodochrosite. Cf. ponite (6th List). 20th List

Ferrorhodonite. I. Kostov, 1968. Kostov, *Mineralogy*, p. 350. An unnecessary name for ferroan rhodonite. 25th List

Ferrorichterite. N. Sundius, 1946. *Årsbok Sveriges Geol. Undersök.*, vol. 40, p. 16. A hypothetical molecule $Na_2CaFe_5''Si_8O_{22}(OH)_2$ to explain the composition of the amphibole group. [*M.A.*, **10**, 70.] 18th List

Ferrorömerite. R. Scharizer, *Zeits. Kryst. Min.*, 1903, vol. xxxvii, p. 546 (Eisenrömerit, ferrorömerit). A synonym of römerite. *See* zinc-römerite. 3rd List

Ferrosalite. H. H. Hess, 1941. *Amer. Min.*, vol. 26, p. 518. A variety of the clinopyroxene sahlite (= salite), rich in iron. [*M.A.*, **8**, 234.] 16th List

Ferrosalites, misprint for Ferrisalites (*M.A.*, **17**, 408), with possible confusion with magnesian Hedenbergite (16th List). 24th List

Ferroschallerite. L. H. Bauer and H. Berman, 1930. *Amer. Min.*, vol. 15, p. 345. A variety of schallerite (10th List) in which the manganese is partly replaced by iron (FeO 17·12%). [*M.A.*, **4**, 345.] 12th List [Is Friedelite; *A.M.*, **38**, 755]

Ferroselite. E. Z. Buryanova and A. I. Komkov, 1955. *Doklady Acad. Sci. USSR*, vol. 105, p. 812 (ферроселит). Iron selenide $FeSe_2$, orthorhombic, resembling marcasite, from Tuva, Siberia. Named from the composition. Cf. achavalite (18th List, *M.A.*, **12**, 236), and eskebornite (19th List). [*M.A.*, **13**, 84.] 21st List

Ferrosilicon. V. Kh. Gevorkyan, 1968. [Докл. акад. наук УССР (*Compt. rend. Acad. sci. Ukrain. RSS*), ser. geol. **2**, 513], abstr. *Zap.*, **99**, 71. Ferrosilicon in drill cores from Konsko-Yalynsk is believed to be natural; later found to consist of two phases, named fersilicite and ferdisilicite (26th List). 27th List

Ferrosilite. 'Iddings and Washington, 1903', according to E. V. Shannon, *Amer. Min.*, 1921, vol. 6, p. 87; H. S. Washington, *Proc. US Nat. Mus.*, 1922, vol. 60, art. 14, p. 6. The ferrous silicate molecule, $FeSiO_3$, used as a 'normative' or standard pyroxene mineral in the calculation of rock analyses. 9th List [Cf. Clino- and Orthoferrosilite]

Ferro-spessartite. L. L. Fermor, 1926. *See* Calc-spessartite. 11th List

Ferrospinel. H. Strunz, 1957. *Min. Tabellen*, 3rd edit., p. 137 (Ferrospinell). Syn. of Hercynite (of Zippe). 22nd List

Ferrostilpnomelane. C. O. Hutton, 1938, *Min. Mag.*, vol. 25, pp. 172, 177. To replace the name stilpnomelane for hydrated silicate of ferrous iron, etc.; the latter name being transferred to the corresponding ferric compound. [The name stilpnomelane was, however, originally applied to the dark-green ferrous silicate, which on oxidation passes to the brown ferric silicate, chalcodite.] 15th List

Ferro-sundiusite. R. Phillips and W. Layton, 1964. *Min. Mag.*, vol. 33, p. 1097. The hypothetical amphibole end-member $Na_2CaFe_3^{2+}Al_4Si_6O_{22}(OH)_2$. 23rd List

Ferrothorite. A Lacroix, 1923. *Minéralogie de Madagascar*, vol. 3, p. 310 (ferrothorite). A variety of thorite containing iron (Fe_2O_3 13·1%) from Madagascar. 11th List

Ferrotine. P. Sasima in O. M. Shubnikova, *Trans. Lomonosov Inst. Acad. Sci. USSR, Ser. Min.*, 1937, no. 10, p. 223 (Ферротин, Ferrotine). An undetermined iron oxide, Fe_nO_{n+1} (FeO 61·30, Fe_2O_3 33·20%), as dark-grey, strongly magnetic scales and spherules from the Ayata river, tributary of the Yenesei, Siberia. (Evidently partially oxidized magnetite.) 15th List [Also 17th List]

Ferrotremolite. A. N. Winchell, 1932. *Amer. Min.*, vol. 17, pp. 114, 472; Optical mineralogy, part 2, 3rd edit., 1933, p. 245. The end-member $H_2Ca_2Fe_5Si_8O_{24}$ of the tremolite-actinolite series, actinolite being really an intermediate member. [*M.A.*, **5**, 216.] 13th List [Ferro-actinolite, *Amph.* (1978)]

Ferrotschermakite. A. N. Winchell, 1945. *Amer. Min.*, vol. 30, p. 29. A hypothetical molecule $Ca_2Fe_3''Fe_2'''Al_2Si_6O_{22}(OH)_2$ to explain the composition of aluminous amphiboles. *See* Tschermakite. [*M.A.*, **9**, 271.] 17th List

Ferrous riebeckite. W. G. Ernst, 1957. *Ann. Rep. Geophysical Lab.*, for 1956–1957, p. 228 (ferrous riebeckite). Artificial $Na_2Fe_3''Fe_2'''Si_8O_{22}(OH)_2$. *See* Magnesian riebeckite. 21st List

Ferrovonsenite. A. A. Brovkin, S. M. Aleksandrov, and I. Ya. Nekrasov, 1963. [Рентгеногр. мин. сырья, Госгеолтехиздат, no. 3, p. 16]; abstr. Зап. Всесоюз. Мин. Общ. (*Mem. All-Union Min. Soc.*), 1965, vol. 94, p. 675 (Ферровонсенит, ferrovonsenite). Syn. of Vonsenite (= Ferroludwigite = Paigeite). 24th List

Ferrowollastonite. S. O. Agrell, 1950. *Amer. Min.*, vol. 35, p. 1080. Synonym of iron-wollastonite (14th List). 19th List

Ferrozincrhodochrosite. H. Strunz, 1957. *Min. Tabellen*, 3rd edit., p. 173 (Ferrozinkrhodochrosit). Syn. of Capillitite (19th List). 22nd List

Ferruccite. G. Carobbi, 1933. *Periodico Min. Roma*, vol. 4, p. 410. Sodium fluoborate, $NaBF_4$, occurring as minute orthorhombic crystals in mixed fumarolic sublimations on Vesuvius. Named after Prof. Ferruccio Zambonini (1880–1932), of Napoli. [*M.A.*, **5**, 390.] 13th List

Ferrutite, error for Ferutite, q.v.

Fersilicite and **Ferdisilicite.** V. Kh. Gevorkyan, 1969. [Геол. журн. Украин (*Geol. Journ. Ukraine*), **29**, 62]; abstr. *Amer. Min.*, **54**, 1737 (1969). Ferrosilicon found in drill cores from sandstones near Zachativsk, Donets region, Ukraine, is believed to be natural; it consists of intergrowths of cubic FeSi, fersilicate, and tetragonal $FeSi_2$, ferdisilicate. Named from the composition. [*M.A.*, 70-747; *Zap.*, **99**, 71.] 26th List

Fersmanite. A. N. Labuntzov, 1929. *Compt. Rend. Acad. Sci. URSS*, Leningrad, ser. A, 1929, p. 296 (Ферсманит, fersmanite). Titano-silicate and fluoride of calcium and sodium, $4RTiO_3.2R_2Si(O,F_2)_3.SiO_2$*, as monoclinic crystals resembling sphene from nepheline-syenite in the Kola peninsula. Named after Prof. Aleksandr Evgenievich Fersman Александр Евгеніевич Ферсман), of Leningrad. [*M.A.*, **4**, 246.] [*Corr. in *M.M.*, **22**.] 12th List [*A.M.*, **16**, 92]

Fersmite. E. M. Bohnstedt-Kupletskaya and T. A. Burova, 1946. *Compt. Rend. (Doklady) Acad. Sci. URSS*, vol. 52, p. 69. Calcium niobate, $CaNb_2O_5$, with some other constituents, as black orthorhombic crystals resembling columbite and euxenite, from the Urals. Named after Aleksandr Evgenievich Fersman, Александр Евгеньевич Ферсман (1883–1945). Not to be confused with fersmanite (12th List). [*M.A.*, **10**, 102.] 18th List [$(Ca,Ce,Na)(Nb,Ti,Fe,Al)_2(O,OH,F)_6$; *A.M.*, **32**, 373]

Ferutite. Y. V. Kazitzyn, 1954. *Mém. Soc. Russ. Min.*, vol. 83, p. 425 (ферутит). The mineral from Mozambique [*Min. Mag.*, **29**, 101, 292], previously referred to davidite, is hexagonal with a distinctive X-ray pattern. Named from the composition, Fe, U, Ti. Abstr. in *Bull. Soc. Franç. Min. Crist.*, 1956, vol. 79, p. 181 (ferrutite); *Amer. Min.*, 1958, vol. 43, p. 382. *See* Mavudzite. 21st List [Is Davidite; *A.M.*, **49**, 447]

Fervanite. E. P. Henderson and F. L. Hess, 1931. *Amer. Min.*, vol. 16, p. 119. F. L. Hess and E. P. Henderson, ibid., p. 273. Hydrous vanadate of iron, $2Fe_2O_3.2V_2O_5.5H_2O$, as a fibrous (monoclinic?) yellow mineral in sandstone from Colorado and Utah. Named from the composition. [*M.A.*, **4**, 498.] 12th List

Finchenite, incorrect transliteration of Фынченит, Fynchenite (22nd List) (I. Kostov, *Mineralogy*, p. 458). 25th List

Finnemanite. G. Aminoff, 1923. *Geol. För. Förh. Stockholm*, vol. 45, p. 160 (Finnemanit). Chloro-arsenite of lead, $3Pb_3(AsO_3)_2.PbCl_2$, found at Långban, Sweden, as hexagonal prisms resembling mimetite (from which it differs in being an arsenite and not an arsenate). Named after Mr. K. J. Finneman, of Långban, by whom it was first noticed. [*M.A.*, **2**, 147.] 10th List

Fischesserite. Z. Johan, P. Picot, R. Pierrot, and M. Kvaček, 1971. Carbonate veins at Předbořice, Bohemia, carry a wide variety of selenides including hakite and permingeatite. The present mineral is cubic, *a* 9·967 Å, $I4_132$, isostructural with petzite. Composition $8[Ag_3AuSe_2]$. Named for R. Fischesser. [*M.A.*, 72-2332.] 27th List

Fizelyite. (F. Krantz's *Circular of January 1914*, pp. 3, 5; 3rd Appendix to 6th edit. of Dana's *System of Mineralogy*, 1915, p. 30.) Black, striated, acicular crystals associated with semseyite, galena, sphaerosiderite, and quartz, from Kisbánya, Torda-Aranyos, Hungary. Said to be monoclinic and with the composition $5PbS.Ag_2S.4Sb_2S_3$. 7th List. J. S. Krenner and J. Loczka, 1926. *Math. Természettud. Értesítö*, Budapest, vol. 42, pp. 18, 21 (Fizélyit). A sulphantimonite of lead and silver, $5PbS.Ag_2S.4Sb_2S_3$, as striated lead-grey prisms from Kisbánya, Hungary. Named after the mining-engineer Sándor Fizély, by whom the mineral was found. *See* 7th List. [*M.A.*, **3**, 8.] 11th List

Flagstaffite. F. N. Guild, 1920. *Amer. Min.*, vol. 5, p. 169; ibid., 1921, vol. 6, p. 133. Colourless, transparent, orthorhombic crystals identical with [cis-]terpin hydrate, $C_{10}H_{20}O_2.H_2O$; found with resin in the radial cracks of buried pine trees near Flagstaff, Arizona. [*M.A.*, **1**, 122, 260.] 9th List [*A.M.*, **50**, 2109]

Flajolotite. A. Lacroix, 1910. *Minéralogie de la France*, vol. iv, p. 509. A compact or earthy mineral of a lemon-yellow colour occurring abundantly as large nodular masses at Hammam N'Baïl, Constantine, Algeria. A hydrated antimonate of iron, $4FeSbO_4.3H_2O$, analysed by — Flajolot in 1871. 6th List [Is Tripuhyite; *A.M.*, **39**, 405]

Fleischerite. G. Gagarin and J. R. Cuomo, 1949, loc. cit., p. 7 (fleischerita). The *6H* polymorph Zn_6S_6 of wurtzite. Named after Dr. Michael Fleischer (1908–) of the United States Geological Survey. [*M.A.*, **10**, 532; **11**, 126, 128.] 19th List [Usage defunct]

—— C. Frondel and H. Strunz, 1960. *Neues Jahrb. Min.*, Monatsh., p. 132; *Amer. Min.*, 1957, vol. 42, p. 747. $Pb_3Ge^{\cdot\cdot}(SO_4)_2(OH)_4.4H_2O$, hexagonal, from Tsumeb, SW Africa. Not to be confused with the fleischerite of G. Gagarin and J. R. Cuomo (19th List), which is an unacceptable name for *6H* polymorph of wurtzite. Named after Dr. Michael Fleischer of the US Geol. Survey. [*M.A.*, **15**, 43.] 22nd List [The accepted usage. Cf. Itoite, Schaurteite; *A.M.*, **45**, 1313]

Fletcherite. J. R. Craig and A. B. Carpenter, 1977. *Econ. Geol.* **72**, 480. Minute crystals in copper ores in the Fletcher mine, Reynolds Co., Missouri, is cubic, *a* 9·520 Å. A thiospinel, $8[Cu(Ni,Co)_2S_4]$, the nickel analogue of carrollite in which Ni replaces Co (in sychnodymite, Ni replaces Cu). Named from the locality. [*A.M.*, **62**, 1057.] 30th List

Flokite. K. Callisen, 1917. *Meddelelser Dansk Geol. For.*, vol. v, no. 9; *Contributions to Mineralogy, Min. Geol. Museum Univ. Copenhagen*, no. 16 (Flokit). A zeolite from Iceland, hitherto regarded as mesolite, forming slender, water-clear or pale yellowish-green, flattened, monoclinic crystals, with the composition $H_8(Ca,Na_2)Al_2Si_9O_{26}.2H_2O$. Named after the viking, Floki Vilgerðarson. 8th List [Is Mordenite, *M.M.*, **31**, 887]

Florencite. E. Hussak and G. T. Prior, *Nature*, 1899, LXI, 119; *Min. Mag.*, 1900, XII, 244, 249. Pale yellow rhombohedral crystals. $AlPO_4.CePO_4.Al_2(OH)_6$. Isomorphous with hamlinite. [Tripuhy, Ouro Preto, Minas Gerais] Brazil. 2nd List [Crandallite group]

Floridin, Floridine. Trade-name for fuller's earth worked by the Floridin Company at Quincy, Florida, and used largely for decolourizing mineral-oils (*6th Ann. Rep. Florida State Geol. Survey*, 1914, p. 34). The form Floridine has been used in Russia (*Min. Abstr.*, vol. 2, p. 465). Floridite (E. T. Cox, 1891) was earlier applied to the phosphate-rock of Florida. 10th List

Fluobaryt. J. F. L. Hausmann, 1847. *Handbuch der Mineralogie*, vol. ii, p. 1441. A compact mixture of fluorite and barytes from Derbyshire analysed by J. Smithson (*Ann. Phil.*, 1820, vol. xvi, p. 48). It was thought to represent a definite compound and was earlier referred to as *Fluss-Schwerspath* (*Journ. für Chemie und Physik*, 1821, vol. xxxi, p. 362), and *Baryt-Flussspath* (Berzelius's *Jahres-Ber.*, 1822, vol. ii, p. 102). 7th List

Fluoborite. P. Geijer*, 1926. *Geol. För. Förh. Stockholm*, vol. 48, p. 85; *Sveriges Geol. Unders.*, 1927, *Årsbok 20* (*for 1926*), no. 4, p. 26. A magnesium fluo-borate, $3MgO.B_2O_3 + 3Mg(F,OH)_2$, as colourless hexagonal prisms from Norberg, Sweden. Named from the composition. [*M.A.*, **3**, 110; **3**, 274.] [*Corr. in *M.M.*, **21**] 11th List [$Mg_3BO_3(F,OH)_3$]

Fluochrysotile. Pu Wang and Wan-chuen Lu, 1965. [*Kexue Tongbao*, no. 9, 822]; abstr. *M.A.*, **19**, 226. An unnecessary name for a fluorian chrysotile. Cf. fluorantigorite, 24th List. 28th List

Fluocollophanite. A Lacroix, 1910. *Minéralogie de la France*, vol. iv, p. 561 (fluocolophanite). A variety of collophanite containing fluorine, the formula being given as

$$\begin{cases} x[Ca_3(PO_4)_2] + y\,CaCO_3 + z\,H_2O \\ x(CaF)_2Ca_8(PO_4)_6 + y\,CaCO_3 + z\,H_2O \end{cases}$$

It is amorphous (colloidal) and optically isotropic, and is an important constituent of the calcium phosphates of sedimentary origin. Colophanite is an error for collophanite (Kollophan of F. von Sandberger, 1870), from κόλλα, glue (Fr. colle) and φαίνεσθαι, to appear. *See* Quercyite. 6th List

Fluor-adelite. *See* tilasite. 1st List

Fluor-amphibole. N. L. Bowen and J. F. Schairer, 1935. *Amer. Min.*, vol. 20, p. 543 (fluor-amphiboles). Artificial $F_2R_7''(Si_4O_{11})_2$ with fluorine replacing hydroxyl of hydroxy-amphibole $[(OH)_2R_7''(Si_4O_{11})_2 = H_2R_7''(SiO_3)_8]$. [*M.A.*, **6**, 353.] 14th List

Fluor-annite, Fluor-biotite, etc. D. P. Grigoriev, 1935. *Mém. Soc. Russ. Min.*, ser. 2, vol. 64, pp. 69, 70, 79. The following names are applied to magnesium-iron-micas:

Fluor-annite (фтор-аннит), $F_2KFe_3''(AlSi_3)O_{10}$.
Fluor-biotite (фтор-биотит), $F_2K(Mg,Fe'')_3(AlSi_3)O_{10}$.
Fluor-phlogopite (фтор-флогопит), $F_2KMg_3(AlSi_3)O_{10}$.
Fluor-siderophyllite (фтор-сидерофиллит), $F_2KFe_3''[(Al,Fe''')Si_3]O_{10}$.
Fluor-lepidomelane (фтор-лепидомелан), $F_2K(Mg,Fe'')_3[(Al,Fe''')Si_3]O_{10}$.
Fluor-meroxene (фтор-мероксен), $F_2KMg_3[(Al,Fe''')Si_3]O_{10}$.

Similarly, Hydroxyl-annite (гидроксил-аннит), $(OH_2KFe_3''(AlSi_3)O_{10}$, Hydroxyl-biotite, Hydroxyl-phlogopite, Hydroxyl-siderophyllite, Hydroxyl-lepidomelane, and Hydroxyl-meroxene, with $(OH)_2$ in place of F_2. 14th List

Fluor-antigorite. Wang Pu and Juan Shou-Tsuen, 1965. [*Scientica Sinica*, vol. 14, p. 327]; abstr. *Amer. Min.* 1965, vol. 50, p. 1506. An unnecessary name for a fluorian Serpentine containing ~2·5% F, from the Shouwangfen magnetite deposit, Hopei, China. 24th List

Fluorapophyllite. P. J. Dunn, R. C. Rouse, and J. A. Norberg, 1978. *Am. Mineral.*, **63**, 196. P. J. Dunn and W. Wilson, *Mineral. Rec.*, **9**, 95 (1978). Apophyllite with F > OH is now named. [*M.A.*, 78-3472.] 30th List

Fluorarfvedsonite. W. G. Ernst, 1962. *Journ. Geol.*, vol. 70, p. 733. A variety of Arfvedsonite rich in fluorine; near $(Na,Ca)_{2\frac{1}{2}}(Fe^{2+},Fe^{3+},Mg)_5Si_8O_{22}(OH)_2$. 23rd List

Fluorbastnäsite. I. V. Aleksandrov, V. I. Ivanov, and L. A. Sinkova, 1965. Зап. Всесоюз. Мин. Общ. (*Mem. All-Union Min. Soc.*), vol. 94, p. 323 (Фторбастнезит). Syn. of Bastnäsite, as distinct from Hydroxyl-bastnäsite (q.v.). [*M.A.*, **17**, 502.] 24th List

Fluor-biotite, *see* Fluor-annite.

Fluor-chlor-hydroxyapatite. P. G. Cooray, 1970. *A.M.*, **55**, 2038. A variety of apatite with OH:F:Cl approaching 1:1:1; the material described, from Ceylon, had OH:F:Cl = 0·76:0·56:0·66, and contained some carbonate. [*M.A.*, 71-2322.] Cf. fluor-chlor-oxy-apatite, 25th List. 28th List

Fluor-chlor-oxy-apatite. E. J. Young and E. L. Munson, 1966. *Amer. Min.*, vol. 51, p. 1476. A chlorian fluorapatite deficient in (F,Cl,OH) from Devils Canyon, Eagle, Colorado. 25th List

Fluor-chondrodite. K. Rankama, 1947. *Amer. Min.*, vol. 32, p. 154. Artificially prepared chondrodite, $MgF_2.2Mg_2SiO_4$, with OH completely replaced by F. [*M.A.*, **10**, 106.] 18th List

Fluor-diopside. H. von Eckermann, 1922. *Geol. För. Förh. Stockholm*, vol. 44, p. 355. *See* Mansjöite. 9th List

Fluor-edenite. J. A. Kohn and J. E. Comeforo, 1955. *Amer. Min.*, vol. 40, p. 410 (fluor-edenite). Artificial $NaCa_2Mg_5(Si_{3.5}Al_{0.5}O_{11})_2F_2$. Cf. fluor-richterite (19th List). [*M.A.*, **13**, 486.] 21st List

Fluorene. R. Rost, 1935. *Věda Přirodni*, Praha, vol. 16, p. 204 (fluoren). An organic compound ($C_{13}H_{10}$) formed by the burning of pyritous shale in Bohemia. [*M.A.*, **6**, 357.] 14th List

Fluore-tremolite, *see* Fluor-tremolite.

Fluorhectorite. J. L. Miller and R. C. Johnson, 1962. *Amer. Min.*, vol. 47, p. 1049. The end-member $K_xMg_{3-x}Li_xSi_4O_{10}F_2$, where x is between $\frac{1}{3}$ and $\frac{2}{3}$. 23rd List

Fluor-herderite. S. L. Penfield, *Amer. J. Sci.*, 1894, XLVII, 330. Herderite containing fluorine with little or no hydroxyl. $Ca(BeF)PO_4$. *See* hydro-herderite. 2nd List

Fluorine-hydroxyl-phlogopite. T. Noda and N. Yamanishi, 1964. [*Kogyo Kogaku Zasshi*, vol. 67, p. 289]; abstr. *Min. Journ.* [*Japan*], 1965, vol. 4, p. 397. An unnecessary name for fluorian phlogopite. 24th List

Fluorichterite, *see* Fluor-richterite.

Fluor-lepidomelane, *see* Fluor-annite.

Fluor-magnesio-richterite. G. V. Gibbs, J. L. Miller, and H. R. Shell, 1962. *Amer. Min.*, vol. 47, p. 75. The synthetic amphibole end-member $Na_2Mg_6Si_8O_{22}F_2$. 23rd List

Fluormanganapatite. H. Laubmann and H. Steinmetz, 1920. *Zeits. Kryst. Min.*, vol. 55, p. 563 (Fluormanganapatit). A variety of apatite containing some manganese (MnO 4·93%) and no chlorine. [*M.A.*, **1**, 125.] 9th List [Cf. Mangan-fluorapatite]

Fluor-meionite. E. V. Shannon, 1920. *Proc. US Nat. Mus.*, vol. 58, p. 482. Fluorine (2·74%) in a scapolite from Trumbull, Connecticut, is assumed to be present as a fluor-meionite in isomorphous mixture. [*M.A.*, **1**, 213.] 9th List

Fluor-meroxene, *see* Fluor-annite.

Fluor-mica. J. E. Comeforo, R. A. Hatch, and W. Eitel, 1950. *Progr. and Abstr. Min. Soc. Amer.*, p. 9 (fluor-micas); *Amer. Min.*, 1951, vol. 36, p. 313. R. A. Hatch, W. Eitel, and R. A. Humphrey, ibid., 1950, p. 13; ibid., 1951, p. 317 (fluorine-micas, F-micas, F-phlogopite, F-muscovite, F-talc, F-pyrophyllite). Cf. Fluor-phlogopite (14th List). 19th List

Fluor-norbergite. K. Rankama, 1947. *Amer. Min.*, vol. 32, p. 154. Artificially prepared norbergite, $MgF_2.Mg_2SiO_4$, with OH completely replaced by F. [*M.A.*, **10**, 106.] 18th List

Fluoromontmorillonite. A.-M. Hecht and E. Geissler, 1972. *Bull.*, **95**, 291. A *synthetic* product of composition $Mg_{5.1}Na_{0.7}Si_8F_4O_{20}Na_{1.1}.nH_2O$. 28th List

Fluoroxyapatite. V. I. Vlodovets, 1933. *Trans. Arctic Inst.*, vol. 12, p. 79 (фтор-окси-апатит), p. 99 (fluorine oxyapatite). O. M. Shubnikova, *Trans. Lomonosov Inst.*, 1936, no. 7, p. 328 (фтороксиапатит, fluor-oxyapatite). 20th List

Fluor-pectolite. P. S. Rogers, 1970. *Min. Mag.*, **37**, 741. $NaCa_2Si_3O_8F$, the fluorine analogue of pectolite; *artificial*, as a devitrification product in a glass. 26th List

Fluor-phlogopite, *see* Fluor-annite.

Fluor-polylithionite. H. Takeda and C. W. Burnham, 1969. *Min. Journ.* [*Japan*], **6**, 102. *Synthetic* $KAlLi_2Si_4O_{10}F_2$; mica family. 27th List

Fluor-richterite. J. E. Comeforo, R. A. Hatch, and W. Eitel, 1950. *Progr. and Abstr. Min. Soc. Amer.*, p. 8 (fluo-richterite); *Amer. Min.*, 1951, vol. 36, p. 312. Artificial fluorine-amphiboles (p. 312), fluoramphibole, fluor-amphiboles (p. 313) containing F in place of OH. Compare fluor-amphibole (14th List) and fluor-tremolite (15th List). 19th List

Fluor-siderophyllite, *see* Fluor-annite.

Fluor-spodiosite. F. K. Cameron and W. J. McCaughey, 1911. *Journn. Physical Chem.*, Ithaca, NY, vol. xv, p. 470. Synonym of spodiosite. *See* Chlor-spodiosite. 6th List

Fluortaeniolite. J. L. Miller and R. C. Johnson, 1962. *Amer. Min.*, vol. 47, p. 1049. Original, incorrect spelling of Fluortainiolite, *q.v.* (Cf. Tainiolite, 2nd List, and Taeniolite, 3rd List.) 23rd List

Fluortainiolite. J. L. Miller and R. C. Johnson, 1962. *Amer. Min.*, vol. 47, p. 1049 (Fluortaeniolite). The mica end-member $KMg_2LiSi_4O_{10}F_2$. 23rd List

Fluortamarite. (G. Murgoci, *Compt. Rend. Acad. Sci. Paris*, 1922, vol. 175, p. 373, in explanation to diagram only.) A soda-iron amphibole. The name, of which no explanation is given, is no doubt an error for Fluotaramite. (*See* Taramite.) [*M.A.*, **2**, 221.] 10th List

Fluor-tremolite. D. P. Grigoriev, 1939. *Compt. Rend.* (*Doklady*) *Acad. Sci. URSS*, vol. 23, p. 71 (fluore-tremolite). Artificially produced tremolite containing fluorine (9·03%) in place of hydroxyl. [*M.A.*, **7**, 477.] 15th List

Fluotaramite. J. Morozewicz, 1923. *See* Taramite. 10th List

Foggite. P. B. Moore, A. J. Irving, and A. R. Kampf, 1975. *A.M.*, **60**, 957 and 965. White plates from the Palermo no. 1 pegmatite, North Groton, New Hampshire; *a* 9·270, *b* 21·324, *c* 5·190 Å, space group $A2_122$; α 1·610 ∥ [001], β 1·610, γ 1·611 ∥ [100], 2Vγ 40 to 45°, cleavage {010} perfect, {100} good. ρ 2·78 g. cm^{-3}. Composition $8[CaAl(OH)_2PO_4.H_2O]$. Named for Mr. F. F. Fogg. 29th List

Foid, foidal. A. Johannsen, 1917. *Journ. Geol.*, Chicago, vol. 25, pp. 69–71; *Descriptive petrology of igneous rocks*, 1931, vol. 1, pp. 141, 174. Abbreviations of felspathoids, felspathoidal. Compare feldspathides (q.v.), feloids (10th List), quarfeloids (10th List). 21st List

Foidal, *see* Foid.

Fojasite, fożasyt. S. J. Thugutt, 1949. *Rocznik Polsk. Tow. Geol.* (*Ann. Soc. Géol. Pologne*), vol. 18 (for 1948), p. 26 (fojasite), p. 35 (fożasyt). Polish spellings of faujasite, named by A. Damour in 1842 after Barthélemy Faujas de Saint-Fond (1741–1819). 19th List

Folgerite. S. H. Emmens, *Journ. Amer. Chem. Soc.*, XIV. 205, 1892; S. L. Penfield, *Amer. Jour. Sci.*, XLV. 494, 1893. Shown by Penfield to be identical with pentlandite. Sudbury, Canada. 1st List

Fontainebleau sandstone. H. A. Miers, 1902. *Mineralogy*, London, p. 215. Synonym of Fontainebleau limestone. 23rd List

Formanite. H. Berman and C. Frondel, 1944. Dana's *System of mineralogy*, 7th edit., vol. 1, p. 757. Tantalate of yttrium, etc., $YtTaO_4$, corresponding with the niobate fergusonite, from Cooglegong, Western Australia, analysed by E. S. Simpson (1909). Named after Francis Gloster Forman, government geologist of Western Australia. 17th List [*A.M.*, **29**, 456]

Fornacite. A Lacroix, 1916. *Bull. Soc. franç. Min.*, vol. xxxix, p. 84. The correct form of Furnacite (A. Lacroix, 1915; cf. 7th list), from the Latin *fornax*. 8th List [$Pb_2CuCrO_4(As,P)O_4(OH)$; *A.M.*, **49**, 447]

Fosfo-escorodita. G. Gagarin and J. R. Cuomo, 1949, loc. cit., p. 14. Spanish form of phospho-scorodite (q.v.). 19th List

Fosfosiderita. G. Gagarin and J. R. Cuomo, 1949, loc. cit., p. 14. Spanish form of phosphosiderite. [*M.A.*, **11**, 129.] 19th List

Foshagite. A. S. Eakle, 1925. *Amer. Min.*, vol. 10, pp. 66, 97. Hydrous calcium silicate, $H_2Ca_5(SiO_4)_3.2H_2O$, as a while, compact, fibrous, orthorhombic mineral filling veins in idocrase, from Crestmore, California. Named after Dr. William F. Foshag, of the United States National Museum at Washington, D.C. [*M.A.*, **2**, 520.] 10th List [Triclinic $Ca_4Si_3O_9(OH)_2$; *A.M.*, **43**, 1]

Foshallassite. P. N. Chirvinsky, 1936. V. I. Vernadsky jubilee volume, *Acad. Sci. USSR*, 1936, vol. 2, p. 757 (Фошалласит), p. 763 (Foschallasit); A. E. Fersman and E. M. Bonshtedt, *Minerals of the Khibina and Lovozero tundras*, 1937, p. 90 (foshallasite). Hydrated calcium silicate, $3CaO.2SiO_2.3H_2O$, from the Kola peninsula, Russia. Related to foshagite and centrallassite and named from a combination of these names. [*M.A.*, **7**, 10.] 15th List [*A.M.*, **23**, 667]

Foskorite. L. Winand, 1963. *Bull. Soc. roy. Sci. Liège*, **32**, 575. A phosphate rock from South Africa; a mixture of apatite and dolomite. [*M.A.*, **17**, 261.] 28th List

Fouchéit, error for Foucherit (Hintze, *Handb. Min.*, Erg.-Bd. II, pp. 544, 707). 22nd List

Foucherite. A. Lacroix, 1910. *Minérologie de la France*, vol. iv, p. 535 (fouchérite). The correct spelling of fucherite (A. Leymerie, 1867), an amorphous hydrated basic phosphate of ferric iron with some aluminium and calcium; from Fouchères, Aube, France. Possibly identical with borickite (J. D. Dana, 1868). 6th List

Fourmarierite. H. Buttgenbach, 1924. *Ann. Soc. Géol. Belgique*, annex to vol. 47 (for 1923–24) (Publ. Congo Belge), p. C 41 (Fourmariérite). A hydrated uranium lead mineral containing perhaps also silica, as minute, red, orthorhombic crystals from Katanga. J. Mélon (*Ann. Soc. Géol. Belgique*, 1924, vol. 47, Bull. p. B 200) gives the formula $PbO.5UO_3.10H_2O$; and A. Schoep (*Bull. Soc. Franç. Min.*, 1924, vol. 47, p. 157) gives $PbO.4UO_3.5H_2O$ or $(UO_2,Pb)O.H_2O$. Named after the Belgian geologist Prof. Paul Fourmarier. [*M.A.*, **2**, 343, 384, 521.] 10th List [Orthorhombic $PbO.4UO_3.4H_2O$; *A.M.*, **45**, 1026; *M.M.*, **41**, 56]

Fozasyt, *see* Fojasite.

Fraipontite. G. Cesàro, 1927. *Ann. Soc. Géol. Belgique*, vol. 50, Bull. p. B 106. Hydrated basic silicate of zinc and aluminium, $Zn_3(AlO)_4(SiO_4)_5.11H_2O$, as a fibrous crust. Named after Julien Jean Joseph Fraipont (1857–1910), a Belgian zoologist, and Charles Fraipont, of Liége. [*M.A.*, **3**, 368.] 11th List [Zinc-bearing Berthierine, $2[(Zn,Al)_3(Si,Al)_2O_5(OH)_4]$; *M.A.*, 76-3582. Cf. Zinalsite]

Framesite. (J. R. Sutton, *Trans. Roy. Soc. South Africa*, 1918, vol. vii, pp. 75, 95; *Chem. News*, 1919, vol. cxviii, pp. 39, 66.) A form of black diamond (bort) from South Africa. Material collected by Mr. P. Ross Frames at the Premier diamond mine, Transvaal, was described in detail by D. P. McDonald, *Trans. Geol. Soc. South Africa*, 1914, vol. xvi, p. 156. 8th List

Francevillite. G. Branche, M. E. Ropert, F. Chantret, B. Morignat, and R. Pouget, 1957. *Compt. Rend. Acad. Sci. Paris*, vol. 245, p. 89; *Bull. Techn. Atomic Energy*, 1957, vol. 7, p. 14. Hydrous vanadate, $(Ba,Pb)(UO_2)_2(VO_4)_2.5H_2O$, orthorhombic, as yellow impregnations in sandstone from Franceville, French Equatorial Africa. Named from the locality. [*M.A.*, **13**, 522.] 21st List [*A.M.*, **43**, 180. Cf. Curienite]

Franckeite. A. W. Stelzner, *Neues. Jahrb. Min.*, II. 114, 1893. $5PbS.2SnS_2.Sb_2S_3$. [Las Animas dist., SE of Chocaya] Bolivia. 1st List [$Pb_5Sn_3Sb_2S_{14}$]

Francoanellite. F. Balenzano, L. Dell'Anna, and M. Dipiero, 1976. *Neues Jahrb. Mineral.*, Monatsh. 49. Yellowish-white nodular aggregates from the Grotte di Castellana, Puglia, Italy; composition $H_6K_3Al_5(PO_4)_8.13H_2O$. Artificial material (*A.M.*, **44**, 138) is rhombohedral, *a* 8·71, *c* 82·8 Å, α 1·510, β 1·515, ρ 2·26 $g.cm^{-3}$. Named for Franco Anelli. [*A.M.*, **61**, 1054; *M.A.*, 76-3676.] 30th List

Frankdicksonite. A. S. Radtke and G. E. Brown, 1974. *A.M.*, **59**, 885. Cubes up to 4 mm in quartz veinlets in the Carlin gold deposit, Eureka County, Nevada, consist of BaF_2; fluorite structure, *a* 6·1964 Å. Named for Dr. Frank W. Dickson. [*Zap.*, **104**, 608; *Bull.*, **98**, 264.] 29th List

Franquenite. R. Van Tassel, 1944. *Bull. Mus. Hist. Nat. Belgique*, vol. 20, no. 16, p. 9. Hydrous sulphate of iron, aluminium, and magnesium, $2(Mg,Fe)SO_4.6(Fe,Al)OHSO_4.41H_2O$, as yellow efflorescences from Franquenies, Belgium. Named from the locality. [*M.A.*, **9**, 125.] 17th List [Is Slavíkite; *A.M.*, **35**, 136]

Franzinite. S. Merlino and P. Orlandi, 1977. *Neues Jahrb. Mineral.*, Monatsh. 163. Squat pearly white prisms in ejected blocks in a pumice deposit at Pitigliano, Tuscany, are hexagonal, *a* 12·884, *c* 26·580 Å. Composition $[Na_{21\cdot5}Ca_{12\cdot5}Si_{31\cdot3}Al_{28\cdot7}O_{1\cdot20}(SO_4)_{7\cdot7}(CO_3)_2(OH)_{3\cdot5}Cl_{0\cdot6}\text{-}4{\cdot}3H_2O]$. A member of the cancrinite group, resembling davyne but with SO_3^{2-} as dominant anion. Afghanite (25th List) and liottite (30th List), also members of the cancrinite group, occur in the same deposit. [*A.M.*, **62**, 1259.] 30th List

Freboldite. H. Strunz, *Min. Tab.*, 3rd edit., 1957, p. 98 (Freboldit). Cobalt selenide, CoSe, hexagonal, pyrrhotine group, artificial. 'Mineral 4' of P. Ramdohr, 1955 [*M.A.*, **13**, 6]. Etymology? 21st List [*A.M.*, **41**, 164; **44**, 907]

Freirinite. W. F. Foshag, 1924. *Amer. Min.*, vol. 9, p. 30. Hydrated copper, sodium, and calcium arsenate, $6(Cu,Ca)O.3Na_2O.2As_2O_5.5H_2O$, as a bluish-green (tetragonal?) mineral from Freirina dept., Chile. Named from the locality. [*M.A.*, **2**, 337.] 10th List [Is Lavendulan, *Bull.*, **79**, 7]

Fremontite. W. T. Schaller, 1914. *Journ. Washington Acad. Sci.*, vol. iv, p. 356; 3rd Appendix to 6th edit. of Dana's *System of Mineralogy*, 1915, p. 31. To replace the earlier name Natramblygonite or Natronamblygonit (W. T. Schaller, 1911; 6th List). Named from the locality, Fremont County, Colorado. *See* Natromontebrasite. 7th List

Fresnoite. J. T. Alfors and M. C. Stinson, 1965. *Min. Inform. Serv., Calif. Div. Mines Geol.*, vol. 18, p. 27. J. T. Alfors, M. C. Stinson, R. A. Matthews, and A. Pabst, *Amer. Min.*, 1965, vol. 50, pp. 279 and 314. Lemon- to canary-yellow tetragonal crystals, fluorescing pale yellow in ultra-violet light, occur in sanbornite–quartz rock in the Big Creek and Rush Creek area of Fresno County, California. Emission spectrographic analysis leads to the formula $Ba_2TiSi_2O_8$, which is confirmed by synthetic experiments. Named from the locality. [*M.A.*, **17**, 400; **17**, 502.] 24th List

Freudenbergite. G. Frenzel, 1961. *Neues Jahrb. Min.*, Monatsh., p. 12 (Freudenbergit). Na_2Fe_2-Ti_7O_{18}, hexagonal, black crystals in the syenite of Katzenbuckel, Odenwald, Germany. Named after Prof. Wilhem Freudenberg, author of a monograph on the Katzenbuckel rocks. [*M.A.*, **15**, 135.] 22nd List [*A.M.*, **46**, 765] [Monoclinic; *M.A.*, **17**, 65, 471.]

Friedrichite. T. T. Chen, E. Kirchner, and W. Paar, 1978. *Can. Mineral.*, **16**, 127. Aggregates in vein-quartz boulders in the Habachtal, Salzburg, Austria, are orthorhombic, *a* 33·84, *b* 11·65, *c* 4·01 Å. Composition $[Cu_{10}Pb_{10}Bi_{14}S_{36}]$, ρ 6·48 g.cm^{-3}. Intermediate between aikinite and hammarite. 30th List

Fringelite. M. Blumer, 1951. *Mikrochem., Mikrochim. Acta*, Wien, vol. 36–37, p. 1052 (Fringelit). A red hydrocarbon dye as minute birefringent crystals in the calcite of fossil crinoids in Jurassic limestone from Fringeli, Bern. Named from the locality. [*M.A.*, **11**, 519.] 19th List

Frohbergite. R. M. Thompson, 1946. *Bull. Geol. Soc. Amer.*, vol. 57, p. 1238; *Univ. Toronto Studies, Geol. Ser.*, 1947, no. 51 (for 1946), p. 35; *Amer. Min.*, 1947, vol. 32, p. 210. Iron telluride, $FeTe_2$, isomorphous with marcasite, found in polished sections of telluride ore from Montbray, Quebec. Named after Dr. Max Hans Frohberg, mining geologist, of Toronto. [*M.A.*, **10**, 99.] 18th List

Frolovite. E. S. Petrova, 1957. *Mém. Soc. Russ. Min.*, vol. 86, p. 622 (фроловит). Hydrous calcium borate, $CaB_2O_4.3\frac{1}{2}H_2O$. With calciborite (q.v.) in limestone skarn from Novo-Frolov copper mine, Turinsk district, northern Urals. 21st List [*A.M.*, **43**, 385]

Frolowit, German transliteration of Фроловит, frolovite (Hintze, *Hand. Min.*, Erg.-Bd. II, p. 708). 22nd List

Frondelite. M. L. Lindberg, 1949. *Amer. Min.*, vol. 34, p. 541. A dufrenite-like mineral, $(Mn'',Fe'')Fe_4'''(PO_4)_3(OH)_5$, isomorphous with rockbridgeite (q.v.) with Mn″ in place of Fe″, orthorhombic, as radially fibrous masses from Brazil. Named after Prof. Clifford Frondel (1907–) of Harvard University. [*M.A.*, **11**, 8.] 19th List [Cf. Mangan-rockbridgeite]

Froodite. J. E. Hawley and L. G. Berry, 1958. *Canad. Min.*, vol. 6, p. 200. A monoclinic palladium bismuthide, identical with artificial α-$PdBi_2$, from the Frood mine, Sudbury, Ontario. Named from the locality. [*M.A.*, **14**, 343.] 22nd List [USSR; *M.A.*, 77-3363]

Fuggerite. E. Weinschenk, *Zeits. Kryst. Min.*, XXVII. 577, 1897. Tetragonal? Between åkermanite and gehlenite in composition, but dimorphous with this group. [Monzonithal] Tryol. 1st List

Fukuchilite. Y. Kajiwara, 1969. *Min. Journ.* (*Japan*), **5**, 399. Cubic Cu_3FeS_8 occurs in the gypsum orebody at the Hanawa mine, Akita Prefecture, Japan. Named for Nobuyo Fukuchi. [*M.A.*, 70-749.] 26th List [*A.M.*, **55**, 1811]

Fülöppite. I. de Finály and S. Koch, 1929. *Min. Mag.*, vol. 22, p. 179; *Magyar Tudom. Akad. Mat. Természett. Értesítő*, 1929, vol. 46, pp. 663, 673 (fülöppit). Lead sulphantimonite, $2PbS.3Sb_2S_3$, as monoclinic crystals; an acid member of the plagionite-semseyite group. From Nagybánya, Hungary (= Baia Mare, Romania). Named after Dr. B. Fülöpp, a Hungarian mineral collector. [*M.A.*, **4**, 244.] 12th List [$Pb_3Sb_8S_{15}$; *M.A.*, **10**, 457]

Fulvite. C. W. Carstens, 1928. *Zeits. Krist.*, vol. 67, p. 272 (Fulvit). Titanium monoxide, TiO, cubic, prepared artificially and present in certain slags and cement clinkers. Named because of its brown colour, from the Latin fulvus, deep yellow (rutile being from rutilus, red). 11th List

Furnacite. A. Lacroix, 1915. *Bull. Soc. franç. Min.*, vol. xxxviii, p. 198. Confused groups of small, dark olive-green crystals (monoclinic?) occurring on dioptase from the French Congo. A preliminary examination suggests that the mineral is a basic chromo-arsenate of lead and copper analogous to vauquelinite (chromo-phosphate of lead and copper). Named from 'furnax' [error for the Latin *fornax*] in honour of the colonial governor, Mr. Lucien Fourneau. 7th List [*See* Fornacite]

Furongite. *Acta Geol. Sinica*, 1976, 203. Anorthic [$Al_2UO_2(PO_4)_2(OH)_2.8H_2O$], *a* 17·87, *b* 14·18, *c* 12·18, α 67·8°, β 77·5°, γ 79·9°; from an undisclosed locality in China. Sp. gr. 2·82–2·90. RIs α 1·543–9, β 1·564–7, γ 1·570–5, $2V_\alpha$ 65°. [*M.A.*, 77-3385; *A.M.*, **63**, 425.] 30th List

Fusain. (C. Grand'Eury, *Ann. des Mines*, 1882, ser. 8, vol. i, Mém., p. 106; A. Lacroix, *Minéralogie de la France*, 1910, vol. iv, p. 651.) A name used by French geologists for a constituent of coal known as 'mineral charcoal' (Ger. Faserkohle). The name, from the Latin *fusus*, a spindle, is applied in French to the spindle-tree, to charcoal (especially that made from the wood of the spindle-tree), to charcoal crayons, and, lastly, to charcoal sketches.

Other visible constituents of banded bituminous coal are distinguished by their external characters and named by M. C. Stopes, *Proc. Roy. Soc. London*, 1919, ser. B, vol. xc, p. 472: *Durain*, dull, hard coal (Ger. Mattkohle); from the Latin *durus*, hard. *Clarain**, bright or glance coal (Ger. Glanzkohle); from the Latin *clarus*, bright. *Vitrain*, similar to the last, with conchoidal fracture and brilliant appearance; from the Latin *vitreus*, glassy, with the same termination as fusain in each case. The chemical characters of these materials have been determined by F. V. Tideswell and R. V. Wheeler, *Journ. Chem. Soc. London*, 1919, Trans. vol. cxv, p. 619. [*Corr. in *M.M.*, **24**.] 8th List [Now, generally, Fusinite]

Fynchenite, standard English transliteration of the Russian transliteration of Fenghuanglite (q.v.). 22nd List

G

Gabrielsonite. P. B. Moore, 1967. *Arkiv Min. Geol.*, vol. 4, p. 401. Black orthorhombic crystals from Långban, Sweden, (Flink's no. 35), are found to be $PbFeAsO_4OH$, a member of the descloizite family. Named for O. Gabrielson. 25th List [*A.M.*, **53**, 1063]

Gagarinite. A. V. Stepanov and E. A. Severov, 1961. Докл. Акад. Наук СССР (*Compt. Rend. Acad. Sci. URSS*), vol. 141, p. 954 (Гагаринит). A white massive mineral from Kazakhstan (exact locality not stated) is near $NaCaLnF_6$; probably hexagonal. Named for Yu. A. Gagarin (*Amer. Min.*, 1962, vol. 47, p. 805; Зап. Всесоюз. Мин. Общ. (*Mem. All-Union Min. Soc.*), 1963, vol. 92, p. 195; *M.A.*, **15**, 459). 23rd List

Gageite. A. H. Phillips, 1910. *Amer. Journ. Sci.*, ser. 4, vol. xxx, p. 283. Radiating agregates of colourless, acicular crystals, associated with leucophoenicite, zincite, willemite, and calcite at Franklin, New Jersey. $8(Mn,Mg,Zn)O.3SiO_2.2H_2O$; related to leucophoenicite (S. L. Penfield and C. H. Warren, 1899). Named after Mr. R. B. Gage, of Trenton, New Jersey, who collected and analysed the material. 6th List [$(Mn,Mg,Zn)_7Si_3O_{10}(OH)_6$; *A.M.*, **53**, 309]

Gahnospinel. B. W. Anderson and C. J. Payne, 1937. *Min. Mag.*, vol. 24, p. 554. A gem magnesium-zinc-spinel containing ZnO up to 18·2%, approaching gahnite in composition. 14th List

Gaidonnayite. G. Y. Chao, 1973. *C.M.*, **12**, 143; G. Y. Chao and D. H. Watkinson, ibid. 316 (1974). An orthorhombic dimorph of catapleiite, $Na_2ZrSi_3O_9.2H_2O$, with *a* 11·740, *b* 12·820, *c* 6·691 Å, occurs in the nepheline syenite of Mont-St.-Hilaire, Quebec. Named for Gabrielle Donnay. [*M.A.*, 74-1446.] 28th List

Gainite. P. N. Chirvinsky, *Materials Geochem. Khibina tundra, Acad. Sci. USSR*, 1935, p. 86 (гаинит), p. 88 (Gainit). Error for hainite (J. Blumrich, 1893; 1st List), due to the transliteration into Russian and back again. *See* Calcium-rinkite. 14th List

Gajite. F. Tućan, 1911. *Centralblatt Min.*, 1911, p. 312 (Gajit). A snow-white, compact mineral from Croatia, resembling magnesite in appearance. It is a hydrated carbonate of calcium and magnesium, similar in composition to the rocks pencatite and predazzite (mixtures of calcite and hydromagnesite), but it is apparently homogeneous. Under the microscope it shows a finely granular texture with the cleavage and optical characters of the rhombohedral carbonates. Named after Ljudevit Gaj (1809–1872), a Croatian leader. 6th List [A mixture; *A.M.*, **46**, 467]

Galafatite. (*Engineering and Mining Journ.*, New York, 1911, vol. xci, p. 261; *Mining Magazine*, London, 1911, vol. iv, p. 229; *Zeits. Kryst. Min.*, 1914, vol. liv, p. 80; 3rd Appendix to 6th edit. of Dana's *System of Mineralogy*, 1915, p. 3.) Error for Calafatite (S. Calderón, 1910; 6th list). 7th List

Galaxite. C. S. Ross and P. F. Kerr, 1932. *Amer. Min.*, vol. 17. p. 15. Manganese aluminate, $MnAl_2O_4$, belonging to the spinel group, as black grains in manganese-ore from North Carolina. Named from the neighbouring town of Galax in Virginia, which itself is named from the local plant galax, milk-wort. Cf. galactite; scarcely an appropriate name for a black mineral. [*M.A.*, **5**, 51.] 13th List

Galchait, Germ. trans. of Галхаит, galkhaite (q.v.). Hintze, 31. 28th List

Galeite. A. Pabst, D. L. Sawyer, and G. Switzer, 1955. *Bull. Geol. Soc. Amer.*, vol. 66, p. 1658 (abstr.). $Na_2SO_4.Na(F,Cl)$, trigonal, the same as schairerite (12th List), but with a difference in the X-ray pattern, the two perhaps intergrown as a polycrystal. From Searles Lake, California. Named after Mr. W. A. Gale. [*M.A.*, **13**, 86.] 21st List [*M.M.*, **40**, 131 (with refs.); *see* Kogarkoite]

Galenobornite. T. A. Satpaeva, G. S. Safargaliev, T. P. Polyakova, M. K. Satpaeva, V. L. Marzuvanov, and M. Z. Fursova, 1964. [Изв. Акад. наук Казах. ССР (*Bull. Acad. Sci. Kazakh SSR, Ser. Geol*), no. 2, p. 29]; abstr. *Amer. Min.*, 1965, vol. 50, p. 809; *M.A.*, **17**, 302; Зап. Всесоюз. Мин. Общ. (*Mem. All-Union Min. Soc.*), 1965, vol. 94, p. 668.

(Галеноборнит, galenobornite). A yellowish brown mineral in prismatic crystals in lead–copper ore of the Dzhezkazgan deposit, Kazakhstan, has a composition near $(Cu,Pb)_{4\cdot7}FeS_4$ (four analyses with rather variable Cu:Pb ratio), and is regarded as a distinct mineral, though all the strongest-lines of the rather diffuse X-ray powder pattern agree with those of galena; eight weak lines are not accounted for. Probably a mixture. 24th List

Galkhaite. V. S. Gruzdev, V. I. Stepanov, N. G. Shumkova, N. M. Chernitsova, R. H. Yudin, and I. A. Bryzgalov, 1972. Докл. **205**, 1194 (Галхаит). $HgAsS_2$, cubic *a* 10·41, from Gal-Khaya, Yakutia, and Khaidarkan, Kirgizia. Named for the type locality. [*M.A.*, 73-1936; *A.M.*, **59**, 208; *Zap.*, **102**, 439.] 28th List

Gallite. H. Strunz, B. H. Geier, and E. Seeliger, 1958. *Neues Jahrb. Min.*, Monatsh., p. 241 (Gallit). Grains and inclusions in germanite, renierite, blende, and other ore minerals from Tsumeb, SW Africa, contain Cu, Ga, and S, and are identical with artificial $CuGaS_2$; tetragonal, related to chalcopyrite. Also from the Kipushi mine, Katanga. The name refers to its being the first mineral with gallium as an essential constituent. [*M.A.*, **14**, 279.] It is the Mineral O of H. Strunz, G. Söhnge, and B. H. Geier, *Neues Jahrb. Min.*, Monatsh., 1958, p. 85. 22nd List [*A.M.*, **44**, 906]

Gallium-albite, gallium-anorthite, gallium-orthoclase, etc. J. R. Goldsmith, 1950. *Journ. Geol. Chicago*, vol. 58, p. 522 (gallium albite, germanium albite, gallium-germanium albite), p. 524 (gallium orthoclase, germanium orthoclase, gallium-germanium orthoclase), p. 527 (gallium anorthite, germanium anorthite, gallium-germanium anorthite). Artificial felspars with gallium in place of aluminium and germanium in place of silicon, $NaGaSi_3O_8$, $NaAlGe_3O_8$, $NaGaGe_3O_8$, etc. [*M.A.*, **11**, 326.] 19th List

Gallium-anorthite, *see* Gallium-albite.

Gallium-orthoclase, *see* Gallium-albite.

Gallium-phlogopite. C. Klingsberg and R. Roy, 1957. *Amer. Min.*, vol. 42, p. 629 (gallium phlogopite), p. 321 (Ga-phlogopite). Artificially prepared $KMg_3GaSi_3O_{10}.(OH)_2$, with Ga in place of Al in phlogopite. 21st List

Gamagarite. J. E. de Villiers, 1943. *Amer. Min.*, vol. 28, p. 329. Vanadate of barium, iron, and manganese, $Ba_4(Fe,Mn)_2V_4O_{15}(OH)_2$, as dark-brown monoclinic needles in manganese ore from Gamagara ridge, Postmasburg, South Africa. Named from the locality. [*M.A.*, **9**, 5.] 17th List

Garibaldite. G. Gagarin and J. R. Cuomo, 1949, loc. cit., p. 5 (garibaldita). Monoclinic β-sulphur = Sulfurite of J. Fröbel, 1845, from the fumaroles of Italian volcanoes. Named after Giuseppe Garibaldi (1807–1882), liberator of Italy. 19th List

Garividite. L. L. Fermor, 1938. *Proc. Nat. Inst. Sci, India*, vol. 4, p. 277. A hypothetical mineral, $3Mn_3O_4.2Fe_2O_3$, which on breaking down gave rise to vredenburgite (5th List; a lamellar intergrowth of hausmannite and jacobsite with excess MnO_2). Named from the locality, Garividi, Vizagapatam, Madras. *See* Devadite. [*M.A.*, **7**, 169.] 15th List

Garnetoid. D. McConnell, 1941. *Amer. Min., Program and Abstracts*, December 1941, p. 18; *Amer. Min.*, 1942, vol. 27, p. 452; *Science*, New York, 1943, vol. 97, p. 99. Substances (silicates, phosphates, etc.) which have structures similar to garnet, including hydrogarnet (q.v.), grossularoid (q.v.), plazolite, griphite, and berzeliite. [*M.A.*, **8**, 343.] 16th List

Garrelsite. C. Milton, J. M. Axelrod, and F. S. Grimaldi, 1955. *Bull. Geol. Soc. Amer.*, vol. 66, p. 1597 (abstr.). Boro-silicate, $(Ba,Ca,Mg)_4B_4(BO_4)_2(SiO_4)_2(OH)_2.2H_2O$. Small monoclinic crystals related to datolite, from an oil boring at Onray, Utah. Named after Robert M. Garrels, of Harvard University. [*M.A.*, **13**, 86.] 21st List [Perhaps $NaBa_3Si_2B_7O_{16}(OH)_4$; *M.A.*, 74-1407; *A.M.*, **59**, 632]

Garronite. G. P. L. Walker, 1960. *Min. Mag.*, vol. 32, p. 505. Tentative name for a zeolite related to phillipsite, occurring in the basalts of the Garron plateau, Co. Antrim. Named from the locality. 22nd List [A definite species; *M.M.*, **33**, 173; Siberia, *M.A.*, 75-1341]

Gaspéite. D. W. Kohls and J. L. Rodda, 1966. *Amer. Min.*, vol. 51. p. 677 (Gaspeite). $(Ni,Mg,Fe)CO_3$, rhombohedral, with Ni dominant, occurs as light green crystals in siliceous dolomite in the Gaspé Peninsula, Quebec. The name, for the locality, is proposed for carbonates with calcite structure in which Ni is dominant, including the end-member $NiCO_3$. 24th List

Gastunite. H. Haberlandt and A. Schiener, 1951. *Tschermaks Min. Petr. Mitt.*, ser. 3, vol. 2, p. 307 (Gastunit Nr. 1). Provisional name for an undetermined mineral of the uranotile group. From the hot springs at Bad Gastein, Salzburg. Named from the locality. Cf. neogastunite (19th List). [*M.A.*, **11**, 433.] 20th List

—— R. M. Honea, 1959. *Amer. Min.*, vol. 44, p. 1047. The names Gastunite 1, Gastunite 1a, Gastunite 1b were given by H. Haberlandt and A. Schiener (1951) to three imperfectly characterized uranium minerals (cf. 20th List); gastunite 1b proves to be β-uranotile, while gastunite 1a has since been described under the name of Haiweeite (q.v.); gastunite 1 appears to be a lower hydrate of haiweeite, but is not identical with the dehydrated haiweeite named Meta-haiweeite (q.v.) (K. Walenta, *Neues Jahrb. Min.*, Monatsh., 1960, p. 37). Unfortunately, R. M. Honea has described another mineral, distinct from any of the three named by Haberlandt and Schiener, under the name of gastunite; this fourth 'gastunite' proves to be identical with the Weeksite (q.v.) (M. Fleischer, priv. comm.; E. Fejer, priv. comm.). In view of the prior uses of gastunite the name weeksite is to be preferred. Honea also describes artificial analogues, Ammonium-, Hydronium-, Potassium-, and Sodium-gastunites. 22nd List [Cf. Ranquilite]

Gatumbaite. O. von Knorring and A.-M. Fransolet, 1977. *Neues Jahrb. Mineral.*, Monatsh. 561. Sheaves and rosettes of a white mineral from Gatumba, Buranga, Rwanda, are monoclinic, *a* 6·907, *b* 5·095, *c* 10·764 Å, β 90° 3′. Composition $2[CaAl_2(PO_4)_2(OH)_2.H_2O]$, sp. gr. 2·92, α 1·610, γ 1·639 || [010], $2V_\alpha$ 60–70°. Named for the locality. [*M.A.*, 78-3471.] 30th List

Gaudefroyite. G. Jouravsky and F. Permingeat, 1964. *Bull. Soc. franç. Min. Crist.*, vol. 87, p. 216. Black hexagonal prisms from Tachgagalt, Morocco, are appreciably altered; after deduction of pyrolusite the probable composition is $Ca_4Mn^{\cdots}_{3-x}(BO_3)_3CO_3(O,OH)_3$, with *x* about 0·17. Named for C. Gaudefroy. 23rd List [*A.M.*, **50**, 806]

Gauslinite. (A. F. Rogers, *Amer. Journ. Sci.*, 1926, ser. 5, vol. 11, p. 473). A name used locally for Burkeite (q.v.). Named after Mr. — Gauslin, superintendent of the American Trona Corporation. 11th List

Gavite. E. Repossi, 1919. *Atti Soc. Ital. Sci. Nat.*, vol. 57, p. 154. A variety of talc from the Gava valley, near Voltri, Genoa, Italy, distinguished by its composition, $H_4(Mg,Fe)_4Si_5O_{16}$, and its solubility in acids. [*M.A.*, **1**, 20.] 9th List

Gaylussacite. J. W. Mellor, 1927. *Compr. Treat. Inorg. Theor. Chem.*, London, vol. 2, p. 711. Variant of Gaylussite. 23rd List

Gearksite. I. F. Grigoriev and E. I. Dolomanova, 1951. *Trudy Min. Mus. Acad. Sci. USSR*, no. 3, p. 93 (геарксит). M. D. Dorfman, ibid., p. 97. Ca-Al hydrofluoride, $CaAl_3(F,OH)_{11}.H_2O$, as white powdery aggregates in hydrothermally metamorphosed sediments, from Transbaikal. Named from analogy with gearksutite ($CaAlF_4OH.H_2O$) and paragearksutite (18th List). 21st List [Is Gearksutite; *A.M.*, **45**, 1135]

Gedroitzite. I. N. Antipov-Karataiev and I. D. Sedletzky, 1937. *Compt. Rend.* (*Doklady*) *Acad. Sci. URSS*, n. ser., vol. 17, p. 251 (gedroitsite). I. D. Sedletzky, ibid., 1939, vol. 23, p. 565 (gedroizite); 1940, vol. 26, p. 241 (gedroitzite). An artificial hydrous alumino-silicate of sodium, $Na_2O.Al_2O_3.3SiO_2.2H_2O$, related to permutite, later identified as a constituent of alkali soil from Ukraine. Named after Konstantin Kaetonovich Gedroitz (Константин Каетонович Гедройц) (1872–1932), professor of pedology in the Forestry Institute of Leningrad. [*M.A.*, **7**, 60, 515.] 15th List

Geikielite. A. B. Dick, *Min. Mag.*, X. 145, 1893; L. Fletcher, *Nature*, XLVI. 620, 1892. $MgO.TiO_2$. [Rakwana] Ceylon. 1st List

Gel-anatase. E. I. Semenov, 1957. Труды Инст. Мин., Геохим., Крист., Редк. Элем. (*Trans. Inst. Min. Geochem. Cryst. Rare Elements*), no. 1, p. 41 (Гельанатаз). Finely divided anatase, a component of leucoxene pseudomorphs after ilmenite in the Lovozero massif, Kola peninsula. An unnecessary name. [*M.A.*, **14**, 278.] 22nd List

Gel-bertrandite. E. I. Semenov, 1957. Труды Инст. Мин., Геохим., Крист., Редк. Элем. (*Trans. Inst. Min. Geochem. Cryst. Rare Elements*), no. 1, p. 64 (Гельбертрандит). A glassy, apparently colloidal mineral (no X-ray data), with the composition of bertrandite plus extra water, from pegmatites in the Khibina and Lovozero tundras, Kola peninsula. [*M.A.*, **14**, 277.] Inadequately characterized. 22nd List

Gel-cassiterite, Gel-cristobalite, Gelgoethite, Gel-thorite, Gel-zircon. E. I. Semenov, 1960. Труды Инст. Мин. Геохим. Крист. Редк. Элем. (*Trans. Inst. Min., Geochem., Cryst.*

Rare Elem.), no. 4, p. 85 (Гелькасситерит, Гелькристобалит, Гельгётит, Гельторит, Гельциркон). Superfluous names for Arandisite, Opal, Limonite, a hydrous Thorite, and Arshinovite respectively. *See also* Зап. Всесоюз. Мин. Общ. (*Mem. All-Union Min. Soc.*), 1963, vol. 92, pp. 198, 211, 212, 217, and *Amer. Min.*, 1962, vol. 47, p. 809; *M.A.*, **16**, 556, 557). 23rd List

Geldiadochite, Gelfischerite, Gelpyrophyllite, Gelvariscite. F. Cornu, 1909. *Zeits. Chem. Indust. Kolloide*, vol. iv, p. 17 (Geldiadochit, Gelfischerit, Gel-Pyrophyllit, Gelvariscit); *Zeits. prakt. Geol.*, 1909, vol. xvii, pp. 84, 144; *Centralblatt Min.*, 1909, p. 330. A. F. Rogers, *Journ. Geol. Chicago*, 1917, vol. xxv, p. 522. The gel-forms equivalent to the crystalloid forms diadochite, etc. *See* Uhligite. 8th List

Geldolomite. K. A. Redlich, 1911. Doelter's *Handbuch d. Mineralchemie*, vol. 1, p. 260 (Geldolomit). A synonym of gurhofian, the amorphous colloidal form of dolomite. 9th List

Gelfischerite, *see* Geldiadochite. 8th List

Gelgoethite, *see* Gel-cassiterite.

Gelmagnesite. K. A. Redlich, 1911. Doelter's *Handbuch d. Mineralchemie*, vol. 1, p. 260 (Gelmagnesit); H. Leitmeier, *Neues Jahrb. Min.*, 1916, Beil.-Bd. 40, p. 678. The amorphous colloidal form of magnesite occurring in serpentine-rocks. 9th List

Gelosite. J. A. Dulhunty, 1939. *Journ. Roy. Soc. New South Wales*, vol. 72, p. 184. Various undetermined microscopical constituents of torbanite from New South Wales are named gelosite, retinosite, humosite, and matrosite. [*M.A.*, **7**, 370.] 15th List

Gel-pristobalite, Geltohorite, errors for Gel-cristobalite, Gel-thorite (Зап. Всесоюз. Мин. Общ. (*Mem. All-Union Min. Soc.*), 1963, vol. 92, pp. 198, 212). 23rd List

Gelpyrophyllite, *see* Geldiadochite.

Gel-rutile. E. I. Semenov, 1957. Труды Инст. Мин., Геохим., Крист., Редк. Элем. (*Trans. Inst. Min. Geochem. Cryst. Rare Elements*), no. 1, p. 41 (Гельрутил). Finely divided rutile, a component of leucoxene pseudomorphs after ilmenite in the Lovozero massif, Kola peninsula. [*M.A.*, **14**, 278.] An unnecessary name. 22nd List

Gel-thorite, -zircon, *see* Gel-cassiterite.

Gelvariscite, *see* Geldiadochite.

Genaruttite. G. Gagarin and J. R. Cuomo, 1949, loc. cit., p. 9 (genaruttita). Cadmium oxide, CdO, cubic (3rd List, p. 380) from Genarutta mine, Monteponi, Sardinia. Syn. of monteponite (18th List). 19th List

Genevite. L. Duparc and M. Gysin, 1927. *Bull. Soc. Franç. Min.*, vol. 50, p. 41 (Genévite). Silicate of calcium, aluminium, etc., as small tetragonal crystals in limestone from Morocco. Previously thought to be dipyre, and perhaps identical with idocrase. Presumably named from Genève, Switzerland, where the authors reside. [*M.A.*, **3**, 368.] 11th List [Is Duparcite (var. of Idocrase); *M.A.*, **5**, 292; **8**, 251]

Genkinite. L. J. Cabri, J. M. Stewart, J. H. G. Laflamme, and J. T. Szymanski, 1977. *Can. Mineral.*, **15**, 389. A specimen from the Onverwacht mine, Transvaal, contains grains up to 165 μm, tetragonal, *a* 7·736, *c* 24·161 Å, composition $8[(Pt,Pd,Rh)_4Sb_3]$. May be identical with the unnamed mineral near $(Pt, Pd)_3Sb_2$ described by Tarkian and Stumpfl (1965) from the Driekop mine, Transvaal. Named for A. D. Genkin. 30th List

Genthelvite. J. J. Glass, R. H. Jahns, and R. E. Stevens, 1944. *Amer. Min.*, vol. 29, p. 178. The zinc end-member, $Zn_4Be_3Si_3O_{12}S$, of the helvine group. Named from helvine and after Prof. Frederick Augustus Louis Charles William Genth (1820–1893), of Philadelphia, who in 1892 described, under the name danalite, a single crystal containing 85% of this component. [*M.A.*, **9**, 62.] 17th List

Gentnerite. Ahmed el Goresy and Joahim Ottemann, 1966. *Zeits. Naturforsch.*, vol. 21*a*, p. 1160. An incompletely described copper iron chromium sulphide near $Cu_8Fe_3Cr_{11}S_{17}$ as veinlets in daubréelite in the Odessa meteorite. Named for W. Gentner. Species not accepted by the I.M.A. Commission on New Minerals and Mineral Names. [*Amer. Min.*, **52**, 559; Зап.. **97**, 67; *Bull.*, **90**, 608.] 25th List

Geolyte. E. A. Wülfing, *Jahreshefte Ver. vaterl. Naturk. Württemberg*, 1900, vol. lvi, p. 35 (Geolyt). To replace the term 'Bodenzeolith' used by agricultural chemists for those constituents of soils which are readily soluble and of indefinite mineralogical composition, but have little in common with zeolites. Named from γῆ, earth, and λύειν, to dissolve. 3rd List

Georgiadesite. A. Lacroix and A. de Schulten, 1907. *Compt. Rend. Acad. Sci. Paris*, vol. cxlv, p. 784; *Bull. Soc. franç. Min.*, 1908, vol. xxxi, p. 86 (georgiadésite). White or brownish-yellow, orthorhombic crystals associated with lead oxychlorides in the ancient lead slags at Laurion, Greece. A chloro-arsenate of lead, $Pb_3(AsO_4)_2.3PbCl_2$, containing much more chlorine than mimetite. Named after Mr. — Georgiades, director of the mines at Laurion. 5th List

Gepherite. I. D. Sedletzky, *Compt. Rend. (Doklady) Acad. Sci. URSS*, 1940, vol. 26, p. 241. Gepherite = гёферит = hoeferite (1st List)! 15th List

Geraesite. O. C. Farrington, 1912. *Bull. Geol. Soc. America*, vol. xxiii, p. 728. A preliminary abstract states this to be 'a hydrous barium aluminium phosphate more acidic than gorceixite', as the result of an analysis of a pebble ('fava') from the diamond washings of Minas Geraes, Brazil. Named after the locality. In the complete account (*Amer. Journ. Sci.*, 1916, ser. 4, vol. xli, p. 356) the name does not appear, the material being evidently impure gorceixite. 8th List

Gerasimovskite. E. I. Semenov, 1957. Труды Инст. Мин., Геохим., Крист., Редк. Элем. (*Trans. Inst. Min. Geochem. Cryst. Rare Elements*), no. 1, p. 41 (Герасимовскит). The niobium analogue of belyankinite (19th List), occurring in an ussingite pegmatite from the Lovozero massif, Kola peninsula. Named after V. I. Gerasimovsky. [*M.A.*, **14**, 278.] 22nd List

Gerassimovskite, spelling variant of or error for Gerasimovskite (22nd List). *Medd. Grønland*, 1967, **181**, no. 5. [*M.A.*, 71-477.] 28th List

Germanate-analcime, -celsian, -leucite, -natrolite, -nepheline, -sodalite. H. Strunz and E. Ritter, 1961. *Neues Jahrb. Min.*, Monatsh., Abt. A, p. 22 (Germanat-Analcim, etc.). Artificial germanium analogues of the silicate minerals. 22nd List

Germanate-pyromorphite. H. Wondratschek, 1963. *Neues Jahrb. Min.*, Abh., vol. 99, p. 116 (Germanatpyromorphit). Artificial $Pb_5(PO_4)_2GeO_4$; apatite family. 23rd List

Germanite. O. Pufahl, 1922. *Metall und Erz*, Halle, vol. 19, p. 324 (Germanit); F. W. Kriesel, ibid., 1923, vol. 20, p. 257; J. Lunt, *South African Journ. Sci.*, 1923, vol. 20, p. 157; J. S. Thomas and W. Pugh, *Journ. Chem. Soc. London*, 1924, vol. 125, p. 816; E. Thomson, *Univ. Toronto Studies, Geol. Ser.*, 1924, no. 17, p. 62. Sulphide of copper, iron, and germanium (Ge 5·10–8·71%), as a massive, reddish-grey mineral intergrown with tennantite at Tsumeb, South-West Africa. E. W. Todd (*in* E. Thomson) suggests the formula $10Cu_2S.4GeS_2.As_2S_3$. Named from the element germanium. [*M.A.*, **2**, 12, 252, 344, 410.] 10th List [$Cu_3(Ge,Fe)(S,As)_4$]

Germanite-(W). B. H. Geier and J. Ottemann, 1970. *Neues Jahrb. Min.*, Abh. **114**, 89. An unnecessary name for a tungstenian germanite; *see* tungsten-germanite. [*A.M.*, **56**, 1487.] 27th List

Germanium-albite, etc. *See* Gallium-albite. 19th List

Germanium-phenakite. A. Van Valkenburg and C. E. Weir, 1957. Bull. Geol. Soc. Amer., vol. 68, p. 1809 (germanium phenacite). Artificial $2BeO.GeO_2$, with Ge in place of Si. 21st List

Gersbyite. L. J. Igelström, *Zeits. Kryst. Min.*, XXVIII. 310, 1897. Near lazulite. [Dicksberg, Wermland] Sweden. 1st List [Is Lazulite; *A.M.*, **49**, 1778; *M.A.*, **17**, 180]

Gerstleyite. C. Frondel and V. Morgan, 1956. *Amer. Min.*, vol. 41, p. 839. Sodium sulphantimonite and sulpharsenite, $Na_4As_2Sb_8S_{17}.6H_2O$, as red spherules, monoclinic(?) with borates in clay from Kramer, California. Named after Mr. J. M. Gerstley, of a borax company. [*M.A.*, **13**, 303.] 21st List

Gerstmannite. E. F. Kashner, 1976. *Rocks and Minerals*, **51**, 124. P. B. Moore and T. Araki. *A.M.*, **62**, 51 (1977). Rosettes of pale pink crystals from the Sterling Hill mine, Ogdensburg, Sussex Co., New Jersey, are orthorhombic, *a* 8·185, *b* 18·65, *c* 6·256 Å, space group *Bbcm*. Composition $8[(Mn,Mg)Mg(OH)_2ZnSiO_4]$, sp. gr. 3·68. RIs α 1·665 ∥ [010], β 1·675 ∥ [001], γ 1·678 ∥ [100], $2V_\alpha$ 50–60°. Named for E. Gerstmann. [*M.A.*, 77-3386, 3387.] 30th List

Getchellite. B. G. Weissberg, 1965. *Amer. Min.*, vol. 50, p. 1817. Dark red monoclinic crystals from the Getchell mine, Humboldt County, Nevada, had the composition $AsSbS_3$. Named from the locality. [*M.A.*, **17**, 696.] 24th List

Geversite. E. F. Stumpfl, 1961. *Min. Mag.*, vol. 32, p. 833. $PtSb_2$, cubic with pyrite-type structure, intergrown with native platinum at the Dreikop mine, Transvaal, South Africa. Named after T. W. Gevers. 22nd List

Gewlekhite, *see* Gyulekhite.

Ghassoulite. G. Millot, 1954. *Compt. Rend. Acad. Sci. Paris*, vol. 238, p. 257. An end-member, $Mg_3Si_4O_{10}(OH)_2$, of the montmorillonite group, previously referred to sepiolite. Named from the locality, gebel Ghassoul, Morocco. [*M.A.*, **12**, 354.] 20th List [Is Hectorite; *A.M.*, **44**, 342]

Ghaussoulith, error for Ghassoulith (Hintze, *Handb. Min.*, Erg.-Bd. II, p. 930). 22nd List

Ghinzburgite, incorrect transliteration of Гинзбургит, Ginzburgite (21st List) (I. Kostov, *Mineralogy*, p. 379). 25th List

Gianellaite. G. Tunell, J. J. Fahey, F. W. Daugherty, and G. V. Gibbs, 1977. *Neues Jahrb. Mineral.*, Monatsh. 119. Yellow rosettes and distorted octahedra in the Mariposa mine, Terlingua district, Brewster Co., Texas, are cubic. *a* 9·5215 Å, space group $F\bar{4}3m$. Composition $4[(NHg_2)_2SO_4]$, the pure sulphate of Millon's base. Mosesite, which contains considerable Cl ($Cl_2 \simeq SO_4$), has *a* 28·618 = 3 × 9·539 Å and a very similar X-ray powder pattern. Named for V. P. Gianella. [*A.M.*, **62**, 1057; *M.A.*, 77-4621.] 30th List

Giannettite. D. Guimarães, 1948. *Bol. Inst. Tecn. Indust. Minas Gerais*, no. 6, p. 22 (giannettita), p. 62 (giannettite). Chloro-zircono-titano-silicate of Ca, Na, Mn, as minute triclinic crystals with aegirine in nepheline-rocks from Poços de Caldas, Brazil. Named after Dr. Americo René Giannetti, mining engineer. [*M.A.*, **10**, 510.] 18th List

Gibbsitogelit. *See* Diasporogelite. 7th List

Giesenherrite. D. P. Grigoriev, *Mém. Soc. Russe Min.*, 1937, ser. 2, vol. 66, p. 264 (гисингерит), p. 300 (*giesenherrite*). Mis-spelling of hisingerite (J. J. Berzelius, 1828), named after Wilhelm Hisinger (1766–1852). Quoted as an example of the many errors that arise through double transliteration. 15th List

Giessenite. S. Graeser, 1963. *Schweiz. Min. Petr. Mitt.*, vol. 43, p. 471. Fine orthorhombic needles of $Pb_8Bi_6S_{17}$ occur in dolomite near Giessen in the Binn valley, Valais, Switzerland. Named from the locality. [*M.A.*, **16**, 546.] 23rd List [$Pb_9CuBi_6Sb_{1.5}S_{30}$(?); *A.M.*, **50**, 264; Norway; *M.A.*, 74-3432]

Gillespite. W. T. Schaller, 1922. *Journ. Washington Acad. Sci.*, vol. 12, p. 7. Silicate of barium and ferrous iron, $Fe''BaSi_4O_{10}$, occurring as red, mica-like scales in a rock from Alaska. Named after the collector, Mr. Frank Gillespie, of Richardson, Alaska. Cf. Taramellite (E. Tacconi, 1908; 5th List). [*M.A.*, **1**, 375.] 9th List [*A.M.*, **14**, 319]

Gilpinite. E. S. Larson and G. V. Brown, 1917. *American Mineralogist*, vol. ii, p. 75. Hydrated sulphate of uranium with some copper, ferrous iron, etc., $RO.UO_3.SO_3.4H_2O$, occurring as minute, monoclinic laths on pitchblende from Gilpin County, Colorado. Differs from zippeite and uranopilite in its optical characters. Named after the locality. 8th List [Is Johannite; *A.M.*, **11**, 1]

Ginorite. G. D'Achiardi, 1934. *Periodico Min. Roma*, vol. 5, p. 22 (ginorite). Hydrous borate of calcium, $H_{12}Ca_4B_{14}O_{29}.2H_2O$, as a compact white mass of minute monoclinic crystals from Tuscany. Named after Prince Piero Ginori Conti, of Firenze (Florence). [*M.A.*, **5**, 484.] 13th List [Syn. Cryptomorphite; *A.M.*, **39**, 406]

Ginzburgite. F. V. Chukhrov, 1955. [*Colloids in the earth's crust*, *Acad. Sci. USSR*, p. 598.] Abstr. in *Mém. Soc. Russ. Min.*, 1956, vol. 85, p. 382 (гинзбургиты, ginsburgites) and *Amer. Min.*, 1957, vol. 42, p. 440. A general term for clay minerals with composition $(Al,Fe)_2O_3{:}SiO_2$ ranging from 1:1 to 1:1·25. Named after I. I. Ginzburg. [*M.A.*, **13**, 628.] 21st List

Giorgiosite. A. Lacroix, 1905. *Compt. Rend. Acad. Sci. Paris*, vol. cxl, p. 1310; *Bull. Soc. franç. Min.*, vol. xxviii, p. 198. A basic magnesium carbonate occurring as a white powder on lava erupted from Santorin in 1866. Under the microscope it is seen to consist of minute, birefringent spherules, and to have the same characters as the artificial compound $4MgCO_3.Mg(OH)_2.4H_2O$. Named from Giorgios, one of the cones formed during the eruption of 1866. 4th List

Girnarite. K. K. Mathur and A. G. Jhingran, 1931. *Quart. Journ. Geol. Mining Metall. Soc.*

India, vol. 3, p. 100. An amphibole of the hastingsite group in nepheline-syenite at Mt. Girnar, Kathiawar, India. Named from the locality. [*M.A.*, **5**, 391.] 13th List [Subsilicic titanian sodian magnesian hastingsite, *Amph.* (1978)]

Gironit. J. Fröbel, 1843. *See* chrysargyrit. 27th List

Giulekhite, variant of Gyulekhite (*Amer. Min.*, 1958, vol. 43, p. 1222). 22nd List

Glacialite. (*18th Ann. Rep. US Geol. Survey for 1896–7*, 1897, part v, p. 1354. G. P. Merrill, *Rep. US National Museum for 1899*, 1901, p. 337.) Trade-name for a white clay from Enid, Oklahoma, put on the American market as a fuller's earth. 6th List

Gladite. K. Johansson, 1924. *Arkiv Kemi, Min. Geol.*, vol. 9, no. 8, p. 17 (Gladit), Sulphobismuthite of lead and copper, $2PbS.Cu_2S.5Bi_2S_3$, as lead-grey to tin-white, prismatic crystals from Gladhammer, Sweden. Named, with hammarite (q.v.), from the locality. [*M.A.*, **2**, 340.] 10th List [C.M. **14**, 194]

Glanzkohle (= Clarain), *see* Fusian.

Glaucamphiboles. H. Rosenbusch, *Sitz.-Ber. Akad. Berlin*, 1898, p. 707 (*Glaukamphibole*). A group-name to include the alkali-amphiboles, glaucophane, gastaldite, and crossite, which are of dynamo-metamorphic origin, as distinct from the arfvedsonite-amphiboles, arfvedsonite, riebeckite, and hastingsite, which occur only as original constituents in igneous rocks. 2nd List

Glaucocerinite. E. Dittler and R. Koechlin, 1932. *Centr. Min.*, Abt. A, 1932, p. 13 (Glaukokerinit). A hydrous ultra-basic sulphate of zinc, aluminium, and copper, $Zn_{13}Al_8Cu_7(SO_4)_2O_{30}.34H_2O$, as sky-blue warty masses of soft waxy material, from Laurion, Greece. Named from *γλαυκός*, blue, and *κήρινος*, waxy. [*M.A.*, **5**, 49.] 13th List [*A.M.*, **17**, 495; **19**, 556. Glaucokerinite, Fleischer's *Gloss.*]

Glaucochroite. S. L. Penfield and C. H. Warren, Dana's *Appendix*, 1899; *Amer. J. Sci.*, 1899, VIII, 343; *Zeits. Kryst. Min.*, 1900, XXXII, 231; Abstr. *Min. Mag.*, XII, 316. A member of the olivine group occurring as embedded prismatic crystals with a bluish-green colour resembling beryl. Rhombic. $CaMnSiO_4$. Franklin Furnace, New Jersey. 2nd List

Glaucodotite. E. T. Wherry. *Journ. Washington Acad. Sci.*, 1920, vol. 10, p. 496. Variant of Glaucodot. 9th List

Glauconie. G. Millot, 1964. *Géol. des Argiles* (Paris, Masson), 238. A variable green mixture, which may contain illite, montmorillonite, chlorite, or various mixed-layer minerals. 29th List

Glaucopargasite. P. R. J. Naidu, 1944. *Quart. Journ. Geol. Mining Metall. Soc. India*, vol. 16, p. 138. A blue amphibole from Mysore intermediate in composition between glaucophane and pargasite. [*M.A.*, **9**, 189.] 17th List

Glaukokerinite, *see* Glaucocerinite.

Glaukosphaerite. M. W. Pryce and J. Just, 1974. *M.M.*, **39**, 737. Green fibrous spherules from Hampton East Location, Kambalda, Western Australia, are monoclinic, *a* 9·34, *b* 11·93, *c* 3·07 Å, *β* 90 to 91°. Composition $4[(Cu,Ni)_2CO_3(OH)_2]$, the nickel analogue of rosasite. Named for the colour and habit. 28th List [Zaire; *M.A.*, 76-2844]

Glendonite. T. W. E. David and T. G. Taylor, 1905. *Records Geol. Survey New South Wales*, vol. viii, p. 161. Large pseudomorphs of granular calcite, probably after glauberite, found in abundance at Glendon on the Hunter River and other localities in New South Wales. They were described by J. D. Dana in 1849, and are similar to thinolite and pseudogaylussite. 4th List

Glimmerton. K. Endell, U. Hofmann, and E. Maegdefrau, 1935. *Zement*, Berlin-Charlottenburg. vol. 24, p. 627. A micaceous clay. Synonym of illite (15th List). 19th List

Glinite. S. V. Potapenko, 1952. [*Publ. Acad. Architecture*, Ukraine, p. 43]. Criticized by I. D. Sedletzky, *Mém. Soc. Russ. Min.*, 1954, ser. 2, vol. 83, p. 289 (глинит). A group name for clay minerals, from глина, clay. Compare Clayite (5th List). *See* Chasovrite. [*M.A.*, **13**, 5.] 21st List

Glucine. N. A. Grigoriev, 1963. Зап. Всесоюз. Мин. Общ. (*Mem. All-Union Min. Soc.*), vol. 92, p. 691 (Глюцин, glucin). $CaBe_4(PO_4)_2(OH)_4.\frac{1}{2}H_2O$, massive and encrusting, with moraesite from a locality in the Urals. Named from the alternative name of beryllium—glucinum. [*M.A.*, **16**, 550.] 23rd List [A.M., **49**, 1152]

Gluschinskit, German transliteration of Глушинскит, glushinskite (H. Strunz, *Min. Tabellen*, 4th edit., 1966, p. 475). 24th List

Glushinskite. Yu. A. Zhemchuznikov and A. I. Ginzburg, 1960. [Основы петрол. углей, Изд. Акад. Наук СССР (*Problems in the Petrology of Coal*), p. 93], abstr. in Зап. Всесоюз. Мин. Общ. (*Mem. All-Union Min. Soc.*), 1962, vol. 91, p. 204 (Глушинскит, glushinskite), and in *Amer. Min.*, 1962, vol. 47, p. 1482. Oxalate of Mg, as orthorhombic plates in chalky clay in an unspecified Arctic locality. [*M.A.*, **16**, 555.] 23rd List

Godlevskite. E. A. Kulagov, T. L. Evstigneeva, and O. E. Yushko-Zakharova, 1969. Геол. рудн. месторожд. (*Geol. Ore-deposits*), **11**, 115 (Годлевскит). β-Ni_7S_6 occurs as grains and aggregates in the Norilsk and Talnakh ore deposits in bornite + chalcopyrite veins. Named for M. N. Godlevskii. [*A.M.*, **55**, 317; *M.A.*, 70-1639; *Zap.*, **99**, 72.] 26th List

Godlewskit, Germ. trans. of Годлевскит, godlevskite (26th List). Hintze, 33. 28th List

Goedkenite. P. B. Moore, A. J. Irving, and A. R. Kampf, 1975. *A.M.*, **60**, 957. Pale yellow crystals, tabular to {001}, from the Palermo no. 1 pegmatite, North Groton, New Hampshire, have *a* 8·45, *b* 5·74, *c* 7·26 Å, β 113·7°, space group $P2_1/m$; α 1·669 ‖ [010], β 1·673, γ 1·692, $2V_\gamma$ 45 to 50°; ρ calc. 3·83 g. cm^{-3}. Composition $2[(Sr,Ca)_2Al(PO_4)_2OH]$; probably isostructural with brackebuschite. Named for Dr. V. L. Goedken. 29th List

Goeschwitzite. I. D. Sedletzky, 1940. *Compt. Rend.* (*Doklady*) *Acad. Sci. URSS*, vol. 26, p. 242 (heshvitcite!). The 'Glimmer von Goeschwitz' of E. Maegdefrau and V. Hofmann (*Zeits. Krist.*, 1937, vol. 98, p. 33), a micaceous clay from Goeschwitz (Göschwitz), Thuringia. Named from the locality. 15th List

Gokaite. T. Tomita, 1936. *Journ. Shanghai Sci. Inst.*, sect. 2, vol. 2, p. 99 (gokaite). A clinohypersthene with small optic axial angle. Named from the locality, Goka, Oki islands, Japan. 14th List

Gold argentide. M. P. Lozhechkin, 1939. *Doklady Acad. Sci. URSS*, vol. 24, p. 454. Synonym of electrum. [*M.A.*, **7**, 515.] 20th List

Gold cupride. M. P. Lozhechkin, 1939. *Doklady Acad. Sci. URSS*, vol. 24, p. 451. Synonym of cuproauride (15th List) and auricupride (19th List). [*M.A.*, **7**, 514.] 20th List

Goldfieldite. F. L. Ransome, 1909. *United States Geol. Survey, Prof. Paper No. 66*, p. 116; *Economic Geology*, 1910, vol. v, p. 453. A massive, dark lead-grey mineral occurring with marcasite, bismuthinite, famatinite, and native gold in the gold ores of Goldfield, Nevada. A cupric sulphantimonite in which part of the antimony is replaced by bismuth (and arsenic) and part of the sulphur by tellurium (17%); $5CuS.(Sb,Bi)_2(S, Te)_3$. Named after the locality. 6th List [tellurian Tetrahedrite; *A.M.*, **32**, 254]

Goldichite. A. Rosenzweig and E. B. Gross, 1954. *Bull. Geol. Soc. Amer.* [*1955*], vol. 65, no. 12 (for 1954), p. 1299; *Amer. Min.*, 1955, vol. 40, pp. 331, 469. Hydrous potassium ferric sulphate, $KFe(SO_4)_2.4H_2O$, as pale-green monoclinic crystals from the decomposition of pyrite. From Utah. Named after Samuel S. Goldich (1909–), professor of geology, University of Minnesota. [*M.A.*, **12**, 511.] 20th List

Goldmanite. R. C. Moench and R. Meyrowitz, 1964. *Amer. Min.*, vol. 49, p. 644. A garnet from the Laguna uranium mining district, Albuquerque, New Mexico, has the composition $Ca_3(V,Fe,Al)_2Si_3O_{12}$, with 60% of the vanadium end-member. Named for M. I. Goldman. 23rd List [Cf. Yamatoite]

Goldschmidtine. M. A. Peacock, 1937. *Amer. Min.*, vol. 22, no. 12, part 2, p. 11 (abstr.); ibid., 1938, vol. 23, p. 176; 1939, vol. 24, p. 227; 1940, vol. 25, p. 372. Described as silver antimonide, Ag_2Sb, but later identified with stephanite (Ag_5SbS_4). [*M.A.*, **7**, 15, 317, 516.] 15th List

Goldschmidtite. W. H. Hobbs, *Amer. J. Sci.*, 1899, VII, 357; *Zeits. Kryst. Min.*, 1899, XXXI, 417. A telluride of gold and silver, Au_2AgTe_6, intermediate between calaverite and sylvanite. Monoclinic. Cripple Creek district, Colorado. 2nd List [Is Sylvanite; *see* Dana *Syst. Min.*, 6th edit., App. II, 46]

Gonnardite. A. Lacroix, *Bull. Soc. fran. Min.*, XIX. 426, 1896. F. Gonnard, *Compt. Rend.*, LXXIII. 1147, 1871. A zeolite. $(Ca,Na_2)_2Al_2Si_5O_{15} + 5\frac{1}{2}H_2O$? Rhombic. Described by Gonnard as mesolite, but it differs from this optically. [Gignat] Puy-de-Dôme. 1st List [$Na_2CaAl_4Si_6O_{20}.7H_2O$]

Gonsogolite. (P. Groth, *Min.-Samml. Strassburg*, 1878, p. 258. *Gonsogolith.*) Probably the same as pectolite. Predazzo, Tyrol. 2nd List

Gonyerite. C. Frondel, 1955. *Amer. Min.*, vol. 40, p. 1090. A chlorite rich in manganese (MnO 33·83%) and poor in aluminium (Al_2O_3 0·58%), from Långban, Sweden. Named after Forest A. Gonyer, analytical chemist, of Harvard University. [*M.A.*, **13**, 87.] 21st List

Goodletite. An unnecessary rock name; the limestone matrix of Burma ruby. R. Webster, *Gems*, 1962, p. 757. 28th List

Goongarrite. E. S. Simpson, 1924. *Journ. R. Soc. Western Australia*, 1924, vol. 20, p. 65. A sulphur-salt of lead and bismuth, $4PbS.Bi_2S_3$, occurring as fibrous to platy masses (monoclinic?) near Lake Goongarrie, Western Australia. Named from the locality. [*M.A.*, **2**, 336.] 10th List [A mixture; *A.M.*, **49**, 1501; confirmed as a mixture; *A.M.*, **62**, 397 (abstr.)]

Gorceixite. E. Hussak, 1906. *Min. Petr. Mitt.* (*Tschermak*), vol. xxv, p. 338 (Gorceixit). Hydrated barium aluminium phosphate, $BaO.2Al_2O_3.P_2O_5.5H_2O$, occurring as brown pebbles ('favas'), with a compact, jaspery structure, in the diamond-bearing sands of Brazil. It is optically uniaxial, and belongs to the hamlinite group of minerals. Named after Professor Henri Gorceix, formerly director of the School of Mines at Ouro Preto, Brazil. 4th List [Cf. Bariohitchcockite]

Gordonite. E. S. Larsen and E. V. Shannon, 1930. *Amer. Min.*, vol. 15, p. 331. Hydrous phosphate of magnesium and aluminium, $MgO.Al_2O_3.P_2O_5.9H_2O$, as colourless monoclinic crystals in variscite nodules from Utah. Related to paravauxite [$FeO.Al_2O_3.P_2O_5.5H_2O$]. Named after 'Mr. S. L. Gordon who first described paravauxite', i.e. Mr. Samuel George Gordon, of the Academy of Natural Sciences of Philadelphia. [*M.A.*, **4**, 344.] 12th List

Görgeyite. H. Mayrhofer, 1953. *Neues Jahrb. Min.*, Monatshefte, p. 35 (Görgeyit). Hydrated sulphate, $K_2SO_4.5CaSO_4.1\text{–}1\frac{1}{2}H_2O$, small tabular monoclinic crystals with glauberite, etc., in salt deposits at Ischl, Upper Austria. Named in memory of Dr. Rolf Görgey von Görgö (1886–1915) of Vienna, who wrote on salt deposits. [*M.A.*, **12**, 132.] Cf. mikheevite. 20th List [*A.M.*, **39**, 403]

Gosseletite. J. Anten, 1923. *Mém.* (*in-4°*) *Acad. Roy. Belgique, Cl. Sci.*, ser. 2, vol. 5, fasc. 3, p. 19. An undetermined mineral, perhaps an orthorhombic manganese silicate, occurring in the phyllites of the Belgian Ardennes. Named after the French geologist, Jules Auguste Alexandre Gosselet (1832–1916). [*M.A.*, **3**, 11.] 11th List [manganian Andalusite; *A.M.*, **22**, 72]

Götzenite. T. G. Sahama and K. Hytönen, 1957. *Min. Soc. Notice*, no. 98; *Min. Mag.*, 1957, vol. 31, p. 503; *M.A.*, **13**, 555. $(Ca,Na,Al)_7(Si,Ti)_5O_{15}F_{3\cdot 5}$, triclinic, rinkite group. In nephelinite from Kivu, Belgian Congo. Named after Count G. A. von Götzen, German traveller in E Africa. 21st List [*M.A.*, 74-1408]

Goumbrine, *see* Gumbrine.

Goureite. A. Lacroix, 1934. *Mém. Acad. Sci. Paris*, vol. 61, p. 320 (gouréite). An undetermined mineral occurring as pale yellow, skeletal, optically uniaxial crystals in microgranite from Gouré (= Gure), west of Lake Chad, French West Africa. Named from the locality. The mineral was first mentioned, but not named, in *Compt. Rend. Acad. Sci. Paris*, 1905, vol. 140, p. 22. [*M.A.*, **6**, 124.] 14th List [Is Narsarsukite; *M.A.*, **15**, 356]

Gouverneurite. F. R. Van Horn, 1926. *Amer. Min.*, vol. 11, p. 54. The brown magnesia tourmaline of Gouverneur, St. Lawrence Co., New York, regarded as a distinct species of the tourmaline group. 11th List [Dravite, q.v.]

Gowerite. R. C. Erd, J. F. McAllister, and H. Almond, 1959. *Amer. Min.*, vol. 44, p. 911. $CaB_6O_{10}.5H_2O$, monoclinic, from Furnace Creek, Death Valley, California. Named after Harrison P. Gower. [*M.A.*, **15**, 501.] 22nd List [*A.M.*, **57**, 381]

Graebeite. A. Treibs and H. Steinmetz, 1933. *Liebigs Ann. Chem.*, vol. 506, p. 171 (Graebeït). Brick-red smears in crevices in shale from a coal mine in Saxony, possessing dyeing properties, and identified as a mixture (graebite a and graebite b) of polyhydroxyanthraquinones. Named after Carl Graebe (1841–1927), formerly Professor of Chemistry at Genève, who did much work on the anthracene series. [Cf. Hoelite, 10th List.] [*M.A.*, **5**, 391.] 13th List

Graemite. S. A. Williams and P. Matter, 1975. *Min. Record*, **6**, 32. Blue-green pleochroic crystals up to 8 mm long from the Cole shaft, Bisbee, Arizona, have space group *Pcmm*, *a*

6·805, *b* 25·613, *c* 5·780 Å, composition 10[$CuTeO_3.H_2O$]. Named for R. Graeme. [*A.M.*, **60**, 486; *Bull.*, **98**, 264.] 29th List

Graftonite. S. L. Penfield, *Amer. J. Sci.*, 1900, IX, 20; *Zeits. Kryst. Min.*, 1900, XXXII, 433. Indistinct monoclinic crystals, consisting of a fine lamellar intergrowth of light coloured graftonite with darker triphylite. $(Fe,Mn,Ca)_3P_2O_8$. Grafton, New Hampshire. 2nd List [Cf. Beusite]

Gralmandite. L. L. Fermor, 1938. *Rec. Geol. Surv. India*, vol. 73, pp. 154, 156; *Indian Assoc. Cultiv. Sci.*, 1938, Spec. Publ. no. 6, p. 20. Garnets intermediate in composition between grossular and almandine. Varieties containing in addition less than 10% of the pyrope and spessartine molecules are called magnesia-gralmandite and manganese-gralmandite. Cf. spandite (4th List), grandite (5th List), and calc-spessartite (11th List). 15th List

Grandidierite. A. Lacroix, *Bull. Soc. franç. Min.*, 1902, vol. xxv, p. 86. A blue, sapphirine-like mineral in granite from Madagascar. Orthorhombic, with strong pleochroism. Sp. gr. = 2·99. A basic silico-aluminate of iron, magnesium, and calcium with 2% of alkalies. Named after Mr. Alfred Grandidier, an authority on the geography and natural history of Madagascar. 3rd List [$MgAl_3(BO_3SiO_4)O_2$; *M.A.*, 70-583]

Grandite. L. L. Fermor, 1909. *Mem. Geol. Survey India*, vol. xxxvii, pp. lxxi, 165. A contraction from grossularite and andradite for garnets of intermediate composition. A manganiferous variety is called 'mangan-grandite'. Compare 'spandite' (L. L. Fermor, 1907; 4th List). 5th List

Grantsite. A. D. Weeks, M. L. Lindberg, and R. Meyrowitz, 1961. *US Geol. Survey Prof. Paper no. 424–B*, p. 293; *Amer. Min.*, 1962, vol. 47, p. 414. $Na_4Ca(V^{4+}O)_2V_{10}O_{30}.8H_2O$ in dark olive-green to black fibrous aggregates at the F–33 mine, Grants, Valencia County, New Mexico, and at the La Salle and Golden Cycle mines, Montrose County, California. Named for the first locality. [*M.A.*, **16**, 551.] 23rd List [*A.M.*, **49**, 1511]

Gränzerite. [J. E. Hibsch, MS.] R. Koechlin, *Min. Taschenbuch Wiener Min. Gesell.*, 2nd edit., 1928, p. 27; *Centr. Min.*, Abt. A, 1933, p. 203. J. E. Hibsch, *Min. Böhm. Mittelgeb.*, 1934, p. 109. Synonym of sanidine. A druse mineral in basalt from Eulenberg, Bohemia. Named after Dr. Josef Gränzer, of Reichenberg, Bohemia. [*M.A.*, **5**, 296, 482.] 13th List

Graphitite. W. Luzi, *Zeits. f. Naturwiss.*, LXIV. 257, 1891, and *Ber. deutsch. chem. Ges.*, XXIV.(2), 4093, 1891 (1892); E. Weinschenk, *Zeits. Kryst. Min.*, XXVIII. 291, 1897. A variety of graphite which does not swell up when moistened with nitric acid and ignited. Weinschenk considers it to be merely graphite. 1st List

Gratonite. C. Palache and D. J. Fisher, 1939. *Amer. Min.*, vol. 24, p. 136; 1940, vol. 25, p. 255. Sulpharsenite of lead, $Pb_9As_4S_{15}$, as rhombohedral crystals from Peru. Named after Louis Caryl Graton (1880–), Professor of Mining Geology at Harvard University. [*M.A.*, **7**, 263, 512.] 15th List

Grayite. S. H. U. Bowie, 1957. *Summ. Progr. Geol. Surv. Great Britain for 1956*, p. 47. Thorium phosphate, hexagonal, related to rhabdophane. Pale yellow, powdery, in lithia-pegmatite from Southern Rhodesia. [*M.A.*, **13**, 494.] 21st List [$(Th,Pb,Ca)PO_4.H_2O$; *A.M.*, **49**, 419; cf. Brockite]

Greenalite. C. K. Leith, 1903. *Monogr. US Geol. Survey*, vol. xliii, pp. 14, 101, 239. A hydrous silicate of iron occurring in the form of green granules and closely resembling glauconite, from which it differs in containing no potassium. The composition varies somewhat; one analysis gives the formula $Fe_2'''(Fe,Mg)_3(SiO_4)_3.3H_2O$. Occurs abundantly as a constituent of sedimentary rocks in the Mesabi iron-bearing district of Minnesota. 4th List

Greenite. John Ruskin, 1884. *Catalogue of a series of specimens in the British Museum (Natural History) illustrative of the more common forms of native silica.* Orpington, Kent (G. Allen), 1884, p. 28. The English equivalent of chlorite. 'Chlorite, which ought to be more simply termed "Greenite" or "Greeny", is a combination of . . .' 4th List

'Green John'. B. Blount and J. H. Sequeira, 1919. *Journ. Chem. Soc. London, Trans.*, vol. 115, p. 707. Green fluorspar, named from analogy to 'Blue John'. [*M.A.*, **1**, 68.] 9th List

Greenolith, error for Greenalith (F. Machatschki, *Spez. Min.*, 1953, pp. 350, 362). 22nd List

Gregorite. J. A. Paris, 1818. *Trans. Roy. Geol. Soc. Cornwall*, vol. 1, p. 226 (Gregorite (Menacchanite)). Synonym of ilmenite. Named after William Gregor (1761–1817), who discovered titanium. Not the gregorite of G. J. Adam, 1869, synonym of agnesite, also from Cornwall and named after William Gregor. 19th List

Greigite. B. J. Skinner, R. C. Erd, and F. S. Grimaldi, 1964. *Amer. Min.*, vol. 49, p. 543. Minute grains and crystals in clays from the Kramer–Four-Corners area, San Bernardino County, California, are the thiospinel of iron, Fe_3S_4; cubic. Named for J. W. Greig. 23rd List

Greinerite. A. K. Boldyrev, 1928. *Kurs opisatelnoi mineralogii*, Leningrad, 1928, part 2, p. 162 (Грейнерит). O. M. Shubnikova and D. V. Yuferov, *Spravochnik po novym mineralam*, Leningrad, 1934, p. 48 (Грейнерит, Greinerite). A 'brown-spar' or mangandolomite, $(Mg,Mn)Ca(CO_3)_2$, from Greiner, Zillerthal, Tyrol. Named from the locality. 14th List

Grenalite. Error for or variant of grenatite (of Saussure), syn. of staurolite. R. Webster, *Gems.*, 1962, p. 757. Not to be confused with greenalite (4th List). 28th List

Griffithite. E. S. Larsen and G. Steiger, 1917. *Journ. Washington Acad. Sci.*, vol. vii, p. 11. A member of the chlorite group occurring as a filling in amygdaloidal cavities in basalt, and differing both optically and chemically from any chlorite previously described. Formula, $4(Mg,Fe,Ca)O.(Al,Fe)_2O_3.5SiO_2.7H_2O$. Named after the locality, Griffith Park, Los Angeles, California. 8th List [Ferroan Saponite; *A.M.*, **40**, 944]

Grimaldiite. C. Milton, D. [E.] Appleman, E. C. T. Chao, F. Cuttitta, J. L. Dinnin, E. J. Dwornik, M. Hall, B. L. Ingram, and H. J. Rose, Jr., 1967. *Geol. Soc. Amer., Progr. Ann. Meeting*, p. 151. CrOOH, isostructural with delafossite, a minor constituent of the mixture of chromium oxides occurring in alluvial gravels of the Merume river, Mazaruni district, Guayana (Merumite, 18th List). Named for F. Grimaldi. 25th List [Cf. Bracewellite, Guyanaite]

Grimselite. K. Walenta, 1972. *Schweiz. Min. Petr. Mitt.*, **52**, 93. Pale yellow crusts from the tunnel between Garstenegg and Sommerloch, Grimsel area, Oberhasli, Switzerland, are hexagonal, *a* 9·30, *c* 8·26 Å, $P\bar{6}2c$. Composition $2[K_3NaUO_2(CO_3)_3.H_2O]$. ω 1·611, yellow, ε 1·480, colourless. Water-soluble. Has been synthesized. Named for the locality. 27th List [*A.M.*, **58**, 139]

Grodnolite. J. Morozewicz, 1924. *Bull. Soc. Franç. Min.*, vol. 47, p. 46; *Spraw. Polsk. Inst. Geol.* (= *Bull. Serv. Géol. Pologne*), 1924, vol. 2, p. 223 (grodnolit). Colloidal calcium phosphate with carbonate, etc., $2Ca_3(PO_4)_2.CaCO_3.Ca(OH)_2 + \frac{1}{4}H_4Al_2Si_2O_9$, occurring as concretions in Cretaceous marls in Grodno, Poland. It differs from the crystalline dahllite and podolite in being optically isotropic and amorphous. Named from the locality. [*M.A.*, **2**, 343, 417.] 10th List [Is Francolite; *A.M.*, **23**, 1]

Grönlandit. M. H. Klaproth, 1809. *Tasch. Min.*, vol. 3, p. 198. A. Breithaupt, 1858. *Berg. Hütt. Zeitung*, vol. 17, p. 61. Original form of Greenlandite (of Klaproth, = almandine) and of Greenlandite (of Breithaupt, = columbite). 22nd List

Grossouvreite. S. Meunier, 1902. *Bull. Soc. Géol. France*, ser. 4, vol. ii, p. 250. A. Lacroix, *Minéralogie de la France*, 1913, vol. v, p. 50 (grossouvréite). To replace the preoccupied name vierzonite of Albert de Grossouvre (1901; 3rd List) for a pulverulent opal from Vierzon, dép. Cher. 7th List

Grosspydite. V. S. Sobolev, 1960. *Internat. Geol. Congr. Rept. 21st Sess.*, part 14, p. 72. A grossular–pyroxene–kyanite (disthene) paragenesis—a rock, not a mineral. 25th List

Grossularoid. D. S. Belyankin and V. P. Petrov, 1941. *Amer. Min.*, vol. 26, p. 450 (grossularoid). Group name for hibschite and plazolite, $3CaO.Al_2O_3.2SiO_2.2H_2O$, related to and associated with grossular, in which $2H_2O$ replaces SiO_2.Cf. garnetoid and hydrogarnet. [*M.A.*, **8**, 145.] 16th List

Grothine. F. Zambonini, 1913. *Atti* (*Rend.*) *R. Accad. Lincei*, Roma, ser. 5, vol. xxii, sem. 1, p. 801 (Grothina). Small, colourless, orthorhombic crystals found in metamorphic limestone at Nocera and Sarno, Campania, Italy. Qualitative tests point to a silicate of aluminium (with a trace of iron) and calcium. Named after Professor Paul Heinrich von Groth, of Munich. Not the Grothite of J. D. Dana, 1867. 7th List

Groutellit, cited without reference, H. Strunz, *Min. Tabellen*, 4th edit., 1966, p. 476. Groutite partially pseudomorphously altered to ramsdellite. 24th List

Groutite. J. W. Gruner, 1945. *Amer. Min.*, vol. 30, p. 169. $HMnO_2$ as orthorhombic crystals of the diaspore group distinct from manganite, from Cuyuna range, Minnesota. Named after

Professor Frank Fitch Grout (1880–) of the University of Minnesota. [*M.A.*, **9**, 126.] 17th List

Grovesite. F. A. Bannister, M. H. Hey, and W. C. Smith, 1955. *Min. Mag.*, vol. 30, p. 645. A chlorite-like mineral from the Benallt manganese mine, Wales. $(Mn,Mg,Al)_3(Si,Al)_2(O,OH)_9$, near pennantite (17th List), but with X-ray pattern similar to berthierine and cronstedtite, suggesting a kaolin- rather than a chlorite-type structure. Named after Dr. Arthur William Groves (1903–), of the Imperial Institute, London. 20th List [Chlorite group; *A.M.*, **59**, 1155]

Grundite. R. E. Grim and W. F. Bradley, *Journ. Amer. Ceramic Soc.*, 1939, vol. 22, p. 157; *Rep. Investig. Geol. Surv. Illinois*, 1939, no. 53, p. 5. Trade-name for a non-bentonitic clay allied to illite (q.v.), from Grundy Co., Illinois. Named from the locality. [*M.A.*, **7**, 423.] 15th List

Grünlingite. W. Muthmann and E. Schröder, *Zeits. Kryst. Min.*, 1897, XXIX, 144; Abstr. *Min. Mag.*, XII, 45. The Cumberland mineral usually referred to tetradymite (or joseite) was found to have the composition Bi_4S_3Te, and therefore to differ from tetradymite. Probably rhombohedral. 2nd List [Status uncertain]

Guadarramite. J. Muñoz del Castillo, 1906. *Bol. Soc. Española Hist. Nat.*, vol. vi, p. 479 (Guadarramita). S. Calderón, *Los Minerales de España*, 1910, vol. i, p. 323. A radioactive variety of ilmenite from the Sierra de Guadarrama, Castille, Spain. 6th List [A mixture; *A.M.*, **37**, 1061]

Guanabacoite. *See* guanabaquite. 1st List

Guanabaquite. F. Vidal y Careta, *Cronica Cientifica, Barcelona*, XIV. 268, 1891. Given to replace the name cubaite (q.v.) for the 'cubical quartz' of Guanabacoa, Cuba. Also includes the 'cubical chalcedony' of the same locality (ibid. XIV. 268, 273, 1891; XIII. 293, 1890); this is shown by L. F. Navarro (*Anal. Soc. Españ. Hist. Nat.*, XXI. *Actas*, p. 120, 1893,) to be pseudomorphous, possibly after fluorite. 1st List

Guanglinite. Hu Tsu-Hsiang, Lin Shu-Jen, Chao Pao, Fang Ching-Sung, and Huang Chi-Shun, 1974. [*Acta Geol. Sinica*, **2**, 202] abstr. *M.A.*, 75-2522. Orthorhombic, *a* 10·83, *b* 3·33, *c* 6·07 Å, composition Pd_3As, from an unnamed locality in China. [Cf. *C.M.*, **13**, 332 (1975); *A.M.*, **61**, 184; the X-ray powder data are very similar to those of tetragonal Pd_3As.] 29th List

Guanine. B. Unger, 1846. *Ann. Chem. Pharm.*, **59**, 58. Unger (*Ann. Chem. Phys.* (*Poggendorff*), 1844, **62**, 158; 1845, **65**, 222) isolated from guano from North Chincha Island, Peru, a compound he believed to be xanthine (xanthic oxide, Harnoxyd), a purine base. P. Einbrodt (*Ann. Chem. Pharm.*, 1846, **58**, 15) showed that this identification could not be correct; Unger (ibid. 18) agreed, and later fully decribed the base, naming it guanine from its source. Formula $C_5H_5N_5O$. Natural occurrence confirmed in bat guano from Murra-el-elevyn Cave, Nullarbor Plain, Western Australia, by P. J. Bridge, *M.M.*, **39**, 467, 889. 28th List

Guayanaite. C. Milton, D. [E.] Appleman, E. C. T. Chao, F. Cuttitta, J. L. Dinnin, E. J. Dwornik, M. Hall, B. L. Ingram, and H. J. Rose, Jr., 1967. *Geol. Soc. Amer.*, *Prog. Ann. Meet.*, p. 151. CrOOH, isostructural with InOOH, a major constituent of the mixture of chromium oxides occurring in alluvial gravels of the Merume river, Mazaruni district, Guayana (Merumite, 18th List). Named from the locality. 25th List [Corrected spelling **Guyanaite**, q.v. Cf. Bracewellite, Grimaldiite]

Gudmundite. K. Johansson, 1928. *Zeits. Krist.*, vol. 68, p. 87 (Gudmundit). Sulphantimonide of iron, FeSbS, as orthorhombic crystals isomorphous with and resembling mispickel [FeAsS], from Gudmundstorp, near Sala, Sweden. Named from the locality. [*M.A.*, **4**, 12.] 12th List

Guerinite. E. I. Nefedov, 1961. [Мат. Всесоюз. Науч.-Иссл. Геол. Инст., vol. 45, p. 113], abstr. in Зап. Всесоюз. Мин. Общ. (*Mem. All-Union Min. Soc.*), vol. 92, p. 196 (Геринит, guerinite), and in *Amer. Min.*, 1962, vol. 47, p. 416. $Ca_5H_2(AsO_4)_4.9H_2O$, in spherulites and rosettes on a specimen labelled Wapplerite from Daniel mine, Schneeberg, Saxony, and one labelled Pharmacolite from Richelsdorf, Hesse. Named for H. Guerin, who synthesized this compound. [*M.A.*, **16**, 553.] 23rd List [*A.M.*, **50**, 812. Nothing to do with Guarinite (*see* Clinoguarinite)]

Guettardite. J. L. Jambor, 1967. *Canad. Min.*, vol. 9, p. 191. A sulphosalt from Madoc,

Ontario, very similar to twinnite and sartorite, but with composition $Pb_6(Sb,As)_{10}S_{21}$. Named for J. E. Guettard. 25th List [*A.M.*, **53**, 1425]

Güggenite. F. Trojer, 1958. *Radex Rundschau*, 365. An *artificial* compound of CuO and MgO, of uncertain formula ($CuMgO_2$ or Cu_2MgO_3 or Cu_3MgO_4), characterized by optical and X-ray data (*Radex Rundschau*, 1963, 383; *Zeits. anorg. Chem.*, 1964, **332**, 230; *Journ. inorg. Chem.*, 1968, **30**, 747). 26th List

Gugiaite. Chi-Jui Peng, Rung-Lung Tsao, and Zu-Rung Zou, 1962. [*Scientia Sinica*, vol. 11, p. 977], abstr. in *Amer. Min.*, 1963, vol. 48, p. 211 and in *Bull. Soc. franç. Min. Crist.*, 1964, vol. 87, p. 113. $Ca_2BeSi_2O_7$, tetragonal, in skarn rocks near Gugia (presumably in China). A member of the Melilite family near Meliphane but containing little Na or F; an unnecessary name. Named from the locality. [*M.A.*, **16**, 548.] 23rd List

Guildite. C. Lausen, 1928. *Amer. Min.*, vol. 13, p. 217. Hydrous iron and copper sulphate, $3(Cu,Fe)O.2(Fe,Al)_2O_3.7SO_3.17H_2O$, as chestnut-brown monoclinic crystals formed by the burning of pyritic ore in a mine in Arizona. Named after Prof. Frank Nelson Guild (1870–), of the University of Arizona. 11th List [$CuFe^{3+}(SO_4)_2(OH).4H_2O$; *A.M.*, **55**, 502]

Guilleminite. R. Pierrot, J. Toussaint, and T. Verbeek, 1965. *Bull. Soc. franç. Min. Crist.*, vol. 88, p. 132. Canary yellow orthorhombic crystals and coatings in the oxidized zone of the copper–cobalt deposit of Musoni, Katanga, have the composition $Ba(UO_2)_3(SeO_3)_2(OH)_4.3H_2O$. Named for C. Guillemin. [*M.A.*, **17**, 400; *A.M.*, **50**, 2103.] 24th List

Guimarãesite. G. Gagarin and J. R. Cuomo, 1949, loc. cit., p. 11 (guimarãesita). Tantalo-niobate of Ti, U, Fe, etc., as orthorhombic crystals from Divino de Ubá, Brazil, resembling ampangabeite but containing more Ti and less U. Named after the Brazilian mineralogist Djalma Guimarães, who described it in 1926 [*M.A.*, **3**, 113; Dana, 7th edit., 1944, vol. 1, p. 807]. 19th List

Gülechit, *see* Gyulekhite.

Gumbrine. A. A. Tvalchrelidze, 1929. [Гумбрин—фуллерова земля из Гумбри. Gumbrine—fuller's earth from Gumbri. Pamphlet issued by the trust for the mining-chemical industries of Georgia, Tiflis]; *Mineral resources of Georgia*, edited by S. A. Godabrelidze, Tiflis, 1933, p. 148 (Гумбрин, Russian), p. ix (გუმბრინი, Georgian), p. xv (goumbrine, French). Fuller's earth similar to floridine (10th List), from Gumbri (Гумбри) near Kutais, Georgia (Грузия), Transcaucasia. Named from the locality. [*M.A.*, **3**, 68; **8**, 106.] 17th List

Gumucionite. R. Herzenberg, 1932. *Revista Minera, Soc. Argentina Mineria y Geol.*, vol. 4, p. 65 (gumucionita); *Centr. Min.*, Abt. A, 1933, p. 77 (Gumucionit). A raspberry-red botryoidal variety of blende containing some arsenic (As 0·64%) from Bolivia. Named after the mining engineer Julio F. Gumucio, by whom it was found. [*M.A.*, **5**, 199; **5**, 295.] 13th List

Gunnarite. G. Landström, *Geol. För. Förh.*, 1887, IX, 368. $3FeS_2.2NiS$. Probably identical with pentlandite (Groth, *Tab. Uebers. Min.*, 4th edit., 1898, p. 20). [Rud, Skedevi parish, Östergötland] Sweden. 2nd List

Gunnbjarnite. O. B. Bøggild, 1951. *Meddel. Grønland*, vol. 142, no. 8, p. 3. $(Fe''',Al)_2O_3.3(Mg,Ca,Fe'')O.6SiO_2.3H_2O$, orthorhombic, as black micaceous plates in basalt, E Greenland. Named after Gunnbjørn Ulfsson, discoverer of Greenland about AD 900 [*M.A.*, **11**, 517]. 19th List [ferrian Sepiolite; *A.M.*, **42**, 920]

Gunningite. J. L. Jambor and R. W. Boyle, 1962. *Canad. Min.*, vol. 7, p. 209; *Amer. Min.*, 1962, vol. 47, p. 1218. $ZnSO_4.H_2O$, the zinc member of the kieserite family, as efflorescences on blende from the Keno Hill and Galena Hill area, Central Yukon. Named for H. C. Gunning. [*M.A.*, **16**, 458.] 23rd List

Gustavite. S. Karup-Møller, 1970. *Canad. Min.*, **10**, 173. Tabular grains with metallic lustre from Ivigtut, Greenland, are orthorhombic, *a* 13·548, *b* 19·449, *c* 4·105 Å. *Bbmm*, *Bb*2*m*, or *Bbm*2. Composition $Ag_3Pb_5Bi_{11}S_{23}$ (microprobe) $Ag_3Pb_6Bi_{11}S_{24}$ fulfils unit-cell requirements. Considered to be an intermediate between $Pb_3Bi_2S_6$ ('lillianite') and $AgPbBi_3S_6$, with the substitution $2Pb \rightleftharpoons AgBi$. The name gustavite, for Gustav A. Hagemann, is assigned to the end-member $AgPbBi_3S_6$ and should include some intermediate members. Gustavite and an unnamed exsolved phase are the 'mineral X' of Karup-Møller (*Canad. Min.*, **8**, 414 (1966)). 27th List [*M.A.*, 74-484; *C.M.*, **13**, 411]

Gutsevichite. E. A. Ankinovich, 1959 [Сб. Науч. Тр. Казахск. Горнометалл Инст., 1959, no. 18, Геология, p. 125]; abstr. Зап. Всесоюз. Мин. Общ. (*Mem. All-Union Min. Soc.*), 1961, vol. 90, p. 104 (Гуцевичит, gutsevichite). Phosphate and vanadate of Al, Fe, etc.; formulated $(Al,Fe)_3\{(P,V)O_4\}_2(OH)_3.7\frac{1}{2}–8\frac{1}{2}H_2O$, but this overlooks some 10% of MgO, CaO, BaO, SO_3, and SiO_2. Named for V. P. Gutsevich. 22nd List [*A.M.*, **46**, 1200]

Guyanaite, corrected spelling of Guayanaite (25th List); *M.A.*, **19**, 127. 28th List

Guyaquilit, another variant of Guayaquilite (H. Strunz, *Min. Tabellen*, 1st edit., 1941, p. 220). 24th List

Gyulekhite. C. M. Khalife-Zade, 1957. [*Doklady Acad. Sci. Azerbaijan SSR*, 1957, vol. 13, p. 647.] Abstr. in *Mém. Soc. Russ. Min.*, 1958, vol. 87, p. 83 (гюлехит, gewlekhite, Gülechit). Hydrous silicate, SiO_2 31·44, Al_2O_3 14·50, Fe_2O_3 29·88, FeO 3·52, MgO 4·00, K_2O 3·71, H_2O + 8·09, H_2O − 3·31%. Named from the locality, Гюлех village, SE Caucasus. 21st List [Doubtful; *A.M.*, **43**, 1223]

H

Haapalaite. M. Huhma, Y. Vuorelainen, T. A. Hakli, and H. Papunen, 1973. *Bull. Geol. Soc. Finland*, **45**, 103. Thin bronze-red scales in the Kokka serpentinite, 33 km NNW of Outokumpu, Finland, are hexagonal, *a* 3·64, *c* 34·02 Å; composition $2(Fe,Ni)S.1{\cdot}61(Mg,Fe)(OH)_2$. Valleriite group. Named for P. Haapala. [*A.M.*, **58**, 1111; *Zap.*, **103**, 355.] 28th List

Habazit, Halkofanit. Double transliterations through Serbian, хабазит and халкофанит, of chabazite and chalcophanite. [*M.A.*, **12**, 483.] 20th List

Hackmanite. L. H. Borgström, *Geol. För. Förh. Stockholm*, 1901, vol. xxiii, p. 563 (Hackmanit). A member of the sodalite group occurring as rhombic dodecahedral crystals of a pale reddish-violet colour in a rock called tawite from the Tawa valley, Kola peninsula, Lapland. It has the composition of sodalite, $Na_4(AlCl)Al_2(SiO_4)_3$, but contains in addition 6·23% of $Na_4[Al(NaS)]Al_2(SiO_4)_3$, a compound which Brøgger and Bäckström (1890) concluded to be a constituent of the artificial product known as 'white ultramarine'. The sulphur is present as monosulphide, and not as polysulphide as in lazurite. Named after Dr. Victor Hackman, of Helsingfors. 3rd List

Haemachatae. *See* discachatae. 3rd List

Haema-ovoid-agates. *See* discachatae. 3rd List

Haematogelite. F. Tućan, 1913. *Centralblatt Min.*, 1913, p. 68 (Hämatogelit). A colloidal form of ferric oxide (Fe_2O_3), occurring as the colouring material of bauxite. *See* Sporogelite. 6th List

Haematophanite. K. Johansson, 1928. *Zeits. Krist.*, vol. 68, p. 102 (Hämatophanit). Oxychloride of lead and iron, $Pb(Cl,OH)_2.4PbO.2Fe_2O_3$, as reddish-brown tetragonal scales, from Jakobsberg, Sweden. Apparently named (from *φαίνεσθαι*, to appear) because of the superficial resemblance to haematite. [*M.A.*, **4**, 13.] 12th List

Haematotokonit, variant of Hematoconite (H. Strunz, *Min. Tabellen*, 3rd (1957) and 4th (1966) edits.). 24th List

Hafnon. J. M. Correia Neves, J. E. Lopez Nunes, and Th. G. Sahama, 1974. *Contr. Min. Petr.*, **48**, 73. Zoned crystals from the Morro, Conca, Moneia, and Muiane mines, Zambezia district, Mozambique, range from hafnian zircon to $(Hf,Zr)SiO_4$ with only 1·2% ZrO_2. Named by analogy with zircon. [*M.A.*, 75-2521; *A.M.*, **61**, 175.] 29th List

Hagatalite. K. Kimura, 1925. *Japanese Journ. Chem.*, vol. 2, p. 81. A variety of zircon containing more rare-earths and less zirconia than naegite. Named from the locality, Hagata, prov. Iyo, Japan. [*M.A.*, **3**, 9.] 11th List

Hagendorfite. H. Strunz, 1954. *Neues Jahrb. Min.*, Monatshefte, 1954, p. 252 (Hagedorfit). $(Na,Ca)(Fe,Mn)_2(PO_4)_2$, triclinic(?), greenish-black. Close to varulite (14th List) and hühnerkobelite (19th List), with Fe > Mn or Na > Ca. Named from the locality, Hagendorf, Bavaria. [*M.A.*, **12**, 462.] 20th List [*A.M.*, **40**, 553. Monoclinic]

Häggite. H. T. Evans and M. E. Mrose, 1958. *Acta Cryst.*, vol. 11, p. 57. Vanadium hydroxide, $V_2O_3.V_2O_4.3H_2O = V_2O_2(OH)_3$, as black monoclinic crystals in sandstone from Wyoming. Named after Prof. Gunnar Hägg of Uppsala University. 21st List [*A.M.*, **45**, 1144]

Haggite, error for Häggite (*Bull. Soc. franç. Min. Crist.*, 1961, vol. 84, p. 209). 22nd List

Hainite. J. Blumrich, *Tsch. Min. Mitth.*, XIII. 472, 1893. Silicate of Na,Ca,Ti,Zr. Anorthic. Related to rosenbuschite, etc. Phonolites of [Hohe Hain, near Mildenau] N Bohemia. 1st List

Haiweeite. T. C. McBurney and J. Murdoch, 1959. *Amer. Min.*, vol. 44, p. 839. $CaU_2^{iv}Si_6O_{17}.5H_2O$, pale yellow, probably monoclinic, from above the Haiwee reservoir, Coso Mts., California. Named from the locality. [*M.A.*, **14**, 415.] Appears to be identical with gastunite

1a (20th List), but though the latter name has priority the description was inadequate and the name haiweeite is preferred (K. Walenta, *Neues Jahrb. Min.*, Monatsh., 1960, p. 37). *See* Gastunite. 22nd List [Cf. Ranquilite]

Hakite. Z. Johan, 1971. *Bull.*, **94**, 45. Grey-brown to grey minute grains with a variety of selenides including permingeatite and fischersserite (qq.v.) in a calcite vein at Přesvorнice, Bohemia, are cubic, *a* 10·88 Å. Composition 8[$(Cu,Hg)_3SbSe_3$]: one analysis shows some sulphur. Tetrahedrite family. Named for J. Hak. [*M.A.*, 71-2332; *Bull.*, **94**, 572.] 27th List [*A.M.*, **57**, 1553]

Halitkainit. *See* Thanite. 7th List

Halkofanit, *see* Habazit.

Hallerite. P. Barbier, 1908. *Compt. Rend. Acad. Sci. Paris*, vol. cxlvi, p. 1221 (Hallérite). A lithium-bearing variety of soda-mica (paragonite) resembling muscovite in appearance. From Mesvres, Autun, France. Named after Albin Haller, professor of organic chemistry at the Sorbonne, Paris. 5th List

Hallimondite. K. Walenta and W. Wimmenauer, 1961. *Jahresheft geol. Landesamt Baden-Württemberg*, vol. 4, p. 21 (Hallimondit). Yellow crystalline crusts from the Michael vein, Weiler bei Lehr, Schwarzwald, Germany, contain Pb, U, and As; they give an X-ray powder photograph (not quoted!) near that of parsonsite, and are presumed to be $Pb_2UO_2(AsO_4)_2.nH_2O$. Named after Dr. A. F. Hallimond of London. Final acceptance of this as a valid species must await publication of the X-ray data, particularly as there is no quantitative chemical analysis (cf. *Bull. Soc. franç. Min. Crist.*, 1961, vol. 84, p. 100, recommendations of the Commission on New Minerals of the International Mineralogical Association). 22nd List [*A.M.*, **50**, 1143; triclinic, no water in formula]

Halogenpyromorphite. H. Wondratschek, 1963. *Neues Jahrb. Min.*, Abh., vol. 99, p. 113 (Halogenpyromorphit). A group name for the artificial compounds $Pb_5(XO_4)_3Z$, where Z is a halogen and X is P, As, or V. 23rd List

Halotri-alunogen. I. Suganuma, 1932. [*Tokyo Buturigakkô-Zassi*, vol. 41, p. 250.] Z. Harada, *Journ. Fac. Sci. Hokkaido Univ.*, Sapporo, Ser. 4, 1936, vol. 3, p. 351 (Halotri-Alunogen). A mixture of halotrichite and alunogen. 14th List

Halurgite. V. V. Lobanova, 1960. Доклады Акад. Наук СССР (*Compt. Rend. Acad. Sci. URSS*), vol. 135, 173 (Галургит); abstr. Зап. Всесоюз. Мин. Общ. (*Mem. All-Union Min. Soc.*), 1961, vol. 90, p. 448 (Галургит, halurgite). Mentioned as a new mineral accompanying strontioborite (q.v.) without any description. 22nd List [$Mg_2B_8O_{10}(OH)_8.H_2O$, monoclinic; *A.M.*, **47**, 1217]

Hämatitogelit. *See* Diasporogelite. 7th List

Hamelite. T. S. Hunt, *Mineral Physiology and Physiography*, 1886, pp. 194, 334; *Trans. Roy. Soc. Canada*, IV.(3), 9, (1886), 1887. Hydrated silicate of Al,Fe,Mg. New Brunswick. 1st List

Hammarite. K. Johansson, 1924. *Arkiv Kemi, Min. Geol.*, vol. 9, no. 8, p. 11 (Hammarit). Sulpho-bismuthite of lead and copper '$5PbS.3Bi_2S_3$', as steel-grey, prismatic (monoclinic?) crystals from Gladhammar, Sweden. Named, with gladite (q.v.), from the locality. [*M.A.*, **2**, 340.] 10th List [$Pb_2Cu_2Bi_4S_9$, Dana's *Syst. Min.* (7th edit.) **1**, 442; a distinct species; *M.A.*, 70-2583; 70-2585]

Hampdenite. A. D. Roe, 1906. *Bull. Minnesota Acad. Sci.*, vol. iv, p. 268. A variety of serpentine from Hampden Co., Massachusetts; it forms the matrix of the pseudomorphs of serpentine after olivine known as hampshirite. C. Palache (*Amer. Journ. Sci.*, 1907, ser. 4, vol. xxiv, p. 494) points out that it does not differ from picrolite. 5th List

Hancockite. S. L. Penfield and C. H. Warren, Dana's *Appendix*, 1899; *Amer. J. Sci.*, 1899, VIII, 339; *Zeits. Kryst. Min.*, 1900, XXXII, 228; Abstr. *Min. Mag.*, XII, 316. A member of the epidote group occurring as brownish red cellular masses of minute lath-shaped crystals. Monoclinic. $R''_2(R'''.OH)R'''_2(SiO_4)_3$; ($R''' = Al,Fe,Mn$; $R'' = Pb,Ca,Mn$). Franklin Furnace, New Jersey. 2nd List

Hanléite. L. L. Fermor, 1952. *Geol. Mag. Hertford*, vol. 89, p. 145 (hanléite). The garnet molecule $Mg_3Cr_2(SiO_4)_3$, represented by a mineral resembling uvarovite, described by F. R. Mallet in 1866, in chromite from near the Hanlé monastery, Kashmir. Named from the locality. [*M.A.*, **11**, 518.] 19th List [Is Uvarovite; *M.M.*, **33**, 508]

Hanuschit. Germanized form of Hanušite (Hintze, *Handb. Min.*, Erg.-Bd. II, p. 142). 22nd List

Hanušite. J. V. Kašpar, 1942. *Chem. Listy*, Praha, vol. 36, no. 6, p. 78 (Hanušit). Hydrated magnesium silicate, $H_2Mg_2(SiO_3)_3.H_2O$, an alteration product of pectolite from Bohemia. Named after Josef Hanuš (1872–), professor of analytical chemistry in the Technical Institute, Praha. [*M.A.*, **9**, 6, 260.] 17th List [Stevensite, or a mixture containing it; *M.A.*, **14**, 339]

Haradaite. T. Watanabe, *et al.*, 1963, in *Appendix to the Second edition of An Index of Mineral Species Arranged Chemically*, M. H. Hey, London, British Museum (Natural Nistory), pp. 57 and 104. T. Yushimura and H. Momoi [*Sci. Rept. Kyushu Univ.*, 1964, Geol. ser. no. 7, p. 85], abstr. *M.A.*, **17**, 183; J. Ito, *Min. Journ.* [*Japan*], 1965, vol. 4, p. 299. A strontium vanadium silicate, $SrVSi_2O_7$, occurring at the Yamato mine, Kagoshima prefecture, Japan. 24th List [*A.M.*, **56**, 1123]

Harbolite. C. E. Taşman, 1946. *Maden Tetkik ve Arama Enstitüsü Mecmuası*, Ankara, vol. 11, p. 50 (harbolit), p. 51 (harbolite); *Bull. Amer. Assoc. Petroleum Geol.*, 1946, vol. 30, p. 1051. A hard lustrous asphalt from Harbol, SE Turkey. Named from the locality. [*M.A.*, **10**, 7, 354.] 18th List

Harbortite. F. Brandt, 1932. *Chemie der Erde*, vol. 7, p. 392 (Harbortit). Hydrous aluminium (and iron) phosphate, $6Al_2O_3.4P_2O_5.17H_2O$, as white to brown spherulites in phosphatized laterite from Brazil. Named after Prof. Erich Harbort (1879–1929), of Berlin-Charlottenburg. [*M.A.*, **5**, 201.] 13th List [A mixture; *A.M.*, **18**, 222]

Hardystonite. J. E. Wolff, *Proc. Amer. Acad. Arts and Sci.*, 1899, XXXIV, 479; *Zeits. Kryst. Min.*, 1899, XXXII, 1; 1900, XXXIII, 147; Abstr. *Min. Mag.*, XII, 315. White granular masses. Tetragonal. $ZnCa_2Si_2O_7$. Franklin Furnace, New Jersey. 2nd List [Melilite group]

Haringtonite. N. W. Wilson, 1945. *Bull. Inst. Mining Metall. London*, no. 470, p. 12. A provisional name for an unidentified mineral resembling guadalcazarite, from Harington Kop, Murchison Range, Transvaal. (Not to be confused with harringtonite of T. Thomson, 1831.) [*M.A.*, **9**, 127.] 17th List [May be Cinnabar; *A.M.*, **32**, 255]

Harkerite. C. E. Tilley, 1948. *Geol. Mag. Hertford*, vol. 85, p. 215; *Min. Mag.*, 1949, vol. 28, p. lxxi. Carbonate-borosilicate of calcium and magnesium in dolomite skarns from Skye. Named after Alfred Harker (1859–1939) of Cambridge. [*M.A.*, **10**, 355.] 18th List [Cubic, complex formula; *M.M.*, **29**, 621 (*A.M.*, **37**, 358)]

Harmotomite. Variant of Harmotome; *A.M.*, **8**, 51. 10th List

Hartsalzkainitit. *See* Thanite. 7th List

Harttite. E. Hussak, 1906. *Min. Petr. Mitt.* (*Tschermak*), vol. xxv, p. 341 (Harttit). Hydrated strontium aluminium sulphato-phosphate, $SrO.2Al_2O_3.P_2O_5.SO_3.5H_2O$, near svanbergite in composition. It is found as flesh-red pebbles ('favas') in the diamond-bearing sands of Brazil. The crystalline structure is rhombohedral. Named in memory of Charles Frederick Hartt (1840–1878), the first director of the Geological Survey of Brazil. 4th List [Is Svanbergite; *M.A.*, **11**, 425]

Hasingtonit, error for Haringtonit (H. Strunz, *Min. Tabellen*, 3rd edit., 1957, p. 378). 22nd List

Hastingsite. F. D. Adams and B. J. Harrington, *Amer. Journ. Sci.*, I. 212, 1896; and *Canadian Record Sci.*, VII. 81, 1896. [*M.M.*, **11**, 244.] A soda hornblende with the orthosilicate formula $(R_2',R'')_3R_2'''Si_3O_{12}$. Monosymmetric. Hastings Co., Ontario. 1st List

Hastite. P. Ramdohr and M. Schmitt, 1955. *Neues Jahrb. Min.*, Monatshefte, no. 6, p. 140 (Hastit). Cobalt selenide, $CoSe_2$, orthorhombic (marcasite group). From Trogtal, Harz. Named after Dr. P. F. Hast, mining director. [*M.A.*, **13**, 5.] 21st List [*A.M.*, **41**, 164]

Hatchite. R. H. Solly and G. F. H. Smith, 1912. *Mineralogical Magazine*, vol. xvi, p. 287. Minute, lead-grey, anorthic crystals of undetermined composition (but suggested to be a sulpharsenite of lead) from the crystalline dolomite of the Binnenthal, Switzerland. Named after Dr. Frederick Henry Hatch. 6th List [$PbTlAgAs_2S_5$; *M.A.*, **19**, 270; 69-2352, 2354]

Hatrurite. S. Gross, 1977. *Bull. Geol. Surv. Israel*, **70**, 82. Pseudohexagonal crystals in the Hatrurim formation, Israel, are near Ca_3SiO_5, and equivalent to alite of Törnebohm (2nd List), an artificial constituent of cement clinker. [*A.M.*, **63**, 425.] 30th List

Hauchecornite. R. Scheibe, *Jahrb. preuss. geol. Landesanst. u. Bergakad.*, for 1892, XII. 91, 1893; H. Laspeyres, *Verh. naturhist. Ver. Bonn*, L. 177, 1893. [*Min. Mag.*, X. 339.] $Ni_{14}SbBi_2S_{13}$ or $(Ni,Co,Fe)_7(S,Bi,As,Sb)_8$. Tetragonal. [Friedrich mine, near Hamm a.d. Sieg] Rhenish Prussia. 1st List [$(Ni,Co)_9(Bi,Sb)_2S_8$; *A.M.*, **35**, 440. Cf. Tučekite]

Hautefeuillite. L. Michel, *Compt. Rend.*, CXVI. 898, 1893; and *Bull. Soc. fran. Min.*, XVI. 38, 1893. [*M.M.*, **11**, 162.] $(Mg,Ca)_3(PO_4)_2 + 8H_2O$. Monosymmetric. [Ödegaarden, Bamle] Norway. 1st List [Is Bobierrite; *A.M.*, **22**, 337]

Hawaiite. C. Elschner, 1906. *Chemiker-Zeitung*, Jahrg. xxx, p. 1119 (Hawaiit). A gem-variety of olivine from the lavas of the Hawaiian Islands. It contains but little iron, and is pale green in colour. 5th List

Hawleyite. R. J. Traill and R. W. Boyle, 1955. *Amer. Min.*, vol. 40, p. 555; *Contr. Canadian Min.*, vol. 5, pt. 7. Cubic CdS, dimorphous with greenockite, as a yellow powder on blende. Named after Prof. James E. Hawley, of Queen's University, Kingston, Ontario. Cf. optically isotropic xanthochroite (8th List). [*M.A.*, **13**, 7.] 21st List

Haxonite. E. R. D. Scott, 1971. *Nature* (*Phys. Sci.*), **229**, 61. A cubic carbide of iron and nickel, near $(Fe,Ni)_{23}C_6$, occurs in the Toluca, Cañon Diablo and other siderites. Named for H. Axon: Not named in the description; the name first appears in *Glossary of Mineral Species*, 1971, by M. Fleischer (*Min. Record*, Bowie, Maryland). [*M.A.*, 72-547.] 27th List

Haycockite. L. J. Cabri and S. R. Hall, 1972. *Amer. Min.*, **57**, 689. Massive sulphides at Mooihoek Farm, Lydenburg District, Transvaal, include a chalcopyrite-yellow orthorhombic, pseudotetragonal mineral, $a \approx b$ 10·71, c 31·56 Å. Composition $12[Cu_4Fe_5S_8]$; about $\frac{1}{2}$% Ni may be essential. Named for M. H. Haycock. [*M.A.*, 72-3345.] 27th List

Headdenite. P. Quensel, 1937. *Geol. För. Förh. Stockholm*, vol. 59, p. 95 (headdenite). A phosphate mineral $Na_2O.5(Fe,Mn,Ca)O.2P_2O_5$, from S Dakota, analogous to varulite (q.v.) but with iron predominating over manganese. Named after William Parker Headden (1850–1932) of Colorado. [*M.A.*, **6**, 486.] 14th List [Is Arrojadite; *M.A.*, **8**, 180]

Heazlewoodite. W. F. Petterd, *Catalogue of Minerals of Tasmania*, 1896. Sulphide of Ni and Fe, related to pentlandite. [Heazlewood, Tasmania.] 1st List [Ni_3S_2; *A.M.*, **32**, 484]

Hebergite. J. Barrington and P. F. Kerr, 1961. *Econ. Geol.*, **56**, 241. Mentioned without a description among secondary uranium minerals from the Midnite mine, Spokane, Washington. [*Zap.*, **97**, 79.] [Possibly an error for liebigite—M.H.H.] 26th List

Hectorite. H. Strese and U. Hofmann, 1941. [*Zeits. Anorg. Chem.*, vol. 247, p. 65.] U. Hofmann and A. Hausdorf, *Zeits. Krist.*, 1942, vol. 104, pp. 266, 274 (Hectorit). Magnesium-bearing bentonite (montmorillonite) from Hector, California. Synonym of magnesium-bentonite (15th List). Named from the locality. Not the hectorite of S. H. Cox, 1882. 16th List

Hedenbergit-Ägirin. *See* Aegirine-hedenbergite. 4th List

Hedleyite. H. V. Warren and M. A. Peacock, 1945. *Univ. Toronto Studies, Geol. Ser.*, no. 49, p. 55. Bismuth-tellurium alloy Bi_7Te_3, a solid solution of Bi_5 in Bi_2Te_3, as rhombohedral cleavage flakes, from Hedley, British Columbia. Named from the locality. [*M.A.*, **9**, 126.] 17th List [*A.M.*, **48**, 435]

Hedroicite. O. M. Shubnikova, *Trans. Inst. Geol. Sci. USSR*, 1940, no. 31 (*Min.-Geochem. Ser. no. 6*), p. 46. Another transliteration of гедройцит, gedroitzite (15th List). 17th List

Heideïte. K. Keil and R. Brett, 1974. *A.M.*, **59**, 465. Rare anhedral grains in the Bustee meteorite, up to 0·1 mm in diameter, are probably monoclinic. Composition $(Fe,Cr)_{1+x}(Ti,Fe)_2S_4$. Named for F. Heide. 28th List

Heidornite. W. v. Engelhardt and H. Füchtbauer, 1956. *Heidelberg. Beitr. Min. Petr.*, vol. 5, p. 177 (Heidornit). Borate, sulphate, and chloride, $Na_2Ca_3Cl(SO_4)_2B_5O_8(OH)_2$, monoclinic, as transparent spear-like crystals (up to 7 cm) with glauberite from a deep boring in anhydrite at Nordhorn, Hannover. Named after Dr. Fritz Heidorn, geologist, of Bentheim. [*M.A.*, **13**, 382.] 21st List [*A.M.*, **42**, 120]

Heikkolite. [Y. Kinosaki, 1935. *Trans. Mining. Eng. Assoc.*, vol. 18, no. 3 (Japanese). *Minerals of Korea*, 2nd edit., *Bull. Geol. Surv. Chosen*, 1941, no. 15, p. 198. *Z. Harada, *Journ. Geol. Soc. Japan*, 1939, vol. 46, p. 290; S. Hori, ibid., 1942, vol. 49, p. 445.] An alkali-amphibole with no extinction between crossed nicols and other optical anomalies. Named from the locality, Heiko, Kogen-do, Korea. Not to be confused with hokutolite (6th List). [*Corr. in *M.M.*, **40**.] 18th List [A trivial variety of crossite, itself near glaucophane]

Heikolite. Correct spelling of heikkolite (18th List). [*M.A.*, **12**, 595.] 20th List

Heinrichite. E. B. Gross, A. S. Corey, R. S. Mitchell, and K. Walenta, 1958. *Amer. Min.*, vol. 43, p. 1134. The arsenic analogue of uranocircite, $Ba(UO_2)_2(AsO_4)_2.10–12H_2O$ and its dehydration product with $8H_2O$ (metaheinrichite, q.v.) occur near Lakeview, Oregon, and in the Black Forest, Germany. Named after Prof. E. William Heinrich. [*M.A.*, **14**, 199.] *See also* Arsenuranocircite. 22nd List

Helictite. [H. C. Hovey, 1882. *Celebrated American caverns*. Cincinnati, p. 186]; Funk and Wagnalls, *Standard Dictionary*, 1893–1895. L. C. Huff, *Journ. Geol. Chicago*, 1940, vol. 48, p. 641. Crooked and branching formations of calcite or aragonite in limestone caverns, as distinct from straight stalactites and stalagmites. Named from ἕλιζ, a spiral, ἑλικτός, twisted. Cf. anemolite (3rd List). [*M.A.*, **8**, 202.] 16th List

Heliodor. (E. Kaiser, *Centralblatt Min.*, 1912, p. 385.) Trade-name for a golden beryl of gem-quality from German SW Africa. 6th List

Hellandite. W. C. Brøgger, *Nyt Mag. Naturvid. Kristiania*, 1903, vol. xli, p. 213 (Hellandit). Monoclinic crystals in the pegmatite-veins near Kragerö, Norway; they are much altered; the freshest contained 7·55% of water and were optically isotropic (like gadolinite, orthite, etc.). Probable formula: $Ca_2R_3'''(R'''O)_3(SiO_4)_4$, where $R''' =$ Ce,Di,La,Al,Fe,Mn. Related to guarinite. Named after Professor Armund Helland, of Christiania. 3rd List [Is a borosilicate; *M.A.*, **16**, 644; **17**, 495; $4[(Ca,Yt)_2(Si,B,Al)_3O_8.H_2O]$; *C.M.*, **11**, 760; but *see A.M.*, **62**, 89]

Hellyerite. K. L. Williams, 1958. [*Australian Inst. Min. Met.*, Stilwell Anniv. vol., p. 263]; abstr. *Amer. Min.*, 1959, vol. 44, pp. 533, 1103. Normal nickel carbonate, $NiCO_3.6H_2O$, occurring with zaratite at the Lord Brassey nickel mine, Heazlewood, Tasmania. Named after Henry Hellyer, surveyor and explorer. [*M.A.*, **14**, 414.] 22nd List

Hematogelite. (3rd Appendix to 6th edit. of Dana's *System of Mineralogy*, 1915, p. 37. A. F. Rogers, *Journ. Geol. Chicago*, 1917, vol. xxv, pp. 523, 528.) Another spelling of Haematogelite (F. Tućan, 1913; 6th List). 7th List [Also 8th List]

Hematophanite. Dana's *System of mineralogy*, 1944, 7th edit., vol. 1, p. 728. Another spelling of haematophanite (12th List). 17th List

Hemiexpandite, *see* expandite.

Hemihedrite. S. A. Williams and J. W. Anthony, 1967. *Canad. Min.*, **9**, 310. Orange to almost black anorthic crystals from the Florence mine, Pinal County, Arizona, and the Pack Rat claim, Wickenburg, Maricopa County, Arizona, are near $Pb_5Zn(CrO_4)_3OF_4$. Named for the symmetry ($P1$). [*A.M.*, **53**, 1427; *Bull.*, **91**, 517.] 26th List [*A.M.*, **55**, 1088]

Hemiopal. H. H. A. Francke, *Min. Nomenklatur*, 1890, p. 81. Syn. of halbopal and semiopal. 1st List

Hemusite. G. Terziev, 1965. [Българ. геол. дружество, **3**, 375 (хемусит); Геол. рудн. месторожд. (*Geol. Ore-deposits*), no. 3, 37 (1966)]; quoted in *Zap.*, **97**, 67 (1968). Cu_3SnS_4, mentioned without description as an accessary mineral in ore-deposits at Chelopech, Bulgaria. [*A.M.*, **53**, 1775; *Bull.*, **92**, 320.] 26th List [Cu_6SnMoS_8, cubic; *A.M.*, **56**, 1847]

Hendersonite. M. L. Lindberg, A. D. Weeks, M. E. Thompson, D. P. Elston, and R. Meyrowitz, 1962. *Amer. Min.*, vol. 47, p. 1252. A black fibrous mineral from the J.J. mine, Paradox Valley, Montrose County, Colorado, and the Eastside mine, San Juan County, New Mexico, has the composition $Ca_2V^{4+}(V^{5+},V^{4+})_8(O,OH)_{24}.8H_2O$, with about 0·1 V^{4+} replacing V^{5+}. Named for E. P. Henderson of the US National Museum. 23rd List

Hendricksite. C. Frondel and J. Ito, 1966. *Amer. Min.*, vol. 51, p. 1107. The zinc analogue of phlogopite, occurring at Franklin, New Jersey. Named for S. B. Hendricks. [*M.A.*, **18**, 48; Зап., **97**, 77.] 25th List

Hengleinite. C. Doelter, 1926. *Handbuch d. Mineralchemie*, vol. 4, part 1, p. 643 (Hengleinit). The cobaltnickelpyrite of M. Henglein, 1914 (7th List), $(Co,Ni,Fe)S_2$ from Müsen, Westphalia. Named after Prof. Martin K. Henglein (1882–), of Karlsruhe. 11th List

Henritemierite, error for Henritermierite (*M.A.*, 69-2890). 28th List

Henritermierite. C. Gaudefroy, M. Orliac, F. Permingeat, and A. Parfenoff, 1969. *Bull. Soc. franç. Min. Crist.*, **92**, 185; ibid. 126. Small brown grains from the Tachgagalt mine, Morocco, are tetragonal, with a deformed hydrogarnet structure and composition $Ca_3(Mn^{3+}_{1\cdot5}Al_{0\cdot5})(SiO_4)_2(OH)_4$. [*A.M.*, **54**, 1739; *M.A.*, 69-3347; *Zap.*, **99**, 81.] 26th List

Heptaphyllite. A. N. Winchell, 1925. *Amer. Min.*, vol. 10, p. 53; *Amer. Journ. Sci.*, ser. 5, vol. 9, pp. 315, 428. The micas fall into two classes, the fundamental units of which contain seven or eight atoms (omitting O, H, F), e.g. KAl_3Si_3 and KMg_3AlSi_3. These are called respectively the heptaphyllite or muscovite-lepidolite system and the octophyllite or biotite system. They correspond roughly with the light-coloured and dark micas. 10th List

Herzenbergite. P. Ramdohr, 1934. *Neues Jahrb. Min.*, Abt. A, Beil.-Bd. 68, p. 293; *Zeits. Krist.*, 1935, vol. 92, p. 186 (Herzenbergit). W. Hofmann, *Fortschr. Min. Krist. Petr.*, 1935, vol. 19, p. 30; *Zeits. Krist.*, 1935, vol. 92, p. 161. To replace the preoccupied name kolbeckine (R. Herzenberg, 1932; 13th List) for tin sulphide from Bolivia, identical with artificial SnS. Named after Dr. Robert Herzenberg, of Oruro, Bolivia. [*M.A.*, **6**, 261, 262, 368.] 14th List [*A.M.*, **20**, 541]

Heshvitcite, error for Goeschwitzite.

Heterobrochantite. H. Buttgenbach, 1926. *Ann. Soc. Géol. Belgique*, vol. 49, Bull. p. B 164 (Hétérobrochantite). A basic copper sulphate, $CuSO_4.2Cu(OH)_2$, orthorhombic, regarded as a variety of antlerite (= stelznerite), from which it differs in optical orientation. Named from ἕτερος, the other, because of its resemblance to brochantite and because the crystals present a heteropolar (hemimorphic) development. [*M.A.*, **3**, 270.] 11th List [Probably identical with antlerite; *M.A.*, **7**, 392]

Heterophyllite. A. Mário de Jesus, [1936]. *Com. Serv. Geol. Portugal*, vol. 19 (for 1933), p. 126 (Heterofilite). A variety of biotite with a formula, $H_8K_4Fe_4Al_6Si_8O_{35}$, differing slightly from siderophyllite. From Mangualde, Portugal. [*M.A.*, **6**, 441.] 14th List

Heulandite baritica. *See* barium heulandite. 1st List

Heulandite-B. F. A. Mumpton, 1960. *A.M.*, **45**, 351. *Syn.* of Metaheulandite (2nd List). [*M.A.*, **15**, 355.] 28th List

Hewettite. W. F. Hillebrand, H. E. Merwin, and F. E. Wright, 1914. *Proc. Amer. Phil. Soc.*, vol. liii, p. 32; *Zeits. Kryst. Min.*, vol. liv, p. 209. An orthorhombic, hydrous vanadate of calcium, $CaO.3V_2O_5.9H_2O$, forming mahogany-red, silky aggregates of minute needles, and occurring as an oxidation product of patronite at Minasragra, Peru. Named after Mr. D. Foster Hewett, of the United States Geological Survey. An isomeric form is called *metahewettite* (q.v.). 7th List [Cf. Rossite]

Hexabolite. H. Strunz, 1949. *Min. Tabellen*, 2nd edit., p. 206. Syn. of basaltic hornblende. 22nd List

Hexacelsian. B. Yoshiki and K. Matsumoto, 1951. *Journ. Amer. Ceramic Soc.*, vol. 34, p. 286. Artificially produced, high-temperature, hexagonal modification of $BaAl_2Si_2O_8$. [*M.A.*, **12**, 307.] 20th List

Hexahydrite. R. A. A. Johnston, 1911. *Summary Rep. Geol. Surv. Canada, for 1910*, p. 256. A hydrated magnesium sulphate containing less water (47·33%) than epsomite, and identical in composition with the artificially prepared hexahydrate $MgSO_4.6H_2O$. As a white, columnar to fibrous material it forms seams in an altered rock in the Lillooet district, British Columbia. 6th List

Hexahydroborite. M. A. Simonov, N. A. Yamnova, E. V. Kazanskaya, Yu. K. Egorov-Tismenko, and N. V. Belov, 1976. *Dokl. akad. nauk SSSR*, **228**, 1337 (Гексагидроьорит). The higher hydrate $CaB_2O_4.6H_2O$ occurs with pentahydroborite (23rd List) at the Solongo deposit, Urals. Monoclinic, *a* 8·006, *b* 8·012, *c* 6·649 Å, *γ* 104·21°, space group *P2/a*. No analysis, optical data, nor X-ray powder data are given. [*A.M.*, **62**, 1259.] [Standard orientation *P2/c*, *a* 8·012, *b* 6·649, *c* 8·006 Å, *β* 104·21° *M.H.H.*] 30th List

Hexastannite. P. Ramdohr, 1960. *Die Erzmineralien und ihre Verwachsungen*, Berlin, p. 514. To replace the name Stannite-I for the hexagonal mineral, believed to approximate to $Cu_3Fe_2SnS_6$. Imperfectly characterized (*Amer. Min.*, 1961, vol. 46, p. 1204; Зап. Всесоюз. Мин. Обı (*Mem. All-Union Min. Soc.*), 1963, vol. 92, p. 195). 23rd List [Cf. Mawsonite, Stannoidite]

Hexastibiopalladite. Anon., 1974. [*Geochimica*, **3**, 169], abstr. *A.M.*, **61**, 182. Syn. of Sudburyite (28th List). [*M.A.*, 75-2529.] 29th List

Hexatestibiopanickelite. Anon. 1974. [*Geochimica*, **3**, 169], abstr. *A.M.*, **61**, 183. $(Ni,Pd)_2SbTe$, hexagonal, *a* 3·98, *c* 5·35 Å; occurs in copper–nickel deposits of SW China. Named from the symmetry and composition. [*M.A.*, 73-2529.] 29th List

Heyite. S. A. Williams, 1973. *M.M.*, **39**, 65. Orange-yellow crystals from the Betty Jo claim, Ely, White Pine County, Nevada, have *a* 8·910, *b* 6·017, *c* 7·734 Å, *β* 111° 53′. Composition $[Pb_5Fe_2(VO_4)_2O_4]$. Cf. also *M.M.*, 1973, **39**, 69. Named for M. H. Hey. [*M.A.*, 73-2943, 2944; *A.M.*, **59**, 382; *Zap.*, **103**, 361.] 28th List

Heyrovskyite. J. Klominsky, M. Rieder, C. Kieft, and L. Mraz, 1971. *Min. Depos.*, **6**, 133. Tin-white orthorhombic crystals, tarnishing black, from Hurky, Czechoslovakia, have *a* 13·705, *b* 31·194, *c* 4·121, D 7·17, *Bbmm* or *Bbm*2. Analyses show some deficiencies from the ideal $4[Pb_6Bi_2S_9]$, with some Ag. [*A.M.*, **57**, 325; *M.A.*, 72-1399.] 27th List [Composition range extended, improperly re-named Goongarrite (q.v.); abstr. in *A.M.*, **62**, 397]

Hibbenite. A. H. Phillips, 1916. *Amer. Journ. Sci.*, ser. 4, vol. xlii, p. 276. A basic zinc phosphate identical in crystalline form, cleavage, and optical properties (so far as determined) with hopeite, occurring in intimate association with spencerite (q.v.) from Salmo in British Columbia. The formula, given as $2Zn_3(PO_4)_2.Zn(OH)_2.6\frac{1}{2}H_2O$, suggests that the material analysed was a mixture of hopeite $[Zn_3(PO_4)_2.4H_2O]$ and spencerite $[Zn_3(PO_4)_2.Zn(OH)_2.3H_2O]$. Named after Dr. John Grier Hibben, President of Princeton University. Hopeite from this locality was subsequently independently described by T. L. Walker (*Journ. Washington Acad. Sci.*, 1916, vol. vi, p. 685). 8th List

Hibonite. H. Curien, C. Guillemin, J. Orcel, and M. Sternberg, 1956. *Compt. Rend. Acad. Sci. Paris*, vol. 242, p. 2845. L. Delbos, ibid., 1957, vol. 244, p. 214. Mixed oxides of aluminium, etc., $(Al,Fe''',Ti,Si,Mg,Fe'')_{12}(Ca,\text{rare-earths})O_{18}$, dark-brown, hexagonal crystals in metamorphic limestone with plagioclase, corundum, spinel, and thorianite, from S Madagascar. Named after P. Hibon, who found the material. [*M.A.*, **13**, 380.] 21st List [*A.M.*, **42**, 119]

Hibschite. F. Cornu, 1905. *Min. Petr. Mitt.* (*Tschermak*), vol. xxiv, p. 327; 1906, vol. xxv, p. 249 (Hibschit). Minute, octahedral, optically isotropic crystals of contact-metamorphic origin, occurring in enclosures of chalk-marl in phonolite at Aussig, Bohemia. It usually forms a colourless layer around a nucleus of green titaniferous melanite. The chemical composition, $H_4CaAl_2Si_2O_{10}$, is the same as of the orthorhombic lawsonite. Named after Professor Josef Emanuel Hibsch, of Tetschen, Bohemia.

The same mineral also occurs in enclosures of limestone in basalt at Aubenas, Ardèche, France, and was described by A. Lacroix in 1893 as colourless garnet with a core of brown melanite. 4th List [*See* Hydrogrossular, Plazolite]

Hidalgoite. R. L. Smith, F. S. Simons, and A. C. Vlisidis, 1953. *Amer. Min.*, vol. 38, p. 1218. Basic sulphate arsenate, $PbAl_3(AsO_4)(SO_4)(OH)_6$, as white, fine-grained masses, hexagonal, from Zimapan, Hidalgo state, Mexico. Named from the locality. Since recorded from France. [*M.A.*, **12**, 302, 571.] 20th List

Higginsite. C. Palache and E. V. Shannon, 1920. *Amer. Min.*, vol. 5. pp. 155, 159. A mineral of the olivenite group with the composition $2CuO.2CaO.As_2O_5.H_2O$, found as green, orthorhombic crystals in the Higgins mine, Bisbee, Arizona. [*M.A.*, **1**, 122.] 9th List [Is Conichalcite; *A.M.*, **36**, 484, 503]

Hilairite. G. Y. Chao, D. H. Watkinson, and T. T. Chen, 1974. *C.M.*, **12**, 237. Small pale-brown trigonal crystals (*a* 10·556, *c* 15·851 Å) from Mont-St.-Hilaire, Quebec, previously listed as ROM 1 and as UK 20, have the composition $Na_2ZrSi_3O_9.3H_2O$, and are named for the locality. 28th List

Hilgardite. C. S. Hurlbut and R. E. Taylor, 1937. *Amer. Min.*, vol. 22, p. 1052. Hydrated chloro-borate of calcium, $Ca_8(B_6O_{11})_3Cl_4.4H_2O$, as colourless monoclinic-domatic crystals in the rock-salt of Louisiana. Named in memory of Prof. Eugene Waldemar Hilgard (1833–1916), who was the first to examine the salt deposits of Louisiana. *See* Parahilgardite. [*M.A.*, **7**, 14, 217, 224, 355.] 15th List $[Ca_2B_5O_8Cl(OH)_2]$

Hilgenstockite. V. A. Kroll, 1911. *Journ. Iron and Steel Inst.*, vol. 84 (no. II for 1911), pp. 128, 185. Tetrabasic calcium phosphate, $4CaO.P_2O_5$, occurring as yellow, orthorhombic plates in the basic slag of the Thomas-Gilchrist process for the dephosphorisation of iron. Named after Gustav Hilgenstock, who was the first to describe these crystals in 1883. *See also* Silico-carnotite, Steadite, and Thomasite. 9th List

Hillebrandite. F. E. Wright, 1908. *Amer. Journ. Sci.*, ser. 4, vol. xxvi, p. 551. J. E. Spurr and G. H. Garrey, *Economic Geology*, 1908, vol. iii, p. 707. Hydrated calcium ortho-silicate, $Ca_2SiO_4.H_2O$. A white, porcellanous mineral with fibrous structure and orthorhombic symmetry. Occurs with spurrite (q.v.) in contact-metamorphic limestone at Velardeña, Durango, Mexico. Named after Dr. William Francis Hillebrand, of Washington, DC. 5th List

Hinsdalite. E. S. Larsen, Junr. and W. T. Schaller, 1911. *Journ. Washington Acad. Sci.*, vol. i, p. 25; *Amer. Journ. Sci.*, ser. 4, vol. xxxii, p. 251; *Bull. US Geol. Survey*, 1912, No. 509, p. 66; *Zeits. Kryst. Min.*, 1912, vol. 1, p. 101. Hydrous sulphate and phosphate of lead (and strontium) and aluminium, $PbSO_4.AlPO_4.2Al(OH)_3$, isomorphous with svanbergite and beudantite. It forms dark-grey, granular masses, and colourless, rhombohedral crystals. Named from the locality, Hinsdale County, Colorado. 6th List

Histrixite. W. F. Petterd, *Papers and Proc. Roy. Soc. Tasmania for 1900–1, 1902*, p. 77; ibid., for 1902, 1903, p. 24; *Rep. Secr. Mines, Tasmania, for 1901–2, 1902*, p. 294. Radiating groups of orthorhombic crystals, associated with chalcopyrite, bismuthinite, etc., at Ringville, Tasmania. Described as a sulphide of antimony and bismuth with the formula $7Bi_2S_3.2Sb_2S_3.5CuFeS_2$. [Probably a mixture.] Named histrixite (porcupine-ore) from the Latin, *hystrix* (*histrix*), a porcupine. 3rd List [A mixture; *M.A.*, **11**, 296]

Hlopinite. *See* Khlopinite. 14th List

Hocartite. R. Caye, Y. Laurent, P. Picot, R. Pierrot, and C. Lévy, 1968. *Bull. Soc. franç. Min. Crist.*, **91**, 383. $Ag_2Sn(Fe,Zn)S_4$, the silver analogue of stannite, as inclusions in blende and wurtzite at the tin mines of Tacama, Hocaya, and Chocaya, Bolivia, and at Fournial, Cantal, France. Named for R. Hocart. [*A.M.*, **54**, 573; *M.A.*, 69-611; *Zap.*, **99**, 73.] 26th List

Hoch-Bassanit. H. Strunz, 1970. *Min. Tab.*, 5th edit., 291. A new name by the same author for β-bassanite (24th List). 28th List

Hoch-Schapbachit. H. Strunz, 1944. *Min. Tab.*, 1st edit., 72 (and in all subsequent editions). The high-temperature cubic modification (α-$AgBiS_2$) of schapbachite. Now a *syn.* of schapbachite (*Min. Tab.*, 1970, 5th edit., pp. 125 and 534), which becomes the high-temperature phase following the redefinition of matildite [*M.A.*, 70-1585] as low-temperature hexagonal β-$AgBiS_2$. [Note: the α- and β-usage is reversed by J. H. Wernick, *A.M.*, **45**, 591.] 28th List

Hochschildite. R. Herzenberg, 1942. *Fac. Nac. Ingen. Univ. Oruro* (*Hochschildita*). Hydrous lead stannate, $PbSnO_3.5$–$6H_2O$, with some Fe and Si, as a yellow, earthy or scaly alteration product of teallite from Bolivia. Named after Dr. Mauricio Hochschild of Oruro. [*M.A.*, **8**, 310.] 16th List

Hodgkinsonite. C. Palache and W. T. Schaller, 1913. *Journ. Washington Acad. Sci.*, vol. iii, p. 474; ibid., 1914, vol. iv, p. 153; *Zeits. Kryst. Min.*, 1914, vol. liii, pp. 529, 675. Hydrous silicate of zinc and manganese, $MnO.2ZnO.SiO_2.H_2O$, occurring as pink, monoclinic crystals at Franklin Furnace, New Jersey. Named after Mr. H. H. Hodgkinson, of Franklin Furnace. 7th List

Hodruschit, Germanized spelling of hodrushite, hodrušit (26th List). Hintze, 37. 28th List

Hodrushite. M. Kodĕra, V. Kupčik, and E. Makovický, 1970. *Min. Mag.*, **37**, 641. Needle-shaped crystals or fine aggregates in the Rosalia vein, Banska Hodruša, near Banska Štiavnica, Czechoslovakia are monoclinic; composition $Cu_8Bi_{12}S_{22}$. Named for the locality. [*M.A.*, 70-2609.] 26th List

Hoeferite. F. Katzer, *Tsch. Min. Mitth.*, XIV, 519, 1895. [*M.M.*, **11**, 161.] $2Fe_2O_3.4SiO_2.7H_2O$. Related to the chloropals. [Křitz, near Rakonitz] Bohemia. 1st List [Topotype material is Chapmanite; *A.M.*, **50**, 2110 (but cf. *M.A.*, **3**, 452)]

—— C. Cipriani and P. Vannuccini, 1961. *Atti* (*Rend.*) *Accad. Nac. Lincei, Cl. sci. fis. mat. nat.*, ser. 8, vol. 30, p. 74. A name proposed for the mineral subsequently named Biringuccite (q.v.), and withdrawn because of prior use for a variety of Nontronite (*see* 1st List). [*M.A.*, **16**, 373.] 23rd List

Hoelite. I. Oftedal, 1922. *Result. Norske Stats. Spitsbergenekspeditioner*, Kristiania, vol. 1, p. 14. Yellow, orthorhombic needles identical with anthraquinone ($C_{14}H_8O_2$) formed by the burning of a coal-seam in Spitsbergen. Named after Mr. Adolf Hoel, the leader of the Norwegian Scientific Expedition to Spitsbergen. [*M.A.*, **2**, 10.] 10th List

Högbohmit, error for Högbomit (Hintze, *Handb. Min.*, Erg.-Bd. II, p. 893). 22nd List

Högbomite. A. Gavelin, 1916. *Bull. Geol. Inst. Univ. Upsala*, vol. xv, p. 289 (Högbomit). Abstr. in *American Mineralogist*, 1919, vol. iv, p. 76, gives the spelling Hoegbomite. A black, rhombohedral mineral occurring intimately intermixed with magnetite, ilmenite, and pleonaste in the iron-ores of Swedish Lapland. It resembles haematite in its physical characters, but approximates to pleonaste (iron-magnesia-spinel) in composition, $MgO.2(Al_2O_3,Fe_2O_3,TiO_2)$. Named after Professor Arvid Gustav Högbom, of Upsala. 8th List [Several polytypes; *M.M.*, **33**, 563 (*A.M.*, **49**, 445). Cf. Taosite]

Högtveitite. A radioactive mineral from a felspar quarry on the farm Högtveit, Evje parish, in Sætersdalen, Norway. A specimen so designated was purchased by the British Museum, in 1910, from Dr. L. Eger, of Vienna. J. Schetelig (*Videnskapsselsk. Skrift. Mat.-naturv. Kl. Kristiania*, 1922, no. 1, p. 144 (Høgtveitit)) considers it to be identical with alvite. [*M.A.*, **2**, 26.] 10th List

Hokutolite. K. Jimbō, 1913. *Bull. Soc. franç. Min.*, 1912 [i.e. 1913], vol., xxxv, p. 471; *Amer. Journ. Sci.*, 1913, ser. 4, vol. xxxv, p. 464. A mixture, in variable proportions, of lead sulphate and barium sulphate deposited as a crystalline crust by the Hokuto hot springs in Taiwan (= Formosa). It was described as a radioactive mineral by Y. Okamoto (first in Japanese, and afterwards in English in *Beiträge zur Mineralogie von Japan*, 1912, no. 4, p. 178), and again by M. Hayakawa and T. Nakano (*Zeits. Anorg. Chem.*, 1912, vol. lxxviii, p. 183), the latter authors applying to it the name *angleso-barate* (p. 183) and, in German, Angleso-Baryt (pp. 184, 190). 6th List

Holdenite. C. Palache, 1921. Title only in *Amer. Min.*, 1921, vol. 6, p. 39. Passing references in *Amer. Min.*, 1923, vol. 8, p. 35; *Proc. Acad. Nat. Sci. Philadelphia*, 1923, vol. 75, p. 271. An undescribed mineral from Franklin Furnace, New Jersey. Named after Albert Fairchild Holden (1866–1913), whose mineral collection, the Holden Collection, is now in Harvard University. C. Palache and E. V. Shannon, *Amer. Min.*, 1927, vol. 12, pp. 82, 144. Basic arsenate of manganese and zinc, $12R''O.As_2O_5.5H_2O$, as red orthorhombic crystals. [*M.A.*, **3**, 365.] 10th & 11th Lists

Hollandite. L. L. Fermor, 1906. *Trans. Mining and Geol. Inst. India*, vol. i, p. 76. A crystallized manganate of barium, manganese, and ferric iron, $m(Ba,Mn)_2MnO_5 + nFe_4(MnO_5)_3$. From Central India. Named after Thomas Henry Holland, Director of the Geological Survey of India. 4th List [$BaMn_8O_{16}$]

Hollingworthite. E. F. Stumpfl and A. M. Clark, 1965. *Amer. Min.*, vol. 50, p. 1068; *Fortschr. Min.*, 1966, vol. 42 (for 1964), p. 213. Minute grains intergrown with Sperrylite, rhodian sulphurian Sperrylite, and Geversite, from the Driekop mine, Transvaal, are isotropic, with composition (Rh,Pd,Pt)AsS, Rh : Pd : Pt ≈ 3·5 : 0·95 : 0·6, and are probably cubic with pyrite structure. Named for S. E. Hollingworth [*M.A.*, **17**, 500]. 24th List

Holmquistite. A. Osann, 1913. *Sitzungsber. Heidelberg. Akad. Wiss.*, Abt. A, Abh. 23 (Holmquistit; *Lithionglaukophan*). A blue amphibole with the optical characters of glaucophane, but containing lithium and little sodium (Li_2O 2·13, Na_2O 1·12%), occurring in a granulitic rock on the island of Utö near Stockholm. Named after Professor Per Johan Holmquist of Stockholm. 7th List [Orthorhombic. Cf. Clinoholmquistite]

Holtite. M. W. Pryce, 1971. *Min. Mag.*, **38**, 21. A fibrous to prismatic mineral with stibiotantalite and tantalite from Greenbushes, Western Australia, is allied to dumortierite; orthorhombic, *a* 11·905, *b* 20·355, *c* 4·690 Å, *Pmcn*, D 3·90. Unit cell contents $[Al_{24\cdot5}Sb^{3+}_{2\cdot56}Ta_{1\cdot36}Nb_{0\cdot16}Sb^{5+}_{0\cdot76}Fe^{3+}_{0\cdot10}Be_{0\cdot05}Ti_{0\cdot03}Mn_{0\cdot02}B_{1\cdot40}Si_{9\cdot09}O_{66\cdot85}]$ or $4[X_{10}O_{17}]$. α 1·744 ‖ [001], β 1·757, γ 1·759. Named for H. E. Holt. [*M.A.*, 71-1385.] 27th List [*A.M.*, **57**, 1556]

Hondurasite. G. Gagarin and J. R. Cuomo, 1949, loc. cit., p. 6 (hondurasita). To replace the name selen-tellurium of E. S. Dana and H. L. Wells, 1890, which has been given in Spanish as teluroselenio (q.v.). Named from the locality, Honduras. 19th List

Honessite. A. V. Heyl, C. Milton, and J. M. Axelrod, 1956. *Bull. Geol. Soc. Amer.*, vol. 67, p. 1706 (abstr.). Hydrous basic nickel-iron sulphate, as a green or brown powdery weathering product of millerite, from Wisconsin. Named after Prof. Arthur Pharaoh Honess (1887–1942). 21st List [*A.M.*, **44**, 995]

Hongquiite. Yu Tsu-Hsiang, Lin Shu-Jen, Chao Pao, Fang Ching-Sung, and Huang Chi-Shun, 1974. [*Acta Geol. Sinica*, **2**, 202], abstr. *A.M.*, **61**, 184. Cubo-octahedra of TiO occur in platinum ores from an undisclosed locality in China; *a* 4·293 Å, *Fm3m*. In *M.A.*, 75-2522 the name is spelt Honquilite. 29th List

Hongshiite. Yu Tsu-Hsiang, Lin Shu-Jen, Chao Pao, Fang Ching-Sung, and Huang Chi-Shun, 1974. [*Acta Geol. Sinica*, **2**, 202], abstr. *M.A.*, 75-2522. Hexagonal PtCuAs, *a* 10·51, *c* 4·59 Å, from an undisclosed locality in China. [*A.M.*, **61**, 185.] 29th List

Honquilite, *see* Hongquiite. 29th List

Hoppingite. G. Gagarin and J. R. Cuomo, 1949, loc. cit., p. 8 (hoppingita). Mercuric iodide as minute scarlet cubes from Broken Hill, New South Wales. Described, as distinct from coccinite, by A. J. Moses, 1901, from material submitted by Roy Hopping, mineral dealer of New York. 19th List

Hormites. R. H. S. Robertson, 1959, in R. C. Mackenzie, *Clay Min. Bull.*, vol. 4, p. 64. An unnecessary group-name for the sepiolite–palygorskite family. 22nd List

Hörnbergite. J. F. Vogl? ([G.J.] Adam, *Tableau Minéralogique*, Paris, 1869, p. 42). An arsenate of uranium. 3rd List

Horobetsuite. K. Hayase, 1955. *Min. Journ.* (*Japan*), vol. 1, p. 189. $(Bi,Sb)_2S_3$, intermediate between stibnite and bismuthinite, with 30 to 70 mol.% Bi_2S_3. Occurs at the Horobetsu mine, Hokkaido, Japan. Named from the locality. 'Nom inutile' (*Bull. Soc. franç. Min. Crist.*, 1959, vol. 82, p. 91). 22nd List

Hoshiite. Yue Chu-Siang, Fuo Kuo-Fun, and S. Chen-Ea 1964. [*Acta Geol. Sinica*, vol. 44, p. 213], abstr. *Bull. Soc. franç. Min. Crist.*, 1965, vol. 88, p. 358; *M.A.*, **17**, 501. An emerald-green mineral in the oxidation zone of a Cu–Ni deposit east of the Yellow River, China, is a nickeloan Magnesite with Mg:Ni ≈ 3:2; the origin of the name is not stated. M. Fleischer (*Amer. Min.*, 1965, vol. 50, p. 2100) comments: 'Should not have been named, but the name may be used if a mineral with Ni > Mg is found.' 24th List

Hovaxite, erroneous transliteration of Ховахсит, khovakhsite (Зап. Всесоюз. Мин. Общ. [*Mem. All-Union Min. Soc.*], 1959, vol. 88, p. 317). 22nd List

Howdenite. *Amer. Journ. Sci.*, 1907, ser. 4, vol. xxiv, p. 184. A local name for the large crystals of chiastolite found by G. R. Howden at Mt. Howden, near Bimbowrie, South Australia, and described by C. Anderson (*Records Australian Museum*, 1902, vol. iv, p. 298). 5th List

Howieite. S. O. Agrell, M. G. Bown, and D. McKie, 1964. *Amer. Min.*, 1965, vol. 50, p. 278 (abstr.) (Howieite). Dark green anorthic blades in metamorphic rocks of the Franciscan formation, Laytonville district, Mendocino County, California, are near $Na(Fe^{2+},Mn)_{11}(Fe^{3+},Al)_2(Si,Ti)_{12}O_{31}(OH)_{13}$. Named for Dr. R. A. Howie. 24th List [*A.M.*, **59**, 86]

Hsiang-hua-shih, *see* Hsianghualite.

Hsianghualite. Wen-Hui Huang, Shao-Hua Tu, K'ung-Hai Wang, Chun-Lin Chao, and Cheng-Chih Yu, 1958. [*Ti-chih-yueh-k'an*, vol. 7, p. 35]; abstr. *Amer. Min.*, 1959, vol. 44, p. 1327 (hsiang-hua-shih); Зап. Всесоюз. Мин. Общ. (*Mem. All-Union Min. Soc.*), 1961, vol. 90, p. 105 (Сянхуанит, hsiang-hua-shih). $Ca_3Be_3Li_2Si_2O_{10}F_2$; cubic, occurring with taaffeite in metamorphosed limestone in Hunan. *See also* A. A. Beus, [Акад. Наук СССР, Инст. Мин., Геохим., Крист., Редк. Элем., 1960, p. 60], abstr. *Amer. Min.*, 1961, vol. 46, p. 244 (hsianghualite). 22nd List $[Ca_3Be_3Li_2(SiO_4)_3F_2]$

Hsihutsunite. C. C. Wang, 1936. *Bull. Geol. Soc. China*, vol. 15, p. 94 (Hsihutsunite). A variety of rhodonite containing 6·24% MgO. Named from the locality, Hsihutsun, prov. Chih-li, China. [*M.A.*, **6**, 442.] 14th List

Huangoite. E. I. Semenov and P'ei-shan Chang, 1961. [*Scientia Sinica*, vol. 10, p. 1007; 1962, vol. 11, p. 251], abstr. in *Amer. Min.*, 1963, vol. 48, p. 1179, and in Зап. Всесоюз. Мин. Общ. (*Mem. All-Union Min. Soc.*), 1963, vol. 92, p. 199. $BaLn(CO_3)_2F$, the barium analogue of Synchysite, in hexagonal platy masses from hydrothermal deposits near the Huang-Ho river. Named from the locality. [*M.A.*, **16**, 181.] 23rd List

Huangtsaoite, alternative transliteration of Hungchaoite (q.v.). 24th List

Huanite. *See* Juanite. 13th List

Huebnerite, variant of Hübnerite; *A.M.*, **9**, 62. 10th List

Huelvite. *See* lacroisite. 3rd List

Huemulite. C. E. Gordillo, E. Linares, R. O. Toubes, and H. Winchell, 1966. *Amer. Min.*, vol.

51, p. 1. Botryoidal masses and infillings in sandstone at the Huemul mine, Malargüe area, Mendoza province, Argentina, have the composition $Na_4MgV_{10}O_{28}.24H_2O$; anorthic. Named from the locality. 24th List

Hügelite. V. Dürrfeld, 1913. *Zeits. Kryst. Min.*, vol. liii, p. 183 (Hügelit). A hydrated vanadate of lead and zinc occurring as yellow, monoclinic needles on corroded galena at Reichenbach, Baden. Previously described, but without a name, by the same author (loc. cit., 1912, vol. li, p. 278). Named after Baron Hügel. 7th List [Redefined as $Pb_2(UO_2)_3(AsO_4)_2(OH)_4.3H_2O$; *A.M.*, **47**, 418]

Hühnerkobelite. M. L. Lindberg, 1950, *Amer. Min.*, vol. 35, pp. 59, 75. Partly oxidized material with the composition $(Na_2,Ca)O.2(Fe,Mn)O.P_2O_5$, from Hühnerkobel, Bavaria, and Norrö, Sweden, previously referred to arrojadite, but differing from this in the X-ray pattern. Named from the locality. [*M.A.*, **11**, 126.] 19th List [*A.M.*, **50**, 713]

Hulsite. A. Knopf and W. T. Schaller, 1908. *Amer. Journ. Sci.*, ser. 4, vol. xxv, p. 323. W. T. Schaller, in 2nd Appendix to Dana's *System of Mineralogy*, 1909, p. 53; *Amer. Journ. Sci.*, 1910, ser. 4, vol. xxix, p. 543. A Knopf and W. T. Schaller, *Zeits. Kryst. Min.*, 1910, vol. xlviii, p. 1. A black, opaque, orthorhombic(?) mineral occurring in a contact-metamorphic limestone in connection with the tin-ores of Alaska. Originally described as a hydrous borate of ferrous and ferric iron and magnesium, but subsequently found to contain some tin. The revised formula is $12(Fe,Mg)O.2Fe_2O_3.SnO_2.3B_2O_3.2H_2O$. Named after Alfred Hulse Brooks, of the United States Geological Survey. *See* Paigeite. 5th List [Cf. Schoenfliesite. A distinct species; *M.A.*, 75-1381. $(Fe,Mg)_2(Fe,Sn)BO_5$]

Humbelite. *Phys. Chem. Min.*, 1977, **1**, 233. Error for gümbelite. 30th List

Humberstonite. G. E. Ericksen. J. J. Fahey, and M. E. Mrose, 1967. *Progr. Ann. Meeting Geol. Soc. Amer.*, p. 59. Colourless hexagonal plates from the Atacama Desert, Chile, have a composition near $Na_7K_3Mg_2(SO_4)_6(NO_3)_2.6H_2O$. Named for J. T. Humberstone. [*A.M.*, **53**, 507; *Bull.*, **91**, 301.] 25th List [*A.M.*, **55**, 1518]

Humite. R. Potonié, 1924. *Kohlenpetrographie*, Berlin, p. 18 (Humite, *pl.*). Coals derived from humic materials. Not the humite of J. L. Bournon, 1813. 14th List

Hummerite. A. D. Weeks, E. A. Cisney, and A. M. Sherwood, 1950. *Progr. and Abstr. Min. Soc. Amer.*, p. 22. Hydrous magnesium vanadate, triclinic, similar in appearance to pascoite. Named from the locality, Hummer mine, Paradox valley, Montrose Co., Colorado. [*M.A.*, **11**, 189.] 19th List [*A.M.*, **36**, 326. $KMgV_5O_{14}.8H_2O$; *A.M.*, **40**, 315]

Humosite. J. A. Dulhunty, 1939. *See* Gelosite. 15th List

Hungchaoite. I-Hua Chün, Hsien-Te Hsieh, Tze-Chiang Chien, and Lai-Pao Liu, 1964. [*Scientia Sinica*, vol. 13, p. 525; Hungtsaoite]; abstr. *Amer. Min.*, 1965, vol. 50, p. 262; Зап. Всесоюз. Мин. Общ. (*Mem. All-Union Min. Soc.*), 1965, vol. 94, p. 675. (Хунчжаоит, Huangtsaoite); *M.A.*, **16**, 648. $MgB_4O_7.9H_2O$, monoclinic, occurring in a Chinese borate deposit (locality not given); identical with synthetic material. Named for Professor Chang Hung-chao. 24th List [Triclinic; *M.A.*, 75-2481]

Hungtsaoite, original transliteration of Hungchaoite (q.v.). 24th List

Huntite. G. T. Faust, 1953. *Amer. Min.*, vol. 38, p. 4. $Mg_3Ca(CO_3)_4$, orthorhombic, as a fine-grained white powder in magnesite deposits in Nevada. Named after Dr. Walter Frederick Hunt (1882–), emeritus professor of mineralogy in the University of Michigan. [*M.A.*, **12**, 132.] 20th List [Trigonal]

Hurlbutite. G. Gagarin and J. R. Cuomo, 1949, loc. cit., p. 7 (hurlbutita). The 4*H* polymorph Zn_4S_4 of wurtzite. Named after Dr. Cornelius Searle Hurlbut, Jr. (1906–) of Harvard University. [*M.A.*, **11**, 126, 128; **10**, 532.] 19th List [Usage defunct]

—— M. E. Mrose, 1951, *Progr. and Abstr., Min. Soc. Amer.*, Nov. 1951, p. 19 (Hurlbutite); *Bull. Geol. Soc. Amer.*, 1951, vol. 62, p. 1464; *Amer. Min.*, 1952, vol. 37, p. 296, 931. Calcium beryllium phosphate, $CaBe_2(PO_4)_2$, orthorhombic, in pegmatite from Newport, New Hampshire. [*M.A.*, **11**, 414.] 19th List [The accepted usage]

Hussakite. E. H. Kraus and J. Reitinger, *Zeits. Kryst. Min.*, 1901, vol. xxxiv, p. 268 (Hussakit); (abstr. *M.M.*, **13**, p. 307) E. Hussak and J. Reitinger, *Zeits. Kryst. Min.*, 1903, vol. xxxvii, p. 563. A synonym of xenotime. Named after Dr. Eugen Hussak, of the Geological Survey of São Paulo, Brazil. 3rd List

Hutchinsonite. R. H. Solly, 1904. *Min. Mag.*, 1905, vol. xiv, p. 72. (A preliminary account of the unnamed mineral appeared in *Nature*, 1903, vol. lxix, p. 142, and the name (Hutchinsonit) was first printed in *Min. Petr. Mitt.* (*Tschermak*), 1904, vol. xxiii, p. 551.) G. T. Prior, *Nature*, 1905, vol. lxxi, p. 534. G. F. H. Smith and G. T. Prior, *Min. Mag.*, 1907, vol. xiv, p. 284. Minute, red, orthorhombic crystals found in the white, crystalline dolomite of the Binnenthal, Switzerland. An analysis made on a small amount of material suggests the formula $(Tl,Cu,Ag)_2S.As_2S_3 + PbS.As_2S_3$; thallium is present to the extent of about 20%. Named after Dr. Arthur Hutchinson, of Cambridge. 4th List [Ag,Cu not essential; $(Tl,Pb)_2As_5S_9$; *M.A.*, **17**, 459]

Huttonite. A. Pabst, 1950. *Nature*, London, vol. 166, p. 157; *Amer. Min.*, 1951, vol. 36, p. 60. C. O. Hutton, *Bull. Geol. Soc. Amer.*, July 1950, vol. 61, p. 678 (uranium-free thorite). Monoclinic thorium silicate, $ThSiO_4$, as minute grains in beach-sand from New Zealand. Named after Professor Colin Osborne Hutton, of Stanford University, California. [*M.A.*, **11**, 188, 209, 310.] 19th List

Hwanghite, variant of Huanghoite, 23rd List. (Докл. 1972, **202**, 422.) 28th List

Hyalithe. Trade name for a red, brown, green, or black glass. R. Webster, *Gems*, 1962, p. 757. Not to be confused with hyalite. 28th List

Hyaloallophane. G. D'Achiardi, *Atti Soc. Toscana Sci. Nat.*, 1898, *Proc. Verb.*, XII, 32 (*Jaloallofane*). Allophane containing an excess of silica, supposed to be due to the presence of admixed hyalite (= jalite in Italian). Sardinia. 2nd List

Hyblite. H. V. Ellsworth, 1927. *Amer. Min.*, vol. 12, p. 368. Alteration products occurring as a white (alpha-hyblite) and yellow (beta-hyblite) skin on uranothorite from Hybla, Hastings Co., Ontario. Microchemical tests suggest hydrous basic sulpho-silicate of thorium with some uranium, iron, and lead. Named after locality. [*M.A.*, **3**, 367.] 11th List [Is Thorogummite; *A.M.*, **38**, 1007]

Hydralsite. D. M. Roy, 1954, *Amer. Min.*, vol. 39, p. 141; R. Roy and E. F. Osborn, ibid., p. 863. Artificial hydrous aluminium silicate, perhaps $2Al_2O_3.2SiO_2.H_2O$, produced by the hydrothermal decomposition of kaolinite. Named from the composition, which would apply to many minerals. [*M.A.*, **12**, 308, 516.] 20th List

Hydrargyrite. J. Fröbel, 1843. *Grundzüge eines Systemes der Krystallologie*. Synonym of moschellandsbergite. Not the hydrargyrite of Bertrand, 1872 (Dana, 6th edit., 159). 27th List

Hydrated metavauxite. M. C. Bandy, 1946. *Min. Llallagua*, La Paz. Provisional name for an undetermined phosphate. [*M.A.*, **10**, 9.] 23rd List

Hydrated paravauxite. M. C. Bandy, 1946. *Min. Llallagua*, La Paz. Provisional name for an undetermined phosphate, subsequently shown to be a valid species and renamed Sigloite (q.v.). [*M.A.*, **10**, 9.] 23rd List

Hydroallanite. E. S. Simpson, 1938. *Journ. Roy. Soc. Western Australia*, vol. 24 (for 1937–1938), p. 113. Hydrated variety of allanite from Cooglegong, Western Australia. 17th List

Hydroamesite. J. Erdélyi, V. Koblencz, and N. S. Varga, 1959. [*Acta Geol. Acad. Sci. Hung.*, vol. 6, p. 95]; abstr. *Amer. Min.*, 1959, vol. 44, p. 1328, and *Bull. Soc. franç. Min. Crist.*, 1960, vol. 83, p. 149. An unnecessary name for a porcellaneous mineral identified as amesite with some excess H_2O^+. 22nd List

Hydro-amphibole. W. Q. Kennedy and B. E. Dixon, 1936. *Zeits. Krist.*, vol. 94, p. 280. An amphibole containing more water (5·78%) than that required by the theoretical formula. [*M.A.*, **6**, 443.] 14th List

Hydro-andradite. R. J. Ford, *Min. Mag.*, **37**, 942. A hydrogarnet, related to andradite by $SiO_4 \rightarrow (OH)_4$ replacement. 26th List

Hydroantigorite. J. Erdélyi, V. Koblencz, and N. S. Varga, 1959. [*Acta Geol. Acad. Sci. Hung.*, vol. 6, p. 65]; abstr. *Amer. Min.*, 1959, vol. 44, p. 1328. An unnecessary name for an antigorite with a small excess of H_2O^+. 22nd List

Hydroascharite. R. von Hodenberg and R. Kühn, 1972. *Kali und Steinsalz*, **6**, 104. An unnecessary name for a szájbelyite with some extra water. [*M.A.*, 74-480.] 28th List

Hydroastrophyllite. Hupei Geologic College, 1974. [*Sci. Geol. Sin.*, **1**, 18], abstr. *A.M.*, **60**, 736. A weathering product in a pegmatite from Szechuan is dark brown, pleochroic, anorthic with *a* 11·86, *b* 11·98, *c* 5·42 Å, α 103° 25′, β 95° 9′, γ 112° 12′; composition near

$(H_3O^+,K)(Ca,H_3O^+,Na)(Fe^{3+},Mn^{4+},Mn^{2+},Mg)_{5\cdot36}(Ti,Nb,Ta)_{1\cdot85}(Si_{4\cdot92}Al_{0\cdot88})(O,OH)_{24}$-(OH,O,F). Named from its relation to astrophyllite. [*M.A.*, 75-555.] 29th List

Hydroauerlite. V. S. Karpenko, N. G. Nazarenko, and O. V. Shchipanova, 1967. [Вопр. прикл. радиогеол., Атомиздат, 100], abstr. *Zap.*, **98**, 330 (Гидроауэрлит). An unnecessary name for a hydrous phosphatian thorite (auerlite). [*A.M.*, **55**, 1070.] 27th List

Hydro-basaluminite. F. A. Bannister and S. E. Hollingworth, 1948. *Nature*, London, vol. 162, p. 565. Hydrous basic aluminium sulphate as a white plastic material occurring with basaluminite (q.v.) and differing from this in its X-ray pattern. When air dried it loses about 50% H_2O, falling to a white powder of basaluminite. [*M.A.*, **10**, 452.] 18th List [Name sometimes unhyphenated; *A.M.*, **33**, 787]

Hydrobismutite. K. A. Nenadkevich, 1917. *Bull. Acad. Sci. Petrograd*, ser. 6, vol. 11, p. 448 (гидробисмутит). Hydrated bismuth carbonate, $Bi_2O_3.CO_2.2–3H_2O$, containing more water than bismutite; from Transbaikal, Siberia. 14th List [Probably Bismutite; *A.M.*, **28**, 251]

Hydrobraunite, Hydroxybraunite, etc. A. K. Boldyrev, 1928. *Kurs opisatelnoi mineralogii*, Leningrad, 1928, part 2, p. 97. O. M. Shubnikova and D. V. Yuferov, *Spravochnik po novym mineralam*, Leningrad, 1934, pp. 44–45. The following are given as members of the psilomelane and wad group with the general formula kMnO.lMnO$_2$.nH$_2$O: Hydrobraunite (Гидробраунит), Hydroxybraunite (Гидроксибраунит), Hydrohausmannite (Гидрогаусманнит), Hydromanganite (Гидроманганит), Hydromanganosite (Гидроманганозит), and Hydropyrolusite (Гидропиролюзит). 14th List

Hydrobritholite. A doubtful hydrated or altered Britholite. [Геохим., мин., генет. типы месторожд. редк. элем., Изд. „Наука", 1964, p. 293]; abstr. Зап. Всесоюз. Мин. Общ. (*Mem. All-Union Min. Soc.*), 1965, vol. 94, p. 683. (Гидробритолит); an unnecessary name. 24th List

Hydrocalcite. H. B. Kosmann, *Zeits. deutsch. geol. Ges.*, XLIV. 155, 1892. $CaCO_3.2H_2O$. [Wolinsdorf, Glatz] Silesia. 1st List

—— H. Marschner, 1969. *Science*, **165**, 1119. A superfluous and preempted name for monohydrocalcite (Semenov, 1964; 24th List). Not the hydrocalcite of Kosmann, 1892 (1st List) or of Dana, 1892 (a synonym of hydroconite). [*A.M.*, **55**, 1069.] 27th List [Cf. Ikaite]

Hydrocalumite. C. E. Tilley, 1934. *Min. Mag.*, 1934, vol. 23, p. 607. Hydrous calcium aluminate, $4CaO.Al_2O_3.12H_2O$, monoclinic, in a chalk-dolerite contact-rock at Scawt Hill, Co. Antrim. Named from the chemical composition. 13th List [*A.M.*, **20**, 316]

Hydrocancrinite. J. Wyart and M. Michel-Lévy, 1949. *Compt. Rend. Acad. Sci. Paris*, vol. 229, p. 131. An artificial hydrothermal product, $NaAlSiO_4.\frac{1}{2}H_2O$, with the same X-ray pattern as cancrinite. Compare lembergite (1st List). [*M.A.*, **11**, 93.] 19th List

Hydrocassiterite. H. Strunz, *Min. Tab.*, 3rd edit., 1957, p. 151 (Hydro-Cassiterit). Synonym of souxite = varlamoffite (18th List). 21st List [Also 22nd List]

Hydrocatapleiite. E. I. Semenov and I. P. Tikhonenkov, 1962. Труды Инст. Мин. Геохим. Редк. Элем. (*Trans. Inst. Min., Geochem., Cryst. Rare Elem.*), no. 9, p. 88 (Гидрокатаплеит). A superfluous name for an insufficiently characterized alteration product of Catapleiite from Mt. Partomchorr, Khibina Tundra, Kola (*Amer. Min.*, 1964, vol. 49, p. 443; Зап. Всесоюз. Мин. Общ. (*Mem. All-Union Min. Soc.*), 1963, vol. 92, p. 575; *M.A.*, **16**, 558). 23rd List

Hydrocatapleite, error for Hydrocatapleiite (23rd List) apparently through straight transliteration of Гидрокатаплеит (*Amer. Min.*, 1964, vol. 49, p. 443; H. Strunz, *Min. Tabellen*, 4th edit., 1966, p. 480). 24th List

Hydrocerite. K. A. Vlasov, M. V. Kuzmenko, and E. M. Eskova, 1959. [The Lovozero Alkaline Massif (Moscow, *Acad. Sci. USSR*), p. 427]; abstr. *Amer. Min.*, 1960, vol. 45, p. 1132. Amorphous, pseudomorphs after manganosteenstrupine (q.v.) in the pegmatite of Mt. Karnasurt, Lovozero massif, Kola peninsula, are formulated (Ln,Th,Ca)(Al,Fe,Ti,Nb)$(Si,P)_2O_7.5H_2O$. The composition is very near that of karnasurtite (q.v.). Not to be confused with the hydrocerite of Glocker (1831, = lanthanite) or of Glocker (1847, = bastnäsite). 22nd List [Cf. *A.M.*, **47**, 420]

Hydrocerusite. Variant of Hydrocerussite (French, hydrocérusite; from Latin, cerussa). *A.M.*, **9**, 62. 10th List

Hydrocervantite. L. B. Shlain, 1950. *Mém. Soc. Russ. Min.*, ser. 2, vol. 79, p. 63 (гидросервантит). O. M. Shubnikova, ibid., 1952, vol. 81, p. 45. $Sb_2O_4.nH_2O$, with 5% H_2O, a variety of cervantite, intermediate between cervantite and stibiconite; alteration product of stibnite. 20th List

Hydrochlorbechilite, apparently an error for Hydrochlorborite (q.v.); cited in *Bull. Soc. franç. Min. Crist.*, 1966, vol. 89, p. 144, and in *Zentr. Min.*, 1964, Teil 1, p. 158. 24th List

Hydrochlorborite. Chien Tzu-Chiang and Chen Shu-Chien, 1965. [*Scientia Sinica*, vol. 14, p. 945]; abstr. in *Amer. Min.*, 1965, vol. 50, p. 2099; *M.A.*, **17**, 501. Massive, colourless, $Ca_4B_8O_{15}Cl_2.22H_2O$, in Tertiary sediments from an unspecified Chinese locality. Named from the composition. [*See also Bull. Soc. franç. Min. Crist.*, 1966, vol. 89, p. 144.] 24th List

Hydroclinohumite. F. Zambonini, 1919. *Bull. Soc. franç. Min.*, vol. 42, pp. 273, 279 (footnote 2). Clinohumite in which fluorine is replaced by hydroxyl. *See* Titanhydroclinohumite (hydroclinohumite titanifère). Abstr. in *Amer. Min.*, 1920, vol. 5, p. 136, proposes the same name as a synonym of 'Titanhydroclinohumite', without regard to the author's previous use. 9th List

Hydrocookeite. A. I. Ginzburg, 1953. *Doklady Acad. Sci. USSR*, vol. 90, p. 871 (гидрокукеит). Hydrated cookeite. [*M.A.*, **12**, 451.] 20th List

Hydrodresserite. J. L. Jambor, A. P. Sabina, and B. D. Sturman, 1977. *Can. Mineral.*, **15**, 399. Compact spherulites of lath-shaped crystals in cavities of a silicocarbonatite sill at St. Michel, Montreal Island, Quebec, are anorthic, *a* 9·79, *b* 10·42, *c* 5·62 Å, α 96·05°, β 92·20°, γ 115·71°, elongation [001]. Composition $2[BaAl_2(CO_3)_2(OH)_4.3H_2O]$, sp. gr. 2·80. RIs α 1·520, β 1·594, γ 1·595, $\gamma' \parallel$ [001]. Named for its relation to dresserite (26th List), to which it gradually dehydrates on exposure. 30th List

Hydro-fluor-herderite. S. L. Penfield, *Amer. J. Sci.*, 1894, XLVII, 330. Herderite containing both fluorine and hydroxyl. $Ca[Be(OH,F)]PO_4$. 2nd List

Hydroforsterite. N. E. Efremov, 1938. *Bull. Acad. Sci. URSS, Cl. Sci. Math. Nat., Sér. Géol.*, 1938, p. 132 (hydroforsterite); *Compt. Rend.* (*Doklady*) *Acad. Sci. URSS*, 1939, vol. 22, p. 432 (Hydrophorsterite!). Hydrated orthosilicate of magnesium, $Mg_2SiO_4.2H_2O$, in the form of asbestos and regarded as an end-member of the serpentine series. [*M.A.*, **6**, 490; **7**, 221, 370.] 15th List

Hydrogadolinite. E. S. Simpson, 1938. *Journ. Roy. Soc. Western Australia*, vol. 24 (for 1937–1938), p. 113. Hydrated variety of gadolinite from Cooglegong, Western Australia. 17th List

Hydrogarnet. E. P. Flint, H. F. McMurdie, and L. S. Wells, 1941. *Journ. Research US Bur. Standards*, vol. 26, p. 13 (garnet-hydrogarnet series), p. 14 (hydrogarnets). Hydrous calcium aluminate ($3CaO.Al_2O_3.6H_2O$) and calcium ferrite ($3CaO.Fe_2O_3.6H_2O$), present in hydrated Portland cement. They are cubic and form a complete series of mixed crystals with grossular ($3CaO.Al_2O_3.3SiO_2$) and andradite ($3CaO.Fe_2O_3.3SiO_2$); plazolite ($3CaO.Al_2O_3.2SiO_2.2H_2O$; 9th List) is an intermediate member. Cf. grossularoid and garnetoid. [*M.A.*, **8**, 101, 146.] 16th List [A group name. Cf. Hydrogrossular]

Hydrogedroitzite. I. D. Sedletzky, 1940. *Compt. Rend.* (*Doklady*) *Acad. Sci. URSS*, vol. 26, p. 241 (Hydrogedroitzite). A hypothetical, hydrated, metastable form of gedroitzite (q.v.). Similarly, Hydromontmorillonite, Hydronontronite, and Hydropyrophyllite, corresponding to montmorillonite, etc. 15th List

Hydrogen-autunite. C. Frondel, 1950. *Amer. Min.*, vol. 35, p. 762 (hydrogen-autunite). V. Ross, ibid., 1955, vol. 40, p. 917. Artificial base-exchange product, $HUO_2PO_4.4H_2O$, tetragonal. [*M.A.*, **13**, 7.] 21st List

Hydrogen-uranospinite. M. E. Mrose, 1950. *Progr. and Abstr. Min. Soc. Amer.*, p. 18 (hydrogen-uranospinite); *Amer. Min.*, 1951, vol. 36, p. 322. Artificial $H_2(UO_2)_2(AsO_4)_2.8H_2O$, isostructural with metatorbernite-I. Replacements of H_2 by Na_2 and $(NH_4)_2$ yield sodium-uranospinite and ammonium-uranospinite. [*M.A.*, **11**, 365.] 19th List

Hydroglauberite. M. N. Slyusareva, 1963. Зап. всесоюз. мин. общ. (*Mem. All-Union Min. Soc.*), **98**, 59. (Гидроглауберит.) Snow-white masses in Tertiary sediments in the Karakalpakii ASSR as an alteration product of glauberite have the composition $Na_{10}Ca_3(SO_4)_8.6H_2O$. Named for the composition. [*A.M.*, **55**, 321; *M.A.*, 70-754; *Zap.*, **99**, 78.] 26th List

Hydro-glockerite. E. Greenly, 1919. *Mem. Geol. Survey, The Geology of Anglesey*, vol. 2, p. 832. An ochreous, hydrated basic ferric sulphate, $2Fe_2O_3.SO_3.8H_2O$, from Parys Mountain, Anglesey, analysed by A. H. Church (*Min. Mag.*, 1895, vol. 11, p. 13), and containing more loosely-held water than the glockerite of Naumann ($2Fe_2O_3.SO_3.6H_2O$). [*M.A.*, **1**, 328.] 9th List

Hydrogoethite. P. Groth, *Tab. Uebers. Min.*, 4th edit., 1898, p. 48. $3Fe_2O_3.4H_2O$. Represents a stage in the alteration of göthite to limonite, as hydrohæmatite (= turgite) is a stage between hæmatite and göthite. 2nd List

Hydrogöthite. P. A. Zemĭatčenskij, *Trav. Soc. Nat. St.-Pétersbourg*, 1889, vol. xx, *Sect. Géol. et Min.*, p. 208 (гидрогетитъ); (abstr., *Zeits. Kryst. Min.*, 1892, vol. xx, p. 185). J. Samojlov, *Zeits. Kryst. Min.*, 1901, vol. xxxv, p. 272. A fibrous (crystalline), cochineal-red hydroxide of iron, $3Fe_2O_3.4H_2O$, occurring as veins in limonite in central Russia. 3rd List

Hydrograndite, error for hydrougrandite (24th List). (H. Strunz, *Min. Tab.*, 1970, 5th edit., 535). 27th List. More probably a direct coinage from grandite (5th List); a group name. 28th List

Hydrogrossular. C. O. Hutton, 1943. *Trans. Roy. Soc. New Zealand*, vol. 73, p. 174 (Hydrogrossular). For members of the series $3CaO.Al_2O_3.3SiO_2$–$3CaO.Al_2O_3.6H_2O$ between grossular and hibschite (= plazolite, $3CaO.Al_2O_3.2SiO_2.2H_2O$). Compare garnetoid, grossularoid, and hydrogarnet (16th List). [*M.A.*, **9**, 61.] 17th List

Hydrohalloysite. I. D. Sedletzky, 1940. *Compt. Rend. (Doklady) Acad. Sci. URSS*, vol. 26, p. 241 (Hydrohalloysite). 'Hydrated halloysite', $Al_2O_3.2SiO_2.4H_2O$, of S. B. Hendricks (*Amer. Min.*, 1938, vol. 23, p. 295). [*M.A.*, **7**, 422.] 15th List [Is Endellite (q.v.)]

—— J. Erdélyi, 1962. *Chemie der Erde*, vol. 21, p. 321. Superfluous name for a Halloysite containing slightly more water than corresponds to the formula $Al_2Si_2O_5(OH)_4.2H_2O$. Cf. Hydrohalloysite of Sedletsky, 1940 (15th List), a synonym of Halloysite. 23rd List [*A.M.*, **48**, 214]

Hydrohausmannite. W. Feitknecht and W. Marti, 1945. *Helvetica Chim. Acta*, vol. 28, p. 133 (Hydrohausmannit). An artificial product obtained, together with hausmannite, by the oxidation of $Mn(OH)_2$. It has the composition $MnO_{1\cdot15}$—$MnO_{1\cdot45}+\frac{1}{2}H_2O$ and gives an X-ray pattern similar to that of hausmannite (Mn_3O_4). Not the hydrohausmannite of A. K. Boldyrev, 1928 (*See* Hydrobraunite). [*M.A.*, **10**, 105.] 18th List [A mixture containing Feitknechtite (q.v.); *A.M.*, **48**, 1420; **50**, 1313]

Hydrohaüyne. G. A. Ilinsky, 1962. Зап. Всесоюз. Мин. Общ. (*Mem. All-Union Min. Soc.*), vol. 91, p. 109 (Гидрогаюин). A variety deficient in Na^+ and SO_4^{2-}, and with some H_2O, believed to be constitutional. 23rd List

Hydro-herderite. S. L. Penfield, *Amer. J. Sci.*, 1894, XLVII, 330. Herderite containing hydroxyl in place of fluorine. [Hebron (and Paris), Maine] $Ca[Be(OH)]PO_4$. *See* fluor-herderite. 2nd List [Hydroxyl-herderite (q.v.)]

Hydrohetaerolite. C. Palache, 1928. *Amer. Min.*, vol. 13, p. 308; *Prof. Paper US Geol. Survey*, 1935, no. 180, pp. 49, 53. A mineral from New Jersey and Colorado previously described as hetaerolite ($ZnO.Mn_2O_3$), but distinct from this in containing some water, the formula being $2ZnO.2Mn_2O_3.H_2O$. It is perhaps tetragonal, like hetaerolite. [*M.A.*, **6**, 261.] 14th List

Hydroilmenite. B. H. Flinter, 1959. *Econ. Geol.*, vol. 54, p. 720. A near-amorphous alteration product of Ilmenite, giving a diffuse pattern of Rutile. [*M.A.*, **15**, 299.] 23rd List

Hydrokaolin. D. S. Belyankin and V. P. Ivanova, 1935. *Zentr. Min.*, Abt. A, 1935, p. 298 (Hydrokaolin). Synonym of halloysite, which is considered to differ from kaolin only in its degree of hydration. [*M.A.*, 7, 103.] 15th List

Hydrokassite. A. A. Kukharenko, 1965. [The Caledonian ultrabasic alkalic rocks and carbonatites of the Kola Peninsula and northern Karelia, Moscow], abstr. *Amer. Min.*, 1967, vol. 52, p. 559. An incompletely described alteration product of kassite (q.v.). Species not accepted by the I.M.A. Commission on New Minerals and Mineral Names. [Зап., **96**, 70.] 25th List

Hydrolepidocrocite. B. P. Krotov, 1936. *Compt. Rend. (Doklady) Acad. Sci. URSS*, 1943, vol. 40, p. 115 (Hydrolepidokrokite). A. E. Fersman and O. M. Shubnikova, *Sputnik geokhimika i mineraloga*, 1937, p. 137 (Гидролепидокрокит). Lepidocrocite with absorbed water, γ-FeOOH.aq, dimorphous with hydrogoethite (α-FeOOH.aq). The hydrogoethite (3rd List) originally described is in reality hydrolepidocrocite. [*M.A.*, **9**, 62.] 17th List

Hydrolepidolite. V. A. Frank-Kamenetsky, 1960. *Clay Min. Bull.*, vol. 4, pp. 162, 170. Name for a group of clay minerals; incompletely defined. 22nd List

Hydrolite. G. Mackenzie (T. Allan, *Min. Nomencl.*, 3rd edit., 1819). The same as siliceous sinter. 2nd List

Hydroloparite. V. I. Gerasimovsky, 1941. Synonym of metaloparite (q.v.). 16th List

Hydromagniolite. N. E. Efremov, 1939. *Compt. Rend.* (*Doklady*) *Acad. Sci. URSS*, vol. 24, p. 287 (hydromagniolites). A collective name for hydrated silicates of magnesium. [*M.A.*, **7**, 419.] 15th List

Hydromanganite, -manganosite, *see* Hydrobraunite.

Hydromelanothallite. F. Zambonini, 1910. *Mineralogia Vesuviana. Mem. R. Accad. Sci. Fis. Mat. Napoli*, ser. 2, vol. xiv, No. 7, p. 57 (idromelanotallite). The black scales of melanothallite (A. Scacchi, 1870) change on exposure to the air to green, and for this alteration-product the name hydromelanothallite is proposed. Scacchi's analyses suggest that the former may be $CuCl_2.CuO.H_2O$ and the latter $CuCl_2.CuO.2H_2O$. 6th List

Hydromelilite. E. M. Epshtein, L. I. Anikeeva, and A. F. Mikhailova, 1961. [Tr. НИИГА, **122**, 116 (Гидромелилит)]; abstr. *Zap.*, **87**, 620. An unnecessary general term for the hydration products of melilite, cebollite, and juanite. 26th List

Hydrometavauxite. M. C. Bandy, 1946. *Mineralogía de Llallagua, Bolivia*, La Paz, p. 57 (Metavauxita Hidratada). *M.A.*, **10**, 9 (hydrated metavauxite); Hintze, *Min.*, 1954, Ergbd. 2, p. 166 (Hydrometavauxit). An undetermined, yellow alteration product of metavauxite. 20th List [Also 23rd List]

Hydromolysite. A. S. Povarennykh, 1962. Мин. Таблицы (translation, with addenda, of H. Strunz, *Min. Tabellen*, 3rd edit., 1957), p. 115 (Гидромолизит). The doubtful unnamed mineral of C. Garavelli (*Period. Min.*, 1958, vol. 27, p. 211); $FeCl_3.6H_2O$, from Rio Marina, Elba. 24th List [*A.M.*, **51**, 1551]

Hydromontmorillonite. I. D. Sedletzky, 1940. *Compt. Rend.* (*Doklady*) *Acad. Sci. URSS*, vol. 26, p. 241. An assumed intermediate stage in the passage from montmorillonite gel into montmorillonite. [*M.A.*, **8**, 146.] 16th List

Hydronasturan. R. V. Getzeva, 1956. [*Atomic Energy*, Moscow, no. 3, p. 135.] Abstr. in *Mém. Soc. Russ. Min.*, 1957, vol. 86, p. 117 (гидронастуран, hydronasturan); *Amer. Min.*, 1957, vol. 42, p. 442. $UO_2.kUO_3.nH_2O$. Compact, amorphous, black, in oxidized ore. Named from the composition. (настуран = pitchblende.) *See* urhite. [*M.A.*, **13**, 385.] 21st List

Hydronatrojarosite. M.-A. Kashkai, 1969. Зап. Всесоюз. мин. общ. (*Mem. All-Union Min. Soc.*), **98**, 153. (Гидронатрояросит.) A superfluous name for a natrojarosite high in water. 26th List

Hydro-naujakasite. O. V. Petersen, 1967. *Medd. Grønland*, **181**, no. 6, 1. An incompletely described alteration product of naujakasite. [*A.M.*, **53**, 1778; *Bull.*, **92**, 320; *Zap.*, **99**, 85.] 26th List

Hydroniojarosite. A. S. Povarennykh, 1962. Мин. Таблицы (translation, with addenda, of H. Strunz, *Min. Tabellen*, 3rd edit., 1957), p. 192 (Гидрониоярозит, Hydroniojarosite). Variant of Hydronium jarosite (22nd List). 24th List

Hydronium gastunite, *see* Gastunite (of Honea). 22nd List

Hydronium jarosite. J. Kubisz, 1960. *Bull. Acad. Polon. Sci., Sér. géol. géogr.*, vol. 8, p. 95. A mineral of the jarosite family from Staszic mine, Holy Cross Mt., Poland, contains only 0·21 (Na,K) to 3 Fe; the author considers the name carphosiderite inappropriate since A. A. Moss has shown (*Min. Mag.*, **31**, 407) that the original carphosiderite and many occurrences subsequently described as carphosiderite are in fact jarosite or natrojarosite. On the other hand, the name carphosiderite is widely accepted for the artificial material. 22nd List

Hydronontronite, *see* Hydrogedroitzite.

Hydroparagonite. H. Strunz, 1957. *Min. Tabellen*, 3rd edit., p. 307. An unnecessary new name for Brammallite (16th List), cf. M. Fleischer, *Amer. Min.*, 1958, vol. 43, p. 1222, and 1959, vol. 44, p. 1329. 22nd List

Hydroparavauxite. M. C. Bandy, 1946. *Mineralogia de Llallagua, Bolivia*, La Paz, p. 57 (Paravauxita hidratada). *M.A.*, **10**, 9 (hydrated paravauxite); Hintze, *Min.*, 1954, Ergbd. 2, p. 167 (Hydroparavauxit). An undetermined, yellow, fibrous, alteration product of paravauxite. 20th List [Also 23rd List] [Re-named Sigloite (q.v.)]

Hydrophlogopite. (W. F. Clarke, *Bull. US Geol. Survey*, 1895, No. 125, p. 49; P. Groth, *Tab. Uebers. Min.*, 4th edit., 1898, p. 131). Contains less alkalies and more loosely combined water than phlogopite. 2nd List

Hydrophorsterite, error for Hydroforsterite (q.v.).

Hydrophyllite, error for Hydrophilite (*Amer. Min.*, **54**, 1021). The confusion of -philite and -phyllite is curiously common. 26th List

Hydropolylithionite. E. I. Semenov, 1959. Труды Мин. Муз. Акад. Наук СССР (*Trans. Min. Mus. Acad. Sci. USSR*), vol. 9, p. 107 (Гидрополилитионит). A brownish-white massive variety of the lepidolite 'polylithionite', high in water. An unnecessary name. 22nd List

Hydropyrochlore. E. I. Semenov, A. N. Spitzyn, and Z. N. Burova, 1963. Докл. Акад. Наук СССР (*Compt. Rend. Acad. Sci. URSS*), vol. 150, p. 1128 (Гидропирохлор). A member of the Pyrochlore group from the Lovozero massif with much Nb, little Ta or Ti, and low in bases (mainly Mg, Sr, and Ca), but with much H_2O. 23rd List [A. A. Ivanov, I. B. Borovskii, and I. A. Yaroshch, 1944. *Zap.*, **73**, 56. Vishnevye Mts., Urals; probably altered metamict pyrochlore; *A.M.*, **62**, 406]

Hydropyrolusite, *see* Hydrobraunite.

Hydropyrophyllite, *see* Hydrogedroitzite.

Hydrorinkite. E. I. Semenov, 1969. [Мин. щелочн. масс. Илимаусак. Изд. 'Наука', p. 49 (*Minerals of the Ilimaussaq alkalic massif*)]; abstr. *Zap.*, **99**, 84 (1970) (Гидроринкит). $NaCa_4Ce_2(Ti,Nb)_2Si_5O_{21}(OH)_{12}F_2$, monoclinic; gives an X-ray pattern close to that of rinkite. Cf. Hydrorinkolite (24th List). 26th List

Hydrorinkolite. Synonym of Mosandrite. [Геохим., мин., генет. типы месторожд. редк. элем. Изд. „Наука", 1964, p. 308], abstr. Зап. Всесоюз. Мин. Общ. (*Mem. All-Union Min. Soc.*), 1965, vol. 94, p. 683 (Гидроринколит). 24th List

Hydroromarchite. R. M. Organ and J. A. Mandarino, 1971. *C.M.*, **10**, 916. White crystals formed on tin pannikins lost about 1801 to 1821 in the Winnipeg River, Ontario, at Boundary Falls, are identical with artificial $5SnO.2H_2O$. Anorthic, *a* 11·5, *b* 6·03, *c* 19·8 Å, α 99°, β 60° 30′, γ 88° 30′. (Cf. *Acta Cryst.*, 1961, **14**, 65.) [*Bull.*, **96**, 241; *A.M.*, **57**, 1555; *Zap.*, **101**, 280.] Cf. romarchite. 28th List [$3SnO.H_2O$; *A.M.*, **58**, 552]

Hydroromeite. G. Natta and M. Baccaredda, 1933. *Zeits. Krist.*, vol. 85, pp. 271, 273 (Idroromeite), p. 295 (Hydroromeit). Calcium-bearing antimony-ochres, $CaO.Sb_2O_5.3H_2O$ and $3CaO.2Sb_2O_5.8H_2O$, from Spain, shown by X-ray analysis to have the same cubic structure as roméite ($2CaO.Sb_2O_5$). [*M.A.*, **5**, 294.] 13th List [Is Stibiconite; *A.M.*, **37**, 982]

Hydroscarbroite. W. J. Duffin and J. Goodyear, 1960. *Min. Mag.*, vol. 32, p. 357. A more highly hydrated phase than scarbroite, $Al_2(CO_3)_3.12Al(OH)_3$, occurring along with scarbroite at South Bay, Scarborough, and dehydrating irreversibly on exposure to air. 22nd List [*A.M.*, **45**, 910]

Hydrosericite. E. V. Dimitriev, 1965. [Геол. рудн. месторож., vol. 7, p. 82]; abstr. Зап. всесоюз. мин. общ. (*Mem. All-Union Min. Soc.*), 1967, vol. 96, p. 79 (Гидросерицит, hydrosericite). A superfluous name for a hydrous mica (H_2O^+ 8·04%) from Krivoi Rog. 25th List

Hydroserizit, literal transliteration of Гидросерицит, Hydrosericite (q.v.) (C. Hintze, *Handb. Min.*, Erg. III, 567). 25th List

Hydroserpentine. V. A. Frank-Kamenetsky, 1960. *Clay Min. Bull.*, vol. 4, pp. 162, 168. $Mg_6Si_4O_{10}(OH)_8.nH_2O$; a 'swelling serpentine'. 22nd List

—— H. Strunz, 1970. *Min. Tab.*, 5th edit., 462. (Hydroserpentin). A group name for the *hypothetical* minerals hydroantigorite and hydroamesite (both 22nd List); not the hydroserpentine (of Frank-Kamenetsky, 1960) (22nd List). 28th List

Hydrosialite. N. E. Efremov, 1939. *Compt. Rend.* (*Doklady*) *Acad. Sci. URSS*, vol. 24, p. 287 (hydrosyalites). A collective name for hydrated silicates of aluminium or clay minerals. *See* Sialite. [*M.A.*, **7**, 419.] 15th List

Hydrosodalite. J. Wyart and M. Michel-Lévy, 1949. *Compt. Rend. Acad. Sci. Paris*, vol. 229, p. 131. An artificial hydrothermal product with the same X-ray pattern as sodalite, but with chlorine replaced by water and CO_2. Compare hydroxyl-sodalite (18th List). [*M.A.*, **11**, 93.] 19th List

—— K. A. Vlasov, M. V. Kuzmenko, and E. M. Eskova, 1959. [The Lovozero Alkaline Massif. (*Moscow, Acad. Sci. USSR*), p. 272]; abstr. *Amer. Min.*, 1960, vol. 45, p. 1131. *See also* V. I. Gerasimovsky, A. I. Polyakov, and L. P. Voronina, Доклады Акад. Наук СССР (*Compt. Rend. Acad. Sci. URSS*), 1960, vol. 131, p. 402. A constituent of the syenites of the Lovozero massif, Kola peninsula, is variously formulated $Na_{10}Al_6Si_6(O,OH)_{25}(O,Cl)_2$ and $2NaAlSiO_4.H_2O$. Perhaps a variety of sodalite with Cl′ partially or wholly replaced by OH′; its relation to lembergite (of Lagorio; 1st List) and to the hydrosodalite of Wyart and Michel-Lévy (19th List), an artificial product containing OH′ and CO_3'' in place of Cl′, remains uncertain. 22nd List

Hydrotenorite. L. De Leenheer, 1937. *Bull. Soc. Belge Géol.*, vol. 47, p. 215 (hydroténorite); ibid., 1938, vol. 48, p. 343. An amorphous black mineral (pitchy copper ore) from Katanga; deducting, chrysocolla, the formula is given as $4CuO{\cdot}H_2O$. V. Billiet and A. Vandendriessche, *Bull. Soc. Belge, Géol.*, 1938, vol. 48, p. 333, prove its identity with tenorite. [*M.A.*, **7**, 10, 226.] 15th List

Hydrothionite. G. Vavrinecz, 1939. *Földtani Közlöny*, Budapest, vol. 69, pp. 82, 98 (Hydrothionit). Hydrogen sulphide, H_2S, as a natural gas. Named from *ὕδωρ*, water, and *θεῖον*, sulphur. [*M.A.*, **7**, 471.] 15th List

Hydrothomsonite. K. D. Glinka, 1906. *Trav. Soc. Nat. Saint-Pétersbourg*, vol. xxxiv, *Sect. Géol. Min.*, p. 61 (Гидротомсонитъ), p. 176 (Hydrothomsonit). A zeolite, $(H_2,Na_2,Ca)Al_2Si_2O_8.5H_2O$, differing from thomsonite in containing more water (29·8%). Small, white or colourless, prismatic crystals with straight optical extinction. Occurs in soil as a product of weathering of augite-andesite in Transcaucasia. 5th List

Hydrothorite. E. S. Simpson, 1928. *Journ. Roy. Soc. Western Australia*, vol. 13 (for 1927), p. 37. Hydrous thorium silicate, $ThSiO_4.4H_2O$, as a pale pink earthy, optically isotropic mineral resulting from the decomposition of mackintoshite; from Western Australia. The composition is that of a hydrated thorite. [*M.A.*, **3**, 544.] 11th List. [Cf. Mozambikite.] E. I. Semenov and M. E. Kazakova, 1961. [Труды Инст. Мин. Геохим. Крист. Редк. Элем. (*Trans. Inst. Min., Geochem., Cryst. Rare Elem.*), no. 7, p. 123], abstr. in Зап. Всесоюз. Мин. Общ (*Mem. All-Union Min. Soc.*), 1964, vol. 93, p. 455 (Гидроторит). Metamict hydrous Thorite (near $ThSiO_4.4H_2O$) from the Lovozero massif. An unnecessary name [for Thorogummite, *A.M.*, **38**, 1007.] 23rd List

Hydrotroilite. M. Sidorenko, 1901. *Mém. Soc. Nat. Nouv.-Russ. Odessa*, vol. xxiv, part i, p. 119 (гидротроилитъ); abstr. in *Neues Jahrb. Min.*, 1902, vol. ii, ref. p. 397 (Hydrotroilit). Hydrated iron sulphide, $FeS.H_2O$, occurring in the black mud of inland seas. 6th List [In part Mackinawite (q.v.), *J. Geol.*, **72**, 293]

Hydrotungstite. P. F. Kerr and F. Young, 1940. *Amer. Min.*, Program and Abstracts, December 1940, p. 9; *Amer. Min.*, 1941, vol. 26, p. 199. Hydrous tungstic oxide, $WO_3.2H_2O$, similar in appearance but distinct from tungstite ($WO_3.H_2O$); from Oruro, Bolivia. [*M.A.*, **8**, 52.] 16th List [*A.M.*, **29**, 192]

Hydrougrandite. Tsao Yung Lung, 1964. [*Acta Geol. Sinica*, vol. 44, p. 219]; abstr. *Bull. Soc. franç. Min. Crist.*, 1965, vol. 88, p. 359; *M.A.*, **17**, 400. A hydrogarnet [q.v.] $(Ca,Mg)_3(Fe,Al)_2[SiO_4,(OH)_4]_3$, with Ca:Mg ≈ 2, Fe $\approx$ Al, and Si $\approx 2\frac{1}{4}$. [Complete miscibility between the four end-members grossular, andradite, $Ca_3Al_2(OH)_{12}$, and $Ca_3Fe_2(OH)_{12}$ appears to be possible artificially (R. H. Boyne, *Chem. Portland Cement*, New York, 1947).] 24th List [*A.M.*, **50**, 2100]

Hydro-wollastonite. A. S. Eakle, 1917. *Bull. Dep. Geol. Univ. California*, vol. x, p. 348. A general term for the hydrated calcium metatsilicates crestmoreite and riversideite (qq.v.). 8th List

Hydroxidsodalith, variant of Hydroxylsodalite (18th List) (H. Strunz, *Min. Tabellen*, 3rd edit., 1957, p. 382). 24th List

Hydroxy-amphibole. N. L. Bowen and J. F. Schairer, 1935. *Amer. Min.*, vol. 20, p. 547 (hydroxy-amphiboles). *See* Fluor-amphibole. 14th List

Hydroxyapatite. W. T. Schaller, 1912. *Bull. US Geol. Survey*, no. 509, p. 100. A hypothetical constituent of apatite, $9CaO.3P_2O_5.CaO.H_2O$, corresponding with fluor-apatite $9CaO.3P_2O_5.CaF_2$. Cf. Voelckerite. 6th List [Also hyphenated; Cf. Hydroxylapatite]

Hydroxyapophyllite. P. J. Dunn, R. C. Rouse, and J. A. Norberg, 1978. *Am. Mineral.*, **63**, 196. P. J. Dunn and W. Wilson, *Mineral Rec.*, **9**, 95 (1978). The pure end-member of the apophyllite series occurs at Great Notch, New Jersey. [*M.A.*, 78-3472.] 30th List

Hydroxybraunite, *see* Hydrobraunite.

Hydroxykeramohalite. A. Dubanský, 1956. *Chem. Listy*, Praha, vol. 50, p. 1350 (hydroxykeramohalit). $Al_4(SO_4)_5(OH)_2.21H_2O$. Derivative of keramohalite (or alunogen), from Dubnik, Slovakia. 21st List

Hydroxylacharite, *see* Hydroxyl-ascharite. 25th List

Hydroxyl-annite, Hydroxyl-biotite, etc. D. P. Grigoriev, 1935. *See* Fluor-annite. 14th List

Hydroxylapatite. P. Niggli, *Lehrbuch der Mineralogie*, 2nd edit., 1926, p. 393 (Hydroxylapatit); A. N. Winchell, 1927, *Elements of optical mineralogy*, 2nd edit., 1927, part 2, p. 128 (Hydroxylapatite). Variant of hydroxyapatite. (6th List), $3Ca_3P_2O_8.Ca(OH)_2$. 15th List [Also hyphenated]

Hydroxyl-ascharite. A. P. Grigoriev*, A. A. Brovkin, and I. Ya. Nekrasov, 1966. Доклады акад. наук СССР (*Compt. Rend. Acad. Sci. URSS*), vol. 166, p. 937 (Гидроксилашарит). An artificial hydrothermal product near szájbelyite (ascharite) is formulated $Mg_2(B,H_3)_2O_5.H_2O$; it also occurs naturally in veins cutting ludwigite in a Siberian skarn deposit. [M. Fleischer, *Amer. Min.*, 1966, vol. 51, p. 1819, considers the substitution $B \rightleftharpoons H_3$ improbable and the evidence inadequate.] [Зап., **96**, 72; hydroxylacharite.] [*Corr. in *M.M.*, **39**.] 25th List

Hydroxyl-bastnäsite. A. S. Kirilov, 1964. Докл. Акад. наук СССР (*Compt. Rend. Acad. Sci. URSS*), vol. 159, p. 1048. [Гидроксил-бастнезит]; Зап. Всесоюз. Мин. Общ. (*Mem. All-Union Min. Soc.*), 1966, vol. 95, p. 51. The hydroxyl analogue of bastnäsite occurs as reniform aggregates in cavities of late carbonatite veins in an unnamed massif of alkalic ultrabasic rocks; $LnCO_3(OH,F)$ with $OH > F$; the lanthanons are mainly La and Ce. [*M.A.*, **17**, 303; *A.M.*, **50**, 805, Hydroxyl-bastnaesite.] 24th List

Hydroxyl-biotite, *see* Fluor-annite.

Hydroxylellestadite. K. Harada, K. Nagashima, K. Nakao, and A. Kato, 1971. *Amer. Min.*, **56**, 1507. An ellestadite from the Chichibu mine, Saitama Prefecture, Japan, has $OH > (Cl + F)$, and is accordingly a hydroxylellestadite. [*M.A.*, 72-1401.] 27th List [Also hyphenated]

Hydroxyl-herderite. Dana's *Mineralogy*, 7th edit., 1951, vol. 2, p. 820. The same as hydroherderite of S. L. Penfield, 1894 (2nd List). 19th List

Hydroxyl-lepidomelane, -meroxene, -phlogopite, *see* Fluor-annite.

Hydroxylpyromorphit. H. Wondratschek, 1963. *Neues Jahrb. Min.*, Abh., vol. 99, p. 147. Synonym of Lead hydroxyapatite. 23rd List

Hydroxyl-siderophyllite, *see* Fluor-annite.

Hydroxyl-sodalite. W. Borchert and J. Keidel, 1947. *Heidelberg. Beitr. Min. Petr.*, vol. 1, p. 3 (Hydroxyd-Sodalith, Hydroxydsodalith). An artificial product $Na_8Al_6Si_6O_{24}(OH)_2$.aq, with the crystal-structure of sodalite (Chloridsodalith). [*M.A.*, **10**, 364.] 18th List

Hydroxyl-szájbelyite, variant of Hydroxyl-ascharite (q.v.). *Min. Abstr.*, 1967, vol. 18, p. 126. 25th List

Hydroxyl-topaz. W. E. Tröger, 1952. *Tabellen opt. Bestim. gesteinsbild. Minerale*, p. 146 (Hydroxyltopas). Synonym of topaz. [*M.A.*, **12**, 4.] 20th List

Hydroxymimetite. C. C. McDonnell and C. M. Smith, 1917. *Journ. Amer. Chem. Soc.*, vol. 39, p. 940 (Hydroxy Mimetite). Artificially prepared basic lead arsenate, $Pb_4(PbOH)(AsO_4)_3.H_2O$, as hexagonal crystals resembling mimetite, but with water of crystallization. 11th List

Hydrozircon. R. G. Coleman and R. C. Erd, 1961. *US Geol. Surv. Prof. Paper, no. 424–C*, p. 297. Another unnecessary name for an altered, hydrous zircon. Also applied (E. Eberhard, *Fortschr. Min.*, 1961, vol. 39, p. 340, Hydrozirkon) to $Zr\{SiO_4,(OH)_4\}$, in analogy with Hydrogrossular. [*M.A.*, **16**, 549.] 23rd List

Hydrozunyite. A. Baumer, H. Gimeney, R. Caruba, and G. Turco, 1974. *Bull.*, **97**, 271. A synthetic zunyite with $(OH)_4$ replacing SiO_4. 29th List

Hyperperthite, Hyperoranite. H. L. Alling, 1921. *See* Hypoperthite. 14th List

Hyperythrin. J. Fröbel, 1843. *See* chrysargyrit. 27th List

Hypo-oranite, *see* Hypoperthite.

Hypoperthite, Hypo-oranite, etc. H. L. Alling, *Journ. Geol. Chicago*, 1921, vol. 21, pp. 234, 253, 254. Hypoperthite, Eutectoperthite, and Hyperperthite (Hyperthite) for intermediate members of the Or-Ab series. Hypo-oranite, Eutecto-oranite, and Hyperoranite for the Or-An (oranite) series. Paraperthite and Para-oranite contain also some An and Ab respectively. 14th List

Hyposiderite. A. Breithaupt, 1847. *Handbuch der Mineralogie*, vol. iii, p. 894 (Picites Hyposiderites; Hyposiderit). A black, shining variety of limonite differing from stilpnosiderite in its lower specific gravity (3·30–3·38) and higher percentage of water (20%). Named from ὑπό, less, and σίδηρος, iron. 6th List

I

Ianthinite. A. Schoep, 1926. *Natuurwetenschappelijk Tijdschrift*, vol. 7 (for 1925), p. 97; 1927, vol. 9, p. 1 (janthiniet); *Ann. Soc. Géol. Belgique*, 1927, vol. 49 (for 1926), *Bull.*, pp. B 188, B 310 (ianthinite). Hydrated uranium dioxide, $2UO_2.7H_2O$, as violet orthorhombic crystals associated with pitchblende from Katanga. Named from ἰάνθινος, violet-coloured. [*M.A.*, **3**, 232–233; **3**, 276; **3**, 370; **3**, 485.] 11th List [$UO_2.5UO_3.10\frac{1}{2}H_2O$; *M.A.*, **14**, 280; *A.M.*, **44**, 1103; cf. Epi-ianthinite]

Idaite. G. Frenzel, 1959. *Neues Jahrb. Min.*, Monatsh., p. 142 (Idait). A sulphide mineral, Cu_5FeS_6, hexagonal and related to covelline, occurring with bornite at the Ida mine, Khan, SW Africa; named from the locality. [*M.A.*, **14**, 279.] Perhaps a ferroan covelline, E. N. Eliseev, Зап. Всесоюз. Мин. Общ. (*Mem. All-Union Min. Soc.*), 1960, vol. 89, p. 128; cf. G. Frenzel, ibid., p. 490, and *Schweiz. Min. Petr. Mitt.*, 1960, vol. 40, p. 243. 22nd List [Cu_3FeS_4, *M.A.*, **19**, 310; 71-2277; 76-2817: $Cu_{5.5}FeS_{6.5}$, Sulphide Mineralogy (M.S.A. 1974), CS-65, CS-73; formerly known as 'orange bornite'. Cf. Mawsonite]

Iddingsite. A. C. Lawson, *Bull. Dept. Geol. Univ. Calif.*, I. 31, 1893. F. L. Ransome, *ibid.*, I. 90, 1893. H. H. Arnold-Bemrose, *Quart. Journ. Geol. Soc.*, L. 617, 1894. [*Min. Mag.*, X. 264.] Hydrated silicate of Fe,Ca,Mg and Na. Rhombic. Possibly a pseudomorph after olivine [Carmelo Bay] California. 1st List [A mixture; *A.M.*, **46**, 92; *M.M.*, **35**, 55]

Idrizite. A Schrauf, *Jahrb. k. k. geol. Reichsanst.*, XLI. 379, 1892. $(Mg_6,Fe)SO_4 + (Al_5,Fe_3)_2S_2O_9 + 16H_2O$. Nearer to quetenite and botryogen than to pickeringite and halotrichite. Idria, Carniola. 1st List

Idromelanotallite. *See* Hydromelanothallite. 6th List

Idroromeite. *See* Hydroromeite. 13th List

Igalikite. O. B. Bøggild, 1933. *Meddelelser om Grønland*, vol. 92, no. 9. Approximate composition $NaKAl_4Si_4O_{15}.2H_2O$, as a polysynthetic aggregate of minute scales. From near Igaliko, S Greenland. Named from the locality. [*M.A.*, **5**, 484.] 13th List [A mixture, *Medd. Grøn.* **162**, 1]

Igdloite. M. Dano and H. Sørensen, 1959. *Medd. Grønland*, vol. 162, no. 5, p. 27; *see also* A. Safiannikoff, *Bull Acad. Roy. Sci. d'Outre-mer* (*Bruxelles*), 1959, new ser., vol. 5, p. 1253. An imperfectly described mineral of perovskite type from Igdlúnguaq, Greenland, formed during the alteration of eudialyte; spectroscopic analysis shows major amounts of Ti, Nb, Na, Ca, and Al. Perhaps essentially $NaNbO_3$, but with some SiO_2, differing in this and its mode of formation from Lueshite (q.v.). The differences seem hardly adequate to justify separate names. 22nd List [*A.M.*, **46**, 1004]

Igmerald. M. Jaeger and H. Espig, 1935. *Deutsche Goldschmiede-Zeitung*, vol. 38, p. 347 (Igmerald). W. F. Eppler, ibid., 1935, vol. 38, no. 15, p. —. B. W. Anderson, *Gemmologist*, London, 1935, vol. 4, pp. 212, 295. H. Espig, *Zeits. Krist.*, 1935, vol. 92, p. 387 (synthetische Smaragd). Trade-name for artificial emerald made by the I. G. [Interessengemeinschaft] Farbenindustrie at Bitterfeld, Germany. Named from I. G. and emerald. [*M.A.* **6**, 200, 497.] 14th List

Iimoriite. A. Kato and K. Nagashima, 1970. *An Introduction to Japanese Minerals*, 39 and 85. Purplish-grey masses from Fusamata, Kawamatamachi, Fukushima Prefecture, Japan, are anorthic, *a* 11·6, *b* 6·65, *c* 26·2 Å, α 94·3°, β 95·0°, γ 93·6°. Composition near $Yt_5(SiO_4)_3(OH)_3$. Named for S. Iimori and T. Iimori. [*Zap.*, **101**, 287.] Cf. Victoria-stone. 28th List [*A.M.*, **58**, 140]

Ikaite. H. Pauly, 1963. [*Naturens. Verden*, June, 1963], abstr. in *Amer. Min.*, 1964, vol. 49, p. 439. Chalky material from skerries in Ika Fjord, Ivigtut, Greenland, prove to be $CaCO_3.6H_2O$, close to the synthetic material. Named for the locality. (The names Lublinite, Hydroconite, Hydrocalcite, Protocalcite are discarded because their true connotation is doubtful.) 23rd List

Ikunolite. A. Kato, 1959. *Min. Journ.* (*Japan*), vol. 2, p. 397. $Bi_4(S,Se)_3$, with S:Se near 12, rhombohedral, from the Ikuno mine, Hyôgo prefecture, Japan; very similar to josëite but contains no Te. Named from the locality. [*M.A.*, **15**, 43.] Cf. Laitakarite. 22nd List [*A.M.*, **45**, 477]

Illidromica, *see* Illite-hydromica, Andreattite.

Ilimaussite. E. I. Semenov, M. E. Kazakova, and V. I. Bukin, 1968. *Medd. Grønland*, **181**, no. 6, 3. Lamellar aggregates of hexagonal crystals from Nakalaq, Ilimaussaq, S Greenland, near $Ba_2Na_4CeFeNb_2Si_8O_{28}.5H_2O$. Named for the locality. [*A.M.*, **54**, 992; *M.A.*, 69-3352; *Zap.*, **99**, 82.] 26th List

Illite. R. E. Grim, R. H. Bray, and W. F. Bradley, 1937. *Amer. Min.*, vol. 22, p. 816. A general term for a micaceous constituent of argillaceous sediments, in particles usually less than a micron across, with the approximate formula $2K_2O.3(Mg,Fe)O.8(Al,Fe)_2O_3.24SiO_2.12H_2O$. Named from the State of Illinois. (Cf. pholidoide and phyllite, 14th List.) [*M.A.*, **7**, 13, 422.] 15th List [A group name]

Illite-chlorite, Illite-montmorillonite. P. Gallitelli, 1959. *Com. Internat. stud. Arcillas*, p. 23. Irregular interstratifications of illite and chlorite or montmorillonite layers. 24th List

Illite-hydromica. C. Andreatta, 1949. *Periodico Min. Roma*, vol. 18, p. 11 (serie illitiidromiche), p. 14 (illidromica). Titles and abstr., *Internat. Geol. Congress*, 18th session, London, 1948, p. 126 (illite-idromica). *Clay Minerals Bull.*, 1949, no. 3, p. 98 (illidromica, illite-hydromica). *Rend. Soc. Min. Ital.*, 1950, vol. 6, p. 40. A micaceous mineral intermediate between illite and hydromica. [*M.A.*, **11**, 172, 347.] 19th List

Illite-montmorillonite, *see* Illite-chlorite.

Ilmaiokite. I. V. Bussen, L. F. Gannibal, E. A. Goiko, A. N. Merkov, and A. P. Nedorezova, 1972. *Zap.*, **101**, 75 (Ильмайокит, Ilmayokite). White crystals in cavities of natrolite in the Karnasurt Mts., Ilmaoik valley, Lovozero Tundra, Kola Peninsula, coated with sodium carbonate, are probably monoclinic. Possibly near $Na_2TiSi_3O_9.6H_2O$. Further study is necessary. 27th List

Ilmajokite, variant (German style) transliteration of Ильмайокит, Ilmaiokite (27th List), *M.A.*, 73-807; *A.M.*, **58**, 139 (1974); *Zap.*, **102**, 453. 28th List

Ilmenitglimmer. H. Rosenbusch, 1905. *Mikroskopische Physiographie d. Mineralien*, 4th edit., vol. i, part 2, p. 81. The same as Titaneisenglimmer (q.v.). 4th List

Ilmeno-corundum. J. de Lapparent, 1937. *Min. Petr. Mitt.* (*Tschermak*), vol. 49, p. 15 (ilméno-corindon, corindon ferro-titané). Corundum in the emery of Samos, previously called taosite (14th List), but not analysed. [*M.A.*, **7**, 150.] 15th List

Ilmenomagnetite. A. F. Buddington, J. [J.] Fahey, and A. [C.] Vlisidis, 1953. *Progr. and Abstr. Min. Soc. Amer.*, 1953, p. 13; *Amer. Min.*, 1954, vol. 39, p. 318 (abstr.). Titaniferous magnetite containing exsolution ilmenite as distinct from any in solid solution. Compare titanomagnetite (2nd List). 20th List

Ilminite, error for Ilmenite (*Nature Phys. Sci.*, **237**, 105 (1972). 29th Lst

Imanite. A. V. Rudneva, 1958. [Акад. Наук СССР, Инст. Геол. Руд. Месторожд., Петрог., Мин., Геохим., Инст. Хим. Силикат., p. 285]: abstr. *Amer. Min.*, 1959, vol. 44, p. 907. The artificial cubic compound $Ca_3Ti_2Si_3O_{12}$. Named from the initials of Inst. Metallurg., Akad. Nauk. [*M.A.*, **15**, 45.] 22nd List

Imerinite. A. Lacroix, 1910. *Minéralogie de la France*, vol. iv, p. 787. A soda-amphibole containing only a small amount of sesquioxides and so allied to soda-richterite. It occurs as colourless to blue needles, resembling tremolite in appearance, in crystalline limestone in the central province Imerina of Madagascar. 6th List

Imgreïte. O. E. Yushko-Zakharova, 1964. Докл. Акад. наук СССР (*Compt. Rend. Acad. Sci. URSS*), vol. 154, p. 613 (Имгреит). A superfluous name for an unanalysed member of the melonite–NiTe solid solution series, possibly near the NiTe end, from the Nittis-Kumuzhya deposit, Monchegorsk. Named for the Institute of Mineralogy, Geochemistry, and Crystal Chemistry of Rare Elements (IMGRE). [*M.A.*, **16**, 647; *Amer. Min.*, **49**, 1151 (Imgreite).] 24th List

Imhofite. G. Burri, S. Graeser, F. Marumo, and W. Nowacki, 1965. [*Chimia*. vol. 19, p. 499); abstr. *Amer. Min.*, 1966, vol. 51, p. 531. Thin translucent copper-red plates from Lengenbach, Binn, Valais, Switzerland, are monoclinic, composition near $Tl_6CuAs_{15}S_{41}$*.

Named for J. Imhof. M. Fleischer comments that while it is probably a valid mineral, the X-ray data should be published. [*Corr. in *M.M.*, **39**] 24th List [X-ray data; *A.M.*, **54**, 1498]

Imogolite. N. Yoshinaga and S. Aomine, 1962. [*Soil Sci. and Plant Nutrition* (*Japan*), vol. 8, pp. 6 and 114], abstr. in *Amer. Min.*, 1963, vol. 48, p. 434. An imperfectly described mineral from Japanese soils, believed 'to have a more ordered structure than allophane'. 23rd List

Impsonite. G. H. Eldridge, 1901. *22nd Ann. Rep. United States Geol. Survey, 1900–1901*, part i, p. 265. Described by J. A. Taff (*Amer. Journ. Sci.*, 1899, ser. 4, vol. viii, p. 219) as an albertite-like asphaltum occurring in the Choctaw Nation, Indian Territory: Impson Valley is one of the localities mentioned. It differs from albertite in being almost insoluble in turpentine. 5th List

Incaite. E. Mackovicky, 1974. *Neues Jahrb. Min.*, Monatsh., 235. Fine lamellae replacing cylindrite from Poopo, Bolivia, are composed of alternating layers, one pseudotetragonal, *a* 86·23, *b* 5·79, *c* 34·98 Å, α 90°, β 90·28°, γ 90°, the other pseudo-hexagonal, *a* 258·7, *b* 3·66, *c* 69·85 Å, α 90°, β 90·28°, γ 90°. Composition near $96[(Pb,Ag)_4Sn_4Sb_2FeS_{14}]$, Named for the Incas. [*A.M.*, **60**, 486; *M.A.*, 75-1391; *Zap.*, **104**, 607.] 29th List

Inderborite. G. S. Gorshkov, 1941. *Compt. Rend.* (*Doklady*) *Acad. Sci. URSS*, vol. 33, p. 255 (inderborite). Hydrous calcium and magnesium borate, $CaMgB_6O_{11}.11H_2O$, as monoclinic crystals from the Inder borate deposits, Kazakhstan. Named from the locality. *See* Metahydroboracite. [*M.A.*, **8**, 341.] 16th List [$CaMg[B_3O_3(OH)_5]_2$; *M.A.*, **17**, 390, 735]

Inderite. A. M. Boldyreva and E. N. Egorova*, 1937. [*Mat. Centr. Sci. Geol. Prosp. Inst. USSR*, General Ser., no. 2.] A. M. Boldyreva, *Mém. Soc. Russe Min.*, 1937, ser. 2, vol. 66, p. 651 (Индерит), p. 672 (Inderite); M. N. Godlevsky, ibid., pp. 327, 355. O. M. Shubnikova, *Trans. Lomonosov Inst. Acad. Sci. USSR*, 1937, no. 10, p. 216. G. B. Boky, *Bull. Acad. Sci. URSS, Cl. Sci. Mat. Nat., Ser. Chim.*, 1937, p. 881 (Индерит), p. 883 (Inderite). Hydrated magnesium borate, $2MgO.3B_2O_3.15H_2O$, as nodular aggregates of acicular crystals (pseudo-orthorhombic). Named from the locality, Inder (Индер), Kazakhstan, SW Siberia. [*M.A.*, **7**, 121–123, 476.] [*Corr. in., *M.M.*, **25**.] 15th List [Syn. Lesserite (q.v.). $Mg[B_3O_3(OH)_5.4H_2O].H_2O$; *M.A.*, **16**, 133; **17**, 20. Dimorph of Kurnakovite (q.v.)]

Indialite. A. Miyashiro and T. Iiyama, 1954. *Proc. Japan Acad.*, vol. 30, p. 746. Artificial $Mg_2Al_4Si_5O_{18}$, hexagonal, distinct from orthorhombic cordierite. Also found in sediments fused by a burning coal seam in India. Named from the locality. *See* Osumilite. [*M.A.*, **12**, 513, 615.] 20th List [*A.M.*, **40**, 787]

Indigirite. L. N. Indolev, Yu. Ya. Zhdanov, K. I. Kashertseva, V. S. Soknev, and K. I. Del'yanidi, 1971. *Zap.*, **100**, 178 (Индигирит)*. Radiating fibrous aggregates from the Sarylakh gold–antimony deposit, Indigirki river, NE Yakutia, have α 1·472, γ 1·502 ∥ elongation, D 1·6. Composition $Mg_2Al_2(CO_3)_4(OH)_2.15H_2O$. [*A.M.*, **57**, 326; *M.A.*, 72-548.] [*Corr. in *M.M.*, **39**.] 27th List

Indite. A. D. Genkin and I. V. Muraveva, 1963. Зап. Всесоюз. Мин. Общ. (*Mem. All-Union Min. Soc.*), vol. 92, p. 445 (Индит). Minute iron-black isotropic grains in cassiterite from the Dzhalind deposit, Little Khingan ridge, Far Eastern Siberia, have a composition near $FeIn_2S_4$. and X-ray data agree closely with synthetic material. Named from the composition. [*M.A.*, **16**, 457.] 23rd List [*A.M.*, **49**, 439]

Indium. V. V. Ivanov, 1964. [Индий самородный. В сб: Хеохим., мин., генет. типы месторожд. редк. элем. Изд. „Наука", vol. 2, p. 568]; abstr. Зап. Всесоюз. Мин. Общ. (*Mem. All-Union Min. Soc.*), 1965, vol. 94, p. 665 (Индий, indium). Native indium is said to occur in granites of E Transbaikal, in association with native lead. 24th List [*A.M.*, **52**, 299]

Innelite. S. M. Kravchenko, 1960. Quoted in Yu. A. Balashov and N. V. Turanskaya, [Геохимия (Geokhimya), 1960, no. 7, p. 618]; abstr. *Amer. Min.*, 1961, vol. 46, p. 769. Mention is made of a new barium silicate from a pegmatite from the Inagli massif, Aldan, containing about 40% BaO, to be described by Kravchenko. Status doubtful pending full description. 22nd List [$Ba_4Ti_{3.5}Si_4O_{18}(OH)_{1.5}.Na_2SO_4$; *M.A.*, **15**, 542; *A.M.*, **47**, 805]

Insizwaite. L. J. Cabri and D. C. Harris, 1972. *Min. Mag.*, **38**, 794. Tiny grains with pentlandite, chalcopyrite, parkerite, and niggliite in the pyrrhotine ore from the Insizwa deposit, Waterfall Gorge, Pondoland, South Africa, are cubic with *a* 6·625 Å, composition $Pt_{1.00}Bi_{1.35}Sb_{0.57}$. This is near the Sb-rich end of the solid-solution series $Pt(Bi,Sb)_2$. The name, from the locality is proposed for $PtBi_2$, the analysed material being an antimonian insizwaite. [*M.A.*, 72-3342.] 27th List [*A.M.*, **58**, 805]

Inyoite. W. T. Schaller, 1914. *Journ. Washington Acad. Sci.*, vol. iv, p. 355; 3rd Appendix to 6th edit. of Dana's *System of Mineralogy*, 1915, p. 41. Hydrated calcium borate, $2CaO.3B_2O_3.13H_2O$, found as large, colourless, monoclinic crystals in the borate deposits in Inyo County, California. Named from the locality. The material is largely altered to Meyerhofferite (q.v.). 7th List

Iocite, erroneous transliteration of Иоцит, Iozite (10th List). *Dokl. Acad. Sci. USSR* (*Earth Sci. Sect.*), 1969, **180**, 133 (translation of Докл. Акад. наук СССР, 1968, **180**, 449). [*M.A.*, 71-1353.] 28th List

Iodatacamite. P. Chirvinsky, 1906. *Bull. Univ. Kiev*, p. 1. (Иодистый атакамитъ); *Zeitschr. Kryst. Min.*, 1909, vol. 46, p. 293 (Jodatacamit). The artificial compound $CuI(OH)_3$, subsequently shown to be an analogue of Botallackite, not of Atacamite, and renamed accordingly. 23rd List

Iodbotallackite. H. R. Oswald, 1961. *Helvet. Chim. Acta*, vol. 44, pp. 2102 (Iodbotallackit) 2103 (Iodobotallackit). The artificial compound $CuI(OH)_3$, formerly termed Iodatacamite (q.v.) is really the iodine analogue of Botallackite and is renamed accordingly. 23rd List

Iodembolite. G. T. Prior and L. J. Spencer, *Centralblatt Min.*, 1902, p. 186 (Jodembolit); *Min. Mag.*, 1902, vol. xiii, p. 176 (iodiferous embolite or iodembolite). To replace the name iodobromite (2AgCl.2AgBr.AgI, which has no existence as a definite compound), and as a general term for minerals of the cerargyrite group (holohedral-cubic silver haloids) containing chlorine, bromine, and iodine, Ag(Cl,Br,I). 3rd List [Iodian Bromargyrite]

Iodomimetite. H. Wondratschek, 1963. *Neues Jahrb. Min.*, Abh., vol. 99, p. 125 (Jodmimetesit). The artificial compound $Pb_5(AsO_4)_3I$; apatite family. 23rd List

Iodopyromorphite. H. Wondratschek, 1959. *Zeits. anorg. Chem.*, vol. 300, p. 41 (Blei-Jod-Apatit). Attempts to synthesize the compound $Pb_5(PO_4)_3I$ failed. 23rd List

Iodovanadinite. H. Wondratschek, 1963. *Neues Jahrb. Min.*, Abh., vol. 99, p. 125 (Jodvanadinit). The artificial compound $Pb_5(VO_4)_3I$; apatite family. 23rd List

Iolanthite. (D. B. Sterrett, *Gems and precious stones in 1914, Mineral Resources, US Geol. Survey, for 1914, 1915*, part II, p. 323; W. T. Schaller, ibid., for 1917, 1918, part II, p. 155.) Trade-name for a jasper-like mineral from Crooked River, Crook Co., Oregon, placed on the gem market by Mr. Don Maguire. 8th List.

Ionite. V. T. Allen, 1927. *Amer. Min.*, vol. 12, p. 78. Hydrous aluminium silicate, $2Al_2O_3.6SiO_2.5H_2O$, occurring in the Ione formation of California. Later (ibid., 1928, vol. 13, p. 145) identified with anauxite. Not the ionite of S. Purnell, 1878. [*M.A.*, **3**, 370; **3**, 487.] 11th List

Iosene. *Brit. Chem. Abstr. A*, 1929, p. 1429; and *Neues Jahrb. Min. Ref. I*, 1930, p. 126 (Iosen); from A. Soltys, *Sitzungsber. Akad. Wiss. Wien, Math.-naturw. Kl.*, Abt. IIb, 1929, vol. 138 (Supplement), p. 175 = *Monatshefte Chem. Wien*, vol. 53–54, p. 175 (Josen). Error for Josen [10th List], a synonym of hartite. The name Josen was used in manuscript by the late Prof. K. B. Hofmann, of Graz, after his wife Josepha, and was first published by F. Machatschki in 1924. [*M.A.*, **2**, 474; **4**, 349.] 12th List

Iosiderite. *See* Iozite. 10th List

Iowaite. D. W. Kohls and J. L. Rodda, 1967. *Amer. Min.*, vol. 52, p. 1261. Bluish-green crystals in drill core, from a Precambrian serpentinite from Sioux County, Iowa, have a hexagonal unit cell containing $4Mg(OH)_2.FeOCl.4H_2O$; the water is zeolitic. Possibily related to the pyroaurite family. Named for the locality. [*Bull.*, **91**, 99.] 25th List

Iozite. A. Brun, 1924. *Arch. Sci. Phys. Nat. Genève*, ser. 5, vol. 6, pp. 253, 263; *Compt. Rend. Soc. Phys. Hist. Nat. Genève*, vol. 41, p. 94; *Schweiz. Min. Petr. Mitt.*, vol. 4, p. 355. Ferrous oxide (FeO) present in a free state as minute black granules in volcanic rocks. On p. 253 (foot-note) the more correct form *Iosidérite* (from ἰὸς σιδήρον, rust of iron) is also mentioned, but preference is given to the shorter form *Iozite*. As a generic term *iozites* includes gradations to magnetite containing an excess of FeO. A German form of the name is Jozit (abstr. in *Chem. Zentr.*, 1924, vol. 2, p. 2519). [*M.A.*, **2**, 383.] 10th List [Cf. Wüstite]

Iranite. P. Bariand and P. Herpin, 1963. *Bull. Soc. franç. Min. Crist.*, vol. 86, p. 133. $PbCrO_4.H_2O$, found in very small amount at the Sebarz mine, NE of Anarak, Iran. Saffron yellow, anorthic. Named for the country. [*M.A.*, **16**, 374; *A.M.*, **48**, 1417.] 23rd List [*Cf. Bull. Brit. Mus. Nat. Hist.* (*Miner.*), **2**, no. 8, 409. Formula uncertain. Cf. Hemihedrite]

Iraqite. A. Livingstone, D. Atkin, D. Hutchinson, and H. M. Al-Hermezi, 1976. *M.M.*, **40**, 441. Pale greenish-yellow massive material from Shakhi-Rash Mtn., Qala-Diza, Iraq, is a member of the ekanite group; space group *P4/mcc*, *a* 7·61, *c* 14·72 Å. Composition $(Ln_{1\cdot23}Th_{0\cdot66}X_{0\cdot15})K_{1\cdot07}(Ca_{3\cdot49}Ln_{0\cdot35}Na_{0\cdot16})(Si_{15\cdot69}Al_{0\cdot27})(O_{39\cdot93}F_{0\cdot07})$, where *X* includes U, Pb, Zr, etc. Named for the country of origin. [*M.A.*, 76-1986.] 29th List [Cf. Kanaekanite.]

Irarsite. A. D. Genkin, N. N. Zhuravlev, N. V. Troneva, and I. V. Muraveva, 1966. Зап. всесоюз. мин. общ. (*Mem. All-Union Min. Soc.*), vol. 95, p. 700 (Ирарсит, Irarcite). Cubic (Ir,Ru,Rh,Pt)AsS occurs in chromite in dunite from Onverwacht, South Africa. Named from the composition (Ir,ars[enic]). [*M.A.*, **18**, 283; *A.M.*, **52**, 1580; Зап. **97**, 67; *Bull.*, **90**, 271.] 25th List

Iraurita. G. Gagarin and J. R. Cuomo, 1949, loc. cit., p. 5. Spanish form of iridium-gold (O. E. Zvyagintzev, 1934) [*M.A.*, **6**, 51], иридистое золото (V. I. Vernadsky, 1908), iridic gold (M. H. Hey, 1950). 19th List

Irhtemite. R. Pierrot and H.-J. Schubnel, 1972. *Bull.*, **95**, 365. Spherulites of fine monoclinic needles from Irhtem, Bou Azzer, Anti-Atlas, Morocco, have *a* 16·73, *b* 9·48, *c* 10·84 Å, *β* 97° 15′; composition $Ca_4MgH_2(AsO_4)_4.4H_2O$. Obtained synthetically by hydrolysis of basic CaMg arsenates and by dehydration of picropharmacolite. Named for the locality [*M.A.*, 73-1397; *Bull.*, **96**, 237; *A.M.*, **59**, 209; *Zap.*, **102**, 450.] 28th List

Iridarsenite. D. C. Harris, 1974. *C.M.*, **12**, 280. Monoclinic $4[IrAs_2]$ (*a* 6·060, *b* 6·071, *c* 6·158 Å, *β* 113° 16′) occurs as inclusions up to 0·06 mm diameter in rutheniridosmine from New Guinea; identical with artificial material. Named from the composition. 28th List

Iridioplatinita. G. Gagarin and J. R. Cuomo, 1949, loc. cit., p. 5. Applied to Ir-Pt alloys rich in platinum. Иридистая платина (V. I. Vernadsky, 1908), iridic platinum (M. H. Hey, 1950). *See* Platinoiridita. 19th List

Iridiumplatin. H. Strunz, 1941. *Min. Tabellen*, 1st edit., p. 58. Synonym of Иридистая платина, Iridic platinum. 24th List

Iriginite. Author? *Guide to USSR exhibit at Conference on Atomic Energy*, Geneva, 1955, p. 9 (priguinite). Later publications, abstr.: *Amer. Min.*, 1956, vol. 41, p. 816 (priguinite), 1958, vol. 43, p. 379 (iriginite); *Mém. Soc. Russ. Min.*, 1953, vol. 87, p. 80 (иригинит, iriginite). $UO_3.2MoO_3.4H_2O$. Etymology? 21st List [$(UO_2)Mo_2O_7.3H_2O$; *A.M.*, **45**, 257]

Irinite. L. S. Borodin and M. E. Kazakova, 1954. *Doklady Acad. Sci. USSR*, vol. 97, p. 725 (иринит). A metamict cubic mineral $(Na,Ce,Th)(Ti,Nb)(O,OH)_3$, near loparite. Named after Dr. Irina (Irene) Dmitrievna Borneman-Starynkevich, Ирина Дмитриевна Борнеман-Старынкевич, of the Geol. Min. Inst., Acad. Sci., Moscow. [*M.A.*, **12**, 462.] 20th List

Iron-åkermanite. A. N. Winchell, 1924. *Amer. Journ. Sci.*, ser. 5, vol. 8, p. 382 (iron-akermanite). The *hypothetical* molecule $Ca_2Fe''Si_2O_7$, corresponding with åkermanite ($Ca_2MgSi_2O_7$), assumed to explain the composition of minerals of the melilite group. The same as Eisen-Åkermanit of K. Hofman-Degen, 1919 (9th List). [*M.A.*, **2**, 427.] 10th List. C. Hlawatsch, 1904. *Tschermaks Min. Petr. Mitt.*, vol. 23, p. 422 (Fe-Åkermannit), p. 449 (Eisen-Åkermannit). N. L. Bowen, J. F. Schairer, and E. Posnjak, *Amer. Journ. Sci.*, 1933, ser. 5, vol. 26, p. 282 (iron-akermanite). $2CaO.FeO.2SiO_2$, tetragonal analogous to åkermanite with Fe in place of Mg. Present in slags and in the system $CaO–FeO–SiO_2$. [*M.A.*, **5**, 454.] 19th List
[The original List entries are confusing. Hlawatsch used Fe-Åkermannit for a slag mineral from Salt Lake City, and Eisen-Åkermannit for the hypothetical end-member. Hofmann-Degen used Eisen-Åkermanit for a slag mineral. Winchell used Iron-Åkermanite for the hypothetical end-member, and Bowen *et al.* used the same form for the slag mineral.]

Iron-akermanite. M. J. LeBas, 1977. *Carbonatite-nephelinite volcanism* (*Wiley*), p. 92. Error for iron-åkermanite (10th List). 30th List

Iron-alabandite. *See* Eisenalabandin. 21st List

Iron-albite. G. T.* Faust, 1936. *Amer. Min.*, vol. 21, p. 762 (Iron-albite). The compound $NaFeSi_3O_8$ with Fe‴ in place of Al in albite. Similarly, Iron-anorthite and Iron-microcline. Cf. Iron-orthoclase (11th List). [*M.A.*, **7**, 145.] [*Corr. in *M.M.*, **25**] 15th List

Iron-andradite. W. Fischer, 1925. *Bol. Acad. Nac. Cienc. Argentina*, vol. 28, p. 153 (andradida ferrugina); *Centr. Min.*, Abt. A, 1926, p. 36 (Eisenandradit). A hypothetical garnet molecule $3FeO.Fe_2O_3.3SiO_2$. *See* Skiagite. [*M.A.*, **3**, 150.] 11th List

Iron-anorthite. G. T.* Faust, 1936. *See* Iron-albite. [*Corr. in *M.M.*, **25**] 15th List

Iron-anthophyllite. C. H. Warren, 1903. *Amer. Journ. Sci.*, ser. 4, vol. xvi, p. 341 (iron anthophyllite). An orthorhombic iron amphibole, $FeSiO_3$, found with fayalite at Rockport, Massachusetts; and later described and analysed (FeO 42·34%) from the rock eulysite at Tunaberg, Sweden (J. Palmgren, *Bull. Geol. Inst. Univ. Upsala*, 1917, vol. xiv, p. 133 (Eisenanthophyllit)). 8th List

Iron-antigorite. H. von Eckermann, 1925. *Geol. För. Förh. Stockholm*, vol. 47, p. 302. Hydrated ferric silicate, $3FeO.2SiO_2 2H_2O$, corresponding with the magnesium compound antigorite (serpentine). Described as iddingsite pseudomorphous after fayalite. *See* Ferro-antigorite. 11th List

Iron-beidellite. C. S. Ross and E. V. Shannon, 1925. *Journ. Washington Acad. Sci.*, vol. 15, p. 467. A variety of beidellite rich in iron (Fe_2O_3 18·54, Al_2O_3 12·22%) and forming a passage to $Fe_2O_3.3SiO_2.4H_2O$. A German translation is Eisenbeidellit. *See* Beidellite. 11th List [Probably Nontronite]

Iron-berlinite. M. A. Gheith, 1953. *Amer. Min.*, vol. 38, p. 621 (iron berlinite). Artificially produced iron phosphate, $FePO_4$, isomorphous with berlinite (Fe in place of Al) and with the quartz-type structure. *See* Polyquartz. [*M.A.*, **12**, 238.] 20th List

Iron chevkinite. Z. Rubai and F. Liangming, 1976. *Geochimica*, **4**, 244. Unnecessary name for a ferroan chevkinite. [*A.M.*, **63**, 424.] 30th List

Iron-chlorite. A. F. Hallimond, 1939. *Min. Mag.*, vol. 25, p. 442 (iron-chlorites). Translation of Eisenchlorite (*pl.*) of J. Holzner, *Neues Jahrb. Min.*, Abt. A, 1938, Beil.-Bd. 73, p. 389. A group name to include various chloritic minerals rich in iron. Eisenchlorit of C. F. Naumann, 1850, was applied only to delessite. 15th List

Iron-copper-chalcanthite. E. S. Larsen and M. L. Glenn, 1920. *See* Zinc-copper-chalcanthite. 9th List

Iron-cordierite. I. L. Fermor, 1923. *Abstr. Proc. Geol. Soc. London*, for 1922–1923, p. 96. A violet-coloured cordierite rich in ferrous oxide and almost devoid of magnesia. The same as Eisencordierit (H. Bücking, 1900; 3rd List). 10th List. [Cf. Ferrocordierite. Now Sekaninaite (q.v.)]

Iron-corundum, Iron mullite. S. O. Agrell and J. M. Langley, 1958. *Proc. Roy. Irish Acad.*, ser. B, **59**, 93. Unnecessary names for ferrian varieties. 30th List.

Iron-dolomite. R. W. G. Wyckoff and H. E. Merwin, 1924. *Amer. Journ. Sci.*, ser. 5, vol. 8, p. 449 (iron dolomite; iron-rich dolomite). A synonym of ferriferous dolomite or of ankerite. 11th List

Iron-epidote. *See* Aluminium-epidote. 10th List

Iron-gedrite. C. Doelter, 1913. *Handbuch d. Mineralchemie*, vol. 2 (i), p. 352 (Eisengedrite). H. von Eckermann, *Geol. För. Förh. Stockholm*, 1922, vol. 44, p. 268 (iron-gedrite). The gedrites (aluminous anthophyllites) richer in iron. 9th List [Ferro-gedrite (q.v.)]

Iron-gehlenite. A. N. Winchell, 1924. *Amer. Journ. Sci.*, ser. 5, vol. 8, p. 382. The hypothetical molecule $Ca_2Fe'''_2SiO_7$, corresponding with gehlenite ($Ca_2Al_2SiO_7$), assumed to explain the composition of minerals of the melilite group. The same as Ferri-gehlenite of P. Niggli, 1922 (9th List). [*M.A.*, **2**, 427.] 10th List

Iron-hornblende. T. Yoshimura, 1939. *Journ. Fac. Sci. Hokkaido Univ.*, ser. 4, vol. 4, p. 421 (ironhornblende). A green hornblende with FeO 17·09, Fe_2O_3 8·05%. 15th List [Oxy-manganoan potassian ferrian ferro-hornblende; *Amph.* (1978)]

Iron-hypersthene. M. Saxén, 1925. *Fennia, Soc. Geogr. Fenniae*, Helsingfors, vol. 45, no. 11, p. 18 (Järnhypersten), p. 39 (Eisen-hypersthen). N. Sundius, *Årsbok Sveriges Geol. Undersök.*, 1932, vol. 26, no. 2, p. 3 (Eisenhypersthen). An iron-rich hypersthene (FeO 42%), $MgSiO_3.3FeSiO_3$. *See* Orthoferrosilite. 14th List

Iron-kaolinite. M. H. Hey, 1936. *Min. Abstr.*, vol. 6, p. 234. 'Many nontronites appear to be ferruginous beidellites, but some seem to be iron-kaolinites.' Cf. iron-beidellite (11th List). 14th List

Iron-knebelite. T. Yoshimura, 1939. *Journ. Fac. Sci. Hokkaido Univ.*, ser. 4, vol. 4, p. 403 (Ironknebelite). The same as järnknebelit, Eisenknebelit (M. Weibull, 1884) and igelströmite (M. Weibull, 1883). An intermediate member (Fe_2SiO_4 60–80 mol.%) of the fayalite-tepheroite series, near to knebelite. 15th List

Iron-lazulite. L. Katz and W. N. Lipscomb, 1951. *Acta Cryst.*, vol. 4, p. 345 (iron lazulite). Artificial $(Fe'',Fe''')_7(PO_4)_4(OH)_4$, tetragonal. [Not in agreement with the lazulite-scorzalite series, *M.A.*, **11**, 244, 426.] 19th List [Now Lipscombite (q.v.)]

Iron-leucite. A. N. Winchell, 1927. *The optic and microscopic characters of artificial minerals, Univ. Wisconsin Studies, Madison*, no. 4, p. 143 (iron-leucite). P. Gaubert, *Compt. Rend. Congr. Soc. Sav.*, 1926, for 1925, p. 408; *Bull. Soc. Franç. Min.*, 1927, vol. 50, p. 514 (leucite ferrique). Pseudo-cubic $KFeSi_2O_6$ analogous to leucite with ferric oxide in place of alumina, prepared artificially by P. G. Hautefeuille (1880). 11th List

Iron-melanterite. A. N. Winchell, 1942. *Elements of mineralogy*, p. 291; *Amer. Min.*, 1949, vol. 34, p. 223 (Iron Melanterite). Syn. of melanterite ($FeSO_4.7H_2O$), classed with boothite ($CuSO_4.7H_2O$) and zinc-copper melanterite ($(Zn,Cu,Fe)SO_4.7H_2O$) as sub-species under a species 'melanterite'. [*M.A.*, **10**, 568.] 18th List

Iron-microcline. G. T.* Faust, 1936. *See* Iron-albite. [*Corr. in *M.M.*, **25**] 15th List

Iron-monticellite. D. S. Belyankin, K. M. Feodotyev, and C. S. Nikogosyan, 1934. *Neues Jahrb. Min.*, Abt. A, Beil.-Bd. 68, p. 337 (Eisen-Monticellit). The molecule $Ca_2SiO_4.Fe_2SiO_4$, as distinct from magnesium-monticellite ($Ca_2SiO_4.Mg_2SiO_4$) and mangan-monticellite ($Ca_2SiO_4.Mn_2SiO_4$). 14th List

Iron mullite *see* Iron Corundum.

Iron-orthoclase. *See* Ferri-orthoclase. 11th List

Iron-pyrochroite. Abstr. in *Amer. Min.*, 1922, vol. 7, p. 214. Translation of Eisenpyrochroit (G. Flink, 1919; 9th List). 10th List

Iron-pyroxene. C. N. Fenner, 1931. *Min. Mag.*, vol. 22, pp. 549, 559. A general term for pyroxenes (hedenbergite, augite, etc.) rich in iron. 14th List

Iron-reddingite. H. H. Woodard, 1951. *Amer. Min.*, vol. 36, p. 881 (Iron reddingite). Synonym of phosphoferrite. [*M.A.*, **11**, 492.] 19th List

Iron-rhodonite. Translation of Järnrhodonit, Eisenrhodonit (M. Weibull, 1884). The original mineral, however, was proved to be sobralite (= pyroxmangite) by N. Sundius, 1931 [*M.A.*, **4**, 527; **5**, 143], who transferred the name to a mineral isomorphous with rhodonite. The slag mineral called iron-rhodonite by J. H. Whiteley and A. F. Hallimond, 1919 [*M.A.*, **1**, 164] has also been identified with pyroxmangite by C. E. Tilley [*M.A.*, **6**, 529] and M. Perutz [*Min. Mag.*, **24**, 573]. 14th List

Iron-richterite. O. B. Bøggild, 1953. *Mineralogy of Greenland, Meddel. om Grønland*, vol. 149, no. 3, p. 282 (Iron Richterite). Translation of Eisenrichterit (q.v.). *See also* ferririchterite. 20th List [Ferro-richterite; *Amph.* (1978)]

Iron-sanidine. D. R. Wones and D. E. Appleman, 1961. *US Geol. Surv. Prof. Paper no. 424-C*, p. 309. Synthetic monoclinic $KFeSi_3O_8$ is believed to be analogous to Sanidine; it is not clear whether this material is distinct from Ferriorthoclase (11th List). [*M.A.*, **17**, 578.] 24th List

Iron-sarcolite. W. T. Schaller, 1916. *See* Soda-sarcolite. 8th List

Iron-sericite. H. Minato and Y. Takano, 1952. *Sci. Papers College Gen. Educ. Univ. Tokyo*, vol. 2, p. 194 (iron sericite). A sericitic clay from Japan containing Fe_2O_3 5·74, FeO 1·50%, MgO trace. *See* Magnesium-sericite. [*M.A.*, **12**, 221.] 20th List

Iron-serpentine. J. W. Gruner, 1936. *Amer. Min.*, vol. 21, p. 453 (iron serpentine). The hypothetical molecule $H_4Fe_3Si_2O_9$ with Fe in place of Mg in serpentine. Cf. Iron-antigorite (11th List), Ferro-chrysotile (14th List). [*M.A.*, **6**, 480.] 15th List

Iron-skutterudite. R. J. Holmes, 1942. *Science*, New York, vol. 96, p. 92. Synonym of arsenoferrite (6th List), which gives the same X-ray pattern as skutterudite, $(Co,Fe)As_3$. [*M.A.*, **8**, 380.] 16th List

Iron-sodium-melilite. M. J. LeBas, 1977. *Carbonatite-nephelinite volcanism* (Wiley), p. 92. A name, later dropped, for natroferrimelilite, the end-member $NaCaFeSi_2O_7$. 30th List

Iron-strigovite. S. Palmqvist, 1935. *Meddel. Lunds Geol.-Min. Inst.*, no. 60, p. 167 (Iron-

Strigovite). The iron silicate constituent of Liassic oolitic iron ores from Scania, Sweden, with the composition $2(Fe,Mg)O.(Fe,Al)_2O_3.2SiO_2.2H_2O$, as calculated from bulk analyses of the ores. This composition is compared with G. Tschermak's hypothetical strigovite molecule $H_4Mg_2Al_2Si_2O_{11}$ of the chlorite group, and the fact that actual analyses of strigovite show the presence of much iron is overlooked. [*M.A.*, **6**, 368.] 14th List

Iron-talc. J. W. Gruner, 1944. *Amer. Min.*, vol. 29, p. 363 (iron talc). *See* Minnesotaite. 17th List

Iron-tephroite. T. Yoshimura, 1939. *Journ. Fac. Sci. Hokkaido Univ.*, ser. 4, vol. 4, p. 406 (Irontephroite). An intermediate member (Fe_2SiO_4 5–20 mol.%) of the fayalite-tephroite series. 15th List

Iron-uranite. d'A. George, 1949. [*US Atomic Energy Comm.*, RMO–563]; C. Frondel, *Min. Mag.*, 1954, vol. 30, p. 343 ('iron uranite'). An incompletely examined mineral from New Mexico, perhaps the same as bassetite or kahlerite (q.v.). 20th List

Iron-wagnerite. A. Henriques, 1957. *Arkiv Min. Geol.*, vol. 2, p. 149. The mineral named Talktriplite by Igelström (1882) is shown to be a ferroan wagnerite and is named accordingly; an international name would have been preferable, though even that is unnecessary. 23rd List

Iron-wollastonite. C. E. Tilley, 1937. *Amer. Min.*, vol. 22, p. 727 (Iron wollastonite); *Min. Mag.*, 1937, vol. 24, p. 572 (iron-wollastonites). Wollastonite with $FeSiO_3$ in solid solution, from Co. Antrim. 14th List

Irosita. G. Gagarin and J. R. Cuomo, 1949, loc. cit., p. 5. Contraction in Spanish form of iridosmine, applied to Os-Ir alloys rich is osmium [*M.A.*, **7**, 162.] *See* Osirita. 19th List

Irvingite. S. Weidman, 1907. *Amer. Journ. Sci.*, ser. 4, vol. xxiii, p. 451. A variety of lithia-mica differing somewhat from other varieties in chemical composition. Occurs with marignacite (q.v.) in pegmatite-veins near Wausau, Wisconsin. Named after Roland Duer Irving (1847–1888), formerly of the Geological Survey of Wisconsin. 4th List

Iscorite. J. Smuts, J. G. D. Steyn, and J. C. A. Boeyens, 1969. *Acta Cryst.*, **B 25**, 1251. An *artificial* iron silicate, $2[Fe_7SiO_{10}]$, monoclinic, found in a steel furnace. [*M.A.*, 71-1751.] 28th List

Ishiganeïte. K. Kami and T. Tanaka, 1937. [*Bull. Electrotechn. Lab. Japan*, vol. 1, pp. 459 and 553; 1938, vol. 2, pp. 19 and 291; *Electrochem.*, Japan, 1938, vol. 6, p. 366; 1939, vol. 7, p. 7], quoted by Y. Hariya,, *Amer. Min.*, 1963, vol. 48, p. 952. A manganese oxide from the Ishigane mine, Aichi prefecture, Japan, subsequently found by M. Nambu and K. Okada [*Journ. Soc. Earth Sci. Amat. Japan*, 1961, vol. 12, p. 249; *Journ. Jap. Ass. Min. Petr. Econ. Geol.*, 1962, vol. 48, p. 76], quoted by Y. Hariya, *loc. cit.*, to be a mixture of Cryptomelane and Birnessite (confirmed by Y. Hariya, *loc. cit.*). 23rd List

Ishikawaite. K. Kimura, 1922. *Journ. Geol. Soc. Tōkyō*, vol. 29, p. 316. Y. Shibata and K. Kimura, *Journ. Chem. Soc. Japan*, 1922, vol. 43, pp. 301, 648; *Japanese Journ. Chem.*, 1923, vol. 2, p. 17. A niobate and tantalate of uranium, iron, etc., $10RO.R_2O_3.6(Nb,Ta)_2O_5$, occurring as black, orthorhombic crystals at Ishikawa, prov. Iwaki, Japan. It differs from samarskite in containing more UO_2 (21·88%) and less rare-earths (8·40%). R. Ōhashi (*Journ. Geol. Soc. Tōkyō*, 1924, vol. 31, p. 166) shows a crystallographic relation to yttrotantalite and samarskite. Named from the locality. [*M.A.*, **2**, 9, 380, 381.] 10th List [Var. of Samarskite; Introduction to Japanese Minerals (*Geol. Surv. Japan*, 1970) p. 105]

Ishkulite. G. P. Barsanov, 1941. [*Compt. Rend.* (*Doklady*) *Acad. Sci. URSS*, vol. 31, p. 468], abstr. in *Amer. Min.*, 1942, vol. 27, p. 62. A variety of magnetite containing Cr_2O_3 11·19% from lake Ishkul (Ишкуль), Ilmen Mts., S Ural. Named from the locality. Not to be confused with ishkyldite (14th List). [*M.A.*, **8**, 230.] 16th List

Ishkyldite. F. V. Syromyatnikov, 1934. *Bull. Soc. Nat. Moscou*, ser. 2, vol. 42 (*Sect. Géol.*, vol. 12), pp. 566, 568 (ишкильдит), p. 574 (ishkyldite); *Amer. Min.*, 1936, vol. 21, p. 48 (ishkyldite). A variety of chrysotile-asbestos differing from α-chrysotile in optical characters and X-ray pattern and with an excess of silica, $H_{20}Mg_{15}Si_{11}O_{47}$. Named from the locality, Ishkyldino, Volga, Russia. [*M.A.*, **6**, 259.] 14th List

Isiganeite. Z. Harada, *Journ. Fac. Sci. Hokkaido Univ.*, ser. 4, 1948, vol. 7, p. 154 (Isigane-isi) from [*Journ. Electrochem. Assoc. Japan*, 1939, vol. 7]. Variety of psilomelane, new mineral from Isigane mine, Sasaoka, E-Kasugai Co., Aiti prefecture. Japanese isi, ishi = stone, -ite. 19th List [Evidently the same as Ishiganeïte (q.v.)]

Isoferroplatinum. L. J. Cabri and C. E. Feather, 1975. *C.M.*, **13**,, 117. Grains in the Pt ores from the Tulameen River, British Columbia, are cubic (primitive), *a* 3·864 Å, composition near Pt_3Fe (ordered). Named from the symmetry and composition. 29th List

Isokite. T. Deans and J. D. C. McConnell, 1955. *Min. Mag.*, vol. 30, p. 681. $CaMgPO_4F$. monoclinic, isomorphous with tilasite with P in place of As. White spherulites from a carbonatite plug near Isoka, Northern Rhodesia. Named from the locality. [*A.M.*, **40**, 776; **41**, 167.] 20th List

Isomertieite. A. M. Clark, A. J. Criddle, and E. E. Fejer, 1974. *M.M.*, **39**, 528. A few grains in the arsenopalladinite concentrates from Itabira, Minas Gerais, Brazil, have a composition near that of mertieite (q.v.), but are pseudo-cubic, *a* 12·283 Å; unit-cell contents $16[(Pd,Cu)_5(Sb,As)_2]$. Metrically cubic but optically uniradial. Named from its chemical similarity to mertieite. [*M.A.*, 74-1444.] 28th List

Isomicrocline. W. Luczizky, 1905. *Min. Petr. Mitt.* (*Tschermak*), vol. xxiv, p. 347 (Isomikroklin). An optically positive microcline, analogous to the isorthose of L. Duparc, 1904 (4th List). 5th List

Iso-orthoclase, Isorthoclase. T. F. W. Barth, *Amer. Min.*, 1933, vol. 18, p. 478 (iso-orthoclase); A. N. Winchell, *Elements of optical mineralogy*, part 2, 2nd edit., 1927, p. 322 (Isorthoclase). Variants of the French Isorthose (L. Duparc, 1904; 4th List). [*M.A.*, **5**, 438.] 13th List

Isoperthite. V. N. Lodochnikov, 1925. *Zap. Ross. Min. Obshch.* (*Mém. Soc. Russe Min.*, ser. 2, vol. 52, p. 107 (Изопертиты, *plur.*). Perthitic intergrowths consisting of the same kind of felspar, e.g. albite-isoperthite альбит-изопертит). 11th List

Isorthoclase, *see* Iso-orthoclase.

Isorthose. L. Duparc, 1904. *Compt. Rend. Acad. Sci. Paris*, vol. cxxxviii, p. 715. A variety of orthoclase (French, orthose) differing from ordinary orthoclase in optical orientation: the acute bisectrix of the optic axes is positive in sign and is perpendicular to the plane of symmetry. It is met with in granite from the northern Urals and in the protogine of Mont Blanc. 4th List

Isostannin, variant of Isostannite (21st List) (H. Strunz., *Min. Tabellen*, 3rd edit., p. 95). 22nd List

Isostannite. G. F. Claringbull and M. H. Hey, 1955. *Min Soc. Notice*, 1955, no. 91 [*M.A.*, **13**, 31]; M. H. Hey, *Index of Minerals* (*Brit. Mus.*) *1955*, pp. 34, 468, 609. Cubic (isometric) modification of Cu_2FeSnS_4, as distinct from tetragonal stannite. 21st List [Is Kësterite (q.v.), *Can. Mineral.*, **13**, 309 (abstr. only)]

Isowolframite. N. P. Senchilo and K. A. Mukhlya, 1972. [Изв. акад. наук Казах. ССР, сер. геол. **4**, 45]; abstr. *A.M.*, **58**, 560. An unnecessary name for members of the hübnerite–ferberite series with $FeWO_4$ 40 to 60 mol.% [*Zap.*, **103**, 362.] 28th List

Istisuite. M. A. Kashkay and A. I. Mamedov, 1955. [*Doklady Acad. Sci. Azerbaijan*, vol. 11, no. 1, p. 21]; abstr. in *Mém. Soc. Russ. Min.*, 1955, ser. 2, vol. 84, p. 347 (истисуит). $(Ca,Na)_7(Si,Al)_8(O,OH)_{24}$, monoclinic. With wollastonite in skarn from Istisu (Истису), a health resort on the Terter river, Azerbaijan republic. [*M.A.*, **13**, 207.] 21st List

Itoite. C. Frondel and H. Strunz, 1960. *Neues Jahrb. Min.*, Monatsh., p. 132 (Itoit). $Pb_3[GeO_2(OH)_2](SO_4)_2$, orthorhombic and isostructural with anglesite. From Tsumeb, SW Africa. Named after Prof. Tei-ichi Ito of Tokyo. [*M.A.*, **15**, 43; *A.M.*, **45**, 1313.] 22nd List

Ivanovite. E. I. Nefedov, 1953. *Mém. Soc. Russ. Min.*, ser. 2, vol. 82, p. 317 (Ивановит). Hydrous chloro-borate of Ca (and K?), monoclinic (pseudo-hexagonal), Inder salt deposits, Kazakhstan. [*M.A.*, **12**, 352.] 20th List

Iwanowit. German transliteration of Ивановит, Ivanovite (20th List) (C. Hintze, *Handb. Min.*, Erg.-Bd. II, p. 732). 24th List

J

Jachymovite. R. Nováček, 1935. *Věstník Král. České Spol. Nauk, Cl. 2*, for 1935, no. 7, p. 28 (jáchymovite); *Čaposis Náradního Musea*, Praha, 1935, vol. 109, p. 100. Hydrated silicate of uranium and copper, $CuO.2UO_3.2SiO_2.6H_2O$, afterwards proved to be identical with cuprosklodowskite (13th List). Named from the locality, Jáchymov (= Joachimsthal), Bohemia. [*M.A.*, **6**, 149, 345.] 14th List

Jade-albite. E. Gübelin, 1965. *Journ. Gemmology*, vol. 9, p. 372. A rock consisting of Albite and Chromojadeite. [*M.A.*, **17**, 377.] 24th List

Jadeite-Aegirite. R. Doht and C. Hlawatsch, 1913. *Verh. Geol. Reichsanst. Wien*, 1913, p. 88 (Jadeit-Ägirin). An aegirite-like pyroxene from Golling, Salzburg, which, in addition to the aegirite molecule $NaFeSi_2O_6$, also contains the jadeite molecule $NaAlSi_2O_6$ in considerable amount. 7th List

Jadeolite. G. F. Kunz, 1908. *Mineral Industry, New York*, for 1907, vol. xvi, p. 810. A deep-green chromiferous syenite cut as a gem-stone and resembling jade in appearance; from the jadeite mine at Bhamo, Burma. Possibly the same as the pseudojadeite (q.v.) of A. W. G. Bleeck, 1908. 5th List

Jadine. Trade-name for an Australian chrysoprase. R. Webster, *priv. comm.* Not to be confused, as presumably intended, with jadeite or jade. 28th List

Jaeneckéite, *see* Jäneckite.

Jaffaite. J. Paclt, 1953. [*Israel Expl. Journ.*, vol. 3, p. 242]; abstr. *M.A.*, **15**, 136. A gum-resin found in calcareous aeolian rock, and believed to originate from a species of Pistacia. From the plain of Sharon, Israel. 22nd List

Jagiite, error, through back-transliteration of Ягиит, for yagiite (26th List). (*Zap.*, **99**, 82.) 27th List

Jagoite. R. Blix, O. Gabrielson, and F. E. Wickman, 1957. *Arkiv. Min. Geol.*, Stockholm, vol. 2, no. 18, p. 215 (jagoite). $(Pb,Ca)_3Fe'''Si_3O_{10}(Cl,OH)$; analysis, SiO_2 22·35, Pb 64·26%, and fractional percentages of Al, Be, K, Mg, Mn, Na, Ti. Trigonal, yellow-green micaceous plates with melanotekite in iron ore. This is the tenth lead silicate from Långban, Sweden. Named after John B. Jago, mineral collector of San Francisco. 21st List [*A.M.*, **43**, 387]

Jagowerite. E. P. Meagher, M. E. Coates, and A. E. Aho, 1973. *C.M.*, **12**, 135. Light green crystalline masses in a Palaeozoic carbonaceous argillite at 63° 35′ N., 132° 48′ 30″ W. in the Yukon Territory, are anorthic, *a* 6·04, *b* 6·964, *c* 4·971 Å, α 116° 51′, β 86° 6′, γ 112° 59′; $BaAl_2(PO_4)_2(OH)_2$. Named for J. A. Gower. 28th List [*A.M.*, **59**, 291]

Jahnsite. P. B. Moore, 1974. *A.M.*, **59**, 48. Short prismatic monoclinic crystals of variable colour (greenish-yellow to purplish-brown) with a wide range of phosphates in the Tip Top pegmatite, Custer, Custer County, S Dakota, have *a* 14·94, *b* 7·14, *c* 9·93 Å, β 110·16°, and composition $2[CaMnMg_2Fe_2{}^{3+}(PO_4)_4(OH)_2.8H_2O]$. Named for R. H. Jahns. 28th List

Jalindite, erroneous transliteration of Джалиндит, dzhalindite (Зап. Всесоюз. Мин. Общ. (*Mem. All-Union Min. Soc.*), 1964, vol. 93, p. 447). 23rd List

Jaloallofane. *See* hyaloallophane. 2nd List

Jamborite. N. Morandi and G. Dalrio, 1973. *A.M.*, **58**, 835. Green pseudomorphs after millerite, occurring in ophiolites near Bologna and Modena, Italy, consist of tiny hexagonal crystals, *a* 3·07, *c* 23·3 Å, with composition $(Ni^{2+},Ni^{3+})(OH)_2(OH,S,H_2O)$ with some Co and Fe. The name, for J. L. Jambor, is intended to include the unnamed nickel hydroxide of Jambor and Boyle (*C.M.*, 1964, **8**, 116.). [*M.A.*, 74-1447.] 28th List

Jäneckeite. L. T. Brownmiller and R. H. Bogue, *Amer. Journ. Sci.*, 1930, ser. 5, vol. 20, p. 251 (Janeckite). F. Krauss and G. Jörns, *Zement*, Charlottenburg, 1931, vol. 20, p. 341 (Jäneckeit). O. F. Honus, *Chimie et Industrie*, Paris, 1931, vol. 26, p. 1011 (Jaeneckéite).

$8CaO.Al_2O_3.2SiO_2$, as a constituent of Portland cement clinkers. Named after E. Jänecke, who first distinguished it from alite (2nd List). [*M.A.*, **4**, 366.] 13 List

Janggunite. S. J. Kim, 1976. *J. Korean Inst. Min. Geol.*, **8**, 117. *M.M.*, **41**, 519. Black aggregates and radiating flaky groups from the Janggun mine, Banghwa, Korea, are orthorhombic, *a* 9·324, *b* 14·05, *c* 7·956 Å. Composition near $4[Mn_{5-x}^{4+}(Mn^{2+},Fe^{3+})_{1+x}O_8(OH)_6]$, with $x = 0{\cdot}2$, sp. gr. 3·59. Named for the locality. [*M.A.*, 78-888; *A.M.*, **63**, 426] 30th List

Janite. S. J. Thugutt, 1933. *Arch. Min. Soc. Sci. Varsovie*, vol. 9, p. 93 (janicie), p. 97 (janite). Hydrated silicate of Fe_2O_3,Al,Mg,Ca, etc., related to chloropal or celadonite, occurring in an altered glassy basalt from the Janowa valley, Volhynia, Poland. Named from the locality. [*M.A.*, **5**, 485.] 13th List [*See* Janowaite]

Janosite. H. Böckh and K. Emszt, 1905, *Földtani Közlöny*, Budapest, vol. xxxv, pp. 76, 139; ibid., vol. xxxvi, pp. 186, 228, 404, 455 (Jánosit). Described as an orthorhombic, hydrated normal ferric sulphate, $Fe_2O_3.3SO_3.9H_2O$, dimorphous with coquimbite; it occurs as a greenish-yellow, powdery, efflorescence at Vashegy, Comitat Gömör, Hungary. Named after Dr. János [= John] Böckh, Director of the Geological Survey of Hungary.

Proved by E. Weinschenk (ibid., 1906, vol. xxxvi, pp. 182, 224, 289, 359) and Z. Toborffy (ibid., 1907, vol. xxxvii, pp. 122, 173; *Zeits. Kryst. Min.*, 1907, vol. xliii, p. 369) to be identical with copiapite, $2Fe_2O_3.5SO_3.18H_2O$, which is orthorhombic and not monoclinic. 4th List

Janovait, variant of Janowait (15th List), a synonym of Janite (13th List). (H. Strunz, *Min. Tabellen,* 3rd edit., 1957, p. 383.) 24th List

Janovite. Still another variation of janite (13th List), janowaite (15th List), and yanite (15th List). [*M.A.*, **13**, 180.] 21st List

Janowaite. C. Hintze, *Handbuch Min.*, Ergänzungsband, 1937, p. 236 (Janowait). A more correct spelling of janite (13th List), from the Janowa valley, Poland. On p. 741 another spelling—Yanit. 15th List

Janthiniet, Janthinit. *See* Ianthinite. 11th List

Japanite. I. Iwasa, 1877. [*Gakugéisirin*, no. 57.] Z. Harada, *Journ. Fac. Sci. Hokkaido Univ.*, Sapporo, ser. 4, 1936, vol. 3, p. 324 (Japanit). An incorrectly determined mineral from Japan, afterwards (Z. Sasamoto, 1895) identified with pennine. 14th List [*See also* Intro. Japanese Minerals (*Geol. Surv. Japan*, 1970), 98]

Jarlite. R. Bøgvad, 1933. *Meddelelser om Grønland*, vol. 92, no. 8, p. 3 (Jarlite). A fluoride, $NaSr_3Al_3F_{16}$, as small colourless monoclinic crystals, from the Greenland cryolite quarry. Named after Mr. C. F. Jarl, of Øresunds Chemiske Fabriker. *See* Metajarlite. [*M.A.*, **5**, 387.] 13th List [Cf. Calcjarlite]

Järnhypersten. M. Saxén, 1925. *See* Iron-hypersthene. 14th List

Jarošite. J. Kokta, 1937. *Sborník Klubu Přírodovědeckého v Brně*, vol. 19 (for 1936), p. 75 (Jarošit). A variety of melanterite containing MgO 5·55%, from Slovakia. Named after Zdeněk Jaroš, keeper of minerals in the museum at Brno. Pronounced Jaroschit, yarroshite. [Not to be confused with jarosite from the Jaroso ravine, Spain.] *See* Cuprojarošite and Kirovite. [*M.A.*, **7**, 316.] 15th List

Jaroslavite, erroneous transliteration of Ярославит (yaroslavite) (Зап. Всесоюз. Мин. Общ. (*Mem. All-Union Min. Soc.*), 1966, vol. 95, p. 43). 24th List

Jaroslawit, German transliteration of Ярославит, Yaroslavite (24th List) (C. Hintze, *Handb. Min.*, Erg. III, 571). 25th List

Jarrowite. [G. A. Lebour, *57th Rept. Brit. Assoc.*, for 1887, 700, 1888]. *British Museum Student's Index*, since 1886. H. A. Miers, *M.M.*, **11**, 264. Syn. of thinolite. Jarrow, Durham. 1st List

Jelinite. J. D. Buddhue, 1938. *Mineralogist*, Portland, Oregon, vol. 6, no. 9, p. 9 (jelinekite, preferably jelinite). To replace the name kansasite (q.v.). Named after Mr. George Jelinek, of Ellsworth, Kansas, who found the material. [*M.A.*, **7**, 265.] 15th List

Jemchuznikovite, erroneous transliteration of Жемчужниковит, Zhemchuzhnikovite (Зап. Всесоюз. Мин. Общ. (*Mem. All-Union Min. Soc.*), 1962, vol. 91, p. 204). 23rd List

Jennite. A. B. Carpenter, R. A. Chalmers, J. A. Gard, K. Speakman, and H. F. W. Taylor,

1966. *Amer. Min.*, vol. 51, p. 56. Fibrous material in a vein in the contact rock at Crestmore, California, are anorthic bladed crystals of composition $Na_2Ca_8Si_5O_{30}H_{22}$*, probably $Na_2Ca_8(SiO_3)_3Si_2O_7(OH)_6.8H_2O$. At about 90°C it loses some water to give Metajennite (q.v.). Named for C. M. Jenni. [*Corr. in *M.M.*, **39**.] 24th List [Near $9CaO.6SiO_2.11H_2O$; *A.M.*, **62**, 365]

Jentschite. *Min. Petr. Mitt.* (*Tschermak*), 1904, vol. xxiii, p. 551; *Min. Mag.*, 1905, vol. xiv, p. 80 footnote. Synonym of lengenbachite (q.v.). Named after Franz Jentsch, mineral dealer of Binn, Switzerland. Not to be confused with the jenzschite of J. D. Dana, 1868. 4th List

Jeromite. C. Lausen, 1928. *Amer. Min.*, vol. 13, p. 227. Sulphide and selenide of arsenic, $As(S,Se)_2$, as black fused globules deposited by hot gases from the burning of pyritic ore in a mine at Jerome, Arizona. Named from the locality. 11th List

Jezekit, variant of Ježekite (7th List) (H. Strunz, *Min. Tabellen*, 4th edit., 1966, p. 483). 24th List

Ježekite. F. Slavík, 1914. [*Memoirs Bohemian Acad. Sci.*]; Doelter's *Handbuch der Mineralchemie*, vol. iii, p. 491; *Bull Soc. franç. Min.*, vol. xxxvii, p. 153. Colourless to white, monoclinic crystals, somewhat resembling epistilbite in appearance, occurring with lacroixite, and roscherite (q.v.) in drusy cavities in lithionite-granite at Greifenstein, near Ehrenfriedersdorf, Saxony. It is a basic fluophosphate of aluminium, calcium, and sodium with a little lithium, $Na_4CaAl(AlO)P_2O_8F_2(OH)_2$, and is closely allied to morinite. Named after Dr. Bohuslav Ježek, of the Bohemian Museum at Prague. 7th List [Is Morinite; *A.M.*, **47**, 398]

Jezekite, spelling variant of Ježekite. *Amer. Min.*, 1963, vol. 48, p. 435. 23rd List

Jimboite. T. Watanabe, A. Kato, T. Matsumoto, and J. Ito, 1963. [*Proc. Japan Acad.*, vol. 39, p. 170], abstr. in *Amer. Min.*, 1963, vol. 48, p. 1416. $(Mn,Mg)_3B_2O_6$, occurring with other manganese minerals at the Kaso mine, Kanuma city, Tochigi prefecture, Japan. Orthorhombic, and the manganese analogue of Kotoite. Named for Prof. K. Jimbo. [*M.A.*, **16**, 547.] 23rd List

Jimthompsonite. D. R. Veblen, P. R. Buseck, and C. W. Burnham, 1977. *Science*, **198**, 359. An orthorhombic mineral, from Chester, Vermont, *a* 18·63, *b* 27·23, *c* 5·30 Å, space group *Pbca*. Composition $(Mg,Fe)_{10}Si_{12}O_{32}(OH)_4$, a member of the biopyribole group and dimorphous with clinojimthompsonite (both 30th List). Named for J. Thompson. [*M.A.*, 78-3473] 30th List

Jiningite. Cheng-Chi Kuo, 1959. [*Kexue Tongbao* (*Scientia*), 1959, no. 6, p. 206]; abstr. *Amer. Min.*, 1960, vol. 45, p. 755 (jiningite); Зап. Всесоюз. Мин. Общ. (*Mem. All-Union Min. Soc.*, 1961, vol. 90, p. 108 (Жинингит, jiningite)). A variety of thorite or thorogummite with 6% P_2O_5, 1% V_2O_5, 7% CaO, 17% Fe_2O_3, found in a muscovite-granite from Inner Mongolia. An unnecessary name. 22nd List

Joaquinite. G. D. Louderback, 1909. *Bull. Dep. Geol. Univ. California*, vol. v, p. 376. Honey-yellow, orthorhombic crystals found associated with benitoite (q.v.): they contain silicon, titanium, calcium, and some iron, but have not yet been completely determined. Named from the Joaquin ridge of the Diablo range, San Benito Co., California. 5th List [*A.M.*, **52**, 1762; **57**, 85; *M.A.*, 73-1291. Formula uncertain]

Jocketan. A. Breithaupt (A. Frenzel, *Min. Lexicon Sachsen*, 1874, p. 160). Hydrated carbonate of iron occurring as concretions on limonite. Jocketa, Saxony. 2nd List

Jodatacamit. German variant of Iodatacamite (q.v.). 23rd List

Jodchromate. *See* dietzeite. 1st List

Jodmimetesit. H. Wondratschek, 1963. Original form of Iodemimetite, (q.v.). 23rd List

Jodvanadinit. H. Wondratschek, 1963. Original form of Iodovanadinite, (q.v.). 23rd List

Joesmithite. P. B. Moore, 1968, *Min. Mag.*, vol. 36, p. 876. A monoclinic mineral, $(Pb,Mn,Ca,Ba)_2Ca_4Fe_2^{3+}(Mg,Fe)_8Si_8O_{24}[Si(O,OH)_4]_4(OH)_8$, from Långban, Sweden, related structurally to the amphiboles. 25th List

Johachidolite. E. Iwase and N. Saito, 1942. [*Sci. Papers Inst. Phys. Chem. Research*, Tokyo, vol. 39, p. 300.] Abstr. in *Amer. Min.*, 1948, vol. 33, p. 98. Hydrous fluoborate, '$H_6Na_2Ca_3Al_4F_5B_6O_{20}$', colourless, optically biaxial. In nepheline veins from Johachido, Korea. Named from the locality. [*M.A.*, **10**, 253.] 18th List [Jōhachidō is the Japanese name for the village Sāngpal-dong, Kilchu Co., N Hamgyong Province.] [Redefined: orthorhombic 4[$CaAlB_3O_7$]; *A.M.*, **62**, 327]

Johannsenite. W. T. Schaller, 1932. *Amer. Min.*, vol. 17, p. 575; 1933, vol. 18, p. 113. N. L. Bowen, J. F. Schairer, and E. Posnjak, *Amer. Journ. Sci.*, 1933, ser. 5, vol. 26, p. 274. A monoclinic pyroxene $MnCaSi_2O_6$, isomorphous with diopside and dimorphous with bustamite. [From Bohemia dist., Oregon; and Schio-Vicentin, N Italy.] Named after Prof. Albert Johannsen, of Chicago. 13th List

Johnsonite. H. D. Richmond and H. Off, 1892. *Chemiker-Zeitung*, Jahrg. xvi, pp. 567, 648 (Johnsonit); *Chem. News*, vol. lxv, p. 257. A fibrous alum, containing the supposed new element masrium, found in Egypt by S. E. Johnson Pasha. The mineral was immediately afterwards re-named masrite (*see* 1st List, *Min. Mag.*, vol. xi, p. 331). 4th List

Johnstonite. E. J. Chapman, *Min.*, 1843, p. 42. The same as vanadanite. 2nd List

Johnstonotite. W. A. Macleod and O. E. White, *Papers and Proc. R. Soc. Tasmania, for 1898–1899*, 1900, p. 74. A variety of garnet. [Port Cygnet, Tasmania.] Named after R. M. Johnston. 2nd List [Calc-spessartine; *A.M.*, **53**, 1065]

Joliotite. K. Walenta, 1976. *Schweiz. Mineral. Petrogr. Mitt.*, **56**, 167. Rare spherulites at Menzenschwand, Schwarzwald, Germany, are orthorhombic, *a* 8·16, *b* 10·35, *c* 6·32 Å. Composition 4[$UO_2CO_3.1\frac{1}{2}–2H_2O$]. RIs α 1·596–1·604, γ 1·637–1·651. Named for J. F. and I. Joliot-Curie. [*M.A.*, 77-2184.] 30th List

Jonesite. W. S. Wise, A. Pabst, and J. R. Hinthorne, 1977. *Mineral. Rec.*, **8**, 453. Rosettes of colourless crystals, fluorescing orange in short-wave U.V. light, occur embedded in natrolite at the Benitoite Gem mine, San Benito Co., California. Orthorhombic, *a* 13·730, *b* 25·904, *c* 10·608, space group $B22_12$. Composition 4[$(K,Na,Ba)_{1-2}Ba_4Ti_4(SiAl)_{12}O_{30}.6H_2O$], ρ 3·25 g.cm^{-3}. RIs α 1·641 ∥ [010], β 1·660 ∥ [100], γ 1·682 ∥ [001], 2V 76–8° [$2V_{\gamma calc.}$ 87°]. Named for F. T. Jones. 30th List

Jordisite. F. Cornu, 1909. *Zeits. Chem. Indust. Kolloide*, vol. iv, p. 190 (Jordisit). A black, powdery, colloidal form of molybdenum sulphide, considered to be distinct from the crystalline mineral molybdenite. It occurs in the Himmelsfürst mine, Freiberg, Saxony, and alters to ilsemannite. 5th List

Josëite C. A. A. Godovikov, K. V. Kochetkova, and Yu. G. Lavrent'ev, 1970. Геол. геофиз. (*Geol. geophys.*), no. 11, 123. (Жозеит C). Small (<10μm) grains of approximate composition $Bi_{16}(TeS_3)_3$ occur with four unnamed and also partially characterized bismuth sulphotellurides at the contact of quartz and native bismuth at Sokhondo, E Transbaikal. Named to follow josëite B (Peacock, 1941) [*A.M.*, **26**, 200] which also needs further study. [*A.M.*, **56**, 1839.] 27th List

Josëite D. A. A. Godovikov, K. V. Kochetkova, and Yu. G. Lavrentev, 1971. *Zap.*, **100**, 425. (Жозеит-D). Material from Ustarasa has the composition $Bi_{3·78}Pb_{0·24}Te_{1·56}S_{1·42}$; perhaps identical with the unnamed mineral of E. A. Dumin-Barkovskii (*Zap.*, **97**, 332). [*Zap.*, **101**, 278.] 28th List

Josen. F. Machatschki, 1924. *Zeits. Krist.*, vol. 60, p. 130 (Josen). A hydrocarbon, $C_{18}H_{30}$ extracted from lignite from Köflach, Styria, and recrystallized as triclinic-pedial (asymmetric) crystals. Identical with hartite, also from Styria. [*M.A.*, **2**, 474.] 10th List

Josenite. A. N. Winchell, *Optical mineralogy*, 1951, part 2, 4th edit., p. 133. Variant of josen (10th List). 19th List

Josephinite. W. H. Melville, *Amer. Journ. Sci.*, XLIII. 509, 1892. A nickel iron, Fe_2Ni_5, possibly meteoric. Josephine Co., Oregon. 1st List [Perhaps Awaruite. Cf. Bobrovkite, Souesite]

Jouravskite. C. Gaudefroy and F. Permingeat, 1965. *Bull. Soc. franç. Min. Crist.*, vol. 88, p. 254. Greenish-yellow to greenish-orange spots on dark manganese minerals on the dumps of the Tachgagalt No. 2 vein, Anti-Atlas, Morocco, consist of minute hexagonal crystals giving an X-ray powder photograph very similar to that of thaumasite. Composition near $Ca_6Mn_2^{4+}\{Si_{1·7}C_{0·3}(O,OH)_4\}(CO_3)_2(OH)_{12}.24H_2O$*. Named for G. Jouravsky. [*M.A.*, **17**, 399; *A.M.*, **50**, 2102.] [*Corr. in *M.M.*, **39**.] 24th List

Joyganit, error for Jogynait (C. Hintze, *Handb. Min.*, Erg. Bd. II, pp. 733, 945). 24th List

Jozit. *See* Iozite. 10th List

Juanite. E. S. Larsen and E. A. Goranson, 1932. *Amer. Min.*, vol. 17, pp. 343, 349, 354 (juanite, pronounced huanite). $10CaO.4MgO.Al_2O_3.11SiO_2.4H_2O$, white fibrous sheaves, perhaps orthorhombic, occurring as an alteration product of melilite. Named after the San Juan Mountains, Colorado. [*M.A.*, **5**, 146.] 13th List [Status uncertain]

Juddite. L. L. Fermor, 1908. *Rec. Geol. Survey India*, vol. xxxvii, p. 211; *Mem. Geol. Survey India*, 1909, vol. xxxvii, pp. lxxi, 159. A manganiferous amphibole associated with the manganiferous pyroxene blanfordite (L. L. Fermor, 1906; 4th List) in a braunite-albite rock at Kácharwáhi, Nágpur district, Central Provinces, India. The pleochroism (carmine, blue or green, and orange) is intense, and the optic axial plane is perpendicular to the plane of symmetry. Named after Professor John Wesley Judd, of London. 5th List [Manganoan magnesio-arfvedsonite; *Amph.* (1978)]

Jujuite. F. Ahlfeld and V. Angelelli, *Las especies minerales de la República Argentina*, Jujuy, 1948, p. 132 (jujuita). G. Gagarin and J. R. Cuomo, 1949, loc. cit., p. 15. Variant of jujuyite (18th List). Named from prov. Jujuy, Argentina. 19th List

Jujuyite. F. Ahlfeld, 1948. *Econ. Geol.*, vol. 43, p. 600. Iron antimonate, as brown to black compact masses intermixed with opal, from prov. Jujuy, Argentina. Named from the locality. [*M.A.*, **10**, 508.] 18th List

Julgoldite. P. B. Moore, 1967. *Canad. Min.*, **9**, 301. Monoclinic $Ca_2Fe^{2+}Fe_2^{3+}SiO_4Si_2O_7(OH)_2.H_2O$, from Långban, Sweden. The iron analogue of pumpellyite. [*A.M.*, **53**, 1426 (Jugoldite); *Bull.*, **91**, 305; *Zap.*, **93**, 330.] 26th List [Scotland, *M.M.*, **40**, 761; review, *M.A.*, 75-3464]

Julienite. A. Schoep, 1928. *Natuurwetenschappelijk Tijdschrift*, Antwerpen, vol. 10, p. 58 (Juliëniet', p. 59 (juliénite). Hydrated chloro-nitrate of cobalt as minute blue needles, presumably hexagonal and isomorphous with buttgenbachite (q.v.); from Katanga. Named after Henry Julien (d. 1920). 11th List [$Na_2Co(SCN)_4.8H_2O$; *M.A.*, **6**, 347; **12**, 337]

Julukulite, erroneous transliteration of Джулукулит, dzhulukulite, q.v. (Зап. Всесоюз. Мин. Общ. [*Mem. All-Union Min. Soc.*], 1959, vol. 88, p. 309). 22nd List

Junitoite. S. A. Williams, 1976. *A.M.*, **61**, 1255. Colourless hemimorphic orthorhombic crystals, *a* 6·309, *b* 12·503, *c* 8·549 Å, space group *Bbm*2, strongly pyroelectric. RIs α 1·656 ∥ [100], β 1·664 ∥ [010], γ 1·672 ∥ [001], $2V_\gamma$ 86°. Composition $4[CaZn_2Si_2O_7.H_2O]$, sp. gr. 3·5. From the Christmas mine, Gila Co., Arizona. Named for Jun Ito. [*M.A.*, 77-3390.] 30th List

Junkerite, variant of Junckerite (T. Thomson, *Outlines Min.*, 1836, vol. 1, p. 448). 22nd List

Junoite. W. G. Mumme, 1975. *Econ. Geol.*, **70**, 369; *A.M.*, **60**, 548. The Juno ore body, Tennant Creek, Australia, carries a mineral with space group *C*2/*m*, *Cm*, or *C*2, *a* 26·66, *b* 4·06, *c* 17·03 Å, β 127·20°; composition near $Cu_2Pb_3Bi_8(S,Se)_8$ with S:Se::5:1. Named for the locality. [*M.A.*, 76-1987; *Bull.*, **98**, 319.] 29th List

Jurbanite. J. W. Anthony and W. J. McLean, 1976. *A.M.*, **61**, 1. Clear colourless monoclinic crystals from the San Manuel copper mine, Pinal County, Arizona, have *a* 8·39, *b* 12·485, *c* 8·154 Å, β 101·90°; α 1·459, β 1·473 ∥ [010], γ 1·483, γ:[100] −5°, $2V_\alpha$ 80°. *D* 1·786 (1·828 calc.). Composition $4[AlSO_4OH.5H_2O]$; identical with synthetic material. Named for Joseph E. Urban. A dimorph of khademite (q.v.). 29th List

Jurupaite. A. S. Eakle, 1921. *Amer. Min.*, vol. 6, p. 107. Hydrated calcium (and magnesium) silicate, $H_2(Ca,Mg)_2Si_2O_7$, forming compact spheres of white, radiating fibres, which are probably monoclinic. It resembles pectolite in appearance, and occurs in metamorphic limestone in the Crestmore Hills, Jurupa Mountains, California. Named from the locality. [*M.A.*, **1**, 253.) 9th List [Magnesian Xonotlite; *A.M.*, **39**, 682, 851]

Jusite. I. Gramling-Mende and G. Leopold, 1943. *Neues Jahrb. Min. Monatshefte*, Abt. A, 1943, p. 178 (Jusit). Hydrated silicate of calcium with some alumina (8·23%) and alkalis (3·61%), '$(Si,Al)_6(Ca,Na,K)_5.5H_2O$', as white radiating needles, probably orthorhombic. Named from the locality, Jus, Schwäbische Alb, Württemberg. [*M.A.*, **9**, 37.] 17th List

Justite. *Min. Petr. Mitt.* (*Tschermak*), 1904, vol. xxiii, p. 97 (Justit); the name also appears on dealers' labels. Synonym of koenenite (*see* 3rd List, *Min. Mag.*, vol. xiii, p. 369). Named from the Justus I salt-mine in Hanover, where the mineral was found. 4th List

—— K. Hofmann-Degen, 1919. *Sitzungsber. Heidelberg. Akad. Wiss. Math.-naturw. Kl.*, 1919, Abt. A. Abh. 14, p. 29 (Justit), p. 99 (Justitfamilie). A polysilicate $(Ca,Mg,Fe,Zn,Mn)_3$-Si_2O_7 present as tetragonal crystals in a furnace-slag. It is allied to melilite, åkermanite, and gehlenite. Named after Mr. W. Just, of Zellerfeld, Harz, who supplied the material. The name justite was earlier used as a synonym of koenenite (4th List). 9th List

Juxporite. *Brit. Chem. Abstr.*, ser. A., 1926, p. 1022; from *Chem. Zentr.*, 1926, vol. 1, p. 1788 (Juxporit). A German spelling of Yuksporite (A. E. Fersman, 1922; 10th List). Another form is Juksporit (A. E. Fersman, *Neues Jahrb. Min.*, Abt. A., 1926, vol. 55, p. 42). 11th List

K

K-bentonite, *see* Potassium-bentonite.

K-priderite, *see* Potassium-priderite.

Kadmoselite. Literal version of Cadmoselite (q.v.). 21st List

Kafehydrocyanite. A. S. Povarennykh and L. D. Rusakova, 1973. Геол. журнал, **33**, part 2, 24 (Кафегидроцианит). Pale yellowish-green stalactites at the Medvezhii Log mine, Olkhovsk, E Sayan, prove to consist of potassium ferrocyanide, $K_4Fe(CN)_6.H_2O$. Named from the composition. [*M.A.*, 74-508; *A.M.*, **59**, 209; *Zap.*, **103**, 353.] Unrelated to hydrocyanite. 28th List [X-ray powder pattern, but not the optical properties, like $K_4Fe(CN)_6.3H_2O$]

Kahlerite. H. Meixner, 1953. *Der Karinthin*, no. 23, p. 277 (Kahlerit, Eisenarsenuranglimmer). Hydrous arsenate of uranyl and iron, $Fe(UO_2)(AsO_4).8H_2O$, as yellow rectangular plates probably monoclinic, from Hüttenberg, Carinthia. Named after Dr. Fritz Kahler, geologist, of the Carinthian Landesmuseum at Klagenfurt. [*M.A.*, **12**, 353; *Min. Mag.*, **30**, 344.] 20th List [*A.M.*, **39**, 1038. Cf. Metakahlerite]

Kalbaite. V. I. Vernadsky, 1913. *See* Elbaite. 7th List

Kalciotalk, variant transliteration of Кальциотальк, calciotalc, q.v. (Hintze, *Handb. Min.*, Erg.-Bd. II, p. 926). 22nd List

Kalgoorlite. E. F. Pittman, *Records Geol. Survey, New South Wales*, 1898, V, 203. Massive, iron-black. $HgAu_2Ag_6Te_6$. Kalgoorlie, Western Australia. 2nd List [A mixture; *M.M.*, **13**, 268]

Kaliägirin. P. Niggli, 1913. *Zeits. Anorg. Chem.*, vol. lxxxiv, p. 42. Artificially prepared potassium ferric metasilicate, $KFe'''(SiO_3)_2$, analogous to aegirite with potassium in place of sodium. 8th List

Kalianorthoklas. Z. Harada, *Journ. Fac. Sci. Hokkaido Univ.*, Sapporo, ser. 4, 1936, vol. 3, p. 265 (Kalianorthoklas). German translation of Potash-anorthoclase (13th List). Identical with hypoperthite (q.v.). 14th List

Kaliastrakanite. A. Naupert and W. Wense, *Ber. deutsch. chem. Ges.*, XXVI. (1), 873, 1893; C. A. Tenne, *Zeits. deutsch. geol. Ges.*, XLVIII. 632. 1896. The name kalium-astrakanite had previously been used by J. K. van der Heide (*Ber. deutsch. chem. Ges.*, XXVI. (1), 414, 1893) for the artificial salt. *See* leonite. $MgSO_4.K_2SO_4 + 4H_2O$. Monosymmetric. [Leopoldshall and Westeregeln] Prussian salt deposits. 1st List [Syn. of Leonite]

Kali-barium-felspar. Z. Harada, 1948. *Journ. Fac. Sci. Hokkaido Univ.*, ser. 4, vol. 7, p. 159 (Kali-Barium-Feldspar). A collective name to include celsian, hyalophane, and kasolite. 19th List

Kaliblödite. Mentioned by C. A. Tenne, *Zeits. deutsch. geol. Ges.*, XLVIII. 632, 1896. Syn. of kaliastrakanite. [= Leonite.] 1st List

Kalicinite. P. Groth, *Tabell. Übersicht der Mineralien*, 4th edit., 1898, p. 57; C. Hintze, *Handbuch der Mineralogie*, 1916, vol. 1, p. 2752 (Kalicinit). Variant of Kalicine (F. Pisani, 1865) for monosymmetric $HKCO_3$. 9th List

Kalicite. C. F. de Landero, *Sinopsis Mineralógica*, Mexico, 1888, p. 255 (Kalicita). Variant of Kalicine. *See* Kalicinite. 9th List

Kali-Harmotom. L. Gmelin, Leonhard's *Zeits. Min.*, 1825, vol. i, p. 11. Synonym of phillipsite. 3rd List

Kali-Klinoptilolith. K. F. Chudoba, 1974. C. Hintze, *Handb. Min.*, Erg.-Bd. **4**, 43. Translation of Potassium-clinoptilolite (q.v.) 28th List

Kali-magnesio-katophorite. R. T. Prider, 1939, *Min. Mag.*, vol. 25, p. 378. *See* Magnophorite. 15th List

Kali-montmorillonit. *See* Potash-montmorillonite. 14th List

Kalioalunite. E. T. Wherry, 1916. *Proc. US Nat. Museum*, vol. li, p. 82 (kalio-alunite), p. 83 (kalioalunite). Synonym of alunite. The potassium (kalium) end-member of the alunite isomorphous group $R_2'[Al(OH)_2]_6(SO_4)_4$, the sodium end-member being natroalunite (W. F. Hillebrand and S. L. Penfield, 1902; 3rd List). 8th List

Kalio-carnotite. E. T. Wherry, 1914. *Science*, New York, new ser., vol. xxxix, p. 576. The original carnotite containing potassium, as distinct from calcio-carnotite (q.v.). In another place (*Bull. United States Geol. Survey*, 1914, no. 580, p. 149) the same author employs the term *potassio-carnotite*. 7th List

Kaliohitchcockite. E. T. Wherry, 1916. *See* Bariohitchcockite. 8th List

Kaliophyllite. Error for kaliophilite. Cf. Corundophyllite (20th List). [*M.A.*, **13**, 493.] 21st List

Kalipyrochlore. I.M.A. Subcom. pyrochlore group, 1978. *A.M.*, **62**, 404. The unnamed mineral of Van Wambeke (1965) is named. 30th List

Kalisaponite. P. N. Chirvinsky, 1939. *Bull. Acad. Sci. URSS, Sér. Géol.*, 1939, no. 4, p. 32 (калисапонит), p. 43 (kalisaponite). A zeolitic mineral resembling saponite in appearance, containing K_2O 6·57%; formula $(R_2O,RO).Al_2O_3.5SiO_2.3H_2O$. [*M.A.*, **7**, 515.] 15th List

Kalistrontite. M. L. Voronova, 1962. Зап. Всесоюз. Мин. Общ. (*Mem. All-Union Min. Soc.*), vol. 91, p. 712 (Калистронцит). Prismatic crystals and hexagonal tablets of $K_2Sr(SO_4)_2$ from Alshtan, Bashkiria. Isostructural with Palmierite. Named from the composition. (*Amer. Min.*, 1963, vol. 48, p. 708; *M.A.*, **16**, 183.) 23rd List

Kalithomsonite. S. G. Gordon, 1924. *Proc. Acad. Nat. Sci. Philadelphia*, vol. 76, p. 261. A variety of thomsonite containing 6·05% K_2O, occurring as a loose aggregate of minute orthorhombic needles in augite-syenite at Narsarsuk, Greenland. [*M.A.*, **2**, 385.] 10th List [Renamed Ashcroftine (q.v.), *M.M.*, **23**, 305]

Kaliumapatit. H. Wondratschek, 1963. *Neues Jahrb. Min.*, Abh., vol. 99, p. 121. Original form of Potassium-apatite, q.v. 23rd List

Kaliumkryolith, *see* Potassium-cryolite.

Kalium-Pharmakosiderit. K. F. Chudoba, 1974. C. Hintze, *Handb. Min.*, Erg.-Bd., **4**, **44**. Syn. of pharmacosiderite. (The name is not used by Walenta in the reference cited; he writes: Der Name Pharmakosiderit hätte nur für das kaliumhaltige Eisenarsenat Gültigkeit.) Cf. Alumopharmacosiderite, 14th List. 28th List

Kaliumpriderit, *see* Potassium-priderite.

Kalium-Richterit. H. Strunz, 1966. *Min. Tabellen*, 4th edit., p. 484. Synonym of Magnophorite (15th List). 24th List

Kaliumstruvit. H. Strunz, 1957. *Min. Tabellen*, 3rd edit., p. 243. The artificial compound $KMgPO_4.6H_2O$, analogous to struvite. 22nd List

Kalkeisencordierit. H. Bücking, *Ber. Senckenberg. Naturf. Ges. Frankfurt-am-Main*, 1900, p. 7. Cordierite with lime and ferrous oxide in place of magnesia. In ejected volcanic blocks from the Celebes. 3rd List

Kalkeisenolivin. C. Doelter, 1913. *Handbuch der Mineralchemie*, vol. 2, part 1, p. 499 (Kalkeisenolivin, Eisenmonticellit). A crystallized slag described by W. von Gümbel (*Zeits. Kryst. Min.*, 1893, vol. 22, p. 269), who compared it with monticellite with MgO replaced by FeO. 11th List

Kalknatronplagioklas. F. Machatschki, 1953. *Spez. Min.*, Wien, p. 321. Syn. of Andesine. 22nd List

Kalk-Olivin. K. Oebbeke, 1877. *Ein Beitrag zur Kenntniss des Palaeopikrits und seiner Umwandlungsproducte. Inaug.-Diss. Würzburg*, 1877, p. 13 (Kalk-Olivin); C. Hintze, *Handbuch d. Mineralogie*, 1889, vol. 2, p. 8 (Kalkolivin). A pale-green olivine, from a palaeopikrite in Nassau, containing CaO 14·09%, and intermediate between olivine and monticellite in composition. The name has also been loosely applied to olivines containing less than this amount of lime, to monticellite, and to calcium orthosilicate. *See* Lime-olivine and Shannonite. 11th List

Kalkorthosilikat. O. Mügge, 1926. H. Rosenbusch's *Mikrosk. Physiogr. Min. u. Gesteine*, 5th edit., vol. 1, part 2, p. 367. The β-modification of Ca_2SiO_4 described by F. P. Paul in 1906. *See* Shannonite. 11th List

Kalkowskyn. E. Rimann, 1925. *Centralblatt Min.*, Abt. A, 1925, p. 18 (Kalkowskyn). Titanate (and silicate) of iron (cerium, etc.), $(Fe,Ce)_2O_3.4(Ti,Si)O_2$, found as minute, black, platy grains in a concentrate from Brazil. Named after Prof. Louis Ernst Kalkowsky, of Dresden. [This termination is also found in Razoumovskyn (J. F. John, 1814)] [*M.A.*, **2**, 419.] 10th List

Kalkthomsonit. C. F. Rammelsberg, 1895. *See* Calciothomsonite. 10th List

Kalkwulfenit, *see* Calcowulfenite.

Kalsilite. F. A. Bannister, 1942. *Min. Mag.*, vol. 26, p. 218. A. Holmes, ibid., p. 197; *Geol. Mag.*, 1942, vol. 79, p. 226. A polymorphous form of potassium aluminosilicate, $KAlSiO_4$, hexagonal, but differing from kaliophilite in its X-ray pattern; occurs in potash-rich volcanic rocks from Uganda and Italy. Named from the chemical formula. [*M.A.*, **8**, 318; *A.M.*, **28**, 62.] 16th List

Kalziotalk, *see* Calciotalc.

Kalzium-Rinkit, *see* Calcium-rinkite.

Kamarezite. K. Busz, *Verh. naturhist. Ver. Bonn*, L. *Sitz.-ber.*, p. 83, 1893; and *Neues Jahrb. Min.*, I, 115, 1895. [*M.M.*, **11**, 108] $3CuO.SO_3.8H_2O$. Rhombic. Kamareza, Greece. 1st List [Is Brochantite; *A.M.*, **50**, 1450; *M.A.*, **17**, 604]

Kamenskite. A. K. Gladkovskii and I. N. Ushatinskii, 1961. [Труды Горно-геол. Инст. Уральск. фил. Акад. наук СССР (*Proc. Mining-geol. Inst. Ural Div. Acad. Sci. USSR*), vol. 56, p. 114], abstr. Зап. Всесоюз. Мин. Общ. (*Mem. All-Union Min. Soc.*), 1966, vol. 95, p. 311 (Каменскит). Merely finely dispersed Diaspore. 24th List

Kamiokalite. K. Sakurai, H. Nagashima, and E. Sorita, 1952. [*Syumi-no-Tigaku* (*Amateur Geologist*), vol. 5, p. 170] abstr. in *Amer. Min.*, 1955, vol. 40, p. 367 (Kamiokalite). Z. Harada, *Journ. Fac. Sci. Hokkaido Univ.*, 1954, ser. 4, vol. 8, p. 317 (Kaniokaite). Blue monoclinic crystals, $3CuO.3ZnO.P_2O_5.7H_2O$. A variety rich in zinc of veszelyite*; compare arakawaite (9th List). Named from the locality, Kamioka mine, Japan. [*Corr. in *M.M.*, **31**.] 20th List

Kanaekanite. A. S. Povarennykh and V. D. Dusmatov, 1970. Инфракрасные спектры поглощения новых минералов из целочных пегматитов Средней Азии. Сб.: Конституция и свойства минералов. Киев, по. 4, стр. 7 (Канаэканит). The 'ekanite' from Central Asia (*M.A.*, **19**, 177; **69**, 1044), which differs from Ceylon (type) ekanite in containing K and Na, is recognized as a separate species and named from its composition. 28th List

Kanasite, erroneous transliteration of Канасит, canasite (q.v.). 22nd List

Kanbaraite. H. Isobe and T. Watanabe, 1930. *Bull. Inst. Phys. and Chem. Research*, vol. 9, p. 440; *Sci. Papers Inst. Phys. and Chem. Research*, Tokyo, 1930, vol. 13, Abstr. p. 40. A hexagonal acid clay (Kanbara clay) from Japan with the composition $H_4Al_2MgSi_6O_{18}.xH_2O$ is called 'kanbaraite A', and the dehydrated (at 800°C) cubic form $H_4Al_2MgSi_6O_{18}$ is called 'kanbaraite B'. [*M.A.*, **4**, 501.] 12th List

Kandite. *Clay Min. Bull.*, 1955, vol. 2, pp. 295, 296. Group name to include kaolinite, nacrite, dickite, halloysite. A portmanteau word, but overlooking halloysite. 21st List

Kanemite. Z. Johan and G. F. Maglione, 1972. *Bull.*, **95**, 371. Small spherules of platy orthorhombic crystals, *a* 7·28, *b* 20·507, *c* 4·956 Å, space group *Pnmb*, occur at the Andjia saline, Kanem, Chad; composition $4[NaHSi_2O_4(OH)_2.2H_2O]$. Named for the locality. [*M.A.*, 73-1938; *Bull.*, **96**, 237; *A.M.*, **59**, 210; *Zap.*, **102**, 454.] 28th List

Kaniokaite, *see* Kamiokalite.

Kaňkite. F. Čech, J. Jansa, and F. Novák, 1976. *Neues Jahrb. Mineral.*, Monatsh., 426. Yellowish-green botryoidal crusts in old dumps near Kaňk, Kutna Hora, Czechoslovakia, are monoclinic, *a* 18·803, *b* 17·490, *c* 7·633 Å, β 92·71°. Composition $16[FeAsO_4.3\frac{1}{2}H_2O]$, sp. gr. 2·70. [*A.M.*, **62**, 594.] 30th List

Kanoite. H. Kobayashi, 1977. *J. geol. Soc. Japan*, **83**, 537. A vein in pyroxmangite–cummingtonite rock at the Tatehira mine, Oshima peninsula, Hokkaido, contains light pinkish-brown monoclinic grains, *a* 9·739, *b* 8·939, *c* 5·260 Å, β 108·56°. A clinopyroxene of composition $(Mn,Mg)_2Si_2O_6$, sp. gr. 3·66. RIs α 1·715, β 1·717 ‖ [010], γ 1·728, γ:[001] 42°, $2V_\gamma$ 40–42°. Named for H. Kano. [*A.M.*, **63**, 598.] 30th List

Kansasite. J. D. Buddhue, 1938. *Mineralogist*, Portland, Oregon, vol. 6, no. 1, p. 8. A fossil resin from Kansas, afterwards named jelinite (q.v.). [*M.A.*, **7**, 265.] 15th List

Kansite. F. H. Meyer, 1957. A.S.T.M. Powder Index Card no. 7–26. Cubic Fe_7S_8, found as a corrosion product on iron pipes in Kansas, Texas. Named from the locality. 23rd List [Probably Mackinawite (q.v.), *Rev. Pure Appl. Chem.*, **20**, 195]

Kaolin-Chamosit. H. Strunz, 1957. *Min. Tabellen*, 3rd edit., p. 323. Syn. of Berthierine. 22nd List

Kaolite. 'Moulded imitation cameos, etc., in baked clay'. R. Webster, *Gems*, 1962, p. 758. A thoroughly objectionable name. 28th List

Karachaite. N. E. Efremov, 1936. *Bull. Acad. Sci. URSS, Cl. Sci. Math. Nat., Sér. Géol.*, 1936, p. 921 (Карачаит), p. 927 (karachaite). An asbestiform variety of chrysotile-serpentine with the composition $MgO.SiO_2.H_2O$. Named from the locality, Karachai, Caucasus. [*M.A.*, **9**, 9, 370.] 15th List

Karelianite. J. V. P. Long, Y. Vuorelainen, and O. Kuovo, 1963. *Amer. Min.*, vol. 48, p. 33. Black grains in boulders from the Outokumpu ore body, Karelia, Finland, are essentially V_2O_3 with a little Fe and Cr. X-ray powder data match synthetic rhombohedral V_2O_3. Named from the locality. [*M.A.*, **16**, 457.] 23rd List

Karibibite. O. von Knorring, Th. G. Sahama, and P. Rehtijärvi, 1973. *Lithos*, **6**, 265. Brownish-yellow fibres in löllingite from the Karibib pegmatite, SW Africa, are orthorhombic (a 27·91, b 6·53, c 7·20 Å); $6[Fe_2^{3+}As_4^{3+}O_9]$, with some replacement of As^{3+} by $3H^+$. Named for the locality. [*A.M.*, **57**, 1315; **59**, 382; *M.A.*, 74-1448.] 28th List

Karnasurtite. M. V. Kuzmenko and S. I. Kozhanov, 1959. [Труды Инст. Мин., Геохим., Крист., Редк. Элем. (trans. *Inst. Min. Geochem. Cryst. Rare Elements*), no. 2, p. 95]; abstr. *Amer. Min.*, 1960, vol. 45, p. 1133; Зап. Всесоюз. Мин. Общ. (*Mem. All-Union Min. Soc.*), 1961, vol. 90, p. 107 (Карнасуртит, karnasurtite). A metamict mineral from Mt. Karnasurt, Lovozero massif, Kola peninsula, approximates to $(Ln,Th)(Ti,Nb)(Al,Fe)(Si,P)_2O_7(OH)_4.3H_2O$. Named from the locality. It had previously been named Kozhanovite without a description. The composition is very near that of hydrocerite (of Vlasov *et al.*) (q.v.). 22nd List

Karpathit, German transliteration of Карпатит, Carpathite (21st List) (H. Strunz, *Min. Tabellen*, 3rd edit., 1957, p. 348). 24th List

Karpatite. Russian карпатит for carpathite (q.v.). 21st List

Karpinskiit, variant transliteration of Карпинскиит, Karpinskyite (21st List) (H. Strunz, *Min. Tabellen*, 3rd edit., 1957, p. 338). [This variant carries an increased risk of confusion with Карпинскит, Karpinskite (21st List).] 24th List

Karpinskite. I. A. Rukavishnikova, 1956. [*Kora vyvetrivaniya, Acad. Sci. USSR*, no. 2, p. 164.] Abstr. in *Mém. Soc. Russ. Min.*, 1957, vol. 86, p. 124 (карпинскит). $(Ni,Mg)_2Si_2O_5(OH)_2$, containing NiO 21·12, MgO 17·56%, compact, greenish-blue, with cerolite minerals (nickel-β-kerolite, ferrikerolite, qq.v.) in crevices in serpentine. Urals. Named after A. P. Karpinsky. Not to be confused with karpinskyite (q.v.). 21st List [*A.M.*, **42**, 584]

Karpinskyite. L. L. Shilin, 1956. *Doklady Acad. Sci. USSR*, vol. 107, p. 737 (Карпинскиит), English table of contents (karpinskyite). Radial aggregates of white hexagonal needles in pegmatite [from Kola peninsula]. $Na_2(Be,Zn,Mg)Al_2Si_6O_{16}(OH)_2$. Named after Alexander Petrovich Karpinsky, Александр Петрович Карпинский (1847–1936), Russian geologist. [*Min. Mag.*, **25**, 293; *M.A.*, **13**, 209.] 21st List [A mixture containing Leifite (q.v.); *A.M.*, **57**, 1006; *M.A.*, 72-2222]

Karrenbergite. E. Walger, 1958. [*Inaug. Diss., Freiburg*, p. 52]; abstr. Hintze, *Handb. Min.*, Erg.-Bd. II, 1959, p. 739 (Karrenbergit). A clay mineral intermediate between nontronite and saponite, from Karrenberg, Reichweiler, Pfalz, Germany. Named from the locality. 22nd List

Karrooite. O. von Knorring and K. G. Cox, 1961. *Min. Mag.*, vol. 32, p. 676. Artificial $MgTi_2O_5$, isomorphous with pseudobrookite. Named from the Karroo volcanic series of Southern Rhodesia, in which there occurs a mineral Kennedyite (q.v.) containing nearly 50% of this end-member. [*M.A.*, **15**, 134.] 22nd List

Karystiolite. J. W. Evans, 1909. *Geol. Mag.*, dec. 5, vol. vi, p. 286. Cf. *Mineralogical*

Magazine, 1906, vol. xiv, p. 143. A modernized form of *καρύστιος λίθος* (Karystian stone), a mineral, identified with chrysotile, from Karystos in Euboea. Suggested as an alternative for chrysotile, owing to the confusion between this name and the name chrysolite. 5th List

Kasoite. T. Yoshimura, 1936. *Journ. Geol. Soc. Japan*, vol. 43, p. 877 (Japanese). A variety of barium-felspar associated with celsian and hyalophane as gangue minerals in the Kaso mine, Tochigi, Japan. Named from the locality. [*M.A.*, **6**, 489.] 14th List

Kasolite. A. Schoep, 1921. *Compt. Rend. Acad. Sci. Paris*, vol. 173, p. 1476. Hydrated silicate of uranium and lead, $3PbO.3UO_3.3SiO_2.4H_2O$, forming ochre-yellow, monoclinic crystals. Named from the locality Kasolo, Katanga, Belgian Congo. [*M.A.*, **1**, 249.] 9th List [$Pb(UO_2)SiO_4.H_2O$]

Kasparit, variant of Kašparite (21st List) (H. Strunz, *Min. Tabellen*, 4th edit., 1966, p. 485). 24th List

Kašparite. A. Dubanský, 1956. *Chem. Listy*, Praha, vol. 50, p. 1352 (Kašparit). (Mg,Co)-$Al_3(SO_4)_5OH.28H_2O$. Variety of pickeringite containing CoO 1·52%, from Dubnik, Slovakia. Named after Prof. Jan V. Kašpar of Praha. [*A.M.*, **42**, 919.] 21st List

Kassite. A. A. Kukharenko, 1965. [*The Caledonian ultrabasic alkalic rocks and carbonatites of the Kola Peninsula and northern Karelia*, Moscow], abstr. *Amer. Min.*, 1967, vol. 52, p. 559. Orthorhombic crystals near $CaTi_2O_4(OH)_2$ in cavities of alkali pegmatites of the Afrikanda massif, Kola peninsula. Not to be confused with kasoite. Named for N. G. Kassin. [Зап., **96**, 70 (Кассит); *Bull.*, **90**, 609.] 25th List

Kassiterolamprit. A. Frenzel. C. Hintze, *Handbuch d. Mineralogie*, 1904, vol. i, p. 1191 footnote. Thought to be a new mineral from Bolivia, but afterwards identified with stannite. 4th List

Katangite. H. Buttgenbach, 1921. *Mém.* (*in-8°*) *Acad. Belgique, Cl. d. Sci.*, ser. 2, vol. 6, fasc. 8, p. 26. A hydrated copper silicate, $CuH_2SiO_4.H_2O$, containing more water than dioptase. Occurs abundantly as a bluish, amorphous mineral with dioptase in Katanga, Belgian Congo. Named from the locality. [*M.A.*, **1**, 250.] 9th List [Var. of Planchéite; *M.A.*, 71-2246]

Katharite. W. Ainsworth, 1834. *An account of the caves of Ballybunian, County of Kerry: with some mineralogical details*, Dublin, 1834, pp. 83, 85. Synonym of alunogen. Named katherite or katharite 'from *χαθαρος* [i.e. *κᾰθᾰρός*], impermixtus, simplex; in allusion to the absence of potash', it being hydrous aluminium sulphate and thus more simple in composition than alum. 4th List

Katoforit. *See* catophorite. 2nd List

Katophorit, variant of Catophorite (H. Strunz, *Min. Tabellen*, 1941 (1st edit.), pp. 194, 249*). [*Corr. in *M.M.*, **33**.] 22nd List [**Katophorite** is the accepted spelling; *Amph.*, Sub.-Comm.]

Katoptrite. G. Flink, 1917. *Geol. För. Förh.*, vol. xxxix, p. 431 (Katoptrit). Abstr. in *Amer. Journ. Sci.*, 1917, vol. xliv, p. 484, and *Amer. Min.*, 1917, vol. ii, p. 129, give the spelling Catoptrite. Silico-antimonate of manganese, aluminium, etc., $14(Mn,Mg,Fe)O.2(Al,Fe)_2O_3.$-$Sb_2O_5.2SiO_2$, forming monoclinic crystals with black colour and metallic lustre. There is a highly perfect cleavage parallel to the orthopinacoid. Occurs with magnetite in granular limestone at Nordmark, Sweden. Named from *κάτοπτρον*, a mirror, because of the brilliant lustre on the cleavage surface. 8th List [$(Mn^{2+}_5Sb^{5+}_2)(Mn^{2+}_8Al_4Si_2)O_{28}$; *A.M.*, **62**, 396; *M.A.*, 76-3602]

Kauaiite. E. Goldsmith, *Proc. Acad. Nat. Sci. Philadelphia*, 105, 1894. [*M.M.*, **11**, 166.] $2Al_2O_3.3(K,Na,H)_2O.SO_3$. [Kauai Island] Hawaii. 1st List

Kawazulite. A. Kato, 1970. *Introduction to Japanese Minerals, Geol. Surv. Japan*, 1970, 87 (provisional description). Silver to tin-white foils with selenian tellurium in vein quartz from the Kawazu mine, Shizuoka Prefecture, Japan, are the Se analogue of tetradymite. a_h 4·24, c_h 29·66 Å, $R\bar{3}m$, $R3m$, or $R32$, D > 7·5 (8·08 calc.), H $1\frac{1}{2}$, R 45 to 50%. Composition $3[BiTe_2Se]$. Named for the locality. [*A.M.*, **57**, 1312.] 27th List

Kayserite. K. Walther, 1922. *Zeits. Deutsch. Geol. Gesell.*, vol. 73 (for 1921), p. 316 (Kayserit). Aluminium hydroxide, $Al_2O_3.H_2O$, stated, on optical grounds to be monoclinic, and therefore dimorphous with diaspore. Occurs as an alteration product of corundum at Cerro

Redondo, Uruguay. Named after Prof. Friedrich Heinrich Emanuel Kayser, of Marburg. [*M.A.*, **2**, 12] 10th List [Is Diaspore; *M.A.*, **8**, 230; **11**, 468; **12**, 340]

Kazakhovite, error for Kazakovite (*M.A.*, 75-1392). 29th List

Kazakovite. A. P. Khomyakov, E. I. Semenov, E. M. Eskova, and A. A. Voronkov, 1974. *Zap.*, **103**, 342 (Казаковит). A rhombohedral member of the lovozerite family, *a* 10·18, *c* 13·06 Å, composition $Na_6H_2TiSi_6O_{18}$, from Mt. Karnasurt, Lovozero massif, is named for M. E. Kazakova. 28th List [*A.M.*, **60**, 161]

Keatite. R. B. Sosman, 1954. *Science* (*Amer. Assoc. Adv. Sci.*), vol. 119, p. 738 (keatite), p. 739 (silica K). A high-pressure tetragonal form of silica prepared hydrothermally by Paul P. Keat, after whom it is named. [*M.A.*, **12**, 410.] 20th List

Keeleyite. S. G. Gordon, 1922. *Proc. Acad. Nat. Sci. Philadelphia*, vol. 74, p. 101. Sulphantimonite of lead, $2PbS.3Sb_2S_3$, occurring as acicular (orthorhombic?) crystals at Oruro, Bolivia. Named after Mr. Frank J. Keeley, of Philadelphia, Pa. [*M.A.*, **2**, 11.] 10th List [Is. Zinckenite; *M.M.*, **25**, 221]

Kegelite. O. Medenbach and K. Schmetzer, 1975. *Naturwiss.*, **62**, 137. Pseudohexagonal plates (30 μm) with hematite and mimetite from Tsumeb, SW Africa, are formulated $Pb_{12}(Zn,Fe)_2Al_4(SO_4)_4Si_{11}O_{38}$. Named for F. W. Kegel. [*A.M.*, **61**, 175.] 29th List

Kehoeite. W. P. Headden, *Amer. Journ. Sci.*, XLVI. 22, 1893. $(Zn_{0{\cdot}75} + Ca_{0{\cdot}25})_3P_2O_8 + 2Al_2(OH)_6 + 21H_2O$. [Merritt mine, Galena] S. Dakota. 1st List [Perhaps related to Analcime; *M.M.*, **33**, 799]

Keldychite, French transliteration of Келдышит, keldyshite (*Bull. Soc. franç. Min. Crist.*, 1963, vol. 86, p. 306). 23rd List

Keldyshite. V. I. Gerasimovsky, 1962. Докл. Акад. Наук СССР (*Compt. Rend. Acad. Sci. URSS*), vol. 142, p. 916 (Келдышит). Irregular grains, optically biaxial, from the Lovozero massif, Kola peninsula, have a composition near $(Na,H)_2ZrSi_2O_7$, with Na ≫ H. Named for M. V. Keldysh (*Amer. Min.*, 1962, vol. 47, p. 1216; Зап. Всесоюз. Мин. Общ. (*Mem. All-Union Min. Soc.*), 1963, vol. 92, p. 209; *M.A.*, **15**, 543). 23rd List [*M.A.*, 73-3351]

Kellerite. G. Gagarin and J. R. Cuomo, 1949, loc. cit., p. 13 (kellerita). $(Mg,Cu)SO_4.5H_2O$, containing MgO 11·36, CuO 12·46, H_2O 38·31%, as bluish-white earthy masses apparently pseudomorphous after chalcanthite, from Copaquire, Chile, described by H. F. Keller, *Proc. Amer. Phil. Soc.*, 1908, vol. 47, p. 81. 19th List [Cuprian Pentahydrite]

Kellyite. D. R. Peacor, E. J. Essene, W. B. Simmons, and W. C. Bigelow, 1974. *A.M.*, **59**, 1153. Yellow plates and laths, up to 1 mm in diameter, in kutnahorite at the Bald Knob manganese deposit, Sparta, North Carolina, have *a* 5·438, *c* 14·04 Å space group $P6_3$, but besides this two-layer polytype there is a six-layer rhombohedral polytype. Composition near $Mn_4Al_4Si_2O_{10}(OH)_8$ with some Mg and Fe; the Mn analogue of amesite: Named for W. C. Kelly. [*M.A.*, 75-2523; *Bull.*, **98**, 265.] 29th List

Kemmlitzite. J. Hak, Z. Johan, M. Kuacek, and W. Liebscher, 1969. *Neues Jahrb. Min.*, *Monatsh.* 201. $(Sr,Ce,La)Al_3AsO_4(P,S)O_4(OH)_6$, the arsenate analogue of svanbergite, was found in the heavy fraction from kaolinized quartz porphyry from Kemmlitz, Saxony. Named for the locality. [*A.M.*, **55**, 320; *Bull.*, **92**, 512; *Zap.*, **99**, 79.] 26th List

Kempite. A. F. Rogers, 1924. *Amer. Journ. Sci.*, ser. 5, vol. 8, p. 145. Preliminary abstr. in *Amer. Min.*, vol. 9, p. 66. Hydrous manganese oxychloride, $MnCl_2.3MnO_2.3H_2O$, as small, emerald-green, orthorhombic crystals in manganese-ore from California. Named after Prof. James Furman Kemp, of Columbia University, New York. [*M.A.*, **2**, 338.] 10th List

Kennedyite. O. von Knorring and K. G. Cox, 1961. *Min. Mag.*, vol. 32, p. 676. $Fe_2MgTi_3O_{10}$, isostructural with pseudobrookite, occurring in the Mateke Hills, Southern Rhodesia. Named after Prof. W. Q. Kennedy of Leeds. [*M.A.*, **15**, 134; *A.M.*, **46**, 766.] 22nd List

Kenngottite. G. Gagarin and J. R. Cuomo, 1949, loc. cit., p. 10 (kenngottita). Vitreous amorphous As_2O_3 (arsenolite) mentioned by J. F. L. Hausmann in 1850 and by Johann Gustav Adolph Kenngott (1818–1897) in 1852. Not the kenngottite (synonym of miargyrite) of W. Haidinger, 1857. 19th List

Kentsmithite. A local name for a black vanadium-bearing sandstone found, on the claim of Mr. J. Kent Smith, in Paradox Valley, Colorado. A specimen so labelled was received at the British Museum in 1914, and the name has appeared in *Amer. Min.*, 1921, vol. 6, p. 171, and *Bull. US Geol. Survey*, 1924, no. 750–D, p. 64. *See* Vanoxite. 10th List [Cf. Montroseite, Paramontroseite]

Kenyaite. H. P. Eugster, 1967. *Science*, vol. 157, p. 1177. Nodular concretions from Lake Magadi, Kenya, approximate to $Na_2Si_{22}O_{41}(OH)_8 6H_2O$; tetragonal. Named from the locality. [*A.M.*, **53**, 510, 2061] 25th List

Keramite. T. S. Hunt, *Mineral Physiology and Physiography*, 1886, p. 371; and *Trans. Roy. Soc. Canada*, III. (3), 76, (1885) 1886. A clay resulting from the alteration of scapolite. Bavaria. 1st List.

—— J. W. Mellor and A. Scott, 1924. *Trans. Ceramic Soc. Stoke-on-Trent*, vol. 23 (for 1923–1924), pp. 328, 336. V. I. Vernadsky, ibid., 1925, vol. 24, p. 18. Aluminium silicate, $3Al_2O_3.2SiO_2$, obtained artificially by heating kaolinite at 1700°. Evidently the same as mullite (10th List). Named from *κέραμος*, potter's earth, pottery. Not the keramite of T. S. Hunt, 1886, nor the ceramite of T. Blount, 1656, and T. Urquhart, 1693. 11th List

Keramsite. J. Dominic, M. Kita-Badak, and A. Manecki, 1975. *Polska Akad. Nauk, Prace Min.*, **40**, 19. A name for light-weight 'bloated' clay aggregates. [*M.A.*, 76-3203.] 30th List

Kerchenite. Another spelling of kertschenite (4th List), being the English transliteration of керченит, from Kerch (Керчь), Crimea. [*M.A.*, **9**, 92, 311.] 17th List

Kernite. W. T. Schaller, 1927. *Amer. Min.*, vol. 12, p. 24. H. S. Gale, *Engin. Mining Journ. New York*, 1927, vol. 123, p. 10. Hydrated sodium borate, $Na_2O.2B_2O_3.4H_2O$, as colourless orthorhombic crystals and cleavage masses from Kern Co., California. Named from the locality. [*M.A.*, **3**, 271.] 11th List

Kersinite, *see* Kerzinite.

Kertisite. Russian кертисит for curtisite. *See* Curtisoids. 21st List

Kertisitoide, Germanized transliteration of Кертиситоиди, Curtisitoids (21st List) (C. Hintze, *Handb. Min.*, Erg.-Bd. II, p. 739). 24th List

Kertschenite. S. P. Popoff, 1906. *Centralblatt Min.*, 1906, p. 113 (Kertschenit). Hydrated phosphate of iron with small amounts of manganese and magnesium. $(Fe,Mn,Mg)O.$-$Fe_2O_3.P_2O_5.7H_2O$. It occurs as radially fibrous aggregates of dark-green, flattened crystals in the limonitic iron-ores of the Kerch (German, Kertsch) peninsula, Crimea. 4th List [Oxidized Vivianite; cf. Paravivianite]

Kerzinite. N. A. Shadlun, 1923. *Commission for the study of the natural products of Russia, Russ. Acad. Sci.*, Petrograd, vol. 4, no. 5 (Nickel), p. 4 (керзинит). Abstr., *Neues Jahrb. Min.*, Abt. A, 1926, vol. 2, p. 113 (Kersinit). A lignite impregnated with a hydrated nickel silicate and used as a nickel ore in the Urals. The ash (6–7%) contains Ni 3–15%. Named after the mining engineer N. A. Kerzin (Н. А. Керзин) who discovered the deposit in 1913. [*M.A.*, **3**, 492.] 11th List

Kësterite. Z. V. Orlova, 1956. [*Trudy Magadan Sci. Research Inst. 1956*, p. 76.] Abstr. in *Mém. Soc. Russ. Min.*, 1958, vol. 87, p. 76 (кёстерит, custerite, Kösterit). (Cu,Sn,Zn)S, containing Cu 30·56, Sn 25·25, Zn 11·16, S 23·40%. In quartz-sulphide ore from Këster, Magadan, Yakutia, NE Siberia. Named from the locality. 21st List [*A.M.*, **43**, 1222; **44**, 1329. Syn. Isostannite (q.v.)]

Kesterite, Kêsterite, further variant transliterations of Кёстерит, kësterite (*Amer. Min.*, 1958, vol. 43, p. 1222). 22nd List

Kettnerite. L. Žak and V. Syneček, 1956. *Časopis Min. Geol.*, vol. 1, p. 195. $CaF(BiO)CO_3$, tetragonal, brown to yellow crystals with bismuth, fluorite, etc., in pegmatite from Krupka, Bohemia. Named after Radim Kettner, professor of geology, Karlova University, Praha. 21st List [*A.M.*, **42**, 121; **43**, 385]

Keweenawite. G. A. Koenig, *Amer. Journ. Sci.*, 1902, ser. 4, vol. xiv, p. 410. A massive, finely-granular mineral, with a pale pinkish-brown colour and metallic lustre, from the Mohawk mine, Keweenaw Co., Michigan. Analyses agree approximately with the formula $(Cu,Ni,Co)_2As$. *See* mohawkite. 3rd List [A mixture; *M.A.*, **3**, 401]

Keyite. P. G. Embrey, E. E. Fejer, and A. M. Clark, 1977. *Mineral. Rec.*, **8**, (3), 87. Deep sky-blue crystals in tennantite ore from Tsumeb, SW Africa, are monoclinic, *a* 11·65, *b* 12·68, *c* 6·87 Å, β 98·95°. Composition $6[(Cu,Zn,Cd)_3(AsO_4)_2]$ with Cu : Zn : Cd 1·19 : 1·11 : 0·55. Pleochroic, α 1·80, pale blue, γ 1·87, deep blue, β greenish-blue, ∥ [010]; α:[001] 10½° (Cd red), $11\frac{1}{2}°$ (Hg yellow), $12\frac{1}{2}°$ (Hg green), $9\frac{1}{2}°$ (Hg violet). Named for C. L. Key. [*A.M.*, **62**, 1259; *M.A.*, 78-3474.] 30th List

Keystoneite. Author and date? Chalcedony coloured blue by chrysocolla [?]. R. Webster, *Gems*, 1962, p. 758. 28th List

Khademite. P. Bariand, J.-P. Berthelon, F. Cesbron, and M. Sadrzadeh, 1973. *Compt. Rend. Acad. Sci. Paris*, **277D**, 1585. Colourless crystals, space group *Pcab*, *a* 11·178, *b* 13·055, *c* 10·887 Å, from Saghand, Iran, have composition $8[AlSO_4OH.5H_2O]$. Named for N. Khadem [*A.M.*, **60**, 486; *M.A.*, 74-3464; *Zap.*, **103**, 625.] A dimorph of jurbanite (q.v.). 29th List

Khakassite. G. A. Bilibin, 1926. *Zap. Ross. Min. Obshch.* (*Mém. Soc. Russ. Min.*), 1928, ser. 2, vol. 57, p. 307 (Хакассит). This name, which appeared in the title of the paper when read in 1926, was afterwards replaced by Alumohydrocalcite (11th List). Named from the locality, Khakassky district in Siberia. 12th List

Khakasskyite, variant of Khakassite (Хакассит, 12th List) (H. Strunz, *Min. Tabellen*, 1st edit., 1941, p. 249). 24th List

Khibinite. *Chem. Abstr.* (*Amer. Chem. Soc.*), 1935, vol. 29, p. 4702; *Amer. Min.*, 1936, vol. 21, p. 269. Erroneously given as a mineral, rather than a rock, name. Chibinit (W. Ramsay, 1899) = Хибинит = Khibinite is a pegmatitic nepheline-syenite with eudialyte, etc., from the Khibina tundra, Kola peninsula. [*M.A.*, **6**, 307, 310, 420; **7**, 196.] 15th List

Khibinskite. A. P. Khomyakov, A. A. Voronkov, S. I. Lebedeva, V. P. Bykov, and K. V. Yurkina, 1974. *Zap.*, **103**, 110 (Хибинскит). Monoclinic pseudorhombohedral crystals (*a* 19·22, *b* 11·10, *c* 14·10 Å, *β* 116° 30′) from the Khibina massif have the composition $16[K_2ZrSi_2O_7]$; named from the locality. Cf. Khibinite, 15th List, a rock name. 28th List [*A.M.*, **59**, 1140]

Khinganite. G. V. Itsikon and D. V. Rundkvist, 1959. [Труды Всесоюз. Геол. Инст., vol. 27, p. 120], abstr. in Зап. Всесоюз. Мин. Общ. (*Mem. All-Union Min. Soc.*), 1962, vol. 91, p. 205 (Хинганит). Synonym of Kësterite (21st List). Named from the locality. 23rd List

Khlopinite. I. E. Starik, 1933. *Problems of Soviet Geology*, vol. 3, no. 7, p. 74 (хлопинит), p. 78 (chlopinite); *Rep. Intern. Geol. Congress*, xvi (1933), USA, 1936, vol. 1, p. 217 (chlopinite). O. M. Shubnikova and D. V. Yuferov, *Report on new minerals*, Leningrad, 1934, p. 104 (Хлопинит, Hlopinite). Titano-niobate of yttrium, uranium, thorium, and iron, $M_2Nb_2TiO_9.H_2O$; black, optically isotropic, from Transbaikal. Named after Professor V. G. Khlopin (В. Г. Хлопин). [*M.A.*, **6**, 258, 518.] 14th List [Is Samarskite; *A.M.*, **57**, 329]

Khoharite. L. L. Fermor, 1938. *Rec. Geol. Surv. India*, vol. 73, p. 145; *Indian Assoc. Cultiv. Sci.*, 1938, Spec. Publ. no. 6, p. 8. The hypothetical garnet molecule $3MgO.Fe_2O_3.3SiO_2$. Named from the Khohar meteorite (India, fell 1910), it being suggested that such a garnet was the original material of enstatite chondrules. [*M.A.*, **7**, 169.] 15th List

Khovakhsite. N. N. Shishkin and V. A. Mikhailova, 1956. [Сборн. Мат. Тех. Инф. (*Collect. Mat. Techn. Inform.*), no. 6, Gipronickel, p. 5]; abstr. Зап. Всесоюз. Мин. Общ. (*Mem. All-Union Min. Soc.*, 1959), vol. 88, p. 317 (Ховахсит, hovaxite). Imperfectly characterized brown oxidation product of smaltite and safflorite from the Khovakhs deposit, Tuva; named from the locality. [*M.A.*, **14**, 278; *A.M.*, **45**, 256.] Cf. Yellow earthy cobalt (Dana, *Syst. Min.*, 6th edit., p. 78). A wholly unnecessary name. 22nd List

Khuniite. D. Adib and J. Ottemann, 1970. *Min. Depos.*, **5**, 86. Brownish-yellow crystals from the T. Khuni mine, Anarak, Iran, have a composition near $Pb_{1\cdot6}Zn_{0\cdot2}Cu_{0\cdot2}CrO_5$; they are stated to be monoclinic, with sp. gr. 5·9; no other data are given. Named for the locality. [This mineral is probably identical with iranite (23rd List), which probably contains Cu and Zn, but without a more adequate description, and in particular without X-ray powder data, this surmise cannot be tested—*M.H.H.*] 26th List [Is probably Iranite; *Bull. Brit. Mus. Hist. Nat.* (*Miner.*), **2**, no. 8, 409; *A.M.*, **59**, 633; **61**, 186. Cf. Hemihedrite]

Kidwellite. P. B. Moore and J. Ito, 1978. *M.M.*, **42**, 137. Mineral 'A' of Moore (*A.M.*, 1970, **55**, 138), from several localities in the Ouachita Mtns., Arkansas, as feathery green-to-yellow crystals, is monoclinic, *a* 20·61, *b* 5·15, *c* 13·75 Å, *β* 112·64°. Composition $2[NaFe^{3}_{9}(OH)_{10}(PO_4)_6.5H_2O]$. RIs α 1·787, *β* 1·800 ∥ [010], *γ* 1·805. Also occurs in Alabama, Virginia, and Germany. Named for A. L. Kidwell. [*M.A.*, 78-2122.] 30th List

Kieselmagnesite. (P. Groth, *Tab. Uebers. Min.*, 3rd edit., 1889, p. 52). A compact mixture of magnesite and quartz. 2nd List

Kieselscheelit, Germanized version of Siliceous scheelite (q.v.) (C. Hintze, *Hand. Min.*, Erg.-Bd. II, p. 739) 24th List

Kietyöite. A. E. Nordenskiöld, *Min. Finland*, 1863, p. 154. Apatite from Kietyö, Finland. 2nd List

Kievite. V. I. Luchitsky, 1912. *Izvyestiya Varshav. Polytechn. Inst.*, vol. 2 (for 1911); also published as a book *Rapakivi Kievskoi gubernii* ... (Russ.), Warsaw, 1912, pp. 73–77 (Кievитъ). A colourless or very pale yellowish-green hornblende, with $c{:}\gamma = 13°$ to $15°$, $\alpha = 1{\cdot}665$, $\gamma = 1{\cdot}710$, negative sign, feeble pleochroism, and prism-cleavage 55° 30′, occurring with biotite, common hornblende, and olivine in the rapakivi-granite of govt. Kiev and of Finland. The grains or fibres are surrounded by green hornblende, the parallel intergrowth sometimes showing a pegmatoid structure. It is allied to cummingtonite and to grünerite, but differs from these in its paragenesis and in some of its characters. Named from the locality. 10th List [Cummingtonite; *Amph.* (1978)]

Kilbreckanite. R. Griffith, *Catalogue of the several localities in Ireland where mines or metalliferous indications have hitherto been discovered*, 1854, p. 2 (reprinted, Dublin, 1884); G. H. Kinahan, *Economic geology of Ireland, Journ. Roy. Geol. Soc. Ireland*, 1899, vol. 8, p. 14. Another spelling of Kilbrickenite (J. Apjohn, 1841) from the Monanoe or Kilbreckan lead mine, near Quin, Co. Clare, Ireland. A synonym of Geocronite. 12th List

Kilchoanite. S. O. Agrell and P. Gay, 1961. *Nature*, vol. 189, p. 743. $Ca_3Si_2O_7$, a polymorph of rankinite, replacing rankinite at Kilchoan, Ardnamurchan, Scotland. Named from the locality. [*M.A.*, **15**, 135.] 22nd List [*A.M.*, **46**, 1203; *M.M.*, **34**, 5]

Killalaite. R. A. Nawaz, 1974. *M.M.*, **39**, 544. Monoclinic colourless crystals (*a* 9·3, *b* 9·9, *c* 7·7 Å, $\beta \approx 105°$) in a thermally metamorphosed limestone at Killala Bay, Inishcrone, Co. Sligo, Ireland, have composition $2[2Ca_3Si_2O_7.H_2O]$. Named from the locality. [*M.A.*, 74-1449.] 28th List [*A.M.*, **59**, 1331]

Kimzeyite. C. Milton and L. V. Blade, 1958. *Science*, vol. 127, p. 1343; *Amer. Min.*, 1961, vol. 46, p. 533. A zirconiferous garnet (ZrO_2 20%) from the Kimzey calcite quarry, Magnet Cove, Arkansas. Named after the Kimzey family. [*M.A.*, **15**, 43.] 22nd List

Kingite. K. Norrish, L. E. R. Rogers, and R. E. Shapter, 1956. *Min. Soc. Notice*, no. 94; *Min. Mag.*, 1957, vol. 31, p. 351. Hydrous aluminium phosphate, $Al_2O_3.Al(OH)_3.P_2O_5.9H_2O$, as white nodules in phosphate deposits in South Australia. Named after Mr. D. King, geologist, Department of Mines, South Australia. (*See* Meta-kingite.) 21st List [*A.M.*, **42**, 580]

Kinoite. J. W. Anthony and R. B. Laughon, 1970. *Amer. Min.*, **55**, 709. Deep blue monoclinic crystals, *a* 6·990, *b* 12·890, *c* 5·654 Å, β 96° 5′, $P2_1/m$, D 3·16, from the northern Santa Rita Mts., Pima Co., Arizona. Composition $2[Cu_2Ca_2Si_3O_{10}.2H_2O]$. α 1·638, pale green-blue; β 1·665, blue; γ 1·676, deep blue. Named for E. F. Kino. [*A.M.*, **56**, 193; *M.A.*, 70-3431; *Zap.*, **100**, 625.] 27th List

Kinoshitalite. M. Yoshii, K. Maeda, T. Kato, T. Watanabe, S. Yui, A. Kato, and N. Nagashima, 1973. *Chigaku Kemkya*, **24**, 181, abstr. *A.M.*, **60**, 486. Small yellow-brown scales in hausmannite–tephroite ore in the Misago ore body, Noda-Tamagawa mine, Iwate prefecture, Japan, have space group $C2/m$, *a* 5·345, *b* 9·250, *c* 10·256 Å, β 99·99°; composition $2[(Ba,K)(Mg,Mn^{2+},Al,Mn^{3+})_3Si_2Al_2O_{10}(OH,F,O)_2]$, the magnesium analogue of anandite. Named for K. Kinoshita. [*M.A.*, 75-476 and 3598; *Zap.*, **104**, 614.] 29th List

Kinradite. (D. B. Sterrett, *Gems and precious stones in 1911, Mineral Resources of the United States*, for 1911, 1912, p. 27 of advance copy.) Local trade-name for a spherulitic jasper-like quartz from California. Named after Mr. J. J. Kinrade, of San Francisco. 6th List

Kipushite. H. Buttgenbach, 1927. *Bull. Cl. Sci. Acad. R. Belgique*, for 1926, p. 905. Hydrated phosphate of copper and zinc, $(Cu,Zn)_3(PO_4)_2.3(Cu,Zn)(OH)_2.3H_2O$, as blue monoclinic crystals from Kipushi, Katanga, Belgian Congo. The same as veszelyite with no arsenate. Named from the locality. [*M.A.*, **3**, 269.] 11th List [Is Arakawaite; *M.A.*, **5**, 94, 137]

Kirchheimerite. K. Walenta, 1964. *Tschermaks Min. Petr. Mitt.*, vol. 9, p. 1. Artificial $Co(UO_2)_2(AsO_4)_2.12H_2O$; possibly occurs naturally. [*M.A.*, **17**, 300.] 24th List [Cf. Metakirchheimerite]

Kirovite. G. N. Vertushkov, 1939. *Bull. Acad. Sci. URSS, Sér. Géol.*, 1939, no. 1, p. 109 (кировит), p. 114 (kirovite). A variety of melanterite, containing MgO 7·45%, $(Fe,Mg)SO_4.7H_2O$, resulting from underground fires in the Kalata mine, Kirovgrad, Ural. Named after Mr. S. M. Kirov (С. М. Киров). *See* Cuprokirovite and Jarošite. [*M.A.*, **7**, 418.] 15th List

Kirschsteinite. T. G. Sahama and K. Hytönen, 1957. *Min. Mag.*, vol. 31, p. 698. $CaFeSiO_4$ (69·4 mol.%, with some Mg and Mn), orthorhombic, monticellite group. With combeite and götzenite (qq.v.) in nephelinite from Kivu, Belgian Congo. Named after the late Dr. Egon Kirschstein, German geologist. 21st List [*A.M.*, **43**, 790]

Kirschteinite, error for Kirschsteinite. *Dokl. Acad. Sci. USSR* (*Earth Sci. Sect.*), 1970, **190**, 136. [*M.A.*, 73-2800.] 28th List

Kiscellite. L. Zechmeister, G. Tóth, and A. Koch, 1934. *Centr. Min.*, Abt. A, 1934, p. 60 (Kiscellit). A brown amber-like resin containing sulphur in place of oxygen, from the Kisczelli tályag (Hung.) = Kleinzeller Tegel (Germ.) beds of Oligocene age at Budapest. Named from the geological horizon, which is named from the locality Kis-Czell near Budapest. [*M.A.*, **5**, 485.] 13th List

Kitkaite. T. A. Häkli, Y. Vuorelainen, and Th. G. Sahama, 1965. *Amer. Min.* vol. 50, p. 581. Carbonate-bearing veinlets in albitite veins in the albite diabase in the valley of the Kitka river, Kuusamo, NE Finland, contain a variety of selenide minerals, including selenian Melonite and the isostructural trigonal Kitkaite, NiSeTe. Named from the locality. [*M.A.*, **17**, 499.] 24th List

Kittlite. Santiago Rivas, 1970. [*Rev. Minera Geol. Mineral. Argentina*, **29**, 56, and **30**, 14], abstr. *A.M.*, **57**, 1313. An inadequately characterized silver-grey cubic mineral, $F\bar{4}3m$, containing major Hg, Ag, Cu, S, and Se; D 5·4, H 5 to $5\frac{1}{2}$; from near Jaguel, Llantenes region, prov. La Rioja, Argentina. May be a selenian metacinnabarite. Named for E. Kittl. 27th List

Kivuïte. L. Van Wambeke, 1958. *Bull. Soc. belge Géol.*, vol. 67, p. 383. *Proc. 2nd UN Internat. Conf. Peaceful Uses Atomic Energy*, vol. 3, p. 541. Minute yellow orthorhombic plates, isostructural with phosphuranylite and renardite, associated with phosphuranylite, renardite, cyrtolite, and columbotantalite in a pegmatite from Kobokobo, Kivu, Congo. Probably formula $ThH_2(UO_2)_4(PO_4)_2(OH)_8.7H_2O$. A plumbian variety is also described. [*M.A.*, **14**, 281.] Needs confirmation and tests for homogeneity (M. Fleischer, *Amer. Min.*, 1959, vol. 44, p. 1326). 22nd List

Kladnoite. R. Rost, 1942. *Rozpravy České Akad.*, 1942, vol. 52, no. 25 (Kladnoit). An organic compound $C_6H_4(CO)_2NH$ (phthalimide) as monoclinic crystals formed in burning waste heaps in the Kladno coal basin, Bohemia. Compare kratochvilite (15th List). Named from the locality. [*M.A.*, **9**, 186.] 17th List [*A.M.*, **31**, 605]

Klebelsbergite. V. Zsivny, 1929. *Magyar Tudom. Akad. Mat. Természett. Értesítő*, Budapest, vol. 46, p. 19 (Klebelsbergit). Hydrous basic antimony sulphate, as monoclinic needles on stibnite from Felsőbánya, Hungary (= Baia Sprie, Romania). Named after Count Kuno Klebelsberg, Hungarian Minister of Education. [*M.A.*, **4**, 150.] 12th List [*A.M.*, **59**, 875]

Kleinite. A. Sachs, 1905. *Sitzungsber. Akad. Wiss. Berlin*, 1905, p. 1091; *Centralblatt Min.*, 1906, p. 200 (Kleinit). An oxychloride of mercury, $Hg_4Cl_2O_3$, containing small amounts of nitrogen and sulphate. It occurs as sulphur-yellow, hexagonal crystals in the mercury-mines at Terlingua, Texas. Named after Professor Carl Klein (1842–1907), of Berlin. 4th List [Perhaps the hexagonal dimorph of Mosesite (q.v.). Cf. Mercurammonite]

Kleit. Polish form of clayite. *See* Cleïte. 18th List

Kliachite. F. Cornu, 1909. *Zeits. Chem. Indust. Kolloide*, vol. iv, p. 90 (Kliachit). Colloidal aluminium hydroxides, α-kliachite, $Al_2O_3.H_2O$, and β-kliachite, $Al_2O_3.3H_2O$, forming, together with iron hydroxides and the crystalloids diaspore ($Al_2O_3.H_2O$) and hydrargillite ($Al_2O_3.3H_2O$), the constituent minerals of bauxite. 5th List

Klinaugite. F. Rinne, *Centralblatt Min.*, 1917, p. 74 (Klinaugite, *plur.*). Variant of Klinoaugite (F. Rinne, 1900; 3rd List). 10th List

Klinoantigorit, original form of Clino-antigorite (q.v.). 24th List

Klinoaugit. *See* Orthoaugite. 3rd List

Klinoberthierin, original form of Clinoberthierine (q.v.). 24th List

Klinobronzit. W. Wahl, 1906. *See* Clinobronzite. 5th List

Klinochrysotil. H. Strunz, 1957. *Min. Tabellen*, 3rd edit., p. 322. German spelling of Clinochrysotile (20th List). 24th List

Klinoëdrit. *See* Clinohedrite. 2nd List

Klinoenstatit. W. Wahl, 1906. *See* Clinoenstatite. 5th List

Klinohypersthen. W. Wahl, 1906. *See* Clinohypersthene. 5th List

Klinoolivin. E. Dittler, 1929. *Sitzungsber. Akad. Wiss. Wien, Math.-naturw. Kl.*, Abt. I, vol. 138, p. 411. A synonym of the monoclinic titanclinohumite. Orthorhombic titanolivine has been prepared artificially, but is not known as a mineral. F. Machatschki, *Centr. Min.*, Abt. A, 1930, p. 200, rightly regards the name as superfluous. [*M.A.*, **4**, 309.] 12th List

Klino-pyroxen. *See* Orthoaugite. 3rd List

Klinotscheffkinit, variant of Clino-chevkinite (20th List). (H. Strunz, *Min. Tabellen*, 3rd edit., 1957, p. 282. C. Hintze, *Handb. Min.*, Erg.-Bd. II, p. 742. 22nd List [Also 24th List]

Klinozoisite. E. Weinschenk, *Zeits. Kryst. Min.*, XXVI. 161, 1896. [*M.M.*, **11**, 237.] Epidote containing only a little iron, and so agreeing with zoisite in composition. $H_2Ca_4Al_6Si_6O_{26}$. Monosymmetric. [Goslerwand, Prägratten] Tyrol. 1st List [Accepted spelling now **Clinozoisite**]

Kljakite. F. Tućan, *Neues Jahrb. Min.*, 1912, Beil.-Bd. xxxiv, p. 411; M. Kišpatić, tom. cit., pp. 519, 534 (Kljakit). An alternative spelling of Cliachite (q.v.) and Kliachite (5th List). *See* Sporogelite. 6th List

Klockmannite. P. Ramdohr, 1928. *Centr. Min.*, Abt. A, 1928, p. 225 (Klockmannit). Copper selenide, CuSe, as metallic slate-blue granular masses, perhaps hexagonal and isomorphous with covellite; from Sierra de Umango, Argentina, and Harz, Germany. Distinct from umangite. Named after Prof. Friedrich Ferdinand Hermann Klockmann (1858–), of Aachen, Germany. 11th List

Kmaite. A. A. Illarionov, 1961. [Вопр. Раз. Мест. Курск. Магнит. Аном., p. 250], abstr. in Зап. Всесоюз. Мин. Общ. (*Mem. All-Union Min. Soc.*), 1962, vol. 91, p. 202 (Кмаит, kmaite) and in *Amer. Min.*, 1962, vol. 47, p. 808. An unnecessary name for a Celadonite rich in Fe_2O_3 and K_2O, from the Midailov iron ores, Kursk. [*M.A.*, **16**, 554.] 23rd List

Knipovichite. E. I. Nefedov, 1953. *Mém. Soc. Russ. Min.*, ser. 2, vol. 82, p. 317 (Книровичит). Hydrous carbonate of Ca,Al,Cr, radiating fibres, optically biaxial, pink. [*M.A.*, **12**, 352.] 20th List [Is chromian Alumohydrocalcite; *A.M.*, **61**, 341]

Knipovitchite. *Bull. Soc. Franç. Min. Crist.*, 1955, vol. 78, p. 218. French transliteration of книповичит, knipovichite (20th List). 21st List

Knipowitschit, German transliteration of Книповичит, Knipovichite (20th List) (C. Hintze, *Handb. Min.*, Erg.-Bd. II, p. 742). 24th List

Knollite. [A. Pelikan, MS.] R. Koechlin, *Centr. Min.*, Abt. A, 1933, p. 204. Synonym of zeophyllite. Named after Mr. — Knoll. [*M.A.*, **5**, 296.] 13th List

Knopite. P. J. Holmqvist, *Geol. För. Förh.*, XVI. 73, 1894. [*M.M.*, **11**, 158.] $CaO.TiO_2$ with some Ce_2O_3. Pseudo-cubic. Alnö, Sweden. 1st List [Formula complex. Cf. *M.A.*, **17**, 296]

Knorringite. P. H. Nixon and G. Hornung, 1968. *Amer. Min.*, **53**, 1833. A bluish-green garnet from the Kao kimberlite pipe, Lesotho, contains a major amount of the end-member $Mg_3Cr_2(SiO_4)_3$; named for Dr. Oleg von Knorring. [*M.A.*, 69-2390; *Bull.*, **92**, 512; *Zap.*, **99**, 81.] 26th List

Koashvite. Yu. L. Kapustin, Z. V. Pudovkina, A. V. Bykova, and G. V. Lyubomilova, 1974. *Zap.*, **103**, 559 (Коашвит). Veinlets replacing lomonosovite in alkali pegmatite on Mt. Koashva, Khibina massif, Kola Peninsula, are near $4[Na_6(Ca,Mn)(Ti,Fe)Si_6O_{18}.H_2O]$. Space group *Pbam*, or *Pba*2, *a* 7·356, *b* 20·950, *c* 10·194 Å. Named for the locality. [*A.M.*, **60**, 487; *M.A.*, 75-2524; *Zap.*, **104**, 613.] 29th List

Kobaltchrysotil, original spelling of Cobalt chrysotile (q.v.). 22nd List

Kobalt-manganspath, original form of Cobalt-manganese-spar (18th List). 24th List

Kobaltnickelpyrit. *See* Cobaltnickelpyrite. 7th List

Kobaltokalzit, variant of Kobaltocalcit (H. Strunz, *Min. Tabellen*, 3rd edit., 1957, p. 386). 24th List

Kobalt-Oligonspat. R. Reissner, 1935. *See* Cobalto-sphaerosiderite. 14th List

Kobaltolivin, German form of Cobalt-olivine (q.v.). 24th List

Kobaltorhodochrosit, German form of Cobaltorhodochrosite (18th List) (H. Strunz, *Min. Tabellen*, 3rd edit., 1957, p. 386). 24th List

Kobaltpentlandit, German form of Cobalt pentlandite (22nd List) (H. Strunz, *Min. Tabellen*, 4th edit., 1966, p. 106). 24th List

Kobalttalkum, German form of Cobalt-talc (q.v.). 24th List

Kobaltullmannit. F. Machatschki, 1953. *Spez. Min.*, Wien, p. 311. Syn. of Willyamite (1st List). 22nd List

Kobeite. J. Takubo, Y. Ukai, and T. Minato, 1950. *Journ. Geol. Soc. Japan*, vol. 56, p. 509; abstr. in *Amer. Min.*, 1951, vol. 36, p. 924. Titanate, niobate, and tantalate (Yt, Fe,U, etc.)(Ti,Nb,Ta, etc.)$_2$ (O,OH)$_6$, differing from euxenite, etc., in the low content of (Nb,Ta)$_2$O$_5$ (4·84–5·45%). Black metamict crystals in pegmatite from Kobe, Kyoto, Japan. Named from the locality. [*M.A.*, **11**, 518.] 19th List [*A.M.*, **42**, 342]

Kobokobite. J. Thoreau, 1957. *Bull. Acad. Roy. Belge, Cl. Sci.*, ser. 5, vol. 43, p. 705. An intermediate member of the rockbridgeite–frondelite series with $Fe^{\cdot\cdot} \approx Mn$, named from the locality, Kobokobo, Congo. [*M.A.*, **14**, 59.] A superfluous name. 'Should not have been named. The formulas of all these minerals are uncertain because of doubt as to the degree of oxidation of FeO' (M. Fleischer, *Amer. Min.*, 1958, vol. 43, p. 795). 22nd List

Kochenite. A. Pichler, *Jahrb. geol. Reichs. Wien*, 1868, XVIII, 47; V. von Zepharovich, *Min. Lexicon Oesterr.*, 1873, II, 170. A fossil amber-like resin from Kochenthal, Tyrol. 2nd List

Kochite. S. Kōzu, K. Seto, and K. Kinoshita, 1922. *Journ. Geol. Soc. Tōkyō*, vol. 29, pp. 1, 148; *Sci. Rep. Tôhoku Univ.*, ser. 3, 1924, vol. 2, p. 1 (kôchite). Hydrated aluminium silicate, $2Al_2O_3.3SiO_2.5H_2O$, occurring as a granular aggregate of minute cubic crystals, at Kōchimura, prov. Rikuchū, Japan. Named from the locality. [*M.A.*, **2**, 51, 521.] 10th List [A mixture containing Zunyite, *Introd. Japanese Minerals* (*Geol. Surv. Japan, 1970*), p. 104]

Koechlinite. W. T. Schaller, 1914. *Journ. Washington Acad. Sci.*, vol. iv, p. 354; 3rd Appendix to 6th edit. of Dana's *System of Mineralogy*, 1915, p. 43. A molybdate of bismuth, $Bi_2O_3.MoO_3$, as minute, greenish-yellow, orthorhombic plates, nearly square in outline with perfect platy cleavage, and resembling torbernite in appearance. From Schneeberg, Saxony. Detected on a single specimen preserved in the mineral collection of the Natural History Museum at Vienna. Named after Dr. Rudolf Koechlin, Curator of that collection. 7th List [Cf. Russellite]

Koenenite. F. Rinne, *Centralblatt Min.*, 1902, p. 493 (Koenenit). A red, scaly mineral with a perfect micaceous cleavage, from the salt deposits of Volpriehausen, Hanover. Rhombohedral, optically uniaxial and positive. $Al_2O_3.3MgO.2MgCl_2.6(\text{or } 8)H_2O$. Named after Professor Adolf von Koenen, of Göttingen. 3rd List [$[Na_4(Ca,Mg)_2Cl_{12}]^{4+}[Mg_7Al_4(OH)_{22}]^{4-}.7H_2O$; *M.A.*, 69-164]

Koesterit, another variant transliteration of Kësterite (21st List) (H. Strunz, *Min. Tabellen*, 4th edit., 1966, p. 487). 24th List

Kogarkoite. A. Pabst and W. N. Sharp, 1973. *A.M.*, **58**, 116. Monoclinic crystals from the Lovozero massif, Kola Peninsula, USSR, and Hortense Hot Spring, Colorado, are identical with synthetic 12[Na_3SO_4F]; *a* 18·073, *b* 6·949, *c* 11·440 Å, *β* 107° 43′. Distinct from schairerite (12th List; $Na_{21}(SO_4)_7F_6Cl$), galeite (21st List; $Na_{15}(SO_4)_5F_4Cl$), and sulphohalite ($Na_6(SO_4)_2FCl$). Named for L. N. Kogarko, who first described the Lovozero material. [*M.A.*, 73-2912; *Zap.*, **103**, 360.] 28th List [*M.A.*, **40**, 131]

Koivinite. A. A. Kukharenko, 1951. *Mém. Soc. Russ. Min.*, ser. 2, vol. 80, p. 238 (койвинит); V. A. Frank-Kamenetzky, A. I. Komkov, and V. V. Nardov, ibid., 1953, vol. 82, p. 297. An incompletely described mineral (locality?) near florencite. Named perhaps from кой (that) and винить (to criticize). [*M.A.*, **12**, 411.] 20th List

Koiwinit, German transliteration of Койвинит, koivinite (20th List). (Hintze, *Handb. Min.*, Erg.-Bd. II, p. 569.) 22nd List

Kokimbit. S. I. Naboko, *Bull. Volcan. Station*, Kamchatka, 1953, no. 18, p. 53 (кокимбит). Double transliteration of coquimbite. [M.A., **12**, 542.] 20th List

Koktaite. J. Sekanina, 1948. *Acta Acad. Sci. Nat. Moravo-Silesicae*, vol. 20, no. 1 (koktait, koktaite, ammonium-syngenite). Hydrated sulphate of calcium and ammonium, $(NH_4)_2Ca(SO_4)_2.H_2O$, as acicular monoclinic crystals pseudomorphous after gypsum, from Moravia. Named after Dr. Jaroslav Kokta of Brno, Moravia. *See* Ammonium-syngenite. [*M.A.*, **10**, 352.] 18th List [*A.M.*, **34**, 618]

Kolbeckine. R. Herzenberg, 1932. *Revista Minera, Soc. Argentina Mineria y Geol.*, vol. 4, p. 33 (Kolbeckina); *Centr. Min.*, Abt. A, 1932, p. 354 (Kolbeckin). Tin sulphide, Sn_2S_3, as minute black scales from Bolivia. Evidently named after Prof. Friedrich Kolbeck, of the Mining Academy, Freiberg, Saxony. [Not the kolbeckite of F. Edelmann, 1926; 11th List.] [*M.A.*, **5**, 199.] 13th List [Renamed Herzenbergite (q.v.)]

Kolbeckite. F. Edelmann, 1926. *Jahrb. Berg- und Hüttenw. Sachsen*, vol. 100, p. A 73 (Kolbeckit). Beryllium phosphate or silicophosphate as cyan-blue monoclinic crystals from Saxony. Named after Prof. Friedrich Kolbeck, of the Mining Academy of Freiberg, Saxony. [*M.A.*, **3**, 472.] 11th List [Contains major Sc and is related to Sterrettite; *A.M.*, **45**, 257]

Kollochrom, error for Kallochrom (H. Strunz, *Min. Tabellen*, 4th edit., 1966, p. 487). 24th List

Kolloid-calcite, Kolloid-magnesite, Kolloid-siderite. A. K. Boldyrev, 1928. *Kurs opisatelnoi mineralogii*, Leningrad, 1928, part 2, p. 163. (Коллоид-кальцит, Коллоид-магнезит, Коллоид-сидерит). O. M. Shubnikova and D. V. Yuferov, *Spravochnik po novym mineralam*, Leningrad, 1934, p. 48 (*коллоид-кальцит*, kolloid-calcite; *коллоид-магнезит*, kolloid-magnesite; *коллоид-сидерит*, kolloid-siderite). Colloidal varieties of minerals of the calcite group. 14th List

Kolloid-magnesite, *see* Kolloid-calcite.

—— -siderite, *see* Kolloid-calcite.

Kolovratite. V. I. Vernadsky, 1922. *Compt. Rend. Acad. Sci. Russie*, 1922, p. 37 (Коловратит). P. N. Chirvinsky, *Min. Mag.*, 1925, vol. 20, p. 290. Vanadate of nickel occurring as amorphous or finely crystalline, yellow or greenish-yellow encrustations and botryoidal crusts on siliceous and carbonaceous slates of Silurian age in Fergana, Russian Turkestan. Named after the late L. S. Kolovrat-Chervinsky (Л. С. Коловрат-Червинский), a Russian radiologist. [*M.A.*, **2**, 417.] 10th List [*A.M.*, **47**, 122]

Kolskite. N. E. Efremov, 1939. *Compt. Rend. (Doklady) Acad. Sci. URSS*, vol. 22, p. 433 (kolskite). V. A. Afanasiev, ibid., 1939, vol. 25, p. 516. A serpentine mineral with the composition $5MgO.4SiO_2.4H_2O$, from the Kola peninsula, Russia. Named from the locality. [*M.A.*, **7**, 370.] 15th List [A mixture; *A.M.*, **59**, 212. Cf. Sungulite]

Komarovite. A. M. Portnov, G. K. Krivokoneva, and T. I. Stolyarova, 1971. *Zap.*, **100**, 599 (Комаровит). Platy rose-coloured aggregates and veinlets in natrolite at Mt. Karnasurt. Lovozero, Kola Peninsula, are orthorhombic, *a* 21·30, *b* 14·00, *c* 17·19 Å; α 1·750 ∥ [010], β 1·766, γ 1·85. The composition is $18[(Ca,Mn)Nb_2(Si_2O_7)O_3.3\frac{1}{2}H_2O]$; the mineral is the end-member of the labuntsovite–nenadkevichite series. Named for V. M. Komarov. [*M.A.*, 72-2334.] 27th List [*A.M.*, **57**, 1315]

Komarowit, Germ. trans. of Комаровит, komarovite (27th List). Hintze, 47. 28th List

Kondrikovite. I. D. Borneman-Starynkevich, 1933. *Khibina Apatite*, Leningrad, 1933, p. 114 (кондрикит); *Materials Geochem. Khibina tundra, Acad. Sci. USSR*, 1935, pp. 54, 56 (кондриковит). Natrolite with microscopic inclusions of a rinkite-like mineral; apparently an alteration product of lovchorrite, which it encrusts, from the Kola peninsula, Russia. Named kondrikite, afterwards corrected to kondrikovite, after Mr. V. I. Kondrikov, President of the Apatite Trust in the Kola peninsula. [*M.A.*, **6**, 341–343.] 14th List

Kordylite. *See* Cordylite. 2nd List

Korea-augite. F. Yamanari, 1926. *Japanese Journ. Geol. Geogr.*, vol. 3 (for 1924), p. 106. A variety of augite, distinguished by its optical properties in micro-sections, present in alkali-trachyte from Korea. [*M.A.*, **3**, 199.] 11th List

Kornerupite, variant of Kornerupine; *A.M.*, **8**, 51. 10th List

Korschinskit, German transliteration of Коржинскит, Korzhinskite (23rd List) (H. Strunz, *Min. Tabellen*, 4th edit., 1966, p. 487). 24th List

Korteite. [C. Prager, 1923?]. *Student's index to the collection of minerals*, British Museum, 27th edit., 1936, p. 21, from a dealer's label. Described as $MgO.Al(OH)_3$, found with wathlingenite (q.v.). Named after G. Korte, director of potash-salt works. Identical with koenenite (3rd List). 18th List

Korunduvite. N. G. D'Ascenzo, 1945. *The gem-table*, Philadelphia, Pennsylvania. To replace the name corundum, because of the confusion with carborundum [and conundrum]. From the Sanskrit kuruvinda. [*M.A.*, **9**, 191.] 17th List

Korzkinskite. S. V. Malinko, 1963. Зап. Всесоюз. Мин. Общ. (*Mem. All-Union Min. Soc.*), vol. 92, p. 555 (Коржинскит). The skarn from the N Urals that contains the calcium borates Calciborite, Frolovite, Nifontovite, Uralborite, Sibirskite, and Pentahydroborite also contains another hydrate, $CaB_2O_4.H_2O$. Named for D. S. Korzhinsky. [*M.A.*, **16**, 547.] 23rd List [*A.M.*, **49**, 441]

Kosmochlor. H. Laspeyres, *Zeits. Kryst. Min.*, XXVII, 592, 1897. Silicate of Cr, etc. Probably monosymmetric. From the Toluca meteoric iron. 1st List [Now, commonly, **Cosmochlore.** Cf. Ureyite]

Kosmochromit. P. Groth, *Tab. Uebers. Min.*, 4th edit., 1898, p. 132. The same as kosmochlor. 2nd List

Kossmatite. O. H. Erdmannsdörffer, 1925. *Centralblatt Min.*, Abt. A, p. 69 (Kossmatit). A mineral of the brittle mica group, with the composition $Si_7O_{42}Al_6Mg_3Ca_7H_{18}F$, occurring as optically positive, white scales in crystalline dolomite in S Serbia. Named after Prof. Franz Kossmat, Director of the Geological Survey of Saxony. [*M.A.*, **2**, 418.] 10th List

Kosterite, Kösterite. Further variant transliterations of Кёстерит, kësterite. (*Amer. Min.*, 1958, vol. 43, p. 1222; Hintze, *Handb. Min.*, Erg.-Bd. II, pp. 743, 932.) 22nd List

Kostovite. G. Terziev, 1966. *Amer. Min.*, vol. 51, p. 29. Small grains in the copper ores of Chelopech, Bulgaria, give X-ray powder data similar to but distinct from those of Calaverite and Sylvanite. Anisotropic, possibly monoclinic. Composition $AuCuTe_4$. Named for I. Kostov. 24th List

Kotoite. T. Watanabe, 1939. *Min. Petr. Mitt.* (*Tschermak*), vol. 50, p. 441; *Fortschr. Min. Krist. Petr.*, vol. 23, p. clxvi (Kotoit). Magnesium borate, $Mg_3B_2O_6$, occurring abundantly as a granular (orthorhombic) constituent of dolomitic marble in Korea. Named after Professor Bundjirô Kotô (1856–1935), Japanese geologist. [*M.A.*, **7**, 315.] 15th List [*A.M.*, **24**, 406. Cf. Jimboite]

Kotoulskite, French transliteration of Котульскит, Kotulskite (*Bull. Soc. franç. Min. Crist.*, 1964, vol. 87, p. 459). 24th List

Kotulskite. A. D. Genkin, N. N. Zhuravlev, and E. M. Smirnova, 1963. Зап. Всесоюз. Мин. Общ. (*Mem. All-Union Min. Soc.*), vol. 92, p. 33 (Котульскит). Minute grains in chalcopyrite from Monchegorsk have the composition $Pd(Te,Bi)_{2-x}$ with $Te:Bi \approx 3$ and $x \approx 0{\cdot}4$. Hexagonal, resembles synthetic PdTe and $PdTe_2$. Preliminary data in [Геол. Рудн. Мест., 1961, no. 5, p. 64], abstr. in *Amer. Min.*, 1962, vol. 47, p. 809. Named for V. K. Kotulsky. [*M.A.*, **16**, 283.] 23rd List [*A.M.*, **48**, 1181. Cf. Yanzhongite]

Koupletskite, *see* Kupletskite.

Koutekite. Z. Johan, 1958. *Nature*, vol. 181, p. 1553; *Chemie der Erde*, 1960, vol. 20, p. 217. Arsenide of copper, near Cu_2As. Distinct from β-Domeykite. Hexagonal, occurring intergrown with an unnamed copper arsenide at Černy Důl, Krkonoše, Bohemia; also obtained artificially. Named after Prof. J. Koutek of Prague. [*M.A.*, **14**, 279.] 22nd List [*A.M.*, **43**, 794; **46**, 467. Cu_5As_2]

Kowalewskit, German transliteration of Ковалевскит, Kovalevskite (C. Hintze, *Handb. Min.*, Erg.-Bd. II, p. 744). 24th List

Kozhanovite. L. L. Shilin, 1956. Доклады Акад. Наук СССР (*Compt. Rend. Acad. Sci. URSS*), vol. 107, p. 739 (Кожановит). Provisional name for an undescribed mineral later named Karnasurtite (q.v.). 22nd List

Kozhanovskite, error for Kozhanovite; *A.M.*, **51**, 1286.

Kôzulite. M. Nambu, K. Tanida, and T. Kitamura, 1969. *Journ. Jap. Ass. Min. Petr. Econ.*

Geol., **62**, 311. An alkali-amphibole from the Tanohata mine, Iwate Prefecture, Japan; occurs in reddish-black prismatic monoclinic crystals, *a* 9·91, *b* 18·11, *c* 5·30 Å, *β* 104·6°, *C*2/*m*. D 3·30, α 1·685, yellow-brown, *β* 1·717, reddish-brown, *γ* 1·720 dark brown. Composition $2[Na_{2\cdot54}K_{0\cdot27}Ca_{0\cdot19}Mn_{3\cdot69}Mg_{0\cdot63}Fe^{3+}_{0\cdot33}Al_{0\cdot31}Si_8O_{21\cdot78}(OH)_{2\cdot22}]$. Named for S. Kôzu. [*A.M.*, **55**, 1815; *M.A.*, 71-1386; *Zap.*, **100**, 90.] 27th List

Kramerite. W. T. Schaller, 1930. *Prof. Paper US Geol. Survey*, no. 158–I, p. 139. The name first appeared, but without description, in *Amer. Min.*, 1928, vol. 13, p. 453. Hydrous borate of sodium and calcium, $Na_2O.2CaO.5B_2O_3.10H_2O$ (i.e. the same as ulexite with $10H_2O$ instead of $16H_2O$), as rosettes of monoclinic prisms in the borate deposits of the Kramer district, Kern Co., California. Named from the locality. Identical with Probertite (q.v.). [*M.A.*, **4**, 245.] 12th List

Kratochvilite. R. Rost, 1937. [*Rozpravy České Akad.*]; *Bull. Internat. Acad. Sci. Bohême*, 1937, vol. —, preprint, p. 6 (Kratochvílite). An organic compound $C_{13}H_{10}$, fluorene (14th List), formed by burning of pyritous shale in Bohemia. Named after Prof. Josef Kratochvíl, of the Karlovy University, Praha. [*M.A.*, **7**, 11.] 15th List [*A.M.*, **23**, 667]

Kratochwilit, Germanized variant of Kratochvílite (14th List). (H. Strunz, *Min. Tabellen*, 3rd edit., 1957, p. 347.) 22nd List

Krausite. W. F. Foshag, 1931. *Amer. Min.*, vol. 16, pp. 115, 352. Hydrous sulphate of iron and potassium, $K_2SO_4.Fe_2(SO_4)_3.2H_2O$, as yellowish-green monoclinic crystals from California. Named after Prof. Edward Henry Kraus, of the University of Michigan. 12th List

Krauskopfite. M. C. Stinson and J. T. Alfors, 1964. *Min. Inform. Serv. Calif. Div. Mines Geol.*, vol. 17, p. 235. J. T. Alfors, M. C. Stinson, R. A. Matthews, and A. Pabst, *Amer. Min.*, 1965, vol. 50, pp. 279, 314. Colourless massive material in veins in sanbornite–quartz rock near Big Creek and Rush Creek, Fresno County, California, are formulated $BaSi_2O_5.3H_2O$ (emission spectrography analysis). Monoclinic. Named for K. B. Krauskopf. [*M.A.*, **17**, 399; **17**, 502.] 24th List [*A.M.*, **50**, 314]

Krautite. F. Fontan, M. Orliac, and F. Permingeat, 1975. *Bull.*, **98**, 78. Thin pale pink lamellae from Nagyag (Săcăramb) and Kapnik (Cavnic), Transylvania, labelled hoernesite manganésifère, prove to consist of $MnHAsO_4.H_2O$; monoclinic, *a* 8·00, *b* 15·93, *c* 6·79 Å, *β* 96° 32′. Named for Dr. F. Kraut. [*M.A.*, 76-1988; *A.M.*, **61**, 503.] 29th List

Kremenchugite. M. N. Dobrokhotov, 1957. Минер. Сборник Львовского Геол. Обш. (*Min. Mag. Lvov Geol. Soc.*), no. 11, p. 295 (Кременчугит). A highly pleochroic chlorite, apparently a thuringite, from the Kremenchug region, Ukraine. Named from the locality. [*M.A.*, **14**, 141.] A superfluous name. 22nd List

Krementschugite. Erratic French transliteration of Кременчугит, kremenchugite (*Bull. Soc. franç. Min. Crist.*, 1958, vol. 81, p. 335). 22nd List

Kreuzbergite. H. Laubmann and H. Steinmetz, 1920. *Zeits. Kryst. Min.*, vol. 55, p. 551 (Kreuzbergit). Small, colourless to pale yellowish, orthorhombic crystals occurring with other phosphates in the pegmatite of the Kreuzberg at Pleystein, Oberpfalz, Bavaria. Qualitative tests show it to be essentially a hydrated aluminium phosphate (near lucinite of W. T. Schaller, 1914). Named from the locality. *See* Pleysteinite. [*M.A.*, **1**, 125.] 9th List [Is Fluellite; *Bull. Soc. Belge Géol.*, **68**, 241]

Kribergite. T. Du Rietz, 1945. *Geol. För. Förh. Stockholm*, vol. 67, p. 78 (Kribergit). Hydrated phosphate and sulphate of aluminium, $2Al_2O_3.2(P_2O_5,SO_3).5H_2O$, as white chalk-like masses from Kristineberg mine, Västerbotten, Sweden. Named from the locality. [*M.A.*, **9**, 188.] 17th List

Krinovite. E. Olsen and L. [H.] Fuchs, 1968. *Science*, **161**, 786. Minute monoclinic grains of $NaMg_2CrSi_3O_{10}$ occur in graphite nodules in the meteoritic irons of Cañon Diablo, Wichita County, and Youndegin. Named for Dr. E. L. Krinov. [*A.M.*, **54**, 578; *M.A.*, 69-3351; *Bull.*, **92**, 513; *Zap.*, **99**, 81.] 26th List

Kroehnkite, variant of Kröhnkite; *A.M.*, **9**, 62. 10th List

Kromspinel. P. de Wijkerslooth, 1943. Alternative spelling of Chromspinell (q.v.). 23rd List

Krupkaite. L. Žak, V. Syneček, and H. Hybler, 1974. *Neues Jahrb. Min.*, Monatsh. 533. Fibrous dark grey aggregates with bismuthinite at Krupka, Teplica, Bohemia, have space group $Pb2_1m$, *a* 11·15, *b* 11·51, *c* 4·01 Å, composition $2[CuPbBi_3S_6]$. Named from the locality. [*A.M.*, **60**, 737; *M.A.*, 75-3599; *Bull.*, **98**, 320.] 29th List [*C.M.*, **14**, 194]

Krutaïte. Z. Johan, P. Picot, R. Pierrot, and M. Kraček, 1972. *Bull.*, **95**, 475. Small (< 1 mm) grains, mostly included in clausthalite from the Petrovice ore deposit, Moravia, Czechoslovakia, of $CuSe_2$ with some Co, Ni, and Fe, are cubic, *a* 6·056 Å, and isostructural with pyrite. Named for Dr. T. Kruťa. [*M.A.*, 73-2945; *Bull.*, **96**, 238; *A.M.*, **59**, 210; *Zap.*, **103**, 356.] 28th List

Krutovite. R. A. Vinogradova, N. S. Rudashevskii, I. A. Budko, L. I. Bochek, P. Kashpar, and K. Padera, 1976. *Zap.*, **105**, 59 (Крутовит). Isotropic grains in quartz veins at Potuchka, Czechoslovakia, are cubic or pseudocubic, *a* 5·79 Å, composition $Ni_{1-x}As_2$ with *x* 0 to 0·1. Named for G. A. Krutov. 29th List

Kruzhanovskite, Kryjanovskite, Kryschanowskit. American (*Amer. Min.*, 1951, vol. 36, p. 382), French (*Bull. Soc. Franç. Min. Crist.*, 1952, vol. 75, pp. 175, 315), and German (Hintze, *Min.*, 1954, Ergbd. 2, pp. 201, 204) transliterations of крыжановскит, kryzhanovskite (19th List). [*M.A.*, **11**, 189.] 20th List

Kryjanovskite, *see* Kruzhanovskite.

Kryohalit, German transliteration of Криогалит, Cryohalite (H. Strunz, *Min. Tabellen*, 3rd edit., 1957, p. 387]. 24th List

Kryptoklas. *See* Cryptoclase. 6th List

Kryptomerite. (P. Groth, *Tab. Uebers. Min.*, 4th edit., 1898, p. 172). A doubtful borate. 2nd List

Kryptonickelmelan, German transliteration of Криптоникелемелан, Cryptonickelemelane (22nd List) (H. Strunz, *Min. Tabellen*, 4th edit., 1966, p. 488). 24th List

Kryshanovskit, another variant German transliteration of Крыжановскит, Kryzhanovskite (19th List) (H. Strunz, *Min. Tabellen*, 4th edit., 1966, p. 488). 24th List

Kryshanowskit, variant of Kryschanowskit (Kryzhanovskite, 19th List). (H. Strunz, *Min. Tabellen*, 3rd edit., 1957, p. 246.) 22nd List

Kryzhanovskite. A. I. Ginzburg, 1950. *Doklady Acad. Sci. USSR*, vol. 72, p. 762 (крыжановскит). A monoclinic mineral, $MnFe_2'''(PO_4)_2(OH)_2.H_2O$, of the dufrenite group. Compare frondelite, laubmannite, and rockbridgeite (qq.v.). Named in memory of Professor Vladimir Ilich Kryzhanovsky Владимир Ильич Крыжановский (1881–1947), curator in the Mineralogical Museum of the Russian Academy of Science. [*M.A.*, **11**, 189.] 19th List [*A.M.*, **36**, 382; **56**, 1]

Ktenasite. P. Kokkoros, 1950. *Tschermaks Min. Petr. Mitt.*, ser. 3, vol. 1, p. 342 (Ktenasit). Hydrous basic sulphate $3(Cu,Zn)O.SO_3.4H_2O$, as blue-green monoclinic crystals from Lavrion, Greece. Named after Constantine A. Ktenas *Κωνσταντιος Α. Κτενας* (1885–1935), professor of mineralogy and petrology in the University of Athens. [*M.A.*, **11**, 125.] 19th List [$(Cu_{3.5}Zn_{1.5})(SO_4)_2(OH)_6.6H_2O$; *M.M.*, **41**, 65; abstr. *A.M.*, **62**, 1262]

Ktypeite. A. Lacroix, *Compt. Rend. Acad. Sci. Paris*, 1898, CXXVI, 602. H. Vater, *Verh. Ges. Deutsch. Naturf. u. Aerzte, 71 Versamml. zu München, 1899*, 1900, II (i), p. 188 (artificial). A form of calcium carbonate. Optically positive and sometimes biaxial; birefringence much less than in calcite and aragonite. Occurs as pisolites at Carlsbad, Bohemia, and in Algeria. (Cf. conchite above). 2nd List [Is Aragonite; *M.A.*, **10**, 536]

Kubeit. L. Darapsky, *Neues Jahrb. Min.*, 1898, I, 163; 1898, II, ref. 366. A misprint for rubrite. 2nd List

Kuckersite. M. D. Zalessky, 1916. *Geol. Vestnik*, Petrograd, vol. 2, p. 227 (Кукерситъ); *Centralblatt Min. Geol.*, 1920, p. 77 (Kuckersit). H. Bekker, *The Kuckers stage of the Ordovician rocks of N.E. Estonia*, Tartu, 1921, p. 5 (kuckersite); *Geol. Mag.*, 1922, vol. 59, p. 361. P. N. Koggerman, *Acta et Comm. Univ. Dorpat*, 1922, vol. 3 (Kukersite). E. H. C. Craig, *Journ. Inst. Petroleum Techn. London*, 1922, vol. 8, p. 349; *Chem. News*, 1922, vol. 125, p. 121 (Kukkersite). An oil-shale occurring in the Kuckers beds (Ordovician) in Esthonia. Named from the locality, Kukruse (= Kuckers). 10th List

Kulanite. J. A. Mandarino and B. D. Sturman, 1976, *C.M.*, **14**, 127. Blue to green plates in an ironstone formation in the region of the Big Fish River and Blow River, Yukon Territory, are anorthic pseudomonoclinic with *a* 9·032, *b* 12·119, *c* 4·936 Å, $\alpha \approx 90°$, β 120° 23′, $\gamma \approx 90°$; sp. gr. 3·91. Composition $2[Ba(Fe,Mn,Mg)_2(Al,Fe)_2(PO_4)_3(OH)_3]$. Named for A. Kulan. 29th List

Kullerudite. Y. Vuorelainen, A. Huhma and A. Häkli, 1964. *Compt. Rend. Soc. Géol. Finland*, vol. 36, p. 113. Orthorhombic $NiSe_2$, mainly as an alteration product of wilkmanite (q.v.) from Kuusamo, NE Finland; probably isostructural with ferroselite. Named for G. Kullerud. [*M.A.*, **17**, 303; *A.M.*, **50**, 520.] 24th List

Kundaite. B. Doss, 1914. *Centralblatt Min.*, 1914, p. 613 (Kundait). A variety of grahamite distinguished by the brown colour of its powder and by its greater solubility in oil of turpentine and in chloroform. Named from the locality, Kunda, Esthonia. 7th List

Kunzite. C. Baskerville, *Science*, New York, 1903, n. ser., vol. xviii, p. 303. G. F. Kunz, *Science*, New York, 1903, vol. xviii, p. 280; *Amer. Journ. Sci.*, 1903, ser. 4, vol. xvi, p. 264. A lilac-coloured gem-variety of spodumene from California. Named after Dr. George Frederick Kunz, of New York. 3rd List

Kupferlovčorrit, Kupferwudjavrit. *See* Cuprolovchorrite. 14th List

Kupfer-Saponit, syn. of Medmontite (H. Strunz, *Min. Tabellen*, 3rd edit., 1957, p. 313). 22nd List

Kupferspießglanze. H. Strunz, 1957. *Min Tabellen*, 3rd edit., pp. 86, 104. Group name, including chalcostibite, emplectite, cuprobismutite, and wittichenite. 22nd List

Kupfer-Vermiculit, translation of Copper vermiculite (22nd List) (H. Strunz, *Min. Tabellen*, 4th edit., 1966, p. 488). 24th List

Kupferwudjavrit, *see* Cuprolovchorrite.

Kupfferite. In addition to its orginal use by Koksharov for a chromian Anthophyllite (the original analysis shows only 3% CaO, which would seem to rule out an Actinolite, suggested by H. Strunz, *Min. Tabellen*, 4th edit., 1966, p. 488), this name has been used for the magnesium end-member of the anthophyllite series (Allen and Clement, 1908), and recently for the *hypothetical* magnesium end-member of the Cummingtonite series (cited by H. W. Jaffe, W. O. J. G. Meijer, and D. H. Sekhow, *Amer. Min.*, 1961, vol. 46, p. 651). A 'kupfferite' analysed by Lorenzen (1884, Dana, 6th edit., p. 347) was a mis-labelled Hypersthene. 24th List [Magnesio-anthophyllite; *Amph.* (1978)]

Kupletskite. E. I. Semenov, 1956. *Doklady Acad. Sci. USSR*, vol. 108, p. 933 (Куплетскит). A variety of astrophyllite rich in manganese (MnO 27·65%), in pegmatite from Kola. Named after Boris Mikhailovich Kupletsky, Борис Михаилович Куплетский, Russian mineralogist. Abstr. in *Bull. Soc. franç. Min. Crist.*, 1957, vol. 80, p. 212 (koupletskite). [*M.A.*, **13**, 384.] 21st List

Kuranakhite. S. V. Yablokova, L. S. Dubakina, A. L. Dmitrik, and G. V. Sokolova, 1975. *Zap.*, **104**, 310 (Куранахит). Minute grains intergrown with gold and electrum in oxidized quartz-hematite ore of the Kuranakh gold deposit, S Yakutia, are orthorhombic, *a* 5·1, *b* 8·9, *c* 5·3 Å. Composition $PbMn^{4+}TeO_6$ (microprobe analysis). Named from the locality. [*M.A.*, 76-881; *A.M.*, **61**, 339.] 29th List

Kurchatovite. S. V. Malinko, A. E. Lisitsyn, K. A. Dorofeeva, I. V. Ostrovskaya, and D. P. Shashkin, 1966. Зап. Всесоюз. Мин. Общ. (*Mem. All-Union Min. Soc.*), vol. 95, p. 203 (Курчатовит, kurchatovite). Orthorhombic crystals from Siberian skarns, composition $Ca(Mg,Mn)B_2O_3$. Named for I. V. Kurchatova. 24th List [*M.A.*, 74-2465]

Kureyt, *see* Kurtzite.

Kurgantaite. Y. Y. Yarzhemski, 1952. *Min. Sbornik, Lvov Geol. Soc.*, no. 6, p. 169 (кургантаит). Hydrous metaborate $2(Sr,Ca)O.2B_2O_3.H_2O$. Named from the locality Kurgan-tau (Курган-тау), Inder district, Kazakhstan. [*M.A.*, **12**, 513.] 20th List [*A.M.*, **40**, 941]

Kurnakite. E. Y. Rode, 1955. *Trans. First Congress of Thermography* (*Kazan, 1953*), *Acad. Sci. USSR*, 1955, p. 217 (курнакит). Two modifications of Mn_2O_3, usually confused with braunite and bixbyite: α-kurnakite is tetragonal and β-kurnakite is cubic. Named after N. S. Kurnakov. [*M.A.*, **13**, 302.] Not to be confused with kurnakovite (16th List). 21st List [Not certainly distinct from Bixbyite. Cf. Patridgeite, Sitaparite]

Kurnakovite. M. N. Godlevsky, 1940. *Compt. Rend.* (*Doklady*) *Acad. Sci. URSS*, vol. 28, p. 638. Hydrous borate of magnesium, $2MgO.3B_2O_3.13H_2O$, as white granular (monoclinic) masses from Inder, Kazakhstan. Named after N. S. Kurnakov (Н. С. Курнаков), member of the USSR Academy of Sciences. [*M.A.*, **8**, 53.] 16th List [$MgB_3O_3(OH)_5.5H_2O$; *M.A.*, 70-3026. Dimorph of Inderite (q.v.)]

Kurskite. P. N. Chirvinsky, 1918. *The phosphorites of Ukraine*. Ukrainian and Russian editions: Kiev, 1918, p. 48 (курскіт); Petrograd, 1919, p. 50 (курскитъ). A phosphorite mineral with the composition $2Ca_3(PO_4)_2.CaF_2.CaCO_3$, occurring as a constituent of the phosphorite nodules of Kursk and elsewhere in Russia. Named from the locality. [*M.A.*, **2**, 54.] 10th List [Is Francolite; *A.M.*, **9**, 118, 155]

Kurtzite. S. J. Thugutt, 1945. *Arch. Min. Tow. Nauk. Warszaw.* (*Arch. Min. Soc. Sci. Varsovie*), vol. 15 (for 1939–1945), p. 182 (courzite), p. 187 (kurcyt). A zeolite from the Crimea, identified by A. E. Fersman (1909) as wellsite, is given a complex structural formula different from that of wellsite. Named from the locality, Kurtzy (Курцы) village near Simferopol. [*M.A.*, **10**, 6.] 18th List

Kurumsakite. E. A. Ankinovich, 1954. (*Izv. Acad. Sci. Kazakhstan*, no. 134, Geol. ser. no. 18, p. 116.] *Abstr. in Mém. Soc. Russ. Min.*, 1955, vol. 84, p. 343 (курумсакит). 8(Zn,Ni,Cu)O. $4Al_2O_3.V_2O_5.5SiO_2.27H_2O$, orthorhombic? In bituminous shale from Kurumsk, Kata-tau Mts., Kazakhstan. Named from the locality. [*M.A.*, **13**, 207.] 21st List

Kushmurunite. A. K. Gladkovskii and I. N. Ushatinskii, 1961. [Труды Горно-геол. Инст. Уралск. Фил. Акад. наук СССР (*Proc. Mining-geol. Inst. Ural Div. Acad. Sci. USSR*), vol. 56, p. 114], abstr. Зап. Всесоюз. Мин. Общ. (*Mem. All-Union Min. Soc.*), 1966, vol. 95, p. 311 (Кушмурунит). Merely finely dispersed Boehmite. 24th List

Kusterite, error for Këstерite [*M.A.*, **14**, 280]. 23rd List

Kusuïte. M. Deliens and P. Piret, 1977. *Bull.*, **100**, 39. Black crystals in the oxidation zone of the Kusu deposit, Kinshasa, Zaïre, are tetragonal, *a* 7·35, *c* 6·56 Å, space group $I4_1/amd$. Composition $4[(Ce^3,Pb^2,Pb^4)VO_4]$. Named for the locality. [*A.M.*, **62**, 1058; *M.A.*, 77-3389.] 30th List

Kutinaite. J. Hak, Z. Johan, and B. J. Skinner, 1970. *Amer. Min.*, **55**, 1083. A silvery-grey mineral with metallic lustre, intimately intergrown with novákite, is cubic, *a* 11·76 Å. D 8·38 on synthetic material. Cell contents $28[Cu_{2\cdot07}Ag_{0\cdot84}As]$. Found in only a few polished sections of the ores from Černy Důl, Bohemia. Named for J. Kutína. [*M.A.*, 71-549; *Zap.*, **100**, 615.] 27th List

Kutnohorite. A. Bukovský, 1901. [*Anz. III Congr. böhm. Naturf. u. Aerzte*, Prag, 1901, p. 293; *Programm d. Oberrealschule in Kuttenberg*, 1902]; *Neues Jahrb. Min.*, 1903, vol. ii, ref. p. 338– (Kutnohorit). A rhombohedral carbonate with the atomic ratios Ca:Mn:Fe:Mg = 7:5:1:2, occurring as reddish-white cleavage-masses at Kutná Hora (German, Kuttenberg), Bohemia. 4th List

Kuttenbergit. R. Koechlin, *Min. Taschenb. Wien. Min. Gesell.*, 2nd edit., 1928, p. 36. Another form of kutnohorite (4th List), named from the locality in Bohemia: Kutná Hora in Czech, Kuttenberg in German. 18th List [Also 20th List]

Kyanophilite. B. R. Rao, 1945. *Current Sci. Bangalore*, vol. 14, p. 196 (kyanophylite). Hydrous silicate of aluminium, as apple-green lumps in association with kyanite-graphite-schist in Mysore. Named from kyanite and *φίλος*, friend. [*M.A.*, **9**, 188.] 17th List [A mixture; *A.M.*, **58**, 807]

Kyanotrichite. A. N. Winchell, *Optical mineralogy*, 1933, part 2, 3rd edit., p. 116. Variant of cyanotrichite, German Kyanotrichit, but the original German spelling (E. F. Glocker, 1839) was Cyanotrichit. 19th List

Kylindrite. A. Frenzel, *Neues Jahrb. Min.*, II. 125, 1893. $6PbS.Sb_2S_3.6SnS_2$. [Santa Cruz mine, Poopó] Bolivia. 1st List [Now, commonly, **Cylindrite**]

Kӯshtӯmite, transliteration of Кыштымит, syn. of Kischtimite (= Bastnäsite), *M.A.*, 1967, **18**, 47. 25th List

L

Labite. N. E. Efremov, 1936. *Mém. Soc. Russ. Min.*, ser. 2, vol. 65, p. 108 (лабит), p. 117 (labite). A yellowish-green pilolite-like mineral, consisting of matted fibres, probably orthorhombic, with a composition, $H_2MgSi_3O_8.H_2O$, near that of the picrocollite (11th List) end-member of the pilolite-palygorskite group. Named from the locality, Laba river, N Caucasus. [*M.A.*, **6**, 439.] 14th List

Labountsovite, Labuntzowit, Labunzovite, Labunzowit, variant transliterations of лабунцовит, Labuntsovite (21st List). 24th List

Labratownite. A. N. Winchell, 1925. *Journ. Geol. Chicago*, vol. 33, p. 726; Elements of optical mineralogy, 2nd edit., 1927, part 2, p. 319. A contraction of labradorite-bytownite for felspars of the plagioclase series ranging in composition from $Ab_{40}An_{60}$ to $Ab_{30}An_{70}$. 11th List

Labuntsovite*. E. I. Semenov and T. A. Burova, 1955. *Doklady Acad. Sci. USSR*, vol. 101, p. 1113 (лабунцовит). Abstr. in *Amer. Min.*, vol. 41, p. 163 (labuntsovite) and *Bull. Soc. Franç. Min. Crist.*, vol. 79, p. 333 (labountsovite). Hydrous titano-silicate, $(K,Na,Ba,Ca,Mn)(Ti,Nb)Si_2(O,OH)_7.\frac{1}{2}H_2O$, orthorhombic, from Kola, Russia. Originally described by A. N. Labuntsov [*M.A.*, **3**, 235] as titaniferous elpidite (titano-elpidite, 11th List), but now shown to contain only a trace of ZrO_2 with TiO_2 25·49, Nb_2O_5 1·45%. Named after Alexander Nikolaevich Labuntsov, Александр Николаевич Лабунцов, formerly of the Lomonosov Institute, Moscow. [*M.A.*, **13**, 4.] [*Corr. in *M.M.*, **35**.] 21st List

Labuntzovite, error for Labuntsovite. 21st List

Labuntzowit(e), variant transliterations of Labuntsovite.

Lacroisite. H. Lienau, *Chemiker-Zeitung*, 1903, Jahrg. xxvii, p. 15 (Lacroisit). Torrensite and viellaurite, from manganese mines in the Pyrénées, previously described as new minerals by Lienau (ibid., 1899, Jahrg. xxiii, p. 418), have been shown on microscopical examination by A. Lacroix (*Bull. Soc. franç. Min.*, 1900, vol. xxiii, p. 251) to be mixtures of various manganese minerals (mainly rhodochrosite, tephroite, and rhodonite). Lienau now describes other similar mixtures under the names lacroisite (a mixture of rhodochrosite and rhodonite; named after Professor Alfred Lacroix, of Paris), Schokaladenstein (chocolate-stone), and huelvite (from Huelva, Spain). 3rd List

Lacroixite. F. Slavík, 1914. [*Memoirs Bohemian Acad. Sci.*]; Doelter's *Handbuch der Mineralchemie*, vol. iii, p. 492; *Bull. Soc. franç. Min.*, vol. xxxvii, p. 157. Imperfect crystals with an orthorhombic aspect and resembling herderite, but probably monoclinic, occurring with ježekite and roscherite (q.v.) in drusy cavities in lithionite-granite at Greifenstein, near Ehrenfriedersdorf, Saxony. It is a hydrated basic fluophosphate of aluminium, calcium, manganese, and sodium, $3AlPO_4.4(Ca,Mn)O.4Na(F,OH).2H_2O$. Named after Prof. Alfred Lacroix, of Paris. Not the Lacroisite of H. Lienau (1903; 3rd List). 7th List [Redefined as $NaAlPO_4(OH)$, the phosphate analogue of Durangite; *A.M.*, **57**, 1914]

Lafittite. Z. Johan, J. Mantienne, and P. Picot, 1974. *Bull.*, **97**, 48. Small grains (up to 0·2 mm) with routhierite in the black dolomites of Jas Roux, Hautes-Alpes, France, are monoclinic, *a* 11·484, *b* 14·020, *c* 6·388 Å, *β* 90·0°. Composition $8[AgHgAsS_3]$. Named for P. Lafitte. 28th List

Lagoriolite. J. Morozewicz, *Tsch. Min. Mitth.*, 1898, XVIII, 147 (*Lagoriolith*; *Natrongranat*); Abstr. *Min. Mag.*, XII, 314. An artificial soda-garnet with the formula $3(Na_2,Ca)O.Al_2O_3.3SiO_2$. Forms a connecting link between the garnet and nosean groups. 2nd List [Repetition of the synthesis yielded nosean; *A.M.*, **57**, 1317]

Laihunite. Guiyang Institute of Geochemistry, 1976. *Geochimica*, **2**, 95. Black crystals in an iron deposit at Lai-He, NE China, are orthorhombic, *a* 4·800, *b* 10·238, *c* 5·857 Å, and essentially oxidized fayalite. It was subsequently recognized as identical with ferrifayalite (23rd List). Composition near $Fe_2Fe^{3}_{2}(SiO_4)_2$. [*A.M.*, **62**, 1058; **63**, 424; *M.A.*, 77-883.] 30th List

Laitakarite. A. Varma, 1959. *Geologi*, Helsinki, vol. 3, no. 11, p. 11 (Laitakariitti); *Bull.*

Comm. géol. Finlande, 1960, vol. 188, p. 1. Bi_4Se_2S, rhombohedral and isostructural with joséite. Named after Aarne Laitakari, who collected the material in 1932 at Orijärvi, Finland. [*M.A.*, **14**, 139; **15**, 134.] Cf. Ikunolite. 22nd List [*A.M.*, **44**, 908]

Lambertite. S. C. Lind and C. W. Davis, 1919. *Science*, New York, new ser., vol. 49, p. 443. An undetermined uranium mineral, perhaps UO_3, occurring as canary-yellow crystals in quartzite at Lusk, Wyoming. Named after Mr. Ross Lambert, of Casper, Wyoming, who discovered this deposit of uranium-ore. [*M.A.*, **1**, 22.] 9th List [Is Uranophane; *M.A.*, **3**, 313]

Lamprobolite. A. F. Rogers, 1940. *Amer. Min.*, vol. 25, p. 826; ibid., 1941, vol. 26, p. 201. To replace the names 'basaltic hornblende', basaltine, and oxyhornblende (13th List) for the black lustrous crystals of hornblende rich in ferric iron, with high refringence and birefringence and strong pleochroism, of volcanic rocks (not only basalts). Named from *λαμπρός*, brilliant, and *βολίς*, a missile. [*M.A.*, **8**, 51.] 16th List

Lamprophyllite. W. Ramsay and V. Hackman [*Fennia, Bull. Soc. Géographie de Finlande*, 1894, XI, no. 2, p. 119] *Neues Jahrb. Min.*, 1896, I, ref. 259; Dana's *Appendix*, 1899. A mineral, related to astrophyllite, occurring in the nepheline-syenite of the Kola peninsula, Russian Lapland. 2nd List [Sr > Ba. Cf. Barytolamprophyllite]

Lamprostibian. L. J. Igelström, *Geol. För. Förh.*, XV. 471, 1893; and *Zeits. Kryst. Min.*, XX. 467, 1894. Antimonate of Fe and Min. Sjö mine, Sweden. 1st List [Probably merely a habit-variety of melanostibian; *Arkiv. Min.*, **4**, 451]

Landauite. A. M. Portnov, L. E. Nikolaeva, and T. I. Stolyarova, 1966. Доклады акад. наук СССР (*Compt. Rend. Acad. Sci. URSS*), vol. 166, p. 1420 (Ландаунт). Fine-grained aggregates in the Burpala massif, N Baikal, approximate to $(Zn, Mn,Fe^{3+})(Ti,Fe^{3+})_3O_7$. Monoclinic, strongly pleochroic in green shades. Named for L. D. Landau. [*M.A.*, **18**, 46; *A.M.*, **51**, 1546; Зап., **96**, 69; *Bull.*, **90**, 116.] 25th List

Landerite. M. M. Villada, 1891. *La Naturaleza*, México, ser. 2, vol. i, p. 502 (Landerita). A beautiful, pink grossularite occurring as large rhombic-dodecahedra embedded in white marble at Xalostoc, Morelos, Mexico. It was analysed by A. Damour in 1891 and by Carlos F. de Landero in 1891, after whom it is named. Also called rosolite and xalostocite (q.v.). 4th List

Landesite. H. Berman and F. A. Gonyer, 1930. *Amer. Min.*, vol. 15, p. 384. Hydrous phosphate of ferric oxide and manganous oxide, $3Fe_2O_3.20MnO.8P_2O_5.27H_2O$, as an alteration product of reddingite, from Maine, USA. Named after Prof. Kenneth K. Landes, of the University of Kansas. [*M.A.*, **4**, 344.] 12th List [$(Mn,FeOH)_3(PO_4)_2.2–3H_2O$, related to Reddingite; *A.M.*, **49**, 1122]

Landevanite. Comte de Limur. A. Lacroix, *Minéralogie de la France*, 1895, vol. i, p. 484. 'Under the name of landevanite,, M. de Limur has distributed to his correspondents a rose-coloured clay coming from Landevan in Morbihan and resulting from the decomposition of a pegmatite rich in albite.' It is placed by Lacroix under montmorillonite. 4th List

Landsbergite. D. R. Hudson, 1943. *Metallurgia*, Manchester, vol. 29, p. 56. Abbreviation of the name moschellandsbergite (15th List). Named from the locality, Landsberg, near Ober-Moschel, Rhenish Bavaria. [*M.A.*, **9**, 55.] 17th List

Laneite. G. Munteanu-Murgoci, 1906. *Bull. Dept. Geol. Univ. California*, vol. iv, p. 384. A variety of amphibole occurring in riebeckite rocks. It is dark coloured with very strong pleochroism. Optically negative and uniaxial or with very small axial angle, the axial plane being perpendicular to the plane of symmetry; c: $\mathfrak{c}$ = 13–26°. Named after Prof. Alfred Church Lane, of Tuft's College, Massachusetts. 7th List [Ferroan or ferro-pargasitic hornblende; *Amph.* Sub-Comm.]

Langbanite, variant of Långbanite; *A.M.*, **9**, 62. 10th List

Langbeinite. S. Zuckschwerdt, *Zeits. angewandte Chem.*, 356, 1891. O. Lüdecke, *Chemiker-Zeitung*, XXI. 264, 1897. $K_2SO_4.2MgSO_4$. Cubic, tetartohedral (Lüdecke). Prussian salt deposits. 1st List [Cf. Manganolangbeinite]

Langisite. W. Petruk, D. C. Harris, and J. M. Stewart, 1969. *Canad. Min.*, **9**, 597. Hexagonal (Co,Ni)As with Co:Ni = 5, isotypic with nickeline, occurs in the ores of the Langis mine, Cobalt–Gowganda area, Ontario. Named for the locality. [*M.A.*, 70-1644; *Zap.*, **99**, 71; *A.M.*, **57**, 1910.] 26th List

Laplandite. E. M. Eskova, E. I. Semenov, A. P. Khomyakov, M. E. Kazakova, and O. V. Sidorenko, 1974. *Zap.*, **103**, 571 (Лапландит). Radiating fibrous aggregates in the natrolite zone of the Jubilee (Юбилей) pegmatite. Mt. Karnasurt, Lovozero massif, Kola Peninsula, are near $4[Na_4CeTiPO_4Si_7O_{18}.5H_2O]$. Space group *Pmmm*, *a* 7·27, *b* 14·38, *c* 22·25 Å. Named for the locality. [*A.M.*, **60**, 487; *M.A.*, 75-2525; *Zap.*, **104**, 616.] 29th List

Laponite, a synthetic hectorite-like clay (B. S. Neumann and K. G. Sansom, [*Israel Journ. Chem.*, **8**, 315 (1970)]); *Clay Min.*, **9**, 231 (1971). [*M.A.*, 71-67, 1680]. 27th List

Lapparentite. H. Ungemach, 1933. *Compt. Rend. Acad. Sci. Paris*, vol. 197, p. 1133; *Bull. Soc. Franç. Min.*, [1934], vol. 56 (for 1933), p. 303. Hydrated basic sulphate of aluminium, $Al_2O_3.2SO_3.10H_2O$, as monoclinic crystals resembling gypsum in appearance. From Chile. Named after Albert Auguste de Lapparent (1839–1908), of Paris. [*M.A.*, **5**, 390.] 13th List [Is Tamarugite; *M.A.*, **8**, 87; the name has also been used by Rost, *M.A.*, **7**, 11, for a different Al sulphate]

Lardite. P. A. Zemĭatčenskij, *Trav. Soc. Nat. St.-Pétersbourg*, 1889, vol. xx, *Sect. Géol. et Min.*, p. 216 (*лярдитъ*); (abstr. *Zeits. Kryst. Min.*, 1892, vol. xx, p. 185). Hydrated silica occurring in clay in central Russia; while still moist it is white and slightly transparent, but on drying it becomes opaque. Named from the Latin, *lardum*, lard. The name lardite was used by J. G. Wallerius in 1788 for steatite. *See* ljardit. 3rd List

Larnite. C. E. Tilley, 1929. *Min. Mag.*, vol. 22, p. 77. Calcium orthosilicate, Ca_2SiO_4, as grains (monoclinic) in a chalk-dolerite contact-rock. Identical with the artificial α-Ca_2SiO_4. Named from the locality, Larne, Co. Antrim, Ireland. 12th List [Now β-Ca_2SiO_4. Cf Bredigite]

Larosite. W. Petruk, 1971. *C.M.*, **11**, 209; ibid. 1972, **11**, 886. Small acicular crystals with chalcosine and stromeyerite from the Foster mine, Cobalt, Ontario, are orthorhombic, *a* 22·15, *b* 24·03, *c* 11·76 Å. Composition $10[(Cu,Ag)_{21}PbBiS_{14}]$. Named for Mr. LaRose. [*M.A.*, 73-3556; 74-1450; *A.M.*, **59**, 382; *Zap.*, **102**, 441.] 28th List

Larsenite. C. Palache, L. H. Bauer, and H. Berman, 1928. *Amer. Min.*, vol. 13, p. 142. Lead and zinc silicate, $PbZnSiO_4$, as colourless orthorhombic prisms from Franklin Furnace, New Jersey. Isomorphous with olivine. Named after Prof. Esper Signius Larsen (1879–), of Harvard University, Cambridge, Massachusetts. [*M.A.*, **3**, 469.] 11th List

Laspeyrit. H. Strunz, 1957. *Min. Tabellen*, 3rd edit., p. 389. A proposed name for Iglesströmite (of Weibull), cf. C. Hintze, *Handb. Min.*, Erg.-Bd. II, p. 748. 24th List

Lassallite. G. Friedel, *Bull. Soc. franç. Min.*, 1901, vol. xxiv, p. 6. A substance resembling matted asbestos, which, under the microscope, is seen to consist of birefringent fibres. $3MgO.2Al_2O_3.12SiO_2.8H_2O$. France. Named after M. Lassalle, a mine proprietor. 3rd List

Lassolatite. F. Gonnard, 1876. *Minérologie du département du Puy-de-Dôme*, p. 9; A. Lacroix, *Minéralogie de la France*, 1901, vol. iii, p. 321. A fibrous, silky variety of opal, identical with fiorite. From the puy de Lassolas, Puy-de-Dôme. 6th List

Lasur-oligoclase. P. von Jereméeff [*Jubilee volume of the St. Petersburg Mining Institute*, 1873, part II, p. 167] *Zeits. Kryst. Min.*, 1900, XXXII, 493 (*Lasur-Oligoklas*). Nordenskiöld's 'Lasur-Feldspath' from Lake Baikal is shown to have the crystal elements of oligoclase. 2nd List

Latiumite. C. E. Tilley and N. F. M. Henry, 1952. *Min. Soc. Notice*, no. 78. Monoclinic sulphatic silicate Ca, K, Al, in ejected blocks from Albano, Latium, Italy. Named from the locality. [*M.M.*, **30**, 39.] 19th List [*A.M.*, **39**, 403]

Latrappite. E. H. Nickel, 1964. *Canad. Min.*, vol. 8, p. 121. The perovskite-family mineral from Oka, Quebec, described by E. H. Nickel and R. C. McAdam (*Canad. Min.*, 1964, vol. 7, p. 683), of composition near $(Ca,Na)(Nb,Ti,Fe)O_3$ with Nb $\sim$ 0·54, is now named for the village of La Trappe, near the deposit in which the mineral was found. [A.M., **50**, 265.] 24th List

Laubmannite. C. Frondel, 1949. *Amer. Min.*, vol. 34, p. 514. A dufrenite-like mineral, $(Fe'',Mn,Ca)_3Fe'''_6(PO_4)_4(OH)_{12}$, orthorhombic?, isostructural with andrewsite; from Polk Co., Arkansas. Named after Dr. Heinrich Laubmann of Munich; the 'dufrenite' from Bavaria which he described in 1923 may perhaps belong here. [*M.A.*, **11**, 8.] 19th List

Laueite. H. Strunz, 1954. *Naturwissenschaften*, vol. 41, p. 256 (Laueit). Hydrous basic phosphate, $MnFe'''_2(PO_4)_2(OH)_2.8H_2O$, triclinic, honey-brown, from Hagendorf, Bavaria. Named after Professor Max (Felix Theodor) von Laue (1879–), of Berlin. [*M.A.*, **12**, 410.] 20th List [*A.M.*, **39**, 1038. Cf. Pseudolaueite, Strunzite]

Launavite, error for launayite (25th List). *C.M.*, **9**, 744. 28th List

Launayite. J. L. Jambor, 1967. *Canad. Min.*, vol. 9, pp. 7, 191. A monoclinic mineral from Madoc, Ontario, with composition $Pb_{22}Sb_{26}S_{61}$. Named for L. de Launay. [*A.M.*, **53**, 1423; *Bull.*, **91**, 302.] 25th List

Lausenite. G. M. Butler, 1928. *Amer. Min.*, vol. 13, p. 594. To replace the name Rogersite (C. Lausen, 1928; 11th List) which was preoccupied (J. L. Smith, 1877). Named after Mr. Carl Lausen, of the Arizona Bureau of Mines. [*M.A.*, **4**, 12.] 12th List

Lavendrine. An absurd trade-name for amethystine quartz. R. Webster, *Gems*, 1962, p. 759. 28th List

Lavendulanite. C. Guillemin, 1956. *Bull. Soc. franç. Min. Crist.*, vol. 79, p. 13. Synonym of Lavendulan [of Breithaupt]. 23rd List

Lavernite, superflous gem-trade name for artificial periclase (R. Webster, *Journ. Gemmology*, **12**, 129 (1970)). 27th List

Lawsonite. F. L. Ransome, *Bull. Dept. Geol. Univ. Calif.*, I. 301, 1895; and F. L. Ransome and C. Palache, *Zeits. Kryst. Min.*, XXV. 531, 1896. [*M.M.*, **11**, 157.] $CaO.Al_2O_3.2SiO_2.2H_2O$. Rhombic. [Tiburon peninsula, Marin Co.] California. Piedmont (*Bull. Soc. fran. Min.*, XX. 5. 1897). Corsica (ibid., p. 110). 1st List

Lazarevićite. C. B. Sclar and M. Drovenik, 1960. *Bull. Geol. Soc. Amer.*, vol. 71, p. 1970 (abstr.). Microscopic grains of a cubic mineral in copper ore from the Tilva Mika deposit, Bor, Serbia, are identified as Cu_3AsS_4, the arsenic analogue of sulvanite, and named after M. Lazarević, pioneer in the study of the Bor deposits. It would have been better to extend the name Arsenosulvanite (20th List) to include all members of the series $Cu_3(As,V)S_4$ with As > V. 22nd List

Lazurquartz. Author and date? A totally unnecessary name for blue quartz (chalcedony). R. Webster, *Gems*, 1962, p. 759. For comments on azur-, lazur-, etc., *see* L. J. Spencer, *A.M.*, **22**, 683. 28th List

Lead-alunite. H. Bassett, 1950, *Journ. Chem. Soc. London*, 1950, p. 1460 (lead alunite). Artificial $Pb[Al_2(OH)_4(H_2O)_2][SO_4]_2$ as mixed crystals with $H_2O[Al_3(OH)_5H_2O][SO_4]_2$, with the alunite X-ray pattern. 19th List

Lead-autunite. J. G. Fairchild, 1929. *Amer. Min.*, vol. 14, p. 265 (lead autunite), p. 273 (lead-autunite). *See* Calcium-autunite. 19th List

Lead-barylite. J. Ito and C. Frondel, 1968. *Arkiv Min. Geol.*, vol. 4, part 4, p. 391. $PbBe_2Si_2O_7$, the synthetic lead analogue of barylite. 25th List

Lead barysilite. W. Petter, A. B. Harnick, and U. Keppler, 1971. *Zeits. Krist.*, **123**, 445 (Blei-Barysilit). The *artificial* compound $18[Pb_3Si_2O_7]$, an analogue of Barysilite. Cf. lead-barylite, 25th List. 28th List

Lead-becquerelite. J. W. Frondel and F. Cuttitta, 1953. *Amer. Min.*, vol. 38, pp. 1019, 1024 (lead-becquerelite). A doubtful variety of becquerelite, perhaps $PbO.6UO_2.11H_2O$. [*M.A.*, **1**, 377; **11**, 109; **12**, 444.] 20th List

Lead feldspar, C. A. Sorrell, 1962. *Amer. Min.*, vol. 47, p. 291. Synthetic $PbAl_2Si_2O_8$ with the Feldspar structure; probably anorthic. 23rd List

Lead-parkerite. J. W. du Preez, 1944. *Ann. Univ. Stellenbosch*, vol. 22, sect. 4, p. 101 (Lead-Parkerite). *See* Bismuth-parkerite. 17th List

Lechatelierite. A. Lacroix, 1915. *Bull. Soc. franç. Min.*, vol. xxxviii, pp. 182, 198 (lechateliérite). Naturally occurring fused (amorphous) silica (silica glass), observed in fulgurites and in quartzose enclosures in volcanic rocks. Named after Professor Henri Le Chatelier, of Paris. 7th List [Cf. Libyanite]

Ledikite. G. Brown and others, 1955. *Clay Min. Bull.*, vol. 2, pp. 299, 300. 'The trioctahedral analogue of illite', as an alteration product of biotite in soil-clay from E Ledikin, Aberdeenshire. Named from the locality. Cf. cardenite (20th List). 21st List

Ledouxite. J. W. Richards, *Amer. Journ. Sci.*, 1901, ser. 4, vol. xi, p. 458; *Chem. News*, 1901, vol. lxxxiv, p. 29. The copper arsenide $(Cu,Ni,Co)_4As$, analysed by Dr. Ledoux, and previously named mohawkite (q.v.). 3rd List [A mixture; Dana *Syst. Min.*, 7th edit., **1**, 170]

Leesbergite. L. Blum, 1908. *Ann. Soc. géol. Belgique*, vol. xxxiv, *Bull.*, p. 118. W. Bruhns, *Mitt. geol. Landesanst. Elsass-Lothringen*, 1908, vol. vi, p. 303. A white, chalky mineral forming a vein in iron-ore (oolitic minette) near Hayingen, Lorraine, and described as a hygroscopic carbonate of calcium and magnesium $Mg_2Ca(CO_3)_3$. Proved by W. Bruhns to be a mixture of hydromagnesite with calcite or dolomite. Named in memory of the late Mining-Captain F. X. H. Leesberg, of Esch on the Alzette, Luxemburg. 5th List

Lefkasbestos. A. Zdarsky, 1910. *Zeits. prakt. Geol.*, vol. xviii, p. 345 (Lefkasbest). A bleached chrysotile occurring in weathered serpentine at Mt. Troodos, Cyprus. Named from λευκός, white, and ἄσβεστος. 6th List

Legrandite. J. Drugman and M. H. Hey, 1932. *Min. Mag.*, vol. 23, p. 175. Hydrous zinc arsenate, approximating to $Zn_3As_2O_8.3H_2O$, but more exactly $Zn_{14}(AsO_4)_9OH.12H_2O$. Bundles of yellow acicular monoclinic crystals with blende from Mexico. Named after the late Mr. — Legrand, a Belgian mining engineer. 13th List [$Zn_2AsO_4(OH).H_2O$; *A.M.*, **48**, 1255]

Lehiite. E. S. Larsen and E. V. Shannon, 1930. *Amer. Min.*, vol. 15, p. 329. Hydrous phosphate of calcium, aluminium, and alkalis, $5CaO.(Na,K)_2O.4Al_2O_3.4P_2O_5.12H_2O$, as white crusts (monoclinic?) in variscite nodules from near Lehi, Utah. Named from the locality. [*M.A.*, **4**, 344.] 12th List

Lehnerite. F. Müllbauer, 1925. *Zeits. Krist.*, vol. 61, p. 331 (Lehnerit). Hydrated basic phosphate of ferrous iron with small amounts of manganese and magnesium, $Fe_7(OH)_2(PO_4)_4.5H_2O$, as apple-green, monoclinic crystals with perfect basal cleavage, from Hagendorf, Bavaria. Named after Mr. Ferdinand Lehner, a mineral collector, of Pleystein, near Hagendorf. Perhaps identical with ludlamite. [*M.A.*, **2**, 417.] 10th List [Is ludlamite; *M.A.*, **3**, 10; **3**, 374; **10**, 538; **12**, 572]

Leifite. O. B. Bøggild, 1915. *Meddelelser om Grønland*, vol. li, p. 427 (Leifit). A highly acidic fluo-silicate of sodium and aluminium, $Na_2Al_2Si_9O_{22}.2NaF$ or $Na_4(AlF)_2Si_9O_{22}$, occurring as colourless hexagonal prisms in the drusy alkali-pegmatite veins at Narsarsuk, Greenland. Named after the Scandinavian explorer Leif Ericsson (fl. 1000 AD). 8th List [Redefined; *M.A.*, 72-2222. $(Na,H_3O)_2(Si,Al,Be,B)_7(O,F,OH)_{14}$. Cf. Karpinskyite]

Leightonite. C. Palache, 1938. *Amer. Min.*, vol. 23, p. 34; M. C. Bandy, ibid., p. 719. A trimetal sulphate, $K_2Ca_2Cu(SO_4)_4.2H_2O$, as pale blue triclinic (pseudo-orthorhombic) crystals from Chile. Named after Dr. Tomas Leighton, Professor of Mineralogy in the University of Santiago, Chile. [*M.A.*, **7**, 59, 223.] 15th List

Leiteite. F. P. Cesbron, R. C. Erd, G. K. Czamanske, and H. Vachey, 1977. *Mineral Rec.*, **8**, (3), 95. Cleavable masses in a specimen of tennantite ore from Tsumeb, SW Africa, are monoclinic, *a* 17·645, *b* 5·019, *c* 4·547 Å, β 90° 59′, space group $P2_1/a$. Composition $4[ZnAs_2O_4]$. RIs α 1·87, β 1·880 ∥ [010], γ 1·98, $2V_\gamma$ $26\frac{1}{2}$°, α:[100] 11°, γ:[001] 10°. Named for L. Teixeira-Leite. [*M.A.*, 78-3476.] 30th List

Lembergite. A. Lagorio, *Trav. Soc. Natural. Varsovie, Ann.*, VI. *Livr.*, XI. 7, 1895. [Abstr. in *Zeits. Kryst. Min.*, XXVIII. 526, 1897.] J. Lemberg, *Zeits. deutsch. geol. Ges.*, **XXXIX**. 562, 1887. [*M.M.*, **11**, 111.] Given to Lemberg's artificial 'Natronnephelinhydrat.' $4Na_2Al_2Si_2O_8 + 5H_2O$. Rhombic. 1st List

—— T. Sudo, 1943. [*Bull. Chem. Soc. Japan*, vol. 18, p.—]; abstr. in *Amer. Min.*, 1947, vol. 32, p. 483. Hydrous silicate of Al, Fe, Mg, Ca, belonging to the montmorillonite group, as a fine-grained green cement in ferruginous sandstones from Japan. Named after Johann Theodor Lemberg (1842–1903) of Dorpat, who described similar material in 1877. Not the lembergite of A. Lagorio, 1895 (1st List). [*M.A.*, **10**, 147.] 18th List [Ferrian Saponite; *A.M.*, **40**, 944]

Lemnäsite. G. Pehrman, 1939. *Acta Acad. Aboensis, Math. et Physica*, vol. 12, no. 6, p. 12 (Lemnäsit). Phosphate of manganese, iron, sodium, and calcium, $3R_3(PO_4)_2.2NaOH$, as black masses, probably monoclinic, from Lemnäs, Kimito, Finland. Named from the locality. [*M.A.*, **7**, 418.] 15th List [Is Alluaudite; *M.A.*, **8**, 180; **9**, 203]

Lemoynite. G. Perrault, E. I. Semenov, A. V. Bikova, and T. A. Capitonova, 1969. *Canad. Min.*, **9**, 585. Colourless monoclinic crystals with elpidite and eudialyte in the St. Hilaire alkaline massif, Quebec; $(Na,Ca)_3Zr_2Si_8O_{22}.8H_2O$. Named for Charles Lemoyne. [*M.A.*, 70-1654; *Zap.*, **99**, 83; *A.M.*, **57**, 1913.] 26th List

Lemuanite, error for Lemoynite (26th List). *C.M.*, **9**, 585. 28th List

Lenad. W. Cross and others, 1903. *Quantitative Classification of Igneous Rocks*, Chicago, 1903, pp. 132, 271. J. P. Iddings, *Igneous Rocks*, New York, 1909, vol. i, p. 411. A contracted form of the names leucite and nephelite, suggested as an alternative group name for the felspathoid minerals. 5th List

Lengenbachite. R. H. Solly, 1904. *Nature*, vol. lxxi, p. 118; *Min. Mag.*, 1905, vol. xiv, p. 78. Thin, blade-shaped crystals, probably anorthic, occurring in the white, crystalline dolomite of the Lengenbach, Binnenthal, Switzerland. Analysis by A. Hutchinson (*Min. Mag.*, 1907, vol. xiv, p. 204) gives the formula $Pb_6(Ag,Cu)_2As_4S_{13}$. *See* Jentschite. 4th List

Lenoblite. F. Cesbron and H. Vachey, 1970. *Bull. Soc. franç. Min. Crist.*, **93**, 235. A sky-blue oxide of vanadium, distinct from duttonite (21st List), with which it occurs at Mounana, Gabon; partially oxidized; the unaltered material is probably $V_2O_4.2H_2O$ (the same composition as duttonite). Named for André Lenoble. [*M.A.*, 70-3426.] 26th List [*A.M.*, **56**, 635]

Leonardite. P. L. Broughton, 1972. *Journ. Sedim. Petr.*, **42**, 356. Weathered lignite from Saskatchewan. Not to be confused with Leonhardite (Blum, 1843) or with Leonhardtite (19th List). 29th List

Leonhardtite. W. Berdesinski, 1952. *Neues Jahrb. Min.*, Monatshefte, p. 29 (Leonhardtit). $MgSO_4.4H_2O$, from the hydration of kieserite. Named after Johannes Leonhardt (1893–), professor of mineralogy and petrography in the University of Kiel. Not to be confused with leonhardite of J. R. Blum, 1843. [*M.A.*, **11**, 517.] 19th List [Renamed Starkeyite (q.v.)]

Leonite. C. A. Tenne, *Zeits. deutsch. geol. Ges.*, XLVIII. 632, 1896. To replace kaliastrakanite (q.v.). 1st List

—— Author and date? 'Tibet stone (eosite)'. R. Webster, *Gems*, 1962, p. 759. Not the leonite (of Tenne), nor the eosite (of Schrauf); *see* eosite. 28th List

Lepidolamprite or Schuppenglanz. A. Breithaupt. C. Hintze, *Handbuch d. Mineralogie*, 1904, vol. i, p. 1198 footnote (Lepidolamprit). A specimen of franckeite (A. W. Stelzner, 1893) in the Freiberg collection was so labelled by Breithaupt in the first half of the nineteenth century. 4th List

Lermontovite. Author? *Guide to USSR exhibit at Conference on Atomic Energy*, Geneva, 1955, p. 8 (lermontovite). Later publications, abstr.: *Amer. Min.*, 1956, vol. 41, p. 816, 1958, vol. 43, p. 379; *Mém. Soc. Russ. Min.*, 1958, vol. 87, p. 81 (лермонтовит, lermontovite). $(U,Ca,\text{rare-earths})_3(PO_4)_4.6H_2O$. Named after Mikhil Yurevich Lermontov (1814–1844), Russian poet of Scottish origin. 21st List

Lesserite. C. Frondel, V. Morgan, and J. L. T. Waugh, 1956. *Amer. Min.*, vol. 41, p. 927. Hydrous magnesium borate, $Mg_2B_6O_{11}.15H_2O$, monoclinic, dimorphous with triclinic inderite, from Kramer, California. Named after Mr. Federico Lesser, of the borate industry. [*M.A.*, **13**, 303.] 21st List [Is Inderite; *A.M.*, **45**, 732]

Lessingite. V. A. Zilbermintz, 1929. *Compt. Rend. Acad. Sci. URSS*, Leningrad, ser. A, 1929, p. 55 (lessingite, лессингит). Hydrous silicate of cerium and calcium, $H_2Ca_2Ce_4Si_3O_{15}$, found as cherry-red glassy pebbles in the gold-washings of the Kyshtymsk district, Ural. Named after Prof. Frantz Yulievich Levinson-Lessing (Франц Юліевич Левинсон-Лессинг), of Leningrad. [*M.A.*, **4**, 150.] 12th List [Is probably Britholite; *M.M.*, **31**, 455]

Letovicite. J. Sekanina, 1932. *Zeits. Krist.*, vol. 83, p. 117 (Letovicit). Triammonium sulphate $H(NH_4)_3(SO_4)_2$ as pseudohexagonal scales from the decomposition of pyrite in coal at Letovice, Moravia. Named from the locality. [*M.A.*, **5**, 145.] 13th List [*A.M.*, **18**, 180]

Leucoglaucite. H. Ungemach, 1933. *Compt. Rend. Acad. Sci. Paris*, vol. 197, p. 1134; *Bull. Soc. franç. Min.*, [1934], vol. 56 (for 1933), p. 303. Hydrated ferric sulphate, $Fe_2O_3.4SO_3.5H_2O$, as pale greenish-blue hexagonal crystals, from Chile. Named from the colour *λευκός*, white, and *γλαυκός*, greenish-blue. [*M.A.*, **5**, 390; **6**, 149.] 13th List [Is Ferrinatrite (q.v.); *A.M.*, **23**, 731]

Leucophœnicite. S. L. Penfield and C. H. Warren, *Amer. J. Sci.*, 1899, VIII, 351; *Zeits. Kryst. Min.*, 1900, XXXII, 239 (*Leukophönicit*); abstr. *Min. Mag.*, XII, 316. Purplish red to raspberry-coloured crystalline masses. A manganese humite with hydroxyl in place of fluorine. $R_5(R.OH)_2(SiO_4)_3$; (R = Mn,Zn,Ca). Probably monoclinic. Franklin Furnace, New Jersey. 2nd List [*A.M.*, **55**, 1146]

Leucophosphite. E. S. Simpson, 1932. *Journ. Roy. Soc. W. Australia*, vol. 18, p. 71. Hydrous phosphate of potassium, iron, and aluminium, perhaps $K_2(Fe,Al)_7(OH)_{11}(PO_4)_4.6H_2O$, as white chalky masses from Western Australia. Named from λευκός, white, and φωσφόρος. [*M.A.*, **5**, 148.] 13th List [$KFe_2^{3+}(PO_4)_2(OH).2H_2O$; *A.M.*, **57**, 397]

Leucosphenite. G. Flink, *Medd. om Grönland*, 1900, XXIV, 137. White prismatic crystals with wedge-shaped terminations. Monoclinic. $BaNa_4(TiO)_2(Si_2O_5)_5$. [Narsarsuk] S Greenland. 2nd List [$BaNa_4Ti_2B_2Si_{10}O_{30}$; *A.M.*, **57**, 1801]

Leukophosphatit. H. Strunz, 1970, *Min. Tab.*, 5th edit., p. 348. A name for the *artificial* 'kubische Modifikation des Leukophosphits' of O. E. Radczewski and E. Wanderer, 1965, *Naturwiss.*, **52**, 426. Note that the formulae of leukophosphatit, $(K,H)Fe_4^{3+}(PO_4)_3(OH)_4.6\text{–}7H_2O$, and of leucophosphite, $K_2(Fe,Al)_7(OH)_{11}(PO_4)_4.6H_2O(?)$ [13th List], now $(K,NH_4)Fe_2^{3+}(PO_4)_2OH.2H_2O$, are not obviously compatible. 28th List

Leukosphenite. G. Flink, *Meddelelser om Grönland*, 1901, Hefte xxiv, plate VII. The same as leucosphenite. 3rd List

Levisite, *see* Livesite.

Levyine, Levyite, variants of Levyne (T. Thomson, *Outlines Min.*, 1836, vol. 1, p. 335). 22nd List

Lewisite. E. Hussak and G. T. Prior, *Min. Mag.*, XI, 80, 1895. $5(Ca,Fe)O.3Sb_2O_5.2TiO_2$. Cubic. [Tripuhy, near Ouro Preto, Minas Gerais] Brazil. 1st List [$(Ca,Fe,Na)_2(Sb,Ti)_2(O,OH)_7$, pyrochlore group; *M.A.*, **5**, 185]

Lewistonite. E. S. Larsen and E. V. Shannon, 1930. *Amer. Min.*, vol. 15, p. 326. Hydrous phosphate of calcium and alkalis, near $15CaO.(K,Na)_2O.4P_2O_5.8H_2O$, as minute hexagonal prisms in variscite nodules from Utah. Named after the locality, Lewiston, Utah. [*M.A.*, **4**, 344.] 12th List [Potassian Hydroxyapatite]

Liberite. Ch'un-Lin Chao. [*Ti Chih Hsueh Pao*, vol. 44, p. 344]; abstr. *Amer. Min.*, 1965, vol. 50, p. 519; *M.A.*, **17**, 399. Li_2BeSiO_4, as pale-yellow monoclinic crystals in veins in tactite from the Nanling range, South China. The name is presumably for the composition. 24th List

Libollite. J. P. Gomes, *Comm. Dir. Trab. Geol. Portugal*, 1898, III, 240, 290. An asphaltum, resembling albertite, from Libollo, Portuguese West Africa. 2nd List

Libyanite. C. Fenner, 1937. Newspaper article in *The Australasian*, Melbourne, October 2, 1937. Silica-glass from the Libyan Desert. (Native silica-glass had already been named lechatelierite, 7th List.) [*Min. Mag.*, **25**, 125.] 17th List

Liddicoatite. P. J. Dunn, D. E. Appleman, and J. [A.] Nelen, 1977. *A.M.*, **62**, 1121. The calcium analogue of elbaite (q.v.) occurs in the detrital soils of Antsirabe, Madagascar. Rhombohedral, a_h 15·867, c_h 7·135 Å. Ideally $Ca(Li,Al)_3Al_6B_3Si_6O_{27}(OH)_3(OH,F)$. ρ 3·02 g.cm^{-3}. Pleochroic, ω 1·637, dark brown, ε 1·621, light brown. Named for R. T. Liddicoat. [*M.A.*, 78-3475.] 30th List

Liebenbergite. S. A. de Waal and L. C. Calk, 1973. *A.M.*, **58**, 733. Minute pale-green remnant grains in a nickel serpentine at the Bon Accord deposit, Barberton Mountain Land, South Africa, approach the end-member Ni_2SiO_4; olivine group. Orthorhombic, *a* 4·727, *b* 10·191, *c* 5·955 Å. Named for W. R. Liebenberg. *Artificial* Ni_2SiO_4 has been called nickel olivine (15th List). [*M.A.*, 74-3465.] 28th List

Li-feldspar. S. Šćavnićar and G. Sabatier, 1957. *Bull. Soc. franç. Min. Crist.*, vol. 80, p. 308 (Feldspath-Li). Artificial $LiAlSi_3O_8$; tetragonal. [*M.A.*, **13**, 637 (Li-felspar).] 22nd List

Likasite. A. Schoep, W. Borchert, and K. Kohler, 1955. *Bull. Soc. franç. Min. Crist.*, vol. 78, p. 84. Nitrate and phosphate of copper, $Cu_{12}(NO_3)_4(PO_4)_2(OH)_{14}$, as sky-blue orthorhombic plates from Likasi, Belgian Congo. Named from the locality. [*M.A.*, **12**, 568.] 20th List [*M.A.*, 74-2461]

Lilianit, error for Lillianit (Hintze, *Handb. Min.*, Erg.-Bd. II, p. 578). 22nd List

Lillehammerit, variant of Lillhammerite (H. Strunz, *Min. Tabellen.*, 1st edit., 1941, p. 253). 24th List

Limaite. J. M. Cotelo Neiva, 1954. *Mem. Notíc. Mus. Laborat. Min. Geol. Univ. Coimbra*, no. 36, pp. 17, 20 (limaíte), p. 55 (limaïte), p. 59 (limaite); Estudos, *Notas e Trab. Serv. Fomento Mineiro* [*1954*], vol. 9, p. 111. Stanniferous variety of gahnite, $(Zn, Sn)Al_2O_4$ with Zn:Sn = 3:1. Octahedra in pegmatite from near Ponte do Lima, Portugal. Named from the locality. Previously mentioned and later described, without name, in *Bull. Soc. franç. Min. Crist.*, 1952, vol. 75, p. xxxvi; 1955, vol. 78, p. 97. [*M.A.*, **12**, 572.] 20th List

Lime-bronzite. A. Poldervaart, 1947. *Min. Mag.*, vol. 28, p. 170. An unstable form of bronzite containing $CaSiO_3$ up to 9%. 18th List

Lime-dravite. F. H. Pough, 1953. *See* Soda-dravite. 20th List

Lime-iron-olivine. R. W. Nurse and H. G. Midgley, 1953. *Journ. Iron and Steel Inst.*, vol. 174, p. 124 (lime-iron olivine, $CaFeSiO_4$). Quoted from N. L. Bowen, J. F. Schairer, and E. Posnjak, *Amer. Journ. Sci.*, 1933, ser. 5, vol. 25, p. 275; vol. 26, p. 195, who, however, have only the form Ca-Fe olivine for artificial material. [*M.A.*, **5**, 252, 454; **12**, 197.] 20th List

Lime-olivine. N. L. Bowen, 1922. *Amer. Journ. Sci.*, ser. 5, vol. 3, p. 30 (Lime olivine). The calcium orthosilicate described by F. P. Paul in 1906. *See* Shannonite and Kalk-Olivin. [*M.A.*, **2**, 77.] 11th List

Limonitogelit. *See* Diasporogelite. 7th List

Lindakerite, error for Lindackerite (*Bull. Soc. franç. Min. Crist.*, 1956, vol. 79, p. 7). 23rd List

Lindesite. L. J. Igelström, *Zeits. Kryst. Min.*, XXIII. 590, 1894. H. Sjögren, *Bull. Geol. Inst. Upsala*, II. 84, 132, 1895. [*M.M.*, **11**, 105, 168.] Shown by Sjögren to be identical with urbanite. [Glakärn mine, Linde Örebro, Sweden] 1st List

Lindgrenite. C. Palache, 1935. *Amer. Min.*, vol. 20, pp. 187, 484. Basic molybdate of copper, $2Cu_2MoO_4.Cu(OH)_2$, as green monoclinic crystals from Chile. Named after Professor Waldemar Lindgren (1860–), of Cambridge, Massachusetts. [*M.A.*, **6**, 54, 147.] 14th List

Lindströmite. K. Johansson, 1924. *Arkiv. Kemi, Min. Geol.*, vol. 9, no. 8, p. 14 (Lindströmit). Sulpho-bismuthite of lead and copper, $2PbS.Cu_2S.3Bi_2S_3$, as lead-grey to tin-white, prismatic crystals from Gladhammar, Sweden. Named after Gustaf Lindström (1838–1916), formerly of the Riksmuseum, Stockholm. [*M.A.*, **2**, 340.] 10th List [Monoclinic $4[Cu_3Pb_3Bi_7S_{15}]$; *C.M.*, **14**, 194 (and refs.)]

Linneite. (E. T. Wherry, *Journ. Washington Acad. Sci.*, 1920, vol. 10, p. 495). Variant of Linnæite. Linneit in German; Linnéite in French. 9th List

Linosite. H. S. Washington, 1908. *Amer. Journ. Sci.*, ser. 4, vol. xxvi, p. 210. A highly titaniferous basaltic hornblende, closely allied to kaersutite, found with anemousite (q.v.) as loose, monoclinic crystals in a volcanic tuff on the Island of Linosa, off the coast of Tunis. 5th List [Ferri or ferrian oxy kaersutite, *Amph.* (1978)]

Lionit. *See* Chillagite. 7th List

Liottite. S. Merlino and P. Orlandi, 1977. *A.M.*, **62**, 321. Flattened hexagonal crystals in ejected blocks in the pumice deposit at Pitigliano, Tuscany, have *a* 12·843, *c* 16·091 Å, space group $P\bar{6}m2$, ρ 2·56 g.cm^{-3}, ω 1·530, ε 1·528. Composition near $(Ca,Na,K)_{24}(Si,Al)_{36}O_{72}(SO_4)_4(CO_3)_2(Cl,OH)_6.2H_2O$. A member of the cancrinite group (cf. franzinite, 30th List), with Ca as dominant cation and a stacking sequence *ABABAC* . . . Named for L. Liotti. [*M.A.*, 78-890.] 30th List

Liparite. This well-known rock-name was earlier applied to three distinct mineral-species—chrysocolla, fluorite, and talc.

(1) F. Casoria, 1846. *Atti della 7ª adunanza degli Scienziati Italiani*, Napoli, vol. 7 (for 1845), part 2, p. 1156. For a hydrous copper silicate from the island of Lipari the formula was deduced as $2CuO.3SiO_2.4\frac{1}{2}H_2O$, but G. Carobbi, *Gazz. Chim. Ital.*, 1928, vol. 58, p. 801, proves the identity of the mineral with chrysocolla. Named from the locality. [*M.A.*, **4**, 139.] 12th List

(2) E. F. Glocker 1847. *Generum et specierum mineralium synopsis*, p. 282 (Liparites, Liparit). Synonym of fluorite. From λιπαρός, splendent. 12th List

(3) A. E. Arppe, 1858. *Acta Soc. Sci. Fennicae*, vol. 5, p. 476 (Jerntalk, Liparit). A ferruginous talc (Fe_2O_3 9·25%) from Finland. From λιπαρός, oily. 12th List

(4) J. Roth, 1861. *Die Gesteins-Analysen*, xxxiv (Liparit). An acid volcanic rock from the Lipari Islands. Synonym of rhyolite of F. von Richthofen, *Jahrb. Geol. Reichsanstalt*, Wien, 1861, vol. 11 (for 1860), p. 156 (Rhyolith). Named from the locality. 12th List

Lipscombite. M. A. Gheith, 1953. *Amer. Min.*, vol. 38, p. 612 (pronounced lips-kum-ite). Artificially produced tetragonal iron phosphate approximating to $Fe''Fe_2'''(PO_4)_2(OH)_2$. To replace the name iron-lazulite (19th List). Named after Professor William Nunn Lipscomb

(1909–) of the University of Minnesota. *See* Barbosalite and Ferro-ferri-lazulite. [*M.A.*, **12**, 238.] 20th List

Listvenit, cited without reference by H. Strunz, *Min. Tabellen*, 3rd edit., 1957, p. 391: 'ein grüner Glimmer, Beresowsk, Ural'. 24th List

Litargita. G. Gagarin and J. R. Cuomo, 1949, loc. cit., p. 9. Spanish spelling of lithargite (8th List). Synonym of litharge. 19th List

Lithargite. E. T. Wherry, 1917. *Amer. Min.*, vol. ii, p. 19. The yellow, orthorhombic modification of lead monoxide, PbO, recognized as a mineral species from Austria and California by E. S. Larsen (ibid., p. 18), as distinct from the red, tetragonal modification called massicot or massicotite. The old name litharge with the mineralogical termination *ite*. 8th List **[Litharge]**

Lithian-muscovite. A. A. Levinson, 1952. *Progr. and Abstr. Min. Soc. Amer.*, p. 30 (lithium muscovite); *Amer. Min.*, 1953, vol. 38, p. 88 (lithian muscovite), p. 93 (lithian-muscovite). A structural variation in the muscovite polymorphs containing Li_2O 3·4–4·0%. Distinct from lithium-muscovite (15th List). [*M.A.*, **12**, 98.] 20th List

Lithia-tourmaline. R. R. Riggs, 1888. *Amer. Journ. Sci.*, ser. 3, vol. 35, p. 50 (Lithia tourmaline). P. Quensel, *Geol. För. Förh. Stockholm*, 1939, vol. 61, p. 63 (lithium tourmaline), p. 76 (Li-tourmaline). J. Sekanina, *Věstník Stat. Geol. Úst. Česk.*, 1946, p. 302 (lithné turmaliny *pl.*), p. 311 (tourmalines lithiques). [*M.A.*, **7**, 335; **10**, 119.] 20th List [Elbaite (q.v.)]

Lithio-mangano-triphylite. E. T. Wherry, 1915. *Proc. United States National Museum*, vol. 49, p. 467. The same as Lithiophilite ($LiMnPO_4$). On the same plan, Triphylite ($LiFePO_4$) is called Lithio-ferro-triphylite, and it is further suggested that the name Natrophilite ($NaMnPO_4$) should be discarded. [*M.A.*, **2**, 471.] 10th List

Lithionglaukophan. *See* Holmquistite. 7th List

Lithiophosphate. V. V. Mathias and A. M. Bondareva, 1957. *Doklady Acad. Sci. USSR*, vol. 112, p. 124 (литиофосфат, lithiophosphate). Lithium phosphate, Li_3PO_4, as white to colourless masses in pegmatite from Kola. It is a hydrothermal alteration product of montebrasite, and weathers to mangan-apatite and davisonite. Named from the composition. The suffix *-ate* is new for mineral names (except agate) and may be confused with chemical and other terms; for example, selenite ($CaSO_4.2H_2O$ and $CaSeO_3$), evaporate, precipitate, separate. [*M.A.*, **13**, 383.] 21st List [*A.M.*, **42**, 585]

Lithiophosphatit, variant of Литиофосфат, Lithiophosphate (21st List) (H. Strunz, *Min. Tabellen*, 4th edit., 1966, p. 273). 24th List

Lithium-amphibole. C. Palache, S. C. Davidson, and E. A. Goranson, 1930. *Amer. Min.*, vol. 15, p. 292. The molecule $Li_2(Mg,Fe)_3Al_2Si_8O_{22}(OH)_2$, or an amphibole containing this in large amount, such as holmquistite or Lithionglaukophan (7th List). [*M.A.*, **4**, 526.] 12th List

Lithium fluor-hectorite. J. L. Miller and R. C. Johnson, 1962. *Amer. Min.*, vol. 47, p. 1049. A synthetic product, $(K,Li)_x Mg_{3-x} Li_x Si_4O_{10}F_2$, obtained as a fine-grained water-swelling phase. 23rd List

Lithium-muscovite. R. E. Stevens, 1938. *Amer. Min.*, vol. 23, pp. 608, 523 (lithium muscovite). A hypothetical molecule $K_4Li_6Al_6Al_4Si_{12}O_{40}(F,OH)_8$ for expressing the composition of lepidolite. [*M.A.*, **7**, 353.] 15th List

Liveingite. R. H. Solly, *Proc. Cambridge Phil. Soc.*, 1901, vol. xi, p. 239; *Min. Mag.*, 1902, vol. xiii, p. 160 footnote; (abstr., *Min. Mag.*, **13**, 206). Crystals resembling in general appearance the several other sulpharsenites of lead from the Binnenthal dolomote. Monoclinic. $5PbS.4As_2S_3$. Named after Professor G. D. Liveing, of Cambridge. 3rd List [Re-defined as identical with Rathite-II, $Pb_9As_{13}S_{28}$; *A.M.*, **54**, 1498]

Liversite. C. F. Barb, *Quart. Colorado Sch. Mines*, 1944, vol. 39, p. 18 ('liversite'). Local name for elaterite from Strawberry river, Utah. 20th List

Livesite. K. Carr, R. W. Grimshaw, and A. L. Roberts, 1951. *Trans. Brit. Ceramic Soc.*, vol. 51, p. 339 (name in table without description). A constituent of Yorkshire fireclay presumably intermediate between kaolinite and halloysite. R. E. Grim, in discussion at symposium, Problems of clay and laterite genesis, *Amer. Inst. Mining and Metall. Engineers*, New York, 1952, p. 221 (levisite); *Clay mineralogy*, New York and London, 1953, p. 49

(levisite). Named after Sir George Thomas Livesey (1834–1908), in whose memory the Livesey professorship of coal gas and fuel was founded in 1910 in Leeds University. [*M.A.*, **12**, 184, 212; cf. *Min. Mag.*, **30**, 139.] 20th List

Livite. L. A. Kulik, 1941. *Meteoritica, Acad. Sci. USSR*, vol. 1, p. 75 (*Ливит* . . . Кварцевое стекло из Ливии) [*sic*], p. 122 (Livit . . . Silica glass). Incorrect form of libyanite (q.v.). [*M.A.*, **9**, 295.] 17th List

Lizardite. E. J. W. Whittaker and J. Zussman, 1955. *Min. Soc. Notice*, no. 91; *Min. Mag.*, 1956, vol. 31, pp. 108, 118; J. Zussman and others, *Amer. Min.*, 1957, vol. 42, p. 134. A platy serpentine mineral with X-ray *c*-axis half that of chrysotile. Named from the locality, Lizard, Cornwall. 21st List [$Mg_3Si_2O_5(OH)_4$]

Ljardit. (*Neues Jahrb. Min.*, 1901, vol. ii, –408–, abstr.). The same as lardite (q.v.). 3rd List

Llallagualite. M. C. Bandy, 1946. *Mineralogía de Llallagua, Bolivia*, La Paz, p. 56 (Llallagualita, Monasita romboédrica). Provisional name for an undescribed rhombohedral mineral which has perhaps the same composition as monazite. Named from the locality Llallagua (pronounced Yayawa). [*M.A.*, **10**, 9.] 18th List [May be Rhabdophane]

Loaisite. R. L. Codazzi, 1905. *Mineralizadores y minerales metálicos de Colombia. Trabajos de la Oficina de Historia Natural, Secc. Min. Geol.*, Bogota, 1905, p. 16 (Loaisita). Scorodite in very pale green, porous masses, from Loaysa, near Marmato, Colombia, analysed by J. B. Boussingault in 1829. 5th List

Lodochnikite. Author? *Guide to USSR exhibit at Conference on Atomic Energy*, Geneva, 1955, p. 9 (lodochnikite). Later publications, abstr.: *Amer. Min.*, 1957, vol. 42, p. 307 (lodochnikite, lodochnikovite), *Amer. Min.*, **43**, 380; *Mém. Soc. Russ. Min.*, 1958, vol. 87, p. 78 (лодочникит, lodochnikite). $2(U,Th)O_2.3UO_3.14TiO_2$. Named after Vladimir (= Wartan) N. Lodochnikov, Владимир Н. Лодочников (1887–1942). Not to be confused with lodochnikovite of E. I. Nefedov, 1953 (20th List). 21st List [Is Brannerite; *A.M.*, **48**, 1419]

Lodochnikovite. E. I. Nefedov, 1953. *Mém. Soc. Russ. Min.*, ser. 2, vol. 82, p. 317 (Лодочниковит). Al,Mg,Ca,Fe oxide, monoclinic, bluish-green, in skarn with spinel, etc. Named after Vladimir (formerly Wartan) N. Lodochnikov, Владимир Н. Лодочников (1887–1942), Petrologist on the Geological Survey of Russia. [*M.A.*, **12**, 352.] 20th List

—— V. I. Gerasimovsky, 1956. [Атом. Энерг. (*Atomic Energy*), no. 4, p. 118], abstr. in *Amer. Min.*, 1957, vol. 42, p. 307. $2(U,Th)O_2.3UO_3.14TiO_2$; synonym of, or probably an error for Lodochnikite (21st List). Not to be confused with Lodochnikovite (of Nefedov; 20th List). 23rd List

Lodotchnikovite, Lodotschnikowit. French (*Bull. Soc. Franç, Min. Crist.*, 1955, vol. 78, p. 219) and German (*Geologie Zeitschr.*, East Berlin, 1955, vol. 4, p. 528) transliterations of лодочниковит, lodochnikovite of E. I. Nefedov, 1953 (20th List). 21st List

Lodotschnikit, German transliteration of Лодочникит, Lodochnikite. (21st List) (C. Hintze, Erg.-Bd. II, p. 753). 24th List

Loesserite. *Soviet Physics—Doklady*, 1962, **7**, 173. Mistransliteration of Лессеоит, Lesserite, now a syn. of inderite [*A.M.*, **45**, 732]. [*M.A.*, **16**, 133.] 28th List

Loewite, error for Löweïte (J. W. Mellor, *Treat. Inorg. Chem.*, **4**, 1060). 26th List

Lohestite. J. Anten, 1923. *Mém.* (*in-4°*) *Acad. Roy. Belgique, Cl. Sci.*, ser. 2, vol. 5, fasc. 3, p. 29. H. Buttgenbach, *Livre Jubilaire Soc. Géol. Belgique*, 1925, vol. 3, p. 29. An almost isotropic material occurring as knots in the phyllites of the Belgian Ardennes and representing a stage in the formation of crystals of andalusite. Named after the Belgian geologist, Marie Joseph Maximin Lohest (1857–). [*M.A.*, **3**, 11.] 11th List

Lok-batanite. S. A. Kovalevsky and A. T. Kochmarev, 1939. [*Trudy Voprosam Neftyanoi Geol.*, 1939, p. 7; *Khim. Referat. Zhur.*, 1940, no. 2, p. 21]; *Chem. Abstr.* (*Amer. Chem. Soc.*), 1942, vol. 36, p. 1874. An organic material from the Lok-Batan mud volcano. 16th List

Lokkaite. V. Perttunen, 1970. *Bull. Geol. Soc. Finland*, **43**, 67. A rare-earth carbonate from Pyörönmaa, Kangasala, SW Finland, originally described as tengerite (*Compt. Rend. Soc. Géol. Finlande*, **38**, 241) proves to be a lower hydrate. Orthorhombic, *a* 39·07, *b* 6·079, *c* 9·19 Å; composition near $(Yt,Ca)_2(CO_3)_3.1{\cdot}58H_2O$. α 1·569, β 1·592, γ 1·620 $\parallel$ [001]. [*A.M.*, **56**, 1838; *M.A.*, 72-3343.] 27th List

Lombaardite. H. J. Nel, C. A. Strauss, and F. E. Wickman, 1949. *Mem. Geol. Surv. South Africa*, no. 43, p. 45. Silicate of Al, Ca, Fe as monoclinic needles, related to pumpellyite and epidote. From Zaaiplaats tin mine, Transvaal. Named after B. V. Lombaard, professor of geology, university of Pretoria. [*M.A.*, **11**, 127.] 19th List [Is probably Allanite; *A.M.*, **48**, 1420]

Lomonosovite. 'A sodium phosphate-titanium silicate' occurring with chinglusuite and nordite in the Kola peninsula, mentioned without description by V. I. Gerasimovsky, *Compt. Rend.* (*Doklady*) *Acad. Sci. URSS*, 1941, vol. 32, p. 498. Named after the pioneer Russian mineralogist Michael Vasilevich Lomonosov, Михаил Васильевич Ломоносов (1711–1765). [M.A., **8**, 279.] 17th List [*M.A.*, 72-901; $Na_5Ti_2(Si_2O_7)(PO_4)O_2$, related to Seidozerite (q.v.)]

β-Lomonosovite. V. I. Gerasimovsky and M. E. Kazakova, 1962, Докл. Акад. Наук СССР (*Compt. Rend. Acad. Sci. URSS*), vol. 142, p. 670 (Беталомоносовит). Differs from Lomonosovite (17th List) by an appreciable water content and a lower alkali content; the X-ray powder photographs show some intensity differences. Identical with Metalomonosovite (q.v.), and the latter name is rejected because there is no evidence for the presence of metaphosphate ions. Formulated $Na_2Ti_2Si_2O_9.(Na,H)_3PO_4$ (Gerasimovsky), $Na_5MnTi_3Si_4O_{21}(OH)_3$ or $Na_2Ti_2Si_2O_9.NaPO_3$ (E. I. Semenov, N. I. Organova, and M. V. Kukharchik, Кристаллография, 1961, vol. 6, p. 925), or $Na_2Ti_2^{3+}Si_2O_8.NaPO_4$ (N. V. Belov and N. I. Organova, Геохимия, 1962, no. 1, p. 6). *See also* Зап. Всесоюз. Мин. Общ. (*Mem. All-Union Min. Soc.*), 1963, vol. 92, pp. 210, 217, and *Amer. Min.*, 1963, vol. 48, p. 1413. Apparently a rather indefinite intermediate stage in the alteration of Lomonosovite to Murmanite by leaching out of phosphate and hydration. [*M.A.*, **16**, 556.] 23rd List

Lonsdaleite. C. Frondel and U. B. Marvin, 1967. *Nature*, vol. 214, p. 587. A hexagonal polymorph of diamond, occurring in the Canyon Diablo and Goalpara meteorites. Named for K. Lonsdale. [*A.M.*, **52**, 1579; Зап., **97**, 64; *Bull.*, **90**, 605.] 25th List

Loparite. A. E. Fersman, 1922. *Compt. Rend. Acad. Sci. Russie*, p. 59; *Trans. Northern Sci. Econ. Exped.*, no. 16 (*Sci. Techn. Dept. Supreme Council of National Economy*, no. 8), Moscow and Petrograd, 1923, pp. 17, 68 (лопарит); [I. G. Kuznetzov, *Dokl. Min. Obshch.*, . . .]. Titanate of rare-earths, sodium, and calcium, as black, cubic crystals with metallic lustre and complex twinning. Related to the perovskite group. From the Kola Peninsula. [*M.A.*, **2**, 264.] 10th List [$(Ce,Na,Ca)_2(Ti,Nb)_2O_6$]

Lopezite. M. C. Bandy, 1937. *Amer. Min.*, vol. 22, p. 929. Potassium dichromate, $K_2Cr_2O_7$, as minute orange-red balls in the soda-nitre of Chile. Named after Dr. Emiliano Lopez, of Iquique. [*M.A.*, **7**, 13.] 15th List

Lorandite. J. A. Krenner, *Math. és term.-tud. Értesitö*, XII. 473, 1894; XIII. 258, 1895; abstr. in *Zeits. Kryst. Min.*, XXVII. 98, 1896. [*M.M.*, **11**, 32, 168.] $Tl_2S.As_2S_3$. Monosymmetric. [Allchar] Macedonia. 1st List

Loranskite. M. P. Melnikoff [separate publication, St. Petersburg]; P. D. Nikolaéeff, *Verh. k. russ. min. Ges.*, 1897 [ii], XXXV, Protocolle, p. 11; K. Flug, ibid., p. 30; abstr. in *Zeits. Kryst. Min.*, 1899, XXXI, 505. A massive black mineral containing Ta_2O_5,ZrO_2,Y earths, etc. [Impilahti, Eräjärvi] Finland. 2nd List

Lorenzenite. G. Flink, *Medd. om Grönland*, 1900, XXIV, 130. Small acicular rhombic crystals with high adamantine lustre; sometimes colourless with black tips. $Na_2O.2(Ti,Zr)O_2.2SiO_2$. [Narsarsuk] S Greenland. 2nd List [*A.M.*, **32**, 59; cf. Ramsayite]

Lorenzit, error for Lorenzenit (Hintze, *Handb. Min.*, Erg.-Bd. II, p. 579). 22nd List

Lorettoite. R. C. Wells and E. S. Larsen, 1916. *Journ. Washington Acad. Sci.*, vol. vi, p. 669. An oxychloride of lead, approximating to $6PbO.PbCl_2$, forming honey-yellow, bladed masses with perfect basal cleavage and probably tetragonal. Named after the locality, Loretto, Tennessee. Cf. Chubutite. 8th List

Lörvite, error for Löweïte (J. W. Mellor, *Treat. Inorg. Chem.*, **2**, 888). 26th List

Loseyite. L. H. Bauer and H. Berman, 1929, *Amer. Min.*, vol. 14, pp. 103, 150. Basic carbonate of manganese and zinc, $2RCO_3.5R(OH)_2$, as bluish-white monoclinic needles from Franklin Furnace, New Jersey. Named after Samuel R. Losey (1833?–1906?), a local collector. [*M.A.*, **4**, 151.] 12th List

Losite. A. Lacroix, 1911. *Les syénites néphéliniques de l'archipel de Los et leurs minéraux.*

Nouv. Arch. Muséum, Paris, ser. 5, vol. iii, p. 37; *Bull. Soc. franç. Min.*, 1912, vol. xxxv, p. 7. An undetermined mineral detected in thin sections of the nepheline-syenite of the Los Islands, W Coast of Africa. It is optically uniaxial and possibly related to cancrinite, but has a lower negative birefringence than this. 6th List [Is Vishnevite; *M.M.*, **26**, 3]

Lossenite. L. Milch, *Zeits. Kryst. Min.*, XXIV. 100, 1894. [*M.M.*, **11**, 106.] $2PbSO_4 + 6(FeOH)_3As_2O_8 + 27H_2O$. Rhombic. Laurion, Greece. 1st List [A mixture of Beudanite and Scorodite; *M.A.*, **2**, 281; **11**, 456]

Lotrite. G. Munteanu-Murgoci, *Bull. Soc. Sci. Bukarest*, 1900, vol. ix, p. 596; 1901, vol. ix, p. 783; *Inaug.-Diss. München*, 1901 (Lotrit). A constituent of an epidote-schist occurring in contact with serpentine in the Lotru valley, S Carpathians. $3(Ca,Mg)O_2.2Al_2O_3.4SiO_2.2H_2O$. Very similar to prehnite, but differs slightly from this in optical characters, as determined under the microscope in rock-sections. 3rd List [Is Pumpellyite; *M.M.*, **30**, 113]

Louderbackite. C. Lausen, 1928. *Amer. Min.*, vol. 13, p. 220. Hydrous ferric and ferrous sulphate, $2FeO.3(Fe,Al)_2O_3.10SO_3.35H_2O$, as pale chestnut-brown crystalline crusts, optically biaxial, formed by the burning of pyritic ore in a mine in Arizona. Named after Prof. George Davis Louderback (1874–), of the University of California. 11th List [Is Römerite; *A.M.*, **35**, 1056]

Loughlinite. J. J. Fahey and J. M. Axelrod, 1947. *Program and Abstracts, 28th Annual Meeting, Min. Soc. Amer.*, p. 9; *Bull. Geol. Soc. Amer.*, vol. 58, p. 1178; *Amer. Min.*, 1948, vol. 33, p. 195 (abstr.). Hydrous magnesium silicate, $MgSi_2O_5$.aq., an asbestiform mineral from Wyoming. Named after Dr. Gerald Francis Loughlin (1880–1946) formerly of the United States Geological Survey. [*M.A.*, **10**, 255.] 18th List [$Na_2Mg_3Si_6O_{16}.8H_2O$; *A.M.*, **45**, 270]

Lovchorrite. E. M. Bonshtedt, 1926. *Bull. Acad. Sci. Russie*, ser. 6, vol. 20, p. 1181 (ловчоррит, lovtchorrite). A. E. Fersman, *Amer. Min.*, 1926, vol. 11, p. 295 (lovchorrite); *Neues Jahrb. Min.*, Abt. A, vol. 55, p. 44 (Lovtschorrit). A colloidal glassy variety of rinkolite (q.v.). Named from the locality, Lovchorr (Ловчорр) plateau in the Khibinsky tundra, Kola peninsula, N Russia. [*M.A.*, **3**, 236; **3**, 275.] 11th List [Is Mosandrite; *A.M.*, **43**, 795]

Lovdarite. Yu. P. Menshikov, A. P. Denisov, E. I. Uspenskaya, and E. A. Lipatova, 1973. Докл., **213**, 429 (Ловдарит). Radiating aggregates and colourless orthorhombic prisms (*a* 38·79, *b* 6·776, *c* 7·012 Å) occur in pegmatoidal deposits on Mt. Karnasurt, Lovozero massif, Kola peninsula. Near $4[(Na,K,Ca)_4(Be,Al)_2Si_6O_{16}.4H_2O]$. Named from Lovozero and дар, a gift. [*M.A.*, 74-2476.] 28th List

Loveringite. B. M. Gatehouse, I. E. Grey, I. H. Campbell, and P. Kelly, 1978. *A.M.*, **63**, 28. Grains up to 50 × 100 μm in the Jimberlana intrusion, Norseman, Western Australia, are rhombohedral, space group $R\bar{3}$, *a* 9·117 Å, α 69·07°, isomorphous with crichtonite and senaite. Composition $[(Ca,Ln)(Ti,Fe,Cr,Mg,Zr,Al)_{20}O_{38}]$. Natural material is metamict, but readily recrystallizes on heating to 800°C. 30th List.

Lovozerite. V. I. Gerasimovsky, 1939. *Compt. Rend. (Doklady) Acad. Sci. URSS*, vol. 25, p. 753. Hydrous silicate of zirconium, etc., $R_2O.RO.ZrO_2.6SiO_2.3H_2O$, as black or pink, optically uniaxial grains in the alkalic rocks of Lovozero, Kola peninsula. Differs from eudialyte in containing less alkalis and more water. Named from locality. [*M.A.*, **7**, 468.] 15th List [$(Na,Ca)_3(Zr,Ti)Si_6(O,OH)_{18}$; *A.M.*, **59**, 633]

Löwite, error for Löweïte (J. W. Mellor, *Treat. Inorg. Chem.*, **2**, 430; **4**, 252). 26th List

Lowtschorrit, German transliteration of Ловчоррит, Lovchorrite (11th List) (C. Hintze, *Handb. Min.*, Erg.-Bd. II, p. 753). 24th List

Lubeckite. J. Morozewicz, 1919. *Bull. Intern. Acad. Sci. Cracovie, Cl. Sci. Math. Nat.*, ser. A, for 1918, p. 185 (Lubeckit). A wad-like mineral, $4CuO.\frac{1}{2}Co_2O_3.Mn_2O_3.4H_2O$, occurring as small spherules in malachite from Poland. Named after Prince Franciszez Xawery Lubecki (1778–1846), a Polish statesman. [*M.A.*, **2**, 52.] 10th List

Lublinite. [J. Morozewicz, 1907 (*see* below).] N. S. Watitsch (Vatič), 1908. *Ann. Géol. Min. Russie*, vol. ix, p. 239 (Люблинитъ), p. 241 (Lublinit). A form of calcite occurring as a mould-like encrustation on chalk-marl in govt. Lublin, Russian Poland. It consists of a matted aggregate of capillary or acicular crystals greatly elongated in the direction of an edge of the primary rhombohedron. Giving oblique optical extinction, these crystals had been previously described as monoclinic or triclinic, and by one author are stated to be hydrated calcium carbonate (L. L. Ivanov, ibid., 1905, vol. viii, p. 23). *See* Pentahydrocalcite and Trihydrocalcite. 5th List. J. Morozewicz, 1907. *Kosmos*, Lemberg, vol. xxxii, p. 487; *Centralblatt Min.*, 1911, p. 229. P. N. Čirvinskij (P. Tschirwinsky), *Mitt. Ges. Naturf. Kiew*, 1910, vol. xxi, p. 285; Doelter's *Handbuch der Mineralchemie*, 1911, vol. i, p. 360. (*See* 5th List, where the name is incorrectly attributed to another author.) 6th List [*See* Ikaite]

Lubumbashite. L. De Leenheer, 1934. *Natuurwet. Tijds. Gent*, vol. 16, pp. 237, 240 (Lubumbashiet). Colloidal hydroxide of cobalt (and copper), earlier compared with heterogenite (A. Schoep, 1921; *M.A.*, **1**, 243). Named from the locality, Lubumbashi, Katanga, Belgian Congo. [*M.A.*, **6**, 52.] 14th List [Clearly the same as Heterogenite]

Lucianite. E. W. Hilgard, 1916. *Proc. National Acad. Sci. U.S.A.*, vol.ii, p. 11. A clayey material consisting mainly of a colloidal hydrated magnesium silicate, which swells up to many times its original volume when immersed in water. Named from the locality, the hacienda Santa Lucia, near the City of Mexico. A. F. Rogers (*Journ. Geol. Chicago*, 1917, vol. xxv, p. 537) suggests that it is identical with stevensite, which he regards as the amorphous equivalent of talc. *See* Auxite. 8th List

Lucinite. W. T. Schaller, 1914. *Journ. Washington Acad. Sci.*, vol. iv, p. 355; 3rd Appendix to 6th edit. of Dana's *System of Mineralogy*, 1915, p. 46. Green, orthorhombic crystals of octahedral habit occurring with tabular orthorhombic crystals of variscite, with which they are identical in composition, $Al_2O_3.P_2O_5.4H_2O$. Named from the locality, Lucin, Utah. 7th List [Is Variscite, *see* Metavariscite]

Ludlockite. R. J. Davis, P. G. Embrey, and M. H. Hey, 1970. *Proc. Int. Min. Ass.*, 7th Gen. Meet., Tokyo (*Min. Soc. Japan Spec. Pap.*, **1**, 264). Red anorthic prisms with zincian chalybite from Tsumeb, SW Africa, have *a* 10·41, *b* 11·95, *c* 9·86 Å, α 113·9°, β 99·7°, γ 82·7°, D 4·40. Composition $9[(Fe,Pb)As_2O_6]$. α 1·96, yellow, β 2·055, deep yellow, γ > 2·11, orange-yellow, near [100]. Named for F. Ludlow Smith III and C. Locke Key. [*A.M.*, **57**, 1003.] [*Min. Record*, **8** (3), 91.] 27th List

Lueneburgite, variant of Lüneburgite; *A.M.*, **9**, 62. 10th List

Lueshite. A. Safiannikoff, 1959. *Bull. Acad. Roy. Sci. d'Outre-mer* (*Bruxelles*), new ser., vol. 5, p. 1251. $NaNbO_3$, cubic, with perovskite structure, occurring with mica at the contact of a carbonatite and a cancrinite-bearing syenite at Lueshe, Goma, Congo. Cf. Igdloite. 22nd List [*A.M.*, **46**, 1004. Cf. Natroniobite]

Luethite. S. A. Williams, 1977. *M.M.*, **41**, 27. Blue monoclinic crystals, *a* 14·743, *b* 5·093, *c* 5·598 Å, β 101° 49′, from Santa Cruz Co., Arizona. RIs α 1·752 ‖ [010], β 1·773, γ 1·796, γ:[001] 10° in obtuse β. Composition $2[Cu_2Al_2(AsO_4)_2(OH)_4.H_2O]$, the Al analogue of chenevixite. Named for R. D. Luethe. [*M.A.*, 77-2186; *A.M.*, **62**, 1058.] 30th List

Luigite. L. Colomba, 1908. *Rend. R. Accad. Lincei*, Roma, ser. 5, vol. xvii, sem. 2, p. 237. The same as aloisiite (q.v.), the latter name being preferred, since the Italian form luigite may be confused with lewisite. 5th List

Lupikkite. M. K. Palmunen, 1939. [*Geol. Toimikunta, Geoteknillisiä julkaisuja*, no. 44.] M. Saksela, *Bull. Comm. Géol. Finlande*, 1951, no. 154, p. 191 (Lupikkit). A mixture of cubanite, pyrrhotine, chalcopyrite, and blende, from Lupikko, Pitkärata, Finland. [*M.A.*, **12**, 70.] 20th List

Lusakite. A. C. Skerl and F. A. Bannister, 1934. *Min. Mag.*, vol. 23, p. 598. Silicate of aluminium, cobalt (CoO 8·48%), iron, magnesium, and nickel, $H_2O.4R''O.9(Al,Fe)_2O_3.8SiO_2$, as deep-blue orthorhombic crystals, from near Lusaka, Northern Rhodesia. A cobaltiferous variety of staurolite. Named after the locality, the new capital of Northern Rhodesia. 13th List

Lusitanite. [T. L. Walker, 1916] *Summary Report, Geol. Survey, Dept. Mines, Canada*, for 1916, 1917, p. 306. Named in memory of the disaster to the steamship *Lusitania*, but later withdrawn. Synonym of spencerite (q.v.). The same name was applied by A. Lacroix (*Compt. Rend. Acad. Sci. Paris*, 1916, vol. clxiii, p. 283) to an igneous rock—a mesocrate form of riebeckite-syenite—from Portugal, being named after the locality, Lùsitania, a Roman province in the Iberian Peninsula. 8th List

Lussatine. F. Laves, 1939. *Naturwiss.*, vol. 27, p. 706 (Lussatin). A form of silica consisting of cryptocrystalline cristobalite as fibres with optically negative elongation, distinct from lussatite with positive elongation, corresponding respectively to the pair chalcedony and quartzine of cryptocrystalline fibrous quartz. [*M.A.*, **7**, 514.] 15th List

Lusterite. A trade-name for *artificial* rutile. M. J. O'Donoghue, *priv. comm. See* under titania. 28th List

Lusungite. L. Van Wambeke, 1958. *Bull. Soc. belge Géol.*, vol. 67, p. 162. A deep-brown coating, associated with limonite, from Kobokobo, Kivu, Congo, contains Fe, Pb, Sr, and P; it is rhombohedral and isostructural with the alunite–beudantite group, and is formulated $(Sr,Pb)Fe_3(PO_4)_2(OH)_5.H_2O$. Named from the river Lusungu, near Kobokobo. [*M.A.*, **14**, 282.] 22nd List

Lutécine. A. Michel-Lévy and Munier-Chalmas, *Bull. Soc. franç. Min.*, 1892, XV, 175. The elements which build up the groupings called lutécite. 2nd List

Lutecite. A. Michel-Lévy and Munier-Chalmas, *Bull. Soc. franç. Min.*, XV. 174, 1892. F. Wallerant, ibid., XX. 57, 1897. [*Min. Mag.*, X. 256]. A fibrous form of silica. 1st List [Cf. Lussatine]

Lybianit, error for Libyanit (Hintze, *Handb. Min.*, Erg.-Bd. II, p. 579). 22nd List

Lyndochite. H. V. Ellsworth, 1927. *Amer. Min.*, vol. 12, p. 212. Niobate and titanate of yttrium, erbium, thorium, and calcium, as orthorhombic crystals resembling euxenite. Named from the locality, Lyndoch township, Renfrew Co., Ontario. [*M.A.*, **3**, 366.] 11th List [Related to Aeschynite and Priorite, not to Euxenite; *M.M.*, **35**, 801]

Lyonite. *See* Chillagite. 7th List

M

Maakite. B. K. Polenov, 1910. *Protok. Obshch. Estestv. Kazan. Univ.*, for 1908–1909, no. 246, p. 8 (маакит). J. Paclt, *Neues Jahrb. Min.*, Monatshefte, 1953, p. 189 (Maakit), Synonym of hydrohalite ($NaCl.2H_2O$), cryohalite (Kryohalit), and bihydrate (a chemical term). Named after Richard Karlovich Maak, P. K. Маакъ (1825–1887), who in 1850 collected material from a salt lake in Yakutsk, Siberia, *See* Cryohalite. [*M.A.*, **12,** 239.] 20th List

Macallisterite. W. T. Schaller, A. C. Vlisidis, and M. E. Mrose, 1965. *Amer. Min.*, vol. 50, p. 629. Small white pellets, consisting of aggregates of minute rhombohedral crystals identical with synthetic $MgB_6O_{10}.7\frac{1}{2}H_2O$, occur with Ginorite, Sassolite, and other boron minerals at the Mott Colemanite prospect, Twenty Mule Team Canyon, Furnace Creek Wash area, Death Valley, Inyo County, California. It is the 'unidentified mineral' of Allen and Kramer (*Amer. Min.*, 1957, vol. 42, p. 56) and the 'magnesium borate of uncertain identity' of Erd, McAllister, and Almond (*Amer. Min.*, 1959, vol. 44, p. 913). Named for J. F. McAllister. [*See also* Trigonomagneborite.] (*M.A.*, **17**, 500.) H. Strunz, *Min. Tabellen*, 4th edit, 1966, p. 253, writes McAllisterit. 24th List [**Mcallisterite** preferred spelling]

Mcconnellite. C. Milton, D. [E.] Appleman, E. C. T. Chao, F. Cuttitta, J. L. Dinnin, E. J. Dwornik, M. Hall, B. L. Ingram, and H. J. Rose, Jr., 1967. *Geol. Soc. Amer. Progr. Ann. Meeting*, p. 151, $CuCrO_2$, isostructural with delafossite, a minor constituent of the mixture of chromium oxides occurring in alluvial gravels of the Merume river, Mazaruni district, Guayana (Merumite, 18th List). Named for D. McConnell. 25th List

Macdonaldite. M. C. Stinson and J. T. Alfors, 1964. *Min. Inform. Serv., Calif. Div. Mines Geol.*, vol. 17, p. 235. J. T. Alfors, M. C. Stinson, R. A. Matthews, and A. Pabst, *Amer. Min.*, 1965, vol. 50, pp. 279, 314. Colourless orthorhombic crystals in veins in sanbornite–quartz rock near Big Creek and Rush Creek, Fresno County, California, are formulated $BaCa_4Si_{15}O_{35}.11H_2O$ (emission spectrography analysis). Named for G. A. Macdonald. [*M.A.*, **17,** 399, **17,** 502.]

Macedonite. D. Radusinović and C. Markov, 1971. *Amer. Min.*, **56,** 387. Small black grains and rare crystals in quartz-syenite veins in Crni Kaman, Prilep, Macedonia, are tetragonal, *a* 3·889, *c* 4·209 Å, D 7·82; composition $PbTiO_3$, identical with synthetic material. Perovskite group. Named from the locality. [*M.A.*, 71-3109.] 27th List

Macgovernite. C. Palache and L. H. Bauer, 1927. *Amer. Min.*, vol. 12, p. 373 (McGovernite). Hydrous basic arsenite, arsenate, and silicate of manganese, magnesium, and zinc, $21(Mn,Mg,Zn)O.3SiO_2.\frac{1}{2}As_2O_3.As_2O_5.10H_2O$, as reddish granular masses with perfect basal cleavage, probably hexagonal. From Stirling Hill, New Jersey. Named after the late Mr. J. J. McGovern (d. 1915), a local collector at Franklin Furnace, New Jersey. [*M.A.*, **3,** 366.] 11th List [$(Mn,Mg,Zn)_{15}As_2Si_2O_{17}(OH)_{14}$; *A.M.*, **45,** 937. **Mcgovernite** preferred spelling]

Machatschkiite. K. Walenta, 1977. *Tschermaks Mineral. Petrogr. Mitt.*, **24**, 125. Crusts on granite in the Anton mine, Heubachtal, Schwarzwald, consist of rhombohedra with a_h 15·10, c_h 22·59 Å. Composition $12[Ca_3(AsO_4)_2.9H_2O]$. RIs ω 1·593, ε 1·585. Named for F. Machatschki. 30th List

Mackayite. C. Frondel and F. H. Pough, 1944. *Amer. Min.*, vol. 29, p. 211. Hydrous tellurite of iron, perhaps $Fe_2(TeO_3)_3.xH_2O$, as green tetragonal crystals from Goldfield, Nevada. Named after John William Mackay (1831–1902), an Irishman who made a great fortune on the Comstock lode in Nevada. [*M.A.*, **9**, 62.] 17th List [$FeTe_2O_5(OH)$; *A.M.*, **55**, 1072]

Mckelveyite. C. Milton, B. Ingram, J. R. Clark, and E. J. Dwornik, 1965. *Amer. Min.*, vol. 50, p. 593. Light to dark green to black trigonal crystals in rocks of the Green River Formation in Sweetwater County, Wyoming, approximate to $Na_2Ba_4(Yt,Ln,Ca,Sr,U^{4+})_3(CO_3)_9.5H_2O$. The crystals include black carbonaceous matter, also Biotite, Acmite, and Quartz. The lanthanons are almost wholly of the yttrium group. Named for V. E. McKelvey. [*M.A.*, **17,** 501.] 24th List

Mackelveyite, variant of Mckelveyite (24th List) (I. Kostov, *Mineralogy*, 1968, p. 541). 25th List

Mckelvyite, error for Mckelveyite (24th List). *Amer. Min.*, **52**, 860 (1967). 26th List

Mackensite. F. Kretschmer, 1918. *Neues Jahrb. Min.*, 1918, pp. 19, 23 (Mackensit (Schwarzeisenerz)); Archiv f. Lagerstättenforschung, Berlin. A chloritic mineral approximating in composition to the hydrated ferric silicate, $Fe_2O_3.SiO_2.2H_2O$, which is regarded as an end-member of the thuringite series. It occurs as iron-black to greenish-black compact masses in diabase and schalstein in N Moravia and S Silesia, where with thuringite, moravite (F. Kretschmer, 1906), and viridite (q.v.), it is mined as an ore of iron. The optical characters of the minute needles indicate monoclinic symmetry. Named after the German Field-Marshal, August von Mackensen. 8th List [Cf. Makensite]

Mackinawite. H. T. Evans, Jr., R. A. Berner, and C. Milton, 1962. *Progr. Ann. Meeting Geol. Soc. Amer.*, p. 47A (abstr.); *Amer. Min.*, 1963, vol. 48, p. 215. A tetragonal polymorph of FeS, from the Mackinaw mine, Snohomish County, Washington. Named from the locality. Compare O. Kouvo, Y. Vuorelainen, and J. V. P. Long, *Amer. Min.*, 1962, vol. 47, p. 511. 23rd List [Non-stoichiometric, widespread occurrence; *see* Sulphide Mineralogy (M.S.A. 1974), CS–25]

Mckinstryite. B. J. Skinner, J. L. Jambor, and M. Ross, 1966. *Econ. Geol.*, vol. 61, p. 1383. A mineral associated with chalcopyrite, stromeyerite, etc., on a specimen from the Foster mine, Cobalt, Ontario, proves to be identical with the synthetic orthorhombic phase $Cu_{1\cdot2}Ag_{0\cdot8}S$ (Djurle, 1958). Named for H. E. McKinstry. [*A.M.*, **52**, 1253; Зап., **97**, 66; *Bull.*, **90**, 605.] 25th List [Kazakhstan, *M.A.*, 76-845; Japan, *M.A.*, 78-852]

Mackinstryite, spelling variant of Mckinstryite (*Amer. Min.*, 1967, vol. 52, p. 1253); [*M.A.*, **18**, 283.] 25th List

Mackintoshite. W. E. Hidden and W. F. Hillebrand, *Amer. Journ. Sci.*, XLVI, 98, 1893. [*Min. Mag.*, X. 341.] $UO_2.3ThO_2.3SiO_2.3H_2O$. Tetragonal. Llano Co., Texas. 1st List [Is Thorogummite; *A.M.*, **38**, 1007. Cf. Maitlandite]

Mackit, cited without reference by H. Strunz, *Min Tabellen*, 3rd edit., 1957, p. 392, as a synonym of Hanksite.

Macrokaolinite. W. C. Isphording and W. Lodding, 1968. *Clays Clay Min.*, **16**, 257. A rock name; illite plus montmorillonite, etc., including large grains ($\geqslant 0{\cdot}2$ mm) of kaolinite. This type of nomenclature invites confusion. [*A.M.*, **58**, 1115.] 28th List

Macrolepidolite. H. Baumhauer, *Eclogae geol. Helvetiae,* 1903, vol. vii, p. 354. Two varieties of lepidolite, one with a large optic axial angle, and the other with a small angle, are distinguished and called respectively macrolepidolite and microlepidolite (Makrolepidolith and Mikrolepidolith); the two varieties also differ in the character of the etched figures on the cleavage planes. 3rd List

Madisonite. R. S. McCaffery and J. F. Oesterle, 1924. *Year Book Amer. Iron and Steel Inst.*, 1924, p. 286. $2CaO.2MgO.Al_2O_3.3SiO_2$(?) as a constituent of iron blast furnace slags. Evidently named after Madison, Wisconsin, where the authors are professors of metallurgy. 13th List

Madocite. J. L. Jambor, 1967. *Canad. Min.*, vol. 9, p. 7. An orthorhombic mineral from Madoc, Ontario, has the composition $Pb_{17}Sb_{16}S_{41}$. Named from the locality. [*A.M.*, **53**, 1421; *Bull.*, **91**, 99.] 25th List

Mafite. A. Johannsen, 1917. *Journ. Geol. Chicago*, vol. 25, p. 70; *Essentials for the microscopical determination of rock-forming minerals*, Chicago, 1922, p. 41. A term to include not only the 'mafic' (i.e. ferromagnesian) minerals, but also some other dark-coloured rock-forming minerals. Metamorphism of ferromagnesian through the stages femag and mafic. 10th List [Also 21st List]

Mafurit. H. Strunz, 1957. *Min. Tabellen*, 3rd edit, p. 392. A proposed name, apparently withdrawn (C. Hintze, *Handb. Min.*, Erg.-Bd. II, p. 755.) 24th List

Magadiite. H. P. Eugster, 1967, *Science*, vol. 157, p. 1177. White massive material from Lake Magadi, Kenya, approximate to $NaSi_7O_{13}(OH)_3.3H_2O$; tetragonal. Named from the locality. 25th List [*A.M.*, **53**, 510, 2061]

Magallanite. G. A. Fester, J. Cruellas, and F. Gargatagli, 1937. *Anal. Soc. Cient. Argentina*, vol. 124, p. 211 (magallanita). A hard asphaltum thrown up by the sea on the coast of Magallanes (Magallan), S America. Named from the locality. [*M.A.*, **7**, 124.] 15th List

Maganthophyllite. A. N. Winchell, 1933. *Optical mineralogy*, part 2, 3rd edit., p. 241. A contraction of Magnesioanthophyllite (9th List). 13th List

Magarfvedsonite. I. V. Ginzburg, G. A. Sidorenko, and D. L. Rogachev, 1961. Труды Мин. Муз. Акад. наук СССР (*Proc. Min. Mus. Acad. Sci. USSR*), vol. 12, p. 3 (Магарфведсонит). An unnecessary name for magnesian Arfvedsonite; synonym of Magnesio-arfvedsonit. 24th List

Magaugite. F. Walker, 1943. *Amer. Journ. Sci.*, vol. 241, p. 518. Alternative name for endiopside [16th List] for augite rich in magnesium. [*M.A.*, **9**, 151.] 17th List

Magbasite. E. I. Semenov, A. P. Khomyakov, and A. V. Bykova, 1965. Докл. Акад. наук СССР (*Compt. Rend. Acad. Sci. URSS*), vol. 163, p. 718 (Магбасит). Colourless or rose-violet, finely fibrous deposits from an unnamed hydrothermal formation have a composition near $KBa(Al,Sc)(Mg,Fe^{2+})_6Si_6O_{20}F_2$. Named from the composition, Mag-ba-si-te. [*M.A.*, **17**, 504; *A.M.*, **51**, 530.] 24th List

Magbassite, variant spelling of Magbasite (*Bull. Soc. franç. Min. Crist.*, 1966, vol. 89, p. 146). 24th List

Maghæmite. P. T. Davey and T. R. Scott, 1957. *Nature*, London, vol. 179, p. 1363. Variant of maghemite (12th List). 21st List

Maghastingsite. I. V. Ginzburg, G. A. Sidorenko, and D. L. Rogachev, 1961. Труды Мин. Муз. Акад. наук СССР (*Proc. Min. Mus. Acad. Sci. USSR*), vol. 12, p. 3 (Маггастингсит). An unnecessary name for magnesian Hastingsite; synonym of Magnesio-hastingsite. 24th List

Maghemite. P. A. Wagner, 1927. *Econ. Geol.*, vol. 22, p. 846; *Mem. Geol. Survey South Africa*, 1928, no. 26, pp. 18, 29. A contraction of magnetite and hematite suggested for magnetic ferric oxide or oxidized magnetite, as distinct from martite. *See* Oxymagnite. [*M.A.*, **4**, 215.] 12th List

—— T. L. Walker (*Univ. Toronto Studies, Geol. Ser.*, 1930, no. 29, p. 19) applies this name to a magnetic and anisotropic mineral consisitng of mixed sesquioxides of iron and titanium, $(Fe,Ti)_2O_3$, from the Bushveld, Transvaal. [*M.A.*, **4**, 348.] 12th List

Maghemo-magnetite. E. Z. Basta, 1959. *Econ. Geol.*, vol. 54, p. 698. A name for intermediate oxides of iron between magnetite and γ-Fe_2O_3 with $Fe_3O_4 > \gamma$-Fe_2O_3. 25th List

Magnalite. S. Richarz, 1920. *Zeits. Deutsch. Geol. Gesell.*, vol. 72, Abh. pp. 36, 41 (Magnalit). A bole- or kerolite-like mineral containing both magnesium and aluminium hydrated silicate (though admittedly a mixture), occurring as an alteration-product of basalt in Oberpfalz, Bavaria. Named from the constituents, in analogy to 'magnalium', an alloy of magnesium and aluminium. [*M.A.*, **2**, 54.] 10th List

Magnalumoid. *Zeits. Geologie*, Berlin, 1952, vol. 1, p. 361. Abbreviation of magnalumoxide (19th List). 20th List

Magnalumoxide. N. A. Bobkov and Y. V. Kazitsyn, 1951, *Mém. Soc. Russe Min.*, vol. 80, p. 108 (магналюмоксид). A spinel mineral with excess sesquioxides (as in artificial spinel), $5(Mg,Fe)(Al,Fe)_2O_4.\ 4(Al,Fe)_2O_3 = (Mg,Fe)_5(Al,Fe)_{18}O_{32}$. Black octahedral crystals from Aldan, SE Siberia. [*M.A.*, **11**, 365.] 19th List

Magnesia-blythite. L. L. Fermor, 1926. *Rec. Geol. Survey India*, vol. 59, pp. 203, 205 (Magnesia-blythite), pl. 10 (Magnesio-blythite). A garnet containing pyrope and blythite (q.v.) molecules. *See* Calc-spessartite. 11th List

Magnesia-cordierite. H. Shibata, 1936. *Japanese Journ. Geol. Geogr.*, vol. 13, p. 227 (magnesia-cordierite). Iron-free cordierite $H_2Mg_4Al_8Si_{10}O_{37}$, as distinct from iron-cordierite $H_2Fe_4Al_8Si_{10}O_{37}$ (10th List). [*M.A.*, **6**, 479.] 14th List

Magnesia-goslarite. C. Milton and W. D. Johnston, 1937. *Amer. Min.*, vol. 22, no. 12, part 2, p. 10; 1938, vol. 23, p. 175. 15th List

Magnesia-gralmandite. L. L. Fermor, 1938. *See* Gralmandite. 15th List

Magnesian chamosite. F. A. Bannister and W. F. Whittard, 1945. *Min. Mag.*, vol. 27, p. 99 (Magnesian chamosite). Variety of chamosite containing MgO 8·75%. 20th List

—— H. S. Yoder, 1952. *Amer. Journ. Sci.*, Bowen volume, p. 588 (Magnesian chamosite), p. 597 (magnesian-chamosite). Artificially produced $3MgO.Al_2O_3.2SiO_2$, corresponding to chamosite with MgO in place of FeO. [*M.A.*, **12**, 83.] 20th List

Magnesian glaucophane. W. G. Ernst, 1957. *Ann. Rep. Geophysical Lab.*, for 1956–1957, p. 228 (magnesian glaucophane). Artificial $Na_2Mg_3Al_2Si_8O_{22}(OH)_2$. [Corr. *M.M.*, **22**.] 21st List

Magnesian riebeckite. W. G. Ernst, 1957. *Ann. Rep. Geophysical Lab.*, for 1956–1957, p. 228 (magnesian riebeckite). Artificial $Na_2Mg_3Fe_2'''Si_8O_{22}(OH)_2$. *See* Ferrous riebeckite. 21st List

Magnesiaspat. F. Machatschki, 1953. *Spez Min.*, Wien, p. 313. Syn. of Magnesite. 22nd List

Magnesio-alumino-katophorite. *Amph.* (1978), **14**, 12. 30th List

Magnesioanthophyllite. E. V. Shannon, 1921. *Proc. US Nat. Museum*, vol. 59, p. 401. *See* Ferroanthophyllite. 9th List

Magnesioarfvedsonite. W. E. Tröger, 1952. *Tabellen opt. Bestim. gesteinsbild. Minerale*, p. 146 (Magnesioarfvedsonit). Variety of arfvedsonite. 20th List [Andreev's usage, *v. inf.*, is the same]

Magnesio-arfvedsonite. A. Miyashiro, 1957. *Journ. Fac. Sci. Tokyo Univ.*, sect. 2, vol. 11, p. 57. Also Yu. K. Andreev, 1957. Труды Инст. Геол. Руд., Месторожд., Петрог., Мин., Геохим. (*Trans. Inst. Geol. Ore-deposits, Petr., Min., Geochem.*, vol. 10, p. 12); abstr. Зап. Всесоюз. Мин. Общ. (*Mem. All-Union Min. Soc.*, 1958, vol. 87, p. 486) (Магнезиоарфведсонит, magnesio-arfvedsonite). The same name is proposed by Miyashiro for the end-members $Na_2Ca_{\frac{1}{2}}Mg_{3\frac{1}{2}}Fe_{1\frac{1}{2}}^{\cdots}Al_{\frac{1}{2}}Si_{7\frac{1}{2}}O_{22}(OH)_2$ of the arfvedsonite series [*M.A.*, **14**, 144], and by Andreev for an amphibole near $Na_3Mg_4Fe^{\cdots}Si_8O_{22}(OH)_2$* (between arfvedsonite and eckermannite) found in a serpentine near a granite intrusion. [*M.A.*, **14**, 281.] [*Corr. *M.M.*, **32**] 22nd List

Magnesio-astrophyllite. Chih-chung Peng and Che-seng Ma, 1964. [*Scientia Sinica*, **13**, 1180]; abstr. *Zap.*, **97**, 76. A superfluous name for magnesian astrophyllite. 26th List

Magnesioaxinite. E. A. Jobbins, A. E. Tresham, and B. R. Young, 1975. *Journ. Gemmology*, **14**, 368. A pale-blue 0·16 g gemstone from Tanzania closely approaches the CaMg axinite end-member. [*M.A.*, 76-882; *A.M.*, **61**, 503.] 29th List

Magnesio-blythite, *see* Magnesia-blythite

Magnesiocarpholite. B. Goffé and P. Saliot, 1977. *Bull.*, **100**, 302. $(Mg,Fe^2)(Fe^3,Al)AlSi_2O_6(OH)_4$, the Mg analogue of carpholite, occurs in the Vanoise region, Savoie, France. 30th List

Magnesiochromite. A. Lacroix, 1910. *Minéralogie de la France*, vol. iv, p. 311 (magnésiochromite). The same as magnochromite (G. M. Bock, 1868). A variety of chromite containing magnesium, $(Fe,Mg)Cr_2O_3$. 6th List

Magnesio-clinoholmquistite. *Amph.* (1978), **6**, 3. 30th List

Magnesiocolumbite. Synonym of magnocolumbite (23rd List). [Геохим., мин., генет. типы месторожд. редк. элем., Изд. „Наука", 1964, p. 456], abstr. Зап. Всесоюз. Мин. Общ. (*Mem. All-Union Min. Soc.*), 1965, vol. 94, p. 683 (Магнезиоколумбит). 24th List

Magnesiocopiapite. L. G. Berry, 1938. *See* Ferricopiapite. 15th List

Magnesiocordierite. W. Schreyer, 1966. *Fortschr. Min.*, vol. 42, p. 213. Variant of Magnesiacordierite (14th List). 24th List

Magnesio-cronstedtite. A. N. Winchell, 1926. *Amer. Journ. Sci.*, ser. 5, vol. 11, p. 284. A hypothetical molecule $H_4Mg_2Fe'''_2SiO_9$ (corresponding with cronstedtite $H_4Fe''_2Fe'''_2SiO_9$) used to explain the composition of the chlorites. [*M.A.*, **3**, 373.] 11th List

Magnesio-cummingtonite. C. E. Tilley, 1939. *Geol. Mag. London*, vol. 76, p. 330. A Mg-Fe monoclinic amphibole, with MgO in excess of FeO, corresponding to the iron member grunerite. 15th List

Magnesiodolomite. A. N. Winchell, 1927. *Elements of optical mineralogy*, part 2, 2nd edit., p. 75. $CaMg(CO_3)_2$ is distinguished as magnesiodolomite; $CaFe(CO_3)_2$ as ferrodolomite [= iron-dolomite, 11th List]; and $CaMn(CO_3)_2$ as mangandolomite [10th List]. These are given as varieties of the mineral dolomite, while in the text it is stated that dolomite is a rock. 14th List

Magnesio-ferri-katophorite. *Amph.* (1978), 14.11. 30th List

Magnesio-ferri-taramite. *Amph.* (1978), 14.16. 30th List

Magnesio-gedrite. *Amph.* (1978), 4.4. 30th List

Magnesiohastingsite. M. Billings, 1928. *See* Ferrohastingsite. 12th List

Magnesio-holmquistite. *Amph.* (1978), **4**, 7. 30th List

Magnesio-hornblende. *Amph.* (1978), **11**, 7. 30th List

Magnesiokatophorite. A. Miyashiro, 1957. *Journ. Fac. Sci. Tokyo Univ.*, sect. 2, vol. 11, p. 57. A name for the end-member, $Na_2CaMg_4FeAlSi_7O_{22}(OH)_2$, of the catophorite series. [*M.A.*, **14**, 144.] 22nd List

Magnesiolaumontite. M. Borcos, 1960. *Stud. Cercet. Geol. Acad. Rouman.*, vol. 5, p. 739 (Magneziolaumontitul), abstr. in Зап. Всесоюз. Мин. Общ. (*Mem. All-Union Min. Soc.*), 1962, vol. 91, p. 203, and in *Amer. Min.*, 1962, vol. 47, p. 1483. $(Ca,Mg)_6(Si,Al)_{36}O_{72}.19H_2O$, with only 0·7% MgO, in monoclinic needles from Musari, Rumania. Relation to Laumontite doubtful, and name unsuitable in view of this and of the low content of MgO. [*M.A.*, **16**, 554.] 23rd List

Magnesioludwigite. B. S. Butler and W. T. Schaller, 1917. *Journ. Washington Acad. Sci.*, vol. vii, p. 29. An ivy-green variety of ludwigite from Utah in which ferrous iron is largely replaced by magnesia, the formula being $MgO.Fe_2O_3.3MgO.B_2O_3$ with only about 15% of the corresponding ferrous compound *ferroludwigite* ($FeO.Fe_2O_3.3MgO.B_2O_3$). 8th List

Magnesio-magnetite. (Author?). J. D. Dana, *Min.*, 5th edit., 1868, p. 150 (Magnesian magnetite). A. E. Fersman and O. M. Shubnikova, *Geochem. Min. Companion*, Moscow, 1937, p. 167 (Магнезиомагнетит). Variety of magnetite containing some MgO. 20th List

Magnesiomargarit. H. Strunz, 1966. *Min. Tabellen*, 4th edit., p. 386. Variant of Магниевий маргарит, Magnesium margarite (21st List). 24th List

Magnesioniobit. H. Strunz, 1966. *Min. Tabellen*, 4th edit., p. 188. Synonym of Магноколумбит, Magnocolumbite (23rd List). 24th List

Magnesioriebeckite. A. Miyashiro, 1957. *Journ. Fac. Sci. Tokyo Univ.*, sect. 2, vol. 11, p. 57; A. Miyashiro and M. Iwasaki, *Journ. Geol. Soc. Japan*, 1957, vol. 63, p. 698. A name for the end-member, $Na_2Mg_3Fe_2Si_8O_{22}(OH)_2$, of the riebeckite series, or for natural minerals near this. [*M.A.*, **14**, 144.] Syn. of Magnesian riebeckite (21st List). 22nd List). [Hyphenated; *Amph.* (1978)]

Magnesioscheelite. E. T. Wherry, 1914. *Proc. United States National Museum*, vol. xlvii, p. 504. A hypothetical magnesium tungstate, $MgWO_4$, isomorphous with scheelite. 7th List

Magnesiospinel. H. Strunz, 1957. *Min. Tabellen,* 3rd edit., p. 137 (Magnesiospinell). Syn. of Spinel. 22nd List

Magnesiosussexite. J. W. Gruner, 1932. *Amer. Min.*, vol. 17, p. 509. Hydrous borate of magnesium and manganese, $2(Mg,Mn)O.B_2O_3.H_2O$, intermediate between camsellite and sussexite, as fibrous veinlets in iron ore from Michigan. [Sussexite also contains a large proportion of magnesia, and camsellite still more.] [*M.A.*, **5**, 201.] 13th List [Perhaps manganoan Szájbelyite]

Magnesiotriplit. H. Strunz, 1966. *Min. Tabellen*, 4th edit., p. 281. An unnecessary variant of Магниотриплит, Magniotriplite (19th List), synonym of Talktriplite. 24th List

Magnesio-wüstite. N. L. Bowen and J. F. Schairer, 1935. *Amer. Journ. Sci.*, ser. 5, vol. 29, pp. 151, 194 (magnesio-wüstites). Solid solutions of MgO and FeO, ranging from pure periclase (cubic MgO) to 78% FeO. But wüstite [13th List] was not defined as FeO. [*M.A.*, **6**, 352.] 14th List

Magnesium-apjohnite. H. Meixner and W. Pillewizer, 1937. *Zentr. Min.*, Abt. A, 1937, p. 265 (Magnesiumapjohnit). Apjohnite with some magnesium replacing manganese. Synonym of bushmanite (= bosjemanite). [*M.A.*, **7**, 12.] 15th List

Magnesium-Astrophyllit, variant of magnesio-astrophyllite (26th List). Hintze, 52. 28th List

Magnesium astrophyllite, *A.M.*, **60**, 737, variant of Magnesio-astrophyllite (26th List). 29th List

Magnesium-axinite. J. Fromme, 1909. *Min. Petr. Mitt.* (*Tschermak*), vol. xxviii, p. 311 (Magnesiumaxinit). A hypothetical axinite with the composition $HMgCa_2BAl_2Si_4O_{16}$. 5th List

Magnesium-beidellite. G. Nagelschmidt, 1938. *Min Mag.*, vol. 25, p. 141. A clay mineral with the composition $Mg_3Si_4O_{11}.nH_2O$, regarded as an end-member of the montmorillonite group. 15th List

Magnesium-bentonite. F. A. Bannister, 1939. *Ann. Rep. Chem. Soc.*, vol. 35 (for 1938), p. 191

(magnesium-bentonite). Described by W. F. Foshag and A. O. Woodford in 1936 [*M.A.*, **6**, 372] as 'bentonitic magnesian clay-mineral from California' with the formula $7MgO.10SiO_2.5H_2O$ or $2MgO.3SiO_2.3H_2O$. 15th List

Magnesium-berzeliite. W. Bubeck, 1934. *Geol. För. Förh. Stockholm*, vol. 56, p. 526 (Magnesium-Berzeliit). W. Bubeck and F. Machatschki. *Zeits. Krist.*, 1935, vol. 90, p. 44 (Mg-Berzeliit). $(Ca,Na)_3Mg_2(AsO_4)_3$, as distinct from $(Ca,Na)_3Mn_2(AsO_4)_3$, manganberzeliite (2nd List). [*M.A.*, **6**, 183.] 14th List [Syn. of Berzeliite, Dana, *Syst. Min.*, 7th edit., **2**, 681]

Magnesium-chalcanthite. A. N. Winchell, 1951. *Optical mineralogy*, part 2, 4th edit., p. 160 (Magnesium chalcanthite). $MgSO_4.5H_2O$, triclinic, the artificial analogue of chalcanthite (*see* copper-chalcanthite, 18th List). Synonym of allenite and pentahydrite (qq.v.). 19th List

Magnesium chloritoid. L. B. Halferdahl, 1961. *Journ. Petrology*, vol. p. 49. A name for the hypothetical end-member $Mg_2Al_4Si_2O_{10}(OH)_4$. [*M.A.*, **15**, 468.] 24th List

Magnesium-chlorophoenicite. C. Palache, 1935. *Prof. Paper US Geol. Survey*, no. 180, p. 123 (Magnesium chlorophoenicite). Basic arsenate of magnesium and manganese, $(Mg,Mn)_2As_2O_8.7(Mg,Mn)(OH)_2$ monoclinic, differing from chlorophoenicite (10th List) in containing magnesium in place of zinc; from Franklin Furnace, New Jersey. [*M.A.*, **6**, 261.] 14th List [$(Mg,Mn)_5(AsO_4)(OH)_7$]

Magnesium chrysotile. W. Noll, H. Kircher, and W. Sybertz, 1960. *Beitr. Min. Petr.*, vol. 7, p. 232 (Magnesiumchrysotil). Syn. of Chrysotile. 22nd List

Magnesium-cordierite. H. M. Richardson and G. R. Rigby, 1949. *Min Mag.*, vol. 28, p. 548. Synonym of cordierite, distinguishing it from iron-cordierite (10th List) and manganese-cordierite (q.v.). 18th List

Magnesium-diopside. H. Rosenbusch, 1905. *Mikroskopische Physiographie d. Mineralien*, 4th edit., vol. i, part 2, pp. 200, 206 (Magnesiumdiopsid). A monoclinic pyroxene near to diopside, but containing only 8–9% CaO. 4th List

Magnesium-glauconite. H. Urgan, 1957. *Tonindustrie-Zeitung*, vol. 81, p. 363; abstr. Hintze, *Handb. Min.*, Erg.-Bd. II, 1960, p. 835 (Magnesium-Glaukonit). Syn. of Celadonite. 22nd List

Magnesium-halotrichite. H. Meixner and W. Pillewizer, 1937. *Zentr. Min.*, Abt. A, 1937, p. 266 (Magnesiumhalotrichit). Halotrichite with [some] MgO in place of FeO. [*M.A.*, **7**, 12.] 15th List

Magnesium-hexahydrite. J. Kubisz, 1958. *Bull. Acad. Polon. Sci., Sér. sci. chim., géol., géogr.*, vol. 6, p. 459. Syn of Hexahydrite (6th List). 22nd List

Magnesium hornblende. W. M. Kowalski, 1967. *Polska Akad. Nauk, Prace Geol.*, **42**, 83. A mistranslation of Hornblenda magnezowa, Magnesian hornblende (ibid 37). [*M.A.*, 71-1312.] 28th List

Magnesium-hydromuscovite. H. Strunz, 1957. *Min. Tabellen*, 3rd edit., p. 308 (Magnesium-Hydromuskovit). Syn. of Gümbelite. 22nd List

Magnesium-Jacobsite. *Zentr. Min.*, 1964, Teil I, p. 152. Apparently an alternative name for Rhombomagnojacobsite (*q.v.*). 24th List

Magnesium kaolinite. N. [E.] Efremov, 1955. *Bull. Geol. Soc. Amer.*, vol. 65, p. 1374. Flaky aggregates from the Elbrus mine, N Caucasus, have a composition near $Mg_2Al_4Si_3O_{10}(OH)_8$, with D.T.A. data and X-ray data nearer Kaolinite than Chlorite. [*M.A.*, **12**, 570.] Evidently a Septechlorite corresponding to Clinochlore; the name is unsuitable if this is the case. 23rd List [Also 20th List]

Magnesium-leonite. W. Schneider, 1961. *Acta Cryst.*, vol. 14, p. 785 (Magnesium-Leonit). Synonym of Leonite (1st List). [*M.A.*, **15**, 419.] 23rd List

Magnesium margarite. N. V. Belov, 1959. Зап. Всесоюз. Мин. Общ. (*Mem. All-Union Min. Soc.*), vol. 88, p. 305 (Магниевый маргарит). Proposed as a more appropriate name for the Calciotalc of D. P. Serdyuchenko (q.v.). 22nd List

Magnesium-melanterite. T. Wieser, 1950. *Rocznik Polsk. Tow. Geol.* (*Ann. Soc. Géol. Pologne*), vol. 19 (for 1949), p. 451 (melanteryt magnezowy), p. 470 (magnesium melanterite). A variety of melanterite containing 3–10% MgO, from Poland. [*M.A.*, **11**, 301.] 19th List

Magnesium-monticellite. *See* Iron-monticellite. 14th List

Magnesium-montmorillonite. *See* Calcium-montmorillonite. 16th List

Magnesium-morenosite. C. O. Hutton, 1947. *Amer. Min.*, vol. 32, p. 559 (magnesian morenosite). Morenosite containing some magnesium, $(Ni,Mg)SO_4.7H_2O$. *See* Nickel-epsomite. 20th List

Magnesium-orthite. P. Geijer, 1927. *Sveriges Geol. Unders. Årsbok 20* (*for 1926*), no. 4, p. 7 (magnesium orthite, Mg-orthite). A variety of orthite from Norberg, Sweden, containing much magnesia (14·15%) and fluorine (3·31%) perhaps present as the molecule MgF_2. [*M.A.*, **3**, 273.] 11th List

Magnesium-pectolite. E. Reuning, 1907. *Centralblatt Min.*, p. 739 (Magnesiumpektolith). Pectolite containing some magnesium (MgO, 5½%). Occurs in crevices in diabase at Burg, Hesse-Nassau. 5th List

Magnesium-pennantite. R. A. H[owie]. *M.A.*, 70-2356. An analogue of pennantite (17th List) with Mg:Mn about 2:1 from the Zhumast deposit, Atasui, Kazakhstan, described by M. M. Kayupova (*Zap.*, 1967, **96**, 155) under the name Магнезальный пеннантит (magnesian pennantite). 28th List

Magnesium-phosphoruranite. H. Strunz, *Min. Tab.*, 1941, p. 165, (Magnesiumphosphoruranit). Syn. of saleite (13th List). Similarly, Kupferphosphoruranit (torbernite), Kupferarsenuranit (zeunerite), Calciumphosphoruranit (autunite), Calciumarsenuranit (uranospinite), Bariumphosphoruranit (uranocircite). 18th List

Magnesium-sericite. H. Minato and Y. Takano, 1952. *Papers College Gen. Educ. Univ. Tokyo*, vol. 2, p. 196 (magnesium sericite). A sericitic clay from Japan containing MgO 2·74, Fe_2O_3 0·46, FeO 1·57%. *See* Iron-sericite. [*M.A.*, **12**, 221.] 20th List

Magnesium-Serizit, variant of Magnesium-sericite (20th List) (H. Strunz, *Min. Tabellen*, 4th edit., 1966, p. 493). 24th List

Magnesium szomolnokite. J. Kubisz, 1960. *Bull. Acad. Polon. Sci., Sér. chim., géol., géogr.*, vol. 8, p. 101. An unnecessary name for magnesian szomolnokite. Not intended as a synonym of kieserite. 22nd List

Magnesium urcilite, error for Magnesium ursilite; *see* Ursilite. 22nd List

Magnesium-vermiculite. G. F. Walker, 1950. *Nature*, London, vol. 166, p. 695 (Magnesium-vermiculite). A. McL. Mathieson and G. F. Walker, *Amer. Min.*, 1954, vol. 39, p. 231 (Magnesium-vermiculite, Mg-vermiculite). Regarded as a clay-mineral occurring in soils and analogous to montmorillonite. [*M.A.*, **12**, 437.] 20th List

Magnesium-wollastonite. C. E. Tilley, 1948. *Amer. Min.*, vol. 33, p. 737 (magnesium wollastonite solid solutions). Cf. iron-wollastonite (14th List). 18th List

Magnesium-zinc-spinel. B. W. Anderson and C. J. Payne, 1937. *See* Gahnospinel. 14th List

Magnesomagnetite. E. Z. Basta, 1959. *Econ. Geol.*, vol. 54, p. 698. Synonym of Magnesiomagnetite. 25th List

Magneso-titanomagnetite. E. Z. Basta, 1959. *Econ. Geol.*, vol. 54, p. 698. An unnecessary name for magnesian titanomagnetite. 25th List

Magnetoilmenite, P. Ramdohr, 1925. *150. Festschr. Bergakad. Clausthal*, p. 324. *Neues Jahrb. Min.*, Abt. A, 1926, vol. 54, p. 345 (Magnetoilmenit). Hexagonal mixed crystals of ilmenite with magnetite, the name titanomagnetite being reserved for cubic mixed crystals of magnetite in ilmenite at the other end of the series with only a small break between. 11th List

Magneto-maghemite. E. Z. Basta, 1959. *Econ. Geol.*, vol. 54, p. 698. A name for intermediate oxides of iron between magnetite and γ-Fe_2O_3, with $Fe_3O_4 < \gamma$-Fe_2O_3. 25th List

Magnetoplumbite. G. Aminoff, 1925. *Geol. För. Förh. Stockholm*, vol. 47, p. 283 (Magnetoplumbit). A double oxide of ferric iron with lead and manganese and some titanium, etc., $2RO.3R_2O_3$, as acute hexagonal pyramids with black colour and metallic lustre, from Långban, Sweden. It is strongly magnetic and contains lead; hence the name. [*M.A.*, **3**, 5.] 11th List [$Pb(Fe^{3+},Mn^{3+})_{12}O_{19}$; *A.M.*, **36**, 512]

Magnetostibian. L. J. Igelström, *Zeits. Kryst. Min.*, XXIII. 212, 1894. Antimonate of Mn and Fe. Sjö mine, [Örebro] Sweden. [Is probably Jacobsite; *A.M.*, **58**, 562.] 1st List

Magneziolaumontitul, original spelling of Magnesiolaumontite. *See, Amer. Min.*, 1962, vol. 47, p. 1483. 23rd List

Magnijmontmorillonit, variant of Magnymontmorillonit (*Zentr. Min.*, 1952, Teil 1, p. 246). 23rd List

Magnioborite. E. I. Nefedov, 1961. [Мат. Всесоюз. Науч. Геол. Инст., ser. 2, vol. 45], referred to in Зап. Всесоюз. Мин. Общ. (*Mem. All-Union Min. Soc.*), 1961, vol. 90, p. 754, and 1962, vol. 91, p. 205. Described as a new mineral but shown to be identical with Suanite (20th List). [*M.A.*, **16**, 647, 648.] 23rd List

Magniophilite. A. A. Beus, 1950. *Doklady Acad. Sci. USSR*, vol. 73, p. 1267 (магниофилит). $(Mn,Fe,Mg)_3(PO_4)_2$ as salmon-pink crystals intergrown with triphylite, in pegmatite from Turkestan. Differs from graftonite in containing Mg and more Mn that Fe. Named from magnesium and *φιλός*, a friend. [*M.A.*, **11**, 190.] 19th List [Is Beusite q.v.]

Magniosiderite. P. V. Zaritsky, 1964. Докл. **155**, 1341 (Магниосидерит). *Syn.* of magnesian chalybite, including sideroplesite and pistomesite. 28th List

Magniotriplite. A. I. Ginzburg, N. A. Kruglova, and V. A. Moleva, 1951. *Doklady Acad. Sci. USSR*, vol. 77, p. 97 (магниотриплит). Triplite rich in magnesium. Synonym of talc-triplite. [*M.A.*, **11**, 311.] 19th List [Magniotriplite preferred, Fleischer's *Gloss.*]

Magnocolumbite. V. V. Matias, L. N. Rossovsky, A. N. Shostsky, and N. M. Kumskova, 1963. Докл. Акад. Наук СССР (*Compt. Rend. Acad. Sci. URSS*), vol. 148, p. 420 (Магноколумбит) Black acicular and tabular orthorhombic crystals from Kugi-Lyal, SW Pamir, are the magnesium analogue of Columbite. Named from the composition. [*M.A.*, **16**, 285.] 23rd List [*A.M.*, **48**, 1182]

Magnodravite. Wang Shu-chang and Hsu Xue-yen, 1966. [*Kexue Tongbao*, vol. 17 (2), p. 91], abstr. *M.A.*, **17**, 606. A superfluous name for magnesian Dravite. 24th List

Magnoferrichromite. A. G. Betekhtin, 1934. *See* Alumochromite. 14th List

Magnofranklinite. *Geol. Survey, New Jersey, Final Rept.*, II. (1), 14, 1892. A. H. Chester, *Dictionary of the Names of Minerals*, 1896, p. 164. A local name for the franklinite of Sterling Hill, NJ, which contains little zinc and is highly magnetic. 1st List

Magnophorite. R. T. Prider, 1939. *Min. Mag.*, vol. 25, pp. 373, 378. An alkalic amphibole in leucite-rich rocks from Western Australia, allied to katophorite but rich in potassium and magnesium and described as kali-magnesio-katophorite, this term being contracted to magnophorite. It had earlier been named simpsonite (q.v.). 15th List [Titanian potassian richterite; *Amph.* (1978)]

Magnostilpnomelane. I. Mincheva-Stefanova and M. Gorova, 1965. [Тр. Върху геол. България, сер. геол., мин., петр., vol. 5, p. 139]; abstr. Зап. всесоюз. мин. общ. (*Mem. All-Union Min. Soc.*, 1968, vol. 97, p. 78. A magnesium-rich variety, with Mg > Fe. 25th List

Magnussonite. O. Gabrielson, 1956. *Arkiv Min. Geol. Stockholm*, vol. 2, p. 133. Arsenite of Mn, etc. $(Mn,Mg,Cu)_5(AsO_3)_3(OH,Cl)$, cubic, as green crusts on dolomite from Långban, Sweden. Named after Nils H. Magnusson, director of the Geological Survey of Sweden. [*M.A.*, **13**, 381.] 21st List [*A.M.*, **42**, 581; **56**, 639]

Magny-monothermite. I. D. Sedletzky and P. S. Samodurov, 1949. *Mém. Soc. Russe Min.*, ser. 2, vol. 78, p. 274 (Магний-монотермит). A clay mineral differing from monothermite (15th List) in containing MgO (2·89%) in place of K_2O. [*M.A.*, **11**, 125.] 19th List

Magnymontmorillonite. I. D. Sedletzky, 1951. *Priroda, Acad. Sci. USSR*, vol. 40, no. 2, p. 61 (Магниймонтмориллонит). Abstr. in *Zentralblatt Min.*, Teil I, 1953, for 1952, p. 246 (Magnijmontmorillonit). Synonym of Magnesium-montmorillonite (16th List). [*M.A.*, **12**, 239.] 20th List

Mahadevite. S. Ramaseshan, 1945. *Proc. Indian Acad. Sci.*, sect. A, vol. 22, p. 177. A bronze-coloured mica intermediate in composition between muscovite and phlogopite, from Warangal, Hyberabad. Named after Professor C. Mahadevan of Andhra University. [*M.A.*, **9**, 189.] 17th List

Maigruen. B. H. Geier and J. Ottemann, 1970. *Min. Depos.*, **5**, 29. Sulphide of Cu and Ga, near Cu_2GaS_3, with sme Zn and V, in small (0·1 mm) grains with germanite and gallite in the Tsumeb ores. Only optical data and an electron-probe analysis given. (The publication of 'working names' is undesirable.) 26th List

Maitlandite. E. S. Simpson, 1930. *Journ. Roy. Soc. W. Australia*, vol. 16, p. 33. Hydrous silicate of thorium, uranium (UO_2), etc., $2(Pb,Ca)O.3ThO_2.4UO_2.8SiO_2.23H_2O$, differing from mackintoshite (to which the mineral had been provisionally referred in 1912) in containing some lead and calcium. Named after Andrew Gibb Maitland, formerly Government Geologist of Western Australia. [*M.A.*, **4**, 346.] 12th List [Is Thorogummite; *A.M.*, **38**, 1007]

Majakite. *Zap.*, **105**, no. 6, cover; *A.M.*, **62**, 1260; *M.A.*, 77-4624. German-style transliteration of Маякит, mayakite. 30th List

Majorite. J. V. Smith and B. Mason, 1970. *Science*, **168**, 832. Minute purple grains in the Coorara meteorite are cubic, *a* 11·524 Å, with a garnet structure and a composition near hypersthene. Named for A. Major. [*A.M.*, **55**, 1815; *M.A.*, 72-3344; *Bull.*, **94**, 573.] 27th List [$Mg_3(Fe,Si,Al)_2(SiO_4)_3$]

Makatite. R. A. Sheppard, A. J. Gude, 3rd, and R. L. Hay, 1970. *Amer. Min.*, **55**, 358. White spherulites from cavities in trona, from drill holes in the Evaporite Series of Lake Magadi, Kenya, have composition $NaSi_2O_3(OH)_3.H_2O$; orthorhombic. Named from the Masai word for soda, emakat, in allusion to its high sodium content. Regrettably close to Magadiite (25th List). [*M.A.*, 70-3430.] 26th List

Makensite. F. Kretschmer, 1917. *Archiv f. Lagerstätten-Forschung, Preuss. Geol. Landesanst.*, Heft 24, pp. 56, 126 (Makensit). The form Makensenit is also mentioned, but rejected on account of length. The same as Mackensite (F. Kretschmer, 1918; 8th List.) Named after the Prussian field-marshal August von Mackensen. [*M.A.*, **1**, 255.] 9th List

Mäkinenite. Y. Vuorelainen,* A. Huhma, and A. Häkli, 1964. *Compt. Rend. Soc. Géol. Finlande*, vol. 36, p. 113. Trigonal γ-NiSe, isostructural with Millerite, occurring with Clausthalite and selenian Melonite at Kuusamo, NE Finland. Named for E. Mäkinen. [*M.A.*, **17**, 303; *A.M.*, **50**, 520.] [*Corr. *M.M.*, **39**] 24th List

Makinthosit. C. Doelter, *Handbuch der Mineralchemie*, 1913, vol. iii, p. 234. Error for Mackintoshite. 7th List

Malachite de plomb. (*Internat. Catal. Sci. Liter.*, 1903, Mineralogy volume for 1901, p. 108.) *See* plumbomalachite. 3rd List

Malanite. Yu Tsu-Hsiang, Lin Shu-Jen, Chao Pao, Fang Ching-Sung, and Huang Chi-Shun, 1974. [*Acta Geol. Sinica*, **2**, 202], abstr. *M.A.*, 75-2522. Cubic $(Cu,Pt,Ir)S_2$, *a* 6·030, from an undisclosed locality in China. [*A.M.*, **61**, 185.] 29th List

Malayaite. J. B. Alexander and B. H. Flinter, 1965. *Min. Mag.*, vol. 35, p. 622. The tin analogue of sphene, mentioned without name in *Malay. Geol. Surv. Mem. 9*, p. 105 (*A.M.*, **46**, 768), found in the valley of the Sungei Lok, Chenderiang, Perak, Malaya, is named for the country; $CaSnSiO_5$. [*A.M.*, **51**, 1551.] 24th List [England, *Econ. Geol.*, **61**, 366; Japan, *M.A.*, 78-761; cf. Zinntitanit]

Malayite, error for Malayaite (24th List). (*Min. Abstr.*, vol. 17, p. 503). 25th List

Malladrite. F. Zambonini and G. Carobbi, 1926. *Atti* (*Rend.*) *R. Accad Lincei, Roma, Cl. Sci. fis. mat. nat.*, ser. 6, vol. 4, p. 173. Sodium fluosilicate, Na_2SiF_6, as hexagonal crystals in deposits from fumaroles on Vesuvius. Named after Prof. Alessandro Malladra, Conservator of the Vesuvian Observatory. Not to be confused with Mallardite of A. Carnot, 1879. [*M.A.*, **3**, 238.] 11th List

Maltesite. J. J. Sederholm, *Geol. För. Förh.*, XVIII. 390, 1896. A variety of andalusite resembling chiastolite. [N of Lake Ladoga] Finland. 1st List

Malthite. W. P. Blake, 1890. *Trans. Amer. Inst. Mining Engin.*, vol. xviii, p. 582. The same as maltha. A group name to include the viscous bituminous hydrocarbons known as maltha, mineral tar, pittasphalt, brea (Spanish), chapapote (Cuban Spanish). 5th List

Manandonite. A. Lacroix, 1912. *Compt. Rend. Acad. Sci. Paris*, vol. clv, p. 446; *Bull. Soc. franç. Min.*, vol. xxxv, p. 223. A borosilicate, $H_{24}Li_4Al_{14}B_4Si_6O_{53}$, occurring as pearly six-sided scales with micaceous cleavage, in corrosion cavities with rubellite in pegmatite near the Manandona river, Madagascar. 6th List

Manasseite. C. Frondel, 1940. *Amer. Min.*, Program and Abstracts, December 1940, p. 6; *Amer. Min.*, 1941, vol. 26, pp. 196, 310. Hydrous basic carbonate of magnesium and aluminium, $Al_2Mg_6(OH)_{16}CO_3.4H_2O$, as hexagonal scales dimorphous with the rhombohedral hydrotalcite. Named after Ernesto Manasse (1875–1922) of Firenze. [*M.A.*, **8**, 51, 100.] 16th List [Cf. Meixnerite]

Mangan-actinolite. T. Yoshimura, 1939. *Journ. Fac. Sci. Hokkaido Univ.*, ser. 4, vol. 4, p. 424 (Manganactinolite). A variety of actinolite containing MnO 5·79%. 15th List

Mangan-alluaudite. Dana's *Mineralogy*, 7th edit., vol. 2, 1951, p. 674. Mn-alluaudite of P. Quensel, 1937. [*M.A.*, **6**, 485.] 19th List

Manganalmandine. J. Palmgren, 1917. *Bull. Geol. Inst. Univ. Upsala*, vol. xiv, p. 171 (Manganalmandin). A manganiferous garnet containing MnO 11·04, FeO 23·48, Al_2O_3 19·37% and thus intermediate between almandine and spessartite in composition. 8th List

Mangan-almandite. L. L. Fermor, 1926. *See* Calc-spessartite. 11th List

Manganandalusite. H. Bäckström, *Geol, För. Förh.*, XVIII. 386, 1896. A variety of andalusite containing 6·91% Mn_2O_3. [Vestanå] Sweden. 1st List

Mangananankerite. S. Koiké, 1935. [*Journ. Japanese Assoc. Min. Petr. Econ. Geol.*, vol. 14, p. 216 (Japanese).] Z. Harada, *Journ. Fac. Sci. Hokkaido Univ.*, Sapporo, ser. 4, 1936, vol. 3, p. 358 (Manganankerit), p. 361 (Mangan-Ankerit). Pink ankerite containing MnO 8.60%, from Japan. 14th List

Mangan-antigorite. H. Strunz, 1957. *Min. Tabellen*, 3rd edit., p. 323 (Mangan-Antigorit). An unnecessary name for manganoan antigorite. 22nd List

Mangan-Arfvedsonit. H. Strunz, 1966. *Min. Tabellen*, 4th edit., p. 494. Synonym of Juddite. 24th List

Manganaxinite. J. Fromme, 1909. *Min. Petr. Mitt.* (*Tschermak*), vol. xxviii, p. 311 (Manganaxinit). Axinite rich in manganese (MnO, 11·54%) from the Harz. $HMnCa_2BAl_2Si_4O_{16}$. 5th List [Cf. Ferroaxinite, Severginite, Tinzenite; *M.A.*, 71-2989]

Manganbabingtonite. R. A. Vinogradova and I. N. Plyusinna, 1967. [Вестн. Москов. унив., сер. 4, Геол. (*Rept. Moscow Univ.*, ser. 4, geol.), 54]; abstr. *Amer. Min.*, **53**, 1064 (1968). The manganese analogue of babingtonite, from the Rudnyi Kaskad deposit, E Sayan, USSR. Named for the composition. [*Zap.*, **98**, 331 (Манганбабингтонит).] 26th List

Mangan-belyankinite. E. I. Semenov, 1957. Труды Инст. Мин., Геохим. Крист. Редк. Элем. (*Trans. Inst. Min. Geochem. Cryst. Rare Elements*), no. 1, p. 41 (Манган-белянкинит). Abstr. Зап. Всесоюз. Мин. Общ. (*Mem. All-Union Min. Soc.*), 1959, vol. 88, p. 312 (Манган-белянкинит, manganbelyankinite); *Amer. Min.*, 1958, vol. 43, p. 1220 (Mangano-belyankinite). A manganiferous variety of belyankinite (19th List), occurring in brownish-black platy crystals, highly pleochroic, from the Lovozero massif, Kola peninsula. [*M.A.*, **14**, 278.] An unnecessary name. 22nd List

Manganberzeliite. L. J. Igelström, *Zeits. Kryst. Min.*, 1894, XXIII, 593. The same as pyrrhoarsenite. 2nd List

Manganboracite. *See* Ericaite. 21st List

Mangnachalcanthite. A. N. Winchell, 1951. *Optical mineralogy,* 4th edit., part 2, p. 10. Variant of manganese-chalcanthite (9th List), artificial $MnSO_4.5H_2O$, triclinic. 19th List

Manganchinglusuite. A. I. Soklakov and M. D. Dorfman, 1964 [Минералы СССР, **15**, 167]; abstr. *M.A.*, **18**, 160. A metamict mineral from Khibina with about twice as much MnO as chinglusuite (15th List), and three or four times as much Fe. Evidence that the mineral is a variety of chinglusuite is not convincing. German variant Mangan-Tschinglusuit, Strunz, *Min. Tab.*, 1970, 5th edit., 551. 28th List

Mangan-chrysotile. H. Strunz, 1957. *Min. Tabellen*, 3rd edit., p. 323 (Mangan-Chrysotil). An unnecessary name for manganoan chrysotile. 22nd List

Mangan-crocidolite. P. de Wijkerslooth, 1943. *Madan Tetkik ve Arama Enstitüsü Mecmuası*, Ankara, vol. 8, p. 99 (manganez krokidolit, Turkish), p. 108 (Mangankrokidolit, German). Variety of crocidolite containing some manganese, from manganese ore deposits in Anatolia. [*M.A.*, **9**, 241.] 20th List [Manganoan riebeckite; *Amph.* Sub-Comm.]

Mangan-Cummingtonit. H. Strunz, 1966. *Min Tabellen*, 4th edit., p. 369. An unnecessary name for manganoan Cummingtonite; synonym of Tirodite (15th List). 24th List

Mangandiaspore. K. Chudoba, 1929. *Centr. Min.*, Abt. A, 1929, p. 11 (Mangandiaspor). A variety of diaspore containing some manganese (Mn_2O_3 4·32%) from South Africa. [*M.A.*, **4**, 148.] 12th List

Mangandickinsonit, error for Manganodickinsonite (21st List) (C. Hintze, *Handb. Min.*, Erg.-Bd. II, p. 758). 24th List

Mangandolomite. C. F. Naumann, 1874. *Elemente der Mineralogie*, 9th edit., p. 289 (Mangandolomit), quoting W. T. Rœpper, 1870, 'Manganesian dolomite' for a mineral intermediate between calcite and rhodochrosite (calciferous rhodochrosite in Dana), to which A. Kenngott in 1872 gave the name Röpperit (not the Rœpperite of J. G. Brush, 1872). In C. Doelter's *Handbuch der Mineralchemie*, 1911, vol. 1, pp. 360, 376, Mangandolomit is given as a synonym of manganiferous dolomite. 10th List

Manganese aluminium chromite. Incorrect translation of Хромит марганцевоалюминиевый, manganoan aluminian chromite (*M.A.*, 70-705). 28th List

Manganese-anorthite. O. Glaser, 1926. *Centralblatt Min.*, Abt. A, p. 86 (Mangananorthit). R. B. Snow, *Journ. Amer. Ceramic Soc.*, 1943, vol. 26, p. 19 (manganese anorthite); J. R. Rait and H. W. Pinder, *Journ. Iron and Steel Inst.*, 1947, vol. 154 (no. 2 for 1946), p. 375 P (manganese felspar, manganese anorthite). Artificial $MnAl_2Si_2O_8$, analogous to anorthite, in the system $MnO–Al_2O_3–SiO_2$. [*M.A.*, **10**, 462.] 18th List

Manganese berzeliite, variant of Mangan-bezeliite (2nd List). *Amer. Min.*, 1968, vol. 53, p. 316. 25th List

Manganese-chalcanthite. E. S. Larsen and M. L. Glenn, 1920. *See* Zinc-copper-chalcanthite. 9th List

Manganese-chlorite. H. von Eckermann, 1927. *Geol. För. Förh. Stockholm*, vol. 49, p. 450; 1944, vol. 66, p. 721. A chlorite from Långban, Sweden, containing 1·06% MnO. (The manganchlorite of A. Hamberg, 1890, contained 2·28% MnO.) [*M.A.*, **3**, 474; **9**, 256.] 17th List [*See also* Manganese-pennine]

Manganese-cordierite. R. B. Snow, 1943. *Journ. Amer. Ceramic Soc.*, vol. 26, p. 15 (manganese compound analogous to cordierite). H. M. Richardson and G. R. Rigby, *Min. Mag.*, 1949, vol. 28, p. 551 (Manganese-cordierite). Artificial $2MnO.2Al_2O_3.5SiO_2$ in the system $MnO–Al_2O_3–SiO_2$. [*M.A.*, **10**, 462.] 18th List

Manganese-gehlenite. O. Glaser, 1926. *Centralblatt. Min.*, Abt. A, p. 86 (Mangangehlenit). Artificial $2MnO.Al_2O_3.SiO_2$, analogous to gehlenite, in the system $MnO–Al_2O_3–SiO_2$. 18th List

Manganese-gralmandite. L. L. Fermor, 1938. *See* Gralmandite. 15th List

Manganese-hörnesite. O. Gabrielson, 1951. *Arkiv Min. Geol. Stockholm*, vol. 1, p. 333 (manganese-hoernesite). $(Mn,Mg)_3(AsO_4)_2.8H_2O$, monoclinic, isomorphous with hörnesite $Mg_3(AsO_4)_2.8H_2O$, and bobierrite $Mg_3(PO_4)_2.8H_2O$. from Långban, Sweden. Named after Moriz Hörnes (1815–1868) of Vienna. [*M.A.*, **12**, 130.] 20th List [*A.M.*, **39**, 159. Cf. Krautite]

Manganese-leonite. W. Schneider, 1961. *Acta Cryst.*, vol. 14, p. 784. Synonym of Manganleonite (19th List). 23rd List

Manganese-merwinite. H. J. Goldschmidt and J. R. Rait, 1943. *Nature*, London, vol. 152, p. 356 (manganese-merwinite). A slag mineral, $3CaO.MnO.2SiO_2$, analogous to merwinite with Mn in place of Mg. [*M.A.*, **9**, 102.] 17th List

Manganese-muscovite, *see* Mangan-muscovite.

Manganese-pennine. G. Aminoff, 1931. H. Eckermann, *Geol. För. Förh. Stockholm*, 1927, vol. 49, p. 450 (manganese-chlorite). G. Aminoff, *Kungl. Svenska Vetenskapsakad. Handl.*, 1931, ser. 3, vol. 9, no. 5, p. 13 (manganese-pennine). A variety of pennine containing MnO 1%, from Långban, Sweden. [*M.A.*, **3**, 474; **4**, 469.] 12th List

Manganese-shadlunite, variant of Mn-shadlunite (Mn-шадлунит). *A.M.*, **58**, 1114 (1973). 28th List

Manganese-sicklerite. P. Quensel, 1937. *Geol. För. Förh. Stockholm*, vol. 59, pp. 85, 96 (Mn-sicklerite). B. Mason, ibid., 1941, vol. 63, p. 139 (Mn-sicklerite). *Amer. Min.*, 1937, vol. 22, p. 876 (manganese sicklerite). The original sicklerite derived from lithiophilite from California (6th List), as distinct from ferri-sicklerite derived from triphylite, from Varuträsk, Sweden (14th List), and containing manganese in excess of iron in the formula $(Li,Mn'',Fe''')PO_4$. [*M.A.*, **6**, 486; **8**, 181.] 16th List

Manganese-zoisite. L. L. Shabynin, 1934. *Mém. Soc. Russ. Min.*, ser. 2, vol. 63, p. 456

(марганцовый цоизит), p. 457 (Mn-цоизит), p. 459 (manganese ziosite), p. 460 (Mn-zoisite). A pink zoisite containing MnO 0·47%. [*M.A.*, **6**, 437.] 14th List [Also Manganzoisit (q.v.)]

Manganez krokidolit, *see* Mangan-crocidolite.

Manganfayalite. C. Hintze, 1897. *Handbuch der Mineralogie*, vol. ii, p. 27 (Manganfayalit). A member of the olivine group between fayalite and knebelite in chemical composition. The name was first applied to an artificial product from a furnace slag; and later (J. Palmgren, *Bull. Geol. Inst. Univ. Upsala*, 1917, vol. xiv, p. 116) to a mineral from Sweden containing MnO 5–26·5%, and with characters more like those of fayalite than knebelite. 8th List

Mangan-fluorapatite. B. Mason, 1941. *Geol. För. Förh. Stockholm*, vol. 63, p. 280 (mangan-fluorapatite), p. 281 (Mn-fluorapatite). Cf. fluormanganapatite (9th List), mangan-hydroxyapatite (16th List). 19th List

Mangangehlenite, *see* Manganese-gehlenite.

Manganglauconite. *See* Marsjatskite. 2nd List

Mangangoslarite. L. J. Lawrence, 1961. *Journ. Proc. Roy. Soc. New South Wales*, vol. 95, p. 13. An unnecessary name for manganoan Goslarite. [*M.A.*, **16**, 144.] 23rd List

Mangan-grandite. L. L. Fermor, 1909, *See* Grandite. 5th List [Also 11th List; *see* Calc-spessartite]

Manganhumite. P. B. Moore, 1978. *M.M.*, **42**, 133. The manganese analogue of humite occurs at the Brattfors mine, Nordmarks Odalfält, Värmland, Sweden. Orthorhombic, *a* 10·54, *b* 21·45, *c* 4·822 Å, space group *Pnma*. Composition $4[(Mn,Mg)_7(OH)_2(SiO_4)_3]$, sp. gr. 3·83. RIs α 1·707, β 1·712, γ 1·732, $2V_\gamma$ 37°. Named from the composition. [*M.A.*, 78-2124.] 30th List

Mangan-hydroxyapatite. B. Mason, 1941. *Geol. För. Förh. Stockholm*, vol. 63, p. 279 (mangan-hydroxyapatite), p. 281 (Mn-hydroxyapatite). Hydroxyapatite (6th List) = hydroxylapatite (15th List) containing manganese (MnO 7·5%), from Varuträsk, Sweden, $(Ca,Mn,Fe)_{10}(PO_4)_6(OH)_2$. [*M.A.*, **8**, 311.] 16th List

Manganilmenite. E. S. Simpson, 1929. *Journ. Roy. Soc. W. Australia*, vol. 15, p. 103. A variety of ilmenite containing MnO 14·40%, from Western Australia. [*M.A.*, **4**, 315.] 12th List

Manganipurpurite. W. T. Schaller, 1907. *See* Ferripurpurite. 5th List

Mangani-sicklerite. P. Quensel, 1952. *Geol. Mag. Hertford*, vol. 89, p. 59. Previously referred to as Mn-sicklerite (P. Quensel, 1937, 14th List, p. 609), manganese sicklerite (W. F. Foshag, *Amer. Min.*, 1937, vol. 22, p. 876), manganese-sicklerite (16th List), and mangan-sicklerite (Dana's *Mineralogy*, 7th edit., 1951, vol. 2, p. 672). 19th List

Mangankiesel. F. Klockmann, *Jahrb. k. preuss. geol. Landesanst. u. Bergakad.*, for 1894, XV, p. xxxii. 1895. A quartz-schist containing much $MnCO_3$. Harz. 1st List

Mangan-knebelite. T. Yoshimura, 1939. *Journ. Fac. Sci. Hokkaido Univ.*, ser. 4, vol. 4, p. 405 (Manganknebelite). An intermediate member (Fe_2SiO_4 20–40 mol.%) of the fayalite-tephroite series, near to knebelite. 15th List

Mangankoninckite. A. A. Beus, 1950. *Doklady Acad. Sci. USSR*, vol. 73, p. 1267 (манганконинкит). $(Fe,Mn)PO_4.3H_2O$, a variety of koninckite containing Mn_2O_3 2·72%, in pegmatite from Turkestan. [*M.A.*, **11**, 190.] 19th List

Manganleonite. H. Anspach, *Zeits. Krist.*, 1939, vol. 101, p. 39 (Mn-Leonit). A. N. Winchell, *Optical mineralogy*, part 2, 4th edit., 1951, p. 167 (manganleonite). Artificial $K_2Mn(SO_4)_2.4H_2O$, monoclinic, isomorphous with leonite ($K_2Mg(SO_4)_2.4H_2O$). [*M.A.*, **7**, 393.] 19th List

Manganmelanterit. H. Strunz, *Min. Tab.*, 1941, p. 134. $(Fe,Mn)SO_4.7H_2O$. Syn. of Luckyit = luckite. 18th List

Mangan-monticellite. *See* Iron-monticellite. 14th List

Mangan-muscovite. P. Eskola, 1914. *Bull. Comm. Géol. Finlande*, vol. 8, no. 40, p. 37 (manganese-muscovite). A compact, fine scaly, manganiferous (MnO 2·30%) variety of muscovite of a deep violet colour, from Kimito, Finland. *Bull. Soc. franç. Min.*, 1933, vol. 56, p. 188 (Manganmuscovite). 13th List

Mangan-neptunite. S. M. Kurbatov, 1923. *Compt. Rend. Acad. Sci. Russie*, 1923, p. 59 (марганцевый нептунит, neptounite manganifère), p. 60 (manganneptounite). A. E. Fersman, *Trans. Northern Sci. Econ. Exped.*, no. 16 (*Sci. Techn. Dept. Supreme Council of National Economy*, no. 8), *Moscow and Petrograd*, 1923, pp. 16, 69, 73 (манган-нептунит), p. 57 (мангано-нептунит). A variety of neptunite (q.v.) containing MnO 9·95 with FeO 5·16%, from the Kola Peninsula. [*M.A.*, **2**, 263–264.] 10th List

Mangan-niobite. H. Strunz, *Min. Tab.*, 1941, p. 106 (Manganniobit). Syn. of manganocolumbite. 18th List

Mangano-alluaudite. D. J. Fisher, 1957. *Amer. Min.*, vol. 42, p. 661. Synonym of Mangan-alluaudite (19th List). 23rd List

Mangano-anthophyllite. C. S. Ross and P. F. Kerr, 1932. *Amer. Min.*, vol. 17, pp. 4, 17. A variety of anthophyllite occurring as a pink alteration product of rhodonite. [*M.A.*, **5**, 51.] 13th List

Mangano-astrophyllite. Chih-chang Peng and Che-sheng Ma, 1964. [*Scientia Sinica*, **13**, 1180]; abstr. *Zap.*, **97**, 76. A superfluous name for manganoan astrophyllite. 26th List

Manganoaxinite. W. T. Schaller, 1909. *See* Ferroaxinite. 5th List

Manganobabingtonite, erroneous translation of Марганцевый бабингтонит, Manganoan babingtonite. *Dokl. Acad. Sci. USSR*, 1966, **169**, 128; translation of Докл. Акад. наук СССР, 1966, **169**, 434. *Syn.* of Manganbabingtonite (26th List). 28th List

Mangano-belyankinite, *see* Mangan-belyankinite.

Manganobrucite. A. Lacroix, 1909. *Minéralogie de la France*, vol. iii, p. 402. The same as manganbrucite (L. J. Iglström, 1882), a variety of brucite containing manganese. 6th List

Manganodickinsonite. D. J. Fisher, 1957. *Amer. Min.*, vol. 42, p. 662 (mangano-dickinsonite). Synonym of dickinsonite (Brush and Dana, 1878). *See* Ferrodickinsonite. 21st List

Manganoferrite. J. Beckenkamp, 1921. *See* Talc-spinel. 9th List

Manganogel. H. Strunz, *Min. Tabellen*, 3rd edit., p. 154. Amorphous MnO_2, present in many occurrences of Wad. 22nd List

Manganolangbeinite. F. Zambonini and G. Carobbi, 1924. *Rend. Accad. Sci. Fis. Mat. Napoli*, ser. 3, vol. 30, p. 126. Manganese and potassium sulphate, $2MnSO_4.K_2SO_4$, as minute, pale-rose tetrahedra from Vesuvius. Named from analogy to langbeinite ($2MgSO_4.K_2SO_4$). [*M.A.*, **2**, 383.] 10th List

Manganomelane. F. Klockmann, 1922. *Lehrbuch der Mineralogie*, 7–8th edit., p. 422 (Manganomelan). Group name for gel forms of MnO_2, including psilomelane, wad, etc. Named from manganese and *μελᾶς*, -*ᾶνος*, black. [*M.A.*, **6**, 53.] 14th List

—— A. A. Levinson, 1961. *Amer. Min.*, vol. 46, p. 355. A portmanteau term for hard, unidentified manganese oxide minerals. [*M.A.*, **15**, 459.] 23rd List

Manganomossite. *Rep. Dept. Mines, Western Australia*, 1923, for 1922. p. 120. An angular pebble from Yinnietharra, Western Australia, which gave an analysis Ta_2O_5 44·53, $Nb_2O_5$34·64, $TiO_2$3·92, MnO 12·02, FeO 4·64, H_2O 0·26 = 100·01, sp. gr. 6·21, is doubtfully referred to manganomossite rather than to manganocolumbite. [Cf. E. S. Simpson, *Min. Mag.*, 1917, vol. 18, p. 108.] 10th List [Metamict Manganocolumbite; *M.A.*, **14**, 274]

Manganonatrolite. C. A. Thomas, 1947. *Rocks and Minerals*, Peekskill, NY, vol. 22, p. 804. Variety of natrolite from Pennsylvania. [*M.A.*, **10**, 294.] 18th List

Manganoniobite. H. Strunz, 1957. *Min. Tabellen*, 3rd edit., p. 155. Syn. of Manganocolumbite. 22nd List

Manganophyllite. T. Yoshimura, 1939. *Journ. Fac. Sci. Hokkaido Univ.*, ser. 4, vol. 4, p. 430 (Manganophyllite). The hypothetical molecule $H_4K_2Mn_5Al_4Si_5O_{24}$, corresponding to siderophyllite, to explain the composition of biotite. Not the manganophyll, manganophyllite of L. J. Igelström (1872) [a manganoan biotite]. 15th List

Manganoplesite. D. P. Bobrovnik, 1959. Мин. Сборник Львов. Геол. Общ. (*Min. Mag. Lvov Geol. Soc.*), vol. 13, p. 343 (Манганоплезит). A manganoan chalybite from the Volhynian coal basin; syn. of Oligon spar. An unnecessary name. 22nd List

Mangan-orthite. L. N. Ovchinnikov and N. N. Tzimbalenko, 1948. *Doklady Acad. Sci.*

USSR, vol. 63, p. 191 (манган-ортит). A variety of orthite containing MnO 5·37%, from Vishnevy Mt., Urals. [*M.A.*, **10**, 453.] 18th List

Manganosideroplesite. D. P. Bobrovnik, 1959. Мин. Сборник Львов. Геол. Общ. (*Min. Mag. Lvov Geol. Soc.*), vol. 13, pp. 340, 348 (Мангансидероплезит, manganosideroplesite). A manganoan sideroplesite from the Volhynian coal basin. An unnecessary name. 22nd List

Manganosphaerite. K. Busz, *Neues Jahrb. Min.*, 1901, vol. ii, p. 129 (Manganosphärit). Resembles sphaerosiderite in appearance, but has the composition of oligonite ($3FeCO_3.2MnCO_3$), from which it only differs in being not in crystals. In cavities in a basalt-vein at Horhausen, Rhenish Prussia. [The name would suggest either a botryoidal manganese carbonate or a manganiferous sphaerite (hydrated basic aluminium phosphate).] 3rd List

Manganosteenstrupine. K. A. Vlasov, M. V. Kuzmenko, and E. M. Eskova, 1959. [*The Lovozero Alkaline Massif* (*Moscow, Acad. Sci. USSR*), p. 421]; abstr. *Amer. Min.*, 1960, vol. 45, p. 1132. Amorphous (metamict?) material in pegmatites on Mt. Karnasurt, Lovozero massif, Kola peninsula, is formulated $(Ln,Th,Ca)MnSiO_3(OH)_3.2H_2O$ and regarded as a manganese analogue of steenstrupine. 22nd List

Manganostibite, variant of Manganostibiite; *A.M.*, **9**, 62. 10th List. [*A.M.*, **55**, 1489]

Manganostilpnomelane. A. N. Winchell, 1951. *Optical mineralogy,* part 2, 4th edit., p. 390. Alternative name for parsettensite [*M.A.*, **2**, 251; *Min. Mag.*, **25**, 189]. 19th List

Manganotantalocolumbite. Ya. A. Kosals, 1967. [Геол. и геофис. no. 2, 116]; abstr. *Zap.*, **98**, 324 (1969) (Манганотанталоколумбит). An unnecessary name for a manganocolumbite with Nb:Ta = 3:2. 26th List

β-Manganous sulphide. G. Baron and J. Debyser, 1957. *Compt. Rend. Acad. Sci. Paris*, vol. 245, p. 1148 (Sulfure manganeux β). Minute black bipyramids in recent sediments from the Baltic Sea were identified by X-ray data as hexagonal β-MnS (Wurtzite group). 22nd List

Manganoxyapatite. B. Mason, 1941. *Geol. För. Förh. Stockholm*, vol. 63, p. 283 (manganoxyapatite), p. 383 (mangan-oxyapatite). [*M.A.*, **8**, 343.] 16th List; also H. Strunz, *Min. Tab.*, 1941, p. 156 (Manganoxyapatit). Syn. of manganvoelckerite (15th List). 18th List

Manganpennine. A. Hamberg, 1890. *Geol. För Förh. Stockholm*, vol. 12, p. 585 (Mangan-Pennin). The terms manganiferous chlorite, manganchlorite, and manganpennine are used alternatively. [*M.A.*, **3**, 474.] 11th List

Mangan-phlogopite. T. Yoshimura, 1939. *Journ. Fac. Sci. Hokkaido Univ.*, ser. 4, vol. 4, p. 431 (Manganphlogopite). A variety of phlogopite containing MnO 18·24%. 15th List

Mangan-pickeringite. H. Meixner and W. Pillewizer, 1937. *Zentr. Min.*, Abt. A, 1937, p. 265 (Manganpickingerit [*sic*]). Pickeringite containing some manganese replacing magnesium. *See* Magnesium-apjohnite. [*M.A.*, **7**, 12.] 15th List

Manganpyrosmalite. C. Frondel and L. H. Bauer, 1953. *Amer. Min.*, vol. 38, p. 755. Variety of pyrosmalite with Mn in excess of Fe. [*M.A.*, **12**, 236.] 20th List [A species]

Mangan-rockbridgeite. M. L. Lindberg, 1949. *Amer. Min.*, vol. 34, p. 341 (manganoan rockbridgeite); *M.A.*, 1950, **11**, 8 (manganiferous rockbridgeite); Hintze, *Min.*, 1954, Ergbd. 2, p. 242 (Mangan-Rockbridgeit). A dufrenite-like mineral $(Fe'',Mn'')Fe_4'''(PO_4)_3(OH)_5$, intermediate between frondelite $Mn''Fe_4'''(PO_4)_3(OH)_5$ and rockbridgeite $Fe''Fe_4'''(PO_4)_3(OH)_5$, (19th List, where these formulae are incorrectly given). 20th List

Mangan-sahlite. V. A. Zharikov and K. V. Podlessky, 1955. Докл. **105**, 1096 (мангансалит). An unnecessary name for a manganoan sahlite, from limestone skarns in W Karamazar. [*M.A.*,,**13**, 56.] 28th List

Manganseverginite. L. D. Kurshakova, 1967. [Геол. и Геофиз., no. 1, p. 118]; abstr. Зап. всесоюз. мин. общ. (*Mem. All-Union Min. Soc.*), 1968, vol. 97, p. 77. Syn of Manganaxinite. 25th List

Mangan-sicklerite. Dana, *Min.*, 7th edit., 1951, vol. 2, p. 672; Hintze, *Min.*, 1954, Ergbd. 2, p. 243 (Mangansicklerit). Variant of Manganese-sicklerite (16th List), Mangani-sicklerite (19th List). 20th List

Manganspinel. P. Groth, 1874. *Tabell. Übers. d. Mineralien*, p. 24 (Manganspinell). A member of the spinel group with the composition $(Mn,Mg)O.(Fe,Mn)_2O_3$, evidently intended as a synonym of jacobsite (but in later editions jacobsite, $MnO.(Fe,Mn)_2O_3$, appears in addition). The same name was later applied by J. S. Krenner (*Magyar Chem. Foly. Budapest*, 1907, vol. 14, pp. 81, 83; *Zeits. Kryst. Min.*, 1907, vol. 43, pp. 473, 571) to brown, octahedral crystals with the composition $(Mn,Mg)O.(Al,Mn)_2O_3$ from an iron-furnace slag. J. Beckenkamp (C. Hintze's *Handbuch d. Min.*, 1921, vol. 1, part 4, p. 30) follows Krenner but simplifies the formula to $MnAl_2O_4$ (*see* Talc-spinel). 9th List

Manganstilpnomelane, variant of Manganostilpnomelane (19th List); syn. of Parsettensite. (Hintze, *Handb. Min.*, Erg.-Bd. II, p. 584.) 22nd List

Mangantapiolite. V. [A.] Khvostova and V. N. Arkhangel'skaya, 1970. Докл. Акад. наук СССР (*Compt. Rend. Acad. Sci. URSS*), **194**, 677 (мангантапиолит). A superfluous name for a manganoan tapiolite. [*A.M.*, **56**, 1122; *Zap.*, **100**, 620.] 27th List

Mangan-tremolite. T. Yoshimura, 1939. *Journ. Fac. Sci. Hokkaido Univ.*, ser. 4, vol. 4, p. 425 (Mangantremolite). A variety of tremolite containing MnO 7·38%. 15th List

Mangan-uralite. S. Kilpady and A. S. Dave, 1958. *Journ. Univ. Geol. Soc. Nagpur*, vol. 1 (for 1955–6), no. 3, p. 4. A pink amphibole formed by alteration of blanfordite, from Ponia, Balaghat District, Madhya Pradesh, India. Near magnesio-arfvedsonite with 3% MnO. 'An unnecessary name . . . somewhat inappropriate' (M. Fleischer, *Amer. Min.*, 1959, vol. 44, p. 692). 22nd List

Manganvoelckerite. P. Quensel, 1937. *Geol. För. Förh. Stockholm*, vol. 59, p. 161 (mangan-voelckerite). Manganapatite with fluorine largely replaced by oxygen, $3(Ca,Mn)_3(PO_4)_2.(Ca,Mn)(O,F_2)$. [*M.A.*, **7**, 10.] 15th List

Mangan-wollastonite. V. M. Goldschmidt, 1911. *Skrifter Videnskap. Kristiania, Mat.-nat. Kl.*, 1911, vol. 1, no. 1, p. 331 (Manganwollastonit). A variety of wollastonite containing manganese (MnO 7%). 14th List

Manganzoisit. Hintze, *Min.*, 1954, Ergbd. 2, p. 244. Variant of Manganese-zoisite (14th List). 20th List

Mangualdite. A. Mário de Jesus, [1936]. *Com. Serv. Geol. Portugal*, vol. 19 (for 1933), p. 141 (Mangualdite). Phosphate of manganese and calcium, $3RO.P_2O_5$, as olive-green orthorhombic (?) crystals. Named from the locality, Mangualde, Portugal. [*M.A.*, **6**, 441.] 14th List [Manganoan Apatite]

Manjak. (R. P. Rothwell's *Mineral Industry*, for 1897, New York, 1898, VI, 54; VII, 72. *Trans. Fed. Inst. Mining Eng.*, 1898, XIV, 539; XVI, 33, 388). A local name for asphaltum from Barbados, West Indies. 2nd List

Manjirōite. M. Nambu and K. Tanida, 1967. [*Journ. Japan. Ass. Min., Petrol. Econ Geol.*, **58**, 39]; abstr. *Amer. Min.*, **53**, 2103 (1968). The sodium analogue of cryptomelane, from the Kohare mine, Iwate Prefecture, Japan; named for Professor Manjiro Watanabe. [*M.A.*, 69-2387; *Zap.*, **97**, 615; *Bull.*, **92**, 513.] 26th List

Mansfieldite. V. T. Allen and J. J. Fahey, 1945. *Program and Abstracts, Min. Soc. Amer.*, 26th Annual Meeting, p. 7; *Amer. Min.*, 1946, vol. 31, p. 189. Hydrated aluminium arsenate, $AlAsO_4.2H_2O$, isomorphous with scorodite, as white cellular crusts in clay from Hobart Butte, Oregon. Named after Dr. George Rogers Mansfield (1875–) formerly of the United States Geological Survey. [*M.A.*, **9**, 262.] 17th List [*A.M.*, **33**, 122]

Mansjöite. H. von Eckermann, 1922. *Geol. För. Förh. Stockholm*, vol. 44, p. 355. A variety of diopside (fluor-diopside) containing fluorine (0·63%). Named from the locality, Mansjö Mt., Sweden. 9th List

Marahuite. O. A. Derby, 1907. *Journ. Geol. Chicago*, vol. 15, p. 231 (Marahuite). O. Stutzer, *Zeits. Deutsch. Geol. Gesell.*, 1935, vol. 87, p. 616 (Marahunit). An earthy bituminous lignite containing algae and corresponding with boghead coal. Named from the locality, Marahú, Bahia, Brazil. 14th List

Marburgite. S. J. Thugutt, 1949. *Rocznik Polsk. Tow. Geol.* (*Ann. Soc. Géol. Pologne*), vol. 18 (for 1948), p. 10 (marburgite), p. 35 (marburgit). The calcium-type of phillipsite, as distinct from the sodium-type ('herschelite'). Named from its locality, Marburg, Hesse. 19th List

Margarosanite. W. E. Ford and W. M. Bradley, 1916. *Amer. Journ. Sci.*, ser. 4, vol. xlii, p. 159. A metasilicate of lead and calcium (with a little manganese replacing calcium), $PbCa_2(SiO_3)_3$, occurring at Franklin, New Jersey, as colourless and transparent lamellar masses with three good cleavages, the lustre being pearly on the best lamellar cleavage. Named from μαργᾰρίτης, a pearl, and σᾰνίς, a board, in reference to the pearly lustre and lamellar structure. Triclinic crystals and large masses of the same mineral were afterwards found at Långban, Sweden (G. Flink, *Geol. För. Förh.*, 1917, vol. xxxix, p. 438). 8th List

Marićite. B. D. Sturman and J. A. Mandarino, 1977. *Can. Mineral.*, **15**, 396. Nodules in ironstones from the Big Fish River area, Yukon Territory, consist largely of elongated anhedral grains, orthorhombic, *a* 6·867, *b* 8·989, *c* 5·049 Å. Composition 4[$NaFe^{2}PO_4$], ρ 3·66 g.cm^{-3}. RIs α 1·676 ‖ [100], β 1·695 ‖ [010], γ 1·698, $2V_\alpha$ 43½°. Named for L. Marić. 30th List

Marignacite. S. Weidman and V. Lenher, 1907. *Amer. Journ. Sci.*, ser. 4, vol. xxiii, p. 287. A variety of pyrochlore occurring as brown octahedra in pegmatite at Wausau, Wisconsin. It differs from other members of the pyrochlore group in containing rather more cerium and yttrium with less calcium and iron, and in the presence of a small amount of silica. Named after Jean Charles Galissard de Marignac (1817–1894). 4th List [Renamed **Ceriopyrochlore**; *A.M.*, **62**, 406; 30th List]

Marokite. G. Gaudefroy, C. Jouravsky, and F. Permingeat, 1963. *Bull. Soc. franç. Min. Crist.*, vol. 86, p. 359. $CaMn_2O_4$, in large black orthorhombic crystals from Tachagalt, Ouarzazate, Morocco. Named from Maroc (= Morocco). [*M.A.*, **16**, 546; *A.M.*, **49**, 817.] 23rd List

Marrite. R. H. Solly, 1904. *Nature*, vol. lxxi, p. 118; *Min. Mag.*, 1905–1906, vol. xiv, pp. 76, 188. Small, steel-grey crystals with brilliant metallic lustre found in the white, crystalline dolomite of the Binnenthal, Switzerland. They are monoclinic with a rich development of faces; their chemical composition is not known. Named after Dr. John Edward Marr, of Cambridge. 4th List [$PbAgAsS_3$; *M.A.*, **19**, 270]

Marshite. A. Liversidge, *Journ. and Proc. Roy. Soc. NS Wales*, XXVI. 328, 1892; abstr. by H. A. Miers, *Zeits. Kryst. Min.*, XXIV. 207, 1894. [*M.M.*, **11**, 236.] Copper iodide. Cubic (Miers). Broken Hill, NSW. 1st List

Marsjatskite. E. von Fedorow and W. Nikitin, *Ann. Géol. Min. Russie*, 1899, III, 90, 102. Glauconite containing much manganese (*Manganglaukonit*); it forms the bulk of a Tertiary sandstone in the Marsjat forest, Urals. 2nd List

Marthozite. F. Cesbron, R. Oosterbosch, and R. Pierrot, 1969. *Bull. Soc. franç. Min. Crist.*, **92**, 278. Yellow-green orthorhombic crystals from the Cu-Co deposit at Musonoi, Katanga, have a composition near $Cu(UO_2)_3(SeO_3)_3(OH)_2.7H_2O$. Named for A. Marthoz. [*M.A.*, 70-751; *A.M.*, **55**, 533.] 26th List

Marvelite. A trade name for *artificial* $SrTiO_3$. M. J. O'Donoghue, *priv. comm. See* under fabulite. 28th List

Mascareignite. A. Lacroix, 1936. *Le volcan actif de l'île de la Réunion*, Paris, 1936, p. 248 (mascareignite). A form of opaline silica (H_2O 13·61%) of vegetable origin consisting largely of the siliceous remains of grasses with some diatoms; from Reunion, Indian Ocean. Named from the Mascarene Islands (*Fr.*, Mascareignes), including Reunion (= Mascarenhas = Bourbon), said to have been discovered in 1513 by the Portuguese navigator, Pedro Mascarenhas. [*M.A.*, **7**, 2.] 15th List

Masicotite. G. Gagarin and J. R. Cuomo, 1949, loc. cit., p. 9 (masicotita). Variant of massicottite (A. D'Achiardi, 1883) and massicotite (A. H. Chester, 1896). Synonym of massicot. 19th List

Masrite. H. D. Richmond and H. Off, *Journ. Chem. Soc. Trans.*, LXI. 491, 1892. A fibrous alum containing 'the element masrium.' Egypt. 1st List [Cf. Johnsonite]

Masutomilite. K. Harada, K. Nagashima, and S. Kanisawa, 1976. *Mineral J.* [*Japan*], **8**, 95. The pink core of a zoned manganoan zinnwaldite from Tanakamiyama, Japan, and a zoned purple hexagonal crystal from Tawara, Gifu Prefecture, Japan, both have over 50% of the Mn analogue of zinnwaldite. The first has 8·12% MnO, 4·45% Li_2O, 1·54% Rb_2O, *a* 5·253, *b* 9·085 *c* 10·107 Å, β 100·15°, α 1·534, colourless, β 1·569, purple, ‖ [010], and γ 1·570, colourless; γ : [100] 3°. The second specimen has 4·27% MnO, 5·78% Li_2O, 1·20% Rb_2O, *a* 5·248, *b* 9·087, *c* 10·090 Å, β 100·10°; α 1·536, colourless, β 1·570, purple, and γ 1·571, colourless, γ:[100] 2–4°. Named for K. Masutomi. [*M.A.*, 78-3478; *A.M.*, **62**, 594; *Zap.*, **105**, 83.] 30th List

Masuyite. J. F. Vaes, 1947. *Ann.* (*Bull.*) *Soc. Géol. Belgique*, vol. 70, p. B 219. Hydrous lead uranate as orange-red orthorhombic scales, from Katanga. Named in memory of Gustave Masuy who died a prisoner of war in 1945. [*M.A.*, **10**, 146.] 18th List [$3PbO.8UO_3.10H_2O$; refs. in *M.M.*, **41**, 51]

Matorolite. Anon., 1967. [*Chamber of Mines, Rhodesia, J.*, **9**, no. 7, 30]; abstr. *Amer. Min.*, **54**, 992 (1969). An unnecessary name for an emerald-green chromian chalcedony; named for the locality. [*Zap.*, **97**, 616; *Bull.*, **92**, 517.] 26th List

Matraite. S. Koch, 1958. [*Acta Univ. Szegediensis*, vol. 11, p. 11]; abstr. *Amer. Min.*, 1960, vol. 45, p. 1131. A mineral occurring at Gyöngyösoroszi with blende and pyrite is identical with the third polymorph of ZnS (γ-ZnS) prepared artificially by Buck and Strock (*Amer. Min.*, 1955, vol. 40, p. 192); this polymorph is termed β'-ZnS by Koch. [*M.A.*, **14**, 279.] 22nd List [*A.M.*, **45**, 1131]

Matrosite. J. A. Dulhunty, 1939. *See* Gelosite. 15th List

Mattagamite. R. I. Thorpe and D. C. Harris, 1973. *C.M.*, **12**, 55. Grains up to 0·1 mm in altaite from the Mattagami Lake mine, Matagami area, Quebec, are orthorhombic *a* 5·305, *b* 6·289, *c* 3·866 Å. Composition $CoTe_2$, with variable amounts of Fe and Sb. Named from the locality. [*A.M.*, **59**, 382.] 28th List

Matteuccite. G. Carobbi and C. Cipriani, 1952. *Atti* (*Rend.*) *Accad. Naz. Lincei, Cl. Sci. fis. mat. nat.*, ser. 8, vol. 12, sem. 1, p. 23. Sodium bisulphate, $NaHSO_4.H_2O$, with mercallite ($KHSO_4$, 14th List) in saline stalactites from Vesuvius. Named after Vittorio Matteucci (1862–1909), a former director of the Vesuvian Observatory. [*M.A.*, **11**, 517.] 19th List

Mattkohle (= Durain), *see* Fusain.

Maucherite. F. Grünling, 1913. *Centralblatt Min.*, 1913, p. 225 (Maucherit). A nickel arsenide, Ni_3As_2, forming reddish silver-white, platy crystals (tetragonal or orthorhombic?) resembling rammelsbergite. It occurs with niccolite in the Kupferschiefer of Eisleben, Thuringia. Named after Mr. Wilhelm Maucher of Munich. 6th List [Near $Ni_{11}S_8$]

Maufite. F. E. Keep, 1930. *Trans. Geol. Soc. S. Africa*, vol. 32 (for 1929), p. 103; *Bull. Geol. Survey Southern Rhodesia*, 1930, no. 16, pp. 26, 33. Hydrous silicate of aluminium, magnesium, and nickel (NiO 4·28%), $(Mg,Fe,Ni)O.2Al_2O_3.3SiO_2.4H_2O$, forming fibrous emerald-green veins in serpentine from Southern Rhodesia. Named after Herbert Brantwood Maufe, Director of the Geological Survey of Southern Rhodesia. [*M.A.*, **4**, 248, 413; *A.M.*, **15**, 275.] 12th List

Mauleonite. A. Leymerie, 1867. *Cours de Minéralogie*, 2nd edit., vol. ii, p. 243 (Mauléonite). A. Lacroix, *Minéralogie de la France*, 1895, vol. i, p. 383, 1910, vol. iv, p. 739. A white chlorite identical with the leuchtenbergite variety of clinochlore. From Mauléon, Basses-Pyrénées. 6th List

Mauritzite. L. Tokody, T. Mándy, and S. Nemes-Varga, 1957. *Neues Jahrb. Min.*, Monatshefte, p. 33 (Mauritzit). Hydrous oxides, $(Fe''',Al)_2O_3.2(Mg,Fe'')O.5H_2O$, as bluish-black rods in altered andesite from Hungary. The structure is similar to that of montmorillonite with $(OH)_4$ in place of SiO_2, but containing much intermixed quartzine. Named after Prof. Béla Mauritz (1881–) of Budapest. [*M.A.*, **13**, 381.] 21st List [*see*, *A.M.*, **42**, 704; **48**, 1183]

Mauzeliite. H. Sjögren, *Geol. För. Förh.*, XVII. 313, 1895. $4(Ca,Pb)O.2Sb_2O_5.TiO_2$. Cubic. [*M.M.*, **11**, 82, 229.] [Jakobsberg, Wermland] Sweden. 1st List [Plumboan Roméite]

Mavinite. B. Rama Rao, 1946. *Rec. Mysore Geol. Dept.*, vol. 43 (for 1944), p. 27. A dark-green brittle mica, $3(Fe,Mg)O.6(Al,Fe)_2O_3.7SiO_2.9H_2O$, intermediate between chloritoid and xanthophyllite. Named from the locality, Mavinhalli, Mysore. [*M.A.*, **10**, 7.] 18th List

Mavudzite. A. V. P. Coelho, 1954. [*Garcia de Orta, Rev. junta missões geogr. e invest. do Ultramar*, vol. 2, p. 209.] *Bibliography of geology, Geol. Soc. Amer.*, 1955, vol. 19, p. 369; *Amer. Min.*, 1956, vol. 41, p. 164. A metamict radioactive mineral from Mavudzi, Tete district, Mozambique. Named from the locality. Previously described [*Min. Mag.*, **29**, 101, 292] as a variety of davidite. *See* Ferutite. 21st List [Is Davidite; *A.M.*, **46**, 700]

Maw-sit-sit. Author and date? A dealers' trade-name for a green decorative jadeite–albite rock, named for the locality: Maw-sit-sit, Tawmaw area, Burma. E. Gübelin, *Journ. Gemmology*, 1965, **9**, 329, 372. Cf. jade-albite, 24th List. 28th List

Mawsonite. N. L. Markham and L. J. Lawrence, 1965. *Amer. Min.*, vol. 50, p. 900. The copper deposits of Mt. Lyell, Tasmania, and of Tingha, New South Wales, carry a mineral corresponding closely to some of the material described by Murdoch (1916) and others as 'orange bornite'; some occurrences of 'orange bornite' have proved to be the germanium mineral, Reniérite, but the Mt. Lyell and Tingha mineral is in fact the tin analogue of Reniérite, and is named Mawsonite. It is pseudocubic, highly anisotropic and pleochroic, composition $Cu_7Fe_2SnS_{10}$. The Mt. Lyell material has been referred to as Reniérite, and also as 'orange stannite'. Some of the other occurrences of 'orange bornite' are probably Mawsonite. Named for Sir Douglas Mawson. [*M.A.*, **17**, 499.] 24th List [Cf. Hexastannite, Stannoidite]

Maxixe-aquamarine, Maxixe-beryl. G. O. Wild, *Centr. Min.*, Abt. A, 1933, p. 38 (Maxixe-Aquamarine, *pl.*); K. Schlossmacher and H. Klang, *Zentr. Min.*, Abt. A, 1935, p. 37; W. Roebling and H. W. Trommau, ibid., p. 134 (Maxixeberyll). A gem beryl of unusual blue colour and pleochroism from Maxixe mine, Minas Geraes, Brazil. The mine, now abandoned, was named from 'machiko' (a gherkin), because of the rugged corroded surface of the beryl, and 'maxixe' is apparently a German modification of this. [*M.A.*, **5**, 295; **6**, 201.] 14th List

Maxixe-beryl, *see* Maxixe-aquamarine.

Mayaite. H. S. Washington, 1922. *Proc. Nat. Acad. Sci. USA*, vol. 8, p. 325. The 'jade' of worked objects left by the ancient Mayas in Central America grades from tuxtlite (q.v.) to nearly pure albite. The name mayaite, from the Maya nation, is applied to this series of rocks. [*M.A.*, **2**, 67.] 10th List

Mayakite. A. D. Genkin, T. L. Evstigneeva, N. V. Troneva, and L. N. Vyal'sov, 1976. *Zap.*, **105**, 698 (Маякит). Small inclusions in polarite and intergrowths with stannopalladinite from the Mayak (Маяк, 'Majak') mine, Talnakh deposit, are hexagonal, *a* 6·066, *c* 7·20 Å. Composition PdNiAs or $(Pd,Ni)As_2$. Named for the locality. [*A.M.*, **62**, 1260; *M.A.*, 77-4624.] 30th List

Mayberyite. S. F. Peckham, *Journ. Franklin Inst.*, 1895, CXL, 382. Named after C. F. Mabery (not Maybery). The names *mayberyite, venturaite* and *warrenite* are proposed in connection with a suggested new classification of petroleum, bitumen, etc. 2nd List

Mayenite. G. Hentschel, 1964. *Neues Jahrb. Min.*, Monatschr., p. 22 (Mayenit). Cubic $Ca_{12}Al_{14}O_{33}$ as grains in metamorphosed limestone from the Bellerberg, Mayen, Eifel, Germany. Named from the locality. This compound has long been known as a constituent of cement clinker. 23rd List [*A.M.*, **50**, 2106]

Mazzite. E. Galli, E. Passaglia, D. Pongiluppi, and R. Rinaldi, 1974. *Contr. Min. Petr.*, **45**, 99. Tiny hexagonal needles (*a* 18·392, *c* 7·646 Å) from Mont Semiol (= Mont Simiouse), Loire, France; a zeolite near $K_2Mg_2Ca_{1\cdot5}Al_{9\cdot5}Si_{25\cdot5}O_{72}.28H_2O$. Named for F. Mazzi. Note that Mont Semiol is the type locality for another hexagonal zeolite, offretite. [*M.A.*, 74-3466.] 28th List

Mboziite. P. W. G. Brock, D. C. Gelatly, and O. von Knorring, 1964. *Min. Mag.*, vol. 33, p. 1057. An amphibole near $Na_2CaFe_3^{2+}Fe_2^{3+}Al_2Si_6O_{22}(OH)_2$, from the Mbozi syenite-gabbro complex in SW Tanganyika. Named from the locality. 23rd List [Potassian taramite; *Amph.* Sub-Comm.]

Mcconnellite and other Mc-names are listed with Mac-names.

Mechernichite. G. Kalb, not published, later replaced by bravoite [*M.A.*, **3**, 154.] Dealer's label (1925) in British Museum. H. Strunz, *Min. Tab.*, 1941, p. 256, P. Ramdohr, *Klockmann's Lehrb. Min.*, 13th edit., 1948, pp. 376, 656 (Mechernichit). Named from the locality, Mechernich, Eifel, Rhineland. Syn. of bravoite. [*M.A.*, **3**, 154, 339; *Min. Mag.*, **25**, 609.] 18th List

Medamaite. Z. Harada, *Journ. Fac. Sci. Hokkaido Univ.*, ser. 4, 1948, vol. 7, p. 153 (Medama-isi) from [*Journ. Japanese Ceramic Assoc.*, 1943, vol. 51, p. 381]. A variety of diaspore from Mutuisi, Wake Co., Okayama prefecture. Japanese isi, ishi = stone, -ite. 19th List

Medfordite. An unnecessary name for a variety of moss agate. R. Webster, *Gems*, 1962, p. 759. 28th List

Medmontite. F. V. Chukhrov and F. Y. Anosov, 1950. *Mém. Soc. Russe Min.*, vol. 79, p. 23 (Медмонтит). A clayey mineral allied to montmorillonite containing CuO 20·96%, from

Kazakhstan. Named from медь (copper) and montmorillonite. *See* Cupromontmorillonite. [*M.A.*, **11**, 124.] 19th List [A mixture; *A.M.*, **54**, 994]

Medziankite. Author?, quoted by C. Guillemin, *Bull. Soc. franç. Min. Crist.*, 1956, vol. 79, p. 29. Error for Miedziankite (10th List), a synonym of sandbergerite (of Breithaupt). 22nd List

Meixnerite. S. Koritnig and P. Süsse, 1975. *Tschermaks Min. Petr. Mitt.*, **22**, 79. Colourless tabular crystals in cracks of a serpentine rock near Ybbs-Persenberg, Lower Austria, have *a* 3·046, *c* 22·93, space group $E\bar{3}m$. Composition $Mg_6Al_2(OH)_{18}.4H_2O$, a CO_2-free member of the hydrotalcite family. Named for Dr. H. Meixner. [*A.M.*, **61**, 176.] 29th List

Melanochalcite. G. A. Koenig, *Amer. Journ. Sci.*, 1902, ser. 4, vol. xiv, p. 404. A pitch-black, amorphous mineral occurring as thin bands in nodules composed of cuprite, malachite, chrysocolla, and quartz at Calumet, near Bisbee, Arizona. A basic silico-carbonate of copper, $Cu_2(Si,C)O_4.Cu(OH)_2$. Named from *μέλας*, *μέλανος*, black, and *χαλκός*, copper. 3rd List [A mixture; *M.A.*, **1**, 263]

Melanostibian. L. J. Igelström, *Geol. För. Förh.*, XIV. 583, 1892; and *Zeits. Kryst. Min.*, XXI. 246, 1893. Antimonite of Mn and Fe. Sjö mine, Sweden. 1st List

Melanostibite. P. B. Moore. 1967. *Arkiv Min. Geol.*, **4**, 449. An unnecessary renaming of melanostibian. [*Zap.*, **99**, 75; *A.M.*, **53**, 1104.] 26th List

Melanovanadite. W. Lindgren, 1921. *Proc. Nat. Acad. Sci. USA*, vol. 7, p. 249; *Amer. Journ. Sci.*, 1922, ser. 5, vol. 3, p. 195. Calcium and vanadium vanadate, $2CaO.2V_2O_4.3V_2O_5$ (hydrated?), occurring as black, monoclinic needles on shale at Minasragra, Peru. Named from *μέλας*, *μέλανος*, black, and vanadium. [*M.A.*, **1**, 250, 376.] 9th List

Melite. F. Zambonini, *Zeits. Kryst. Min.*, 1899, XXXII, 161. Hydrated silicate of Al,Fe. Massive. [Saalfeld] Thuringia. 2nd List

Melkovite. B. L. Egorov, A. D. Dara, and V. M. Senderova, 1968. Зап. всесоюз. мин. общ. (*Mem. All-Union. Min. Soc.*), **98**, 207 (Мелковит). Minutely crystalline films and veinlets in the oxidation zone of a molybdenite deposit in the Shunak Mts. Kazakhstan, have a composition near $CaFeH_6(MoO_4)_4PO_4.6H_2O$. Named for Vracheslav Gavrilovich Melkov. [*A.M.*, **55**, 320; *M.A.*, 70-1648; *Bull.*, **92**, 516; *Zap.*, **99**, 80.] (Cf. the molybate-arsenate betpakdalite, 22nd List.) Not to be confused with Melnikovite. 26th List

Melkowit, Germ. trans. of Мелковит, melkovite (26th List). Hintze, 58. 28th List

Mellahite. E. Niccoli, 1925. *Giornale di Chimica Industriale ed Applicata*, Milano, vol. 7, p. 189; 1926, vol. 8, pp. 309, 604. The mixed salts (containing $MgSO_4$ 32, $MgCl_2$ 3, NaCl 19, KCl 20%) deposited from sea-water in the large natural salt-pans Mellahet Brigá and Mellahet Bu-Numa near Bu-Kammash on the coast of Tripolitania. The name is given in connection with a process for the extraction of potash from the mixed salts. From the Arabic, ملاحة mallāhat = saltpan, salina. 11th List

Mellonite. L. Palmieri, 1873. *Rend. Accad. Sci. Fis. Mat. Napoli*, Anno XII, p. 94. F. Zambonini, *Mineralogia Vesuviana. Mem. Accad. Sci. Fis. Mat. Napoli*, 1910, ser. 2, vol. xiv, no. 7, p. 50. Impure pseudocotunnite, or a mixture of chlorides and sulphates of sodium, potassium, copper, and lead. Named after Professor Macedonio Melloni (1798–1854) of Naples. 6th List

Mellorite. W. Hugill, 1939. [*Bull. British Refractories Research Association*, no. 49, p. 15]; *Journ. Iron and Steel Inst.*, 1941, vol. 144 (1941, no. 2), p. 257 P; *Trans. Brit. Ceramic Soc.*, 1942, vol. 41, p. 50. Silicate of ferric iron, calcium, etc., approaching garnet in composition, but with optical properties similar to those of an orthorhombic pyroxene. Formed by the action of basic slag on silica brick in a steel furnace. Named after Dr. Joseph William Mellor (1869–1938). [*M.A.*, **8**, 279.] 16th List

Melnikovite. B. Doss, 1911. *Annuaire Géol. Min. Russie*, vol. xiii, p. 130 (мельниковитъ), p. 139 (Melnikowit); *Neues Jahrb. Min.*, 1912, Beil.-Bd. xxxiii, p. 683; *Zeits. prakt. Geol.*, 1912, vol. xx, p. 453. A labile form of iron disulphide (FeS_2) occurring as minute, black, magnetic grains in Miocene clay on the estates of the Brothers Melnikov in govt. Samara, Russia. It is more readily attacked by chemical reagents than is iron pyrites, and it is regarded as having been derived from a colloidal form of iron sulphide. 6th List [Now named Greigite (q.v.)]

Melnikovite-marcasite. H. Rechenberg, 1950. *Neues Jahrb. Min.*, Monatshefte, 1950, p. 141

(Melnikovit-Markasit). Colloform marcasite containing some water and crystallized from a gel. Analogous to melnikovite-pyrite (14th List). [*M.A.*, **11**, 353.] 19th List

Melnikovite-pyrite. H. Schneiderhöhn and P. Ramdohr, 1931. *Lehrbuch der Erzmikroskopie*, Berlin, vol. 2, p. 170 (Melnikovit-Pyrit), p. 173 (Melnikovtpyrit). Shelly concentric mixtures of pyrite and marcasite crystallized from a gel, which is assumed to be identical with melnikovite (6th List). 14th List

Melonjosephite. A.-M. Fransolet, 1973. *Bull.*, **96**, 135. Dark-green fibrous masses in the pegmatite of Angarf-Sud, Zenaga, Anti-Atlas, Morocco, are orthorhombic, *a* 9·548, *b* 10·847, *c* 6·380 Å; composition $4[CaFe^{2+}Fe^{3+}(PO_4)_2OH]$, with some Mg. Named for Joseph Mélon. 28th List [*A.M.*, **60**, 946]

Melosark. [A. Breithaupt, 1841, MS.] R. Koechlin, *Min. Taschenbuch Wiener Min. Gesell.*, 2nd edit., 1928, p. 41; *Centr. Min.*, Abt. A, 1933, p. 203 (Melosark). Synonym of melopsite. Named from *μῆλον*, apple, and *σάρξ*, flesh (*ὄψον*, meat). [*M.A.*, **5**, 296.] 13th List

Mendeleevite. Standard English transliteration of Менделеевит, replacing Mendeleyevite [and Mendelyeevite]. 23rd List

Mendelyeevite. V. I. Vernadsky, 1914. *Bull. Acad. Sci. St.-Pétersbourg*, ser. 6, vol. 8 (part 2), pp. 1368, 1366 (менделѣевитъ). Also mentioned by Vernadsky in P. G. Mezernitsky *Physical Therapeutics*, vol. 3, p. 177, foot-note 1, and coloured plate at p. 193, Petrograd (published by the journal *Practical Medicine*), 1915. A calcium urano-titano-niobate (U_3O_8 25, CaO over 15%) referred to the betafite group and occurring as black rhombic-dodecaheda in pegmatite near Slyudianka, Lake Baikal. Named after the Russian chemist Dmitri Ivanovicn Mendelyeev (1834–1907). 9th List [Syn. of Betafite; *A.M.*, **62**, 406]

Mendiffit. J. Fröbel, 1843. *Grundzüge eines Systems der Krystallologie*, 32. Composition given as $PbCl_2 + 2\dot{P}b$, i.e. $PbCl_2.2PbO$. *Syn.* of mendipite. 28th List

Mendocita. G. Gagarin and J. R. Cuomo, 1949, loc. cit., p. 15. Variant of mendozite. Named after the province of Mendoza, but in Spanish *z* before *e* or *i* is pronounced as *c*. [*M.A.*, **11**, 129.] 19th List

Mercallite. G. Carobbi, 1935. *Rend. R. Accad. Lincei, Cl. Sci. fis. mat. nat. Roma*, ser. 6, vol. 21, sem. 1, p. 385 (mercallite). Potassium hydrogen sulphate, $KHSO_4$, as minute orthorhombic crystals in a saline efflorescence from Vesuvius. Named after Prof. Giuseppe Mercalli (1850–1914), a former Director of the Vesuvian Observatory. [Misenite, described by A. Scacchi in 1849 as $KHSO_4$, has been proved to be monoclinic $K_2SO_4.6KHSO_4$.] [*M.A.*, **6**, 148.] 14th List

Mercurammonite. W. F. Hillebrand and W. T. Schaller, 1909. *Bull. United States Geol. Survey*, no. 405, p. 18; *Zeits. Kryst. Min.*, 1910, vol. xlvii, p. 444 (Merkurammonit). A name suggested by the chemical composition of the mineral previously called kleinite (A. Sachs, 1905; 4th List), but regarded by the authors as a synonym. 5th List

Mercurarsite, Меркурарсит, *syn.* of aktashite (26th List). *Zap.*, **102**, 440, 457 (1973) apparently a provisional name, referring to the composition. 28th List

Merenskyite. G. A. Kingston, 1966. *Min. Mag.*, vol. 35, p. 815. Small grains, often intergrown with Kotulskite, in platinum ore from the Rustenburg mine, Pretoria, Transvaal, is hexagonal, composition $(Pd,Pt)(Te,Bi)_2$, and isostructural with Moncheïte. Named for H. Merensky. [*M.A.*, **17**, 697; *A.M.*, **52**, 526.] 24th List

Merlinoite. E. Passaglia, D. Pongiluppi, and R. Rinaldi, 1977. *Neues Jahrb. Mineral.*, Monatsh. 355. Fibrous aggregates from Cupaello, Santa Rufina, Rieti, Italy, are orthorhombic, *a* 14·116, *b* 14·229, *c* 9·946 Å, space group *Immm*, *ρ* 2·14 g.cm^{-3}, *n* 1·494, elongation +. Composition near $(K,Na)_5(Ca,Ba)_2Al_9Si_{23}O_{64}.23H_2O$. Named for S. Merlino. [*M.A.*, 78-891; *A.M.*, **63**, 598.] 30th List

Merrihueite. R. T. Dodd, Jr., W. R. Van Schmus, and U. B. Marvin, 1965. *Science*, vol. 149, p. 972. Rare minute inclusions in the Mezö-Madaras meteorite prove to be a member of the Osmulite group: $(K,Na)_2(Fe,Mg)_5Si_{12}O_{30}$, with K > Na and Fe > Mg; hexagonal. Named for C. M. Merrihue. [*A.M.*, **50**, 2096.] 24th List

Merrillite. E. T. Wherry, 1917. *American Mineralogist,* vol. ii, p. 119. A calcium phosphate belonging to the apatite group occurring in meteoric stones and identified by G. P. Merrill (*Amer. Journ. Sci.*, 1917, ser. 4, vol. xliii, p. 322) as francolite, from which, however, it differs somewhat in its optical characters. Named after Dr. George Perkins Merrill, of the United States National Museum, Washington. 8th List [Cf. Whitlockite; *Min. Record* **2**, 277]

Mertieite. G. A. Desborough, J. J. Finney, and B. F. Leonard, 1973. *A.M.*, **58**, I. Brassy-coloured grains from Goodnews Bay, Alaska, are pseudohexagonal, a_h 15·04, c_h 22·41 Å; composition $Pd_5(Sb,As)_2$. Named for J. B. Mertie, *Jr.* [*M.A.*, 73-2946, *Zap.*, **103**, 353.] 28th List [Cf. Isomertieite]

Merumite. S. Bracewell, 1946. Geology and mineral resources, sect. 4 in *Handbook of natural resources of British Guiana*, Georgetown, 1946, p. 36. Hydrous chromic oxide (Cr_2O_3 81·30, Al_2O_3 6·55, H_2O 8·18%, etc.), from the Merume river, Mazaruni district. Named from the locality. [*M.A.*, **10**, 292.] 18th List [A mixture containing the new species Bracewellite, Grimaldite, Guyanaite, and Mcconnellite (qq.v)]

Merwinite. E. S. Larsen and W. F. Foshag, 1921. *Amer. Min.*, vol. **6**, p. 143. Calcium-magnesium orthosilicate, $Ca_3Mg(SiO_4)_2$, forming with gehlenite, etc., a granular aggregate in contact-metamorphic limestone at Crestmore, California. The optical characters show it to be monoclinic, thus distinguishing it from monticellite. Named after Dr. Herbert E. Merwin, of the Geophysical Laboratory, Washington. [*M.A.*, **1**, 254.] 9th List [Ireland; *A.M.*, **59**, 1117]

Mesabite. H. V. Winchell, *Trans. Amer. Inst. Mining Engineers*, XXI. 660, 1893. The ochreous göthite abundant at Mesabi, Minnesota. 1st List

Mesodialyte. A. E. Fersman, 1922. *Compt. Rend. Acad. Sci. Russie*, p. 59; *Trans. Northern Sci. Econ. Exped.*, no. 16 (*Sci. Techn. Dept. Supreme Council of National Economy*, no. 8), *Moscow and Petrograd*, 1923, pp. 16, 69 (мезодиалит). An optically-isotropic, intermediate member of the eudialyte-eucolite series. [*M.A.*, **2**, 262, 264.] 10th List

Mesoenstatite. E. Thilo and G. Rogge, 1939. *Ber. Deutsch. Chem. Gesell.*, Abt. B, vol. 72, p. 352 (Mesoenstatit). E. Thilo, *Forschungen und Fortschritte*, 1939, vol. 15, p. 171. An enantiotropic modification of $MgSiO_3$ stable between 900° and 1270°C. Enstatite $\rightleftharpoons$ mesoenstatite (900°) $\rightleftharpoons$ clinoenstatite (1270°). [*M.A.*, **7**, 410.] 15th List [Identical with Protoenstatite (18th List) = Metatalc (15th List)]

Mesolitine. J. Apjohn, *Journ. Geol. Soc. Dublin*, 1844, vol. iii, p. 77. Identical with thomsonite. Named mesolitine because of a supposed relation to mesolite. 3rd List

Mesomicrocline. Author? H. Strunz, *Min. Tab.*, 3rd edit., 1957, p. 334 (Mesomikroklin). Triclinic pseudo-monoclinic, $K(Al,Si)_2Si_2O_8$. Greek *μεσος*, middle. 21st List

Mesquitelite. A. Mário de Jesus, [1936]. *Com. Serv. Geol. Portugal*, vol. 19 (for 1933), p. 136 (Mesquitelite). A clayey alteration product of felspar with a formula $(Mg,Ca)O.2Al_2O_3.9SiO_2.5H_2O$, slightly different from that of montmorillonite. Named from the locality, Mesquitela, near Mangualde, Portugal. [*M.A.*, **6**, 441.] 14th List

Meta-allanite. S. C. Robinson, 1955. *Bull. Geol. Surv. Canada,* no. 31, p. 70. Undescribed; presumably a term for metamict Allanite. 23rd List

Meta-aluminite. C. Frondel. 1968. *Amer. Min.*, **53**, 717. White, microcrystalline, monoclinic, with basaluminite and gypsum in sandstone at the Fuemrole mine, Temple Mountain, Emery County, Utah; $Al_2(SO_4)(OH)_4.5H_2O$; formed when aluminite is heated in air to 55°C. [*M.A.*, 69-619; *Bull.*, **92**, 317; *Zap.*, **99**, 78.] 26th List

Meta-alunogen. S. G. Gordon, 1942. *Notulae Naturae, Acad. Nat. Sci. Philadelphia*, no. 101, p. 6. Partly dehydrated alunogen (triclinic, $Al_2O_3.3SO_3.16H_2O$) yielding monoclinic $Al_2O_3.3SO_3.13\frac{1}{2}H_2O$. [*M.A.*, **8**, 278; *A.M.*, **28**, 61.] 16th List

Meta-ankoleïte. [M.V.]. Gallagher and [D.]. Atkin, 1963. Cited, without reference, by H. Strunz, *Min. Tabellen*, 4th edit., 1966, p. 312 (Meta-Ankoleit). $KUO_2PO_4.3H_2O$; synonym of Potassium autunite. 24th List

Meta-arsenuranocircite, *see* Arsenuranocircite. 22nd List

Metaautunite. P. Gaubert, 1904. *Bull. Soc. franç, Min.*, xxvii, p. 224 (Métaautunite). The same as Metakalkuranit (F. Rinne, 1901; 3rd List, *Min. Mag.*, xiii, p. 371). 4th List [Preferably hyphenated]

Metabasaluminite. F. A. Bannister, 1950. *Min. Mag.*, vol. 29, p. 9. $Al_4SO_4(OH)_{10}$ = $2Al_2O_3.SO_3.5H_2O$, produced by heating hydrobasaluminite (18th List) at 150°C. 19th List [Cf. Meta-aluminite]

Meta-bassetite. H. Strunz, 1957. *Min. Tabellen*, 3rd edit., p. 253 (Meta-Bassetit). The lower hydrate of $Fe(UO_2)_2(PO_4)_2$, with $8H_2O$ or less. The name appears to be attributed by Strunz (loc. cit., p. 254) to C. Frondel (*Min. Mag.*, 1954, vol. 30, p. 343), but was not used by Frondel. 22nd List

Metabayleyite. T. W. Stern and A. D. Weeks, 1952. *Amer. Min.*, vol. 37, p. 1060. Artificially dehydrated bayleyite (18th List). [*M.A.*, **12**, 281.] 20th List

Meta-bentonite. C. S. Ross, 1928. *Bull. Amer. Assoc. Petroleum Geol.*, vol. 12, p. 164 (meta-bentonites). V. T. Allen, *Journ. Geol. Chicago*, 1932, vol. 40, p. 259. Metamorphosed bentonite as altered volcanic tuffs in Palaeozic rocks. 14th List

Metaberyllite. E. I. Semenov, 1972. [Мин. Ловозерск. щелочн. масс., Изд. Наука]; abstr. *Zap.*, **102**, 75 (Метобериллит). $Be_3SiO_5.2H_2O$, from the Lovozero massif; a lower hydrate of beryllite (20th List). 28th List

Metabiotite. F. Rinne, 1921. *Die Kristalle als Vorbilder des feinbaulichen Wesens der Materie*, Berlin, p. 79 (Metabiotit); English translation, London, 1924, p. 152. Synonym of bauerite (F. Rinne, 1911; 6th List). 10th List

Metaboracite. E. S. Fedorov, 1892. *Zeits. Kryst. Min.*, vol. xx, p. 74. The names metaboracite, metaleucite, and metaperovskite (Metaboracit, Metaleucit, Metaperowskit) are applied to those dimorphous forms possessing lower* symmetry met with at ordinary temperatures, whilst the names boracite, leucite, and perovskite respectively, are restricted to the cubic, optically isotropic modifications of the same substances which are stable only above certain temperatures. [*Corr. in *M.A.*, **15**.] 4th List

Metaborite. V. V. Lobanova and N. P. Avrova, 1964. Зап. Всесоюз. Мин. Общ. (*Mem. All-Union Min. Soc.*), vol. 93, p. 329 (Метаборит, metaborite). White crusts were found to be the cubic modification of metaboric acid, HBO_2; named from the composition. [*A.M.*, **50**, 261.] 23rd List

Metabrucite. F. Rinne, 1913. *Fortschritte Min. Krist. Petr.*, vol. iii, p. 160 (Metabrucit); O. Westphal, *Inaug.-Diss.*, Leipzig, 1913. Brucite, $Mg(OH)_2$, artificially dehydrated, but retaining its original crystalline structure. 7th List

Metacalciouranoite. V. P. Rogova, L. N. Belova, G. P. Kiziyarov, and N. N. Kusnetsova, 1973. *Zap.*, **102**, 75 (Метакальцураноит, metacaltsuranoite). Fine-grained orange aggregates replacing pitchblende at an unnamed Russian locality have a composition $(Ca,Ba,Na_2)O.2{\cdot}1UO_3.1{\cdot}7H_2O$. Named from its composition. [*M.A.*, 74-505; *Bull.*, **96**, 239; *A.M.*, **58**, 1111; *Zap*, **103**, 358.] 28th List

Metacalciowardite. H. G. Clinton, 1929. *Amer. Min.*, vol. 14, pp. 435, 436. An undescribed mineral, presumably a calciferous variety of wardite (1st List) occurring with other aluminium phosphates at Manhattan, Nevada. [*M.A.*, **4**, 296.] 13th List

Metacaltsuranoite, *see* Metacalciouranoite.

Metaceinerite, error for Metazeunerite. (L. N. Belova, *Proc. 2nd UN Conf. Internat. Uses Atomic Energy*, 1958, vol. 2, p. 294; *M.A.*, **14**, 344.) 22nd List

Metachabazite. F. Rinne, 1921. *Die Kristalle als Vorbilder des feinbaulichen Wesens der Materie*, Berlin, p. 81 (Meta-Chabasit); 2nd edit., 1922, p. 166 (Metachabasit); English translation, London, 1924, p. 153 (metachabasite). Artificially dehydrated chabazite showing a change in the optical characters. 10th List

Metachalcolite. *Bull. Soc. franç. Min.*, 1902, vol. xxv, p. 372 (abstr. of Rinne's paper). Synonym of metakupferuranit (q.v.). 3rd List [Now **Metatorbernite**]

Metachalcophyllite. P. Gaubert, 1904. *Bull. Soc. franç. Min.*, vol. xxvii, p. 224 (Métachalcophyllite). Following F. Rinne (1901), this name is given to chalcophyllite which has been partly dehydrated by heating artificially until only the water of constitution remains; the loss of water is accompanied by changes in the optical characters. 4th List

Meta-chamosite. H. Jung, 1932. *Chem. Erde*, vol. 7, p. 598 (Meta-Chamosit). Artificially dehydrated chamosite. [*M.A.*, **5**, 284.] 13th List

Metacinnabar. W. H. Cropp, 1923. *Proc. Austr. Inst. Min. Met.*, new ser., no 52, p. 259. Variant of Metacinnabarite. Another recent and unnecessary change in this name is Metazinnabarit. (q.v.). [*M.A.*, **2**, 491.] 10th List

Metacristobalite. A. Lacroix, 1909. *Minéralogie de la France*, vol. iii, p. 806 (métacristobalite). The optically isotropic phase of cristobalite stable above 175°C. 5th List

Metadelrioïte. M. L. Smith, 1970. *Amer. Min.*, **55**, 185. Topotype delrioïte (22nd List) is shown to be a mixture of two phases. Delrioïte, the more hydrous phase, $8[CaSrV_2O_6(OH)_2.3H_2O]$, is monoclinic, *a* 17·170, *b* 7·081, *c* 14·644 Å, *β* 102° 29′, *Ia* or *I2/a*; metadelroïte, $2[CaSrV_2O_6(OH)_2]$, is anorthic, *a* 7·343, *b* 8·382, *c* 5·117 Å, *α* 119° 39′, *β* 90° 16′, *γ* 102° 49′. The fibrous intergrowths have *a* of delrioïte parallel to *c* of metadelrioïte. [*M.A.*, 70-2573; *Zap.*, **100**, 623.] 27th List

Metadesmine. F. Rinne, *Neues Jahrb. Min.*, I, 57, 1897. [*M.M.*, **11**, 343.] Given to the substances of definite chemical composition and optical properties produced when water is artificially expelled from stilbite (= desmine). 1st List

Metagreenalite. F. Jolliffe, 1935. *Amer. Min.*, vol. 20, pp. 406, 411. The crystalline equivalent of the amorphous greenalite (4th List), with the composition $3FeO.4SiO_2.2H_2O$, occurring as green granules in the iron ores of Minnesota. J. W. Gruner, *Amer. Min.*, 1936, vol. 21, p. 449, considers this to be coarser grained greenalite with the composition $9FeO.Fe_2O_3.8SiO_2.8H_2O$. [*M.A.*, **6**, 152, 480; *Min. Mag.*, **24**, 434.] 14th List

Meta-haiweeite. T. C. McBurney and J. Murdoch, 1959. *Amer. Min.*, vol. 44, p. 839. K. Walenta, *Neues Jahrb. Min.*, Monatsh., 1960, p. 37. Artificial, produced by dehydration of Haiweeite (q.v.) at 300°C or higher [*M.A.*, **14**, 415]. Walenta considers that the name would more appropriately have been given to the partially hydrated phase formed by dehydration of haiweeite at 100°C, which occurs naturally and was described by Haberlandt and Schiener (1951) as Gastunite 1 (*see* Gastunite). 22nd List

Metahalloysite. M. Mehmel, 1935. *Zeits. Krist.*, 1935, vol. 90, p. 35; *Chemie der Erde*, 1937, vol. 11, p. 9 (Metahalloysit). Halloysite ($H_4Al_2Si_2O_9 + 2H_2O$) when partially dehydrated at 50°C loses $2H_2O$, then having the kaolin formula $H_4Al_2Si_2O_9$ but with a distinct crystal-structure. [*M.A.*, **6**, 181.] 14th List [*A.M.*, **40**, 1110]

Metaheinrichite. E. B. Gross, A. S. Corey, R. S. Mitchell, and K. Walenta, 1958. *Amer. Min.*, vol. 43, p. 1134. K. Walenta, *Jahresheft geol. Landesamt Baden-Württemberg*, 1958, vol. 3, p. 23. The arsenic analogue of meta-uranocircite, $Ba(UO_2)_2(PO_4)_2.8H_2O$, occurs along with the fully hydrated mineral near Lakeview, Oregon, and in the Schwarzwald, Germany. Named after Prof. E. William Heinrich. [*M.A.*, **14**, 199.] L. N. Belova, 1958, gave the name Arsenuranocircite to this lower hydrate; M. Fleischer, *Amer. Min.*, 1959, vol. 44, p. 466, points out that Meta-arsenuranocircite would be more appropriate. It is not clear which name has priority. 22nd List

Metaheulandite. F. Rinne, *Neues Jahrb. Min.*, 1899, I, 13, 30. Heulandite from which part of the water has been artificially expelled, causing a variation in the optical characters (Abstr. *Min. Mag.*, XII, 312). 2nd List

Metahewettite. W. F. Hillebrand, H. E. Merwin, and F. E. Wright, 1914. *Proc. Amer. Phil. Soc.*, vol. liii, p. 33; *Zeits. Kryst. Min.*, vol. liv, p. 209. An orthorhombic, hydrous vanadate of calcium, $CaO.3V_2O_5.9H_2O$, occurring as a dark-red powdery impregnation in sandstone in Colorado and Utah. It differs slightly from hewettite (q.v.) in its behaviour during dehydration. 7th List [Much 'metahewettite' is Barnesite (of Ross). (q.v.)]

Metahohmannite. M. C. Bandy, 1938. *Amer. Min.*, vol. 23, p. 748. Hydrated basic ferric sulphate, $Fe_2(SO_4)_2(OH)_2.3H_2O$, as an orange-coloured powder from the partial dehydration of hohmannite, from Chile. [*M.A.*, **7**, 223.] 15th List

Metahydroboracite. N. Y. Ikornikova and M. N. Godlevsky, 1941. *Compt. Rend.* (*Doklady*) *Acad. Sci. URSS*, vol. 33, p. 257 (metahydroboracite). Hydrous calcium and magnesium borate, $CaMgB_6O_{11}.11H_2O$, like hydroboracite but with more water. Synonym of inderborite (q.v.). [*M.A.*, **8**, 341.] 16th List

Meta-jarlite. R. Bøgvad, 1933. *Meddelelser om Grønland*, vol. 92, no. 8, p. 7 (Meta-Jarlite). A fluoride, $NaSr_3Al_3F_{16}$, as small grey crystals enclosed in chiolite, from the Greenland cryolite quarry. Supposed to differ from jarlite (q.v.) in its optical characters. [*M.A.*, **5**, 388.] 13th List [Is Jarlite; *A.M.*, **34**, 386]

Meta-jennite. A. B. Carpenter, R. A. Chalmers, J. A. Gard, K. Speakman, and H. F. W. Taylor, 1966. *Amer. Min.*, vol. 51, p. 56. An *artificial* product of partial dehydration of Jennite; monoclinic $Na_2Ca_8Si_5O_{26}H_{14}$. 24th List

Metakahlerite. K. Walenta, 1958. *Jahresheft geol. Landesamt Baden-Württemberg*, vol. 3, p. 17 (Meta-Kahlerit). $Fe(UO_2)_2(AsO_4)_2.8H_2O$, the lower hydrate of Kahlerite (20th List), found in the Sophia shaft, Baden. [*A.M.*, **45**, 254.] 22nd List

Metakalkuranit. F. Rinne, *Centralblatt Min.*, 1901, p. 709 (Metakalkuranite, *pl.*). Calco-uranite (autunite) from which part of the water has been artificially expelled by heating, thus causing an important series of changes in the optical characters. 3rd List [Is **Meta-autunite**]

Metakalzuranoit, Germ. trans. of Метакальцураноит, metacalciouranoite (q.v.). Hintze, 59. 28th List

Metakamacite. E. A. Owen, 1940. *Phil. Mag.*, ser. 7, vol. 29, p. 561. Nickel-iron of the metastable α_2-form occurring as the granular plessite in meteorites, as distinct from the plessite consisting of fine crystals of kamacite and taenite. [*M.A.*, **7**, 538.] 15th List

Metakaolin, Metanacrite. F. Rinne, 1925. *Zeits Krist.*, vol. 61, p. 119 (Metakaolin, Metanakrit). Artificially dehydrated kaolin and nacrite. Cf. Anhydro-. [*M.A.*, **2**, 505.] 10th List

Metakaolinite. Chang-Ling Liu, Te-Yeh Liu, Fuchang, Chin-Cheng Li, Mo-Chun Sun, and Wen-Han Lu, 1963. [*K'o Hsueh T'ung Pao*, no. 10, p. 59]; abstr. in *Amer. Min.*, 1964, vol. 49, p. 1777. The name Metakaolinite, previously used (Johns, 1953, *Min. Mag.*, vol. 30, p. 186) as a synonym of Metakaolin (Rinne, 1925) for an artificial dehydration product of Kaolinite, is now applied to an inadequately described mineral containing somewhat less water than Kaolinite. 24th List

Metakernite. H. Menzel, H. Schulz, and H. Deckert, 1935. *Die Naturwissenschaften*, vol. 23, p. 832 (Metakernit). The amorphous dihydrate, $Na_2B_4O_7.2H_2O$, produced artificially by the partial dehydration of kernite ($Na_2B_4O_7.4H_2O$). 14th List

Meta-kingite. K. Norrish, L. E. R. Rogers, and R. E. Shapter, 1957. *Min. Mag.*, vol. 31, p. 351. An artificially, partially dehydrated form, $Al_2O_3.Al(OH)_3.P_2O_5.4H_2O$, of kingite (q.v.). 21st List

Meta-kirchheimerite. K. Walenta, 1958. *Jahresheft geol. Landesamt Baden-Württemberg*, vol. 3, p. 34 (Meta-Kirchheimerit). $Co(UO_2)_2(AsO_4)_2.8H_2O$, a member of the meta-torbenite group, from the Sophia shaft, Baden, on pitchblende. Named after Dr. F. Kirchheimer. 22nd List [*A.M.*, **44**, 466]

Metakoenenite. F. Rinne, *Centralblatt Min.*, 1902, p. 498 (Metakoenenit). Koenenite (q.v.) when heated in water leaves a residue having the composition $Al_2O_3.MgO.H_2O$; with ammonium chloride solution it yields $Al_2O_3.2H_2O$, and this on ignition gives anhydrous alumina. These artificially produced secondary products ('metakoenenites') retain the scaly form of the original mineral, and are optically uniaxial and negative. 3rd List

Metakupferuranit. F. Rinne, *Centralblatt Min.*, 1901, p. 618 (Metakupferuranite, *pl.*). Cupro-uranite (torbernite) from which part of the water has been artificially expelled by heating, thus causing an important series of changes in the optical characters. *See* metachalcolite. 3rd List [Is Metatorbernite]

Metalaumontite. H. Strunz, 1957. *Min. Tabellen*, 3rd edit., p. 342 (Metalaumontit). Syn. of Leonhardite. 22nd List

Metaleucite. E. S. Fedorov, 1892. *See* Metaboracite. 4th List

Metaliebigite. J. Kiss, 1966. [*Ann. Univ. Sci. Budapest Rolando Eötvös Naminat., Sec. Geol.*, vol. 9 (for 1965) p. 139]; abstr. *Amer. Min.*, 1968, vol. 53, p. 509. A premature 'provisional' name for a uranium mineral occurring with liebigite on pitchblende in the Mecsek Mts., Hungary; contains Ca, Mg, and U. 25th List

Meta-lodevite. H. Agrinier, F. Chantret, J. Geffroy, B. Hery, B. Bachet, and H. Vachey, 1972. *Bull.*, **95**, 360 (Méta-lodèvite). Tetragonal platy crystals from Lodève, Hérault, France, have $a = b = 7{\cdot}16$, c 17·20 Å, space-group $P4_2/m$, though the optical properties ($2V_\alpha$ 27 to 37°) suggest orthorhombic symmetry. Composition $Zn(UO_2)_2(AsO_4)_2.10H_2O$. Named for the locality. [*M.A.*, 73-1940; *Bull.*, **96**, 239; *A.M.*, **59**, 210; *Zap.*, **102**, 450.] 28th List

Metalomonosovite. E. I. Semenov, N. I. Organova, and M. V. Kukharchik, 1961. Кристаллография (Crystallography), vol. 6, p. 925 (Металомоносовит). A member of the Lomonosovite family, to which the formula $Na_2Ti_2Si_2O_9.NaPO_3$ was attributed; named from the supposed presence of a metaphosphate. The name was later changed to β-Lomonosovite (q.v.). *See also* Зап. Всесоюз. Мин. Общ. (*Mem. All-Union Min. Soc.*), 1963, vol. 92, p. 210, and *Amer. Min.*, 1963, vol. 48, p. 1413. [*M.A.*, **16**, 182, 645.] 23rd List

Metalomonossowit, variant transliteration of Металомоносовит, Metalomonosovite (23rd List) (H. Strunz, *Min. Tabellen*, 4th edit., 1966, p. 497). 24th List

Metaloparite. V. I. Gerasimovsky, 1941. *Compt. Rend.* (*Doklady*) *Acad. Sci. URSS*, vol. 33, p. 61 (metaloparite), p. 63 (hydroloparite). Hydrous titanoniobate of rare-earths pseudomorphous after loparite (10th List), from which it differs in containing water in place of alkalis. [*M.A.*, **8**, 341.] 16th List

Metamesolite. A. Cavinato, 1927. *Mem. R. Accad. Lincei, Cl. Sci. fis. mat. nat. Roma*, ser. 6, vol. 2, p. 339. Artificially dehydrated mesolite showing a change in the optical characters. Analogous to the metascolecite, etc., of F. Rinne. [*M.A.*, **4**, 321.] 12th List

Metamilarite. F. Rinne, 1927. *Centr. Min., Abt. A*, 1927, p. 1 (Metamilarit). Milarite dehydrated by artificial heating with an accompanying change in the optical characters. [*M.A.*, **3**, 451.] 11th List

Metamontmorillonite. H. Strunz, 1957. *Min. Tabellen*, 3rd edit., pp. 310, 325 (Metamontmorillonit). The product of dehydration of montmorillonite at 400°C., with *c* about 10 Å. 22nd List

Metamurmanite. E. I. Semenov, N. I. Organova, and M. V. Kukharchik, 1961. Кристаллография (Crystallography), vol. 6, p. 925 (Метамурманит). A superfluous name for a weathering product intermediate between Metalomonosovite (= β-Lomonosovite, q.v.) and Murmanite (Зап. Всесоюз. Мин. Общ. (*Mem. All-Union Min. Soc.*), 1963, vol. 92, p. 211; *Amer. Min.*, 1963, vol. 48, p. 1413; *M.A.*, **16**, 182, 645). 23rd List

Metanacrite, *see* Metakaolin.

Meta-Natriumautunit. Original form of Metanatro-autunite, (q.v.). 24th List

Meta-Natrium-uranospinit. The original form of Meta-sodium-uranospinite (q.v.). 24th List

Metanatro-autunite. K. Walenta, 1965. *Chemie der Erde*, vol. 24, p. 263 (Meta-Natrium-autunit). Natro-autunite (Sodium autunite, 9th List, internationalized Fersman and Shubnikova, 1937) probably belongs to the meta series, and if so should be named accordingly. 24th List

Metanatrolite. *See* Epinatrolite. 6th List

Meta-Na-Uranospinit. H. Strunz, 1966. *Min. Tabellen*, 4th edit., p. 313. Synonym of Meta-sodium-uranospinite (q.v.). 24th List

Metanhydrite. E. Sommerfeldt, 1907. *Neues Jahrb. Min.*, vol. i, p. 140 (Metanhydrit). A modification of calcium sulphate which in external form appears to be isomorphous with barytes, though in physical characters (sp. gr. and cleavage) it is identical with anhydrite. The crystals examined were prepared artificially: the same substance probably occurs as a volcanic product at Santorin. 4th List

Metanocerite. F. v. Sandberger, *Neues Jahrb. Min.* I. 221, 1892. A mineral resembling nocerite. Arendal, Norway. 1st List

Metanováčekite. C. Frondel, 1956. *Prof. Paper US Geol. Surv.*, no. 300, p. 578 (metanovacekite). $Mg(UO_2)_2(AsO_4)_2.8H_2O$, partly dehydrated form of nováčekite (19th List). H. Strunz, *Min. Tab.*, 3rd edit., 1957, p. 253 (Meta-Novačekit), $Mg(UO_2)_2(AsO_4)_2.4H_2O$, tetragonal. 21st List

Meta-otenite, error of transliteration for Meta-autunite (Доклады Акад. Наук СССР [*Compt. Rend. Acad. Sci. URSS*], 1960, vol. 132, p. 493). 22nd List

Metaparisite. F. Rinne, 1921. *Zeits. Metallkunde*, Berlin, vol. 13, p. 405 (Metaparisit). Parisite from which carbon dioxide has been artificially expelled by heat without the destruction of the crystalline structure. 9th List

Metaperovskite. E. S. Fedorov, 1892. *See* Metaboracite. 4th List

Metaquartz. I. D. Sedletzky, 1940. *Compt. Rend.* (*Doklady*) *Acad. Sci. URSS*, vol. 26, p. 241. An assumed intermediate stage in the passage of amorphous silica gel to chalcedony. [*M.A.*, **8**, 146.] 16th List

Metaranquilite. Author? *Intro. Min. Japan*, 39 (1970), Geol. Surv. Japan. A lower hydrate of ranquilite (22nd List) from the Togo mine, Tottori Prefecture, Japan. $(CaO)_{1.5}(UO_3)_2$-$(SiO_2)_5.10H_2O$; no further details. [*Zap.*, **101**, 286.] 28th List

Metarossite. W. F. Foshag and F. L. Hess, 1927. *Proc. United States Nat. Mus.*, vol. 72, art. 11, p. 1. Hydrated vanadate of calcium, $CaO.V_2O_5.2H_2O$, occurring in Colorado as a dehydration product of rossite (q.v.). [*M.A.*, **3**, 470; *A.M.*, **13**, 160.] 11th List

Meta-saléeite. M. E. Mrose, 1950. *Amer. Min.*, vol. 35, p. 529 (meta-saléeite). Further alteration of name, saléite (13th List) and saléeite (15th List), for the Katanga mineral $Mg(UO_2)_2(PO_4)_2.8H_2O$ corresponding to meta-autinite, the name saléeite being transferred to the fully hydrated mineral with $10H_2O$. [*M.A.*, **11**, 230.] 19th List

Metasandbergerite. K. Walenta, 1958. *Techn. Hochschule*, Stuttgart; abstr. *Bull. Soc. franç. Min. Crist.*, 1958, vol. 61, p. 68. $Ba(UO_2)_2(PO_4)_2.8H_2O$; later renamed metaheinrichite (q.v.), the name sandbergerite having twice been used before (1866 and 1883). 22nd List

Metasandbergit, error for Meta-Sandbergerit (= Metaheinrichite). Hintze, *Handb. Min.*, Erg.-bd. II, p. 765. 23rd List

Metasanidine. C. Oftedahl, 1949. *Skrifter Norske Vidensk.-Akad. Oslo, I. Mat.-Naturv. Kl.*, for 1948, no. 3, p. 58 (meta-sanidine), p. 68 (metasanidine). An alkali-felspar showing partial exsolution of sanidine to cryptoperthite. [*M.A.*, **11**, 11.] 19th List

Metascarbroite. W. J. Duffin and J. Goodyear, 1960. *Min. Mag.*, vol. 32, p. 358. An artificial product obtained by heating scarbroite $[Al_2(CO_3)_3.12Al(OH)_3]$ to between 130° and 230°C.; still contains carbonate. 22nd List

Metaschoderite. D. M. Hausen, 1960. *Bull. Geol. Soc. Amer.*, vol. 71, p. 1883. Al_2PO_4-$VO_4.3H_2O$, a dehydration product of Schoderite (q.v.). [*M.A.*, **15**, 44.] 22nd List [*A.M.*, **47**, 637; $6H_2O$, not 3. Cf. Rusakovite]

Metaschoepite. C. L. Christ and J. R. Clark, 1960. *Amer. Min.*, vol. 45, p. 1059. Schoepite (= Schoepite–I) is unstable in air, and dehydrates spontaneously to two lower hydrates, Metaschoepite (= Schoepite–II) and Paraschoepite (18th List; also called Schoepite–III). 22nd List

Metascolezite. F. Rinne, *Neues Jahrb. Min.* II. 60, 1894. A dehydrated form of scolecite. $CaAl_2Si_3O_{10} + 2H_2O$. Monosymmetric. 1st List

Metasideronatrite. M. C. Bandy, 1938. *Amer. Min.*, vol. 23, p. 733. Hydrated basic sulphate of sodium and ferric iron, $Na_4Fe_2(SO_4)_4(OH)_2.3H_2O$, as yellow, fibrous, orthorhombic crystals from Chile. A partly dehydrated form of sideronatrite. [*M.A.*, **7**, 223.] 15th List

Metasimpsonite. E. S. Simpson, 1938. *Rep. Dept. Mines Western Australia*, for 1937, p. 88 (Metasimpsonite). L. E. R. Taylor, *Journ. Roy. Soc. Western Australia*, 1939, vol. 25 (for 1938–1939), p. 93 (meta-simpsonite). An alteration product of simpsonite (q.v.), later identified with microlite. 15th List [*A.M.*, **62**, 407]

Metaskolezit, variant of Metascolecite. (H. Strunz, *Min. Tabellen*, 1st edit., 1941, p. 257). 24th List

Meta-sodium-uranospinite. K. Walenta, 1965. *Chemie der Erde*, vol. 24, p. 254 (Meta-Natrium-uranospinit). The mineral described as sodium uranospinite (19th List; *M.A.*, **14**, 53; **15**, 364) belongs to the meta series and is re-named accordingly. 24th List

Metastrengite. Dana's *Mineralogy*, 7th edit., 1951, vol. 2, p. 769. Monoclinic $FePO_4.2H_2O$ dimorphous with orthorhombic strengite; named to correspond with metavariscite and variscite $AlPO_4.2H_2O$. Synonym of phosphosiderite (Bruhns and Busz, 1890) and clinostrengite (q.v.). 19th List

Metataenite. J. D. Buddhue, 1936. *See* Orthotaenite. 14th List

Metatalc. E. Thilo, 1937. *Ber. Deutsch. Cham. Gesell.*, Abt. B, vol. 70, p. 2373 (Metatalk). A modification of magnesium metasilicate, $MgSiO_3$, with crystal-structure different from that of enstatite and clinoenstatite, obtained by dehydrating talc. [*M.A.*, **7**, 410.] 15th List [Identical with Protoenstatite]

Metathenardite. A. Lacroix, 1910. *Minèralogie de la France*, vol. iv, p. 32 (métathénardite). At temperatures above 200°C. thenardite (Na_2SO_4) becomes transformed into one or more different crystalline modifications; the latter on cooling may exist in a metastable state for some days at the ordinary temperature. Clear crystals of metathenardite collected from the hot fumaroles of Mt. Pelée, Martinique, gradually became cloudy and white, with a change in their optical characters, due to the transformation into an aggregate of differently orientated crystals of thenardite. 6th List

Metathomsonite. M. H. Hey, 1932. *Min. Mag*., vol. 23, pp. 93–97, 118, 120. Partially dehydrated thomsonite showing a change in phase; probably a high-temperature form identical with gonnardite. *See* Epithomsonite. 19th List

Meta-thuringite. H. Jung, 1932. *Chem. Erde*, vol. 7, p. 596 (Meta-Thuringit). Artificially dehydrated thuringite. [*M.A.*, **5**, 284.] 13th List

Meta-torbernite. A. F. Hallimond, 1916. *Mineralogical Magazine*, vol. xvii, p. 333. Synonym of Metakupferuranit (F. Rinne, 1901) and Metachalcolite (3rd List). 7th List [Now the accepted name]

Metratriplite. A. Mário de Jesus, [1936]. *Com. Serv. Geol. Portugal*, vol. 19 (for 1933), p. 146 (Metatriplite). A black alteration product of triplite with the composition $6MnO.3Fe_2O_3.3P_2O_5.2RF_2.4H_2O$. From Mangualde, Portugal. [*M.A.*, **6**, 442.] 14th List

Metatujamunit, variant of Metatyuyamunite (20th List) (H. Strunz, *Min. Tabellen*, 4th edit., 1966, p. 316). 24th List

Metatyuyamunite. A. D. Weeks, M. E. Thompson, and R. B. Thompson, 1953. *Progr. and Abstr. Min. Soc. Amer*., p. 43; *Amer. Min*., 1954, vol. 39, p. 348 (abstr.). A. D. Weeks and M. E. Thompson [*Bull. US Geol. Surv*., 1954, no. 1009–B, p. 37]; *Amer. Min*., 1954, vol. 39, p. 1037 (abstr.). A lower hydrate $5–7H_2O$ of tyuyamanunite $Ca(UO_2)(VO_4)_2.9H_2O$ (6th List). [*M.A.*, **12**, 511, 566.] 20th List [*A.M.*, **41**, 187]

Metauramphite. Z. A. Nekrasova, 1957. [Вопр. геол. уран., p. 67], cited by H. Strunz, *Min. Tabellen*, 4th edit., 1966, p. 312. The lower hydrate of Uramphite (22nd List). 24th List

Metauranocircite. P. Gaubert, 1904. *Bull. Soc. franç. Min*., vol. xxvii, p. 226 (Méta-uranocircite). *See* Metachalcophyllite. 4th List [Usually hyphenated]

Meta-uranopilite. C. Frondel, 1951. Dana's *Mineralogy*, 7th edit., vol. 2, p. 582. Partially dehydrated uranopilite, β-uranopilite of R. Nováček, 1935 [*M.A.*, **6**, 148]. 19th List [*A.M.*, **37**, 950]

Meta-Uranosandbergit. K. Walenta, quoted in Hintz, *Handb. Min*., Erg.-bd. II, pp. 765, 878. Synonym of Metaheinrichite. 23rd List

Meta-uranospinite. H. Strunz, 1957. *Min. Tabellen*, 3rd edit., p. 254 (Meta-Uranospinit). The lower hydrate of Uranospinite; occurs naturally at the Sophia shaft, Baden (K. Walenta, *Jahresheft geol. Landesamt Baden-Württemberg*, 1958, vol. 3, p. 30). [*A.M.*, **45**, 1054.] 22nd List

Metavandendriesscheite. C. L. Christ and J. R. Clark, 1960. *Amer. Min*., vol. 45, p. 1059. Vandendriesscheite (= Vandendriesscheite–I) is unstable in air, and dehydrates spontaneously to a lower hydrate, Metavandendriesscheite (= Vandendriesscheite–II). 22nd List

Meta-vanuralite. F. Cesbron, 1970. *Bull. Soc. franç, Min. Crist*., **93**, 242. A reversible dehydration product, $Al(UO_2)_2(VO_4)_2(OH).8H_2O$, of vanuralite (23rd List); anorthic; occurs with vanuralite at Mounana, Gabon. [*M.A.*, 70-3425; *A.M.*, **56**, 637.] 26th List

Metavariscite. E. S. Larsen and W. T. Schaller, 1925. *Amer. Min*., vol. 10, p. 23. Hydrated aluminium phosphate, $Al_2O_3.P_2O_5.4H_2O$, orthorhombic and dimorphous with variscite; the distinction depending apparently on the optical data as determined under the microscope. This is the 'crystallized variscite' from Lucin, Utah, described by W. T. Schaller in 1911, 1912, and 1916; his 'lucinite' (1914 and 1916; 7th List) is now referred to variscite. [*M.A.*, **2**, 421.] 10th List [Cf. Clinovariscite]

Metavauxite. S. G. Gordon, 1927. *Amer. Min*., vol. 12, p. 264. Hydrated phosphate of ferrous iron and aluminium, $FeO.Al_2O_3.P_2O_5.4H_2O$, as monoclinic crystals from Bolivia. Named after Mr. George Vaux, junior (1863–1927), of Bryn Mawr, Pennsylvania, *See* Vauxite and Paravauxite, 10th List. [*M.A.*, **3**, 370.] 11th List

Metavermiculite. H. Strunz, 1957. *Min. Tabellen*, 3rd edit., pp. 311, 325 (Metavermiculit). The product of dehydration of vermiculite at 400°C., with c about 10 Å. 22nd List

Metavivianite. C. Ritz, E. J. Essene, and D. R. Peacor, 1974. *A.M.*, **59**, 896. A polymorph of vivianite, isomorphous with symplesite, has a 7·81, b 9·08, c 4·65 Å, α 94·77°, β 97·15°,

γ 107·37°; it occurs in the Big Chief pegmatite near Glendale, South Dakota. Named for the relation to vivianite, [*M.A.*, 75-1393; *Zap.*, **104**, 611; *Bull.*, **98**, 265.] 29th List

Metavoltite. Variant of Metavoltine, *A.M.*, **9**, 62. [T. Egleston, 1892, and A. H. Chester, 1896, give the more correct form Metavoltaite; cf. Voltaite]. 10th List

Metazellerite. R. G. Coleman, D. R. Ross, and R. Meyrowitz, 1966. *Amer. Min.*, vol. 51, p. 1567. Zellerite (23rd List) is shown to be $CaUO_2(CO_3)_2.5H_2O$; it readily loses water to give metazellerite $CaUO_2(CO_3)_2$; $3H_2O$. Named for Howard D. Zeller. [Зап., **97**, 71; *Bull.*, **90**, 273.] 25th List

Metazeolites. R. Brauns, 1892. *Neues Jahrb. Min.*, vol. ii, p. 240 (Metazeolithe), in abstr. of F. Rinne, 1890, though not given in the original paper. The names metadesmine (1st List), metascolecite (1st List), metanatrolite (6th List), metaepistilbite, were given by F. Rinne, *Sitz.-ber. Akad. Wiss. Berlin*, 1890, pp. 1203–1204, for partially dehydrated zeolites showing changes in optical properties and in some cases of crystal system. 19th List

Metazeolith. H. Strunz, 1957. *Min. Tabellen*, 3rd edit., p. 325. A general name for partially dehydrated zeolites. 22nd List

Metazeunerite. A. E. Fersman and O. M. Shubnikova. *Geochemist's and mineralogist's companion*, 1937, p. 173 (Метацейнерит), $Cu(UO_2)_2(AsO_4)_2.8H_2O$; the same formula is given (p. 306) for zeunerite. J. W. Frondel, *Amer. Min.*, 1951, vol. 36, p. 255 (meta-zeunerite); artificial zeunerite contains $5–16H_2O$, and to be consistent with meta-autunite (4th List) and meta-torbernite (7th List) the natural mineral with $8H_2O$ might be called meta-zeunerite. Dana's *Mineralogy*, 7th edit., 1951, vol. 2, p. 993 (metazeunerite). [*M.A.*, **11**, 323, 432.] 19th List [*A.M.*, **42**, 222]

Metazinnabarit. P. Groth and K. Mieleitner, 1921. *Mineralogische Tabellen*, p. 21. Variant of Metacinnabarite. *See* Metacinnabar. 10th List

Meta-zircon. J. Lietz, 1937. *Zeits. Krist.*, vol. 98, p. 209 (Meta-Zircon). Zircon of medium density consisting of normal crystalline material with amorphous material. [*M.A.*, **7**, 131.] 15th List

Meyerhofferite. W. T. Schaller, 1914. *Journ. Washington Acad. Sci.*, vol. iv, p. 355; 3rd Appendix to 6th edit. of Dana's *System of Mineralogy*, 1915, p. 50. Hydrated calcium borate, $2CaO.3B_2O_3.7H_2O$, occurring as colourless, triclinic crystals or as a white, fibrous alteration product of Inyoite (q.v.) in the borate deposits of Mt. Blanco district, Inyo Co., California. Named after Wilhelm Meyerhoffer (1864–1906), of Berlin, who, in association with J. H. van't Hoff, prepared the mineral artificially. 7th List

Meyersite. C. Elschner, 1922. *Kolloid-Zeits.*, vol. 31, p. 95 (Meyersit). Hydrated aluminium phosphate containing a little more water than variscite, and forming banded, agate-like masses in lava on a guano island of the Hawaii group. Named after Mr. H. H. Meyers, of Pittsburgh, Pennsylvania. [*M.A.*, **2**, 11.] 10th List

Meyamacite. R. Pierrot and R. Van Tassel, 1965. *Bull.*, **88**, 613. The Meymacite of Carnot (1874) is shown to be Ferritungstite, and the name is reinstated for amorphous $WO_3.2H_2O$. [*M.A.*, **17**, 695; *A.M.*, **53**, 1065.] 28th List

Mg-berzeliit, *see* Magnesium-berezeliite.

Mg-blödite. M. Giglio, 1958. *Acta Cryst.*, vol. 11, p. 789. Synonym of Blödite (*M.A.*, **14**, 103). 23rd List

Mg-orthite, *see* Magnesium-orthite.

Mg-ursilite, variant of Magnesium-ursilite (22nd List). (I. Kostov, *Mineralogy*, 1968, p. 327.) 25th List

Mg-vermiculite, *see* Magnesium-vermiculite.

Mg-wollastonite. I. Shinno, 1970. *Journ. Jap. Assoc. Min. Petr. Econ. Geol.*, **63**, 146. An *artificial* product with the composition of diopside but the crystal-structure of wollastonite. [*M.A.*, 71-1002.] Cf. Magnesium-wollastonite, 18th List. 28th List

Micatite. A misleading trade-name for a phenolic resin plastic. R. Webster, *Gems*, 1962, p. 759. Nothing to do with micaultite (error for micaultlite). 28th List

Micaultlite. Comte de Limur, 1883. *Catalogue raisonné des minéraux du Morbihan*. Vannes, 1883, p. 38. An earthy, brick-red decomposition product of rutile, which is suggested may be an aluminous hydrorutile. Presumably named after Victor Micault, of Paris. 4th List

Micheewit. E. I. Nefedov, *Geologie, Zeits. Geol.*, East Berlin, 1955, vol. 4, p. 526. German spelling of mikheevite (20th List). Named after Viktor Ivanovich Mikheev, Виктор Иванович Михеев (1912–1956), X-ray crystallographer. [*M.A.*, **13**, 85.] 21st List

Michejewit, German transliteration of Михеевит, mikheevite (Hintze. *Handb. Min.*, Erg.-Bd. II, p. 588). 22nd List

Michelottin. J. Fröbel, 1843. *See* chrysargyrit. 27th List

Michenerite. J. E. Hawley and L. G. Berry, 1958. *Canad. Min.*, vol. 6, p. 200. A palladium bismuthide, cubic with pyrite structure, probably $PdBi_2$, occurring in the ores of the Frood mine, Sudbury, Ontario. Named after C. E. Michener, who discovered the mineral in 1940. [*M.A.*, **14**, 343; *A.M.*, **44**, 207.] 22nd List [*A.M.*, **48**, 1184. Cf. Testibiopalladinite]

Microantigorite. W. N. Lodochnikov, 1933. *Problems of Soviet Geology*, vol. 2, p. 121 (микроантигорит). p. 145 (microantigorite). A minutely crystalline antigorite. 13th List

Microantiperthite. A. F. Buddington, J. [J.] Fahey, and A. [C.] Vlisidis, 1955. *Amer. Journ. Sci.*, vol. 253, p. 503. A fine intergrowth of potash-felspar and plagioclase with the former predominating. Cf. Antiperthite (4th List). 21st List

Micro-dunhamite. E. E. Fairbanks, 1947. *Amer. Min.*, vol. 32, p. 683. Synonym of dunhamite (q.v.). The prefix 'micro' to indicate that the determination was made only in polished sections under the microscope. [*M.A.*, **10**, 255.] 18th List

Microlepidolite. *See* macrolepidolite. 3rd List

Miedziankite. J. Morozewicz, 1923. *Spraw. Polsk. Inst. Geol.* (= *Bull. Serv. Géol. Pologne*), 1923, vol. 2, p. 1 (miedziankite), p. 3 (miedziankit); [*Compt. Rend. Congrès Géol. Intern.*, Sess. XIII, 1922, Bruxelles]; abstr. in *Bull. Soc. Franç. Min.*, [1923], vol. 45 (for 1922), p. 255. A sulpho-salt of copper and zinc, $2Cu_3AsS_3.ZnS$, allied to tennantite. Occurs as granular masses at Miedzianka, Poland, where by its alteration it gives rise to staszicite (q.v.). Named from the locality. 10th List

Miersite. L. J. Spencer, *Nature*, 1898, LVII, 574. A cubic-tetrahedral modification of silver iodide containing some copper iodide; isomorphous with marshite. Broken Hill, New South Wales. 2nd List

Mikheevite. E. I. Nefedov, 1953. *Mém. Soc. Russ. Min.*, ser. 2, vol. 82, p. 317 (Михеевит). $K_2Ca_4(SO_4)_5.H_2O$, triclinic, in salt deposits. Named after V. I. Mikheev, В. И. Михеев. [*M.A.*, **12**, 352.] 20th List [Is Görgeyite (q.v.); *A.M.*, **41**, 816]

Mikroantigorit. Hintze, *Min.*, Ergbd. 2, p. 264. Variant of Microantigorite (13th List). 20th List

Millisite. E. S. Larsen and E. V. Shannon, 1930. *Amer. Min.*, vol. 15, p. 329. Hydrous phosphate of calcium, sodium, and aluminium, $2CaO.Na_2O.6Al_2O_3.4P_2O_5.17H_2O$, as white crusts (monoclinic?) in variscite nodules from Utah. Named after Mr. F. T. Millis, of Lehi, Utah. [*M.A.*, **4**, 344.] 12th List

Millosevichite. U. Panichi, 1913. *Rend. R. Accad. Lincei*, Roma, ser. 5, vol. xxii, sem. 1, p. 303. Normal ferric and aluminium sulphate occurring as an encrustation with a delicate violet colour and vitreous lustre in the Alum Grotto, on the Island of Vulcano, Lipari Islands. Named after Professor Federico Millosevich, of Florence. 6th List [$(Al,Fe)_2(SO_4)_3$; *A.M.*, **59**, 1140]

Milowite (T. H. Barry, *The Industrial Chemist*, London, 1928, vol. 4, p. 227). Trade-name for a white powdery, 'amorphous silica' (SiO_2 98·12, Al_2O_3 0·85, MgO trace, alk. 0·41, H_2O 0·62%) from the island of Milos, Greece. 11th List

Miltonite. G. Gagarin and J. R. Cuomo, 1949, loc. cit., p. 13 (miltonita). Calcium sulphate hemihydrate, $CaSO_4.\frac{1}{2}H_2O$, obtained in the preparation of thin sections of gypsum. [*M.A.*, **8**, 280.] Named after Dr. Charles Milton (1896–) of the United States Geological Survey. Found as a natural mineral in deserts in Central Asia, V. I. Popov and A. L. Vorobiev, 1947 [*M.A.*, **11**, 366]. Cf. bassanite, 6th List. 19th List

Minasite. O. C. Farrington, 1912. *Bull. Geol. Soc. America,* vol. xxiii, p. 728. Analysis of a pebble ('fava') from the diamond washings of Minas Geraes, Brazil, gave results suggesting at first a new aluminium hydroxide, $2Al_2O_3.3H_2O$. In the complete account (*Amer. Journ. Sci.*, 1916, ser. 4, vol. xli, p. 360) this name is omitted and the material admitted to be impure. W. E. Ford (*Amer. Journ. Sci.*, 1916, ser. 4, vol. xli, p. 569) incorrectly describes it as a hydrous aluminium phosphate *See also* Geraesite. The two names from the locality Minas Geraes. 8th List

Minasragrite. W. T. Schaller, 1915. *Journ. Washington Acad. Sci.*, vol. v, p. 7; 3rd Appendix to 6th edit. of Dana's *System of Mineralogy*, p. 51. A hydrated acid vanadyl sulphate, $(V_2O_2)H_2(SO_4)_3.15H_2O$, probably monoclinic, occurring as a blue efflorescence on patronite at Minasragra, Peru. 7th List [$VOSO_4.5H_2O$; *M.A.*, 74-491]

Mindigite. L. De Leenheer, 1934. *Natuurwet. Tijdschrift*, Gent, vol. 16, pp. 237, 240 (mindigiet), p. 241 (mindigite). Hydroxide of cobalt and copper, $9Co_2O_3.2CuO.16H_2O$, as pitch-black colloidal material, from Mindigi, Katanga, Belgian Congo. Named from the locality. [*M.A.*, **6**, 52.] 14th List [Is Heterogenite; *M.M.*, **33**, 253]

Minervite. A. Gautier, *Compt. Rend.* CXVI. 1173, 1893. $Al_2O_3.P_2O_5.7H_2O$. A. Carnot (*Ann. des Mines*, VIII. 311, 1895) gives it as a phosphate of Al and K. Grotte de Minerve, France. 1st List [Is Taranakite; *M.M.*, **28**, 31. Cf. Palmerite]

Minette. A mining term applied to oolitic brown-iron-ores (limonite) of Lorraine and Luxemburg. A diminutive of the French *mine*, ore, in allusion to the low content of metal in the ore. 4th List

Minguetite. A. Lacroix, 1910. *Bull. Soc. franç. Min.*, vol. xxxiii, p. 273; *Minéralogie de la France*, 1910, vol. iv, p. 738 (minguétite). A black micaceous mineral, previously described as biotite, from the Minguet iron-mine, near Segré (Maine-et-Loire). In composition, $17SiO_2.4(Fe,Al)_2O_3.8(Fe,Mg)O.(K,Na)_2O.8H_2O$, it is intermediate between the iron-mica lepidomelane and the iron-chlorite stilpnomelane. 6th List [Is Stilpnomelane; *A.M.*, **54**, 1223]

Minguzzite. C. L. Garavelli, 1955. *Rend. Accad. Lincei, Cl. Sci. fis. mat. nat.*, ser. 8, vol. 18, p. 400. Oxalate of potassium and ferric iron, $K_3Fe(C_2O_4)_3.3H_2O$, green monoclinic crystals, with oxalite = humboldtine, $FeC_2O_4.2H_2O$, in limonite from Elba. Named in memory of Carlo Minguzzi (1910–1953), Italian mineralogist. [*M.A.*, **13**, 86.] 21st List [*A.M.*, **41**, 370]

Miniumite. Variant of Minium, *A.M.*, **8**, 51. 10th List

Minnesotaite. J. W. Gruner, 1944. *Amer. Min.*, vol. 29, p. 363. Hydrous silicate of ferrous iron (magnesium, etc.), $(OH)_{5\cdot5}(Fe'',Mg)_{5\cdot5}(Si,Al,Fe''')_8O_{18\cdot5}$, with the crystal-structure of talc and regarded as an iron-talc, of abundant occurrence in iron ores in Minnesota. Named from the locality. [*M.A.*, **9**, 88.] 17th List

Minyulite. E. S. Simpson and C. R. LeMesurier, 1933. *Journ. Roy. Soc. Western Australia*, vol. 19, p. 13. Hydrated basic phosphate of aluminium and potassium, $KAl_2(OH,F)(PO_4)_2.3\frac{1}{2}H_2O$, as radiating groups of white orthorhombic fibres, from Minyulo Well, Dandaragan, Western Australia. Named from the locality. [*M.A.*, **5**, 293.] 13th List

Miormirite. V. Vujanovič, 1969. [*Rad. Inst. geol. rud. istraž. ispit. nukl. min. sirov*, **4**, 147], abstr. *Zap*, **100**, 619. Plumbian (?) davidite from Nežilova, Skopl, Macedonia. The title of the paper (Miomirit – olovni davidit iz Nežilova) would refer to a stannian davidite, but the analysis shows PbO and no SnO_2. 27th List

Miromirite, error for miomirite (27th List). *A.M.*, **58**, 560. 28th List

Mirupolskite. G. A. Yurgenson, N. G. Smirnova, and L. A. Karenina, 1968. [Вестн. научн. информ. Забайкал. фил. Геогр. общ. СССР, no. 9, 3]; abstr. *Zap.*, **99**, 78 (1970) (Мирупольскит), $2CaSO_4.H_2O$; synonym of bassanite (6th List). 26th List

Mirzaanite. L. D. Melikadze and T. A. Eliava, 1948. [*Kolloid. Zhurnal*, Voronezh, vol. 10, p. 115.] Abstr. in *Amer. Chem. Abstr.*, 1949, vol. 43, col. 5350 (mirzaanite); *Brit. Chem. Abstr.*, 1949, B II, col. 356; *Bull. Soc. Franç. Min. Crist.*, 1954, vol. 77, p. 1270 (mirsaanite). An elastic bitumen, variety of elaterite, from the Mirzaani (Мирзаани) oil-field, Georgia, Transcaucasia. [*M.A.*, **12**, 511.] 20th List

Miserite. W. T. Schaller, 1950. *Amer. Min.*, vol. 35, p. 911. $KCa_4Si_5O_{13}(OH)_3$, a pink, fibrous (orthorhombic?) alteration product of wollastonite in metamorphosed shale from Arkansas. Previously described as natroxonotlite (J. F. Williams, 1891). Named after Dr. Hugh Dinsmore Miser (1884–) of the United States Geological Survey. [*M.A.*, **11**, 187.] 19th List

Miskeyite. Trade-name for a compact chlorite (pseudophite) used as an ornamental stone from St. Gallenkirch, Vorarlberg. Named after J. von Miskey, director of the Miskeyitwerksgesellschaft. F. Berwerth, Tschermak's *Min. Petr. Mitt.*, 1912, vol. 31, p. 112. W. C. Smith, *Min. Mag.*, 1924, vol. 20, p. 242. 18th List

Mitchellite. J. H. Pratt, *Amer. J. Sci.*, 1899, VII, 286. A variety of chromite containing much $MgO.Al_2O_3$. [Webster] North Carolina. The same as magnochromite. 2nd List

Mitridatite. P. A. Dvoichenko, 1914. [*Zap. Krym. Obshch. Est.* (*Bull. Soc. Nat. Crimée*), vol. 4, p. 114.] O. M. Shubnikova, *Trans. Lomonosov Inst. Acad. Sci. USSR*, 1936, no. 7, p. 327; F. V. Chukhrov, ibid., 1937, no. 10, p. 139 (митридатит), p. 148 (mitridatite). Hydrated phosphate of calcium and ferric iron, $3CaO.2Fe_2O_3.2P_2O_5.5H_2O.n$aq, as earthy yellowish-green nodules and veinlets in iron ore in the Kerch peninsula, Crimea. Named from the hill of Mithridat at Kerch, which was called after Mithridates or Mithradates VI (died *c.* 63 BC), king of Pontus. [*M.A.*, **7**, 60.] 15th List [*A.M.*, **59**, 48; $Ca_3Fe_4(PO_4)_4(OH)_6.3H_2O$]

Mitscherlichite. F. Zambonini and G. Carobbi, 1925. *Ann. R. Osservatorio Vesuviano*, ser. 3, vol. 2, p. 9. Hydrated potassium cupric chloride, $K_2CuCl_4.2H_2O$, as minute greenish-blue tetragonal crystals on saline sublimations in the crater of Vesuvius. Named after the German chemist, Eilhardt Mitscherlich (1794–1863), who prepared this salt artificially in 1840. 11th List

Miyashiroite. R. Phillips and W. Layton, 1964. *Min. Mag.*, vol. 33, p. 1097. The hypothetical amphibole end-member $Na_3Mg_3Al_3Si_7O_{22}(OH)_2$. Named for A. Miyashiro. 23rd List

Mn-alluaudite, *see* Mangan-alluaudite

Mn-alumochromite. E. I. Nefedov and N. I. Shuvalova, 1968. *Zap.*, **97**, 90 (Mn-алюмохромит). An unnecessary name for manganoan alumochromite. *See* alumochromite, 14th List. 28th List

Mn-Beljankinit. H. Strunz, 1966. *Min. Tabellen*, 4th edit., p. 182. Another variant German transliteration of Манган-Белянкинит, Manganbelyankinite (22nd List). 24th List

Mn-feldspat. E. Eberhard, 1963. *Fortschr. Min.*, vol. 40, p. 52. Synonym of Manganese-anorthite. 23rd List

Mn-ferripalygorskite, *see* Mn-palygorskite. 28th List

Mn-ferrisepiolite, *see* Mn-sepiolite. 28th List

Mn-fluorapatite, *see* Mangan-fluorapatite.

Mn-hydroxyapatite, *see* Mangan-hydroxyapatite.

Mn-leonite, *see* Manganleonite.

Mn-palygorskite. E. I. Semenov, 1969. [Инст. мин. геохим. крист. редк. элем. 1969, 100]; abstr. *A.M.*, **55**, 2139 (1970). A red pleochroic mineral, α 1·56, yellowish, γ 1·58, red-brown, from Siorarsuit, Ilimaussaq, Greenland, is regarded as identical with material (α 1·545, γ 1·58) from Karnasurt, Lovozero, Kola Peninsula; the latter has a composition near $NaMgMnFe_2AlSi_7O_{20}(OH)_2.10H_2O$. X-ray data are near those of palygorskite. The name Mn-ferripalygorskite is also used, apparently for the same mineral. Cf. ferripalygorskite, 22nd List. 28th List

Mn-pyroxene, Mn-pyroxmangite, Mn-rhodonite. H. Narita, K. Koto, and N. Morimoto, 1977. *Mineral. J.* [*Japan*], **8**, 329. The *artificial* polymorphs of $MnSiO_3$ with the pyroxene, pyroxmangite, and rhodonite structures respectively. 30th List

Mn-proxmangite, *see* Mn-pyroxene.

Mn-rhodonite, *see* Mn-pyroxene.

Mn-Schadlunit, Germ. trans. of Mn-шадлунит, Mn-shadlunite (q.v.). Hintze, 60. 28th List

Mn-sepiolite. E. I. Semenov, 1969. [Инст. мин. геохим. крист. редк. элем. 1969, 100] abstr. *A.M.*, **55**, 2138 (1970). A dark reddish-brown pleochroic mineral, α 1·72 yellowish, $\gamma \parallel [001]$ 1·74, red-brown, from Nakalak Mountain, Ilimaussaq, Greenland, is formulated as $Mn_4Fe_4Si_{12}O_{30}(OH)_6.8H_2O$ and as $(Fe,Mn)_9Si_{12}O_{30}(OH)_6.10H_2O$ (no analysis). X-ray powder data resemble those of sepiolite. The name Mn-ferrisepiolite is also used, apparently for the same mineral. Cf. ferrisepiolite, 22nd List. 28th List

Mn-shadlunite. T. L. Evstigneeva, A. D. Genkin, N. V. Troneva, A. A. Filimonova, and A. I. Tsepin, 1973. *Zap.*, **102**, 63 (Mn-шадлунит). A manganoan variety of shadlunite (q.v.), with Mn replacing most of the Pb, from the Oktyabr copper deposit, Norilsk, Siberia. Cubic, *a* 10·73 Å. [*M.A.*, 73-4082; *A.M.*, **58**, 1114.] 28th List [Also Manganese-shadlunite (q.v.). Perhaps a species]

Mn-sicklerite, *see* Mangani-sicklerite.

Mn-zoisite, *see* Manganese-zoisite.

Moctezumite. R. V. Gaines. 1965. *Amer. Min.*, vol. 50, p. 1158. Orange blades and rosettes in the oxidized zone of the Te-Au deposit at the Moctezuma mine, Sonora, Mexico, are monoclinic, with composition $PbUO_2(TeO_3)_2$. Named from the locality. [*M.A.*, **17**, 607.] 24th List

Modderite. R. A. Cooper, 1924. *Journ. Chem. Metall. Mining Soc. South Africa*, vol. 24, p. 265. Monoarsenide of cobalt, CoAs as a bluish-white (incompletely determined) mineral, occurring with niccolite (NiAs) on the Witwatersrand, Transvaal. Evidently named from the Modderfontein mine. [*M.A.*, **3**, 115.] 11th List

Mofettite. G. Vavrinecz, 1939. *Földtani Közlöny*, Budapest, vol. 69, pp. 82, 98 (Mofettit). Carbon dioxide, CO_2, as a natural gas. Named from *mofeta* (Span.), mephitis, noxious emanation; and *mofetta* (Ital.), the fissure from which the gas, mainly CO_2, escapes. [*M.A.*, **7**, 471.] 15th List

Mogensenite. A. F. Buddington, J. [J.] Fahey, and A. [C.] Vlisidis, 1955. *Amer. Journ. Sci.*, vol. 253, p. 409. A titaniferous magnetite containing exsolution ulvöspinel. Named after Fredrik Mogensen, of Djursholm, Sweden, who described ulvöspinel (Fe_2TiO_4, 18th List). *See* Titanmagnetite. 21st List

Mohavite. W. T. Schaller, 1928. *Amer. Min.*, vol. 13, p. 453. Synonym of tincalconite (C. U. Shephard, 1878) or 'octahedral borax', rhombohedral $Na_2B_4O_7.5H_2O$. Named from the Mohave desert, California. [*M.A.*, **4**, 246.] 13th List

Mohawk-algodonite. G. A. Koenig, *Amer. Journ. Sci.*, 1902, ser. 4, vol. xiv, pp. 414, 416. Massive copper arsenides from the Mohawk mine, Michigan, of variable composition, but approximating to that of algodonite. *See* mohawkite. 3rd List

Mohawkite. J. R. Stanton, *Engin. and Mining Journ. New York*, 1900 (April), vol. lxix, p. 413. A massive copper arsenide, with some cobalt and nickel, from the Mohawk mine, Michigan. Ledoux deduced from his analysis the formula $(Cu,Ni,Co)_4As$. This was afterwards re-named ledouxite (q.v.). 3rd List

—— G. A. Koenig, *Amer. Journ. Sci.*, 1900 (December), ser. 4, vol. x, p. 440; *Zeits. Kryst. Min.*, 1901, vol. xxxiv, p. 67. A nickeliferous and cobaltiferous domeykite, $(Cu,Ni,Co)_3As$, from the Mohawk mine. The copper arsenides from this locality show wide variations in composition, and, as pointed out by Koenig, are mixtures, or more of the nature of alloys rather than definite mineral species; nevertheless this author has introduced for them several other new names, viz.:—keweenawite, mohawk-algodonite, mohawk-whitneyite and semi-whitneyite (q.v.). 3rd List [A mixture, Dana *Syst. Min.*, 7th edit., **1**, 170; *A.M.*, **56**, 1319]

Mohawk-whitneyite. G. A. Koenig, *Amer. Journ. Sci.*, 1900, ser. 4, vol. x, p. 466; *Zeits. Kryst. Min.*, 1901, vol. xxxiv, p. 75 (Mohawkit-Whithneyit); *Amer. Journ. Sci.*, 1902, ser. 4, vol. xiv, pp. 414, 416. Intimate mixtures of mohawkite and whitneyite from the Mohawk mine, Michigan. *See* mohawkite. 3rd List

Mohelnite. R. Dvořák, 1943. [*Příroda*, Brno, vol. 35, p. 190.] J. Paclt, *Neues Jahrb. Min.*, Monatshefte, 1953, p. 189 (Mohelnit). An incompletely described chlorite in the serpentine of Mohelno, Moravia, perhaps identical with clinochlore or moravite (4th List). Named from the locality. [*M.A.*, **12**, 240.] 20th List

Mohrite. C. L. Garavelli, 1964. *Atti (Rend.) Accad. Naz. Lincei, Cl. sci. fis. mat. nat.*, vol. 36, p. 524. Pale green incrustations collected by A. Pelloux in 1927 from the boriferous soffioni of Travale, Val de Cecina, Tuscany, prove to consist of $(NH_4)_2Fe(SO_4)_2.6H_2O$ and $(NH_4)_2$-$(Fe,Mg)(SO_4)_2.6H_2O$ with Fe $\geqslant$ Mg. The Fe end-member has long been known as Mohr's salt after K. F. Mohr. [*M.A.*, **17**, 505; *A.M.*, **50**, 805.] 24th List

Moissanite. G. F. Kunz, 1905. *Centralblatt Min.*, 1905, p. 154; *Amer. Journ. Sci.*, ser. 4, vol. xix, p. 396. Silicide of carbon, CSi, identical with the artificial product known as carborundum (q.v.), found by H. Moissan (*Compt. Rend. Acad. Sci. Paris*, 1904, vol. cxxxix, p. 778; 1905, vol. cxl. p. 405) as small, green, hexagonal plates in the meteoric iron of Cañon Diablo, Arizona. Named after Professor Henri Moissan (1852–1907). 4th List [The alpha phase]

Moldavite. P. Poni, *Ann. Sci. Univ. Jassy*, 1900, vol. i, p. 144; *Anal. Acad. Române*, Bukarest, 1900, vol. xii. Earlier used by Cobalcescu [reference?] as a specific name for ozocerite, and later by C. Istrati (1897) [who, however, spells the word moldovite] for a variety of ozocerite from Moldavia (Moldova). 3rd List [Not the moldavite of Zippe, a tektite.]

Moldovite. C. [I.] Istrati, *Bult. Soc. Sci. Bucarest*, 1897, VI, 91. A variety of ozocerite from Moldova (Moldavia). 2nd List

Molengraaffite. H. A. Brouwer, 1911. *Centralblatt Min.*, p. 129 (Molengraaffit). A titanosilicate of calcium, sodium, etc. forming yellowish-brown, monoclinic prisms with a perfect orthopinacoidal cleavage, and resembling astrophyllite in appearance. It occurs as a constituent of the rock lujaurite associated with the nepheline-syenites of the Pilandsberg, Transvaal. Named after Professor Gustaaf Adolf Frederik Molengraaff, of Delft, Holland. 6th List [Identical with lamprophyllite; *M.A.*, **7**, 463]

Molibdomenita. G. Gagarin and J. R. Cuomo, 1949, loc. cit., p. 15. Variant of molybdomenite. 19th List

Molibdosodalite, *see* Molybdosodalite.

Moluranit. Author? *Guide to USSR exhibit on Atomic Energy*, Geneva, 1955, p. 9 (moluranite). Later publications, abstr.: *Amer. Min.*, 1956, vol. 41, p. 816, *Amer. Min.*, **43**, 381, $UO_2.2UO_3.5MoO_3.12H_2O$; *Mém. Soc. Russ. Min.*, 1958, vol. 87, p. 79 (молуранит), $UO_2.2UO_3.5MoO_3.12H_2O$? Named from the composition. 21st List [*A.M.*, **45**, 258]

Molybdophyllite. G. Flink, *Bull. Geol. Inst. Univ. Upsala,* 1901, vol. v (1900), p. 91 (Molybdophyllit). Platy masses with a perfect basal cleavage, and having quite the appearance of white mica. Found with hausmannite in granular limestone or dolomite at Långban, Sweden. Hexagonal. $2(Pb,Mg)O.SiO_2.H_2O$. Differs from barysilite in containing water. Named from *μόλυβδος*, lead, and *φύλλον*, a leaf. 3rd List

Molybdoscheelite. H. Strunz, *Min. Tab.*, 1941, p. 144 (Molybdoscheelite, *pl.*), p. 258 (Molybdoscheelit). O. H. Ödman, *Årsbok Sveriges Geol. Undersök.*, 1947, vol. 41, no. 6, p. 47 (molybdo-scheelite). Calcium tungstate and molybdate, $Ca(W,Mo)O_4$, as mixed crystals intermediate between scheelite and powellite. Syn. of seyrigite (15th List). [*M.A.*, **10**, 492.] 18th List

Molybdosodalite. F. Zambonini, 1910. *Mineralogia Vesuviana. Mem. R. Accad. Sci. Fis. Mat. Napoli*, ser. 2, vol. xiv, no. 7, pp. 214, 358 (molibdosodalite). The green sodalite from Monte Somma was found by G. Freda in 1878 to contain molybdenum (MoO_3, 2·87%) and a deficiency of chlorine (2·28%). 6th List

Monalbite. T. R. Schneider and F. Laves, 1957. *Zeits. Krist.*, vol. 109, p. 241 (Monalbit). Since re-examination of the type barbierite shows it to be a microcline, the name monalbite is proposed for the monoclinic modification of $NaAlSi_3O_8$, not yet found in nature. 22nd List

Moncheïte. A. D. Genkin, N. N. Zhuravlev, and E. M. Smirnova, 1963. Зап. Всесоюз. Мин. Обш. (*Mem. All-Union Min. Soc.*), vol. 92, p. 33 (Мончеит). Minute steel-grey grains in Chalcopyrite from Monchegorsk have the composition $(Pt,Pd)(Te,Bi)_2$, with Pt : Pd $\approx$ 2 to 4 and Te : Bi 2 to 10. Hexagonal and isostructural with $PtTe_2$. Named from the locality. Preliminary data in [Геол. Рудн. Мест., 1961, no. 5, p. 64], abstr. in *Amer. Min.*, 1962, vol. 47, p. 809 (*Amer. Min.*, 1963, vol. 48, p. 1181; *M.A.*, **16**, 283, 550). 23rd List

Monocerotite. G. Arrhenius, 1975. *Meteoritics*, **10**, 354. 'proposed as an appropriate zoomorphic term for this characteristic pseudomorph' of olivine in epitaxic replacement of enstatite. 29th List

Monohydrallite. H. Harrassowitz, 1927. *See* Allite. 11th List

Monohydrocalcite. E. I. Semenov, 1964. Кристаллография, vol. 9, p. 109 (Моногидрокальцит). The unnamed $CaCO_3.0{\cdot}65H_2O$ of Sapozhnikov and Tsvetkov (Докл. Акад. наук СССР (*Compt. Rend. Acad. Sci. URSS*), 1959, vol. 124, p. 402) gives an X-ray powder pattern very near that of $CaCO_3.H_2O$ (Brooks, Clark, and Thurston, *Phil. Trans.*, Ser. A, vol. 243, p. 145; Baron and Pesneau, *Compt. Rend. Acad. Sci. Paris*, 1956, vol. 243, p. 1217; Lippmann, *Naturwiss.*, 1959, vol. 46, p. 553). Sapozhnikov and Tsvetkov report γ 1·590, α and β 1·545 for the natural mineral, while Lippmann found the synthetic material to be uniaxial negative, ω 1·590, ε 1·543. Named from the composition. [*M.A.*, **16**, 648; *AM.*, **49**, 1151.] 24th List [Australia, *A.M.*, **60**, 690; bladderstone, *A.M.*, **62**, 273]

Monothermite. D. S. Belyankin and V. P. Ivanova, 1936. *Vernadsky jubilee volume*, vol. 1, p. 561 (монотермит), p. 562 (Monothermit). D. S. Belyankin, *Compt. Rend. (Doklady) Acad. Sci. URSS*, 1938, vol. 18, p. 673 (monothermite), p. 674 (monothermite or endothermite). A finely scaly clay mineral, $0{\cdot}2K_2O.Al_2O_3.3SiO_2.1{\cdot}5H_2O + 0{\cdot}5aq$, differing from kaolin in showing only one thermal effect on the heating curve, viz. an endothermal effect at 550° and no exothermal effect at 900–950°C. [*M.A.*, **7**, 170.] 15th List

Monrepite. W. Wahl, 1925. *Fennas* (*Bull. Soc. Géogr. Finlande*), Helsingfors, vol. 45, no. 20, p. 87 (Monrepit). A ferro-ferri-mica, $H_2KFe'''Fe_3''(SiO_4)_3$, forming a series parallel to the phlogopite-lepidomelane series. Occurs as a constituent of rapakivi-granite at Monrepos, Viborg, Finland. Named from the locality. [*M.A.*, **3**, 500.] 11th List

Monsmedite. V. Manilici, D. Giuscă, and V. Stiopol, 1965. [*Mem. Com. Geol. Repub. Soc. Romània*, vol. 7, p. 46]; abstr. *Min. Abstr.*, 1967, vol. 18, pp. 246 and 285. Dark green to black orthorhombic crystals from the Rotmundi vein at Baia Sprie (= Felsobánya) have a composition given as $Tl_2O_3.K_2O.8SO_3.15H_2O$. Named for the Latin name (Mons Medius) of the locality. The cited composition seems highly improbable and needs confirmation. 25th List [*A.M.*, **54**, 1496]

Montasite. (A. L. Hall, *Mem. Geol. Survey South Africa*, 1930, no. 12 (2nd edit.), pp. 27, 40.) Trade name for a fine quality fibre asbestos from the Montana mine, Pietersburg, Transvaal; identical with amosite, a variety of anthophyllite. 12th List

Montbrayite. M. A. Peacock and R. M. Thompson, 1945. *Program and Abstracts, Min. Soc. Amer.*, 26th Annual Meeting, p. 21; *Amer. Min.*, 1946, vol. 31, p. 204. Gold telluride, Au_2Te_3, as tin-white triclinic crystals from Montbray, Quebec. Named from the locality. [*M.A.*, **9**, 262.] 17th List [*A.M.*, **31**, 515; **57**, 146]

Monteponite. E. E. Fairbanks, 1946. *Econ. Geol.*, vol. 41, p. 767. Cadmium oxide, CdO, as minute black octahedra described by B. Neumann and E. Wittich (*Natürliches Cadmiumoxyd, Chem.-Ztg.*, 1901, vol. 25, p. 561; E. Wittich and B. Neumann, *Cbl. Min.*, 1901, p. 549) from Monteponi, Sardinia. Named from the locality. [*M.A.*, **10**, 100.] 18th List

Montesite. R. Herzenberg, 1949. *Publ. Techn. Inst. Boliviano Ingen. Minas y Geol. Mineria Boliviana*, no. 45 (montesita). Lead sulphostannite, $PbSn_4S_5$, between teallite ($PbSnS_2$) and herzenbergite SnS) in composition, and similar to them in appearance. From Bolivia. Named after the late Ismael Montes, founder of the School of Mines at Oruro. [*M.A.*, **11**, 10.] 19th List [Var. of Herzenbergite; *A.M.*, **60**, 163]

Montgomeryite. E. S. Larsen, 3rd, 1940. *Amer. Min.*, vol. 25, p. 315. Hydrous phosphate of calcium and aluminium, $Ca_4Al_5(PO_4)_6(OH)_5.11H_2O$, as green to colourless monoclinic crystals in variscite nodules from Utah. Named after Mr. Arthur Montgomery, of New York City. [*M.A.*, 7, 513.] 15th List [$Ca_4Mg(H_2O)_{12}Al_4(OH)_4(PO_4)_6$; *A.M.*, **59**, 843]

Montmorillonoids. D. M. C. MacEwan, 1951. *X-ray identification and mineral structures of clay minerals, Min. Soc.*, London, p. 86; *Clay Min. Bull.*, 1951, vol. 1, p. 195. Group name for minerals related to the original montmorillonite from Montmorillon, France. [*M.A.*, **11**, 253.] 20th List

Montroseite. A. D. Weeks, E. A. Cisney, and A. M. Sherwood, 1950. *Progr. and Abstr. Min. Soc. Amer.*, p. 22. $2FeO.V_2O_3.7V_2O_4.4H_2O$, as black orthorhombic blades, from Paradox valley, Montrose Co., Colorado. Named from the locality. [*M.A.*, **11**, 189; *A.M.*, **36**, 327.] 19th List [(V,Fe)O(OH), *A.M.*, **38**, 1235; Cf. Paramontroseite, Kentsmithite, Vanoxite]

Montroydite. A. J. Moses, *Amer. Journ. Sci.*, 1903, ser. 4, vol. xvi, p. 259. Orthorhombic mercuric oxide, HgO, occurring as velvety encrustations of orange-red needles on mercury-ores at Terlingua, Texas. Named after Mr. Montroyd Sharpe, one of the owners of the mines. 3rd List

Mooihoekite. L. J. Cabri and S. R. Hall, 1972, *Amer. Min.*, **57**, 689. Massive sulphides at Mooihoek Farm, Lydenburg District, Transvaal, include a chalcopyrite-yellow tetragonal mineral, *a* 10·58, *c* 5·37 Å, $P\bar{4}2m$. Composition [$Cu_9Fe_9S_{16}$]. Obtained synthetically. Named for the locality. [*M.A.*, 72-3345.] 27th List

Mooraboolite. G. B. Pritchard, *Victorian Naturalist*, 1901, vol. xviii, p. 63. A zeolite occurring as white, radial aggregates in the decomposed basalt of the Moorabool valley, Victoria. In chemical composition, and in the crystallographic and physical characters as far as determined, it agrees with natrolite. 3rd List

Mooreite. L. H. Bauer and H. Berman, 1929. *Amer. Min.*, vol. 14, pp. 103, 165. Hydrous basic sulphate of magnesium, zinc, and manganese, $7R(OH)_2.RSO_4.4H_2O$ (for δ-mooreite, $6R(OH)_2.RSO_4.4H_2O$), as tabular monoclinic crystals from [Sterling Hill, Sussex Co.] New Jersey. Named after Gideon Emmet Moore (1842–1895), mineral chemist of New York City. [*M.A.*, **4**, 151.] 12th List [δ-Mooreite renamed Torreyite, q.v.; $(Mn,Zn,Mg)_8(SO_4)(OH)_{14}.3H_2O$, *A.M.*, **54**, 973]

Moorhouseite. J. L. Jambor and R. W. Boyle, 1965. *Canad. Min.*, vol. 8, p. 166. A pink efflorescence, with Aplowite (q.v.) on sulphates at the Magnet Cove Barium Corporation mine, Walton, Nova Scotia, have the composition $(Co,Ni,Mn)SO_4.6H_2O$, with Co : Ni : Mn ≈ 11 : 5 : 2, with a little Cu, monoclinic and isomorphous with hexahydrite. The name is given for W. W. Moorhouse, and defined to include all members of the hexahydrite family with Co as principal cation. [*A.M.*, **50**, 808.] 24th List

Moraesite. M. L. Lindberg, W. T. Pecora, and A. L. de M. Barbosa, 1953. *Amer. Min.*, vol. 38, p. 1126. Hydrous beryllium phosphate, $Be_2PO_4(OH).4H_2O$, as white fibrous masses, monoclinic; from Sapucaia pegmatite mine, Brazil. Named after Dr. Luciano Jacques de Moraes, Brazilian geologist. [*M.A.*, **12**, 301.] 20th List

Moravite. F. Kretschmer, 1906. *Centralblatt Min.*, 1906, p. 293 (Moravit). A finely scaly chloritic mineral, of the leptochlorite group, occurring abundantly as iron-black masses with sub-metallic lustre in association with iron-ores at Gobitschau in Moravia. It resembles thuringite in appearance, but differs chemically from this, the formula being $H_4(Al,Fe)_4(Fe,Mg)_2Si_7O_{24}$. Named from the locality. 4th List

Morencite. W. Lindgren and W. F. Hillebrand, 1904. *Amer. Journ. Sci.*, ser. 4, vol. xviii, p. 455; *Bull. US Geol. Survey*, 1905, no. 262, p. 49; *Profess. Paper US Geol. Survey*, 1905, no. 43, p. 115. A brownish-yellow mineral occurring in finely fibrous seams, the minute fibres having straight extinction. Hydrated ferric silicate with some magnesia, lime, etc., and a little alumina; $H_{22}R_2''Fe_6'''(SiO_4)_{11}$. Occurs in calcareous shale at Morenci, Arizona, and is probably an alteration product of some contact-metamorphic mineral. 4th List [Is Nontronite; *M.A.*, **3**, 452]

Morganite. G. F. Kunz, 1911. *Amer. Journ. Sci.*, ser. 4, vol. xxxi, p. 81; *Mineral Industry*, for 1910, 1911, vol. xix, p. 583. Trade-name for a rose-coloured beryl of gem-quality occurring with lithium-bearing minerals in the pegmatites of Madagascar and California. Named after Johy Pierpont Morgan (1837–1913) of New York City. The same name had earlier been used for the manufactured products of the Morgan Crucible Company in London. Further, the variety-name vorobyevite (5th List) had been previously applied to the pink alkali-bearing beryl from Madagascar. 6th List

Moronite. S. Calderon and M. Paul, *Anal. Soc. Españ. Hist. Nat.*, 1886, XV, 477. S. Calderon, *ibid.*, 1894, XXIII, 21. A mixed siliceous and calcareous deposit formed of radiolaria, diatoms, foraminifera, etc., Moron, Spain. 2nd List

Moschellandsbergite. H. Berman and G. A. Harcourt, 1938: *Amer. Min.*, vol. 23, p. 764. The body-centred cubic γ-phase of silver-amalgam, Ag_2Hg_3 (Hg 73%), as distinct from the face-centred cubic α-phase of mercurial silver (var. arquerite). Named from the locality, Moschellandsberg, Rhenish Bavaria. [*M.A.*, **7**, 224.] 15th List [Cf. Landsbergite, Schachnerite, Paraschachnerite]

Mosesite. F. A. Canfield, W. F. Hillebrand, and W. T. Schaller, 1910. *Amer. Journ. Sci.*, ser. 4, vol. xxx, p. 202; *Zeits. Kryst. Min.*, 1911, vol. xlix, p. 1; *Bull. US Geol. Survey*, 1912, no. 509, p. 104; F. A. Canfield, *School of Mines Quart.*, New York, 1913, vol. xxxiv, p. 276. A mercurous-ammonium compound containing chlorine and sulphate. It was found at Terlingua, Texas, as minute, canary-yellow octahedra and spinel-twins; and is very similar in composition and appearance to kleinite (4th List; = mercurammonite, 5th List) from the same locality. Named after Professor Alfred Joseph Moses, of Colombia University, New York. 6th List [Chloride, molybdate, and sulphate of Millon's base, Hg_2NOH; *A.M.*, **38**, 1225]

Mossite. W. C. Brøgger, *Skrifter Vid.-Selsk. Christiania*, I, *Math.-natur. Kl.*, 1897, no. 7; Abstr. *Min. Mag.*, XII, 130. Black tetragonal crystals twinned so as to simulate simple rhombic crystals of prismatic habit. $FeNb_2O_6.FeTa_2O_6$. Isomorphous with tapiolite. Moss, Norway. 2nd List

Mouchketovite. *Ann. Géol. et Min. Russie*, 1901, vol. iv, sect. iii, p. 107 (Mouchkétovite, Мушкетовитъ). The same as muschketowit (E. S. Fedorov and W. Nikitin, 1899). 3rd List

Mounanaïte. F. Cesbron and J. Fritsche, 1969. *Bull Soc. franç. Min. Crist.*, **92**, 196. $PbFe_2^{3+}(VO_4)_2(OH)_2$ occurs in very small amounts at Mounana, Gabon; composition ascertained by synthesis. Anorthic. Named for the locality. [*A.M.*, **54**, 1738; *M.A.*, 69-3348; *Zap.*, **99**, 80.] 26th List

Mountainite. J. A. Gard and H. F. W. Taylor, 1957. *Min. Soc. Notice*, 1957, no. 97; *Min. Mag.*, 1957, vol. 31, p. 611. A fibrous zeolitic mineral containing no Al_2O_3, unit-cell contents

$(Ca,Na_2,K_2)_{16}Si_{32}O_{80}.24H_2O$, monoclinic. Closely associated and confused with rhodesite (q.v.) from Bultfontein mine, Kimberley, South Africa. Named after Prof. Edgar Donald Mountain (1901–) of Rhodes University, Grahamstown. 21st List [*A.M.*, **43**, 624]

Mounténite, *see* Muntenite.

Mourite. E. V. Kopchenova, K. V. Skvortsova, N. I. Silanteva, G. A. Sidorenko, and L. V. Mikhailova, 1962. Зап. Всесоюз. Мин. Общ. (*Mem. All-Union Min. Soc.*), vol. 91, p. 67 (Моурит). Violet nodules, crusts, and plates from an unnamed locality have a composition near $(UO_2,UO_3).5\frac{1}{2}MoO_3.5\frac{1}{3}H_2O$. Named from the composition (*Amer. Min.*, 1962, vol. 47, p. 1217; *M.A.*, **15**, 541). 23rd List [$UMo_5O_{12}(OH)_{10}$; *A.M.*, **56**, 163]

Mourmanite. N. N. Gutkova, *Dokl. Akad. Nauk SSSR* (*Compt. Rend. Acad. Sci. URSS*), Ser. A, 1930, p. 730 (мурманит, mourmanite). French spelling of murmanite (A. E. Fersman, 1923; 10th List). Hydrated silico-titanate of sodium, etc. Named from the Murman Coast, Russian Lapland. [*M.A.*, **5**, 198.] 12th List [Also 13th List]

Mozambikite. J. M. Cotelo Neiva and J. M. Correia Neves, 1960. *Internat. Geol. Congr.*, Rept. 21st session, Norden, part. 17, p. 53. Yellow-brown octahedra, near $ThSiO_4$ but hydrated; cubic, X-ray pattern unaltered by heating at 1000°C. From the pegmatites of Alto-Ligonha, Mozambique. Named from the locality. 22nd List [Cf. Hydrothorite]

Mozarkite. Author unknown. *Lapidary J.* **31**, 160 (1977). A term for a multi-coloured chert from the Ozark Mts., Missouri. 30th List

Mpororoite. O. von Knorring, Th. G. Sahama, and M. Lehtinen, 1972. *Bull. Geol. Soc. Finland*, **44**, 107. A fine-grained greenish-yellow alteration product of scheelite in the Mpororo tungsten deposit, Kigezi, Uganda, is monoclinic, *a* 8·27, *b* 9·32, *c* 16·40 Å, *β* 92° 29′; composition 5 $[(Al,Fe)_2W_2O_9.6H_2O]$. Named for the locality. [*A.M.*, **58**, 1112; *Zap.*, **103**, 362.] 28th List

Mrazekite. Gh. Neacşu, 1970. [*Rev. Roum. Geol. Geophys. Geogr., Ser. Geol.*, **14**, 24], abstr. *A.M.*, **57**, 595. White earthy masses, turning brown on exposure, occur with murgocite (q.v.); and are considered to be a 'montmorillonoide magnésien'. Analytical, X-ray, and d.t.a. data are given. 'The mineral resembles saponite, stevensite, and talc. . . . The data are inadequate to characterize a new phase'—*M.F.* [*Zap.*, **100**, 625.] 27th List

Mroseite. J. A. Mandarino, R. S. Mitchell, and R. G. V. Hancock, 1975. *Geol. Soc. Amer. Abstr. Progr.*, **7**, 814; *C.M.*, **13**, 286. White massive material from Mina la Moctezuma, Sonora, Mexico, has space group *Pbca*, *a* 6·93, *b* 11·16, *c* 10·54 Å, composition $8[CaTeO_2CO_3]$. Named for M. E. Mrose. [*A.M.*, **60**, 946; **61**, 339.] 29th List

Mtorodite. E. H. Rutland, 1970. [*Zeits. deut. gemm. Ges.*, **19**, 141]; quoted in *C. Hintze, Handb. Min.* (K. F. Chudoba, editor), *Erg.-Bd.*, **4**, 62. A green chromian chalcedony. Named for its locality in Rhodesia. A variant of matorolite, 26th List. 28th List

Muchinite, non-standard transliteration of Мухинит, mukhinite (*Zap.*, **99**, 80). 27th List

Muirite. J. T. Alfors and M. C. Stinson, 1965. *Min. Inform. Serv., Calif. Div. Mines Geol.*, vol. 18, p. 27. J. T. Alfors, M. C. Stinson, R. A. Matthews, and A. Pabst, *Amer. Min.*, 1965, vol. 50, pp. 279, 314, and 1500. Orange grains and tetragonal crystals in sanbornite–quartz rock near Big Creek and Rush Creek, Fresno County, California, are near $Ba_{10}Ca_2MnTiSi_{10}O_{30}(OH,Cl,F)_{10}$ (emission spectrography analysis, Cl by X-ray spectrography). Named for J. Muir. [*M.A.*, **17**, 400; **17**, 502.] 24th List

Mukhinite. A. B. Shepel and M. V. Karpenko, 1969. Докл. Акад. наук СССР (*Compt. Rend. Acad. Sci. URSS*), **185**, 1342 (Мухинит). A vanadian clinozoisite (V_2O_3 11·29%) occurring in marble from the roof of the Tashelginsk iron ore deposit, Gornaya Shoriya, W Siberia, is named for A. S. Mukhin. [*A.M.*, **55**, 322; *M.A.*, 70-746; *Zap.*, **99**, 80.] 26th List [*M.A.*, 76-732]

Mullanite. E. V. Shannon, 1918. *Amer. Journ. Sci.*, ser. 4, vol. xlv, p. 66. A sulphantimonite of lead, $5PbS.2Sb_2S_3$ (corresponding with the silver-lead diaphorite), occurring as acicular (orthorhombic?) crystals and felted masses near Mullan in Idaho, and in W Montana. Named after Captain John Mullan, a pioneer road-maker in this region. 8th List [Is Boulangerite; *M.A.*, **8**, 6]

Müllerite. F. Zambonini, *Zeits. Kryst. Min.*, 1899, XXXII, 157. Yellowish green, soft and compact. $Fe_2O_3.3SiO_2 + 2H_2O$. Nontron, France. Differs from chloropal (nontronite) in having $2H_2O$ instead of $5H_2O$. 2nd List [Is Nontronite, *M.A.*, **3**, 452; Cf. Zamboninite]

Mullite. N. L. Bowen, J. W. Greig and E. G. Zies, 1924. *Journ. Washington Acad. Sci.*, vol. 14, p. 183; *Journ. Amer. Ceramic Soc.*, vol. 7, p. 244. Aluminium silicate, $3Al_2O_3.2SiO_2$, as orthorhombic crystals with optical constants differing only slightly from those of sillimanite. It is the artificially-prepared 'sillimanite' and it occurs naturally in the 'sillimanite-buchite' (inclusions of fused phyllite in igneous intrusions) of the island of Mull, Scotland. Named from the locality. [*M.A.*, **2**, 303, 377.] 10th List

Munkforssite. L. J. Igelström. *Zeits. Kryst. Min.*, XXVIII. 601. 1897. Spelt munkforsite, *ibid.*, XXVIII. 310, 1897. Phosphate and sulphate of Ca and Al; resembles svanbergite. Sweden. 1st List [Probably manganoan apatite, *A.M.*, **49**, 1778; *M.A.*, **17**, 180]

Munkrudite. L. J. Igelström. *Zeits. Kryst. Min.*, XXVIII. 310, 1897. Phosphate and sulphate of Fe and Ca; resembles svanbergite. Munkerud, Sweden. 1st List [Is Kyanite, *A.M.*, **49**, 1778; *M.A.*, **17**, 180]

Muntenite. (G. G. Longinescu, *Bull. Sect. Sci. Acad. Roumaine*, 1925, vol. 9, p. 170 [= no. 9–10, p. 20] (Mounténite); L. J. Spencer, *Min. Mag.*, 1927, vol. 21, p. 247 (muntenite).) A fossil resin from Eocene beds at Olăneşti, in Oltenia, Romania, related to copalite and gedanite, and differing from romanite (which occurs in the Oligocene). It was described by G. Murgoci (Georges Munteanu-Murgoci, 1872–1925) in 'Gisements du succin de Roumaine . . . et un nouvelle résine-fossile d'Olăneşti' (Bucarest, 1903; Romanian edition in 1902). The name was first mentioned to me [*L.J.S.*] by Prof. Murgoci in 1907, when he said that it would be published by C. I. Istrati and M. A. Mihăilescu, but he was never able to give me the bibliographical reference. 11th List

Murataite. J. W. Adams, T. Botinelly, W. N. Sharp, and K. Robinson, 1974. *A.M.*, **59**, 172. Black anhedral grains or poorly developed crystals in a pegmatite in the St. Peters Dome area, Colorado, are cubic, *a* 14·863 Å; composition $8[(Na,Yt,Ln)_4(Zn,Fe)_3(Ti,Nb)_6O_{18}$-$(F,OH)_4]$, with Er > Dy > Gd the predominant lanthanons. Named for K. J. Murata. [*M.A.*, 74-3467.] 28th List

Murdochite. J. J. Fahey, 1953. *Progr. and Abstr. Min. Soc. Amer.*, p. 21; *Amer. Min.*, 1954, vol. 39, p. 327. C. L. Christ and J. R. Clark, ibid., pp. 15, 321. Cu_6PbO_8 as tiny black octahedra with NaCl structure. Named after Prof. Joseph Murdoch (1890–) of the University of California at Los Angeles. [*M.A.*, **12**, 304.] 20th List [*A.M.*, **40**, 905]

Murgocite. V. Ianovici and Gh. Neacşu, 1970. [*Rev. Roum. Geol. Geophys. Geogr., Geol. Ser.*, **14**, 3], abstr. *A.M.*, **57**, 594. Soft earthy masses, greasy to the touch, from Moldava Noua, Banat, Romania, are regarded as a regular 1:1 interstratified saponite and swelling chlorite mineral with 20% of stevensite sheets. Analytical, X-ray, and d.t.a. data are given. The origin of the name is not given. 'The data are inadequate to characterize a new species'—*M.F.* [*Zap.*, **100**, 625.] 27th List

Murmanite. A. E. Fersman, 1923. *Compt. Rend. Acad. Sci. Russie*, p. 63 (мурманит). An undescribed mineral from the Kola Peninsula. Named from the Murman Coast. [*M.A.*, **2**, 263.] 10th List [$Na_2MnTi[Ti_2(OH)_4(Si_2O_7)_2].4H_2O$, *M.A.*, **19**, 268; *see also* under Mourmanite and Metamurmanite]

β-Murmanite. Synonym of Metamurmanite (23rd List). [Геохим., мин., генет. типы месторожд. редк. элем., Изд. „Наука", 1964, p. 546]; abstr. Зап. Всесоюз. Мин. Общ. (*Mem. All-Union Min. Soc.*), 1965, vol. 94, p. 683 (Бетамурманит). 24th List

Muschketowite. E. von Federow and W. Nikitin, *Ann. Géol. Min. Russie*, 1899, III, 87, 99. A pseudomorph of magnetite after hæmatite. Urals. 2nd List

Musgravite. D. R. Hudson, A. F. Wilson, and I. M. Threadgold, 1967. *Min. Mag.*, vol. 36, p. 305. A provisional name, withdrawn before publication, for a polytype of taaffeite, but listed by I. Kostov, *Mineralogy*, 1968, p. 213. 25th List

Muskoxite. J. L. Jambor, 1969. *Amer. Min.*, **54**, 684. Hydrous oxide of Mg and Fe^3. occurring as trigonal crystals, dark reddish-brown, in the Muskox Intrusion, Coppermine River, Canada. Named for the locality. [*M.A.*, 70-748.] 26th List

Mussolinite. A Serra, 1936. [*Osservazioni sul giacimento «Orantalco» in provincia di Nuoro (Sardegna), Studi Sassaresi*, vol. 14, p. 5], *Periodico Min. Roma*, 1937, vol. 8, p. 88 (Mussolinite). A variety of talc from Sardinia. Presumably named after Signor Benito Mussolini (1883–). 15th List

Mutabilite. G. Mueller, 1964. *Rept. 22nd Internat. Geol. Congr., India*, part 1, 47. A highly oxygenated olefinic bitumen. 26th List

Muthmannite. F. Zambonini, 1911. *Zeits. Kryst. Min.*, vol. xlix, p. 246 (Muthmannit). C. Gastaldi, *Rend. Accad. Sci. Fis. Mat. Napoli*, 1911, ser. 3, vol. xvii, p. 26. Monotelluride of silver and gold, (Ag,Au)Te, hitherto confused with krennerite (a ditelluride, $AuTe_2$, containing but little silver). Named after Dr. Wilhelm Muthmann, Professor of Inorganic Chemistry in the Technical High School at Munich. 6th List

Motukoreaite. K. A. Rodgers, J. E. Chisholm, R. J. Davis, and C. S. Nelson, 1977. *M.M.*, **41**, 389 and M21. A white clay-like mineral in beach rock at Brown's Island (Motukorea), Waitemata Harbour, Auckland, New Zealand, consists of microscopic hexagonal crystals, *a* 9·336, *c* 44·72 Å. Composition near $[NaMg_{19}Al_{12}(CO_3)_{6\cdot5}(SO_4)_4(OH)_{54}.28H_2O]$, sp. gr. 1·43. Named from the locality. [*M.A.*, 77-4625; *A.M.*, **63**, 598.] 30th List

Myrickite. (D. B. Sterrett, *Gems and precious stones in 1911, Mineral Resources of the United States*, for 1911, 1912, p. 7 of advance copy.) Local trade-name for a variety of chalcedony from San Bernardino Co., California. It shows red spots on a grey ground and resembles 'St. Stephen's stone'. Named after Mr. F. M. Myrick, of Randsburg, California. 6th List

Myrmeki-perthitoid. P. Quensel, 1939. *See* Perthitoid. 15th List

Mysite. H. Moissan, 1905. *Traité de chemie minérale*, vol. 4, p. 372. Error for misy. (On the same page the names of other iron sulphates are incorrectly spelt.) 20th List

N

Na-Autunit. H. Strunz, 1966. *Min. Tabellen*, 4th edit., p. 311. Synonym of Natro-autunite. 24th List.

Na-heterosite, *see* Soda-heterosite.

Na-Meta-Autunit. H. Strunz, 1966. *Min. Tabellen*, 4th edit., p. 312. Synonym of Metanatro-autunite (q.v.). 24th List

Na-purpurite, *see* Soda-purpurite.

Na-spodumene, *see* Soda-spodumene.

Naegite. T. Wada, 1904. *Minerals of Japan, Tōkyō*, p. 49 (Naëgite). A tetragonal mineral occurring as small, rather indistinct crystals with angles near those of zircon. Described as a silicate of uranium, thorium, etc., and considered to be isomorphous with zircon; in a later analysis (*Beiträge zur Mineralogie von Japan*, 1906, no. 2, p. 23) zirconium figures largely. [Further investigation will probably prove its identity with zircon.] Found in alluvial tin-washings at Naëgi, near Takayama, Prov. Mino, Japan. 4th List [*M.A.*, 74-2364]

Nafalapatite. D. J. Fisher and A. Volborth, 1960. *Amer. Min.*, vol. 45, p. 645. Artificial sodian aluminian apatite produced by heating Morinite to 400°C. An unnecessary name given from the composition. 23rd List

Nafalwhitlockite. D. J. Fisher and A. Volborth, 1960. *Amer. Min.*, vol. 45, p. 645. Artificial sodian aluminian Whitlockite produced by heating Morinite to 800°C. An unnecessary name given from the composition. 23rd List

Nagatelite. S. Iimori, J. Yoshimura, and S. Hata, 1931. *Sci. Papers Inst. Physical and Chemical Research*, Tokyo, vol. 15, p. 87; *Chem. News*, London, 1931, vol. 142, p. 211. Hydrous phosphosilicate of cerium earths, aluminium, iron, and calcium, $4R^{II}O.3R^{III}{}_2O_3.6(SiO_2,P_2O_5).2H_2O$, as black monoclinic prisms, belonging to the epidote group and close to allanite. Named from the locality, Nagatejima, Ishikawa, Japan. [*M.A.*, **4**, 500.] 12th List [Phosphatian Allanite]

Nagelschmidtite. R. L. Barrett and W. J. McCaughey, 1941. *Amer. Min., Program and Abstracts*, December 1941, p. 4; *Amer. Min.*, 1942, vol. 27, p. 689. Calcium silico-phosphate $Ca_7Si_2P_2O_{16}$ as optically biaxial grains in basic slag. Named after Dr. Gunther Nagelschmidt (1906–) of Harpenden, Hertfordshire, who described the material in 1937. [*M.A.*, **7**, 147; **8**, 312.] 16th List

Nagolnite. G. Vavrinecz, 1936. *Földtani Közlöny*, Budapest, vol. 66, p. 250; 1937, vol. 67, p. 60 (Nagolnit). Hypothetical molecule $H_4Al_2SiO_7$ as a constituent of many orthochlorites, corresponding to V. I. Vernadsky's chlorite acid $H_2Al_2SiO_6(+H_2O)$, and approximating in composition to Y. V. Samoilov's α-chloritite (7th List) from Nagolno, Donetz, Russia. Named from the locality. [*M.A.*, **7**, 361.] 17th List

Nahcolite. F. A. Bannister, 1928. *Nature*, London, vol. 122, p. 866; *Min. Mag.*, 1929, vol. 22, p. 60. Sodium bicarbonate, as minute [monoclinic] crystals in a mixture of salts found in a Roman conduit near Naples. Named from the chemical formula $NaHCO_3$ and λίθος, stone. 12th List

Nakalifite. A. V. Stepanov, 1961. Зап. Всесоюз. Мин. Общ. (*Mem. All-Union Min. Soc.*), 1963, vol. 92, p. 195. A provisional name for the mineral subsequently named Gagarinite (q.v.); derived from натрий, кальций, фтор (Na,Ca,F). 23rd List

Nakaséite. Tei-Ichi Ito and H. Muraoka, 1960. *Zeits. Krist.*, vol. 113, p. 93. Near $Pb_4Ag_3Sb_{12}S_{24}$; X-ray powder data are essentially identical with those of andorite, and the mineral is admitted to be 'a structural variety of andorite', but a new name is proposed 'because, apart from the difference in the chemical compositions, it has a very complicated twinned lattice essentially different from that of andorite'. Named from the locality, the Nakasé mine, Japan. A new name should not have been proposed while the status of webnerite and fizelyite remains uncertain. M. Fleischer (*Amer. Min.*, 1960, vol. 45, p. 1314) suggests that the material, which has a *c*-axis 24 × 4·26 Å., should be provisionally called andorite–XXIV. [*M.A.*, **15**, 44.] 22nd List

Nakauriite. J. Suzuki, M. Ito, and T. Sugiura, 1976. *J. Japan Assoc. Mineral. Petrol. Econ. Geol.*, **71**, 183. Clear sky-blue fibres from Nakauri, Achi Prefecture, Japan, are orthorhombic, *a* 14·585, *b* 11·47, *c* 16·22 Å. Composition $2[Cu_8(SO_4)_4CO_3(OH)_6.48H_2O]$, sp. gr. 2·39. RIs α 1·585, colourless, β 1·604, light greenish-blue, γ 1·612, light blue. Named from the locality. [*A.M.*, **62**, 594.] 30th List

Nambulite. M. Yoshi, Y. Aoki, and K. Maeda, 1972. *Min. Journ.* [*Japan*], **7**, 29. Reddish-brown veinlets in braunite ores at the Tunakozowa mine, Kitakami Mts., Japan, are anorthic, *a* 7·621, *b* 11·761, *c* 6·731 Å, α 92° 46′, β 95° 5′, γ 106° 52′; composition near $LiNaMn_8Si_{10}O_{28}(OH)_2$. Named for M. Nambu. Differs from hydrorhodonite (Engström, 1875) in containing much less water. [*A.M.*, **58**, 1112.] 28th List

Nanlingite. Gu Xiongfei, Ding Kuishou, and Xu Yingnian, 1976. *Geochimica*, **2**, 107. Brownish-red grains in a granite-limestone contact zone in the Nan Ling area, China, are rhombohedral, *a* 10·42 Å, α 58° 50′, space group *R3m or R*$\bar{3}$*m*, ρ 3·927, ω 1·82, ε 1·78. Named from the locality. [The composition is given as near $5[CaMg_3F_3(AsO_3)_2]$, but the cell contents derived from the analysis, cell-size, and density approximate to $4[Ca_{0\cdot96}Na_{0\cdot27}Mg_{2\cdot84}Fe_{0\cdot41}Li_{0\cdot37}Al_{0\cdot09}Mn_{0\cdot01}(AsO_3)_{2\cdot04}(OH)_{0\cdot30}F_{3\cdot28}O_{0\cdot02}]$, or ideally $4[CaMg_4(AsO_3)_2F_4]$—*G.Y.C.* Determination of FeO is not mentioned, and part of the iron may be ferrous, leading to a better balance—*M.H.H.*] [*A.M.*, **62**, 1058; *M.A.*, 77-885.] 30th List

Naphthine, Naphteine, or Naphtine. A. N. Desvaux, 1834. *Mém. Soc. Agric. Sci. and Arts*, Angers, 1834, vol. ii, p. 139 (Naphtéine); *Bull. Soc. Géol. France*, 1835, vol. vi, p. 139 (Naphthine). A. Lacroix, *Minéralogie de la France*, 1910, vol. iv, p. 619 (Naphtine). A mineral-wax, identical with hatchettite, found in cavities in limestone at Beaulieu, dép. Maine-et-Loire. So named because it has an odour like that of naphtha. 6th List

Naphtolithe. P. Barbier, 1911. *Bull. Soc. Hist. Nat. Autun*, vol. xxiv, p. 115. A. Lacroix, *Minéralogie de la France*, 1913, vol. v, p. 24. A bituminous shale from Thelots, Saône-et-Loire. 7th List

Napoleonite. An obsolete synonym of orthoclase. The name appears in T. Thomson's *Outlines of Mineralogy*, 1836, vol. i, p. 291, and in the 1st to 4th edits. (1837–1854) of Dana's *System of Mineralogy*. The same name was later applied to the orbicular diorite, or corsite, of Corsica. In T. Egleston's *Catalogue of Minerals and Synonyms*, 1892, it appears as a synonym of amphibole. 6th List

Narsarsukite. G. Flink, *Medd. om Grönland*, 1900, XXIV, 154. Honey-yellow tabular or cube-shaped tetragonal crystals. $3Na_2O.Fe'''F.2TiO_2.12SiO_2$. Narsarsuk, S Greenland. 2nd List [$Na_2TiOSi_4O_{10}$; *A.M.*, **47**, 539. Cf. Gouréite]

Narsasukite. G. Flink, *Meddelelser om Grönland*, 1901, Hefte xxiv, plate VIII, and p. 154 footnote. Error for narsarsukite. 3rd List

Nasinite. C. Cipriani, 1961. *Atti* (*Rend.*) *Accad. Naz. Lincei, Cl. sci. fis. mat. nat.*, ser. 8, vol. 30, pp. 74, 235; vol. 31, p. 141. $Na_4B_{10}O_{17}.7(\text{or } 5)H_2O$, in recent incrustations at Larderello, Tuscany; monoclinic. Named for R. Nasini. Identical with 'Auger's borate' (*Amer. Min.*, 1963, vol. 48, p. 709; *M.A.*, **16**, 373). 23rd List

Nasledovite. M. R. Enikeev, 1958. [Доклады Акад. Наук Узбек. ССР (*Compt. Rend. Acad. Sci. Uzbek SSR*), no. 5, p. 13]; abstr. Зап. Всесоюз. Мин. Общ. (*Mem. All-Union Min. Soc.*), 1958, vol. 88, p. 313 (Наследовит. nasledovite). A hydrous carbonate and sulphate of Pb, Mn, Mg, and Al, from the Altyn-Topken mining field, Sardob, Central Asia. Named after B. N. Nasledov. [*M.A.*, **14**, 278; *A.M.*, **44**, 1325.] 22nd List

Nasonite. S. L. Penfield and C. H. Warren, Dana's *Appendix*, 1899; *Amer. J. Sci.*, 1899, VIII, 346; *Zeits. Kryst. Min.*, 1900, XXXII, 234; abstr. *Min. Mag.*, XII, 316. White, massive. $Pb_4(PbCl)'_2Ca_4(Si_2O_7)_8$. Probably tetragonal and isomorphous with ganomalite. Franklin Furnace, New Jersey. 2nd List [*A.M.*, **56**, 1174]

Natisite. Yu. P. Menshikov, Ya. A. Pakhomovskii, E. A. Goiko, I. V. Bussen, and A. N. Merkov, 1975. *Zap.*, **104**, 314 (Натисит). Yellowish-green platy crystals in natrolite–ussingite veinlets on Karnasurt Mt., Lovozero massif, Kola Peninsula, have space group *P4/nmm*, *a* 6·50, *c* 5·07 Å. Composition $2[Na_2TiSiO_5]$; identical with the synthetic compound. Named from the composition. [*M.A.*, 76-883; *A.M.*, **61**, 339.] 29th List

Natramblygonite. W. T. Schaller, 1911. *Amer. J. Sci.*, ser. 4, vol. xxxi, p. 48; *Zeits. Kryst. Min.*, 1911, vol. xlix, p. 233; 1912, vol. li, p. 246 (Natronamblygonit); *Bull. US Geol. Survey*, 1912, no. 509, p. 101. A soda-amblygonite occurring as greyish-white cleavage-masses in pegmatite near Canyon City, Colorado. It differs from ordinary amblygonite in that the lithium is largely replaced by sodium (Na_2O, 11·23%), the formula being $(Na,Li)Al(OH,F)PO_4$. 6th List [Renamed Natromontebrasite, q.v.]

Natrion-betpakdalite, *see* Sodium betpakdalite.

Natrium-Betpakdalit, German form of sodium betpakdalite (27th List). Hintz, 63. 28th List

Natrium-Hewettit. H. Strunz, 1957. *Min. Tabellen*, 3rd edit., p. 257; ibid., 4th edit., 1966, p. 318. A *hypothetical* sodium analogue of Hewettite, formulated $Na_2V_6O_{16}.9H_2O$; the only hydrate of $Na_2V_6O_{16}$ known is the trihydrate, Barnesite (22nd List), the analogue of Metahewettite. 24th List

Natriummimetesit. H. Wondratschek, 1963. Original form of Natromimetite (q.v.). 23rd List

Natriummordenit, German form of sodium mordenite (26th List). Hintze, 64. 28th List

Natro-alumobiotite. D. P. Serdyuchenko, 1954. *Doklady Acad. Sci. USSR*, vol. 97, p. 317 (натровый алюмобиотит). A composition variety of biotite containing Al_2O_3 22·85, Na_2O 2·76%. [*M.A.*, **12**, 535.] 20th List

Natroalunite. W. F. Hillebrand and S. L. Penfield, *Amer. Journ. Sci.*, 1902, ser. 4, vol. xiv, pp. 218, 220; *Zeits. Kryst. Min.*, 1902, vol. xxxvi, pp. 552, 554. Alunite with part of the potash replaced by soda; in material from Colorado $K_2O:Na_2O = 4:7$. $(Na,K)_2[Al(OH)_2]_6(SO_4)_4$. Rhombohedral. 3rd List

Natro-autunite. A. E. Fersman and O. M. Shubnikova, 1937. *Geochemist's and mineralogist's companion*, Moscow, p. 177 (Натро(вый) отунит). Variant of sodium-autunite (q.v.). 19th List

Natrochalcite. C. Palache and C. H. Warren, 1908. *Amer. Journ. Sci.*, ser. 4, vol. xxvi, p. 342; *Zeits. Kryst. Min.*, vol. xlv, p. 534. Hydrous sulphate of copper and sodium, $Na_2SO_4.Cu_4(OH)_2(SO_4)_3.2H_2O$, forming bright emerald-green, monoclinic crystals with an acute pyramidal habit. It occurs with several other species of sulphates in copper-veins in the mining district of Chuquicamata, Antofagasta, Chile. 5th List

Natrodavyne. F. Zambonini, 1910. *Mineralogia Vesuviana. Mem. R. Accad. Sci. Fis. Mat. Napoli*, ser. 2, vol. xiv, no. 7, p. 188 (natrodavyna). Hexagonal crystals, rich in faces, hitherto referred to nepheline or to davyne, but on analysis found to differ from the members of the davyne-microsommite group in containing no potassium and much carbon dioxide. 6th List

Natrofairchildite. Yu. L. Kapustin, 1971. [Мин. карбонатитов, Изд. наука, 1971, 181]; abstr. *Zap.*, **103**, 359. $Na_2Ca(CO_3)_2$, from the Vuoriyarvi massif, Karelia, ω 1·525, ε 1·459, strongest X-ray powder lines 3·18, 2·64, 6·71, 2·20, 1·891 Å. [Probably identical with nyerereite, 24th List, which has not been adequately described—*M.H.H.*] 28th List

Natro-ferrophlogopite. D. P. Serdyuchenko, 1954. *Doklady Acad. Sci. USSR*, vol. 97, p. 317 (натровый феррофлогопит). A composition variety of phlogopite containing FeO 15·56, Na_2O 3·18%. [*M.A.*, **12**, 535.] 20th List

Natrohisingerite. A. I. Soklakov and M. D. Dorfman, 1964. [Минерал ı СССР, **15**, 167]; abstr. *M.A.*, **18**, 160. A metamict mineral (locality not given) formulated $R_{0\cdot26}Mn_{0\cdot02}Fe_{2\cdot68}Si_3O_{10}.7{\cdot}1H_2O$, where R is largely Na. The evidence that it is a variety of hisingerite is not convincing. 28th List

Natrohitchcockite. E. T. Wherry, 1916. *See* Bariohitchcockite. 8th List

Natrojarosite. W. F. Hillebrand and S. L. Penfield, *Amer. Journ. Sci.*, 1902, ser. 4, vol. xiv, p. 211; *Zeits. Kryst. Min.*, 1902, vol. xxxvi, p. 545. A yellowish-brown, glistening powder consisting wholly of minute, perfectly developed rhombohedra with basal planes. $Na_2[Fe(OH)_2]_6(SO_4)_4$. From Soda Springs Valley, Nevada. 3rd List

Natro-melilite. O. M. Shubnikova and D. V. Yuferov, 1934. *New minerals 1922–1932*, Leningrad, Moscow, 1934, p. 71 (натро-мелилит). Synonym of soda-melilite (12th List). 19th List

Natromimetite. H. Wondratschek, 1963. *Neues Jahrb. Min., Abh.*, vol. 99, p. 116 (Natrium-mimetesit). The artificial compound $Pb_4Na(AsO_4)_3$; apatite family. 23rd List

Natromontebrasite. F. Gonnard, 1913. *Bull. Soc. franç. Min.*, vol. xxxvi, p. 120. To replace the name Natramblygonite or Natronamblygonit of W. T. Schaller (1911; 6th List), since the mineral is a hydrofluophosphate rather than a fluophosphate. *See* Fremontite. 7th List

Natronamblygonit. *See* Natramblygonite. 6th List

Natronanorthite. S. J. Thugutt, *Neues Jahrb. Min.*, 1895. *Beil.-Bd.*, IX, 561. Abstr. *Min. Mag.*, XI, 111, *soda-anorthite*. An artificial compound, $Na_2Al_2Si_2O_8$. 2nd List [Syn. of Carnegieite]

Natron-berzeliite. H. Bäckström, *Zeits. Kryst. Min.*, XXVI. 102, 1896 (abstr. of H. Sjögren's paper). Syn. of soda-berzeliite (q.v.). 1st List [Var. of Berzeliite, Dana, *Syst. Min.*, 7th edit., **2**, 681]

Natronbiotite. H. Strunz. *Min. Tab.*, 1941, p. 199 (Natronbiotit). Biotite with potassium partly replaced by sodium. 18th List

Natron-Carnotit. H. Strunz, 1966. *Min. Tabellen*, 4th edit., p. 316. Synonym of Sodium carnotite (*q.v.*). 24th List

Natroncatapleiite. G. Flink, *Geol. För. Förh.*, 1893, XV, 206; *Zeits. Kryst. Min.*, 1894, XXIII, 359 (*Natronkatapleït*). Catapleiite containing soda in place of lime. 2nd List

Natrongranat. *See* lagoriolite. 2nd List

Natron-Heulandit. H. Strunz, 1966. *Min. Tabellen*, 4th edit., p. 430. Synonym of Clinoptilolite (10th List). 24th List

Natroniobite. A. G. Bulakh, A. A. Kukharenko, Yu. N. Knipovich, V. V. Kondrateva, K. A. Baklanova, and E. N. Baranova, 1960. [Матер. Всесоюз. Науч. Геол. Инст.]. abstr. in Зап. Всесоюз. Мин. Общ. (*Mem. All-Union Min. Soc.*), 1962, vol. 91, p. 190. The monoclinic dimorph of Lueshite (22nd List), $NaNbO_3$, from the Lesnaya Baraka and Sallanlatvi massifs, Kola peninsula. Named from the composition. [*M.A.*, **16**, 552; *A.M.*, **47**, 1483.] 23rd List

Natronite. Variant of Natron; *A.M.*, **8**, 51. A. H. Chester, 1896, lists natronite as a variant of natrolite. 10th List

Natronkalisimonyit. R. Köchlin, *Min. petr. Mitt.* (*Tschermak*), 1902, vol. xxi, p. 356. A provisional name for crystals agreeing with blödite (simonyite) in form, but apparently differing slightly in chemical composition: in addition to soda, a little potash (0·43%) is present, and more water is expelled at 100°C than from typical blödite. From the salt deposits of Kalusz, Galicia. 3rd List

Natronkatapleit, variant of Natronkatapleiit (H. Strunz, *Min. Tabellen*, 1st edit., 1941, p. 260). 24th List

Natronmanganwollastonit. F. Machatschki, 1953. *Spez. Min.*, Wien, p. 314. Syn. of Schizolite (2nd List). 22nd List

Natronmargarit. C. Hintze, *Handbuch der Mineralogie*, Ergänzungsband, 1936, p. 390. A German form of Soda-margarite (12th List). 14th List

Natronmelilith. W. C. Brøgger, 1898. *Die Eruptivgesteine des Kristianiagebietes*, III. *Vidensk.-Selsk. Skrifter, Math.-Naturv. Kl.*, Kristiania, for 1897, no. 6, p. 69. A melilite-like mineral occurring in the rock farrisite; afterwards (loc. cit., p. 367) suggested to belong to the scapolite group. 6th List [Also 2nd List]

Natron-mesomicrocline. H. Strunz, 1957. *Min. Tabellen*, 3rd edit., p. 334 (Natron-Mesomikroklin). Syn. of Natronorthoklas. 22nd List

Natronmikroklin. W. C. Brøgger, *Die Silur. Etag. Kristianiagebiet*, 1882, p. 262; *Skrifter Vid.-Selsk., Christiania*, I, *Math.-natur. Kl.*, 1898, no. 6, p. 11. Dana's *Appendix*, 1899, p. 5, *soda-microcline*. The same as anorthoclase. 2nd List

Natronphlogopit. E. Weinschenk, *Abhandl. Akad. Wiss. München*, 1901, vol. xxi, p. 272. A white mica containing magnesia and soda (and potash) in crystalline limestone from Styria. 3rd List

Natronrichterite. H. Sjögren, *Geol. För. Förh.*, XIV. 253, 1892. *See* soda-richterite. 1st List

Natronsanidine. F. von Wolff, 1904. *Centralblatt Min.*, 1904, p. 208 (Natronsanidin). A monoclinic felspar with the habit of sanidine, but containing much soda, the composition being $Or_1Ab_1 = K_{\frac{1}{2}}Na_{\frac{1}{2}}AlSi_3O_8$. 5th List

Natronsarkolith. Abst. in *Zeits. Krist.*, 1922, vol. 57, p. 100 (Natronsarkolith), p. 105 (Natron-Sarkolith). A German form of Soda-sarcolite (W. T. Schaller, 1916; 8th List). 9th List

Natronthomsonit. C. F. Rammelsberg, 1895. *See* Calciothomsonite. 10th List

Natronwollastonit. F. Machatschki, 1953. *Spez. Min.*, Wien, p. 314. Syn. of Pectolite. 22nd List

Natropal. V. I. Gerasimovsky, 1956. [Геокимия, no. 5 (Натропал)], translated in Geochemistry, no. 5, p. 494. Unnecessary name for a sodian Opal. 23rd List

Natrophosphate. Yu. L. Kapustin, A. V. Bykova, and V. I. Bukin, 1972. *Zap.*, **101**, 80 (Натрофосфат, Natrophosphate). An irregular monomineralic mass from the Yukspor Mts., Khibina massif, is isotropic, *n* 1·460–1·462. Analysis corresponds to $Na_6H(PO_4)_2F.17H_2O$. X-ray powder data are close to those of artificial $Na_7(PO_4)_2F.19H_2O$ (E. W. Neuman, 1933; *M.A.*, **5**, 473). Cubic, *a* 27·79 Å. Further study is desirable. [*A.M.*, **58**, 139.] 27th List

Natrosilite. I. M. Timoshenkov, Yu. P. Menshikov, L. F. Gannibal, and I. V. Bussen, 1975. *Zap.*, **104**, 317 (Натросилит). Irregular grains intergrown with analcime, natrolite, and lomonosovite in nepheline syenite on Mt. Karnasurt, Lovozero massif, are identical with synthetic β-$Na_2Si_2O_5$. Monoclinic, *a* 12·30, *b* 4·88, *c* 8·27 Å, β 104° 14′, space group $P2_1/a$. [*M.A.*, 76-884; *A.M.*, **61**, 339.] 29th List

Naujakasite. O. B. Bøggild, 1933. *Meddelelser om Grønland*, vol. 92, no. 9. Approximate composition $HNa_3Al_2Si_4O_{13}$, as micaceous scales, from Naujakasik, S Greenland. Named from the locality. [*M.A.*, **5**, 484.] 13th List [Contains essential Fe, *M.A.*, 71-478; *A.M.*, **53**, 1780]

Naurodite. W. von Knebel, 1903. [*Sitz.-Ber. Phys.-Med. Soc. Erlangen*, vol. 35, p. 213]; abstr. in *Neues Jahrb. Min.*, 1907, vol. i, p. 229 (Naurodit). A blue hornblende differing from glaucophane, riebeckite, crossite, and arfvedsonite in optical extinction-angle and pleochroism, occurring in naurodite-schist at Naurod, NW of Wiesbaden, Germany. Named from the locality. 16th List

Nauruite. C. Elschner, 1913. *Corallogene Phosphat-Inseln Austral-Oceaniens und ihre Produkte*, Lübeck, 1913, p. 54 (Nauruit). A resinous, colloidal calcium phosphate, probably $3Ca_3P_2O_8.Ca(OH,F)_2$, encrusting the phosphate rock of Nauru or Pleasant Island, in the Pacific. Evidently a kind of phosphorite. 7th List [Is Francolite; *M.A.*, **9**, 33]

Navajoite. A. D. Weeks, M. E. Thompson, and A. M. Sherwood, 1954. *Science* (*Amer. Assoc. Adv. Sci.*), vol. 119, p. 326; *Amer. Min.*, 1955, vol. 40, p. 207. $V_2O_5.3H_2O$, as brown fibres, probably monoclinic, from Arizona. Named after the Navajo Indians. [*M.A.*, **12**, 408, 567.] 20th List

Neighborite. E. C. T. Chao, H. T. Evans, jr., B. J. Skinner, and C. Milton, 1961. *Amer. Min.*, vol. 46, p. 379. $NaMgF_3$, orthorhombic and isostructural with perovskite, occurs in dolomitic oil-shale at South Ouray, Uintah Co., Utah, as rounded grains and as octahedral crystals. Named after Mr. Frank Neighbor, District Geologist of the Sun Oil Co. 22nd List

Nekoite. J. A. Gard and H. F. W. Taylor, 1955. *Min. Soc. Notice*, no. 90; *Min. Mag.*, 1956, vol. 31, p. 5. A dimorphous triclinic form of $CaO.2SiO_2.2H_2O$, applied to the 'okenite' from Crestmore, California, which differs in optical and X-ray data from okenite from other localities. Anagrammatic name. 21st List [*A.M.*, **40**, 933; **41**, 958]

Nemaphyllite. F. Focke, 1902. *Min. Petr. Mitt.* (*Tschermak*), vol. xxi, p. 323 (Nemaphyllit). A chlorite-like mineral occurring as greenish scales with a fibrous structure: it has the composition of serpentine ($H_4Mg_3Si_2O_9$), containing also 2·11% Na_2O. It is from the Zillerthal, Tyrol, where it sometimes forms regular intergrowths with dolomite. Named from *νῆμα*, a thread, and *φύλλον*, a leaf. 4th List

Němecite. J. V. Kašpar, 1941. *Rozpravy České Akad.*, 1941, vol. 51, no. 14 (němecit); *Bull. Internat. Acad. Sci. Bohême*, 1943, vol. 43 (for 1942), p. 276. Hydrous ferric silicate, $H_4Fe_2Si_2O_9.5H_2O$, as a limonite-like encrustation on pyrrhotine from Chiuzbaia (= Kisbánya), Romania. Named after Alois Němec, mineral collector and engineer, of Přerov, Moravia. Probably identical with hisingerite and canbyite (10th List). [*M.A.*, **9**, 186.] 17th List

Nenadkevichite. M. V. Kuzmenko and M. E. Kazakova, 1955. *Doklady Acad. Sci. USSR*, vol. 100, p. 1159 (Ненадкевичит). Niobo-titano-silicate $(Na,Ca)(Nb,Ti)Si_2O_7.2H_2O$, orthorhombic, in alkalic rock from Kola. Named after Konstantin Avtonomovich Nenadkevich, Константин Автономович Ненадкевич, Russian mineralogist and geochemist. [*M.A.*, **12**, 569; *A.M.*, **40**, 1154.] 20th List

Nenadkevite. V. A. Polikarpova, 1956. [*Atomic Energy*, Moscow, no. 3, p. 132.] Abstr. in *Mém. Soc. Russ. Min.*, 1957, vol. 86, p. 123 (ненадкевит). Earlier note in guide to USSR exhibit at Conference on Atomic Energy, Geneva, 1955, p. 7 (nenadkevite). Abstr. in *Amer. Min.*, **41**, 816; **42**, 441. Composition variously given as $(U^{iv},Yt,Ce,Th)U^{vi}(Ca,Mg,Pb)(SiO_4)_2(OH)_4.nH_2O$ and $U^{vi}(U^{iv},Th)(Mg,Ca,Pb)_3(SiO_4)_2(OH)_8.nH_2O$. Named after Konstantin Avtonomovich Nenadkevich, Константин Автономович Ненадкевич, Russian mineralogist and geochemist. Not to be confused with Nenadkevichite of M. V. Kusmenko and M. E. Kazakova, 1955 (20th List). [*M.A.*, **13**, 385.] 21st List [Is probably Coffinite, *A.M.*, **42**, 442; a mixture, *A.M.*, **62**, 1261]

Nenadkewitschit, German transliteration of Ненадкевичит, nenadkevichite (H. Strunz, *Min. Tabellen*, 3rd edit., 1957, p. 278). 22nd List

Neocolemanite. A. S. Eakle, 1911. *Bull. Dep. Geol. Univ. California*, vol. vi, no. 9, p. 179. Described as a variety of colemanite, $2CaO.3B_2O_3.5H_2O$, differing slightly in the angles of the crystals and in optical orientation; but proved by A. Hutchinson (*Mineralogical Magazine*, 1912, vol. xvi, p. 239) to be identical with colemanite. 6th List

Neodigenite. P. Ramdohr, 1943. *Zeits. Prakt. Geol.*, vol. 51, pp. 1, 9 (Neodigenit). Cubic Cu_9S_5 in the system Cu_2S–CuS, and in copper ores. The original digenite (A. Breihaupt, 1844) of approximately this composition is a mixture of chalcosine and covelline, but N. W. Buerger [*M.A.*, **7**, 482; **8**, 254, 364) has transferred the name digenite to cubic Cu_9S_5. [*M.A.*, **9**, 263.] 17th List [Cf. Djurleïte]

Neogastunite. H. Haberlandt and A. Schiener, *Tschermaks Min. Petr. Mitt.*, 1951, ser. 3, vol. 2, pp. 311, 315 (Neogastunit). Presumably a local name for a uranium mineral, later identified as schröckingerite, from the hot springs at Bad Gastein, Salzburg. [*M.A.*, **11**, 433.] 19th List

Neoglauconite [A. V. Kazakov, 1938. *Trans. Sci. Inst. Fertilizers*, Moscow, no. 140, p. 146.] O. M. Shubnikova, *Trans. Inst. Geol. Sci. USSR*, 1940, no. 31 (*Min.-Geochem. Ser.*, no. 6), p. 45 Неоглауконит, Neoglauconite). A variety of glauconite of distinctive form and colour, from phosphorite mines in the Moscow region. 17th List

Neokaolin. A. E. Fersman, 1935. *Scientific study of Soviet mineral resources*, New York, p. 76 (neokaolin). Kaolin produced artificially from nepheline. 14th List

Neomesselite. C. Frondel, 1955. *Amer. Min.*, vol. 40, p. 828. $(Ca,Fe,Mn,Mg)_3(PO_4)_2.2H_2O$, triclinic, from Palermo mine, North Groton, New Hampshire. Synonym of messelite (W. Muthmann, 1889) from Messel, Hesse. [*M.A.*, **13**, 8.] 21st List [*A.M.*, **44**, 469. Cf Parbigite]

Neo-Permutit. *Zeits. Krist.*, 1937, vol. 98, pp. 33–34. Trade-name of the Permutit Company for glauconite used as a water softener. 15th List

Neopurpurite. A Mário de Jesus, [1936]. *Com. Serv. Geol. Portugal*, vol. 19 (for 1933), p. 149 (Neopurpurite). An alteration product of lithiophilite, with a formula, $7(Fe,Mn)_2O_3.5P_2O_5.4H_2O$, slightly different from that of purpurite. From Mangualde, Portugal. [*M.A.*, **6**, 442.] 14th List [Is Heterosite; *M.A.*, **8**, 180]

Neotantalite. P. Termier, *Bull. Soc. franç. Min.*, 1902, vol. xxv, p. 34 (Néotantalite). Minute, regular octahedra resembling pyrochlore, in kaolin from dép. Allier, France. Hydrated tantalate (and niobate) of iron, manganese, and alkalies. Considered to be a dimorphous form of tantalite, but differs considerably from this in composition. 3rd List [Is metamict Microlite, *M.A.*, 73-2887; *A.M.*, **62**, 407]

Nephediewit. (*Chem. Zentralblatt*, 1913, vol. ii, p. 1250. *Fortschritte Min. Krist. Petr.*, 1914, vol. iv, p. 167.) Error for nefedieffite, nefedevite, нефедьевитъ (P. Puzuirevskij, 1872). 7th List

Nephritoid. J. Fromme, 1909. *Min. Petr. Mitt.* (*Tschermak*), vol. xxviii, p. 306. A nephrite-like mineral occurring as a vein in weathered harzburgite in the Radauthal, Harz. It is also similar to nephrite in chemical composition, but it differs, as seen under the microscope, in having the fibres parallel, instead of matted as in true nephrite; it may therefore be described as a compact actinolite. 5th List

—— G. P. Barsanov, 1933. *Trav. Inst. Lomonossoff, Acad. Sci. Leningrad*, no. 2, p. 5 (нефритоид, nephritoïde). A variety of serpentine similar to bowenite. [Not the nephritoid of J. Fromme, 1909; 5th List.] 13th List

Nepouite. E. Glasser, 1906. *Compt. Rend. Acad. Sci. Paris*, vol. cxliii, p. 1173; *Bull. Soc. franç. Min.*, 1907, vol. xxx, p. 17 (Népouite). A hydrated silicate of nickel (NiO,18·1–50·7%) and magnesium (MgO,3·0–30·0%) with the formula $3(Ni,Mg)O.2SiO_2.2H_2O$, occurring as a finely crystalline, bright green to pale yellowish-green powder in niccoliferous peridotite at Népoui in New Caledonia. It differs from connarite, garnierite, etc., and in its optical characters it resembles the chlorites. 4th List [Cf. Lizardite]

Neptunite. G. Flink, *Geol. För. Förh.*, XV. 196, 1893; and *Zeits Kryst. Min.*, XXIII. 346, 1894. O. Sjöström, *Geol. För. Förh.*, XV. 393, 1893. [*M.A.*, **11**, 100.] $(Fe,Mn)O.(Na,K)_2O.4SiO_2.TiO_2$. Monosymmetric. [Narsarsuk] Greenland. 1st List [*A.M.*, **57**, 85. $Na_2KLi(Fe,Mn)_2Ti_2Si_8O_{24}$. Cf. Mangan-neptunite]

Neslite. A. Leymerie, 1846. *Statistique géologique et minéralogique du département de l'Aube*, Troyes, 1846, pp. 73, 117, 669. A. Lacroix, *Minéralogie de la France*, 1901, vol. iii, p. 335. A variety of opal, closely resembling menilite, occurring as reniform nodules in marl at Nesle, dep. Marne. 5th List

Neuquenite. A. W. Allen, 1932. *Engin. and Mining Journ. New York*, vol. 133, p. 566. A variety of asphalt from Neuquen Territory, Argentina, differing from rafaelite [11th List]. Named from the locality. 13th List

Neyite. A. D. Drummond, J. Trotter, R. M. Thompson, and J. A. Gower, 1970. *Canad. Min.*, **10**, 90. $Pb_7(Cu,Ag)_2Bi_6S_{17}$ occurs as monoclinic needles with other sulphosalts in quartz veins of the Lime Creek molybdenum deposit, Kitsault, Alice Arm, British Columbia. Named for C. S. Ney. [*A.M.*, **55**, 1444.] 26th List

N'hangellite. (L. A. Boodle, *Bull. Roy. Botanic Gardens*, Kew, 1907, p. 145; Sir Boverton Redwood, ibid., p. 151.) An elastic bitumen derived from a gelatinous alga, and resembling coorongite; from the neighbourhood of Lake N'hangella in Portuguese E Africa. 4th List [Var. of Coorongite; *M.A.*, 69-2188]

Ni-Chlorit, variant of Nickelchlorit (H. Strunz, *Min. Tabellen*, 4th edit., 1966, p. 398). 24th List

Ni-phlogopite, *see* Nickel-phlogopite.

Ni-Skutterudit, variant of Nickel-skutterudite (H. Strunz, *Min. Tabellen*, 3rd edit., 1957, p. 400). 24th List

Ni-serpentine. D. M. Roy and R. Roy, 1954. *A.M.*, **39**, 957–975. *Artificially* prepared serpentine minerals with Ni substituting for Mg. Cf. nickel-antigorite (22nd List) and nickel-chrysotile (q.v.). 28th List

Nichellinneite, *see* Nickel-linnæite.

Nicholsonite. G. M. Butler, 1913. *Economic Geology*, vol. viii, p. 8. A variety of aragonite containing up to 10% of zinc, from Leadville, Colorado. Named after Mr. S. D. Nicholson of the Western Mining Company. 6th List [Zn concentrated in inclusions; *M.A.*, **18**, 123]

Nichtspießglanze. H. Strunz, 1957. *Min. Tabellen*, 3rd edit., p. 111. A term covering proustite, pyrargyrite, xanthoconite, and pyrostilpnite. 22nd List

Nickel. P. Ramdohr, 1967. [Геол. рудн. месторожд. (Geol. ore-deposits), no. 2, p. 32]; abstr. *Amer. Min.*, vol. 53, p. 348. Native nickel (98% Ni) occurs as cubic grains up to 0·1 mm enclosed in heazlewoodite from Bogota, Canala, New Caledonia. 25th List

Nickel-antigorite. H. Strunz, 1957. *Min. Tabellen*, 3rd edit., pp. 323, 385 (Nickel-Antigorit). An unnecessary name for nickelian antigorite. Karpinskite (21st List) is 'vielleicht ein Nickel-Saponit oder Nickel-Antigorit'. 22nd List

Nickel-asbolane. S. Dimitrov, 1942. *Ann. Univ. Sofia, Fac. Phys.-Mat., Sci. Nat.*, vol. 38, p. 207 (никеловъ асболанъ, Bulgarian), p. 225 (Nickel-Asbolan, German). A variety of asbolane containing NiO 3·58%, from Bulgaria [*M.A.*, **10**, 300.] 18th List

Nickelbleipyrit. F. Machatschki, 1953. *Spez. Min.*, Wien, p. 306. A sulphide of nickel and lead with some Ag, Cu, and Co, formulated $(Ni,Pb,Ag,Cu,Co)S_2$, is mentioned; its specific gravity is given as 6·01, but no reference, note of origin, or other data are given. Doubtful. Cf. Penroseite (11th List). 22nd List

Nickelblödite. E. H. Nickel and P. J. Bridge, 1977. *M.M.*, **41**, 37. Magnesian $Na_2Ni(SO_4)_2.4H_2O$, the Ni analogue of blödite, occurs at Kambalda and Carr Boyd Rocks, Western Australia. The Kambalda material, with Ni:Mg = 5·6, has α 1·513, γ 1·520; the Carr Boyd material has Ni:Mg = 1·2, and α 1·504, γ 1·509. [*M.A.*, 77-2187; *A.M.*, **62**, 1059.] 30th List

Nickelbloedite. *A.M.*, **62**, 1059. An improperly-formed name for the Ni analogue of blödite [Cf. Dana, *Syst. Min.*, 5th edit., 1868, pp. xxxxii and xxxxiii, sect. 13h.] 30th List

Nickel-cabrerite. H. Meixner, 1951. *Neues Jahrb. Min.*, Monatshefte, 1951, p. 19 (Nickel-cabrerit). Synonym of cabrerite, $(Ni,Mg)_3(AsO_4)_2.8H_2O$. *See* Cobalt-cabrerite. [*M.A.*, **11**, 312.] 19th List

Nickelchlorite. H. Strunz, 1957. *Min. Tabellen*, 3rd edit., p. 317 (Nickelchlorit). Some of the natural nickel silicates may be members of the chlorite group (cf. *M.A.*, **8**, 14). 22nd List

Nickel-chrysotile. H. Strunz, 1957. *Min. Tab.*, 3rd edit., 323 (Nickel-Chrysotil). An unnecessary name for nickeloan chrysotile. Cf. nickel-antigorite (22nd List) and Ni-serpentine (q.v.). 28th List

Nickel-cobaltomelane. K. K. Nikitin, 1960. Кора выветривания (Crust of weathering), vol. 3, p. 41 (Никель-кобальтомелан). *See* Alumocobaltomelane. 22nd List [A mixture]

Nickeleisenkies. F. Machatschki, 1953. *Spez. Min.*, Wien, p. 310. Syn. of Violarite. 22nd List

Nickeleisenpyrit. F. Machatschki, 1953. *Spez. Min.*, Wien, p. 306. Syn. of Bravoite. 22nd List

Nickelemelane. I. I. Ginzburg and I. A. Rukavishnikova, 1951 (никелемелан). *See* Alumocobaltomelane. 22nd List

Nickel-epsomite. C. O. Hutton, 1947. *Amer. Min.*, vol. 32, p. 553 (nickelian epsomite); Hintz, *Min.*, 1954, Ergbd. 2, p. 284 (Nickelepsomit). A mineral (NiO 12·22, MgO 9·81%) of the epsomite group, approaching morenosite in composition. *See* Magnesium-morenosite. [*M.A.*, **10**, 388.] 20th List

Nickelhexahydrite. B. V. Oleinikov, S. L. Shvartsev, N. T. Mandrikova, and H. N. Oleinikova, 1965. Зап. Всесоюз. Мин. Общ. (*Mem. All-Union Min. Soc.*), vol. 94, p. 534 (Никельгексагидрит, Nickelhexahydrite). Crusts and coatings in the Severnaya mine, Norilsk, USSR, are a magnesian ferroan variety of the nickel analogue of Hexahydrite (monoclinic, dimorphous with Retgersite). Named from the composition. [*M.A.*, **17**, 697; *A.M.*, **51**, 529.] 24th List

Nickel iodine boracite. Author? Mentioned in *The New Scientist*, 1966, vol. 30, p. 314. *Artificial* cubic $Ni_3B_7O_{13}I$, analogous to Boracite. 24th List

Nickel-β-kerolite. I. A. Rukavishnikova, 1956. [*Kora vyvetrivaniya, Acad. Sci. USSR*, no. 2, p. 141.] Abst. in *Mém. Soc. Russ. Min.*, 1957, vol. 86, p. 125 (никелевый β-керолит, nickel-β-kerolite). $(Mg,Ni)_3Si_4O_{10}(OH)_2.H_2O$, containing MgO 20·05, NiO 18·06%, compact, bright green, with karpinskite (q.v.) in serpentine. Urals. 21st List

Nickelkobaltkies. F. Machatschki, 1953. *Spez. Min.*, Wien, p. 310. Syn. of Siegenite. 22nd List

Nickel-linnæite. F. Zambonini, 1916. *Riv. Min. Crist. Italiana*, vol. 47, p. 48 (nichellinneite). An alternative name for polydymite, which is regarded as belonging to the linnæite group. [*M.A.*, **1**, 259] 9th List

Nickel-magnetite. C. Doelter, 1926. *Mineralchemie*, vol. 3, part 2, p. 666 (Nickelmagnetit). Synonym of trevorite $NiFe_2O_4$ (10th List). 19th List [Also 20th List]

Nickelmagnetkies, a mixture of pentlandite and pyrrhotine (H. Strunz, *Min. Tabellen*, 3rd edit., 1957, p. 400, without reference). 24th List

Nickel-montmorillonite. K. Kinoshita, N. Tanaka, and T. Honda, 1958. *Journ. Min. Soc. Japan*, vol. 3, pp. 468, 791. (Nickeliferous montmorillonite, pl. 3 and p. 792; nickel-montmorillonite, p. 792.) A pale blue mineral previously classed as garnierite, from the Niu mine, Oita prefecture, Japan. [*M.A.*, **14**, 281.] Unnecessary name for a nickelian montmorillonite. 22nd List

Nickel-olivine. D. P. Grigoriev, 1937. *Bull. Soc. Nat. Moscou*, vol. 45, (*Sect. Géol.*, vol. 15), p. 152 (никелевый оливин), p. 153 (nickel olivine). Artificial crystals of nickel orthosilicate, Ni_2SiO_4. [*M.A.*, **7**, 142.] 15th List

Nickel-Palladium Arsenide. L. V. Razin, V. D. Begizov, and V. I. Meshchankina, 1973.

[*Trudy Ts NIGRI*, **108**, 96], abstr. *A.M.*, **61**, 180. Two minerals from the Talnakh deposit, USSR, are formulated $(Ni,Pd)_{5\pm x}As_2$ and β-$(Ni,Pd)_{2+x}As$. The powder pattern of the former is indexed as hexagonal, *a* 9·910, *c* 6·601 Å. [Not enough data to differentiate between the two; synthetic Ni_5As_2 is hexagonal, *a* 6.70, *c* 12·41 Å.—*L.J.C.*] 29th List

Nickel-phlogopite. C. Klingsberg and R. Roy, 1957. *Amer. Min.*, 1957, vol. 42, p. 629 (nickel phlogopite), p. 631 (Ni-phlogopite). Artificially prepared $KNi_3AlSi_3O_{10}(OH)_2$, with Ni in place of Mg (Mg-phlogopite). 21st List

Nickel-pimelite. C. W. F. T. Pistorius, 1963. *Neues Jahrb. Min.*, Monatsh., p. 30. Synonym of Alipite (= Pimelite of Schmidt, as distinct from Pimelite of Karsten). 24th List

Nickel-pyrite. H. Strunz, *Min. Tab.*, 1941, p. 80; P. Ramdohr, Klockmann's *Lehrb. Min.*, 13th edit., 1948, p. 376 (Nickelpyrit). Syn. of bravoite. 18th List

Nickel-skutterudite. E. Waller and A. J. Moses, *School of Mines Quart.*, XIV. 49, 1892. $(Ni,Co,Fe)As_3$. [Bullard's Peak dist., Grant Co.] New Mexico. 1st List [Included in chloanthite field; *M.A.*, **14**, 471]

Nickelspeise. *See* Placodine. 7th List

Nickelspinel. H. Strunz, 1957. *Min. Tabellen*, 3rd edit., p. 137 (Nickelspinell). A name for artificial $NiAl_2O_4$. 22nd List

Nickel-talc. C. W. F. T. Pistorius, 1963. *Neues Jahrb. Min.*, Monatsh., p. 30. *Artificial* $Ni_3Si_4O_{10}(OH)_2$, isostructural with Talc. 24th List

Nickel-vermiculite. C. S. Ross and E. V. Shannon, *Amer. Min.*, 1926, vol. 11, p. 90 (nickeliferous vermiculite), NiO 11·25%, North Carolina [*M.A.*, **3**, 352]. A. P. Nikitina, 1956. [*Kora vyvetrivaniya, Acad. Sci. USSR*, no. 2, p. 188 (никелевый вермикулит).] Variety of vermiculite containing NiO 8·6%, from weathering of ultra-basic rocks, Ukraine. [*M.A.*, **13**, 514.] 21st List

Nicolatie, error for Niocalite.

Nicolayite. E. S. Simpson, 1930. *Journ. Roy. Soc. W. Australia*, vol. 16, p. 33. Hydrous silicate of thorium, uranium (UO_3), etc., $2(Pb,Ca)O.3ThO_2.4UO_3.8SiO_2.21H_2O$, differing from thorogummite (to which the mineral had been provisionally referred in 1912) in containing some lead and calcium. Named after the late Rev. Charles G. Nicolay, who formed the first collection of Western Australian minerals. [*M.A.*, **4**, 346.] 12th List [Is Thorogummite; *A.M.*, **38**, 1007]

Nifesite. H. Löfquist and C. Benedicks, 1940. *Jernkontorets Ann. Stockholm*, vol. 124, pp. 658, 682; *K. Svenska Vetenskapsakad. Handl.*, 1941, ser. 3, vol. 19, no. 3, p. 43 (nifesit), p. 91 (nifesite). Nickel-iron sulphide consisting of a very fine-grained aggregate of bravoite and pentlandite, occurring in native iron and in pyrrhotine from Ovifak, Greenland. Named from the chemical symbols Ni, Fe, S. [*M.A.*, **8**, 312.] 16th List [A mixture]

Nifontovite. S. V. Malinko and A. E. Lisitsyn, 1961. Доклады Акад. Наук СССР (*Compt. Rend. Acad. Sci. URSS*), vol. 139, p. 188 (Нифонтовит). $CaB_2O_4.3\frac{1}{2}H_2O$, small anhedral grains in skarn deposits in the Urals. Named after the geologist R. V. Nifontov (Р. В. Нифонтов.)* 22nd List [*Corr. in *M.M.*, **33**] [$2{\cdot}3H_2O$, not $3\frac{1}{2}$; error in 22nd List. *A.M.*, **47**, 172, 1482. Cf. Frolovite]

Nigerite. R. Jacobson and J. S. Webb, 1947. *Min. Mag.*, vol. 28, p. 118, F. A. Bannister, M. H. Hey, and H. P. Stadler, ibid., p. 129. Aluminate of tin (SnO_2 25·33%), iron, zinc, etc. $(Zn,Mg,Fe'')(Sn,Zn)_2(Al,Fe''')_{12}O_{22}(OH)_2$, as dark brown hexagonal plates from Nigeria. Named from the country. [*M.A.*, **10**, 276.] 18th List [*A.M.*, **52**, 864.] R. Jacobson and J. S. Webb, 1946. *Bull. Geol. Surv. Nigeria*, no. 17, p. 27. Earlier reference than that given in 18th List. 20th List

Niggliite. D. L. Schlotz, 1936. *Publ. Univ. Pretoria*, 1936, ser. 2, no. 1, p. 184; preprint from *Trans. Geol. Soc. South Africa*, 1937, vol. 39 (for 1936), p. 184. Platinum telluride ($PtTe_3$?) as anisotropic silver-white grains from South Africa. Named after Professor Paul Niggli (1888–) of Zürich. [*M.A.*, **6**, 440.] 14th List [Composition PtSn; *M.M.*, **38**, 796]

Nigrite. C. A. Peterson, *20th Ann. Rept. US Geol. Survey*, 1899, part VI, p. 257. A kind of asphaltum from Utah. 2nd List

Nimesite. Z. Maksimović, 1972. [*Bull. Sci. Cons. Acad. Sci. Arts. R.S.F. Yugoslav., sect. A*, **17**, 224]; abstr. *A.M.*, **58**, 1112 (1973). The nickel analogue of amesite, in green fibrous aggregates from the Marmara bauxite, Greece. Not to be confused with nimite, 26th List, a chlorite. 28th List [Renamed Brindleyite (q.v.)]

Nimite. S. A. Hiemstra and S. A. de Waal, 1968. *Nat. Inst. Met.* (*South Africa*) *Res. Rept.*, **344**, 1; *Amer. Min.*, **55**, 18 (1970). $(Ni,Mg,Fe,Al)_{12}(Si,Al)_8O_{20}(OH)_{16}$, a chlorite with Ni as dominant cation; yellowish-green, monoclinic; from rocks near the Scotia talc mine, Barberton Mountain Land, Transvaal. Named from the initials of the Nat. Inst. Metallurgy. [*A.M.*, **54**, 1740; *M.A.*, 70-2605.] 26th List

Ningyoite. T. Muto, R. Meyrowitz, A. M. Pommer, and T. Murano, 1959. *Amer. Min.*, vol. 44, p. 633. $CaU(PO_4)_2.1\frac{1}{2}H_2O$, with some replacement of Ca and U by lanthanons; orthorhombic. Occurs in an unoxidized zone of the Ningyo-toge mine, Tottori prefecture, Japan. Named from the locality. [*M.A.*, **14**, 415.] 22nd List

Niningerite. K. Keil and K. G. Snetsinger, 1967. *Science*, vol. 155, p. 451. (Mg,Fe,Mn)S, cubic, occurs in six enstatite chondrites. Named for H. H. Nininger. [*A.M.*, **52**, 925; Зап., **97**, 66; *Bull.*, **90**, 271.] 25th List

Niobanatase. E. I. Semenov, 1957. Труды Инст. Мин., Геохим., Крист., Редк. Элем. (*Trans. Inst. Min. Geochem. Cryst. Rare Elements*), no. 1, p. 41 (Ниобоанатаз, Nb-анатаз). Abstr. Зап. Всесоюз. Мин. Общ. (*Mem. All-Union Min. Soc.*), 1959, vol. 88, p. 311 (Ниобанатаз, niobanatase). An incompletely analysed mineral (TiO_2 31·1, Nb_2O_5 21·6%) from natrolite-lined cavities in the Lovozero massif, Kola peninsula. X-ray powder photographs resemble those of anatase. [*M.A.*, **14**, 278.] The name is premature with only 53% of the composition accounted for. 22nd List

Niobo-aeschynite. A. G. Zhabin, G. N. Mukhitdinov, and M. E. Kazakova, 1960. Труды Инст. Мин. Геохим. Крист. Редк. Элем. (*Trans. Inst. Min., Geochem., Cryst. Rare Elem.*), vol. 4, p. 51 (Ниобоэшинит). Near $Ln(Nb,Ti)_2O_6$ with Nb > Ti (Aeschynite has Ti > Nb), in black prisms from Vishnevye Gor. Named from the composition [*M.A.*, **16**, 556; *A.M.*, **47**, 417.] 23rd List

Niobobelyankinite. E. I. Semenov, 1957. Труды Инст. Мин., Геохим., Крист., Редк. Элем. (*Trans. Inst. Min., Geochem., Cryst. Rare Elements*), no. 1, p. 41 (Ниобобелянкинит, Nb-белянкнит). Syn. of Gerasimovskite (q.v.). 22nd List

Niobochevkinite. A superfluous name for niobian Chevkinite [Геохим., мин., генет. типы месторожд. редк. элем., Изд. „Наука", 1964, p. 308]; abstr. Зап. Всесоюз. Мин. Общ. (*Mem. All-Union Min. Soc.*), vol. 94, p. 683 (Ниобочевкинит, Niobochevkinite). 24th List

Niobo-eschynite, undesirable variant spelling of Niobo-aeschynite., Зап. Всесоюз. Мин. Общ. (*Mem. All-Union Min. Soc.*), 1963, vol. 92, p. 199; *Amer. Min.*, 1962, vol. 47, p. 417. 23rd List

Niobolabuntsovite. A superfluous name for niobian Labuntsovite. [Геохим., мин., генет. типы месторожд. редк элем., Изд. „Наука", 1964, p. 533]; abstr. Зап. Всесоюз. Мин. Общ. (*Mem. All-Union Min. Soc.*), 1965, vol. 94, p. 683 (Ниоболабунцовит, niobolabunzovite). 24th List

Nioboloparite. I. P. Tikhonenkov and M. E. Kazakova, 1957. *Mém. Soc. Russ. Min.*, vol. 86, p. 641 (ниоболопарит). Variety of loparite (10th List) with niobium predominating over tantalum, $(Na,Ce,Ca)(Ti,Nb)O_3$, as black octahedra in alkali-pegmatite from Kola. 21st List

Niobophyllite. E. H. Nickel, J. F. Rowland, and D. J. Charette, 1964. *Canad. Min.*, vol. 8, p. 40. An anorthic mineral from Seal Lake, Labrador, has a unit-cell containing approximately $(K,Na)_3(Fe,Mn)_{6\cdot4}(Nb,Ti)_2(Si,Al)_{7\cdot7}(O,OH,F)_{31}$, and is regarded as the niobium analogue of Astrophyllite for which a unit-cell containing twice $(K,Na)_3(Fe,Mn)_7Ti_2Si_8(O,OH,F)_{31}$ has recently been found (*M.A.*, **16**, 611, 615). [*A.M.*, **50**, 263.] 24th List

Niobotapiolite. A superfluous name for niobian Tapiolite. [Геохим., мин., генет. типы месторожд. редк. элем., Изд. „Наука", 1964, p. 438]; abstr. Зап. Всесоюз. Мин. Общ. (*Mem. All-Union Min. Soc.*), 1965, vol. 94, p. 683 (Ниоботапиолит, niobotapiolite). 24th List

Niobotschewkinit, German transliteration of Ниобочевкинит, Niobochevkinite (24th List) (C. Hintze, *Handb. Min.*, Erg. III, 605). 25th List

Niobozirconolite. L. S. Borodin, A. V. Bykova, T. A. Kapitonova, and Yu. A. Pyatenko, 1960. Доклады Акад. Наук СССР (*Compt. Rend. Acad. Sci. URSS*), vol. 134, p. 1188 (Ниобоцирконолит). An unnecessary name for a niobian zirconolite from the Vuorijarvi massif, Kola peninsula. 22nd List [Syn. of Zirkelite, refs. in *A.M.*, **62**, 408]

Niobpyrochlore. *See* under Antimonpyrochlore. Syn. of pyrochlore, *A.M.*, **62**, 407.

Niocalite. E. H. Nickel, 1956. *Amer. Min.*, vol. 41, p. 785, p. 969 (nicolatie). K. Kern, A. Rimsky, and J. C. Monier, *Compt. Rend. Acad. Sci. Paris*, 1957, vol. 245, p. 2063. Niobo-silicate of calcium, $Ca_4NbSi_2O_{10}(OH,F)$, orthorhombic, as pale yellow prismatic crystals in metamorphic limestone from Oka, Quebec. Named from the composition. [*M.A.*, **13**, 211.] 21st List

Nisaite. J. de Lencastre and M. M. Varinho, 1970. *Estudo de um mineral urainífero da região de Nisa* (*Portugal*): [*Presidência do Conselho, Junta de Energia Nuclear, Direcção Geral dos Serviços de Prospecção e Exploração Mineiras*]. Thin yellow films and crusts associated with saléeite at Nisa, (39°30′N, 7°40′W), Portugal, consist of a phosphate of uranium and calcium; X-ray powder data (strongest lines 8·00, 4·00, 3·09 Å) are distinct from those of phosphuranylite and of autunite [also from those of pseudo-autunite (22nd List) and ningyoite (24th List)]. RIs α 1·67‖elongation, β 1·72, γ 1·77, straight extinction; weak yellow-green fluorescence in U.V. light. Named for the locality. [May well be a valid species, but adequate chemical data are lacking—*M.H.H.*] 30th List

Nisbite. L. J. Cabri, D. C. Harris, and J. M. Stewart, 1970. *Canad. Min.*, **10**, 240. Small (<20 μm) grains with metallic lustre in two polished sections from Mulcahy Township, Kenora District, Ontario, have the composition $NiSb_2$. X-ray powder data agree with those of synthetic $NiSb_2$, and lead to a cell with *a* 5·162, *b* 6·303, *c* 3·839 Å. Named for the composition. Occurs in the same deposit as paracostibite (q.v.). [*A.M.*, **56**, 631; *M.A.*, 71-2333; *Zap.*, **100**, 615.] 27th List

Nissonite. M. E. Mrose, R. Meyrowitz, J. T. Alfors, and C. W. Chesterman, 1966. *Progr. Ann. Meeting Geol. Soc. Amer.*, p. 145. Blue-green monoclinic crystals of $CuMgPO_4(OH).2\frac{1}{2}H_2O$ occur in metamorphic rocks in Panoche Valley, California. Named for W. H. Nisson. [*A.M.*, **52**, 927; *Bull.*, **91**, 99.] 25th List

Nitrate-apatite. D. McConnell, 1938. *Amer. Min.*, vol. 23, p. 8. 'The fact that a nitrate-apatite is not known does not weaken the hypothesis' that NO_3 groups may enter into the apatite structure. 16th List

Nitrate-hydrotalcite. D. M. Roy, R. Roy, and E. F. Osborn, 1953. *Amer. Journ. Sci.*, vol. 251, p. 355 (nitrate-hydrotalcite). Artificially produced $Mg_6Al_2(N_2O_5)(OH)_{16}.4H_2O$, analogous to hydrotalcite ('carbonate-hydrotalcite') with N_2O_5 in place of CO_2. [*M.A.*, **12**, 196.] 20th List

Nitratite. Variant of Nitratine, *A.M.*, **8**, 52; **9**, 62. 10th List

Nitrokalite. H. Strunz, *Min. Tab.*, 1941, p. 115 (Nitrokalit, Kalisalpeter). Syn. of nitre, KNO_3. 18th List

Nitronatrite. E. F. Glocker, 1847. *Generum et specierum mineralium . . . synopsis*, p. 292 (Nitronatrites, Nitronatrit). Syn. of nitratine, soda-nitre, $NaNO_3$. 18th List

Nobleite. R. C. Erd, J. F. McAllister, and A. C. Vlisidis, 1961. *Amer. Min.*, vol. 46, p. 560. $CaB_6O_{10}.4H_2O$, monoclinic crystals with other borates from Furnace Creek, Death Valley, California. Named after Dr. Levi F. Noble of the US Geol. Survey. 22nd List

Nogisawaite. Z. Harada, *Journ. Fac. Sci. Hokkaido Univ.*, 1954, sect. 4, vol. 8, p. 322 [*M.A.*, **12**, 477]. Another spelling of nogizawalite (19th List), perhaps a variety of beckelite. [*M.A.*, **11**, 311.] 20th List

Nogizawalite. T. Kawai, 1949. *Journ. Chem. Soc. Japan, Pure Chem. Sect.*, vol. 70, p. 268. Silicate and phosphate of rare-earths, Zr, Al, Fe, Mg, Ca, tetragonal. Named from the locality, Nogizawa village, Ishikawa district, Japan. [Perhaps a mixture of zircon, xenotime, etc. Cf. oyamalite, 11th List, and yamagutilite, 14th List.] [*M.A.*, **11**, 311.] 19th List

Nolanite. S. C. Robinson, 1955. *Bull. Geol. Surv. Canada*, no. 31, p. 67. W. H. Barnes and M. M. Qurashi, *Amer. Min.*, 1952, vol. 37, p. 420 (iron vanadate [*M.A.*, **12**, 210]). S. C. Robinson, H. T. Evans, W. T. Schaller, and J. J. Fahey, *Amer. Min.*, 1957, vol. 42, p. 619. Iron vanadate, $3FeO.V_2O_3.3V_2O_4$ or $4FeO.V_2O_3.4V_2O_4$, as minute, black hexagonal plates with uranium ore from Beaverlodge (= Goldfields), Saskatchewan. Intergrown with a second phase ($FeO.2V_2O_4$). Named after Thomas Bennan Nolan (1901–), Director, US Geological Survey. 21st List

Noonkanbahite. R. T. Prider, 1965. *Min. Mag.*, vol. 34, p. 403. A highly pleochroic accessory mineral in coarse-grained lamproite at Wolgidee Hills, West Kimberley, Western Australia, has a composition near $NaKBaTi_2Si_4O_{14}$, and is related to Batisite ($Na_2BaTi_2Si_4O_{14}$) and Shcherbakovite ($Na\check{K}(Ba,K)(Ti,Nb)_2Si_4O_{14}$). The X-ray powder data differ from those of Batisite except for the four strongest lines. Named from the Noonkanbah sheep station, on which most of the lamproites occur. [*A.M.*, **50**, 2105.] 24th List

Norbergite. P. Geijer, 1926. *Geol. För. Förh. Stockholm*, vol. 48, p. 84; *Sveriges Geol. Unders.*, 1927, Årsbok 20 (for 1926), no. 4, p. 16. A magnesium silicate with fluorine and hydroxyl, $Mg_2SiO_4.Mg(F,OH)_2$, as pink, massive aggregates from Norberg, central Sweden. It is optically distinct from 'prolectite' for which the same formula was suggested. Named from the locality. [*M.A.*, **3**, 110; **3**, 273.] 11th List

Nordenskioeldite. Variant of Nordenskiöldine [of W. C. Brögger, 1887; not the nordenskiöldite of A. Kenngott, 1854]. *A.M.*, **8**, 51. 10th List

Nordite. V. I. Gerasimovsky, 1941. *Compt. Rend. (Doklady) Acad. Sci. URSS*, vol. 32, p. 496. Silicate of rare-earths, Na, Sr, Ca, Mn, as pale-brown orthorhombic crystals from the Kola peninsula, Russia. So named because of its northern origin. [*M.A.*, **8**, 279.] 16th List

Nordstrandite. D. Papée, R. Tertian, and R. Biais, 1958. *Bull. Soc. Chim. France*, p. 1301. The polymorph of $Al(OH)_3$ synthesized by R. A. van Nordstrand, W. P. Hettinger, and C. D. Keith, 1956. Since found occurring as a mineral in Sarawak (J. R. D. Wall, E. B. Wolfenden, E. H. Beard, and T. Deans, *Nature*, 1962, vol. 196, p. 264; *Min. Mag.*, vol. 33, p. lxix) and in Guam (J. C. Hathaway and S. O. Schlanger, *Nature*, 1962, vol. 196, p. 265). [*M.A.*, **16**, 283, 284; *A.M.*, **48**, 214.] 23rd List [Colorado; *M.A.*, 75-3553]

Norilskite. O. E. Zvyagintsev, 1940. *Compt. Rend. (Doklady) Acad. Sci. URSS*, vol. 26, p. 790. An alloy of platinum with iron, nickel, copper, and palladium, found as grains and cubic crystals in alluvial deposits at Norilsk, N Siberia. Named from the locality. [*M.A.*, **8**, 53, 341.] 16th List [A mixture; *A.M.*, **55**, 1067]

Normannite. A. Weisbach [1833–1901], MS. A. Tetzner and F. Edelmann, *Jahrb. Berg- u. Hüttenw. Sachsen.*, 1927, vol. 101, p. A 102. Basic bismuth carbonate, $3Bi_2O_3.CO_2$, as brown globular aggregates from Schneeberg, Saxony. Named after Dr. — Normann (d. 1883 in Zwickau). [*M.A.*, **3**, 540.] 11th List [Is Bismutite; *A.M.*, **28**, 521]

Norsethite. C. Milton, M. E. Mrose, E. C. T. Chao, and J. J. Fahey, 1959. *Bull. Geol. Soc. Amer.*, vol. 70, p. 1646 (abstr.); *Amer. Min.*, 1961, vol. 46, p. 420. $BaMg(CO_3)_2$, in rhombohedral crystals in the Westvaco trona mine, Wyoming. Named after Mr. Keith Norseth, engineering geologist. [*M.A.*, **14**, 343.] 22nd List

Northetit, error for Norsethit. (C. Hintze, *Handb. Min.*, Erg. III, p. 232.) 25th List

Northupite. W. M. Foote, *Proc. Acad. Nat. Sci.*, Philadelphia, 1895, 408, 1896; and *Amer. Journ. Sci.*, L. 480, 1895. J. H. Pratt, *Amer. Journ. Sci.*, II. 124, 1896; and *Zeits. Kryst. Min.*, XXVII. 418, 1896. [*M.M.*, **11**, 159, 226.] $MgCO_3.Na_2CO_3.NaCl$. Cubic. [Borax Lake, San Bernardino Co.] California. 1st List [Cf. Tychite]

Nováčekite. C. Frondel, 1951. *Amer. Min.*, vol. 36, p. 680 (Novacekite). $Mg(UO_2)_2(AsO_4)_2.9H_2O$, pseudo-tetragonal, isomorphous with saléeite, the corresponding phosphate. From Schneeberg, Saxony. Named after Radim Nováček (1905–1942), Czech mineralogist. [*M.A.*, **11**, 413.] 19th List

Novacikit, error for Nováčekit (F. Machatschki, *Spez. Min.*, Wien, 1953, pp. 348, 370). 22nd List

Novákite. Z. Johan and J. Hak, 1959. *Chemie der Erde*, vol. 20, p. 49 (Novákit). Arsenide of Cu and Ag, occurring in carbonate gangue at Černy Důl (Schwarzenthal), Czechoslovakia. Named after Prof. Jiři Novák of Prague. The formula is given as $(Cu,Ag)_4As_3$ or $(Cu,Ag)_{11}As_8$, but the constants given lead to a cell content of $(Cu,Ag)_{26.5}As_{19.9}$ (M. Fleischer, *Amer. Min.*, 1959, vol. 44, p. 1321). 22nd List [$(Cu,Ag)_4As_3$; *A.M.*, **46**, 885]

Novoelpidite. Mentioned without description by V. I. Gerasimovsky, *Compt. Rend. (Doklady) Acad. Sci. URSS*, 1939, vol. 22, p. 263, as an associate of chkalovite in the Kola peninsula. *M.A.*, **7**, 314.] 17th List

Nowackiite. F. Marumo and G. Burri, 1965. [*Chimia*, vol. 19, p. 500]; abstr. *Amer. Min.*, 1966, vol. 51, p. 532. About ten grey to black crystals on honey-yellow blende from Lengenbach, Binn, Valais, Switzerland, are rhombohedral, composition near $Cu_6Zn_3As_4S_{12-13}$. X-ray powder data (not published!) show a strong relation to the Blende structure. Named for W. Nowackii. 24th List [*M.A.*, 76-1967]

Nsutite. W. K. Zwicker, W. O. J. Groenveld Leijer, and H. W. Jaffe, 1962. *Amer. Min.*, vol. 47, p. 246. The natural occurrence of γ-MnO_2 at Nsuta, Ghana, is confirmed; it is named from the locality, and is shown to be a fairly common mineral. It had earlier been described and named Yokosukaite (q.v.) by K. Kami and T. Tanaka (1937). [*M.A.*, **16**, 65.] 23rd List

Nuevite. J. Murdoch, 1946. *Bull. Geol. Soc. Amer.*, vol. 57, p. 1219; *Amer. Min.*, 1947, vol. 32, p. 204. Titano-niobate of yttrium and iron as black grains and orthorhombic (?) crystals in quartz from Nuevo, Riverside Co., California. Named from the locality. [*M.A.*, **10**, 100.] 18th List [Is Samarskite; *A.M.*, **36**, 358]

Nuffieldite. P. W. Kingston, 1968. *Canad. Min.*, **9**, 439. Steel-grey orthorhombic crystals of $Cu_4Pb_{10}Bi_{10}S_{27}$ occur in a quartz vein in the Lime Creek stock near Alice Arm, British Columbia. Named for Professor E. W. Nuffield. [*A.M.*, **54**, 574; *Bull.*, **92**, 317; *M.A.*, 70-1641; *Zap.*, **98**, 321.] 26th List [*A.M.*, **57**, 319]

Nuissierite. The same as nussierite (G. Barruel, 1836). A calciferous mimetite from the Nuissière lead-mine near Chenelette, dep. Rhône, France. The new spelling (nuissiérite) is given by A. Des Cloizeaux, *Manuel de Minéralogie*, 1893, vol. ii, p. 519. 4th List

Nuolaite. L. Lokka, 1928. *Bull. Comm. Géol. Finlande*, no. 82, p. 21 (Nuolait). A black mineral resembling wiikite but consisting of an intimate intergrowth of an opaque crystalline material and a colourless isotropic material. The bulk analysis indicates a hydrous meta-niobate (and tantalate) and titanate of yttrium, iron, thorium, etc. Named from the locality, Nuolainniemi, Impilahti parish, Lake Ladoga, Finland. [*M.A.*, **4**, 250.] 12th List [A mixture, *A.M.*, **47**, 812; **62**, 407]

Nyerereïte. C. Milton and B. Ingram, 1963. Cited by H. Strunz, *Min. Tabellen*, 4th edit., 1966, p. 214, without reference or details of occurrence, etc. $Na_2Ca(CO_3)_2$. 24th List [Cf. Natrofairchildite]

O

Oakermanite. *Amer. Min.*, 1920, vol. 5, p. 81. An Anglicized form of the Swedish name åkermanite. 9th List

Oakite. Mentioned by J. W. Gruner (*Amer. Min.*, 1943, vol. 28, pp. 174, 615) as due to W. E. Richmond. Tentative name for a supposed new manganese mineral, afterwards identified with lithiophorite, from White Oak Mountain, Tennessee. 17th List

Oborite. T. L. Ho, 1935. *Bull. Geol. Soc. China*, vol. 14, p. 279 (Oborite). An undetermined mineral presumed to contain rare-earths (La,Ce,Yt,Er), as minute grains (rhombohedral) in fluorite from Beiyin Obo, Inner Mongolia. With beiyinite (q.v.), named from the locality. [*M.A.*, **6**, 151.] 14th List

Obruchevite. Author? *Guide to USSR exhibit at Conference on Atomic Energy*, Geneva, 1955, p. 9 (obruchevite), contains U, Ti, Nb, Ta. Later publications, abstr.: *Amer. Min.*, 1957, vol. 42, p. 307, 1958, vol. 43, p. 380, $(Yt,U,Na_2)Ta_2O_6(OH,F)$; *Mém. Soc. Russ. Min.*, 1958, vol. 87, p. 79 (обручевит, obruchevite), $3Na_2O.4(Ca,Fe)O.3Yt_2O_3.(U,Th)O_2$. [A. P. Kalita], *Doklady Acad. Sci. USSR*, 1957, vol. 117, p. 117, обручевит, variety of pyrochlore in pegmatite from Karelia. Named after the veteran geologist Vladimir Afanasevich Obruchev, Владимир Афанасьевич Обручев (1863–1956). 21st List [Later shown to be two phases, one related to Samiresite and another now to be re-named Yttropyrochlore, *A.M.*, **62**, 407 (contains refs.)]

Obvenite, error for Olivenite (*Amer. Min.*, 1959, vol. 44, p. 1321). 22nd List

Ocrite. H. von Philipsborn, 1953. *Tafeln zum Bestimmen der Mineralien*, Stuttgart, p. xi (Bi-ocrit, Mo-ocrit), p. 72 (Wismutocker, Molybdänocker). Collective name for powdery ochres. [*M.A.*, **12**, 297.] 20th List

'Octahedral borax'. The conditions necessary for the formation of the artificial salt long known as 'octahedral borax' (sodium borate, $Na_2B_4O_7.5H_2O$, crystallizing in rhombohedral forms *r c* with the appearance of regular octahedra) have been investigated by J. H. van 't Hoff and W. C. Blasdale (*Sitzungsber. Akad. Wiss. Berlin*, 1905, p. 1086), who point out that this substance was observed by E. Bechi in 1854 as an incrustation at the Tuscan lagoons. [= Tincalconite.] 4th List

Octobolite. H. Strunz, 1957. *Min. Tabellen*, 3rd edit., p. 291 (Oktobolit). Syn. of Augite. 22nd List

Octophyllite. A. N. Winchell, 1925. *See* Heptaphyllite. 10th List

Odenite. J. J. Berzelius, MS. 1815. *Jac. Berzelius brev* (*Jac. Berzelius lettres*) by H. G. Söderbaum, Uppsala, 1921, vol. 4, part 1, pp. 45, 49, 89, 99 (odenit, odeniten); N. Zenzén, *Geol. För. Förh. Stockholm*, 1926, vol. 48, p. 95 (Odenit). Berzelius in his letters (1815, 1817) to W. Hisinger applied this name to a black mica (biotite) from Finbo, Sweden, which he believed to contain a new chemical element 'odenium'. Other forms of the name, odinite, odite, oderite, have appeared (A. H. Chester's *Dictionary of the names of minerals*, 1896, p. 192); cf. also wodanite (9th List). Named from Odin, the chief deity of Norse mythology. 11th List

Oehrnite. E. S. Fedorov, 1905. *Gornyi Zhurnal*, St. Petersburg, year lxxxi, vol. iii, p. 264 (эрнитъ, Oehrnit). A rock-forming mineral from the Caucasus resembling diallage in appearance: it has three rectangular cleavages, but is shown by its optical characters to be monoclinic. The formula is given as $6(Mg,Fe,Ca)O.6SiO_2.H_2O$, but the analysis shows also Al_2O_3, 6·74%. Named after A. G. Ern, a Russian mining engineer. 4th List [A more correct transliteration is Ernite]

Okermanite. Variant of Åkermanite, *A.M.*, **9**, 62; cf. Oakermanite. 10th List

Olefinite. G. Mueller, 1964. *Rept. 22nd Internat. Geol. Congr.*, India, part 1, 46. A bitumen containing 'according to the results of fractionation, I.R., U.V. spectra, determination of I values, etc., mainly olefinic molecules'. 26th List

Oligoexpandite, *see* Expandite.

Oligonsiderite. F. Ulrich and R. Munk, 1936. *Schlägel und Eisen, Teplitz-Schönau*, reprint p. 10 (Oligonsiderit). Manganese concretions containing $FeCO_3$ 43·73, $MnCO_3$ 33·31% from Slovakia. Synonym of oligon-spar = oligonite. Not to be confused with oligosiderite in A. Daubrée's (1867) classification of meteorites. 17th List

Oliveiraite. T. H. Lee, 1917. *Revista Soc. Brasileira Sci.*, no. 1, p. 31; *Amer. Journ. Sci.*, 1919, vol. xlvii, p. 126; *Chem. News*, 1919, vol. cxviii, p. 125. Hydrated titanate of zirconium, $3ZrO_2.2TiO_2.2H_2O$, occurring as greenish-yellow, radially-fibrous masses with euxenite in Brazil. Named after the Brazilian geologist Dr. Francisco de Paula Oliveira. 8th List

Olmsteadite. P. B. Moore, T. Araki, A. R. Kampf, and I. M. Steele, 1976. *A.M.*, **61**, 5. Deep brown to black prismatic crystals elongated along [010] from the Big Chief pegmatite, Black Hills, South Dakota, have *a* 7·512, *b* 10·000, *c* 6·492 Å, space group $Pb2_1m$; α 1·725, dark blue, ∥ [001], absorption ≫ β 1·755, light brown ∥ [100], and γ 1·815, brown; *D* 3·31. Composition $[K_2Fe_4^{2+}(Nb,Ta)_2(PO_4)_4.4H_2O]$. Also found at the Hesnard pegmatite, Custer, Custer County, South Dakota. Named for Milo Olmstead. 29th List

Olovotantalite, direct transliteration of Оловотанталит; synonym of Stannotantalite. *Amer. Min.*, 1961, vol. 46, p. 1514. 23rd List [Cf. Wodginite]

Olsacherite. C. S. Hurlbut, Jr., and L. F. Aristarain, 1969. *Amer. Min.*, **54**, 1519. $Pb_2SO_4SeO_4$ occurs as an alteration product of penroseite at the Pacajake mine, Colquechaca, Bolivia; the crystals are isostructural with anglesite and $PbSeO_4$, but have a doubled *b*-axis, and hence probably an ordered structure. Named for Professor Juan A. Olsacher. [*M.A.*, 70-2611.] 26th List

Olschanskit, Germ. trans. of Ольшанскит, olshanskyite (26th List). Hintze, 67. 28th List

Olshanskyite. M. A. Bogomolov, I. B. Nikitina, and N. N. Pertsev, 1969. Докл. Акад. наук СССР (*Compt. Rend. Acad. Sci. URSS*), **184**, 1398 (Ольшанскит, olshanskyte). Veinlets in sakhaite from E Siberia contain fibrous, monoclinic or anorthic $3CaO.2B_2O_3.9H_2O$; as the I.R. spectrum shows OH bands but not H_2O bands, the formula is written $Ca_3[B(OH)_4]_4(OH)_2$. Named for Yakov Iosifovich Olshanskii. [*A.M.*, **54**, 1737 (olshanskyite); *M.A.*, 70-755 (olshanskyite); *Zap.*, **99**, 77.] 26th List

Ondřejite. J. V. Kašpar, 1944. *Věda Přírodní*, Praha, vol. 23, p. 132 (Ondřejit). Hydrated carbonate and silicate of magnesium, calcium, and sodium, as a white powder on aragonite from Zbrašov caves, Hranice, Moravia. Named after Prof. Augustin Ondřej of the Technical High School of Praha. [*M.A.*, **9**, 261.] 17th List [A mixture; *A.M.*, **49**, 1502]

Ondrschejit, variant of Ondřejite (C. Hintze, *Handb. Min.*, Erg.-Bd. II, p. 292). 24th List

Onoratoite. G. Belluomini, M. Fornaseri, and M. Nicoletti, 1968. *Min. Mag.*, vol. 36, p. 1037. $Sb_8O_{11}Cl_2$, as anorthic crystals from Cetine di Cotorniano, Rosia, Siena, Italy, and synthetic. Named for E. Onorato. [*A.M.*, **54**, 1219.] 25th List [Monoclinic, *Min. Record*, **8**, 285]

Ooguanolite. A. A. Hayes, 1855. *Proc. Boston Soc. Nat. Hist.*, vol. 5, p. 167 (Oöguanolite). G. E. Hutchinson, *Bull. Amer. Mus. Nat. Hist.*, 1950, vol. 96, p. 94 (ooguanolite). Sulphate of potassium and ammonium, $K_2SO_4.(NH_4)_2SO_4$, in a fossilized egg from guano deposits on an island off the coast of Peru. Named from ᾠόν, egg, and guano. Cf. taylorite $(5K_2SO_4.(NH_4)_2SO_4)$ and guanapite. [*M.A.*, **11**, 245, 302, 551.] 19th List

Oonachatae. *See* discachatae. 3rd List

Oosterboschite. Z. Johan, P. Picot, R. Pierrot, and T. Verbeek, 1970, *Bull.*, **93**, 476. A yellowish metallic-looking mineral with trogtalite (21st List) in the Musonoi Cu–Co deposit, Katanga, is orthorhombic, *a* 10·42, *b* 10·60, *c* 14·43 Å. Composition $8[(Pd,Cu)_7Se_5]$. Named for M. R. Oosterbosch. [*M.A.*, 71-1387; *Zap.*, **100**, 617; *A.M.*, **57**, 1553.] 27th List

Oranite. H. L. Alling, 1921. *Journ. Geol. Chicago*, vol. 29, p. 237. Intergrowths of potash-felspar (orthoclase or microcline) and a plagioclase (near anorthite), analogous to perthite (in which the plagioclase is near albite). A contraction of *or*thoclase-*an*orth*ite*. 10th List

Orcelite. S. Caillère, J. Avias, and J. Falgueirettes, 1959. *Compt. Rend. Acad. Sci. Paris*, vol. 249, p. 1771 (Orcélite). *Bull. Soc. franç. Min. Crist.*, 1961, vol. 84, p. 9. A vein in serpentinized harzburgite in the Tiebaghi massif, New Caledonia, consists almost wholly of a new mineral, Ni_2As, distinct from maucherite and niccolite. Named after Prof. J. Orcel. [*M.A.*, **14**, 342.] Probably identical with the artificial phase $Ni_{5-x}As_2$ (M. Fleischer, *Amer. Min.*, 1960, vol. 45, p. 753). 22nd List

Ordite. Yu. M. Abramovich, 1956. [Вопросы мин. осад. образ. (*Problems Min. Sedim. Formations*), Lvov Univ., Book 3–4, p. 80]; abstr. Зап. Всесоюз. Мин. Общ. (*Mem. All-Union Min. Soc.*), 1959, vol. 87, p. 481 (Ордит.) A name for a pseudomorph of gypsum from Orda, Perm region; named for the locality. [*M.A.*, **14**, 277.] An unnecessary name; confusion with Орлит (Orlite, 21st List) is likely. 22nd List

Ordoñezite. G. Switzer and W. F. Foshag, 1953. *Progr. and Abstr. Min. Soc. Amer.*, p. 40; *Amer. Min.*, 1954, vol. 39, p. 346; 1955, vol. 40, p. 64. Zinc antimonate, $ZnSb_2O_6$, tetragonal, brown crystals in tin ore from Guanajuato, Mexico. Named after the late Ezequiel Ordoñez, Mexican geologist. [*M.A.*, **12**, 303, 512.] 20th List

Oregonite. P. Ramdohr and M. Schmitt, 1959. *Neues Jahrb. Min.*, Monatsh., p. 239 (Oregonit). Probably Ni_2FeAs_2; hexagonal. From Josephine Creek, Josephine Co., Oregon. Named from the locality. [*M.A.*., **14**, 500; *A.M.*, **45**, 1130.] 22nd List

Oriental jasper. R. Jameson, 1800. *Min. Scot. Isles*, vol. 2, p. 68. An early name for Heliotrope. 23rd List

Orientite. D. F. Hewett and E. V. Shannon, 1921. *Amer. Journ. Sci.*, ser. 5, vol. 1, p. 491. Hydrous silicate of manganese and calcium, $4CaO.2Mn_2O_3.5SiO_2.4H_2O$, occurring as small, brown, orthorhombic crystals in manganese-ores in Oriente province, Cuba. 9th List [*M.A.*, **1**, 201]

Orlandinite. F. Ahlfeld and J. Muñoz Reyes, *Mineralogie von Bolivien*, 1938, p. 31. Local name for a lead-grey compact ore abundant in the Porvenir mine, Huanuni, Bolivia. It is presumably boulangerite. 15th List

Orlite. Author? *Guide to USSR exhibit at Conference on Atomic Energy*, Geneva, 1955, p. 7 (orlite). Later publications, abstr., *Amer. Min.*, 1956, vol. 41, p. 816, 1957, vol. 42, p. 307, 1958, vol. 43, p. 381; *Mém. Soc. Russ. Min.*, 1958, vol. 87, p. 83 (орлит, orlite). Formula variously given as $3PbO.3UO_3.4SiO_2.6H_2O$ and '$Pb_3(UO_2)_3Si_2O_7.6H_2O$'. Etymology? 21st List

Oroseite. A. Amstutz, 1925. *Schweiz. Min. Petr. Mitt.*, vol. 5, p. 308. A red-brown alteration product of olivine in basalt from Orosei, Sardinia. Named from the locality. It differs from traversite (q.v.) in its optical orientation with respect to the olivine. [Evidently the same as iddingsite.] [*M.A.*, **3**, 369.] 11th List

Orpheite. I. Kostov, 1968. *Mineralogy* (Oliver and Boyd, London), 474. B. Kolkovski, [Годишник Софийск, унив., Геол.-геогр. фак., геол., 1971/2, **64**, 107], abst. *Zap.*, **102**, 451. $H_6Pb_{10}Al_{20}(PO_4)_{12}(SO_4)_5(OH)_{40}.11H_2O$, from the Modjarovo deposit, Rhodope Mts, Bulgaria. 28th List

Orthamphibole. F. Rinne, *Centralblatt Min.*, 1917, p. 80 (Orthamphibole, *plur.*). A collective name for the orthorhombic amphiboles, analogous to Orthaugite (q.v.). 10th List

Orthaugite. F. Rinne, *Centralblatt Min.*, 1917, p. 76 (Orthaugite, *plur.*). Variant of Orthoaugite (F. Rinne, 1900; 3rd List). 10th List

Ortho-antigorite. G. W. Brindley and O. von Knorring, 1954. *Amer. Min.*, vol. 39, p. 794 (ortho-antigorite). A variety of antigorite based on an ortho-hexagonal cell; from Unst, Shetland. [*M.A.*, **12**, 463.] 20th List [*A.M.*, **42**, 666. Is lizardite-$6T_1$; *C.M.*, **14**, 314]

Ortho-armalcolite. S. E. Haggerty, 1973. *Nature* (*Phys. Sci.*), **242**, 123. A premature name for a colour variety of armalcolite (27th List). [*A.M.*, **59**, 632.] 28th List

Orthoaugite. F. Rinne, *Sitz.-ber. Akad. Wiss. Berlin*, 1900, p. 482 (Orthoaugit). A collective name for the orthorhombic pyroxenes, enstatite, bronzite, and hypersthene, which cannot always be distinguished from each other in rock-sections. Similarly, monoclinic pyroxenes are called clinoaugites (Klinoaugite).

Following Rinne, E. Düll (*Zeits. Kryst. Min.*, 1902, vol. xxxvi, p. 654 footnote, abstr.) proposes the terms ortho-pyroxene and klino-pyroxene, abbreviated as o-pyroxene and kl-pyroxene. 3rd List

Orthoberthierine. H. Strunz, 1966. *Min. Tabellen*, 4th edit., p. 403. To replace Orthochamosite (22nd List), since the mineral is not a chlorite. 24th List

Orthobromite. J. V. Samojlov, 1906. *Materialien zur Geologie Russlands*, vol. xxiii, p. 146 (Ортобромитъ); abstr. in *Neues Jahrb. Min.*, 1907, vol. ii, ref. p. 194 (Orthobromid [*sic*]). A variety of embolite with the composition AgCl.AgBr, as distinct from A. Breithaupt's megabromite (4AgCl.5AgBr), and microbromite (3AgCl.AgBr). From the Donetz Basin, S Russia. 5th List

Orthochamosite. F. Novák, J. Vtelensky, J. Losert, F. Kupa, and Z. Valcha, 1957. [*Czech Acad. Sci., F. Slavík Memorial Vol.*, p. 315]; abstr. *Amer. Min.*, 1958, vol. 43, p. 792. The orthorhombic polymorph described by Brindley (*Min. Mag.*, **29**, 502). 22nd List

Ortho-chevkinite. S. Bonatti and G. Gottardi, 1953. *Rend. Soc. Min. Ital.*, vol. 9, p. 242 (ortochevkinite). Orthorhombic chevkinite from Madagascar [*M.A.*, **1**, 376; **2**, 269], as distinct from clino-chevkinite (q.v.). [*M.A.*, **12**, 240, 498.] 20th List

Ortho-chrysotile. E. J. W. Whittaker, 1951. *See* Clino-chrysotile. 20th List

Orthoenstatite. N. Morimoto and K. Koto, 1969. *Zeits. Krist.*, **129**, 65. *Syn.* of enstatite. 28th List

Orthoericssonite. P. B. Moore, 1971. *Lithos*, **4**, 137. Ericssonite (26th List), from Långban, Sweden, is shown to consist of two intergrown phases, monoclinic ericssonite, *a* 20·46, *b* 7·03, *c* 5·34 Å, *β* 95° 30′, *C*2/*m*, and orthorhombic orthoericssonite, *a* 20·37, *b* 7·03, *c* 5.34 Å. Both have *α* 1·807, pale greenish tan, ||[010], *β* 1·833, red-brown, *γ* 1·89, deep brown. Analysis of material containing both phases, believed to be polymorphous, leads to $4[BaMn_2Fe^{3+}Si_2O_7OOH]$. [*A.M.*, **56**, 2157; *M.A.*, 71-3104.] 27th List

Orthoferrosilite. N. F. M. Henry, 1935. *Min. Mag.*, vol. 24, p. 266; 1937, vol. 24, p. 528. Hypothetical end-member $FeSiO_3$ of the enstatite-hypersthene series of orthorhombic pyroxenes. Named from analogy with ferrosilite (H. S. Washington, 1903; 9th List) and clinoferrosilite (q.v.). Synonym of iron-hypersthene (q.v.). 14th List

Orthoguarinite. G. Cesàro, 1932. *See* Clinoguarinite. 13th List

Orthokalsilite. Th. G. Sahama, 1961. *Fortschr. Min.*, vol. 39, p. 25. The artificial orthorhombic high-temperature polymorph of $KAlSiO_4$. 23rd List

Ortholomonosovite. N. V. Belov and N. I. Organova, 1962. Геохимия (Geochemistry), no. 1, p. 6 (Ортоломоносовит). An unnecessary name for Lomonosovite (17th List) as distinct from Metalomonosovite (= *β*-Lomonosovite; q.v.), given to emphasize the presence of orthophosphate ions. *See also* Зап. Всесоюз. Мин. Общ. (*Mem. All-Union Min. Soc.*), 1963, vol. 92, p. 211, and *Amer. Min.*, 1963, vol. 48, p. 1413. 23rd List

Ortholomonossowit. German transliteration of Ортоломоносовит, Ortholomonosovite (23rd List) (H. Strunz, *Min. Tabellen*, 4th edit., 1966, p. 503). 24th List

Orthomimic Felspars. V. Souza-Brandão, 1909. *Communic. Commissão do Serviço Geologico de Portugal*, vol. vii, p. 136 (feldspaths orthomimiques). Triclinic felspars which by repeated twinning (orthomimie, orthomimicry) simulate a higher degree of symmetry with rectangular cleavages. They include orthoclase, anorthoclase, and cryptoclase (q.v.). 6th List

Orthopinakiolite. R. Randmets, 1961. *Arkiv Min. Geol.*, vol. 2, p. 551. Orthorhombic polymorph of pinakiolite, $Mg_3Mn^{\cdot\cdot}Mn_2^{\cdot\cdot\cdot}B_2O_{10}$, as black needles in dolomite from Långban, Sweden. The 'pinakiolite' of Bäckström (1895) was in fact orthopinakiolite. [*A.M.*, **46**, 768.] 22nd List

Orthorhombic låvenite. A. M. Portnov, V. I. Simonov, and G. P. Sinyugina, 1966. Доклады акад. наук СССР (*Compt. Rend. Acad. Sci. URSS*), vol. 166, p. 1199 (Ромбический ловенит). Låvenite from the Burpala massif, N Baikal, gave X-ray data indicating orthorhombic symmetry and is regarded as a distinct polymorph. E. H. Nickel (*Amer. Min.*, 1966, vol. 51, p. 1549; orthorhombic Lavenite) considers the evidence inadequate, as the material is polysynthetically twinned. [*M.A.*, **18**, 48; *Bull.*, **90**, 117.] 25th List

Orthoriebeckite. K. Willmann, 1937. *Neues Jahrb. Min.*, Abt. A, Beil.-Bd. 72, p. 390 (Orthoriebeckit). The darkest coloured and least birefringent member of the riebeckite-laneite series, i.e. riebeckite proper. 15th List

Orthotaenite. J. D. Buddhue, 1936. *Popular Astronomy*, Northfield, Minnesota, vol. 44, p. 512 (orthotaenite). Taenite of the composition Fe_2Ni is called orthotaenite; while material ranging up to Fe_7Ni, due to admixture with kamacite, is called metataenite. [*M.A.*, **6**, 390.] 14th List

Orthotorbernite. H. Strunz, 1961. *Der Aufschluss*, p. 25 (Orthotorbernit). Syn. of Torbernite. 22nd List

Orthotscheffkinit, variant of Orthochevkinite (20th List) (H. Strunz, *Min. Tabellen*, 4th edit., 1966, p. 503). 24th List

Orthozoisite syn. of zoisite. K.-H. Nitsch and H. G. F. Winkler, *Beitr. Min. Petr.*, 1964, vol. 11, p. 470. *M.A.*, **17**, 671.] 25th List

Orto-chevkinite, *see* Ortho-chevkinite.

Oruetite. S. Pìña de Rubies, 1919. *Anal. Soc. Española Fis. Quím.*, vol. xvii, p. 83 (Oruetita). A bismuth sulphotelluride, Bi_8TeS_4, occurring in the Serranía de Ronda, Spain, as brilliant steel-grey lamellar masses very like tetradymite in appearance. All the naturally occurring bismuth sulphotellurides (tetradymite, joseite, grünlingite, and oruetite) are regarded as eutectic mixtures of $Bi_2Te_3.Bi_2S_3$ with Bi_2Te_3 and Bi_2S_3 and bismuth. Named after Domingo de Orueta, who collected the material. 8th List [Is Josëite; *M.A.*, **8**, 286]

Orvillite. T. H. Lee, 1917. *Revista Soc. Brasileira Sci.*, no. 1, p. 31; *Amer. Journ. Sci.*, 1919, vol. xlvii, p. 126; *Chem. News*, 1919, vol. cxviii, p. 125. Hydrated zirconium silicate, $8ZrO_2.6SiO_2.5H_2O$, occurring with zircon in one variety of the zirconia-ore (*see* Caldasite and Zirkite) of the Caldas district, Brazil. It is evidently an altered zircon similar to those already described under various names. Named after Orville A. Derby (1851–1915), formerly director of the Geological Survey of Brazil. 8th List

Osannite. C. Hlawatsch, 1906. *Festschrift Harry Rosenbusch*, Stuttgart, 1906, p. 76 (Osannit). Those soda-amphiboles, intermediate between riebeckite and arfvedsonite, in which the optic axial plane is perpendicular to the plane of symmetry and the acute, negative bisectrix nearly coincides with the vertical prism-axis. A constituent of amphibole-gneiss at Cevadaes, Portugal. Named after Professor Alfred Osann, of Freiberg in Baden. 4th List [Riebeckite, *Amph.* Sub-Comm.]

Osarizawaite. Y. Taguchi, 1961. *Min. Journ.* (*Japan*), vol. 3, p. 181. A yellow powdery crust from the Osarizawa mine, Akita prefecture, Japan, has a composition near $PbCuAl_2(SO_4)_2(OH)_6$, and is the aluminium analogue of Beaverite. Named from the locality. *See also Amer. Min.*, 1962, vol. 47, pp. 1079 and 1216; *M.A.*, **16**, 69. 23rd List

Osarsite. K. G. Snetsinger, 1972. *Amer. Min.*, **57**, 1029. Grains from placer gravels at Gold Bluff, Humboldt Co., California, contain aggregated curved monoclinic laths (100 to 150 × 30 μm) with irarsite (25th List) of composition 4(?)[Os,Ru)AsS]; *a* 5·933, *b* 5·916, *c* 6·009 Å, β 112° 21′. Analysis shows Os > Ru and As > S. Arsenopyrite family. Named for the composition, by analogy with irarsite. 27th List

Osirita. G. Gagarin and J. R. Cuomo, 1949, loc. cit., p. 5. Contraction in Spanish form of osmiridium, applied to Os-Ir alloys rich in iridium. *See* Irosita. [*M.A.*, **7**, 162.] 19th List

Osmiridin. J. Fröbel, 1843. *Grundzüge eines Systemes der Krystallologie*. Synonym of nevyanskite. 27th List

Osmiridium. C. Lévy and P. Picot, 1961. *Bull. Soc. franç. Min. Crist.*, vol. 84, p. 312. This name is here applied to the hexagonal Os-Ir alloys instead of to the cubic (compare M. H. Hey, *Chem. Index Min.*, 1950; F. V. Chukhrov, *Min.*, vol. 1, pp. 47 and 53; and the review in *Min. Mag.*, 1963, vol. 33, p. 712). 23rd List

Osmite. R. Hermann, *Bull. Soc. Nat. Moscou*, 1836, IX, 228. Iridosmine with 40·83% of osmium. 2nd List. V. I. Vernadsky, 1909. *Opuit Opisatelnoĭ Mineralogii*, St. Petersburg, vol. i, pp. 248, 249 (Осмитъ); translation in *Mining Journ. London*, 1912, vol. xcviii, p. 851. Native osmium, perhaps present amongst the grains of crude iridosmine from Brazil and the Urals. In the Supplement, 1914, p. 752, the name is applied to an iridosmine from Borneo containing Os 80, Ir 10, Rh 5%. 8th List

Ostwaldite. F. Cornu, 1909. *Zeits. Chem. Indust. Kolloide*, vol. iv, p. 187 (Ostwaldit). The colloidal (hydrogel) form of silver chloride obtained by precipitation, and represented in nature by 'buttermilk silver', cerargyrite being the crystalloidal form. Named after Professor Wilhelm Ostwald, of Leipzig. 5th List

Osumilite. A. Miyashiro, 1953. *Proc. Japan Acad.*, vol. 29, p. 321 (osumilite). A mineral resembling cordierite, but hexagonal, $(K,Na,Ca)(Mg,Fe'')_2(Al,Fe''',Fe'')_3(Si,Al)_{12}O_{30}.H_2O$, in volcanic rock from prov. Ôsumi, Japan. Named from the locality. [*M.A.*, **12**, 304, 616.] 20th List [*A.M.*, **41**, 104; **59**, 383]

Otavite. O. Schneider, 1906. *Centralblatt Min.*, 1906, p. 389 (Otavit). A basic carbonate of cadmium (Cd, 61·5%) occurring as minute, curved rhombohedra and forming white to reddish crystalline crusts on copper-ores at Otavi in German SW Africa. 4th List

Otaylite. (H. S. Spence, *Mines Branch*, Canada, 1924, no. 626, p. 11: C. S. Ross and E. V. Shannon, *Journ. Amer. Ceramic Soc.*, 1926, vol. 9, pp. 88–89; J. Melhase, *Engin. Mining Journ.-Press*, New York, 1926, vol. 121, p. 837.) Trade-name for a bentonite clay from Otay, San Diego Co., California. *See* Amargosite, Ardmorite. [*M.A.*, **3**, 143.] 11th List

Ottemannite. G. H. Moh, 1966. *Fortschr. Min.*, vol. 42, p. 211. β-Sn_2S_3, orthorhombic, described from Cerro de Potosi, Bolivia (G. H. Moh and F. Berndt, *Neues Jahrb. Min.*, Monatsh., 1964, p. 94) is named. [*A.M.*, **50**, 2107; **51**, 1551.] 24th List

Otwayite. E. H. Nickel, B. W. Robinson, C. E. S. Davis, and R. D. MacDonald, 1977. *A.M.*, **62**, 999. Occurs as narrow veinlets in serpentine and ore minerals at the Otway prospect, Nullagine, Western Australia. Orthorhombic, *a* 10·18, *b* 27·4, *c* 3·22 Å. Sp. gr. 3·41. Composition $8[Ni_2CO_3(OH)_2.H_2O]$, with some Mg. Pale green, weakly pleochroic, α 1·65, ‖ elongation, γ' 1·72. Named for C. Otway. [*M.A.*, 78-2125.] 30th List

Ouralborite, French transliteration of Уралборит, Uralborite (*Bull. Soc. franç. Min. Crist.*, 1964, vol. 87, p. 460). 24th List

Ourayite. S. Karup-Møller, 1977. *Bull. Geol. Soc. Denmark*, **26**, 41. Orthorhombic, *a* 13·457, *b* 44·042, *c* 4·100 Å, composition $[Ag_{12\cdot5}Pb_{15}Bi_{20\cdot5}S_{52}]$. From the Old Lout mine, Ouray, Colorado. Named from the locality. [*M.A.*, 78-899.] 30th List

Ousbékite. *See* Uzbekite. 11th List

Overite. E. S. Larsen, 3rd, 1938. *Amer. Min.*, vol. 23, no. 12, part 2, p. 9; 1939, vol. 24, p. 188; 1940, vol. 25, p. 315. Hydrous phosphate of calcium and aluminium, $Ca_3Al_8(PO_4)_8(OH)_6.15H_2O$, as pale green to colourless orthorhombic crystals in variscite nodules from Utah. Named after Mr. Edwin Over, of Colorado Springs, Colorado. [*M.A.*, **7**, 224, 513.] 15th List [$CaMgAl(PO_4)_2(OH).4H_2O$; *A.M.*, **59**, 48]

Ovulite. W. H. Bucher, 1918. *Journ. Geol. Chicago*, vol. 26, p. 593. L. Déverin, *Schweiz. Min. Petr. Mitt.*, 1940, vol. 20, p. 102. An individual grain of oolite. From Latin ovulum, dim. of ovum. To replace the name ooide (Ooid, E. Kalkowsky, *Zeits. Deutsch. Geol. Gesell.*, 1908, vol. 60, p. 72). The name ovulite was in use before 1848 for a fossil egg. 16th List

Owyheeite. E. V. Shannon, 1921. *Amer. Min.*, vol. 6, p. 82. Sulphantimonite of lead and silver, $5PbS.Ag_2S.3Sb_2S_3$, forming indistinctly fibrous masses and orthorhombic (?) needles. The needles are brittle with basal cleavage and the mineral was in consequence previously described as 'silver jamesonite' (q.v.). Named from the locality, Owyhee County, Idaho. [*M.A.*, **1**, 150.] 9th List

Oxidapatite. A. N. Winchell, *Elements of optical mineralogy*, 2nd edit., 1927, p. 128 (oxidapatite). Variant of oxy-apatite (6th List), synonym of voelckerite (6th List), $3Ca_3P_2O_8.CaO$. 15th List

Oxide-meionite. L. H. Borgström, 1914. *See* Carbonate-marialite. 7th List

Oxide-pearlite. C. Benedicks, 1912. *Compt. Rend. XI Congr. Géol. Internat.* (*Stockholm, 1910*), 1912, vol. 2, p. 888 (perlit à oxyde). M. Schwarz, *Zentr. Min.*, Abt. A, 1937, p. 85 (Oxyd-Perlit). Pearlite in which the lamellae of ferrite (Schwarz says those of cementite) have been altered to iron oxide, occurring as a constituent of the native iron in basalt at Ovifak in Greenland and at Bühl near Kassel in Germany. [*M.A.*, **6**, 528.] 14th List

Oxoferrite. R. Schenck and T. Dingmann, 1927. *Zeits. Anort. Chem.*, vol. 166, p. 140 (Oxoferrit). Metallic iron with some FeO in solid solution. 13th List

Oxonic pyrochlore, for Oxonium pyrochlore (q.v.). 24th List

Oxonio-alunite. M.-A. Kashkai, 1969. Зап. Всесоюз. мин. общ. (*Mem. All-Union Min. Soc.*), **98**, 153 (Оксониоалунит). A name for the end-member $(H_3O)Al_3(SO_4)_2(OH)_6$. 26th List

Oxonium pyrochlore. A. S. Sergeev, 1961. Зап. Всесоюз. Мин. Общ. vol. 90, p. 400 (Оксониевый пирохлор, Oxonic pyrochlore). Amber yellow to pale brown Pyrochlore in fenites from the Kola peninsula is very low in Na, low in total cations, and high in H_2O. It is formulated with 0·98 H_3O^+ in the *A* positions of the general pyrochlore formula $A_2B_2X_7$. [*M.A.*, **16**, 645.] 24th List

Oxyallanite. W. A. Dollase, 1973. *Zeits. Krist.*, **138**, 41. An *artificial* product; allanite loses hydrogen on heating and forms oxyallanite. 28th List

Oxyannite. H. P. Eugster and D. R. Wones, 1962. *Journ. Petrology*, **3**, 83. The trioctahedral oxidized annite (of Winchell) 'molecule' $K[Fe^{2+}Fe^{3+}][AlSi_3]O_{12}$. *Hypothetical* end-member. 27th List

Oxy-apatite. A. F. Rogers, 1912. *See* Voelckerite. 6th List

Oxybiotite. H. P. Eugster and D. R. Wones, 1962. *Journ. Petrology*, **3**, 83. General name for trioctahedral oxidized biotites. 27th List

Oxychildrenite. A. I. Ginzburg and N. V. Voronkova, 1950. *Doklady Acad. Sci. USSR*, vol. 71, p. 145 (оксичильдренит). An oxidized form of childrenite with (Fe,Mn)O changed to $(Fe,Mn)_2O_3$, from Kalbin Mts., Kazakhstan. [*M.A.*, **11**, 124.] 19th List

Oxydhydratmarialith and Oxydhydratmejonit. R. Brauns, 1915. *See* Carbonate-marialite. 7th List

Oxyd-Perlit, *see* Oxide-pearlite.

Oxyferropumpellyite. P. B. Moore, 1971. *Lithos*, **4**, 93. A name for a pumpellyite-group mineral in which MgOH is replaced by $Fe^{3+}O$ (cf. Ferropumpellyite). *Hypothetical* end-member. [*A.M.*, **56**, 2158.] 27th List

Oxyhornblende. A. N. Winchell, 1932. *Amer. Min.*, vol. 17, pp. 114, 472; *Optical mineralogy*, part 2, 3rd edit., 1933, p. 252. 'Basaltic hornblende' in which the ferrous oxide has been oxidized to ferric. [*M.A.*, **5**, 216.] 13th List

Oxyjulgoldite. P. B. Moore, 1971. *Lithos*, **4**, 93. A member of the pumpellyite family, in which the $Fe^{2+}OH$ of julgoldite (26th List) is replaced by $Fe^{3+}O$. *Hypothetical* end-member. [*A.M.*, **56**, 2158.] 27th List

Oxykaersutite. K. Aoki and H. Matsumoto, 1959. *Journ. Japan Assoc. Min. Petr. Econ. Geol.*, vol. 43, p. 248. A variety of kaersutite from Iki island, Nagasaki prefecture, Japan, characterized by a high Fe_2O_3:FeO ratio, high refractive indices, and strong pleochroism. [*M.A.*, **15**, 44.] 22nd List

Oxykerchenite. Another spelling of oxykertschenite (5th List). *See* kerchenite. [*M.A.*, **9**, 92, 311.] 17th List

Oxykertschenite. S. P. Popov, 1907. *Bull. Acad. Sci. Saint-Pétersbourg*, ser. 6, vol. i, p. 138 (Оксикерченитъ). Hydrated ferric phosphate with small amounts of manganese, etc., $(Mn,Mg,Ca)O.4Fe_2O_3.3P_2O_5.21H_2O$. Occurs as a brown alteration product of paravivianite and kertschenite (S. P. Popov, 1906; 4th List) in the Kerch (German, Kertsch) peninsula, Crimea. 5th List

Oxymagnite. A. N. Winchell, 1931. *Amer. Min.*, vol. 16, p. 270. A contraction of oxidized magnetite, to replace the name maghemite (q.v.) for magnetic ferric oxide. [*M.A.*, **4**, 502.] 12th List

Oxymimetesit. H. Wondratschek, 1963. Original form of Oxymimetite (q.v.). 23rd List

Oxymimetite. H. Wondratschek, 1963. *Neues Jahrb. Min., Abh.*, vol. 99, p. 145 (Oxymimetesit). The hypothetical compound $Pb_{10}(AsO_4)_6O$; attempts at synthesis failed. Cf H. P. Rooksby, Analyst, 1952, vol. 77, p. 759. 23rd List

Oxypyromorphite. H. Wondratschek, 1960. *Zeits. anorg. Chem.*, vol. 306, p. 25 (Oxypyromorphit). Artificial $Pb_{10}(PO_4)_6O$; apatite family. *See also Analyst*, 1952, vol. 77, p. 759; *Naturwiss.*, 1956, vol. 43, p. 494; and *Neues Jahrb. Min., Abh.*, 1963, vol. 99, p. 135. 23rd List

Oxytourmaline. C. Frondel, A. Beidl, and J. Ito, 1966. *Amer. Min.*, vol. 51, p. 1501. A term for tourmalines in which an excess cationic charge is balanced by replacement of some (OH,F) by O^{2-}. 25th List

Oxytschildrenit. K. F. Chudoba, Hintze, *Min.*, 1954, Ergbd. 2, p. 294 (Oxychildrenit, richtiger Oxytschildrenit?). Superfluous synonym of oxychildrenite (19th List). Named after John George Children (1777–1852) of the British Museum. 20th List

Oxy-Turmalin, German variant of oxytourmaline (25th List). Hintze, 69. 28th List

Oxyvanadinite. H. Wondratschek, 1963. *Neues Jahrb. Min., Abh.*, vol. 99, p. 113 (Oxyvanadinit). The hypothetical compound $Pb_{10}(VO_4)_6O$. 23rd List

Oyamalite. K. Kimura, 1925. *Japanese Journ. Chem.*, vol. 2, p. 81. A variety of zircon containing rare-earths and phosphoric acid. Named from the locality, Oyama, prov. Iyo, Japan [*M.A.*, **3**, 9.] 11th List

P

Pabstite. E. B. Gross, J. E. N. Wainwright, and B. W. Evans, 1965. *Amer. Min.*, vol. 50, p. 1164. Colourless grains fluorescing blueish-white in short-wave ultraviolet light, occurring in recrystallized siliceous limestone at Santa Cruz, California, are trigonal and have the composition $Ba(Sn,Ti)Si_3O_8$ with Sn:Ti $\approx$ 7:2; they are a titanian variety of the tin analogue of Benitoite, to which the name is assigned for A. Pabst. [*M.A.*, **17**, 606.] 24th List

Padparadschah. A. K. Coomaraswamy, *Administration Reports*, Ceylon, for 1904, part 4, Mineralogical Survey, 1905, p. E16 (Padmaragaya). M. Bauer, *Edelsteinkunde*, Leipzig, 2nd edit., 1909, p. 363 (patparachan). R. Brauns, *Künstliche Schmucksteine, Handwörterbuch der Naturwissenschaften*, 1913, vol. 8, p. 968 (Padparadschah). German corruptions (with other variations) of the Sinhalese padmaragaya, from padma, lotus, and raga, colour. A trade-name for reddish-yellow gem corundum, now used more especially for the artificially produced material. [*M.A.*, **6**, 198.] 14th List

Paewelite, error for parwelite (26th List) (*Zap.*, **99**, 83). 27th List

Pageit, error for Paigeite (Hintze, *Handb. Min.*, Erg.-Bd. II, 1959, p. 673). 22nd List

Paigeite. A. Knopf and W. T. Schaller, 1908. *Amer. Journ. Sci.*, ser. 4, vol. xxv, p. 324. W. T. Schaller, in 2nd Appendix to Dana's *System of Mineralogy*, 1909, p. 78; *Amer. Journ. Sci.*, 1910, ser. 4, vol. xxix, p. 543. A. Knopf and W. T. Schaller, *Zeits. Kryst. Min.*, 1910, vol. xlviii, p. 1. A coal-black, lustrous and opaque aggregate of matted fibres and long needles, with a foliated appearance, occurring in a contact-metamorphic limestone in connexion with the tin-ores of Alaska. Originally described as a hydrous borate of ferrous and ferric iron, but subsequently found to contain a considerable amount of tin, the probable formula being $30FeO.5Fe_2O_3.SnO_2.6B_2O_3.5H_2O$. It is suggested, however, that the mineral may possibly be a mixture of hulsite (q.v.) and an iron borate. Named after Sidney Paige, of the United States Geological Survey. 5th List [Is Vonsenite (q.v.), *M.A.*, 75-1381; Cf. Schoenfliesite]

Painite. G. F. Claringbull, M. H. Hey, and C. J. Payne, 1956. *Min. Soc. Notice*, 1956, no. 95; *Min. Mag.*, 1957, vol. 31, p. 420. Mixed oxide, near $Al_{20}Ca_4BSiO_{37\cdot5}$. A single, small, transparent, dark-red, hexagonal crystal from gem-gravel at Mogok, Burma. Named after Arthur Charles Davy Pain, who collected the material. Cf. Hibonite (q.v.). 21st List [$CaZrB(Al_9O_{18})$; *A.M.*, **61**, 88]

Palacheite. A. S. Eakle, *Bull. Dep. Geol. Univ. California*, 1903, vol. iii, p. 231. Loosely coherent aggregates of brick-red, monoclinic crystals, found as a recent formation in the old workings of the Redington mercury mine, Knoxville, California. $2MgO.Fe_2O_3.4SO_3.15H_2O$. It is not proved to differ essentially from the imperfectly determined rubrite. Named after Dr. Charles Palache, of Harvard University. 3rd List [Var. of Botryogen]

Palaeo-albite, etc. T. Scheerer, 1854. *Ann. Phys. Chem.* (*Poggendorff*), vol. xci, p. 379 (Paläo-Albit). Following Haidinger (*see* Palaeo-amphibole), the following additional names are given for the original minerals of what were believed to be paramorphs: Paläo-Albit, Paläo-Oligoklas-Albit (p. 379), Paläo-Krokydolith (p. 383), Paläo-Epidot (p. 387). Scheerer regarded albite and scapolite as dimorphous, and the pseudomorphous crystals from Norway as paramorphs of albite after scapolite. His palaeo-albite is therefore a synonym of scapolite. V. M. Goldschmidt (*Vidensk.-Selsk. Skrifter, Math.-Naturv. Kl.*, Kristiania, 1911, no. 1, pp. 309, 312) incorrectly applies the name 'Paläoalbit' to the pseudomorphous crystals themselves. 6th List

Palaeo-amphibole, etc. W. Haidinger, 1854. *Sitzungsber. Akad. Wiss. Wien*, vol. xi (Jahrg. 1853), p. 399 (Paläo-Amphibol). This and other similar names (Paläo-Krystalle (p. 397), Paläo-Natrolith (p. 398), Paläo-Calcit (p. 399), Paläo-Uralit (p. 399)) are suggested for the original minerals of paramorphs; palaeo-amphibole and palaeo-calcite are therefore synonyms of pyroxene and aragonite, respectively. 6th List

Palaeoleucite. H. Rosenbusch, 1905. *Mikroskopische Physiographie d. Mineralien*, 4th edit., vol. i, part 2, p. 33 (Paläoleucit). The original mineral of pseudoleucite. *See* Soda-leucite. 4th List

Palaite. W. T. Schaller, 1912. *Journ. Washington Acad. Sci.*, vol. ii, p. 144. A flesh-coloured hydrous manganese phosphate, $5MnO.2P_2O_5.4H_2O$, probably monoclinic, resulting from the alteration of lithiophilite in the gem-tourmaline mines at Pala, San Diego Co., California. Named after the locality. 6th List [Is Hureaulite; *M.A.*, **8**, 180]

Palermoite. M. E. Mrose, 1952. *Progr. and Abstr. Min. Soc. Amer.*, p. 35; *Bull. Geol. Soc. Amer.*, vol. 63, p. 1283; *Amer. Min.*, 1953, vol. 38, p. 354 (abstr.). $(Li,Na)_4SrAl_9(PO_4)_8(OH)_9$, orthorhombic prisms in pegmatite from Palermo mine, North Groton, New Hampshire. Named from the locality. [*M.A.*, **12**, 131.] 20th List [$(Li,Na)_2(Sr,Ca)Al_4(PO_4)_4(OH)_4$; *A.M.*, **50**, 777]

Palladite. V. I. Vernadsky, 1914. *Essay descrip. Min.*, St. Petersburg, pp. 128, 232, 262 (палладит). Variant of palladinite, PdO (C. U. Shepard 1857). 20th List

Palladium-amalgam. J. B. Harrison, 1925. *Official Gazette,* British Guiana, no. 71; *Min. Mag.*, 1928, vol. 21, p. 398 (palladium amalgam). A. Cissarz, *Zeits. Krist.*, 1930, vol. 74, p. 510 (Palladium-amalgam). H. Berman and G. A. Harcourt, *Amer. Min.*, 1938, vol. 23, p. 764 (Palladium-Amalgam). Synonym of potarite (11th List). [*M.A.*, **3**, 4; **7**, 224.] 15th List

Palladium-bismuthide. O. E. Yushko-Zakharova and L. A. Chernyaev*, 1966. Доклады акад. наук СССР (*Compt. Rend. Acad. Sci. URSS*), vol. 170, p. 183 (Визмутид палладия)*. A mineral in the Monchegorsk ores, Kola Peninsula, is distinct from michernerite and froodite, and has the composition $PdBi_3$. [*M.A.*, **18**, 125; Зап., **97**, 65.] [*Corr. in *M.M.*, **39**.] 25th List

—— L. V. Razin, V. D. Begizov, and V. I. Meshchankina, 1973. [*Trudy Ts NIGRI*, **108**, 96], abstr. *A.M.*, **61**, 180. A single grain, near PdBi, from the Talnakh deposit, USSR, is indexed as hexagonal, *a* 4·20, *c* 5·64 Å. Not to be confused with Palladium bismuthide of Yushko-Zakharova and Cherayev, $PdBi_3$ (24th List). Apparently identical with Sobolevskite (q.v.). 20th List

Palladium-Copper-Platinum Stannide. L. V. Razin, V. D. Begizov, and V. I. Meshchankina, 1973. [*Trudy Ts NIGRI,* **108**, 96], abstr. *A.M.*, **61**, 180. A premature name for a tetragonal mineral from the Talnakh deposit, USSR with *a* 4·072, *c* 3·720 Å, space group *P4/mmm*. [May be the same as some of the minerals grouped under Unnamed Z of Cabri, 1972. The essential nature of all the elements found is yet to be proved.—L.J.C.] 29th List

Palladium Platinum Arsenoplumbostannide. L. V. Razin, V. D. Begizov, and V. I. Meshchankina, 1973. [*Trudy Ts NIGRI*, **108**, 96], abstr. *A.M.*, **61**, 180. A premature name for a cubic mineral (*a* 3·99 Å) from the Talnakh deposit, USSR. [Not enough data to determine whether the analysis represents a new species or a Pb-rich ordered Pd_3Sn with some As.—L.J.C.] 29th List

Palladium-Platinum Arsenostannide and Palladium-Platinum Plumbostannoarsenide. L. V. Razin, V. D. Begizov, and V. I. Meshchankina, 1973. [*Trudy Ts NIGRI*, **108**, 96], abstr. *A.M.*, **61**, 180. Premature names assigned on the basis of single probe analyses of two grains; both are tetragonal (*a* 3·99, *c* 3·655 Å, *P4/mmm*) from the Talnakh deposit, USSR. [Not enough data to determine whether either mineral is a unique species, and the second appears to be merely a plumbian variety of the first.—L.J.C.] 29th List

Palladium-Platinum Stannide. L. V. Razin, V. D. Begizov, and V. I. Meshchankina, 1973. [*Trudy Ts NIGRI*, **108**, 96], abstr. *A.M.*, **61**, 180. An inappropriate name for a platinian ordered form of Pd_3Sn from the Talnakh deposit, USSR, *a* 3·984 Å, space group *Pm3m*. 29th List

Palladium Plumboarsenide. L. V. Razin, V. D. Begizov, and V. I. Meshchankina, 1973. [*Trudy Ts NIGRI*, **108**, 96], abstr. *A.M.*, **61**, 181. A mineral from the Talnakh deposit, USSR is near PdPbAs; orthorhombic, *a* 7·180, *b* 8·619, *c* 10·662 Å. 29th List

Palladium Stibiostannoarsenide. L. V. Razin, V. D. Begizov, and V. I. Meshchankina, 1973. [*Trudy Ts NIGRI*, **108**, 96], abstr. *A.M.*, **61**, 181. A mineral from the Talnakh deposit, USSR is formulated $Pd_{2\pm x}(As,Sn,Sb)$ and its ten-line X-ray patterns is indexed as orthorhombic, *a* 8·107, *b* 5·625, *c* 4·360 Å. [The powder pattern is very similar to that of the monoclinic Palladoarsenide Pd_2As (28th List)—L.J.C.] 29th List

Palladium-Wismutid, Germ. trans. of Визмутид палладия, palladium bismuthide (25th List). Hintze, 69. 28th List

Palladoarsenide. V. D. Begizov, V. I. Meshchankina, and L. S. Dubakina, 1974. *Zap.*, **103**, 104 (Палладоарсенид). Monoclinic Pd_2As (*a* 9·25, *b* 8·47, *c* 10·44 Å, *β* 94·0°), from the Oktyabr copper-nickel deposit. Named from the composition. 28th List

Palladobismutharsenide. L. J. Cabri, T. T. Chen, J. M. Stewart, and J. H. G. Laflamme, 1976. *Can. Mineral.*, **14**, 410. Grains from the Stillwater Complex, Montana, are orthorhombic, *a* 7·504, *b* 18·884, *c* 6·841 Å. Composition $4[Pd_{10}As_2Bi]$. Named from the composition. [*M.A.*, 78-892.] 30th List

Palladseïte. R. J. Davis, A. M. Clark, and A. J. Criddle, 1977. *M.M.*, **41**, 123 and M10. Rare grains among concentrates from gold washings at Itabira, Minas Gerais, Brazil, are cubic, *a* 10·635. Composition $Pd_{15\cdot47}Cu_{1\cdot85}Hg_{0\cdot24}Se_{14\cdot43}$ (on a basis of thirty-two atoms), agrees closely with synthetic $Pd_{17}Se_{15}$. Named from the composition. [*M.A.*, 77-2188; *A.M.*, **62**, 1059.] 30th List

Pallite. L. Capdecomme and R. Pulou, 1954. *Compt. Rend. Acad. Sci. Paris*, vol. 239, p. 288. A hypothetical type of Ca-Al phosphate in phosphorites from Senegal. Derivation of name not given. [*M.A.*, **12**, 440.] 20th List [Is ferrian Millisite; *A.M.*, **45**, 256]

Palmerite. E. Casoria, 1904. *Atti. R. Accad. Georgofili*, Firenze, ser. 5, vol. i, p. 293; *Ann. R. Scuola Sup. Agric. Portici*, 1904, vol. vi. Hydrated aluminium potassium phosphate, $HK_2Al_2(PO_4)_3.7H_2O$, occurring as a white powder under bat-guano in a cave on Monte Alburno, Salerno, Italy. Named after Paride Palmeri, professor of chemistry in the Royal School of Agriculture at Portici, near Naples. 4th List [Is Taranakite; *M.M.*, **28**, 31. Cf. Minervite]

Palmierite. A. Lacroix, 1907. *Compt. Rend. Acad. Sci. Paris*, vol cxliv, p. 1400 (palmiérite). Minute hexagonal scales which are optically uniaxial with strong negative birefringence; found enclosed in aphthitalite amongst the products of the Vesuvian eruption of April 1906. Anhydrous sulphate of lead, potassium, and sodium, $PbSO_4.(K,Na)_2SO_4$. Named after Luigi Palmieri (1807–1896). 4th List

Pamirite. Labuntsov, 1930. Quoted in Yu. K. Vorobev, *Zap.*, 1967, **96**, 333. Described as a distinct species intermediate between forsterite and clinohumite, but is shown to be merely forsterite. [*M.A.*, **19**, 44.] 28th List

Pandaite. E. Jäger, E. Niggli, and A. H. van der Veen, 1959. *Min. Mag.*, vol. 32, p. 10. A member of the pyrochlore group containing Ba and Sr, with only small amounts of other bases, and much water, occurring in a carbonatite at Panda Hill, Mbeya, Tanganyika. Named from the locality. 22nd List [Renamed **Bariopyrochlore**, *A.M.*, **62**, 406; 30th List]

Panethite. L. H. Fuchs, E. Olsen, and E. P. Henderson, 1966. *Abstr. 29th Meeting Meteoritical Soc.*, p. 12; *Geochimica Acta*, 1967, vol. 31, p. 1711. Tiny grains with brianite, whitlockite, etc. in pockets in the Dayton meteorite (a siderite) prove to be monoclinic, near $(Na,Ca,K)_{1-x}(Mg,Fe,Mn)PO_4$; also obtained synthetically. Named for F. A. Paneth. [*A.M.*, **52**, 309; **53**, 509; *Bull.*, **91**, 302.] 25th List

Pantellarite. (A. Dufrénoy, *Min.*, 1859, IV, 59 (*pontellarite*); A. Des Cloizeaux, *Min.*, 1862, I, 323). The felspar of Pantellaria analysed by Abich in 1840; since called anorthoclase. 2nd List

Paolovite. A. D. Genkin, T. L. Evstigneeva, L. N. Vyalsov, I. P. Laputina, and N. V. Groneva, 1974. *Geol. Rudn. Mestorzhd.*, **16**, 98 (Паоловит). Polysynthetically twinned grains intergrown with sperrylite, from the Oktyabr deposit, Talnakh ore field, USSR, have *a* 8·11, *b* 5·662, *c* 4·324 Å, space group *Pbnm*; composition Pd_2Sn, with small amounts of Pt. Named for *Pa*lladium and *olovo* (tin). [*M.A.*, 75-558; *A.M.*, **59**, 1331; *Zap.*, **103**, 612; *Bull.*, **98**, 321.] 29th List

Pao-t'ou-k'uang, *see* Baotite. 22nd List

Papagoite. C. O. Hutton and A. C. Vlisidis, 1960. *Amer. Min.*, vol. 45, p. 599. Near $CaCuAlSi_2O_6(OH)_3$; monoclinic blue crystals from Ajo, Pima Co., Arizona. Named from the tribe that formerly inhabited the region. [*M.A.*, **15**, 44.] 22nd List

Para-alumohydrocalcite. D. and B. L. Srebrodolskii, 1977. *Zap.*, **106**, 336 (Параалюмогидрокальцит). White radiating aggregates from the Vodinsk and Gaurdok sulphide ore deposits have composition $CaAl_2(CO_3)_2(OH)_4.6H_2O$, sp. gr. 2·0, positive elongation, $2V_\alpha$ 69°. Named from the relation to alumohydrocalcite, of which it is a higher hydrate. [*M.A.*, 78-3480.] 30th List

Para-armalcolite. S. E. Haggerty, 1973. *Nature* (*Phys. Sci.*), **242**, 123. A premature name for a colour variety of armalcolite (27th List). [*A.M.*, **59**, 632.] 28th List

Para-autunite. Author? H. Strunz, *Min. Tab.*, 3rd edit., 1957, p. 254 (Para-Autunit). $Ca(UO_2)_2(PO_4)_2$, orthorhombic (pseudo-tetragonal). Meta-autunite-II of J. Beintema, 1938 [*M.A.*, **7**, 237]. Completely dehydrated in the para-uranite (Para-Uranit) series. In other mineral names (e.g. 18th List) the prefix para- is used in different senses. [*artificial*] 21st List [Also 22nd List]

Parabayldonite. F. K. Biehl, 1919. *Inaug.-Diss. Münster* (*Westf.*), pp. 47, 57 (Parabayldonit). Basic copper-lead arsenate, $2R_3As_2O_8.R(OH)_2.\frac{1}{2}H_2O$, differing from bayldonite in containing slightly less water, occurs as green pseudomorphous crusts of Tsumeb, SW Africa. Compare Cuproplumbite and Duftite. [*M.A.*, **1**, 203.] 9th List [Is Plumboan Conichalcite; *A.M.*, **42**, 123]

Paraboleïte. A. Mücke, 1972. *Fortschr. Min.*, **50**, Beiheft 1, 67 (Paraboleit). A name proposed for all minerals intermediate between boléite and pseudoboleïte, and particularly for the material with composition $2[28PbCl_2.24Cu(OH)_2.6AgCl.8H_2O]$ and tetragonal cell dimensions *a* 15·249, *c* 30·831 Å from Mina Santa Ana, Caracoles, Sierra Gorda, Chile [Should not have been named—*M.F.*] [*A.M.*, **59**, 221.] 28th List

Parabutlerite. M. C. Bandy, 1938. *Amer. Min.*, vol. 23, p. 742. Hydrated basic ferric sulphate, $Fe(SO_4)(OH).2H_2O$, orange-coloured, orthorhombic, as an alteration product of copiapite, from Chile. Dimorphous with butlerite (11th List). [*M.A.*, **7**, 223.] 15th List

Paracancrinite. J. Wyart and M. Michel-Lévy, 1949. *Compt. Rend. Acad. Sci. Paris*, vol. 229, p. 133. An artificial hydrothermal product with the same X-ray pattern as cancrinite, but containing no calcium. [*M.A.*, **11**, 93.] 19th List

Paracelsian. E. Tacconi, 1905. *Rend. R. Istit. Lombardo*, ser. 2, vol. xxxviii, p. 642 (Paracelsiana). A mineral with the composition $Ba_3Al_8Si_8O_{31}$, occurring as pale yellow granules in veins in crystalline schists at Candoglia, Piedmont. It is near to celsian ($BaAl_2Si_2O_8$) in composition and optical characters, but it seems to differ from this in possessing only poor cleavage. 4th List [Wales, *M.M.*, **26**, 231; $BaAl_2Si_2O_8$. Cf. Slawsonite]

Para-chrysotile. E. J. W. Whittaker and J. Zussman, 1955. *Min. Soc. Notice*, 1955, no. 91; *Min. Mag.*, 1956, vol. 31, p. 116. A variety of chrysotile with X-ray fibre-axis 9·2 Å., distinct from clino- and ortho-chrysotile (20th List). 21st List

Paracoquimbite. J. Klvaňa, 1882. *Sitzungsber. bohm. Gesell. Wiss. Prag*, 1882, vol. for 1881, p. 272 (Paracoquimbit). A hydrated ferric sulphate forming minute rhombohedral crystals and occurring as a siskin-green encrustation on phyllite at Troja near Prague. It differs from coquimbite in the proportions of its constituents ($Fe_2O_3$21·70, Al_2O_3 2·02, MgO 3·44, SO_3 33·22, H_2O 36·65%). 9th List [Is Slavíkite; *M.A.*, **9**, 204]

—— H. Ungemach, 1933. **Compt. Rend. Acad. Sci. Paris**, vol. 197, p. 1133; **Bull. Soc. Franç Min.** [*1934*], vol. 56 (for 1933), p. 303. Violet rhombohedral crystals dimorphous with the hexagonal coquimbite, $Fe_2(SO_4)_3.9H_2O$, the two often grown in parallel position. From Chile. [*M.A.*, **5**, 390.] 13th List [The accepted usage]

Paracostibite. L. J. Cabri, D. C. Harris, and J. M. Stewart, 1970. *Canad. Min.*, **10**, 232. Small (<130 μm) grains with metallic lustre in drill cores from Mulcahy Township, Kenora District, Ontario, give an X-ray powder pattern near that of pararammelsbergite, indexable as *a* 5·764, *b* 5·952, *c* 11·635 Å; composition CoSbS. Occurs in the same deposit as nisbite (q.v.). Named from the composition (cf. costibite, 26th List) and probable structural relation to pararammelsbergite. [*A.M.*, **56**, 631; *M.A.*, 71-2333; *Zap.*, **100**, 617.] 27th List

Paradamin, variant of Paradamite (21st List). (H. Strunz, *Min. Tabellen*, 3rd edit., 1957, p. 228.) 22nd List

Paradamite. G. Switzer, 1956. *Science* (*Amer. Assoc. Adv. Sci.*), vol. 123, p. 1039. Zinc arsenate, Zn_2AsO_4OH, triclinic, dimorphous with adamite and isomorphous with tarbuttite. From [Ojuela mine, Mapimi, Durango] Mexico. [*M.A.*, **13**, 380; *A.M.*, **41**, 958.] 21st List [*A.M.*, **51**, 1218]

Paradeweylite. N. E. Efremov, 1939. *Compt. Rend.* (*Doklady*) *Acad. Sci. URSS*, vol. 22, p. 423 (paradeveilite). A hypothetical hydrated silicate of magnesium, $4MgO.3SiO_2.3H_2O$, differing from deweylite in containing less water. [*M.A.*, **7**, 370.] 15th List

Paradocrasite. B. F. Leonard and C. W. Mead, 1971. *Amer. Min.*, **56**, 1127. Silvery-white prisms (0·5 mm) with stibarsen (16th List) and antimonian löllingite, replacing calcite, are monoclinic, *a* 7·252, *b* 4·172, *c* 4·431 Å, *β* 123° 8·4′, *C*2, composition [$Sb_{2\cdot93}As_{1\cdot07}$]; *R* 66·75%. From Broken Hill, New South Wales. Named from *παράδοξος* and *κρᾶσις*, unexpected alloy, by analogy with dyscrasite, which it resembles. [*M.A.*, 72-549.] 27th List [Perhaps $Sb_2(Sb,As)_2$]

Paradokrasit, German variant of paradocrasite (27th List). Hintze, 70. 28th List

Paraduttonite. H. T. Evans jr., 1960. *US Geol. Surv. Prof. Paper 400–B*, p. 444. A black oxidation product of Duttonite ($V_2O_4.H_2O$; 21st List), probably near $V_2O_5.H_2O$, found at Monument Valley, Arizona, USA. 23rd List

Paraffinite. G. Mueller, 1964. *Rept. 22nd Internat. Geol. Congr., India,* part 1, 47. A group term for ozocerite and petroleum. 26th List

Paragearksutite. N. A. Smolyaninov and E. N. Isakov, 1946. *D.S. Belyankin Jubilee vol., Acad. Sci. USSR*, p. 147 (парагеарксутит), Hydrous calcium aluminium fluoride $4CaF_2.4Al(F,OH)_3.3H_2O$, differing from gearksutite in containing slightly less water, from Transbaikalia. [*M.A.*, **10**, 453.] 18th List

Paragite. V. von Zepharovich, *Min. Lexicon Oesterr.*, 1873, II, 350. The same as Korallenerz (corallinerz, Dana) or hepatic cinnabar, an ore of mercury in Idria. 2nd List

Paraguanajuatite. P. Ramdohr, 1948. Klockmann's *Lehrb. Min.*, 13th edit., p. 360. Rhombohedral paramorphs after natural and artificial orthorhombic guanajuatite. $Bi_2(Se,S)_3$. [*A.M.*, **34**, 619.] 18th List

Parahalloysite. D. P. Serdyuchenko, 1953. *Problems* (вопросы) *Petr. Min.* (*Acad. Sci. USSR*), vol. 2, p. 100 (парагаллузит). Clays similar to halloysite in composition but to silica-poor montmorillonite in structure. Mg-, Fe-Mg-, and Fe-Cr- varieties are distinguished. [*M.A.*, **12**, 513.] 20th List

Parahilgardite. C. S. Hurlbut, 1938. *Amer. Min.*, vol. 23, p. 765. Triclinic-pedial crystals dimorphous with hilgardite (q.v.). [*M.A.*, **7**, 224, 355.] 15th List [Cf. Hilgardite–3Tc; *A.M.*, **44**, 1102]

Parahopeite. L. J. Spencer, 1907. *Nature*, London, vol. lxxvii, p. 143; *Mineralogical Magazine*, 1908, vol. xv, p. 18. Hydrated zinc phosphate, $Zn_3P_2O_8.4H_2O$, dimorphous with hopeite, forming divergent groups of colourless, platy, anorthic crystals, resembling hemimorphite in appearance. It is found with tarbuttite ($Zn_3P_2O_8.Zn(OH)_2$; L. J. Spencer, 1907; 4th List at Broken Hill, NW Rhodesia. 5th List

Parajamesonite. V. Zsivny and I. v. Náray-Szabó, 1947. *Schweiz. Min. Petr. Mitt.*, vol. 27, p. 183 (Parajamesonit). Dimorphous with jamesonite, $4PbS.FeS.3Sb_2S_3$, distinguished by the X-ray pattern. [*M.A.*, **10**, 254.] 18th List

Parakalinepheline, Parakaliophilite. F. A. Bannister, 1942. *Min. Mag.*, vol. 26, p. 221. Alternative, but rejected, names for kalsilite (q.v.). 16th List

Parakaolinite. D. P. Serdyuchenko, 1945. *Compt. Rend.* (*Doklady*) *Acad. Sci. URSS*, vol. 46, p. 117. End-member of the series serpentine-parakaolinite, $H_4(Mg_3,Al_2)Si_2O_8$. *See* Alumino-chrysotile. [*M.A.*, **9**, 185.] 17th List

Para-Kupferglanz. H. Strunz, *Min. Tab.*, 1941, p. 65. Paramorphs of low-temperature orthorhombic Cu_2S after high-temperature cubic [hexagonal?] α-Cu_2S. Syn. of chalcosine. [*M.A.*, **2**, 506; **10**, 202.] 18th List

Parakutnohorite. V. I. Pavlishin and M. M. Slivko, 1962. Мин. Сборн. Львовск. Геол. Общ. (*Min. Mag. Lvov Geol. Soc.*), no. 16, p. 445 (Паракутнагорит). $(Ca,Mn)CO_3$, near the 1:1 ratio but with calcite (disordered) rather than dolomite (ordered) structure. (Зап. Всесоюз. Мин. Общ. (*Mem. All-Union Min. Soc.*), 1964, vol. 93, p. 450 Паракутнагорит, parakutnahorite). The use of separate names for ordered and disordered phases is, in general, undesirable; the suffixes *-o* and *-d* are to be preferred (cf. *M.A.*, **14**, 21). 23rd List

Paralaurionite. G. F. H. Smith, *Min. Mag.*, 1899, XII, 108, 183; *Zeits. Kryst. Min.*, 1900 XXXII, 217. Dimorphous with laurionite, PbClOH. Monoclinic. In Roman lead slags from Laurion, Greece. *See* rafaelite. 2nd List

Paramontmorillonite. A. Fersmann, 1908. *Bull. Acad. Sci. Saint-Pétersbourg*, ser. 6, vol. ii, p. 264 (Paramontmorillonit), p. 656 (Парамонтмориллонитъ). A fibrous mineral placed with pilolite ('mountain-leather') in the palygorskite group; its composition, $H_{12}Al_2Si_4O_{17}$, is near to that of montmorillonite. 5th List

Paramontroseite. H. T. Evans and M. E. Mrose, 1955, *Amer. Min.*, vol. 40, p. 861. Vanadium dioxide V_2O_4, orthorhombic, as an oxidation product of montroseite, (V,Fe)O(OH) (19th List), from Paradox Valley, Colorado. [*M.A.*, **13**, 9.] 21st List [Cf. Kentsmithite]

Paranatrolite. A. A. Maier, N. S. Manuilova, and B. G. Varshai, 1964. Докл., **154**, 363 (Паранатролит). A supposed polymorph of natrolite, produced by dehydration and rehydration, differing only in dielectric constant and ease of dehydration. Is merely natrolite with some lattice breakdown (E. E. Senderov and G. V. Yukhnevich, 1964. Геохимия, 849). 28th List

Para-oranite, *see* Hypoperthite.

Para-orthose. (V. Souza-Brandão, *Communic. Commissão do Serviço Geologico de Portugal*, 1909, vol. vii, p. 136.) A French form of parorthoclase (German, Parorthoklas, F. Zirkel, 1893; 2nd List). A synonym of anorthoclase (French, anorthose). 6th List

Parapechblende, *see* Parapitchblende. 22nd List

Parapectolite. W. F. Müller, 1976. *Z. Kristallogr.*, **144**, 401. An unnecessary name for pectolite-M2abc. [*A.M.*, **63**, 427.] 30th List

Paraperthite, Para-oranite. H. L. Alling, 1921. *See* Hypoperthite. 14th List

Paraphane. G. A. Sidorenko, 1960. [Рентген. опред. уран мин., Госгеолиздат., p. 46] abstr. Зап. всесоюз. мин. общ. (*Mem. All-Union Min. Soc.*) 1967, vol. 96, p. 79. An inadequately characterized hydrous silicate of uranium. [*Bull.*, **90**, 610.] 25th List

Parapitchblende. J. Geffroy and J. A. Sarcia, 1954. [*Sciences de la terre*, Univ. Nancy, vol. 2, p. 1]; abstr. in *Amer. Min.*, 1958, vol. 43, p. 792; Hintze, *Handb. Min.*, Erg.-Bd. II, p. 810 (Parapechblende). A black, isotropic alteration product of pitchblende from Bauzot and Ruaux, Saône-et-Loire, France. Inadequately described, and probably not a valid species. 22nd List

Pararammelsbergite. M. A Peacock, 1939. *Amer. Min.*, vol. 24, no. 12, part 2, p. 10; 1940, vol. 25, pp. 211, 561. Material recently described as rammelsbergite, $NiAs_2$, from Ontario is found to give X-ray data differing from those for rammelsbergite from Germany, and it is now named pararammelsbergite. [*M.A.*, **7, 469, 507**.] 15th List

Para-schachnerite. E. Seeliger and A. Mücke, 1972. *Neues Jahrb. Min. Abh.*, **117**, 1 (Para-Schachnerit). $2[Ag_{1\cdot2}Hg_{0\cdot8}]$, orthorhombic pseudohexagonal, formed by loss of Hg from moschellandsbergite at the Vertrauen Gott mine, Obermoschel, Pfalz, Germany. *a* 2·96, *b* 5·13, *c* 4·83 Å. Named for Dr. D. Schachner. [*M.A.*, 73-1941; *Bull.*, **96**, 240; *A.M.*, **58**, 347.] 28th List

Paraschoepite. A. Schoep and S. Stradiot, 1947. *Amer. Min.*, vol. 32, p. 344. Uranic hydroxide, $5UO_3.9\frac{1}{2}H_2O$, as yellow orthorhombic crystals, differing from schoepite (10th List) in its optical properties; from Katanga, [*M.A.*, **10**, 145, 544.] 18th List [*A.M.*, **45**, 1026]

Parasepiolite. A. Fersmann, 1908. *Bull. Acad. Sci. Saint-Pétersbourg*, ser. 6, vol. ii, p. 263 (Parasepiolit), p. 657 (Парасепіолитъ). A fibrous mineral placed in the palygorskite group; its composition, $H_8Mg_2Si_3O_{12}$, is near to that of sepiolite. 5th List

Paraserandite. W. F. Müller, 1976. *Z. Kristallogr.*, **144**, 401. An unnecessary name for a *hypothetical* polymorph of serandite. [*A.M.*, **63**, 427.] 30th List

Para-Silberglanz. H. Strunz, *Min. Tab.*, 1941, p. 66. Paramorphs of low-temperature orthorhombic Ag_2S (acanthite) after high-temperature cubic α-Ag_2S. Syn. of argenite. [*M.A.*, **3**, 338.] 18th List

Paraspurrite. A. A. Colville and P. A. Colville, 1977. *A.M.*, **62**, 1003. A monoclinic polymorph of spurrite occurring near Darwin. Inyo Co., California. *a* 10·473, *b* 6·706, *c* 27·78 Å, *β* 90·58°, space group $P2_1/a$. RIs α 1·650 ‖ [010], *β* 1·672, *γ* 1·677, *γ* : [001] 30°, $2V_\alpha 47°$. Named from its relation to spurrite. [*M.A.*, 78-2126.] 30th List

Parastrengite. S. V. Gevorkyan, L. N. Egorova, and A. S. Povarennykh, 1974. *Geol. Zhurn.*, **34**, no. 3, 27 (Парастренгит). Name proposed for material giving I.R. absorption patterns analogous to those of paravariscite (q.v.), but with the composition and orthorhombic symmetry of strengite. [*A.M.*, **60**, 340.] 29th List

Parasymplesite. T. Ito, H. Minato, and K. Sakurai, 1954. *Proc. Japan Acad.*, vol. 30, p. 318. Monoclinic $Fe_3(AsO_4)_2.8H_2O$, dimorphous with triclinic symplesite; from Japan. [*M.A.*, **12**, 412; *A.M.*, **40**, 368.] 20th List [Mexico; *C.M.*, **14**, 437]

Paratacamite. G. F. H. Smith, 1905. *Nature*, vol. lxxi, p. 574; *Min. Mag.*, 1906, vol. xiv, p. 170; *Zeits. Kryst. Min.*, 1907, vol. xliii, p. 28. Hydrated oxychloride of copper, $Cu_2Cl(OH)_3$, dimorphous with atacamite. The bright green crystals are rhombohedral but with optical anomalies. From Chili. 4th List [*M.M.*, **29**, 34]

Paratellurite. G. Switzer and H. E. Swanson, 1960. *Amer. Min.*, vol. 45, p. 1272. The tetragonal modification of TeO_2, found at Cananea, Sonora, Mexico. 22nd List

Paratenorite. H. Strunz, *Min. Tab.*, 1941, p. 95 (Paratenorit). Syn. of paramelaconite. 18th List

Parathenardite. P. N. Chirvinsky, 1906. P. Tchirwinsky, *Reproduction artificielle de minéraux au XIXe siècle* (*Russ.*), Kieff, 1903–1906, p. 583 (паратенардитъ, parathénardite). The same as Metathenardite (A. Lacroix, 1910; 6th List). 9th List

Parathuringite. O. Koch, 1884. *Inaug.-Diss. Jena*, p. 37 (Parathuringit). A mineral from Vogtland very similar to thuringite, but with the composition $Al_8Fe_5''Mg_2Si_6O_{31}.9aq$. 17th List

Paratooite. D. Mawson and W. T. Cooke, 1907. *Trans. Roy. Soc. South Australia*, vol. xxxi, p. 68. An insoluble residue of bird-guano, consisting mainly of hydrated aluminium and ferric phosphate, but of variable and indefinite composition. Found on the top of an isolated rock near Paratoo railway siding in South Australia. 5th List

Para-uranite. H. Strunz, 1957. *Min. Tabellen*, 3rd edit., pp. 252, 254 (Para-Uranit). The fully dehydrated, anhydrous series of compounds corresponding to the autunite group. 22nd List

Paraurichalcite. F. K. Biehl, 1919. *Inaug.-Diss. Münster* (*Westf.*), pp. 24, 28, 34 (Paraurichalcit). Basic carbonate of copper and zinc, $3(Cu,Zn)CO_3.4(Cu,Zn)(OH)_2$ or $4(Cu,Zn)CO_3.5(Cu,Zn)(OH)_2$, approximating to aurichalcite in composition. It is formed by the alteration of malachite and is perhaps a mixture of malachite and hydrozincite. From Tsumeb, SW Africa. Compare Cuprozincite. [*M.A.*, **1**, 203] 9th List

Paravariscite. S. V. Gevorkyan, L. N. Egorova, and A. S. Povarennykh, 1974. *Geol. Zhurn.*, **34**, no. 3, 27 (Паравариcкит). A name for the 'Messbach' type of orthorhombic variscite, which gives I.R. spectra appreciably different from those of the 'Lucin' type. Synonym of Redondite (Shepard, 1869), *see* D. McConnell, *M.M.*, **40**, 609. [*A.M.*, **60**, 340.] 29th List

Paravauxite. S. G. Gordon, 1922. *Science*, New York, vol. 56, p. 50; *Amer. Min.*, vol. 7, p. 108; *Proc. Acad. Nat. Sci. Philadelphia*, 1923, vol. 75, p. 265. Hydrated phosphate of iron and aluminium, $5FeO.4Al_2O_3.5P_2O_5.26H_2O + 21H_2O$, occurring as colourless, triclinic crystals of prismatic habit on wavellite from the tin mines of Llallagua, Bolivia. *See* Vauxite. [*M.A.*, **2**, 148.] 10th List [$FeAl_2(PO_4)_2(OH)_2.10H_2O$]

Paraveatchite, error for *p*-Veatchite. J. R. Clark, *in* J. Murdoch and R. W. Webb, *Calif. Div. Mines Geol. Bull.*, **173** (1964), suppl. 2, 16; cf. *A.M.*, **55**, 1936. The prefixed *p* (for primitive) was mistaken for *p* for para. 28th List

Paravivianite. S. P. Popoff, 1906. *Centralblatt Min.*, 1906, p. 112. A variety of vivianite with part of the iron replaced by small amounts of manganese and magnesium, $(Fe,Mn,Mg)_3P_2O_8.8H_2O$. Occurs as blue, acicular crystals in limonitic iron-ore in S Russia. 4th List [Cf. Kertschenite]

Parawollastonite. M. A. Peacock, 1935. *Amer. Journ. Sci.*, ser. 5, vol. 30, pp. 495, 525. Monoclinic calcium metasilicate, $CaSiO_3$, the name wollastonite being reserved for the more common triclinic modification of this compound. [*M.A.*, **6**, 260.] 14th List [Cf. Wollastonite-2M; *A.M.*, **49**, 224]

Parbigite. Y. V. Mirtov, 1958. [Вес. Зап.-Сиб. Новосиб. Геол. Упр. (*Bull. West-Siberian and New Siberian Geol. Dept.*), no. 1, p. 72]; abstr. Зап. Всесоюз. Мин. Общ. (*Mem. All-Union Min. Soc.*), 1959, vol. 88, p. 318 (Парбигит, Parbighite). Incompletely characterized; a phosphate of the collinsite-fairfieldite family, but optically negative. Found in sandstone from a borehole near Parbig, Tomsk, Siberia. Named from the locality. [*M.A.*, **14**, 278.] Probably messelite; evidence for a new name quite inadequate (M. Fleischer, *Amer. Min.*, 1960, vol. 45, p. 256). 22nd List

Paredrite. O. C. Farrington, 1916. *Amer. Journ. Sci.*, ser. 4, vol. xli, p. 356. A black, compact form of titanium dioxide differing from rutile in containing a little water, 0.6% [not sufficient, however, to form a hydrate, and no doubt present as an impurity]. Named from *πάρεδρος*, an associate, on account of its association as 'favas' (bean-shaped pebbles) with the diamonds of Brazil. Cf. Doelterite. 8th List

Parianite. S. F. Peckham, *Journ. Franklin Inst.*, 1895, CXL, 381. Asphaltum from the Pitch Lake of Trinidad. 2nd List

Parkerite. D. L. Scholtz, 1936. *Publ. Univ. Pretoria*, 1936, ser. 2, no. 1, p. 186; preprint from *Trans. Geol. Soc. South Africa*, 1937, vol. 39 (for 1936), p. 186. Nickel sulphide, Ni_2S_3 or NiS_2, monoclinic, from South Africa. Named after Professor Robert Luling Parker of Zürich. [*M.A.*, **6**, 440.] 14th List [Redefined as $Ni_3Bi_2S_2$; *A.M.*, **28**, 343. Cf. Bismuthparkerite, Shandite]

Parorthoclase. F. Zirkel, *Lehrb. d. Petrogr.*, 1893, I, 238 (*Parorthoklas*). The same as anorthoclase. 2nd List

Parryite. A. F. Williams, 1932. *The genesis of the diamond*, London, vol. 1, p. 172. An undescribed hydrous calcium silicate from the diamond mines at Kimberley, South Africa. Named after John Parry (1863–1931), chemist to the De Beers Consolidated Mines at Kimberley. [*M.A.*, **5**, 97.] 13th List

Parsettensite. J. Jakob, 1923. *Schweiz. Min. Petr. Mitt.*, vol. 3, p. 227 (Parsettensit) Hydrated manganese silicate, $3MnO.4SiO_2.4H_2O$, forming copper-red masses with platy basal cleavage (pseudo-hexagonal), allied to friedelite. Named from the locality, Alp Parsettens, Val d'Err, Grisons. [*M.A.*, 2, 251.] 10th List

Parsonsite. A. Schoep, 1923. *Compt. Rend. Acad. Sci. Paris*, vol. 176, p. 171; *Bull. Soc. Belge Géol. Bruxelles*, 1924, vol. 33 (for 1923), p. 195. Hydrated phosphate of lead and uranyl, $2PbO.UO_3.P_2O_5.H_2O$, occurring as a brownish, crystalline powder (monoclinic or triclinic) on torbenite from Kasolo, Katanga. Named after Prof. Arthur Leonard Parsons, of Toronto, Canada. [*M.A.*, **2**, 50, 342.] 10th List [*A.M.*, **35**, 245]

Partridgeite. J. E. de Villiers, 1943. *Amer. Min.*, vol. 28, pp. 336, 468. Manganese and ferric oxide $(Mn,Fe)_2O_3$, differing from sitaparite in colour and etching reactions; from Postmasburg, South Africa. Named in memory of Francis Chamberlain Partridge (1903–1939), formerly of the Geological Survey of South Africa. [*M.A.*, **9**, 4, 5.] 17th List [Not certainly distinct from Bixbyite, Cf. Kurnakite]

Parweelite, error for Parwelite (*Bull. Soc. franç. Min. Crist.*, **91**, 305). 26th List

Parwelite. P. B. Moore, 1967. *Canad. Min.*, **9**, 301; *Arkiv Min. Geol.*, **4**, 467 (1969). Yellowish-brown crystals in manganoan carbonate from Långban, Sweden, have the composition $(Mn,Mg)_2Sb(Si,As)_2O_{10-11}$; monoclinic. Named for Alexander Parwel, who analysed the mineral. [*A.M.*, **53**, 1426; **55**, 323; *M.A.*, 69-2393; *Bull.*, **91**, 305; *Zap.*, **98**, 320; **99**, 83.] 26th List

Pascoite. W. F. Hillebrand, H. E. Merwin, and F. E. Wright, 1914. *Proc. Amer. Phil. Soc.*, vol. liii, p. 49; *Zeits. Kryst. Min.*, vol. liv, p. 209. A monoclinic (?), hydrous vanadate of calcium, $2CaO.3V_2O_5.11(?)H_2O$, occurring as an orange-yellow, powdery efflorescence on vanadium ores at Minasragra, Cerro de Pasco, Peru. Named from the locality. 7th List [$Ca_3V_{10}O_{28}.16H_2O$; *A.M.*, **40**, 315. Cf. Hummerite]

Patagosite. S. Meunier, 1917. *Compt. Rend. Somm. Soc. Géol. France*, 1917, p. 84. A variety of calcite forming the material of fossil crinoids, etc., which shows some indications of organic structure and contains some black organic matter. Named from *πάταγος*, a clattering, because the material decrepitates when heated. [*M.A.*, **1**, 257.] 9th List

Paternoite. F. Millosevich, 1920. *Rend. R. Acad. Lincei*, Roma, ser. 5, vol. 29, sem. 2, p. 286. Hydrated magnesium tetraborate, $MgO.4B_2O_3.4H_2O$, occurring as white nodules, composed of minute orthorhombic or monoclinic scales, with bloedite in salt deposits in Sicily. Named after the Italian chemist Emanuele Paternò. [*M.A.*, **1**, 149.] 9th List [Is Kaliborite; *M.A.*, **17**, 178]

Patiñoite. M. C. Bandy, 1946. *Mineralogia de Llallagua*, Bolivia, La Paz, p. 57 (Patiñoita). Provisional name for an undescribed mineral as yellow tetragonal crystals probably a phosphate or arsenate. Named after Simon Irtubi Patiño, who was the first to work the Llallagua mines. [*M.A.*, **10**, 9.] 18th List

Patparachan, *see* Padparaschah.

Patronite. F. Hewett. *Engineering and Mining Journ. New York*, 1906, vol. lxxxii, p. 385. C. Matignon, *Revue Scientifique*, Paris, 1906, ser. 5, vol. vi, p. 597. A dark green mineral containing much vanadium, perhaps vanadium sulphide, but not yet completely determined. It occurs with vanadiferous asphaltum near Cerro de Pasco in Peru. Named after Antenor Rizo Patrón, of the Huaraucaca smelting works near Cerro de Pasco, who detected the vanadium. 4th List

Paucilithionite. A. N. Winchell, 1941. *Amer. Min.*, Program and Abstracts, December 1941, p. 26; *Amer. Min.*, 1942, vol. 27, pp. 117, 235. A hypothetical end-member $K_2Li_3Al_5Si_6O_{20}F_4$ which together with polylithionite and protolithionite, enters into the composition of lepidolite. Named from the Latin *paucus*, few, little. 16th List

Paulingite. W. B. Kamb and W. C. Oke, 1960. *Amer. Min.*, vol. 45, p. 79. A cubic zeolite, forming rhombic dodecahedra, from the Columbia River at Rock Island Dam, Wenatchee, Washington. Chemical analyses unsatisfactory. Named after Prof. Linus Pauling. [*M.A.*, **15**, 135.] 22nd List

Paulite. H. W. Bültemann, 1960. [*Der Aufschluss*, vol. 11, no. 11, p. 281]; abstr. *Amer. Min.*, 1961, vol. 46, p. 465. Thin light yellow tablets from uranium deposits at Bühlskopf, Ellweiler, Birkenfeld, Germany, contain Al, U, and As; it is suggested that they are the As analogue of sabugalite, and the name Paulite is proposed after Hans Paul. The name paulite has been used before (Werner, 1812), and the material is not adequately characterized. 22nd List

Pavonite. E. W. Nuffield, 1953. *Progr. and Abstr. Min. Soc. Amer.*, p. 33; *Amer. Min.*, 1954, vol. 39, pp. 338, 409. *Contrib. Canadian Min.*, 1954, vol. 5, part 6, $AgBi_3S_5$, monoclinic, for a Bolivian mineral previously referred to alaskaite and to benjaminite. Named from the Latin pavo, pavonis, peacock, in honour of Prof. Martin Alfred Peacock (1898–1950) of the University of Toronto. [*M.A.*, **12**, 304, 410.] 20th List [Variable comp.; *M.A.*, 74-484]

Paxite. Z. Johan, 1962. *Acta Univ. Carolinae*, Geol. no. 2, p. 77. Cu_2As_3, probably orthorhombic and isostructural with Sb_2S_3, in intergrowths with Novákite, Koutekite, and Arsenic, from Černy Důl, Krkonoše, Bohemia. (*Amer. Min.*, 1962, vol. 47, p. 1484; *M.A.*, **14**, 279). Previously referred to as Mineral X (*Nature*, 1958, vol. 181, p. 1553; *Chemie der Erde*, 1960, vol. 20, p. 217; *Amer. Min.*, 1961, vol. 46, p. 885). Named from the Latin *pax*. [*M.A.*, **16**, 557.] 23rd List

Pb-Dolomit, variant of Plumbodolomite (14th List) (H. Strunz, *Min. Tabellen,* 4th edit., 1966, p. 213). 24th List

Pearceite. S. L. Penfield, *Amer. Journ. Sci.*, II. 17, 1896; and *Zeits. Kryst. Min.*, XXVII. 65, 1896. [*M.M.*, **11**, 224.] A name for the arsenical varieties of polybasite. $9Ag_2S.As_2S_3$. Monosymmetric. [Drumlummon mine, Marysville, Lewis and Clark Co., Montana] 1st List

Pearcit. P. Groth. *Tabell. Uebersicht d. Mineralien*, 4th edit., 1898, p. 38. Error for pearceite: named after Dr. Richard Pearce. 4th List

Pecoraite. G. T. Faust, J. J. Fahey, B. Mason, and E. J. Dwornik, 1969. *Science*, **165**, 59. The nickel analogue of clinochrysotile occurs filling cracks in the oxidized Wolf Creek meteoritic iron. Named for William T. Pecora. [*M.A.*, 70-1653; *A.M.*, **54**, 1740.] 26th List [$Ni_6Si_4O_{10}(OH)_8$; *M.A.*, 75-1320]

Peligonit, error for Peligotite (original Пелигоит, a misprint; named for E. M. Péligot) (C. Hintze, *Handb. Min.*, Erg.-Bd. II, p. 599). 24th List

Peligotite. V. G. Melkov, 1942. *Mém. Soc. Russ. Min.*, vol. 71, p. 9 (пелигоит, *sic*), p. 11 (peligotite). A secondary uranium sulphate, '$CuO.2UO_3.2SO_3.3H_2O$.aq', from Tadzhikstan. Named after Eugène Melchior Péligot (1811–1890), French chemist, who isolated metallic uranium in 1840. [Not sufficiently distinguished from johannite.] [*M.A.*, **12**, 461.] 20th List

Pelinite. A. B. Searle, 1912. *The Natural History of Clay, Cambridge Manuals of Science and Literature, 1912*, pp. 83, 148, 149. An unidentified hydrated aluminium silicate, which is highly plastic and partly colloidal. The essential clay substance of the widely distributed, secondary (i.e. transported from their place of origin) plastic clays. The term is analogous to the term clayite (J. W. Mellor, 1909; 5th List), which is here restricted to the corresponding essential constituent of the less plastic, primary (i.e. not transported) clays, such as china-clay (kaolin). Named from *πήλῖνος*, made of clay. 8th List

Pelionite. W. F. Petterd, *Catalogue of the Minerals of Tasmania, Papers and Proc. R. Soc. Tasmania, for 1893*, 1894, p. 13. Cannel coal from Mt. Pelion. 2nd List

Pellouxite. G. Gagarin and J. R. Cuomo, 1949, loc. cit., p. 9 (pellouxita). Lime, calcium oxide, CaO, reported from Vesuvius (calce, A. Scacchi, 1883). Named after the Italian mineralogist Alberto Pelloux (1868–1948). 19th List

Pellyite. E. P. Meagher, 1971. *Prog. Abstr. Geol. Soc. Amer. Ann. Meet.*, 644. J. H. Montgomery, R. M. Thompson, and E. P. Meagher, 1972. *C.M.*, **11**, 444. A colourless to pale yellow mineral in contact metamorphic skarns near the headwaters of the Pelly and Ross Rivers, Yukon Territory, is orthorhombic, *Cmcm*, *a* 15·677, *b* 7·151, *c* 14·209 Å, composition $4[Ba_2Ca(Fe,Mg)_2Si_6O_{17}]$. Named from the locality. [*A.M.*, **57**, 597; **58**, 806; *Zap.*, **102**, 453.] Unrelated to pumpellyite, 11th List. 27th and 28th Lists

Pendletonite. J. Murdoch and T. A. Geissman, 1967. *Amer. Min.*, vol. 52, p. 611. The aromatic hydrocarbon coronene, $C_{24}H_{12}$, has been found in monoclinic crystals in a small mercury deposit near the New Idria mine, San Benito county, California. Named for N. H. Pendleton. [*Bull.*, **90**, 605.] 25th List [Is Carpathite, *A.M.*, **54**, 329; **61**, 1055]

Penfieldite. F. A. Genth, *Amer. Journ. Sci.*, XLIV. 260, 1892. S. L. Penfield, *ibid.*, XLVIII. 114, 1894. [*M.M.*, **11**, 43.] $PbO.2PbCl_2$. Hexagonal. Laurion, Greece. 1st List [Pb_2Cl_3OH; *A.M.*, **26**, 293]

Penikisite. J. A. Mandarino, B. D. Sturman and M. I. Corlett, 1977. *Can. Mineral.*, **15**, 393. Crystals of kulanite (29th List) from about 1 km from the type locality (an ironstone formation in the region of Big Fish River and Blow River, Yukon Territory) proved to be strongly zoned. The rims of some crystals had Mg > Fe and so constitute a distinct species. Material with Fe:Mg:Ca 1·01:0·83:0·16 had ρ 3·79 g.cm^{-3}, α 1·684, grass-green, β 1·688, blue-green, γ 1·705, pale pink; γ:[001] −6°, β:[010] 0° to 19°. Anorthic, pseudo-monoclinic, *a* 8·999, *b* 12·069, *c* 4·921 Å, α and γ:*c* 90°, β 100° 31′. Composition $2[Ba(Mg,Fe,Ca)_2Al_2(PO_4)_3(OH)_3]$. Named for G. Penikis. 30th List

Peniskisite. *Can. Mineral.*, **15**, no. 3, contents. Error for penikisite. 30th List

Penkvilskite. I. V. Bussen, Yu. P. Menshikov, A. N. Merkov, A. P. Nedorezova, E. I. Uspenskaya, and A. P. Khomyakov, 1974. *Dokl. akad. nauk. SSSR*, **217**, 1161 (Пенквилскит). White massive material from the Lovozero massif, Kola Peninsula, are monoclinic or orthorhombic, *a* 7·48, *b* 8·77 Å, γ 90° (from electron-diffraction data). Composition that of a hydrated narsarsukite $(Na,Ca)_4(Ti,Zr)_2Si_8O_{32}.5H_2O$. Named for пенк, curly and вилкис, white. [*A.M.*, **60**, 340; *M.A.*, 75-1394; *Zap.*, **104**, 615.] 29th List

Pennaite. D. Guimarães, 1948. *Bol. Inst. Tecn. Indust. Minas Gerais*, no. 6, p. 19 (pennaita), p. 59 (pennaite). Chloro-zircono-titano-silicate of Na, Ca, Fe, as minute triclinic crystals with aegirine in nepheline-rocks from Poços de Caldas, Brazil. Named after Dr. José Moreira dos Santos Penna, director of the Instituto de Tecnologia Industrial. [*M.A.*, **10**, 510.] 18th List

Pennantite. W. C. Smith, F. A. Bannister, and M. H. Hey, 1946. *Min Mag.*, vol. 27, p. 217. A manganese-rich chlorite, $(Mn,Al)_6(Si,Al)_4O_{10}(OH)_8$, as orange-coloured optically uniaxial scales from Benallt manganese mine, Wales. Analogous to thuringite with MnO(39%) in place of FeO. Named after Thomas Pennant (1726–1798), Welsh zoologist and mineralogist. 17th List

Penroseite. S. G. Gordon, 1926. *Proc. Acad. Nat. Sci. Philadelphia*, vol. 77 (for 1925), p. 317; *Amer. Min.*, vol. 11, p. 42. Selenide of lead, copper, and nickel, perhaps $PbSe.Cu_2S.(Ni,Co)Se_3$, as lead-grey, radiating columnar masses with perfect orthorhombic cleavages. From Bolivia. Named after Dr. Richard Alexander Fullerton Penrose, junior (1863–), mining geologist, of Philadelphia, Pennsylvania. [*M.A.*, **3**, 112.] 11th List [Cf. Blockite]

Pentagonite. L. W. Staples, H. T. Evans, *Jr.*, and J. R. Lindsay, 1973. *A.M.*, **58**, 405. Prismatic orthorhombic crystals, *a* 10·298, *b* 13·999, *c* 8·891 Å, from Owyhee Dam, Oregon, have composition $4[CaVOSi_4O_{10}.4H_2O]$, and are dimorphous with cavansite (25th List). [*M.A.*, 73-4079, 4080.] 28th List

Pentrahydrite. C. Frondel, 1951. Dana's *Mineralogy*, 7th edit., vol. 2, p. 492. Magnesium sulphate pentahydrate, $MgSO_4.5H_2O$, triclinic; artificial, but occurring naturally in mixed crystals (*see* comstockite, kellerite). Named analogously to hexahydrite ($MgSO_4.6H_2O$). The same as allenite and magnesium-chalcanthite (qq.v.). 19th List

Pentahydroborite. S. V. Malinko, 1961. Зап. Всесоюз. Мин. Общ. (*Mem. All-Union Min. Soc.*), vol. 90, p. 673 (Пентагидроборит). $CaB_2O_4.5H_2O$ in granular masses from a skarn deposit in the Urals. (*Amer. Min.*, 1962, vol. 47, p. 1482; *M.A.*, **16**, 556). 23rd List

Pentahydrocalcite. P. N. Čirvinskij (Tschirwinsky), 1906. *Ann. Géol. Min. Russie*, vol. viii, p. 241 (Пентагидрокальцитъ), p. 245 (Pentahydrocalcit). Hydrated calcium carbonate, $CaCO_3.5H_2O$, occurring as a mould-like encrustation on chalk-marl near Nova-Alexandria, govt. Lublin, Russian Poland. *See* Lublinite and Trihydrocalcite. 5th List

Percivalite. S. Weidman, 1907. *Bull. Wisconsin Geol. Nat. Hist. Surv.*, no. 16, p. 283 (Percivalite). A green soda-pyroxene intergrown with blue amphibole (crocidolite) consisting mainly of $NaAlSi_2O_6$ (jadeite) and $NaAlSiO_4$, from Wisconsin. Named after James Gates Percival (1795–1856), at one time State Geologist of Wisconsin. 15th List

Perdell, variant of Peredell (var. of Topaz) (H. Strunz, *Min. Tabellen*, 4th edit., 1966, p. 506). 24th List

Perhamite. P. J. Dunn and D. Appleman, 1977. *M.M.*, **41**, 437. Rare brown spherulitic masses at the Bell Pit, Newry, Maine, and at the Dunton Gem mine, Newry Hill, consist of platy hexagonal crystals, *a* 7·02, *c* 20·21 Å. Unit cell contents approximately $3CaO.3{\cdot}5Al_2O_3.3SiO_2.2P_2O_5.18H_2O$, ρ 2·64 g.cm^{-3}. ω 1·564, ε 1·577. Named for F. C. Perham. [*M.A.*, 78-893.] 30th List

Perite. M. Gillberg, 1961. *Arkiv Min., Geol.*, vol. 2, p. 565. $PbBiO_2Cl$, small orthorhombic plates with hausmannite, calcite, etc., from Långban, Sweden, and artificial. Named after Prof. Per Geijer. [*A.M.*, **46**, 765.] 22nd List

Perloffite. A. R. Kampf, 1977. *Mineral. Rec.*, **8**(2), 112. Brown crystals from the Big Chief pegmatite, Glendale, South Dakota, are monoclinic, *a* 9·223, *b* 12·422, *c* 4·995 Å, β 100·39°, space group $P2_1/m$. Composition (microprobe analysis only) $2[Ba(Mn,Fe)_2Fe_2(OH)_3(PO_4)_3]$, the Fe analogue of bjarebyite. RIs α 1·793, dark greenish-brown, $\parallel$ [010], β 1·803, light greenish-brown, β:[001] *c*. 42°, γ 1·808, dark greenish-brown. Named for L. Perloff. [*A.M.*, **62**, 1059.] 30th List

Permingeatite. Z. Johan, P. Picot, R. Pierrot, and M. Kravček, 1971. *Bull.*, **94**, 162. Microscopic grains with a variety of selenides, including hakite and fischesserite (qq.v.), in a calcite vein at Předbořice, Bohemia, are tetragonal, *a* 5·63, *c* 11·23 Å, $I\bar{4}2m$. Composition $2[Cu_3SbSe_4]$, a member of the luzonite-germanite family. Named for F. Permingeat. [*M.A.*, 72-1402; *A.M.*, **57**, 1554.] 27th List

Permutite. (*Chem. Zentralblatt*, 1907, vol. ii, pp. 363, 1664; 1908, vol. ii, p. 988; 1909, vol. i, p. 2031, etc.) Trade-name for an artificial zeolite-like substance approximating to $Na_2O.Al_2O_3.2SiO_2.6H_2O$ in composition; employed in water purificiation and sugar refining. Evidently named from the Latin permuto, to change completely. 6th List

Perplexite. P. Marshall, 1946. *New Zealand Journ. Sci. Techn.*, Sect. B, vol. 28, p. 51. A compact zeolitic mineral not definitely identified in the groundmass of phonolitic rocks. [*M.A.*, **10**, 295.] 18th List

Perrierite. S. Bonatti and G. Gottardi, 1950. *Atti* (*Rend.*) *Accad. Naz. Lincei, Cl. Sci. fis. mat. nat.*, ser. 8, vol. 9, sem. 2, p. 361. Titanosilicate of Yt, Ce, with some Th, Fe, Ca, P_2O_5 (near chevkinite); monoclinic (orthite habit). Found in beach-sand at Nettuno, Roma, Named in memory of the Italian mineralogist Carlo Perrier (1886–1948). [*M.A.*, **11**, 310.] 19th List [Structure; *A.M.*, **45**, 1]

Perryite. Kurt Fredriksson and E. P. Henderson, 1965. *Trans. Amer. Geophys. Union*, vol. 46, p. 121 (abstr.). A silicide of nickel, approximately $Ni_5(Si,P)_2$, occurring in the Horse Creek iron meteorite and the St. Mark's enstatite chondrite. Named for S. H. Perry. [*Amer. Min.*, **52**, 559; Зап., **97**, 64; *Bull.*, **99**, 610; *Min. Mag.*, **36**, 850.] 25th List [*A.M.*, **54**, 579; **56**, 1123]

Perthitoid. P. Quensel, 1939. *Geol. För. Förh. Stockholm*, vol. 60 (for 1938), p. 626. A perthitic texture shown by non-felspathic minerals, e.g. symplektic intergrowths of chondrodite-calcite, epidote-quartz, plagioclase-hornblende. Myrmeki-perthitoid when one of the components is vermicular. [*M.A.*, **7**, 335.] 15th List

Peruvite. L. Pflücker y Rico, 1883. *Anales de la Escuela de Construcciones Civiles y de Minas del Perú*, Lima, vol. 3, no. 1, p. 62 (Peruvita). I. Domeyko, *Mineralogía*, 3rd Appendix, Santiago de Chile, 1884, p. 31 (Perulita [*sic*]). C. Hintze, *Handb. Min.*, 1902, vol. 1, part 1, p. 991 (Peruvit). Synonym of matildite ($AgBiS_2$) which was first described from the Matilde vein near Morococha, Peru, under the name Silberwismuthglanz (C. F. Rammelsberg, 1877; *Monatsber. Akad. Wiss. Berlin*, for 1876, p. 700). The names matildite, morocochite, and peruvite were independently given in 1883 to replace this German chemical name. 11th List

Petamene. Local name for a spodumene-quartz mixture chemically equivalent to petalite. (E. W. Heinrich, *Indian Journ. Earth Sci.*, 1975, **2**, 18.) 29th List

Petersberg-Illite. R. A. Koch, 1958. *Neues Jahrb. Min.*, Monatsh., p. 168 (Petersberg-Illit). A variety of illite distinguished by birefringence, exchange capacity, staining reactions, d.t.a. curve, etc., from typical illite. 22nd List

Petricichite. P. Poni, *Ann. Sci. Univ. Jassy*, 1900, vol. i, p. 143; *Anal. Acad. Române*, Bukarest, 1900, vol. xxii. Error (?) for pietricikite, which is given by C. Istrati (*Bull. Soc. Sci. Bukarest*, 1897, vol. vi, p. 65) as the correct spelling of Dana's zietrisikite. A variety of ozocerite from Mt. Pietricica, Moldavia. 3rd List

Petrovicite. Z. Johan, M. Kvaček, and P. Picot, 1976. *Bull.*, **99**, 310. Tabular crystals in hydrothermal dolomite-calcite veins of the Petrovice deposit, Czechoslovakia, are orthorhombic, *a* 16·176, *b* 14·684, *c* 4·331 Å. Composition near $4[Cu_3HgPbBiSe_5]$. Named from the locality. [*A.M.*, **62**, 594; *M.A.*, 77-2189.] 30th List

Petterdite. W. H. Twelvetrees, *Rep. Secr. Mines*, Tasmania, for 1900–1901, 1901, p. 356; *Papers and Proc. Roy. Soc. Tasmania*, for 1900–1901, 1902, p. 51. An oxychloride of lead occurring in Tasmania as white, thin hexagonal plates. Named after Mr. W. F. Petterd, of Launceston, Tasmania. 3rd List [Is Adamite; *Cat. Mins. Tasmania* (*Tasmania Dept. Mines, 1969*), p. 8]

Pharaonite. E. M. El Shazly and G. S. Saleeb, 1972. *Rept. 24th Int. Geol. Congr.*, *Montreal, sect.* 14, 192. Prismatic crystals up to 14 cm long, from St. John's Island, Egypt, are hexagonal, *a* 12·74, *c* 5·35 Å, composition $(Na,Ca,K)_{5\cdot76}Mg_{2\cdot87}(Al,Si)_{12}O_{26}Cl_{1\cdot49}(SO_4)_{0\cdot80}\cdot 2{\cdot}3H_2O$. An unnecessary name for a magnesian cancrinite [or a magnesium-rich Davyne]. [*A.M.*, **58**, 1113.] 28th List

Phenaksite, error of transliteration for Fenaksite (Фенаксит; 22nd List). *Soviet Physics—Doklady*, 1964, **9**, 80. [*M.A.*, **17**, 178.] 28th List

Phenaxite, error of transliteration for Fenaksite (Фенаксит; 22nd List). *Soviet Physics—Doklady*, 1971, **15**, 902 [*M.A.*, 71-2672.] 28th List

Phenicochroite. Variant of Phoenicochroite, *A.M.*, **9**, 62. 10th List

Phianite. Trade-name for cubic ZrO_2, stabilized by Yt_2O_3(?), produced by the Lebedov Physical Institute, Moscow, as a gem diamond substitute. Similar material is marketed through MSB Industries, New York City, under the name 'diamonesque'. Cf. djevalite, 30th List. (K. Nassau, *Lapidary J.*, 1977, p. 904.) 30th List

Philipstadite, R. A. Daly, *Proc. Amer. Acad. Arts and Sci.*, 1899, XXXIV, 433; *Bull. Soc. franç. Min.*, 1899, XXII, 144. A variety of hornblende from Philipstad, Sweden. 2nd List [Ferrian ferro-hornblende; *Amph.* Sub-Comm.]

Pholidite. C. Hintze, 1892. *Handbuch d. Mineralogie*, vol. ii, p. 835 footnote (Pholidit). The more correct derivation of pholerite, from *φολίς*, *φολίδος*, a scale; a scaly variety of kaolinite. 4th List

Pholidoide. K. Smulikowski, 1936. *Arch. Min. Tow. Nauk. Warszaw.* (*Arch. Min. Soc. Sci. Varsovie*), vol. 12, pp. 165 (pholidoïde), p. 180 (folidoid). The group of aluminous glauconites grading into normal (ferruginous) glauconite and occurring in sedimentary rocks. To replace the name phyllite of J. L. Thiébaut (q.v.) and including skolite (q.v.). Named from *φολιδοειδης*, scale-like. Not the pholidolite of G. Nordenskiöld, 1890, although near to this in composition. [*M.A.*, **6**, 345.] 14th List

Phosinaite. Yu. L. Kapustin, A. P. Khomyakov, E. I. Semenov, E. M. Eskova, A. V. Bykova, and Z. V. Pudovkina, 1974. *Zap.*, **103**, 567 (Фосинаит). Colourless to brownish rose crystals from ussingite veinlets at Mt. Karnasurt, Lovozero massif, and infillings between anorthoclase from Mt. Koashva, Khibina (both in the Kola Peninsula) are near $H_2Na_3(Ca,Ce,Nd,La)SiO_4PO_4$. Space group $P22_12$ or $P22_12_1$, *a* 12·23, *b* 14·62, *c* 7·21 Å. Named for *Pho*sphorus, *si*licon, and *na*trium. [*A.M.*, **60**, 488; *M.A.*, 75-2526; *Zap.*, **104**, 614.] 29th List

Phosphate-allophane. S. G. Gordon, 1929. *Amer. Min.*, 1929, vol. 14, p. 105 (phosphate allophane); *Proc. Acad. Nat. Sci. Philadelphia*, 1944, vol. 96, p. 355 (Phosphate allophane, Phosphate-Allophane). A variety of allophane containing P_2O_5 7·97%. [*M.A.*, **9**, 209.] 17th List

Phosphate-belovite. *See* Arsenate-belovite. 21st List

Phosphate-schultenite. H. Strunz, 1957. *Min. Tabellen*, 3rd edit., p. 224 (Phosphat-Schultenit). Artificial $PbHPO_4$, the phosphate analogue of schultenite. 22nd List

Phosphate-Walpurgine. H. Strunz, 1966. *Min. Tabellen,* 4th edit., p. 310. The phosphate analogue of Walpurgite; natural occurrence attributed to V. G. Melkov, 1946. $(BiO)_2UO_2(PO_4)_2.3H_2O$. 24th List

Phosphocristobalite. R. J. Manly, *Jr.*, 1950. *A.M.*, **35**, 111. *Artificial.* A polymorph of berlinite ($AlPO_4$) with the cristobalite structure; formed by heating evansite and some other Al phosphates. Named from the analogy with cristobalite. 28th List

Phosphoferrite. H. Laubmann and H. Steinmetz, 1920. *Zeits. Kryst. Min.*, vol. 55, p. 569 (Phosphoferrit). An acid phosphate of ferrous iron (magnesium, calcium, etc.), $4\frac{1}{2}R''.3PO_4.H_3PO_4$, occurring as cloudy-white or greenish, crystalline masses with greasy lustre in pegmatite at Hagendorf, Bavaria. [*M.A.*, **1**, 125.] 9th List [Redefined as $(Fe,Mn)_3(PO_4)_2.3H_2O$; *see* Dana's *Syst. Min.*, 7th edit., **2**, 727]

Phosphophyllite. H. Laubmann and H. Steinmetz, 1920. *Zeits. Kryst. Min.*, vol. 55, p. 566 (Phosphophyllit). Hydrated phosphate and sulphate of ferrous iron, magnesium, calcium, potassium and aluminium, $3R_3P_2O_8.2AlOHSO_4.9H_2O$, forming colourless or pale-blue, monoclinic crystals with perfect micaceous cleavage (hence the name). Occurs in pegmatite at Hagendorf, Bavaria. [*M.A.*, **1**, 125.] 9th List [$Zn_2(Fe,Mn)(PO_4)_2.4H_2O$; Mn substitution, earlier refs., *A.M.*, **62**, 818]

Phosphoralunogen. H. Strunz, *Min. Tab.*, 1941, p. 135. Alunogen containing some P_2O_5 replacing SO_3, $Al_2[PO_3OH,(SO_4)_2].16H_2O$. The formula $Al_2(SO_4)_2(HPO_4).11\frac{1}{2}H_2O$, as originally given [*M.A.*, **7**, 470], however, points to meta-alunogen [*M.A.*, **8**, 278] rather than to alunogen. 18th List

Phosphormimetesit. H. Strunz, *Min. Tab.*, 1941, p. 156. Mimetite with some arsenic replaced by phosphorus. Identified with campylite (Kampylit). 18th List

Phosphorogummit, error for Phosphor-Gummit (C. Hintze, *Handb. Min.*, Erg.-Bd. II, p. 814). 24th List

Phosphoro-orthite. F. Machatschki, 1931. *Centr. Min.*, Abt. A, 1930, p. 347 (Phosphoroorthit, Phosphorerdenepidot). A variety of orthite in which some phosphorus takes the place of silicon. Synonym of nagatelite (12th List). [*M.A.*, **5**, 52.] 13th List

Phosphor-rösslerite. O. M. Friedrich and J. Robitsch, 1939. *Zentr. Min.*, Abt. A, 1939, p. 142 (Phosphorrößlerit), p. 143 (Phosphor-Rößlerit). Hydrous acid magnesium phosphate, $MgHPO_4.7H_2O$, as monoclinic crystals isomorphous with rösslerite ($MgHAsO_4.7H_2O$). From Salzburg. *See* Arsen-rösslerite. [*M.A.*, **7**, 316, 495.] 15th List

Phosphorus. Evidence of the existence of native phosphorus in a meteoric stone is adduced by O. C. Farrington, *Amer. Journ. Sci.*, 1903, ser. 4, vol. xv, p. 71; *Chem. News*, 1903, vol. lxxxvii, p. 66. 4th List

Phosphoscorodite. T. N. Shadlun and Y. S. Nesterova, 1947. *Mém. Soc. Russe Min.*, vol. 66, p. 212 (Фосфоскородит). A white powdery, optically biaxial mineral midway in composition between scorodite ($FeAsO_4.2H_2O$) and strengite ($FePO_4.2H_2O$). [*M.A.*, **11**, 11.] 19th List

Phosphothorogummite. V. S. Karpenko, N. G. Nazarenko, and O. V. Shchipanova*, 1957. [Сборн. вопрос. прик. радио. Атомиздат, 100; M. S. Filippov and L. V. Komlev, Труды Радиев. инст. Акад. наук СССР **8** (1958)]; abstr. *Bull.*, **92**, 517 (1969) and *Zap.*, **98**, 330 (1967) (Фосфоторогумит); in the latter the date is given as 1967. Unnecessary name for a phosphatian thorogummite. [*Corr. in *M.M.*, **39**.] 26th List

Phosphotridymite. R. J. Manly, *Jr.*, 1950. *A.M.*, **35**, 111. *Artificial.* A polymorph of berlinite ($AlPO_4$) with the tridymite structure; formed by heating wavellite and some other Al phosphates. Named from the analogy with tridymite. 28th List

o-Phthalic acid. M. Louis, J.-C. Guillemin, J.-C. Goñi, and P.-P. Ragot, 1968. *Proc. 4th*

Intern. Meeting on Organic Geochem., Amsterdam, 1968, 535. A constituent of the oil included in the calcite matrix of quincite. [*M.A.*, 72-1768.] 28th List

Phurcalite. M. Deliens and P. Piret, 1878. *Bull.*, **101**, 356. A specimen labelled bergenite (= barium-phosphuranylite, *see* 22nd List), from Bergen an der Trieb, Vogtland, Saxony, was found to be non-fluorescent, so was re-examined. It occurs as yellow orthorhombic plates, *a* 17·426, *b* 16·062, *c* 13·592 Å, space group *Pbca*. Composition $8[Ca_2(UO_2)_3(PO_4)_2(OH)_4.4H_2O]$. RIs α 1·690, β 1·730, γ 1·749, $2V_\alpha$ 68°. Named for *ph*osphate, *ur*anium, *ca*lcium. 30th List

Phyllite. A general term used by some French authors for the scaly minerals micas, chlorites, and clays (J. de Lapparent, *Leçons de pétrographie*, Paris, 1923, p. 255), and more recently applied to minerals with a layered crystal-structure. J. L. Thiébaut (*Contribution à l'étude des sédiments argilo-calcaires du bassin de Paris*, Nancy, 1925, p. 46) found the phyllitic constituent of marls to be near bravaisite or glauconite in composition. This has since been named pholidoide (q.v.). Phyllite of T. Thomson (1828) is a synonym of ottrelite. Phyllite (C. F. Naumann, 1849), a variant of the French phyllade (A. Brongniart, 1813), is in common use for a slaty-schistose rock. Also a general term for fossil plant leaves. 14th List

Picotpaulite. Z. Johan, R. Pierrot, H.-J. Schubnel, and F. Permingeat, 1970. *Bull.*, **93**, 545. A thallium mineral, only observed as patches (<0·5 mm) in polished sections, associated with raguinite (26th List) and pyrite in realgar at Allchar, Macedonia, is orthorhombic, *a* 5·40, *b* 10·72, *c* 9·04 Å. Composition $4[TlFe_2S_3]$. Pseudohexagonal by interpenetration twinning on {120}. Named for Paul Picot. [*A.M.*, **57**, 1909.] 27th List

Picroamosite. D. P. Serdyuchenko, 1936. *Bull. Acad. Sci. URSS, Cl. Sci. Math. Nat., Sér. Géol.*, 1936, p. 689 (пикроамозит), p. 695 (picroamosite). A fibrous orthorhombic amphibole analogous to amosite (8th List; *M.A.*, **4**, 92), with MgO (29·26%) in place of FeO; from Caucasus. Named from πικρός, bitter, and amosite. [*M.A.*, **7**, 9.] 15th List [Ferrian anthophyllite; *Amph.* (1978)]

Picrochromite. E. S. Simpson, 1920. *Min. Mag.*, vol. 19, pp. 100, 104. The hypothetical molecule $MgO.Cr_2O_3$ of the spinel–chromite series of minerals: it is present in predominating amount in a 'chromite' from Quebec. Named from the composition (πικρός, bitter, alluding to the magnesia). 9th List [Cf. Magnesiochromite]

Picrocollite. E. S. Simpson, 1928. *Journ. Roy. Soc. Western Australia*, vol. 13 (for 1927), p. 43. A hypothetical molecule $H_4MgSi_3O_8.2H_2O$ containing the same number of atoms as halloysite ($H_4Al_2Si_2O_8.2H_2O$), the two being regarded as end-members of the pilolite-palygorskite group. [*M.A.*, **3**, 545.] 11th List

Picrocrichtonite. (*Collection de Minéralogie du Muséum d'Histoire Naturelle, Paris, Guide du Visiteur*, 2nd edit., 1900, p. 24.) A. Lacroix, *Minéralogie de la France*, 1901, vol. iii, p. 284. Synonym of picroilmenite $(Fe,Mg)TiO_3$ (cf. *Min. Mag.*, vol. xiv, p. 165). Lacroix distinguishes between crichtonite ($FeTiO_3$) and ilmenite ($FeTiO_3.xFe_2O_3$); hence the above name for the magnesian variety. 4th List [Crichtonite refs., *see A.M.*, **63**, 36]

Picrogalaxite. T. Yoshimura, 1936. [*Journ. Geol. Soc. Japan*, **43**, 444], quoted in *Intro. Jap. Min.*, 116 (1970), Geol. Surv. Japan. A magnesian galaxite from the Oashi mine, Tochigi Prefecture, Japan. 28th List

Picroilmenite. E. Hussak*, 1895. *Min. Petr. Mitt.* (*Tschermak*), **14**, 408. (*Pikroilmenit*). The same as picrotitanite, a variety of ilmenite rich in magnesium. 2nd List [*Corr. in. *M.M.*, **14**.]

Picroknebelite. T. Yoshimura, 1939. *Journ. Fac. Sci. Hokkaido Univ.*, ser. 4, vol. 4, p. 404 (Picroknebelite). A variety of knebelite containing MgO 4%. [*M.A.*, **7**, 412.] W. F. Foshag, *Amer. Min.*, 1939, vol. 24, p. 659, gives this as 'Picrotephroite'. 15th List

Picrophengite. A. N. Winchell, 1949. *Amer. Min.*, vol. 34, p. 223. A hypothetical molecule $K_2MgAl_3(OH)_4Si_7AlO_{20}$, classed as a sub-species of muscovite. Compare ferrophengite (q.v.). [*M.A.*, **10**, 568.] 18th List

Picrourbanite. T. Yoshimura, 1937. [*Journ. Geol. Soc. Japan*, **44**, 561], quoted in *Intro. Jap. Min.* 117 (1970), Geol. Surv. Japan. A clinopyroxene from the Nodo-Tamagawa mine, Iwate Prefecture, Japan. [*Zap.*, **101**, 286.] 28th List

Piddintonite, error for Piddingtonite (H. Strunz, *Min. Tabellen*, 3rd edit., 1957, p. 404; 4th edit., 1966, p. 506). 24th List

Pierrepontite. F. R. Van Horn, 1926. *Amer. Min.*, vol. 11, p. 54. The black iron tourmaline of Pierrepont, St. Lawrence Co., New York, regarded as a distinct species of the tourmaline group. 11th List

Pierrotite. C. Guillemin, Z. Johan, C. Laforêt, and P. Picot, 1970. *Bull. Soc. franç. Min Crist.*, **93**, 66. $Tl_2(Sb,As)_{10}S_{17}$, grey-black, metallic lustre, massive, in quartz veins from Jas-Roux, Hautes Alpes. Orthorhombic. Named for R. Pierrot. [*M.A.*, 70-3428; *A.M.*, **57**, 1909.] 26th List

Pietersite, a variety of agate from SW Africa (H. Strunz, *Min. Tabellen*, 4th edit., 1966, p. 507). 24th List

Pietricikite. C. [I.] Istrati, *Bult. Soc. Sci. Bucarest*, 1897, VI, 65, 93. The correct spelling of Dana's zietrisikite. A variety of ozocerite from Mt Pietricica, Moldavia. 2nd List

Piezotite. P. A. Vaughan and R. Berman, 1963. *Acta Cryst.*, vol. 26, suppl., p. A.13. *Artificial* $Al_3Si_2O_7(OH)_3$, obtained by hydrothermal decomposition of Spessartine; anorthic. 24th List

Pigeonite. A. N. Winchell, *Amer. Geologist*, 1900, vol. xxvi, pp. 204, 368; *Thèse Fac. des Sci. Paris*, 1900. A pyroxene, with optic axial angle small and variable (2E = 13° 16′–62° 24′), occurring as a constituent of olivine-diabase at Pigeon Point, Minnesota. 3rd List [*See Rock-forming Minerals* (Longmans, 1963), **2**, 143]

Pigeonite-augite. [W.] E. Tröger, 1951. *Neues Jahrb. Min.*, Monatshefte, 1951, pp. 136, 137 (Pigeonitaugit). A clinopyroxene intermediate between pigeonite and augite. [*M.A.*, **11**, 390.] 19th List

Pikrophyll, variant of Picrophyll, Pikrophyllit (H. Strunz, *Min. Tabellen*, 1st edit., 1941, p. 264). 24th List

Pikrourbanit, German variant of Picrourbanite (q.v.). Hintze, 73. 28th List

Pilbarite. E. S. Simpson, 1910. *Chem. News*, vol. cii, p. 283; *Journ. Nat. Hist. Sci. Soc. Western Australia*, 1911, vol. iii, p. 130. A canary-yellow, amorphous (and pseudomorphous) mineral consisting of hydrated silicate of thorium, uranium, and lead, $ThO_2.UO_3.PbO.2SiO_2.2H_2O + 2H_2O$, and differing from thorogummite and mackintoshite in the relative proportions of these constituents. It is found as earthy nodules in the Pilbara goldfield, Western Australia. 6th List [A mixture; *M.A.*, **13**, 635]

Pilite. E. Schulze, *Lithia Hercynica*, Leipzig, 1895, p. 29. The same as tinder-ore (Zundererz). Formula given as $Pb_2Sb_3S_8$, Harz. 2nd List [i.e. impure Jamesonite]

Pinchite. B. D. Sturman and J. A. Mandarino, 1974. *C.M.*, **12**, 417. Brown to blackish crystals from Terlingua, Brewster County, Texas, have space group *Ibam*, *a* 11·6, *b* 6·07, *c* 11·7 Å; composition $4[Hg_5O_4Cl_2]$. Named for W. W. Pinch. [*A.M.*, **61**, 340; *M.A.*, 75-3602; *Bull.*, **97**, 508; *Zap.*, **104**, 608.] 29th List

Pintadoite. F. L. Hess and W. T. Schaller, 1914. *Journ. Washington Acad. Sci.*, vol. iv, p. 576. A hydrous vanadate of calcium, $2CaO.V_2O_5.9H_2O$, occurring as a thin, green efflorescence on sandstone at Cañon Pintado, San Juan Co., Utah. Named from the locality. 7th List

Pirssonite. J. H. Pratt, *Amer. Journ. Sci.*, II. 126, 1896; and *Zeits. Kryst. Min.*, XXVII. 420, 1896. [*M.M.*, **11**, 226.] $CaCO_3.Na_2CO_3.2H_2O$. Rhombic. [Borax Lake, San Bernardino Co.] California. 1st List

Pisekite. A. Krejčí, 1923. *Časopis Min. Geol. Prague*, vol. 1, p. 2 (Pisekit). B. Ježek, ibid., p. 69. An optically isotropic mineral containing Nb, Ta, Ti, U, Ce, Yt, Yb, Th, as monoclinic crystals resembling monazite. It is perhaps isomorphous with or pseudomorphous after monazite. Named from the locality, Písek in Bohemia. [*M.A.*, **2**, 335, 336.] 10th List

Pitankite. (Author?). A. K. Boldyrev, *Kurs opisatelnoi min.*, Leningrad, 1926, vol. 1, p. 259 (питанкит, index only); A. E. Fersman and O. M. Shubnikova, *Sputnik geochim. min.*, Moscow, 1937, p. 187. '$2(Ag,Pb,Cu)S.Bi_2S_3$'. (Locality?, derivation?) 20th List

Pizit. (A. Rzehak, *Verh. naturforsch. Ver. Brünn*, 1920, vol. 57, p. 142). Alternative spelling of Picite (E. Bořický, 1869, *Sitzungsber. Math.-Naturwiss. Cl. Akad. Wiss. Wien*, vol. 59, Abth. I, p. 591; A. Nies, 1880; from 'Picites resinaceus' of A. Breithaupt, 1847). 9th List

Placodine. A. Breithaupt, 1841. *Ann. Phys. Chem. (Poggendorff)*, vol. liii, p. 631 (Plakodin, Placodinus niccoleus). A nickel arsenide described as tabular (πλᾰκώδης) crystals from the

Jungfer mine, Müsen, Westphalia, but afterwards considered to be an artificial *Nickelspeise*. (For history and literature *see* C. Hintze, *Handbuch der Mineralogie*, 1900, vol. i, p. 621.) A. Rosati (*Atti* (*Rend.*) *Accad. Lincei*, Roma, 1913, ser. 3, vol. xxii, sem. 2, p. 243; *Zeits. Kryst. Min.*, 1914, vol. liii, p. 389) proves it to be identical with the recently-described mineral maucherite, tetragonal Ni_3As_2 (F. Grünling, 1913; 6th List). *See* Temiskamite. 7th List

Plaffeiite. A. Tschirch and [T.] Kato, 1926. *Mitt. Naturfor. Gesell. Bern*, for 1925, p. 13 (Plaffeiit). A fossil resin occurring in the Flysch at Plaffeien, Switzerland. Named from the locality. [*M.A.*, **3**, 475.] 11th List

Plancheite. A. Lacroix, 1908. *Compt. Rend. Acad. Sci. Paris*, vol. cxlvi, p. 724; *Bull. Soc. franç. Min.*, vol. xxxi, p. 250 (planchéite). A blue, fibrous copper silicate, $H_{10}Cu_{15}Si_{12}O_{44}$ or $H_2(CuOH)_8Cu_7(SiO_3)_{12}$ occurring with dioptase as botryoidal masses or fibrous veins in limestone at Mindouli, French Congo. Named after Mr. — Planche. 5th List [A copper amphibole; refs. in *Chem. Index*, App. II, 14.2.2. $Cu_8(Si_4O_{11})_2(OH)_4.nH_2O$; *A.M.*, **62**, 491]

Planoferrite. L. Darapsky, *Verh. deutsch. wiss. Ver. Santiago, Chili*, 1897, III, 423; *Bol. Soc. Nac. Mineria, Santiago*, 1898 [iii], X, 106; *Zeits. Kryst. Min.*, 1898, XXIX, 213. Probably rhombic (F. Grünling). $Fe_2O_3.SO_3.15H_2O$. Groth (*Tab. Uebers. Min.*, 4th edit., 1898, p. 74) writes the formula as $SO_4Fe_2(OH)_4 + 13H_2O$. Chili. 2nd List

Platarsite. L. J. Cabri, J. H. G. Laflamme, and J. M. Stewart, 1977. *Can. Mineral.*, **15**, 385. A specimen from the Onverwacht platinum deposit, Transvaal, contains grains of a cubic mineral, *a* 5·790–5·824 Å, space group *Pa*3. Composition 4[(Pt,Rh,Ru)AsS]. This is the Rh-sperrylite of Stumpfl and Clark (30th List), which is inappropriate because Pt > Rh. Named for the composition. 30th List

Platinoiridita. G. Gagarin and J. R. Cuomo, 1949, loc. cit., p. 5. Variant of platiniridium, applied to Ir-Pt alloys rich in iridium. *See* Iridioplatinita. 19th List

Platinum-Palladium Stannide. L. V. Razin, V. D. Bergizov, and V. I. Meshchankina, 1973. [*Trudy Ts NIGRI*, **108**, 96], abstr. *A.M.*, **61**, 180. A palladian ordered form of Pt_3Sn from the Talnakh deposit, USSR. *a* 3·984 space group *Pm*3*m*. 29th List

Plattnerite, error for Planerite (Hintze, *Handb. Min.*, Erg.-Bd. 11, 1960, p. 815). 22nd List

Platynite. G. Flink, 1910. *Arkiv Kemi, Min. Geol.*, vol. iii, no. 35 (*Bidrag till Sveriges mineralogi*), p. 5 (Platynit). An iron-black metallic mineral found as thin plates at Falun, Sweden. It shows basal and rhombohedral cleavages. Formula $PbS.Bi_2S_2$. Named from πλατύνειν, to flatten. 6th List [$PbBi_2(Se,S)_3$]

Playfairite. J. L. Jambor, 1967. *Canad. Min.*, vol. 9, pp. 7, 191. A monoclinic mineral from Madoc, Ontario, with composition $Pb_{16}Sb_{18}S_{43}$. Named for J. Playfair. [*A.M.*, **53**, 1424; *Bull.*, **91**, 303.] 25th List

Plazolite. W. F. Foshag, 1920. *Amer. Min.*, vol. 5, p. 183. Hydrated silicate (and carbonate) of calcium and aluminium, $3CaO.Al_2O_3.2(SiO_2,CO_2).2H_2O$, occurring as small, colourless rhombic-dodecahedra in metamorphic limestone at Crestmore, California. Named from πλάζω, to perplex [and λίθος, stone], in allusion to the difficulty of interpreting the chemical composition. A. S. Eakle (*Amer. Min.*, 1921, vol. 6, p. 109) points out a similarity to garnet. [*M.A.*, **1**, 151, 254] 9th List [*See* Hibschite, **Hydrogrossular** (preferred name)]

Pleistoexpandite, *see* Expandite.

Pleochroite. H. G. Midgley, 1968. *Trans. Brit. Ceram. Soc.*, **67** (1), I. An *artificial*, highly pleochroic (deep-blue to colourless) monoclinic (pseudo-orthorhombic) phase occurring in high-alumina cements. Composition $Ca_{22}Fe_3Al_{14}(Al_2O_7)_8(AlO_4)_4(SiO_4)_2$. [*M.A.*, 69-289.] 28th List

Pleysteinite. P. Groth, 1916. *Zeits. prakt. Geol.*, vol. 24, p. 190. Mentioned in a preliminary announcement as 'ein wohlkristallisiertes neues Phosphat (Pleysteinit)'. Evidently the same mineral as that later described under the name Kreuzbergite (q.v.), From the Kreuzberg (or Kreuzstein) at Pleystein, Oberpfalz, Bavaria. 9th List [i.e., Fluellite]

Plioexpandite, *see* Expandite.

Plumalsite. G. Ya. Gornyi, M. G. Dyadchenko, and T. A. Kudykina, 1967. [Доповиди акад. наук Укр. ССР, сер. Б, геол. геофиз. хим., no. 6, p. 514]; abstr. *Amer. Min.*, 1968, vol. 53, p. 349. Colourless, yellow, green, and black angular platy fragments in weathered crystalline rocks of the Ukrainian shield consist of a silicate of lead and aluminium, near $(Pb,Ca,Mg)_4(Al,Fe)_2(SiO_3)_7$; orthorhombic. Named from the composition (Plum[bum],Al,Si). The name is uncomfortably near the mineral plumosite and the rock plumasite. 25th List

Plumangite. D. Adib and J. Ottemann, 1970. *Min. Depos.*, **5**, 86. A greyish mineral replacing murdochite along fractures, from the T. Khuni mine, Anarak, Iran, is formulated $Cu_{0\cdot85}Zn_{0\cdot15}PbMn_4O_{11}$ on the basis of electron-probe analysis. Some optical but no X-ray data are given. Named for the composition. [It is difficult to see how the cited formula, with some Mn in a valency state higher than 4, is arrived at; if it is assumed that the manganese is all Mn^{4+}, the composition is close to that of an analogue of coronadite with the Mn^{2+} replaced by Cu. The name is unfortunate in suggesting relation to plumosite or umangite—*M.H.H.*] [*A.M.*, **55**, 1812.] 26th List

Plumboallophane, variant of Plumballophane (Bombicci, 1868). [*Godishn. Univ. Sofia, Fak. Geol.-Geogr.*, Geol. **63**, 217]; abstr. *A.M.*, **58**, 348. An unnecessary name for an allophane with a little PbO [*Zap.*, **103**, 365.] 28th List

Plumboalunite. M.-A. Kashkai, 1969. Зап. Всесоюз. мин. общ. (*Mem. All-Union Min. Soc.*), **98**, 153 (свинцовый алунит). A name for the end-member $PbAl_6(SO_4)_4(OH)_{12}$. 26th List

Plumbobetafite. A. A. Ganzeev, A. F. Efimov, and G. V. Lyubomilova, 1969. Труды Мин. муз. Акад. наук СССР (*Trav. Mus. Min. Acad. Sci. URSS*), **19**, 135 (Плюмбобетафит). Metamict yellowish grains (up to 2 to 3 mm) and octahedra, cubic after heating to 800°C (*a* 10·33 Å), D 4·64, from a dyke in the Burpala massif, N Baikal, analyse as $(Pb_{0\cdot44}U_{0\cdot25}Ca_{0\cdot18}Na_{0\cdot12}Ln_{0\cdot12})Nb_{1\cdot12}Ti_{0\cdot78}Fe_{0\cdot07}Ta_{0\cdot02}O_6(OH)_{0\cdot58}F_{0\cdot42}$. [*A.M.*, **55**, 1068; *Zap.*, **100**, 83.] 27th List [*A.M.*, **62**, 403]

Plumbobinnite. A. Weisbach, *Char. Min.*, 1880, p. 42. The same as dufrenoysite. 2nd List

Plumbodolomite. W. Siegl, 1936. *Min. Petr. Mitt.* (*Tschermak*), vol. 48, p. 288 (Plumbodolomit). A variety of dolomite containing some lead, from Kreuth, Carinthia. [*M.A.*, **6**, 529.] 14th List

Plumbocolumbite. H. Strunz, *Min. Tab.*, 1941, p. 108 (Plumbocolumbit). Syn. of plumboniobite (5th List). 18th List

Plumbojarosite. W. F. Hillebrand and S. L. Penfield, *Amer. Journ. Sci.*, 1902, ser. 4, vol. xiv, p. 213; *Zeits. Kryst. Min.*, 1902, vol. xxxvi, p. 548. A dark-brown, glistening powder consisting wholly of minute, perfectly developed rhombohedra with basal planes. $Pb[Fe(OH)_2]_6(SO_4)_4$. From Cook's Peak, New Mexico. 3rd List

Plumbolimonite. G. N. Vertushkov* and Yu. A. Sokolov, *Mém. Soc. Russ. Min.*, 1958, vol. 87, p. 96 (плумболимонит). PbO 15·38, Fe_2O_3 45·20, MnO 18·37, SiO_2 5·00, H_2O 12·51%. With pyromorphite from Verkhne Ufaleye, Urals. [*Corr. in *M.M.*, **31**.] 21st List

Plumbomalachite. S. F. Glinka and I. A. Antipov, St. Petersburg, *Devn. XI Sjezda russ. jest. vrač.*, 1901, p. 468 (свинцовомъ малахитъ). *See* Bleimalachit, malachite de plomb. 3rd List

Plumbomatildite. A. A. Godovikov, 1972. Abstr. in *Zap.*, **103**, 617 (1974) (Плюмбоматилдит). Unnecessary, invalid name based solely on an old analysis near $Ag_6PbBi_6S_{13}$. 29th List

Plumbomicrolite. A. Safiannikoff and L. Van Wambeke, 1962. *Bull. Soc. franç. Min. Crist.*, vol. 84, p. 384. Greenish-yellow and orange crystalline masses and octahedra from Kivu, Congo, have a composition near $(Pb,Na,Ca)_{2-x}(Ta,Nb,Ti)_2(O,OH)_7$, with x about 0·9. The authors do not definitely class the mineral as a microlite, but refer to it as a plumboan member of the Microlite family; yet in a few places they use the name Plumbomicrolite. 23rd List [*A.M.*, **62**, 406]

Plumbonacrite. J. K. Olby, 1966. *Journ. Inorg. Nucl. Chem.*, **28**, 2507. A *synthetic* phase, $Pb_{10}(CO_3)_6(OH)_6O$; hexagonal, *a* 9·076, *c* 24·96 Å; previously mis-named hydrocerussite (cf. Cowley, *Acta Cryst.*, 1956, **9**, 391, also Reed, ibid., 1957, **10**, 142). Not the plumbonacrite of Heddle (*M.M.*, 1889, **8**, 203), a name proposed to replace hydrocerussite (Nordenskiöld, 1877), '*Not* being a hydrated cerussite.' [*A.M.*, **52**, 563.] 28th List

Plumboniobite. O. Hauser and L. Finckh, 1909. *Ber. Deutsch. Chem. Ges.*, xlii, p. 2270; O. Hauser, ibid., 1910, xliii, p. 417 (Plumboniobit). A niobate resembling samarskite in composition, but containing some lead (PbO, 7·55%): formula $R''_2Nb_2O_7.R'''_4(Nb_2O_7)_3$, where

R″ = Fe,Pb,UO,Ca, and R‴ = Gd,Sm,Y,Al. Occurs as dark brown to black, imperfectly crystalline (optically isotropic) masses, associated with pitchblende, in the mica mines at Morogoro, Uluguru Mts. German East Africa. 5th List

Plumbopalladinite. A. D. Genkin, T. L. Evstigneeva, L. N. Vyalsov, I. P. Laputina, and N. V. Troneva, 1970. Геол. Рудн. Месторожд. **5**, 63 (Плюмбопалладинит). Aggregates of minute grains, in veinlets of cubanite in talnakhite (25th List) in the Talnakh Ni–Cu deposits, are hexagonal, *a* 4·470, *c* 5·719 Å. Composition Pd_3Pb_2 with small amounts of Ag, Cu, Bi, Sn, Sb. Named from the composition. [*A.M.*, **56**, 1121; *M.A.*, 71-2335; *Zap.*, **100**, 614.] 27th List

Plumbopyrochlore. H. V. Skorobogatova, G. A. Sidorenko, K. A. Dorofeeva, and T. I. Stolyarova, 1966. [Геол. месторожд. редк. элем., no. 30, p. 84]; abstr. Зап. всесоюз. мин. общ. (*Mem. All-Union Min. Soc.*), 1968, vol. 97, p. 69 (Плюмбопирохлор, plumbopyrochlore). A variety with Pb predominant in the *A* cations. 25th List [*A.M.*, **62**, 406]

Plumbosvanbergite. A superfluous name for plumbian Svanbergite. [Геохим., мин., генет. типы месторожд. редк. элем., Изд. „Наука“, 1964, p. 194]; abstr. Зап. Всесоюз. Мин. Обш. (*Mem. All-Union Min. Soc.*), 1965, vol. 94, p. 683 (Плюмбосванбергит). 24th List

Plumbosynadelphite. C. S. Hurlbut, 1937. *Amer. Min.*, vol. 22, p. 531. A variety of synadelphite containing some lead (PbO 3·24%), forming a red outer zone on colourless crystals of synadelphite. [*M.A.*, **6**, 488.] 14th List

Plumbozincocalcite. M. Z. Kantor, 1964. [Изв. высш. учебн. завед., геол. разв. no. 3, 61]; abstr. *Zap.*, **97**, 70 (1968) (Плюмбоцинкокальцит). An unnecessary name for a plumbian zincian calcite. [*A.M.*, **53**, 1776; *Bull.*, **92**, 321.] 26th List

Plumbozinkocalcit, Germ. trans. of Плюмбоцинкокальцит, plumbozincocalcite (26th List). Hintze, 75. 28th List

Plusinglanz. A. Breithaupt, *Vollst. Char. Min.-Syst.*, 1823, p. 277. Shown by A. Frenzel (*Tsch. Min. Mitth.*, 1900, XIX, 244) to be identical with argyrodite. 2nd List

Podolite. W. Tschirwinsky, 1907. *Centralblatt Min.*, 1907, p. 279 (Podolit). A mineral closely related to apatite and staffelite, but with the composition $3Ca_3(PO_4)_2.CaCO_3$, occurring as minute hexagonal crystals in phosphorite nodules in govt. Podolia, S Russia. It had earlier been called carbapatite (q.v.). 4th List

Poechite. F. Katzer, 1911. *Oesterreich. Zeits. Berg- und Hüttenwesen*, vol. lix, p. 229 (Poechit). A massive, reddish-brown manganese-iron-ore from Vareš, Bosnia. Analysis gives the formula—$H_{16}Fe_8Mn_2Si_3O_{29}$ or $(MnO)_2SiO_3.2(FeO)_2SiO_3.5H_2O + 2Fe_2O_3.3H_2O$. Named after Franz Poech, Chief of the Department of Mines of Bosnia-Herzegovina. 6th List

Poitevinite. J. L. Jambor, G. R. Lachance, and S. Courville, 1964. *Canad. Min.*, vol. 8, p. 109. The unnamed $(Cu,Fe)SO_4.H_2O$ of Jambor, *Canad. Min.*, 1962, vol. 7, p. 245, from Bonaparte River, Lillooet District, British Columbia, a copper analogue of Szomolnokite, has been restudied, and named for Dr. E. Poitevin. Poitevinite is defined as the Cu-rich half of the series. [*A.M.*, **50**, 263.] 24th List

Polarite. A. D. Genkin, T. L. Evstigneeva, N. V. Troneva, and L. N. Vyalsov, 1969. Зап. Всесоюз. мин. общ. (*Mem. All-Union Min. Soc.*), **98**, 708. The mineral Pd(Bi,Pb) described by L. J. Cabri and R. J. Traill (*Canad. Min.*, **8**, 541, 1966)* from Norilsk, W Siberia, has also been found in the Talnakh deposits; further study confirms its species status, and it is named polarite (Полярит). [*Corr. in *M.M.*, **39**.] 26th List

Polialite. Italian spelling of polyhalite. 20th List

Poliophane. F. Machatschki, 1953. *Spezielle Mineralogie,* Wien, p. 226 (Fahle, Poliophane, *pl.*). Group name for fahlore (grey copper) and bournonite. Named from *πολιός*, grey. 20th List

Polipyrites. E. F. Glocker, 1839. *Grundriss der Mineralogie,* 1839, p. 321. An obsolete synonym of marcasite. Named from *πολιός*, grey, and pyrites; German, Graueisenkies. 6th List

Poly-, prefix in some thirty mineral names (polybasite, polycrase, polyhalite, polyaugite, etc.) also in the terms polymorph, polymer, polysynthetic twinning, polyhedron, etc. It has now been extended with other significations, giving rise to polyonymous complexities. 20th List

Polyamphibole. B. I. Goroshnikov and L. D. Yur′ev, 1965. Докл. Акад. наук СССР (*Compt.*

Rend. Acad. Sci. URSS), **163**, 143 (Полиамфибол). The term polyamphibole rocks is used for rocks containing two distinct amphiboles. [*M.A.*, 69-521.] 28th List

Polyaugite. A. N. Winchell, 1949. *Amer. Min.*, vol. 34, p. 224. To replace clinopyroxene as a species, rather than a group, name, with diopside, augite, jadeite as sub-species, excluding clinoenstatite and spodumene [*M.A.*, **10**, 568.] 18th List

Polybrookite. F. Machatschki, *Spezielle Mineralogie*, Wien, 1953, p. 319 (Polybrookite, *pl.*), p. 371 (Polybrookit). Columbite and tantalite isotypic with brookite. [*M.A.*, **12**, 232.] 20th List

Polycrasite. Variant of Polycrase; *A.M.*, **8**, 52. 10th List

Polycrystal. [G. Donnay, *Amer. Cryst. Assoc.*, June 1953], G. Donnay and J. D. H. Donnay, *Amer. Min.*, 1953, vol. 38, p. 941. An apparently single crystal consisting of a regular intergrowth of different minerals. [*M.A.*, **12**, 329.] 20th List

Polygorskite. C. E. Marshall, *Colloid chemistry of silicate minerals*, New York, 1949, p. 54. Error for palygorskite. 20th List

Polyhydrate. V. I. Popov and A. L. Vorobiev, 1947. *Mém. Soc. Russ. Min.*, vol. 76, p. 268 (полугидрат). Hintze, *Min.*, 1954, Ergbd. 2, p. 315 (Polyhydrat). Error for hemi(polu) hydrate $CaSO_4.\frac{1}{2}H_2O$. [*M.A.*, **11**, 366.] *See* Miltonite (19th List), Bassanite (6th List). 20th List

Poly-irvingite. P. Quensel, 1937. *Geol. För. Förh. Stockholm*, vol. 59, p. 467 (poly-irvingite). A lithia-mica much richer in silica than irvingite [4th List], occurring as an alteration product of amblygonite in Sweden. [*M.A.*, **7**, 108.] 15th List

Polymigmite. F. Machatschki, 1953, ibid., p. 319 (Polymigmit). Error for polymignite (*μιγνύω*, to mix, not *μίγμα*, a mixture). 20th List

Polymignyte. Dana's *System of mineralogy*, 1944, 7th edit., vol. 1, p. 764. Another spelling of polymignite (J. J. Berzelius, 1824). 17th List

Polymineralic rocks. F. Machatschki, 1953, ibid., p. 1 (Polyminerale Gesteine), as distinct from monomineralic rocks. 20th List

Polynite. E. A. Yarilova and E. I. Parfenova, 1957. Pochvovedenie (Pedology), 1957, no. 9, p. 37 (полынит). A montmorillonoid clay mineral in soils. Named after B. B. Polynov, Б. Б. Полынов (1877–1952). [*M.A.*, **13**, 580.] 21st List

Polyophane. F. Machatschki, ibid., p. 371. Error for poliophane (q.v.). 20th List

Polyosmin. J. Fröbel, 1843. *Grundzüge eines Systems der Krystallologie*. Synonym of syserts-kite. 27th List

Polyphant stone. Greyish-green potstone flecked with white and brown, from Polyphant, Cornwall, used as an ornamental stone in churches since Norman times. 20th List

Polyplatinum. F. Machatschki, 1953, ibid., p. 302 (Polyplatin). Platinum containing Fe, Ir, Os, Rh, Pd. 20th List

Polyquartz. F. Machatschki, 1953, ibid., p. 82 (Polyquarze, *pl.*), p. 371 (Polyquarz). Compounds $AlPO_4$ (berlinite), $AlAsO_4$, $FePO_4$, BPO_4, isotypic with quartz ($SiSiO_4$). (The corresponding polymorphs have been named Al-phosphorotridymite and Al-phosphorocristobalite, R. L. Manly, 1950, *M.A.*, **11**, 180.) 20th List

Polyrutile. F. Machatschki, 1953, ibid., p. 118 (Polyrutile, *pl.*), p. 371 (Polyrutil). Tapiolite and mossite isotypic with rutile. (Cf. polyrutiles and trirutiles of V. M. Goldschmidt, 1923, *M.A.*, **3**, 182.) 20th List

Polywurtzite. F. Machatschki, 1953, ibid., p. 371 (Polywurtzit), p. 305 (Polywurtzite, *pl.*). Hexagonal polymorphs of ZnS. 20th List

Ponite. V. C. Buţureanu, 1912. *Ann. Sci. Univ. Jassy*, vol. vii, p. 185. A ferriferous variety of rhodochrosite, $5MnCO_3.FeCO_3$, from Roumania. Named after Petru Poni, Professor of Chemistry at Jassy. 6th List

Pontellarite, *see* Pantellarite.

Porcelainite. A. L. Roussin, 1932. *In* W. H. Taylor, *Journ. Soc. Glass Techn. Sheffield*, 1932, vol. 16, p. 118. Later replaced by Porzite (q.v.). Porcelainite has long been used as a trade-name for certain kinds of white stoneware. Not the porcellanite (= porcelain-jasper) of J. T. A. Peithner, 1794. 13th List

Porcupine-ore. *See* Histrixite. 3rd List

Porpecita. G. Gagarin and J. R. Cuomo, 1949, loc. cit., p. 4. Spanish pronunciation of porpezite. *See* Mendocita. 19th List

Portlandite. C. E. Tilley, 1933. *Min. Mag.*, vol. 23, p. 419. Calcium hydroxide, $Ca(OH)_2$, as hexagonal plates in a chalk-dolerite contact-rock at Scawt Hill, Co. Antrim. So named because crystals of this substance had been earlier observed in Portland cement. (Portland cement was named in 1824 because of its resemblance in colour to the oolitic limestone—Portland stone—from the Isle of Portland in Dorsetshire.) 13th List

Porzite. A. J. Bradley and A. L. Roussin, 1932. *Trans. Ceramic Soc. Stoke-on-Trent*, vol. 31, p. 426. A fibrous constituent of porcelain, belonging to the fibrolite-mullite series. The X-ray pattern shows a closer relation to that of 'pink mullite' than to that of 'grey mullite'. E. Posnjak and J. W. Greig (*Journ. Amer. Ceramic Soc.*, 1933, vol. 16, p. 579) suggest it is identical with mullite. [*M.A.*, **5**, 323, 473.] 13th List

Posnjakite. A. I. Komkov and E. I. Nefedov, 1967. Зап. всесоюз. мин. общ. (*Mem. All-Union Min Soc.*), vol. 96, p. 58 (Posnjakite, Познякит)*. Dark blue crystals, pseudomorphous after langite, from the Nura–Talinsk tungsten deposits, Kazakhstan, have a composition $Cu_4SO_4(OH)_6.H_2O$. Named for E. W. Posnjak. [*M.A.*, **18**, 285; *A.M.*, **52**, 1582; Зап., **97**, 72; *Bull.*, **91**, 303.] [*Corr. in *M.M.*, **39**.] 25th List

Potarite. Sir John B. Harrison, MS. 1925. *Bull. Soc. franç. Min.*, 1926, vol. 49, p. 5. L. J. Spencer, *Min. Mag.*, 1928, vol. 21, p. 397. Palladium amalgam or mercuride, PdHg, cubic, as white grains from the diamond-washings on the Potaro river, British Guiana. Named from the locality. [*M.A.*, **3**, 4.] 11th List

Potash-aegirine. G. T. Faust, *Amer. Min.*, 1936, vol. 21, p. 737 (potassium-aegerite). Translation of Kaliägirin (8th List). 14th List

Potash-albite, etc. H. L. Alling, 1921. *Journ. Geol. Chicago*, vol. 29, p. 252 (potash albite, etc.). T. Tomita, *Journ. Shanghai Sci. Inst.*, sect. 2, 1933, vol. 1, p. 7; 1934, vol. 1, pp. 112, 116 (potash-albite, etc.). Potash-soda-lime felspar series, including potash-albite, potash-oligoclase (5th List), potash-andesine, potash-labradorite, potash-bytownite, and potash-anorthite, containing more than 10% of the potash component $KAlSi_3O_8$. 14th List

Potash-analcime. E. S. Larsen and B. F. Buie, 1938. *Amer. Min.*, vol. 23, p. 837 (potash analcime). A potash-rich (K_2O 4·48%) variety of analcime in analcime-basalt from Montana. [*M.A.*, **7**, 448.] 15th List

Potash-anorthoclase. K. Kimizuka, 1932. *Japanese Journ. Geol. Geogr.*, vol. 9, p. 225. A triclinic felspar of the composition $Or_{66}Ab_{31}An_3$. [*M.A.*, **5**, 363.] 13th List

Potash-bentonite. P. F. Kerr and P. K. Hamilton, *Glossary of clay mineral names*, New York, 1949, p. 52 (potash bentonite). Syn. of meta-bentonite (C. S. Ross, 1928, 14th List), meta-bentonite (J. G. Kay, *Journ. Geol. Chicago*, 1931, vol. 39, p. 361). Cf. potash-montmorillonite (14th List). 19th List

Potash-margarite. F. C. Phillips, 1931. *Min. Mag.*, vol. 22, p. 485. Synonym of lesleyite, for a margarite in which calcium is largely replaced by potassium. 12th List

Potash-montmorillonite. V. T. Allen, 1932. *Journ. Geol. Chicago*, vol. 40, p. 263 (potash-montmorillonite); abst. in *Neues Jahrb. Min.*, Ref. III, 1933, p. 1174 (Kali-Montmorillonit). A clay mineral of the meta-bentonites (q.v.) with high potash (K_2O 4·60%) and low water, from Missouri. 14th List

Potash-nepheline. A. Holmes, *Min. Mag.*, 1936, vol. 24, p. 413 (potash-nepheline). A potash-rich nepheline. Not the same as Kalinephelin (Dana), Kali-Nephelin (Hintze), a synonym of kaliophilite. 14th List

Potash-oligoclase, J. P. Iddings, 1906. *Rock Minerals*, New York, p. 232. A lime-soda-microcline (anorthoclase) in which lime appears to form part of the potash-soda-felspar, the composition being nearly that of oligoclase with part of the soda replaced by potash. Occurs in the rhomb-porphyry of S Norway and in the kenyte of Kilimanjaro, E. Africa. 5th List

Potash-richterite. H. Sjögren, *Bull. Geol. Inst. Upsala*, II. 77, 1895. The original richterite analysed by Michaelson in 1863. 1st List

Potash-scapolite. D. P. Serdyuchenko, 1955. *Problems of geology of Asia, Acad. Sci. USSR*, vol. 2, p. 742 (калиевые скаполиты). Scapolites containing up to 5·88% K_2O from Yakutia. [*M.A.*, **13**, 186.] 21st List

Potassalumite. A. N. Winchell, 1927. *Elements of optical mineralogy*, 2nd edit., part 2, p. 114. Cubic potash-alum; it being suggested that the fibrous kalinite is monoclinic. 11th List

Potassio-carnotite. *See* Kalio-carnotite. 7th List

Potassium allevardite. Yu. M. Korolev, 1965. Докл., **162**, 650 (Калиевый аллевардит), transl. as *Dokl. Acad. Sci. USSR* (*Earth Sci. Sect.*), 1965, **162**, 143 (Potassium allevardite). The potassium analogue of allevardite (19th List), with 5% K_2O, from Kuli-Kolon, USSR. [*M.A.*, **19**, 179.] But allevardite is now considered to be rectorite [*A.M.*, **49** 446]. 28th List

Potassium-apatite. H. Wondratschek, 1963. *Neues Jahrb. Min.*, Abh., vol. 99, p. 121 (Kaliumapatit). Artificial $Ca_4K(PO_4)_3$; apatite family (*Zeits. anorg. Chem.*, 1938, vol. 237, p. 49). 23rd List

Potassium-bentonite. C. E. Weaver and T. F. Bates, 1951. *Bull. Geol. Soc. Amer.*, vol. 62, p. 1488 (potassium bentonites, K-bentonites). C. E. Weaver, *Amer. Min.*, 1953, vol. 38, p. 698. A montmorillonite clay containing potassium, from Ordovician limestone, Pennsylvania. Same as metabentonite (q.v.). [*M.A.*, **12**, 489.] 20th List

Potassium boltwoodite, syn. of Boltwoodite (21st List). *See* Ammonium boltwoodite. 22nd List

Potassium-clinoptilolite. H. Minato and Y. Takano, 1964. [*Nendo Kagaku* (*Journ. Clay Sci. Soc. Japan*)], abstr. *M.A.*, **19**, 130. A variety in which the principal cation is potassium. 28th List

Potassium-cryolite. A. Duboin, 1892. *Bull. Soc. Franç. Min.*, vol. 15, p. 191 (cryolithe potassique). P. Groth, *Chem. Kryst.*, 1906, vol. 1, p. 416 (Kaliumkryolith). C. Brosset, *Diss. Stockholm*, 1942, p. 119 (potassium-cryolite). Artificial potassium fluo-aluminate near K_3AlF_6, analogous to cryolite. *See* Ammonium-cryolite. [*M.A.*, **10**, 16, 208.] 18th List

Potassium gastunite, *see* Gastunite (of Honea). 22nd List

Potassium-melilite. R W. Nurse and H. G. Midgley, 1953. *Journ. Iron and Steel Inst.*, vol. 174, p. 121 (Potassium-melilite). Hypothetical end-member, $KCaAlSi_2O_7$, of the melilite series. Present in small amount in natural melilite and up to 20% in solid solution in artificial gehlenite. [*M.A.*, **12**, 197.] 20th List

Potassium-priderite. K. Norrish, 1951. *Min. Mag.*, vol. 29, p. 500 (K-priderite). Hintze, *Min.*, 1954, Ergbd. 2, p. 31 (Kaliumpriderit), p. 189 (Kalium-Priderit). *See* Barium-priderite. 20th List

Potassium pseudo-edingtonite, *see* Pseudo-edingtonite.

Poterite, error for Potarite (Mellor, *Treatise Inorg. Chem.*, vol. 15, p. 649). 22nd List

Poubaite. F. Čech and I. Vavřin, 1978. *Neues Jahrb. Mineral.*, Monatsh. 9. Lath-shaped crystals from hydrothermal quartz-carbonate veins at Oldřichov, Tachov, Bohemia, are rhombohedral, a_h 4·252, c_h 40·094 Å. Composition $3[PbBi_2(Se,Te,S)_4]$. Named for Z. Pouba. [*M.A.*, 78-3481.] 30th List

Poughite. R. V. Gaines, 1968. *Amer. Min.*, **53**, 1075. Yellow orthorhombic crystals of $Fe_2(TeO_3)_2SO_4.3H_2O$ from the Moctezuma mine, Sonora, Mexico. Named for Dr. Frederick H. Pough. [*M.A.*, 69-620; *Bull.*, **92**, 317; *Zap.*, **98**, 325.] 26th List

Pouzacite. — Frossard, 1890. [*Bull. Soc. Ramond, Bagnères-de-Bigorre.*] A. Lacroix, *Minéralogie de la France*, 1895, vol. i, pp. 383, 384. A variety of clinochlore, identical with leuchtenbergite, occurring in metamorphic limestones at Pouzac, Hautes-Pyrénées. 6th List

Pozzuolite. G. Gagarin and J. R. Cuomo, 1949, loc. cit., p. 7 (pozzuolita). Solfuro arsenicale (Arsenschwefel), $As_2S_3.H_2O$, of E Monaco 1903, from the solfatara of Pozzuoli, Italy. Not to be confused with pozzolana, pozzuolana, a volcanic tuff used for the manufacture of Roman cement. 19th List

Pragit, German transliteration of Прагит, Praguite (22nd List) (H. Strunz, *Min. Tabellen*, 4th edit., 1966, p. 333). 24th List

Praguite, R. Barta and C. Barta, 1956. [Журн. прикл. хим. (*Journ. Appl. Chem.*), vol. 29, p. 341]; referred to by S. O. Agrell and J. V. Smith, *Acta Crist.*, 1957, vol. 10, p. 761 (Praguite). Syn. of β-Mullite. 22nd List

Prasemalachite. Author and date? Prase enclosing malachite; a useless name. R. Webster, *Gems*, 1962, p. 761. 28th List

Prasiolite. C. Hintze, *Handbuch d. Mineralogie*, 1892, vol. ii, p. 940 (Prasiolith). Variant of praseolite (Praseolit, A. Erdmann, 1842). Hintze gives as derivation *πράσιος*, leek-green, and *λίθος*, stone; whilst Erdmann gives *πράσον*, leek, and *λίθος*. 4th List

Prasochrome. X. Landerer, *Neues Jahrb. Min.*, 1850, 313, 682. A dark green alteration product coating chromite. Grecian Archipelago. 2nd List

Pravdite. A. N. Nurlbaev, 1962. Дкол. Акад. Наук СССР (*Compt. Rend. Acad. Sci. URSS*), vol. 147, p. 689). A metamict mineral from the Ishim pegmatites, Central Kazakhstan, near $Ca_4Ln_2Al_{10}Si_5O_{32}$, with about 5% ThO_2. Named for the newspaper *Pravda*. (*Amer. Min.*, 1963, vol. 48, p. 709). Discredited, being shown to be an aluminian Britholite, comparable to Alumobritholite (this List), see Зап. Всесоюз. Мин. Общ. (*Mem. All-Union Min. Soc.*), 1964, vol. 93, p. 106. [*M.A.*, **16**, 558.] 23rd List

Prawdit, German transliteration of Правдит, Pravdite (23rd List) (C. Hintze, *Handb. Min.*, Erg. III, p. 501). 25th List

Pregibbsite. M.-C. Gastuche and A. Herbillon, 1962. *Bull. Soc. Chim. France*, 1404. An aluminium hydroxide gel giving a diffuse X-ray pattern related to that of gibbsite, and crystallizing to gibbsite; *artificial*. 26th List

Prelaumontite. N. Katayama. 1958. [*Journ. Geol. Soc. Japan*, **45**, 458], quoted in *Intro. Jap. Min.*, 120 (1970), Geol. Surv. Japan. *Syn.* of laumontite; fully hydrated material from the Ashio mine, Tochigi Prefecture, Japan, and three other localities was compared with partially dehydrated specimens accepted as laumontite. 28th List

Preobratschenskit, German transliteration of Преображенскит, Preobrazhenskite. *Der Aufschluss*, 1962, vol. 13, p. 63. 23rd List

Preobrazhenskite. Y. Y. Yarzhemski, 1956. *Doklady Acad. Sci. USSR*, vol. 111, p. 1087 (преображенскит, preobrazhensquite). Magnesium borate, $3MgO.5B_2O_3.4{\cdot}5H_2O$, monoclinic (?). From the salt deposits at Inder, Kazakhstan. Named after Paul Ivanovich Preobrazhensky, Павл Иванович Преображенский (1874–1944), investigator of Russian salt deposits. Abstr. in *Bull. Soc. Franç. Min. Crist.*, 1957, vol. 80, p. 217 (preobrajenskite). [*M.A.*, **13**, 300.] 21st List [$Mg_3B_{11}O_{15}(OH)_9$; *A.M.*, **55**, 1071]

Preslite. V. Rosický, 1912. *Zeits. Kryst. Min.*, vol. li, p. 521 (Preslit). Synonym of Tsumebite (q.v.). Named in memory of the Bohemian naturalist Jan Swatopluk Presl (1791–1849). 6th List

Priasowit, German transliteration of Приазовит, Priazovite (24th List) (C. Hintze, *Handb. Min.*, Erg. III, p. 502). 25th List

Priazorit, error for Priazovit (H. Strunz, *Min. Tabellen*, 4th edit., 1966, pp. 175, 509). 24th List

Priazovite. Yu. Yu. Yurk, 1956. [Редк. Мин. Пегматит. Приазовья] cited in A. I. Ginzburg, S. A. Gorzhevskaya, E. A. Erofeeva, and G. A. Sidorenko [Геохимия, 1958, p. 486], translated as *Geochemistry*, 1958, p. 615. A cation-deficient yttrian uranian Pyrochlore from the Azov region. Named for the region. An unnecessary name. In an abstr. of a paper by V. S. Dzhun [Доповиди Акад. наук УССР, 1963, vol. 10, p. 1379], abstr. *Amer. Min.*, 1965, vol. 50, p. 268, the date of Yurk's name is given as 1941. 24th List [Samarskite group mineral; *A.M.*, **62**, 407 gives refs., including Yurk (1941)]

Priderite. K. Norrish, 1951. *Min. Mag.*, vol. 29, p. 496. $(K,Ba)_{1\cdot3}(Ti,Fe''')_8O_{16}$, minute red tetragonal crystals, previously mistaken for rutile in leucite-rocks from Kimberley, Western Australia. Related structurally to cryptomelane. Synthetic K-priderite and Ba-priderite (p. 500). Named after Rex Tregilgas Prider, professor of geology in the University of Western Australia. 19th List

Priguinite. Error for Iriginite (q.v.). 21st List

Přilepite. E. Bořický, Zepharovich's *Min. Lexicon Oesterr.*, 1873, II, 246. A resinous substance occurring as reniform crusts on coal shales at Přilep, Bohemia. 2nd List

Priorite. W. C. Brøgger, 1906. *Videnskabs-Selsk. Skrifter*, Kristiania, 1906, no. 6, p. 110 (Priorit). A titano-niobate of yttrium, cerium-earths, etc., found as indistinct orthorhombic crystals in the tin-gravels of Swaziland, South Africa. Named after Dr. George Thurland

Prior, of the British Museum, who analysed it (*Min. Mag.*, 1899, vol. xii, p. 96). It is isomorphous with blomstrandine (q.v.) and dimorphous with euxenite, the four minerals euxinite-polycrase and priorite-blomstrandine forming an isodimorphous group. 4th List [Aeschynite-(Y); *A.M.*, **51**, 153]

Prjevalskite, *see* Przhevalskite.

Proarizonite. A. D. Bykov, 1964. Докл. Акад. Наук СССР (*Compt. Rend. Acad. Sci. URSS*), vol. 156, p. 567 (Проаризонит). An intermediate stage in the alteration of Ilmenite to arizonite. 23rd List [A mixture]

Probertite. A. S. Eakle, 1929. *Amer. Min.*, vol. 14, p. 427. Hydrous borate of sodium and calcium, '$Na_2CaB_6O_{11}.6H_2O$', as rosettes of monoclinic prisms embedded in kernite from California. Named after Prof. Frank Holman Probert, Dean of the Mining College, University of California. Identical with Kramerite (q.v.). [*M.A.*, **4**, 245.] 12th List

Proglauconite. I. Y. Mikei, 1936. *V. I. Vernadsky jubilee volume, Acad. Sci. USSR*, vol. 2, pp. 818, 825 (проглауконит), p. 826 (Proglaukonit). A hypothetical molecule, an alumoferrisilicate $R_2O_3.4SiO_2$, forming the basis of glauconite. [*M.A.*, **7**, 162.] 15th List

Prokaolin. R. Schwarz and G. Trageser, 1932. *Chemie der Erde*, vol. 7, p. 566 (Prokaolin). Precipitated amorphous $Al_2O_3.2SiO_2.nH_2O$, from which kaolin can be prepared artificially. [*M.A.*, **5**, 361.] 13th List

—— I. D. Sedletzky, 1940. *Compt. Rend.* (*Doklady*) *Acad. Sci. URSS*, vol. 26, p. 241. An assumed amorphous weathering product passing progressively into kaolinite. [*M.A.*, **8**, 146.] 16th List

Prokoenenite. W. Berdesinski, 1952. *Neues Jahrb. Min.*, Abh., vol. 84, p. 147 (Prokoenenit). An early stage in the artificial production of koenenite. [*M.A.*, **12**, 77.] 20th List

Prolectite. H. Sjögren, *Bull. Geol. Inst. Upsala*, II. 99, 1895. [*M.M.*, **11**, 139, 161.] Probably $Mg[Mg(F,OH)]_2SiO_4$. Monosymmetric. [Ko mine, Nordmark] Sweden. 1st List [Is Chondrodite. *M.A.*, **3**, 153]

Promontmorillonite. I. D. Sedletzky, 1937. *Compt. Rend.* (*Doklady*) *Acad. Sci. URSS*, vol. 17, p. 375 (Promontmorillonit); ibid., 1940, vol. 26, p. 241 (Promontmorillonite). An artificial preparation with the chemical composition of montmorillonite, but amorphous (showing no X-ray pattern). On long standing it passes into typical montmorillonite. 15th List

Promullite. J. E. Comeforo, R. B. Fischer, and W. F. Bradley, 1948. *Journ. Amer. Ceramic Soc.*, vol. 31, p. 259 (pro-mullite). An amorphous stage in the dehydration of kaolin, yielding mullite at a higher temperature. Synonym of metakaolin (10th List). Cf. prokaolin (16th List) and promontmorillonite (15th List). [*M.A.*, **11**, 452.] 19th List

Proto-amphibole. G. V. Gibbs, F. D. Bloss, and H. R. Shell, 1960. *Amer. Min.*, vol. 45, p. 974. A name for a series of artificial orthorhombic fluor-amphiboles having only half the *a*-dimension of anthophyllite. The presence of Li and absence of Ca appears to be essential to their formation. Named because of a structural relation to proto-enstatite. 22nd List

Protoastrakhanite. B. Friedel, 1976. *Neues Jahrb. Min.*, Abh., **126**, 187 (Protoastrakanit). $Na_2Mg(SO_4)_2.5H_2O$, a metastable phase found in salt soils and efflorescences on soils (localities not stated). 29th List

Protocalcite. E. Balogh, 1937. *Erdélyi Múzeum*, Kolozsvár, 1937, vol. 42, p. 147 (Protokálcit, Protocalcit). A mould-like encrustation of calcium carbonate in a limestone cavern, consisting of fine needles with oblique optical extinction. Evidently identical with lublinite (5th and 6th Lists). [*M.A.*, **7**, 515.] 15th List [*See* Ikaite]

Protodolomite. D. L. Graf and J. R. Goldsmith, 1955. *Bull. Geol. Soc. Amer.*, vol. 66, p. 1566 (abstr.); *Journ. Geol. Chicago*, 1956, vol. 64, p. 173. Imperfectly crystallized, artificial material approaching $CaCO_3.MgCO_3$ in composition. [*M.A.*, **13**, 88, 270.] 21st List

Protodoloresite. H. T. Evans, Jr., 1960. *US Geol. Surv. Prof. Paper 400–B*, p. 433. A tentative name for the mineral from Carlile, Wyoming, described in *Acta Cryst.*, 1958, vol. 11, p. 56 as Phase B; formulated $V_2O_3.2V_2O_4.5H_2O$. 23rd List

Protoenstatite. W. Büssem and C. Scheusterius, 1938. *Wiss. Veröff. Siemens-Werken*, vol. 17, p. 62 (Protoenstatit). Replacing the name metatalc (15th List) for an artificially produced modification of $MgSiO_3$. 18th List

Protokálcit, *see* Protocalcite.

Protomelane. L. L. Fermor, 1945. *Proc. South Wales Inst. Engin.*, vol. 61, p. 30. Barium-bearing psilomelane as distinct from potassium-bearing psilomelane (cryptomelane, 16th List). [*M.A.*, **9**, 189.] 17th List

Protopartzite. S. Koritnig, 1967. *Mitt. blatt. Mus. Bergbau, Geol., Paleont. Landesmus. Joanneum, Abt. Min.*, no. 1–2, p. 51. A name for an ill-defined material from Veitsch, Styria, referred by Cornu (1908) to thrombolite. The electron-probe analyses vary widely, and both Schrauf's analysis (1880) of thrombolite from the type locality (Rezbanya) and Mason and Vitaliano's formula for partzite (1930) fall within their range. Because Californian partzite gives an X-ray powder pattern while the Styrian material is X-ray amorphous, the name protopartzite is proposed. The evidence for a new mineral, distinct from either thrombolite or partzite, is inadequate and the name is superfluous. [*A.M.*, **52**, 1581.] 25th List

Protowollastonite. T. Ito, 1950. *X-ray studies on polymorphism*, Tokyo, 1950, p. 105 (protowollastonite). Hypothetical monoclinic wollastonite, which by different modes of twinning has given rise to monoclinic parawollastonite (14th List) and the more common form of triclinic wollastonite. [*M.A.*, **11**, 308.] 20th List

Proudite. W. G. Mumme, 1976. *A.M.*, **61**, 839. Monoclinic, *a* 31·96, *b* 4·12, *c* 36·69 Å, *β* 109·52°. Composition near $CuPb_{7\cdot5}Bi_{9\cdot33}(S,Se)_{22}$. From the Juno mine, Tennant Creek, Northern Territory, Australia, with junoite and other sulphobismuthites. 30th List

Przhevalskite. Author? *Guide to USSR exhibit at Conference on Atomic Energy*, Geneva, 1955, p. 8 (prjevalskite). Later publications, abstr.: *Amer. Min.*, vol. 41, p. 816, vol. 43, p. 381; *Mém. Soc. Russ. Min.*, 1958, vol. 87, p. 81 (пржевальскит, przhewalskite). $PbO.2UO_3.P_2O_5.4H_2O$. Named after Nikolai Mikhailovich Przhevalsky, Н. М. Пржевальский (1839–1888), Russian explorer in Central Asia. 21st List

Pseudo-aenigmatite. A. A. Kukharenko *et al.*, 1965. [*The Caledonian ultrabasic alkalic rocks and carbonatites of the Kola Peninsula and northern Karelia. Izd. 'Neda'*, Moscow. pp. 501–502]; abstr. *Amer. Min.*, 1967, vol. 52, p. 561. An incompletely characterized aluminosilicate of Fe, Ti, and alkalis near $(Na,K,Ca)(Fe,Ti,Mg)(Si,Al,Ti)_3O_8$, giving an X-ray pattern resembling that of aenigmatite. Probably a distinct species but needs further study. From the Turii peninsula, Kola. [Зап., **96**, 76, Псевдоэнигматит; *Bull.*, **90**, 610.] 25th List

Pseudo-apatelite. A. Magne, 1942. *Bull. Soc. Franç. Min.*, vol. 65, p. 41 (pseudo-apatélite). Abstr. in *Amer. Min.*, 1945, vol. 30, p. 86 (Pseudoapatelite). $(Fe,Al)_2(OH)_4SO_4.H_2O$ or $5Fe_2O_3.2Al_2O_3.7H_2O.21H_2O$, described by A. Lacroix in 1910 as apatelite, but differing from the apatelite $(3Fe_2O_3.4SO_3.6H_2O)$ of A. Meillet (1841) in containing Al_2O_3 11·0%. [*M.A.*, **9**, 126.] 17th List

Pseudo-autunite. A. S. Sergeev, 1964. [Мин. геохим., Ленинград унив., Сборн. статеи, no. 1, p. 31]; abstr. *Amer. Min.*, 1965, vol. 50, p. 1505; *M.A.*, **17**, 400; Зап. Всесоюз. Мин. Общ. (*Mem. All-Union Min. Soc.*), 1965, vol. 94, p. 679 (Псевдоотенит, pseudoautunite). Crusts of small platy hexagonal crystals in cavities of Albite–Acmite veins in an alkalic ultrabasic massif of N Karelia are formulated $(H_3O)_2(UO_2,Ca)_2(PO_4)_2.2\frac{1}{2}H_2O$, with $UO_2 \sim 1{\cdot}14$. M. Fleischer notes that there is no evidence that alkalis or ammonium were looked for, and the oxonium formulation is therefore not proven. 24th List

Pseudobarthite. Author?; quoted by C. Guillemin, *Bull. Soc., franç. Min. Crist.*, 1956, vol. 79, p. 71. Synonym of *β*-Duftite. 23rd List

Pseudoboehmite. D. Papée, R. Tertian, and R. Biais, 1958. *Bull. Soc. Chim. France*, p. 1301. An imperfectly crystallized or highly disordered form of Boehmite. 23rd List

Pseudoboleite. A. Lacroix, *Bull. Mus. d'Hist. Nat. Paris*, p. 39, 1895. Between boleite and cumengeite. Boleo, Lower California. 1st List [A valid species, formula uncertain; *Min. Record*, **5**, 283]

Pseudo-chalcedonite. A. Lacroix, *Compt. Rend. Acad. Sci. Paris*, 1900, CXXX, 430. A form of fibrous anhydrous silica; optically biaxial and negative. 2nd List

Pseudochlorite. V. A. Frank-Kamenetsky, 1958. Referred to by R. C. Mackenzie, *Clay Min. Bull.*, 1959, vol. 4, p. 61. A term for the cronstedtite–amesite–berthierine group; syn. of septechlorite (q.v.). 22nd List

—— R. F. Youell, 1960. *Clay Min. Bull.*, vol. 4, p. 191. An artificial product obtained by adsorbing Mg salts on Montmorillonite and precipitating $Mg(OH)_2$ between the layers of the mineral. Not to be confused with the Pseudochlorite of Frank-Kamenetsky (1958; 22nd List), a synonym of Septechlorite. 23rd List

Pseudo-copiapite. H. Ungemach, 1935. *Bull. Soc. Franç. Min.*, vol. 58, p. 154. A crystallographically aberrant form of copiapite from Chile. [*M.A.*, **6**, 149.] 14th List

Pseudo-crocidolite. (F. W. Rudler, in T. E. Thorpe's *Dictionary of Applied Chemistry*, 1890, vol. i, p. 620. L. J. Spencer, *The World's Minerals*, London and Edinburgh, 1911, p. 156, New York, 1911, p. 202.) Quartz pseudomorphous after crocidolite: the well-known 'tiger-eye' and 'hawk's-eye', from the Asbestos Mountains, Orange River, South Africa, used for ornamental purposes. To replace the misleading trade-name 'crocidolite' often applied to this material. 6th List

Pseudodeweylite. F. Zambonini, 1908. *Atti. R. Accad. Sci. Fis. Mat. Napoli*, ser. 2, vol. xiv, no. 1, p. 84; *Rend. R. Accad. Sci. Fis. Mat. Napoli*, ser. 3, vol. xiv, p. 148. A hydrated magnesium silicate, $Mg_3Si_2O_7.3H_2O$, from Chester Co., Pennsylvania, closely resembling deweylite, but differing slightly in composition (deweylite being $Mg_4Si_3O_{10}.6H_2O$). 5th List

Pseudo-edingtonite. M. H. Hey, 1934. *Min. Mag.*, vol. 23, pp. 491, 493. Potassium and sodium pseudo-edingtonites are base-exchange products of edingtonite with a crystal-structure different from that of edingtonite. 13th List

Pseudo-eucryptite. F. M. Jaeger and A. Šimek, 1914. [*Verslagen k. Akad. Wetensch. Amsterdam, Wis- en Natuurk. Afd.*, vol. xxiii]; abstr. in *Neues Jahrb. Min.*, 1915, vol. ii, ref. p. 146 (*pseudo-* or *β-Eukryptit*); *Proc. Roy. Acad. Sci. Amsterdam*, vol. xvii, p. 242. An artificial form of $LiAlSiO_4$ dimorphous with eucryptite. 7th List

Pseudogaylussite. F. J. P. van Calker, *Zeits. Kryst. Min.*, 1897, XXVIII, 556. The 'barley-corn' pseudomorphs of calcite after gaylussite (or celestite?). 2nd List

Pseudoglaucophane. L. Duparc, 1927. *Compt. Rend. Soc. Phys. Hist. Nat. Genève*, vol. 44, p. 49. An amphibole differing from ordinary glaucophane in its optical characters. 11th List

Pseudogymnite. P. Groth and K. Mieleitner, 1921. *Mineralogische Tabellen*, p. 100 (Pseudogymnit). The same as Pseudodeweylite (of F. Zambonini, 1908; 5th List). 10th List

Pseudoheterosite. A. Lacroix, 1910. *Minéralogie de la France,* vol. iv, p. 469 (pseudohétérosite). Probably represents a stage in the alteration of triphylite ($Li[Fe,Mn]PO_4$) to heterosite ($[Fe,Mn]PO_4 + H_2O$ (?)). It occurs as a zone between these two minerals, and is detected only by the difference in optical character as shown in thin sections. From Huréaux, Haute-Vienne. 6th List [Is Ferri-sicklerite; *M.M.*, **8**, 180]

Pseudo-ixiolite. E. H. Nickel, J. F. Rowland, and R. C. McAdam, 1963. *Amer. Min.*, vol. 48, p. 975. $(Fe,Mn,Nb,Ta)O_2$, a disordered structure corresponding to Tantalite and Columbite in composition. Differs from Ixiolite in the absence of Sn and in transforming to Tantalite or Columbite on heating. 23rd List

Pseudo-jade. A name which may be applied to any mineral resembling jade in appearance; for example, bowenite (C. A. McMahon, *Min. Mag.*, 1890, vol. ix, p. 187). 4th List

Pseudo-jadeite. F. W. Clarke, 1906. H. R. Bishop, *Investigations and Studies in Jade*, New York, 1906, vol. 1, p. 157 (pseudo-jadeite). H. S. Washington, *Proc. US Nat. Museum*, 1922, vol. 60, art. 14, p. 6 (pseudojadeite). For the chemical molecule (Ca,Mg,Fe)-$Al_2(SiO_3)_4$ assumed to be sometimes present in isomorphous replacement with the normal jadeite molecule $NaAl(SiO_3)_2$. The same name was also used independently by A. W. G. Bleeck in 1907 in a different sense (5th List). 9th List

—— A. W. G. Bleeck, 1907. *Zeits. prakt. Geol.*, Jahrg. xv, p. 353; *Rec. Geol. Survey India*, 1908, vol. xxxvi, p. 267. Some of the material collected as jadeite from the jadeite quarry in Upper Burma was afterwards found on examination to be albite. Cf. Jadeolite. 5th List

Pseudo-kaliophilite. E. Gruner, 1935. *Zeits. Anorg. Chem.*, vol. 224, p. 366 (Pseudokaliophilit). W. Borchert and J. Keidel, *Heidelberger Beitr. Min. Petr.*, 1947, vol. 1, p. 11. $KAlSiO_4$ prepared by desulphurizing ultramarine in fused KCN, differing in X-ray pattern from kaliophilite. [*M.A.*, **10**, 364.] 18th List

Pseudokrokydolith. C. Hintze, *Handbuch Min.*, Ergänzungsband, 1937, p. 514 (Pseudokrokydolith). German form of pseudo-crocidolite (6th List). 17th List

Pseudolaueite. H. Strunz, 1956. *Naturwissenschaften*, vol. 43, p. 128. Orange-yellow monoclinic crystals from Hagendorf, Bavaria, with the same composition, $MnFe_2'''(PO_4)_2(OH)_2.8H_2O$, as laueite (20th List). [*M.A.*, **13**, 210.] 21st List

Pseudo-laumontite. F. F. Grout, 1910. *Journ. Geol. Chicago*, vol. xviii, p. 654. Hydrous silicate of aluminium, iron, magnesium, and potassium, forming green pseudomorphs after red laumontite in the amygdaloidal diabase of the Keweenawan area, Minnesota. 6th List

Pseudo-låvenite. A. Lacroix, 1911. *Les syénites néphéliniques de l'archipel de Los et leurs minéraux. Nouv. Arch. Muséum*, Paris, ser. 5, vol. iii, p. 60 (Pseudo-låvénite). An undetermined mineral detected in a thin section of the nepheline-syenite of the Los Islands, W Coast of Africa. It resembles låvenite, but differs from this in its optical orientation. 6th List

Pseudolussatine. O. Braitsch, 1957. *Heidelberger Beitr. Min. Petr.*, vol. 5, p. 331 (Pseudolussatin). A structural modification of low-cristobalite differing from lussatine. 22nd List

Pseudo-manganite. L. L. Fermor, 1909. *Mem. Geol. Survey India*, vol. xxxvii, p. lxviii (pseudomanganite), p. 84 (pseudo-manganite). Crystals of manganite altered wholly or partially to pyrolusite. 6th List

Pseudomeionite. H. Rosenbusch, 1902. *Mitt. Grossherz. Badisch. geol. Landesanst.*, vol. iv, p. 391 (Pseudomejonit). A name provisionally given to a felspathoid mineral occurring with bytownite in a para-amphibole-gneiss from the Black Forest; it has the microscopical characters of meionite, except in possessing a good basal cleavage. 4th List

Pseudo-mendipite. E. Rimann, 1918. *Anal. Soc. Quím. Argentina*, vol. 6, p. 326 (Pseudo-mendipita). Defined as an orthorhombic lead oxychloride, $3PbO.PbCl_2$, but this formula is based on a misquoted old analysis. [*M.A.*, **1**, 121] 9th List

Pseudomesolite. A. N. Winchell, *Amer. Geologist*, 1900, vol. xxvi, pp. 275, 372; *Thèse Fac. des Sci. Paris*, 1900. A zeolite agreeing with mesolite in chemical composition, but differing in optical characters. Occurs as colourless to white, radially fibrous masses in plagioclasite in Minnesota. 3rd List

Pseudonocerite, variant of Pseudonocerina (*B.M. Index*, 27th edit., 1936). 23rd List

Pseudo-orthoclase. A. Cathrein, 1915. *Neues Jahrb. Min.*, 1915, vol. 1, p. 32 (Pseudo-Orthoklas). Crystals of felspar with the appearance of orthoclase, but found on examination to be anorthoclase. [*M.A.*, **1**, 238.] 9th List

Pseudo-ozocerite. Förtsch, *Zeits. f. Naturwiss. Ver. f. Sachsen u. Thüringen*, 1898, LXX, 322. Ozocerite from Central Persia, differing in some respects from Galician ozocerite. 2nd List

Pseudopolaite. A Mário de Jesus, [1936]. *Com. Serv. Geol. Portugal*, vol. 19 (for 1933), p. 151 (Pseudopalaite). An alteration product of lithiophilite, with a formula, $6(Mn,Fe)O.2P_2O_5.5H_2O$, slightly different from that of palaite. From Mangualde, Portugal. [*M.A.*, **6**, 442.] 14th List [Is Hureaulite; *M.A.*, **8**, 180]

Pseudoparisite. G. Flink, *Medd. om Grönland*, 1898, XIV, pp. 236 and iii; 1900, XXIV, 179. The same as cordylite (*see* above). 2nd List

Pseudophillipsite. F. Zambonini, *Neues Jahrb. Min.*, 1902, vol. ii, p. 73 (Pseudophillipsit). A zeolite, from the leucitic lavas near Rome, agreeing with phillipsite in form and twinning (though octahedral in habit, like gismondite), but containing less silica (38%) than typical phillipsite: formula, $(Ca,Na_2)_2Al_4Si_5O_{18}.9H_2O$. 3rd List

Pseudo-pirssonite. E. Stolley, 1909. *Medd. Dansk Geol. For.*, no. 15, p. 361 (Pseudo-Pirssonit). Pseudomorphs, resembling pseudogaylussite, found in the alum-shales of Cambrian age on the island of Bornholm, Denmark. In a postscript (p. 368) it is suggested that the original mineral of these pseudomorphs may have been struvite rather then pirssonite, and the name pseudo-struvite (Pseudo-Struvit) is added. 5th List

Pseudopyrochroite. G. Aminoff, 1918. *Geol. För. Förh. Stockholm*, vol. 40, p. 427 (Pseudopyrochroit). Later described under the name Bäckströmite. (q.v.). 9th List

Pseudopyrophyllite. F. Löwinson-Lessing, *Verh. russ. min. Ges.*, XXXIII. 283, 1895. $3MgO.4Al_2O_3.9SiO_2.8H_2O$. Rhombic [Pyshminsk] Urals. 1st List [A mixture; *M.A.*, **12**, 285]

Pseudo-quartzine. O. Braitsch, 1957. *Heidelberger Beitr. Min. Petr.*, vol. 5, p. 331 (Pseudoquarzin). A structural modification of quartz. differing from quartzine. 22nd List

Pseudorutile. G. Teufer and A. K. Temple, 1966. *Nature*, vol. 211, p. 189. $Fe_2Ti_3O_9$, a distinct intermediate stage in the topochemical transformation of ilmenite to rutile; hexagonal; observed in 'Ilmenites' from Florida, New Jersey, India, and Brazil. Distinct from 'Arizonite' (5th List) and 'Proarizonite' (23rd List). 24th List

Pseudo-sarcolite. C. E. Tilley, 1929. *Geol. Mag. London*, vol. 66, pp. 349, 353 (pseudo-sarcolite), pp. 349, 352 (pseudosarcolite). The artificial product $3(Ca,Na_2)O.Al_2O_3.3SiO_2$, tetragonal and optically negative, prepared by A. F. Buddington (1922), differing from sarcolite in its optical properties. It is regarded as one of the constituent molecules of the minerals of the melilite group. [*M.A.*, **4**, 218; *Min. Mag.*, **22**, 463.] 12th List

Pseudosillimanite. J. de Lapparent, 1920. *Compt. Rend. Congrès Soc. Sav. Sci. Paris*, 1920, p. 77; *Bull. Serv. Carte Géol. Alsace Lorraine*, Strasbourg, 1923, vol. 1, p. 54 (pseudosillimanite). An undetermined mineral occurring as minute prisms resembling fibrolite (sillimanite) in a metamorphic phthanite containing radiolaria from Alsace. A similar mineral is recorded from Tierra del Fuego by E. H. Kranck, *Acta Geogr. Helsingfors*, 1932, vol. 4, no. 2, p. 146 (pseudosillimannite), p. 150 (pseudo-sillimannite). 13th List

Pseudo-struvite. E. Stolley, 1909. *See* Pseudo-pirssonite. 5th List

Pseudo-succinite. J. D. Buddhue, 1938. *Mineralogist*, Portland, Oregon, vol. 6, no. 1, p. 21. Amber from Equilleres, Basses-Alpes, France, differing from Baltic amber in its reaction to solvents [*M.A.*, **7**, 265.] 15th List

Pseudotetrahedrite. K. Tatsuka and N. Morimoto, 1973. *A.M.*, **58**, 425. Tetrahedrite is shown to have a variable composition, and may be Cu-rich, near $Cu_{14}Sb_4S_{13}$, or Cu-poor, near $Cu_{12}Sb_4S_{13}$; while material near Cu_3SbS_3 has a superstructure, with a cell-edge double that of normal tetrahedrite, and is named. [*M.A.*, 73-3709.] 28th List

Pseudothuringite. O. Koch, 1884. *Inaug.-Diss. Jena*, p. 27 (Pseudothuringit). A mineral from Vogtland very similar to thuringite, but with the composition $Al_4Fe_4''Mg_2Si_3O_{18}.5aq$. 17th List

Pseudo-topaz. A. Sachs, 1910. *Centralblatt Min.*, 1910, p. 498 (Pseudotopas). A crystal from the granite of Striegau, Silesia, which had been described by A. Sachs (ibid., 1909, p. 438) as topaz, was proved by V. Goldschmidt to be merely quartz with a prismatic habit in the direction of one of the horizontal axes. As so orientated there is a remarkable agreement between the interfacial angles of quartz and topaz. In correcting his error, the author adds to the confusion by using the above name. 6th List

Pseudowavellite. H. Laubmann, 1922. *In* F. Henrich, *Ber. Deutsch. Chem. Gesell.*, vol. 55, Abt. B, p. 3016 (Pseudo-wavellit); *Geognost. Jahresh. München*, 1923, vol. 35 (for 1922), p. 203 (Pseudowavellit). Hydrated phosphate of aluminium with lime, ferric iron, and rare-earths; occurring as white encrustations (trigonal needles) on limonite and wavellite at Amberg, Bavaria. So named because of its resemblance to wavellite, of which it is perhaps an alteration product. [*M.A.*, **2**, 13, 522.] 10th List [Is Crandallite; Dana *Syst. Min.*, 7th edit., **2**, 837]

Pseudo-willemite. G. Sabatier, 1952. *Bull. Soc. Franç. Min. Crist.*, vol. 75, p. 521 (pseudo-willémite). Cubic modification of Zn_2SiO_4, artificially produced. [*M.A.*, **12**, 81.] 20th List

Pseudowollastonite. E. T. Allen and W. P. White, 1906. *Amer. Journ. Sci.*, ser. 4, vol. xxi, p. 89 (Pseudo-Wollastonite). An artificial product usually known as hexagonal calcium metasilicate: the crystals are, however, pseudo-hexagonal and probably monoclinic. Wollastonite when heated to 1180°C. becomes changed, without melting, into this dimorphous modification. 4th List. In 4th List incorrectly attributed to E. T. Allen and W. P. White, 1906; apparently first used by A. Lacroix, *Minéralogie de la France*, 1895, vol. i, p. 624 (pseudowollastonite); 1910, vol. iv, p. 777 (pseudo-wollastonite). The artificially-produced, pseudo-hexagonal calcium metasilicate dimorphous with wollastonite. Earlier called *Bourgeoisite* (R. Breñosa, 1885; 3rd List). C. Doelter, *Handbuch der Mineralchemie*, 1913, vol. ii, p. 450, refers to it as *β-Wollastonite*, in preference to α-Wollastonite of other authors, since it is the modification stable at the higher temperature. 7th List [Cf. Cyclowollastonite]

Pseudo-zircon. B. W. Anderson and C. J. Payne, 1937. *Gemmologist*, London, vol. 7, p. 298. Zircons of low specific gravity (3·95–4·05), which consist of amorphous SiO_2 and ZrO_2, and increase in density when heated. *See* Meta-zircon and Zirconoid. 15th List

Psilomelanite. (T. Egleston, *Catalogue of Minerals and Synonyms, Bull. US Nat. Mus.*, 1887

(1889), no. 33, p. 137. E. T. Wherry, *Proc. US Nat. Mus.*, 1916, vol. li, p. 84.) Synonym of Psilomelane (W. Haidinger, 1828) with the termination *ite*.

Crystalline psilomelane is distinguished by L. L. Fermor (*Rec. Geol. Survey India*, 1917, vol. xlviii, p. 120) as X-psilomelane [perhaps a misprint for χ-psilomelane], and is considered to be identical with hollandite. The colloidal form he describes as κ-hollandite or simply as psilomelane. (*See* 7th List under Diasporogelite.) 8th List [*See* Romanechite]

Pteochroite, error for Pleochroite (q.v.) (*New Scientist*, **63**, 235). 28th List

Pufahlite. [F.] Ahlfeld, 1925. *Metall u. Erz*, Halle, vol. 22, p. 135 (Pufahlit). Sulpho-stannate of lead (and zinc) as black (orthorhombic or monoclinic) scales from Bolivia. [No doubt identical with Teallite.] Named after Prof. Otto Pufahl (1855–1924), of Berlin. [*M.A.*, **2**, 520.] 10th List

Pulleite. J. W. H. Adam, 1909. *Zeits. prakt. Geol.*, Jahrg. xvii, p. 500 (Pulleit). Apatite in violet crystals of peculiar habit resembling sheaf-like crystals of stilbite; from the pegmatite-veins at San Piero in Campo, Elba. Earlier described as a variety of apatite by R. Görgey (*Centralblatt Min.*, 1909, p. 337). Named after Count J. G. Pullé of Elba. 5th List

Pulszkyite. J. S. Krenner, 1948. *Schweiz. Min. Petr. Mitt.*, vol. 28, p. 707 (Pulszkyit). Sulphate of copper and magnesium as green hexagonal crystals with herrengrundite from Hungary (now Slovakia). Named after Ferencz Pulszky (1814–1897) formerly director of the Hungarian National Museum. [*M.A.*, **10**, 510.] 18th List

Pumpellyite. C. Palache and H. E. Vassar, 1925. *Amer. Min.*, vol. 10, p. 412. Hydrous alumino-silicate of calcium, $6CaO.3Al_2O_3.7SiO_2.4H_2O$, as minute, bluish-green, orthorhombic fibres and plates in the copper-bearing amygdaloidal rocks of the Keweenaw Peninsula, Michigan. It has some resemblances to zoisite. Named after the American geologist, Raphael Pumpelly (1837–1923). [*M.A.*, **3**, 8.] 11th List [Cf. Julgoldite; review, *M.A.*, 75-3464]

Pungernite. Bulgarine, *Quart. Journ. Geol. Soc.*, 1851, VII, 66. Yellowish brown organic matter from the Silurian of Russia. 2nd List

Purpurite. L. C. Graton and W. T. Schaller, 1905. *Amer. Journ. Sci.*, ser. 4, vol. xx, p. 146; *Zeits. Kryst. Min.*, 1906, vol. xli, p. 433. A hydrous manganic ferric phosphate, $2(Mn,Fe)PO_4.H_2O$, with a purple or dark reddish colour; hence the name, from the Latin, purpura, purple. It is probably orthorhombic, and occurs in small irregular masses as an alteration product of lithiophilite and triphylite in the lithium-bearing pegmatite-veins of North Carolina and of San Diego Co., California. 4th List

Pyknochlorite. J. Fromme, *Min. petr. Mitt.* (*Tschermak*), 1903, vol. xxii, p. 62 (Pyknochlorit). A greyish-green, compact chlorite occurring in a quartz and calcite vein in the gabbro of the Radauthal, Harz. It has the same general formula (Sp_3At_4, Tschermak) as clinochlore, but differs from this in containing much more ferrous iron and in its compact (πυκνός) texture. 3rd List

Pyralmandite. L. L. Fermor, 1926. *Rec. Geol. Survey India*, vol. 59, p. 205. A contraction of pyrope and almandite for garnets of intermediate composition. Compare 'spandite' and 'grandite' (L. L. Fermor, 1907, 1909; 4th and 5th Lists) and 'spalmandite' below. *See* Calcspessartite. 11th List

Pyralspite. A. N. Winchell, 1927. *Elements of optical mineralogy*, 2nd edit., part 2, p. 257. A contraction of the names pyrope, almandine, and spessartine for this series of garnets in which there is complete isomorphous replacement. Two species of garnet are recognized—'pyralspite' and 'ugrandite' (q.v.). 11th List

Pyrandine. B. W. Anderson, 1947. *Journ. Gemmology,* London, vol. 1, no. 2, p. 15. A contraction of pyrope and almandine for gem garnets of intermediate composition. (Cf. rhodolite, 2nd List; pyralmandite and pyralspite, 11th List.) [*M.A.*, **10**, 102.] 18th List

Pyraphrolith. (P. Groth, *Tab. Uebers. Min.*, 3rd edit., 1889, p. 156). A mixture of felspar and opal. 2nd List

Pyribole. A. Johannsen, 1911. *Journ. Geol. Chicago,* vol. xix, p. 319. A contraction of the words pyroxene and amphibole. *See* Biopyribole. 6th List

Pyritogelit. *See* Diasporogelite. 7th List

Pyrobelonite. G. Flink, 1919. *Geol. För. Förh. Stockholm*, vol. 41, p. 433 (Pyrobelonit). Vanadate of lead and manganese, $4PbO.7MnO.2V_2O_5.3H_2O$, found at Långban, Sweden, as fire-red, needle-shaped (orthorhombic) crystals. Named from πῦρ, fire, and βελόνη, needle. [*M.A.*, **1**, 124.] 9th List [$MnPbVO_4(OH)$; Wales, *M.M.*, **41**, 85 (with refs.)]

Pyrochlore-wiikite. H. Strunz, 1957. *Min. Tabellen*, 3rd edit., p. 147 (Pyrochlor-Wiikit). Wiikite (3rd List) is subdivided into pyrochlore-wiikite, samarskite-wiikite, and silicate-wiikite (qq.v.). 22nd List [A mixture; *A.M.*, **62**, 407, 408]

Pyrochlorite. Variant of Pyrochlore; *A.M.*, **8**, 52 [not a Chlorite]. 10th List

Pyroxene-perthite, Pyroxene-microperthite, Pyroxene-cryptoperthite. W. Wahl, 1908. *Öfvers. Finska Vet. Soc. Förh.*, 1906–1907, no. 2, p. 19 (Pyroxenperthite, Pyroxenmikroperthite, Pyroxenkryptoperthite, plur.). Lamellar intergrowths of pyroxenes of different kinds, as with the felspars. 5th List

Pyroxenoid. H. Berman, 1937. *Amer. Min.*, vol. 22, pp. 360, 389 (Pyroxenoid family). Group name for minerals of the rhodonite and wollastonite series, as distinct from the pyroxene group. 15th List

Pyroxferroite. E. C. T. Chao, J. A. Minkin, C. Frondel, C. Klein, Jr., J. C. Drake, L. [H.] Fuchs, B. Tani, J. V. Smith, A. T. Anderson, P. B. Moore, G. R. Zechman, Jr., R. J. Traill, A. G. Plant, J. A. V. Douglas, and M. R. Dence, 1970. *Proc. Apollo XI Lunar Sci. Conf.*, **1**, 65. Yellow anorthic grains in lunar rocks from Tranquillity Base, composition $(Fe_{0\cdot85}Ca_{0\cdot15})SiO_3$, have *a* 6·62, *b* 7·54, *c* 17·35 Å, α 114·4°, β 82·7°, γ 94·5°, D 3·68, 3·76; α 1·748–1·756, β 1·750–1·758, γ 1·768–1·767. The mineral is the iron analogue of pyroxmangite, and is named accordingly. [*A.M.*, **55**, 2137; *M.A.*, 71-1388; *Zap.*, **100**, 624.] 27th List

Pyroxmangite. W. E. Ford and W. M. Bradley, 1913. *Amer. Journ. Sci.*, ser. 4, vol. xxxvi, p. 169; *Zeits. Kryst. Min.*, vol. liii, p. 225. A triclinic manganese pyroxene, $(Fe,Mn)SiO_3$, found as brown cleavage masses in South Carolina. Alters to Skemmatite (q.v.). 7th List

Pyrrhochrysit. J. Fröbel, 1843. *See* Chrysargyrit. 27th List

Q

Quarfeloids. A. Johannsen, 1917. *Journ. Geol. Chicago*, vol. 25, p. 70; *Essentials for the microscopical determination of rock-forming minerals*, Chicago, 1922, p. 41. A portmanteau-word from quartz, felspar, and felspathoids. Cf. Feloid. 10th List

Quartzine. A. Michel-Lévy and Munier-Chalmas, *Bull. Soc. fran. Min.*, XV. 166, 1892; F. Wallerant, *ibid.*, XX. 52, 1897. [*Min. Mag.*, X. 254.] A fibrous form of silica. 1st List

Quenselite. G. Flink, 1926. *Geol. För. Förh. Stockholm,* vol. 47 (for 1925), p. 377 (Quenselit). Basic lead manganite, $2PbO.Mn_2O_3.H_2O$, as black monoclinic crystals from Långban, Sweden. Named after Prof. Percy Dudgeon Quensel (1881–), of Stockholm. [*M.A.*, **3**, 110.] 11th List

Quercyite. A. Lacroix, 1910. *Compt. Rend. Acad. Sci. Paris*, vol. cl, pp. 1217, 1388; *Minéralogie de la France,* vol. iv, p. 579. W. T. Schaller, *Bull. US Geol. Survey*, 1912, no. 509, p. 92. A type of phosphorite consisting of an intimate interbanded mixture of amorphous (colloidal), optically isotropic collophanite (*see* Fluocollophanite), and a finely fibrous constituent which is optically uniaxial or nearly so. When the crystalline element is optically negative the mixture is called quercyite α, and when optically positive quercyite β. The optically negative constituent corresponds with dahllite ($2Ca_3(PO_4)_2.CaCO_3.\frac{1}{2}H_2O$) and staffelite ($(CaF)_2Ca_8(PO_4)_6.CaCO_3.H_2O$), but the optically positive constituent does not correspond with any known species. The name quercyite, from Quercy, an ancient district in France noted for its deposits of phosphorite, is thus a rock name rather than a mineral name. Not to be confused with quercite, the chemical name for 'sugar of acorns'. 6th List

Quetzalcoatlite. S. A. Williams, 1973. M.M. **39**, 261. Greenish-blue hexagonal crystals, *a* 10·097, *c* 4·944 Å, from the Bambollita mine, Moctezuma, Sonora, Mexico, have the composition [$Cu_4Zn_8(TeO_3)_3(OH)_{18}$]. Named for the Toltec god Quetzalcoatl. [*M.A.*, 74-510.] 28th List

Quiroguite. L. F. Navarro, *Anal. Soc. Españ. Hist. Nat.*, XXIV. *Actas*, p. 96, 1895. Spelt quirogite in abstr. in *Zeits. Kryst. Min.*, XXVIII. 202, 1897; *Neues Jahrb. Min.*, I. 452, ref. 1897; *Bull. Soc. fran. Min.*, XX. 163, 1897. [*M.M.*, **11**, 241.] $23PbS.3Sb_2S_3$. Tetragonal. [San Andres and other mines, Sierra Almagrera] Spain. Probably only galena. 1st List

Quisqueite. D. F. Hewett, 1907. Quoted by W. F. Hillebrand, *Amer. Journ. Sci.*, 1907, ser. 4, vol. xxiv, p. 141; *Journ. Amer. Chem. Soc.*, 1907, vol. xxix, p. 1019. D. F. Hewett, *Trans. Amer. Inst. Min. Engin.*, 1910, vol. xl, p. 286. A lustrous, black, brittle substance very like asphaltum in appearance, but containing much sulphur (S, $46\frac{1}{2}\%$ with C, 43%) and only little hydrogen. It occurs with vanadium-ore (patronite) in the Quisque district, near Cerro de Pasco, Peru. 5th List

R

Rabbittite. M. E. Thompson, A. D. Weeks, and A. M. Sherwood, 1954. [*Trace Elements Rep. US Geol. Surv.*, no. 405, p. —.] A. D. Weeks and M. E. Thompson. [*Bull. US Geol. Surv.*, 1954, no. 1009–B, p. —.] Abst. in *Amer. Min.*, 1954, vol. 39, p. 1037 (rabbitite). *Amer. Min.*, 1955, vol. 40, p. 201 (rabbittite). Uranyl carbonate with Ca and Mg, $Ca_3Mg_2(UO_2)_2(CO_3)_6(OH)_4$. Monoclinic, minute greenish-yellow needles as an efflorescence on mine walls in Utah. Named after John Charles Rabbitt (1907–) of the US Geological Survey. [*M.A.*, **12**, 511, 566–567.] 20th List

Rabdopissite. A. N. Krishtofovich, 1927. *Bull. Com. Géol. URSS*, vol. 44 (for 1925), p. 57 (рабдописсит). O. M. Shubnikova, *Trans. Inst. Geol. Sci. USSR*, 1940, no. 31 (*Min. Geochem. Sér.*, no. 6), p. 61. Rods of brown bituminous material (C 66·72, H 7·49, N 0·49, O + S 25·30%) in coal from Siberia. Named from ῥάβδος, a rod, πίσσα, pitch. 17th List

Racewinite. A. N. Winchell, 1918. *Economic Geology*, vol. xiii, p. 611. Hydrated silicate of aluminium and iron (Fe_2O_3 7·37%) approximating to the formula $2(Al,Fe)_2O_3.5SiO_2.9H_2O$. The bluish green, coarsely crystalline masses change on exposure to brownish-black, and the material undergoes certain other remarkable changes in colour. It occurs in metamorphic limestone at Bingham, Utah. Named from Racewin, the cable address of H. V. Winchell, who collected the material examined. 8th List

Radiobaryt. J. Knett, 1904. *Sitzungsber. Akad. Wiss. Wien*, vol. cxiii, Abt. IIa, p. 761. Crystals of barytes from the hot springs of Carlsbad, Bohemia, but not crystals from other localities, were found to be radio-active (as demonstrated by their action on a photographic plate). 4th List

Radiofluorite. F. L. Hess, 1931. *Amer. Journ. Sci.*, ser. 5, vol. 22, p. 220; 1933, ser. 5, vol. 25, p. 426. Strongly radioactive fluorite, suggested to be $(Ca,Ra)F_2$. W. L. Brown (*Univ. Toronto Studies, Geol. Ser.*, 1932, no. 32, p. 56) shows that the supposed radioactive effect on a photographic plate is due to phosphorescence. [*M.A.*, **5**, 52, 235, 330.] 13th List

Radiophyllite. A. Brauns and R. Brauns, 1924. *Centralblatt Min.*, p. 551 (Radiophyllit). Hydrated calcium meta-silicate, $CaSiO_3.H_2O$, as small, white globules sometimes showing a radial-scaly structure. Belongs to the group of micaceous zeolites, near gyrolite. Named on account of the structure [from Latin *radius*, and Gr. *φύλλον*, a leaf]. [Cf. Crestmoreite, A. S. Eakle, 1917; 8th List.] [*M.A.*, **2**, 341.] 10th List [Is Zeophyllite; *M.A.*, **15**, 290]

Radiotine. R. Brauns, 1904. *Neues Jahrb. Min.*, Beil.-Bd. xviii, p. 314 (Radiotin). Small spheres with radially fibrous structure occurring in a serpentine which has resulted from the alteration of picrite near Dillenburg, Nassau. It has the same composition as serpentine ($H_4Mg_3Si_2O_9$), but differs from this in not being attacked by hydrochloric acid and in its higher specific gravity of 2·70. So named because of its radial structure and its relation to serpentine. 4th List

Rafaelite. A. Arzruni, *Zeits. Kryst. Min.*, 1899, XXXI, 243 (*Rafaëlit*); Abstr. *Min. Mag.*, XII, 308. G. F. H. Smith, *Min. Mag.*, 1899, XII, 183; *Zeits. Kryst. Min.*, 1900, XXXII, 217. Identical with paralaurionite (q.v.). San Rafael mine, Chili. 2nd List

—— A. Windhausen and P. T. Vignau, 1912. *Informes Preliminares de la Dirección General de Minas, Geología é Hidrología*, Buenos Aires, no. 1 (rafaelita). G. Fester and F. Bertuzzi, *Zeits. angew. Chem.*, 1925, vol. 38, p. 364 (Rafaelit). A vanadiferous asphaltum (the ash, $\frac{1}{4}$–$\frac{10}{2}$%, containing 21–44% V_2O_5) found in 1890 near San Rafael, Argentina. [Not the Rafaelite of A. Arzruni, 1899, 2nd List.] 11th List

Raguinite. Y. Laurent, P. Picot, F. Permingeat, and T. Ivanov, 1969. *Bull. Soc. franç. Min. Crist.*, **92**, 38; ibid., 237. Hexagonal plates from Allchar, Macedonia, are pseudomorphs consisting of fibres of $TlFeS_2$; brilliant bronze, orthorhombic. Named for Professor E. Raguin. [*A.M.*, **54**, 1495, 1741; *M.A.*, 69-2382; *Zap.*, **99**, 73.] 26th List

Raite. A. N. Merkov, Y. P. Menshikov, and A. P. Nedorezova, 1973. *Zap.*, **102**, 54 (Раит). Radiating fibrous needles from pegmatite in the Karnasurt Mts., Ilmaoik valley, Lovozero tundra, Kola Peninsula, approximate to $(Na,Ca,K,Ln)_4(Mn,Ti,Fe^{3+},Mg,Fe^{2+})_3$-$(Si_2\{O_4(OH)_3\}_7)_4.5H_2O$. Named for the papyrus boat Ra. [*M.A.*, 73-4081; *Bull.*, **96**, 240; *A.M.*, **58**, 1113; *Zap.*, **103**, 363.] 28th List

Ramdohrite. F. Ahlfeld, 1930. *Centr. Min.*, Abt. A, 1930, p. 365 (Ramdohrit). Sulphantimonite of lead and silver, $Ag_2S.3PbS.3Sb_2S_3$, as grey-black prisms from Bolivia. Named after Prof. Paul Ramdohr, of Aachen. [*M.A.*, **4**, 341.] 12th List [Mixture with andorite; redefined as $AgSbS_2.Sb_2S_3.2PbS$, *M.A.*, 77-861]

Rameauite. F. Cesbron, W. L. Brown, P. Bariand, and J. Geffroy, 1972. *Min. Mag.*, **38**, 781. Orange monoclinic crystals, *a* 13·97, *b* 14·26, *c* 14·22 Å, β 121° 1′, *C*2/*c*, D 5·55, occur with uranophane on pitchblende at Margnac, France. Composition $4[K_2O.CaO.6UO_3.9H_2O]$. $\alpha \parallel [010]$, γ:[001] 4 to 6°. Named for J. Rameau. [*M.A.*, 72-3346; *A.M.*, **58**, 805.] 27th List

Ramsayite. A. E. Fersman, 1922. *Compt. Rend. Acad. Sci. Russie*, p. 59; *Trans. Northern Sci. Econ. Exped.*, no. 16 (*Sci. Techn. Dept. Supreme Council of National Economy*, no. 8), Moscow and Petrograd, 1923, pp. 16, 36, 73. E. E. Kostyleva, *Compt. Rend. Acad. Sci. Russie*, 1923, p. 55 (рамзаит, ramsayite). Sodium titanosilicate, $Na_2O.2SiO_2.2TiO_2$, occurring as brown, orthorhombic crystals resembling sphene in nepheline-syenite-pegmatite in the Kola Peninsula. Named after Prof. Wilhelm Ramsay, of Helsingfors. [*M.A.*, **2**, 250, 262–264, 399.] 10th List [*M.A.*, 70-2107. Is Lorenzenite (q.v.), *A.M.*, **32**, 59]

Ramsdellite. M. Fleischer and W. E. Richmond, 1943. *Econ. Geol.*, vol. 38, p. 278. Orthorhombic MnO_2 dimorphous with tetragonal pyrolusite (= polianite). Named after Prof. Lewis Stephen Ramsdell (1895–) of the University of Michigan, who described the material in 1932. [*M.A.*, **5**, 180; **9**, 4.] 17th List [*A.M.*, **47**, 47]

Rancieite. (*Collection de Minéralogie du Muséum d'Histoire Naturelle, Paris, Guide du Visiteur*, 2nd edit., 1900, p. 29 (Ranciéite).) The correct spelling of rancierite (A. Leymerie, 1857), a variety of wad (hydrated manganese oxide) from the iron-mines of Rancié, Ariège, France. 4th List [$(Ca,Mn)O.4MnO_2.3H_2O$; *A.M.*, **54**, 1741. Cf. Takanelite]

Randomite. R. A. Van Nordstrand, W. P. Hettinger, and C. D. Keith, 1956. *Nature*, London, vol. 177, p. 714. Preliminary name for an artificial form of $Al(OH)_3$, with X-ray pattern distinct from those of associated gibbsite and bayerite. At first thought to have a random stacking structure. Later named bayerite-II, [and, finally, Nordstrandite]. 21st List

Rankamaite. O. von Knorring, A. Vorma, and P. H. Nixon, 1969. *Bull. Geol. Soc. Finland*, **41**, 47. Water-worn pebbles in alluvial deposits from the Mumba area, Kivu, Congo, are probably an alteration product of simpsonite. Composition near $(Na,K,Pb,Li)_6(Ta,Nb,Al)_{22}(O,OH)_{60}$. Probably orthorhombic and related to $K_2Nb_8O_{21}$,$PbNb_2O_6$, $SrTa_4O_{11}$, and the tungsten bronzes. Named for Professor Kalervo Rankama. [*M.A.*, 70-758; *A.M.*, **55**, 1814.] 26th List

Rankinite. C. E. Tilley, 1942. *Min. Mag.*, vol. 26, p. 190. Tricalcium disilicate, $3CaO.2SiO_2$, as monoclinic crystals in the dolerite-chalk contact at Scawt Hill, Co. Antrim, and in blast-furnace slag. Named after George Atwater Rankin (1884–), of Washington, DC, formerly of the Geophysical Laboratory. [*M.A.*, **8**, 229, 244.] 16th List [Cf. Kilchoanite]

Ranquilite. M. Jimenez de Abeledo, M. Rodriguez de Benyacar, and E. E. Galloni, 1960. *Amer. Min.*, vol. 45, p. 1078. Near $3CaO.4UO_3.10SiO_2.24H_2O$, microcrystalline, orthorhombic, with gypsum from the Ranquil–C6 area, Portezuelo Hill, Malargue Dept., Mendoza Prov., Argentina. Named from the locality. 22nd List [Is Haiweeite (q.v.), Fleischer's *Gloss.* Cf. Gastunite]

Ransätite. L. J. Igelström, *Geol. För. Förh.*, XVIII. 43, 1896; and *Zeits. Kryst. Min.*, XXVII. 604, 1897. '$3(Mn,Ca,Mg)SiO_3 + (Fe,Al)_4Si_3O_{12}$.' 'Cubic.' [Bliaberg, Ransät parish, Wermland] Sweden. 1st List

Ransomite. C. Lausen, 1928. *Amer. Min.*, vol. 13, p. 221. Hydrous copper ferric sulphate, $CuO.Fe_2O_3.4SO_3.7H_2O$, as sky-blue orthorhombic crystals formed by the burning of pyritic ore in a mine in Arizona. Named after Prof. Frederick Leslie Ransome (1868–), of the California Institute of Technology. 11th List

Raphite. H. How. [T. Egleston, *Catalogue of Minerals*, (1887), 1889, p. 144. A. H. Chester, *Dictionary of the Names of Minerals*, 1896, p. 227. In the British Museum since 1879.] Synonym of ulexite. 1st List

Rashleighite. A. Russell, 1948. *Min. Mag.*, vol. 28, p. 353. Hydrous phosphate of aluminium, ferric iron, and copper, intermediate between turquoise and chalcosiderite, as blue to green encrustations from Cornwall. Named after the Cornish mineralogist Philip Rashleigh (1729–1811). 18th List

Rasorite. L. A. Palmer, 1927. *Engin. and Mining Journ. New York*, vol. 123, p. 494, 1928, vol. 125, p. 207; H. S. Gale, ibid., 1928, vol. 125, p. 702. *The Mineral Industry*, New York and London, 1927, vol. 35 (for 1926), p. 91. E. H. Kraus and W. F. Hunt, *Mineralogy*, 2nd edit., 1928, pp. 303, 380. Synonym of kernite (11th List). Named after Mr. Clarence M. Rasor, field engineer to the Pacific Coast Borax Company. [*M.A.*, **4**, 244.] 12th List

Raspite. C. Hlawatsch, *Ann. naturh. Hofmus, Wien*, 1897, XII, 33; *Zeits. Kryst. Min.*, 1897, XXIX, 130; 1899, XXXI, 1; *Records Geol. Survey, New South Wales*, 1898, VI, 51; abstr. *Min. Mag.*, XII, 47. Small yellow or brownish crystals with high adamantine lustre. Dimorphous, with stolzite, $PbWO_4$. Monoclinic. Broken Hill, New South Wales. 2nd List

Rassoulite, error for or syn. of Ghassoulite (20th List) (*Bull. groupe franç. argiles*, vol. 8, p. 37). 22nd List

Rasvumite. M. N. Sokolova, M. G. Dobrovol'skaya, N. I. Organova, and A. L. Dmitrik, 1970. *Zap.*, **99**, 712 (Расвумит). Grains with metallic lustre, consisting of fine orthorhombic fibres, *a* 9·12, *b* 11·08, *c* 5·47 Å, D 3·1, occur in the Rasvumchorr and Kukisvumchorr pegmatites of the Khibina massif. Composition $[K_3Fe_9S_{14}]$, with a little Na and Mg. Named for the locality. [*A.M.*, **56**, 1121; *M.A.*, 71-2337; *Zap.*, **100**, 615.] 27th List

Raswumit, Germ. trans. of Расвумит, rasvumite (27th List). Hintze, 78. 28th List

Rathite. H. Baumhauer, *Zeits. Kryst. Min.*, XXVI. 593, 1896. [*M.A.*, **11**, 225.] A sulpho-salt of Pb,As (and Sb). Rhombic. [Lengenbach quarry] Binnenthal, Switzerland. 1st List $[(Pb,Tl)_3As_5S_{10}]$

Rathite–I, Rathite–II, Rathite–III. Marie-Thérèse Le Bihan, 1959. *Compt. Rend. Acad. Sci. Paris*, vol. 249, p. 719; ibid. 1960, vol. 251, p. 2196. Three distinct structures: Rathite–I 'parait s'identifier au minéral nommé "Rathite" ou plutôt au minéral "sans nom" . . . decrit par Solly (1919)'; it has formula $Pb_7As_9S_{20}$, *a* 8·43, *b* 25·80, *c* 7·91 Å, *β* 90° 15′. Rathite–II is also $Pb_7As_9S_{20}$ but has *a* 8·43, *b* 72, *c* 7·91 Å, *β* 90° 15′. Rathite–III has *a* 8·43, *b* 7·91, *c* 24·40 Å, *β* 90° 15′, and composition $Pb_6As_{10}S_{20}$. 22nd List [Rathite–II is now Liveingite (q.v.); *see* review in *Chemical Index of Minerals*, App. II, p. 18 (1974)]

Rathite–IV, Rathite–V. W. Nowacki and C. Bahezre, 1964, *Schweiz. Min. Petr. Mitt.*, **44**, 7. Doubtful sulphosalts from the Lengenbach quarry, Binnatal, Switzerland. No formulae given. Rathite–IV (longest cell edge 45·96 Å) is later (*Jahrb. Naturhist. Mus. Bern*, 1963–1965, 296) withdrawn as 'weitgehend identisch' with sartorite, and rathite–V (longest cell edge 17 × 8·26 = 140·42 Å) becomes the new rathite–IV. 28th List [*M.A.*, 75-2500]

Rauenthalite. R. Pierrot, 1964. *Bull. Soc. franç. Min. Crist.*, vol. 87, p. 169. White spherules and minute crystals, monoclinic or anorthic, of $Ca_3(AsO_4)_2.10H_2O$ from the Rauenthaler vein system at Sainte-Marie-aux-Mines, Alsace. Named from the vein system. [*M.A.*, **17**, 79; *A.M.*, **50**, 805.] 24th List

Rauvite. F. L. Hess, 1922. *Engin. Mining Journ. New York*, vol. 114, p. 274; *Bull. US Geol. Survey*, 1924, no. 750-D, p. 68 (rauvite; p. 70, raw′vite; on plate X, raubite). Hydrous vanadate of uranium and calcium, $CaO.2UO_3.6V_2O_5.20H_2O$, as a purplish impregnation in sandstone from Utah. Named from the chemical symbols Ra, U, and V. [*M.A.*, **2**, 420.] 10th List $[Ca(UO_2)_2V_{10}O_{28}.16H_2O$, *Bull. US Geol. Surv. 1009-B*, pp. 13–62 (1954)]

Reaumurite. A. Lacroix, 1908. *La Montagne Pelée après ses éruptions*, Paris, 1908, p. 134; *Minéralogie de la France*, 1910, vol. iv, p. 778 (réaumurite). Glass when kept for some time at the softening temperature becomes converted into an opaque, white material, the 'porcelain of Réaumur' (René Antoine Ferchault de Réaumur, 1683–1757). This fibrous crystalline material is shown by its optical characters to be orthorhombic, and it has nearly the same composition as glass (approximately $(Ca,Na_2)O.3SiO_2$). The same substance has been produced under partly natural conditions by the action of volcanic heat on glass vessels in the houses at St. Pierre, Martinique (eruption of Mt. Pelée, 1902), and at Boscotrecase (eruption of Vesuvius, 1906). 6th List [A mixture; *A.M.*, **7**, 64]

Rectorite, *see* Allevardite.

Redledgeite. H. Strunz, 1961. *Neues Jahrb. Min.*, Monats., p. 107; *Amer. Min.*, 1961, vol. 46, p. 1201. The mineral from the Red Ledge mine, Nevada County, California, formerly regarded as a variety of Rutile and named Chromrutile (11th List), is a distinct species, of composition near $Mg_4Cr_6Ti_{24}Si_2O_{65}$, and is named from the locality. [*M.A.*, **15**, 211.] 23rd List

Reedmergnerite. C. Milton, J. M. Axelrod, and F. S. Grimaldi, 1954. *Bull. Geol. Soc. Amer.* [*1955*], vol. 65, no. 12 (for 1954), p. 1286; *Amer. Min.*, 1955, vol. 40, p. 326 (abstr.). $Na_2O.B_2O_3.6SiO_2$, triclinic, small colourless crystals from oil wells in Utah. Named after Frank S. Reed and John L. Mergner, technicians of the United States Geological Survey. [*M.A.*, **12**, 511.] 20th List [*A.M.*, **45**, 188. Related to Albite]

Reevesite. J. S. White, E. P. Henderson, and B. Mason, 1967. *Amer. Min.*, vol. 52, p. 1190. $Ni_6Fe_2CO_3(OH)_{16}.4H_2O$, the nickel analogue of pyroaurite, occurring among the weathering products of the Wolf Creek sideritic meteorite. [*Bull.*, **91**, 100.] 25th List

Reinerite. B. H. Geier and K. Weber, 1958. *Neues Jahrb. Min.*, Monatsh., p. 160 (Reinerit). Pale yellow-green orthorhombic crystals from Tsumeb, SW Africa, have the composition $Zn_3(AsO_3)_2$; a sea-blue variety contains 2% CuO. Named after Mr. Willy Reiner. [*M.A.*, **14**, 282.] An unfortunate name, easily confused with renierite (18th List, also from Tsumeb). 22nd List

Reitingerite. G. Gagarin and J. R. Cuomo, 1949, loc. cit., p. 11 (reitingerita). Zirconium oxide, ZrO_2, as nodular radially fibrous masses from Brazil, supposed by E. Hussak and J. Reitinger to be dimorphous with baddeleyite, but later shown to be identical. [*Min. Mag.*, **13**, 398; **21**, 171.] Named after J. Reitinger of Munich. 19th List

Renardite. A. Schoep, 1928. *Bull. Soc. Franç. Min.*, vol. 51, p. 247 (renardite); *Ann. Musée Congo Belge, Ser. I, Minéralogie, Tervueren* (*Belgique*), 1930, vol. 1, fasc. 2, p. 34. Hydrous phosphate of uranium and lead, $PbO.4UO_3.P_2O_5.9H_2O$, as yellow orthorhombic crystals from Katanga. Named after Prof. Alphonse François Renard (1842–1903) of Gand (= Gent). [*M.A.*, **4**, 15, 313.] 12th List [*A.M.*, **39**, 448]

Renierite. J. F. Vaes, 1948. *Ann.* (*Bull.*) *Soc. Géol. Belgique*, vol. 72 (for 1948–1949), p. B 19 (reniérite). Sulphide of Cu, Fe, Ge (7·75%), Zn, As, near RS, tetrahedral-cubic, from Katanga. Named after Prof. Armand Renier, Director of the Geological Survey of Belgium. Near germanite. [*M.A.*, **10**, 454.] 18th List [*A.M.*, **38**, 794. $Cu_3(Fe,Ge,Zn)(S,As)_4$. Cf. Mawsonite]

Reniforite. K. Kawai, 1925. [*Journ. Geol. Soc. Tokyo*, 1925, vol. 32, p. 106]; abstr. in *Japanese Journ. Geol. Geogr.*, '1924', vol. 3, Abstr. p. (15). Sulpharsenite of lead, $5PbS.As_2S_3$, as reniform aggregates from Japan. Evidently named from reniform. The analysis given agrees with those of jordanite. [*M.A.*, **3**, 114.] 11th List

Reniformite. O. M. Shubnikova and D. V. Yuferov, *Report on new minerals*, Leningrad, 1934, p. 30 (Рениформит, Reniformite). The correct spelling of reniforite (11th List), since the name evidently has reference to the reniform structure of the mineral. The mineral has since been proved to be identical with jordanite (M. Watanabé and N. Nakano, *Journ. Japanese Assoc. Min. Petr. Econ. Geol.*, 1936, vol. 15, p. 216). 14th List

Reposit, variant of or error for Repossite (14th List) (H. Strunz, *Min. Tabellen*, 4th edit., 1966, p. 512). 24th List

Repossite. E. Grill, 1935. *Periodico Min. Roma*, vol. 6, p. 23. Phosphate of iron, manganese, and calcium, $(Fe,Mn,Ca)_3(PO_4)_2$, in pegmatite from Lake Como. Evidently identical with graftonite. Named after Emilio Repossi (1876–1931), Professor of Mineralogy, University of Torino. [*M.A.*, **6**, 52.] 14th List

Restite. Author and date? A petrological term for the remainder after mobile constituents have moved out of a rock during metamorphism. 29th List

Retgersite. C. Frondel and C. Palache, 1948. *Program and Abstracts Min. Soc. Amer.*, 29th Annual Meeting, p. 6; *Bull. Geol. Soc. Amer.*, vol. 59, p. 1323. *Amer. Min.*, 1949, vol. 34, pp. 188, 276. Nickel sulphate hexahydrate, $NiSO_4.6H_2O$, tetragonal trapezohedral, long known as an artificial salt, has been recognized as a natural mineral at five localities, in association with morenosite (orthorhombic, $NiSO_4.7H_2O$). Named after Jan Willem Retgers (1856–1896), Dutch chemical crystallographer. [*M.A.*, **10**, 452, 506.] 18th List

Retinosite. J. A. Dulhunty, 1939. *Journ. Roy. Soc. New South Wales*, vol. 72, p. 184. *See* Gelosite. 15th List

Retzian. H. Sjögren, *Bull. Geol. Inst. Upsala*, II. 54, 1895; *Geol. För. Förh.*, XIX. 106, 1897.

[*M.M.*, **11**, 167.] Hydrated basic arsenate of Mn and Ca; near flinkite. Rhombic. [Moss mine, Nordmark] Sweden. 1st List [$Mn_2YtAsO_4(OH)_4$; *A.M.*, **52**, 1603]

Revdinite. E. N. Kortkov, *Bull. Soc. Ouralienne d'Amateurs des Sci. Nat.*, 1891–1892, vol. xiii, pp. 83, 76 (ревдинитъ). The same as rewdanskit (R. Herman, 1867); other forms of spelling are: refdanskite, revdanskite, revdinskite, rewdjanskit. An impure hydrous nickel silicate from Revda (= Revdinsk), Urals. 3rd List

Revoredite. G. C. Amstutz, P. Ramdohr, and F. De Las Casas, 1957. [*Soc. Geol. Peru, Ann.*, part 2, p. 25]; abstr. *Amer. Min.*, 1958, vol. 43, p. 794. Cf. C. Milton and B. Ingram, *Amer. Min.*, 1959, vol. 44, p. 1070. A natural Pb–S–As glass from Cerro de Pasco, Peru. There appears to be a continuous amorphous series from As_2S_3 to $2As_2S_3.PbS$ and probably beyond. The name was proposed by Amstutz, Ramdohr, and De Las Casas, after Dr. J. F. A. Revoredo, 'in case a crystalline sample should be found'; Milton and Ingram consider the name should be rejected. 22nd List

Reyerite. F. Cornu, 1906. *Min. Petr. Mitt.* (*Tschermak*), 1906–1907, vol. xxv, pp. 211, 519 (Reyerit). A 'micaceous zeolite' indistinguishable in appearance from gyrolite and zeophyllite. The thin plates with perfect cleavage and pearly lustre are rhombohedral and form radial aggregates. Sp. gr. 2·499–2·578; refractive index (ω) 1·564. Hydrous silicate of calcium with a little aluminium. The specimens examined were collected in Greenland by C. L. Giesecke in 1807–1813, and the mineral is probably identical with his 'Glimmerzeolith' (*see Min. Mag.*, vol. xiv, p. 95). Named after Professor Eduard Reyer, of Vienna. 4th List [*A.M.*, **58**, 517; *M.M.*, **33**, 839; **35**, 1]

Rezhikite. M. V. Soboleva and N. D. Sobolev, 1959. Совет. Геол., no. 9, p. 94. A deep blue amphibole asbestos near magnesioriebeckite and magnesioarfvedsonite but differing somewhat in optics. A premature name. 22nd List

Rh-sperrylite. E. F. Stumpfl and A. M. Clark, 1965. *A.M.*, **50**, 1068. Rare grains in platinum concentrates from the Driekop mine, Transvaal, have the composition $Pt_{0\cdot33}Rh_{0\cdot30}Ir_{0\cdot25}Pd_{0\cdot05}As_{0\cdot90}S_{1\cdot10}$. Name inappropriate, since Pt > Rh and As $\simeq$ S, and is replaced by platarsite (30th List). 30th List

Rhatite, error for Rathite. *Bull. Soc. franç. Min. Crist.*, 1961, vol. 85, p. 316. 23rd List

Rhenanite. H. H. Franck, M. A. Bredig, and R. Frank, 1936. *Zeits. Anorg. Chem.*, vol. 230, p. 2 (Rhenaniaphosphat, Rhenanit). An artificial fertilizer prepared by the Rhenania process, said to have a composition near $Ca_4Na_6(PO_4)_4CO_3$, but shown by R. Klement and P. Dihn (ibid., 1938, vol. 240, p. 40) to be a mixture of $CaNaPO_4$ and Na_2CO_3. Potassium-rhenanite, Kalium-Rhenanit (H. H. Franck, M. A. Bredig, and E. Kanert, ibid., 1938, vol. 237, p. 49) is the corresponding compound $CaKPO_4$. Named from Latin, *Rhenus*, *Rhenanus* = Rhine, Rhenish. [*M.A.*, **7**, 553.] 15th List [*A.M.*, **28**, 600. *See* Buchwaldite]

Rhodanite. G. Vavrinecz, 1939. *Földtani Közlöny*, Budapest, vol. 69, pp. 82, 98 (Rhodanit). Sulphocyanic (rhodanic) acid, HCNS, as a natural gas. Named from *ῥόδον*, rose. [*M.A.*, **7**, 471.] 15th List

Rhodesite. E. D. Mountain, 1956. *Min. Soc. Notice*, 1957, no. 97; *Min. Mag.*, 1957, vol. 31, p. 607. J. A. Gard and H. F. W. Taylor, *Min. Soc. Notice*, 1956, no. 95, 1957, no. 97; *Min. Mag.*, 1957, vol. 31, p. 611. A fibrous zeolitic mineral containing Al_2O_3 only 0·29%, $4(Ca,Na_2,K_2)O.10SiO_2.7H_2O$ unit cell contents $(Ca,Na_2,K_2)_8Si_{16}O_{40}.11H_2O$, orthorhombic. Closely associated with mountainite (q.v.) from Bultfontein mine, Kimberley, South Africa. Named after Cecil John Rhodes (1853–1902) and Rhodes University, Grahamstown. [*A.M.*, **43**, 624.] 21st List

Rhodium. L. J. Cabri and J. H. G. Laflamme, 1974. *C.M.*, **12**, 399. A single grain from the platinum ores of the Stillwater complex, Montana, has composition $Rh_{0\cdot57}Pt_{0\cdot43}$ and is a platinian rhodium. Cubic, *a* 3·856 A. [*A.M.*, **61**, 340; *Bull.*, **97**, 509.] 29th List

Rhodoarsenian. L. J. Igelström, *Zeits. Kryst. Min.*, XXII. 469, 1892. Hydrated basic arsenate of Mn,Ca and Mg. Sjö mine, [Örebro] Sweden. 1st List [May be Rhodonite, *A.M.*, **58**, 562]

Rhodolite. W. E. Hidden and J. H. Pratt, *Amer. J. Sci.*, 1898, V, 294; VI, 463; *Min. Mag.*, 1899, XII, 133, 145. Garnet ('*rose garnet*') of a rose-red or rhododendron-red colour used as a gem-stone. $2Mg_3Al_2(SiO_4)_3 + Fe_3Al_2(SiO_4)_3$. North Carolina. 2nd List

Rhodomacon. I. C. C. Campbell, 1972. *Journ. Gemmology*, **13**, 53. Unnecessary gem name, synonym of rhodolite (2nd List), suggested because of possible confusion with rhodonite;

derived from the colour and the type locality, Macon County, North Carolina. [*M.A.*, 72-3216]. 27th List

Rhodophosphite. L. J. Igelström, *Zeits. Kryst. Min.*, XXV. 433, 1895. Phosphate of Ca, (Mn and Fe) with chloride and sulphate. [Hörrsjöberg Mts., Wermland] Sweden. [Part of the analysis given for this has been published (*Bull. Soc. fran. Min.*, V. 303, 1882) for mangan-apatite.] 1st List [Is manganoan Apatite; *A.M.*, **44**, 910]

Rhodostannite. G. Springer, 1968. *Min. Mag.*, **36**, 1045. $Cu_2FeSn_3S_8$, hexagonal, with stannite, in comparison with which it is reddish, from Vila Apacheta, Bolivia. Named from its reddish colour. [*A.M.*, **54**, 1218; *M.A.*, 69-1530.] 26th List

Rhodusite. H. v. Foullon, *Ber. Akad. Wien*, C. (1), 144, 1891. Asbestiform glaucophane. Island of Rhodes. 1st List [Magnesio-riebeckite; *Amph.*, Sub-Comm.]

Rhomboclase. J. A. Krenner, 1891. *Akadémiai Értesítö*, Budapest, vol. ii, p. 96 (rhomboklas); *Földtani Közlöny*, Budapest, 1907, vol. xxxvii, pp. 204, 205; private letter from Professor Krenner, 1910. Hydrated acid ferric sulphate, $Fe_2O_3.4SO_3.9H_2O$, forming colourless rhombic plates with basal cleavage (hence the name, from ῥόμβος and κλάω), and occurring together with szomolnokite (q.v.) and other iron sulphates (kornelite, copiapite, coquimbite, etc.) at Szomolnok, Hungary.

The same compound, $(HO)_6Fe'''_2S_4O_{12}.6H_2O$, has been prepared artificially by R. Scharizer, *Zeits. Kryst. Min.*, 1901, vol. xxxv, p. 345; 1907, vol. xliii, p. 113. 5th List [$HFe^{3+}(SO_4)_2.4H_2O$]

Rhombomagnojacobsite. Fan De Lian, 1964. [*Acta Geol. Sinica*, vol. 44, p. 343]; abstr. *Bull. Soc. franç. Min. Crist.*, 1965, vol. 88, p. 361; *M.A.*, **17**, 398. A black orthorhombic mineral, composition given as $(Mn^{2+},Mg)(Mn^{3+},Fe)_2O_4$ with $Mn^{2+}:Mg \approx 3:2$ and $Mn^{3+} \approx Fe$; locality of origin not stated. M. Fleischer (*Amer. Min.*, 1965, vol. 50, p. 2101, comments: 'A badly chosen and confusing name, even if correct, for an orthorhombic analogue of magnesian Hausmannite (not jacobsite). Further study of the symmetry and of the chemistry, including determination of active oxygen, is essential.' 24th List

Rhönite. J. Soellner, 1907. *Neues Jahrb. Min.*, Beil.-Bd. xxiv, p. 475 (Rhönit). X. Galkin, ibid., 1910, Beil.-Bd. xxix, p. 715. A dark brown, triclinic amphibole resembling aenigmatite in its microscopical characters, but differing from this in chemical composition, $(Ca,Na_2K_2)_3Mg_4Fe''_2Fe'''_2Al_4(Si,Ti)_6O_{30}$. Occurs as a constituent of basaltic rocks in the Rhön Mountains and other localities. A. Lacroix (*Bull. Soc. franç. Min.*, 1909, vol. xxxii, p. 325) describes an amphibole of the same kind, occurring in doleritic nephelinite at Puy de Barneire, Puy-de-Dôme, for which he gives the formula $(Na,K,H)_2Ca_8(Fe,Mg)_{15}(Al,Fe)_{16}(Si,Ti)_{21}O_{90}$. 5th List [*M.A.*, 70-2106, 3358; 73-1825; *A.M.*, **55**, 864]

Ribeirite. W. Florencio, 1952. *Anais Acad. Brasil. Cienc.*, vol. 24, p. 259 (ribeirita). An altered zircon containing yttrium earths 7·45%, etc., from Brazil. Named after Prof. Joaquim Costa Ribeiro. [*M.A.*, **12**, 305.] 20th List

Richetite. J. F. Vaes, 1947. *Ann.* (*Bull.*) *Soc. Géol. Belgique*, vol. 70, p. B 221. Hydrous (?) lead uranate as black monoclinic plates, from Katanga. Named in memory of Emile Richet, died 1939, chief geologist of the Union Minière du Haut-Katanga. [*M.A.*, **10**, 146.] 18th List [Remains unconfirmed; *M.M.*, **41**, 51]

Rickardite. W. E. Ford, *Amer. Journ. Sci.*, 1903, ser. 4, vol. xv, p. 69; *Engin. and Min. Journ. New York*, 1903, vol. lxxv, p. 113; *Chem. News*, 1903, vol. lxxxvii, p. 56; *Zeits. Kryst. Min.*, 1903, vol. xxxvii, p. 609. A massive telluride of copper, Cu_4Te_3, with a rich purple colour resembling an iridescent tarnish. From Vulcan, Gunnison Co., Colorado. Named after Mr. T. A. Rickard, of New York. *See* sanfordite. 3rd List [Cu_7Te_5. Cf. Vulcanite]

Ricolite. A trade-name for an impure serpentinous rock from New Mexico. Named from *rico* (Spanish), rich, in allusion to its rich green colour. (G. P. Merrill, *Stones for Building and Decoration*, 2nd edit., 1897, p. 64.) 5th List

Riebeckite-arfvedsonite, W. G. Ernst, 1962. *Journ. Geol.*, **70**, 689. Used for solid solutions intermediate between riebeckite and arfvedsonite. 29th List

Riebeckrichterite. T. Yoshimura, 1937. [*Journ. Geol. Soc. Japan*, **44**, 561], quoted in *Intro. Jap. Min.*, 118 (1970), Geol. Surv., Japan. A variety of richterite from the Noda-Tamagawa mine, Iwate Prefecture, Japan. [*Zap.*, **101**, 286.] 28th List

Rijkeboerite. A. H. van der Veen, 1963. [*Verhandl. Kon. Ned. Geol. Mijnb. Gen.*, geol. ser., vol. 22, p. 1], abstr. *M.A.*, **16**, 373. A member of the Pyrochlore family, with Ta > Nb, much H_2O, and Ba as principal cation; the tantalum analogue of Pandaite (22nd List); near $Ba_{0\cdot 3}Ta_2O_{5\cdot 3}.2H_2O$. Cf. Hydropyrochlore. 23rd List [Renamed **Bariomicrolite**, *A.M.*, **62**, 406; 30th List]

Rilandite. E. P. Henderson and F. L. Hess, 1933. *Amer. Min.*, vol. 18, p. 202. Hydrous basic silicate of chromium and aluminium (Cr_2O_3 47·59%), as brownish-black pitch-like masses in sandstone from Colorado. Named after Mr. J. L. Riland, of Meeker, Colorado. [*M.A.*, **5**, 293.] 13th List

Rimpylite. G. Murgoci, 1922. *Compt. Rend. Acad. Sci. Paris*, vol. 175, p. 429 (rimpylites). Certain green or brown amphiboles, very rich in sesquioxides, and which fall outside the group of those poorer in magnesia, are named rimpylites, from one of the localities, Rimpy [not stated where]. [*M.A.*, **2**, 221.] 10th List [Hornblende; *Amph.*, Sub-Comm.]

Ringwoodite. R. A. Binns, R. J. Davis, and S. J. B. Reed, 1969. *Nature*, **221**, 943. Rounded grains and pseudomorphs after olivine in the Tenham meteorite have the composition of olivines with Fa 26–34%, but with the structure of spinel. The name, for Professor A. E. Ringwood, is applied to the whole range of $(Mg,Fe)_2SiO_4$ spinels. [*A.M.*, **54**, 1219; *M.A.*, 70-745.] 26th List

Rinkolite. E. M. Bonshtedt, 1926. *Bull. Acad. Sci. Russie*, ser. 6, vol. 20, p. 1181 (ринколит, rinkolite). A. E. Fersman, *Amer. Min.*, 1926, vol. 11, p. 295; *Neues Jahrb. Min.*, Abt. A, 1926, vol. 55, p. 44 (Rinkolit). Titano-silicate of cerium, calcium, strontium, and sodium, as large yellowish-green monoclinic crystals in the nepheline-syenites of the Kola peninsula. Related to rinkite. [*M.A.*, **3**, 236; **3**, 275.] 11th List [Is Mosandrite, *A.M.*, **43**, 795]

Rinneite. H. E. Boeke, 1908. *Chemiker-Zeitung*, Jahrg. xxxii, p. 1228; *Centralblatt Min.*, 1909, p. 72; *Neues Jahrb. Min.*, 1909, vol. ii, p. 19; *Sitzungsber. Akad. Wiss. Berlin*, 1909, p. 632 (Rinneit). O. Schneider, *Centralblatt Min.*, 1909, p. 503. An anhydrous chloride of ferrous iron, potassium, and sodium, $FeCl_2.3KCl.NaCl$, occurring as large, lenticular masses in the Prussian salt deposits at Nordhausen and Hildesheim. It is clear and colourless, but quickly becomes yellow on exposure to the air. Rhombohedral crystals are occasionally found in the coarsely granular aggregates, and have also been prepared artificially. Named after Professor Fritz Rinne, of Kiel. 5th List

Riolith*. J. Fröbel, 1843. *Grundzüge eines Systemes der Krystallologie*. The riolite of Brooke (1836) being commonly discredited, Fröbel transfers the name to the supposed $AgSe_2$ from Tasco described by Del Rio (1827) and later named tascine. 27th List [*Corr. in *M.M.*, **39**]

Ripidolite, *see* Angaralite.

Risörite. O. Hauser, 1908. *Zeits. Anorg. Chem.*, vol. lx, p. 230 (Risörit). A preliminary description was given in *Ber. Deutsch. Chem. Ges.*, 1907, vol. xl, p. 3118. A niobate (and titanate) of yttrium, occurring as glassy, optically isotropic masses at Risör, Norway. 5th List

Rivadavite. C. S. Hurlbut, Jr., and L. F. Aristarian, 1967, *Amer. Min.*, vol. 52, p. 326. $Na_6MgB_{24}O_{40}.22H_2O$, monoclinic crystals in borax from Tincalayu, Salta, Argentina. Named for B. Rivadavia. [*M.A.*, **18**, 284; Зап., **97**, 71; *Bull.*, **90**, 606.] 25th List

Rivaite. F. Zambonini, 1912. *Rend. Accad. Sci. Fis. Mat. Napoli*, ser. 3, vol. xviii, p. 223; *Riv. Min. Crist. Ital.*, xli, p. 94. A Vesuvian mineral with the formula $(Ca,Na_2)Si_2O_5$. Named after the mineralogist, Dr. Carlo Riva (1872–1902) of Pavia. 6th List [Is Wollastonite; *M.A.*, **2**, 43]

Riversideite. A. S. Eakle, 1917. *Bull. Dep. Geol. Univ. California*, vol. x, p. 347. Hydrated calcium silicate, perhaps $2CaSiO_3.H_2O$, occurring as a white, fibrous material filling small veins in massive idocrase and blue calcite at Crestmore, Riverside Co., California. Named from the locality. 8th List [Unstable. $Ca_5Si_6O_{16}(OH)_2.2H_2O$; *M.M.*, **30**, 293 (*A.M.*, **39**, 1038)]

Rizopatronite. J. J. Bravo, 1906. [*Informaciones y Memorias Soc. Ingen. Lima*, vol. viii, p. 171; *Bol. Soc. Ingen. Lima*, August 1906], quoted by W. F. Hillebrand, *Amer. Journ. Sci.*, 1907, ser. 4, vol. xxiv, p. 144; *Journ. Amer. Chem. Soc.*, 1907, vol. xxix, p. 1022 (Rizo-Patronita). A synonym of patronite (D. F. Hewett, 1906; 4th List). A dark greenish-black vanadium sulphide, VS_4 or V_2S_9, occurring in vanadium-ore at Minasragra, near Cerro de Pasco, Peru. D. F. Hewett (*Trans. Amer. Inst. Min. Engin.*, 1910, vol. xl, p. 287) suggests the formula $V_2S_5 + nS$. Named after Antenor Rizo Patrón, of the Huaraucaca smelting works near Cerro de Pasco. 5th List

Robellazite. E. Cumenge, *Bull. Soc. franç. Min.*, 1900, XXIII, 17. Black concretionary masses occurring with carnotite in Colorado. Contains V,Nb,Ta,W,Al,Fe,Mn. 2nd List

Robertsite. P. B. Moore, 1974. *A.M.*, **59**, 48. Blood-red to black, highly pleochroic plates and fibres with other phosphates in the Tip Top pegmatite, Custer, Custer County, South Dakota, are monoclinic, *a* 17·36, *b* 19·53, *c* 11·30 Å, *β* 96·0°; composition $8[Ca_3Mn_4^{3+}(PO_4)_4(OH)_6.3H_2O]$. Named for Willard L. Roberts. 28th List

Robertsonite. F. V. Chukhrov, 1936. [*Colloids in the earth's crust* (*Russ.*), *Acad. Sci. USSR*, 1936, p. 83.] O. M. Shubnikova, *Trans. Inst. Geol. Sci., Acad. Sci. USSR*, 1938 [i.e. 1940], no. 11 (*Min. Geochem. Ser.*, no. 3), p. 3 (Робертсонит, Robertsonite). Colloidal zinc sulphide from Cherokee County, Kansas. Named after James D. Robertson, who described the material (*Amer. Journ. Sci.*, 1890, ser. 3, vol. 40, p. 160). *See* Brunckite (15th List). 16th List

The same name, Robertsonite (E. Poitevin, *Ann. Rep. Dept. Mines*, Canada, 1931 for 1929–1930, pp. 19, 20), was provisionally given to a mineral from British Columbia, at first thought to be a new species. 16th List

Robinsonite. L. C. Berry, J. J. Fahey, and E. H. Bailey, 1951. *Program and Abstracts, Min. Soc. Amer.*, November 1951, p. 8; *Bull. Geol. Soc. Amer.*, 1951, vol. 62, p. 1423; *Amer. Min.*, 1952, vol. 37, pp. 285, 438. Still another lead sulphantimonite, $7PbS.6Sb_2S_3$, triclinic, from Nevada, and as artificial crystals. Named after Dr. S. C. Robinson, of the Geological Survey of Canada. [*M.A.*, **10**, 457–458 ('mineral X'), **11**, 414.] 19th List [May be $Pb_4Sb_6S_{13}$; *C.M.*, **13**, 415]

Rockbridgeite. C. Frondel, 1949. *Amer. Min.*, vol. 34, p. 513. Dana's *Mineralogy*, 7th edit., 1951, vol. 2, p. 867. A dufrenite-like mineral, $(Fe'', Mn'')Fe'''_4(PO_4)_3(OH)_5$, orthorhombic, isomorphous with frondelite (q.v.). Named from one of the localities, Rockbridge County, Virginia. [*M.A.*, **11**, 7, 187.] 19th List [Cf. Mangan-rockbridgeite]

Rodalquilarite. J. Sierra Lopez, G. Leal, R. Pierrot, Y. Laurent, J. Protas, and Y. Dusausov, 1965. *Bull. Soc. franç. Min. Crist.*, vol. 91, p. 28. Small emerald green anorthic crystals from Rodalquilar, Almeria, Spain, have the composition $Fe_2^{3+}TeO_3(TeO_3H)_3Cl.0{\cdot}5H_2O$. Named from the locality. [*A.M.*, **53**, 2104.] 25th List

Rodita. G. Gagarin and J. R. Cuomo, 1949, loc. cit., p. 4. Spanish spelling of rhodite. Not the rodite of A. Brezina, 1885, applied to the Roda (Spain) meteorite and a class of meteorites. 19th List

Rœblingite. S. L. Penfield and H. W. Foote, *Amer. Journ. Sci.*, III. 413, 1897. [*M.M.*, **11**, 343.] $5(H_2CaSiO_4) + 2(CaO.PbSO_3)$. [Parker shaft, Franklin Furnace] New Jersey. 1st List [$Pb_2Ca_7Si_6O_{14}(OH)_{10}(SO_4)_2$; *A.M.*, **51**, 504]

Roedderite. L. H. Fuchs, C. Frondel, and C. Klein, 1966. *Amer. Min.*, vol. 51, p. 949. $(Na,K)_2(Mg,Fe)_5Si_{12}O_{30}$, hexagonal, the sodium magnesium analogue of merrihueite, occurs as an accessory mineral in the Indarch meteorite. Named for E. W. Roedder who synthesized $K_2Mg_5Si_{12}O_{30}$. [*M.A.*, **18**, 47; Зап., **97**, 77.] 25th List

Roentgenite, *see* Röntgenite.

Roewolfeite. *Zap.*, **105**, 78 (1978). Error for wroewolfeite. 30th List

Rogersite. C. Lausen, 1928. *Amer. Min.*, vol. 13, p. 225. Hydrous ferric sulphate, $Fe_2O_3.3SO_3.6H_2O$, as aggregates of monoclinic fibres formed by the burning of pyritic ore in a mine in Arizona. Named after Prof. Austin Flint Rogers (1877–), of Stanford University, California. 11th List [Renamed Lausenite (q.v.). Not the Rogersite (of Smith), which is Churchite; *A.M.*, **48**, 1168]

Roggianite. E. Passaglia, 1969. *Rend. Soc. Ital. Min. Petr.*, **25**, 105; *Clay Min.*, **8**, 107 and 112. Fibrous aggregates of composition $(Na,K)_2Ca_{12}Al_{16}(Si,Al)_{28}O_{52}(OH)_{80}$ occur at Alpe Rosso, Val Vigezzo, Novare, Italy; tetragonal. Named for Aldo G. Roggiani. [*A.M.*, **54**, 1741; **55**, 322; *M.A.*, 69-3345, 3346; *Zap.*, **99**, 83.] 26th List

Rogueite. Another unnecessary name for a variety of jasper, this time greenish and from Oregon. R. Webster, *Gems*, 1962, p. 762. Not to be confused with roquesite (23rd List). 28th List

Romanechite. (*Collection de Minéralogie du Muséum d'Histoire Naturelle, Paris, Guide du*

Visiteur, 2nd edit., 1900, p. 29 (Romanéchite).) Psilomelane from Romanèche, France, with the composition (Ba,Mn)O.$3MnO_2.H_2O$; analysed by A. Gorgeu (1890). 4th List [Now adopted as the species name; Psilomelane becomes a group name]

Romanite. The same as rumänite (O. Helm, 1891), rumanite, roumanite (2nd List). An amber-like resin from Roumania; (or Rumania; in Roumanian, România). (G. Munteanu-Murgocĭ, *Gisements du succin de Roumanie, Asoc. Română Sci. Mem. Congres.*, Bucarest, 1903.) 5th List

Romarchite. R. M. Organ and J. A. Mandarino, 1971. *C.M.*, **10**, 916. Black crystals formed on tin pannikins lost about 1801–1821 in the Winnipeg River at Boundary Falls, Ontario, are identical with synthetic SnO. Tetragonal, *a* 3·79, *c* 4·83 Å, *P4/nmm*. Named for the *R*oyal *O*ntario *M*useum of *Arch*aeology. [*Bull.*, **96**, 241; *A.M.*, **57**, 1555; *Zap.*, **101**, 280.] 28th List [Cf. Hydroromarchite]

Röntgenite. G. Donnay, 1953. *Amer. Min.*, vol. 38, p. 868 (roentgenite). Minute wax-yellow to brown, trigonal pyramidal crystals, intergrown with synchysite, parisite, and bastnäsite, from Narsarsuk, Greenland. From X-ray and optical data the composition is deduced as $3CeFCO_3.2CaCO_3$. Named after Wilhelm Conrad von Röntgen (1845–1923) of München, discoverer of X-rays (Röntgen rays). [*M.A.*, **12**, 238, 329.] 20th List [*A.M.*, **38**, 932]

Rooseveltite. R. Herzenberg, 1946. *Boletín Técnico, Fac. Nac. Ingeniería, Univ. Técnica, Oruro*, no. 1, p. 10 (Rooseveltita). Bismuth arsenate, $BiAsO_4$, as a white coating on wood-tin from Santiaguillo, Bolivia. Named after Franklin Delano Roosevelt (1882–1945), President of the United States of America. [*M.A.*, **10**, 9; *A.M.*, **32**, 372.] 18th List

Roquesite. P. Picot and R. Pierrot, 1963. *Bull. Soc. franç. Min. Crist.*, vol. 86, p. 7. Sulphide of copper and indium, $CuInS_2$, isostructural with Chalcopyrite, as small (0·2 × 0·3 mm) inclusions in Bornite from Charrier, Allier, France. Named for Prof. M. Roques. [*M.A.*, **16**, 372; *A.M.*, **48**, 1178.] 23rd List [Kazakhstan; *M.A.*, 75-1359]

Rosaline. *Syn.* of thulite, *var.* of zoisite. R. Webster, *Gems*, 1962, p. 762. 28th List

Rosasite. D. Lovisato, 1908. *Rend. R. Accad. Lincei*, Roma, ser. 5, vol. xvii, sem. 2, p. 723. A fibrous, pale green, basic carbonate of zinc and copper, $5ZnCO_3.3CuCO_3.2CuO$, resembling aurichalcite, but differing from this in composition. From Rosas, Sardinia. 5th List [Cf. Glaukosphaerite]

Roscherite. F. Slavík, 1914. [*Memoirs Bohemian Acad. Sci.*]; Doelter's *Handbuch der Mineralchemie*, vol. iii, p. 499; *Bull. Soc. franç. Min.*, vol. xxxvii, p. 162. Dark-brown, monoclinic crystals occurring with ježekite, lacroixite (q.v.), and childrenite in drusy cavities in lithionite-granite at Greifenstein, near Ehrenfriedersdorf, Saxony. It is a hydrated basic phosphate of aluminium, manganese, calcium, and iron, $(Mn,Ca,Fe)_2.Al(OH)P_2O_8.2H_2O$. Named after Mr. Walter Roscher, apothecary and mineral collector, of Ehrenfriedersdorf. 7th List [$(Ca,Mn,Fe)_3Be_3(PO_4)_3(OH)_3.2H_2O$; *A.M.*, **43**, 824]

Rose garnet. *See* Rhodolite. 2nd List

Roseite. J. Ottemann and S. S. Augustithis, 1967. *Mineralium Deposita*, vol. 1, p. 269. A platinum nugget from Ethiopia contains several possibly new minerals. One, containing Os, Ir, and S, and is formulated (Os,Ir)S though the sulphur content (15%) is open to some doubt. It is named for Hermann Rose. The name roseite is pre-empted (Stubb, 1879), and in any case more data are needed to establish a new species. [*A.M.*, **52**, 1579; *Bull.*, **90**, 610.] 25th List

Roselite. Error for or variant of rosolite (4th List). R. Webster, *Gems*, 1962, p. 762. Not to be confused with the roselite of (Lévy), β-roselite (beta-roselite, 21st List), nor rosellite (*syn.* of rosellan). Cf. xalostocite (4th List). 28th List

Rosenhahnite. A. Pabst, E. B. Gross, and J. T. Alfors, 1967. *Amer. Min.*, vol. 52, p. 336. $CaSiO_3.\frac{1}{3}H_2O$, as anorthic crystals in veins in a metamorphosed sediment, found as boulders in Russian River, Cloverdale, Mendocino County, California. Named for the finder, L. Rosehahn. [*M.A.*, **18**, 284; Зап., **97**, 76; *Bull.*, **90**, 606.] 25th List [$Ca_3Si_3O_8(OH)_2$; *M.M.*, **41**, 394 (with refs.)]

Rosenite. C. Zincken, *Pogg. Ann.*, 1835, XXXV, 357. The same as plagionite. 2nd List

Rosickyite. J. Sekanina, 1931. *Zeits. Krist.*, vol. 80, p. 174 (Rosickýit). Monoclinic γ-sulphur as minute crystals from the decomposition of pyrite nodules in Cretaceous clay in Moravia. Named after Prof. Vojtěch* Rosický, of Brno (= Brünn), Moravia. [*M.A.*, **5**, 49.] 13th List [*Corr. in *M.M.*, **25**]

Rosieresite. A. Lacroix, 1910. *Minéralogie de la France*, vol. iv, pp. 532, 399 (rosiérésite). A hydrated phosphate of aluminium containing some lead (PbO 10%) and copper (CuO 3%), analysed by P. Berthier in 1841. Found as stalactites in the Rosières copper-mine near Carmaux, Tarn. 6th List

Rosinca, German variant of Inca rose, a massive rhodochrosite used as an ornamental stone (H. Strunz, *Min. Tabellen*, 2nd edit., 1949, p. 290). 24th List

Rosolite. (*Trans. Amer. Inst. Mining Engin.*, 1902, vol. xxxii, pp. 55, 57.) A rose-pink garnet from Mexico. *See* Landerite. 4th List [Erroneous variant Roselite (q.v.)]

Rössingite. J. W. von Backström, 1970. [*Uranium exploration geol., Proc. of a panel held in Vienna*, 143]; abstr. *M.A.*, 72-1028. Mentioned, without details, as a mineral from the Rössing uranium deposit, Swakopmund, SW Africa. 28th List

Rossite. F. L. Hess and W. F. Foshag, 1926. *Amer. Min.*, vol. 11, p. 66; W. F. Foshag and F. L. Hess, *Proc. United States Nat. Mus.*, 1927, vol. 72, art. 11, p. 1. Hydrated vanadate of calcium, $CaO.V_2O_5.4H_2O$; triclinic. From [Bull Pen Canyon, San Miguel Co.] Colorado. Named after Dr. Clarence Samuel Ross (1880–), of the United States Geological Survey. *See* Metarossite. [*M.A.*, **3**, 239; **3**, 470.] 11th List

Rosstrevorite. (Greg and Lettsom, *Min. of Great Brit.*, 1858, p. 105.) Fibrous stellated epidote from near Rosstrevor, Co. Down. 2nd List

Roubaultite. F. Cesbron, R. Pierrot, and T. Verbeek, 1970. *Bull.*, **93**, 550. Rosettes of green platy crystals with other uranium minerals at Shinkolobwe, Katanga, are anorthic, *a* 7·73, *b* 6·87, *c* 10·87 Å, α 86° 29′, β 134° 12′, γ 93° 10′. Composition $Cu_2(UO_2)_3(OH)_{10}.5H_2O$. α' 1·700, colourless, β' 1·800, colourless, nearly ‖ [100], γ' 1·84, greenish yellow. Named for M. Roubault. [*M.A.*, 71-2338; *Bull.*, **94**, 573; *Zap.*, **100**, 619; *A.M.*, **57**, 1912.] 27th List

Roumanite. C. [I.] Istrati [*Bult. Soc. Sci. Fizice*, 1895, IV, 59; *Anal. Acad. Române*, 1895, XVI]. *Bult. Soc. Sci. Bucarest*, 1897, VI, 55; 1898, VII, 272. The same as rumänite. 2nd List

Routhierite. Z. Johan, J. Mantienne, and P. Picot, 1974. *Bull.*, **97**, 48. Dark-red veins in the black dolomites of Jas Roux, Hautes-Alpes, France, are tetragonal, *a* 9·977, *c* 11·290 Å; composition $8[TlHgAsS_3]$, with some Cu and Ag replacing Tl and some Zn replacing Hg. Named for P. Routhier. 28th List

Roweite. H. Berman and F. A. Gonyer, 1937. *Amer. Min.*, vol. 22, p. 301. Hydrated borate $H_2(Mn,Mg,Zn)Ca(BO_3)_2$ as light-brown orthorhombic crystals, differing from sussexite in containing calcium. Named after Mr. George Rowe, a mine captain at Franklin Furnace, New Jersey, where he collected the mineral. [*M.A.*, **6**, 488.] 14th List [$4[Ca_2(Mg,Mn,Zn)_2(OH)_4B_4O_7(OH)_2]$; *A.M.*, **59**, 66]

Royalite. Trade-name for a purplish-red glass. R. Webster, *Gems*, 1962, p. 762. 28th List

Royite. N. L. Sharma, 1940. *Proc. Indian Acad. Sci.*, Sect. B, vol. 12, p. 215. Flattened prisms of quartz of blade-like habit occurring on joint-planes of sandstones and shales in the Jhaira coal-field. Named after Professor S. K. Roy, of the Indian School of Mines, Dhanbad. [*M.A.*, **8**, 230.] 16th List

Rozenite. J. Kubisz, 1960. *Bull. Acad. Polon. Sci., Sér. sci. géol. géogr.*, vol. 8, p. 107. $FeSO_4.4H_2O$, isomorphous with ilesite and leonhardtite, from Ornak, W High Tatra, Poland, and from the Staszic mine, Rudki, Poland. Optical data show that it is identical with siderotile, the water content of which has hitherto been uncertain. An unnecessary name. 22nd List [Distinct from Siderotile. *C.M.*, **7**, 751 (*A.M.*, **49**, 820)]

Rozhkovite. L. V. Razin, 1975. *Trudy Mineral. Muz. akad. nauk SSSR*, **24**, 93 (Рожковит). This unnecessary name is proposed for a mineral from the Talnakh deposit, described by Razin *et al.* in 1971 (*Zap.*, **100**, 66) as palladian cuproauride. X-ray powder data (24 lines) indexed as orthorhombic with *a* 3·86, *b* 39·00, *c* 3·84 Å, as for artificial CuAu(II), and also with *a* 3·88, *b* 42·68, *c* 3·84 Å, as for artificial Cu_3Au_2. The composition (5 grains) is near (Cu,Rh)(Au,Pd), with a little Ag, Ni, and Bi. [In view of the lack of single-crystal data . . . its indexing must be considered tentative, but if any likeness to synthetic phases is to be made, it must be to CuAu(II)—*L.J.C.* Cuproauride (Lozhechkin, 1939), from Karabash, Urals, had a composition near Cu_3Au_2, but was regarded as a solid solution, probably extending to Cu_3Au, not a definite compound—*M.H.H.*] [*A.M.*, **62**, 595.] 30th List

Rozircon. An objectionable and misleading trade-name for a pink *artificial* spinel. R. Webster, *Gems*, 1962, p. 762. *See* under corundolite. 28th List

Rubber-sulphur. T. Wada, 1916. *Minerals of Japan* (*in Japanese*), 2nd edit., 1916, p. 21 (Rubber-Sulphur). Amorphous, plastic sulphur from the Kobui sulphur mine, prov. Oshima, Japan. [*M.A.*, **1**, 63.] 9th List

Rubidium-microcline. V. I. Vernadsky, 1913. *Bull. Soc. franç. Min.*, vol. xxxvi, p. 263 (microcline rubidifère, Rubidiummikroklin, rubidievyj mikroklin). The microcline (Amazonstone) of the Ilmen Mts., Urals, was found to contain 3·12% of rubidium oxide corresponding to 10·89% of the silicate $RbAlSi_8O_8$. 7th List

Rubiesite. C. Doelter, 1926. *Mineralchemie*, vol. 4, part 1, p. 838 (Rúbiesit). A mineral described in 1920 by S. Piña de Rúbies and evidently regarded by him as a mixture of sulphide, selenide, and telluride of bismuth and antimony. Doelter's only information was from [*M.A.*, **1**, 201]. 19th List

Rubolite. An unnecessary name for a red variety of common opal. R. Webster, *Gems*, 1962, p. 762. Not to be confused with rubellite. 28th List

Rucklidgeite. E. N. Zav'yalov and V. D. Begizov, 1977. *Zap.*, **106**, 62 (Раклиджит, Racklidgite). Foliated aggregates in the gold ores of Zod, Armedia, and of Kokchar, Urals, are rhombohedral a_h 4·416, c_h 41·45 Å, space group *R*3*m*. Composition $3[(Bi,Pb)_3Te_4]$. This agrees well with the unnamed mineral of J. Rucklidge (*M.A.*, 70-1605), from the Robb Montbray mine, Montbray Township, Quebec, and is therefore named for him. [*M.A.*, 77-4626; *A.M.*, **63**, 599.] 30th List

Ruffite. W. B. Blumenthal, 1958. *Chem. Behavior Zirconium*, Princeton, 1962, p. 161. A name for artificial tetragonal ZrO_2. 23rd List

Ruizite. S. A. Williams and M. Duggan, 1977. *M.M.*, **41**, 429. Orange to brown spherules from the Christmas mine, Gila Co., Arizona, are monoclinic, *a* 11·95, *b* 6·17, *c* 9·03 Å, β 91° 22½′. Composition $4[CaMn^3(SiO_3)_2.2H_2O]$. RIs α 1·663, β 1·715 ∥ [010], γ 1·734, γ:[001] 44° in obtuse β, $2V_\alpha$ 60·2° with inclined dispersion, $\rho > v$ strong. Named for J. A. Ruiz. [*M.A.*, 78-894.] 30th List

Rumongite. S. Bracewell, 1950. *Rep. Geol. Surv. British Guiana*, for 1949, pp. 38, 40 ('Rumongite'). Provisional name for a mineral afterwards identified as ilmenorutile, from the Rumong-Rumong river, British Guiana. [*M.A.*, **11**, 251.] 19th List

Rusacovite, variant of Rusakovite (Русаковит). *Bull. Soc. franç. Min. Crist.*, 1961, vol. 84, p. 107. 23rd List

Rusakovite. E. A. Ankinovich, 1960. Зап. Всесоюз. Мин. Общ. (*Mem. All-Union Min. Soc.*), vol. 89, p. 440 (Русаковит). Abstr. ibid., 1961, vol. 90, p. 443 (Русаковит, rusacovite), and *Bull. Soc. franç. Min. Crist.*, 1961, vol. 84, p. 1075 (Rusacovite). $(Fe,Al)\{(V,P)O_4\}_2$-$(OH)_9.3H_2O$, from Balasauskandyk, Karatau, Kazakhstan. Named after M. P. Rusakova (Михаила Петровича Русакова). 22nd List [*A.M.*, **45**, 1315. Cf. Schoderite]

Russellite. M. H. Hey and F. A. Bannister, 1938. *Min. Mag.*, vol. 25, p. 41. Yellow pellets approximating in composition to $Bi_2O_3.WO_3$. X-ray examination suggests that it is a mixed (tetragonal) crystal of Bi_2O_3 and WO_3, rather than a bismuth tungstate. Occurs as an alteration product with native bismuth and wolframite in Cornwall. Named after Mr. Arthur Edward Ian Montagu Russell (1878–), of Swallowfield Park, Berkshire. 15th List

Rustenbergite. P. Mihalik, S. A. Hiemstra, and J. P. R. de Villiers, 1975. *C.M.*, **13**, 146. Two grains in concentrates from the Rustenberg and Atok mines in the Bushveld Complex, S Africa, gave $Pt_{1.52}Pd_{1.48}Sn_{0.83}$ (palladian rustenbergite) and $Pd_{1.85}Pt_{1.15}Sn_{0.81}$ (platinian atokite). Space group *Fm*3*m*, *a* 3·991 Å. Synthetic study shows a complete solid solution series from Pt_3Sn_{1-x} (rustenbergite) to Pd_3Sn_{1-x}, (atokite). Named from the localities. [*A.M.*, **61**, 340.] 29th List

Rustenite, error for Ruténite (Hintze, *Handb. Min.*, Erg.-Bd. II, 1958, p. 612). 22nd List

Rustite. J. D. Buddhue, 1939. *See* Ayasite. 15th List

Rustonite. K. F. Chudoba, 1958. Hintze, *Handb. Min.*, Erg.-Bd. II, p. 612. Said to be a synonym of Ruthenosmiridium. 22nd List

Rustumite. S. O. Agrell, 1965. *Min. Mag.*, vol. 34, p. 1. Crudely tabular monoclinic crystals in metamorphosed limestone at Kilchoan, Ardnamurchan, Scotland. Composition $Ca_4Si_2O_7(OH)_2$. Named for Rustum Roy. [*A.M.*, **50**, 2104.] 24th List

Ruthenarsenite. D. C. Harris, 1974. *C.M.*, **12**, 280. Orthorhombic 4[RuAs] (*a* 5·628, *b* 3·239, *c*

6·184 Å) occurs as inclusions up to 0·1 mm diameter in rutheniridosmine from New Guinea; identical with artificial material. Named from the composition. 28th List

Ruthen-Iridosmium. H. Strunz, 1966. *Min. Tabellen*, 4th edit., p. 93. A name to replace Ruthenosmiridium (15th List), because the mineral, being hexagonal, is a ruthenian Iridosmine rather than a ruthenian Osmiridium; neither name is necessary. 24th List

Ruthenium. Y. Urashima, T. Wakabayashi, T. Masaki, and Y. Terasaki, 1974. *Min. Journ. (Japan)*, **7**, 438. A tabular crystal (7 × 35 mm) of iridian ruthenium was found in platy ruthenosmiridium from the Horokanai placer, Hokkaido, Japan. [*A.M.*, **61**, 177.] 29th List

Ruthenosmiridium. S. Aoyama, 1936. *Sci. Rep. Tôhoku Univ.*, ser. 1, K. Honda anniv. vol., p. 527. A variety of iridosmine containing ruthenium (Ru 21·08%); hexagonal, RuOsIr. From Japan. [*M.A.*, **7**, 315.] 15th List

Rutherfordine. W. Marckwald, 1906. *Centralblatt Min.*, 1906, p. 763 (Rutherfordin). A yellow uranyl carbonate, $UO_2.CO_3$, resembling uranochre in appearance and resulting by the alteration of uraninite; from German East Africa. It is strongly radio-active, and is named after a prominent worker in radio-activity, Professor Ernest Rutherford, of Manchester, formerly of the McGill University, Montreal. Not to be confused with rutherfordite (C. U. Shepard, 1851) from Rutherford Co., North Carolina. 4th List [Orthorhombic UO_2CO_3; *A.M.*, **41**, 127]

Rutilohematite. A. F. Buddington, J. [J.] Fahey, and A. [C.] Vlisidis, 1953. *Progr. and Abstr. Min. Soc. Amer.*, 1953, p. 13; *Amer. Min.*, 1954, vol. 39, p. 319 (rutilohematite). The most oxidized facies of Fe-Ti oxides. *See* Ilmenomagnetite. 20th List

Rutosirita. G. Gagarin and J. R. Cuomo, 1949, loc. cit., p. 5. Contraction in Spanish form of ruthenosmiridium (15th List). 19th List

S

Saamite. M. I. Volkova and B. N. Melentiev, 1939. *Compt. Rend.* (*Doklady*) *Acad. Sci. URSS*, vol. 25, p. 122 (strontium apatite, saamite). A variety of apatite containing SrO (5·58–11·42%) together with rare-earths (1·75–4·90%) from the Kola peninsula, Russia. (Not the strontium-apatite of A. N. Winchell, 1927, 15th List.) [*M.A.*, **8**, 52.] 16th List [Cf. Belovite (of Borodin and Kazakova), Strontium-apatite]

Sabalite. (D. B. Sterrett, *Mineral Resources, US Geol. Survey*, for 1914, 1915, part II, p. 334; W. T. Schaller, ibid., for 1917, 1918, part II, p. 161.) Trade-name for a banded variscite, from Utah, used as a gemstone. Cf. Trainite. 8th List

Sabugalite. C. Frondel, 1951. *Amer. Min.*, vol. 36, p. 671. $HAl(UO_2)(PO_4)_2.16H_2O$, pseudo-tetragonal, isostructural with fully hydrated autunite. Crusts of minute platy yellow crystals in pegmatite. Named from the locality, Sabugal, Beira, Portugal. Syn. aluminium-autunite, aluminum-autunite (q.v.). [*M.A.*, **11**, 412.] 19th List

Sachait, German transliteration of Сахаит, Sakhaite (24th List) (C. Hintze, *Handb. Min.*, Erg. III, 625). 25th List

Sacharowait, German transliteration of Сахаровaиг, Sakharovaite (22nd List) (C. Hintze, *Handb. Min.*, Erg. III, 627). 25th List

Saffronite. Variant of safranite (13th List), *syn.* of citrine. R. Webster, *Gems*, 1962, p. 762. 28th List

Safranite. *Gemmologist*, London, 1933, vol. 2, pp. 346, 375; 1933, vol. 3, p. 61; 1934, vol. 3, p. 328. *Deutsche Goldschmiede-Zeitung*, 1933, vol. 36, p. 207. *Neues Jahrb. Min.*, Abt. A, ref. I, 1934, p. 154. Safranite or Topaz-safranite, trade-names for yellow gem quartz, sometimes sold as topaz. Apparently named from safran, French for saffron. 13th List

Sahamalite. H. W. Jaffe, R. Meyrowitz, and H. T. Evans, 1953. *Amer. Min.*, vol. 38, p. 741. $(Mg,Fe)(Ce,La,Nd)_2(CO_3)_4$, monoclinic, as minute colourless crystals from the alteration of bastnäsite. From Mountain Pass, San Bernardino Co., California. Named after Professor Thure Georg Sahama (1910–) of Helsinki, Finland. [*M.A.*, **12**, 237.] 20th List

Sahlinite. G. Aminoff, 1934. *Geol. För. Förh. Stockholm*, vol 56, p. 493 (Sahlinite). Very basic chloro-arsenate of lead, $12PbO.As_2O_5.2PbCl_2$, as yellow monoclinic scales, from Långban, Sweden. Named after Dr. Carl Sahlin, formerly manager of the iron works at Laxå, Sweden. [*M.A.*, **6**, 51; *A.M.*, **20**, 315.] 14th List

Sainfeldite. R. Pierrot, 1964. *Bull. Soc. franç. Min. Crist.*, vol. 87, p. 169. Colourless to light pink rosettes and monoclinic crystals of $Ca_5H_2(AsO_4)_4.4H_2O$ from Sainte-Marie-aux-Mines, Alsace, identical with synthetic material. Named for P. Sainfeld. [*M.A.*, **17**, 79; *A.M.*, **50**, 806.] 24th List

Sakhaite. I. V. Ostrovskaya, N. N. Pertsev, and I. B. Nikitina, 1966. Зап. Всесоюз. Мин. Общ. (*Mem. All-Union Min. Soc.*), vol. 95, p. 193 (Сахаит, Sakhaite). A cubic borate-carbonate, with chloride, of Ca and Mg, closely related to Harkerite (18th List); the composition is nearly that of Harkerite with Si completely replaced by B—near $Ca_{12}Mg_4(CO_3)_4$-$(BO_3)_7Cl(OH)_2.H_2O$. Named for the place of find, in Siberia. [*A.M.*, **51**, 1817.] 24th List

Sakharovaite. Ivan Kostov, 1959. Труды Мин. Муз. Акад. Наук СССР (*Trans. Min. Mus. Acad. Sci. USSR*), vol. 10, p. 148 (Сахароваит). The mineral named bismuth-jamesonite by M. S. Sakharova (21st List) is regarded as a distinct species and named after its describer. [*M.A.*, **13**, 164; **14**, 500.] 22nd List [$(Pb,Fe)(Bi,Sb)_2S_4$]

Sakharowit, variant German transliteration of Sakharovaite (22nd List) (H. Strunz, *Min. Tabellen*, 4th edit., 1966, p. 514). 24th List

Sakiite. N. S. Kurnakov and B. L. Ronkin, 1931. [*Priroda, Acad. Sci. USSR*, 1931, no. 7, p. 619.] O. M. Shubnikova, *Trans. Lomonosov Inst. Acad. Sci. USSR*, 1936, no. 7, p. 333 (Сакиит, Sakiite). Synonym of hexahydrite, $MgSO_4.6H_2O$, from the Saki salt lakes, Crimea. Named from the locality. [*M.A.*, **4**, 378.] 15th List

Sakuraiite. A. Kato, 1965. [*Chigaku Kenkyu* (*Earth Science Studies*), Sakurai vol., 1]; abstr. *Amer. Min.*, **53**, 1421. Metallic grey, tetragonal $(Cu,Zn,Fe)_3(In,Sn)S_4$, the indium analogue of kësterite, with stannite, etc., in a vein of the Ikuno mine, Hyogo Prefecture, Japan. Named for Dr. Kin-ichi Sakurai. [*Zap.*, **98**, 321.] 26th List

Salammonite. *Committee on Nomenclature, Amer. Min.*, 1923, vol. 8, p. 52; 1924, vol. 9, p. 61. An adaptation of sal-ammoniac. 10th List

Saléeite. A. Schoep, *Meded. Kl. Wetens. Kon. Vlaamsche Acad. Wetens. Lett.*, 1939, p. 65 (saléeiet), p. 70 (saléeite). The correct form of saleite (13th List), named after Prof. Achille Salée, of Louvain, Belgium. 15th List

Saleite. J. Thoreau and J. F. Vaes, 1932. *Bull Soc. Belge Géol.*, vol. 42, p. 96 (saléite). Hydrous phosphate of uranium and magnesium, $MgO.2UO_3.P_2O_5.8H_2O$, as yellow square plates (orthorhombic) from Katanga. Named after Prof. Achille Salée, of Louvain, Belgium. [*M.A.*, **5**, 292.] 13th List [Now **Saléeite**]

Salesite. C. Palache and O. W. Jarrell, 1939. *Amer. Min.*, vol. 24, p. 388. Copper iodate, $CuIO_3(OH)$, as bluish-green orthorhombic crystals from Chile. Named after Mr. Reno H. Sales, chief geologist of the Anaconda Copper Mining Company. [*M.A.*, **7**, 369.] 15th List [*A.M.*, **63**, 172]

Salmoite. E. S. Larsen [1921]. *Bull. US Geol. Survey*, no. 679, p. 135. An undetermined mineral (the optical characters only being stated) occurring as colourless grains with spencerite from Salmo, British Columbia. Presumably the 'new basic zinc phosphate', mentioned by A. H. Phillips, *Amer. Journ. Sci.*, 1916, ser. 4, vol. 42, p. 278. 10th List

Salmonsite. W. T. Schaller, 1912. *Journ. Washington Acad. Sci.*, vol ii, p. 144. Hydrous manganese and iron phosphate, $Fe_2O_3.9MnO.4P_2O_5.14H_2O$, forming buff-coloured cleavable masses, and resulting from the alteration of hureaulite. From the gem-tourmaline mines at Pala, San Diego Co., California. Named after Mr. Frank A. Salmons, formerly of Pala. 6th List [A mixture containing Jahnsite (q.v.); *M.M.*, **42**, 318]

Saltspar. P. M. Murzaev, 1941. *Compt. Rend.* (*Doklady*), vol. 33, p. 306 (saltspar). Coarsely crystallized and cleavable halite. [*M.A.*, **9**, 127.] 17th List

Salvadorite. W. Herz, *Zeits. Kryst. Min.*, XXVI. 16, 1896. [*M.M.*, **11**, 240.] $FeCu_2(SO_4)_3 + 21H_2O$. Monosymmetric. [Salvador mine, Quetana, near Calama] Chili. 1st List

Samarskite-wiikite. H. Strunz, 1957. *Min. Tabellen*, 3rd edit., pp. 147, 158 (Samarskit-Wiikit). *See* Pyrochlore-wiikite. 22nd List

Samiresite. A. Lacroix, 1912. *Compt. Rend. Acad. Sci. Paris*, vol. cliv, p. 1042 (samiresite); *Bull. Soc. franç. Min.*, 1912, vol. xxxv, p. 89 (samirésite). Titano-niobate of uranium, lead, etc., occurring as yellow (altered and hydrated) octahedra in pegmatite at the hill of Samiresy, near Antsirabe, Madagascar. Named after the locality. *See* Betafite. 6th List [The name should be dropped; *A.M.*, **62**, 407 (contains refs.)]

Sampleite. C. S. Hurlbut, 1942. *Amer. Min.*, vol. 27, p. 586. Hydrous phosphate and chloride of copper, calcium, and sodium, $NaCaCu_5(PO_4)_4Cl.5H_2O$, as blue crusts of minute orthorhombic crystals. Named after Mr. Mat Sample of Chuquicamata, Chile, where the mineral was found. [*M.A.*, **8**, 309.] 16th List

Samsonite. – Werner and – Fraatz, 1910. *Centralblatt Min.*, 1910, p. 331 (Samsonit). Steel-black (red by transmitted light), 'monoclinic' crystals resembling miargyrite in appearance. In composition, $2Ag_2S.MnS.Sb_2S_3$, resembles pyrargyrite with part of the silver replaced by manganese (Mn, 5·86%). Found with pyrargyrite and pyrolusite in the Samson mine, St. Andreasberg, Harz. The name samsonite has been used for a mining explosive. 5th List

Samuelsonite. P. B. Moore and T. Araki, 1975. *Geol. Soc. Amer. Abstr. Progr.*, **7**, 825; P. B. Moore, A. J. Irving, and A. R. Kampf, *A.M.*, **60**, 957 (1975). Colourless prisms up to 1 mm long from the Palermo no. 1 pegmatite, North Groton, New Hampshire, space group *C2/m*, have *a* 18·495, *b* 6·805, *c* 14·000 Å, *β* 112·75°; α 1·645, β 1·650, γ 1·655. Composition $(Ca,Ba)Fe_2^{2+}Mn_2Ca_8Al_2(PO_4)_{10}(OH)_2$, with considerable cation vacancies; occurs with whitlockite and hydroxyapatite. Named for P. B. Samuelson. [*A.M.*, **60**, 947.] 29th List

Sanbornite. A. F. Rogers, 1932. *Amer. Min.*, vol. 17, pp. 117, 161. Barium metadisilicate, $BaSi_2O_5$, as white pearly cleavages, triclinic, from California. Named after Mr. Frank Sanborn, of the Division of Mines, California. [*M.A.*, **5**, 145.] 13th List

Sandbergerite. K. Walenta, 1958. [Techn. Hochschule, Stuttgart]; abstr. *Bull. Soc. franç. Min. Crist.*, 1958, vol. 81, p. 69. The mineral later re-named heinrichite (q.v.), the name sandbergerite having been used twice before. 22nd List

Sandbergit, error for Sandbergerit (= Heinrichite). Hintze, *Handb. Min.*, Erg.-bd. II, p. 765. 23rd List

Sanderite. W. Berdesinski, 1952. *Neues Jahrb. Min.*, Monatshefte, p. 28 (Sanderit). $MgSO_4.2H_2O$, from the hydration of kieserite. Named after Bruno (Hermann Max) Sander (1884–), professor of mineralogy and petrography in the University of Innsbruck. [*M.A.*, **11**, 517.] 19th List

Sanfordite. *Ores and Metals*, Denver, Colorado, April, 1903 (Sandfordite in title). A synonym of the earlier name rickardite (q.v.). Named after Mr. Albert B. Sanford, of Denver, who first noticed the mineral in 1901. 3rd List

Sangarite. V. A. Drits and A. G. Kossovskaya, 1963. Докл. Акад. Наук СССР (*Compt. Rend. Acad. Sci. URSS*), vol. 156, p. 934 (Сангарит). An alteration product of Biotite from the Vilui depression, Sangar region, proves to be a regularly alternating layer structure. Named for the locality. 'The introduction of a new name to apply to one so carefully analyzed example of a structure which will apparently emerge to be infinitely variable is unwise' (W. F. Bradley, *Amer. Min.*, 1946, vol. 49, p. 444). 23rd List

Sanidine-anorthoclase. K. Chudoba, 1930. *Centr. Min.*, Abt. A, 1930, p. 145 (Sanidinanorthoklas), p. 150 (Sanidin-Anorthoklas, Anorthoklas-Sanidin). Felspar from the Drachenfels, Rhine, with the tabular habit of sanidine but with the optical extinction of anorthoclase. [*M.A.*, **4**, 389.] 12th List

Sanjuanite. M. E. J. de Abeledo, V. Angelelli, M. A. R. de Benyacar, and C. Gordillo, 1968. *Amer. Min.*, vol. 53, p. 1. White compact masses in slates from the Sierra Chica de Zonda, Dept. Pocito, San Juan Province, Argentina, gave X-ray powder data resembling those of kribergite (17th List), but the ratios $Al:P:S:H_2O$ are quite different. The composition approximates to $Al_2PO_4SO_4(OH).9H_2O$. Named for the province. [*M.A.*, **19**, 314.] 25th List

Sanmartinite. V. Angelelli and S. G. Gordon, 1948. *Notulae Naturae, Acad. Sci. Philadelphia*, no. 205. Zinc tungstate, near $ZnWO_4$ with small amounts of Fe, Mn, Ca, as minute monoclinic crystals very similar to wolframite; from San Martin, prov. San Luis, Argentina. Named from the locality, which in turn was named after the liberator José de San Martin (1778–1850). [*M.A.*, **10**, 353.] 18th List

Santafeite. M. S. Sun and R. H. Weber, 1957. *Bull. Geol. Soc. Amer.*, vol. 68, p. 1802 (abstr.). Hydrous vanadate, $Na_2O.3MnO_2.6(Mn,Ca,Sr)O.3(V,As)_2O_5.8H_2O$, orthorhombic, black needles, on limestone from [Grants dist., McKinley Co.] New Mexico. Named after the Santa Fe Railroad Company. [*M.A.*, **13**, 624.] 21st List [*A.M.*, **43**, 677]

Santanaite. A. Mücke, 1972. *Neues Jahrb. Min.*, Monatsh. 455. Hexagonal yellow platelets from the Santa Ana mine, Caracoles, Sierra Gorda, Chile, gave *a* 9·03, *c* 39·84 Å. Composition Pb 88·0, Cr 1·9%, from which the formula $Pb_{11}CrO_{16}$ is deduced. [In fact, allowing for the probable analytical error, the formula could be anywhere between $Pb_{11}CrO_{13}$ and $Pb_{11}CrO_{19}$—*M.H.H.*]. [*M.A.*, 73-2948; *Bull.*, **96**, 245; *A.M.*, **58**, 966; *Zap.*, **103**, 357.] 28th List [Cf. Scheibeite]

Santite. S. Milano and F. Satori, 1970. *Contr. Min. Petr.*, **27**, 159. Small grains intergrown with larderellite and sassolite, from Larderello, Italy, agree closely in X-ray and other physical data with synthetic orthorhombic $KB_5O_8.4H_2O$. Named for Giorgio Santi. [*A.M.*, **56**, 636.] 26th List

Sapperite. R. Potonié, 1924. *Kohlenpetrographie*, Berlin, 1924, p. 220 (Sapperit). H. Meixner, *Zentr. Min.*, Abt. A, 1938, p. 208; *Fortschr. Min. Krist. Petr.*, 1939, vol. 23, p. xlvi. Pure white cellulose occurring in brown coal of Miocene age from Saxony, and in fossil wood in basalt-tuff from Styria. Named after mining director – Sapper, of Klettwitz, Saxony. [*M.A.*, 7, 170.] 15th List

Sapphirite. Variant of Sapphirine; *A.M.*, **8**, 52. [Not sapphire.] 10th List

Sapromyxite. M. D. Zalessky, 1915. *See* Tomite. 10th List

Sapropelite. H. Potonié, 1906. *Abhand. Preuss. Geol. Landesanst.*, no. 49, pp. 12, 21 (Sapropelite, *pl.*). R. Potonié, *Kohlenpetrographie*, Berlin, 1924, p. 4. Coals derived from algal materials. Named from σαπρός, putrid, and πηλός, mud. 14th List

Sarabauite. I. Nakai, K. Koto, K. Nagashima, and N. Morimoto, 1977. *Chem. Lett.* (*Japan*), 275. Crystal structure data are given for a monoclinic mineral (*a* 25·33, *b* 5·655, *c* 16·88 Å, *β* 117° 51′) from the Sarabau mine, Sarawak, Malaysia. Formula $CaSb_{10}O_{10}S_6$, given without analysis or other data. [*A.M.*, **62**, 1260.] 30th List

Sarmientite. V. Angelelli and S. G. Gordon, 1941. *Notulae Naturae, Acad. Nat. Sci. Philadelphia*, no. 92; *Science*, New York, September 26, 1941, vol. 94, suppl. p. 9. Hydrous arsenate and sulphate of ferric iron, $FeAsO_4.Fe(OH)SO_4.5H_2O$, as minute, lemon-yellow, monoclinic crystals, isomorphous with destinezite; in sulphate deposits from Argentina. Named after Domingo Faustino Sarmiento (1811–1888), President of the Argentine Republic. [*M.A.*, **8**, 187.] 16th List [*A.M.*, **53**, 2077]

Sarospatakite. I. D. Sedletzky, 1940. *Compt. Rend.* (*Doklady*) *Acad. Sci. URSS*, vol. 26, p. 242 (sarospatakite). The 'Glimmer von Sarospatak' of E. Maegdefrau and U. Hofmann (*Zeits. Krist.*, 1937, vol. 98, p. 33), a micaceous clay from Sarospatak, Hungary. Named from the locality. 15th List

Sartorite-II. H. Rösch and E. Hellner, *Naturwiss.*, 1959, **46**, 72. An *artificial* product in the system $PbS–As_2S_3$ with a *b*-axis double that of sartorite. 28th List

Sary-arkite. O. F. Krol, V. I. Chernov, Yu. V. Shapovalov, and G. A. Khan, 1964. Зап. Всесоюз. Мин. (*Mem. All-Union Min. Soc.*), vol. 93, p. 147 (Сарыаркит). Silicate and phosphate of Al, rare earths, Th, and Ca; tetragonal. 23rd List

Saryarkite, erroneous transliteration of Сарыаркит, Sary-arkite (Зап. Всесоюз. Мин. Общ. (*Mem. All-Union Min. Soc.*), 1964, vol. 93, no. 2, p. ii, *Amer. Min.*, 1964, vol. 49, p. 1775). 24th List

Sasaite. J. Martini, 1978. *M.M.*, **42**, 401. Soft white chalky nodules from West Driefontein Cave, Carlstonville, Transvaal, consist of microscopic plates, probably orthorhombic with *a* 21·50, *b* 30·04, *c* 92·06 Å. Composition $10[(Al,Fe^3)_{14}(PO_4)_{11}(OH)_7SO_4.83H_2O]$, sp. gr. 1·75. RIs α 1·465, *β* 1·473, *γ* 1·477. Named for the South African Speleological Association. 30th List

Satelite. Trade-name for a serpentine cat's-eye from Tulare Co., California. Resembles chrysotile, but is harder and has a coarse splintery fracture. (D. B. Sterrett, *Mineral Resources of the United States*, for 1908, 1909, part ii, p. 839.) 5th List

Satimolite. V. M. Bocharov, I. I. Khal'turina, N. P. Avrova, and Yu. V. Shipovalov, 1969. Труды Мин. муз. Акад. наук СССР (*Trav. Mus. Min. URSS*), **19**, 121 (Сатимолит). I. V. Ostrovskaya, ibid., 202. White powdery aggregates or small orthorhombic crystals (*a* 12·62, *b* 18·64, *c* 6·97 Å); α 1·535, *β* 1·552, *γ* 1·552. Composition $4[KNa_2Al_4(B_2O_5)_3Cl_3.13H_2O]$. Named for the locality—which is not stated. [*A.M.*, **55**, 1069; *Zap.*, **100**, 86.] 27th List

Satpaevite. E. A. Ankinovich, 1959. Зап. Всесоюз. Мин. Общ. (*Mem. All-Union Min. Soc.*), vol. 88, p. 157 (Сатпаевит). Abstr. Hintze, *Handb. Min.*, Erg.-Bd. II, p. 934 (Satpajewit); *M.A.*, **14**, 280 (Satpayevite). A yellow aluminium vanadyl vanadate near $6Al_2O_3.V_2O_4.3V_2O_5.30H_2O$ in minute weakly pleochroic flakes in the argillaceous anthraxolitic vanadiferous deposits of Kurumsak and Balasanskandyk, Karatau, Kazakhstan. Named after K. I. Satpaev. [*M.A.*, **14**, 280.] The colour is unexpected for a mineral containing both V^{4+} and V^{5+} (M. Fleischer, *Amer. Min.*, 1959, vol. 44, p. 1326). 22nd List

Sauconite. W. T. Roepper, F. A. Genth's *Mineralogy of Pennsylvania*, 1875, p. 120. A zinciferous clay from Saucon Valley, Pennsylvania. (Analysis in Dana's 5th edit. 1868, p. 409.) 2nd List [*A.M.*, **31**, 411. Cf. Zinalsite, Zincsilite]

——. D. M. Roy and F. A. Mumpton, 1955. *Bull. Geol. Soc. Amer.*, vol. 66, p. 1610 (abstr.). Artificial $Zn(OH)_2$ as various polymorphs, the original sauconite (W. T. Roepper, 1875, 2nd List) being a montmorillonoid phase. 21st List [Delete this entry; the use of the name sauconite is restricted to the original montmorillonoid phase (D. M. Roy). Corr. in *M.M.*, **31**]

Saukovite. V. I. Vasilev, 1966. Доклады акад. наук СССР (*Compt. Rend. Acad. Sci. URSS*), vol. 168, p. 182 (Сауковит). An unnecessary name for cadmian metacinnabarite, found in the Altai and named for A. A. Saukov. [*M.A.*, **18**, 45; *A.M.*, **51**, 1818; Зап., **97**, 65. Saukowit, *Germ.* Not to be confused with Sauconite (2nd and 21st Lists).] 25th List

Sazhinite. E. M. Eskova, E. I. Semenov, A. P. Khomyakov, M. E. Kazakova, and N. G. Shumyatskaya, 1974. *Zap.*, **103**, 338 (Сажинит). Orthorhombic $2[Na_3CeSi_6O_{15}.6H_2O]$, *a* 7·35, *b* 7·50, *c* 15·62 Å, from Karnasurt Mt., Lovozero massif, is named for N. P. Sazhina. [*A.M.*, **60**, 162.] 28th List

Sborgite. C. Cipriani, 1957. *Atti* (*Rend.*) *Accad. Naz. Lincei, Cl. Sci. fis. mat. nat.*, ser. 8, vol. 22, p. 524. Hydrous sodium borate, $Na_2O.5B_2O_3.10H_2O$, triclinic. Deposit in steam pipes at the hot springs of Larderello, Tuscany. Named for Umberto Sborgi (1883–1955), professor of chemistry in the University of Milano. [*A.M.*, **43**, 378.] 21st List

Sc-beryl. G. Bergerhoff and W. Nowacki, 1955. *Schweiz. Min. Petr. Mitt.*, vol. 35, p. 410 (Sc-Beryll). Syn. of Bazzite (7th List). 22nd List

Sc-perrierite. E. I. Semenov, M. P. Kulakov, L. P. Kostynina, M. E. Kazakova, and A. S. Dudykina, 1966. Геохимия, **2**, 244 (Sc-Перьерит); transl. as *Geochem. Internat.*, 1966, **3**, 160 (Sc-perrierite). An unnecessary name for scandian perrierite. [*M.A.*, **19**, 53.] 28th List

Scawtite. C. E. Tilley, 1929. *Nature*, London, vol. 124, p. 896. *Min. Mag.*, 1930, vol. 22, p. 222; 1931, vol. 22, p. 457. Silicate and carbonate of calcium, $6CaO.4SiO_2.3CO_2$ (or perhaps $2CaCO_3.Ca_2Si_3O_8$), as minute monoclinic crystals in vesicles at the contact of chalk and dolerite at Scawt Hill, near Larne in Co. Antrim. Named from the locality. 12th List [$Ca_7Si_6(CO_3)O_{18}.2H_2O$; *A.M.*, **40**, 505]

Schachnerite. E. Seeliger and A. Mücke, 1972. *Neues Jahrb. Min.*, Abh. **117**, 1. $2[Ag_{1\cdot1}Hg_{0\cdot9}]$, hexagonal, *a* 2·978, *c* 4·842 Å, formed by loss of Hg from moschellandsbergite at the Vertrauen Gott mine, Obermoschel, Pfalz, Germany. Named for Dr. D. Schachner. [*M.A.*, 73-1941; *Bull.*, **96**, 242; *A.M.*, **58**, 347; *Zap.*, **102**, 437.] 28th List [Cf. Paraschachnerite]

Schadeite. M. Lazarevič, 1909. *Zeits. Chem. Indust. Kolloide*, vol. iv, p. 306 (Schadeit). The colloidal equivalent of plumbogummite; the material from Huelgoat, Brittany, being partly optically isotropic and in part showing a chalcedonic structure. Named after Dr. Heinrich Schade, of Kiel. 8th List

Schadlunit, Germ. trans. of Шадлунит, shadlunite (q.v.). Hintze, 83. 28th List

Schafarzikite. J. A. Krenner, 1921. *Zeits. Krist.*, 1921, vol. 56, p. 198 (Schafarzikit). Red, tetragonal crystals found with kermesite (and resembling this in appearance) at Pernek, Hungary, contain iron and phosphorus, and, since they are isomorphous with trippkeite, the composition suggested is a ferrous phosphite, nFeO.P_2O_3. Named after Professor Ferencz Schafarzik, of Budapest. [*M.A.*, **1**, 200.] 9th List [Near $FeSb_2O_4$; *A.M.*, **37**, 136]

Schairerite. W. F. Foshag, 1931. *Amer. Min.*, vol. 16, p. 133. Sulphate and fluoride of sodium, $Na_2SO_4.Na(F,Cl)$, as colourless rhombohedral crystals from Searles Lake, California. Named after Dr. John Frank Schairer, of the Geophysical Laboratory, Washington. [*M.A.*, **4**, 498.] 12th List [*A.M.*, **56**, 174; *M.M.*, **40**, 131 (with refs.); *see* Kogarkoite]

Schallerite. R. B. Gage, E. S. Larsen, and H. E. Vassar, 1925. *Amer. Min.*, vol. 10, p. 9. Hydrated arseno-silicate of manganese, $9MnSiO_3.Mn_3As_2O_8.7H_2O$, as light-brown masses; optically uniaxial with perfect basal cleavage. From Franklin Furnace, New Jersey. Named after Dr. Waldemar Theodore Schaller, of the United States Geological Survey, Washington, DC. [*M.A.*, **2**, 419.] 10th List

Schaniawskit, *see* Shanyavskite.

Schanjawskit. (*Zeits. Kryst. Min.*, 1915, vol. 55, p. 178; *Fortsch. Min. Krist. Petr.*, 1920, vol. 6, p. 89). Another spelling of Shanyavskite, Шанявскитъ (T. A. Nikolaevsky, 1912; 6th List). 9th List

Schapbacite, error for Schapbachite (*Econ. Geol.*, 1960, vol. 55, pp. 762, 763, 782). 22nd List

Scharizerite. J. Schadler, 1926. *Anz. Akad. Wiss. Wien, Math.-naturw. Kl.*, vol. 62 (for 1925), p. 180. A nitrogenous hydrocarbon found as black patches in the phosphatic earth in a cave in Styria. Named after Prof. Rudolf Scharizer (1859–), of Graz, Styria. [*M.A.*, **3**, 474.] 11th List

Schaumopal. O. Hauser, 1911. *Centralblatt Min.*, 1911, p. 436. A German name for porous opal. Synonym of Float-stone (Ger. Schwimmstein). 7th List

Schaurteïte*. H. Strunz and C. L. Tennyson, 1965. *Min. Tabellen*, 4th edit., 1966, p. 515 (Schaurteit). Hexagonal $Ca_3Ge(SO_4)_2(OH)_4.4H_2O$, analogous to Fleischerite, from Tsumeb, SW Africa. 24th List [*Corr. in *M.M.*, **39**] [*A.M.*, **52**, 926]

Scheibeite. A. Mücke, 1970. *Neues Jahrb. Min.*, Monatsh. 276. A deep red basic lead chromate from the Santa Ana mine, Sierra Gorda, Caracoles, Chile, gives an identical powder pattern to those of a synthetic product formulated $Pb_8(CrO_4)_3O_5$ [but which may be contaminated by lead hydroxide] and of a specimen from Berezovsk labelled phoenicochroite; the author accepts Hermann's very doubtful 1833 analysis and density as adequate evidence that this specimen was not phoenicochroite, and assigns the pre-empted name Scheibeite for Dr. R. Scheibe (the name was given, in honour of the same person, by O. von Linstow, *Neues Jahrb. Min.*, Beil.-Bd., **33**, 814 (1912), to a resin). [It seems probable that Mücke's mineral is a new occurrence of phoenicochroite. Further study is needed—*M.H.H.*] 26th List [*A.M.*, **56**, 359, 1840. Cf. Santanaite]

Schemtschuschnikovite, German transliteration of Жемчужниковит, Zhemchuzhnikovite (23rd List) (H. Strunz, *Min. Tabellen*, 4th edit., 1966, p. 444). 24th List

Scherbakovit, variant German transliteration of Щербаковит, Shcherbakovite (21st List), alternative to Schtscherbakowit (22nd List) (H. Strunz, *Min. Tabellen*, 4th edit., 1966, p. 347). 24th List

Scherbakowit, erroneous German transliteration of Щербаковит, **Shcherbakovite (20th List)** (H. Strunz, *Min. Tabellen*, 3rd edit., 1957, p. 278). 22nd List

Scherbinaite, error for shcherbinaite (*Zap.*, **102**, 445). 28th List

Schernikite. D. S. Martin, 1912. *Annals New York Acad. Sci.*, vol. xxi, p. 189. A pink, fibrous variety of muscovite occurring intergrown with lepidolite at Haddam Neck, Connecticut. Named after Mr. Ernest Schernikow, of New York, who presented to the Oxford Museum the specimens described by H. L. Bowman, *Mineralogical Magazine*, 1902, vol. xiii, p. 98. A special name is here clearly superfluous. 8th List

Schertalite. R. W. E. MacIvor, *Chem. News*, 1902, vol. lxxxv, p. 217. To replace the name muellerite for a phosphate in a bat guano from Skipton Caves, Ballarat, Victoria. Small, indistinct crystals with the composition $Mg(NH_4)_2H_2(PO_4)_2.4H_2O$. Named after Professor Arnulf Schertal, late of Freiberg, Saxony. 3rd List [Now **Schertelite**]

Schertelite. *Zeits. Kryst. Min.*, 1906, vol. xlii, p. 386 (Schertelit). The correct spelling of schertalite (R. W. E. MacIvor, 1902; 3rd List, *Min. Mag.*, vol. xiii, p. 376). Named after Arnulf Schertel (1841–1902), formerly professor in the Bergakademie of Freiberg, Saxony. 4th List

Scheteligite. H. Bjørlykke, 1937. *Norsk Geol. Tidskrift*, vol. 17, p. 47 (scheteligite). $(Ca,Fe,Mn,Sb,Bi,Yt)_2(Ti,Ta,Nb,W)_2(O,OH)_7$, as black orthorhombic (?) crystals in pegmatite from Iveland, Norway. Named after Professor Jakob Grubbe Cock Schetelig (1875–1935) of Oslo. [*M.A.*, **6**, 487.] 14th List [Perhaps in betafite group; *A.M.*, **62**, 407]

Schilkinit. *See* Chilkinite. 20th List

Schizolite. C. Winther, *Medd. om Grönland*, 1900, XXIV, 196. A variety of pectolite, containing much manganese, occurring as pink prismatic crystals. Monoclinic. [Siorarssuit North] S Greenland. Cf. manganopectolite. 2nd List [*A.M.*, **40**, 1022]

Schmeiderite. J. Olsacher, 1962. Appendix to 2nd edit., *Chem. Index Min.*, p. 84. A selenate of Pb and Cu, probably $(Pb,Cu)_2SeO_4OH$ and isostructural with Linarite, from Condor mine, La Rioja, Argentina. 23rd List [*A.M.*, **49**, 1498]

Schmitterite. R. V. Gaines, 1971. *Amer. Min.*, **56**, 411. Pale straw-yellow rosettes (1 mm) of orthorhombic plates from the 'Moctezuma mine', Sonora, Mexico, have *a* 7·860, *b* 10·089, *c* 5·363 Å, *Pmab*, D 6·878; composition $4[UO_2TeO_3]$. Optically biaxial, negative, $n > 2{\cdot}0$. Named for E. Schmitter Villada. [*M.A.*, 71-3111.] 27th List [Zaire; *M.A.*, 77-3361]

Schmöllnitzit. R. Koechlin, *Min. Taschenb. Wien. Min. Gesell.*, 2nd edit., 1928. Another form of szomolnokite (5th List), named from the locality in Slovakia: Szomolnok in Hungarian, Schmöllnitz in German, and Smolník in Slovak. Evidently identical with ferropallidite, $FeSO_4.H_2O$ (3rd List). 18th List

Schneiderhöhnite. J. Ottemann, B. Nuber, and B. H. Geier, 1973. *Neues Jahrb. Min.*, Monatsh. 517. Brown crystal aggregates from Tsumeb, SW Africa, are anorthic, *a* 8·940, *b* 9·998, *c* 9·145 Å, α 63·00°, β 116·20°, γ 81·79°. Composition $8FeO.5As_2O_3$. Named for H. Schneiderhöhn. [*M.A.*, 74-3468; *A.M.*, **59**, 1139.] 28th List

Schoderite. D. M. Hausen, 1960. *Bull. Geol. Soc. Amer.*, vol. 71, p. 1883. Orange microcrystalline coatings on sandstone from Eureka, Nevada, have the composition $2Al_2O_3.V_2O_5.P_2O_5.12H_2O$. Named after William P. Schoder. [*M.A.*, **15**, 44.] 22nd List [*A.M.*, **47**, 637. Cf. Metaschoderite, Rusakovite]

Schoenfliesite. G. T. Faust and W. T. Schaller, 1971. *Zeits. Krist.*, **134**, 116. A dark reddish-brown mass of 'altered hulsite', with a little fluorite, from Brooks Mt., Seward Peninsula, Alaska, contains minute particles of $MgSn(OH)_6$ as a major constituent (19·5%), with goethite, maghemite, hematite, and hulsite. Cubic, *Pn*3, *a* 7·759 Å, Z = 4. D 3·36 (adjusted from mixture); *n* 1·590. Properties partly derived from synthetic material. Related to wickmanite (25th List). Named for A. M. Schoenflies. 27th List

Schoepite. T. L. Walker, 1923. *Amer. Min.*, vol. 8, pp. 55, 67. A. Schoep, *Bull. Soc. Franç. Min.*, [1924], vol. 46 (for 1923), p. 9; ibid., 1924, vol. 47, p. 147. At first suggested to be a carbonate of uranium, but later stated by A. Schoep to have the same composition, $UO_3.2H_2O$, as becquerelite (A. Schoep, 1922; 9th List). Both minerals are found in Katanga as minute, yellow, orthorhombic crystals; they are distinguished by their optical characters. Named after Prof. Alfred Schoep, of Ghent. [*M.A.*, **2**, 147, 249, 383–384.] 10th List [*A.M.*, **45**, 1026. Syn. Epi-ianthinite (q.v.). Cf. Ianthinite]

Schoepite–I, Schoepite–II, Schoepite–III. C. L. Christ and J. R. Clark, 1960. *Amer. Min.*, vol. 45, pp. 1027, 1028. Synonyms respectively of Schoepite (10th List), Metaschoepite (22nd List), and Paraschoepite (18th List). 22nd List

Schokoladenstein. *See* Lacroisite. 3rd List

Scholzite. H. Strunz, 1949. *Min. Tab.*, 2nd edit., p. 164 (Scholzit); *Fortschr. Min.*, 1950, vol. 27 for 1948, p. 31. $Ca_3Zn(OH)_2(PO_4)_2.H_2O$, as colourless monoclinic crystals with blende and triplite in pegmatite from Hagendorf, Bavaria. Named after Dr. Adolf Scholz of Regensburg. [*M.A.*, **11**, 189.] 19th List [$CaZn_2(PO_4)_2.2H_2O$; *A.M.*, **46**, 1519]

Schoonerite. P. B. Moore and A. R. Kampf, 1977. *A.M.*, **62**, 246. Rare thin pale-brown laths in the Palermo no. 1 pegmatite, North Groton, New Hampshire, are orthorhombic, *a* 11·119, *b* 25·546, *c* 6·437 Å, space group *Pmab*. Composition $4[ZnMn_2Fe^{2}_{2}Fe^{3}(OH)_2(PO_4)_3.2H_2O]$. RIs α 1·618, pale yellow, ∥ [010], *β* 1·652, pale brown, ∥ [001], *γ* 1·682, brown, ∥ [100], $2V_\alpha$ 70–80°. [*M.A.*, 77-4627.] Named for R. Schooner. 30th List

Schorsuit, German transliteration of Шорсуит, shorsuite (21st List) (Hintze, *Handb. Min.*, Erg.-Bd. II, p. 614). 22nd List

Schreyerite. O. Medenbach and K. Schmetzer, 1976. *Naturwiss.*, **63**, 293. Twinned grains up to 30 μm in vanadian rutile from the Kwale district, Kenya, are monoclinic, *a* 7·06, *b* 5·01, *c* 18·74 Å, *β* 119·4°. Composition $V_2Ti_3O_9$. Named for W. Schreyer. [*A.M.*, **62**, 395.] 30th List

Schroetterite. Variant of Schrötterite; *A.M.*, **9**, 62. 10th List

Schtscherbakowit, German transliteration of Щербаковит, shcherbakovite (21st List) (Hintze, *Handb. Min.*, Erg.-Bd. II, p. 833). 22nd List

Schtscherbinait, Germ. trans. of Щербинаит, shcherbinaite (q.v.). Hintze, 84. 28th List

Schubnelite. F. Cesbron, 1970. *Bull.*, **93**, 470. Black crystals (0·5 mm) with fervanite in the oxidation zone of the Mounana uranium deposit, Gabon, are anorthic, *a* 6·59, *b* 5·43, *c* 6·62, α 125°, *β* 104°, *γ* 84° 43′, $P\bar{1}$. Composition $[Fe_2V_2O_8.2H_2O]$. Named for H.-J. Schubnel. [M.A., 71-1389; *Zap.*, **100**, 622; *A.M.*, **57**, 1556.] 27th List

Schubnikowit. *Geologie Zeits.*, East Berlin, 1955, vol. 4, p. 528. German transliteration of щубниковит, shubnikovite (20th List). 21st List

Schuetteïte. E. H. Bailey, F. A. Hildebrand, C. L. Christ, and J. J. Fahey, 1959. *Amer. Min.*, vol. 44, p. 1026 (Schuetteite). Basic mercuric sulphate, $HgSO_4.2HgO$, yellow, hexagonal, from Nevada, California, Oregon, and Idaho. Named after Curt Nicolaus Schuette. [*M.A.*, **14**, 501.] 22nd List

Schuilingite. J. F. Vaes, 1947. *Ann.* (*Bull.*) *Soc. Géol. Belgique*, vol. 70, p. B 233. Carbonate of lead and copper as blue orthorhombic needles from Katanga. Named after H. J. Schuiling, geologist to the Union Minière du Haut-Katanga. [*M.A.*, **10**, 147; *A.M.*, **33**, 385.] 18th List [$Pb_3Cu_2Ca_6(CO_3)_8(OH)_6.6H_2O$; *A.M.*, **43**, 796]

Schultenite. L. J. Spencer, 1926. *Nature*, London, vol. 118, pp. 412, 754: *Min. Mag.*, vol. 21, p. 149. Lead hydrogen arsenate, $PbHAsO_4$, as colourless monoclinic* plates from Tsumeb,

SW Africa. Named after Baron August Benjamin de Schulten (1856–1912) of Helsingfors and Paris, who prepared and described artificial crystals of this compound. [*M.A.*, **3**, 232.] [*Corr in *M.M.*, **23**.] 11th List [*Min. Record*, **8**(3), 98]

Schulzenite. P. Martens, *Actes Soc. Sci. Chili*, V. 87, 1895. $CuO.2CoO.Co_2O_3.4H_2O$. Resembles wad. Chili? 1st List [Var. of Heterogenite; *M.M.*, **33**, 253]

Schuppenglanz. *See* Lepidolamprite. 4th List

Scleroclasite. E. T. Wherry and W. F. Foshag, 1921. *Journ. Washington Acad. Sci.*, vol. 11, p. 3. Variant of scleroclase (Skleroklas, W. Sartorius von Waltershausen, 1855); synonym of sartorite. 9th List

Sclerospathite. W. F. Petterd, *Rep. Secr. Mines, Tasmania*, for 1901–1902, 1902, p. 297; *Papers and Proc. Roy. Soc. Tasmania*, for 1902, 1903, p. 27. A hydrated sulphate of iron and chromium, perhaps allied to knoxvillite, occurring as tough (*σκληρός*), felted masses of short fibres, at Salisbury, Tasmania. 3rd List

Scorzalite. W. T. Pecora and J. J. Fahey, 1947. *Program and Abstracts, 28th Annual Meeting, Min. Soc. Amer.*, p. 18; *Bull. Geol. Soc. Amer.*, vol. 58, p. 1216; *Amer. Min.*, 1948, vol. 33, p. 205 (abstr.); *Mineração e Metalurgia*, 1948, vol. 13, p. 53; *Amer. Min.*, 1949, vol. 34, p. 83. Hydrous phosphate, $Al_2O_3.(Fe,Mg)O.P_2O_5.H_2O$, blue, monoclinic, differing from lazulite in having FeO in excess of MgO. From Divino, Brazil. Also from Custer, South Dakota (W. T. Pecora and J. J. Fahey, *Amer. Min.*, 1949, vol. 34, p. 282). Named after Dr. Evaristo Pena Scorza, of the Mineral Survey of Brazil. [*M.A.*, **10**, 254, 456, 507.] 18th List [*A.M.*, **34**, 685]

Scotite, error for Scawtite (12th List) (Crystallography, a translation of Кристаллография, 1961, vol. 5, p. 659). 22nd List

Scyelite. M. F. Heddle, 1883, in J. Young, *Trans. Geol. Soc. Glasgow*, vol. 7, p. 419. A silicate 'similar to bastite' from Loch Scye, Caithness, later described as a rock (*Quart. Journ. Geol. Soc.*, 1885, vol. 41, p. 401). 23rd List

Seamanite. E. H. Kraus, W. A. Seaman, and C. B. Slawson, 1930. *Amer. Min.*, vol. 15, p. 220. Hydrous phospho-borate of manganese, $3MnO.(B_2O_3,P_2O_5).3H_2O$, as pale-yellow orthorhombic needles from Michigan. Named after Prof. Arthur E. Seaman, of the Michigan College of Mining and Technology. [*M.A.*, **4**, 342.] 12th List [$Mn_3(PO_4)B(OH)_6$; *A.M.*, **56**, 1527]

Searlesite. E. S. Larsen and W. B. Hicks, 1914. *Journ. Washington Acad. Sci.*, vol. iv, p. 397; *Amer. Journ. Sci.*, ser. 4, vol. xxxviii, p. 437. A hydrated borosilicate of sodium, $NaB(SiO_3)_2.H_2O$, occurring as small white spherulites (probably monoclinic) in clay at Searles Lake, San Bernardino Co., California. Named after Mr. John W. Searles, pioneer at this locality. 7th List

Sebkhainite. L. Berthon, 1922. *L'industrie minérale en Tunisie*, p. 176 (Sebkhaïnite). Mixed salts containing carnallite (60%), epsomite, and halite from salt-pans (sebkhas) in Tunisia. Cf. mellahite (11th List). 17th List

Sederholmite. Y. Vuorelainen, A. Huhma, and A. Häkli, 1964. *Compt. Rend. Soc. Géol. Finlande*, vol. 36, p. 113. Hexagonal β-NiSe, often nickel-deficient, with Penroseite and Wilkmanite (q.v.) and as grains in Clausthalite, from Kuusamo, NE Finland. Named for J. J. Sederholm. [*M.A.*, **17**, 303; *A.M.*, **50**, 519.] 24th List

Sedovite. K. V. Skvortsova and G. A. Siderenko, 1965. Зап. Всесоюз. Мин. Общ. (*Mem. All-Union Min. Soc.*), vol. 94, p. 548 (Седовит, Sedovite). Brown powdery deposits or radiating fibrous on altered Pitchblende and Femolite (23rd List) at an unnamed uranium-molybdenum deposit consist of uranous molybdate, $U(MoO_4)_2$; probably orthorhombic. [*A.M.*, **51**, 530; *M.A.*, **17**, 607.] 24th List

Seelandite. A. Brunlechner, *Jahrb. naturh. Landes-Museums, Klagenfurt*, XXII. 192, 1893; '*Carinthia*,' *Klagenfurt*, no. 2, 1891. $MgO.Al_2O_3.4SO_3.27H_2O$. Near pickeringite. [Lölling] Carinthia. 1st List

Seeligerite. A. Mücke, 1971. *Neues Jahrb. Min.*, Monatsh. 210. Thin bright yellow plates with schwartzembergite at Santa Ana mine, Caracoles, Sierra Gorda, Chile, are identical with synthetic $8[Pb_3IO_3Cl_3O]$. $a = b = 7{\cdot}964$, c 27·88 Å, $C222_1$, D 6·83; α 2·12, β and γ 2·32. Named for E. Seeliger. [*A.M.*, **57**, 327; *M.A.*, 71-3112.] 27th List

Sefströmite. D. Mawson. Although unpublished by the author, the name appeared in 1907 on the labels and lists of dealers, being incorrectly spelt seffstromite. Supposed to be a vanadiferous variety of ilmenite. T. Crook (*Mineralogical Magazine*, 1910, vol. xv, p. 281) proves 'sefströmite' and 'davidite' (D. Mawson, 1906; 4th List) to be merely mixtures. Named after Nils Gabriel Sefström (1787–1845), the discoverer of vanadium. 5th List

Segelerite. P. B. Moore, 1974. *A.M.*, **59**, 48. Pale-green prismatic orthorhombic crystals with a wide range of phosphates in the Tip Top pegmatite, Custer, Custer County, South Dakota, have *a* 14·826, *b* 18·751, *c* 7·307 Å, and composition $8[CaMgFe^{3+}(PO_4)_2OH.4H_2O]$. Named for Curt G. Segeler. 28th List

Seidoserit, German transliteration of Сейдозерит, Seidozerite (22nd List) (C. Hintze, *Handb. Min.*, Erg.-Bd. II, p. 834). 24th List

Seidozerite. E. I. Semenov, M. E. Kazakova, and V. I. Simonov, 1958. Зап. Всесоюз. Мин. Общ. (*Mem. All-Union Min. Soc.*), vol. 87, p. 590 (Сейдозерит). V. I. Simonov and N. V. Belov, Доклады Акад. Наук СССР (*Compt. Rend. Acad. Sci. URSS*), 1958, vol. 122, p. 473. Abstr. *Bull. Soc. franç. Min. Crist.*, 1959, vol. 82, p. 93 (Seidoserite). Fan-like clusters of brownish-red needles embedded in microcline in an alkali pegmatite from near Lake Seidozero, Lovozero tundra, Kola peninsula, near $Na_8Zr_3Ti_3Mn_2Si_8O_{32}F_4$. Named from the locality. [*M.A.*, **14**, 198.] A member of the wöhlerite family, and in need of full examination (M. Fleischer, *Amer. Min.*, 1959, vol. 44, p. 468). 22nd List [Cf. Lomonosovite; *M.A.*, 72-901]

Seinäjokite. N. N. Mozgova, Yu. S. Borodaev, N. A. Ozerova, V. Paakonen, O. L. Sveshnikova, V. S. Balitskii, and B. A. Dorogovin, 1976. *Zap.*, **105**, 617 (Сейняйокит). Inclusions in native antimony from near Seinäjoki, Vaasa, Finland, are orthorhombic, *a* 3·19, *b* 5·81, *c* 6·49 Å. Composition $2[(Fe,Ni)(Sb,As)_2]$. Named for the locality. [*A.M.*, **62**, 1059; *M.A.*, 77-3392.] 30th List

Sekaninaite. J. Stanek and J. Miskovsky, 1968. Quoted by H. Strunz, *Min. Tab.*, 5th edit., 1970, 406. The iron analogue of cordierite, from Dolní Bory, Moravia (cf. *Časopis Min. Geol.*, 1964, **9**, 191). A name proposed to replace Eisencordierit (3rd List) or Iron-cordierite (10th List). [*Zap.*, **102**, 453.] 28th List

Selenide-spinel. F. Machatschki, 1958. *Tschermaks Min. Petr. Mitt.*, ser. 3, vol. 6, p. 402 (Selenidspinell). A name proposed for the unnamed cobalt–nickel–copper selenide of S. C. Robinson and E. J. Brooker (1952). *See* Tyrellite. 22nd List

Selenio-melonite. Y. Vuorelainen and A. Häkli, 1964. *Geologi* (*Helsinki*), vol. 5, p. 53. A highly selenian Melonite from Kuusamo, NE Finland. [*M.A.*, **16**, 647.] An unnecessary name. 24th List

Selenio-polydymite. Y. Vuorelainen and A. Häkli, 1964. *Geologi* (*Helsinki*), vol. 5, p. 53. Selenian polydymite from Kuusamo, NE Finland. [*M.A.*, **16**, 547.] An unnecessary name. 24th List

Selenio-siegenite. J. F. Vaes, 1947. *Ann.* (*Bull.*) *Soc. Géol. Belgique*, vol. 70, p. B 231 (Sélénio-Siegenite). A variety of siegenite containing Se 11·65, Te 3·80%, $(Ni,Co,Cu)_3(S,Se,Te)_4$, from Katanga. [*M.A.*, **10**, 147.] 18th List

Selenio-vaesite. J. F. Vaes, 1947. *Ann.* (*Bull.*) *Soc. Géol. Belgique*, vol. 70, p. B 229 (Selenio-Vaesite), p. B 230 (Sélénio-Vaesite). A variety of vaesite (17th List) containing Se13·70–19·70%, $Ni(S,Se)_2$, from Katanga. [*M.A.*, **10**, 146.] 18th List

Selenjoseïte. L. G. Berry and R. M. Thompson, 1962. *Mem. Geol. Soc. Amer.*, no. 85. Material from Falun, described under this name, was later found by L. G. Berry (*Canad. Min.*, 1963, vol. 7, p. 677) to be identical with Laitakarite (22nd List). 24th List

Selenobismutite. V. I. Vernadsky, 1918. *Opuit Opisatelnoĭ Mineralogii*, Petrograd, vol. 2, p. 34 (Селенобисмутитъ). Synonym of guanajuatite; a translation of the German Selenwismuthglanz. The same name has also been used by E. T. Wherry (*Journ. Washington Acad. Sci.*, 1920, vol. 10, p. 490) in a restricted sense for the orthorhombic Bi_2Se_3 as distinct from guanajuatite, $Bi_2(Se,S)_3$. 9th List

Selenocosalite. O. H. Ödman, 1941. *Årsbok Sveriges Geol. Undersökning*, vol. 35, no. 1, p. 87. A variety of cosalite containing Se 6·43%, $Pb_2Bi_2(S,Se)_5$, from Boliden, Sweden. [*M.A.*, **8**, 311.] 16th List

Selenojarosite. A. E. Fersman and O. M. Shubnikova, *Sputnik geokhimika i mineraloga, Acad. Sci. USSR*, 1937, p. 197 (селен(о)ярозит). B. K. Breshenkov, *Compt. Rend.* (*Doklady*)

Acad. Sci. URSS, 1946, vol. 52, p. 329 (selenojarosite). Jarosite with some SO_3 replaced by SeO_3 (0·20%) from Altai and Kazakhstan. [*M.A.*, **2**, 113; **7**, 6; **10**, 248.] 18th List

Selenokobellite. O. H. Ödman, 1941. *Årsbok Sveriges Geol. Undersökning*, vol. 35, no. 1, p. 89. A variety of kobellite containing Se 4·78–5·74%, $Pb_2(Bi,Sb)_2(S,Se)_5$, from Boliden, Sweden. [*M.A.*, **8**, 311.] 16th List

Selenolinnaeite. V. Cuvelier, 1929. *Natuurwetenschappelijk Tijdschrift*, Antwerpen, vol. 11, p. 176 (Selenolinneïet). A seleniferous variety of linnaeite from Katanga. [*M.A.*, **3**, 362; **4**, 248.] 12th List

Selensulfur. Variant of Selensulphur; *A.M.*, **9**, 61. 10th List

Selenwismuthglanz, *see* Selenobismutite.

Seligmannite. H. Baumhauer, *Sitz.-ber. Akad. Wiss. Berlin*, 1901, p. 110 (abstr., *M.M.*, **13**, 205); ibid., 1902, p. 611. R. H. Solly, *Min. Mag.*, 1903, vol. xiii, p. 336. Crystals resembling in general appearance the several sulpharsenites of lead from the Binnenthal dolomite. Orthorhombic, with angles, habit, and twinning similar to those of bournonite, and probably therefore with the composition $Cu_2S.2PbS.As_2S_3$. Named after Herr Gustav Seligmann, of Coblenz. 3rd List

Semenovite. O. V. Petersen and J. G. Ronsbo, 1972. *Lithos*, **5**, 163. Twinned tetragonal crystals, *a* 13·866, *c* 9·892 Å, in albitites of the Ilimaussaq alkalic intrusive, S Greenland, have a composition $2[(Ca,Ln,Yt,Mn,Na)_{12}(Si,Be)_{20}O_{40}(O,OH,F)_8.H_2O]$. Named for E. I. Semenov. [*A.M.*, **58**, 1114; *Zap.*, **102**, 454.] 28th List

Seminephrite. F. J. Turner, 1935. *Trans. Roy. Soc. New Zealand*, vol. 65, p. 190 (Seminephrites). The New Zealand 'greenstones' include besides interfelted fibrous nephrite also more coarsely crystalline tremolite as acicular prisms and sheaves of parallel fibres. The term is introduced rather as a rock-name for forms intermediate between nephrite and tremolite-schist. [*M.A.*, **6**, 501.] 14th List

Semi-whitneyite. G. A. Koenig, *Amer. Journ. Sci.*, 1902, ser. 4, vol. xiv, p. 416. Massive copper arsenides in which the ratio of Cu:As is variable and very high (up to 30:1). *See* Mohawkite. 3rd List

Senaite. E. Hussak and G. T. Prior, *Min. Mag.*, 1898, XII, 30. Black crystals with the same symmetry as ilmenite (rhombohedral with parallel-faced hemihedrism), to which the mineral is related. $(Fe,Pb)O.2(Ti,Mn)O_2$. [Diamantina, Minas Gerais] Brazil. 2nd List [$Pb(Ti,Fe,Mn)_{24}O_{38}$; *A.M.*, **53**, 869]

Senegalite. Z. Johan, 1976. *Lithos*, **9**, 165. Colourless crystals in the oxidation zone of the Kouroudiako iron deposit, Senegal, have *a* 9·678, *b* 7·597, *c* 7·668 Å, space group *Pna*2; composition $4[Al_2PO_4(OH)_3.H_2O]$. Named for the locality. 29th List

Sengierite. J. F. Vaes and P. F. Kerr, 1949. *Amer. Min.*, vol. 34, p. 109. (The name first appeared in newspapers in October 1948.) $2CuO.2UO_3V_2O_5.10H_2O$, as small green orthorhombic crystals from Katanga, Belgian Congo. Related to carnotite and tyuyamunite. Named after Edgard Sengier, Director of the Union Minière du Haut-Katanga. [*M.A.*, **10**, 507.] 18th List [Monoclinic $Cu(UO_2)_2(VO_4)_2.6\text{–}8H_2O$; *A.M.*, **39**, 323]

Septeamesite. W. R. Phillips, 1954. Syn. of Amesite; *see* Septechlorite. 22nd List

Septeantigorite. W. R. Phillips, 1954. [Ph.D. Thesis, Univ. Utah]; quoted by W. M. Tuddenham and R. J. P. Lyon, *Anal. Chem.*, 1959, vol. 31, p. 377. A member of the amesite–cronstedtite–berthierine family having a composition near that of antigorite. *See* Septechlorite. 22nd List

Septechamosite. B. W. Nelson and R. Roy, 1958. *Amer. Min.*, vol. 43, p. 721. Syn. of Berthierine. *See* Septechlorite. 22nd List

Septechlorite. B. W. Nelson and R. Roy, 1954. [*Clays and clay minerals. Publ. Nat. Acad. Sci. and Nat. Res. Council*, Washington, no. 327, p. 335]; abstr. *Amer. Min.*, 1958, vol. 43, p. 707. A group name for the amesite–cronstedtite–berthierine family, dimorphous with the chlorites; the name refers to the 7 Å *c*-spacing characteristic of this family. 22nd List

Septekämmererite. W. R. Phillips, 1954. [Ph.D. Thesis, Univ. Utah]; quoted by W. M. Tuddenham and R. J. P. Lyon, *Anal. Chem.*, 1959, vol. 31, p. 377. A member of the amesite–cronstedtite–berthierine (septechlorite) family having the same composition as kämmererite. 22nd List

Serandite. A. Lacroix, 1931. *Compt. Rend. Acad. Sci. Paris*, vol. 192, pp. 189, 193, 1324 (sérandite). Acid metasilicate of manganese, calcium, and sodium, $(Mn,Ca)_{7.5}Na_3H_2(SiO_3)_{10}$, as peach-blossom-red monoclinic crystals in sodalite-syenite from the Los Islands, W Africa. Named after Mr. J. M. Sérand, a local collector. [*M.A.*, **4**, 497.] 12th List [$Na(Mn,Ca)_2Si_3O_8(OH)$, series with Pectolite; *A.M.*, **40**, 1022]

Serendibite. G. T. Prior and A. K. Coomáraswámy, *Nature*, 1902, vol. lxv, p. 383; *Quart. Journ. Geol. Soc.*, 1902, vol. lviii, p. 420; *Min. Mag.*, 1903, vol. xiii, p. 224. Blue, embedded crystals in the contact-zones between crystalline limestone and granulite. Probably anorthic; with polysynthetic twinning and strong pleochroism.

$$10(Fe,Ca,Mg)O.5Al_2O_3.6SiO_2.B_2O_3.$$

Named from 'Serendib', an old Arab name for Ceylon, where the mineral was found. 3rd List [Aenigmatite group, *M.A.*, 75-2420]

Serpentin-Asbest. C. F. Naumann, *Elem. d. Min.*, 2nd edit., 1850, p. 284. Synonym of chrysolite. 3rd List

Serpentine-jade. R. Brauns, 1929. *Deutsche Goldschmiede-Zeitung*, vol. 32, nos. 3, 13; *Handwörterbuch der Naturwissenschaften*, 2nd edit., 1933, vol. 8, p. 1094 (Serpentin-Jade). A serpentine from China resembling jade, used as an ornamental stone [cf. bowenite]. Other jade-like minerals are distinguished as Garnet-jade (Granat-Jade, Transvaal-Jade) and Vesuvian-Jade (= Californite, 4th List). 14th List

Serpentine-talc. H. Füchtbauer and H. Goldschmidt, 1956. *Heidelberg. Beitr. Min. Petr.*, vol. 5, p. 195 ('Serpentintalk'). Between serpentine and talc in composition, $Mg_6Si_6O_{15}(OH)_6$, and physical characters. From silicification of dolomite in anhydrite bed of the Werra salt-deposits, Thuringia. 21st List

Serpentite. (C. U. Shepard, *Contributions to Mineralogy*, Amherst College, 1877.) Variant of Serpentine. 7th List

Serpochlorite or Chloritoserpentine. [A. A. Menyailov, 1935. *Mat. Petr. Geol. Kuznetz Alatau, Acad. Sci. USSR, Ser. Siberia*, no. 19, p. 64.] O. M. Shubnikova, *Trans. Inst. Geol. Sci. USSR*, 1938, no. 11 (*Geochem. Ser.*, no. 3), p. 14 (Серпохлорит, Serpochlorite, Хлоритосерпентин, Chlorite-serpentine), p. 35 (Chloritoserpentine). Undefined variety of blue-green chlorite occurring in serpentine rocks. 17th List

Serpophite. W. N. Lodochnikov, 1933. *Problems of Soviet Geology*, vol. 2, p. 120 (Серпофит), p. 145 (Serpophite, serpentinophite). Suggested for the compact varieties of serpentine. A combination of the synonyms serpentine and ophite. 13th List

Severginite. G. P. Barsanov, 1951. *Trudy Min. Mus. Acad. Sci. USSR*, no. 3, p. 10 (севергинит). A variety of axinite containing MnO 14·79%. Synonym of manganaxinite (5th List) and tinzenite (10th List, *M.A.*, **12**, 340). Named after Vasilii Mikhailovich Severgin, Василий Михайлович Севергин (1765–1826), Russian mineralogist. [*M.A.*, **13**, 353.] 21st List [Cf. Manganseverginite]

Sewerginit, German transliteration of Севергинит, severginite (21st List) (Hintze, *Handb. Min.*, Erg.-Bd. II, p. 838). 22nd List

Seyrigite. A. Lacroix, 1940. *Compt. Rend. Acad. Sci. Paris*, vol. 210, p. 276 (seyrigite). Tungstate and molybdate of calcium (MoO_3 24%) intermediate between scheelite and powellite. Named after Mr. Seyrig, manager of the phlogopite mine in Madagascar where the mineral was found. [*M.A.*, **7**, 469.] 15th List [Cf. Molybdoscheelite]

Shadlunite. T. L. Evstigneeva, A. D. Genkin, N. V. Trobeva, A. A. Filimonova, and A. I. Tsepin, 1973. *Zap.*, **102**, 63 (Шадлунит). Veinlets in cubanite in the Talnakh and Oktyabr copper deposits, Norilsk, Siberia, are cubic with *a* 10·91 Å and composition $(Fe,Cu)_8(Pb,Cd)S_8$, with Pb:Cd ≈ 2. Named for T. Shadlun. [*M.A.*, 73-4082; *Bull.*, **96**, 242; *A.M.*, **58**, 1113; *Zap.*, **103**, 355.] 28th List

Shandite. P. Ramdohr, 1950. *Die Erzmineralien und ihre Verwachsungen*, Berlin, p. 820 (Shandit, in locality index only); *Sitz.-ber. Akad. Wiss. Berlin*, 1950, no. 6 (for 1949); *Fortschr. Min.*, 1950, vol. 28 (for 1949), p. 70. M. A. Peacock and J. McArthur, *Amer. Min.*, 1950, vol. 35, p. 425. Rhombohedral $Ni_3Pb_2S_2$ with X-ray pattern distinct from parkerite and from the artificial β-phase in the system $Ni_3Bi_2S_2$–$Ni_3Pb_2S_2$. From Trial Harbour, Tasmania. Named after Professor Samuel James Shand (1882–), Scottish petrologist. [*M.A.*, **11**, 186, 187, 312, 466.] 19th List

Shannonite. C. E. Tilley, 1927. *Geol. Mag.*, vol. 64, p. 144; also independently a few months later by A. N. Winchell, *Elements of optical mineralogy*, part 2, 1927, p. 165. Supposed to be calcium orthosilicate, β-Ca_2SiO_4; but afterwards (C. E. Tilley, *Geol. Mag.*, 1928, vol. 65, p. 29) proved to be monticellite, and the name was withdrawn. Named from the locality, Shannon Tier, Tasmania. The same mineral has also been referred to as Kalkorthosilikat and Lime-olivine (q.v.). [*M.A.*, **3**, 273; **3**, 474.] 11th List

Shanyavskite. T. A. Nikolaevskij, 1912. *Bull. Acad. Sci. St.-Pétersbourg*, ser. 6, vol. vi, pp. 717, 724 (Шанявскитъ). Abstr. in *Chem. Zentralblatt*, 1912, vol. ii, p. 630 (Schaniawskit). A transparent, amorphous, glassy material consisting of almost pure hydrated alumina ($Al_2O_3.4H_2O$), occurring with allophanoid minerals in crevices in dolomite near Moscow. Named after A. L. Shanyavskij, of the University of Moscow. 6th List

Sharpite. J. Mélon, 1938. *Bull. Séan. Inst. Roy. Colon. Belge*, vol. 9, p. 333. Hydrated carbonate of uranyl, $6UO_3.5CO_2.8H_2O$, as yellowish-green fibrous crusts, perhaps orthorhombic, from Shinkolobwe, Katanga. Named after Major R. R. Sharp, who discovered the uranium deposit at Shinkolobwe. [*M.A.*, **7**, 225.] 15th List

Shattuckite. W. T. Schaller, 1915. *Journ. Washington Acad. Sci.*, vol. v, p. 7; 3rd Appendix to 6th edit. of Dana's *System of Mineralogy*, p. 72. A massive, fibrous, blue hydrated silicate of copper, $2CuSiO_3.H_2O$. From the Shattuck Arizona Copper Company's mine at Bisbee, Arizona. *See* Bisbeeite. 7th List [$Cu_5(SiO_3)_4(OH)_2$, *A.M.*, **52**, 782; **62**, 491]

Shcherbakovite. E. M. Eskova and M. E. Kazakova, 1954. *Doklady Acad. Sci. USSR*, vol. 99, p. 837 (шербаковит). Silicate and titanoniobate $Na(K,Ba)_2(Ti,Nb)_2(Si_2O_7)_2$, monoclinic. Named after D. I. Shcherbakov, Д. И. Щербаков. [*M.A.*, **12**, 569; *A.M.*, **40**, 788.] 20th List

Shcherbinaite. L. F. Borisenko, 1972. *Zap.*, **101**, 464 (Щербинаит). Crystalline V_2O_5 occurs on the walls of fissures on the Berzymyanny volcano, Kamchatka. Named for V. V. Shcherbina. Cf. *A.M.*, **56**, 1487. [*A.M.*, **58**, 560; *Zap.*, **102**, 445.] 28th List

Shentulite. Peng, Ch'i-Jui, 1959. [*Ti-chih K'o-hsueh*, vol. 10, p. 289]; abstr. *Amer. Min.*, 1960, vol. 45, p. 755 (Shen-t'u-shih). (Th,Fe,Ca,Ce){(Si,P,As)O_4,CO_3OH}; metamict. An unnecessary name for a variety of thorite or thorogummite. *See* Arsenothorite. 22nd List

Shen-t'u-shih, *see* Shentulite.

Sheridanite. J. E. Wolff, 1912. *Amer. Journ. Sci.*, ser. 4, vol. xxxiv, p. 476. A pale greenish, talc-like chlorite (near to leuchtenbergite) containing much alumina and very little iron, $H_6Mg_2Al_2Si_2O_{13}$. It occurs as foliated masses in Sheridan Co., Wyoming. 6th List

Sherwoodite. M. E. Thompson, C. H. Roach, and R. Meyrowitz, 1958. *Amer. Min.*, vol. 43, p. 749. Near $Ca_3V_8O_{22}.15H_2O$, blue-black tetragonal prisms from numerous vanadium mines on the Colorado plateau. Named after Dr. Alexander M. Sherwood of the US Geological Survey. [*M.A.*, **14**, 141.] 22nd List

Shilkinite. G. V. Merkulova, 1939. [*Mém. Soc. Russe Min.*, vol. 68, p. 559.] Abstr. in *Amer. Min.*, 1943, vol. 28, p. 62. $K_2O.4Al_2O_3.8SiO_2.4H_2O$, as green, fibrous and fan-shaped aggregates in pegmatite from Shilka river, Transbaikal. Named from the locality. 16th List [Perhaps Illite]

Shinkolobwite. H. Buttgenbach, 1925. *Mém. Soc. Roy. Sci. Liége*, ser. 3, vol. 13, no. –, p. 72 (*see also* pp. 13, 62, 182). Another spelling of Chinkolobwite (A. Schoep, 1923; 10th List), named from the Shinkolobwe or Chinkolobwe copper mine. The mineral, however, comes from the neighbouring Kasolo hill in Katanga, Belgian Congo. 11th List [Is Sklodowskite (q.v.)]

Shishimskite. L. L. Shilin, 1940. *Compt. Rend.* (*Doklady*) *Acad. Sci. URSS*, 1940, vol. 28, p. 346. An ore mixture of perovskite, spinel, magnetite, and haematite from the Shishim Mts., Urals. [*M.A.*, **8**, 174.] 18th List

Shorsuite. N. T. Vinnichenko, 1955. [*Trudy Central Asiatic State Univ.*, no. 63, p. 19.] Abstr. in *Mém. Soc. Russ. Min.*, 1956, vol. 85, p. 377 (шорсуит, shorsuite); *Amer. Min.*, 1957, vol. 42, p. 441. White, fibrous, $(Fe,Mg)Al_2(SO_4)_4.19{\cdot}6H_2O$, between halotrichite and pickeringite. Named from the locality, Шор-Су, in Turkestan. [*M.A.*, **13**, 521.] 21st List

Shortite. J. J. Fahey, 1939. *Amer. Min.*, vol. 24, p. 515. Double carbonate of sodium and calcium, $Na_2CO_3.2CaCO_3$, as hemimorphic orthorhombic crystals from Wyoming. Named after Prof. Maxwell Naylor Short (1889–), of the University of Arizona. [*M.A.*, **7**, 370.] 15th List

Shubnikovite. E. I. Nefedov, 1953. *Mém. Soc. Russ. Min.*, ser. 2, vol. 82, p. 317 (Шубниковит). Hydrous chloro-arsenate of Cu, Ca, and K, orthorhombic?, blue plates. Named after Aleksei Vasilievich Shubnikov, Алексей Васильевич Шубников (1887–), Director of the Crystallographic Institute, Acad. Sci., Moscow. [*M.A.*, **12**, 352.] 20th List

Sialite. N. E. Efremov, 1939. *Compt. Rend.* (*Doklady*) *Acad. Sci. URSS*, vol. 22, p. 434 (sialites). A collective name for hydrous aluminium (Al) silicates (Si) or clay minerals. Later changed to Hydrosialite (q.v.). Cf. Siallite (11th List). [*M.A.*, **7**, 370.] 15th List

Siallite. H. Harrassowitz, 1926. *See* Allite. 11th List

Sibirskite. N. N. Vasilkova, 1962. Зап. Всесоюз. Мин. Общ. (*Mem. All-Union Min. Soc.*), vol. 91, p. 455 (Сибирскит). Minute crystals, probably orthorhombic, from limestone from an undisclosed area have a composition near $CaHBO_3$. M. Fleischer (*Amer. Min.*, 1963, vol. 48, p. 433) notes that the X-ray powder data are inadequate for comparison with the synthetic $CaHBO_3$ of Lehmann (1958). *See also Bull. Soc. franç. Min. Crist.*, 1963, vol. 86, p. 96; *M.A.*, **16**, 66. 23rd List

Sicklerite. W. T. Schaller, 1912. *Journ. Washington Acad. Sci.*, vol. ii, p. 145. Hydrous manganese and iron phosphate, $Fe_2O_3.6MnO.4P_2O_5.3(Li,H)_2O$, forming dark brown cleavable masses, and resulting from the alteration of lithiophilite. From the gem tourmaline mines at Pala, San Diego Co., California. Named after the Sickler family, formerly of Pala. 6th List [$Li(Mn,Fe)PO_4$; *A.M.*, **26**, 681]

Siderazotite. E. T. Wherry, *Journ. Washington Acad. Sci.*, 1920, vol. 10, p. 492. Variant of Siderazot or Siderazote. 9th List

Siderogel. H. Strunz, 1941. *Min. Tab.*, p. 111 (Siderogel). Colloidal FeO(OH), in bog-iron-ore. 19th List

Siderotil. A. Schrauf, *Jahrb. k. k. geol. Reichsanst.*, XLI. 380, 1892. $FeSO_4+5H_2O$. Idria, Carniola. 1st List [Redefined; *A.M.*, **49**, 820]

Sigloite. C. S. Hurlbut and R. Honea, 1962. *Amer. Min.*, vol. 47, p. 1. The alteration product of Paravauxite described by M. C. Bandy under the name Hydrated paravauxite (q.v.) is shown to be a valid mineral and is named from the locality, Siglo XX mine, Llallagua, Bolivia. $(Fe^{3+},Fe^{2+})Al_2(PO_4)_2(O,OH).8H_2O$; anorthic. [*M.A.*, **16**, 68.] 23rd List

Silberspießglanze. H. Strunz, 1957. *Min. Tabellen*, 3rd edit., pp. 86, 105. A group name, including smithite, trechmannite, pavonite, bolivian, tapalpite, stephanite, pearceite, and polybasite. Not to be confused with Silberspießglanz of early German authors (Dana, *Syst. Min.*, 6th edit., p. 42), a synonym of dyscrasite. 22nd List

Silesite. A. Pauly, 1926. *Centr. Min.*, Abt. A, p. 43 (Silesit). Stated to be a silicate of tin, as concretionary forms resembling chalcedony. Named after Dr. Hernando Siles, President of Bolivia. [*M.A.*, **3**, 112; **3**, 370.] 11th List

Silfbergite. Author?, date?. *Zeits. Krist.*, 1924, vol. 60, p. 337 (abstr., no ref.); Dana's *Mineralogy*, 7th edit., 1944, vol. 1, p. 702. Presumably the 'Mangan-Magnetit' (MnO 3·80%) from Vester Silfberg, Sweden, analysed by M. Weibull, 1886. Not the silfbergite of M. Weibull, 1883. 19th List

Silhydrite. A. J. Gude III and R. A. Sheppard, 1972. *Amer. Min.*, **57**, 1053. Soft white masses (up to 4 cm) of minute (<4 μm) orthorhombic crystals remain after Na has been leached from magadiite (25th List) at Trinity County, California, by spring water. Composition $3SiO_2.H_2O$. *a* 14·519, *b* 18·80, *c* 15·938 Å; D 2·141; *n* 1·466. Named for the composition. 27th List

Silicate-pyromorphite. H. Wondratschek, 1963. *Neues Jahrb. Min.*, Abh., vol. 99, p. 112 (Silikatpyromorphit). The artificial compound $Pb_5(PO_4)_2SiO_4$; apatite family. (*Ber. Glastechn.*, 1956, vol. 29, p. 345.) 23rd List

Silicate-wiikite. H. Strunz, 1957. *Min. Tabellen*, 3rd edit., pp. 147, 158 (Silikat-Wiikit). *See* Pyrochlore-wiikite. 22nd List [A mixture; refs. in *A.M.*, **62**, 407, 408.]

Siliceous scheelite. N. N. Kohanowski, 1953. *Mines Mag.*, Denver, Colorado, vol. 43, p. 17. An extremely doubtful and inadequately described mineral found as coatings on scheelite from various localities, formulated $2CaO.SiO_2.12WO_3.24H_2O$. [*M.A.*, **12**, 306; *A.M.*, **39**, 160.] 24th List

Silico-apatite. I. D. Borneman-Starynkevich, 1938. *Compt. Rend.* (*Doklady*) *Acad. Sci. URSS*, vol. 19, p. 225 (Silico-apatite). A hypothetical molecule $Ca_5Si_3O_9(OH)_4$ substituting $Ca_5P_3O_{12}F$ in apatite (SiO_2 5·95%) from Shishim mines, Ural. [*M.A.*, **7**, 352.] 15th List

Silico-Carnotite. V. A. Kroll, 1911. *Journ. Iron and Steel Inst.*, vol. 84 (no. II for 1911), pp. 126, 185; on pp. 172–175 the name appears simply as Carnotite. Calcium silico-phosphate, $3CaO.P_2O_5 + 2CaO.SiO_2$, occurring as blue, orthorhombic crystals with vivid pleochroism in the basic slag of the Thomas-Gilchrist process for the dephosphorisation of iron. Named after Professor Adolphe Carnot (1839–1920), of Paris, who first described the material in 1883. Not the Carnotite of C. Friedel and E. Cumenge, 1899 (2nd List). 9th List

Silicoglaserite. H. Strunz, 1957. *Min. Tabellen*, 3rd edit., p. 263 (Silicoglaserit). A name for high-temperature α-Ca_2SiO_4 (Strunz inverts the usual designations, calling the low-temperature form α-Ca_2SiO_4 and the high γ). 22nd List

Silicoilmenite. P. P. Pilipenko, 1930. *Mineralnoe Syre*, Moskva, 1930, vol. 5, p. 981 (Силикоильменит). A red-brown mineral intergrown with ilmenite from the Ilmen Mts., Ural, thought to represent a solid solution of silicate or silica in ilmenite. [*M.A.*, **4**, 499.] 12th List

Silicomagnesiofluorite. P. A. Zemĭatčenskij, 1906. *Zeits. Kryst. Min.*, vol. xlii, p. 209 (Silicomagnesiofluorit). Ash-grey, greenish, or bluish, radially-fibrous aggregates associated with quartz and serpentine, in a boulder at Luppiko, Finland. The fibres are optically positive and give straight extinction. Formula, $H_2Ca_4Mg_3Si_2O_7F_{10}$. So named on account of its composition. 4th List

Silicomanganberzeliite. M. M. Kayupova, 1963. [Изв. Акад. наук Казах. ССР, сер. геол. (*Bull. Acad. Sci. Kazakh. SSR, ser. geol.*), vol. 6, p. 57]; abstr. Зап. Всесоюз. Мин. Общ. (*Mem. All-Union Min. Soc.*), 1965, vol. 94, p. 678 (Силикоманганберцелиит, silicomanganberzeliite). An unnecessary name for a manganoan and possibly silicatian Berzeliite. 24th List

Silicomonazite. I. Ya. Nekrasov, 1972. Докл., **204**, 491 (Силикомонацит). An unnecessary name for a silicatian monazite from the Kular district. [*A.M.*, **58**, 348; *Zap.*, **102**, 451.] Cf. silicorhabdophane, 22nd List (addenda), and sulphate-monazite, 24th List. 28th List [*M.A.*, 78-759]

Silicorhabdophane. E. I. Semenov, 1959. [Матер. мин. Кольск. полуостр., no. 1, p. 102]; abstr. Зап. Всесоюз. Мин. Общ. (*Mem. All-Union Min. Soc.*), 1961, vol. 90, p. 444 (Силикорабдофанит, silicorhabdophane). Spherulites in a pegmatite from the Lovozero massif give a powder photograph near that of rhabdophane but contain much SiO_2; near $(Ce,Al,Fe)(P,Si)(O,OH)_4.H_2O$. [*A.M.*, **47**, 419.] 22nd List

Silicosmirnovskite. E. I. Semenov, 1959. [Матер. мин. Кольск. полуостр., no. 1, p. 102]; abstr. Зап. Всесоюз. Мин. Общ. (*Mem. All-Union Min. Soc.*), 1961, vol. 90, p. 446 (Силикосмирновскит, silicosmirnovskite). Massive and metamict material in a pegmatite from the Lovozero massif is formulated $(Th,Ln,Ca)_3[(P,Si)(O,OH)_4]_4.4H_2O$, but contains only 3% P_2O_5, and is not obviously related to smirnovskite (21st List); a doubtful species. [*A.M.*, **47**, 419.] 22nd List

Silikatapatit. G. Trömel and W. Eitel, 1957. *Zeits. Krist.*, vol. 109, p. 231. Group name for Britholite, Abukumalite, and other silica-bearing members of the Apatite family. 23rd List

Silikatpyromorphit. H. Wondratschek, 1963. Original form of Silicate-pyromorphite (q.v.). 23rd List

Silikatsulfatapatit. Synonym of wilkeite. *See* Bleiapatit. 20th List [Also 22nd List]

Silikomanazit, Germ. trans. of Силикомонацит, silicomonazite (q.v.). Hintze, 85. 28th List

Sillenite. C. Frondel, 1943. *Amer. Min.*, vol. 28, p. 525. Body-centred cubic modification of Bi_2O_3 as greenish waxy masses from Durango, Mexico. Named after Dr. Lars Gunnar Sillén of Stockholm, who prepared the material artificially. [*M.A.*, **7**, 234; **9**, 6.] 17th List

Silver-analcite, Silver-chabazite. G. Steiger, *Amer. Journ. Sci.*, 1902, ser. 4, vol. xiv, p. 31. Artificial derivatives of analcite and chabazite with silver in place of sodium or calcium. 3rd List

Silver Jamesonite. E. V. Shannon, 1920. *Proc. US Nat. Museum*, vol. 58, p. 603. At first described as a variety of jamesonite, but later named owyheeite (q.v.). [*M.A.*, **1**, 151.] 9th List

Silivialite. R. Brauns, 1914. *Neues Jahrb. Min.*, Beil.-Bd. xxxix (Festschrift Max Bauer), p. 121; *Neues Jahrb. Min.*, 1915, vol. ii, ref. p. 141 (*Silvialith*). In the 3rd Appendix to 6th edit. of Dana's *System of Mineralogy* (1915, p. 70) the name is spelt *Sylvialite*. A hypothetical molecule of the scapolite group, identical with L. H. Borgström's Sulphate-meionite, $CaSO_4.3CaAl_2Si_2O_8$. *See* Carbonate-marialite. Named after Dr. (Mrs.) Silvia Hillebrand, of Vienna, daughter of Gustav Tschermak. 7th List

Simanite, error for Seamanite, through back-transliteration of Симанит. *Soviet Physics—Doklady*, 1971, **16**, 272 (translation of Докл. Акад. наук СССР, 1971, **197**, 1070). [*M.A.*, 72-1860.] 28th List

Simonellite. G. Boeris, 1919. *Rend. Accad. Sci. Ist. Bologna*, n. ser., vol. 23 (for 1918–1919), p. 87. R. Ciusa and A. Galizzi, *Gazzetta Chim. Italiana*, 1921, vol. 51, part i, p. 57. A hydrocarbon, $C_{15}H_{20}$, found as a white crystalline (orthorhombic) encrustation on lignite from Fognano, Tuscany. Named after the geologist, Professor Vittorio Simonelli, who collected the material. [*M.A.*, **1**, 202, 376.] 9th List [1,1-dimethyl-7-isopropyl-1,2,3,4–tetrahydrophenanthrene; *A.M.*, **55**, 1818]

Simplotite. M. E. Thompson, C. H. Roach, and R. Meyrowitz, 1956. *Science* (*Amer. Assoc. Adv. Sci.*), vol. 123, p. 1078; *Amer. Min.*, 1958, vol. 43, p. 16. Hydrous tetravanadite of calcium, $CaV_4O_9.5H_2O$, monoclinic, as dark-green plates and warty aggregates with U-V ores in crevices in sandstone from Peanut mine, Montrose Co., Colorado, and other localities in Colorado and Utah. Named after J. R. Simplot, former owner of the Peanut mine. [*M.A.*, **13**, 379.] 21st List

Simpsonite. H. Bowley, 1938. *Rep. Dept. Mines Western Australia*, for 1937, pp. 93, 88; *Journ. Roy. Soc. Western Australia*, 1939, vol. 25 (for 1938–1939), p. 89; L. E. R. Taylor, ibid., p. 93. Hydrous tantalate of aluminium and calcium, $CaO.5Al_2O_3.4Ta_2O_5.2H_2O$, hexagonal, from Western Australia. Named after Dr. Edward Sydney Simpson (1875–1939), Government Mineralogist and Analyst of Western Australia. [*M.A.*, **7**, 369.] [$Al_4(Ta,Nb)_3O_{13}(F,OH,O)$; *M.A.*, **17**, 561]

The name simpsonite has also been applied to an alkalic amphibole (A. Wade and R. T. Prider, *Rep. Brit. Assoc. Adv. Sci.*, Cambridge, 1938, p. 419. R. T. Prider, *Abstr. Diss. Univ. Cambridge*, 1939, for 1937–1938, p. 93; *Min. Mag.*, 1939, vol. 25, p. 378) which was afterwards renamed magnophorite (q.v.). 15th List [Titanian potassian richterite; *Amph.* (1978)]

Sincosite. W. T. Schaller, 1922. *Journ. Washington Acad. Sci.*, vol. 12, p. 195. Hydrous phosphate of calcium and vanadyl, $CaO.V_2O_4.P_2O_5.5H_2O$, occurring as green tetragonal plates at Sincos, Peru. Named from the locality. [*M.A.*, **1**, 375.] 9th List

Sinhalite. G. F. Claringbull and M. H. Hey, 1952. *Min. Mag.*, vol. 29, p. 841. Borate $MgAlBO_4$, orthorhombic. Hitherto mistaken for brown gem olivine. Named from Sinhala, the Sanskrit name for Ceylon. [*A.M.*, **37**, 700, 1072.] 19th List

Sinicite. Ho Chen-Tsi and Chun Chi-Chen, 1957. [*Kexue Tongbao* (*Scientia*), no. 12, p. 378]; abstr. Зап. Всесоюз. Мин. Общ. (*Mem. All-Union Min. Soc.*), 1958, vol. 87, p. 479 (Синисит sinicite) and *Amer. Min.*, 1959, vol. 44, p. 467. A tantaloniobate of rare earths and Th from an undefined locality in China. A species of doubtful validity. 22nd List

Sinnerite. F. Marumo and W. Nowacki, 1964. *Schweiz. Min. Petr. Mitt.*, vol. 44, p. 439; abstr. with correction, *Amer. Min.*, 1965, vol. 50, p. 1504. A small steel-grey crystal found with tennantite in the Lengenbach quarry near Binn, Switzerland, is pseudocubic with a pseudocell content $Cu_{1.4}As_{0.9}S_{2.1}$, but is probably anorthic and twinned. Named for R. von Sinner. In the abstract E. Seeliger notes that the X-ray powder pattern is very near that of luzonite, tetragonal Cu_3AsS_4, and suggests that it might be twinned luzonite or an oriented intergrowth of luzonite and tennantite. [*M.A.*, **17**, 74.] 24th List [$Cu_6As_4S_9$; *A.M.*, **57**, 824]

Sinoite. C. A. Andersen, K. Keil, and B. Mason, 1964. *Science*, vol. 146, p. 256. Silicon oxynitride, Si_2N_2O, identical with the synthetic material (Brosset and Idrestedt, *Nature*, 1964, vol. 201, p. 1211), occurs in the Jajh deh Kot Lalu meteorite and some other enstatite chondrites. It appears to have been first observed as an unidentified mineral in the Pillistfer and Hvittis falls by Lacroix in 1905 (Mason, *Geochimica Acta*, 1966, vol. 30, p. 23). Named from the composition. [*M.A.*, **17**, 302; *A.M.*, **50**, 521.] 24th List

Sitaparite. L. L. Fermor, 1908. *Rec. Geol. Survey India*, vol. xxxvii, p. 207; *Mem. Geol. Survey India*, 1909, vol. xxxvii, pp. lxvii, 49. A dark bronze-grey, crystalline, and cleavable mineral resembling vredenburgite (q.v.) in appearance, but differing from this in being only slightly magnetic. Formula perhaps $9Mn_2O_3.4Fe_2O_3.MnO_2.3CaO$. Occurs with manganese-ores at Sitapár, Chhindwára district, Central Provinces, India. 5th List [Is Bixbyite, *A.M.*, **27**, 653; *see also A.M.*, **28**, 468, **29**, 66. Cf. Kurnakite, Partridgeite]

Sjanchualinit, German form of Hsianghualite (22nd List). *Zentr. Min.*, 1964, Teil 1, p. 151. 24th List

Sjögrenite. C. Frondel, 1940. *Amer. Min.*, Program and Abstracts, December 1940, p. 6 (Sjogrenite); *Amer. Min.*, 1941, vol. 26, p. 196 (Sjogrenite), p. 295* (Sjögrenite). Hydrous basic carbonate of magnesium and ferric iron, $Fe_3Mg_6(OH)_{16}CO_3.4H_2O$, as hexagonal scales dimorphous with the rhombohedral pyroaurite. Named after Sten Anders Hjalmar Sjögren (1856–1922) of Stockholm. [*M.A.*, **8**, 51, 99.] [*Corr. in *M.M.*, **28**.] 16th List

—— J. S. Krenner, 1910. *Compt. Rend. Congrès Géol. Intern.*, XI Session, 1910, Stockholm, 1912, vol. 1, p. 129 (Sjögrenit); *Földtani Közlöny*, Budapest, 1913, vol. 43, pp. 10, 121. P. Quensel, *Geol. För. Förh. Stockholm*, 1946, vol. 68, p. 110. Hydrous ferric phosphate, $5Fe_2O_3.3P_2O_5.8H_2O$, probably triclinic. First described from Cornwall by E. Kinch, F. H. Butler, and H. A. Miers (*Min. Mag.*, 1886, vol. 7, p. 85) as distinct from dufrenite. Named after Carl Anton* Hjalmar Sjögren (1822–1893)* of Stockholm. Not the sjögrenite of C. Frondel, 1940 (16th List). [*M.A.*, **9**, 263.] [*Corr. in *M.M.*, **28**.] 17th List [Is chalcosiderite; *A.M.*, **34**, 521]

Sjögrufvite. L. J. Igelström, *Geol. För. Förh.*, XIV. 309, 1892; *Zeits. Kryst. Min.*, XXII. 471, 1894. Hydrated arsenate of Mn and Fe. Sjö mine, Sweden. 1st List [A mixture, *A.M.*, **49**, 447; *M.A.*, **17**, 180; is Caryinite, *A.M.*, **58**, 562]

Skemmatite. W. E. Ford and W. M. Bradley, 1913. *Amer. Journ. Sci.*, ser. 4, vol. xxxvi, p. 169; *Zeits. Kryst. Min.*, vol. liii, p. 225. Hydrated oxide of manganese and iron, $3MnO_2.2Fe_2O_3.6H_2O$, occurring as a black alteration product of pyroxmangite (q.v.). Named for *σκέμμα*, a question, because of the doubt whether the mineral really represents a distinct species. 7th List

Skiagite. L. L. Fermor, 1926. *Rec. Geol. Survey India*, vol. 59, p. 202. A hypothetical garnet molecule $3FeO.Fe_2O_3.3SiO_2$ present (nearly 20%) in garnet from Glen Skiag in Scotland, and in some Indian garnets. *See* Iron-andradite. 11th List

Skinnerite. S. Karup-Møller and E. Makovicky, 1974. *A.M.*, **59**, 889. Aggregates intergrown with tetrahedrite, chalcostibite, etc., in analcime–natrolite veins in the Ilimaussaq intrusion, S Greenland, have space group $P2_1/c$, *a* 7·81, *b* 10·25, *c* 13·27 Å, *β* 90° 21′; composition $8[Cu_3SbS_3]$. Undergoes a reversible transition to an orthorhombic form at 122°C. Named for B. J. Skinner. [*M.A.*, 75-1397; *Zap.*, **104**, 606; *Bull.*, **98**, 166.] 29th List

Sklodowskite. A. Schoep, 1924. *Compt. Rend. Acad. Sci. Paris*, vol. 179, p. 413; *Bull. Soc. Franç. Min.*, 1924, vol. 47, p. 162. Hydrated silicate of uranium and magnesium, $MgO.2UO_3.2SiO_2.7H_2O$, as citron-yellow, orthorhombic needles from Katanga. Named after Madame Marie Curie, *née* Sklodowska. [*M.A.*, **2**, 341, 384.] 10th List

Skolite. K. Smulikowski, 1936. *Arch. Min. Tow. Nauk. Warszaw.* (*Arch. Min. Soc. Sci. Varsovie*), vol. 12, p. 145 (Skolite), p. 179 (Skolit). A dark green, finely scaly mineral, $H_4K(Mg,Fe'',Ca)(Al,Fe''')_3Si_6O_{20}.4H_2O$, of the glauconite group and close to bravaisite, occurring in sandstone at Skole, E Carpathians, Poland. Named from the locality. [*M.A.*, **6**, 345.] 14th List

Skorzalith. W. E. Tröger, 1952. *Tabellen opt. Bestim. gesteinsbild. Minerale*, p. 147. German form of scorzalite (18th List), named after E. P. Scorza. 20th List

Skunolite (Скунолит), error for Ikunolite (Икунолит), Зап. Всесоюз. Мин. Общ. (*Mem. All-Union Min. Soc.*), 1962, vol. 92, p. 187. 23rd List

Skupit, variant of Schoepite (H. von Philipsborn, *Erzkunde*, 1965, p. 65). 25th List

Slavikite. R. Jirkovský and F. Ulrich, 1926. *Věstnik Státního Geol. Ústavu Československ.*, vol. 2, p. 345 (Slavíkit). Hydrated sodium and ferric sulphate, $(Na,K)_2SO_4.2Fe_5(OH)_3$-$(SO_4)_6.63H_2O$, as minute greenish-yellow rhombohedral crystals on weathered pyritic shales from Bohemia. Named after Prof. František Slavík (1876–), of Prague. [*M.A.*, **3**, 365; *A.M.*, **13**, 492.] 11th List [Mg essential, *M.A.*, **11**, 246; **13**, 369]

Slavyanskite. B. V. Dolishnii, 1977. *Zap.*, **106**, 331 (Славянскит). Tetragonal crystals from a drill core in the Slavyansk salt deposit, Dneprovsk-Donets basin, have *a* 11·26, *c* 6·56 Å. Composition near $4[CaAl_2O_4.8\frac{1}{2}H_2O]$, sp. gr. 2·52. RIs ω 1·571, ε 1·600. Named from the locality. [*M.A.*, 78-896; *A.M.*, **63**, 599.] 30th List

Slawsonite. D. T. Griffin, P. H. Ribbe, and G. V. Gibbs, 1977. *A.M.*, **62**, 31. The Sr analogue of paracelsian occurs in Wallowa Co., Oregon. Monoclinic, *a* 8·888, *b* 9·344, *c* 8·326 Å, β 90·33°. [*M.A.*, 77-3393.] 30th List

Smirnovite. V. G. Melkov and L. Ch. Pukhalsky, 1957. [*Minerals of the Urals*, Moscow, 1957, p. 35.] Abstr. in *Mém. Soc. Russ. Min.*, 1958, vol. 87, p. 79 (смириовит, smirnovite = торутит, thorutite). $2[(Th,U,Ca)Ti_2O_6].H_2O$. Presumably named after S. S. Smirnov (*see* smirnovskite). Thorutite = thoria + rutile (?). 21st List [**Thorutite**]

Smirnovskite. I. F. Grigoriev and E. I. Dolomanova, 1957. *Mém. Soc. Russ. Min.*, vol. 86, p. 607 (смирновскит). Hydrous fluo-silico-phosphate of thorium, etc. (Th,Ce,Ca...)OH. $(P,Si,Al)(O,F,OH)_4$, Metamict. With cassiterite in quartz veins from Trans-Baikal. Named after academician Sergei Sergesich Smirnov, Сергей Сергесич Смирнов. 21st List

Smirnowit, German transliteration of Смирновит, smirnovite (21st List) (Hintze, *Handb. Min.*, Erg.-Bd. II, p. 842). 22nd List

Smirnowskit, German transliteration of Смирновскит, smirnovskite (21st List) (Hintze, *Handb. Min.*, Erg.-Bd. II, p. 843). 22nd List

Smithite. R. H. Solly, 1905. *Min. Mag.*, vol. xiv, p. 74. (A preliminary description of the unnamed mineral appeared in *Nature*, 1903, vol. lxix, p. 142.) G. F. H. Smith and G. T. Prior, *Min. Mag.*, 1907, vol. xiv, p. 293. Minute, red, monoclinic crystals found in the white, crystalline dolomite of the Binnenthal, Switzerland. Sulpharsenite of silver, $AgAsS_2$. Named after Dr. George Frederick Herbert Smith, of the British Museum. 4th List

Smoljaninowit, German transliteration of Смоляниновит, Smolyaninovite (21st List) (Hintze, *Handb. Min.*, Erg.-Bd. II, p. 844). 22nd List

Smolyaninovite. L. K. Yakhontova, 1956. *Doklady Acad. Sci. USSR*, vol. 109, p. 849 (смольяниновит, smolianinovite). Hydrous arsenate, $2As_2O_5.(Fe,Al)_2O_3.4(Co,Ni,Mg,Ca)O.11H_2O$. Yellow, earthy oxidation product of Ni-Co ores from Bou-Azzer, Morocco. Named after Prof. N. A. Smolyaninov, Н. А. Смольянинов. 21st List [*A.M.*, **59**, 1141; Australia, *M.A.*, 77-4609]

Smythite. R. C. Erd and H. T. Evans, 1956. *Journ. Amer. Chem. Soc.*, vol. 78, p. 2017. R. C. Erd, H. T. Evans, and D. H. Richter, *Amer. Min.*, 1957, vol. 42, p. 309. Iron sulphide, Fe_3S_4, rhombohedral; X-ray pattern similar to but distinct from pyrrhotine. Named after Prof. Charles Henry Smyth, Jr. (1866–1937), economic geologist. [*M.A.*, **13**, 380, 523.] 21st List [Redefined as (pseudorhombohedral) $(Fe,Ni)_9S_{11}$; *A.M.*, **57**, 1571]

Sobolevskite. T. L. Evstigneeva, A. D. Genkin, and V. A. Kovalenker, 1974. *Zap.*, **104**, 568 (Соболевскит). Small grains in the Oktyabr ore deposit, E Siberia, are hexagonal, $P6_3/mmc$, *a* 4·23, *c* 5·69 Å, composition PdBi. Named for P. G. Sobolevskii. 29th List

Sobotkite. C. Haranczyk and K. Prchazka, 1974. [*Prace Muzeum Ziemi*, **22**, 3], abstr. *A.M.*, **61**, 177. A trioctahedral member of the montmorillonite group, $(Ca_{0\cdot13}K_{0\cdot015})(Mg_{1\cdot91}Al_{0\cdot95})$-$(Si_{3\cdot06}Al_{0\cdot94})(OH)_2O_{10}.5{\cdot}18H_2O$, occurs in the serpentinites of the Gogolow–Jordanow massif, Lower Silesia, at Wiryin. Named from Mt. Sobotka. [Could be considered an aluminian saponite—*M.F.*] 29th List

Sobralite. J. Palmgren, 1917. *Bull. Geol. Inst. Univ. Upsala*, vol. xiv, p. 173 (Sorbalit). A triclinic manganese pyroxene, $4MnSiO_3.2FeSiO_3.CaSiO_3.MgSiO_3$, from Tunaberg, Sweden, differing from pyroxmangite (7th List) in the position of the optic axial plane. Named after Dr. J. M. Sobral, of Buenos Aires. Cf. Vogtite. 8th List [Is Pyroxmangite; *M.A.*, **6**, 528]

Soda-adularia, Soda-sanidine. A. N. Winchell, 1927. *Optical mineralogy*, part II, 2nd edit., pp. 318, 322 (soda-adularia, soda-sanidine); 4th edit., 1951, p. 302 (sodian adularia, sodian sanidine). Varieties of soda-orthoclase (q.v.). *See* Natronsanidine (5th List). 20th List

Soda-alunite. J. D. Laudermilk, 1935. *Amer. Min.*, vol. 20, p. 57 (Soda-alunite). Alunite containing 4·62% Na_2O from Hawaii. Synonym of Natroalunite (3rd List). [*M.A.*, **6**, 188.] 15th List

Soda-amblygonite. *See* Natramblygonite. 6th List

Soda-anorthite. *See* natronanorthite. 2nd List

Soda-augite. K. Yagi, 1953. *Bull. Geol. Soc. Amer.*, vol. 64, p. 781. Augite containing Na_2O up to 1·7%, intermediate between augite and aegirine-augite, as zoned crystals and single crystals. 20th List

Soda-beryl. A. A. Beus and N. E. Zalashkova, 1936. *Min. Sbornik Lvov Geol. Soc.*, no. 10, p. 273 (натриевый берилл). An alkali-beryl (12th List) containing up to 1·44% Na_2O. [*M.A.*, **13**, 184.] 21st List

Soda-berzeliite. H. Sjögren, *Bull. Geol. Inst. Upsala*, 11. 92, 1895. [*M.M.*, **11**, 163.] 10RO.-$3As_2O_5$ or $3RO.As_2O_5$. Cubic [Långban] Sweden. 1st List [Var. of Berzeliite, Dana *Syst. Min.*, 7th edit., **2**, 681]

Soda-catapleiite. G. Flink. *Medd. om Grönland*, 1900, XXIV, 100. *See* Natroncatapleiite. 2nd List

Sodaclase. A. Johannsen, 1926. *Journ. Geol. Chicago*, vol. 34, p. 841. Members of the plagioclase series between pure albite and $Ab_{90}An_{10}$. *See* Calciclase. 11th List

Soda-dehrnite. E. S. Larsen and E. V. Shannon, 1930. *See* Dehrnite. 12th List

Soda-dravite. F. H. Pough, 1953. *Field guide to rocks and minerals*, Boston, p. 280 (Soda-dravite). $NaMg_3B_3Al_3(Al_3Si_6O_{27})(OH)_4$, a brown variety of tourmaline. Similarly, lime-dravite, $CaMg_3B_3Al_3(Al_3Si_6O_{27})(O,OH)_4$, a white variety. 20th List

Soda-garnet. *See* Lagoriolite. 2nd List

Soda-glauconite. A. F. Hallimond, 1922. *Min. Mag.*, vol. 19, p. 333. A variety of glauconite in which part of the potash is replaced by soda. 9th List

Soda-heterosite. P. Quensel, 1937. *Geol. För. Förh. Stockholm*, vol. 59, p. 96 (Na-heterosite; 15th List, p. 645). G. L. English, *Descriptive list of new minerals*, 1939, p. 208 (Soda-heterosite). B. Mason, *Geol. För. Förh. Stockholm*, 1941, vol. 63, p. 161 (Na-heterosite). [*M.A.*, **6**, 486; **7**, 313; **8**, 180.] 16th List [A mixture]

Soda-jadeite. H. S. Washington, 1922. *Proc. US Nat. Museum*, vol. 60, art. 14, pp. 7, 9 (soda jadeite). Synonym of jadeite. *See* Diopside-jadeite. 9th List

Soda-killinite. P. Quensel, 1938. *Geol. För. Förh. Stockholm*, vol. 60, pp. 205, 215 (soda-killinite). An alteration product of spodumene, similar to killinite but rich in soda, and apparently consisting of a mixture of soda-spodumene (q.v.), cimolite, halloysite, and illite (q.v.). [*M.A.*, **7**, 120.] 15th List

Soda-leucite. An analysis of pseudoleucite, consisting of a mixture of orthoclase, nephelite, scapolite, etc., showed the presence of 7·08% Na_2O with 8·49% K_2O; this suggests that the original mineral was a soda-leucite (C. W. Knight, *Amer. Journ. Sci.*, 1906, ser. 4, vol. xxi, p. 292). A soda-leucite (Natronleucit) was prepared artificially by J. Lemberg in 1876. *See* Palaeoleucite. 4th List

Sodalumite. A. N. Winchell, 1927. *Elements of optical mineralogy*, 2nd edit., part 2, p. 114; *The optic and microscopic characters of artificial minerals*, Madison, 1927, pp. 100, 101. Sodium-alum, $Na_2Al_2(SO_4)_4.24H_2O$, the cubic modification prepared artificially and not known with certainty as a mineral. Mendozite, with the same chemical composition, is optically birefringent and probably uniaxial. 11th List

Soda-margarite. F. C. Phillips, 1931. *Min. Mag.*, vol. 22, pp. 482, 485. Synonym of ephesite, for a margarite in which calcium is largely replaced by sodium. 12th List

Soda-melilite. H. Berman, 1929. *Amer. Min.*, vol. 14, p. 398 (soda-melilite), pp. 400, 405 (soda melilite). A hypothetical molecule $Na_2Si_3O_7$ which is regarded as an end-member of the melilite group. Similarly, sub-melilite is $CaSi_3O_7$. [*M.A.*, **4**, 204.] 12th List

Soda-microcline. *See* Natronmikroklin. 2nd List

Soda-orthoclase. J. P. Iddings, 1906. *Rock minerals*, pp. 203, 232. 'Those apparently mono-symmetric feldspars with notable amount of soda may be called soda-orthoclase. . . . When the soda equals or exceeds the potash the crystals exhibit triclinic symmetry and are soda-microcline'. *See* Soda-microcline = Natronmikroklin = anorthoclase (2nd List). 20th List

Soda-purpurite. P. Quensel, 1937. *Geol. För. Förh. Stockholm*, vol. 59, p. 96 (Na-purpurite); ibid., 1939, vol. 61, p. 69 (soda-purpurite). The end-product, $(Mn,Fe)PO_4$, of alteration of the soda mineral varulite (14th List); as distinct from purpurite, $(Mn,Fe)PO_4$, an alteration product of the lithium mineral lithiophilite. Similarly, Na-heterosite and heterosite, both also $(Mn,Fe)PO_4$, alteration products of headdenite (14th List) and triphylite. [*M.A.*, **7**, 120.] 15th List

Soda-richterite. H. Sjögren, *Bull. Geol. Inst. Upsala*, II. 71, 1895. To replace the name astochite (*Geol. För. Förh.*, XIII. 604, 1891). *See* Natronrichterite. 1st List [Manganoan richterite; *Amph.*, Sub-Comm.]

Soda-sanidine, *see* Soda-adularia.

Soda-sarcolite. W. T. Schaller, 1916. *Bull. US Geol. Survey*, no. 610, p. 112 (soda-sarcolite), p. 115 (soda sarcolite). A hypothetical molecule, $3Na_2O.Al_2O_3.3SiO_2$, corresponding with sarcolite but containing sodium in place of calcium, assumed to explain the composition of minerals of the melilite group. It is present to the extent of 12·2% in a melilite from Vesuvius. A ferric iron sarcolite, $3CaO.Fe_2O_3.3SiO_2$, is also suggested (p. 119). 8th List

Soda-spodumene. P. Quensel, 1938. *Geol. För. Förh. Stockholm*, vol. 60, p. 214 (Na-spodumene). A presumed base-exchange product of spodumene with Na in place of Li, as distinct from the mixture β-spodumene of Brush and Dana, 1880. Not the Natron-Spodumen of J. J. Berzelius, 1824, = soda-spodumene (J. D. Dana, 1844) = oligoclase. [*M.A.*, **7**, 120.] 15th List

Soda-tremolite. H. Berman and E. S. Larsen, 1931. *Amer. Min.*, vol. 16, p. 143. E. S. Larsen and E. A. Goranson, *Amer. Min.*, 1932, vol. 17, pp. 351–353. A variety of tremolite with the formula $Na_2CaMg_5Si_8O_{22}(OH)_2$. [*M.A.*, **5**, 216.] 13th List

Soda-triphylite. V. Ziegler, 1914. *Bull. South Dakota School of Mines*, no. 10, p. 192 (soda-triphylite), p. 193 (soda triphylite). A mineral from Black Hills, South Dakota, analysed by W. P. Headden in 1891 and described by him as being near triphylite. It was later named arrojadite by D. Guimarães (1925, 11th List) and headdenite by P. Quensel (1937, 14th List). B. Mason (*Geol. För. Förh. Stockholm*, 1941, vol. 63, p. 132) points out that the earliest name soda-triphylite would suggest identity of the mineral with natrophilite, and he adopts the name arrojadite: soda-triphylite = headdenite = arrojadite. [*M.A.*, **3**, 113; **6**, 486; **8**, 180.] 16th List

Soddite. A. Schoep, 1922. *Compt. Rend. Acad. Sci. Paris*, vol. 174, p. 1066. Hydrated silicate of uranium $12UO_3.5SiO_2.14H_2O$, as pale-yellow, orthorhombic crystals forming with curite (q.v.) fine-grained aggregates at Kasolo, Katanga. Named after Prof. Frederick Soddy, F.R.S., of Oxford. [*M.A.*, **1**, 377.] 9th List

Soddyite. (V. Billiet, *Natuurwetenschappelijk Tijdschrift*, Antwerpen, 1926, vol. 7 (for 1925), p. 112 (Soddyiet); A. Schoep, ibid., 1927, vol. 9, p. 25 (Soddyiet), p. 29 (Soddyite)). A more correct form of soddite (A. Schoep, 1922; 9th List). Named after Prof. Frederick Soddy (1877–), of Oxford. [*M.A.*, **3**, 371.] 11th List [$(UO_2)_5Si_2O_9.6H_2O$; *A.M.*, **37**, 386]

Sodium-anthophyllite. *Amph.* (1978), **3**, 3 (Sodium anthophyllite). 30th List

Sodium-autunite. J. G. Fairchild, 1929. *Amer. Min.*, vol. 14, p. 265 (sodium autunite), p. 269 (sodium-autunite). *See* Calcium-autunite, natro-autunite. 19th List [*A.M.*, **43**, 383]

Sodium-bentonite. J. W. Jordan, 1949. *Min. Mag.*, vol. 28, p. 598. A highly swelling clay from Wyoming. 19th List

Sodium betpakdalite. K. V. Skvortsova*, G. A. Sidorenko, Yu. S. Nesterova, G. A. Arapova, A. D. Dara, and L. I. Rybakova, 1971. *Zap.*, **100**, 603 (натриевый бетпакдалит, Natrion betpakdalite). Lemon-yellow crystals in the oxidation zone of a molybdenum deposit, with α 1·792, γ 1·810, are monoclinic, *a* 11·28, *b* 19·30, *c* 17·67 Å, β 94° 30′, D 2·92, and have a composition corresponding to that of betpakdalite (22nd List) with rather more than half the Ca replaced by Na_2. The name as originally given employs an adjectival modifier, and would have been better rendered Sodian (or Natrian) betpakdalite. [*M.A.*, 72-2335; *A.M.*, **57**, 1312.] [*Corr. in *M.M.*, **39**.] 27th List

Sodium boltwoodite, *see* Ammonium boltwoodite. 22nd List

Sodium carnotite. P. B. Barton, 1958. *Amer. Min.*, vol. 43, p. 799. *Synthetic* $NaUO_2VO_4$, the sodium analogue of carnotite; several other analogues were also synthesized (Rb,Cs,Tl,NH_4) but not formally named. 24th List

Sodium gastunite, *see* Gastunite (of Honea). 22nd List

Sodium-gedrite. *Amph.* (1978), **3**, 6 (Sodium gedrite). 30th List

Sodium-illite. F. A. Bannister, 1943. *Min. Mag.*, vol. 26, p. 304. Synonym of brammallite (q.v.). 16th List

Sodium-jarosite. F. V. Chukhrov, P. E. Arest-Yakubovich, and N. A. Kozlova, 1940. *Compt. Rend.* (*Doklady*) *Acad. Sci. URSS*, vol. 28, p. 829 (sodium-jarosites). Synonym of natrojarosite (3rd List). [*M.A.*, **8**, 142.] 16th List

Sodium-melilite. R. W. Nurse and H. G. Midgley, 1953. *Journ. Iron and Steel Inst.*, vol. 174, p. 121. Artificially produced end-member, $NaCaAlSi_2O_7$, of the melilite series. Distinct from the hypothetical end-member soda-melilite, $Na_2Si_3O_7$ (12th List). [*M.A.*, **12**, 197.] 20th List

Sodium-mordenite. I. M. King, W. J. King, and R. Wallis, 1968. *Nature*, **217**, 1968. A *synthetic* preparation having the composition of the sodium end-member of the mordenite series. [*M.A.*, 69-314.] 26th List

Sodium-nepheline. G. Donnay, 1957. *Ann. Rep. Geophysical Lab.*, for 1956–1957, p. 237 (sodium nepheline). Artificial $NaAlSiO_4$, as distinct from the usual nepheline $(Na,K)AlSiO_4$. Also, as deduced from X-ray data, natural at Monte Somma, Vesuvius. 21st List

Sodium pseudo-edingtonite, *see* Pseudo-edingtonite.

Sodium-uranospinite. *See* Hydrogen-uranospinite. 19th List [*A.M.*, **43**, 383]

Sogdianite. V. D. Dusmatov*, A. F. Efimov, Z. T. Kataeva, L. A. Khoroshilova, and K. P. Yanulov, 1968. Докл. Акад. наук СССР (*Compt. Rend. Acad. Sci. URSS*), **182**, 1176 (Согдианит). Violet platy masses occur in an alkalic intrusive in the Alai range, Tadzhik SSR; hexagonal, composition $(K,Na)_2Li_2(Li,Fe,Al,Ti)_{1\cdot8}(Zr,Ti)(Si_2O_5)_6$. Named for the ancient Central Asian state of Sogdiana. [*A.M.*, **54**, 1221, where a contradiction in the optical data is noted; *M.A.*, 69-1539; *Zap.*, **99**, 83.] [*Corr. in *M.M.*, **39**.] 26th List

Sogdianovit, error for Sogdianite (26th List). *A.M.*, **54**, 1221; **55**, 1073; **57**, 1914. 28th List

Sogrenite. Author? [Russian textbooks, 1956, 1957.] Abstr. in *Amer. Min.*, 1958, vol. 43, p. 382. A black organo-uranium complex, perhaps identical with thucholite. Locality? etymology? 21st List

Söhngeïte. H. Strunz, 1965. *Naturwiss.*, vol. 52, p. 493 (Söhngeit). $Ga(OH)_3$, as light-brown crystal aggregates on germanite from Tsumeb, SW Africa; cubic. Named for G. Söhnge. [*A.M.*, **51**, 1815.] 24th List [Orthorhombic; *A.M.*, **56**, 355]

Sokolovite. A. K. Sharova and A. K. Gladkovsky, 1958. [Отдел геол.-геогр. Наук, Акад. Наук СССР, 1958, p. 70]; abstr. Зап. Всесоюз. Мин. Общ. (*Mem. All-Union Min. Soc.*), 1961, vol. 90, p. 103 (Соколовит, sokolovite) and *Amer. Min.*, 1961, vol. 46, p. 243. Near $2(Ca,Sr)O.4Al_2O_3.P_2O_5.11H_2O$, in cavities in the Sokolov bauxite deposits, middle Urals. Named from the locality. Doubtful; the composition is near to goyazite and crandallite, but the X-ray data agree rather closely with svanbergite and woodhouseite. It is not stated whether SO_4'' was tested for. 22nd List

Solanite. Huang Yun-Hue, 1965. [*Dizhi lumping, geological review*, vol. 23, p. 7], abstr. in Зап. Всесоюз. Мин. Общ. (*Mem. All-Union Min. Soc.*), 1966, vol. 95, p. 321 (Соланит, solanite); Tseng Jo-ku, Hsueh Chi-yueh, and Peng Chih-chung, [*Kexue Tongbao*, 1966, vol. 17 (1), p. 45], abstr. *M.A.*, **17**, 605 (suolunite). Orthorhombic $Ca_2H_2Si_2O_7.H_2O$. 24th List [*A.M.*, **52**, 560; **53**, 349: **Suolunite** is now the preferred spelling (not to be confused with Sulunite). Bosnia, *M.A.*, 75-1306]

Sollyite. G. Gagarin and J. R. Cuomo, 1949, loc. cit., p. 7 (sollyita). A fibrous lead sulpharsenite, $Pb_3As_4S_9$, from Binn, Switzerland, near to rathite in composition, but with interfacial angles near to dufrenoysite. Described (*Min. Mag.*, 1919, **18**, 360) by, and now named after, Richard Harrison Solly (1851–1925) of Cambridge. 19th List [Probably a mixture, *Jahrb. Naturhist. Mus. Bern*, 1966–1968, p. 71]

Solongoite. S. V. Malinko, 1974. *Zap.*, **113**, 117 (Солонгоит). Monoclinic crystals (*a* 7·93, *b* 7·26, *c* 12·54 Å, *β* 94°) from the Solongo deposit, Buryat ASSR, have the composition $2[Ca_4B_6O_8(OH)_9Cl]$. Named from the locality. [*A.M.*, **60**, 162.] 28th List

Sommairite. (*British Museum* (*Natural History*), *Students' Index to the Collection of Minerals*, 1897. *Collection de Minéralogie du Muséum d'Histoire Naturelle, Paris, Guide du Visiteur*, 2nd edit., 1900, p. 32.) Zinciferous melanterite from Laurion, Greece; analysed, but not named sommairite, by L. Michel (*Bull. Soc. franç. Min.*, 1893, vol. xvi, p. 204). 4th List

Sonolite. M. Yoshinaga, 1963. [*Mem. Fac. Sci. Kyushi Univ.*, ser. D, Geol., vol. 14, p. 1], abstr. in *Amer. Min.*, 1963, vol. 48, p. 1413. $4Mn_2SiO_4.Mn(OH,F)_2$, the manganese analogue of Clinohumite, occurring as monoclinic prisms and anhedral crystals at the Sono, Hanawa, and other manganese mines in Japan; named from the locality at which it was first found. [*M.A.*, **16**, 549]. 23rd List

Sonoraite. R. V. Gaines, G. Donnay, and M. H. Hey, 1968. *Amer. Min.*, **53**, 1828. Yellowish-green monoclinic crystals of $Fe_2^{3+}Te_2^{4+}O_6(OH)_2.2H_2O$ from the Moctezuma mine, Sonora, Mexico. Named from the locality. [*M.A.*, 69-2396; *Bull.*, **92**, 514; *Zap.*, **99**, 76.] 26th List

Sorbalite, error for Sobralite (q.v.).

Sorbite. *See* Cochranite. 8th List

Sorbyite. J. L. Jambor, 1967. *Canad. Min.*, vol. 9, pp. 7 and 191. A monoclinic mineral from Madoc, Ontario, with composition $Pb_{17}(Sb,As)_{22}S_{50}$. Named for H. C. Sorby. Not to be confused with Sorbite (of Osmond or of Howe; 8th List). [*A.M.*, **53**, 1425; *Bull.*, **91**, 303.] 25th List

Sørensenite. E. I. Semenov, N. V. Maksimova, and O. V. Petersen 1965. *Medd. Grønland*, vol. 181, no. 1. Colourless monoclinic crystals in hydrothermal veins at Nakalaq in the Ilinaussaq alkalic intrusive, S Greenland, have the composition $Na_4SnBe_2Si_6O_{16}(OH)_4$. Named for H. Sørensen. [*M.A.*, **17**, 766; *A.M.*, **51**, 1547; **52**, 928.] 25th List

Soretite. L. Duparc and F. Pearce, 1903. *Arch. Sci. Phys. Nat. Genève*, ser. 4, vol. xvi, p. 599 (Soretite); *Bull. Soc. franç. Min.*, vol. xxvi, p. 126 (sorétite). A variety of common aluminous hornblende, defined by its optical characters, which vary, however, in different specimens. It occurs in anorthite-diorite at Koswinsky Kamen, N Urals. Named after the late Professor Charles Soret (1854–1904), of Geneva. 4th List

Sorit, Germ. trans. of Зорит, Zorite (q.v.). Hintze. 86. 28th List

Sosmanite. H. Schneiderhöhn and P. Ramdohr, 1931. *Lehrbuch der Erzmikroskopie*, vol. 2, p. 537 (Sosmanit). Synonym of maghemite (P. A. Wagner, 1927; 12th List) for magnetic ferric oxide (γ-Fe_2O_3). Named after Dr. Robert Browning Sosman (1881–), formerly of the Geophysical Laboratory, Washington, DC, who described the mineral in 1925. [*M.A.*, **3**, 217.] 14th List

Souesite. G. C. Hoffmann, 1905. *Amer. Journ. Sci.*, ser. 4, vol. xix, p. 319. A native nickel-iron alloy occurring as small, rounded grains in the auriferous gravels of the Fraser River, British Columbia. Named after F. Soues, who sent the material for identification.

G. S. Jamieson (*Amer. Journ. Sci.*, 1905, ser. 4, vol. xix, p. 413; *Zeits. Kryst. Min.*, 1905, vol. xli, p. 157) points out that awaruite, josephinite, souesite, and other terrestrial nickel-iron alloys have no definite composition, though they all lie between $FeNi_3$ and $FeNi_2$; and he proposes that they should all be referred to the earliest name awaruite (W. Skey, 1885). 4th List

Soumansite. A. Lacroix, 1910. *Minéralogie de la France*, vol. iv, p. 541. A fluo-phosphate of aluminium and sodium with some water or hydroxyl; found in small amount as colourless tetragonal crystals on corroded amblygonite from Montebras in Soumans, Creuse. 6th List [Is Wardite; *M.A.*, **3**, 343]

Souxite. R. Herzenberg, 1946. *Boletín Técnico, Fac. Nac. Ingeniería, Univ. Técnica, Oruro*, no. 1, p. 5. Hydrous tin oxide, $SnO_2.xH_2O$, as a yellow colloidal powder from Potosí, Bolivia. Named after the late Louis Soux of Potosí. Cf. Varlamoffite (q.v.). [*M.A.*, **10**, 9.] 18th List

Souzalite. W. T. Pecora and J. J. Fahey, 1947. *Program and Abstracts, 28th Annual Meeting, Min. Soc. Amer.*, p. 18; *Bull. Geol. Soc. Amer.*, vol. 58, p. 1216; *Amer. Min.*, 1948, vol. 33, p. 205 (abstr.); *Mineração e Metalurgia*, 1948, vol. 13, p. 53; *Amer. Min.*, 1949, vol. 34, p. 83. Hydrous phosphate, $(Al,Fe)_2O_3.3(Mg,Fe)O.2P_2O_5.5H_2O$, green, fibrous, monoclinic(?); an alteration product of scorzalite (q.v.) from Divino, Brazil. Named after Dr. Antonio José Alves de Souza, formerly director of the Mineral Survey of Brazil. [*M.A.*, **10**, 254, 456, 507.] 18th List

Spalmandite. L. L. Fermor, 1926. *Rec. Geol. Survey India*, vol. 59, p. 205. A contraction of spessartite and almandite for garnets of intermediate composition. *See* Pyralmandite. 11th List

Spandite. L. L. Fermor, 1907. *Records Geol. Survey India*, vol. xxxv, p. 22. The term spessart-andradite, contracted to spandite, is applied to garnets intermediate in chemical composition between spessartite and andradite. 4th List

Sparklite. A trade-name for colourless zircon. R. Webster, *Gems*, 1962, p. 763. 28th List

Speculite. E. H. Liveing, *Engin. and Mining Journ.*, New York, 1903, vol. lxxv, p. 814. A specular gold and silver telluride from Kalgoorlie, Western Australia, resembling sylvanite in colour and perfect cleavage, but differing from this in chemical composition (Au 36, Ag 4%) and sp. gr. (8·64). 3rd List

Spencerite. T. L. Walker, 1916. *Nature*, London, vol. xcvii, p. 375; *Mineralogical Magazine*, vol. xviii, p. 76; *Journ. Washington Acad. Sci.*, 1917, vol. vii, p. 456; *Univ. Studies, Geol. Ser. No. 10*, Toronto, 1918. A. H. Phillips, *Amer. Journ. Sci.*, 1916, ser. 4, vol. xlii, p. 275. A hydrated basic zinc phosphate, $Zn_3(PO_4)_2.Zn(OH)_2.3H_2O$, occurring in some abundance as pearly white, scaly cleavage masses, and small monoclinic crystals, forming large stalactites thinly encrusted with hemimorphite, from Salmo, British Columbia. Named after Leonard James Spencer, of the Mineral Department, British Museum. (*See* Lusitanite.) [Cf. Salmoite]

The same name (Spencerit) had been used earlier (C. Hlawatsch, *Min. Petr. Mitt.*, 1903, vol. xxii, p. 498) for an artificial furnace product, the rhombic modification of iron carbide, Fe_3C, or $(Fe,Mn)_3(C,Si)$ corresponding with Spiegeleisen, as distinct from the anorthic modification corresponding with ferro-manganese (cf. *Mineralogical Magazine*, 1903, vol. xiii, p. 296). 8th List

Spencite. C. Frondel, 1961. *Canad. Min.*, vol. 6, p. 576. A dark brown metamict borosilicate of Ca and Y, near $(Ca,Fe)_2Y_3B_3(Si,Al)_5(O,OH)_{20}$, from Cardiff township, Haliburton Co., Ontario. Named after Hugh S. Spence who collected the material in 1934. 22nd List [Perhaps derived from Hellandite (q.v.) and related to tritomite, *C.M.*, **12**, 66; Tritomite-(Y), Fleischer's *Glossary* (1975); USSR, *M.A.*, 76-2736]

Speziaite. L. Colomba, 1914. *Atti Accad. Sci. Torino*, vol. xlix, p. 625. A member of the amphibole group occurring as dark-green fibres and acicular (monoclinic) crystals at Traversella, Piedmont, and with the ortho-silicate formula $5Fe_4'''(SiO_4)_3$ + $12(Ca,Mg,Fe'',Na_2H_2)_2SiO_4$. Named after the late Professor Giorgio Spezia (1842–1911), of Turin. 7th List [Hornblende; *Amph.*, Sub-Comm.]

Sphaerobertrandite, *see* Spherobertrandite.

Sphærodialogite. A. W. Woodland, 1939. *Quart. Journ. Geol. Soc. London*, vol. 95, p. 34. Rhodochrosite (= dialogite) as minute granular globules in Welsh manganese ore. [*M.A.*, **7**, 437.] 16th List

Sphaeromagnesite. A. Sigmund, 1909. *Die Minerale Niederösterreichs*, p. 71 (Sphäromagnesit); K. A. Redlich, *Zeits. prakt. Geol.*, 1909, vol. 17, p. 307. Radial aggregates of magnesite crystals forming large (10 cm diam.) spheres in a crystalline magnesite on the Eichberg near Gloggnitz, Austria. 9th List

Sphenomanganite. G. Flink, 1919. *Geol. För. Förh. Stockholm*, vol. 41, p. 329 (Sphenomanganit). Crystals of manganite from Långban, Sweden, belong to the sphenoidal class of the orthorhombic system. Until this is proved to be a constant character of the species, the name sphenomanganite is provisionally applied to these crystals. [*M.A.*, **1**, 123.] 9th List

Spherite. W. H. Bucher, 1918. *Journ. Geol. Chicago*, vol. 26, p. 593. Spherical grains, including ovulite (q.v.) with concentric structure and spherulite with radial structure. 16th List [Not the Sphaerite (of Zepharovich, 1867) [Sphärit], changed to Spherite, *A.M.*, **9**, 62. 10th List]

Spherobertrandite. E. I. Semenov, 1957. Труды Инст. Мин., Геохим., Крист., Редк. Элем. (*Trans. Inst. Min. Geochem. Rare Elements*), no. 1, p. 64 (Сферобертрандит). Abstr. Зап. Всесоюз. Мин. Общ. (*Mem. All-Union Min. Soc.*), 1958, vol. 87, p. 485 (Сферобертрандит, spherobertrandite), *Amer. Min.*, 1958, vol. 43, p. 1219 (spherobertrandite), and *M.A.*, **14**, 277 (sphaerobertrandite). Bertrandite high in BeO and low in SiO_2, from pegmatites in the Khibina and Lovozero tundras, Kola peninsula. 'L'individualité de la sphérobertrandite n'est pas démontrée' (*Bull. Soc. franç. Min. Crist.*, 1959, vol. 82, p. 91). 22nd List

Spherocobaltite. Variant of Sphaerocobaltite; *A.M.*, **9**, 61. 10th List. [Cf. Cobaltocalcite (of Frondel)]

Spinellide. A. Lacroix, 1910. *Minéralogie de la France*, vol. 4, p. 297 (Groupe des spinellides). Group name for the spinel group. 9th List

Spiroffite. J. A. Mandarino, S. J. Williams, and R. S. Mitchell, 1962. *Canad. Min.*, vol. 7, p. 340; *Min. Soc. Amer. Spec. Paper no. 1 (1963)*, p. 305. Red to purple cleavable masses from Moctezuma, Sonora, Mexico, have the composition $(Mn,Zn)_2Te_3O_8$; monoclinic. Named for Prof. K. Spiroff. (*Amer. Min.*, 1962, vol. 47, p. 196; 1964, vol. 49, p. 444.) 23rd List

Spodiophyllite. G. Flink, *Medd. om Grönland*, 1900, XXIV, 85. Rough, ash-grey, hexagonal prisms with the physical characters of chlorite, but containing no water. $(Al,Fe)_2(Mg,Fe,Mn)_8(Na_2,K_2)_2Si_8O_{24}$. [Narsarsuk] S Greenland. 2nd List

Spodulite. Local name for a spodumene–quartz mixture chemically equivalent to petalite (E. W. Heinrich, *Indian Journ. Earth Sci.*, 1975, **2**, 18). 29th List

γ-Spodumene. C. T. Li, 1972. *A.M.*, **57**, 321. The *artificial* compound $LiAlSi_2O_6$-III, having a structure based on that of β-quartz. 28th List

Sporogelite. M. Kišpatić, 1912. *Neues Jahrb. Min.*, Beil.-Bd. xxxiv, p. 519; F. Tućan, tom. cit., p. 411 (Sporogelit). C. Doelter and E. Dittler, *Centralblatt Min.*, 1913, p. 193. To replace F. Cornu's name α-kliachite (5th List) for the colloidal form of $Al_2O_3.H_2O$, which occurs as a constituent of bauxite and 'terra rossa'. So named as being the gel corresponding with the crystalloid diaspore. *See* Cliachite and Kljakite. 6th List

Spurrite. F. E. Wright, 1908. *Amer. Journ. Sci.*, ser. 4, vol. xxvi, p. 547. J. E. Spurr and G. H. Garrey, *Economic Geology*, vol. iii, p. 707. A silicate and carbonate of calcium, $2Ca_2SiO_4.CaCO_3$, forming granular masses with glistening surfaces and resembling crystalline limestone in appearance. It effervesces in dilute hydrochloric acid with the separation of gelatinous silica. The optical characters suggest monoclinic symmetry. Occurs with hillebrandite (q.v.) in contact-metamorphic limestone at Valardeña, Durango, Mexico. Named after Josiah Edward Spurr, of New York. 5th List

Staalerts. C. A. Münster, *Nyt Mag. f. Naturv.*, 1892, XXXII, 269; *Skrifter Vid.-Selsk. Christiania*, I, *Math.-natur. Kl.*, 1894, no. 1, p. 54. A massive steel-grey ore which may be 'arsenical silver' or argentiferous mispickel. Kongsberg, Norway. *Staalerts* (Norwegian) = *Stahlerz* (Germ.) = *Steel-ore* (Engl.). 2nd List

Stagmalite. O. C. Farrington, *Field Columbian Museum*, Chicago, Publication 53, Geol. ser., 1901, vol. i, p. 261 (Stagmalites). A general term, to include both stalactite and stalagmite, for formations produced by dropping water. From στάγμα, a drop, and λίθος, a stone. 3rd List

Stainierite. V. Cuvelier, 1929. *Natuurwetenschappelijk Tijdschrift*, Antwerpen, vol. 11, p. 177 (Stainieriet). A. Schoep and V. Cuvelier, *Bull. Soc. Belge Géol. Pal. Hydrol.*, 1930, vol. 39 (for 1929), p. 74. W. F. de Jong, *Natuurwet. Tijds.*, 1930, vol. 12, p. 69. A. Schoep, *Ann. Service Mines, Katanga*, 1930, vol. 1, p. 55. Hydrous sesquioxide of cobalt with some iron and aluminium, $(Co,Fe,Al)_2O_3.H_2O$, regarded as the crystalline equivalent of the colloidal heterogenite. From Katanga, Belgian Congo. Named after Xavier Stainier, Professor of Geology in the University of Gent (= Gand). [*M.A.*, **4**, 248, 347, 501.] 12th List [Is Heterogenite; *M.A.*, **8**, 86; *M.M.*, **33**, 253]

Stanfieldite. L. H. Fuchs, 1967. *Science*, vol. 158, p. 190. A mineral occurring in the Estherville mesosiderite and several pallasites as irregular grains or veinlets is monoclinic, with composition $Ca_4(Mg,Fe,Mn)_5(PO_4)_6$. Named for Stanley Field. [*A.M.*, **53**, 508.] 25th List

Stanierit, error for Stainierite (12th List) (C. Hintze, *Handb. Min.*, Erg.-Bd. II, pp. 619, 954). 24th List

Stannoenargite. G. H. Moh and J. Ottemann, 1962. *Neues Jahrb. Min.*, Abh., vol. 99, p. 1 (Stannoenargit). An unnecessary name for a stannian Enargite. 23rd List

Stannoidite. A. Kato, 1969; *Bull. Nat. Sci. Mus.* (Tokyo), **12**, 165: *Min. Journ.* (*Japan*), **5**, 417. A stannite-like mineral from the Konjo mine, Okayama Prefecture, Japan, has the composition $Cu_5(Fe,Zn)_2SnS_8$; orthorhombic, and distinct from hexastannite. [*A.M.*, **54**, 1495; *M.A.*, 70-1642; *Bull.*, **92**, 514; *Zap.*, **99**, 73.] 26th List [Cf. Mawsonite]

Stannoluzonite. G. H. Moh and J. Ottemann, 1962. *Neues Jahrb. Min.*, Abh., vol. 99, p. 1 (Stannoluzonit). An unnecessary name for a stannian Luzonite. 23rd List

Stanniomicrolite. I.M.A. Subcomm. pyrochlore group, 1978. *A.M.*, **62**, 406. A systematic name to replace sukulaite. 30th List

Stannopalladinite. I. N. Maslenitzky, P. V. Faleev, and E. V. Iskyul, 1947. *Doklady Acad. Sci. USSR*, vol. 58, p. 1137 (станнопалладинит). Cubic alloy of tin and palladium, Pd_3Sn_2, with some Pt and Cu. [*M.A.*, **10**, 453.] 18th List [*A.M.*, **56**, 360]

Stannotantalite. V. V. Matias, 1961. [Геол. Мест. Редк. Элем. (*Geol. Depos. Rare Elem.*), vol. 9, p. 30], abstr. in Зап. Всесоюз. Мин. Общ. (*Mem. All-Union Min. Soc.*), 1962, vol. 91, p. 190 (Оловотанталит), in *Amer. Min.*, 1961, vol. 46, p. 1514 (Tin-tantalite, Olovotantalite), and in *M.A.*, **16**, 552 (Stannotantalite). Black crystals, possibly monoclinic, with a composition near $(Mn,Sn)(Ta,Nb,Sn)_2O_6$. Possibly identical with the unnamed phase obtained by heating Ixiolite, or may have the same cell-dimensions as Wodginite (*Amer. Min.*, 1963, vol. 48, p. 216). 23rd List

Stantienite. E. Pieszczek [*Archiv f. Pharmacie*, [iii], XIV, 433–436]. *Journ. Chem. Soc. Abstracts*, 1881, XL, 687. A black resin occurring with Prussian amber. 2nd List

Staringite. E. A. J. Burke, C. Kieft, R. O. Felius, and M. S. Adusumilli, 1969. *Min. Mag.*, **37**, 447. Small inclusions of $Fe_{0.5}Sn_{4.5}TaO_{12}$, with some Mn, Ti, and Nb, occur as inclusions in tapiolite from Seridózinho and Pedra Lavreda, Paraiba State, Brazil; tetragonal, probably with a trirutile structure. Named for Dr. W. C. H. Staring. [*M.A.*, 70-759; *A.M.*, **55**, 1446.] 26th List

Starkeyite. O. R. Grawe, 1945. *Missouri Geol. Surv.*, ser. 2, vol. 30, p. 209; *Amer. Min.*, 1956, vol. 41, p. 662. First described from the X-ray pattern as $FeSO_4.4H_2O$, and later shown to be $MgSO_4.4H_2O$ [leonhardtite, 19th List, not to be confused with leonhardite*]. Named from the locality, Starkey mine, Madison Co., Missouri. [*M.A.*, **13**, 302.] 21st List [*Corr. in *M.M.*, **31**] [Name replaces Leonhardtite]

Starlite. G. F. Kunz, 1927. *Jewelers' Circular*, New York, vol. 94, p. 65 (Starlite). *Amer. Min.*, 1927, vol. 12, p. 265 (starlight), p. 294 (starlite). Trade-name for 'artificially coloured' blue zircon from Siam. Named from star and *λίθος*, stone. 11th List

Starolite. (J. B. Sterrett, *Mineral Resources, US Geol. Survey*, for 1914, 1915, part II, p. 324; W. T. Schaller, ibid., for 1917, 1918, part II, p. 162.) Jeweller's trade-name for asteriated quartz showing a six-rayed star by reflected light. 8th List

—— A trade-name for a rose-quartz doublet showing asterism. R. Webster, *Gems*, 1962, p. 763. Not, apparently, the starolite of the 8th List, nor to be confused with staurolite nor starlite (11th List). 28th List

Stasite. A. Schoep, 1922. *Compt. Rend. Acad. Sci. Paris*, vol. 174, p. 875. Hydrated phosphate of uranium and lead, $8UO_3.4PbO.3P_2O_5.12H_2O$, dimorphous with dewindtite (q.v.), from which it differs in density, colour, and form of the minute crystals. From Kasolo, Katanga. Named after the Belgian chemist, Jean Servais Stas (1813–1891). [*M.A.*, **1**, 377.] 9th List [The same as Dewindtite (Schoep, 1923, ref. *see* Parsmsite)]

Staszicite. J. Morozewicz, 1918. *Bull. Intern. Acad. Sci. Cracovie, Cl. Sci. Math. Nat.*, ser. A, for 1918, p. 4 (Staszycyt, Staszicit). Basic arsenate of calcium, copper, and zinc, $R_3(AsO_4)_2.2R(OH)_2$, forming yellowish-green botryoidal masses with radially-fibrous (orthorhombic?) structure. From Miedzianka, Poland. Named after Stanisław Staszic (1755–1826), a Polish statesman. [*M.A.*, **2**, 51.] 10th List

Stcherbakovite. *Bull. Soc. Franç. Min. Crist.*, 1955, vol. 78, p. 352. French transliteration of шербаковит, shcherbakovite (20th List). 21st List

Steadite. A. Sauveur, 1902. *Journ. Iron and Steel Inst.*, vol. 61 (no. I for 1902), p. 118; *The Metallography of Iron and Steel*, New York, 1912, lesson xx, p. 3; 2nd edit., 1916, p. 391. A eutectic consisting of iron phosphide (Fe_3P about 61%) with iron, the latter containing some phosphorus in solution. As a constituent of grey cast-iron it was first observed in 1900 by Dr. J. E. Stead, FRS, of Middlesbrough.

The same name was also later applied by V. A. Kroll (*Journ. Iron and Steel Inst.*, 1911, vol. 84 (no. II for 1911), pp. 130, 186) to a basic calcium silico-phosphate, $3(3CaO.P_2O_5).2CaO.(2CaO.SiO_2)$, occurring as yellow, hexagonal needles in the basic slag of the Thomas-Gilchrist process for the dephosphorisation of iron, and first described by

J. E. Stead and C. H. Ridsdale (*Journ. Chem. Soc. London*, 1887, vol. 51, p. 605). *See also* Hilgenstockite, Silico-Carnotite, and Thomasite. 9th List

Steenstrupite. O. B. Boeggild, *Meddelelser om Grönland*, 1901, vol. xxiv, p. 203. The same as steenstrupine. 3rd List

Steigerite. E. P. Henderson, 1935. *Amer. Min.*, vol. 20, p. 769. Hydrated vanadate of aluminium, $Al_2O_3.V_2O_5.6\frac{1}{2}H_2O$, as a canary-yellow amorphous powder in sandstone from [Gypsum Valley, San Miguel Co.] Colorado. Named after Dr. George Steiger (1869–), formerly chief chemist of the United States Geological Survey. [*M.A.*, **6**, 260.] 14th List

Stellarite, error for Stellerite (*Bull. Soc. franç. Min. Crist.*, **91**, 307) (not the Stellarite of How, 1869). 26th List

Stellerite. J. A. Morozewicz, 1909. *Bull. Intern. Acad. Sci. Cracovie*, 1909, p. 344 (stellerycie, Stellerit). Calcium alumo-hepta-silicate, $CaAl_2Si_7O_{18}.7H_2O$. An orthorhombic zeolite, resembling stilbite, found in diabase-tuff from the Komandor Islands, Bering Sea. Named after Georg Wilhelm Steller (1709–1746), the discoverer of the Komandor Islands. 5th List [*A.M.*, **53**, 511]

Stelznerite. A. Arzruni, *Zeits. Kryst. Min.*, 1899, XXXI, 232; abstr. *Min. Mag.*, XII, 308. Green, transparent, prismatic crystals resembling brochantite in appearance. Rhombic. $CuSO_4.2Cu(OH)_2$. [Remolinos, Vallinar] Chili. 2nd List [Is Antlerite]

Stenhuggarite. P. B. Moore, 1967. *Canad. Min.*, **9**, 301. Orange tetragonal crystals of $CaFe^{3+}SbO(AsO_3)_2$ from Långban, Sweden, are named for Brian Mason (Swedish *stenhuggar*, stonemason). [*A.M.*, **53**, 1427; *Bull.*, **81**, 305; *Zap.*, **98**, 329.] 26th List [*A.M.*, **56**, 636]

Stenonite. H. Pauly, 1962. *Medd. om Grønland*, vol. 169, no. 9. $Sr_2AlCO_3F_5$, in monoclinic grains in the Cryolite mine at Ivigtut, S Greenland. Named for N. Steensen (= Steno), 1638–1686 (*Amer. Min.*, 1963, vol. 48, p. 1178). [*M.A.*, **16**, 648.] 23rd List

Stepanovite. E. I. Nefedov, 1953. *Mém. Soc. Russ. Min.*, ser. 2, vol. 82, p. 317 (Степановит). Oxalate $NaMgFe'''(C_2O_4)_3.8\text{–}9H_2O$, trigonal, yellowish-green, granular, in coal. [*M.A.*, **12**, 353.] 20th List [*A.M.*, **49**, 442]

Stepanowit, German transliteration of Степановит, stepanovite (20th List) (Hintze, *Handb. Min.*, Erg.-Bd. II, p. 848). 22nd List

Sterretit, error for Sterrettite (16th List) (C. Hintze, *Handb. Min.*, Erg.-Bd. II, pp. 620, 955). 24th List

Sterrettite. E. S. Larsen, 3rd, and A. Montgomery, 1940. *Amer. Min.*, vol. 25, p. 513. Hydrous basic aluminium phosphate, $Al_6(PO_4)_4(OH)_6.5H_2O$, as minute, colourless, orthorhombic crystals in variscite nodules from Fairfield, Utah. Named after Dr. Douglas Bovard Sterrett. Shown by F. A. Bannister (*Min. Mag.*, **26**, 131) to be identical with eggonite. [*M.A.*, **8**, 3.] 16th List [$ScPO_4.2H_2O$, *A.M.*, **45**, 257; related to Kolbeckite]

Sterryite. J. L. Jambor, 1967. *Canad. Min.*, vol. 9, pp. 7, 191. An orthorhombic mineral from Madoc, Ontario, with composition $Pb_7(Sb,As)_8S_{19}$. Named for T. Sterry Hunt. [*A.M.*, **53**, 1423; *Bull.*, **91**, 304.] 25th List

Stevensite. A. R. Leeds [1873(?).] [A. H. Chester, *Dictionary of the Names of Minerals*, 1896, p. 257.] Talc pseudomorphous after pectolite. [Bergen Hill, New Jersey.] 1st List [$Mg_3Si_4O_{10}(OH)_2$. *M.A.*, **16**, 541; *A.M.*, **38**, 973; **44**, 343]

Stewartite. J. R. Sutton, 1911. *Nature*, London, vol. lxxxvii, p. 550. A variety of bort, or iron-bort, possessing magnetic properties; from the diamond mines of Kimberley, South Africa. Named after Mr. James Stewart, the manager of the pulsator at the diamond mines. 6th List

—— W. T. Schaller, 1912. *Journ. Washington Acad. Sci.*, vol. ii, p. 144. A hydrous manganese phosphate, probably triclinic, occurring as an alteration-product of lithiophilite in the Stewart gem-tourmaline mine, Pala, San Diego Co., California. Named after the locality. 6th List [$MnFe_2(PO_4)_2(OH)_2.8H_2O$; *A.M.*, **59**, 1272]

Stibarsen. P. E. Wretblad, 1941. *Geol. För. Förh. Stockholm*, vol. 63, pp. 38, 46 (Stibarsen, Allemontite II). Allemontite from Varuträsk, Sweden, of the composition AsSb. Allemontite I consists of an eutectoid mixture of this with the Sb phase, and allemontite III with the As phase. [*M.A.*, **8**, 98.] 16th List [Cf. *A.M.*, **21**, 202]

Stiberite. H. How. [T. Egleston, *Catalogue of Minerals*, 3rd edit., 1892, p. 328. A. H. Chester, *Dictionary of the Names of Minerals*, 1896, p. 257. In the British Museum since 1879.] Syn. of ulexite. 1st List

Stibiobaumhauerite. W. Nowacki, 1964. *Schweiz. Min. Petr. Mitt.*, vol. 44, p. 459. The antimony analogue of baumhauerite is stated to occur naturally but no details are given. 25th List

Stibiobismutantalit, error for Stibiobismutotantalite (19th List) (C. Hintze, *Handb. Min.*, Erg.-Bd. II, pp. 375, 452, 955). 24th List

Stibiobismuthinite. G. A. Koenig, *Journ. Acad. Nat. Sci. Philadelphia*, ser. 2, vol. xv, p. 424. Large, cleavable prisms resembling stibnite rather than bismuthinite in appearance, from Nacozari, Sonora, Mexico, consisting of bismuth sulphide with Sb 8·12%; formula $(Bi,Sb)_4S_7$. *See* Aurobismuthinite. 7th List

Stibiobismutotantalite. M. C. Bandy, 1951. *Rocks and Minerals*, vol. 26, p. 521. A variety of stibiotantalite ($Sb(Ta,Nb)O_4$) containing Bi_2O_3 3·98%, from Mozambique. [*M.A.*, **11**, 477.] 19th List

Stibiocolumbite. W. T. Schaller, 1915. 3rd Appendix to 6th edit. of Dana's *System of Mineralogy*, p. 74. Name proposed for the 'stibiotantalite' from Mesa Grande, California, described by S. L. Penfield and W. E. Ford (1906), because [in one analysis, cf. W. E. Ford, *Amer. Journ. Sci.*, 1911, vol. xxxii, p. 287] the amount of niobium (columbium) is greatly in excess of the tantalum. 7th List

Stibio-domeykite. G. A. Koenig, *Amer. Journ. Sci.*, 1900, ser. 4, vol. x, p. 445; *Zeits. Kryst. Min.*, 1901, vol. xxxiv, p. 67. Massive domeykite containing a small amount of antimony (0·78 to 1·29%). From the Mohawk mine, Lake Superior copper region. 3rd List

Stibiodufrenoysite. W. Nowacki, 1964. *Schweiz. Min. Petr. Mitt.*, vol. 44, p. 459; I. Burkart-Baumann, J. Ottemann, and G. C. Amstutz, *Neues Jahrb. Min.*, 1966, Monatsh., 353. Crystalline inclusions in amorphous Pb–As–Sb sulphides from Cerro de Pasco, Peru, gave an X-ray diagram corresponding to dufrenoysite; electron-probe analysis gave a composition $Pb_2(Sb,As)_2S_5$ with Sb > As. Similar material had been stated by Nowacki to occur naturally, but no details were given. Named from the composition. Probably identical with veenite (q.v.). 25th List

Stibioenargite. Author? H. Strunz, *Min. Tab.*, 3rd edit., 1957, p. 97 (Stibioenargit). $Cu_3Sb.S_4$, orthorhombic. Artificial(?). Natural enargite $Cu_3(As,Sb)S_4$ contains up to 6% Sb. 21st List

Stibio-luzonite. *See* Antimon-luzonite. 3rd List

Stibioluzonite. H. Strunz, 1957. *Min. Tabellen*, 3rd edit., p. 97 (Stibioluzonit). A synonym of Famatinite. Not to be confused with the Stibio-luzonite of S. Stevanovic (3rd List), an antimonian Luzonite. 23rd List

Stibiomicrolite. P. Quensel and T. Berggren, 1938. *Geol. För. Förh. Stockholm*, vol. 60, p. 216 (Stibiomicrolite). O. Rosén and A. Westgren, ibid., p. 226 (Stibio-Microlite). An unknown mineral with the assumed composition $(Sb,Ca)(Ta,Nb)(O,OH)_4$, which by alteration has given rise to a mixture of stibiotantalite and microlite, together with some antimony and senarmontite. [*M.A.*, **7**, 120.] 15th List [*A.M.*, **62**, 407]

Stibioniobite. R. Koechlin, *Min. Taschenb., Wien. Min. Gesell.*, 2nd edit., 1928, p. 60 (Stibioniobit). Syn. of stibiocolumbite (7th List). 18th List

Stibiopalladinite. P. A. Wagner, 1929. *The platinum deposits and mines of South Africa*, Edinburgh and London, 1929, pp. 12, 15, 18, 227, 237. H. Schneiderhöhn, *Centr. Min.*, Abt. A, 1929, p. 193; *Chem. Erde*, 1929, vol. 4, pp. 268, 275. Palladium antimonide, Pd_3Sb, as minute white grains (cubic?) in the platiniferous norite of the Bushveld, Transvaal. First described as 'a new palladium mineral' by H. R. Adam, *Journ. Chem. Metall. Mining Soc. South Africa*, 1927, vol. 27, p. 249, and later named from the chemical composition. [*M.A.*, **3**, 369; **4**, 145, 149.] 12th List [*A.M.*, **58**, 1; **61**, 1249]

Stibiopearceite. A. D. Genkin and M. G. Dobrovolskaya, 1965. [Труды Мин. Муз. Акад. наук СССР (*Proc. Min. Mus. Acad. Sci. USSR*), vol. 16, p. 90], abstr. Зап. Всесоюз. Мин. Общ. (*Mem. All-Union Min. Soc.*), 1966, vol. 95, p. 310 (Стибиопирсит). Synonym of Antimonpearceite (23rd List). 24th List

Stibioskleroklas. W. Nowacki, 1964. *Schweiz. Min. Petr. Mitt.*, vol. 44, p. 459. The antimony analogue of sartorite, $Pb(Sb,As)_2S_4$, with Sb:As ≈ 2, is stated to occur naturally, but no details are given. Probably identical with twinnite (q.v.). 25th List

Stibiotantalite. G. A. Goyder, *Journ. Chem. Soc. Trans.*, LXIII. 1076, 1893. $(Ta,Nb)_2O_5.Sb_2O_3$[?] Rhombic? [Greenbushes] Western Australia. 1st List [Cf. Stibiocolumbite]

Stibio-tellurobismutite. I. G. Magakyan, 1956. [*Doklady Acad. Sci. Armenian SSR*, 1956, vol. 23, p. 215.] Abstr. in *Mém. Soc. Russ. Min.*, 1958, vol. 87, p. 76 (стибио-теллуровисмутит, stibio-tellurobismutite). $(Bi,Sb)_2Te_3$, rhombohedral, containing Sb 2·7%. From Zodsk, Armenia. 21st List [Stibian Tellurbismuth]

Stichtite. W. F. Petterd, 1910. *Catalogue of the Minerals of Tasmania*, 3rd edit., Hobart, 1910, p. 167; *Papers Roy. Soc. Tasmania*, 1910, p. 167. A foliated mineral of a bright lilac colour, which previous to analysis had been mistaken for kaemmererite. A hydrated carbonate of magnesium, chromium, and iron, $(Cr,Fe)_2O_3.6MgO.CO_2.13H_2O$. It occurs with chromite as spots and veins in serpentine at Dundas, Tasmania. Named after Mr. Robert Sticht, of Tasmania. Cf. chrom-brugnatellite. 6th List [Cf. Barbertonite]

Stiepelmannite. P. Ramdohr and E. Thilo, 1940. *Zentr. Min.*, Abt. A, 1940, p. 1 (Stiepelmannit). Basic phosphate of yttrium, ytterbium, and aluminium, $(Yt,Yb)PO_4.AlPO_4.2Al(OH)_3$, as rhombohedral crystals isomorphous with the cerium mineral florencite. Named after Mr. – Stiepelmann, owner of the gem mine in SW Africa where the mineral was found. [*M.A.*, **7**, 514.] 15th List [Is Florencite; *A.M.*, **32**, 485]

Stilleite. P. Ramdohr, 1956. *Geotektonisches Symposium zu Ehren von Hans Stille*, Stuttgart, 1956, p. 481 (Stilleit). Abstr. in *Amer. Min.*, vol. 42, p. 584. Microscopical examination of an ore sample from Katanga showed a cubic mineral with X-ray pattern similar to that of artificial ZnSe. Named after Hans Stille (1876–), German geologist. 21st List

Stillwellite. J. McAndrew and T. R. Scott, 1955. *Nature*, London, vol. 176, p. 509. P. Gay, *Min. Mag.*, 1957, vol. 31, p. 465. Boro-silicate of lanthanons, $(Ln,Ca)BSiO_5$, rhombohedral, in radioactive ore from Queensland. Named after Dr. Frank Leslie Stillwell, of Melbourne. [*M.A.*, **13**, 7; *A.M.*, **41**, 370.] 21st List

Stilpnochlorane. F. Krestschmer, 1905. *Centralblatt Min.*, 1905, p. 203; 1907, p. 292 (Stilpnochloran). A chloritic mineral occurring as shining yellow scales in the iron-mines at Gobitschau, Moravia. It is an alteration product of thuringite, and has the composition $H_{24}(Al,Fe)_{10}(Ca,Mg)Si_9O_{46}$. Named from στιλπνός, glistening, and χλωρός, pale-green, yellow. 4th List [Is Nontronite; *A.M.*, **20**, 485]

Stipoverite. D. P. Grigoriev, 1962. [Вест. Акад. Наук СССР, vol. 4, p. 21], abstr. in *Amer. Min.*, 1963, vol. 48, p. 434. A name proposed for Stishovite, but discarded (Зап. Всесоюз. Мин. Общ. (*Mem. All-Union Min. Soc.*), 1962, vol. 91, p. 635). 23rd List

Stishovite. E. C. T. Chao, J. J. Fahey, J. Littler, and D. J. Milton, 1962. *Journ. Geophys. Research*, vol. 67, p. 419. A dense polymorph of silica isolated from the coesite-bearing Coconino sandstone at Meteor Crater, Arizona, proves to be identical with the high-pressure polymorph described by S. M. Stishov (С. М. Стишов) and S. V. Popova (С. В. Попова), and is named for the former. (*Amer. Min.*, 1962, vol. 47, p. 807; Зап. Всесоюз. Мин. Общ. (*Mem. All-Union Min. Soc.*), 1963, vol. 92, p. 197.) 23rd List [Cf. Stipoverite]

Stistaite. E. P. Nikolaeva, V. A. Grigorenko, and P. E. Tsypkina, 1970. *Zap.*, **99**, 68 (Стистаит). Light-grey cubic crystals with metallic lustre, *a* 4·15 Å, from placer samples near the Elkiaidan river, Uzbekistan, associated with native tin and Cu(Sn,Sb), have the composition SnSb, and are identical with the synthetic product. Named for the composition (STIbium, STAnnum). [*A.M.*, **56**, 358; *M.A.*, 71-550; *Zap.*, **100**, 78.] 27th List

Stoffertite. C. Klein, *Sitz.-ber. Akad. Wiss. Berlin*, 1901, p. 722 (Stoffertit). Differs from brushite ($HCaPO_4.2H_2O$) only in containing a little more water ($HCaPO_4.2\frac{3}{4}H_2O$). This difference is probably not essential, but, in case it should at some future time be proved to be so, the new name is provisionally given. From the guano deposits of the island of Mona, Porto Rico, West Indies. Named after Dr. Adolph Stoffert, of Hamburg, who sent the material for examination. 3rd List [Is Brushite; *M.A.*, **9**, 33]

Stokesite. A. Hutchinson, *Phil. Mag.*, 1899, [v], XLVIII, 480; *Proc. Cambridge Phil. Soc.*, 1900, X, 216; *Min. Mag.*, 1900, XII, 274. A colourless crystal with the habit of gypsum. Rhombic. $CaO.SnO_2.3SiO_2.2H_2O$. Cornwall. 2nd List [*M.M.*, **41**, 411 (with refs.)]

Stottite. H. Strunz, G. Söhnge, and B. H. Geier, 1958. *Neues Jahrb. Min.*, Monatsh., p. 85

(Stottit). Ferrous germanate, $Fe^{\cdot\cdot}Ge(OH)_6$, tetragonal, occurring with tennantite and renierite at Tsumeb, SW Africa. Named after Mr. Charles E. Stott. [*M.A.*, **14**, 281.] *See also* J. Zemann, *Neues Jahrb. Min.*, Monatsh., 1959, p. 67, and H. Strunz and M. Giglio, *Fortschr. Min.*, 1960, vol. 38, p. 40. [*A.M.*, **43**, 1006.] 22nd List

Stranskiite. H. Strunz, 1960. *Naturwiss.*, vol. 47, p. 376 (Stranskiit). $Zn_2Cu(AsO_4)_2$, anorthic, blue crystals, from Tsumeb, SW Africa, and synthetic. Named after Prof. I. N. Stranski of Berlin. [*A.M.*, **45**, 1315.] 22nd List [$(Zn,Cu)_3(AsO_4)_2$; *A.M.*, **63**, 213]

Straschimirit, Germ. trans. of Страшимирит, strashimirite (26th List). Hintze, 89. 28th List

Strashimirite. I. Mincheva-Stefanova, 1968. Зап. Всесоюз. мин. общ. (*Mem. All-Union Min. Soc.*), **97**, 470 (Страшимирит). Pale green spherulites in the Zapachitsa copper deposit, Stara-Planina, Bulgaria, are monoclinic, composition near $Cu_2AsO_4OH.1{\cdot}25H_2O$, with some Zn replacing Cu. Named for Strashimir Dimitrov of Bulgaria. [*A.M.*, **54**, 1221; *M.A.*, 69-1541; *Bull.*, **92**, 318; *Zap.*, **98**, 328.] 26th List

Strashmirite, error for Strashimirite (*M.A.*, **69**, 1541). 26th List [Also 28th List]

Strätlingite. G. Hentschel and H.-J. Kuzel, 1976. *Neues Jahrb. Min.*, Monatsh. 326. $Ca_2Al_2SiO_7.8H_2O$ ('gehlenite hydrate') occurs naturally in a limestone inclusion in basalt at Bellerberg, Mayen, Germany. Uniaxial negative plates, space group *R*3 or $R\bar{3}$, *a* 5·737, *c* 37·59. Named for W. Strätling, who described the artificial compound in 1938. 29th List

Strelkinite. M. A. Alekseeva, A. A. Chernikov, D. P. Shashkin, E. A. Komkov, and I. N. Gavrilova, 1974. *Zap.*, **103**, 576 (Стрелкинит). Yellow powdery crusts and small plates from an unnamed Russian locality have space group *Pnmm* or *Pnm*2, *a* 10·64, *b* 8·36, *c* 32·72 Å. Composition $8[NaUO_2VO_4.3H_2O]$, 'the Na analog of carnotite'. Named for M. F. Strelkin. (Cf. Sodium carnotite (artificial), 24th List.) [*A.M.*, **60**, 488; *M.A.*, 75-2527; *Zap.*, **104**, 612.] 29th List

Striegovite. P. Groth, *Tab. Uebers. Min.*, 4th edit., 1898, p. 135. The same as strigovite. 2nd List

Stillwaterite. L. J. Cabri, J. H. G. Laflamme, J. M. Stewart, J. F. Rowland, and T. T. Chen, 1975. *Can. Mineral.*, **13**, 321. Small anhedral grains from the Stillwater Complex, Montana, are hexagonal, *a* 7·399, *c* 10·311 Å, space group *P*3 or $P\bar{3}$. Composition $3[Pd_8As_3]$. Named from the locality. [*A.M.*, **62**, 1060; *Bull.*, **99**, 344; *M.A.*, 77-886; *Zap.*, **105**, 71.] 30th List

Stringhamite. J. R. Hindeman, 1976. *A.M.*, **61**, 189. Deep-blue crystals with thaumasite in a diopside–magnetite skarn at the Bwana mine, Milford, Utah, are monoclinic, *a* 5·028, *b* 16·07, *c* 5·303 Å, *β* 102·58°, space group $P2_1/c$; α 1·709, light grey-blue, ∥ [010], *β* 1·717, light blue, *γ* 1·729, dark blue; *β*:[001] 2·5°. Composition $CuCaSiO_4.2H_2O$. Named for B. F. Stringham. Appears to be identical with the unnamed 'Mineral F' of Woodford, 1943. A polytype with doubled *a*-axis occurs at the Christmas mine. 29th List

Strongite. A trade-name for *artificial* (synthetic) spinel. R. Webster and M. J. O'Donoghue, both *priv. comm. See* under corundolite. 28th List

Strontian-apatite. E. S. Larsen, M. H. Fletcher, and E. A. Cisney, 1952. *Amer. Min.*, vol. 37, p. 656 (Strontian apatite, Strontian fluorapatite, Strontian hydroxylaptite). Variety of apatite containing SrO 11·6%. [*M.A.*, **12**, 225.] 20th List

Strontiapatite. D. McConnell, 1973. *Apatite: its crystal chemistry, mineralogy, utilization, and geologic and biologic occurrences.* Vienna and New York (Springer), 90. The name strontium apatite having been used in four different senses (15th, 16th, 22nd, and 23rd Lists), $(Sr,Ca)_5(PO_4)_3(OH,F)$, the strontium apatite of Efimov, Kravchenko, and Vasileva, is renamed. A useful clarification. [*M.M.*, **39**, 617.] 28th List

Strontioborite. V. V. Lobanova, 1960. Доклады Акад. Наук СССР (*Compt. Rend. Acad. Sci. URSS*), vol. 135, p. 173 (Стронциоборит). Near $(Sr,Ca)_2MgB_{12}O_{21}.4\frac{1}{2}H_2O$, in soils from the Caspian region. 'Requires verification. Some of the data could be construed as indicating a mixture of strontioginorite, boracite, and anhydrite' (M. Fleischer, *Amer. Min.*, 1961, vol. 46, p. 768). 22nd List [Unit cell and formula disagree; *A.M.*, **50**, 1508]

Strontiodresserite. J. L. Jambor, A. P. Sabina, A. C. Roberts, and B. D. Sturman, 1977. *Can. Mineral.*, **15**, 405. Coatings on specimens of a silicocarbonatite sill at St. Michel, Montreal Island, Quebec, are the Sr analogue of dresserite (26th List). Orthorhombic, *a* 9·14, *b* 15·91, *c* 5·59 Å, *ρ* 2·71 g.cm^{-3}. Composition $4[(Sr,Ca)Al_2(CO_3)_2(OH)_4.H_2O]$. RIs α 1·510, in plane of platy crystals, *β* 1·583, ∥ elongation, $2V_\alpha$ $42\frac{1}{2}$°. Named from the composition and analogy with dresserite. 30th List

Strontiogehlenite. C. Brisi and F. Abbatista, 1960. [*Ann. Chim.* (*Roma*), vol. 50, p. 1061], cited by H. Strunz, *Min. Tabellen*, 4th edit., 1966, p. 346. *Artificial* tetragonal $Sr_2Al_2SiO_7$, the strontium analogue of gehlenite. 24th List. P. S. Dear, 1969. *Lithos*, **3**, 13. 26th List

Strontioginorite. O. Braitsch, 1959. *Beitr. Min. Petr.*, vol. 6, p. 366. A strontian variety of ginorite, $(Sr,Ca)_2B_{14}O_{23}.8H_2O$, from the 'Old Halite' bed of the Königshall-Hindenburg mine, Reyershausen, Germany. The unit cell and refractive indices are near those of volkovite (20th List). Cf. Strontium ginorite. 22nd List

Strontiohilgardite-1Tc. O. Braitsch, 1959. *Beitr. Min. Petr.*, vol. 6, p. 233 (1Tc-Strontiohilgardit). $(Ca,Sr)_2B_5O_8(OH)_2Cl$, from the Königshall-Hindenburg mine, Reyershausen, Germany. Apart from a higher strontium content (not in itself a good reason for a new name), this anorthic mineral differs from parahilgardite (calciumhilgardite-3Tc) by having a *c*-axis of 6·61 Å instead of 17·50 Å. [*A.M.*, **44**, 1102.] 22nd List

Strontiohitchcockite. E. T. Wherry, 1916. *See* Bariohitchcockite. 8th List [Is Goyazite]

Strontium-anorthite. E. Dittler and H. Lasch, 1931. *Sitzungsber. Akad. Wiss. Wien, Math.-naturw. Kl.*, Abt. I, p. 659 (Strontium-anorthit). Artificially prepared $SrAl_2Si_2O_8$, triclinic and isomorphous with anorthite. [*M.A.*, **5**, 102.] 13th List

Strontium-apatite. A. N. Winchell, 1927. *Elements of optical mineralogy*, 2nd edit., part 2, p. 128 (Strontiumapatite). Synonym of fermorite (5th List). 15th List

—— H. Strunz, 1957. *Min. Tabellen*, 3rd edit., p. 235 (Strontium-Apatit). An unnecessary name for strontian apatite. 22nd List

—— A. F. Efimov, S. M. Kravchenko, and Z. V. Vasileva, 1962. Докл. Акад. Наук СССР (*Compt. Rend. Acad. Sci. URSS*), vol. 142, p. 439 (Стронцийапатит). The use of this name for all phosphatic members of the Apatite family with > 50 atomic % Sr is proposed. (*Amer. Min.*, 1962, vol. 47, p. 808; Зап. Всесоюз. Мин. Общ. (*Mem. All-Union Min. Soc.*), 1963, vol. 92, p. 206 (Стронцийапатит, strontiumapatite); *M.A.*, **15**, 542). Not to be confused with the Strontiumapatite of Winchell (15th List), the Strontium apatite of Volkova and Melentiev [*see* Saamite, 16th List], or the Strontium-Apatit of Strunz (22nd List). The use of this name to replace Phosphate-belovite (21st List) has also been suggested; with all these alternatives the term is best dropped altogether. 23rd List [See Strontiapatite]

Strontium-aragonite. C. O. Hutton, 1936. *Trans. Roy. Soc. New Zealand*, vol. 66, p. 36 (Strontium-aragonite). A variety of aragonite containing $SrCO_3$ 5·51%, from Otago. Synonym of mossotite. [*M.A.*, **6**, 364.] 14th List

Strontium-arsenapatite. H. Strunz, *Min. Tab.*, 1941, p. 156 (Strontiumarsenapatit). Syn. of fermorite (5th List). 18th List

Strontium-barylite. J. Ito and C. Frondel, 1968. *Arkiv. Min. Geol.*, vol. 4, p. 391. $SrBe_2Si_2O_7$, the synthetic strontium analogue of barylite. 25th List

Strontium ginorite. O. Braitsch, 1959. *Beitr. Min. Petr.*, vol. 6, p. 370 (Strontiumginorit). The strontium analogue of ginorite, $Sr_2B_{14}O_{23}.8H_2O$, as distinct from Strontioginorite (q.v.). 22nd List

Strontiumolivine. H. Hayashi, N. Yamamoto, M. Yoshida, M. Mizuno, T. Noguchi, K. Yamamoto, 1964. [*Rept. Govt. Indust. Res. Inst.*, Nagoya, vol. 13, p. 318]; abstr. *Min. Journ.* [*Japan*], 1965, vol. 4, p. 323. *Artificial* Sr_2SiO_4. 24th List

Strontiumthomsonite. A. F. Efimov, S. M. Kravchenko, and E. M. Vlasova, 1963. [Труды Инст. Мин., Геохим., Крист. Редк. Элем. (*Trans. Inst. Min., Geochem., Cryst. Rare Elem.*), no. 16, p. 141]; abstr. *Amer. Min.*, 1965, vol. 50, p. 2100; *M.A.*, **17**, 402; Зап. Всесоюз. Мин. Общ. (*Mem. All-Union Min. Soc.*), 1965, vol. 94, p. 199 (Стронцийтомсонит). A rose-coloured mineral from the Inagli massif gave on analysis: $(Na_{1.5}K_{0.09})(Ca_{0.63}Sr_{0.44}Ba_{0.01})(Al_{4.14}Fe_{0.02}Mg_{0.60}Mn_{0.01})Si_{5.86}O_{20.46}.4{\cdot}28H_2O$. It is interpreted as a substituted thomsonite. 24th List

Strontium weilite. M. A. Nobar and A. P. Dalvi, 1977. *Bull.*, **100**, 353. *Artificial* $SrHAsO_4$, the Sr analogue of weilite (23rd List), is anorthic, *a* 7·43, *b* 6·97, *c* 7·40 Å, α 96° 0′, β 103° 31′, γ 86° 52′. 30th List

Strunzite. C. Frondel, 1957. *Neues Jahrb. Min.*, Monatshefte, 1957, p. 222 (Strunzit). Monoclinic polymorph of laueite (20th List) and pseudo-laueite (q.v.), $MnFe'''_2(PO_4)_2(OH)_2.8H_2O$, as straw-yellow radiating fibres from alteration of triphyline, at Hagendorf, Bavaria, and twelve United States localities. Named after Prof. Hugo Strunz of Regensburg and Berlin. [*A.M.*, **43**, 793.] 21st List

Strüverite. F. Zambonini, 1907. *Rend. Accad. Sci. Napoli*, ser. 3, vol. xiii, p. 35. A black, tetragonal mineral, with angles near those of rutile and tapiolite, found in pegmatite at Craveggia, Piedmont. In chemical composition it is near to ilmenorutile. Named after Professor Giovanni Strüver, of Rome. Not the strüverite of A. Brezina, 1876. 4th List [Trirutile structure; *A.M.*, **49**, 792]

Studtite. J. F. Vaes, 1947. *Ann.* (*Bull.*) *Soc. Géol. Belgique*, vol. 70, p. B 223. Hydrous uranium carbonate as yellow flexible fibres, orthorhombic, from Katanga. Named after F. E. Studt, geologist, Tanganyika Concessions. [*M.A.*, **10**, 146.] 18th List [*A.M.*, **59**, 166. Claimed to be $UO_4.4H_2O$]

Stuetzite. E. T. Wherry, *Journ. Washington Acad. Sci.*, 1920, vol. 10, p. 491. Variant of Stützite. 9th List

Stumpflite. Z. Johan and P. Picot, 1972. *Bull.*, **95**, 610. Small, highly reflecting grains in platinum concentrates from Driekop, Transvaal, are hexagonal, *a* 4·175, *c* 5·504 Å; 2[Pt(Sb,Bi)], with the nickeline structure. Named for E. F. Stumpfl, who first described (but did not name) the mineral (*M.M.*, 1961, **32**, 833). [*Bull.*, **96**, 243; *A.M.*, **59**, 211; *Zap.*, **103**, 354.] 28th List

Sturtite. T. Hodge-Smith, 1930. *Rec. Australian Museum*, vol. 17, p. 410. Hydrous silicate of manganous oxide and ferric oxide, $H_3FeMn_3Si_4O_{14}.10H_2O$, as amorphous jet-black masses from Broken Hill, New South Wales. Named after Captain Charles Sturt (1795–1869), who visited the locality in 1844. [*M.A.*, **4**, 345.] 12th List

Suanite. T. Watanabe, 1952. [*Geology and mineral resources of the Far East*]; *Min. Journ.* (*Min. Soc. Japan*), 1953, vol. 1, p. 54. Pyroborate of magnesium, $Mg_2B_2O_5$, monoclinic, from Suan, Korea. Named from the locality. [*M.A.*, **12**, 411; *A.M.*, **40**, 941.] 20th List

Subglaucophane. A. Miyashiro, 1957. *Journ. Fac. Sci. Univ. Tokyo*, sect. 2, vol. 11, p. 57. A name for the clino-amphibole end-member $Na_2Mg^{\cdot\cdot}{}_{1\frac{1}{2}}Fe(Al,Fe^{\cdot\cdot\cdot})Si_8O_{22}(OH)_2$ with $Fe^{\cdot\cdot\cdot}$ 0·3 to 0·7. 22nd List

Subhydrocalcite. M. Copisarow, 1923. *Journ. Chem. Soc. London*, trans. vol. 123, p. 792. The same as Trihydrocalcite, $CaCO_3.3H_2O$, [5th List]. 10th List

Sub-melilite. H. Berman, 1929. *See* Soda-melilite. 12th List

Sudburyite. L. J. Cabri and J. H. Gilles Laflamme, 1974. *C.M.*, **12**, 275. Small inclusions, up to 0·1 by 0·018 mm, in cobaltite or maucherite from Copper Cliff South mine, Sudbury, Ontario, are hexagonal, *a* 4·06, *c* 5·59 Å, have the composition 2[PdSb], with or without some Ni replacing Pd; nickeline structure. Named from the locality. 28th List

Sudoite. G. Müller, 1962, quoted in W. von Engelhardt, G. Müller, and H. Kromer, *Naturwiss.*, 1962, vol. 49, p. 205. A name for the dioctahedral series of phyllosilicates corresponding to the Chlorites, the name Chlorite being restricted to the trioctahedral series. [*A.M.*, **53**, 2103.] 23rd List

Sugilite. N. Murakami, T. Kato, Y. Miura, and F. Hirowatari, 1976. *Mineral. J.* [*Japan*], **8**, 110. Subhedral grains in aegirine syenite at Iwagi Islet, Ehime Prefecture, Japan, are brownish-yellow, hexagonal, *a* 10·007, *c* 14·000 Å, space group *P6/mcc*. Composition near $K_{0\cdot81}Na_{1\cdot42}Li_{2\cdot12}Fe_{0\cdot03}{}^{2}Fe_{1\cdot60}{}^{3}Al_{0\cdot59}Ti_{0\cdot06}Si_{12}O_{29\cdot63}.0{\cdot}45H_2O$, sp. gr. 2·74. RIs ω 1·610, ε 1·607. Named for K. Sugi. [*A.M.*, **62**, 596.] 30th List

Sukulaite. Atso Vorma and J. Siivola, 1967. *Bull. Comm. Géol. Finlande*, **229**, 173. Probably $Sn_2Tl_2O_7$, isostructural with microlite, as yellowish-brown inclusions in cassiterite from Sukula, Tammela, Finland; named from the locality. [*A.M.*, **53**, 2103; *Zap.*, **98**, 325.] 26th List [Renamed **Stannomicrolite**, *A.M.*, **62**, 406; 30th List]

Sulfatapatit, *see* Sulphate-apatite.

Sulfate-monazite, Sulfat-Monazit, American and German forms of Сулфат-монацит, Sulphate-monazite (22nd List) (*Amer. Min.*, 1962, vol. 47, p. 417; H. Strunz, *Min. Tabellen*, 4th edit., 1966, p. 522). 24th List

Sulfatskapolith, *see* Sulphate-scapolite.

Sulfoborite. A. Naupert and W. Wense, *Ber. deutsch. chem. Ges.*, XXVI. (1), 874, 1893. H.

Bücking, *Ber. Akad. Berlin*, 967, 1893. K. Thaddéeff, *Zeits. Kryst. Min.*, XXVIII. 264, 1897. [*M.M.*, **11**, 103.] $3MgSO_4.2Mg_3B_4O_9.12H_2O$. According to Thaddéeff—$MgSO_4.Mg_2B_2O_5.4\frac{1}{2}H_2O = 4MgHBO_3 + 2MgSO_4 + 7H_2O$. Rhombic (Bücking). [Westeregeln] Prussian salt deposits. 1st List [$Mg_3B_2(SO_4)(OH)_{10}$]

Sulfohalite. Variant of Sulphohalite; *A.M.*, **9**, 62 [German, Sulfohalit]. 10th List

Sulfur. Variant of Sulphur; *A.M.*, **9**, 61. 10th List

Sulfurite. F. Rinne, *Centralblatt Min.*, 1902, p. 500 (Sulfurit). Naturally occurring amorphous sulphur. *See* arsensulfurite. 3rd List [*See also* Sulphurite]

Sulfurosite. G. Vavrinecz, 1939. *Földtani Közlöny*, Budapest, vol. 69, pp. 82, 98 (Sulfurosit). Sulphur dioxide, SO_2, as a natural gas. [*M.A.*, **7**, 471.] 15th List

Sulphate-apatite. R. Brauns, 1916. *Neues Jahrb. Min.*, Beilage-Band xli, p. 60 (Sulfatapatit). Small, colourless crystals of apatite occurring in the sanidinite bombs of the Laacher See district, Rhine, contain some sulphuric acid (SO_3 1·13–1·35%) in addition to fluorine and chlorine. The molecule $3Ca_3P_2O_8.CaSO_4$ therefore enters into their composition, the general formula for apatite being $3Ca_3P_2O_8.Ca(F_2,Cl_2SO_4,CO_3,O,[OH]_2)$. 8th List

Sulphate ferrithorite. E. V. Sveshnikova, D. N. Knyazeva, and M. T. Dmitrieva, 1964. [Минералы СССР, **15**, 239], abstr. in *M.A.*, **18**, 280. A sulphur-bearing (1·35% SO_3, 4·19% S) fluorian ferrian thorite from the Enisei region. Cf. ferrithorite, 16th List. 28th List

Sulphate-marialite and Sulphate-meionite. L. H. Borgström, 1914. *See* Carbonate-marialite. 7th List

Sulphate-meionite, *see* Sulphate-marialite.

Sulphate-monazite. A. A. Kukharenko, A. G. Bulakh, and K. A. Baklanova, 1961. Зап. Всесоюз. Мин. Общ. (*Mem. All-Union Min. Soc.*), vol. 90, p. 373 (Сулфат-монацит). $(Ce,Ca)(P,S)O_4$, with some Fe, Th, and H_2O, occurring as small nodules in a carbonatite in the Kola peninsula; the powder pattern is close to that of monazite, the refractive indices considerably lower. Named from the composition. An unnecessary name; there is only 3% SO_3, and sulphatic monazite would have been a better term. 22nd List

Sulphate-scapolite. R. Brauns, 1914. *Neues Jahrb. Min.*, Beil.-Bd. xxxix (Festschrift Max Bauer), pp. 83, 120 (Sulfatskapolith). Also named Silvialite (q.v.). 7th List

Sulphatic Cancrinite. E. S. Larsen and G. Steiger, 1916. *Amer. Journ. Sci.*, ser. 4, vol. xlii, p. 332. A cancrinite from Colorado with about half of the CO_3 isomorphously replaced by SO_4. 8th List [*See* Vishnevite, Wischnewite]

Sulphoborite. *Journ. Chem. Soc. Abstracts*, 1893, LXIV (i), 325. The same as sulfoborite (q.v.). 2nd List

Sulphojosëite. E. A. Dunin-Barkovskaya, V. V. Lider, and V. N. Rozhansky, 1968. *Zap.*, **97**, 332 (Сульфожозеит). *Syn.* of Josëite-A. [*M.A.*, 69-569.] 28th List

Sulphurin. J. Fröbel, 1843. *Grundzüge eines Systemes der Krystallologie*. Synonym of sulphur. 27th List

Sulphurite. (E. T. Wherry, *Journ. Washington Acad. Sci.*, 1917, vol. vii, p. 447 (sulfurite); *Chem. News*, 1917, vol. cxvi, p. 251 (sulphurite).) A mineralogical name for native sulphur. The name Sulfurit has been previously used (F. Rinne, 1902; 3rd List) for naturally-occurring amorphous sulphur. 8th List

Sulunite. A. A. Nyrkov, 1959. Зап. Всесоюз. Мин. Общ. (*Mem. All-Union Min. Soc.*), vol. 88, p. 571 (Сулунит). Hydrous aluminosilicate of alkalis and ferric iron, described as a new mineral of the chlorite group, from the Sulin region of the Donetz basin; named from the locality. The X-ray data are said to be comparable to gümbelite, a hydromuscovite. [*M.A.*, **14**, 501; *A.M.*, **45**, 478.] The analysis is unsatisfactory, and the material, which is high in alkalis, may well be a mixture. Cf. D. P. Serdyuchenko and N. V. Belov, Зап. Всесоюз. Мин. Общ. (*Mem. All-Union Min. Soc.*), 1960, vol. 89, p. 367. 22nd List

Sulvanite. G. A. Goyder, *Trans. R. Soc. South Australia*, 1900, XXIV, 69; *Journ. Chem. Soc. Trans.*, 1900, LXXVII, 1094. Sulphovanadate of copper, $3Cu_2S.V_2S_5$. Massive; bronze-yellow with metallic lustre. [Near Burra] South Australia. 2nd List [Cf. Arsenosulvanite, Lazarevićite]

Sundiusite. R. Phillips and W. Layton, 1964. *Min. Mag.*, vol 33, p. 1097. The hypothetical amphibole end-member $Na_2CaMg_3Al_4Si_6O_{22}(OH)_2$. 23rd List

Sundtite. W. C. Brøgger, *Forh. Vid.-Selsk. Cristiania*, for 1892, no. 18, 1893. [*M.M.*, **11**, 286.] Syn. of andorite. [Itos mine, Oruro, Bolivia.] 1st List [Is Andorite-IV; *A.M.*, **39**, 161]

Sungulite. G. A. Sokolov, 1925. *Bull. Sci.-Techn. Circle Metall. and Chem.*, Leningrad Polytechn. Inst., no. 1(2), p. 74; *Trans. Geol. Prospecting Service USSR*, 1931, no. 56, p. 32 (сунгулит). A variety of serpentine from Sungul lake, Kyshtym, Urals. Named from the locality. [*M.A.*, **6**, 219, 436.] 14th List [A mixture; *A.M.*, **59**, 212. Cf. Kolskite]

Suolunite, alternative transliteration for Solanite (q.v.). 24th List [Now the preferred spelling; not to be confused with Sulunite]

Suomite. G. Gagarin and J. R. Cuomo, 1949, loc. cit., p. 11 (suomita). Tantalic ochre (tantalochra of A. E. Nordensiöld, 1855) believed to be Ta_2O_5 as a brown coating on crystals of tantalite from Finland = Suomi. 19th List

Surinamite. E. W. F. de Roever, C. Kieft, E. Murray, E. Klein, and W. H. Drucker, 1976. *A.M.*, **61**, 193. A mineral closely resembling sapphirine, from the Bakhuis Mts., W Surinam, is monoclinic, *a* 9·64, *b* 11·36, *c* 4·95 Å, *β* 109·0°, space group $P2_1/a$; *α* 1·738, *β* 1·743 ∥ [010], violet, γ_{Na} 1·746; *γ* to a cleavage ⊥ (010) 31° for violet, 44° for yellow; on (010), blue-green for vibrations ∥ the cleavage, pale greenish-brown ⊥ it; $2V_\alpha$ 67°. Composition $(Al_{1\cdot38}Mg_{1\cdot12}Fe_{0\cdot46}Mn_{0\cdot04})_{\Sigma3}(Si_{1\cdot51}Al_{0\cdot49})_{\Sigma2}\{O_{7\cdot36}(OH)_{0\cdot64}\}_{\Sigma8}$. 29th List

Sursassite. J. Jakob, 1926. *Schweiz. Min. Petr. Mitt.*, vol. 6, p. 376 (Sursassit). Hydrous manganese and aluminium silicate, $5MnO.2Al_2O_3.5SiO_2.3H_2O$, as copper-red fibrous aggregates, from Oberhalbstein, Switzerland. Named from Sursass, the name for this locality in the Rhæto-Romanic dialect. [*M.A.*, **3**, 272.] 11th List [*A.M.*, **49**, 168]

Svetlozarite. M. N. Maleev, 1976. *Zap.*, **105**, 449 (Светлозарит). Spherulites in chalcedony veinlets from Zvezdel, Rhodope Mts., Bulgaria, are orthorhombic, *a* 19·482, *b* 20·963, *c* 7·554 Å. Composition near $[(Ca,Na,K)_{6\cdot85}(Si,Al)_{55\cdot6}O_{112}.25H_2O]$. RIs *α* 1·481, *β* 1·482, *γ* 1·485. Named for Svetlozar I. Bontev. [*A.M.*, **62**, 1060; *M.A.*, 77-3394.] 30th List

Svidneïte. V. Mincheva-Stefanova, 1951. [*Izv. Geol. Inst. Bulg. Akad. Nauk*, 1951, pp. 41–62]; abstr. *Amer. Min.*, 1967, vol. 52, p. 562. An oxyamphibole near magnesioriebeckite (Miyashiro, 1957; 22nd List) from Svidnya, near Sofia; named from the locality. [Зап. **96**, 79; Swidneit, *Germ.*] 25th List [Oxy magnesio-riebeckite; *Amph.*, Sub-Comm.]

Svitalskite. A. P. Nikolsky and A. N. Efimov, 1960. [Труды ВСЕГЕИ, vol. 37, p. 142], abstr. in Зап. Всесоюз. Мин. Общ. (*Mem. All-Union Min. Soc.*), 1963, vol. 92, p. 213 (Свитальскит) and in *Amer. Min.*, 1963, vol. 48, p. 1181. An unnecessary name for a green mica from Krivoi Rog, formulated $(K,Na,H_3O)_{0\cdot98}(Mg,Fe)_{0\cdot78}(Fe,Al)_{1\cdot14}(Si,Al)_4O_{10}(OH)_2$ and classed with the phlogopite–biotite series. It is a dioctahedral mica and very near some Celadonites; 'the name is therefore unnecessary' (M. Fleischer). Named for N. I. Svitalsky. [*M.A.*, **16**, 557.] 23rd List

Swartzite. J. Axelrod, F. [S.] Grimaldi, C. Milton, and K. J. Murata, 1948, *Program and Abstracts, Min. Soc. Amer.*, 29th Annual Meeting, p. 4; *Bull. Geol. Soc. Amer.*, vol. 59, p. 1310; *Amer. Min.*, 1949, vol. 34, p. 274. Hydrous calcium-magnesium-uranyl carbonate. $CaMgUO_2(CO_3)_3.nH_2O$, from Arizona. Named after Charles Kephart Swartz (1861–), Emeritus Professor of Geology, Johns Hopkins University. [*M.A.*, **10**, 452.] 18th List [*A.M.*, **36**, 1]

Swedenborgite. G. Aminoff, 1924. *Zeits. Krist.*, vol. 60, p. 262 (Swedenborgit). Antimonate of aluminium and sodium, $NaAl_2SbO_6$, as colourless, hexagonal crystals, from Långban, Sweden. Named after the Swedish philosopher, Emanuel Swedenborg (1688–1772). [*M.A.*, **2**, 338.] 10th List [$NaBe_4SbO_7$; *A.M.*, **20**, 492]

Swidneit, *see* Svidneïte.

Swinefordite. Pei-lin Tien, P. B. Leavens, and J. A. Nelen, 1975. *A.M.*, **60**, 540. A swelling clay from the Foote Mineral Co. mine, Kings Mountain, Cleveland County, North Carolina, has d_{001} 13 Å at relative humidity 60%, and composition approximately $Li_{0\cdot72}K_{0\cdot04}Na_{0\cdot11}Ca_{0\cdot23}Mg_{0\cdot05}(Al_{1\cdot87}Fe^{3}_{0\cdot15}Fe^{2+}_{0\cdot09}Mg_{1\cdot31}Li_{1\cdot76})(Si_{7\cdot86}Al_{0\cdot34})O_{20}\{F_{0\cdot65}(OH)_{3\cdot35}\}$. Named for A. Swineford. [*M.A.*, 76-887.] 29th List

Switzerit, variant of Schweizerite (H. Strunz, *Min. Tabellen*, 3rd edit., 1957, p. 419, without reference). 24th List

Switzerite. P. B. Leavens and J. S. White, Jr., 1967. *Amer. Min.*, vol. 52, p. 1595. Pale pink to brown monoclinic crystals from the Foote Mineral Company's spodumene mine, King's Mountain, North Carolina, have the composition $(Mn,Fe)_3(PO_4)_2.4H_2O$. Named for G. Switzer. 25th List

Swjaginzewit, German transliteration of Звягинцевит, Zvyagintsevite (q.v.) (C. Hintze, *Handb. Min.*, Erg. III, 634). 25th List

Sylvialite. *See* Silvialite. 7th List

Sylvinite. (C. Hintze, *Handbuch der Mineralogie*, 1911, vol. i, p. 2155 (Sylvinit).) A mining term for the mixtures of sylvite and rock-salt occurring in the Prussian salt-deposits. 6th List

Symant. A trade-name for *artificial* $SrTiO_3$. H. Strunz, *Min. Tabellen*, 4th edit., 1966, p. 522; 5th edit., 1970, 579. *See* Fabulite (q.v.). 24th and 28th Lists

Synchysite. G. Flink, *Bull. Geol. Inst. Univ. Upsala*, 1901, vol. v (1900), p. 81 (Synchysit); abstr., [*M.M.*, **13**], p. 207. The mineral from Narsarsuk, Julianehaab, Greenland, previously described as parisite, has characters sufficiently distinct from those of the Colombian parisite to be regarded as a distinct species. Hexagonal. $CeFCa(CO_3)_2$. Named from *σύγχῠσις*, confounding. 3rd List [Cf. Röntgenite. *See A.M.*, **45**, 92]

Synkysit. O. B. Boeggild, *Mineralogia Groenlandica*, Copenhagen, 1905, p. 167. Variant of synchysite (G. Flink, 1901). 4th List

Syntagmatite (of Tröger). W. E. Tröger, 1952. A name given by A. Breithaupt (*Berg.-Hütten. Zeitung*, 1865, vol. 24, p. 428) to a hornblende from Vesuvius analysed by C. F. Rammelsberg (*Ann. Phys. Chem.* (*Pogg.*), 1858, vol. 103, p. 451) because the crystals occurred in stellate groups ('*συνταγμα*, d. i. das Gruppirte'). Relegated to the synonymy by J. D. Dana (*Syst. Min.*, 5th edit., 1868, p. 235), was revived by R. Scharizer (*Neues Jahrb. Min.*, 1884, vol. 2, p. 147) as a name for the hornblende end-member $(R_2{}^{+}, R^{2+})_3R_2{}^{3+}Si_3O_{12}$, the other end-member being actinolite, $Ca(Mg,Fe)_3Si_4O_{12}$. It was revived by Tröger (*Tab. opt. Best. gesteinsbild. Min.*, 1952, pp. 65, 76) as the $MgFe^{3+}$ end-member of a series of titaniferous hornblendes, $NaCa_2(Mg,Fe,Ti)_4Fe_3Si_6O_{22}(OH)_2$*, the other (MgAl) end-member being kaersutite. Tröger later came to regard the name as superfluous (*Opt. Best. gesteinsbild. Min.*, Teil 2, 1967, p. 465), but it has had some usage in this third sense. [*Corr. in *M.M.*, **39**.] 25th List [Titanian hastingsite; *Amph.* (1978)]

Syntholite. Trade-name for *artificial* corundum simulating alexandrite. R. Webster, *Gems*, 1962, p. 763. *See* under zircolite. 28th List

Szamozyt. Polish spelling of chamosite. [*M.A.*, **12**, 350.] 20th List

Szaskaite. (T. A. Readwin, *Index to Mineralogy*, 1867.) The same as calamine ($ZnCO_3$). Szaska, Hungary. 2nd List

Szechenyiite. J. A. Krenner, *Scientific Results of Count Béla Széchenyi's Travels in Eastern Asia*, Budapest, 1897, III, p. 290 (Hungarian); abstr. in *Zeits. Kryst. Min.*, 1899, XXXI, 503. A soda-amphibole resembling diallage in appearance. $2Na_2O.10MgO.3CaO.Al_2O_3.16SiO_2$. Occurs embedded in jadeite from Burma. 2nd List

Szomolnokite. J. A. Krenner, 1891. *Akadémiai Értesítö*, Budapest, vol. ii, p. 96 (szomolnokit); *Földtani Közlöny*, Budapest, 1907, vol. xxxvii, pp. 204, 205; private letter from Professor Krenner, 1910. Hydrated ferrous sulphate, $FeSO_4.H_2O$, forming yellowish or brownish, monoclinic pyramids isomorphous with kieserite ($MgSO_4.H_2O$). Occurs with rhomboclase (q.v.) and other iron sulphates (kornelite, copiapite, coquimbite, etc.) at Szomolnok (German, Schmöllnitz), Hungary. 5th List

T

Taaffeite. B. W. Anderson, 1951. *Gemmologist*, London, vol. 20, p. 76. B. W. Anderson, C. J. Payne, and G. F. Claringbull, *Min. Mag.*, 1951, vol. 29, p. 765. B. W. Anderson, *Gemmologist*, 1952, vol. 21, p. 23 (pronounced 'tarfite'; other names provisionally suggested, 'bemagalite' and 'berinel', refer to the chemical composition and to the relation to spinel). $BeMgAl_2O_8$, hexagonal. Known only as two small faceted gemstones resembling mauve-coloured spinel. Named after Count Edward Charles Richard Taaffe (1898–) of Dublin, who first detected it. [*M.A.*, **11**, 309, 369, 485; *A.M.*, **37**, 360.] 19th List

Tabbyite. (*Mineral Resources, US Geol. Survey*, for 1907, 1908, part ii, p. 726.) Trade-name for a solid elastic bitumen from the Uinta basin, Utah. It is closely related to wurtzilite and elaterite. 6th List

Tablite. *Bull. Soc. Franç. Min. Crist.*, 1950, vol. 73, p. 151. 'Un antimonioarséniate de cuivre et de cobalt d'un gisement métallifère d'Afrique du Sud'. *See* Allevardite and tabulite. 19th List

Tabulite. E. Lemoine, 1950. *Bull. Soc. Franç. Min. Crist.*, vol. 73, p. 146. Alternative name suggested for allevardite (q.v.). Named from the locality, La Table (mountain) in Savoie, near Allevard in Isère. 19th List [Now Rectorite]

Tacharanite. J. M. Sweet, 1961. *Min. Mag.*, vol. 32, p. 745. $(Ca,Mg,Al)(Si,Al)O_3.H_2O$, with gyrolite, tobermorite, and mesolite in dolerite from Portree, Isle of Skye. Readily changes to a mixture of tobermorite and gyrolite. Named from the Gaelic *tacharan*, a changeling. 22nd List [Near $Ca_{12}Al_2Si_{18}O_{69}H_{36}$, *M.M.*, **40**, 113; Tasmania, *M.M.*, **40**, 887]

Tacherite. German transliteration of Тахерит, takherite (Hintze, *Handb. Min.*, Erg.-Bd. II, p. 858). 22nd List

Tadshikit, German transliteration of Таджикит, tadzhikite (27th List). Hintze, 91. 28th List

Tadzhikite. A. F. Efimov, V. D. Dusmatov, V. Yu. Alkhazov, Z. G. Pudovkina, and M. E. Kazakova, 1970. Докл. Акад. наук СССР (*Compt. Rend. Acad. Sci. URSS*), **195**, 1190 (Таджикит). Platy crystals and prisms, with somewhat different compositions, occur in pegmatite dykes in the Turkestana alkalic province of Tadzhikistan. Anorthic, *a* 17·93, *b* 4·71, *c* 10·39 Å, β 100° 45′, α and $\gamma \approx 90°$; D 3·73, 3·86. Composition $2[Ca_3Ln_2(Ti,Al,Be)B_4Si_4O_{22}]$. Similar to but distinct from hellandite (3rd List). [*A.M.*, **56**, 1838; *M.A.*, 71-3113; *Zap.*, **100**, 613.]

Taeniolite. *Amer. Journ. Sci.*, 1900, ser. 4, vol. x, p. 324, abstr. (Tæniolite). The same as tainiolite (Flink, 1901). 3rd List

Tageranite, non-standard transliteration of Тажеранит, tazheranite (*Zap.*, **99**, 75). 27th List

Tainiolite. G. Flink, *Medd. om Grönland*, 1900, XXIV, 115. A mica occurring in the form of strips. Monoclinic. $(MgOH)_2(K,Na,Li)Si_3O_8 + H_2O$. [Narsarsuk] S Greenland. 2nd List [Commonly **Taeniolite** (q.v.)]

Taiyite. Chi Ling-Yi, 1974 [*Acta. Geol. Sinica*, 91], abstr. *A.M.*, **61**, 178. An unnecessary name for a priorite (aeschynite-(Yt)) from S China. 29th List

Tajikite, non-standard transliteration of Таджикит, tadzhikite (*Zap.*, **100**, 623). 27th List

Takanelite. M. Nambu and K. Tanida, 1971. *Journ. Jap. Ass. Min. Petr. Econ. Geol.*, **65**, 1. Grey to black grains in the oxidation zone of the Nomura mine, Ehime Prefecture, Japan, are the manganous analogue of ranciéite (4th List). Powder data are indexed on a hexagonal unit cell, *a* 8·68, *c* 9·00 Å, D 3·41, containing $3[(Mn^{2+},Ca)Mn_4^{4+}O_9.1{\cdot}3H_2O]$. [*A.M.*, **56**, 1487; *M.A.*, 72-1404.] 27th List

Takizolite. S. Iimori and J. Yoshimura, 1929. *Bull. Chem. Soc. Japan*, vol. 4, p. 2; *Sci. Papers Inst. Physical and Chemical Research*, Tokyo, vol. 10, p. 225. A pink clay from Tanokami, Japan, intermediate in composition, $2Al_2O_3.7SiO_2.7H_2O$, between catlinite and montmorillonite. Named after the late Takizo Ueno, of Tanokami, who collected the material. [*M.A.*, **4**, 247.] 12th List

Takovite. Z. Maksimović, 1957. [*Zapis. srp. geol. drustva zo 1955 god.* (*Compt. Rend. Soc. serbe Géol.*, ann. 1955), p. 219]; abstr. *M.A.*, **13**, 624 and Hintze, *Handb. Min.*, Erg.-Bd. II, p. 858 (Takowit). A blue-green massive clay-like mineral, $Ni_5Al_4O_2(OH)_{18}.6H_2O$, from Takovo, Serbia. Named from the locality. 22nd List [$Ni_6Al_2(OH)_{16}(CO_3,OH).4H_2O$, pyroaurite group; *A.M.*, **62**, 458]

Takowit, Germanized form of Takovite (22nd List) (C. Hintze, *Handb. Min.*, Erg.-Bd. II, p. 858). 24th List

Talasskite. V. D. Nikitin, 1936. *Mém. Soc. Russ. Min.*, ser. 2, vol. 65, p. 282 (таласскит), p. 288 (talasskite). A variety of fayalite containing Fe_2O_3 12·07%. Named from the locality, the Talassa valley, Kirghiz republic, Siberia. [*M.A.*, **6**, 439.] 14th List

Talc-knebelite. *See* Talkknebelite. 1st List

Talc-spinel. J. Beckenkamp, 1921. C. Hintze's *Handbuch d. Min.*, vol. 1, part 4, pp. 3 et seq. Synonym of spinel; named from Talkerde (Ger.) = magnesia. Similar chemical names are given for other members of the spinel group; the following are new (*see also* Manganspinel, above):

Talc-spinel (Talkspinell), $Mg(AlO_2)_2$, synonym of spinel. Ferroferrite (Ferroferrit), $Fe(FeO_2)_2$, synonym of magnetite. Manganoferrite (Manganoferrit), $Mn(FeO_2)_2$, synonym of jacobsite. Zincoferrite (Zinkoferrit), $Zn(FeO_2)_2$, synonym of franklinite. Ferrochromite (Ferrochromit), $Fe(CrO_2)_2$, synonym of chromite. 9th List

Talenite, error for Thalenite, through back-transliteration of Таленит. *Soviet Physics—Doklady*, 1972, **17**, 88. 28th List

Talkknebelite. L. J. Igelström, *Neues Jahrb. Min.*, I. 248, 1890. Knebelite with 4·7% MgO. [Hilläng mine, Dalecarlia] Sweden. 1st List

Talknebeit, error for Talkknebelit (C. Hintze, *Handb. Min.*, Erg.-Bd. II, p. 955). 24th List

Talmessite. P. Bariand and P. Herpin, 1960. *Bull. Soc. franç. Min. Crist.*, vol. 83, p. 120. $Ca_2(Mg,Ba)(AsO_4)_2.2H_2O$, anorthic and isomorphous with β-Roselite. Named from the locality, the Talmessi mine, Anarak, Iran. [*M.A.*, **15**, 45.] 22nd List [Cf. Belovite (of Nefedov), and Arsenate-belovite]

Talnachit, German transliteration of Талнахит, talnakhite (25th List). Hintze, 92. 28th List

Talnakhite. I. A. Buko and E. A. Kulagov, 1968. Зап. всесоюз. Мин. Общ.(*Mem. All-Union Min. Soc.*), vol. 97, p. 63 (Талнахит, talnakhite; p. 63, talnachite). A cubic polymorph of chalcopyrite, from the Norilsk- and Talnakh ores (cf. Доклады акад. наук СССР (*Compt. Rend. Acad. Sci. URSS*), 1963, vol. 152). 25th List [$Cu_9(Fe,Ni)_8S_{16}$. *A.M.*, **55**, 2135; **56**, 2159]

Talspat, error for Talkspat (C. Hintze, *Handb. Min.*, Erg.-Bd. II, p. 955). 24th List

Tamanite. S. P. Popoff, *Zeits. Kryst. Min.*, 1903, vol. xxxvii, p. 267. The same as anapaite (q.v.). Professor Groth, in an editorial footnote (l.c.), calls attention to this identity, but he gives no reason for admitting the later name as a synonym. Named from the locality, Taman peninsula, Black Sea. 3rd List

Tanatarite. O. A. Petrushkevich, 1926. *Bull. Geol.-Min. Circle Ekaterinoslav Mining Inst.*, no. 2, p. 19 (танатарит); I. I. Tanatar, ibid., 1927, p. 9. Aluminium hydroxide, $Al_2O_3.H_2O$, stated on optical grounds to be monoclinic and therefore dimorphous with diaspore. Occurs in chromite in Russian Central Asia. Named after Prof. Josef. I. Tanatar (Иосиф И. Танатар) of the Ekaterinoslav (= Dnepropetrovsk) Mining Institute. Kayserite (K. Walther, 1922; 10th List) was distinguished from diaspore on the same grounds. [*M.A.*, **3**, 237; **3**, 473.] 11th List [Is Diaspore; *M.A.*, **8**, 230]

Tangaite. D. McKie, 1958. *Records Geol. Surv. Tanganyika*, vol. 5 (for 1955), p. 81. $AlPO_4.2H_2O$ (with 3½% Fe_2O_3, 1% Cr_2O_3) from Gerevi Hill, Tanga district, Tanganyika, is essentially a low-iron redondite and not a variscite. Named from the locality. The name is liable to confusion with tangeïte. 22nd List [*A.M.*, **49**, 445]

Tangeite. A. E. Fersman, 1925. *Priroda*, Leningrad, 1925, no. 7–9, col. 238 (тангеит). K. A. Nenadkevich and P. A. Volkov, *Compt. Rend. Acad. Sci. URSS*, 1926, ser. A, p. 43 (тангеит, tanguéite). I. Kurbatov, *Centr. Min.*, Abt. A, 1926, p. 346 (Tangeit). Hydrated vanadate of copper and calcium, $2CuO.2CaO.V_2O_5.H_2O$, as dark-green radially fibrous botryoidal masses from Tyuya-Muyun, Fergana. Named from the Tange gorge. The colloidal form is termed 'Turkestan-volborthite' (q.v.), and is compared with calciovolborthite. [*M.A.*, **3**, 234.] 11th List [Is Calciovolborthite]

Tangenite. G. Gagarin and J. R. Cuomo, 1949, loc. cit., p. 10 (tangenita). A betafite-like mineral containing TiO_2 32·27–35·05% from Tangen near Kragerø, Norway, described by H. Bjørlykke, 1931 [*M.A.*, **5**, 425.] 19th List [A mixture; *A.M.*, **62**, 407]

Tangiwaite. (R. Koechlin, *Mineralogisches Taschenbuch der Wiener Mineralogischen Gesellschaft*, 1911, p. 55 (Tangawait).) A variety of serpentine, identical with bowenite, from New Zealand. The Maori name, tangiwai, for the stone means 'tear-water', since polished specimens have the appearance of drops of water. For a description of the mineral and its occurrence *see* F. Berwerth, *Sitz.-Ber. Akad. Wiss. Wien, Math.-naturw. Cl.*, 1880, vol. lxxx (for 1879), Abt. I, p. 116; A. M. Finlayson, *Quart. Journ. Geol. Soc. London*, 1909, vol. lxv, pp. 361, 369. 8th List

Tangueite. *See* Tangeite. 11th List

Tanjeloffite. J. Tanjeloff, 1971, in L. Zara, *Min. Digest*, **2**, 21. Synonym of blue zoisite (tanzanite). 27th List

Tansanit, German variant of tanzanite (26th List). Hintze, 92. 28th List

Tantal-aeschynite. V. A. Kornetova, V. B. Aleksandrov, and M. E. Kazakova, 1963. Труды Мин. муз. Акад. наук СССР, **14**, 108 (Тантал-эшинит). Cf. M. S. Adusumilli, *Jorn. Mineralogia, Recife, Ed. Espl.*, **6**, 11 (1968) and M. S. Adusumilli, C. Kieft, and E. A. J. Burke, *M.M.*, **39**, 571 (1974). This name was applied by Kornetova *et al.* to a mineral with Ti:Nb:Ta = 0·89:0·52:0·57, metamict but giving an X-ray powder pattern similar to that of aeschynite after heating, from a Siberian pegmatite; Adusumilli *et al.* regard this as merely a tantalian aeschynite, and reserve the name for material with Ta > Ti or Nb, such as occurs as idiomorphic metamict crystals at São Jose do Sabugí, Paraíba State, Brazil (this also gives an X-ray pattern similar to that of aeschynite after heating, orthorhombic with a 5·34, b 10·97, c 7·38 Å). [*M.A.*, 71-2325; *A.M.*, **59**, 1331.] 28th List

Tantalbetafite. A. P. Kalita and A. V. Bykova, 1961. Труды Инст. Мин. Геохим. Крист. Редк. Элем. (*Trans. Inst. Min., Geochem., Cryst. Rare Elem.*), no. 7, p. 104 (Танталбетафит). Betafite with Ta > Nb, in contrast to normal Betafite, which has Nb > Ta (Зап. Всесоюз. Мин. Общ. (*Mem. All-Union Min. Soc.*), 1963, vol. 92, p. 199; *M.A.*, **16**, 555). 23rd List [Syn. or var. of Betafite; *A.M.*, **62**, 407]

Tantalcarbid. H. Strunz, 1966. *Min. Tabellen*, 4th edit., p. 93. The doubtful mineral Tantalum carbide (q.v.), formerly supposed to be native Tantalum. 24th List [*A.M.*, **47**, 786]

Tantallyndochite. S. A. Gorzhevskaya, G. A. Sidorenko, and A. I. Ginzburg, 1974. *Abstr. Zap.*, **105**, 76 (Танталлиндокит). Unnecessary name for tantalian lyndochite. 30th List

Tantalobetafite, variant of Tantalbetafite (23rd List). A. H. van der Veen, *Verhandl. Kon. Ned. Geol. Mijnbouw. Genootsch.*, geol. ser., 1963, no. 22. 24th List

Tantalohatchettolite. J. E. de Villiers, 1941. *Amer. Min.*, vol. 26, p. 505. Microlite containing much uranium. [*M.A.*, **8**, 188.] 16th List [Syn. of Uranmicrolite; *A.M.*, **62**, 407. Cf. Djalmaite]

Tantalo-obruchevite. A. H. van der Veen, 1963. *Verhandel. Kon. Nederland. geol. mijnbouw. Genootschap.*, geol. ser., no. 22. A microlite in which the principal A cation in the formula $A_2B_2X_7$ is yttrium. 24th List [Remains hypothetical; *A.M.*, **62**, 407]

Tantalopolycrase. E. S. Simpson, 1938. *Journ. Roy. Soc. W. Australia*, vol. 24, p. 112; *Rep. Dept. Mines, W. Australia*, for 1937, p. 88. Titano-tantalate of yttrium, $YtTi_2TaO_8$, differing from polycrase in containing tantalum in place of niobium. From Western Australia. Previously included under tanteuxenite [$YtTi_2TaO_8 + Yt_2Ta_2O_8(+ CaTiTa_2O_8$, etc.)] (12th List). 15th List

Tantalo-rutile. A. B. Edwards, 1940. *Proc. Austral. Inst. Mining and Metall.*, no. 120, p. 735 (tantalo-rutile). To replace the name ilmenorutile for rutile containing tantalum and niobium. [*Min. Mag.*, **15**, 85; *M.A.*, **8**, 307.] 16th List

Tantalotitanocolumbite. Ya. A. Kosals, 1967. [Геол. и геофис. no. 2, 116]; abstr. *Zap.*, **98**, 324 (1969) (Танталотитаноколумбит). An unnecessary name for a titanian columbite with Nb:Ti:Ta 1·05:0·57:0·35. 26th List

Tantalpolykras, variant of Tantalopolycrase (15th List) (H. Strunz, *Min. Tabellen*, 3rd edit., 1957, p. 419). 24th List

Tantalum. P. Walther, 1909. *Nature*, London, vol. lxxxi, p. 335. W. von John, ibid., 1910, vol. lxxxiii, p. 398. A bright, greyish-yellow crystalline powder from gold-washings in the Urals, containing Ta 98·5%, Nb 1·5%. The crystalline grains have been determined to be cubic by C. O. Trechmann. Found also in the Alta Mts. 5th List [Is Tantalum carbide; *A.M.*, **47**, 786]

Tantalum carbide. C. Frondel, 1962. *Amer. Min.*, vol. 47, p. 786. Type material of 'native Tantalum' is shown to be tantalum carbide; it is uncertain whether the material was native or of artificial origin. 24th List

Tanteuxenite. E. S. Simpson, 1928. *Journ. Roy. Soc. Western Australia*, vol. 14, p. 45; *Amer. Min.*, 1928, vol. 13, p. 467. Titanotantalate of yttrium, etc., $YtTi_2TaO_8$, differing from euxenite in containing tantalum in place of niobium; as tabular orthorthombic crystals from Western Australia. [*M.A.*, **4**, 9, 184.] 12th List [*A.M.*, **45**, 756. Cf. Delorenzite]

Tantpolycrase, synonym of Tantalopolycrase (15th List). [Геохим., мин., генет. типы месторожд. редк. элем., Изд. „Наука", 1964, p. 470]: abstr. Зап. Всесоюз. Мин. Общ. (*Mem. All-Union Min. Soc.*), 1965, vol. 94, p. 683 (Тантполикраз). 24th List

Tanzanite. H. B. Platt, 1967. *Wall Street Journal*, October 14. *See also Lapidary Journ.*, **22**, 636, 736, 738, 740 (1968); *Rocks and Minerals*, **43**, 332 (1968); *Zeits. Deut. Gesell. Edelsteinkunde*, **61**, 27 (1967); *Amer. Min.*, **54**, 702 (1969). Deep violet-blue vanadiferous zoisite, highly pleochroic, from the Gerevi Hills, Tanzania, has been named from the locality. [*M.A.*, 69-320, 321.] 26th List [*M.A.*, 74-3345]

Taosite. J. de Lapparent, 1935. *Compt. Rend. Acad. Sci. Paris*, vol. 201, p. 156. A form of alumina distinct from corundum, occurring in the emery of Samos. Named from ταώς, a peacock, the emblem of Hera (Juno) at Samos. [*M.A.*, **6**, 150.] 14th List [Is Högbomite; *A.M.*, **37**, 136]

Taramellite. E. Tacconi, 1908. *Rend. R. Accad. Lincei*, Roma, ser. 5, vol. xvii, sem. 1, p. 810; *Centralblatt Min.*, 1908, p. 506; *Riv. Min. Crist. Italiana*, 1909, vol. xxxix, p. 26. A basic meta-silicate, $Ba_4Fe''Fe'''_4Si_{10}O_{31}$ or $Ba_4Fe''(Fe'''O)Fe'''_3(SiO_3)_{10}$. Occurs as brownish-red, radially fibrous aggregates or as thin veins penetrating magnetite and iron-pyrites in the granular limestone of Candoglia, Piedmont. It is optically biaxial, strongly pleochroic, and probably orthorhombic. Named after Torquato Taramelli, professor of geology in the University of Pavia. 5th List [$Ba_4(Fe,Mg)Fe_2^{3+}TiSi_8O_{24}(OH)_4$; *A.M.*, **44**, 470]

Taramite. J. Morozewicz, 1923. *Spraw. Polsk. Inst. Geol.* (= *Bull. Serv. Géol. Pologne*), vol. 2, p. 6 (Taramit). A soda-iron amphibole occurring in nepheline-syenite (mariupolite) at Walitarama, Mariupol, Ukraine. A fluotaramite (p. 8, fluotaramit) contains fluorine 1·75–2·40%. Named from the locality. 10th List

Tarankite, error for Taranakite (*Amer. Min.*, 1959, vol. 44, p. 138). 22nd List

Tarasovite. E. K. Lazarenko and Yu. M. Korolev, 1970. Зап. Всесоюз. мин. общ. (*Mem. All-Union Min. Soc.*), **99**, 214 (Тарасовит). A disordered dioctahedral phyllosilicate, $Na_{1\cdot00}K_{1\cdot18}(H_3O)_{0\cdot61}(Ca_{0\cdot18}Na_{0\cdot24})_{exch.}(Si_{12\cdot65}Al_{3\cdot35})O_{40}Al_8(OH)_8.2H_2O$; X-ray photographs show basal spacings only, with d_{001} 44 Å. Occurs in the Tarasov region. Named from the locality. [*A.M.*, **56**, 1123.] 26th List

Tarasowit, German transliteration of Тарасовит, tarasovite (26th List). Hintze, 92. 28th List

Taraspite. C. v. John. *Verh. k. k. geol. Reichsanst.*, p. 67, 1891. Syn. of miemite (var. of dolomite). Tarasp, Switzerland. 1st List

Tarbuttite. L. J. Spencer, 1907. *Nature*, vol. lxxvi, p. 215. Hydrous phosphate of zinc occurring as colourless or faintly coloured, anorthic crystals with descloizite at the Rhodesia Broken Hill mines, NW Rhodesia. The crystals have a perfect cleavage in one direction, through which emerges obliquely the acute, negative bisectrix of the optic axes. Named after Mr. Percy Coventry Tarbutt, a director of the mining company, who furnished the material for determination. 4th List [*M.M.*, **15**, 22; triclinic $Zn_2PO_4(OH)$]

Tartarkaite. (3rd Appendix to 6th edit. of Dana's *System of Mineralogy*, 1915, p. 76.) Error for Tatarkaite (q.v.). 7th List

Tartufite. Venetian mineralogists, before 1812. T. A. Catullo, *Manuale Mineralogico*, Belluno, 1812, p. 19 (Tartufite). J. Desnoyers, *Mém. Soc. Hist. Nat. Paris*, 1823, vol. i, p. 179; *Ann. Sci. Nat. Paris*, 1824, vol. i, p. 58 (Tartuffite). R. Koechlin, *Mineralogisches Taschenbuch der Wiener Mineralogischen Gesellschaft*, 1911, p. 56 (Tartüffit). A fetid, fibrous calcite from Monte Viale, Venetia, which when struck emits an odour like that of truffles (Italian, tartufo). *See* Truffite (5th List). 6th List

Tasheranit, Germ. trans. of Тажеранит, tazheranite (26th List). Hintze, 93. 28th List

Tatarkaite. A. Meister, 1910. *Explorations géologiques dans les régions aurifères de la Sibérie, Région aurifère d'Iénisséi, Livraison IX*, 1910, p. 498 (татаркаитъ); abstr. in *Zeits. Kryst. Min.*, 1914, vol. liii, p. 597 (Tatarkait). A hydrated silicate of aluminium, magnesium, etc., $R_2O.11RO.13R_2O_3.30SiO_2.19H_2O$, occurring as dark-grey, elongated plates (optically uniaxial and positive) in metamorphosed limestone near the Tatarka river, a tributary of the Angara, Siberia. *See* Angaralite. 7th List [A chlorite near ripidolite; *A.M.*, **50**, 2111]

Tatarskite. V. V. Lobanova, 1963. Зап. Всесоюз. Мин. Общ. (*Mem. All-Union Min. Soc.*), vol. 92, p. 697 (Татарскит). $Ca_3Mg(SO_4)(CO_3)Cl_2(OH)_2.3\frac{1}{2}H_2O$, massive, in Anhydrite. Named for Prof. V. B. Tatarsky. [*M.A.*, **16**, 548; *A.M.*, **49**, 1152.] 23rd List

Tavorite. M. L. Lindberg and W. T. Pecora, 1954. *Science* (*Amer. Assoc. Adv. Sci.*), vol. 119, p. 739. Hydrous lithium ferric phosphate, $LiFe(PO_4)(OH)$, as yellow fine-grained aggregates from Brazil. Named after Prof. Elysiario Tavora, University of Brazil, Rio de Janeiro. [*M.A.*, **12**, 408; *A.M.*, **40**, 952.] 20th List

Tawmawite. A. W. G. Bleeck, 1907. *Zeits. prakt. Geol.*, Jahrg. xv, p. 354 (Tawmawit); *Rec. Geol. Survey India*, 1908, vol. xxxvi, p. 269. A dark green, chromiferous (Cr_2O_3, 11·16%) variety of epidote, found as massive fragments in the jadeite quarry at Tawmaw in Upper Burma. 5th List

Taylorite. W. C. Knight, *Engineering and Mining Journ.*, New York, 1897, LXIII, 600. An impure clay from Wyoming. The name afterwards changed to *bentonite* (*see* above). [Not the Taylorite (of Dana), 1868] 2nd List

Tazheranite. A. A. Konev, Z. F. Ushchapovskaya, A. A. Kashaev, and V. S. Lebedeva, 1969. Докл. Акад. наук СССР (*Compt. Rend. Acad. Sci. URSS*), **186**, 917 (Тажеранит). Irregular grains and cubic crystals in calciphyres from the Tazheran massif, W of Lake Baikal, Siberia, have a unit-cell content $Zr_{2\cdot36}Ca_{0\cdot77}Ti_{0\cdot7}{}^{3+}Ti_{0\cdot13}{}^{4+}Al_{0\cdot06}Fe_{0\cdot06}O_{6\cdot07}$; they are essentially the cubic modification of ZrO_2, stabilized by foreign cations. Named from the locality. [The name is unfortunately close to tacharanite (22nd List).] [*A.M.*, **55**, 318; *M.A.*, 70-1638; *Zap.*, **99**, 75.] 26th List

Tchinglusuite, Tschinglusuit. V. I. Gerasimovsky, 1938. French (*Bull. Soc. franç. Min. Crist.*, 1951, vol. 74, p. 191) and German (Hintze, *Min.*, 1954, Ergbd. 2, p. 399) transliterations of чинглусуит, chinglusuite (15th List). 20th List [Also 15th List]

Teallite. G. T. Prior, 1904. *Min. Mag.*, vol. xiv, p. 21. Sulphostannite of lead, $PbSnS_2$, occurring as black, metallic, soft, and flexible folia resembling graphite, which are orthorhombic with a perfect basal cleavage. Named after Dr. J. J. Harris Teall, Director of the Geological Survey of Great Britain. From Bolivia; R. Koechlin (*Min. Petr. Mitt.* (*Tschermak*), 1905, vol. xxiv, p. 114) gives a more definite locality. [Santa Rosa mine, Bolivia] 4th List

Teepleite. (C. Doelter, *Handbuch d. Mineralchemie*, 1927, vol. 4, part 2, p. 663; Teepleit). Synonym of burkeite and gauslinite (11th List), described by J. E. Teeple in 1921. 12th List

—— W. A. Gale, W. F. Foshag, and M. Vonsen, 1939. *Amer. Min.*, vol. 24, p. 48. (Earlier mentioned in *Amer. Min.*, 1938, vol. 23, p. 90; and A. Pabst, *Minerals of California*, 1938, p. 167.) Hydrated borate and chloride of sodium, $Na_2B_2O_4.2NaCl.4H_2O$, as tetragonal crystals from the dried-up Borax Lake, California. Named after the late Dr. John E. Teeple, who described the artificial salt in 1929. [*M.A.*, **7**, 263.] 15th List [The accepted usage]

Teineite. T. Yoshimura, 1939. *Journ. Fac. Sci. Hokkaido Univ.*, ser. 4, vol. 4, p. 465. Hydrous tellurate and sulphate of copper, $Cu(Te,S)O_4.2H_2O$, as blue orthorhombic crystals from Teine mine, Japan. Named from the locality. [*M.A.*, **7**, 316.] 15th List [$Cu(TeO_3)_2.2H_2O$; *A.M.*, **46**, 466]

Telargpalite. V. A. Kovalenker, A. D. Genkin, T. L. Evstigneeva, and I. P. Laputina, 1974. *Zap.*, **103**, 595 (Теларгпалит). Rounded grains 5 to 200 μm in chalcopyrite in the Oktyabr deposit, Norilsk region USSR, are cubic, *a* 12·60 Å. Composition near $(Pd,Ag)_{4+x}Te$. Named for *Tel*lurium, *arg*entum, *pal*ladium. [*A.M.*, **60**, 489; *M.A.*, 75-2528; *Zap.*, **104**, 607.] 29th List

Telegdite. L. Zechmeister and V. Vrabély, 1927. *Centr. Min.*, Abt. A, p. 287. A fossil resin from Transylvania, containing sulphur 1·73% and no succinic acid. Named after Dr. Karl Roth von Telegd (telegdi Róth Károly dr.) of the Geological Survey of Hungary. [*M.A.*, **3**, 369.] 11th List

Tellite. G. Gagarin and J. R. Cuomo, 1949, loc. cit., p. 8 (tellita). A single steel-grey crystal of unknown composition, from Binn, Switzerland, described crystallographically (triclinic?) by G. F. H. Smith (*Min. Mag.*, **19**, 40). Named after the Swiss national hero William Tell. 19th List [Syn. of Sinnerite]

Tellurantimony. R. I. Thorpe and D. C. Harris, 1973. *C.M.*, **12**, 55. Lath-shaped crystals in altaite, up to 0·17 × 0·35 mm, are rhombohedral (*a* 4·258, *c* 30·516 Å) and isostructural with tellurbismuth. Named from the composition, Sb_2Te_3. [*A.M.*, **59**, 382.] 28th List

Tellurobismuthite. E. T. Wherry, *Journ. Washington Acad. Sci.*, 1920, vol. 10, p. 490. Translation of the German Tellurwismuth. Applied to rhombohedral Bi_2Te_3, as distinct from tetradymite ($Bi_2S_3.2Bi_2Te_3$). 9th List [i.e. **Tellurbismuth**]

Tellurobismutite. M. N. Short, 1931. *Bull. US Geol. Surv.*, no. 825, p. 73 (telluro-bismutite). Synonym of tellurobismuthite (9th List) Bi_2Te_3, as distinct from tetradymite (Bi_2Te_2S). The earlier names Tellurwismuth (J. J. Berzelius, 1823) and tellurbismuth (D. M. Balch, 1863) covered both. [*M.A.*, **8**, 8, 109, 183, 311; **9**, 213, 262; **10**, 126, 446.] 20th List

Teluroselenio. F. Pardillo, 1947. *Tratado de mineralogía*, translation of 12th edit. of F. Klockmann and P. Ramdohr, Barcelona, p. 318; G. Gagarin and J. R. Cuomo, 1949, loc. cit., 1949, p. 6. Spanish form of selen-tellurium. *See* Hondurasite. 19th List

Temagamite. L. J. Cabri, J. H. G. Laflamme, and J. M. Stewart, 1973. *C.M.*, **12**, 193. Small grains (to 0·1 mm) in chalcopyrite from the Temagami deposit, Ontario, are probably orthorhombic (*a* 11·57, *b* 12·16, *c* 6·76 Å); weakly anisotropic. Composition Pd_3HgTe_3. Named for the deposit (formerly spelt Timagami). [*A.M.*, **60**, 947.] 28th List

Temiskamite. T. L. Walker, 1914. *Amer. Journ. Sci.*, ser. 4, vol. xxxvii, p. 170. A nickel arsenide, Ni_4As_3, occurring as silver-white, radially-fibrous masses in the Temiskaming district, Ontario. According to C. Palmer (*Economic Geology*, 1914, vol. ix, p. 671; *Chem. News*, 1915, vol. cxi, p. 219; *Zeits. Kryst. Min.*, 1915, vol. liv, p. 437) it is identical with maucherite. *See* Placodine. 7th List

Teremkovite. D. A. Timofeevskii, 1967. Зап. Всесоюз. мин. общ. (*Mem. All-Union Min. Soc.*), **96**, 30 (Теремковит). Sulphantimonite of lead and silver, very near owyheeite, occurring in the Ust-Terenki deposit of E Transbaikal. Named for the locality. The distinction from owyheeite is very slight, and the name seems unnecessary. [*A.M.*, **54**, 990; *Zap.*, **97**, 613; *M.A.*, 69-1533.] 26th List

Teremkowit, Germ. trans. of Теремковит, teremkovite (26th List). Hintze, 93. 28th List

Terlinguaite. W. H. Turner, *Mining and Scientific Press*, San Francisco, July 21, 1900. Yellowish-green crystals on the mercury-ores of Terlingua, Brewster Co., Texas, determined by S. L. Penfield to be an oxychloride of mercury [Hg_4Cl_2O]. A. J. Moses (*Amer. Journ. Sci.*, 1903, ser. 4, vol. xvi, p. 253) distinguishes two oxychlorides of mercury from this locality, which he names eglestonite (q.v.) and terlinguaite. The latter name is applied to monoclinic crystals having the composition Hg_2ClO; these are sulphur-yellow, and alter to olive-green on exposure to light. 3rd List

Termierite. G. Friedel, *Bull. Soc. franç, Min.*, 1901, vol. xxiv, p. 6 (Termiérite). A clay, resembling halloysite in appearance, with the formula $Al_2O_3.6SiO_2.18H_2O$. Associated with kaolinite, etc., France. Named after Professor Pierre Termier, of Paris. 3rd List [A mixture; *A.M.*, **42**, 586]

Ternovskite. Y. I. Polovinkina, 1924. *Zap. Ross. Min. Obshch.* (*Mém. Soc. Russe Min.*), ser. 2, vol. 53, p. 216 (*Терновскит*), p. 233 (*ternovskit*). A variety of alkali-amphibole distinguished from other members of the group by its optical characters. Named from the locality, Ternovsky mine, Krivoy Rog, Ukraine. [*M.A.*, **3**, 196.] 11th List [Magnesio-riebeckite; *Amph.* (1978)]

Terpitzite. – Dürr [Pastor Dürr died in 1828; this name had been used by him since 1811, but apparently not published]. J. C. Freiesleben, *Magazin für die Oryktographie von Sachsen*, 1829, Heft iii, p. 168 (Terpizit). A. Frenzel, *Mineralogisches Lexicon für das Königreich Sachsen*, 1874, p. 274 (Terpitzit). A siliceous sinter (Kieselsinter) passing into hornstone or chalcedony, occurring in crevices in porphyry at Terpitz, Saxony. 6th List

Tertschite. H. Meixner, 1953. *Fortschr. Min.*, vol. 31 (for 1952), p. 39 (Tertschit). Hydrous calcium borate, $Ca_4B_{10}O_{19}.20–21H_2O$, finely fibrous, probably monoclinic, from Turkey. Named after Prof. Hermann Tertsch (1880–) of Vienna. [*M.A.*, **12**, 353; *A.M.*, **39**, 849.] 20th List

Teruelite. A. Maestre, *Anales de Minas, Madrid*, 1845, III, 264; F. Quiroga, *Anal. Soc. Españ. Hist. Nat.*, 1873, II, 249; F. Chaves, *Anal. Soc. Españ. Hist. Nat.*, 1891, XX, Actas p. 9. Black, acute rhombohedra (with basal planes) of dolomite from Teruel, Aragon. 2nd List

Teruggite. L. F. Aristarain and C. S. Hurlbut Jr., 1968. *Amer. Min.*, **53**, 1815. Colourless well-formed monoclinic crystals of $Ca_4MgB_{12}O_{20}(AsO_4)_2.18H_2O$ from the Loma Blanca borate deposit, Jujuy Province, Argentina. Named for Professor Mario E. Teruggi. [*M.A.*, 69-2394; *Bull.*, **92**, 514; *Zap.*, **99**, 77.] 26th List

Teshirogilite. Local name for ilmenorutile from Teshirogi, Fukushima-ken, Japan. T. Ito and K. Sakurai, Wada's *Minerals of Japan*, 3rd edit., 1947, vol. 1, p. 211 (figs. 3 and 4), p. 212 (two chemical analyses). [*M.A.*, **10**, 351.] 18th List

Testibiopalladite. Anon., 1974 [*Geochimica*, **3**, 169], abstr. *M.A.*, 75-2529. Cubic 4[Pd(Sb,Bi)Te], *a* 6·56 to 6·58 Å occurs in copper-nickel sulphide deposits in SW and NE China. Named from the composition. This is the Sb analogue of michenerite. [*A.M.*, **61**, 183.] 29th List

Tetraferribiotite. O. M. Rimskaya-Korsakova and E. P. Sokolova, 1964. Зап. Всесоюз. Мин. Общ. (*Mem. All-Union Min. Soc.*), vol. 93, p. 411 (Тетраферрибиотит). A term for biotites having some ferric iron in tetrahedral co-ordination. [*M.A.*, **17**, 504.] 24th List

Tetraferriphlogopite. O. M. Rimskaya-Korsakova and E. P. Sokolova, 1964. Зап. Всесоюз. Мин. Общ. (*Mem. All-Union Min. Soc.*), vol. 93, p. 411. (Тетраферрифлогопит). A term for phlogopites having some ferric iron in tetrahedral co-ordination. [*M.A.*, **17**, 504.] 24th List

Tetraferroplatinum. L. J. Cabri and C. E. Feather, 1975. *C.M.*, **13**, 117. Tetragonal ordered PtFe occurs at Mooihoek, Transvaal. *a* 3·850, *c* 3·693 Å. [*A.M.*, **61**, 341.] 29th List

Tetragophosphite. L. J. Igelström, *Zeits. Kryst. Min.*, XXV. 433, 1895. Slightly more basic than lazulite. [Hörrsjöberg] Sweden. 1st List [Is Lazulite; *M.A.*, **14**, 523]

Tetrahydrite. I. Kostov, 1968. *Mineralogy*, p. 494. Syn. of Leonhardtite (19th List). 25th List

Tetra-kalsilite. J. V. Smith and O. F. Tuttle, 1957. *Amer. Journ. Sci.*, vol. 255, p. 286. T. G. Sahama and J. V. Smith, *Amer. Min.*, vol. 42, p. 286. A form of kalsilite, $(K,Na)AlSiO_4$, with the hexagonal *a*-axis 20 Å, in the composition range Na_{20-30}. *See* Tri-kalsilite. [*M.A.*, **13**, 524.] 21st List

Tetrawickmanite. J. S. White, Jr., and J. A. Nelen, 1973. *Min. Record*, **4**, 24. A tetragonal polymorph of the cubic mineral wickmanite, $MnSn(OH)_6$, (25th List) occurs in honey-yellow crystals at the Foote Mineral Company's spodumene mine, Kings Mountain, Cleveland County, North Carolina. *a* 7·7870, *c* 7·797 Å. (Cf. *M.A.*, **9**, 209.) [*M.A.*, 73-2949; *A.M.*, **58**, 966; *Zap.*, **103**, 358.] 28th List

Texasite. W. W. Crook III, 1977. *A.M.*, **62**, 1006. An alteration crust between crystals of allanite, gadolinite, and yttrofluorite at the Clear Creek pegmatite, Burnet Co., Texas, is almost pure $Pr_2(SO_4)O_2$, identical with artificial material. Orthorhombic, *a* 4·139, *b* 4·243, *c* 13·431 Å. RIs α 1·826, colourless, ‖ [001], β 1·917, pale green, ‖ [010], γ 1·921, pale green, ‖ [100], $2V_\alpha$ 26–31°. Named for the State of Texas. [This name was used by Kenngott (1853) for zaratite (Casares, 1851) but has not been generally accepted. Though it has occasionally been suggested that a name discredited and unused for over fifty years can properly be re-used, this is undesirable—*M.H.H.*] [*M.A.*, 78-2127.] 30th List

Thalcusite. V. A. Kovalenker, I. P. Laputina, T. L. Evstigneeva, and V. M. Izoitko, 1976. *Zap.*, **105**, 202 (Талкусит). Platy grains up to 150 μm along galena–chalcopyrite boundaries in ores of the Talnakh deposit are tetragonal, *a* 3·882, *c* 13·25 Å, composition $Cu_{3-x}Tl_2Fe_{1+x}S_4$, the S analogue of bukovite (28th List). X-ray data match those of synthetic $Cu_3Tl_2FeS_4$. Named from the composition. 29th List

Thalenite. C. Benedicks, *Geol. För. Förh.*, 1898, XX, 308; *Bull. Geol. Inst. Univ. Upsala,* 1899 (for 1898), IV, 1 (*Thalénit*). Flesh-red; massive and in dull monoclinic crystals. $H_2O.2Y_2O_3.4SiO_2$. Österby, Sweden. 2nd List [$Yt_3Si_3O_{10}(OH)$; *M.A.*, 73-1295]

Thanite. M. Rózsa, 1914. *Zeits. Anorg. Chem.*, vol. lxxxviii, p. 328 (Thanit; *Hartsalzkainitit*), p. 332 (*Halitkainit*). A mixture of kainite and halite occurring in the salt deposits of the Werra district, Prussia. Named after the late Professor Karl von Than. 7th List

—— G. Vavrinecz, 1939. *Földtani Közlöny*, Budapest, vol. 69, pp. 82, 98 (Thanit). Carbon oxysulphide, COS, as a natural gas. Named after Karl Than, who discovered it in 1867. [*M.A.*, **7**, 471.] 15th List

Thellite. ([— Phipson] *Chem. News*, 1870, XXI, 13, *thelline*, *thellite*). Damour's yttrium silicate from Brazil (Dana's *Min.*, 6th edit., p. 512). 2nd List

Thelotite. C. E. Bertrand and B. Renault, *Bull. Soc. Hist. Nat. Autun*, 1892, vol. v, pp. 163, 240 (thélotite). An undetermined carbonaceous constituent of Boghead coal from the Thélots pits at Autun. 3rd List

Theophrastite. A. Breithaupt, 1849. *Die Paragenesis der Mineralien*, p. 216 (Theophrastit). Synonym of Nickelwismuthglanz (F. von Kobell, 1835) = grünauite (J. Nicol, 1849) = saynite (F. von Kobell, 1853). Named after Theophrastus (*c.* 372–286 BC). 16th List

Thermitocorundum. R. L. Pevzner, 1947. *Compt. Rend.* (*Doklady*) *Acad. Sci. URSS*, vol. 55, p. 233. Artificial corundum prepared by the thermite process. [*M.A.*, **10**, 259.] 18th List

Thermitospinel. R. L. Pevzner, 1947. *Compt. Rend.* (*Doklady*) *Acad. Sci. URSS*, vol. 55, p. 233. Artificial spinel prepared by the thermite process. [*M.A.*, **10**, 259.] 18th List

Thermokalite. H. J. Johnston-Lavis, MS. K. W. Earle, *Proc. Geol. Assoc. London*, 1928, vol. 39, p. 96. An undescribed 'member of the haloid group' in the Johnston-Lavis collection of Vesuvian minerals. Perhaps named from analogy with thermonatrite [$Na_2CO_3.H_2O$], in which case it would belong to the carbonate group. 11th List [A mixture; *M.M.*, **22**, 59]

Thermospinell, error for Thermitospinel (18th List) (C. Hintze, *Handb. Min.*, Erg.-Bd. II, pp. 388, 955). 24th List

Thioelaterite. B. L. Dunicz, 1936. *Arch. Min. Tow. Nauk. Warszaw.* (*Arch. Min. Soc. Sci. Varsovie*), vol. 12, p. 90 (tioelateryt), p. 95 (tioélaterite, tioélatérite). An elastic bitumen containing 3% of sulphur (θεῖον), from Bolivia: a natural vulcanized caoutchouc or elaterite. [*M.A.*, **6**, 344.] 14th List

Thiolaterit, error for Thioelaterit (14th List) (C. Hintze, *Handb. Min.*, Erg.-Bd. II, p. 955). 24th List

Thiospinels. J. Flahaut, L. Domange, and M. Ourmitchi, 1960. *Compt. Rend. Acad. Sci. Paris*, vol. 250. p. 134 (thiospinelles). A general term for compounds AB_2S_4 with spinel-type structures. 22nd List

Thomasite. V. A. Kroll, 1911. *Journ. Iron and Steel Inst.*, vol. 84 (no. II for 1911), pp. 163, 187. Basic calcium silico-phosphate, $(6CaO.P_2O_5).(2FeO.SiO_2)$, as minute, bluish-green, hexagonal pyramids aggregated as needles and larger rough crystals. A constituent of the basic slag of the Thomas-Gilchrist process for the dephosphorisation of iron. Named after Sidney Gilchrist Thomas (1850–1885). 9th List

Thorbastnäsite. A. S. Pavlenko, L. P. Orlova, M. V. Akhmanova, and K. I. Tobelko, 1965. Зап. Всесоюз. Мин. Общ. (*Mem. All-Union Min. Soc.*), vol. 94, p. 105 (Торбастнезит, Thorbastnesite). Brown cryptocrystalline material from an unnamed nepheline syenite intrusion in E Siberia has a composition near $(Ca,Ln)Th(CO_3)_2F_2.3H_2O$. X-ray powder data indicate hexagonal symmetry. The water may not be essential. [*M.A.*, **17**, 398; *A.M.*, **50**, 1505, Thorbastnaesite.] 24th List

Thorchevkinite, a superfluous name for the thorian Chevkinite (?) analysed by Hermann (1866). [Геохим., мин., генет. типы месторожд. редк. элем., Изд. „Наука", 1964, p. 305]; abstr. Зап. Всесоюз. Мин. Общ. (*Mem. All-Union Min. Soc.*), 1966*, vol. 94, p. 683 (Торчевкинит). 24th List [*Corr. in *M.M.*, **39**]

Thoreaulite. H. Buttgenbach, 1933. *Ann. Soc. Géol. Belgique*, vol. 56, *Bull.*, p. B 327. Tantalate of tin, perhaps $SnO_2.Ta_2O_5$, as brown crystals, perhaps monoclinic, in pegmatite from Katanga. Named after Prof. Jacques Thoreau, of Louvain, Belgium. [*M.A.*, **5**, 389.] 13th List [*A.M.*, **59**, 1026]

Thorgadolinite. L. B. Zubkov, V. I. Paribok, and A. B. Cheryakhovskii, 1970. Докл. Акад. наук СССР (*Compt. Rend. Acad. Sci. URSS*), **192**, 633 (Торгадолинит). An unnecessary name for a thorian (4·65% ThO_2) gadolinite. [*A.M.*, **56**, 2156.] 27th List

Thorianite. W. R. Dunstan, 1904. *Nature*, vol. lxix, p. 510; *Bull. Imp. Inst. London*, 1904, vol. ii, p. 13; 1905, vol. iii, p. 155. W. R. Dunstan and G. S. Blake, *Proc. Roy. Soc. London*, 1905, ser. A, vol. lxxvi, p. 253. W. R. Dunstan and B. M. Jones, *Proc. Roy. Soc. London*, 1906, ser. A, vol. lxxvii, p. 546. A heavy, black mineral found as water-worn, cubic crystals in the gem-bearing gravels of Ceylon. It was first described as uraninite (A. K. Coomaraswamy, *Spolia Zeylanica*, 1904, vol. i, p. 112; *Administration Reports, Ceylon*, for 1903, 1904, part iv, p. L 11), but it differs from this in that the uranium oxide (UO_2) is largely replaced isomorphously by thoria (58·8–78·8% ThO_2). So named because of the large amount of thorium which it contains. 4th List

Thoro-aeschynite. E. M. Eskova, A. G. Zhabin, and G. N. Mukhitdinov, 1964. [Мин. геохим. редк. элем. Вишнев. Гор (*Min. geochem. rare elements in the Vishnevaya Mts., Urals*), Moscow (Izdatel Nauka)]; abstr. *Amer. Min.*, 1965, vol. 50, p. 2101; Зап. Всесоюз. Мин. Общ. (*Mem. All-Union Min. Soc.*), 1966, vol. 95, p. 312 (Тороэшинит, thoroaeshynite). An unnecessary name for a thorian Aeschynite from the alkalis complex at the Vishnevaya Mts., Urals. 24th List

Thorobritholite. I. I. Kupriyanova and G. A. Sidorenko, 1963. Докл. Акад. Наук СССР (*Compt. Rend. Acad. Sci. URSS*), vol. 148, p. 212 (Торобритолит). Thorian britholite. 'A superfluous name' (Зап. Всесоюз. Мин. Общ. (*Mem. All-Union Min. Soc.*), 1964, vol. 93, p. 456). 23rd List

Thorolite, error for Thoreaulite, due to back-transliteration of Торолит (*Soviet Physics—Crystallography*, 1967, vol. 12, p. 105; transl. of Кристаллография, 1967, vol. 92, p. 133.) 25th List

Thoromelanocerite. I. I. Kupriyanova and G. A. Sidorenko, 1963. Докл. Акад. Наук СССР (*Compt. Rend. Acad. Sci. URSS*), vol. 148, p. 212 (Торомеланоцерит). Thorian melanocerite. 'A superfluous name' (Зап. Всесоюз. Мин. Общ. (*Mem. All-Union Min. Soc.*), 1964, vol. 93, p. 456). 23rd List

Thorosteenstrupine. I. I. Kupriyanova, T. I. Stolyarova, and G. A. Sidorenko, 1962. Зап. Всесоюз. Мин. Общ. (*Mem. All-Union Min. Soc.*), vol. 91, p. 325. (Торостенструпин). A dark brown metamict mineral from an undisclosed locality in E Siberia has a composition near $CaMnThSi_4O_{11}(OH)F.5\frac{1}{3}H_2O$, 'similar to Steenstrupine with rare earths almost entirely replaced by Th and Ca'. Named from the composition. Relation to Steenstrupine uncertain, and name therefore inappropriate (*Amer. Min.*, 1963, vol. 48, p. 433; *M.A.*, **16**, 66). 23rd List

Thorotungstite. J. B. Scrivenor and J. C. Shenton, 1927. *Amer. Journ. Sci.*, ser. 5, vol. 13, p. 487; *Geologist's Ann. Rep. Federated Malay States*, 1927, for 1926, p. 2. Hydrated oxide of tungsten and thorium $2WO_3.H_2O + ThO_2.H_2O$, as yellow masses of minute (perhaps orthorhombic) crystals, from Perak. It shows a relation to tungstite, hence the name. [*M.A.*, **3**, 367.] 11th List [Renamed **Yttrotungstite**, as it contains Yt, not Th; *M.A.*, **11**, 189]

Thortveitite. J. Schetelig, 1911. *Centralblatt Min.*, 1911, p. 721 (Thortveitit). Silicate of scandium (yttrium, dysprosium, and erbium), $(Sc,Y)_2O_3.2SiO_2$, occurring as long, greyish-green, orthorhombic crystals, somewhat resembling epidote in appearance, in pegmatite near Iveland, Sætersdalen, Norway. Named after Mr. Olaus Thortveit, of Iveland, who sent the material for determination. Remarkable in being the only mineral containing scandium in large amount [then known]. 6th List [Cf. Thalenite]

Thorutite. *See* Smirnovite. 21st List [*A.M.*, **43**, 1007; **48**, 1419]

Thucholite. H. V. Ellsworth, 1928. *Amer. Min.*, vol. 13, p. 66. 'A remarkable carbon mineral' from Parry Sound, Ontario. 11th List. *Amer. Min.*, vol. 13, pp. 419, 438 (thū′-chō-lite). H. S. Spence, *Amer. Min.*, 1930, vol. 15, p. 499. A. Faessler, *Centr. Min.*, Abt. A, 1931, p. 10 (Thucholith). A brittle jet-black carbonaceous material with variable amount of ash which contains thorium, uranium, etc. It occurs as small nodules intimately associated with uraninite in pegmatite in Canada. Named from the chemical symbols of its chief components, Th, U, C, H, O, and λίθος, stone. [*M.A.*, **4**, 11, 502.] 12th List

Tialite, Tielite, Tieilite. Various renderings of the Japanese name. H. Strunz, *Min. Tab.*, 3rd edit., 1957, p. 155 (Tialit). K. F. Chudoba, Hintze, *Min.*, Erg.-Bd. II, p. 390 (Tielit). *M.A.*,

11, 415; 19th List (Tieilite). Artificial aluminium titanate, Al_2TiO_5, isomorphous with pseudobrookite. 21st List

Tibergite. N. H. Magnusson, 1930. *Avh. Sveriges Geol. Undersök., Ser. Ca*, no. 23, p. 46 (tibergit), p. 107 (tibergite). A brownish-purple variety of amphibole rich in alkalis, Al_2O_3 and Fe_2O_3, containing $R_2(Al,Fe)_4Si_2O_{12}$ in addition to the richterite molecule; from Långban, Sweden. Named after Mr. H. V. Tiberg, formerly manager of the Långban mines. [*M.A.*, **9**, 4.] 17th List [Manganoan sodian magnesio-hastingsite; *Amph.* (1978)]

Tieilite. G. Yamaguchi, 1944. [*Journ. Japanese Ceramic Assoc.*, vol. 52, p. 6]; *Chem. Abstr.*, 1951, vol. 45, col. 7925. Al_2TiO_5, orthorhombic, isomorphous with pseudobrookite (Fe_2TiO_5); prepared artificially, and present in abrasives. [*M.A.*, **11**, 415.] 19th List [Also *see* Tialite]

Tielite, *see* Tialite.

Tienshanite. V. D. Dusmatov, A. F. Efimov, V. Yu. Alkhazov, M. E. Kazakova, and N. G. Mumyatskaya, 1967. Докл. Акад. наук СССР (*Compt. Rend. Acad. Sci. URSS*), **177**, 678. (Тяншанит.) Green, hexagonal $Na_2BaMnTiB_2Si_6O_{20}$, from a pegmatitic vein of a massif of the Turkestan-Alai province, S Tienshan. Named for the locality. [*A.M.*, **53**, 1426; *Zap.*, **97**, 618.] 26th List

Tiffanyite. G. F. Kunz, *Trans. N.Y. Acad. Sci.*, XIV. 260, 1895. [*M.M.*, **11**, 241.] A hydrocarbon to which is attributed the phosphorescence of diamonds. 1st List

Tikhonenkovite. A. P. Khomyakov, V. I. Stepanov, V. A. Moleva, and Z. V. Pudovkina, 1964. Докл. Акад. Наук СССР (*Compt. Rend. Acad. Sci. URSS*), vol. 156, p. 345 (Тихоненковит). $SrAlF_4OH.H_2O$, from Karasug, Tannu-Ola, Tuva in small monoclinic crystals; named for I. P. Tikhonenkov. The strontium analogue of Gearksutite. [*A.M.*, **49**, 1774.] 23rd List

Tikhvinite. O. M. Ansheles and N. I. Vlodavets, 1927. *Zap. Ross. Min. Obshch.* (*Mém. Soc. Russe Min.*), ser. 2, vol. 56, p. 53 (тихвинит), p. 60 (tikhvinite). Phosphate and sulphate of aluminium and strontium, $2SrO.3Al_2O_3.P_2O_5.SO_3.6H_2O$ (or $7H_2O$), occurring as nodules in bauxite in the Tikhvin district, Russia. Named after the locality. [*M.A.*, **3**, 473.] 11th List

Tilasite. H. Sjögren, *Geol. För. Förh.*, XVII. 291, 1895. [*M.M.*, **11**, 229.] Fluor-adelite. $2CaO.MgO.MgF_2.As_2O_5$. Anorthic. [Långban] Sweden. 1st List [$CaMg(AsO_4)F$]

Tilleyite. E. S. Larsen and K. C. Dunham, 1933. *Amer. Min.*, vol. 18, p. 469. Silicate and carbonate of calcium, $Ca_2SiO_4.CaCO_3$, monoclinic (?), in a contact metamorphic rock at Crestmore, California. Named after Prof. Cecil Edgar Tilley, of Cambridge. [*M.A.*, **5**, 387.] 13th List [$Ca_5Si_2O_7(CO_3)_2$; *M.A.*, **12**, 337]

Timazit, listed by H. Strunz (*Min. Tabellen*, 1st edit., 1941, p. 278) as a synonym of Gamsigradite, is a rock name (A. Breithaupt, *Berg. Hüttenmann. Zeit.*, 1861, p. 51). Timagite or Timazite, included in T. A. Readwin's list of minerals 'of very doubtful character' (*Chem. News*, 1874, vol. 30, p. 164; 1875, vol. 32, p. 18) is probably the same. 24th List

Tinaksite. Yu. G. Rokov, V. P. Rogova, A. A. Voronkou, and V. A. Moleva, 1965. Докл. Акад. наук СССР (*Compt. Rend. Acad. Sci. URSS*), vol. 162, p. 658. (Тинаксит). Pale yellow prismatic anorthic crystals in potash-feldspar metasomatic rocks in the border of the Murunsk massif, NW Aldan. The unit cell contains $2[NaK_2(Ca,Fe^{3+})_2(Ti,Fe^{3+})_2Si_7O_{11}OH]$. Named for the composition (Ti,Na,K,Si). [*M.A.*, **17**, 502; *A.M.*, **50**, 2098.] 24th List

Tinaxite, erroneous transliteration of Тинаксит (Tinaksite, 24th List). *Soviet Physics—Doklady*, 1971, **16**, 338 (translation of Докл. 1971, **198**, 575). [*M.A.*, 72-1803.] 28th List

Tinkalite. G. Gagarin and J. R. Cuomo, 1949, loc. cit., p. 12 (tinkalita). Synonym of tincal = borax. 19th List

Tin-tantalite, translation of Оловотанталит; synonym of Stannotantalite. *Amer. Min.*, 1961, vol. 46, p. 1514. 23rd List [Cf. Wodginite]

Tinticite. B. Stringham, 1946. *Amer. Min.*, vol. 31, p. 395. Hydrated basic ferric phosphate, $2FePO_4.Fe(OH)_3.3\frac{1}{2}H_2O$, as a compact creamy-white wall-coating in a limestone cave at Tintic, Utah. Named from the locality. [*M.A.*, **10**, 5.] 18th List

Tintinaite. D. C. Harris, J. L. Jambor, G. R. Lachance, and R. I. Thorpe, 1968. *Canad. Min.*, vol. 9, p. 371. A mineral from the Tintina silver mines, Yukon, proves to be essentially $Pb_5Sb_8S_{17}$, and to be the antimony analogue of kobellite. [*A.M.*, **54**, 573.] 25th List

Tinzenite. J. Jakob, 1923. *Schweiz. Min. Petr. Mitt.*, vol. 3, p. 234 (Tinzenit). Silicate of manganese, aluminium, and calcium, $Mn_2O_3.Al_2O_3.2CaO.4SiO_2$, forming yellow, monoclinic crystals. Named from the locality, Tinzen, Grisons. [*M.A.*, **2**, 251.] 10th List [Cf. Ferroaxinite, Manganaxinite, Severginite; proposed redefinition, *M.A.*, 71-2989]

Tioelateryt, Tioélatérite, *see* Thioelaterite.

Tirodite. J. A. Dunn and P. C. Roy, 1938. *Rec. Geol. Surv. India*, vol. 73, p. 295 (tirodite). A manganese amphibole (MnO 8·25%) differing from dannemorite and richterite, from Tirodi, Central Provinces, India. Named from the locality. [*M.A.*, **7**, 317; *A.M.*, **25**, 380.] 15th List

Titanaugite. A. Knop, 1892. *Der Kaiserstuhl im Breisgau. Leipzig*, p. 72 (Titan-Augit). Titaniferous augite. 4th List

Titanbetafit, variant of Titanobetafite (H. Strunz, *Min. Tabellen*, 4th edit., 1966, p. 175). 24th List

Titanbiotite. W. Freudenberg, 1919. *See* Wodanite. 9th List

Titanclinohumite. F. Machatschki, 1930. *Centralblatt Min.*, Abt. A, 1930, p. 194 (Titanklinohumit). F. de Quervain, *Schweiz. Min. Petr. Mitt.*, 1938, vol. 18, p. 591. Synonym of titanolivine (A. Damour, 1879), titanhydroclinohumite (9th List), and Klinolivin (12th List). [*M.A.*, **8**, 85.] 16th List

Titaneisenglimmer. H. Rosenbusch, 1885. *Mikroskopische Physiographie d. Mineralien*, 2nd edit., vol. i, p. 331. Ilmenite as thin, transparent scales of a clove-brown colour by transmitted light; described by K. Hofmann (1879) in basaltic rocks from Hungary. *See* Ilmenitglimmer. 4th List

Titanelpidite. A. N. Winchell, 1951. *Optical mineralogy*, part 2, 4th edit., p. 455. Variant of titano-elpidite (11th List). 19th List

Titangarnet, Titanhornblende, Titanmica, Titantourmaline. W. Kunitz, 1936. *Neues Jahrb. Min.*, Abt. A, Beil.-Bd. 70, p. 392 (Titangranate, *pl.*), p. 385 (Titanhornblenden), p. 397 (Titan-Hornblenden), p. 399 (Titanglimmer), p. 401 (Titan-Glimmer), p. 385 (Titanturmaline, *pl.*). Titaniferous silicates analogous to titanaugite (4th List), titanbiotite (9th List), titanolivine (A. Damour, 1879). 14th List

Titanhaematite. A. B. Edwards, 1938. *Proc. Australasian Inst. Mining and Metall.*, n. ser., no. 110, p. 42 (titanohematite). C. O. Hutton, *New Zealand Journ. Sci. Techn.*, sect. B, 1945, vol. 26, p. 299 (titanhaematite). Haematite with TiO_2 in solid solution, with dark brown to black streak. When more than about 10% of TiO_2 is present the excess separates out as intergrowths of ilmenite. [*M.A.*, **9**, 263.] 17th List

Titanhornblende, Titan-Hornblende, *see* Titangarnet.

Titanhydroclinohumite. F. Zambonini, 1919. *Bull. Soc. fanç. Min.*, vol. 42, p. 279 (titanhydroclinohumite or hydroclinohumite titanifère). The 'titanolivine' of A. Damour (1879) is proved to be identical with clinohumite, but containing some titanium (TiO_2 1·92%) and hydroxyl in place of fluorine. *See* Hydroclinohumite. [*M.A.*, **1**, 106.] 9th List

Titania. The old chemical name for titanium dioxide has now been used as a trade-name for *artificial* rutile. H. Strunz, *Min. Tab.*, 1970, 5th edit., 582. Other trade-names include: astryl, brilliante, diamothyst, kimberlite gem, lusterite (q.v.), miridis, tirum gem, titanium [*sic*], and titanstone (from a collection of 26 names supplied by M. J. O'Donoghue, priv. comm.). 28th List

Titanium lueshite, a mistranslation of титановый луешит, titanian lueshite (*Dokl. Acad. Sci. USSR* (*Earth Sci. Sect.*), **171**, 160). 28th List

Titan-låvenite. E. I. Kutukova, 1940. [*Trans. Inst. Geol. Sci., Acad. Sci. USSR*, no. 31, *Min. Geochem. Ser.*, no. 6, p. 23.] Abstr. in *Amer. Min.*, 1941, vol. 26, p. 135 (Titan-lovenite). A variety of låvenite containing TiO_2 11·30%, from the Kola peninsula. 16th List

Titan-Lueshit. K. F. Chudoba, 1974. Hintze, 95. A mistranslation of титановый луешит, titanian lueshite Докл. **171**, 956 (1966)). 28th List

Titanmagnetite. A. F. Buddington, J. Fahey, and A. Vlisidis, 1955. *Amer. Journ. Sci.*, vol. 253, p. 409. A titaniferous magnetite (titanomagnetite, P. Groth, 2nd List) containing TiO_2 in solid solution, as distinct from ilmenomagnetite (20th List) with exsolution ilmenite. *See* Mogensenite. 21st List

Titanmelanite. A. Knop, 1892. *Der Kaiserstuhl im Breisgau. Leipzig*, p. 145 (Titanmelanit). Titaniferous melanite approaching schorlomite in composition. 4th List

Titanmica, *see* Titangarnet.

Titanmikrolith. H. Strunz, 1966. *Min. Tabellen*, 4th edit., p. 175. A name for a member of the pyrochlore family in which Ti > Ta > Nb. Since there appears to be no break in the Ti ⇌ (Nb + Ta) substitution, but a distinct break in the Nb ⇌ Ta substitution in natural minerals, van der Veen's division of the Nb:Ta:Ti triangle of pyrochlore minerals into three (Pyrochlore, Microlite, and Titanopyrochlore) appears preferable. 24th List [Remains hypothetical; *A.M.*, **62**, 407]

Titano-aeschynite. A. G. Zhabin, G. N. Mukhidtinov, and M. E. Kazakova, 1960. Труды Инст. Мин. Геохим. Крист. Редк. Элем. (*Trans. Inst. Min., Geochem., Cryst. Rare Elem.*), no. 4, p. 51 (Титанозшинит). An unnecessary name for Aeschynite, which normally has Ti > Nb (Зап. Всесоюз. Мин. Общ. (*Mem. All-Union Min. Soc.*), 1963, vol. 92, p. 199; *Amer. Min.*, 1962, vol. 47, p. 417; *M.A.*, **15**, 556). 23rd List

Titanobetafite. A. I. Ginzburg, S. A. Gorzhevskaya, E. A. Erofeeva, and G. A. Sidorenko, 1958. [Геохимия, p. 486], translated as Geochemistry, 1958, p. 615. A name for members of the pyrochlore family in which Ti > (Nb + Ta). 24th List [Syn. of betafite; *A.M.*, **62**, 407]

Titanochondrodite. K. Aoki, K. Fujino, and M. Akaogi, 1976. *Contr. Min. Petr.*, **56**, 243. Unnecessary name for titanian chondrodite (9% TiO_2), occurring in the Buell Park Kimberlite, Arizona. 29th List

Titanochromite. E. N. Cameron, 1970. *Science*, **167**, 623. An unnecessary name for a titanian chromite occurring in Apollo XI lunar rocks. [*A.M.*, **55**, 2135; *Zap.*, **100**, 618.] 27th List

Titanoclinohumite. W. T. Huang, 1957. *Amer. Min.*, vol. 42, p. 686. A titanian variety of Clinohumite containing appreciable F (1%), as distinct from the fluorine-free Titanhydroclinohumite (9th List). The variant Titanclinohumite has been used (16th List) as a synonym of Titanhydroclinohumite, but should properly include both fluorine-free and fluorine-bearing varieties, and has priority in this sense. 23rd List

Titano-elpidite. A. E. Fersman, 1926. *Amer. Min.*, vol. 11, p. 295. *Neues Jahrb. Min.*, Abt. A, 1926, vol. 55, p. 43 (Titanoelpidit). A. N. Labuntzov, *Compt. Rend. Acad. Sci. URSS*, ser. A, 1926, p. 39 (титановый эльпидит, elpidite titanifère). A variety of elpidite containing more titanium than zirconium. [*M.A.*, **3**, 235; **3**, 236.] 11th List [Renamed Labuntsovite (q.v.)]

Titano-eschynite, undesirable spelling variant of Titano-aeschynite. *Amer. Min.*, 1962, vol. 47, p. 417; Зап. Всесоюз. Мин. Общ. (*Mem. All-Union Min. Soc.*), 1963, vol. 92, p. 199. 23rd List

Titano-Euxenite. Chan Pei-Shan, 1963. [*Scientia Sinica*, vol. 12, p. 237], quoted in C. Hintze, *Handb. Min.*, Erg. III, p. 641. Syn. of Euxenite. 25th List

Titano-haematite. R. Chevallier, J. Bolfa, and S. Mathieu, 1955. *Bull. Soc. franç. Min. Crist.*, vol. 78, p. 310 (titano-hématites rhomboédriques). Variant of titanhaematite (17th List). *See* Ferri-ilmenite. 21st List

Titanohaematite. H. Strunz, 1957. *Min. Tabellen*, 3rd edit., p. 144. Another variant of Titanohematite. The attribution of this variant to L. J. Spencer in Hintze, *Handb. Min.*, Erg.-Bd. II, p. 867, is in error. 22nd List

Titanoludwigite. A. A. Konev, V. S. Lebedeva, A. A. Kashaev, and Z. F. Ushchapovskaya, 1970. Зап. Всесоюз. мин. общ. (*Mem. All-Union Min. Soc.*), **99**, 225 (Титанолюдвигит). A titanian lugwigite from Tazheran, $(Mg_{6\cdot55}Fe_{1\cdot18}^{2+}Mn_{0\cdot03}Ca_{0\cdot06}Na_{0\cdot16})(Fe_{2\cdot28}^{3+}Ti_{0\cdot72}^{4+}Mg_{0\cdot72}^{2+}Al_{0\cdot20})B_{4\cdot12}O_{19\cdot97}$. An unnecessary name. Cf. Azoproite. 26th List

Titanomaghemite. E. Z. Basta, 1959. *Econ. Geol.*, vol. 54, p. 698. An unnecessary varietal name for titanian maghemite. 22nd List

Titanomagnetite. (P. Groth, *Tab. Uebers. Min.*, 4th edit., 1898, p. 79.) Titaniferous magnetite. $[(Fe,Ti)O_2]_2Fe$. 2nd List

Titanoenadkevichite. A superfluous name for titanian Nenadkevichite. [Геохим., мин., генет. типы месторожд. редк. элем., Изд. „Наука", 1964, p. 533]; abstr. Зап. Всесоюз. Мин. Общ. (*Mem. All-Union Min. Soc.*), 1965, vol. 94, p. 683 (Титаноненадкевичит). 24th List

Titano-obruchevite. A. H. van der Veen, 1963. *Verhandel. Kon. Nederland, geol. mijnbouw. genootschap*, geol. ser. no. 22. A Titanopyrochlore in which the principal *A* cation in the formula $A_2B_2X_7$ is yttrium. 24th List [Syn. of Yttrobetafite, *A.M.*, **62,** 407]

Titanopriorite. Cited from *Scientia Sinica*, 1963, vol. 12, p. 237 in Зап. Всесоюз. Мин. Общ. (*Mem. All-Union Min. Soc.*), 1965, vol. 94, p. 672 (Титаноприорит, titanopriorite). Syn. of blomstrandine. 24th List

Titanopyrochlore. A. H. van der Veen, 1963. *Verhandel. Kon. Nederland. geol. mijnbouw. genootschap*, geol. ser. no. 22. F. Machatschki (1932) proposed the name Titanpyrochlore (13th List, *see* under Antimonpyrochlore) for a *hypothetical* end-member of the pyrochlore group; a spelling variant is now proposed for all members of the Pyrochlore–Microlite family in which Ti > Nb and Ti > Ta. 24th List [Possible related specimens, but name to be dropped, *A.M.*, **62**, 407]

Titanorhabdophane. E. I. Semenov, 1959. [Матер. мин. Кольск. полуостр., no. 1, p. 102]; abstr. Зап. Всесоюз. Мин. Общ. (*Mem. All-Union Min. Soc.*), 1961, vol. 90, p. 445 (Титанорабдофанит, titanorhabdophane). Spherulites from a pegmatite near Nepkha in the Lovozero massif contain 48% Ln_2O_3, 13% TiO_2, 9% SiO_2, 15% H_2O, and only 3·6% P_2O_5, but are regarded as a member of the rhabdophane group on the ground of a certain resemblance in the powder patterns. An unsatisfactory name. 22nd List [Renamed Tundrite (q.v.)]

Titano-spinel. J. de Lapparent, 1937. *Min. Petr. Mitt.* (*Tschermak*), vol. 49, p. 15 (titano-spinelle). Spinel in the emery of Samos assumed to contain titanium, but not analysed. [*M.A.*, **7**, 150.] 15th List

Titano-thucholite. G. Aminoff, 1943. *Geol. För. Förh. Stockholm*, vol. 65, p. 31 (Titano-Thucholite). O. H. Ödman, *Årsbok Sveriges Geol. Unders.*, 1944, vol. 38, no. 6, pp. 3, 7. A radioactive bituminous mineral differing from thucholite (12th List) in containing titanium (TiO_2 4·5%), from Boliden, Sweden. [*M.A.*, **9**, 37, 257.] 17th List

Titanpigeonite. T. Tomita, 1933. *Journ. Shanghai Sci. Inst.*, sect. 2, vol. 1, p. 3 (titan-pigeonite); 1934, vol. 1, p. 120 (titanpigeonite). A variety of titanaugite related to pigeonite. [*M.A.*, **6**, 119.] 14th List

Titanspinel. F. Mogensen, 1943. *See* Ulvöspinel. 18th List

Titantourmaline, Titanturmalin, *see* Titangarnet.

Titanvesuvianite. L. L. Shilin, 1940. *Compt. Rend.* (*Doklady*) *Acad. Sci. URSS*, vol. 29, p. 325 (titanvesuvianite). V. S. Myasnikov, ibid., 1940, vol. 28, p. 446 (titaniferous vesuvianite). A variety of idocrase containing 4·59% TiO_2. [*M.A.*, **8**, 100, 101.] 16th List

Tĭujamunite, Tjuiamunit, *see* Tyuyamunite.

Tjanschanit, Germ. trans. of Тяншанит, tienshanite (26th List). Hintze, 95. 28th List

Tlalocite. S. A. Williams, 1975. *M.M.*, **40**, 221. Blue velvety crusts on fracture surfaces in ore are probably orthorhombic but X-ray powder data could not be indexed. Composition $Cu_{10}Zn_6TeO_3(TeO_4)_2Cl(OH)_{25}.27H_2O$. Named for Tlaloc, the god of rain in Toltec and Aztec mythology. [*M.A.*, 75-3606; *A.M.*, **61**, 504.] 29th List

Tlapallite. S. A. Williams and M. Duggan, 1978. *M.M.*, **42**, 183. Thin green paint-like films on rock-fracture surfaces at the Bambollita mine, Moctezuma, Sonora, Mexico, and at the Lucky Cuss mine, Tombstone, Arizona, are monoclinic, *a* 11·94, *b* 9·11, *c* 15·66 Å, *β* 90° 36′. Composition $H_6(Ca,Pb)_2(Cu,Zn)_3SO_4(TeO_3)_4TeO_6$. RIs α 1·815, $\beta = \gamma$ 1·960 (Bambollita); α 1·955, $\beta = \gamma$ 2·115 (Tombstone). Named from the Nahua word *tlapalli*, paint. 30th List

Tochilinite. N. I. Organova, A. D. Genkin, V. A. Drits, S. P. Molotkov, O. V. Kuzmina, and A. L. Dmitrik, 1971. *Zap.*, **100**, 477 (Точилинит). Bronze-black grains and fibrous aggregates from the Lower Maman intrusive, Voronezh, USSR, are monoclinic, *a* 5·4, *b* 15·7, *c* 10·7 Å, *β* 95°; composition $6Fe_{1-x}S.5(Mg,Fe)(OH)_2$, the FeS analogue of valleriite. Named for M. S. Tochilin. [*M.A.*, 73-1943; *A.M.*, **57**, 1552; *Zap.*, **101**, 277.] 28th List [Canada and USA; *M.A.*, 77-866]

Toddite. H. V. Ellsworth, 1926. *Amer. Min.*, vol. 11, p. 332. Niobate (and tantalate) of uranium, iron, etc., as rounded masses, optically isotropic, in pegmatite from Ontario. Named after Mr. E. W. Todd, of the Ontario Department of Mines. [*M.A.*, **3**, 271.] 11th List [A mixture; *A.M.*, **47**, 1363]

Todorokite. T. Yoshimura, 1934. *Journ. Fac. Sci. Hokkaido Univ.*, Sapporo, ser. 4, vol. 2, p. 289 (todorokite). Hydrated manganese oxide or manganate with some Ca,Ba,Mg, formula $2(RO.MnO_2.2H_2O).3(Mn_2O_3.3MnO_2.2H_2O)$, probably monoclinic; from Todoroki mine, Hokkaido, Japan. Named from the locality. [*M.A.*, **6**, 53.] 14th List [$(Mn,Ca,Mg)Mn_3^{4+}O_7.H_2O$; *A.M.*, **45**, 1167. Cf. Woodruffite]

Toernebohmite. *See* Törnebohmite. 9th List

Tohdite. I. Yamaguchi, H. Yanagida, and S. Ono, 1964. [*Bull. Chem. Soc. Japan*, **37**, 752 and 1555], quoted *Zap.*, **99**, 333. *Synthetic* $5Al_2O_3.H_2O$, identical with the Al_2O_3–KI of Torkar and Krischmer [*Monatschr. Chem.*, **91**, 638 (1960)] and the AS(H)–1 of S. Aramaki and R. Roy, *Amer. Min.*, **48**, 1322 (1962). Synonym of akdalaite (Shpanov, 1970; q.v.). 27th List

Tolypite. A. Uhlemann, 1909. *Min. Petr. Mitt.* (*Tschermak*), vol. xxviii, p. 461 (Tolypit). A name applied to a peculiar structure shown by chlorite from Saxon Vogtland. The mineral has the form of small balls which are built up of irregularly arranged fibres. Named from *τολύπη*, a ball of thread. 6th List

Tomazit, error for Tombazit (H. Strunz, *Min. Tabellen*, 3rd edit., p. 421). 22nd List

Tombarthite. H. Neumann and B. Nilssen, 1968. *Lithos*, **1**, 113. Brownish-black masses, partly metamict but giving measurable X-ray powder patterns similar to those of monazite, occur with thalenite at Høgetveit, Evje, Norway. A water-rich silicate of rare earths, principally Yt. Named for Professor Tom Barth. [*M.A.*, 69-618; *A.M.*, **54**, 327; *Bull.*, **92**, 515; *Zap.*, **98**, 330.] 26th List

Tomite. M. D. Zalessky, 1915. *Geol. Vestnik*, Petrograd, vol. 1 (for 1914–1915), p. 234; *Mém. Comité Géol. Petrograd*, 1915, n. ser., livr. 139, p. 6 (*томитъ*), p. 44 (*Tomite*). A variety of coal showing algae in its micro-structure, from the river Tom, Tomsk, Siberia. In the second paper, the name is altered to *sapromyxite* (p. 6, *сапромикситъ*; p. 44, *sapromyxite*), evidently derived from *σαπρός*, putrid, and *μύξα*, slime. 10th List

Topaz-safranite. *See* Safranite. 13th List

Torendrikite. A. Lacroix, 1920. *Compt. Rend. Acad. Sci. Paris*, vol. 171, p. 596; *Minéralogie de Madagascar*, 1922, vol. 1, p. 541. A bluish-black, strongly pleochroic, monoclinic amphibole containing ferric (and ferrous) iron, magnesium, and sodium, and occurring as a constituent of syenite in Madagascar. Named from the locality Itorendrika (the definite article being omitted). [*M.A.*, **1**, 376.] 9th List [Magnesio-riebeckite; *Amph.*, Sub-Comm.]

Törnebohmite. P. Geijer, 1921. *Sveriges Geol. Undersökning*, Årsbok 14 (for 1920), no. 6, p. 16. Abstr. in *Amer. Min.*, 1921, vol. 6, p. 118, gives the spelling Toernebohmite. Silicate of cerium, lanthanum, and didymium with some aluminium, $R_2(ROH)(SiO_4)_2$, observed as green, pleochroic grains in micro-sections of the cerite ore from Bastnäs, Sweden. It differs from cerite in its optical characters. Named after Alfred Elis Törnebohm (1838–1911), a former Director of the Geological Survey of Sweden. [*M.A.*, **1**, 251.] 9th List

Torniellite. E. Dittler and F. Kirnbauer, 1937. *Zeits. Prakt. Geol.*, vol. 45, p. 120 (Torniellit). An amorphous clay mineral, $Al_2Si_2O_5(OH)_4.H_2O$, near to halloysite. Named from the locality, Torniella, Toscana. [*M.A.*, **7**, 12.] 15th List

Torrensite. H. Lienau, *Chem.-Zeit.*, 1899, XXIII, 418. A compact reddish grey to sepia coloured mineral occurring with rhodonite at Torrens mine, Hautes-Pyrénées. $MnSiO_3.MnCO_3.\frac{1}{2}H_2O$. 2nd List

Torreyite. J. Prewitt-Hopkins, 1949. *Amer. Min.*, vol. 34, p. 589. $(Mn,Mg,Zn)_8(SO_4)(OH)_{14}.4H_2O$, monoclinic, from Sterling Hill, New Jersey. Previously named δ-mooreite, but giving an X-ray pattern distinct from that of mooreite (12th List). Named after John Torrey (1796–1873), American botanist, chemist, and mineralogist. Not the torrelite of T. Thomson, 1836, or of J. Renwick, 1824. [*M.A.*, **11**, 9.] 19th List

Toryanit. (*Rozpr. Akad. Kraków*, 1911, vol. xi, ser. A, p. 475; *Bull. Intern. Acad. Cracovie*, 1911, ser. A, p. 558.) A Polish form of Thorianite (W. R. Dunstan, 1904; 4th List). 7th List

Tosalite. T. Yoshimura, 1967. [*Sci. Rept. Fac. Sci. Kyushu Univ.*, ser. D, **9**, special issue no. 1], abstr. *Min. Abstr.* An intermediate member of the bementite–greenalite series, having Mn : Fe ≈ 1, from mines in the Tosa province, Kochi prefecture, Japan. Serpentine group. Named for the locality. 25th List

Tosudite. V. A. Frank-Kamenetsky, N. V. Logvinenko, and V. A. Drits, 1963. Зап. Всесоюз. Мин. Общ. (*Mem. All-Union Min. Soc.*), vol. 92, p. 560 (Тосудит). One constituent of Alushtite is a mixed-layer phase composed of Dickite and Montmorillonite [*M.A.*, **16**, 279]; this proves to be a regularly alternating structure and is named for Prof. Toshio Sudo, who described a similar assemblage [*M.A.*, **14**, 96; **16**, 549; *A.M.*, **49**, 816.] 23rd List

Totschilinit, Germ. trans. of Точилинит, tochilinite (q.v.). Hintze, 96. 28th List

Tozalite, error for tosalite (25th List). *Bull.*, **95**, 164. 28th List

Trachyaugite. F. Yamanari, 1926. *Japanese Journ. Geol. Geogr.*, vol. 3 (for 1924), p. 107. A variety of augite, distinguished by its optical properties in micro-sections, present in alkali-trachyte from islands in the Sea of Japan. [*M.A.*, **3**, 199.] 11th List

Trainite. (W. T. Schaller, *Mineral Resources, US Geol. Survey*, 1918, for 1917, part II, p. 163.) An impure banded variscite from Manhattan, Nevada. Named after the collector, Mr. Percy Train, of Manhattan. A specimen so named, and described on the dealer's label as aluminium phosphosilicate, was acquired by the British Museum mineral collection in 1913. The material has been described, although not under this name, by E. T. Wherry (*Journ. Washington Acad. Sci.*, 1916, vol. vi, p. 105; and *Chem. News*, 1916, vol. cxiii, p. 290), who determines it to be a lamellar intergrowth of colloidal vashegyite ($4Al_2O_3.3P_2O_5.30H_2O$; 5th List) and a hydrated calcium aluminium silicate near the zeolite laubanite. Cf. Sabalite. 8th List

Tranquillitite, error for tranquillityite (*Geotimes*, **17**, 46 (1972)). 27th List

Tranquillityite. J. F. Lovering, D. A. Wark, A. F. Reid, N. G. Ware, K. Keil, M. Prinz, T. E. Bunch, A. El Goresy, P. Ramdohr, G. M. Brown, A. Peckett, R. Phillips, E. N. Cameron, J. A. V. Douglas, and A. G. Plant, 1971. *Proc. Second Lunar Sci. Conf.* (*suppl. to Geochim. Acta*, **35**), **1**, 39. Rare semi-opaque dark foxy-red laths ($<65 \times 15$ μm) in basaltic rocks from Apollo XI and XII missions are hexagonal pseudocubic, *a* 11·69, *c* 11·11 Å. Composition $3[Fe_8^{2+}(Zr,Yt)_2Ti_3Si_3O_{24}]$. Apparently unrelated to any terrestrial species. This is the 'phase A' of Ramdohr and El Goresy (1970), the 'unnamed Y–Zr-silicate' of Cameron (1970), and the 'Fe, Ti, Zr silicate' of Dence *et* al. (1970). [*M.A.*, 72-3349; *A.M.*, **58**, 140.] 27th List

Transvaalite. H. Strunz, 1957. *Min. Tabellen*, 3rd edit., p. 51 (Transvaalit). Cobaltous hydroxide, $Co(OH)_2$; no natural occurrence mentioned. Not to be confused with transvaalite of McGhie and Clark (Dana, *Syst. Min.*, 6th edit., p. 260), a hydrous cobaltic oxide near heterogenite. 22nd List

Traskite. J. T. Alfors and M. C. Stinson, 1965. *Min. Inform. Serv., Calif. Div. Mines Geol.*, vol. 18, p. 27. J. T. Alfors, M. C. Stinson, R. A. Matthews, and A. Pabst, *Amer. Min.*, 1965, vol. 50, pp. 279, 314, and 1500. Brownish-red grains and hexagonal crystals in Sanbornite–Quartz rock near Big Creek and Rush Creek, Fresno County, California, are near $Ba_9Fe_2Ti_2Si_{12}O_{30}(OH,Cl,F)_{18}$ (emission spectrography analysis). Named for J. B. Trask. [*M.A.*, **17**, 400; **17**, 502.] 24th List

Traversite. A. Amstutz, 1925. *Schweiz. Min. Petr. Mitt.*, vol. 5, p. 302. A red-brown alteration product of olivine in basalt from Traversa, near Orosei, Sardinia. Named from the locality. *See* Oroseite. [Evidently the same as iddingsite. Not to be confused with traversoite (10th List), also from Sardinia.] [*M.A.*, **3**, 369.] 11th List

Traversoite. A. D'Ambrosio, 1924. *Ann. Museo Civ. Stor. Nat. Genova*, vol. 51, p. 253. A variety of chrysocolla containing alumina (and lime), and admitted to be a mixture of chrysocolla and gibbsite; found as bright blue, amorphous masses at Arenas, Sardinia. Optically isotropic (*Amer. Min.*, 1925, vol. 10, p. 108). Named in memory of the mining-engineer Giovanni Battista Traverso (1843–1914) of Genova. [*M.A.*, **2**, 521.] 10th List

Treanorite. A. O. Woodford, J. D. Laudermilk, and E. H. Bailey, 1940. *Bull. Geol. Soc. Amer.*, vol. 51, p. 1965. A. O. Woodford, R. A. Crippen, and K. B. Garner, *Amer. Min.*, 1941, vol. 26, p. 375. An undescribed mineral similar to allanite, from Crestmore, California. [*M.A.*, **8**, 145.] 16th List [Is Allanite; *M.A.*, **10**, 148]

Treasurite. S. Karup-Møller, 1977. *Bull. Geol. Soc. Denmark*, **26**, 41. Monoclinic, *a* 13·349, *b* 26·538, *c* 4·092 Å, *β* 92·77°. Composition $[Ag_7Pb_6Bi_{15}S_{32}]$. From the Treasury mine, Colorado. Named from the locality. [*M.A.*, 78-899.] 30th List

Trechmannite. R. H. Solly, 1904. *Min. Mag.*, 1905, vol. xiv, pp. 75, 189. (A preliminary description of the unnamed mineral appeared in *Nature*, 1903, vol. lxix, p. 142, and the name (Trechmannit) was first printed in *Min. Petr. Mitt.* (*Tschermak*), 1904, vol. xxiii, p. 552.) G. F. H. Smith and G. T. Prior, *Min. Mag.*, 1907, vol. xiv, p. 300. Minute, red crystals associated with tennantite, etc., in the white, crystalline dolomite of the Binnenthal, Switzerland. The crystals are rhombohedral with parallel-faced hemihedrism, and were found to contain silver and arsenic: probably a sulpharsenite of silver. Named after Dr. Charles O. Trechmann, of Castle Eden, Co. Durham. 4th List [$AgAsS_2$; *M.A.*, 70-183. Trechmannite-alpha (*M.M.*, **18**, 363) is Nowackiite (q.v.), *M.A.*, 76-1967]

Trevolite, error for Trevorite (*Journ. Min. Soc. Japan*, 1959, vol. 4, p. 137). 22nd List

Trevorite. A. F. Crosse, 1921. *Journ. Chem. Metall. and Mining Soc. S. Africa*, vol. 21, p. 126. T. L. Walker, *Univ. Toronto Studies, Geol. Ser.*, 1923, no. 16, p. 53. Oxide of nickel and ferric iron, $NiFe_2O_4$. An opaque black mineral with metallic lustre and strongly magnetic, from Barberton, Transvaal. Named after Major Tudor G. Trevor, Inspector of Mines for the Pretoria district. [*M.A.*, **2**, 249.] 10th List

Triamond. A trade-name for *artificial* yttrium aluminate. M. J. O'Donoghue, *priv. comm. See* under cirolite. 28th List

Trieuite. L. De Leenheer, 1935. *Natuurwet. Tijdschrift*, Gent, vol. 17, p. 91 (trieuiet), p. 95 (trieuite). Hydrated oxide of cobalt and copper, $2Co_2O_3.CuO.6H_2O$; a black amorphous mineral differing from heterogenite in containing no CoO. From Katanga. Named after Robert du Trieu de Terdonck, chief geologist of the Union Minère du Haut Katanga. Cf. Mindigite. [*M.A.*, **6**, 152.] 14th List [Is Heterogenite; *M.M.*, **33**, 253]

Trigonite. G. Flink, 1920. *Geol. För. Förh. Stockholm*, vol. 42, p. 436 (Trigonit). Acid arsenite of lead and manganese, $Pb_3MnH(AsO_3)_3$, occurring as sulphur-yellow to brownish, monoclinic (domatic class) crystals at Långban, Sweden. Named from *τρίγωνον*, a triangle, because of the triangular, wedge-shaped habit of the crystals. [*M.A.*, **1**, 149.] 9th List

Trigonomagneborite. I-Hua Ch'u, Yu-T'en Han, Tzu-chiang Chen, Lai-Pao Liu, and Ling-Sheng Min, 1965. [*Scientia Sinica*, vol. 14, p. 1246]; abstr. *Amer. Min.*, 1965, vol. 50, p. 2110. An independent discovery, as a white efflorescence in salts of an unnamed Chinese lake, of $MgB_6O_{10}.7\frac{1}{2}H_2O$; clearly identical with Macallisterite, which name has priority. 24th List

Trihydrallite. H. Harrassowitz, 1927. *See* Allite. 11th List

Trihydrocalcite. P. N. Čirvinskij (Tschirwinsky), 1906. *Ann. Géol. Min. Russie*, vol. viii, p. 241 (Тригидрокальцитъ), p. 245 (Trihydrocalcit). Hydrated calcium carbonate, $CaCO_3.3H_2O$, occurring as a mould-like encrustation on chalk-marl near Nova-Alexandria, govt. Lublin, Russian Poland. *See* Lublinite and Pentahydrocalcite. 5th List

Tri-kalsilite. T. G. Sahama and J. V. Smith, 1957. *Amer. Min.*, vol. 42, p. 286. A form of kalsilite, $(K,Na)AlSiO_4$, in parallel intergrowth with nepheline, $(Na,K)AlSiO_4$. The hexagonal *a*-axis 15·4 Å, is three times longer than that of kalsilite (*a* 5·15 Å). *See* Tetra-kalsilite. [*M.A.*, **13**, 524.] 21st List

Trilithionite. M. D. Foster, 1960. *US Geol. Surv. Prof. Paper 354–E*, p. 115. Proposed as a more appropriate name for the mica end-member $K_2Li_3Al_5Si_6O_{20}(F,OH)_4$, in place of the name Lithium-muscovite (15th List). 23rd List

Trilliumite. Author?, 1975. *Lapidary Journ.*, **29**, 1463. A yellowish-green gem quality apatite. 29th List

Trimontite. I. Iwasa, 1877. [*Gakugéisirin*, no. 57.] Z. Harada, *Journ. Fac. Sci. Hokkaido Univ.*, Sapporo, ser. 4, 1936, vol. 3, p. 357 (Trimonit, Trimontit). Analysis gave the composition $5CaO.3WO_3$, but the mineral was afterwards (T. Wada, 1904) proved to be scheelite, $CaWO_4$. Named from the locality, San-no-také (= three-of-mountains), Hukuoka, Japan. 14th List

Trinitatin, J. Fröbel, 1843. *See* Chrysargyrite. 27th List

Tripletine. A thoroughly objectionable trade-name for emerald-coloured beryl triplet cut stones. R. Webster, *Gems*, 1962, p. 764. Not to be confused with triplite, triploidite, nor even tripestone. 28th List

Tripuhyite. E. Hussak and G. T. Prior, *Min. Mag.*, XI. 302, 1897. $2FeO.Sb_2O_5$. Tripuhy, [near Ouro Preto, Minas Gerais] Brazil. 1st List

Troegerite. Variant of Trögerite; *A.M.*, **9**, 62. 10th List

Trogtalite. P. Ramdohr and M. Schmitt, 1955. *Neues Jahrb. Min.*, Monatshefte, no. 6, p. 139 (Trogtalit). Cobalt selenide, $CoSe_2$, cubic (pyrite group). Named from the locality, Trogtal, Harz. [*M.A.*, **13**, 5; *A.M.*, **41**, 164.] 21st List

Trömelite. W. L. Hill, G. T. Faust, and D. S. Reynolds, 1944. *Amer. Journ. Sci.*, vol. 242, p. 467, 545. An artificial calcium phosphate, perhaps $7CaO.5P_2O_5$. Named after Gerhard Trömel of Düsseldorf, who made the first systematic study of the system $CaO–P_2O_5$. [*M.A.*, **9**, 92.] 17th List

Tronite. Variant of Trona; *A.M.*, **8**, 52. 10th List

Trudellite. S. G. Gordon, 1926. *Proc. Acad. Nat. Sci. Philadelphia*, vol. 77 (for 1925), p. 317; *Amer. Min.*, vol. 11, p. 42. Hydrous basic chloride, and sulphate of aluminium, $Al_2(SO_4)_3.4AlCl_3.4Al(OH)_3.30H_2O$, as compact (trigonal), amber-yellow masses from Chile. Named after Mr. Harry William Trudell (1884–), of Philadelphia, Pennsylvania. [*M.A.*, **3**, 112.] 11th List [A mixture; *A.M.*, **57**, 1317]

Truffite. E. Dumas, 1876. *Statistique géologique, minéralogique, etc., du Département du Gard, Paris, etc.*, 1876, vol. ii, pp. 431, 433 (not p. 491). A fibrous lignite which when struck emits an odour like that of truffles (French, truffe); it occurs as large, nodular masses, associated with ordinary lignite, in Cretaceous (Turonian) limestone at Pont-Saint-Esprit, dep. Gard. A. Lacroix (*Minéralogie de la France*, 1909, vol. iii, pp. 436, 509) describes it as a fetid, fibrous calcite. 5th List

Truscottite. P. Hövig, 1914. *Jaarboek van het Mijnwezen in Nederlandsch Oost-Indië*, Batavia, 1914, vol. 41 (for 1912), Verhand., p. 202 (p. 116 in reprint) (Truscottiet). A hydrous calcium silicate containing SiO_2 62·52, CaO 30·20, H_2O 3·85% [near $CaO.2SiO_2.\frac{1}{2}H_2O$], as pearly white scales from the Lebong Donok gold-silver mine, Benkulen, Sumatra. Named after Prof. Samuel John Truscott, of the Royal School of Mines, London, formerly manager of the mine where the mineral was found. [Material presented to the British Museum by Prof. Truscott in 1925 shows micaceous cleavage, and is optically uniaxial and negative with $\omega < 1{\cdot}55$; sp. gr. 2·48.] 10th List [Identity with Reyerite uncertain, *M.M.*, **33**, 821; *A.M.*, **53**, 511]

Trustedtite. Y. Vuorelainen, A. Huhma, and A. Häkli, 1964. *Compt. rend. Soc. géol. Finlande*, vol. 36, p. 113. Cubic Ni_3Se_4, isostructural with Linnaeite, as crystals in Clausthalite, associated with Penroseite and Sederholmite (q.v.) at Kuusamo, NE Finland. Named for O. Trustedt. [*M.A.*, **17**, 303; *A.M.*, **50**, 520.] 24th List

Tsavolite. H. Bank, 1975. [*Zeits. deut. gemmol. Gesell.*, **24**, 13], abstr. in *A.M.*, **61**, 178. An unnecessary name for a green chromian vanadian grossular from the Tsavo National Park, Kenya. E. G. Gübelin and M. Weibel (*Lapidary Journ.*, **29**, 402, 424) spell the name Tsavorite and give the locality as Laulenyi, Voi, Kenya. [*M.A.*, 76-549.] 29th List

Tsavorite, *see* Tsavolite. 29th List

Tschelkarit, Germ. trans. of Челкарит, chelkarite (27th List). Hintze, 97. 28th List

Tschermakite. A. N. Winchell, 1945. *Amer. Min.*, vol. 30, pp. 29, 44. A hypothetical 'Tschermak molecule' $Ca_2Mg_3Al_4Si_6O_{22}(OH)_2$ to explain (together with the corresponding iron molecule $Ca_2Fe_3Fe_2'''Al_2Si_6O_{22}(OH)_2$, named ferrotschermakite) the composition of aluminous amphiboles. Named after Gustav Tschermak (1836–1927) of Vienna. Not the tschermakite of F. Kobell, 1873, a synonym of albite. [*M.A.*, **9**, 271.] 17th List

Tschernichewite. L. Duparc and F. Pearce, 1907. *Compt. Rend. Acad. Sci. Paris*, vol. cxliv, p. 763 (tschernichéwite). A variety of amphibole occurring with magnetite in a quartzite in the N Urals. It is defined by its optical characters: the plane of the optic axes is perpendicular to the plane of symmetry and the axial angle is near 90°; $\gamma{:}\acute{c} = 4°$; the pleochroism is very strong. These characters point to a soda-iron-amphibole near riebeckite or arfvedsonite; cf. osannite. Named after Theodosij Nikolajevič Černyšev (Tschernyschew), Director of the Geological Survey of Russia. 4th List

Tschernowit, Germ. trans. of Черновит, chernovite (25th List). Hintze, 97. 28th List

Tschernychit, Germ. trans. of Черныхит, chernykhite (q.v.). Hintze, 98. 28th List

Tschevkinit, another variant of Chevkinite (H. Strunz, *Min. Tabellen*, 4th edit., 1966, p. 355). 24th List

Tschinglusuit, Germ. trans. of Chinglusuite (q.v.). *See* Tchinglusuite.

Tschirwinskite. G. Gagarin and J. R. Cuomo, 1949, loc. cit., p. 13 (tschirwinskita). Hydrous ferric phosphate, $Fe_2O_3.P_2O_5.2\frac{1}{2}H_2O$, from Kerch, Crimea, described in 1904 by Petr Nikolaevich Chirvinsky Петр Николаевич Чирвинский (1880–). Not the chirvinskite of N. K. Platonov 1941 (17th List). 19th List

Tschkalowit, German transliteration of Чкаловит, chkalovite (H. Strunz, *Min. Tabellen*, 3rd edit., p. 149). 22nd List

Tsilaisite. W. Kunitz, 1929. *Chemie der Erde*, vol. 4, p. 225 (Tsilaisit). A hypothetical molecule, $H_8Na_2Mn_6Al_{12}Si_{12}B_6O_{62}$, of the tourmaline group, to explain the composition of a manganiferous tourmaline (MnO about 6%) from Tsilaisina (Tsilaizina), Madagascar. Named from the locality. [*M.A.*, **4**, 204.] 12th List

Tsumcorite. B. H. Geier, K. Kautz, and G. Müller, 1971. *Neues Jahrb. Min. Monatsh.*, 305. Yellow-brown crusts and red-brown crystals from the Tsumeb mine oxidation zone are monoclinic, *C*2/*m*, *a* 9·131, *b* 6·326, *c* 7·583 Å, *β* 115·2°. Composition $2[PbZnFe(AsO_4)_2.H_2O]$. *n* 1·9, 2V 90°, D 5·2 (all approx.); H 4½. May be related to brackebuschite. Named for the TSUMeb CORporation. [*M.A.*, 72-1405; *A.M.*, **57**, 1558.] 27th List

Tsumebite. K. Busz, 1912. *Festschrift Gesell. Deutsch. Naturf. u. Aerzte*, Münster, p. 182 (Tsumebit). Hydrated basic phosphate of lead and copper, $5(Pb,Cu)O.P_2O_5.8H_2O$, found as small, emerald-green, monoclinic crystals on white calamine at Tsumeb, Otavi, German SW Africa. *See* Preslite. 6th List [$Pb_2Cu(PO_4)(SO_4)(OH)$; *A.M.*, **51**, 258]

Tućanite. M. Karšulin, *Yugoslav. Akad. Znan. Umjet.*, 1960, p. 113; *Acad. Yougoslav. Sci. Arts, Sympos. Bauxites*, 1964, vol. 2, p. 37. White chalky material from bauxite at Carev Most, Niksic, Montenegro, has a composition near $Al_2O_3.5H_2O$; at 150°C the composition is $Al_2O_3.3H_2O$ but the powder photograph is quite different from that of gibbsite. The powder data are indexed on a monoclinic cell containing $8[Al_2(OH)_6.H_2O]$ and the mineral is named for F. Tućan. M. Fleischer (*Amer. Min.*, 1965, vol. 50, p. 1504) points out that the X-ray powder data, dehydration curves, and electron diffraction photographs match those of Scarbroïte (*Min. Mag.*, 1960, vol. 32, pp. 353, 363); there can be no doubt that CO_2 (8%) has been overlooked and that Tućanite is Scarbroïte (R. C. Mackenzie, priv. comm.). 24th List

Tučekite. J. Just and C. E. Feather, 1978. *M.M.*, **42**, 278, M21. Microscopic grains, partly replacing millerite, at Kanowna, Western Australia, and at several gold deposits of the Witwatersrand system, South Africa, are the antimony analogue of hauchecornite. Tetragonal, *a* 7·174, *c* 5·402 Å. Composition $Ni_9Sb_2S_8$. Named for K. Tuček. 30th List

Tugtupite. H. Sørensen, 1962. *Medd. om Grønland*, vol. 169, no. 1. Aluminosilicate and chloride of Be and Na, from the Ilimaussaq massif, SW Greenland. Very near to and perhaps identical with Beryllosodalite (22nd List). *Amer. Min.*, 1963, vol. 48, p. 1178; 1961, vol. 46, p. 241; *M.A.*, **15**, 212. 23rd List

Tuhualite. P. Marshall, 1932. *New Zealand Journ. Sci. Techn.*, vol. 13, p. 202. A violet-coloured variety of alkali-amphibole with pleochroism colourless, purplish violet, and deep violet, occurring in comendite from Mayor (= Tuhua) Island, Bay of Plenty, New Zealand. Named from the locality. [*M.A.*, **5**, 295.] 13th List [$H_9(Na,K,Mn)_{12}Fe_6(Al,Fe,Mg,Ti)_9(Si_3O_8)_{15}$; not an amphibole; *M.M.*, **31**, 96]

Tulameenite. L. J. Cabri, D. R. Owens, and J. H. G. Laflamme, 1973. *C.M.*, **12**, 21. Rounded grains up to 0·4 mm occurring with platinum minerals in placers of the Tulameen River area, British Columbia, have composition Pt_2CuFe; tetragonal, *a* 3·891, *c* 3·577 Å; closely matches synthetic Pt_2CuFe. Named for the locality. [*M.A.*, **59**, 383.] 28th List

Tundrite. E. I. Semenov, 1963. [Мин. редк. элем. (*Mineralogy of rare earths*), Издат. Акад. наук СССР, p. 209]; abstr. *Amer. Min.*, 1965, vol. 50, p. 2097: *M.A.*, **17**, 401. The mineral previously described as Titanorhabdophane (22nd List) is found to be anorthic, and is renamed for the Lovozero tundra; composition $Ce_2Ti(Si,P)(O,OH)_7.4H_2O$. 24th List [Contains essential carbonate; new (complex) formula, *A.M.*, **59**, 633]

Tunellite. R. C. Erd, V. Morgan, and J. R. Clark, 1961. *US Geol. Surv. Prof. Paper 424–C*, p. 294. $SrB_6O_{10}.4H_2O$, monoclinic and isostructural with the calcium borate Nobleite. In monoclinic crystals from Kramer and from Furnace Creek, Death Valley, California. Named for Prof. G. Tunell (*Amer. Min.*, 1962, vol. 47, p. 416; *M.A.*, **16**, 547). 23rd List

Tungomelane. P. F. Kerr, 1940. *Bull. Geol. Soc. Amer.*, vol. 51, p. 1379. Tungsten-bearing psilomelane (WO_3 1·54–2·78%) from Golconda, Nevada. [*M.A.*, **8**, 310.] 16th List

Tungsten-germanite. B. H. Geier and J. Ottemann, 1970. *Neues Jahrb. Min. Abh.*, **114**, 89. Unnecessary name for a tungstenian germanite. [*A.M.*, **56**, 1487; *M.A.*, 71-1391.] 27th List

Tungstenite. R. C. Wells and B. S. Butler, 1917. *Journ. Washington Acad. Sci.*, vol. vii, p. 596. Tungsten sulphide, probably WS_2, occurring as minute graphite-like scales intimately intermixed with other minerals in a compact ore from the Little Cottonwood district, Utah. Named from analogy to molybdenite. [*A.M.*, **3**, 30.] 8th List

Tungsten-powellite. O. H. Ödman, 1950. *Årsbok Sveriges Geol. Undersök.*, vol. 44, no. 2, p. 23. A variety of powellite ($CaMoO_4$) containing WO_3 9·6–14·0%. [*M.A.*, **11**, 473.] 19th List

Tungsto-powellite, variant of Tungsten-powellite (19th List) (H. Strunz, *Min. Tabellen*, 3rd edit., 1957, p. 216, Tungsto-Powellit). 22nd List

Tungusite. V. I. Kudryashova, 1966. Доклады акад. наук СССР (*Compt. Rend. Acad. Sci. URSS*), vol. 171, p. 1167 (Тунгуст). Platy green aggregates from amygdales in lavas on the Lower Tunguska river, Siberia, are near $Ca_4Fe_2^{2+}Si_6O_{15}(OH)_6$ in composition. Named for the locality. [*A.M.*, **52**, 927; Зап., **97**, 77.] 25th List

Tunisite. Z. Johan, P. Povondra, and E. Slánsky, 1969. *Amer. Min.*, **54**, 1. Tetragonal crystals and fine-grained aggregates found on the dumps of the Sakiet Sidi Tousseff ore deposit, Tunisia, have the composition $NaHCa_2Al_4(CO_3)_4(OH)_{10}$. Named for the country of origin. [*M.A.*, 69-2389; *Bull.*, **92**, 515; *Zap.*, **99**, 76.] 26th List

Tunnerite. F. Cornu, 1909. *Zeits. Chem. Indust. Kolloide*, vol. 4, p. 297 (Tunnerit). The 'Zinkmanganerz' of A. Brunlechner (1893, 1st List) regarded as a colloidal psilomelane or wad with adsorbed zinc oxide. P. Groth (*Tab. Übers. Min.*, 1898, p. 64; P. Groth and K. Meileitner, *Min. Tab.*, 1921, p. 47) regarded it is perhaps identical with chalcophanite. 10th List [May be Woodruffite; *A.M.*, **56**, 1840]

Turanite. K. A. Nenadkevich, 1909. *Bull. Acad. Sci. Saint-Petersbourg*, ser. 6, vol. iii, p. 185 (Туранитъ). Hydrated copper vanadate, $5CuO.V_2O_5.2H_2O$, forming compact, spongy, or radially fibrous aggregates and reniform crusts of an olive-green colour. Occurs with alaite (q.v.) in cavities in malachite and limestone near the Alai Mountains in the Turan district, Russian Central Asia. 5th List [May be Tangeïte; *Bull.*, **79**, 222]

Turite. (J. A. Samoilov, *Bull. Soc. Nat. Moscou*, 1900, vol. xiii (for 1899), p. 142, Турьитъ). The correct form of the name Turgite (R. Hermann, 1845) or Turjit (R. Hermann, 1860), after the Turya (Turia, Turja, Турья) river, N Urals. [*M.M.*, **18**, 341.] 8th List

—— A. A. Kukharenko *et al.*, 1965. [*The Caledonian ultrabasic alkalic rocks and carbonatites of the Kola Peninsula and northern Karelia. Izd. 'Neda'*, Moscow, pp. 418–423]; abstr. *Amer. Min.*, 1967, vol. 52, p. 561. An unnecessary name for a cerian götzenite. The name (from the locality, Turii Peninsula, Kola) has previously been used for hydrohematite (*see* 8th List). [Зап., **96**, 76; Турит.] 25th List

Turkestan-volborthite. I. A. Antipov, 1908. *Gornyi Zhurnal*, St. Petersburg, year 84, vol. 4, p. 261; abstr. in *Neues Jahrb. Min.*, 1909, vol. 2, ref. p. 39 (turkestanischer Volborthit). K. A. Nenadkevich and P. A. Volkov, *Compt. Rend. Acad. Sci. URSS*, ser. A, 1926, p. 46 (туркестанский фольбортит). I. Kurbatov, *Centr. Min.*, Abt. A, 1926, p. 346. A greenish-black mineral from Turkestan differing somewhat from volborthite in chemical composition, and regarded by Nenadkevich and Volkov as the colloidal equivalent of tangeite (q.v.). [*M.A.*, **3**, 234.] 11th List

Turyite. E. T. Wherry, 1920. *Amer. Min.*, vol. 5, p. 18. An alternative spelling of Turgite = Turite, Турьитъ (cf. 8th List). This only adds to the existing confusion. It would be better to replace the name turgite (R. Hermann, 1845) by its synonym hydrohaematite (A. Breithaupt, 1847). 9th List

Tuscanite. P. Orlandi, L. Leoni, M. Mellini, and S. Merlino, 1977. *A.M.*, **62**, 1110. Colourless crystals in ejected blocks in a pumice from Pitigliano, Tuscany, Italy, are monoclinic, *a* 24·036, *b* 5·110, *c* 10·888 Å, *β* 106·95°, space group $P2_1/a$. Composition $2[K_{0\cdot 9}(H_2O)_{1\cdot 1}(Ca_{5\cdot 25}Na_{0\cdot 5}Fe_{0\cdot 1}{}^3Mg_{0\cdot 1})(Si_{6\cdot 34}Al_{3\cdot 66})O_{22}(SO_4)_{1\cdot 38}(CO_3OH)_{0\cdot 55}(OH)_{0\cdot 44}]$, sp. gr. 2·83. RIs α 1·581, β 1·590, γ 1·591, α:[001] 40°, $2V_\alpha$ 40°. Closely related to latiumite. Named from the locality. [*M.A.*, 78-3482.] 30th List

Tusiite. M. A. Kashkai and R. M. Aliev, 1960. [Труды Азербайдж. Геогр. Общ. (*Trans. Azerbaidzhan Geogr. Soc.*), p. 49], abstr. in Зап. Всесоюз. Мин. Общ. (*Mem. All-Union Min. Soc.*), 1962, vol. 91, p. 196 and in *Amer. Min.*, 1962, vol. 47, p. 807. An alternative name proposed for the mineral finally named Calciocopiapite (q.v.). 23rd List

Tuvite. N. N. Shishkin and V. A. Mikhailova, 1956. [Сборн. Мат. Тех. Инф. (*Collect. Mat. Tech. Inform.*), no. 6, Gipronickel, p. 5]; abstr. Зап. Всесоюз. Мин. Общ. (*Mem. All-Union Min. Soc.*), 1959, vol. 88, p. 317 (Тувит, tuvite). Imperfectly characterized yellow oxidation products of smaltite and safflorite from the Khovaks deposit, Tuva. Named from the locality. [*M.A.*, **14**, 278; *A.M.*, **45**, 256.] Cf. Yellow earthy cobalt, Dana, *Syst. Min.*, 6th edit., p. 78. A wholly unnecessary name. 22nd List

Tuwit, German transliteration of Тувит, Tuvite. Hintze, *Handb. Min.*, Erg.-Bd. II, p. 873. 23rd List

Tuxtlite. H. S. Washington, 1922. *Proc. Nat. Acad. Sci. USA*, vol. 8, p. 321. A pyroxene ('jade' from Tuxtla, Mexico) with the diopside and jadeite molecules in about equal amounts; previously referred to as diopside-jadeite (9th List). Named from the locality. *See* Mayaite. [*M.A.*, **2**, 67.] 10th List

Tuyamunite, *see* Tyuyamunite.

Tvalchrelidzeite. V. S. Gruzdev, N. M. Mchedlishvili, G. A. Terekhova, Z. Ya. Tsertsvadze, N. M. Chernitsova, and N. G. Shumkova, 1975. *Dokl. akad. nuak SSSR*, **225**, 911 (Твалчрелидзеит). Granular aggregates from the Gomi deposit, Georgia, USSR, are monoclinic, *a* 11·51, *b* 4·39, *c* 14·62 Å, *β* 92° 14′. Composition near $Hg_{12}(Sb,As)_8S_{15}$. Named for A. A. Tvalchrelidze. [*A.M.*, **62**, 174; *M.A.*, 77-889; *Zap.*, **105**, 77.] 30th List

Tveitite. S. Bergstøl, B. B. Jensen, and H. Neumann, 1977. *Lithos*, **10**, 81. Massive, pale yellow, from a pegmatite dyke at Høydalen, Telemark, Norway. X-ray powder data are indexed on a pseudocubic monoclinic pseudocell with *a* 3·924, *b* 3·893, *c* 5·525 Å, *β* 90·26°. Composition $Ca_{1-x}(Yt,Ln)_xF_{2+x}$, with $x \simeq 0{\cdot}3$. Named for J. Tveit. [The heating product, giving a fluorite pattern, is presumably yttrofluorite. Chatterjee (*Z. Kristallogr.*, 1940, **102**, 245; *M.A.*, **7**, 73) reported that artificial $(Ca,Yt)F_{2+x}$ has the fluorite structure with interstitial F when $YtF_3 < 50\%$, but for $YtF_3 > 50\%$ the structure becomes more complicated—*M.H.H.*] [*A.M.*, **62**, 1060; *M.A.*, 78-2128.] 30th List

Twinnite. J. L. Jambor, 1967. *Canad. Min.*, vol. 9, pp. 4, 7, 191. A lead sulphosalt from Madoc, Ontario, proves to have the composition $Pb(Sb,As)_2S_4$, analogous to sartorite. The name twinnite, in honour of R. M. Thompson, is proposed for all members of the series with Sb > As. [*A.M.*, **53**, 1424.] 25th List

Tyanshanite, standard English transliteration of Тяншанит (Tienshanite, q.v.). 26th List

Tychite. S. L. Penfield and G. S. Jamieson, 1905. *Amer. Journ. Sci.*, ser. 4, vol. xx, p. 217; *Zeits. Kryst. Min.*, vol. xli, p. 235. Octahedral crystals indistinguishable in appearance from northupite, and differing from this in containing sodium sulphate in place of sodium chloride, the formula being $2MgCO_3.2Na_2CO_3.Na_2SO_4$. Amongst several thousand crystals of northupite from Borax Lake, San Bernardino Co., California, only four crystals of tychite were found, and the first one only by the merest chance, hence the name, from *τύχη*, luck. The formula was determined by the analysis of artificially prepared crystals. Mixed crystals containing the northupite and tychite molecules in varying proportions have been prepared by A. de Schulten (*Compt. Rend. Acad. Sci. Paris*, 1906, vol. cxliii, p. 403). 4th List

Tynite. L. N. Ovchinnikov, 1960. [Труды Горно-геол. Инст.Урал. фил. Акад. наук СССР, vol. 39, p. 297], abstr. in Зап. Всесоюз. Мин. Общ. (*Mem. All-Union Min. Soc.*), 1962, vol. 91, p. 203 (Тынит, tynite) and in *Amer. Min.*, 1962, vol. 47, p. 1483. An inadequately characterized aluminosilicate of Fe, Ca, and Mg, near $Ca_3(Fe,Mg)(Fe,Al)_2Si_3O_{12}(OH)_2.H_2O$ from the Urals. Named from the Tyny river. [*M.A.*, **16**, 554.] 23rd List

Tyretskite. V. V. Kondrateva, 1964. [Рентгеногр. мин. сырья, Moscow, no. 4, p. 10]; abstr. *M.A.*, **17**, 500. A. A. Ivanov and Ya. Ya. Yarzhemskii, [Труды Всесоюз. науч. исслед. Инст. Галурги,1954, vol. 29, p. 210]; abstr. *M.A.*, **17**, 500; Зап. Всесоюз. Мин. Общ. (*Mem. All-Union Min. Soc.*), 1966, vol. 95, p. 315 (Тыретскит). Fibrous aggregates with Sylvine, Carnallite, Halite, and Anhydrite in dolomite from near Tyret on the E Siberian railway are anorthic, near $Ca_3B_8O_{13}(OH)_4$. 24th List [*A.M.*, **53**, 2084]

Tyrrellite. I. D. Sindeeva, 1959. [Минералогия (Mineralogy), publ. *Acad. Sci. USSR*, p. 58]; quoted in Зап. Всесоюз. Мин. Общ. (*Mem. All-Union Min. Soc.*), 1961, vol. 90, p. 92 (Тирреллит, tyrrellite); Index to the X-ray powder data file, 1960, p. 298. The unnamed $(Cu,Co,Ni)_3Se_4$ of S. C. Robinson and E. J. Brooker, 1952 [*A.M.*, **37**, 542] named by S. C.

Robinson after Joseph Burr Tyrrell. Д. У. Тиррелл in Bonshtedt-Kupletskaya's list is an error. *See also* Selenide-spinel. 22nd List

Tyuyamunite. K. A. Nenadkevič, 1912. *Bull. Acad. Sci. St.-Pétersbourg*, ser. 6, vol. vi, p. 945 (тюямунитъ. The French translation of the title on the wrapper gives the form Tĭujamunite). Abstr. in *Chem. Zentralblatt*, 1913, vol. i, p. 326 (Tjuiamunit). A hydrated urano-vanadate of calcium, $V_2O_5.2UO_3.CaO.4H_2O$, occurring with ferganite and taranite (5th List) at Tyuya-Muyun, Fergana, Russian Central Asia. W. F. Hillebrand (*Amer. Journ. Sci.*, 1913, ser. 4, vol. xxxv, p. 440) suggests that it is a calcium carnotite; he spells the name tuyamunite. 6th List [Cf. Metatyuyamunite]

U

U-galena. P. F. Kerr, 1935. *Amer. Min.*, vol. 20, p. 443. Galena containing uranium lead of isotope Pb^{206}. [*M.A.*, **6**, 152.] 14th List

U-wulfenite. O. B[radley], 1962. *Min. Abstr.*, vol. 15, p. 460. A name for the uraniferous wulfenite of A. A. Chernikov, T. L. Pokrovskaya, Yu. S. Nesterova, and N. I. Organova. 23rd List

Udokanite. G. A. Yurgenson, N. G. Smirnova, and L. A. Kerenina, 1968. [Вестн. научн. информ. Забайкал. фил. Геогр. общ. СССР, 1968, no. 9], 3; abstr. *Zap.*, **99**, 78 (1970) (Удоканит). A basic copper sulphate from the Udokan Mts. in Transbaikal is formulated $Cu_8(SO_4)_3(OH)_{10}.H_2O$ and named from the locality. The description is inadequate, and the material is probably antlerite. 26th List

Uduminelite. Author? *Introd. Japanese Min.* (1970), 126. White acicular crystals, α 1·623, β 1·626, γ 1·621 (?) in cracks of perthite in a pegmatite at Udumine, Fukushima Prefecture, Japan. Composition $Ca_3Al_8(PO_4)_2O_{12}.2H_2O$. Named from the locality. [*A.M.*, **58**, 806; *Zap.*, **101**, 284.] 28th List

Ufertite. Author? *Guide to USSR exhibit at Conference on Atomic Energy*, Geneva, 1955, p. 9 (ufertite). Later publications, abstr.: *Amer. Min.*, 1957, vol. 42, p. 307, 1958, vol. 43, p. 378 (ufertite, uferite); *Mém. Soc. Russ. Min.*, 1958, vol. 87, p. 78 (уфертит, ufertite). $20FeO.8Fe_2O_3$.4(rare-earths).$74TiO_2$. Named from the composition U, Fe, Ti, 21st List [Is Davidite; *A.M.*, **49**, 447]

Ugandite. E. J. Wayland and L. J. Spencer, 1929. *Min. Mag.*, vol. 22, pp. 186, 187. Synonym of bismutotantalite, from Uganda. 12th List

Ugrandite. A. N. Winchell, 1927. *Elements of optical mineralogy*, 2nd edit., part 2, p. 257. A contraction of the names uvarovite, grossular, and andradite for this series of garnets in which there is complete isomorphous replacement. Two species of garnet are recognized—'ugrandite' and 'pyralspite' (q.v.). 11th List

Uhligite. O. Hauser, 1909. *Zeits. Anorg. Chem.*, vol. lxiii, p. 342 (Uhligit). Bright, black octahedra, at first thought to be perovskite, but found to have the composition $5Ca(Zr,Ti)_2O_5.Al_2TiO_5$; and thus regarded as an aluminous zirkelite with titanium predominating over zirconium: [zirkelite is, however, rhombohedral and not cubic]. Collected by Dr. Carl Uhlig from a metamorphosed nepheline-rock in the Great Rift Valley in German East Africa. 5th List

—— F. Cornu, 1909. *Zeits. Chem. Indust. Kolloide*, vol. iv, p. 17 (Uhligit). Apparently given as an alternative name for gelvariscite (q.v.) for the amorphous variscite of Leoben in Styria; but it also appears as an alternative name for gelfischerite (q.v.) for the amorphous fischerite of Roman-Gladna in Hungary. Probably named after Viktor Karl Uhlig (1857–1911), Professor of Geology at Vienna. Not the uhligite of O. Hauser, 1909 (5th List). 8th List

Uklonskovite. M. N. Shyusareva, 1964. Докл. Акад. наук СССР (*Compt. Rend. Acad. Sci. URSS*), vol. 158, p. 1093 (Уклонсковит). I. M. Rumanova and E. P. Popova, Кристаллография, 1964, vol. 9, p. 275. Monoclinic $NaMgSO_4OH.1\frac{1}{2}$ or $2H_2O$, in cavities in clay at Kara-Kalpakii, lower Amu-Darya river, with Glauberite and Polyhalite. Named for A. S. Uklonskii. [*M.A.*, **17**, 304; *A.M.*, **50**, 520.] 24th List

Ulmite. T. Steel, 1921. *Proc. Linnean Soc. New South Wales*, vol. 46, p. 213; *Chem. News*, London, vol. 123, p. 293. A form of humus coating the grains of a black, friable sandstone from New South Wales. Evidently named from ulmin, a constituent of humus. [*M.A.*, **1**, 257.] 9th List

Ulrichite. G. Kirsch, 1925. *Tschermaks Min. Petr. Mitt.*, vol. 38, p. 227 (Ulrichit). Cubic uranium oxide, UO_2, representing the original mineral which by 'radioactive transformation pseudomorphism' gave rise to uraninite. (Pitchblende is regarded as a distinct mineral

—uranium uranate.) Named after Dr. Carl Ulrich (–1925), radio-chemist, of Vienna. [*M.A.*, **3**, 106.] 11th List [Syn. of Uraninite]

Ultrabasite. V. Rosický and J. Štěrba-Böhm, 1916. *Rozpr. České Akad.*, class 2, vol. 25, no. 45; *Zeits. Kryst. Min.*, 1920, vol. 55, p. 430 (Ultrabasit). Black metallic, orthorhombic crystals with the ultrabasic formula $Sb_4Ag_{22}Pb_{28}Ge_3S_{53}$; from Freiberg, Saxony. [*M.A.*, **1**, 149.] 9th List

Ultralite. Trivial trade-name for red-violet *artificial* sapphire. R. Webster, *Gems*, 1962, p. 764. 28th List

Ulvite. Author? H. Strunz, *Min. Tab.*, 3rd edit., pp. 138, 147 (Ulvit). Fe_2TiO_4, cubic. Abbreviated form of ulvöspinel (18th List). 21st List [Also 23rd List]

Ulvöspinel. F. Mogensen, 1943. *Blad för Bergshandteringens Vänner*, Stockholm, vol. 26, p. 135 (Ulvöspinellen), p. 134 (titanspinell). A titaniferous iron-ore in which a large excess of FeO is assumed to be present as Fe_2TiO_4 with a spinel structure. Named from the locality, Ulvö islands, Sweden. In a later paper, F. Mogensen, *Geol. För. Förh. Stockholm*, 1946, vol. 68, p. 578, these two names are replaced by 'ferro-ortho-titanate'. [*M.A.*, **10**, 6, 101.] 18th List [*A.M.*, **40**, 138]

Umanguita. G. Gagarin and J. R. Cuomo, 1949, loc. cit., p. 15. Variant of umangite to accord with Spanish pronunciation. Named from the locality, Sierra de Umango, La Rioja, Argentina. 19th List

Umbozerite. E. M. Eskova, A. P. Khomyakov, A. N. Merkov, S. I. Lebedina, and L. S. Dubakina, 1974. *Dokl. akad. nauk. SSR*, **216**, 169 (Умбозерит). Bottle-green to greenish-brown translucent metamict material from Umbozera, Kola Peninsula, have composition near $Na_3Sr_4ThSi_8(O,OH)_{24}$. [*A.M.*, **60**, 341; *M.A.*, 75-1398; *Zap.*, **104**, 617.] 29th List

Umohoite. P. F. Kerr and G. P. Brophy, 1953. *Rocks and Minerals*, Peekskill, NY, vol. 28, p. 480. Hydrous molybdate of uranium, black, flaky, hexagonal, from Utah. Named from the composition. [*M.A.*, **12**, 239.] 20th List [$(UO_2)MoO_4.4H_2O$; *A.M.*, **42**, 657]

Ungemachite. M. A. Peacock and M. C. Bandy, 1936. *Amer. Min.*, 1936, vol. 21, no. 12, part 2, p. [2]; 1937, vol. 22, p. 207. Hydrated basic sulphate of sodium, potassium, and ferric iron, $Na_4(K,Fe''')_2(OH)(SO_4)_3$*$.5H_2O$, as colourless rhombohedral crystals, from Chile. Named after Dr. Henri Léon Ungemach (1879–1936) of Strasbourg. *See* Clino-ungemachite. [*M.A.*, **6**, 443.] 14th List [*Corr. in *M.M.*, **25**.]

Ungvarite. Variant of Unghwarite (of E. F. Glocker, 1837), a variety of chloropal from Ungvár, Hungary, now Užhorod in Carpathian Ruthenia, Czechoslovakia. V. Zepharovich (*Min. Lex. Oesterr.*, 1859) has the form Unghvarit; *British Museum Index*, since 1863, Unghvarite; M. Tóth (*Min. Hungary*, 1882), Ungvárit. [*Min. Mag.*, **21**, 459.] 11th List

Uralborite. S. V. Malinko, 1961. Зап. Всесоюз. Мин. Общ. (*Mem. All-Union Min. Soc.*), vol. 90, p. 673 (Уралборит). $CaB_2O_4.2H_2O$, in radiating fibrous aggregates from a skarn deposit in the Turinsk area, Urals (*Amer. Min.*, 1962, vol. 47, p. 1482; *M.A.*, **16**, 556). 23rd List

Uralolite. N. A. Grigoriev, 1964. Зап. Всесоюз. Мин. Общ. (*Mem. All-Union Min. Soc.*), vol. 93, p. 156 (Уралолит) $CaBe_3(PO_4)_2(OH)_2.4H_2O$; monoclinic. Named from its occurrence in the Urals. 23rd List [*A.M.*, **49**, 1776]

Uramphite. Z. A. Nekrasova, 1957. [Вопросы Геол. Ур.... (*Problems Geol. Uranium*), Атомиздат., p. 67]; abstr. Зап. Всесоюз. Мин. Общ. (*Mem. All-Union Min. Soc.*), 1958, vol. 87, p. 483 (Урамфит, uramphite). *Proc. 2nd UN Internat. Conf. Peaceful Uses Atomic Energy*, 1958, vol. 2, p. 286. $NH_4UO_2PO_4.3H_2O$, occurring as bottle-green flakes in the oxidation zone of a uranium–coal deposit; locality not given. Named from the composition, *ur*anium *am*monium *ph*osphate. [*M.A.*, **14**, 277, 344; *A.M.*, **44**, 464.] 22nd List

Uran-apatite. I. G. Chentzov, 1956. [Атом. Энерг. (*Atomic energy*), no. 5, p. 113]; abstr. Зап. Всесоюз. Мин. Общ. (*Mem. All-Union Min. Soc.*), 1958, vol. 87, p. 482 (Уран-апатит, uran-apatite). An unnecessary name for a uranian apatite with 1 to 4% UO_2. 22nd List

Uranelain. R. Hermann, *Bull. Soc. Nat. Moscou*, 1832, V, 45; *Pogg. Ann.*, 1833, XXVIII, 566. A transparent, yellow, gummy substance like rancid oil, containing C, H, and O. So-called inflammable snow which fell in gov. Moscow in 1832. 2nd List

Uran-galena. O. M. Shubnikova, *Trans. Lomonosov Inst. Acad. Sci. USSR*, 1937, no. 10, p. 171 (Урановый галенит, Uran-galena). Variant of U-galena (14th List). 15th List

Uran-microlite. Author? H. Strunz, *Min. Tab.*, 3rd edit., 1957, p. 147 (Uran-Mikrolith). Variety of microlite containing some uranium. 21st List [Name replaces Djalmaite (q.v.); *A.M.*, **62**, 406]

Uranoanatase. Y. Vuorelainen, A. Huhma, A. Häkli, 1964. *Bull. Comm. Géol. Finlande*, vol. 33, no. 215, 116. Mentioned without description among the uranium minerals from Kuusamo, NE Finland. Presumably a uranian anatase. [Зап., **96**, 71.] 25th List

Uranoflorescite. H. von Philipsborn, 1953. *Tafeln zum Bestimmen der Minerale*, Stuttgart, pp. xi, 74 (Uranoflorescit, Uranblüte). A collective name for efflorescences of secondary uranium minerals. 20th List

Uranohydrothorite. V. A. Khvostova, 1969. Геохимия, 328 (Ураногидроторит). An ill-defined mineral, presumably a hydrous uranian thorite. An yttrian variety occurs in Precambrian conglomerates in the Urals. [*Zap.*, **100**, 624.] 27th List

Uranolepidite. J. Thoreau, 1933. *Ann. Soc. Géol. Belgique, Publ. Congo Belge*, annex to vol. 55, p. c3 (uranolépidite), p. c5 (uranolepidite). Hydrous uranate of copper $CuO.UO_3.2H_2O$, as dark green lamellar masses, monoclinic or triclinic, from Katanga. Identical with Vandenbrandeite (q.v.). Apparently named from uranium and *λεπίς, λεπίδος*, scale. [*M.A.*, **5**, 389.] 13th List

Uranopissinit, error for Uranopissite (6th List). Hintze, *Handb. Min.*, Erg.-bd. II, p. 877. 23rd List

Uranopissite. E. F. Glocker, 1847. *Generum et specierum mineralium synopsis*, p. 74 (Uranopissites, Uranopissit). An obsolete synonym of uraninite or pitchblende. Named from uranium and *πίσσα*, pitch. 6th List

Uranosandbergit. K. Walenta, quoted in Hintze, *Handb. Min.*, Erg.-bd. II, pp. 765, 878. Synonym of Heinrichite (22nd List). 23rd List

Uranospathite. A. F. Hallimond, 1915. *Mineralogical Magazine*, vol. xvii, p. 221. A yellow or pale-green 'uranium-mica' from Redruth, Cornwall, hitherto classed as autunite. It is orthorhombic (pseudotetragonal), with the form of thin, rectangular plates; and is probably a hydrated uranyl salt. Named from uranium and *σπάθη*, a broad blade. Cf. Bassetite. 7th List [$Cu(UO_2)_2(AsO_4,PO_4)_2.11H_2O$; *M.M.*, **30**, 343]

Uranospherite. Variant of Uranosphaerite; *A.M.*, **9**, 62. 10th List

Uranothorianite. R. C. Wells, J. G. Fairchild, and C. S. Ross, 1933. *Amer. Journ. Sci.*, ser. 5, vol. 26, p. 47 (footnote). Suggested for a mineral intermediate between uraninite and thorianite, $(U,Th)O_2$. 13th List

Uranothorogummite(s). L. van Wambeke, 1967. *Bull. Soc. belge geol.*, **76**, 7. Unnecessary varietal name for uranian thorogummite. [*Zap.*, **100**, 624.] 27th List

Uran-pyrochlore. Author? H. Strunz, *Min. Tab.*, 3rd edit., 1957, p. 147 (Uran-Pyrochlor). Variety of pyrochlore containing some uranium. 21st List [**Uranpyrochlore** (Holmquist, 1896); name now replaces Hatchettolite; *A.M.*, **62**, 406]

Uranspat. *Neues Jahrb. Min.*, 1920, ref. p. 146; *Fortsch. Min. Krist. Petr.*, 1922, vol. 7, p. 170. Variant of uranospathite (A. F. Hallimond, 1915; 7th List). 9th List

Urbanite. H. Sjögren, *Geol. För. Förh.*, XIV. 251, 1892; and *Bull. Geol. Inst. Upsala*, II, 77, 106, 1895. [*M.M.*, **11**, 167.] A monosymmetric pyroxene. $(Ca,Mg)O.SiO_2 + Na_2O.Fe_2O_3.4SiO_2$. [Långban, Sweden] 1st List

Urea. P. J. Bridge, 1973. *M.M.*, **39**, 346. The organic compound urea, $CO(NH_2)_2$, occurs as tetragonal bipyramids in guano from a cave or rock shelter near Toppin Hill, Western Australia. [*M.A.*, 74-511; *A.M.*, **59**, 874.] 28th List

Ureyite. C. Frondel and C. Klein, 1965. *Science*, vol. 149, p. 742. An emerald-green mineral found in the Coahuila, Toluca, and Hex River Mts. siderites, is $NaCrSi_2O_6$, the chromium analogue of Jadeite; monoclinic. Named for H. C. Urey. It is clearly identical with Cosmochlore, which name has priority (Laspeyres, 1897; 1st List). [*A.M.*, **50**, 2096.] 24th List

Urhite. R. V. Getzeva, 1956. [*Atomic energy*, Moscow, no. 3, p. 135.] Abstr. in *Mém. Soc. Russ. Min.*, 1957, vol. 86, p. 118 (ургит, urhyte); *Amer. Min.*, 1957, vol. 42, p. 442 (urgite);

Bull. Soc. Franç, Min. Crist., 1957, vol. 88, p. 538 (urgite). Hydrous oxide, $UO_3.2{\cdot}3$–$3{\cdot}1H_2O$, amorphous, yellow, in oxidized ore. Named from the composition. *See* Hydronasturan. [*M.A.*, **13**, 385 (urhyte).] 21st List

Uricite. P. J. Bridge, 1974. *M.M.*, **39**, 889. Anhydrous uric acid, occurring in bird-guano deposits in Dingo Donga Cave, Western Australia. Named for the composition. 28th List

Urquhartite. M. F. Heddle, 1878. *Trans. Roy. Soc. Edinburgh*, vol. 28, p. 310. A name for a wholly undefined mineral in the gneiss of Milton, Glen Urquhart, Scotland. Named from the locality. 22nd List

Ursilite. A. A. Chernikov, O. V. Krutetzkaya, and V. D. Sidelnikova, 1957. [Вопросы Геол. Уран. (*Problems Geol. Uranium*), Атомиздат., p. 73]; abstr. Зап. Всесоюз. Мин. Общ. (*Mem. All-Union Min. Soc.*), 1958, vol. 87, p. 486 (Урсилит, ursilite). $2(Ca,Mg)O.2UO_3.5SiO_2.9H_2O$, earthy masses or spherulites in quartz-porphyry (locality not given). Two varieties occur, calcium ursilite (кальцийурсилит) with Ca $\gg$ Mg and magnesium ursilite (магнийурсилит) with Mg $\gg$ Ca. Named from the composition, *ur*anium *sil*icate. *See also Proc. 2nd UN Internat. Conf. Peaceful Uses Atomic Energy*, vol. 2, p. 295 (Calcium urcilite, magnesium urcilite). [*M.A.*, **14**, 277, 344.] Calcium ursilite is very near haiweeite (q.v.) in composition but has a different powder pattern (K. Walenta, *Neues Jahrb. Min.*, Monatsh., 1960, p. 37). 22nd List

Urvantsevite. N. S. Rudoshevskii, V. N. Makarov, E. M. Mededeva, V. V. Ballakh, Y. I. Permyakov, G. A. Mitenkov, A. M. Karpenkov, I. D. Budk'ko, and N. N. Shishkin, 1976. *Zap.*, **105**, 704 (Урванцевит). Intergrowths with other platinum-group minerals in sulphide ores from the Mayak (Маяк, 'Majak') mine, Talnakh deposit, are hexagonal, *a* 13·82, *c* 6·53 Å. Composition $Pd(Bi,Pb)_2$. Named for N. N. Urvantsev. [*A.M.*, **62**, 1260; *M.A.*, 77-4628.] 30th List

Usbekit. *See* Uzbekite. 11th List

Usigit, error for Usihite (Усигит, 21st List). Hintze, *Handb. Min.*, Erg.-bd. II, p. 879. 23rd List

Usihite. V. G. Melkov and L. Ch. Pukhalsky, 1957. [*Minerals of the Urals*, Moscow, 1957, p. 67.] Abstr. in *Mém. Soc. Russ. Min.*, 1958, vol. 87, p. 84 (усигит, usihyte). $R(UO_2)_2Si_2O_7.nH_2O$. Named from the composition. 21st List

Usovite. A. D. Nozhkin, V. A. Gavrilenko, and V. A. Moleva, 1967. Зап. всесоюз. мин. общ. (*Mem. All-Union Min. Soc.*), vol. 96, p. 63 (Усовит, Usovite). A brown mineral occurring in a fluorite vein in the Upper Noiby river area, Yenisei region, Siberia, is probably orthorhombic. Composition, after deduction of Ca as fluorite (11·7%), $Ba_2MgAl_2F_{12}$. Named for M. A. Usov. [*A.M.*, **52**, 1582; Зап., **97**, 67.] 25th List [Monoclinic, Ca essential; *A.M.*, **60**, 739]

Ussingite. O. B. Bøggild, 1914. *Meddelelser om Grønland*, vol. li, p. 103; *Zeits. Kryst. Min.*, vol. liv, p. 120 (Ussingit). A reddish-violet mineral from the pegmatite veins in the naujaite (sodalite-syenite) at Kangerdluarsuk, Greenland. It is a silicate, $HNa_2Al(SiO_3)_3$, showing certain relations to the zeolites. Triclinic, with one perfect and two poor cleavages. Named after the late Professor Niels Viggo Ussing (1864–1911), of Copenhagen. 7th List [*A.M.*, **59**, 335]

Ustarasite. M. S. Sakharova, 1955. *Trudy Min. Mus. Acad. Sci. USSR*, no. 7, p. 116 (устарасит). Sulphosalt $PbS.3(Bi,Sb)_2S_3$, as grey prismatic crystals in bismuth ore from the Ustarasaisk deposit in W Tyan-shan, Siberia. Named from the locality. [*M.A.*, **13**, 164; *A.M.*, **41**, 814.] 21st List

Utahlite. G. F. Kunz, *16th Ann. Rept. US Geol. Survey*, for 1894–1895, Part IV. 602, 1895. The compact nodular variscite from [Lewiston] Cedar Valley, [Tooele Co.] Utah. 1st List

Uvanite. F. L. Hess and W. T. Schaller, 1914. *Journ. Washington Acad. Sci.*, vol. iv, p. 576; 3rd Appendix to 6th edit. of Dana's *System of Mineralogy*, 1915, p. 81. An orthorhombic, hydrous vanadate of uranium, $2UO_3.3V_2O_5.15H_2O$, occurring as a brownish-yellow powder disseminated in sandstone in [Temple Rock] Utah. The name is a contraction of the words uranium and vanadium with the termination *ite*. 7th List

Uvite. W. Kunitz, 1929. *Chemie der Erde*, vol. 4, p. 221 (Uvit). A hypothetical molecule, $H_8Ca_2Mg_8Al_{10}Si_{12}B_6O_{62}$, of the tourmaline group, to explain the composition of magnesia-lime-tourmaline (CaO 5·35%) from province Uva, Ceylon. Named from the locality. [*M.A.*, **4**, 204.] 12th List [*See* Elbaite]

Uxporite. O. M. Shubnikova and D. V. Yuferov, *Spravochnik po novym mineralam*, Leningrad, 1934, p. 57. Still another spelling of yuksporite, юкспорит (10th List), juxporite and juksporite (11th List). 14th List

Uzbekite. A. E. Fersman, 1925. *Priroda*, Leningrad, 1925, no. 7–9, col. 238 (узбекит), *Musée de Minéralogie, Acad. Sci. Leningrad*, 1925, p. 4 (ousbékite). I. Kurbatov, *Centr. Min.*, Abt. A, 1926, p. 234 (Usbekit). Hydrous copper vanadate, $3CuO.V_2O_5.3H_2O$, as thin dark-green crusts. Named from the locality, Uzbekistana district in Fergana (from the Uzbek or Uzbeg race in Central Asia). [*M.A.*, **3**, 345.] 11th List [Is Volborthite: *A.M.*, **50**, 2111]

V

Vabanite. Author and date? There seems to be no end to the trivial names for dirty quartz; this one is a jasper from California. R. Webster, *Gems*, 1962, p. 764. Not to be confused with vrbaite (6th List), nor with urbanite. 28th List

Vaesite. P. F. Kerr, 1945. *Amer. Min.*, vol. 30, pp. 483, 498. Nickel disulphide, NiS_2, cubic with pyrite structure, from Kasompi mine, Belgian Congo. Named after Johannes Vaes, mineralogist to the Union Minière du Haut Katanga. [*M.A.*, **9**, 188, 224.] 17th List

Valahite, spelling variant of Vallachite [*M.A.*, **17**, 138]. 24th List

Vallachite. G. Gita and E. Gita, 1962. [*Ann. Pedology Sect., Centr. Res. Inst. Agric.*, Bucharest, vol. 30, p. 279]; abstr. *M.A.*, **17**, 138. An undesirable name for a mixed-layer clay mineral from a Rumanian soil. 'Belongs to the open illite type or ammersooite' (21st List). Presumably named for Wallachia. 24th List

Valleite. G. Cesàro, *Bull. Acad. Belg.*, XXXII. 536, 1896; XXIX. 508, 1895. [*M.M.*, **11**, 228.] A rhombic amphibole near anthophyllite. $(Mg,Ca)O.SiO_2$. [Edwards] New York. 1st List [Calcian manganoan anthophyllite; *Amph.* (1978)]

Vanadine. [G.J.] Adam, 1869. *Ann. Mines*, Paris, ser. 6, Mém. vol. 15, p. 488; *Tableau Minéralogique*, Paris, 1869, p. 33. Synonym of vanadic ochre (J. D. Dana, 1868) = vanadic acid (J. E. Teschemacher, 1851). The existence of this as a mineral has been questioned by W. T. Schaller, *Amer. Journ. Sci.*, 1915, ser. 4, vol. 39, p. 404 [*M.A.*, **1**, 262]. Cf. Alaite, 5th List. 12th List

Vanadin-Spinelle. H. Strunz, 1970. *Min. Tab.*, 5th edit., 177. An unnecessary group name for coulsonite (14th List). Cf. Vanadiumspinell (24th List). 28th List

Vanadioardennite. F. Zambonini, 1922. *See* Arsenioardennite. 10th List

Vanadiochrome spinel. C. O. Mathiesen, 1970. *Norges Geol. Undersög.*, **266**, 86. An unnecessary name for a vanadian chromian magnetite. [*A.M.*, **57**, 1004.] 28th List

Vanadio-laumontite. A. E. Fersman, 1922. *Trav. Musée Géol. Min. Pierre le Grand, Russ. Acad. Sci.* [*ser. 2*], vol. 2 (for 1916), p. 311 (ванадіо-ломонтитъ). A vanadiferous laumontite, containing V_2O_5 2·5%, as yellowish-red, botryoidal masses from Fergana. [*M.A.*, **2**, 299.] 10th List

Vanadiorutile. C. O. Mathiesen, 1970. *Norges Geol. Undersög.*, **266**, 86. Unnecessary name for a vanadium rutile. [*A.M.*, **57**, 1004.] 28th List

Vanadium-Arsen Germanit. K. F. Chudoba, 1974. C. Hintze, *Handb. Min.*, Erg.-Bd., **4**, 101. An unnecessary name for the vanadian arsenian germanite (Ванадиевомышьяковый германит) of I. M. Mitryaeva, M. A. Yarenskaya, E. A. Kosyak, and A. N. Muratova (*Zap.*, **97**, 325). 28th List

Vanadium-garnet. S. T. Badalov, 1951. *Mém. Soc. Russ. Min.*, vol. 80, p. 212 (ванадиевый гранат). Variety of grossular garnet containing V_2O_3 4·52%. 21st List

Vanadium-germanite. B. H. Geier and J. Ottemann, 1970. *Neues Jahrb. Min. Abh.*, **114**, 89. Unnecessary name for a vanadian germanite. [*A.M.*, **56**, 1487; *M.A.*, 71-1391.] 27th List

Vanadiumgranat, German form of Vanadium-garnet (21st List). Hintze, *Handb. Min.*, Erg.-bd. II, p. 881. 23rd List

Vanadiumspinell. H. Strunz, 1966. *Min. Tabellen*, 4th edit., p. 528. Synonym of Coulsonite (14th List). 24th List

Vanadium-tourmaline. S. T. Badalov, 1951. *Mém. Soc. Russ. Min.*, vol. 80, p. 212 (ванадиевый турмалин). Variety of tourmaline containing V_2O_3 5·76%. 21st List

Vanadiumturmalin, German form of Vanadium-tourmaline (21st List). Hintze, *Handb. Min.*, Erg.-bd. II, p. 881. 23rd List

Vanado-magnetite. A. M. Heron, 1936. *Rec. Geol. Surv. India*, vol. 71, p. 44 (vanado-magnetite). G. H. Tipper, *Bull. Imp. Inst. London*, 1936, vol. 34, p. 451. A magnetic iron ore containing variable amounts of vanadium (V_2O_5 up to 8%) and titanium (up to 25%), from Bihar, India. Later named coulsonite (q.v.). [*M.A.*, **6**, 489.] 14th List

Vanalite. E. A. Ankinovich, 1962. Зап. Всесоюз. Мин. Общ. (*Mem. All-Union Min. Soc.*), vol. 91, p. 307 (Ваналит). $NaAl_8V_{10}O_{38}.30H_2O$, in bright yellow incrustations on weathered shales from NW Kara-Tau, Kazakhstan. Named from the composition. [*M.A.*, **16**, 68.] 23rd List [*A.M.*, **48**, 1180; **57**, 597]

Vandenbrandeite. A. Schoep, 1932. *Ann. Musée Congo Belge, A, Ser. I Minéralogie*, Tervueren (Belgique), vol. 1, fasc. 3, p. 22. Hydrous uranate of copper, $2CuO.2UO_3.5H_2O$, perhaps triclinic, as a dark-green alteration product of uraninite and chalcopyrite; from Katanga. Named after P. Van den Brande, of the Geological Survey of Katanga. *See* Uranolepidite. [*M.A.*, **5**, 292; *A.M.*, **18**, 179.] 13th List [*A.M.*, **36**, 394]

Vandendriesscheite. J. F. Vaes, 1947. *Ann.* (*Bull.*) *Soc. Géol. Belgique*, vol. 70, p. B 217. Hydrous lead uranate as amber-orange orthorhombic crystals resembling fourmarierite, from Katanga. Named in memory of Adrien Vandendriessche (1914–1940) of Ghent. [*M.A.*, **10**, 146, 255.] 18th List [$PbO.7UO_3.12H_2O$; *M.M.*, **41**, 51 (with refs.)]

Vandendriesscheite–I, Vandendriesscheite–II. C. L. Christ and J. R. Clark, 1960. *Amer. Min.*, vol. 45, p. 1031. Synonyms respectively of Vandendriesscheite (18th List) and Metavandendriesscheite (q.v.). 22nd List

Vandiestite. (Author?) Described as probably a new species, but without name, by R. Pearce (*Colorado Sci. Soc. Bull.*, 1898, no. 6, p. 4; *Proc.* [*1902*], vol. vi, p. 163). The information here given was reproduced, under the name Von Diestite, in a note by E. Cumenge (*Bull. Soc. franç. Min.*, 1899, vol. xxii, p. 25 *bis*): in abstracts of this note the name Diestit (q.v.) has been used. The form vandiestite is given in a dealer's advertisement (*Amer. Journ. Sci.*, November 1901). A massive telluride of silver, bismuth, gold, and lead, from Sierra Blanca, Colorado. Named after Mr. P. H. van Diest, of San Luis, Colorado, by whom the mineral was found. 3rd List

Vanoxite. F. L. Hess, 1924. *Bull. US Geol. Survey*, no. 750–D, p. 63 (vanoxite; p. 67, van-ox′ite). Vanadate of vanadyl, $2V_2O_4.V_2O_5.8H_2O$, as a black impregnation in sandstone from Colorado. Derivation presumably from vanadium and oxygen. *See* Kentsmithite. [*M.A.*, **2**, 420.] 10th List

Vanthoffite. K. Kubierschky, *Sitz.-ber. Akad. Wiss. Berlin*, 1902, p. 407 (Vanthoffit). The results of several analyses of an apparently homogeneous specimen, from the salt deposits of Wilhelmshall near Stassfurt, were plotted as curves, and the compound $3Na_2SO_4.MgSO_4$ inferred to be present in the mixture of salts. This is colourless, with a vitreous lustre, and coarsely crystalline structure. Pure material was isolated from another specimen by means of a heavy liquid. The salt $3Na_2SO_4.MgSO_4$ was afterwards prepared artificially by Professor J. H. van't Hoff, of Berlin, after whom the mineral is named. 3rd List

Vanuralite. G. Branche, P. Bariand, F. Chantret, R. Pouget, and A. Rimsky, 1963. *Compt. Rend. Acad. Sci. Paris*, vol. 256, p. 5374 $(UO_2)_2Al(VO_4)_2OH.8H_2O$; yellow, monoclinic, occurring at Mounana, Gabon Republic. Named from the composition. [*M.A.*, **16**, 374.] 23rd List [*A.M.*, **48**, 1415; **56**, 639]

Vanuranilite. E. Z. Buryanova, G. C. Strokova, and V. A. Shitov, 1965. Зап. Всесоюз. Мин. Общ. (*Mem. All-Union Min. Soc.*), vol. 94, p. 437 (Ванураналит, vanuranilite). A yellow microcrystalline mineral of composition $H_3O^+UO_2VO_4.2H_2O$ with some replacement of H_2O^+ by Ba,Ca,Pb, and K. The composition is near that of Vanuralite ($AlOH(UO_2)_2(VO_4)_2.8H_2O$; 23rd List), and the optics, X-ray powder data, and unit-cell dimensions are close to those of Vanuralite; the very similar names may lead to confusion. 24th List [**Vanuranylite** preferred]

Vanuranylite, variant of Vanuranilite (Ванураналит; 24th List) presumably to recognize the presence of UO_2 groups. *Amer. Min.*, 1968, vol. 51, p. 1548. 25th List

Variolite. An old rock name (author and date?) denoting 'stones that have rounded protuberances, of a different nature from the common mass of the stone, from their resemblance to variolae'. Localities Durance and the river Drae, France. R. Kirwan, *Mineralogy*, 1794 (2nd edit.), **1**, 368. The name was dropped from Dana's *Syst. Min.*, 1892 (6th edit.), after appearing in earlier editions, but has re-appeared in R. Webster, *Gems*, 1962, p. 764, as a *var.* of orthoclase. 28th List

Varlamoffite. R. De Dycker, 1947. H. Buttgenbach, *Les minéraux de Belgique et du Congo Belge*, Liège, 1947, p. 182. N. Varlamoff, *Ann.* (*Bull.*) *Soc. Géol. Belgique*, 1948, vol. 71 (for 1947–1948), pp. B 224, B 226, B 232; ibid., 1948, vol. 72 (for 1948–1949), p. B 41. Metastannic acid, H_2SnO_3, as a yellow earthy material from alteration of stannite, from Kalima, Belgian Congo. Named after Nicolas Varlamoff, mining engineer, Belgian Congo, who found the mineral. Cf. Souxite (q.v.). [*M.A.*, **10**, 354, 454, 495.] 18th List

Varulite. P. Quensel, 1937. *Geol. För. Förh. Stockholm*, vol. 59, p. 95 (varulite). A phosphate mineral $Na_2O.5(Mn,Fe,Ca).2P_2O_5$, presumably orthorhombic, occurring as olive-green granular masses at Varuträsk, N Sweden. Named from the locality. [*M.A.*, **6**, 486; *A.M.*, **26**, 861.] 14th List

Vashegyite. K. Zimányi, 1909. *Math. és természettudományi Értesítö*, Budapest, vol. xxvii, p. 64 (Vashegyit); *Zeits. Kryst. Min.*, xlvii, p. 53. A basic aluminium phosphate, $4Al_2O_3.3P_2O_5.30H_2O$, occurring as compact, dull white masses, resembling meerschaum in appearance, in the iron mine of Vashegy, comitat Gömör, Hungary. 5th List [Redefinition proposed; *M.A.*, **18**, 203]

Vashegyrite. Error for vashegyite (5th List). R. Webster, *Gems*, 1962, p. 764. 28th List

Vaterite. W. Meigen, 1911. *Verh. Ges. Deutsch. Naturforscher u. Ärzte, 82. Versamml. zu Königsberg*, 1910, II. Teil, 1. Hälfte, p. 124 (Vaterit). W. Diesel, *Zeits. Kryst. Min.*, 1911, vol. xlix, p. 272. G. Linck, Doelter's *Handbuch der Mineralchemie*, 1911, vol. i, p. 113. Vater's third modification of calcium carbonate prepared artificially in the form of minute spherules with specific gravity 2·6 and less stable than aragonite and calcite. Named after Professor Heinrich Vater, of Tharandt, Saxony. 6th List [*M.M.*, **32**, 535]

Vauxite. S. G. Gordon, 1922. *Science*, New York, vol. 56, p. 50; *Amer. Min.*, vol. 7, p. 108; *Proc. Acad. Nat. Sci. Philadelphia*, 1923, vol. 75, p. 261. Hydrated phosphate of iron and aluminium, $4FeO.2Al_2O_3.3P_2O_5.24H_2O + 3H_2O$, occurring as radial aggregates of sky-blue, strongly pleochroic, triclinic crystals on wavellite from the tin mines of Llallagua, Bolivia. Named after Mr. George Vaux, junior, of Bryn Mawr, Pennsylvania. *See* Paravauxite. [*M.A.*, **2**, 148.] 10th List [*A.M.*, **53**, 1025]

Väyrynenite. A. Volborth and E. Stradner, 1954. *Anz. Math.-naturwiss. Kl. Österreich. Akad. Wiss.*, Wien, vol. 92, p. 21. A. Volborth, *Ann. Acad. Sci. Fennicae*, 1954, ser. A, III Geol. Geogr., no. 39, p. 66 (Väyrynenit). $MnBe(PO_4)(OH,F?)$, similar to herderite with Mn in place of Ca. Monoclinic, rose-red, in Li-pegmatite from Finland. Named after Dr. Heikki Väyrynen, professor of mineralogy and geology in the Technical Highschool, Helsinki. [*M.A.*, **12**, 354, 568.] 20th List [*M.A.*, 77-2175]

Veatchite. G. Switzer, 1938. *Amer. Min.*, vol. 23, p. 409; J. Murdoch, ibid., 1938, vol. 23, no. 12, part 2, p. 11. Hydrous borate of calcium, $Ca_2B_6O_{11}.2H_2O$, monoclinic, occurring in cross-fibre veins; from California. Named after Dr. John A. Veatch, who first detected borates in California in 1856. [*M.A.*, **7**, 169, 263.] 15th List [$Sr_2B_{11}O_{16}(OH)_5.H_2O$; *A.M.*, **56**, 1934]

***p*-Veatchite.** O. Braitsch, 1959. *Beitr. Min. Petr.*, vol. 6, p. 352 (*p*-Veatchit); C. A. Beevers and F. H. Stewart, *Min. Mag.*, 1960, vol. 32, p. 500. A polymorph of veatchite, crystallizing in space-group $P2_1/m$ or possibly $P2_1$; the space-group of veatchite is $A2/a$ or perhaps Aa. The prefixed *p* is for 'einfach-*p*rimitivem Raumgitter'. It would have been preferable to use a suffix and avoid any danger of displacement in an index. 22nd List

Vedrite. Error for verdite. 5th List

Veenite, J. L. Jambor, 1967. *Canad. Min.*, vol. 9, p. 7. A lead sulphantimonate giving X-ray data near those of dufrenoysite, but with Sb > As, composition $Pb_2(Sb,As)_2S_5$, occurs at Madoc, Ontario. It is probably identical with stibiodufrenoysite of Burkart-Baumann, Ottemann, and Amstutz (q.v.), but as it has not been shown to be isostructural with dufrenoysite, a new name is assigned, in honour of R. W. van der Veen. [*A.M.*, **53**, 1422.] 25th List

Vegasite. A. Knopf, 1915. *Journ. Washington Acad. Sci.*, vol. v, p. 501. A hydrated lead ferric sulphate, $PbO.3Fe_2O_3.3SO_3.6H_2O$, differing slightly from plumbojarosite ($PbO.3Fe_2O_3.4SO_4.6H_2O$, W. F. Hillebrand and S. L. Penfield, 1902; 3rd List) in composition and optical characters. Occurs as straw-yellow, ochreous masses, consisting of minute fibres and six-sided plates (optically uniaxial), in Clark Co., Nevada. Named from Las Vegas, the principal town of the county. 7th List [Probably Plumbojarosite]

Velardeñite. W. T. Schaller, 1914. *Journ. Washington Acad. Sci.*, vol. iv, p. 355; 3rd Appendix to 6th edit. of Dana's *System of Mineralogy*, 1915, p. 82. A member of the melilite group with the composition $2CaO.Al_2O_3.SiO_2$. It has been produced artificially, and this molecule occurs in large amount in 'gehlenite' from the Velardeña mining district, Mexico. 7th List

Velikhovite. [I. V. Lopukhov, *Mineralnoe Syre*, Moscow, 1931, no. 2, pp. 152, 154.] O. M. Shubnikova, *Trans. Inst. Geol. Sci. USSR*, 1938, no. 11 (*Min. Geochem. Ser.*, no. 3), p. 32 (Велиховит, Velikhovite). A bitumen (C 77·34, H 7·71, N 2·00, O 10·59, S 2·36) with shining conchoidal fracture. Named from the locality, Velikhov (Велихов), Urals. 17th List

Velikite. L. N. Kaplunnik, E. A. Pobedimskaya, and N. V. Belov, 1977. *Kristallografiya*, **22**, 175 (Великит). A dark-grey mineral from the Khaidarkan deposit is tetragonal, *a* 5·542, *c* 10·908 Å, space group *I*4/*mmm*. The structure is stated to be that of stannite, with Hg in 7/8 of the Fe positions, and the unit cell contents are cited as $[Cu_{3\cdot18}Hg_{1\cdot53}Zn_{0\cdot44}Sn_{1\cdot90}As_{0\cdot08}Sb_{0\cdot02}S_{9\cdot94}]$, (analysis not given). [*A.M.*, **62**, 1260; *M.A.*, 78-1506.] 30th List

Venaite. I. Kostov, 1968. *Mineralogy* (Oliver and Boyd, London), 167. Pb_3BiSbS_3; no details are given. 28th List

Vendeennite. A. E. A. Rivière, 1840. *Compt. Rend. Acad. Sci. Paris*, vol. xi, p. 208. A. Lacroix, *Minéralogie de la France*, 1910, vol. iv, p. 640 (vendéennite). A fossil resin from the coal-measures of Vendée. 7th List

Venturaite. *See* under Mayberyite. 2nd List

Verdelite. P. Quensel and O. Gabrielson, 1939. *Geol. För. Förh. Stockholm*, vol. 61, p. 67. Green tourmaline, completing the series of colour varieties—achroite, rubellite, and indicolite. From *verd* (Old French), *verde* (Italian), and *λίθος*. [*M.A.*, **7**, 336.] 15th List [*See* Elbaite]

Verdite*. Trade-name for an ornamental stone of a rich chrome-green colour, found as large blocks on the S bank of the North Kaap river, near Kaap Station, South Africa. The stone contains chrome-muscovite (fuchsite) and some argillaceous material. (G. F. Kunz, *Mineral Industry*, New York, for 1907, 1908, vol. xvi, p. 810.) 5th List [*Corr. in *M.M.*, **16**]

Vermilite. Author and date? Cinnabar in opal. Presumably derived from vermilion. R. Webster, *Gems*, 1962, p. 764. 28th List

Vernadite. A. G. Betekhin, 1944. *Bull. Acad. Sci. URSS, Sér. Géol.*, no. 4, p. 35 (Вернадит), p. 45 (vernadite). Colloidal hydrated MnO_2 as a weathering product of manganese ores; considered to be a manganic acid H_2MnO_3, giving rise to salts of the psilomelane group. Named after Vladimir Ivanovich Vernadsky, Владимир Иванович Вернадский (1863–1945), who had predicted its existence. Distinct from vernadskite (6th List). [*M.A.*, **9**, 185.] 17th List

Vernadskite. F. Zambonini, 1910. *Mineralogia Vesuviana. Mem. R. Accad. Sci. Fis. Mat. Napoli*, ser. 2, vol. xiv, no. 7, p. 337 (Vernadskijte, Vernadskyte). A green, basic sulphate of copper, $4CuO.3SO_3.5H_2O$, produced by the action of acid vapours from fumaroles on dolerophanite (Cu_2SO_5). Named after Professor Vladimir Ivanovič Vernadsky, of St. Petersburg. 6th List [Is Antlerite pseudom. after Dolerophane; *A.M.*, **46**, 146]

Verplanckite. J. T. Alfors and M. C. Stinson, 1965. *Min. Inform. Serv., Calif. Div. Mines Geol.*, vol. 18, p. 27. J. T. Alfors, M. C. Stinson, R. A. Matthews, and A. Pabst, *Amer. Min.*, 1965, vol. 50, pp. 279, 314, 1500. Brownish-orange to brownish-yellow hexagonal crystals in sanbornite-quartz rock near Big Creek and Rush Creek, Fresno County, California, are near $Ba_2(Mn,Fe,Ti)Si_2O_6(O,OH,Cl,F)_2.3H_2O$ (emission spectrography analysis, Cl by X-ray spectroscopy). Named for W. E. Ver Planck. [*M.A.*, **17**, 400; **17**, 502.] 24th List

Vertumnite. E. Passaglia and E. Galli, 1977. *Tschermaks Mineral. Petrogr. Mitt.*, **24**, 57. Flattened hexagonal prisms on tobermorite from Campo Morto, Montalto di Castro, Viterbo, Tuscany, Italy, are monoclinic pseudohexagonal with *a* 5·744, *b* 25·12, *c* 5·766 Å, *β* 119·72°. Composition $[Ca_4Al_4Si_4O_6(OH)_{24}.3H_2O]$, sp. gr. 2·15. RIs *α* 1·531 ∥ [010], *β* 1·535, *γ* 1·541, *γ*:[001] 16° in obtuse *β*. Named for the Etruscan god Vertumnus. [*A.M.*, **62**, 1061; *M.A.*, 78-2129.] 30th List

Vesignieite. C. Guillemin, 1955. *Compt. Rend. Acad. Sci. Paris*, vol. 240, p. 2331 (vésigniéite); *Bull. Soc. Franç. Min. Crist.*, 1956, vol. 79, p. 250. Copper-barium vanadate $Cu_3Ba(VO_4)_2(OH)_2$, with X-ray pattern similar to bayldonite $[Cu_3Pb(AsO_4)_2(OH)_2]$. Named after Colonel Jean Paul Louis Vésignié (1870–1954) of Paris, who supplied the material from his private collection. [*M.A.*, **13**, 6; *A.M.*, **40**, 942.] 21st List [**Vesigniéite**]

Vestanite. (C. Hintze, *Handb. d. Min.*, 1892, II, 832). The same as westanite. 2nd List

Vibertite. N. R. Goodman, 1957. [*The geology of Canadian industrial mineral deposits*, p. 110]; abstr. *Amer. Min.*, 1958, vol. 43, p. 791. Syn. of Bassanite (6th List). 22nd List

Victoria-stone. S. Iimori, before 1969. Also called Iimori-stone. An *artificial* material, radially fibrous and 'mineralogically similar to the nephrite of the amphibole group', produced in a variety of colours for cutting as cabochons. *Lapidary Journal*, 1969, p. 696. Not iimoriite, (q.v.). 28th List

Viellaurite. H. Lienau, *Chem.-Zeit.*, 1899, XXIII, 418. A massive, dark grey mineral from near Vielle-Aure, Hautes-Pyrénées. $5MnCO_3.2Mn_2SiO_4$. 2nd List [A mixture of tephroite and rhodocrosite; Lacroix, *Min. de la France*, **3**, 626]

Vierzonite. A. de Grossouvre, *Bull. Soc. géol. France*, 1901, ser. 4, vol. i, p. 432. A. Lacroix, *Min. de France*, 1901, vol. iii, p. 329. A deposit ('argile à silex') consisting essentially of pulverulent opal, which has been produced by the weathering of Cretaceous beds in the neighbourhood of Vierzon, dép. Cher. The same name has been previously applied to a yellow ochreous clay from the same locality (H. W. Bristow, *Glossary of Mineralogy*, 1861). 3rd List

Vigorite. Trade-name for a phenolic resin plastic. R. Webster, *Gems*, 1962, p. 764. 28th List

Vikingite. S. Karup-Møller, 1977. *Bull. Geol. Soc. Denmark*, **26**, 41. Monoclinic, *a* 13·603, *b* 25·248, *c* 4·112 Å, *β* 95·55°. Composition $[Ag_5Pb_8Bi_{13}S_{30}]$. From Ivigtut, Greenland. Named for the colonizers of Greenland. [*M.A.*, 78-899.] 30th List

Vilateite. A. Lacroix, 1910. *Minéralogie de France*, vol. iv, pp. 477, 501 (vilatéite). The violet crystals described by A. Des Cloizeaux (1858) as type I of hureaulite are shown to differ essentially from that species: they give the same qualitative chemical reactions as strengite (hydrated phosphate of iron) with a little manganese. From La Vilate, near Chanteloube, Haute-Vienne. 6th List

Villamaninite. W. R. Schoeller and A. R. Powell, 1919. *Nature*, London, vol. 104, p. 326; *Min. Mag.*, 1920, vol. 19, p. 14; 1921, vol. 19, p. 273. Disulphide of copper, nickel, cobalt, and iron, $(Cu,Ni,Co,Fe)S_2$, with some selenium (1·5%), occurring as small, black, cubic crystals and as nodular aggregates in crystalline dolomite near Villamanín, prov. León, Spain. Named from the locality. E. Thomson (*Univ. Toronto Studies, Geol. Ser.*, 1921, no. 12, p. 39) suggests that it is a mixture of two undetermined minerals (but the original material is definitely crystallized). [*M.A.*, **1**, 24, 260.] 9th List [*M.M.*, **33**, 169; redefined, *M.A.*, 69-2343]

Villiaumite. A. Lacroix, 1908. *Compt. Rend. Acad. Sci. Paris*, vol. cxlvi, p. 215. *Bull. Soc. franç, Min.*, vol. xxxi, p. 50. Sodium fluoride, [NaF], occurring as a primary constituent in nepheline-syenite from the Los Islands, W coast of Africa. The small crystals are tetragonal (pseudo-cubic) with three perfect cleavages at right angles, a deep carmine colour, and strong pleochroism. The mineral is soluble in water. Named after Mr. – Villiaume, who collected the material. 5th List

Villiersite. G. Gagarin and J. R. Cuomo, 1949, loc. cit., p. 14 (villiersita). A nickel-bearing silicate (NiO 30·6%), $(Ni,Mg,Fe,Co)_7Si_8O_{22}(OH)_2$ resembling talc, described by F. C. Partridge from Bon Accord, Transvaal [*M.A.*, **9**, 189]. Named after Dr. J. E. de Villiers of the Geological Survey of South Africa, who prepared the posthumous paper for publication. 19th List

Vimsite. D. P. Shashkin, M. A. Simonov, N. I. Chernova, S. V. Malinko, T. I. Stolyarova, and N. V. Belov, 1968. Докл. Акад. наук СССР (*Compt. Rend. Acad. Sci. URSS*), **182**, 821 (Вимсит); ibid., 1402. Colourless transparent monoclinic crystals of composition $CaB_2O_2(OH)_4$ occur with uralborite in skarn from the Urals. Named from the initials of the All-Union Research Institute of Mineral Resources (Всесоюз. инст. мин. сырья) [*A.M.*, **54**, 1220; *M.A.*, 69-2388; *Zap.*, **99**, 77]. 26th List

Vincentite. E. F. Stumpfl and M. Tarkian, 1974. *M.M.*, **39**, 525. Minute grains (7 to 10 μm, rarely up to 40 μm) in iron-bearing platinum from the Riam Kanan River, SE Borneo, have composition $(Pd,Pt)_3(As,Sb,Te)$ with Pd > Pt and As ≈ (Sb + Te); the X-ray pattern cannot be matched by that of arsenopalladinite, mertieite, isomertieite, stibiopalladinite, or atheneïte. Named for E. A. Vincent. [*M.A.*, 74-1452; *A.M.*, **59**, 1332.] 28th List

Vinogradovite. E. I. Semenov, E. M. Bonshtedt-Kupletskaya, V. A. Moleva, and N. N. Sludskaya, 1956. *Doklady Acad. Sci. USSR*, vol. 109, p. 617 (виноградовит). Titanosilicate, $Na_5Ti_4AlSi_6O_{24}.3H_2O$, monoclinic (pseudo-orthorhombic), as white to colourless crystals and radial spherical aggregates in nepheline-syenite pegmatite from Kola. Named after Alexander Pavlovich Vinogradov, Александр Павлович Виноградов. [*M.A.*, **13**, 300; *A.M.*, **42**, 308.] 21st List

Violaite. E. S. Fedorov, *Annales de l'Institut agronomique de Moscou*, 1901, vol. vii, p. 43 (Віолаитъ). A strongly pleochroic pyroxene which forms an essential constituent of a rock called kedabekite from the Kedabek copper-mine, Caucasus. Differs from fedorovite (Viola, 1899) in having a considerable portion of the magnesia replaced by ferrous oxide. Named after Dr. Carlo Viola, of Rome. 3rd List

Violarite. W. Lindgren and W. M. Davy, 1924. *Econ. Geol.*, vol. 19, p. 318; A. F. Buddington, ibid., p. 527. 'Polydymite' from Key West mine in Nevada and from Vermilion mine in Ontario differs from the original polydymite from Westphalia. Deducting iron as admixed pentlandite, the composition approximates to NiS_2. Named from the Latin *violaris*, of violets, in allusion to the violet-grey colour of the mineral on polished sections. [*M.A.*, **2**, 338, 447.] 10th List [Ni_2FeS_4]

Violite. Trade-name for a purple *artificial* corundum. R. Webster, *Gems*, 1962, p. 764. Not to be confused with violan, violaite (3rd List), violarite (10th List), nor with violite (of Melville), a *var.* of copiapite. *See* under zircolite (q.v.). 28th List

Vioralite, error for Violarite. *Intro. Jap. Min.*, 13 (1970), Geol. Surv. Japan. 28th List

Virgilite. B. M. French, P. A. Jezek, and D. E. Appleman, 1978. *A.M.*, **63**, 461. Hexagonal bipyramids up to 50 μm, and fibrous overgrowths on quartz and other minerals in the volcanic glass of Macusani, Peru, have *a* 5·13, *c* 5·44 Å, and a β-quartz type structure. Composition $[(Li,Al)_{0\cdot6}Si_{2\cdot4}O_6]$. *n* 1·520, uniaxial negative, ω–ε 0·005. There is no ordering of Al and Si. Named for Virgil E. Barnes. 30th List

Viridine. G. Klemm, 1912 [*Notizblatt Ver. Erdkunde*, Darmstadt, for 1911, p. 4]; abstr. in *Chem. Zentralblatt*, 1913, vol. i, p. 54 (Viridin). A green variety of andalusite containing some iron (Fe_2O_3 4·16%) and manganese (Mn_2O_3 4·77%) replacing aluminium. It occurs as small grains in a contact-hornfels near Darmstadt. Cf. manganandalusite (H. Bäckström, 1896; 1st List). The name viridin has been previously used for chlorophyll; viridine for a coal-tar colour, and viridite for a mineral. 6th List

Viridite. F. Kretschmer, 1918. *Neues Jahrb. Min.*, 1918, p. 19 (Viridit); Archiv f. Lagerstättenforschung, Berlin. A chloritic mineral approximating in composition to hydrated ferrous silicate, $4FeO.2SiO_2.3H_2O$, occurring as compact, leek-green masses in diabase and schalstein in N Moravia and S Silesia, where, with mackensite (q.v.), etc., it is mined as an ore of iron. The optical characters of the minute needles and scales indicate monoclinic symmetry. Named from *viridis*, green. Not the viridite of H. Vogelsang, 1872. 8th List

Viseite. J. Mélon, 1943. *Ann. Soc. Géol. Belgique*, vol. 66, *Bull.*, p. 53 (viséite). Hydrous silicophosphate of aluminium and calcium, $5Al_2O_3.5CaO.3SiO_2.3P_2O_5.25\text{–}30H_2O$, as white, wart-like, optically isotropic masses. Named after the locality, Visé, Belgium. [*M.A.*, **9**, 88, 188.] 17th List [$NaCa_5Al_{10}(SiO_4)_3(PO_4)_5(OH)_{14}.16H_2O$(?); *A.M.*, **37**, 609]

Vishnevite. D. S. Belyankin, 1931. *Bull. Geol. Prosp. Service USSR*, vol. 50. no. 47, p. 751 (вишневит), p. 752 (vishnevite). The original spelling of Wischnewite (12th List). Named from the locality, Vishnevy Mountains (Вишневые горы), S Ural. [*M.A.*, **4**, 499.] 13th List [Sulphatian Cancrinite]

Viterbite. R. L. Codazzi, 1925. *Notas mineralógicas y petrográficas*, Bogotá, 1925, p. 26; *Los minerales de Colombia*, Bogotá, 1927, p. 124 (viterbita). A compact chocolate-coloured or white and powdery mineral consisting of hydrated silicate and phosphate of aluminium, and explained as a mixture of allophane and wavellite. Named from the locality, Santa Rosa de Viterbo, Boyacá, Colombia. [*M.A.*, **3**, 129.] 11th List

Viterite. D. Siftar and I. Jurkovic, 1961. [*Geol. Vjesnik*, vol. 14, p. 89], abstr. in *M.A.*, **16**, 540. Croatian spelling of Witherite. 23rd List

Vitrain. M. C. Stopes, 1919, *See* Fusain. 8th List

Vittinkite. M. Saxén, 1925. *Fennia, Soc. Geogr. Fenniae*, Helsingfors, vol. 45, no. 11, p. 24 (Vittinkit). A variant of wittingite (N. Nordenskiöld, 1849), a synonym of neotocite. Named from the locality, Vittinki (Wittingi), SE Bothnia, Finland. 14th List

Vladimirite. E. I. Nefedov, 1953. *Mém. Soc. Russ. Min.*, ser. 2, vol. 82, p. 317 (Владимирит). $3CaO.As_2O_5.4H_2O$, monoclinic?, colourless radiating needles. [*M.A.*, **12**, 352.] 20th List [$Ca_5H_2(AsO_4)_4.5H_2O$; *A.M.*, **50**, 813; **56**, 639]

Vlasovite. R. P. Tikhonenkova and M. E. Kazakova, 1961. Доклады Акад. Наук СССР (*Compt. Rend. Acad. Sci. URSS*), vol. 137, p. 944 (Власовит, vlasovite). $Na_2ZrSi_4O_{11}$, colourless monoclinic crystals from the contact zone of the Lovozero massif, Kola peninsula. Named after K. A. Vlasov. 22nd List [*A.M.*, **46**, 1202; also triclinic, *M.M.*, **36**, 233; *C.M.*, **12**, 211]

Voelckerite. A. F. Rogers, 1912. *Amer. Journ. Sci.*, ser. 4, vol. xxxiii, p. 479 (Vœlckerite); *Zeits. Kryst. Min.*, 1913, vol. lii, p. 209 (Voelckerit). A basic calcium phosphate, $3Ca_3(PO_4)_2.CaO$, or oxyapatite, being a hypothetical member of the apatite group. Named after John Augustus Voelcker, agricultural chemist, whose analyses of Norwegian apatite led him, in 1883, to the formula $3Ca_3(PO_4)_2.Ca(F_2,Cl_2,O)$. 6th List

Vogtite. C. Hlawatsch, 1907. *Zeits. Kryst. Min.*, vol. xlii, p. 593 (Vogtit). A crystallized slag of unknown origin and chemical composition: the triclinic crystals possess a certain resemblance to rhodonite and were found to contain much manganese. The material is assumed to belong to the triclinic group of slags with the composition $(Mg,Fe,Mn)SiO_3$, studied by Professor Johan H. L. Vogt, of Christiania, after whom it is named. The angles of the crystals correspond with those of a triclinic metasilicate $(Fe,Ca,Mn,Mg)SiO_3$ formed in acid steel-furnace slags and described by A. F. Hallimond, *Mineralogical Magazine*, 1919, vol. xviii, p. 368. Cf. Sobralite. 8th List

Volkonkoit, error for Volkonskoit (C. Hintze, *Handb. Min.*, Erg.-Bd. II, pp. 412, 957). 24th List

Volkovite. E. I. Nefedov, 1953. *Mém. Soc. Russ. Min.*, ser. 2, vol. 82, p. 317 (Волковит). Hydrous borate of Sr and K, monoclinic, Inder salt deposits, Kazakhstan. [*M.A.*, **12**, 352.] 20th List [May be Ginorite; *A.M.*, **40**, 551]

Volkovskite. V. V. Kondrateva, I. V. Ostrovskaya, and Ya. Ya. Yarzhemski, 1966. Зап. Всесоюз. Мин. Обш. (*Mem. All-Union Min. Soc.*), vol. 95, p. 45 (Волковскит, volkovskite). A calcium borate, $CaB_6O_{10}.3H_2O$. [*A.M.*, **51**, 1550.] 24th List

Volynskite. M. S. Bezsmertnaya and L. N. Soboleva, 1965. [Экспер. метод. исслед. рудн. мин., Акад. наук СССР, p. 129]; abstr. *Amer. Min.*, 1966, vol. 51, p. 531. The unnamed telluride of Bi and Ag, reported by the authors in 1963 (Труды Инст. мин., геохим., крист. редк. элем., vol. 18, p. 70; abstr. *Amer. Min.*, 1964, vol. 49, p. 818) from gold ores of Armenia is fully described and named. Analyses lead to the composition $AgBi_{1\cdot16}Te_2$, while X-ray patterns match those of artificial orthorhombic $AgBiTe_2$. Named for I. S. Volynskii. 24th List

Von Diestite. E. Cumenge, *Bull. Soc. franç. Min.*, 1899, XXII, 25. Telluride of silver, bismuth, etc. Colorado. 2nd List [Also 3rd List. *See* Vandiestite]

Vonsenite. A. S. Eakle, 1920. *Amer. Min.*, vol. 5, p. 141. A coal-black, lustrous mineral from a granite-limestone contact at Riverside, California. It has the ludwigite formula $(3Fe,Mg)O.B_2O_3 + FeO.Fe_2O_3$ with ferrous oxide largely in excess of magnesia, but it appears to differ crystallographically from ludwigite, being orthorhombic or monoclinic. Named after Mr. M. Vonsen, of Petaluma, California, who collected the material. (Cf. Ferroludwigite of B. S. Butler and W. T. Schaller, 1917; 8th List.) [*M.A.*, **1**, 122.] 9th List [Cf. Paigeite]

Vorobyevite. V. I. Vernadsky, 1908. *Trav. Musée. Géol. Pierre-le-Grand*, St.-Pétersbourg, vol. ii, p. 81; *Bull. Acad. Sci. St.-Pétersbourg*, ser. 6, vol. ii, p. 975 (Воробьевитъ, vorobyevite); abstr. in *Neues Jahrb. Min.*, 1909, vol. ii, p. 21 (Worobieffit, Vorobyevite). A variety of beryl containing caesium (Cs_2O, 3·1%, also Li_2O, 1·39%), in composition very like the beryl from Hebron, Maine, analysed by H. L. Wells (1892). White crystals of short prismatic habit from the Urals. Named in memory of Viktor Ivanovič Vorobyev (В. И. Воробьевъ) (1875–1906), by whom the material had been examined crystallographically. The name (worobewite) is also applied by A. Lacroix (*Bull. Soc. franç. Min.*, 1910, vol. xxxiii, p. 44) to tabular crystals of rose-coloured and colourless beryl, rich in alkalis, from Madagascar. 5th List

Vrbaite. B. Ježek, 1912. *Rozpravy České Akad.*, vol. xxi, no. 26; *Bull. Intern. Acad. Sci. Prague*, vol. xvii, p. 130; *Zeits. Kryst. Min.*, vol. li, p. 364 (Vrbait). A thallium sulpho-salt, $TlAs_2SbS_5$, found as small orthorhombic crystals embedded in realgar and orpiment from Allchar, Macedonia. Named after Professor Karel Vrba, of Prague. 6th List [Contains essential Hg: $Tl_4Hg_3Sb_2As_8S_{20}$; *M.A.*, **19**, 57]

Vredenburgite. L. L. Fermor, 1908. *Rec. Geol. Survey India*, vol. xxxvii, p. 200; *Mem. Geol. Survey India*, 1909, vol. xxxvii, pp. lxvi, 42. A bronze-grey, crystalline, and cleavable mineral, which is strongly magnetic and sometimes with polarity. $3Mn_3O_4.2Fe_2O_3$. From the manganese-ore deposits of Central Provinces and Madras, India. Named after Ernest Watson Vredenburg, of the Geological Survey of India. 5th List [*See* Alpha-vredenburgite]

Vuagnatite. H. Sarp, J. Bertrand, and E. McNear, 1976. *A.M.*, **61**, 825. Orthorhombic, *a* 7·055, *b* 8·542, *c* 5·683 Å, space group $P2_12_12_1$. Composition $CaAlSiO_4OH$, isostructural with conichalcite. RIs α 1·700, β 1·725, γ 1·730, $2V_\alpha$ 48°. In rodingite near Bögürtlencik Tepe, Taurus Mts., Turkey. Named for M. Vuagnat. [*M.A.*, 77-2191.] 30th List

Vudyavrite. I. D. Borneman-Starynkevich, 1933. *Khibina Apatite*, Leningrad, 1933, p. 114 (вудъяврит); *Materials Geochem. Khibina tundra, Acad. Sci. USSR*, 1935, pp. 43, 62; P. N. Chirvinsky, ibid., p. 89 (Wudjavrit). A. E. Fersman, *The scientific study of Soviet mineral resources*, New York, 1935, p. 49 (vudiavrite). Hydrated titano-silicate of cerium earths, $Ce_4(Ti_2O_6)_3.nSiO_2.mH_2O$, as a yellowish, amorphous glassy alteration product of lovchorrite from Vudyavrchorr (Вудъяврчорр), Kola peninsula, Russia. Named from the locality. [*M.A.*, **6**, 341–343.] 14th List

Vulcanite. E. N. Cameron and I. M. Threadgold, 1961. *Amer. Min.*, vol. 46, p. 258. CuTe, with rickardite and native tellurium, as coatings on rock fragments from the Good Hope mine, Vulcan, Gunnison Co., Colorado; also synthetic. Orthorhombic. Named from the locality. 22nd List

Vuonnemite. I. V. Bussen, A. P. Denisov, N. I. Zabavnikova, L. V. Kozyreva, Yu. P. Menshikov, and E. A. Lipatova, 1973. *Zap.*, **102**, 423 (Вуоннемит, wuonnemit). A member of the lomonosovite–murmanite family from the Khibina massif; anorthic, *a* 7·02, *b* 14·15, *c* 5·38 Å, α 93° 40′, β 89° 30′, γ 87° 30′. Composition $Na_4TiNbSi_4O_{17}.2Na_3PO_4$. [*Zap.*, **103**, 362; *A.M.*, **59**, 875.] 28th List

Vysotskite. A. D. Genkin and O. E. Zvyagintsev, 1962. Зап. Всесоюз. Мин. Общ. (*Mem. All-Union Min. Soc.*), vol. 91, p. 718 (Высоцкит, Vysozkite); ibid., 1963, vol. 93, p. 567 (Высоцкит vyssotskite). (Pd,Ni)S as minute grains or prismatic tetragonal crystals, isomorphous with Braggite, from Norilsk. Named for N. K. Vysotsky (*Amer. Min.*, 1961, vol. 46, p. 464; 1963, vol. 48, p. 708; *M.A.*, **16**, 180). 23rd List

W

Wadeite. A. Wade and R. T. Prider, 1938. *Rep. Brit. Assoc. Adv. Sci.* (*Cambridge, 1938*), 1938, p. 419. R. T. Prider, *Abstr. Diss. Univ. Cambridge*, 1939, for 1937–1938, p. 93; *Min. Mag.*, 1939, vol. 25, pp. 373, 382. Silicate of potassium, zirconium, etc., approximately $K_2CaZrSi_4O_{12}$, as hexagonal plates from Western Australia. Named after Dr. Arthur Wade, who collected the material. 15th List [Near $K_2ZrSi_3O_9$. *M.A.*, **17**, 493; **18**, 199]

Wairakite. A. Steiner, 1955. *Min. Mag.*, vol. 30, p. 691. D. S. Coombs, ibid., p. 699. A zeolite, $CaO.Al_2O_3.4SiO_2.2H_2O$, the calcium analogue of analcime, but optically biaxial, monoclinic (pseudo-cubic). From hot springs at Wairaki, New Zealand. Named from the locality. *See* Calcium-analcime. [*A.M.*, **41**, 166.] 20th List

Wairauite. G. A. Challis and J. V. P. Long, 1964. *Min. Mag.*, vol. 33, p. 942. Minute grains of CoFe, occurring with Awaruite in the Red Hills serpentinites of Wairau Valley, South Island, New Zealand. Named from the locality. [*A.M.*, **50**, 521.] 23rd List

Wakabayashilite. A. Kato, K. I. Sakurai, and K. Ohsumi, 1970. *Introduction to Japanese Minerals, Geol. Surv. Japan,* 1970, 92 (provisional description), golden to lemon-yellow fibres [010] (5 mm) in quartz druses at the Nishinomaki mine, Gumma Prefecture, Japan, are monoclinic $P2_1$ or $P2_1/m$, *a* 25·17, *b* 6·48, *c* 25·24 Å, *β* 120° 0′. Composition $6[(As,Sb)_{11}S_{18}]$. Also occurs at White Caps, Nevada (<2 cm); associated with realgar and orpiment at both localities. Named for Y. Wakabayashi. [*A.M.*, **57**, 1311.] 27th List [Nevada, $Sb_2As_{20}S_{36}$; *C.M.*, **13**, 418]

Wakefieldite. D. D. Hogarth and N. Miles, 1969. *Ann. Meet. Geol. and Min. Ass. Canada*, 1969, abstr. 20. $YtVO_4$ occurs in a Precambrian pegmatite near Wakefield, Quebec; tetragonal. Named from the locality. [*M.A.*, 70-1650; *A.M.*, **55**, 1446.] 26th List

Walaite. E. Bořický, *Lotos*, 1869, XIX, 19. The same as valaite. 2nd List

Walderite. Superfluous gem-trade name for colourless artificial corundum. L. H. Benson, *Highlights at the Gem Trade Laboratory in Los Angeles, 1958/59*, **9**, no. 8, 231 and 254; R. Webster, *Gems*, 1962, 764. 27th List

Wallisite. W. Nowacki, 1965. [*Eclogae Geol. Helvet.*, vol. 58, p. 403]; abstr. *Amer. Min.*, 1966, vol. 51, p. 532. A preliminary account of an overgrowth on rathite (rathite-I) from Lengenbach, Binn, Valais, Switzerland. The material is anorthic, with composition near $(Cu,Ag)TlPbAs_2S_4$. Named for the Canton (Wallis *Germ.*, Valais *Fr.*). [Publication of a new name with preliminary, inadequate data is undesirable; no density or X-ray powder data are given.] 24th List [S_5, not S_4; *A.M.*, **54**, 1497]

Walstromite. M. C. Stinson and J. T. Alfors, 1964. *Min. Inform. Serv., Calif. Div. Mines Geol.*, vol. 17, p. 235. J. T. Alfors, M. C. Stinson, R. A. Matthews, and A. Pabst, 1965. *Amer. Min.*, 1965, vol. 50, pp. 279, 314. Colourless grains and anorthic crystals, fluorescing pink in ultraviolet light, in sanbornite-quartz rock near Big Creek and Rush Creek, Fresno County, California are formulated $BaCa_2Si_3O_9$ from emission spectrography analysis; X-ray and optical data show that the mineral is identical with synthetic $BaCa_2Si_3O_9$ (Glasser and Glasser, *Zeits. Krist.*, 1961, vol. 116, p. 263). Named for R. E. Walstrom. [*M.A.*, **17**, 399; **17**, 502.] 24th List [*A.M.*, **53**, 9]

Wardite. J. M. Davison, *Amer. Journ. Sci.*, II. 154, 1896. [*M.M.*, **11**, 226.] $P_2O_5.2Al_2O_3.4H_2O$. [Cedar Valley, Tooele Co.] Utah. 1st List [$NaAl_3(PO_4)_2(OH)_4.2H_2O$; *M.M.*, **37**, 598]

Wardsmithite. R. C. Erd, J. F. McAllister, and A. C. Vlisidis, 1970. *Amer. Min.*, **55**, 349. Hexagonal plates of $5CaO.MgO.12B_2O_3.30H_2O$ occur in the Furnace Creek area, Death Valley, California. Named for Ward C. Smith. [*M.A.*, 70-3429.] 26th List

Warrenite. A. K. Boldyrev, 1928. *Kurs opisatelnoi mineralogii*, Leningrad, 1928, part 2, p. 162 (Варренит). O. M. Shubnikova and D. V. Yuferov, *Spravochnik po novym mineralam*, Leningrad, 1934, p. 48 (Варренит, Warrenite). A pink variety of smithsonite containing cobalt (CoO 10·25%), from Boleo, Lower California, Mexico. Named after Professor Charles Hyde Warren (1876–), of Yale University, who analysed the mineral in 1898. Synonym of Cobaltsmithsonite (12th List). Not the warrenite of L. G. Eakins (1888) or of S. F. Peckham (1895; 2nd List). [*See* under Mayberyite.] 14th List

Warthaite. J. S. Krenner, 1909. *Akadèmiai Értesítö*, Budapest, vol. 20, p. 105, 'Warthait egy új magyar ásvány' (title only). J. S. Krenner, *Math. Természettud. Értesitö*, Budapest, 1926, vol. 42, pp. 4, 5; J. Loczka, ibid., pp. 10, 20 (Warthait). A sulphobismuthite of lead, $4PbS.Bi_2S_3$, as steel-grey fibrous aggregates from Hungary. Named after Professor Vince Wartha, of the József Polytechnic in Budapest. Evidently identical with goongarrite (q.v.) (E. S. Simpson, 1924; 10th List). [*M.A.*, **3**, 7.] 11th List [A mixture; *A.M.*, **49**, 1501]

Warthite. *Mem. Geol. Survey India*, 1913, vol. 43, p. 124. Synonym of blödite. Named after Dr. Henry Warth, formerly of the Geological Survey of India, who discovered the mineral in the Punjab salt mines. 10th List

Wathlingenite. [C. Prager, 1923]. *Student's index to the collection of minerals*, British Museum, 27th edit., 1936, p. 37 (Wathlingite, Wathlingenite), from dealers' labels in 1923 and 1930. Described as $CaSO_4.MgSO_4.H_2O$, from salt deposits at Wathlingen, near Celle in Hanover. Probably a mixture of anhydrite and kieserite. 18th List [Is Kieserite; *M.A.*, **15**, 540]

Wawellit, error for Wavellite (Hintze, *Handb. Min.*, Erg.-Bd. II, p. 543). 22nd List

Waylandite. O. von Knorring and M. E. Mrose, 1962. *Progr. Ann. Meeting Geol. Soc. Amer.*, p. 156A; *Amer. Min.*, 1963, vol. 48, p. 216. A white mineral replacing Bismutotantalite at Wampiro Hill, Busiro County, Buganda, Uganda, is a member of the Plumbogummite family, $(Bi,Ca)Al_3\{(P,Si)O_4\}_2(OH)_6$. Named for E. J. Wayland, sometime Director of the Uganda Geol. Survey. 23rd List

Weberite. R. Bøgvad, 1938. *Meddel. om Grønland*, vol. 119, no. 7 (Weberite). Fluoride of sodium, magnesium, and aluminium, Na_2MgAlF_7, as pale grey monoclinic grains in cryolite from Greenland. Named after Theobald Weber, who in 1857 founded the Øresund cryolite works. [*M.A.*, **7**, 225; *A.M.*, **24**, 278.] 15th List

Webnerite. A. W. Stelzner, *Zeits. Kryst. Min.*, XXIV. 125, 1894. [*M.M.*, **11**, 286.] Syn. of andorite. [Itos mine, Oruro, Bolivia; *see* Sundtite] 1st List

Weddellite. C. Frondel and E. L. Prien, 1942. *Science* (*Amer. Assoc. Adv. Sci.*), vol. 95, p. 431; E. L. Prien and C. Frondel, *Journ. Urology*, Baltimore, 1947, vol. 57, p. 965. Calcium oxalate, $CaC_2O_4.2H_2O$, described by F. A. Bannister and M. H. Hey [*M.A.*, **6**, 341] in deep-sea deposits from the Weddell Sea, Antarctic, and also present in urinary calculi. [*M.A.*, **10**, 40, 215.] 18th List

Weeksite. W. F. Osterbridge, M. H. Staatz, R. Meyrowitz, and A. M. Pommer, 1960. *Amer. Min.*, vol. 45, p. 39. $K_2(UO_2)_2(Si_2O_5)_3.4H_2O$, orthorhombic, from the Thomas range, Juab Co., Utah, and elsewhere. Resembles uranophane in appearance. Named after Dr. Alice D. Weeks. 22nd List

Wegscheiderite. J. J. Fahey, K. P. Yorks, and D. E. Appleman, 1961. *Progr. Ann. Meeting Geol. Soc. Amer.*, p. 48A; *Amer. Min.*, 1962, vol. 47, p. 415. $Na_5H_3(CO_3)_4$, in fibrous aggregates from Perkins Wells nos. 1 and 2 and Grierson Well no. 1, Sweetwater County, Wyoming; anorthic. Named for R. Wegscheider, who synthesized this compound in 1913. [*M.A.*, **15**, 358; **16**, 284.] [*A.M.*, **48**, 400.] 23rd List

Weibullite. G. Flink, 1910. *Arkiv Kemi, Min. Geol.*, vol. iii, no. 35 (*Bidrag till Sveriges mineralogi*), p. 4 (Weibullit). This mineral, from Falun, Sweden, was described by M. Weibull in 1885 as a seleniferous variety of galenobismutite, $2PbS.Bi_2S_3.Bi_2Se_3$. It differs from galenobismutite in possessing distinct cleavages. Named after Professor Mats Weibull, of Alnarp, Sweden. 6th List [Redefined as $Pb_5Bi_8Se_7S_{10-11}$; *A.M.*, **62**, 397. Cf. Wittite]

Weidgerite, *see* Wiedgerite.

Weilerite. K. Walenta and W. Wimmenauer, 1961. *Jahresheft geol. Landesamt Baden-Württemberg*, vol. 4, p. 29 (Weilerit). White earthy crusts of microscopic rhombohedra with mimetite and adamite contain Ba, Al, SO_4'', AsO_4'', and H_2O; they give a powder pattern (not quoted!) similar to that of the beudantite group, and are presumed to be $BaAl_3AsO_4SO_4(OH)_6$, the barium analogue of svanbergite. Named from the locality, the Michael vein, Weiler bei Lehr, Schwarzwald, Germany. Cf. note under Hallimondite. 22nd List [*A.M.*, **47**, 415; **52**, 1588]

Weilite. P. Herpin and R. Pierrot, 1963. *Bull. Soc. franç. Min. Crist.*, vol. 86, p. 368. $CaHAsO_4$, the arsenic analogue of Monetite, has been found on museum specimens labelled Pharmacolite and Haidingerite from various French and German localities. Named for R. Weil (*Amer. Min.*, 1964, vol. 49, p. 816; *M.A.*, **16**, 551). 23rd List

Weinbergerite. F. Berwerth, 1906. *Min. Petr. Mitt.* (*Tschermak*), vol. xxv, p. 181 (Weinbergerit). Radially-fibrous aggregates occurring with diopside and bronzite in the meteoric iron of Kodaikanal, India. The optical characters agree with orthorhombic symmetry. $(Na,K)_2O.6(Fe,Ca,Mg)O.Al_2O_3.8SiO_2$ or $R'AlSiO_4 + 3R''SiO_3$. Named after J. Weinberger, of Vienna. 4th List

Weinbergit, error for Weinbergerit (4th List) (H. Strunz, *Min. Tabellen*, 4th edit., 1966, p. 529; C. Hintze, *Handb. Min.*, Erg.-Bd. II, p. 957). 24th List

Weinschenkite. (1) H. Laubmann, 1922. *In* F. Henrich, *Ber. Deutsch. Chem. Gesell.*, vol. 55, Abt. B, p. 3013; *Geognost. Jahresh. München*, 1923, vol. 35 (for 1922), p. 201 (Weinschenkit). Hydrated phosphate of rare-earths, $(Er,Y)PO_4.2H_2O$ occurring as white spherules and radiating needles (monoclinic) on limonite at Auerbach, Bavaria. Named after Prof. Ernst Heinrich Oskar Kasimir Weinschenk (1865–1921) of Munich. [*M.A.*, **2**, 12, 522.] 10th List [Is Churchite; *M.M.*, **30**, 211]

—— (2) G. Murgoci, 1922. *Compt. Rend. Acad. Sci. Paris*, vol. 175, pp. 373, 428. A dark-brown hornblende of the magnesium-calcium group, but rich in sesquioxides and water, and poor in ferrous oxide. [*M.A.*, **2**, 221.] 10th List [Ferri-magnesio-hornblende, or magnesio-hastingsite; *Amph.* (1978)]

Weisbachite. F. Kolbeck, 1907. C. F. Plattner's *Probierkunst mit dem Lötrohre*, 7th edit., Leipzig, 1907, pp. 241, 253; 8th edit., 1927, pp. 235, 246 (Weisbachit). K. Hlawatsch, *Ann. Naturhist. Museums Wien*, 1925, vol. 38, p. 19. A variety of anglesite containing barium sulphate, $5PbSO_4.BaSO_4$; from Chile. Named after Professor Julius Albin Weisbach (1833–1901), of Freiberg, Saxony. Evidently the same as hokutolite (of K. Jimbō, 1913; 6th List). 11th List

Weissbergite. F. W. Dickson and A. S. Radtke, 1977. This mineral is mentioned as occurring in the Carlin gold deposit, Elko, Nevada, and a formula $TlSbS_2$ is given, but no further description, in Radtke *et al.*, *A.M.*, **62**, 421 (1977). 30th List

Weissite. W. P. Crawford, 1927. *Amer. Journ. Sci.*, ser. 5, vol. 13, p. 345. Copper telluride, Cu_5Te_3, as a massive bluish-black mineral from Vulcan, Colorado. Named after the late Dr. Loui Weiss, owner of the Good Hope mine where the mineral was found. [Evidently the same as rickardite; not the weissite of H. G. Trolle-Wachtmeister, 1828.] [*M.A.*, **3**, 367.] 11th List [A distinct species; *A.M.*, **34**, 357]

Weldite. F. M. Krausé, *Proc. Roy. Soc. Tasmania*, for 1884, 1885, p. lxxv. A silicate of aluminium and sodium, from the Weld river, upper Huon district, Tasmania. 3rd List. W. F. Petterd, *Catalogue of Minerals of Tasmania*, 1896. 1st List

Welichowit, German transliteration of Велиховит, Velikhovite (17th List) (H. Strunz, *Min. Tabellen*, 4th edit., 1966, p. 530; C. Hintze, *Handb. Min.*, Erg.-Bd. II, p. 416). 24th List

Welinite. P. B. Moore, 1967. *Arkiv Min. Geol.*, vol. 4, p. 407. A deep red-brown uniaxial mineral from Långban, Sweden, (Flink's no. 100) has a composition near $2[Mn^{2+},W,Mg)_{3-x}(Mn^{4+},W)_{1-x}Si(O,OH)_7]$*. Named for W. Welin. 25th List [*Corr. in *M.M.*, **39**.]

Wellsite. J. H. Pratt and H. W. Foote, *Amer. Journ. Sci.*, III. 443. 1897. $(Ba,Ca,K_2)O.Al_2O_3.3SiO_2 + 3H_2O$. Monosymmetric. Related to phillipsite. [Buck Creek mine, Clay Co.] North Carolina. 1st List [Cf. Kurtzite; review, *M.A.*, 77-3334]

Weloganite. A. P. Sabina, J. L. Jambor, and A. G. Plant, 1968. *Canad. Min.*, **9**, 468. Yellow hexagonal (but optically biaxial) crystals from a sill in the Trenton limestone at St. Michel, Montreal Island, Quebec, have the composition $Sr_5Zr_2(CO_3)_9.4H_2O$. Named for Sir William E. Logan. [*A.M.*, **54**, 576; *Bull.*, **92**, 319; *M.A.*, 70-1651; *Zap.*, **98**, 325.] 26th List [Triclinic, $(Sr,Ca)_3ZrNa_2(CO_3)_6.3H_2O$; *C.M.*, **13**, 209]

Welshite. P. B. Moore, 1970. *Min. Record*, **1**, 161. $Ca_2Mg_4SbFeO_2(Be_3Si_3O_{18})$. Rare, deep-brown prisms occurring with adelite at Långban, Sweden. Published without chemical analysis, crystallographic, or optical data in a review article; etymology not stated. 28th List

Wenkite. J. Papageorgakis, 1962. *Schweiz. Min. Petr. Mitt.*, vol. 42, p. 269. Hexagonal crystals in the marbles of Candoglia, Italy, are possibly related to Cancrinite. Formulated $(Ba,Ca)_7Al_9Si_{12}(SO_4)_2O_{42}(OH)_5$, or (M. Fleischer, *Amer. Min.*, 1963, vol. 48, p. 213) $(Ba,Ca,Na,K)_9Al_9Si_{12}O_{42}(SO_4)_2(OH)_4$. Named for Prof. E. Wenk. [*M.A.*, **15**, 360; **16**, 67.] 23rd List

Wenzelite. F. Müllbauer, 1925. *Zeits. Krist.*, vol. 61, p. 333 (Wentzelit). Hydrated phosphate of manganese, etc. $(Mn,Fe,Mg)_3(PO_4)_2.5H_2O$, as rosettes of pale rose-red, monoclinic crystals from Hagendorf, Bavaria. Named after *Pater* Hieronymus Wenzel (not Wentzel), of Münnerstadt, Bavaria, formerly of Pleystein, who discovered the phosphate minerals at Pleystein. [*M.A.*, **2**, 418.] 10th List [Is Hureaulite; *A.M.*, **40**, 370]

Wermlandite. P. B. Moore, 1971. *Lithos*, **4**, 213. Greenish-grey hexagonal plates on museum specimens from Långban, Wermland, Sweden, have *a* 9·260, *c* 22·52 Å, D 1·93; ε 1·482, ω 1·493. Idealized formula (based on a 1938 analysis that may not relate to the material here described) $2[(CaMg(OH)_4(Mg_6(Al,Fe)_2(OH)_{16})(CO_3)_{0\cdot5}OH.15H_2O]$. The mineral is related to the pyroaurite group. Named for the province. [*M.A.*, 71-3115; *A.M.*, **57**, 327.] 27th List

Weslienite. G. Flink, 1923. *Geol. För. Förh. Stockholm*, vol. 45, p. 567 (Weslienit). Antimonate of calcium, ferrous iron, and sodium, $Na_2O.FeO.3CaO.2Sb_2O_5$ or $5RO.2Sb_2O_5$, occurring as small cubic crystals with optical anomalies, at Långban, Sweden. Named after Mr. J. G. H. Weslien, manager of the Långban mines. Near atopite. [*M.A.*, **2**, 252.] 10th List [Var. of Roméite]

Westerveldite. I. S. Oen, E. A. J. Burke, C. Kieft, and A. B. Westerhof, 1972. *Amer. Min.*, **57**, 354. An orthorhombic mineral (*a* 3·46, *b* 5·97, *c* 5·33 Å) occurring as intergrowths with maucherite in chromite–nickeline ores from La Gallego, Spain, has the composition (Fe,Ni)As with Fe:Ni around 2:1, near the Ni-rich end of the synthetic solid-solution series; the name, for J. Westerveld, is applied to the series. 27th List [Greenland; *M.M.*, **41**, 77]

Westgrenite. O. von Knorring and M. E. Mrose, 1962. *Progr. Ann. Meeting Geol. Soc. Amer.*, p. 156A; *Amer. Min.*, 1963, vol. 48, p. 215. The pink alteration product of Bismutotantalite from Uganda, mentioned by L. J. Spencer (*Min. Mag.*, 1929, vol. 22, p. 191), is shown to be a distinct mineral of the Pyrochlore group, $BiTa_2O_6OH$; the powder pattern closely resembles that of $BiTa_2O_6F$, which was synthesized by Prof. A. Westgren, for whom the mineral is named. 23rd List [Renamed **Bismutomicrolite**; *A.M.*, **62**, 406]

Wetherilite. (A. Danby, *Natural Rock Asphalts*, London, 1913, p. 5.) A difficultly fusible bitumen from Canada. 7th List

Wetherillite. W. S. Ward in MS., quoted by W. E. Ford and W. M. Bradley, *Zeits. Kryst. Min.*, 1913, vol. liii, p. 219 (Wetherillit). Synonym of hetaerolite and wolftonite (6th List). Named after the late Mr. W. C. Wetherill. 7th List

Whartonite. S. H. Emmens, *Journ. Amer. Chem. Soc.*, XIV. 209, 1892. S. L. Penfield, *Amer. Journ. Sci.*, XLV. 496, 1893. Shown by Penfield to be nickeliferous pyrites. Sudbury, Canada. 1st List

Wherryite. J. J. Fahey, E. B. Daggett, and S. G. Gordon, 1950. *Amer. Min.*, vol. 35, p. 93. $PbCO_3.2PbSO_4.Pb(Cl,OH)_2.CuO$. Pale green, finely granular, optically biaxial, from Mammoth mine, Arizona. Named after Edgar Theodore Wherry (1885–), professor of plant ecology, University of Pennsylvania. [*M.A.*, **11**, 127.] 19th List [*A.M.*, **55**, 505]

White clinohumite. Labuntsov, 1930, quoted in Yu. K. Vorobev, *Zap.*, 1967, **96**, 333. *Syn.* of Pamirite (q.v.). [*M.A.*, **19**, 44.] 28th List

Whiteite. P. B. Moore and J. Ito, 1978. *M.M.*, **42**, 309. The aluminium analogue of jahnsite occurs as tan crystals, flattened on {001}, at Ilha de Taquaral, Minas Gerais, Brazil, and at Blow River, Yukon Territory, Canada. Monoclinic, *a* 14·90, *b* 6·98, *c* 10·13 Å, β 113° 7′, space group *P2/a*. Composition $2[Ca(Fe,Mn)Mg_2Al_2(PO_4)_4(OH)_2.8H_2O]$, sp. gr. 2·58. RIs α 1·580, β 1·585, γ 1·593. Named for J. S. White, Jr. 30th List

Whitlockite. C. Frondel, 1940. *Amer. Min., Program and Abstracts*, December 1940, p. 7; *Amer. Min.*, 1941, vol. 26, pp. 145, 197. Calcium phosphate, $Ca_3(PO_4)_2$, with small amounts of Mg and Fe, as colourless rhombohedra from North Groton, New Hampshire. Named after Mr. Herbert Percy Whitlock (1868–) of the American Museum of Natural History, New York. [*M.A.*, **8**, 52, 99, 232.] 16th List [Perhaps the same as Merrillite; *Min. Record*, **2**, 277; further refs. in *Chem. Index of Minerals*, App. II, p. 86]

Whitmanite. J. Murdoch, 1948. *Progr. and Abstr. Min. Soc. Amer.*, 1948, p. 11. A preliminary spectroscopic analysis suggested Mg-Mn titanate. A later chemical analysis (*Amer. Min.*, 1949, vol. 34, p. 835 [*M.A.*, **11**, 243]) showed only 0·4% MnO, identifying the supposed new mineral with geikielite. Named after the late Alfred Russell Whitman, of the University of California at Los Angeles. 19th List

Whitmoreite. P. B. Moore, A. R. Kampf, and A. J. Irving, 1974. *A.M.*, **59**, 900. Crystals up to 2 mm long in cavities of altered triphylite at the Big Chief pegmatite, Glendale, South Dakota, have space group $P2_1/c$, *a* 10·00, *b* 9·73, *c* 5·47 Å, *β* 93·8°; composition $2[Fe^{2+}Fe_2^{3+}(PO_4)_2(OH)_2.4H_2O]$. Named for R. W. Whitmore. [*M.A.*, 75-1399; *Bull.*, **98**, 266.] 29th List

Wickelkamazite. A. Brezina and E. Cohen, *Die Structur d. Meteoreisen*, Stuttgart, 1886; E. Cohen, *Meteoritenkunde*, 1894, p. 94. The kamacite surrounding accessory constituents in meteoric irons. 2nd List

Wickenburgite. S. A. Williams, 1968. *Amer. Min.*, **53**, 1433; *Canad. Min.*, **9**, 582. Colourless hexagonal crystals, $CaPb_3Al_2Si_{10}O_{24}(OH)_6$, from Wickenburg, Arizona. Named from the locality. [*M.A.*, 69-1538; *Bull.*, **92**, 319; *Zap.*, **98**, 331.] 26th List

Wickmanite. P. B. Moore, 1967. *Arkiv. Min. Geol.*, vol. 4, p. 395. Yellow octahedra from Långban, Sweden (Flink's nos. 161, 234, 374) prove to be $MnSn(OH)_6$. Named for F. E. Wickman. [*A.M.*, **53**, 1063.] 25th List [*A.M.*, **56**, 1488. Cf. Tetrawickmanite]

Wickmannite, error for Wickmanite. *Norsk. geol. Tidsskr.*, 1972, **52**, 193. 28th List

Widenmannite. K. Walenta and W. Wimmenauer, 1961. *Jahresheft geol. Landesamt Baden-Württemberg*, vol. 4, p. 22 (Widenmannit). Yellow tabular orthorhombic crystals with cerussite and galena from the Michael vein, Weiler bei Lehr, Schwarzwald, Germany; a carbonate of Pb and U. X-ray powder data (not quoted!) differ from all known uranium carbonates. Named after J. F. Widenmann (1764–1798), who first discovered uranite in the Schwarzwald. Cf. note under Hallimondite. [*A.M.*, **47**, 415.] 22nd List

Wiedgerite. (*Mineral Resources, US Geol. Survey*, for 1909, 1911, part ii, p. 733; for 1910, 1911, part ii, p. 836 (wiedgerite); *Mineral Industry*, for 1910, 1911, vol. xix, p. 56 (weidgerite).) Trade-name for a soft bitumen resembling elaterite, but containing much sulphur and water. 6th List

Wightmanite. J. Murdoch, 1962. *Amer. Min.*, vol. 47, p. 718. Colourless, roughly hexagonal prisms from Commercial quarry, Crestmore, California; anorthic, cell-contents $Mg_9B_2O_{12}.8H_2O$. Named for R. H. Wightman. [*M.A.*, **16**, 65.] 23rd List [$Mg_5(BO_3)O(OH)_5.2H_2O$; *A.M.*, **59**, 985]

Wiikite. W. Ramsay. Mentioned, but without name, by W. Ramsay and A. Zilliacus, *Öfvers. Finska Vetenskaps-Soc. Förhandl.*, 1897, vol. xxxix, p. 58. The name wiikite appears in a dealer's advertisement (*Amer. Journ. Sci.*, December 1899). An euxenite-like mineral found with monazite in a felspar-quarry at Impilaks, Lake Lagoda, Finland. Named after Professor F. J. Wiik, of Helsingfors. 3rd List [Is Euxenite or Obruchevite; *A.M.*, **47**, 812; a mixture, refs. in *A.M.*, **62**, 407]

Wilconite. Error for or variant of wilsonite, *var.* of scapolite. R. Webster, *Gems*, 1962, p. 764. 28th List

Wilkeite. A. S. Eakle and A. F. Rogers, 1914. *Amer. Journ. Sci.*, ser. 4, vol. xxxvii, p. 262. A member of the apatite group containing calcium silicate, sulphate, carbonate, and oxide in addition to phosphate, $3Ca_2(PO_4)_2.CaCO_3 + 3Ca_3(SiO_4)(SO_4).CaO$; occurring as pink, hexagonal crystals in blue crystalline limestone in Riverside Co., California. Named after Mr. R. M. Wilke, a mineral collector and dealer, of Palo Alto, California. 7th List

Wilkinite. S. D. Wells, *Paper*, New York, 1920, vol. 27, no. 14, p. 19 (wilkinite); P. F. Kerr and P. K. Hamilton, *Glossary of clay mineral names*, New York, 1949, p. 68 (wilkonite). Trade-name for a highly colloidal bentonitic clay ('jelly rock') used in paper making. 19th List [Also 20th List]

Wilkmanite. Y. Vuorelainen, A. Huhma, and A. Häkli, 1964. *Compt. Rend. Soc. Géol. Finlande*, vol. 36, p. 113. Monoclinic Ni_3Se_4, as a primary mineral and as an alteration product of sederholmite (q.v.), from Kuusamo, NE Finland. Named for W. W. Wilkman. [*M.A.*, **17**, 303; *A.M.*, **50**, 519.] 24th List

Wilkonite, variant of Wilkinite (19th List) (C. Hintze, *Handb. Min.*, Erg.-Bd. II, p. 421). 24th List

Willemseite. S. A. Hiemstra and S. A. de Waal, 1968. *Nat. Inst. Met.* (*South Africa*) *Res. Rept.*, **352**, 1; *Amer. Min.*, **55**, 31 (1970). $(Ni,Mg)_3Si_4O_{10}(OH)_2$, the nickel analogue of talc, occurs near Barberton, Transvaal; light green, monoclinic. Named for Professor Johannes Willemse. [*A.M.*, **54**, 1740; *M.A.*, 70-2606.] 26th List

Willyamite. E. F. Pittman, *Journ. and Proc. Roy. Soc. N. S. Wales*, XXVII. 366, 1893. [*M.M.*, **11**, 236.] $CoS_2.NiS_2.CoSb_2.NiSb_2$. Cubic. Broken Hill, New South Wales. 1st List [Redefined as pseudocubic (Co,Ni)SbS; *M.A.*, 73-4063]

Wiltschireit, Germanized form of Wiltshireite (6th List) (C. Hintze, *Handb. Min.*, Erg.-Bd. II, pp. 447, 957). 24th List

Wiltshireite. W. J. Lewis, 1910. *Nature*, London, vol. lxxxiv, p. 203; *Phil. Mag.*, 1910, ser. 6, vol. xx, p. 474; *Zeits. Kryst. Min.*, 1910, vol. xlviii, p. 514; *Mineralogical Magazine*, 1912, vol. xvi, p. 197. R. H. Solly, *Mineralogical Magazine*, 1911, vol. xvi, p. 121. An incompletely determined mineral occurring as small, lead-grey, monoclinic crystals in the crystalline dolomite of the Binnenthal, Switzerland. Probably identical with the rathite of H. Baumhauer (1896). Named in memory of the Rev. Dr. Thomas Wiltshire (1826–1902), formerly Professor of Geology and Mineralogy in King's College, London. 6th List

Wimsit, Germ. trans. of Вимсит, vimsite (26th List). Hintze, 104. 28th List

Winchellite. D. S. Martin, 1912. *Annals New York Acad. Sci.*, vol. xxi, p. 190. Synonym of lintonite [a var. of Thomsonite, *M.M.*, **23**, 114]. Named after Professor Newton Horace Winchell (1839–1914), who examined the material and found it to be distinct in optical characters from both mesolite and thomsonite (*Amer. Geol.*, 1898, vol. xxii, p. 349). 8th List

Winchite. L. L. Fermor, 1906. *Trans. Mining and Geol. Inst. India*, vol. i, p. 79. A blue amphibole closely allied to tremolite in chemical composition, but containing also iron, sodium, potassium, and manganese. Found by Howard J. Winch with manganese-ores in Central India. A preliminary description of the mineral was given by L. L. Fermor in 1904 (*Records Geol. Survey India*, vol. xxxi, p. 235). 4th List [Redefined, 'essentially a magnesio-arfvedsonite'; *M.M.*, **40**, 395. But *see* report of *Amph.* (1978)]

Winogradowit, German transliteration of Виноградовит (Vinogradovite, 21st List). Hintze, *Handb. Min.*, Erg.-Bd. II, p. 883. 23rd List

Wisaksonite. J. H. Druif, 1948. *Mededeel. Algem. Proefstation Landbouw.* (*Communic. General Agricultural Experiment Station*), Buitenzorg, Java, 1948, no. 69, p. 8. Clear green metamict zircon as minute crystals in river sands from Celebes. Named after Wisaksono Wirjodihardjo, of the Institute for Soil Research, Buitenzorg. [Metamict zircon has been previously referred to as α-zircon, meta-zircon, pseudo-zircon, zirconoid, low-zircon, low-density zircon. *Min. Mag.*, **14**, 48; *M.A.*, **7**, 130–131, 522–523; **8**, 123.] [*M.A.*, **10**, 455.] 18th List [Is Uranothorite; *A.M.*, **39**, 825]

Wischnewite. D. S. Belyankin, 1931. *Centr. Min.*, Abt. A, 1931, p. 196 (Wischnewit). To replace the name sulphate-cancrinite or sulphatic cancrinite (E. S. Larsen and G. Steiger, 1916; 8th List) for $3Na_2Al_2Si_2O_8.Na_2SO_4.3H_2O$, of the davyne-cancrinite group. Named from the locality, Wischnewy Gory [Vishnevy Mts.], S Ural. [*M.A.*, **4**, 499.] [= Vishnevite (*q.v.*).] 12th List

Wismutantimon. A. Volborth, 1960. *Neues Jahrb. Min.*, Abh., vol. 94, p. 140. An unnecessary name for a bismuthian antimony from Viitaniemi, Eräjäarvi, Finland. [*M.A.*, **15**, 134.] 22nd List

Wismutblüte. H. Strunz, 1957. *Min. Tabellen*, 3rd edit., p. 426. Synonym of Bismite. 23rd List

Wismut-Jamesonit, German form of Bismuthjamesonite (21st List), the original name given to Sakharovaite (22nd List). Hintze, *Handb. Min.*, Erg.-Bd. II, p. 884. 23rd List

Wismutmikrolith, German form of Bismuthmicrolite (22nd List). Hintze, *Handb. Min.*, Erg.-Bd. II, p. 885. 23rd List

Wismutparkerit, German form of Bismuth-parkerite (17th List) H. Strunz, *Min. Tabellen*, 4th edit., 1966, p. 530). 24th List

Wittite. K. Johansson, 1924. *Arkiv Kemi, Min. Geol.*, vol. 9, no. 9, p. 2 (Wittit). Sulpho- and seleno-bismuthite of lead, $5PbS.3Bi_2(S,Se)_3$, lead-grey with a good cleavage in one direction (orthorhombic or monoclinic?). Named after Mr. Th. Witt, a former engineer of the Falu mine, Fahlun, Sweden, where the mineral was found. [*M.A.*, **2**, 340.] 10th List [Orthorhombic $Pb_9Bi_{12}Se_7S_{20}$, *A.M.*, **62**, 397; distinct from Weibullite (q.v.) and the new mineral proudite (q.v.), *A.M.*, **61**, 839]

Wladimirit. E. I. Nefedov, *Geologie, Zeits. Geol.*, East Berlin, 1955, vol. 4, p. 528. German transliteration of владимирит, vladimirite (20th List). [*M.A.*, **13**, 85.] 21st List

Wodanite. W. Freudenberg, 1919. *Mitt. Badisch. Geol. Landesanst.*, vol. 8, p. 317 (Titanbiotit (Wodanit)). A biotite (meroxene) rich in titanium (TiO_2 12·5%) from the mica-nepheline porphyry of the Katzenbuckel, Odenwald, Baden. Evidently named from the Teutonic deity Wodan = Woden = Odin. 9th List

Wodginite. E. H. Nickel, J. F. Rowland, and R. C. McAdam, 1963. *Canad. Min.*, vol. 7, p. 390. Black grains from Bernic Lake, Manitoba, have the composition $(Ta,Sn,Mn)O_2$. X-ray powder patterns resemble those of Manganotantalite, but with some major differences, and the unit cell is monoclinic and contains 32 oxygen. Material described by E. S. Simpson in 1909 as Ixiolite proves to be Wodginite. Named from the locality, Wodgina, Western Australia. Cf. Stannotantalite. [*A.M.*, **48**, 1417.] 23rd List

Wodingite, error for Wodginite (23rd List). *Dokl. Acad. Sci. USSR* (*Earth Sci. Sect.*), 1966, **167**, 109 (translation of Докл. 1966, **167**, 1135). [*M.A.*, 69-1485.] 28th List

Woehlerite. Variant of Wöhlerite (of Scheerer, 1843); *A.M.*, **9**, 62. 10th List

Wöhlerite. G. P. Vdovykin, 1960. Метеоритика (*Meteoritika*), no. 18, p. 78. An unsatisfactory name for the carbon compounds in carbonaceous chondrites. Not to be confused with Wöhlerite of Scheerer, 1843 (*Amer. Min.*, 1961, vol. 46, p. 244). 23rd List

Wolfeite. C. Frondel, 1949. *Amer. Min.*, vol. 34, p. 692. $(Fe'',Mn'')_2(PO_4)(OH)$, with Fe > Mn, monoclinic; previously included in triploidite $(Mn'',Fe'')_2(PO_4)(OH)$. Named after Prof. Caleb Wroe Wolfe, of Boston University, Massachusetts, who first noticed the mineral (as triploidite) at North Groton, New Hampshire. [*M.A.*, **11**, 9.] 19th List

Wolframoixiolite*. A. I. Ginzburg, S. A. Gorzhevskaya, G. A. Sidorenko, and T. A. Ukhina, 1969. Зап. Всесоюз. мин. общ. (*Mem. All-Union Min. Soc.*), **98**, 63 (Вольфрамоиксиолит). Prismatic grains intergrown with microcline, quartz, ilmenite, and fluorite on a specimen from an unknown locality have a composition $(Nb,W,Fe,Mn,Ta, etc.)_{1\cdot97}O_4.0\cdot84H_2O$, and are regarded as a variety of ixiolite. Named from its relation to both ixiolite and wolframite. [*A.M.*, **55**, 318; *M.A.*, 70-2614; *Zap.*, **99**, 76.] [*Corr. in *M.M.*, **39**.] 26th List [Not certainly a single phase; *A.M.*, **62**, 1262]

Wolfram-Powellit, Germanized version of Tungsten-powellite (19th List) (H. Strunz, *Min. Tabellen*, 3rd edit., 1957, p. 216). 22nd List

Wolftonite. G. M. Butler, 1913. *Economic Geology*, vol. viii, p. 8. An oxide of manganese and zinc from the Wolftone mine at Leadville, Colorado. Identical with hetaerolite, $2ZnO.2Mn_2O_3.H_2O$ (W. E. Ford and W. M. Bradley, *Amer. Journ. Sci.*, 1913, ser. 4., vol. xxxv, p. 600). 6th List

Wölkérite. A. Lacroix, *Minéralogie de Madagascar*, 1922, vol. 1, p. 364. Error for Voelckerite (named after J. A. Voelcker; 6th List). 10th List

Wolkowit. E. I. Nefedov, *Geologie. Zeits. Geol.*, East Berlin, 1955, vol. 4, p. 528. German transliteration of волковит, volkovite (20th List). [*M.A.*, **13**, 85.] 21st List

β-Wollastonite. *See* Pseudowollastonite. 7th List

Wölsendorfite. J. Protas, 1957. *Compt. Rend. Acad. Sci. Paris*, vol. 244, p. 2942. (Pb,Ca)O.-$2UO_3.2H_2O$, orthorhombic, as bright-red crusts in fluorite from Wölsendorf, Bavaria. Named from the locality. [*M.A.*, **13**, 520.] 21st List [*M.M.*, **41**, 51; USSR, *M.A.*, 75-3551]

Woodfordite. J. Murdoch and R. A. Chalmers, 1958. *Bull. Geol. Soc. Amer.*, vol. 69, p. 1620 (abstr.); *Amer. Min.*, 1960, vol. 45, p. 1275. A variety of ettringite containing significant amounts of SiO_2 and CO_2, with afwillite and calcite from the Commercial quarry, Crestmore, California. Named after Prof. A. O. Woodford who found the mineral. The name was later withdrawn. 22nd List

Woodhouseite. D. M. Lemmon, 1937. *Amer. Min.*, vol. 22, p. 939. Hydrated sulphate and phosphate of calcium and aluminium, $2CaO.3Al_2O_3.P_2O_5.2SO_3.6H_2O$, as small colourless rhombohedral crystals from California. Named after Mr. C. D. Woodhouse, of Santa Barbara, California. [*M.A.*, **7**, 13.] 15th List

Woodruffite. C. Frondel, 1953. *Amer. Min.*, vol. 38, p. 761. A psilomelane-like mineral with the composition $2(Zn,Mn)O.5MnO_2.4H_2O$, from Sterling Hill, New Jersey. Named after the late Samuel Woodruff, for many years a miner with the New Jersey Zinc Company and a keen collector of minerals. [*M.A.*, **12**, 237.] 20th List [Cf. Todorokite]

Worobieffite. *See* Vorobyevite. 5th List

Wotanite. W. Kunitz, 1936. *Neues Jahrb. Min.*, Abt. A., Beil.-Bd. 70, p. 399 (Wotanite, *pl.*). Variant of wodanite (9th List). 14th List

Wretbladite. G. Gagarin and J. R. Cuomo, 1949, loc. cit., p. 5 (wretbladita). The compound AsSb in the system As–Sb. To replace the name stibarsen = allemontite II (P. E. Wretblad, 1941, 16th List) which has been given in Spanish as arsenoestibio (q.v.). Named after P. Erik Wretblad of Fagersta, Sweden. 19th List

Wroewolfeite. P. J. Dunn and R. C. Rouse, 1975. *M.M.*, **40**, 1. Blue platy crystals from the Loudville lead mine, Massachusetts, have space group *Pc* or *P2/c*, *a* 6·058, *b* 5·654, *c* 14·360, *β* 93° 28′; composition $2[Cu_4SO_4(OH)_6.2H_2O]$. Named for C. Wroe Wolfe. [*M.A.*, 75-2530; *Bull.*, **98**, 322.] 29th List

Wudjavrit. *See* Vudyavrite. 14th List

Wuonnemite, German-style transliteration of Вуоннемит, Vuonnemite (*Zap.*, **102**, 423). 28th List

Wüstite. R. Schenck and T. Dingmann, 1927. *Zeits. Anorg. Chem.*, vol. 166, p. 141 (Wüstit). Ferrous oxide containing excess of oxygen due to the presence of Fe_3O_4 in solid solution. Named after Geheimrat F. Wüst, of Düsseldorf. [*M.A.*, **5**, 252, 470.] 13th List [Cf. Iozite]

Wyartite. C. Guillemin and J. Protas, 1959. *Bull. Soc. franç. Min. Crist.*, vol. 82, p. 80. The 'ianthinite' of Bignand [*M.A.*, **12**, 587] is distinct from the type ianthinite of Schoep, and is named wyartite after Prof. J. Wyart. It is a calcium uranium carbonate, near $3CaO.UO_2.6UO_3.2CO_2.12\text{–}14H_2O$, and occurs with ianthinite on uraninite from Shinkolobwe, Katanga. [*M.A.*, **14**, 280; *A.M.*, **44**, 908.] 22nd List

Wyllieite. P. B. Moore and J. Ito, 1973. *Min. Record*, **4**, 131. Deep greenish-black crystalline masses from the Victory mine, Custer, Custer County, South Dakota, formerly identified as triphylite, are monoclinic, *a* 11·868, *b* 12·382, *c* 6·35 Å, *β* 114·52°. Composition near $4[Na_2Fe_2^{2+}Al(PO_4)_3]$, with some $(OH)_4$ replacing PO_4, and some Mn and Mg. Related to arrojadite. Named for P. J. Wyllie. [*A.M.*, **59**, 211; *M.A.*, 74-3469.] [*A.M.*, **59**, 280.] 28th List

X

Xalostocite. (*Collection de Minéralogie du Muséum d'Histoire Naturelle, Paris, Guide du Visiteur*, 2nd edit., 1900, p. 49.) The pink grossularite occurring in white marble at Xalostoc, Morelos, Mexico. *See* Landerite. 4th List

Xanthochroite. A. F. Rogers, 1917. *Journ. Geol. Chicago*, vol. xxv, p. 524. Amorphous cadmium sulphide, probably containing some adsorbed water, $CdS.xH_2O$, occurring as a thin, powdery, yellow coating on blende. It has hitherto been referred to the hexagonal species greenockite, but it is now to be regarded as the amorphous equivalent of this. Named from ξανθός, yellow, and χρῶμα, colour. It is possible, however, that, being optically isotropic, this material may be the cubic modification of cadmium sulphide, corresponding with zincblende. 8th List [Cf. Hawleyite]

Xanthotitane. P. Groth, *Tab. Uebers. Min.*, 3rd edit., 1889, p. 156 (*Xanthotitan*). The correct form of xanthitane. 2nd List

Xanthoxenite. H. Laubmann and H. Steinmetz, 1920. *Zeits. Kryst. Min.*, vol. 55, p. 580 (Xanthoxen). A basic ferric phosphate with some manganese and calcium forming small, wax-yellow, monoclinic crystals. Occurs with cacoxenite and dufrenite in pegmatite at Rabenstein, Bavaria. Named from ξανθός, yellow, and ξένος, a stranger, on account of its relation to cacoxenite. [*M.A.*, **1**, 125.] 9th List [Probably Stewartite; *see Chem. Index of Minerals*, App. II, p. 90; *M.M.*, **42**, 309, 316]

Xanthus. R. Jameson, 1800. *Min. Scot. Isles*, vol. 2, p. 68. An early name for Heliotrope. 23rd List

Xenotimite. Variant of Xenotime; *A.M.*, **8**, 52. 10th List

Xiangjiangite. *Acta Geol. Sinica*, 183 (1978). A light-yellow microcrystalline mineral from Xiangjiang (Hsiang) River, China, is pseudotetragonal $a = b$ 7·17, c 22·22 Å. Composition $[(Fe,Al)(UO_2)_4(PO_4)_2(SO_4)_2OH.22H_2O]$. RIs α 1·558, β 1·576, γ 1·593. 30th List

Xingzhongite. Yu Tsu-Hsiang, Lin Shu-Jen, Chao Pao, Fang Ching-Sung, and Huang Chi-Shun, 1974. [*Acta Geol. Sinica*, **2**, 202], abstr. *M.A.*, 75-2522. (Ir,Cu,Rh)S, cubic or pseudocubic, a 8·72 Å, from an undisclosed locality in China. [*A.M.*, **61**, 185.] 29th List

Xiphonite. G. Platania, *Atti Accad. Sci. Acireale*, V. 55, 1893. [*M.M.*, **11**, 168.] A variety of amphibole. Monosymmetric. Etna. 1st List

Xocomecatlite. S. A. Williams, 1975. *M.M.*, **40**, 221. Emerald-green spherules on altered rhyolite at the Bambollita mine, Moctezuma, Sonora, Mexico, are probably orthorhombic with a 12·140, b 14·318, c 11·662 Å and composition $12[Cu_3TeO_4(OH)_4]$. Named from the Nahua word for grapes, xocomecatl. [*M.A.*, 75-3606; *A.M.*, **61**, 504.] 29th List

Y

Yagiite. T. E. Bunch and L. H. Fuchs, 1969. *Amer. Min.*, **54**, 14. The sodium–magnesium analogue of osumilite, occurring interstitially to pyroxene in silicate inclusions of the Colomera iron meteorite. Named for Dr. Kenzo Yagi. [*M.A.*, 69-2392; *Zap.*, **99**, 82.] 26th List

Yagite, error for yagiite (26th List). *Bull.*, **95**, 163. 28th List

Yamaguchilite. S. Hata, *Sci. Papers Inst. Phys. Chem. Research*, Tokyo, 1938, vol. 34, p. 622. Another spelling of yamagutilite (14th List). [*M.A.*, **7**, 263, 264.] 15th List

Yamagutilite. K. Kimura, 1933. [*Rep. Japanese Assoc. Adv. Sci.*, vol. 8, p. 157 (Japanese).] Z. Harada, *Journ. Fac. Sci. Hokkaido Univ.*, Sapporo, ser. 4, 1936, vol. 3, p. 298 (Yamagutilith). A variety of zircon containing P_2O_5 4·23, rare-earths 15·89, HfO_2 *ca.* 3·4%, etc., from Yamaguti, Nagano, Japan. Named from the locality. 14th List

Yamatoite. T. Yoshimura, and H. Momoi, 1964. [*Sci. Rep. Fac. Sci. Kyushu Univ.*, geol. ser., no. 7, p. 85]; abstr. *Amer. Min.*, 1965, vol. 50, p. 810; *M.A.*, **17**, 183. A manganoan goldmanite, $(Ca,Mn)_3(V,Al)_2Si_3O_{12}$ with Ca:Mn ≈ 3:2 and V:Al ≈ 1·4 : 1, has been found at the Yamato mine, Kagoshima prefecture, Japan. The name Yamatoite, from the locality, is given to the hypothetical garnet end-member $Mn_3V_2Si_3O_{12}$. 24th List

Yamazuchilith, error for Yamuguchilit (15th List) (C. Hintze, *Handb. Min.*, Erg.-Bd. II, p. 957). 24th List

Yanshainshynite, erroneous transliteration of Яаншайншит, Yanshynshite (22nd List). *Amer. Min.*, 1961, vol. 46, p. 1200. 23rd List

Yanshynshite. Cheng-Chi Kuo, 1959. [*Kexue Tongbao* (*Scientia*), no. 6, p. 206]; abstr. Зап. Всесоюз. Мин. Общ. (*Mem. All-Union Min. Soc.*), 1961, vol. 90, p. 108 (Яншайншит, yanshynshite). An unnecessary name for a variety of thorogummite with 12% P_2O_5, 14% CaO, and 4% Fe_2O_3. 22nd List

Yanzhongite. Yu Tsu-Hsiang, Lin Shu-Jen, Chao Pao, Fang Ching-Sung, and Huang Chi-Shun, 1974. [*Acta Geol. Sinica,* **2**, 202], abstr. *M.A.*, 75-2522. Synonym of Kotulskite (23rd List). [*A.M.*, **61**, 185.] 29th List

Yaroslavite. M. I. Novikova, G. A. Sidorenko, and N. N. Kuznetsova, 1966. Зап. Всесоюз. Мин. Общ. (*Mem. All-Union Min. Soc.*), vol. 95, p. 39 (Ярославит, Jaroslavite). An orthorhombic fibrous mineral, in spherules from an unnamed Siberian locality, has the composition $Ca_3Al_2F_{10}(OH)_2.H_2O$. [*A.M.*, **51**, 1546, 1820.] 24th List

Yavapaiite. C. O. Hutton, 1959. *Amer. Min.*, vol. 44, p. 1105. $KFe(SO_4)_2$, monoclinic crystals on one specimen from the United Verde copper mine, Jerome, Arizona. Named after the Yavapai tribe, who inhabit the region around Jerome. [*M.A.*, **14**, 502.] 22nd List

Yeatmanite. C. Palache, L. H. Bauer, and H. Berman, 1937. *Amer. Min.*, vol. 22, no. 12, part 2, p. 11 (abstr.); ibid., 1938, vol. 23, pp. 176, 527. Silico-antimonate of manganese and zinc, $(Mn,Zn)_{16}Sb_2Si_4O_{29}$, as brown triclinic crystals from Franklin Furnace, New Jersey. Named after Mr. Pope Yeatman, mining engineer at Franklin Furnace. [*M.A.*, **7**, 14, 168.] 15th List [*A.M.*, **51**, 1494. Perhaps $(Mn^{2+}_5Sb^{5+}_2)(Mn^{2+}_2Zn_8Si_4)O_{28}$, and related to Katoptrite; *A.M.*, **62**, 396]

Yedlinite. W. J. McLean, R. A. Bideaux, and R. W. Thomssen, 1974. *A.M.*, **59**, 1157. M. M. Wood, W. J. McLean, and R. B. Laughon, ibid., 1160. Prismatic red-violet crystals, space group $R\bar{3}$, *a* 12·868, *c* 9·821, occur with diaboleïte, quartz, wulfenite, etc., in a complex paragenesis on a few specimens from the Mammoth mine, Tiger, Arizona. Composition $Pb_6Cl_6CrX_6Y_2$, where *X* and *Y* are O, OH, or H_2O; the Cr is 6-coordinated, but it could not be decided whether it is present as Cr^{3+} or as Cr^{6+}. Named for N. Yedlin. [*M.A.*, 75-2531; *Bull.*, **98**, 267.] 29th List

Yenerite. J. Steiger and O. Bayramgil, 1943. *Schweiz. Min. Petr. Mitt.*, vol. 23, p. 616

(Yenerit). O. Bayramgil, ibid., 1945, vol. 25, p. 46. Sulphantimonite of lead, $11PbS.4Sb_2S_3$, **between boulangerite and falkmanite (15th List); from Işıkdağ, Turkey. Named after** Hadi Yener, formerly director of MTA (Maden Tetkik ve Arama Enstitüsü, Turkish Mining Research Institute). [*M.A.*, **9**, 37, 262.] 17th List [Is Boulangerite; *A.M.*, **33**, 716]

Yenshanite. Jen Ying-Chen and Huang Wan-Kang, 1973. [*Geochimica*, **1**, 23], abstr. *A.M.*, **60**, 737. An unnecessary name for a nickel-free Vysotskite (23rd List) from an unnamed Chinese locality. 29th List

Yftisite. Yu. M. Shipalov, 1965. A rare-earth titanium silicate with fluoride from Kazakhstan, orthorhombic with *a* 14·90, *b* 10·60, *c* 7·08 Å, was given this name. It has been applied by V. P. Balko and V. V. Bakakin (*Zp. Strukt. Khim.*, 1975, **16**, 837) to a mineral from the Kola peninsula described, with partial analysis, by N. I. Pletneva, A. P. Denison, and N. A. Elina (*Mater. Mineral. Kol'sk Poluostr.*, 1971, **8**, 176), for which Balko and Bakakin give *a* 15·04, *b* 10·63, *c* 7·052 Å, sp. gr. 3·96, RIs α 1·690, β 1·705, γ 1·710–2. The formula $(Yt,Ln)_4(F,OH)_6TiO(SiO_4)_2$ is assigned. [But is probably incorrect, since the partial analysis (total 87·3%), taken with the cell-dimensions and density cited, lead to cell contents including 11·3(Yt,Ln,Th).3·3Ti.5·9Si.12·6F.21·1OH and 18·5 oxygen—plus, of course, the 13% of undetermined elements—*M.H.H.*] Presumably named from the composition. [*A.M.*, **62**, 396.] 30th List

Yixunite. Yu Tsu-Hsiang, Lin Shu-Jen, Chao Pao, Fang Ching-Sung, and Huang Chi-Shun, 1974. [*Acta Geol. Sinica*, **2**, 202], abstr. *M.A.*, 75-2522. Cubic PtIn, *a* 3·948, from an undisclosed locality in China. [*A.M.*, **61**, 185.] 29th List

Yoderite. D. McKie, 1959. *Min. Mag.*, vol. 32, p. 282. Near $Mg_2Al_6Si_4O_{18}(OH)_2$; a monoclinic, highly pleochroic purple mineral occurring as a quartz–yoderite–kyanite–talc schist at Mautia Hill, Kongwa, Tanganyika. Named after Dr. Hatten Schuyler Yoder of the US Geological Survey. 22nd List [*M.A.*, **16**, 249; **18**, 116. Near $(Al,Mg,Fe)_6Al_2(Si,Al)_4O_{18}$-$(OH)_2$] **[*for* U.S. Geol. Survey, *read* Geophysical Laboratory, Washington, D.C.]**

Yofortierite. G. Perrault, Y. Harvey, and R. Pertowsky, 1975. *C.M.*, **13**, 68. The manganese analogue of palygorskite occurs as pink to violet fibres in pegmatite veins at Mt. St. Hilaire, Quebec. Composition $Mn_5Si_8O_{20}(OH)_2(H_2O)_4$.4 to $5H_2O$. Named for Y. O. Fortier. [*A.M.*, **61**, 341; *Bull.*, **98**, 322.] 29th List

Yokosukaite. K. Kami and T. Tanaka, 1937. [*Bull. Electrotechn. Lab. Japan*, vol. 1, pp. 459, 553; 1938, vol. 2, pp. 19, 21; *Electrochem., Japan*, 1938, vol. 6, p. 366; 1939, vol. 7, p. 7], quoted by Y. Hariya, *Amer. Min.*, 1963, vol. 48, p. 952. A manganese oxide from Yokosuka, Aichi prefecture, Japan, gave a distinctive X-ray pattern, recognized by M. Nambu and K. Okada ([*Journ. Soc. Earth Sci. Amat. Japan*, 1961, vol. 12, p. 249; *Journ. Jap. Ass. Min. Petr. Econ. Geol.*, 1962, vol. 48, p. 76], quoted by Y. Hariya, loc. cit.), and by Y. Hariya (loc. cit.) as that of artificial γ-MnO_2, subsequently named Nsutite (q.v.) by W. K. Zwicker *et al.* The name Yokosukaite has priority. 23rd List

Yokosukalite, variant of Yokosukaite (23rd List). *Intro. Jap. Min.*, 120 (1970), Geol. Surv. Japan. 28th List

Yoshimuraite. T. Watanabe, 1959. *Min. Journ.* (*Japan*), vol. 2, p. 408; T. Watanabe, Y. Takéuchi, and J. Ito, ibid., 1961, vol. 3, p. 156. Orange-brown tabular crystals or stellate groups in an alkali-pegmatite from the Noda-Tamagawa mine, Iwate prefecture and the Taguchi mine, Aichi prefecture, Japan; anorthic, near $(Ba,Sr)_2Mn_2(Ti,Fe)\{(P,S)O_4\}$-$Si_2O_8OH$. Named after Prof. Toyofumi Yoshimura. 22nd List [*A.M.*, **45**, 479; **46**, 1515]

Ytro-Columbo-Tantalite, Ytrocolumbite, *see* Yttro-columbo-tantalite.

Yttererdensilikatapatit. F. Machatschki, 1953. *Spez. Min.*, Wien, p. 330. Syn. of Abukumalite. 22nd List

Yttrianite. J. Takubo *et al.*, 1953. *Journ. Geol. Soc. Japan*, vol. 59, p. 58. Error for yttrialite. [*M.A.*, **12**, 279.] 20th List

Yttrium-apatite. G. Flink, *Medd. om Grönland*, 1900, XXIV, 173. Apatite containing Y_2O_3 (3·36%), cerium, etc. [Narsarsuk] S Greenland. 2nd List

Yttrium-Bastnäsit, undesirable variant of Bastnaesite-(Yt) (27th List). Hintze, 105. 28th List

Yttrium-Granat, German form of Yttrium garnet (Bergemann, 1854, Yttergranat) (H. Strunz, *Min. Tabellen*, 4th edit., 1966, p. 329). 24th List

Yttrium-orthite. J. W. Frondel and M. Fleischer, 1950. *US Geol. Surv. Circular 74*, p. 10 (yttrium orthite), p. 20 (Yttriumorthite). Synonym of yttro-orthite (13th List). 19th List

Yttroalumite. H. S. Yoder and M. L. Keith, 1951. *Amer. Min.*, vol. 36, p. 522. The high-temperature tetragonal modification of $Yt_3Al_2(AlO_4)_3$. *See* Yttrogarnet. [*M.A.*, **11**, 364.] 19th List

Yttroapatite. K. Omori and H. Konno, 1962. *Amer. Min.*, vol. 47, p. 1191. A name for the hypothetical Apatite end-member $Y_{10/3}(PO_4)_3(F,OH)$. 23rd List

Yttrobetafite. A. P. Kalita, A. V. Bykova, and M. V. Kukharchik, 1962. Труды Инст. Мин. Геохим. Крист. Редк. Элем. (*Trans. Inst. Min., Geochem., Cryst. Rare Elem.*), no. 8, p. 210 (Иттробетафит); 1959, no. 2, p. 164. A superfluous name for intermediate members of the Pyrochlore-Obruchevite series from Alakurti, NW Karelia (Зап. Всесоюз. Мин. Общ. (*Mem. All-Union Min. Soc.*), 1963, vol. 92, p. 570; *Amer. Min.*, 1963, vol. 49, p. 440; *M.A.*, **16**, 558). 23rd List

Yttrobritholite. I. I. Kupriyanova and G. A. Sidorenko, 1963. Докл. Акад. Наук СССР (*Compt. Rend. Acad. Sci. URSS*), vol. 148, p. 212 (Иттробритолит) Yttrian britholite. 'A superfluous name' (Зап. Всесоюз. Мин. Общ. (*Mem. All-Union Min. Soc.*), 1964, vol. 93, p. 456). 23rd List

Yttrocalcite. E. S. Fedorov, 1905. *Gornyi Zhurnal*, St. Petersburg, year lxxxi, vol. iii, p. 264 (иттрокальцитъ). Described as a fluoride of calcium, yttrium, and cerium, but found on further examination to be identical with fluor-apatite (*Ann. Rep. Progr. Chem. Soc. London*, 1907, for 1906, vol. iii, p. 311). 4th List

Yttro-columbo-tantalite. C. Lepierre, 1937. *Mem. Acad. Ciên. Lisboa, Cl. Ciên.*, vol. 1, p. 374 (Ytro-Columbo-Tantalite), p. 369 (Ytrocolumbite). A variety of yttrotantalite with niobium in excess of tantalum, from Mozambique. The name yttrocolumbite appears in the index to Dana's 'System' (1892) but not in the text. [*M.A.*, **7**, 470.] 15th List

Yttrocrasite. W. E. Hidden and C. H. Warren, 1906. *Amer. Journ. Sci.*, ser. 4, vol. xxii, p. 515; *Zeits. Kryst. Min.*, 1907, vol. xliii, p. 18. Hydrous titanate of yttrium-earths, thorium, uranium, etc., found in Burnet Co., Texas, as black, orthorhombic crystals with a pitchy lustre and uneven fracture. Apparently so named because of its resemblance in appearance to polycrase (from *κρᾶσις*, a mixing), though it contains no more yttrium than does polycrase. 4th List

Yttroepidote. B. G. Lutta and D. A. Mineev, 1967. [Редкие элементы в породах различных метаморфических фаций, p. 93]; abstr. *Zap.*, **97**, 620 (Иттроэпидот). An unnecessary name for an epidote with 1·36% Ln_2O_3. 26th List

Yttrofluorite. T. Vogt, 1911. *Centralblatt Min.*, 1911, p. 373 (Yttrofluorit). A variety of fluorite containing yttrium earths (17%); formula $(Ca_3,Y_2)F_6$. Occurs in pegmatite in N Norway. 6th List

Yttrogarnet. H. S. Yoder and M. L. Keith, 1951. *Amer. Min.*, vol. 36, p. 522. Artificial $3Yt_2O_3.5Al_2O_3 = Yt_3Al_2(AlO_4)_3$ forming a continuous series of cubic mixed crystals with spessartine $(Mn_3Al_2(SiO_4)_3)$. It is presumably present in natural yttrium-bearing spessartine (yttergarnet, ytter-garnet, yttriogarnet, emildine, erinadine). At about 1970°C. it is transformed to yttroalumite (q.v.). [*M.A.*, **11**, 364.] 19th List

Yttrogranat, German form of Yttrogarnet (19th List) (H. Strunz, *Min. Tabellen*, 4th edit., 1966, p. 531; C. Hintze, *Handb. Min.*, Erg.-Bd. II, p. 435). 24th List

Yttromelanocerite. I. I. Kupriyanova and G. A. Sidorenko, 1963. Докл. Акад. Наук СССР (*Compt. Rend. Acad. Sci. URSS*), vol. 148, p. 212 (Иттромеланоцерит). Yttrian melanocerite. 'A superfluous name' (Зап. Всесоюз. Мин. Общ. (*Mem. All-Union Min. Soc.*), 1964, vol. 93, p. 456). 23rd List

Yttroniobite. H. Strunz, *Min. Tab.*, 1941, p. 107 (Yttroniobit). Syn. of yttrocolumbite. 18th List

Yttro-orthite. A. E. Fersman, 1931. *Pegmatites*, Leningrad, vol. 1, p. 310 (иттро-ортит). A variety of orthite containing 8% Yt_2O_3, from Kareliya. 13th List

Yttroparisite. E. I. Nefedov, 1941. *Compt. Rend. (Doklady) Acad. Sci. URSS*, 1941, vol. 32, p. 363. A variety of parisite containing much yttria, from Adun-Chalon, Transbaikalia. [*M.A.*, **8**, 279.] 16th List

Yttropyrochlore. I.M.A. Sub. Comm. pyrochlore group, 1978. *A.M.*, **62**, 407. A systematic name to replace obruchevite (q.v.). 30th List

Yttrosynchysite. Synonym of Doverite (21st List). [Геохим., мин., генет. типы месторожд. редк. элем, Изд. „Наука", 1964, p. 272]; abstr. Зап. Всесоюз. Мин. Общ. (*Mem. All-Union Min. Soc.*), 1965, vol. 94, p. 683 (Иттросинхизит). 24th List

Yttrotungstite. E. H. Beard, 1950. *Colonial Geol. and Min. Resources*, London, vol. 1, p. 51. To replace the name 'thorotungstite' [11th List] for a Malay mineral which contains rare-earths (Yt, Ce) and no Th. [*M.A.*, **11**, 189.] 19th List [Cf. Cerotungstite]

Yugawaralite. K. Sakurai and A. Hayashi, 1952. [*Sci. Rep. Yokohama Nat. Univ.*, sect. 2, no. 1, p. 60]; abstr. in *Amer. Min.*, 1953, vol. 38, p. 426. A monoclinic zeolite, '$Ca_4Al_7Si_{20}O_{54}.14H_2O$', in altered andesite tuffs near the Yugawara hot spring, Kanagawa, Japan. Named from the locality. [*M.A.*, **12**, 133.] 20th List [$CaAl_2Si_6O_{16}.4H_2O$; *A.M.*, **54**, 306]

Yukonite. J. B. Tyrrell and R. P. D. Graham, 1913. *Trans. Roy. Soc. Canada*, ser. 3, vol. vii, sect. 4, p. 13. A brownish-black, resinous, amorphous mineral, resembling pitticite in appearance, consisting of hydrous arsenate of calcium and ferric iron, $(Ca_3,Fe_2''')As_2O_8.Fe_2'''(OH)_6.5H_2O$. Named after the locality, Yukon, Canada. 7th List

Yuksporite. A. E. Fersman, 1922. *Compt. Rend. Acad. Sci. Russie*, 1922, p. 60 (юкспорит); *Trans. Northern Sci. Econ. Exped.*, no. 16 (*Sci. Techn. Dept. Supreme Council of National Economy*, no. 8), Moscow and Petrograd, 1923, pp. 16, 52, 68, 73; [E. E. Kostyleva, *Acad. Sci. Russie* . . .]. Hydrous silicate of calcium, potassium, sodium, and ferric iron, as rose-red, fibrous and lamellar masses from Yukspor in the Kola Peninsula, Russian Lapland. It belongs to the pectolite group. Named from the locality. [*M.A.*, **2**, 262, 264.] 10th List [New analysis, perhaps a valid species; *A.M.*, **62**, 1262]

Z

Zaherite. A. P. Ruotsala and L. L. Babcock, 1977. *A.M.*, **62**, 1125. Veinlets in kaolinite–boehmite rock in the Salt Range, Pakistan, consist of aggregates of small grains. ρ 2·007 g.cm^{-3}, n 1·4981, almost isotropic. Composition $Al_{12}(SO_4)_5(OH)_{26}.20H_2O$. Dehydrates reversibly at room temperature with change in the intense X-ray diffraction at d 18·1 to 15·5 Å. Named for M. A. Zaher. [*M.A.*, 78-3483.] 30th List

Zaïreite. L. van Wambeke, 1975. *Bull.*, **98**, 351. Small greenish masses in the weathering zone of quartz veins at Eta-Eta, Kivu, Zaïre, are rhombohedral, a_h 7·015, c_h 16·365, space group $R\bar{3}m$. Uniaxial negative, $n \approx$ 1·82–1·83; ρ 4·37 g. cm^{-3}. Ideal composition $[Bi(Fe,Al)_3(PO_4)_2(OH)_6]$; some replacement of Bi by Ba, Ca, and Cu, and of P by Si, S, Te, and H_4. Crandallite family. Named for the country of origin. 29th List

Zamboninite. M. Bauer, 1901. *Neues Jahrb. Min.*, 1901, vol. i, ref. p. 200 (Zamboninit). Synonym of the müllerite of F. Zambonini (1899; 2nd List). The name müllerite (also müllerine and millerite) being preoccupied, the alternative name zamboninite is proposed in an abstract of F. Zambonini's paper. 6th List [Is Nontronite; *M.A.*, **3**, 452]

—— F. Stella Starrabba, 1930. *Boll. Soc. Geol. Italiana*, vol. 48 (for 1929), p. 259. Fluoride of magnesium and calcium. $CaF_2.2MgF_2$, as white radially fibrous masses from Etna, Sicily. Named after Professor Ferruccio Zambonini, of Napoli. The alternative name zambonina (= zambonine) is mentioned, but not adopted. Not the zamboninite of M. Bauer (1901, 6th List). [*M.A.*, **4**, 249.] 12th List [A mixture; *M.A.*, **5**, 392]

Zapatalite. S. A. Williams, 1972. *Min. Mag.*, **38**, 541. Pale blue crusts with libethenite, chenevixite, beaverite, etc., in a small prospect near Agua Prieta, Sonora, Mexico, are tetragonal. a 15·22, c 11·52 Å, D 3·02; composition $6[Cu_3Al_4(PO_4)_3(OH)_9.4H_2O]$. ε 1·635, ω 1·646. Named for E. Zapata. [*M.A.*, 72-1406.] 27th List

Zavaritskite. E. I. Dolomanova, V. M. Senderova, and M. T. Yanchenko, 1962. Докл. Акад. Наук СССР (*Compt. Rend. Acad. Sci. URSS*), vol. 146, p. 680 (Заварицкит, zawaryzkite). BiOF, in pseudomorphs after Bismuthinite, from Sherlova Gory, E. Transbaikal. Named for A. N. Zavaritsky (*Amer. Min.*, 1963, vol. 48, p. 210; *M.A.*, **16**, 181). 23rd List

Zawaryzkite, erroneous transliteration of Заварицкит, Zavaritskite (q.v.). 23rd List

Zeathite. A trade-name for *artificial* $SrTiO_3$. R. Webster, priv. comm. *See* under fabulite. 28th List

Zebedassite. A. Brusoni, 1917. *Rend. R. Ist. Lombardo*, ser. 2, vol. 1, p. 646; *Riv. Min. Crist. Italiana*, 1918, vol. 1, p. 74. Hydrated silicate of magnesium and aluminium, $H_8Al_2Mg_5(SiO_4)_6$, occurring as silky white, fibrous (orthorhombic?) aggregates in fissures of an altered serpentinous rock. Named from the locality, Zebedassi, near Volpedo in Piedmont. 8th List [Is Saponite; *M.A.*, **12**, 54]

Zeiringit. *See* Zeyringite. 4th List

Zektzerite. P. J. Dunn, R. C. Rouse, B. Cannon, and J. A. Nelen, 1977. *A.M.*, **62**, 416. Colourless to pink crystals in a riebeckite-granite from the Golden Horn batholith near Washington Pass, Okanogan Co., Washington, are orthorhombic, pseudohexagonal, a 14·306, b 17·330, c 10·140 Å. Composition $8[LiNaZrSi_6O_{15}]$ with some Ti and Hf, ρ 2·79 g. cm^{-3}. RIs α 1·582 ∥ [100], β 1·584 ∥ [010], γ 1·584 ∥ [001], $2V_\alpha$ near 0°. [*M.A.*, 78-898.] 30th List

Zellerite. H. G. Granger, 1963. [*New Mexico Bur. Mines Min. Resources Mem. 15*, p. 21], abstr. in *Amer. Min.*, 1964, vol. 49, p. 439. An inadequately described mineral, said to be a carbonate, from Gas Hills, Wyoming, and Alta mine, Grants District, New Mexico. 23rd List [$Ca(UO_2)(CO_3)_2.5H_2O$; *A.M.*, **51**, 1567]

Zemannite. J. A. Mandarino, E. Matzat, and S. A. Williams, 1969. *Canad. Min.*, **10**, 139. Brown hexagonal crystals, a 9·41, c 7·64 Å, $P6_3/m$, from Moctezuma, Sonora, Mexico.

Uniaxial, ω 1·85, ε 1·93. $2[Na_xH_{2-x}Zn_2(TeO_3)_3.yH_2O]$. Named for J. Zemann. [*A.M.*, **55**, 1448; *Zap.*, **100**, 88.] 27th List

Zeolite mimetica. G. D'Achiardi, 1906. *Atti Soc. Toscana Sci. Nat., Mem.*, vol. xxii, p. 160. A zeolitic mineral from Elba was determined to have the form of octagonal prisms built up of sectors with different optical orientation, and to have the composition $(Ca,K_2,Na_2)_3Al_4(Si_2O_5)_9.14H_2O$. Had the author been in no doubt as to this being really a new mineral he would not have hesitated to call it 'dachiardite' (p. 164) in memory of his father, Antonio D'Achiardi (1839–1902); for the present, however, he prefers to call it 'zeolite mimetica'. 4th List [Later named Dachiardite (q.v.)]

Zeophyllite. A. Pelikan, *Anzeiger Akad. Wiss. Wien, Math.-naturw. Cl.*, 1902, Jahrg. xxxix, p. 113; Sitz.-ber., 1902, vol. cxi, Abth. I, p. 336 (Zeophyllit). A zeolite containing fluorine; $H_4Ca_4F_2Si_3O_{11}$. It occurs at Gross-Priesen, Bohemia, as hemispherical and spherical aggregates of small plates with a perfect cleavage and pearly lustre. Named from *ξεῖν*, to boil, and *ωύλλον*, a leaf. 3rd List [Cf. Radiophyllite]

Zeraltite, error for Ceraltite, *Doklady Acad. Sci. USSR*, 1961, Earth Sci. Sect., vol. 134, p. 1025 (translation of Докл. Акад. Наук СССР (*Compt. Rend. Acad. Sci. URSS*)). 23rd List

Zeyringite. – Pantz, 1811. *Taschenb. Min.* (*Leonhard*), vol. v, p. 373 (Zeyringit). A finely fibrous, greenish-white or sky-blue calcareous sinter containing nickel, from Zeyring, Styria. It is placed under aragonite by V. von Zepharovich (*Min. Lexicon Österreich*, 1859, vol. i, p. 28) and E. Hatle (*Die Minerale des Herzogthums Steiermark*, 1885, p. 68); the latter spells Zeiringit. 4th List [Colour due to aurichalcite; *Fortschr. Min.*, 1962, **40**, 60]

Zhemchuzhnikovite. Yu. A. Zhemchuzhnikov and A. I. Ginzburg, 1960. [Основы петрол. углей, Изд. Акад. наук СССР (*Problems of coal petrology*), p. 93], abstr. in Зап. Всесоюз. Мин. Общ. (*Mem. All-Union Min. Soc.*), 1962, vol. 91, p. 206 (Жемчужниковит, Jemchuznikovite) and in *Amer. Min.*, 1962, vol. 47, p. 1482; [Труды Всесоюз. Науч.-Иссл. Геол. Инст. (*Trans. All-Union Geol. Inst.*), 1963, vol. 96, p. 131], abstr. in *Amer. Min.*, 1964, vol. 49, p. 442 and with corrections in Зап. Всесоюз. Мин. Общ., 1964, vol. 93, p. 457. $NaMg(Fe,Al)(C_2O_4)_3.8H_2O$, in green trigonal crystals (violet in artificial light) from veinlets in coal in the Chaitumusuk deposits, 200 km from the estuary of the Lena river, Siberia. The unit cell is given as *a* 16·70, *c* 12·50, while Stepanovite ($NaMgFe(C_2O_4)_3.8$ to $9H_2O$; 20th List) is given as *a* 9·84 ($= 17{\cdot}04/\sqrt{3}$), *c* 36·77; confirmation of this difference is desirable. Named for Yu. A. Zhemchuzhnikov (1885–1957). [*M.A.*, **16**, 551, 555.] 23rd List

Ziegelite. (*Comm. Dir. Trab. Geol. Portugal*, 1898, III, 203.) The same as Ziegelerz = tile-ore (variety of cuprite). 2nd List

Zillerite. (A. Fersmann, *Rozpr. České Akad.*, 1912, vol. xxi, no. 15, p. 2; *Bull. Internat. Acad. Sci. Prague*, 1912, vol. xvii, p. 118 (Zillerit).) A variant of zillerthite (J. C. Delamétherie, 1795). The name is here applied to the matted-fibrous asbestos (mountain-cork) belonging to the species tremolite-actinolite. 7th List

Zinalsite. F. V. Chukhrov, 1956. [Кора быветривания, vol. 2, p. 107]; abstr. Зап. Всесоюз. Мин. Общ (*Mem. All-Union Min. Soc.*), 1958, vol. 87, p. 487 (Цинальсит, zinalsite). A zinciferous clay, much higher in Zn than sauconite (q.v.) and apparently not of the montmorillonite family. Named from the composition, Zn, Al, Si. 'Parait être identique à la moresnetite' (*Bull. Soc. franç. Min. Crist.*, 1958, vol. 81, p. 337). 22nd List [Cf. Fraipontite; *A.M.*, **44**, 208. Also, Zincsilite]

Zincalunite. K. Omori and P. F. Kerr, 1963. *Bull. Geol. Soc. Amer.*, no. 74, p. 709. R. C. Erd, R. J. P. Lyon, and B. C. Madsden, ibid., 1965, no. 76, p. 271; K. Omori and P. F. Kerr, ibid., 1965, vol. 74, p. 283. A mineral referred to as zincalunite from Laurion, Greece, gave an infra-red adsorption spectrum resembling that of alunite, while X-ray fluorescence data indicate the presence of zinc. The X-ray powder data, however, do not resemble those of the alunite group. The name can be accorded no standing without proper chemical analysis. [*A.M.*, **50**, 810.] 24th List

—— M.-A. Kashkai, 1969. Зап. Всесоюз. мин. общ. (*Mem. All-Union Min. Soc.*), **98**, 153 (Цинкалунит). A name for the end-member $ZnAl_6(SO_4)_4(OH)_{12}$. 26th List

Zinc-aragonite. A. K. Boldyrev, *Kurs opisatelnoi mineralogii*, Leningrad, 1928, part 2, p. 202 (цинкарагонит). H. Strunz, *Min. Tab.*, 1941, p. 118 (Zinkaragonit). Syn. of nicholsonite (6th List). 18th List

Zincbotryogen, error for Zincobotryogen (q.v.), *Zentr. Min.*, 1964, Teil 1, p. 159. 24th List

Zinc chkalovite. *Soviet Physics—Cryst.*, 1970, **14**, 918. *Syn.* of Zn-chkalovite. 28th List

Zinc-chromium spinel. C. Milton, D. [E.] Appleman, E. C. T. Chao, F. Cuttitta, J. L. Dinnin, E. J. Dwornik, M. Hall, B. L. Ingram, and H. J. Rose, Jr., 1967. *Progr. Abstr. Ann. Meet. Geol. Soc. Amer.*, 151. Minute violet-coloured crystals of a spinel high in zinc and chromium occur with grimaldite and mcconnellite in the 'merumite' [18th List] from Guyana. [*M.A.*, **19**, 127.] 28th List

Zinc chrysotile. W. Noll, H. Kircher, and W. Sybertz, 1960. *Beitr. Min. Petr.*, vol. 7, p. 240 (Zinkchrysotil). The hypothetical zinc analogue of chrysotile. 22nd List

Zinc-copper-chalcanthite. E. S. Larsen and M. L. Glenn, 1920. *Amer. Journ. Sci.*, ser. 4, vol. 50, p. 228 (Zinc-copper chalcanthite, etc.). Abstr. in *Amer. Min.*, 1922, vol. 7, p. 74 (zinc-copper-chalcanthite). Members of the triclinic chalcanthite group, $RSO_4.5H_2O$, in which copper is replaced by other metals. Zinc-copper-chalcanthite results by the partial dehydration (on exposure to dry air) of zinc-copper-melanterite (q.v.); iron-copper-chalcanthite from pisanite; cobalt-chalcanthite from bieberite; and manganese-chalcanthite was prepared artificially. [*M.A.*, **1**, 121.] 9th List

Zinc-copper-melanterite; **Zinc-melanterite.** E. S. Larsen and M. L. Glenn, 1920. *Amer. Journ. Sci.*, ser. 4, vol. 50, p. 225 (Zinc-copper melanterite, etc.). Abstr. in *Amer. Min.*, 1922, vol. 7, p. 74 (zinc-copper-melanterite, etc.). Members of the monoclinic melanterite group, $RSO_4.7H_2O$, in which iron is replaced by other metals, are named zinc-, zinc-copper-, copper-zinc-, and cobalt-melanterites, according to the relative preponderance of the replacing metals. Zinc-copper-melanterite is described as a secondary mineral from Colorado; the remainder are hypothetical. [*M.A.*, **1**, 121.] 9th List

Zinc-dibraunite. K. Nenadkevič, 1911. *Trav. Mus. Géol. Pierre le Grand, Acad. Sci. St.-Pétersbourg*, vol. v, p. 37 (Цинкдибраунитъ); abstr. in *Zeits. Kryst. Min.*, 1914, vol. liii, p. 609 (Zinkdibraunit). A soft, earthy, chocolate-coloured mineral occurring in cavities in calamine ores at Olkush, Russia. Composition $ZnO.2MnO_2.2H_2O$; this being a salt of a meta-manganese acid is called a braunite, whilst salts of the ortho-acid are called manganites. There is here some confusion between chemical and mineralogical names. 7th List [A specimen from Upper Silesia, not type material, contains Hydrohetaerolite; *A.M.*, **60**, 739]

Zinc-epsomite. M. Yu. Fishkin and V. S. Melnikov, 1965. Мин. Сборн. Львов. Геол. Общ. (*Min. Mag. Lvov Geol. Soc.*), vol. 19, p. 495 (Zn-эпсомит). Synonym of goslarite. 24th List

Zinc-fauserite. L. Tokody, 1949. *Földtani Közlöny*, Budapest, vol. 79, p. 68 (Cinkfauserit, Hungarian), p. 78 (Zinkfauserint, misprint for Zinkfauserit, German). (Mn,Zn,Mg)-$SO_4.5H_2O$, orthorhombic. A variety of fauserite containing ZnO 5·08%, from Felsöbánya. [*M.A.*, **11**, 10.] 19th List

Zinc-ferro-hexahydrite. J. Kubisz, 1958. *Bull. Acad. Polon. Sci., Sér. sci. chim., géol., géogr.*, vol. 6, p. 459. Syn. of Bianchite (12th List). 22nd List

Zinc-ferro-magnesio-hexahydrite. J. Kubisz, 1958. *Bull. Acad. Polon. Sci., Sér. sci. chim., géol., géogr.*, vol. 6, p. 459. An unnecessary name for magnesian bianchite. 22nd List

Zinc-hexahydrite. J. Kubisz, 1958. *Bull. Acad. Polon. Sci., Sér. sci. chim., géol., géogr.*, vol. 6, p. 459. A name for monoclinic $ZnSO_4.6H_2O$, not known in nature. 22nd List

Zinc-högbomite. V. A. Moleva and V. S. Myasnikov, 1952. *Doklady Acad. Sci. USSR*, vol. 83, p. 733 (цинк-хёгбомит). Variety of högbomite containing ZnO 11·12%. [*M.A.*, **12**, 13.] 20th List

Zinc-lavendulan. H. Strunz, 1959. *Fortschr. Min.*, vol. 37, p. 89 (Zinklavendulan). An unnecessary name for zincian lavendulan from Tsumeb, $(Ca,Na)_2(Cu,Zn)_5(AsO_4)_4Cl.4–5H_2O$. 22nd List

Zinc-magnesia-chalcanthite. C. Milton and W. D. Johnston, 1937. *Amer. Min.*, vol. 22, no. 12, part 2, p. 10; 1938, vol. 23, p. 175. 15th List

Zinc-manganese-cummingtonite. L. H. Bauer and H. Berman, 1930. *Amer. Min.*, vol. 15, p. 341. An amphibole containing ZnO 10·46, MnO 13·79%, as green prisms from Franklin Furnace, New Jersey. [*M.A.*, **4**, 345.] 12th List

Zinc-manganocalcite. G. Gagarin, 1936. *Ann. Géol. Péninsule Balkan.*, vol. 13, p. 74 (Zink-manganokalcit), p. 78 (Zinc-manganocalcite). A variety of calcite containing MnO 3·71% and ZnO 0·78%, from Serbia. [*M.A.*, **7**, 256.] 15th List

Zinc-melanterite, *see* Zinc-copper-melanterite.

Zinc mica. B. W. Evans and R. G. J. Strens, 1966. *Nature*, **211**, 619. *Syn.* of Hendricksite (25th List). [*M.A.*, **18**, 39.] 28th List

Zincnontronite. M. R. Enikeev, 1970. [Труды Ташкент. унив. 1970, **371**, 56]; abstr. *Zap.*, **103**, 365 (Цинкнонтронит). An unnecessary name for a zincian nontronite. 28th List

Zincobotryogen. Kuang Tu-Chih, Hsi-Lin Li, Hsien-Te Hsieh, and Shu-Shen Yin, 1964. [*Ti Chih Hsueh Pao*, vol. 44, p. 99]; abstr. *Amer. Min.*, 1964, vol. 49, p. 1776; *M.A.*, **17**, 504. The unnamed zinc analogue of Botryogen (Zeman, *Fortschr. Min.*, 1961, vol. 39, p. 84) has been found on the N border of the Tsadam basin, China, and is named from its composition. 24th List

Zincocopiapite. Kuang Tu-Chih, Hsi-Lin Li, Hsien-Te Hsieh, and Shu-Shen Yin, 1964. [*Ti Chih Hsueh Pao*, vol. 44, p. 99]; abstr. *Amer. Min.*, 1964, vol. 49, p. 1777; *M.A.*, **17**, 504. The zinc member of the Copiapite family, previously known from several localities (Scharizer, *Zeits. Kryst. Min.*, 1913, vol. 52, p. 372) has been found on the N border of the Tsadam basin, China, and is named from its composition. 24th List

Zincoferrite. J. Beckenkamp, 1921. *See* Talc-spinel. 9th List

Zincorhodochrosite. E. Manasse, 1911. *Mem. Soc. Toscana Sci. Nat.*, vol. xxvii, p. 79 (zincorodocroisite). A variety of rhodochrosite containing much zinc (ZnO 31·03%). From [Rosseto] Elba. 6th List

Zincosite. Variant of Zinkosite; *A.M.*, **9**, 62. 10th List

Zinc-pisanite. H. Strunz, 1957. *Min. Tabellen*, 3rd edit., p. 428 (Zinkpisanit). An unnecessary name for zincian pisanite. 22nd List

Zinc-rockbridgeite. M. L. Lindberg and C. Frondel, 1950. *Amer. Min.*, vol. 35, p. 1028 (Zincian rockbridgeite). H. Strunz, *Neues Jahrb. Min.*, Abh., 1952, vol. 84, p. 89 (Zinkrockbridgeit). A zinciferous variety of rockbridgeite (ZnO 5·20%), from Portugal and Bavaria. [*M.A.*, **11**, 187.] 20th List

Zinc-römerite. R. Scharizer, *Zeits. Kryst. Min.*, 1903, vol. xxxvii, p. 546 (Zinkrömerit). Römerite in which ferrous iron is replaced by zinc. In the natural mineral, from the Harz, this replacement is only partial, but it is more complete in an artificial product. 3rd List

Zincrosasite. H. Strunz, 1959. *Fortschr. Min.*, vol. 37, p. 89 (Zinkrosasit). A variety of rosasite having Zn > Cu; from Tsumeb, SW Africa. [*A.M.*, **44**, 1323.] 22nd List [But Zincorosasite in Fleischer's *Glossary* (1975)]

Zinc-saponite. C. E. Marshall, 1949. *The colloidal chemistry of silicate minerals*, p. 62 (zinc saponite). Syn. of sauconite [*M.A.*, **10**, 27]. 19th List

Zinc-schefferite. J. E. Wolff, *Zeits. Kryst. Min.*, 1900, XXXIII, 148. Schefferite containing some zinc. Franklin Furnace, New Jersey. 2nd List

Zincselenide. H. Rose, 1927, *Fortschr. Min. Krist. Petr.*, vol. 12, p. 73 (Zinkselenid). W. Geilmann and H. Rose, *Neues Jahrb. Min. Geol.*, Abt. A, 1928, pp. 797, 803, 807. Dana's *System of mineralogy*, 7th edit., 1944, p. 215 (Zincselenide). An incompletely determined mineral, supposed to be selenide of zinc, occurring in selenium ores from Andreasberg, Harz. [*M.A.*, **4**, 76.] 17th List

Zincsilite. N. N. Smolyaninova, V. A. Moleva, and N. I. Organova, 1960. [*Rept. Meeting Internat. Comm. Study Clays*, p. 45]; abstr. *Amer. Min.*, 1961, vol. 46, p. 241. The aluminium-free end-member of the montmorillonite–sauconite series, near $Zn_3Si_4O_{10}(OH)_2.nH_2O$; from Batystau, Kazakhstan. Named from *zinc sili*cate. 22nd List [Cf. Zinalsite]

Zinc-stottite. B. H. Geier and J. Ottemann, 1970. *Neues Jahrb. Min. Abh.*, **114**, 89. Secondary minerals at Tsumeb include a stottite with Fe 13·4, Zn 12·6%; since Fe is still >Zn, this is a zincian stottite and a new name is unnecessary. [*A.M.*, **56**, 1488; *M.A.*, 71-1391 ('Mineral A').] 27th List [Also 28th List]

Zinc-teallite. F. Ahlfeld, 1926. *Centr. Min.*, Abt. A, 1926, p. 388 (Zinkteallit). Teallite with some zinc replacing lead, $(Pb,Zn)SnS_2$, from Bolivia. Previously described under the name pufahlite (F. Ahlfeld, 1925; 10th List). [*M.A.*, **3**, 272.] 11th List

Zinkbotryogen, German form of Zincobotryogen (q.v.) *Zentr. Min.*, 1964, Teil 1, p. 159. 24th List

Zink-Chrom-Spinell, undesirable variant of zinc-chromium spinel. Hintze, 106. 28th List

Zinkchrysotil, *see* Zinc chrysotile.

Zinkcopiapit, German form of Zincocopiapite (q.v.) *Zentr. Min.*, 1964, Teil 1, p. 159. 24th List

Zinkdibraunit, *see* Zinc-dibraunite.

Zinkfauserit, *see* Zinc-fauserite.

Zinkferrit. J. T. Ebelmen, 1851. *Journ. prakt. Chem.*, vol. 54, p. 143. In a translation of Ebelmen's paper in *Ann. Chim. Phys.*, 1851, ser. 3, vol. 33, p. 34, the term ferrite de zinc for artificial $ZnFe_2O_4$ is translated Zinkferrit; it is a chemical, not a mineral name, and its essential identity with Franklinite was recognized by Ebelmen (cf. C. Hintze, *Handb. Min.*, Erg.-Bd. II, p. 646). 24th List

Zink-Hogbomit, German transliteration of Цинк-хёгбомит, Zinc-högbomite (20th List) (H. Strunz, *Min. Tabellen*, 4th edit., 1966, p. 532). 24th List

Zinklavendulan, *see* Zinc-lavendulan.

Zinkmanganerz. A. Brunlechner, *Jahrb. naturh. Landes-Museums, Klagenfurt*, XXII. 194, 1893. Hydrated manganate of zinc. [Bleiberg] Carinthia. 1st List

Zink-manganokalcit, *see* Zinc-manganocalcite.

Zinkmontmorillonit. F. Machatschki, 1953. *Spez. Min.*, Wien, p. 349. Syn. of Sauconite (2nd List). 22nd List

Zinkpisanit, *see* Zinc-pisanite.

Zinkrockbridgeit, *see* Zinc-rockbridgeite.

Zinkrömerit, *see* Zinc-römerite.

Zinkrosasit, *see* Zincrosasite.

Zinksaponit, Germanized form of Zinc-saponite (19th List) (C. Hintze, *Handb. Min.*, Erg.-Bd. II, p. 439). 24th List

Zinkselenid, *see* Zincselenide.

Zinkstottit, German variant of zinc-stottite (q.v.). Hintze, 107. 28th List

Zinkteallit, *see* Zinc-teallite.

Zinkvredenburgit. H. Strunz, 1949. *Min. Tabellen*, 2nd edit., p. 106. An unnecessary name for the zincian vredenburgite of B. Mason (1946). 22nd List

Zinkwolframit. F. Machatschki, 1953. *Spez. Min.*, Wien, p. 315. Syn. of Sanmartinite (18th List). 22nd List

Zinnfahlerz. G. H. Moh and J. Ottemann, 1962. *Neues Jahrb. Min.*, Abh., vol. 99, p. 1. Synonym of Fredricite. 23rd List

Zinntitanit. P. Ramdohr, 1935. *Neues Jahrb. Min.*, Abt. A, Beil.-Bd. 70, p. 15 (Zinntitanit). A variety of sphene containing 10% of tin, from Arandis, SW Africa. [*M.A.*, **6**, 368.] 14th List

Zircolite. A misleading trade-name for a white *artificial* sapphire. R. Webster, *Gems*, 1962, p. 765. Contains no zirconium, and not to be confused with zirkelite (1st List), zirconite, nor zirconolite (21st List). Trivial trade-names that have been used for artificial corundum include adamite (8th List); aloxite (7th List); alundum (5th List); [corindite (9th List)] damburite, diamantin(e), diamondite, diamontine, syntholite, ultralite, violite (28th List); walderite (27th List); and 'Crown Jewels' (M. J. O'Donoghue, priv. comm.). 28th List

Zirconoid. E. E. Kostyleva, 1936. *Trans. Lomonosov Inst., Min. Ser., Acad. Sci. USSR*, no. 7, p. 223 (цирконоиды), p. 224 (zirconoids). Zircons of low specific gravity and optically isotropic, consisting of a mixture of ZrO_2 and SiO_2 in a metamict state. The name zirconoid is already in use for the tetragonal form (*hkl*) [e.g. Dana's *Textbook Min.*]. [*M.A.*, **7**, 130.] 15th List

Zirconolite. L. S. Borodin, I. I. Nazarenko, and T. L. Richter, 1956. *Doklady Acad. Sci. USSR*, vol. 110, p. 845 (цирконолит). A metamict mixed oxide $CaZrTi_2O_7$, containing also Nb, Fe, U, Ce, etc. In nepheline-pyroxenite [from Kola]. Similar to zirkelite [*Min. Mag.*, **11**, 86; **16**, 309] from Brazil and Ceylon. [*M.A.*, **13**, 383.] 21st List [Is Zirkelite (q.v.), *A.M.*, **60**, 341; **62**, 408]

Zircon-pyroxenes. W. C. Brøgger, 1890. *Zeits. Kryst. Min.*, vol. xvi, p. 363 (Zirkonpyroxene). C. Hintze, *Handbuch d. Mineralogie*, 1894, vol. ii, pp. 1140, 1176 (Zirkon-Pyroxene, *plur.*). A group-name for the zircono-silicates rosenbuschite, låvenite, wöhlerite, hiortdahlite, etc. Following Brøgger, these minerals are usually classified with the pyroxenes, although they do not exhibit prismatic cleavages. 6th List

Zircopal. V. I. Gerasimovsky, 1956. *Geochemistry* (translation of Геохимия), no. 5, p. 494. Unnecessary name for a zirconian Opal. 23rd List

Zircophyllite. Yu. L. Kapustin, 1972. *Zap.*, **101**, 459 (Циркофиллит). $(K,Na)_3(Fe,Mn)_7$-$(Zr,Ti)_2Si_8O_{24}(O,OH,F)_7$, with Zr : Ti ≈ 4 : 1. The Zr analogue of astrophyllite, from the Korgeredabinsh massif. Tuva. [*M.A.*, 73-2951; *Bull.*, **96**, 244; *A.M.*, **58**, 967; *Zap.*, **102**, 454.] 28th List

Zircosulfate, variant of Zircosulphate (q.v.). 24th List

Zircosulphate. Yu. L. Kapustin, 1965. Зап. Всесоюз. Мин. Общ. (*Mem. All-Union Min. Soc.*), vol. 94, p. 530 (Циркосулбфат, Zircosulphate). Compact powdery masses in a cavity in nepheline-syenite pegmatite of the Korgeredabin massif, SE Tuva agree with synthetic orthorhombic $Zr(SO_4)_2.4H_2O$ in X-ray pattern and composition. Named from the composition. [*A.M.*, **51**, 529, Zircosulfate; *M.A.*, **17**, 607.] 24th List

Zirfesite. E. E. Kostyleva, 1945. *Compt. Rend.* (*Doklady*) *Acad. Sci. URSS*, vol. 48, p. 502. Hydrous silicate of zirconium, iron, etc., $(ZrO_2,Fe_2O_3)SiO_2.nH_2O$, as a powdery clay-like mineral from the alteration of eudialyte in the Kola peninsula. Named from the composition Zr,Fe,Si. [*M.A.*, **9**, 261.] 17th List [Cf. Zirsite]

Zirkelite. E. Hussak and G. T. Prior, *Min. Mag.*, XI. 86, 1895; XI. 180, 1897. (Ca,Fe)O. $2(Zr,Ti,Th)O_2$. Cubic. [Jacupiranga, São Paulo] Brazil. 1st List [*A.M.*, **60**, 341. $(Ca,Th,Ce)Zr(Ti,Nb)_2O_7$. Cf. Zirconolite]

Zirkite. *Monthly Price List*, Foote Mineral Company, Philadelphia, September 1916, p. 26; November 1916, p. 29; *Mineral Foote-Notes*, March 1917, p. 2. Trade-name for a zirconia-ore from Brazil containing 73–97% ZrO_2. It is stated to be a mechanical mixture of baddeleyite, zircon, and a new zirconium silicate (*see* Orvillite). [It includes the 'favas' of zirconia described by E. Hussak in 1899, and the large mamillated lumps with radially-fibrous structure described by E. Hussak and J. Reitinger in 1903; the latter was regarded as possibly a modification of zirconia distinct from baddeleyite.] *See* Caldasite. 8th List

Zirklerite. E. Harbort, 1928. *Kali*, Halle, vol. 22, p. 157 (Zirklerit). Hydrous chloride of ferrous iron, magnesium, and calcium with aluminium hydroxide, $9\{(Fe,Mg,Ca)Cl_2.2H_2O\} + 2\{Al_2O_3.H_2O\}$, rhombohedral; from the N German potash-salt deposits. Named after Bergrat Dr. – Zirkler, Director of the Aschersleben potash works. [*M.A.*, **4**, 14.] 12th List

Zirkoneuxenite. F. Machatschki, 1953. *Spezielle Mineralogie*, Wien, p. 319 (Zirkoneuxenit = Polymigmit, q.v.). Synonym of polymignite. 20th List

Zirkonoid, German transliteration of Цирконоид, Zirconoid (15th List) (C. Hintze, *Handb. Min.*, Erg.-Bd. II, p. 440). 24th List

Zirkonolith, German transliteration of Цирконолит, Zirconolite (21st List) (H. Strunz, *Min. Tabellen*, 4th edit., 1966, p. 175). 24th List

Zirkonpyroxen, *see* Zirconpyroxene.

Zirkophyllitt, German-style transliteration of Циркофиллит, Zircophyllite (*Zap.*, **102**, 454). 28th List

Zirsinalite. Yu. L. Kapustin, Z. V. Pudovkina, and A. V. Bykova, 1974. *Zap.*, **103**, 551 (Цирсиналит). Colourless irregular deposits on Mt. Koashva, Khibina massif, Kola Peninsula, have space group $R3m$, $R\bar{3}m$, or $R32$, a_h 10·29, c_h 13·11 Å, a_{rh} 7·38, α 88° 28′. Composition $Na_6(Ca,Mn,Fe)ZrSi_6O_{18}$. Named for the composition. [*A.M.*, **60**, 489; *M.A.*, 75-2532; *Zap.*, **104**, 612.] 29th List

Zirsite. I. D. Dorfman, 1962, quoted by E. M. Bonshtedt-Kupletskaya, Зап. Всесоюз. Мин. Общ. (*Mem. All-Union Min. Soc.*), 1963, vol. 92, pp. 214, 217. An unnecessary name for a low-iron variety of Zirfesite (17th List) from the Yukspor Mts., Khibina massif, Kola peninsula. Named from the composition (*Amer. Min.*, 1963, vol. 48, p. 1182; *M.A.*, **16**, 557). 23rd List

Zittavite. F. Glöckner, 1912. *Zeits. Deutsch. Geol. Ges.*, vol. lxiii, Monatsber., p. 418 (Zittavit). A black, lustrous variety of lignite resembling dopplerite, but more brittle and harder than this. It occurs in the lignite deposits of Zittau, Saxony. 6th List

Zn–blödite. M. Giglio, 1958. *Acta Cryst.*, vol. 11, p. 789. Artificial $Na_2Zn(SO_4)_2.4H_2O$, the zinc analogue of Blödite. [*M.A.*, **14**, 103.] 23rd List

Zn,Cd-chkalovite. M. A. Simonov and N. V. Belov, 1965. Докл., **164**, 406. (Zn,Cd-чкаловит). *Artificial* $Na_4ZnCd(Si_2O_6)_2$ an analogue of Chkalovite. (15th List). In N. V. Belov. *Min. Sborn. Lvov*, 1966, **20**, 143, this compound is misnamed Zn-chkalovite (Zn-чкаловит). [*M.A.*, **19**, 177.] 28th List

Zn-chkalovite. E. L. Belokoneva, Yu. K. Egorov-Tismenko, M. A. Simonov, and N. V. Belov, 1970. Кристаллография, **14**, 1060 (Zn-чкаловит); *Soviet Physics—Cryst.*, 1970, **14**, 918, (Zn-chkalovite, Zinc chkalovite). *Artificial* $Na_2ZnSi_2O_6$, analogous to Chkalovite (15th List). Also, in error, for $Na_4ZnCd(Si_2O_6)_2$ (N. V. Belov, *Min. Sborn. Lvov*, 1966, **20**, 143). [*M.A.*, 70-3011.] 28th List

Zn-Dolomit. H. Strunz, 1966. *Min. Tabellen*, 4th edit., p. 213. An unnecessary name for a zincian Dolomite. 24th List

Zn-Fahlerz. B. M. Honnorez-Guerstein, 1971. *Min. Depos.*, **6**, 111. A zincian tetrahedrite is named Zn-Fahlerz in the summaries only. [*A.M.*, **57**, 326.] 28th List

Zodite. I. G. Magakyan, 1956. [Доклады Акад. Наук Арм. ССР (*Compt. Rend. Acad. Sci. Armen. SSR*), vol. 23, p. 215]; abstr. Зап. Всесоюз. Мин. Общ. (*Mem. All-Union Min. Soc.*), 1958, vol. 87, p. 76 (Зодит). A provisional name for the mineral later named Stibiotellurobismutite (21st List). Named from the locality. 22nd List

Zoesite. S. Meunier, 1911. *Compt. Rend. Acad. Sci. Paris*, vol. clii, p. 1879 (zoésite). A variety of fibrous silica (sp. gr. 2·59) isolated by dissolving in acid the fossil shells from the Chalk formation. It differs in its optical orientation from the other forms of fibrous silica (viz. chalcedonite, quartzine, and lutecite). Named from *ζωή*, life, in allusion to the influence of the living organism in producing the fibrous structure. Not to be confused with zoisite. 6th List

Zonite. Trade or lapidary trivial name for a chert or jasper, from Arizona. R. Webster, *Gems*, 1962, p. 765. Not related to zonolite (11th List). 28th List

Zonolite. E. N. Alley, *Engin. Mining Journal-Press*, New York, 1925, vol. 120, p. 819. Trade-name for a light flaky material obtained by roasting vermiculite (altered phlogopite), from near Libby, Montana. It is used as a packing material for heat insulation and in the manufacture of wall-paper. [*M.A.*, **3**, 78.] 11th List

Zonotlite. E. S. Larsen and H. Berman, 1934. *Bull. US Geol. Surv.*, no. 848 (2nd edit.), pp. 108, 266. Error for xonotlite, from Tetela de Xonotla (= Tetela del Oro), Puebla, Mexico. 20th List

Zorite. A. N. Merkov, I. V. Bussen, E. A. Goiko, E. A. Kulitskaya, Yu. P. Menshikov, and A. P. Nedorezova, 1973. *Zap.*, **102**, 63 (Зорит). Rosy crystals from the same pegmatite as raite (q.v.) and ilmaiokite (27th List) are orthorhombic, *a* 23·9, *b* 7·23, *c* 14·25 Å, with composition $Na_2Ti(Si,Al)_3O_9.nH_2O$. Named from зори, зорка, the rosy tint of dawn. [*M.A.*, 73-4081; *Bull.*, **96**, 244; *A.M.*, **58**, 1113; *Zap.*, **103**, 364.] 28th List

Zussmanite. S. O. Agrell, M. G. Bown, and D. McKie, 1964. *Amer. Min.*, 1965, vol. 50, p. 278 (abstr.). Pale green tabular rhombohedral crystals in metamorphic rock of the Franciscan formation, Laytonville, Mendocino County, California, are near $K(Fe^{2+},Mg,Mn,Al)_{13}(Si,Al)_{18}O_{42}(OH)_{14}$. Named for Dr. J. Zussman. 24th List

Zvyagintsevite. A. D. Genkin, I. V. Muraveva, and N. V. Troneva, 1966. [Геол. рудн. месторожд. (Geol. ore-deposits), no. 8, p. 94]; abstr. in *Amer. Min.*, 1967, vol. 52, p. 299. 'Mineral 5' of Borovskii, Deev, and Marchukova (abstr. *A.M.*, **46**, 464) is shown to be near $(Pd,Pt)_{5.5}PbSn$ or $Pd_3(Pb,Sn)$ and is named for O. E. Zvyagintsev. *See also Canad. Min.*, 1966, vol. 8, p. 541 [Зап. **97**–64; Звягинцевит.] 25th List [*A.M.*, **52**, 299, 1587]

Zvyagintsivite, error for Zvyagintsevite (25th List). *Canad. Min.*, **8**, 541 (1966). 26th List

AUTHOR INDEX

Abbatista, F., *v.* Brisi, C.
Abel, O.
  Chiropterite (1922)
Abeledo, M. Jiminez de, *v.* de Abeledo, M. J.
Abraham, K., *v.* Schreyer, W.
Abramovich, Yu. M.
  Ordite (1956)
Acheson, E. G.
  Carborundum (1893)
Achiardi *v.* D'Achiardi
Adam, [G. J.]
  Vanadine (1869)
Adam, J. W. H.
  Pulleite (1909)
Adams, F. D., & Harrington, B. J.
  Hastingsite (1896)
Adams, J. W., Botinelly, T., Sharp, W. N., & Robinson, K.
  Murataite (1974)
Adamson, O. J.
  Eckermannite (1942)
Adib, D., & Ottemann, J.
  Anarakite (1972)
  Chrominium (1970)
  Khuniite (1970)
  Plumangite (1970)
Adusumilli, M. S., *v.* Burke, E. A. J.
Agrell, S. O.
  Dellaite (1965)
  Ferrowollastonite (1950)
  Rustumite (1965)
——, Bown, M. G., & McKie, D.
  Deerite (1965)
  Howieite (1964)
  Zussmanite (1964)
——, & Gay, P.
  Kilchoanite (1961)
——, & Langley, J. M.
  Iron-corundum (1958)
  Iron mullite (1958)
Agrinier, H., Chantret, F., Geffroy, J., Hery, B., Bachet, B., & Vachey, H.
  Meta-lodevite (1972)
Ahlfeld, F.
  Jujuyite (1948)
  Pufahlite (1925)
  Ramdohrite (1930)
  Zinc-teallite (1926)
——, & Angelelli, V.
  Jujuite (1948)
——, & Muñoz Reyes, J.
  Orlandinite (1938)
——, *v.* Ramdohr, P.
Aho, A. E., *v.* Meagher, E. P.
Ainsworth, W. F.
  Katharite (1843)
Akaogi, M., *v.* Aoki, K.
Akhmanova, M. V., *v.* Pavlenko, A. S.
Akhvlediani, R. A., *v.* Ivanitskii, R. V.
Aleksandrov, I. V., Ivanov, V. I., & Sinkova, L. A.
  Fluorbastnäsite (1965)
Aleksandrov, S. M., *v.* Brovkin, A. A.
——, *v.* Pertsev, N. N.
Aleksandrov, V. B., *v.* Kornetova, V. A.
Alekseeva, E. F., & Godlevsky, M. N.
  Ferrigarnierite (1937)
Alekseeva, M. A., Chernikov, A. A., Shashkin, D. P., Komkov, E. A., & Gavrilova, I. N.
  Strelkinite (1974)
Alexander, J. B., & Flinter, B. H.
  Malayaite (1965)
Alexander, L. T., Faust, G. T., Hendricks, S. B., Insley, H., & McMurdie, H. F.
  Endellite (1943)
Alfors, J. T., & Stinson, M. C.
  Fresnoite (1965)
  Muirite (1965)
  Traskite (1965)
  Verplanckite (1965)
——, *v.* Mrose, M. E.
——, *v.* Pabst, A.
——, *v.* Stinson, M. C.
Al-Hermezi, H. M., *v.* Livingstone, A.
Aliev, R. M., *v.* Kashkai, M. A.
Alkhazov, V. Yu., *v.* Dusmatov, V. D.
——, *v.* Efimov, A. F.
Allen, A. W.
  Nequenite (1932)
Allen, E. T., *v.* Wright, F.E.
Allen, R. D., *v.* Muessig, S.
Allen, V. T.
  Ionite (1927)
  Potash-montmorillonite (1932)
——, & Fahey, J. J.
  Mansfieldite (1945)
Alley, E. N.
  Zonolite (1925)
Alling, H. L.
  Analbite (1936)
  Eutecto-oranite (1921)
  Eutectoperthite (1921)
  Hyperoranite (1921)
  Hyperperthite (1921)
  Hypo-oranite (1921)
  Hypoperthite (1921)

Alling, H. L.—*cont.*
Oranite (1921)
Para-oranite (1921)
Paraperthite (1921)
Potash-albite (1921)
Almond, H., *v.* Erd, R. C.
Alvir, A. D.
Antamokite (1928)
Amelandov, A. S., & Ozerov, K. N.
Buldymite (1934)
Aminoff, G.
Arsenoklasite (1931)
Bäckströmite (1919)
Bromellite (1925)
Finnemanite (1923)
Magnetoplumbite (1925)
Manganese-pennine (1931)
Pseudopyrochroite (1918)
Sahlinite (1934)
Swedenborgite (1924)
Titano-thucholite (1943)
——, & Mauzelius, R.
Armangite (1920)
Amor, I. Asensio, *v.* Asensio Amor, I.
Amstutz, A.
Oroseite (1925)
Traversite (1925)
Amstutz, G. C., Ramdohr, P., & De Las Casas, F.
Revoredite (1957)
Andersen, C. A., Keil, K., & Mason, B.
Sinoite (1964)
Anderson, A. T., Bunch, T. E., Cameron, E. E., Haggerty, S. E., Boyd, F. R., James, O. B., Keil, K., Prinz, M., Ramdohr, P., & Goresy, A. E.]
Armalcolite (1970)
——, *v.* Chao, E. C. T.
Anderson, B. W.
Pyrandine (1947)
——, Claringbull, G. F., Davis, R. J., & Hill, D. K.
Ekanite (1961)
——, & Payne, C. J.
Gahnospinel (1937)
Magnesium-zinc-spinel (1937)
Pseudo-zircon (1937)
——, ——, & Claringbull, G. F.
Taaffeite (1951)
Andreatta, C.
Bianchite (1930)
Daunialite (1943)
Illite-hydromica (1949)
Andreeva, O. V., *v.* Filonenko, N. E.
Angelelli, V.
Capillitite (1948)
——, & Gordon, S. G.
Sanmartinite (1948)
Sarmientite (1941)
——, & Trelles, R. A.
Calingastite (1938)
——, *v.* Ahlfeld, F.
——, *v.* de Abeledo, M. E. J.
Anikeeva, L. I., *v.* Epshtein, E. M.
Ankinovich, E. A.
Alvanite (1959)
Bokite (1963)
Chromsteigerite (1963)
Gutsevichite (1959)
Kurumsakite (1954)
Rusakovite (1960)
Satpaevite (1959)
Vanalite (1962)
——, Gekht, I. I., & Zaiteseva, R. I.
Carbonate-cyanotrichite (1963)
——, *v.* Ankinovich, G.
Ankinovich, G., Ankinovich, E. A., Rozhdestvenskaya, I. V., & Frank-Kamenetskii, V. A.
Chernykhite (1972)
Anosov, F. Y., *v.* Chukhrov, F. V.
Ansheles, O. M., & Vlodavets, N. I.
Tikhvinite (1927)
Anspach, H.
Manganleonite (1939)
Anten, J.
Gosseletite (1923)
Lohestite (1923)
Anthony, J. W., & Laughon, R. B.
Kinoite (1970)
——, & McLean, W. J.
Jurbanite (1976)
——, *v.* Williams, S. A.
Antipov, I. A.
Ferganite (1908)
Turkestan-volborthite (1908)
——, *v.* Glinka, S. F.
Antipov-Karataiev, I. N., & Sedletzky, I. D.
Gedroitzite (1937)
Anzai, T., *v.* Sudo, T.
Aoki, K., Fujino, K., & Akaogi, M.
Titanochondrodite (1976)
——, & Matsumoto, H.
Oxykaersutite (1959)
Aoki, Y., *v.* Yoshi, M.
Aomine, S., *v.* Yoshinaga, N.
Aoyama, S.
Ruthenosmiridium (1936)
Apjohn, J.
Mesolitine (1844)
Appleman, D. E., *v.* Dunn, P. J.
——, *v.* Fahey, J. J.
——, *v.* French, B. M.
——, *v.* Milton, C.
——, *v.* Wones, D. R.
Araki, T., *v.* Moore, P. B.
Arapova, G. A., *v.* Skvortsova, K. V.
Arem, J. E., *v.* White, J. S.
Arest-Yakubovich, P. E., *v.* Chukhrov, F. V.
Aristarain, L. F., & Hurlbut, C. S.
Ameghinite (1967)
Teruggite (1968)
——, *v.* Hurlbut, C. S.
Arkangel'skaya, V. N., *v.* Khvostova, V. A.

Arppe, A. E.
Liparite (1858)
Arrhenius, G.
Monocerotite (1975)
Artini, E.
Bavenite (1901)
Bazzite (1915)
Brugnatellite (1909)
Arzruni, A.
Rafaelite (1899)
Stelznerite (1899)
——, *v.* Cossa, A.
Asensio Amor, I.
Cardosonite (1955)
Atanassov, V. A., & Kirov, G. N.
Balkanite (1973)
Atkin, D., *v.* Gallagher, M. J.
——, *v.* Livingstone, A.
Augustithis, S. S., *v.* Ottemann, J.
Autenreith, H., & Braune, G.
D'Ansite (1958)
Avias, J., *v.* Caillère, S.
Avrova, N. P., Bocharov, V. M., Khalturina, I. I., & Yunosova, Z. R.
Aldzhanite (1968)
Chelkarite (1968)
——, *v.* Bocharov, V. M.
——, *v.* Lobanova, V. V.
Aweng, E., *v.* Tschirch, A.
Axelrod, J., Grimaldi, F. S., Milton, C., & Murata, K. J.
Andersonite (1948)
Bayleyite (1948)
Swartzite (1948)
——, *v.* Fahey, J. J.
——, *v.* Heyl, A. V.
——, *v.* Milton, C.
Aymé, V.
Carbite (1901)

Babcock, L. L., *v.* Ruotsala, A. P.
Babkine, J., Conquéré, F., & Vilminot, J.-C.
Ferripléonaste (1968)
Baccaredda, M., *v.* Natta, G.
Bachet, B., *v.* Agrinier, H.
Backenkamp, J.
Talc-spinel (1921)
Backet, B., *v.* Cesbron, F.
Backström, H.
Manganandalusite (1896)
Natron-berzeliite (1896)
——, *v.* Brøgger, W. C.
Backström, J. W. von. *See* under von Backström
Badalov, S. T.
Vanadium-garnet (1951)
Vanadium-tourmaline (1951)
——, & Golovanov, I. M.
Birunite (1957)
Bahezre, C., *v.* Nowacki, W.
Bailey, E. H., Hildebrand, F. A., Christ, C. L., & Fahey, J. J.
Schuetteïte (1959)
——, *v.* Berry, L. C.
——, *v.* Woodford, A. O.
Bailey, S. W., & Tyler, S. A.
Al-lizardite (1960
Baklanova, K. A., *v.* Bulakh, A. G.
——, *v.* Kukharenko, A. A.
Balenzano, F., Dell'Anna, L., & Dipiero, M.
Francoanellite (1976)
Balitskii, V. S., *v.* Mozgova, N. N.
Ballakh, V. V., *v.* Rudashevskii, N. S.
Balogh, E.
Protocalcite (1937)
Bandy, M. C.
Cuprocopiapite (1938)
Hydrated metavauxite (1946)
Hydrated paravauxite (1946)
Hydrometavauxite (1946)
Hydroparavauxite (1946)
Llallagualite (1946)
Lopezite (1937)
Metahohmannite (1938)
Metasideronatrite (1938)
Parabutlerite (1938)
Patiñoite (1946)
Stibiobismutotantalite (1951)
——, *v.* Peacock, M. A.
Bank, H.
Tsavolite (1975)
Bannister, F. A.
Braggite (1932)
Brammallite (1943)
Earlandite (1936)
Kalsilite (1942)
Magnesium-bentonite (1939)
Metabasaluminite (1950)
Nahcolite (1928)
Parakalinepheline, Parakaliophilite (1942)
Sodium-illite (1943)
——, Claringbull, G. F., & Hey, M. H.
Arsenopalladinite (1955)
——, Hey, M. H., & Smith, W. C.
Grovesite (1955)
——, & Hollingworth, S. E.
Basaluminite (1948)
Hydro-basaluminite (1948)
——, & Whittard, W. F.
Magnesian chamosite (1945)
——, *v.* Hey, M. H.
——, *v.* Skerl, A. C.
——, *v.* Smith, W. C.
Banno, S., *v.* Kanehira, K.
Baranova, E. N., *v.* Bulakh, A. G.
Barb, C. F.
Liversite (1944)
Barbier, P.
Hallerite (1908)
Naphtolithe (1911)
Barbosa, A. L. de M., *v.* Lindberg, M. L.
Barea, P. Castro, *v.* Castro Barea, P.
Baret, C.
Dubuissonite (1904)
Bariand, P., Berthelon, J.-P., Cesbron, F., & Sadrzadeh, M.

Belyankin, D. S.
Askanite (1934)
Endothermite (1938)
Vishnevite (1931)
Wischnewite (1931)
——, Feodotyev, K. M., & Nikogosyan, C. S.
Iron-monticellite (1934)
——, & Ivanova, V. P.
Hydrokaolin (1935)
Monothermite (1936)
——, & Petrov, V. P.
Grossularoid (1941)
Benedicks, C.
Oxide-pearlite (1912)
Thalenite (1898)
——, *v.* Löfquist, H.
Beneslavsky, S. I.
Alumogoethite (1957)
Alumohematite (1957)
Alumomaghaemite (1957)
Alumomaghemite (1957)
Benson, L. H.
Walderite (1958/9)
Benson, W. N.
Ferropigeonite (1944)
Benyacar, M. Rodriguez de, *v.* de Abeledo, M. Jimenez
Benyacar, M. A. R., de, *v.* de Benyacar, M. Rodriguez de
Berdesinski, W.
Leonhardtite (1952)
Prokoenenite (1952)
Sanderite (1952)
Berendsen, P., *v.* Foord, E. E.
Bergemann, C.
Cobalt-manganese-spar (1857)
Bergerhoff, G., & Nowacki, W.
Sc-beryl (1955)
Berggren, T., *v.*, Quensel, P.
Bergstøl, S., Jensen, B. B., & Neumann, H.
Tveitite (1977)
Berman, H.
Pyroxenoid (1937)
Soda-melilite (1929)
Sub-melilite (1929)
——, & Frondel, C.
Formanite (1944)
——, & Gonyer, F. A.
Landesite (1930)
Roweite (1937)
——, & Harcourt, G. A.
Moschellandsbergite (1938)
——, & Larsen, E. S.
Soda-tremolite (1931)
——, & Wolfe, C. W.
Bellingerite (1940)
——, *v.* Bauer, L. H.
——, *v.* Larsen, E. S.
——, *v.* Palache, C.
Berman, R., *v.* Vaughan, P. A.
Berndt, F., *v.* Ramdohr, P.
Berner, R. A., *v.* Evans, H. T. Jr
Berry, L. G.
Aluminocopiapite (1947)
Ferricopiapite (1938)
Ferrocopiapite (1938)
Magnesiocopiapite (1938)
——, Fahey, J. J., & Bailey, E. H.
Robinsonite (1951)
——, & Thompson, R. M.
Selenjosëite (1962)
——, *v.* Hawley, J. E.
——, *v.* Radcliffe, D.
Berthelon, J.-P., *v.* Bariand, P.
Berthon, L.
Sebkhainite (1922)
Bertrand, C. E., & Renault, B.
Thelotite (1892)
Bertrand, J., *v.* Sarp, H.
Berwerth, F.
Weinbergerite (1906)
Berzelius, J. J.
Odenite (1815)
Bespalov, M. M.
Aldanite (1941)
Betekhtin, A. G.
Alumochromite (1934)
Arsenosulvanite (1941)
Ferrichrompicotite (1934)
Ferrichromspinel (1934)
Magnoferrichromite (1934)
Vernadite (1944)
Beus, A. A.
Magniophilite (1950)
Mangankoninckite (1950)
——, & Zalashkova, N. E.
Soda-beryl (1936)
Bezsmertnaya, M. S., & Soboleva, L. N.
Volynskite (1965)
——, *v.* Semenov, E. I.
——, *v.* Sørensen, C.
Biais, R., *v.* Papée, D.
Bianchini, G.
Cupro-mangano-aphthitalite (1937)
Bideaux, R. A., *v.* McLean, W. J.
——, *v.* Williams, S. A.
Biehl, F. K.
Cuproplumbite (1919)
Cuprozincite (1919)
Parabayldonite (1919)
Paraurichalcite (1919)
Bigelow, W. C., *v.* Peacor, D. R.
Bikova, A. V., *v.* Perrault, G.
Bilgrami, S. A.
Chiklite (1955)
Ferririchterite (1955)
Bilibin, G. A.
Alumohydrocalcite (1926)
Alumolimonite (1928)
Cobaltsmithsonite (1927)
Khakassite (1926)
Billiet, V.
Soddyite (1926)
Billings, M.
Alkali-femaghastingsite (1928)

Borgström, L. H.—*cont.*
Oxide-meionite (1914)
Sulphate-marialite (1914)
Sulphate-meionite (1914)
Bořický, E.
Přilepite (1873)
Walaite (1869)
Borisenko, L. F.
Shcherbinaite (1972)
Borneman-Starynkevich, I. D.
Calcium-rinkite (1935)
Carbocer (1933)
Kondrikovite (1933)
Silico-apatite (1938)
Vudyavrite (1933)
Borodaev, Yu. S., *v.* Mozgova, N. N.
Borodin, L. S., Bykova, A. V., Kapitonova, T. A., & Pyatenko, Yu. A.
Niobozirconolite (1960)
——, & Kazakova, M. E.
Belovite (1954)
Irinite (1954)
——, Nazarenko, I. I., & Richter, T. L.
Zirconolite (1956)
Boronkov, A. A., *v.* Semenov, E. I.
Borovskii, I. B., *v.* Ivanov, A. A.
Botinelly, T., *v.* Adams, J. W.
Bournon, Comte J. L. de
Bardiglione (1811)
Bowen, N. L.
Clinoferrosilite (1935)
Echellite (1920)
Lime-olivine (1922)
——, Greig, J. W., & Zies, E. G.
Mullite (1924)
——, & Schairer, J. F.
Fluor-amphibole (1935)
Hydroxy-amphibole (1935)
Magnesio-wüstite (1935)
——, *v.* Morey, G. W.
Bowie, S. H. U.
Grayite (1957)
——, & Horne, J. E. T.
Cheralite (1953)
Bowley, H.
Simpsonite (1938)
Bown, M. G., *v.* Agrell, S. O.
——, *v.* Gittins, J.
Boyd, F. R., *v.* Anderson, A. T.
Boyle, R. W., *v.* Jambor, J. L.
——, *v.* Traill, R. J.
Bracewell, S.
Merumite (1946)
Rumongite (1950)
Bradley, A. J., & Roussin, A. L.
Porzite (1932)
B[radley], O.
U-wulfenite (1962)
Bradley, W. F., *v.* Comeforo, J. E.
——, *v.* Grim, R. E.
Bradley, W. M.
Empressite (1914)
——, *v.* Ford, W. E.
Braitsch, O.
Calciumhilgardite-3Tc, Calciumhilgardite-2M(Cc) (1959)
Pseudolussatine (1957)
Pseudo-quartzine (1957)
Strontioginorite (1959)
Strontiohilgardite-1Tc (1959)
Strontium ginorite (1959)
ρ-Veatchite (1959)
Branche, G., Bariand, P., Chantret, F., Pouget, R., & Rimsky, A.
Vanuralite (1963)
——, Ropert, M. E., Chantret, F., Morignat, B., & Pouget R.
Francevillite (1957)
Brandt, F.
Harbortite (1932)
Braun, H., *v.* Borchert, H.
Brauns, A., & Brauns, R.
Radiophyllite (1924)
Brauns, R.
Carbonate-apatite (1916)
Carbonate-sodalite (1916)
Garnet-jade (1929)
Metazeolites (1892)
Oxydhydratmarialith (1915)
Oxydhydratmejonit (1915)
Radiotine (1904)
Serpentine-jade (1929)
Silvialite (1914)
Sulphate-apatite (1916)
Sulphate-scapolite (1914)
Transvaal-jade (1929)
Vesuvian-jade (1929)
——, *v.* Brauns, A.
Bravo, J. J.
Rizopatronite (1906)
Bray, R. H., *v.* Grim, R. E.
Bredig, M. A., *v.* Franck, H. H.
Breithaupt, A.
Melosark (1841)
Theophrastite (1849)
Breithaupt, J. F. A.
Basitom-Glanz (1832)
Cliachite (1847)
Euban (1823)
Hyposiderite (1847)
Jocketan
Lepidolamprite
Placodine (1841)
Plusinglanz (1823)
Breñosa, R.
Bourgeoisite (1885)
Brett, R., *v.* Keil, K.
Breusing, E.
Angolite (1900)
Bricker, O.
Feitknechtite (1965)
Bridge, P. J.
Archerite (1977)
Urea (1973)
Uricite (1974)
——, & Pryce, M. W.

Bridge, P. J.—*cont.*
Clinobisvanite (1974)
——, *v.* Nickel, E. H.
Brindley, G. W., & von Knorring, O.
Ortho-antigorite (1954)
——, & Youell, R. F.
Ferric chamosite (1953)
Ferro-chamosite (1953)
Brisi, C., & Abbatista, F.
Strontiogehlenite (1960)
Brock, P. W. G., Gelatly, D. C., & von Knorring, O.
Mboziite (1964)
Brøgger, W. C.
Aegirine-diopside (1898)
Blomstrandine (1906)
Catophorite (1894) (Katophorite)
Cergadolinite (1922)
Hellandite (1903)
Mossite (1897)
Natronmelilith (1898)
Natronmikroklin (1882)
Priorite (1906)
Sundtite (1893)
Zircon-pyroxenes (1890)
——, & Backström, H.
Alkali-garnet (1890)
Brophy, G. P., *v.* Kerr, P. F.
Broughton, P. L.
Leonardite (1972)
Brouwer, H. A.
Molengraaffite (1911)
Brovkin, A. A., Aleksandrov, S. M., & Nekrasov, I. Ya.
Ferrovonsenite (1963)
——, *v.* Grigoriev, A. P.
Brown, G., & others
Ledikite (1955)
Brown, G. E., *v.* Radtke, A. S.
Brown, G. M., *v.* Lovering, J. F.
Brown, G. V., *v.* Larsen, E. S.
Brown, K. L., *v.* Radtke, A. S.
Brown, R.
Collieite (1927)
Brown, W. L., *v.* Cesbron, F.
Brownmiller, L. T., & Bogue, R. H.
Jäneckeite (1930)
Brugnatelli, L.
Artinite (1902)
Brun, A.
Iozite (1924)
Brunlechner, A.
Seelandite (1891)
Zinkmanganerz (1893)
Brusoni, A.
Zebedassite (1917)
Bryzgalov, I. A., *v.* Gruzdev, V. S.
Bubeck, W.
Magnesium-berzeliite (1934)
Bucher, W. H.
Ovulite (1918)
Spherite (1918)
Buchwald, V. F., & Scott, E. R. D.
Carlsbergite (1971)
Bücking, H.
Eisencordierite (1900)
Buckle, E. R., & Taylor, H. F. W.
Calcio-chondrodite (1958)
Buddhue, J. D.
Ayasite (1939)
Bacalite (1935)
Jelinite (1938)
Kansasite (1938)
Metataenite (1936)
Orthotaenite (1936)
Pseudo-succinite (1938)
Rustite (1939)
Buddington, A. F., Fahey, J. J., & Vlisidis, A. C.
Ilmenomagnetite (1953)
Microantiperthite (1955)
Mogensenite (1955)
Rutilohematite (1953)
Titanmagnetite (1955)
Budko, I. A., *v.* Rudashevskii, N. S.
——, *v.* Vinogradova, R. A.
——, *v.* Yalovai, A. A.
Buie, B. F., *v.* Larsen, E. S.
Bukin, V. I., *v.* Kapustin, Yu. L.
——, *v.* Semenov, E. I.
Buko, I. A., & Kulagov, E. A.
Talnakhite (1968)
Bukovský, A.
Kutnohorite (1901)
Bulakh, A. G., Kondrateva, V. V., & Baranova, E. N.
Carbocernaite (1961)
——, Kukharenko, A. A., Knipovich, Yu. N., Kondrateva, V. V., Baklanova, K. A., & Baranova, E. N.
Natroniobite (1960)
——, *v.* Kukharenko, A. A.
Bulgarine
Pungernite (1851)
Bültemann, H. W.
Ellweilerite (1960)
Paulite (1960)
——, & Moh, G. H.
Bergenite (1959)
Bunch, T. E., & Fuchs, L. H.
Brezinaite (1969)
Yagiite (1969)
——, *v.* Anderson, A. T.
——, *v.* Lovering, J. F.
——, *v.* Olsen, E.
Burke, E. A. J., Kieft, C., Felius, R. O., & Adusumilli, M. S.
Staringite (1969)
——, *v.* Oen, I. S.
Burnham, C. W., *v.* Rapoport, P. A.
——, *v.* Takeda, H.
——, *v.* Veblen, D. R.
Burns, J. H., *v.* Beck, C. W.
Burova, T. A., *v.* Bonstedt-Kupletskaya, E. M.
——, *v.* Dorfman, M. D.

Burova, T. A.—*cont.*
——, *v.* Semenov, E. I.
Burova, Z. N., *v.* Semenov, E. I.
Burri, G., Graeser, S., Marumo, F., & Nowacki, W.
Imhofite (1965)
——, *v.* Marumo, F.
Buryanova, E. Z.
Bilibinite (1958)
——, & Komkov, A. I.
Ferroselite (1955)
——, Kovalev, G. A. & Komkov, A. I.
Kadmoselite (1957)
——, Strokova, G. C., & Shitov, V. A.
Vanuranilite (1965)
Buseck, P. R.
Barringerite (1969)
——, *v.* Veblen, D. R.
Bussell, M., *v.* Barker, W. W.
Büssem, W., & Scheusterius, C.
Protoenstatite (1938)
Bussen, I. V., Denisov, A. P., Zabavnikova, N. I., Kozyreva, L. V., Menshikov, Yu. P., & Lipatova, E. A.
Vuonnemite (1973)
——, Gannibal, L. F., Goiko, E. A., Merkov, A. N., & Nedorezova, A. P.
Ilmaiokite (1972)
——, Menshikov, Yu. P., Merkov, A. N., Nedorezova, A. P., Uspenskaya, E. I., & Khomyakov, A. P.
Penkvilskite (1974)
——, *v.* Menshikov, Yu. P.
——, *v.* Merkov, A. N.
——, *v.* Timoshenkov, I. M.
Busz, K.
Manganosphaerite (1901)
Tsumebite (1912)
Butler, B. S., & Schaller, W. T.
Beaverite (1911)
Ferroludwigite (1917)
Magnesioludwigite (1917)
——, *v.* Wells, R. C.
Butler, G. M.
Lausenite (1928)
Nicholsonite (1913)
Wolftonite (1913)
Buttgenbach, H.
Berthonite (1923)
Bialite (1929)
Cornetite (1917)
Cuprosklodowskite (1933)
Droogmansite (1925)
Fourmarierite (1924)
Heterobrochantite (1926)
Katangite (1921)
Kipushite (1927)
Shinkolobwite (1925)
Thoreaulite (1933)
——, & Gillet, C.
Cesarolite (1920)
Buţureanu, V. C.
Ponite (1912)
Bykhover, N. A., *v.* Ozerov, K. N.
Bykov, A. D.
Proarizonite (1964)
Bykov, V. P., *v.* Khomyakov, A. P.
Bykova, A. V., *v.* Borodin, L. S.
——, *v.* Dusmatov, V. D.
——, *v.* Kalita, A. P.
——, *v.* Kapustin, Yu. L.
——, *v.* Mineev, D. A.
——, *v.* Semenov, E. I.
——, *v.* Zdorik, T. B.

Cabri, L. J., Chen, T. T., Stewart, J. M., & Laflamme, J. H. G.
Palladobismutharsenide (1976)
——, & Feather, C. E.
Isoferroplatinum (1975)
Tetraferroplatinum (1975)
——, & Hall, S. R.
Haycockite (1972)
Mooihoekite (1972)
——, & Harris, D. G.
Insizwaite (1972)
——, ——, & Stewart, J. M.
Costibite (1970)
Nisbite (1970)
Paracostibite (1970)
——, & Laflamme, J. H. G.
Rhodium (1974)
Sudburyite (1974)
——, ——, & Stewart, J. M.
Platarsite (1977)
Temagamite (1973)
——, ——, ——, Rowland, J. F., & Chen, T. T.
Stillwaterite (1975)
——, Owens, D. R., & Laflamme, J. H. G.
Tulameenite (1973)
——, Stewart, J. M., Laflamme, J. H. G., & Szymanski, J. T.
Genkinite (1977)
——, *v.* Harris, D. C.
Caillère, S., Avias, J., & Falgueirettes, J.
Orcelite (1959)
——, & Hénin, S.
Allevardite (1950)
Calderón, S.
Almeriite (1910)
Calafatite (1910)
——, & Paul, M.
Moronite (1886)
Calk, L. C., *v.* de Waal, S. A.
Callisen, K.
Flokite (1917)
Cameron, E. N.
Titanochromite (1970)
——, & Threadgold, I. M.
Vulcanite (1961)
——, *v.* Anderson, A. T.
——, *v.* Lovering, J. F.
Cameron, F. K., & McCaughey, W. J.
Chlor-spodiosite (1911)
Fluor-spodiosite (1911)

Claringbull, G. F.—*cont.*
Sinhalite (1952)
——, ——, & Russell, A.
Cornubite (1958)
——, ——, & Payne, C. J.
Painite (1956)
——, *v.* Anderson, B. W.
——, *v.* Bannister, F. A.
Clark, A. H.
α-Arsenic sulphide (1970)
η′-bronze (1972)
——, & Sillitoe, R. H.
Chalcocite, Tetragonal (1971)
Clark, A. M., Criddle, A. J., & Fejer, E. E.
Atheneïte (1974)
Isomertieite (1974)
——, *v.* Davis, R. J.
——, *v.* Embrey, P. G.
——, *v.* Fejer, E. E.
——, *v.* Stumpfl, E. F.
Clark, J. R., *v.* Christ, C. L.
——, *v.* Erd, R. C.
——, *v.* Milton, C.
Clarke, F. W.
Pseudo-jadeite (1906)
Clarke, W. B.
Baddeleyite (1882)
Clinton, H. G.
Metacalciowardite (1929)
Coates, M. E., *v.* Meagher, E. P.
Codazzi, R. L.
Codazzite (1925)
Loaisite (1905)
Viterbite (1925)
Coelho, A. V. P.
Mavudzite (1954)
Coleman, R. G., & Erd, R. C.
Hydrozircon (1961)
——, Ross, D. R., & Meyrowitz, R.
Metazellerite (1966)
Collins, B. J. S.
Allokite (1955)
Colomba, L.
Aloisiite (1908)
Luigite (1908)
Speziaite (1914)
Colville, A. A. & P. A.
Paraspurrite (1977)
Comeforo, J. E., Fischer, R. B., & Bradley, W. F.
Promullite (1948)
——, Hatch, R. A., & Eitel, W.
Fluor-mica (1950)
Fluor-richterite (1950)
——, *v.* Kohn, J. A.
Conquéré, F., *v.* Babkine, J.
Cooke, W. T., *v.* Mawson, D.
Coomaraswamy, A. K., *v.* Prior, G. T.
Cooper, R. A.
Modderite (1924)
Cooray, P. G.
Carbonate-fluor-chlor-hydroxyapatite (1970)
Fluor-chlor-hydroxyapatite (1970)
Copisarow, M.
Subhydrocalcite (1913)
Corey, A. S., *v.* Gross, E. B.
Corlett, M. I., *v.* Mandarino, J. A.
Cornu, F.
Ehrenwerthite (1909)
Geldiadochite (1909)
Gelfischerite (1909)
Gelpyrophyllite (1909)
Gelvariscite (1909)
Hibschite (1905)
Jordisite (1909)
Kliachite (1909)
Ostwaldite (1909)
Reyerite (1906)
Tunnerite (1909)
Uhligite (1909)
Correia Neves, J. M., Lopez Nunes, J. E., & Sahama, Th. G.
Hafnon (1974)
——, *v.* Cotelo Neiva, J. M.
Corti, H.
Chubutite (1918)
Cossa, A., & Arzruni, A.
Chrome-tourmaline (1883)
Cotelo Neiva, J. M.
Limaite (1954)
——, & Correia Neves, J. M.
Mozambikite (1960)
Couper, A. G., *v.* Fejer, E. E.
Courville, S., *v.* Jambor, J. L.
Cox, K. G., *v.* von Knorring, O.
Craig, J. R., & Carpenter, A. B.
Fletcherite (1977)
Crawford, W. P.
Weissite (1927)
Criddle, A. J., *v.* Clark, A. M.
——, *v.* Davis, R. J.
Crook, W. W. III
Texasite (1977)
——, & Marcotty, L.-A.
Albrittonite (1978)
Cross, W., & others
Lenad (1903)
Crosse, A. F.
Trevorite (1921)
Cruellas, J., *v.* Fester, G. A.
Cumenge, E.
Bouglisite (1895)
Robellazite (1900)
Von-Diestite (1899)
——, *v.* Friedel, C.
Cuomo, J. R., *v.* Gagarin, G.
Curien, H., Guillemin, C., Orcel, J., & Sternberg, M.
Hibonite (1956)
Cuttitta, F., *v.* Frondel, J. W.
——, *v.* Milton, C.
——, *v.* Raup, O. B.
Cuvelier, V.
Selenolinnaeite (1929)
Stainierite (1929)

Czamanske, G. K., *v.* Cesbron, F. P.

D'Achiardi, G.
Ginorite (1934)
Hyaloallophane (1898)
Zeolite-mimetica (1906)
Daggett, E. B., *v.* Fahey, J. J.
Dake, H. C.
Brodrickite (1941)
Dalrio, G., *v.* Morandi, N.
Daly, R. A.
Philipstadite (1899)
Dalvi, A. P., *v.* Nobar, M. A.
D'Ambrosio, A.
Traversoite (1924)
Dannenberg, A.
Arzrunite (1899)
Dano, M., & Sørensen, H.
Igdloite (1959)
D'Ans, J.
Ammonium-syngenite (1906)
Dara, A. D., *v.* Egorov, B. L.
——, *v.* Skvortsova, K. V.
Darapsky, L.
Planoferrite (1897)
Darbyshire, D. P. F., *v.* Hawkes, J. R.
D'Ascenzo, N. G.
Korunduvite (1945)
Daugherty, F. W., *v.* Tunell, G.
Dave, A. S., *v.* Kilpady, S.
Davey, P. T., & Scott, T. R.
Maghæmite (1957)
David, T. W. E., & Taylor, T. G.
Glendonite (1905)
Davidson, S. C., *v.* Palache, C.
Davis, C. E. S., *v.* Nickel, E. H.
Davis, C. W., *v.* Lind, S. C.
Davis, R. J., Clark, A. M., & Criddle, A. J.
Palladseïte (1977)
——, Embrey, P. G., & Hey, M. H.
Ludlockite (1970)
——, & Hey, M. H.
Arthurite (1964)
——, *v.* Anderson, B. W.
——, *v.* Binns, R. A.
——, *v.* Rodgers, K. A.
Davison, J. M.
Wardite (1896)
Davy, W. M., *v.* Lindgren, W.
de Abeledo, M. E. J., Angelelli, V., de Benyacar, M. A. R., & Gordillo, C.
Sanjuanite (1968)
——, *v.* Jimenez de Abeledo, M.
Deans, T., & McConnell, J. D. C.
Isokite (1955)
Dear, P. S.
Strontio-gehlenite (1969)
de Areitio y Larrinaga, A.
Ciempozuelite (1873)
de Benyacar, M. A. R., *v.* de Abeledo, M. E. J.
de Béthune, P.
Borgniezite (1956)
de Bournon, Count J. L. *See* under Bournon, Count J. L. de
Debyser, J., *v.* Baron, G.
Deckert, H., *v.* Menzel, H.
De Dycker, R.
Varlamoffite (1947)
Deer, W. A., & Wager, L. R.
Ferrohortonolite (1939)
Deferne, J., *v.* Sarp, H.
De Fiore, O., *v.* Zambonini, F.
de Jesus, Mario A., *v.* Mario de Jesus, A.
de Grossouvre, A.
Vierzonite (1901)
de Lapparent, J.
Attapulgite (1935)
Boehmite (1927)
Ilmeno-corundum (1937)
Pseudosillimanite (1920)
Taosite (1935)
Titano-spinel (1937)
De Las Casas, F., *v.* Amstutz, G. C.
De Leenheer, L.
Boodtite (1936)
Cupro-asbolane (1938)
Hydrotenorite (1937)
Lubumbashite (1934)
Mindigite (1934)
Trieuite (1935)
de Lencastre, J., & Vairinho, M. M.
Nisaite (1970)
De Lian, Fan
Rhombomagnojacobsite (1964)
Deliens, M., & Piret, P.
Kusuïte (1977)
Phurcalite (1978)
Dell'Anna, L., *v.* Balenzano, F.
Del'yanidi, K. I., *v.* Indolev, L. N.
de Mello, E. Z. Vieira, *v.* Vieira de Mello, E. Z.
de Montreuil, L. A.
Bellidoite (1975)
de Moraes, L. F., *v.* Lee, T. H.
Denaeyer, M. E., & Ledent, D.
Camermanite (1950)
Dence, M. R., *v.* Chao, E. C. T.
Denisov, A. P., *v.* Bussen, I. V.
——, *v.* Menshikov, Yu. P.
de Paiva Netto, J. E.
Eunicite (1955)
Derby, O. A.
Caldasite (1917)
Marahuite (1907)
de Roever, E. W. F., Kieft, C., Murray, E., Klein, E., & Drucker, W. H.
Surinamite (1976)
de Roever, W. P.
Ferrocarpholite (1951)
de Rubies, S. Piña, *v.* Piña de Rubies, S.
Desborough, G. A., Finney, J. J., & Leonard, B. F.
Mertieite (1973)
de Schulten, A.
Bromcarnallite (1897)
——, *v.* Lacroix, A.

Dubanský, A.—*cont.*
Hydroxykeramohalite (1956)
Kašparite (1956)
Duboin, A.
Potassium-cryolite (1892)
Dudkin, O. B.
Barium-lamprophyllite (1959)
Dudykina, A. S., *v.* Semenov, E. I.
Dufet, H.
Ceruleite (1900)
Duffin, W. J., & Goodyear, J.
Hydroscarbroite (1960)
Metascarbroite (1960)
DuFresne, E. R., & Roy, S. K.
Farringtonite (1961)
Duggan, M., *v.* Williams, S. A.
Dulhunty, J. A.
Gelosite (1939)
Humosite (1939)
Matrosite (1939)
Retinosite (1939)
Düll, E.
Klino-pyroxene (1902)
Ortho-pyroxene (1902)
Dumas, E.
Truffite (1876)
Dunham, K. C., *v.* Larsen, E. S.
——, *v.* Smythe, J. A.
Dunicz, B. L.
Thioelaterite (1936)
Dunin-Barkovskaya, E. A., Lider, V. V., & Rozhansky, V. N.
Sulphojoseïte (1968)
Dunn, J. A.
Coulsonite (1937)
——, & Roy, P. C.
Tirodite (1938)
Dunn, P. J., & Appleman, D. E.
Perhamite (1977)
——, ——, & Nelen, J. A.
Liddicoatite (1977)
——, & Rouse, R. C.
Wroewolfeite (1975)
——, ——, Cannon, B., & Nelen, J. A.
Zektzerite (1977)
——, ——, & Norberg, J. A.
Fluorapophyllite (1978)
Hydroxyapophyllite (1978)
Dunstan, W. R.
Thorianite (1904)
Duparc, L.
Isorthose (1904)
Pseudoglaucophane (1927)
——, & Gysin, M.
Genevite (1927)
——, & Pearce, F.
Soretite (1903)
Tschernichewite (1907)
——, & Tikonowitch, M. N.
Bobrovkite (1920)
Du Rietz, T.
Kribergite (1945)
Dürr
Terpitzite (1811)
Dürrfeld, V.
Hügelite (1913)
du Preez, J. W.
Bismuth-parkerite (1944)
Lead-parkerite (1944)
Dusmatov, V. D., Efimov, A. F., Alkhazov, V. Yu., Kazakova, M. E., & Mumyatskaya, N. G.
Tienshanite (1967)
——, ——, Kataeva, Z. T., Khoroshilova, L. A., & Yanulov, K. P.
Sogdianite (1968)
——, Semenov, E. I., Khomyakov, A. P., Bykova, A. V., & Dzhafarov, N. Kh.
Baratovite (1975)
——, *v.* Efimov, A. F.
——, *v.* Povarennykh, A. S.
——, *v.* Semenov, E. I.
Dusausov, Y., *v.* Sierra Lopez, J.
Duvornik, E. J., *v.* Milton, C.
Dvoichenko, P. A.
Bosphorite (1914)
Mitridatite (1914)
Dvořák, R.
Mohelnite (1943)
Dwornik, E. J., *v.* Faust, G. T.
——, *v.* Milton, C.
——, *v.* Pabst, A.
——, *v.* Raup, O. B.
Dyadchenko, M. G., *v.* Gornyi, G. Ya.
Dybinchik, V. T., *v.* Portnov, A. M.
Dzhafarov, N. Kh., *v.* Dusmatov, V. D.

Eakle, A. S.
Crestmoreite (1917)
Erionite (1898)
Esmeraldaite (1901)
Foshagite (1925)
Hydro-wollastonite (1917)
Jurupaite (1921)
Neocolemanite (1911)
Palacheite (1903)
Probertite (1929)
Riversideite (1917)
Vonsenite (1920)
——, & Rogers, A. F.
Wilkeite (1914)
Ebelmen, J. T.
Zinkferrit (1851)
Eberhard, E.
Mn-feldspat (1963)
Eckermann, H.
Barium-phlogopite (1925)
Iron-antigorite (1925)
——, *v.* Quensel, P.
Eckhardt, F. J., & Heimbach, W.
Chromatite (1963)
Edelmann, F.
Kolbeckite (1926)
Edwards, A. B.
Tantalo-rutile (1940)
Titanhaematite (1938)

Efimov, A. F., Dusmatov, V. D., Alkhazov, V. Yu., Pudovkina, Z. G., & Kazakova, M. E.
Tadzhikite (1970)
——, ——, Ganzeev, A. A., & Kataeva, Z. T.
Caesium astrophyllite (1969)
Caesium kupletskite (1971)
——, Kravchenko, S. M., & Vasileva, Z. V.
Strontium-apatite (1962)
——, ——, & Vlasova, E. V.
Strontiumthomsonite (1963)
——, *v.* Dusmatov, V. D.
——, *v.* Ganzeev, A. A.
Efimov, A. N., *v.* Nikolsky, A. P.
Efremov, N. E.
Abkhazite (1940)
Adigeite (1939)
Alumoantigorite (1951)
Alumochrysotile (1951)
Alumodeweylite (1939 and 1951)
Azovskite (1937)
Bedenite (1935)
Ferrihalloysite (1936)
Hydroforsterite (1938)
Hydromagniolite (1939)
Hydrosialite (1939)
Karachaite (1936)
Kolskite (1939)
Labite (1936)
Magnesium kaolinite (1955)
Paradeweylite (1939)
Sialite (1939)
Egorov, B. L., Dara, A. D., & Senderova, V. M.
Melkovite (1968)
Egorova, E. N., *v.* Boldyreva, A. M.
Egorova, L. N., *v.* Gevorkyan, S. V.
Egorov-Tismenko, Yu. K., *v.* Belokoneva, E. L.
——, *v.* Simonov, M. A.
Ehlmann, A. J., & Mitchell, R. S.
Behoite (1970)
Eitel, W., *v.* Comeforo, J. E.
——, *v.* Trömel, G.
Eldrige, G. H.
Impsonite (1901)
El Goresy, A., *v.* Lovering, J. F.
Eliava, T. A., *v.* Melikadze, L. D.
Elliott, C. J., *v.* Fejer, E. E.
Ellsworth, H. V.
Calciosamarskite (1928)
Hyblite (1927)
Lyndochite (1927)
Thucholite (1928)
Toddite (1926)
——, & Poitevin, E.
Camsellite (1921)
Elschner, C.
Hawaiite (1906)
Meyersite (1922)
Nauruite (1913)
El Shazly, E. M., & Saleeb, G. S.
Pharaonite (1972)
Elston, D. P., *v.* Lindberg, M. L.
Embrey, P. G., Fejer, E. E., & Clark, A. M.
Keyite (1977)
——, *v.* Davis, R. J.
Emmens, S. H.
Folgerite (1892)
Whartonite (1892)
Emszt, K., *v.* Böckh, H.
Endell, K., Hofmann, U., & Maegdefrau, E.
Glimmerton (1935)
Engelhardt, W. V., & Füchtbauer, H.
Heidornite (1956)
Enikeev, M. R.
Nasledovite (1958)
Zincnontronite (1970)
Epshtein, E. M., Anikeeva, L. I., & Mikhailova, A. F.
Hydromelilite (1961)
Erd, R. C., & Evans, H. T.
Smythite (1956)
——, Foster, M. D., & Proctor, P. D.
Faustite (1953)
——, McAllister, J. F., & Almond, H.
Gowerite (1959)
——, ——, & Vlisidis, A. C.
Nobleite (1961)
Wardsmithite (1970)
——, Morgan, V., & Clark, J. R.
Tunellite (1961)
——, White, D. E., Fahey, J. J., & Lee, D. E.
Buddingtonite (1964)
——, *v.* Cesbron, F. P.
——, *v.* Coleman, R. G.
——, *v.* Hurlbut, C. S.
——, *v.* Skinner, B. J.
Erdélyi, J.
Hydrohalloysite (1962)
——, Koblencz, V., & Varga, N. S.
Hydroamesite (1959)
Hydroantigorite (1959)
Erdmannsdörffer, O. H.
Kossmatite (1925)
Eremeev, P. V.
Lasur-oligoclase (1873)
Ericksen, G. E., Fahey, J. J., & Mrose, M. E.
Humberstonite (1967)
——, *v.* Mrose, M. E.
Erlichman, J., *v.* Olsen, E.
Ermilova, L. P., Moleva, V. A., & Klevtsova, R. F.
Chukhrovite (1960)
——, & Senderova, V. M.
Betpakdalite (1961)
——, *v.* Chukhrov, F. V.
Ernst, W. G.
Ferrous riebeckite (1957)
Fluorarfvedsonite (1962)
Magnesian glaucophane (1957)
Magnesian riebeckite (1957)
Riebeckite-arfvedsonite (1962)
Erofeeva, E. A., *v.* Ginzburg, A. I.
Eskola, P.
Chrome-epidote (1933)
Chrome-tremolite (1933)

Feit, W.
Ascharite (1891)
Feitknecht, W., & Marti, W.
Hydrohausmannite (1945)
——, *v.* Giovanoli, R.
Fejer, E. E., Clark, A. M., Couper, A. G., & Elliott, C. J.
Claringbullite (1977)
——, *v.* Clark, A. M.
——, *v.* Embrey, P. G.
Felius, R. O., *v.* Burke, E. A. J.
Fenner, C.
Libyanite (1937)
Fenner, C. N.
Iron-pyroxene (1931)
Feodotyev, K. M., *v.* Belyankin, D. S.
Feraud, J., *v.* Johan, Z.
Fermor, L. L.
Beldongrite (1909)
Blanfordite (1906)
Blythite (1926)
Calc-pyralmandite (1938)
Calc-spessartite (1926)
Devadite (1938)
Ferro-calderite (1926)
Ferro-spessartite (1926)
Garividite (1938)
Gralmandite (1938)
Grandite (1909)
Hanléite (1952)
Hollandite (1906)
Iron-cordierite (1923)
Juddite (1908)
Khoharite (1938)
Magnesia-blythite (1926)
Magnesia-gralmandite (1938)
Mangan-almandite (1926)
Manganese-gralmandite (1938)
Mangan-grandite (1909)
Protomelane (1945)
Pseudo-manganite (1909)
Pyralmandite (1926)
Sitaparite (1908)
Skiagite (1926)
Spalmandite (1926)
Spandite (1907)
Vredenburgite (1908)
Winchite (1906)
Fernández Navarro, L.
Quiroguite (1895)
——, & Castro Barea, P.
Bolivarite (1921)
Ferraris, G., Franchini-Angela, M., & Orlandini, P.
Canavesite (1978)
Fersman, A. E.
Alushtite (1914)
Belomorite (1925)
Calcioancylite (1922)
Cerapatite (1926)
Elatolite (1922)
Ferronemalite (1911)
Loparite (1922)
Mesodialyte (1922)
Murmanite (1923)
Neokaolin (1935)
Ramsayite (1922)
Tangeite (1925)
Titano-elpidite (1926)
Uzbekite (1925)
Vanadio-laumontite (1922)
Yttro-orthite (1931)
Yuksporite (1922)
——, & Shubnikova, O. M.
Cryohalite (1937)
Ferrorhodochrosite (1937)
Metazeunerite (1937)
Natro-autunite (1937)
Selenojarosite (1937)
Fersmann, A.
Caciopalygorskite (1908)
Paramontmorillonite (1908)
Parasepiolite (1908)
Fester, G. A., & Cruellas, J.
Broggite (1935)
——, *et al.*
Magallanite (1937)
Fiala, J., *v.* Bauer, J.
Filimonova, A. A., *v.* Evstigneeva, T. L.
Filonenko, N. E., Lavrov, I. V., Andreeva, O. V., & Pevzner, R. L.
Aluminium spinel (1957)
Finály, I., & Koch, S.
Fülöppite (1929)
Finckh, L., *v.* Hauser, O.
Finkelman, R. B., & Mrose, M. E.
Downeyite (1977)
——, *v.* Pabst, A.
Finney, J. J., *v.* Desborough, G. A.
Fischer, F., *v.* Giovanoli, R.
Fischer, M., *v.* Strunz, H.
Fischer, R. B., *v.* Comeforo, J. E.
Fischer, W.
Iron-andradite (1925)
Fisher, D. J.
Bastinite (1945)
Ferro-alluaudite (1957)
Ferrodickinsonite (1954)
Ferrofillowite (1955)
Mangano-alluaudite (1957)
Manganodickinsonite (1957)
——, & Volborth, A.
Nafalapatite (1960)
Nafalwhitlockite (1960)
——, *v.* Palache, C.
Fisher, F. G., & Meyrowitz, R.
Brockite (1962)
Fishkin, M. Yu., & Melnikov, V. S.
Zinc-epsomite (1965)
Fishman, M. V., *v.* Goldin, B. A.
Flahaut, J., Domange, L., & Ourmitchi, M.
Thiospinels (1960)
Fleischer, M.
Beta-vredenburgite (1944)
Cupromontmorillonite (1951)
——, & Richmond, W. E.

Fleischer, M.—*cont.*
Ramsdellite (1943)
——, *v.* Faust, G. T.
——, *v.* Frondel, J. W.
——, *v.* Richmond, W. E.
Fletcher, A. B., *v.* Barker, W. W.
Fletcher, M. H., *v.* Larsen, E. S.
Flink, G.
Akrochordite (1922)
Ancylite (1900)
Chalcolamprite (1900)
Cordylite (1900)
Dixenite (1920)
Eisenpyrochroit (1919)
Ektropite (1917)
Endeiolite (1900)
Katoptrite (1917)
Leucosphenite (1900)
Molybdophyllite (1901)
Narsarsukite (1900)
Natroncatapleiite (1893)
Neptunite (1893)
Platynite (1910)
Pseudoparisite (1898)
Pyrobelonite (1919)
Quenselite (1926)
Soda-catapleiite (1900)
Sphenomanganite (1919)
Spodiophyllite (1900)
Synchysite (1901)
Tainiolite (1900)
Trigonite (1920)
Weibullite (1910)
Weslienite (1923)
Yttrium-apatite (1900)
Flint, E. P., McMurdie, H. F., & Wells, L. S.
Hydrogarnet (1941)
Flinter, B. H.
Hydroilmenite (1959)
——, *v.* Alexander, J. B.
Florencio, W.
Alvarolite (1952)
Ribeirite (1952)
Focke, F.
Nemaphyllite (1902)
Folch, J., *v.* Tomás, L.
Fölster, H., *v.* Scheffer, F.
Fong, D. G., *v.* Jambor, J. L.
Fontan, F., Orliac, M., & Permingeat, F.,
Krautite (1975)
——, ——, ——, Pierrot, R., & Stahl, R.
Brassite (1974)
Foord, E. E., Berendsen, P., & Storey, L. O.
Corderoite (1974)
Foote, H. W., *v.* Penfield, S. L.
——, *v.* Pratt, J. H.
Foote, W. M.
Northupite (1895)
Zirkite (1916)
Ford, R. J.
Hydro-andradite
Ford, W. E.
Rickardite (1903)
——, & Bradley, W. M.
Margarosanite (1916)
Pyroxmangite (1913)
Skemmatite (1913)
Fornaseri, M., *v.* Belluomini, G.
Förtsch
Pseudo-ozocerite (1898)
Foshag, W. F.
Freirinite (1924)
Krausite (1931)
Plazolite (1920)
Schairerite (1931)
——, & Gage, R. B.
Chlorophoenicite (1924)
——, & Hess, F. L.
Metarossite (1927)
——, *v.* Gale, W. A.
——, *v.* Hess, F. L.
——, *v.* Larsen, E. S.
——, *v.* Palache, C.
——, *v.* Switzer, G.
Foster, M. D.
Trilithionite (1960)
——, *v.* Erd, R. C.
Fournier, R. O., *v.* Morey, G. W.
Fraatz, *v.* Werner
Franchini-Angela, M., *v.* Ferraris, G.
Franck
Ernite (1911)
Franck, H. H., Bredig, M. A., & Frank, R.
Rhenanite (1936)
Francotte, J., Moreau, J., Ottenburgs, R., & Lévy, C.
Briartite (1965)
Frank, R., *v.* Franck, H. H.
Frankel, J. J.
Chrome-magnetite (1942)
Frank-Kamenetsky, V. A.
Ferrophlogopite (1958)
Hydrolepidolite (1960)
Hydroserpentine (1960)
Pseudochlorite (1958)
——, Logvinenko, N. V., & Drits, V. A.
Tosudite (1963)
——, *v.* Ankinovich, G.
Fransolet, A.-M.
Melonjosephite (1973)
——, *v.* von Knorring, O.
Fredriksson, K., & Henderson, E. P.
Perryite (1965)
French, A. G.
Canadium (1911)
French, B. M., Jezek, P. A., & Appleman, D. E.
Virgilite (1978)
Frenzel, A.
Bismutosmaltite (1897)
Kassiterolamprite (1904)
Frenzel, G.
Blaubleibender Covellin (1959)
Bismutoniobite (1955)
Calcium-gümbelite (1971)
Freudenbergite (1961)

Ginzburg, A. I.
Ferroferrimargarite (1955)
Hydrocookeite (1953)
Kryzhanovskite (1950)
——, Gorzhevskaya, S. A., Erofeeva, E. A., & Sidorenko, G. A.
Titanobetafite (1958)
——, ——, Sidorenko, G. A., & Ukhina, T. A.
Wolframoixiolite (1969)
——, Kruglova, N. A., & Moleva, V. A.
Magniotriplite (1951)
——, & Voronkova, N. V.
Oxychildrenite (1950)
——, *v*. Gorzhevskaya, S. A.
——, *v*. Zhemchuznikov, Yu. A.
Ginzburg, I. I.
Buryktalskite (1960)
——, & Ponomarev, A. I.
Ferromontmorillonite (1939)
——, & Rukavinshnikova, I. A.
Alpha-cerolite (1950)
Alumocobaltomelane (1951)
Beta-cerolite (1950)
Cobaltomelane (1951)
Cobalto-nickelemelane (1951)
Nickelemelane (1951)
Ginzburg, I. V.
Clinoholmquistite (1965)
——, Lisitsina, G. A., Sadikova, A. T., & Sidorenko, G. A.
Ferrifayalite (1936)
——, Sidorenko, G. A., & Rogachev, D. L.
Magarfvedsonite (1961)
Maghastingsite (1961)
Giovanoli, R., Feitknecht, W., & Fischer, F.
Buserite (1971)
Giraud, R., *v*. Bariand, P.
Gita, E., *v*. Gita, G.
Gita, G., & Gita, E.
Vallachite (1962)
Gittins, J., Bown, M. G., & Sturman, D.
Agrellite (1976)
Giusca, D., *v*. Manilici, V.
Giuseppetti, G., *v*. Ramusino, C. C.
Givens, D. B., *v*. Beck, C. W.
Gladkovskii, A. K., & Ushatinskii, I. N.
Ayatite (1961)
Kamenskite (1961)
Kushmurunite (1961)
——, *v*. Sharova, A. K.
Glaser, O.
Manganese-anorthite (1926)
Manganese-gehlenite (1926)
Glass, J. J., Jahns, R. H., & Stevens, R. E.
Genthelvite (1944)
Glasser, E.
Nepouite (1906)
Glebov, A. V., *v*. Serdyuchenko, D. P.
Glenn, M. L., *v*. Larsen, E. S.
Glinka, K. D.
Hydrothomsonite (1906)
Glinka, S. F., & Antipov, I. A.
Plumbomalachite (1901)
Glocker, E. F.
Liparite (1847)
Nitronatrite (1847)
Poliopyrites (1839)
Uranopissite (1847)
Glöckner, F.
Zittavite (1912)
Gmelin, L.
Kali-harmotom (1825)
Godlevsky, M. N.
Aidyrlite (1934)
Kurnakovite (1940)
——, *v*. Alekseeva, E. F.
——, *v*. Ikornikova, N. Y.
Godovikov, A. A.
Argentoaikinit, etc. (1972)
Bismutodiaphorite (1972)
Cuprocosalite (1972)
Plumbomatildite (1972)
——, Kochetkova, K. V., & Lavrent'ev, Yu. G.
Joseïte C (1970)
Joseïte D (1971)
Goffé, B., & Saliot, P.
Magnesiocarpholite (1977)
Goiko, E. A., *v*. Bussen, I. V.
——, *v*. Menshikov, Yu. P.
——, *v*. Merkov, A. N.
Goldin, B. A., Yuskin, N. P., & Fishman, M. V.
Chernovite (1967)
Goldschlag, M.
Aluminiumepidot (1916)
Eisenepidot (1915)
Goldschmidt, H., *v*. Füchtbauer
Goldschmidt, H. J., & Rait, J. R.
Manganese-merwinite (1943)
Goldschmidt, V. M.
Epidote-orthite (1911)
Mangan-wollastonite (1911)
Goldsmith, J. R.
Cadmium-dolomite (1958)
Gallium-albite (1950)
Gallium-anorthite (1950)
Gallium-orthoclase (1950)
——, *v*. Graf, D. L.
Golovanov, I. M., *v*. Badalov, S. T.
Gomes, J. P.
Libollite (1898)
Goñi, J.-C., *v*. Louis, M.
Gonnard, F.
Lassolatite (1876)
Natromontebrasite (1913)
Gonyer, F. A., *v*. Berman, H.
Goodman, N. R.
Vibertite (1957)
Goodyear, J., *v*. Duffin, W. J.
Goranson, E. A., *v*. Palache, C.
——, *v*. Schaller, W. T.
Gordillo, C., *v*. de Abeledo, M. E. J.
Gordillo, C. E., Linares, E., Toubes, R. O., & Winchell, H.

Gross, E. B.—*cont.*
Metaheinrichite (1958)
——, Wainwright, J. E. N., & Evans, B. W.
Pabstite (1965)
——, *v.* Pabst, A.
——, *v.* Rosenzweig, A.
——, *v.* Thompson, M. E.
Gross, S.
Hatrurite (1977)
Grosspietsch, O.
Eichbergite (1911)
Groth, P.
Blätterserpentin (1898)
Faserserpentin (1898)
Manganspinel (1847)
Picroilmenite (1898) (error in 2nd List)
Pleysteinite (1916)
——, & Mieleitner, K.
Pseudogymnite (1921)
Grout, F. F.
Pseudo-laumontite (1910)
Grum-Grzhimailo, S. V., *v.* Gritsaenko, G. S.
Gruner, E.
Pseudo-kaliophilite (1935)
——, *v.* Jahn, A.
Gruner, J. W.
Ammonium-mica (1939)
Groutite (1945)
Iron-serpentine (1936)
Iron-talc (1944)
Magnesiosussexite (1932)
Minnesotaite (1944)
Oakite (1943)
Grünling, F.
Maucherite (1913)
Gruzdev, V. S., Mchedlishvili, N. M., Terekhova, G. A., Tsertsvadze, Z. Ya., Chernitsova, N. M., & Shumkova, N. G.
Tvalchrelidzeite (1975)
——, Stepanov, V. I., Shumkova, N. G., Chernitsova, N. M., Yudin, R. H., & Bryzgalov, I. A.
Galkhaite (1972)
Gübelin, E.
Chrome-jadeite (1965)
Jade-albite (1965)
Gude, A. J. III, & Sheppard, R. A.
Silhydrite (1972)
——, *v.* Raup, O. B.
——, *v.* Sheppard, R. A.
Guild, F. N.
Flagstaffite (1920)
Guillemin, C.
α-Duftite, β-Duftite (1956)
Lavendulanite (1956)
Vesigniéite (1955)
——, Johan, Z., Laforêt, C., & Picot, P.
Pierrotite (1970)
——, & Protas, J.
Wyartite (1959)
——, *v.* Curien, H.
——, *v.* Louis, M.
Guimarães, C. P.
Calogerasite (1944)
Djalmaite (1939)
Guimarães, D.
Arrojadite (1925)
Eschwegeite (1926)
Giannettite (1948)
Pennaite (1948)
Guiyang Institute of Geochemistry
Laihunite (1976)
Gvakhariya, G. A., & Nazarov, Yu. I.
Ferroalunite (1963)
Gysin, M., *v.* Duparc, L.

Haberlandt, H., & Schiener, A.
Gastunite (1951)
Neogastunite (1951)
Hackman, V., *v.* Ramsay, W.
Hägele, G., & Machatschki, F.
Alumopharmacosiderite (1937)
Haggerty, S. E.
Ortho-armalcolite (1973)
Para-armalcolite (1973)
——, *v.* Anderson, A. T.
Hahn, F. V.
Cornuite (1925)
Haidinger, W.
Palaeo-amphibole (1854)
Palaeo-calcite (1854)
Haji-Vassiliou, A., & Puffer, J. H.
Attapulgite-palygorskite (1975)
Hak, J., Johan, Z., Kuacek, M., & Liebscher, W.
Kemmlitzite (1969)
——, ——, & Skinner, B. J.
Kutinaite (1970)
——, *v.* Johan, Z.
Häkli, A., *v.* Vuorelainen, Y.
Häkli, T. A., Vuorelainen, Y., & Sahama, Th. G.
Kitkaite (1965)
——, *v.* Huhma, M.
Halferdahl, L. B.
Magnesium chloritoid (1961)
Hall, A. L.
Amosite (1918)
Hall, M. *v.* Milton, C.
Hall, S. R., *v.* Cabri, L. J.
Hallimond, A. F.
Bassetite (1915)
Iron-chlorite (1939)
Meta-torbernite (1916)
Soda-glauconite (1922)
Uranospathite (1915)
Hamberg, A.
Manganpennine (1890)
Hamilton, P. K., *v.* Kerr, P. F.
Han, Yu-T'en, *v.* Ch'u, I-Hua
Hancock, R. G. V., *v.* Mandarino, J. A.
Handmann, R.
Aglaurite (1907)
Hanks, H. G.
Arizonite (before 1878)

Harada, K., Nagashima, K., & Kanisawa, S.
Masutomilite (1976)
——, ——, Nakao, K., & Kato, A.
Hydroxylellestadite (1971)
Harada, Z.
Heikkolite (1939)
Isiganeite (1948)
Kalianorthoclas (1936)
Kali-barium-felspar (1948)
Medamaite (1948)
Nogisawaite (1954)
Haranczyk, C.
Boleslavite (1961)
——, & Prchazka, K.
Sobotkite (1974)
Harbort, E.
Zirklerite (1928)
Harcourt, G. A., *v.* Berman, H.
Harding, R. R., *v.* Hawkes, J. R.
Harnick, A. B., *v.* Petter, W.
Harrassowitz, H.
Allite (1926)
Monohydrallite (1927)
Siallite (1926)
Trihydrallite (1927)
Harrington, B. J., *v.* Adams, F. D.
Harris, D. C.
Iridarsenite (1974)
Ruthenarsenite (1974)
——, Cabri, L. J., & Kaiman, S.
Athabascaite (1970)
——, Jambor, J. L., Lachance, G. R., & Thorpe, R. I.
Tintinaite (1968)
——, & Nuffield, E. W.
Bambollaite (1972)
——, *v.* Cabri, L. J.
——, *v.* Nuffield, E. W.
——, *v.* Petruk, W.
——, *v.* Thorpe, R. I.
Harrison, J. B.
Palladium-amalgam (1925)
Potarite (1925)
Harvey, Y., *v.* Perrault, G.
Hata, S.
Abukumalite (1938)
Yamaguchilite (1938)
——, *v.* Iimori, S.
Hatch, R. A., Humphrey, R. A., & Worden, E. C.
Boron-phlogopite (1956)
——, *v.* Comeforo, J. E.
Hatle, E.
Erzbergite (1892)
Hausen, D. M.
Metaschoderite (1960)
Schoderite (1960)
Hauser, O.
Risörite (1908)
Schaumopal (1911)
Uhligite (1909)
——, & Finckh, L.
Plumboniobite (1909)
Hausmann, J. F. L.
Fluobaryt (1847)
Hawkes, J. R., Merriman, R. J., Harding, R. R., & Darbyshire, D. P. F.
Bazirite (1975)
Hawkins, A. C., & Shannon, E. V.
Canbyite (1924)
Hawley, J. E., & Berry, L. G.
Froodite (1958)
Michenerite (1958)
Hay, R. L., *v.* Sheppard, R. A.
Hayase, K.
Horobetsuite (1955)
Hayashi, A., *v.* Sakurai, K.
Hayashi, H., Nakayama, N., Yoshida, M., Kozuka, T., Mizuno, M., Yamamoto, K., Yamamoto, T., & Noguchi, T.
Cadmium olivine (1964)
——, Yamamoto, N., Yoshida, M., Mizuno, M., Noguchi, T., & Yamamoto, K.
Strontiumolivine (1964)
Hayes, A. A.
Ooguanolite (1855)
Hayton, J. D., *v.* Rowledge, H. P.
Headden, W. P.
Doughtyite (1905)
Heberdey, P. P.
Bleizinkcrysolith (1892)
Hecht, A.-M., & Geissler, E.
Fluoromontmorillonite (1972)
Heddle, M. F.
Discachatae (1901)
Haemachatae (1901)
Haema-ovoid-agates (1901)
Oonachatae (1901)
Scyelite (1883)
Urquhartite (1878)
Heimbach, W., *v.* Eckhardt, F. J.
Heixner, H.
Nickel-cabrerite (1951)
Hellner, E., *v.* Rösch, H.
Hemi, C., Kusachi, I., Henni, K., Sabine, P. A., & Young, B. R.
Bicchulite (1973)
Henderson, E. P.
Steigerite (1935)
——, & Hess, F. L.
Corvusite (1933)
Fervanite (1931)
Rilandite (1933)
——, *v.* Fredriksson, K.
——, *v.* Fuchs, L. H.
——, *v.* Pough, F. H.
——, *v.* Ross, C. S.
——, *v.* White, J. S.
Hendricks, S. B., *v.* Alexander, L. T.
Henglein, M.
Cobaltnickelpyrite (1914)
——, & Meigen, W.
Barthite (1914)
Hénin, S., *v.* Caillère, S.
Henni, K., *v.* Hemi, C.

Henriques, A.
Iron-wagnerite (1957)
Henry, N. F. M.
Orthoferrosilite (1935)
——, *v.* Tilley, C. E.
Hentschel, G.
Mayenite (1964)
——, & Kuzel, H.-J.
Strätlingite (1976)
Herbillon, A., *v.* Gastuche, M.-C.
Herbstein, F. H., *v.* de Villiers, P. R.
Hermann, H. R.
Osmite (1836)
Uranelain (1832)
Heron, A. M.
Vanado-magnetite (1936)
Herpin, P., & Pierrot, R.
Weilite (1963)
——, *v.* Bariand, P.
Herrmann, V., *v.* Götz, W.
Hery, B., *v.* Agrinier, H.
Herz, W.
Salvadorite (1896)
Herzenberg, R.
Ahlfeldite (1935)
Blockite (1935)
Brunckite (1938)
Gumucionite (1932)
Hochschildite (1942)
Kolbeckine (1932)
Montesite (1949)
Rooseveltite (1946)
Souxite (1946)
Hess, F. L.
Radiofluorite (1931)
Rauvite (1922)
Vanoxite (1924)
——, & Fahey, J. J.
Caesium-biotite (1932)
——, & Foshag, W. F.
Rossite (1926)
——, & Schaller, W. T.
Pintadoite (1914)
Uvanite (1914)
——, & Wells, R. C.
Brannerite (1920)
——, *v.* Foshag, W. F.
——, *v.* Henderson, E. P.
Hess, H. H.
Endiopside (1941)
Ferroaugite (1941)
Ferrosalite (1941)
——, & Phillips, A. H.
Ferrohypersthene (1940)
Hettinger, W. P., *v.* Van Nordstrand, R. A.
Hewett, D. F.
Patronite (1906)
Quisqueite (1907)
——, & Shannon, E. V.
Orientite (1921)
——, *v.* Radtke, A. S.
Hey, M. H.
Iron-kaolinite (1936)
Metathomsonite (1932)
Pseudo-edingtonite (1934)
——, & Bannister, F. A.
Ashcroftine (1932)
Russellite (1938)
——, *v.* Bannister, F. A.
——, *v.* Claringbull, G. F.
——, *v.* Davis, R. J.
——, *v.* Donnay, G.
——, *v.* Drugman, J.
——, *v.* Gaines, R. V.
——, *v.* Leavens, P. B.
——, *v.* Smith, W. C.
Heyl, A. V., Milton, C., & Axelrod, J.
Honessite (1956)
Hezner, L.
Chrom-brugnatellite (1912)
Hibsch, J. E.
Gränzerite (1928)
Hicks, W. B., *v.* Larsen, E. S.
Hidden, W. E., & Pratt, J. H.
Rhodolite (1898)
——, & Warren, C. H.
Yttrocrasite (1906)
Hiemstra, S. A., & de Waal, S. A.
Nimite (1968)
Willemseite (1968)
——, *v.* Mihalik, P.
Higgins, D. F.
Collbranite (1918)
Hildebrand, F. A., *v.* Bailey, E. H.
Hilgard, E. W.
Auxite (1916)
Lucianite (1916)
Hill, D. K., *v.* Anderson, B. W.
Hill, P. G., *v.* Upton, B. G. J.
Hill, R. E. T., *v.* Barker, W. W.
Hill, W. L., Faust, G. T., & Reynolds, D. S.
Trömelite (1944)
Hillebrand, W. F.
Bravoite (1907)
——, Merwin, H. E., & Wright, F. E.
Hewettite (1914)
Metahewettite (1914)
Pascoite (1914)
——, & Penfield, S. L.
Natroalunite (1902)
Natrojarosite (1902)
Plumbojarosite (1902)
——, & Schaller, T. W.
Mercurammonite (1909)
——, *v.* Canfield, F. A.
——, *v.* Lindgren, W.
Hinderman, J. R.
Stringhamite (1976)
Hinthorne, J. R., *v.* Wise, W. S.
Hintze, C.
Ammonium-cryolite (1913)
Analcidite (1897)
Manganfayalite (1897)
Pseudokrokydolith (1937)
Hirowatari, F., *v.* Murakami, N.

Hlawatsch, C.
Iron-åkermanite (1904)
Osannite (1906)
Raspite (1897)
Spencerite (1903)
Vogtite (1907)
——, *v.* Doht, R.
Ho, T. L.
Beiyinite (1935)
Oborite (1935)
Hobbs, W. H.
Goldschmidtite (1899)
Hodge, E. T.
Feloid (1924)
Hodge-Smith, T.
Sturtite (1930)
Hoffmann, G. C.
Baddeckite (1898)
Souesite (1905)
Hofmann, U., *v.* Endell, K.
——, *v.* Strese, H.
——, *v.* Weiss, A.
Hofmann-Degen, K.
Eisen-Åkermanit (1919)
Justite (1919)
Hofmeister, W., & Tilmanns, E.
Arsenbrackebuschite (1976)
Hogarth, D. D., Chao, G. Y., Plant, A. G., & Steacy, H. R.
Caysichite (1974)
——, & Miles, N.
Wakefieldite (1969)
Holden, E. F.
Ceruleofibrite (1922)
Hollingworth, S. E., *v.* Bannister, F. A.
Holmes, A.
Chrome-acmite (1937)
Potash-nepheline (1936)
Holmes, R. J.
Cobalt-löllingite (1942)
Iron-skutterudite (1942)
Holmquist, P. J.
Knopite (1894)
Holroyd, W. F., *v.* Barnes, J.
Honda, T., *v.* Kinoshita, K.
Honea, R. M.
Ammonium boltwoodite (1961)
Gastunite (1959)
Sodium boltwoodite (1961)
——, & Beck, F. R.
Cambersite (1962)
——, *v.* Frondel, C.
——, *v.* Hurlbut, C. S.
Honnorez-Guerstein, B. M.
Zn-Fahlerz (1971)
Horne, J. E. T., *v.* Bowie, S. H. U.
Hornung, G., *v.* Nixon, P. H.
Hough, G. J.
Cocinerite (1919)
Hovey, H. C.
Helictite (1882)
Hövig, P.
Truscottite (1914)
How, H.
Raphite
Stiberite
Howe, H. M.
Cementite (1890)
Ferrite (1890)
Sorbite (1890)
H[owie], R. A.
Fe-pennantite (1970)
Magnesium-pennantite
Hřichová, R., *v.* Bauer, J.
Hsieh, Hsien-Te; Chien, Tze-Chiang; & Liu, Lai-Pau
Carboborite (1964)
——, *v.* Chün, I-Hua
——, *v.* Tu-Chih, Kuang
Hsien-Chueh, Wang, *v.* Chi-Ti, Kuo
Hsien-Hsien, Yeh, *v.* Van-Kang, Huang
Huang, Wen-Hui; Tu, Shao-Hua; Wang, K'ung-Hai; Chao, Chun-Lin; & Yu, Cheng-Chih
Hsianghualite (1958)
Huang, W. T.
Titanoclinohumite (1957)
Huang-Chuan, Hou, *v.* Chi-Ti, Kuo
Huckenholz, H. G., Schairer, J. F., & Yoder, H. S. Jr.
Ferri-diopside (1969)
Hudson, D. R.
Argental (1943)
Landsbergite (1943)
——, Wilson, A. F., & Threadgold, I. M.
Musgravite (1967)
——, *v.* Barker, W. W.
Hugill, W.
Mellorite (1939)
Huhma, A., *v.* Vuorelainen, Y.
Huhma, M., Vuorelainen, Y., Hakli, T. A., & Papunen, H.
Haapalaite (1973)
——, *v.* Kuovo, O.
Humphrey, R. A., *v.* Hatch, R. A.
Hupei Geologic College
Hydroastrophyllite (1974)
Hurlbut, C. S.
Aminoffite (1937)
Bermanite (1936)
Bikitaite (1957)
Parahilgardite (1938)
Plumbosynadelphite (1937)
Sampleite (1942)
——, & Aristarain, L. F.
Beusite (1968)
Olsacherite (1969)
Rivadavite (1967)
——, & Erd, R. C.
Aristarainite (1974)
——, & Honea, R.
Sigloite (1962)
——, & Taylor, R. E.
Hilgardite (1937)
——, *v.* Aristarain, L. F.

Hunt, T. S.
Hamelite (1886)
Keramite (1886)
Hussak, E.
Chalmersite (1902)
Gorceixite (1906)
Harttite (1906)
Picroilmenite (1895)
——, & Prior, G. T.
Derbylite (1897)
Florencite (1899)
Lewisite (1895)
Senaite (1898)
Tripuhyite (1897)
Hutchinson, A.
Stokesite (1899)
Hutchison, D., *v.* Livingstone, A.
Hutton, C. O.
Ferrostilpnomelane (1938)
Hydrogrossular (1943)
Magnesium-morenosite (1947)
Nickel-epsomite (1947)
Strontium-aragonite (1936)
Yavapaiite (1959)
——, & Vlisidis, A. C.
Papagoite (1960)
Hybler, J., *v.* Zak, L.
Hytönen, K., *v.* Sahama, T. G.

I.M.A. amphibole subcommittee (1978)
Alumino-barroisite
Alumino-ferro-hornblende
Alumino-katophorite
Alumino-magnesio-hornblende
Alumino-taramite
Alumino-tschermakite
Alumino-winchite
Ferri-barroisite
Ferri-katophorite
Ferri-taramite
Ferri-winchite
Ferro-alumino-barroisite
Ferro-alumino-tschermakite
Ferro-alumino-winchite
Ferro-clinoholmquistite
Ferro-ferri-barroisite
Ferro-ferri-tschermakite
Ferro-ferri-winchite
Ferro-holmquistite
Ferro-hornblende
Ferro-kaersutite
Ferro-pargasite
Magnesio-alumino-katophorite
Magnesio-clinoholmquistite
Magnesio-ferri-katophorite
Magnesio-ferri-taramite
Magnesio-gedrite
Magnesio-holmquistite
Magnesio-hornblende
Sodium-anthophyllite
Sodium-gedrite
I.M.A. pyrochlore subcommittee
Bariomicrolite (1978)
Bariopyrochlore (1978)
Ceriopyrochlore (1978)
Kalipyrochlore (1978)
Stannomicrolite (1978)
Yttropyrochlore (1978)
Ianovici, V., & Neacşu, Gh.
Murgocite (1970)
Iddings, J. P.
Potash-oligoclase (1906)
Soda-orthoclase (1906)
Igelström, L. J.
Bliabergite (1897)
Dicksbergite (1896)
Elfstorpite (1893)
Gersbyite (1897)
Manganberzeliite (1894)
Munkforssite (1897)
Munkrudite (1897)
Ransätite (1896)
Rhodophosphite (1895)
Talkknebelite (1890)
Tetragophosphite (1895)
I-Hsien, Wang, *v.* Chi-Ti, Kuo
Iimori, S.
Victoria-stone (before 1969)
——, & Yoshimura, J.
Takizolite (1929)
——, ——, & Hata, S.
Nagatelite (1931)
Iiyama, T., *v.* Miyashiro, A.
Ikornikova, N. Y., & Godlevsky, M. N.
Metahydroboracite (1941)
Ilinsky, G. A.
Hydrohaüyne (1962)
Illarionov, A. A.
Kmaite (1961)
Ilyukhin, V. V., *v.* Dorfman, M. D.
——, *v.* Gan'ev, R. M.
Indolev, L. N., Zhdanov, Yu. Ya., Kashertseva, K. I., Soknev, V. S., & Del'yanidi, K. I.
Indigirite (1971)
Ingamells, C. O., *v.* Donnay, G.
Ingram, B., *v.* Milton, C.
——, *v.* Thompson, M. E.
Insley, H., *v.* Alexander, L. T.
Irving, A. J., *v.* Moore, P. B.
Isakov, E. N., *v.* Smolyaninov, N. A.
Iskyul, E. V., *v.* Maslenitzky, I. N.
Isobe, H., & Watanabe, T.
Kanbaraite (1930)
Isphording, W. C., & Lodding, W.
Macrokaolinite (1968)
Istrati, C. I.
Moldovite (1897)
Pietricikite (1897)
Roumanite (1895)
——, & Mihailescu, M. A.
Albanite (1912)
Ito, J., & Frondel, C.
Lead-barylite (1968)
Strontium-barylite (1968)
——, *v.* Frondel, C.

Karup-Møller, S.—*cont.*
Eskimoite (1977)
Gustavite (1970)
Ourayite (1977)
Treasurite (1977)
Vikingite (1977)
——, & Makovicky, E.
Skinnerite (1974)
Kashaev, A. A., *v.* Konev, A. A.
——, *v.* Vladykin, N. V.
Kashertseva, K. I., *v.* Indolev, L. N.
Kashima, N.
Anthodite (1965)
Kashkai, M.-A.
Hydronatrojarosite (1969)
Oxonio-alunite (1969)
Plumboalunite (1969)
Zincalunite (1969)
——, & Aliev, R. M.
Calciocopiapite (1960)
Tusiite (1960)
——, & Mamedov, A. I.
Istisuite (1955)
Kashner, E. F.
Gerstmannite (1976)
Kashpar, P., *v.* Vinogradova, R. A.
Kašpar, J. V.
Hanušite (1942)
Němecite (1941)
Ondřejite (1944)
Kataeva, Z. T., *v.* Dusmatov, V. D.
——, *v.* Efimov, A. F.
Katayama, N.
Prelaumontite (1958)
Kato, A.
Ikunolite (1959)
Kawazulite (1970)
Sakuraiite (1965)
Stannoidite (1969)
——, & Nagashima, K.
Iimoriite (1970)
——, Sakurai, K. I., & Ohsumi, K.
Wakabayashilite (1970)
——, *v.* Harada, K.
——, *v.* Watanabe, T.
——, *v.* Yoshii, M.
Kato, T., *v.* Murakami, N.
——, *v.* Tschirch, A.
——, *v.* Yoshii, M.
Katz, L., & Lipscomb, W. N.
Iron-lazulite (1951)
Katzer, F.
Hoeferite (1895)
Poechite (1911)
Kautz, K., *v.* Geier, B. H.
Kawai, K.
Reniforite (1925)
Kawai, T.
Aluminobetafite (1960)
Nogizawalite (1949)
Kayupova, M. M.
Silicomanganberzeliite (1963)
Kazakov, A. V.
Neoglauconite (1938)
Kazakova, M. E.
Beryllium-felspar (1946)
——, *v.* Borodin, L. S.
——, *v.* Dusmatov, V. D.
——, *v.* Efimov, A. F.
——, *v.* Eskova, E. M.
——, *v.* Gerasimovsky, V. I.
——, *v.* Kornetova, V. A.
——, *v.* Kuzmenko, M. V.
——, *v.* Semenov, E. I.
——, *v.* Tikhonenkov, I. P.
——, *v.* Tikhonenkova, R. P.
——, *v.* Zhabin, A. G.
Kazanskaya, E. V., *v.* Simonov, M. A.
Kazitzyn, Y. V.
Bobkovite (1955)
Ferutite (1954)
——, *v.* Bobkov, N. A.
Keep, F. E.
Maufite (1930)
Keester, K. L., *v.* Moore, P. B.
Keidel, J., *v.* Borchert, W.
Keil, K., & Brett, R.
Heideïte (1974)
——, & Snetsinger, K. G.
Niningerite (1967)
——, *v.* Andersen, C. A.
——, *v.* Anderson, A. T.
——, *v.* Lovering, J. F.
Keith, C. D., *v.* Van Nordstrand, R. A.
Keith, M. L., *v.* Yoder, H. S.
Kelly, A.
Conchite (1900)
Kelly, C. J.
Eicosyl alcohol (1970)
Kelly, P., *v.* Gatehouse, B. M.
Kennedy, W. Q., & Dixon, B. E.
Hydro-amphibole (1936)
Keppler, U., *v.* Petter, W.
Kerr, J. H., *v.* Pecora, W. T.
Kerr, P. F.
Berkeyite (1926)
Cattierite (1945)
Tungomelane (1940)
Vaesite (1945)
——, & Brophy, G. P.
Umohoite (1953)
——, & Hamilton, P. K.
Potash-bentonite (1949)
——, & Young, F.
Hydrotungstite (1940)
——, *v.* Barrington, J.
——, *v.* Omori, K.
——, *v.* Ross, C. S.
——, *v.* Vaes, J. F.
Khalezova, E. B., *v.* Sørensen, C.
Khalife-Zade, C. M.
Gyulekhite (1957)
Khalturina, I. I., *v.* Avrova, N. P.
——, *v.* Bocharov, V. M.
Khan, G. A., *v.* Krol, O. F.

Kharitonov, Yu. A., *v.* Gan'ev, R. M.
Khmaruk, T. E., & Shcherbakov, I. B.
Ferrisalites (1963)
Khomyakov, A. P., Semenov, E. I., Eskova, E. M., & Voronkov, A. A.
Kazakovite (1974)
——, Stepanov, V. I., Moleva, V. A., & Pudovkina, Z. V.
Tikhonenkovite (1964)
——, Voronkov, A. A., Lebedeva, S. I., Bykov, V. P., & Yurkina, K. V.
Khibinskite (1974)
——, *v.* Bussen, I. V.
——, *v.* Dusmatov, V. D.
——, *v.* Eskova, E. M.
——, *v.* Kapustin, Yu. L.
——, *v.* Menshikov, Yu. P.
——, *v.* Semenov, E. I.
Khoroshilova, L. A., *v.* Dusmatov, V. D.
Khvostova, V. A.
Uranohydrothorite (1969)
——, & Arkhangel'skaya, V. N.
Mangantapiolite (1970)
Kieft, C., *v.* Burke, E. A. J.
——, *v.* de Roever, E. W. F.
——, *v.* Klominsky, J.
——, *v.* Oen, I. S.
Kilpady, S., & Dave, A. S.
Mangan-uralite (1958)
Kim, S. J.
Janggunite (1976)
Kimizuka, K.
Potash-anorthoclase (1932)
Kimura, K.
Hagatalite (1925)
Ishikawaite (1922)
Oyamalite (1925)
Yamagutilite (1933)
——, & Miyake, Y.
Enalite (1932)
Kimura, M., *v.* Yoshimura, T.
Kinahan, G. H.
Alumyte (1889)
King, I. M., King, W. J., & Wallis, R.
Sodium-mordenite (1968)
King, W. J., *v.* King, I. M.
Kingston, G. A.
Merenskyite (1966)
Kingston, P. W.
Nuffieldite (1968)
Kinosaki, Y.
Heikkolite (1935)
Kinoshita, K.
Bunkolite (1927)
——, Tanaka, N., & Honda, T.
Nickel-montmorillonite (1958)
——, *v.* Kōzu, S.
Kirchner, E., *v.* Chen, T. T.
Kircher, H., *v.* Noll, W.
Kirilov, A. S.
Hydroxyl-bastnäsite (1964)
Kirnbauer, F., *v.* Dittler, E.
Kirov, G. N., *v.* Atanassov, V. A.
Kirsch, G.
Ulrichite (1925)
Kisenezova, N. N., *v.* Rogova, V. P.
Kišpatić, M.
Sporogelite (1912)
Kiss, J.
Metaliebigite (1966)
Kissin, S. A., Owens, D. R., & Roberts, W. L.
Černýite (1978)
Kita-Badak, M., *v.* Dominic, J.
Kitamura, T., *v.* Nambu, M.
Kiziyarov, G. P., *v.* Rogova, V. P.
Klaproth, M. H.
Grönlandit (1809)
Klebs, R.
Cedarite (1897)
Klein, C.
Chromocyclite (1892)
Stoffertite (1901)
——, *v.* Frondel, C.
——, *v.* Fuchs, L. H.
Klein, C. Jr., *v.* Chao, E. C. T.
Klein, E., *v.* de Roever, E. W. F.
Klemm, G.
Viridine (1912)
Klevtsova, R. F., *v.* Ermilova, L. P.
Klingsberg, C., & Roy, R.
Gallium-phlogopite (1957)
Nickel-phlogopite (1957)
Klockmann, F.
Mangankiesel (1895)
Manganomelane (1922)
Klominsky, J., Rieder, M., Kieft, C., & Mraz, L.
Heyrovskyite (1971)
Klvaňa, J.
Paracoquimbite (1882)
Knett, J.
Radiobaryt (1904)
Knight, W. C.
Bentonite (1898)
Taylorite (1897)
Knipovich, Yu. N., *v.* Bulakh, A. G.
Knop, A.
Titanaugite (1892)
Titanmelanite (1892)
Knopf, A.
Vegasite (1915)
——, & Schaller, W. T.
Hulsite (1908)
Paigeite (1908)
Knorring, O. von, *v.* von Knorring, O.
Knyazeva, D. N., *v.* Lapin, V. V.
——, *v.* Sveshnikova, E. V.
Kobayashi, H.
Kanoite (1977)
Koblencz, V., *v.* Erdélyi, J.
Koch, G., *v.* Weiss, A.
Koch, O.
Parathuringite (1884)
Pseudothuringite (1884)

Koch, R. A.
Petersberg-Illite (1958)
Koch, S.
Csiklovaite (1948)
Matraite (1958)
——, *v*. Finály, I.
——, *v*. Zechmeister, L.
Kochetkova, K. V., *v*. Godovikov, A. A.
Kochmarev, A. T., *v*. Kovalevsky, S. A.
Koděra, M., Kupčik, V., & Makovický, E.
Hodrushite (1970)
Koechlin, R.
Cuproarquerite (1928)
Kuttenbergit (1928)
Natronkalisimonyit (1902)
Schmöllnitzit (1928)
Stibioniobite (1928)
Tangiwaite (1911)
——, *v*. Dittler, E.
Koenig, G. A.
Argento-domeykite (1903)
Aurobismuthinite (1912)
Keweenawite (1902)
Melanochalcite (1902)
Mohawk-algodonite (1902)
Mohawk-whitneyite (1900)
Semi-whitneyite (1902)
Stibiobismuthinite
Stibio-domeykite (1900)
Kohanowski, N. N.
Farallonite (1953)
Siliceous scheelite (1953)
Kohler, K., *v*. Schoep, A.
Kohls, D. W., & Rodda, J. L.
Gaspéite (1966)
Iowaite (1967)
Kohn, J. A., & Comeforo, J. E.
Boron-edenite (1955)
Fluor-edenite (1955)
Koiké, S.
Mangananankerite (1935)
Kokkoros, P.
Ktenasite (1950)
Kokta, J.
Cuprojarošite (1937)
Jarošite (1937)
Kolaczkowska, M.
Chacaltaite (1936)
Kolbeck, F.
Weisbachite (1907)
Komada, K., *v*. Wakabayashi, Y.
Komkov, A. I., & Nefedov, E. I.
Posnjakite (1967)
——, *v*. Buryanova, E. Z.
Komkov, E. A., *v*. Alekseeva, M. A.
Kondrateva, V. V.
Tyretskite (1964)
——, Ostrovskaya, I. V., & Yarzhemski, Ya. Ya.
Volkovskite (1966)
——, *v*. Blazko, L. N.
——, *v*. Bulakh, A. G.
——, *v*. Kukharenko, A. A.
Konev, A. A., Lebedeva, V. S., Kashaev, A. A., & Ushchapovskaya, Z. F.
Azoproite (1970)
Titanoludwigite (1970)
——, Ushchapovskaya, Z. F., Kashaev, A. A., & Lebedeva, V. S.
Tazheranite (1969)
Konkova, E. A., *v*. Karpova, Kh. N.
Konno, H., *v*. Omori, K.
Kopchenova, E. V., & Sidorenko, G. A.
Bearsite (1962)
——, Skvortsova, K. V., Silanteva, N. I., Sidorenko, G. A., & Mikhailova, L. V.
Mourite (1962)
Korin, I. Z.
Ferrimontmorillonite (1939)
Koritnig, S.
Protopartzite (1967)
——, & Süsse, P.
Meixnerite (1975)
Kornetova, V. A., Aleksandrov, V. B., & Kazakova, M. E.
Tantal-aeschynite (1963)
Korolev, K. G., *v*. Razumnaya, E. G.
Korolev, Yu. M.
Potassium allevardite (1965)
——, *v*. Lazarenko, E. K.
Korovashkin, V. V., *v*. Chukhrov, F. V.
Kosals, Ya. A.
Manganotantalocolumbite (1967)
Tantalotitanocolumbite (1967)
Kosmann, H. B.
Hydrocalcite (1892)
Kossovskaya, A. G., *v*. Drits, V. A.
Kostov, I.
Bonchevite (1958)
Ferrorhodonite (1968)
Orpheite (1968)
Sakharovaite (1959)
Tetrahydrite (1968)
Venaite (1968)
Kostyleva, E. E.
Zirconoid (1936)
Zirfesite (1945)
Kostynina, L. P., *v*. Semenov, E. I.
Koto, K., *v*. Morimoto, N.
——, *v*. Nakai, I.
——, *v*. Narita, H.
Kovalenker, V. A., Genkin, A. D., Evstigneeva, T. L., & Laputina, I. P.
Telargpalite (1974)
——, Laputina, I. P., Evstigneeva, T. L., & Izoitko, V. M.
Thalcusite (1976)
——, *v*. Evstigneeva, T. L.
Kovalenko, V. I., *v*. Vladykin, N. V.
Kovalev, G. A., *v*. Buryanova, E. Z.
Kovalevsky, S. A., & Kochmarev, A. T.
Lok-batanite (1939)
Kovyazina, V. M., *v*. Kukharenko, A. A.
Kowalski, W. M.
Magnesium hornblende (1967)
Kozhanov, S. I., *v*. Kuzmenko, M. V.

Kunz, G. F.
Azurchalcedony (1907)
Azurlite (1907)
Azurmalachite (1907)
Californite (1903)
Jadeolite (1908)
Moissanite (1905)
Morganite (1911)
Starlite (1927)
Tiffanyite (1895)
Utahlite (1895)
Kuo, Cheng-Chi
Jinigite (1959)
Yanshynshite (1959)
Kuo-Cheng, Lin, *v.* Pao-Kwei, Chang
Kuo-Fun, Fuo, *v.* Chu-Siang, Yue
Kuovo, O., Huhma, M., & Vuorelainen, Y.
Cobalt pentlandite (1959)
——, & Vuorelainen, Y.
Eskolaite (1958)
——, *v.* Long, J. V. P.
Kupa, F., *v.* Novák, F.
Kupčik, V., *v.* Koděra, M.
Kupriyanova, I. I., & Sidorenko, G. A.
Alumomelanocerite (1963)
Alumospencite (1963)
Thorobritholite (1963)
Thoromelanocerite (1963)
Yttrobritholite (1963)
Yttromelanocerite (1963)
——, Stolyarova, T. I., & Sidorenko, G. A.
Thorosteenstrupine (1962)
Kurbatov, S. M.
Chrome-vesuvian (1922)
Mangan-neptunite (1923)
Kurnakov, N. S., & Ronkin, B. L.
Sakiite (1931)
Kurshakova, L. D.
Manganseverginite (1967)
Kusachi, I., *v.* Hemi, C.
Kusnestova, N. N., *v.* Rogova, V. P.
Küstel, G.
Belmontite (1911)
Kutukova, E. I.
Titan-låvenite (1940)
Kuzel, H.-J., *v.* Hentschel, G.
Kuzmenko, M. V.
Beryllite (1954)
——, & Kazakova, M. E.
Nenadkevichite (1955)
——, & Kozhanov, S. I.
Karnasurtite (1959)
——, *v.* Vlasov, K. A.
Kuzmina, O. V., *v.* Organova, N. I.
Kuznetsova, N. N., *v.* Nazarova, A. S.
——, *v.* Novikova, M. I.
——, *v.* Rogova, V. P.
Kvaček, M., *v.* Johan, Z.
Kvalheim, A., *v.* Barth, T. F. W.
Kyrtseva, H. H., *v.* Lapin, V. V.

Labuntzov, A. N.
Carburan (1934)
Fersmanite (1929)
Lachance, G. R., *v.* Harris, D. C.
——, *v.* Jambor, J. L.
Lacroix, A.
Ambatoarinite (1916)
Ampangabeite (1912)
Befanamite (1923)
Betafite (1912)
Bityite (1908)
Chizeuilite (1910)
Chromohercynite (1920)
Chromojadeite (1930)
Cobaltoadamite (1910)
Cokeite (1910)
Cuproadamite (1910)
Doelterite (1913)
Eugeiite (1910)
Faratsihite (1915)
Ferrobrucite (1909)
Ferropicotite (1910)
Ferrorhabdite (1909)
Ferrothorite (1923)
Flajolotite (1910)
Fluocollophanite (1910)
Fornacite (1916)
Foucherite (1910)
Furnacite (1915)
Giorgiosite (1905)
Gonnardite (1896)
Goureite (1934)
Grandidierite (1902)
Imerinite (1910)
Ktypeite (1898)
Lechatelierite (1915)
Losite (1911)
Manandonite (1912)
Manganobrucite (1909)
Mascareignite (1936)
Metacristobalite (1909)
Metathenardite (1910)
Minguetite (1910)
Palmierite (1907)
Picrocrichtonite (1900)
Plancheite (1908)
Pseudoboleite (1895)
Pseudochalcedonite (1900)
Pseudoheterosite (1910)
Pseudo-låvenite (1911)
Pseudowollastonite (1895)
Quercyite (1910)
Reaumurite (1908)
Rosieresite (1910)
Samiresite (1912)
Serandite (1931)
Seyrigite (1940)
Soumansite (1910)
Torendrikite (1920)
Vilateite (1910)
Villiaumite (1908)
——, & de Schulten, A.
Georgiadesite (1907)
Ladieva, V. D.
Chromspinellids (1964)

Mellor, J. W.
Clayite (1909)
Gaylussacite (1927)
——, & Scott, A.
Keramite (1924)
Melnikov, M. P.
Loranskite (1896)
Melnikov, V. S., *v.* Fishkin, M. Yu.
Mélon, J.
Sharpite (1938)
Viseite (1943)
Menshikov, Yu. P., Bussen, I. V., Goiko, E. A., Zabavnikova, N. I., Merkov, A. N., & Khomyakov, A. P.
Bornemanite (1975)
——, Denisov, A. P., Uspenskaya, E. I., & Lipatova, E. A.
Lovdarite (1973)
——, Pakhomovskii, Ya. A., Goiko, E. A., Bussen, I. V., & Merkov, A. N.
Natisite (1975)
——, *v.* Bussen, I. V.
——, *v.* Merkov, A. N.
——, *v.* Timoshenkov, I. M.
Menyailov, A. A.
Chloritoserpentine (1935)
Serpochlorite (1935)
Menzel, H., Schulz, H., & Deckert, H.
Metakernite (1935)
Merchukova, I. D., *v.* Poplavko, E. M.
Merkov, A. N., Bussen, I. V., Guiko, E. A., Kulitskaya, E. A., Menshikov, Yu. P., & Nedorezova, A. P.
Zorite (1973)
——, Menshikov, Yu. P., & Nedorezova, A. P.
Raite (1973)
——, *v.* Bussen, I. V.
——, *v.* Eskova, E. M.
——, *v.* Menshikov, Yu. P.
Merkulova, G. V.
Shilkinite (1939)
Merlino, S., & Orlandi, P.
Franzinite (1977)
Liottite (1977)
——, *v.* Orlandi, P.
Merriman, R. J., *v.* Hawkes, J. R.
Merwin, H. E., *v.* Hillebrand, W. F.
——, *v.* Palache, C.
——, *v.* Wyckoff, R. W. G.
Meshchankina, V. I., *v.* Begizov, V. B.
——, *v.* Razin, L. V.
Meunier, S.
Grossouvreite (1902)
Patagosite (1917)
Zoesite (1911)
Meyer, B., *v.* Scheffer, F.
Meyer, F. H.
Kansite (1957)
Meyrowitz, R., *v.* Coleman, R. G.
——, *v.* Fisher, F. G.
——, *v.* Jaffe, H. W.
——, *v.* Lindberg, M. L.
——, *v.* Moench, R. C.
——, *v.* Mrose, M. E.
——, *v.* Muto, T.
——, *v.* Osterbridge, W. F.
——, *v.* Thompson, M. E.
——, *v.* Weeks, A. D.
——, *v.* Young, E. J.
Michel-Lévy, A.
Feldspathides (1889)
Michel-Lévy, M., *v.* Wyart, J.
Midgley, H. G.
Pleochroite (1968)
——, *v.* Nurse, R. W.
Mieleitner, K., *v.* Groth, P.
Miers, H. A.
Fontainebleau sandstone (1902)
Mihailescu, M. A., *v.* Istrati, C. I.
Mihalik, P., Hiemstra, S. A., & de Villiers, J. P. R.
Atokite (1975)
Rustenbergite (1975)
Mikei, I. Y.
Elbrussite (1930)
Mikhailova, A. F., *v.* Epshtein, E. M.
Mikhailova, L. V., *v.* Kopchenova, E. V.
Mikhailova, V. A., *v.* Shishkin, N. N.
Mikheev, V. I.
Elizavetinskite (1957)
Milano S., & Satori, F.
Santite (1970)
Miles, N., *v.* Hogarth, D. D.
Miller, J. L., & Johnson, R. C.
Fluorhectorite (1962)
Fluortaeniolite (1962)
Fluortainiolite (1962)
Lithium fluor-hectorite (1962)
——, *v.* Gibbs, G. V.
Miller, S. L.
Craigite (1970)
Millosevich, F.
Cobaltocalcite (1910)
Paternoite (1920)
Millot, G.
Ghassoulite (1954)
Glauconie (1964)
Milne, A. A., *v.* Jones, L. H. P.
Milton, C., Appleman, D. E., Chao, E. C. T., Cuttitta, F., Dinnin, J. L., Duvornik, E. J., Hall, M., Ingram, B., & Rose, H. J. Jr.
Bracewellite (1967)
Grimaldiite (1967)
Guayanaite (1967)
Mcconnellite (1967)
Zinc-chromium spinel (1967)
——, & Axelrod, J.
Buetschliite (1946)
Fairchildite (1946)
——, ——, & Grimaldi, F. S.
Eitelite (1954)
Garrelsite (1955)
Reedmergnerite (1954)
——, & Blade, L. V.
Kimzeyite (1958)

Milton, C., & Ingram, B.
Nyerereïte (1963)
——, ——, Clark, J. R., & Dwornik, E. J.
Mckelveyite (1965)
——, & Johnston, W. D.
Copper-zinc-epsomite (1937)
Magnesia-goslarite (1937)
Zinc-magnesia-chalcanthite (1937)
——, Mrose, M. E., Chao, E. C. T., & Fahey, J. J.
Norsethite (1959)
——, *v.* Axelrod, J.
——, *v.* Chao, E. C. T.
——, *v.* Evans, H. T. Jr.
——, *v.* Heyl, A. V.
——, *v.* Pabst, A.
Milton, D. J., *v.* Chao, E. C. T.
Min, Ling-Sheng, *v.* Ch'u, I-Hua
Minato, H., & Takano, Y.
Iron-sericite (1952)
Magnesium-sericite (1952)
Potassium-clinoptilolite (1964)
——, *v.* Ito, T.
Minato, T., *v.* Takubo, J.
Mincheva-Stefanova, I.
Strashimirite (1968)
——, & Gorova, M.
Ferroankerite (1967)
Magnostilpnomelane (1965)
Mineev, D. A., Lavrishcheva, T. I., & Bykova, A. V.
Bastnaesite-(Yt) (1970)
——, *v.* Lutta, B. G.
Minguzzi, C.
Cuprorivaite (1938)
Minkin, J. A., *v.* Chao, E. C. T.
Mira, C.
Corencite (1939)
Miropolsky, L. M.
Al-chamosite (1936)
Mirtov, Y. V.
Parbigite (1958)
Miskovsky, J., *v.* Stanek, J.
Mitchell, R. S., *v.* Ehlmann, A. J.
——, *v.* Gross, E. B.
——, *v.* Mandarino, J. A.
Mitenkov, G. A., *v.* Rudashevskii, N. S.
Miura, Y., *v.* Murakami, N.
Miyachiro, A.
Ferroglaucophane (1957)
Miyaka, Y., *v.* Kimura, K.
Miyashiro, A.
Magnesio-arfvedsonite (1957)
Magnesiokatophorite (1957)
Magnesioriebeckite (1957)
Osumilite (1953)
Subglaucophane (1957)
——, & Iiyama, T.
Indialite (1954)
Mizuno, M., *v.* Hayashi, H.
Moench, R. C., & Meyrowitz, R.
Goldmanite (1964)
Mogensen, F.
Ferro-ortho-titanate (1946)
Titanspinel (1943)
Ulvöspinel (1943)
Moh, G. H.
Berndtite (1966)
Ottemannite (1966)
——, & Ottemann, J.
Stannoenargite (1962)
Stannoluzonite (1962)
Zinnfahlerz (1962)
——, *v.* Bültemann, H. W.
Mohseni-Koutchesfehani, Samad, & Montel, Gerard
Barim carbonate-apatite (1961)
Moissan, H.
Mysite (1905)
Mokretsova, A. V., *v.* Dorfman, M. D.
Moleva, E. M., *v.* Semenov, E. I.
Moleva, V. A., & Myasnikov, V. S.
Zinc-högbomite (1952)
——, *v.* Ermilova, L. P.
——, *v.* Ginzburg, A. I.
——, *v.* Khomyakov, A. P.
——, *v.* Nozhkin, A. D.
——, *v.* Rokov, Yu. G.
——, *v.* Smolyaninova, N. N.
Molotkov, S. P., *v.* Organova, N. I.
Momoi, H., *v.* Yoshimura, T.
Montel, Gerard, *v.* Mohseni-Koutchesfehani, Samad
Montgomery, A., *v.* Larsen, E. S.
Montgomery, J. H., Thompson, R. M., & Meagher, E. P.
Pellyite (1972)
Moore, P. B.
Ericssonite (1967)
Eveite (1967)
Ferripumpellyite (1971)
Ferri-reddingite (1964)
Ferropumpellyite (1971)
Gabrielsonite (1967)
Jahnsite (1974)
Joesmithite (1968)
Julgoldite (1967)
Manganhumite (1978)
Melanostibite (1967)
Orthoericssonite (1971)
Oxyferropumpellyite (1971)
Oxyjulgoldite (1971)
Parwelite (1967)
Robertsite (1974)
Segelerite (1974)
Stenhuggarite (1967)
Welinite (1967)
Welshite (1970)
Wermlandite (1971)
Wickmanite (1967)
——, & Araki, T.
Bahianite (1976)
Samuelsonite (1975)
——, ——, Kampf, A. R., & Steele, I. M.
Olmsteadite (1976)
——, Irving, A. J., & Kampf, A. R.

Moore, P. B.—*cont.*
Foggite (1975)
Goedkenite (1975)
——, & Ito, J.
Kidwellite (1978)
Whiteite (1978)
Wyllieite (1973)
——, & Kampf, A. R.
Schoonerite (1977)
——, ——, & Irving, A. J.
Whitmoreite (1974)
——, Lund, D. H., & Keester, K. L.
Bjarebyite (1973)
——, & Ribbe, P. H.
Esperite (1965)
——, *v.* Chao, E. C. T.
——, *v.* Olsen, E.
Moraes, L. F. de, *v.* de Moraes, L. F.
Morandi, N., & Dalrio, G.
Jamborite (1973)
Morawiecki, A.
β-Alumohydrocalcite (1961)
Moreau, J., *v.* Francotte, J.
Morey, G. W., & Bowen, N. L.
Devitrite (1931)
——, Rowe, J. J., & Fournier, R. O.
Calcium-langbeinite (1964)
Morgan, F.
Fairbanksite (1965)
Morgan, V., *v.* Erd, R. C.
——, *v.* Frondel, C.
Morignat, B., *v.* Branche, G.
Morimoto, N., & Koto, K.
Orthoenstatite (1969)
——, ——, & Shimazaki, Y.
Anilite (1969)
——, *v.* Nakai, I.
——, *v.* Narita, H.
——, *v.* Roseboom, E. H., Jr.
——, *v.* Tatsuka, K.
Morin, N., *v.* Cesbron, F.
Morozewicz, J.
Bardolite (1924)
Beckelite (1904)
Fluotaramite (1923)
Grodnolite (1924)
Lagoriolite (1898)
Lubeckite (1919)
Lublinite (1907)
Miedziankite (1923)
Staszicite (1918)
Stellerite (1909)
Taramite (1923)
Morris, R. C.
Edgarite (1962)
Morton, R. D.
Chloroxyapatite (1962)
Moses, A. J.
Eglestonite (1903)
Montroydite (1903)
Mountain, E. D.
Bismoclite (1935)
Boksputite (1935)
Rhodesite (1956)
Mozgova, N. N., Borodaev, Yu. S., Ozerova, N. A., Paakonen, V., Sveshnikova, O. L. Balitskii, V. S., & Dorogovin, B. A.
Seinäjokite (1976)
Mraz, L., *v.* Klominsky, J.
Mrose, M. E.
Calcium-uranospinite (1953)
Hurlbutite (1951)
Hydrogen-uranospinite (1950)
Meta-saléeite (1950)
Palermoite (1952)
——, Ericksen, G. E., & Marinenko, J. W.
Brüggenite (1971)
——, Meyrowitz, R., Alfors, J. T., & Chesterman, C. W.
Nissonite (1966)
——, *v.* Ericksen, G. E.
——, *v.* Evans, H. T.
——, *v.* Finkelman, R. B.
——, *v.* Milton, C.
——, *v.* Schaller, W. T.
——, *v.* von Knorring, O.
Mücke, A.
Paraboleïte (1972)
Santanaite (1972)
Scheibeite (1970)
Seeligerite (1971)
——, *v.* Seeliger, E.
Mueller, G.
Aromite (1964)
Bernalite (1964)
Carbonite (1964)
Cyclite (1964)
Mutabilite (1964)
Olefinite (1964)
Paraffinite (1964)
Muessig, S., & Allen, R. D.
Ezcurrite (1957)
Mügge, O.
Kalkorthosilikat (1926)
Mukhitdinov, G. N., *v.* Eskova, E. M.
——, *v.* Zhabin, A. G.
Mukhlya, K. A., *v.* Senchilo, N. P.
Müllbauer, F.
Baldaufite (1925)
Lehnerite (1925)
Wenzelite (1925)
Müller, G.
Sudoite (1962)
——, *v.* Geier, B. H.
——, *v.* Puchelt, H.
Müller, W. F.
Parapectolite (1976)
Paraserandite (1976)
——, *v.* Schreyer, W.
Mumme, W. G.
Junoite (1975)
Proudite (1976)
Mumpton, F. A.
Heulandite-B (1960)
——, Jaffe, H. W., & Thompson, C. S.
Coalingite (1965)

Neiva, J. M. C., *v.* Cotelo Neiva, J. M.
Neiva, J. M. Cotelo, *v.* Cotelo Neiva, J. M.
Nekrasov, I. Ya.
Silicomonazite (1972)
——, *v.* Brovkin, A. A.
——, *v.* Grigoriev, A. P.
Nekrasova, Z. A.
Metauramphite (1957)
Uramphite (1957)
Nel, H. J., Strauss, C. A., & Wickman, F. E.
Lombaardite (1949)
Nelen, J. A., *v.* Dunn, P. J.
——, *v.* Tien, Pei-Lin
——, *v.* White, J. S.
Nelson, B. W., & Roy, R.
Septechamosite (1958)
Septechlorite (1954)
Nelson, C. S., *v.* Rodgers, K. A.
Nemes-Varga, S., *v.* Tokody, L.
Nenadkevich, K. A.
Alaite (1909)
Basobismutite (1917)
Hydrobismutite (1917)
Turanite (1909)
Tyuyamunite (1912)
Zinc-dibraunite (1911)
Nesterova, Yu. S., *v.* Shadlun, T. N.
——, *v.* Skvortsova, K. V.
Netto, J. E. de Paiva, *v.* de Paiva Netto, J. E.
Neumann, B. S., & Sanson, K. G.
Laponite (1970)
Neumann, H.
Armenite (1939)
——, & Nilssen, B.
Tombarthite (1968)
——, *v.* Bergstøl, S.
Neves, J. M. Correia, *v.* Correia Neves, J. M.
Niccoli, E.
Mellahite (1925)
Nickel, E. H.
Cuprospinel (1973)
Latrappite (1964)
Niocalite (1956)
——, & Bridge, P. J.
Nickelblödite (1977)
——, Robinson, B. W., Davis, C. E. S., & MacDonald, R. D.
Otwayite (1977)
——, Rowland, J. F., & Charette, D. J.
Niobophyllite (1964)
——, ——, & McAdam, R. C.
Pseudo-ixiolite (1963)
Wodginite (1963)
——, *v.* Barker, W. W.
Nicol, A. W., *v.* Tarney, J.
Nicolau, T.
Calciodialogite (1910)
Calciorhodochrosite (1910)
Nicolet, S. E.
Duparcite (1932)
Nicoletti, M., *v.* Belluomini, G.
Niedźwiedzki, J.
Delatynite (1908)
Niggli, E., *v.* Jäger, E.
Niggli, P.
Ferri-gehlenite (1922)
Ferri-sarcolite (1922)
Hydroxylapatite (1926)
Kaliägirin (1913)
Nikitin, A. V., *v.* Rudneva, A. V.
Nikitin, K. K.
Cryptonickelemelane (1960)
Nickel-cobaltomelane (1960)
Nikitin, V. D.
Talasskite (1936)
Nikitin, W., *v.* Fedorov, E. S.
Nikitina, I. B., *v.* Bogomolov, M. A.
——, *v.* Ostrovskaya, I. V.
——, *v.* Pertsev, N. N.
Nikogosyan, C. S., *v.* Belyankin, D. S.
Nikolaeva, E. P., Grigorenko, V. A., & Tsypkina, P. E.
Stistaite (1970)
Nikolaeva, L. E., *v.* Portnov, A. M.
Nikolaevsky, F. A.
Ferriallophane (1914)
Shanyavskite (1912)
Nikol'skaya, N. V., Novozhilov, A. I., & Samioilovich, M. I.
Charoite (1976)
Nikolsky, A. P., & Efimov, A. N.
Svitalskite (1960)
Nilssen, B., *v.* Neumann, H.
Nixon, P. H., & Hornung, G.
Knorringite (1968)
——, *v.* von Knorring, O.
Nobar, M. A., & Dalvi, A. P.
Strontium weilite (1977)
Nockolds, S. R., & Zies, E. G.
Barium-anorthite (1933)
Barium-plagioclase (1933)
Noda, T., & Yamanishi, N.
Fluorine-hydroxyl-phlogopite (1964)
Noguchi, T., *v.* Hayashi, H.
Noll, W.
Alkali-montmorillonite (1936)
Calcium-montmorillonite (1936)
——, Kircher, H., & Sybertz, W.
Cobalt chrysotile (1958)
Magnesium chrysotile (1960)
Zinc chrysotile (1960)
Norberg, J. A., *v.* Dunn, P. J.
Norrish, K.
Barium-priderite (1951)
Potassium-priderite (1951)
Priderite (1951)
——, Rogers, L. E. R., & Shapter, R. E.
Kingite (1956)
Meta-kingite (1957)
——, *v.* Rogers, L. E. R.
Nose, K. W.
Deodatite (1790)
Dolomian (1797)
Nováček, R.
Jachymovite (1935)

Novák, F., Povondra, P., & Vtělenský, J.
Bukovskýite (1967)
——, & Valcha, Z.
Ferri-Orthochamosit (1964)
——, Vtelensky, J., Losert, J., Kupa, F., & Valcha, Z.
Orthochamosite (1957)
——, *v.* Čech, F.
Novikova, M. I., Sidorenko, G. A., & Kuznetsova, N. N.
Yaroslavite (1966)
Novorossova, L. E., *v.* Semenov, E. I.
Novotný, M., & Staněk, J.
Cyrilovite (1953)
Novozhilov, A. I., *v.* Nikol'skaya, N. V.
Nowacki, W.
Baumhauerite-I (1966)
Stibiobaumhauerite (1964)
Stibiodufrenoysite (1964)
Stibioskleroklas (1964)
Wallisite (1965)
——, & Bahezre, C.
Rathite-IV, Rathite-V (1964)
——, *v.* Bergerhoff, G.
——, *v.* Burri, G.
——, *v.* Marumo, F.
Nozhkin, A. D., Gavrilenko, V. A., & Moleva, V. A.
Usovite (1967)
Nuber, B., *v.* Ottemann, J.
Nuffield, E. W.
Pavonite (1953)
——, & Harris, D. C.
Berryite (1965)
——, *v.* Harris, D. C.
Nunes, J. E. Lopez, *v.* Lopez Nunes, J. E.
Nurlbaev, A. N.
Pravdite (1962)
Nurse, R. W., & Midgley, H. G.
Lime-iron-olivine (1953)
Potassium-melilite (1953)
Sodium-melilite (1953)
Nyrkov, A. A.
Sulunite (1959)

Ödman, O., *v.* Ramdohr, P.
Ödman, O. H.
Selenkobellite (1941)
Selenocosalite (1941)
Tungsten-powellite (1950)
Oebbeke, K.
Kalk-Olivin (1877)
Oen, I. S., Burke, E. A. J., Kieft, C., & Westerhof, A. B.
Westerveldite (1972)
Oesterle, J. F., *v.* McCaffery, R. S.
Off, H., *v.* Richmond, H. D.
Oftedahl, C.
Apoanalcite (1947)
Metasanidine (1949)
Oftedal, I.
Hoelite (1922)
Ohsumi, K., *v.* Kato, A.
Oke, W. C., *v.* Kamb, W. B.
Olby, J. K.
Plumbonacrite (1966)
Oleinikov, B. V., Shvartsev, S. L., Mandrikova, N. T., & Oleinikova, H. N.
Nickelhexahydrite (1965)
Oleinikova, H. N., *v.* Oleinikov, B. V.
Olsacher, J.
Achavalite (1939)
Schmeiderite (1962)
Olsen, E., Erlichman, J., Bunch, T. E., & Moore, P. B.
Buchwaldite (1977)
——, & Fuchs, L. H.
Krinovite (1968)
——, *v.* Fuchs, L. H.
Omori, K., & Kerr, P. F.
Zincalunite (1963)
——, & Konno, H.
Yttroapatite (1962)
Ono, S., *v.* Yamaguchi, I.
Oosterbosch, R., *v.* Cesbron, F.
Orcel, J., *v.* Curien, H.
Organ, R. M., & Mandarino, J. A.
Hydroromarchite (1971)
Romarchite (1971)
Organova, N. I., Genkin, A. D., Drits, V. A., Molotkov, S. P., Kuzmina, O. V., & Dmitrik, A. L.
Tochilinite (1971)
——, *v.* Belov, N. V.
——, *v.* Semenov, E. I.
——, *v.* Smolyaninova, N. N.
——, *v.* Sokolova, M. M.
Orlandi, P., Leoni, L., Mellini, M., & Merlino, S.
Tuscanite (1977)
——, *v.* Ferraris, G.
——, *v.* Merlino, S.
Orliac, M., *v.* Dietrich, J. E.
——, *v.* Fontan, F.
——, *v.* Gaudefroy, C.
Orlova, L. P., *v.* Pavlenko, A. S.
Orlova, Z. V.
Kësterite (1956)
Osann, A.
Holmquistite (1913)
Osborn, E. F., *v.* Roy, D. M.
Ossaka, J., *v.* Torii, T.
Osterbridge, W. F., Staatz, M. H., Meyrowitz, R., & Pommer, A. M.
Weeksite (1960)
Ostrovskaya, I. V., Pertsev, N. N., & Nikitina, I. B.
Sakhaite (1966)
——, *v.* Kondrateva, V. V.
——, *v.* Malinko, S. V.
——, *v.* Pertsev, N. N.
Oswald, H. R.
Brombotallackite (1961)
Iodbotallackite (1961)
Ottemann, J., & Augustithis, S. S.
Roseite (1967)

Parsons, A. L., *v.* Walker, T. L.
Partridge, F. C.
Arandisite (1930)
Parwell, A., *v.* Gabrielson, O.
Passaglia, E.
Roggianite (1969)
——, & Galli, E.
Vertumnite (1977)
——, & Pongiluppi, D.
Barrerite (1975)
——, ——, & Rinaldi, R.
Merlinoite (1977)
——, *v.* Galli, E.
Pattiaratchi, D. B., Saari, E., & Sahama, Th. G.
Anandite (1967)
Paul, M., *v.* Calderón, S.
Pauly, A.
Bolivianite (1923)
Silesite (1926)
Pauly, H.
Chalcopentlandite (1958)
Ikaite (1963)
Stenonite (1962)
Pavlenko, A. S., Orlova, L. P., & Akhmanova, M. V.
Cerphosphorhuttonite (1965)
——, Orlova, L. P., Akhmanova, M. V., & Tobelko, K. I.
Thorbastnäsite (1965)
Pavlishin, V. I., & Slivko, M. M.
Parakutnohorite (1962)
Pavlov, V. A., *v.* Serdyuchenko, D. P.
Payne, C. J., *v.* Anderson, B. W.
——, *v.* Claringbull, G. F.
Peacock, M. A.
Goldschmidtine (1937)
Pararammelsbergite (1939)
Parawollastonite (1935)
——, & Bandy, M. C.
Clino-ungemachite (1936)
Ungemachite (1936)
——, & Thompson, R. M.
Montbrayite (1945)
——, *v.* Warren, H. V.
Peacor, D. R., Essene, E. J., Simmons, W. B., & Bigelow, W. C.
Kellyite (1974)
——, *v.* Ritz, C.
Pearce, F., *v.* Duparc, L.
Peckett, A., *v.* Lovering, J. F.
Peckham, S. F.
Mayberyite (1895)
Parianite (1895)
Pecora, W. T., & Fahey, J. J.
Scorzalite (1947)
Souzalite (1947)
——, & Kerr, J. H.
Burbankite (1953)
Calkinsite (1953)
——, *v.* Lindberg, M. L.
Pehrman, G.
Lemnäsite (1939)
Pei-Shan, Chan
Titano-Euxenite (1963)
Pekun, Yu. F.
Ferri-metahalloysite (1956)
Pelikan, A.
Knollite (1933)
Zeophyllite (1902)
Penfield, S. L.
Alkali-beryl (1884)
Arsenpolybasite (1896)
Caesium-beryl (1888)
Fluor-herderite (1894)
Graftonite (1900)
Hydro-fluor-herderite (1894)
Pearceite (1896)
——, & Foote, H. W.
Bixbyite (1897)
Clinohedrite (1898)
Rœblingite (1897)
——, & Jamieson, G. S.
Tychite (1905)
——, & Warren, C. H.
Glaucochroite (1899)
Hancockite (1899)
Leucophœnicite (1899)
Nasonite (1899)
——, *v.* Hillebrand, W. F.
Peng, Chih-chang, & Ma, Che-sheng
Magnesio-astrophyllite (1964)
Mangano-astrophyllite (1964)
Peng, Chi-Jui
Fenghuanglite (1959)
——, & Liu, Yuan-Lung
Fenghuangite (1962)
——, Tsao, Rung-Lung, & Zou, Zu-Rung
Gugiaite (1962)
Peng, Tze-Chung, & Chang, Chien-hung
Barytolamprophyllite (1965)
Permingeat, F., *v.* Dietrich, J. E.
——, *v.* Fontan, F.
——, *v.* Gaudefroy, C.
——, *v.* Johan, Z.
——, *v.* Jouravsky, G.
——, *v.* Laurent, Y.
Permyakov, Y. I., *v.* Rudashevskii, N. S.
Perrault, G., Harvey, Y., & Pertowsky, R.
Yofortierite (1975)
——, Semenov, E. I., Bikova, A. V., & Capitonova, T. A.
Lemoynite (1969)
Pertowsky, R., *v.* Perrault, G.
Pertsev, N. N., & Aleksandrov, S. M.
Alumoludwigite (1964)
——, Ostrovskaya, I. V., & Nikitina, I. B.
Borcarite (1965)
——, *v.* Bogomolov, M. A.
——, *v.* Ostrovskaya, I. V.
Perttunen, V.
Lokkaite (1970)
Petersen, O. V.
Hydro-naujakasite (1967)
——, & Ronsbo, J. G.
Semenovite (1972)
——, *v.* Semenov, E. I.

Polyakova, T. P., *v*. Satpaeva, T. A.
Pommer, A. M., *v*. Muto, T.
——, *v*. Osterbridge, W. F.
Pongiluppi, D., *v*. Galli, E.
——, *v*. Passaglia, E.
Poni, P.
Badenite (1900)
Brostenite (1900)
Moldavite (1900)
Ponomarev, A. I., *v*. Ginzburg, I. I.
Poplavko, E. M., Merchukova, I. D., & Zak, C. S.
Dzhezkazganite (1962)
Popov, S. P.
Kertschenite (1906)
Oxykertschenite (1907)
Paravivianite (1906)
Tamanite (1903)
Popov, V. I., & Vorobiev, A. L.
Polyhydrate (1947)
Portnov, A. M.
Calcium catapleiite (1964)
——, Dybinchik, V. T., & Solntseva, L. S.
Calciocatapleiite (1972)
——, Krivokoneva, G. K., & Stolyarova, T. I.
Komarovite (1971)
——, Nikolaeva, L. E., & Stolyarova, T. I.
Landauite (1966)
——, Simonov, V. I., & Sinyugina, G. P.
Orthohombic låvenite (1966)
Posnjak, E., *v*. Ross, C. S.
Potapenko, S. V.
Chasovrite (1952)
Glinite (1952)
Potonié, H.
Caustobiolites (1910)
Caustolites (1910)
Denhardtite (1905)
Sapropelite (1906)
Potonié, R.
Clarite (1924)
Humite (1924)
Sapperite (1924)
Pouget, R., *v*. Bariand, P.
——, *v*. Branche, G.
Pough, F. H.
Lime-dravite (1953)
Soda-dravite (1953)
——, & Henderson, E. P.
Brasilianite (1945)
——, *v*. Frondel, C.
Povarennykh, A. S.
Calcjarlite (1973)
Calcurmolite (1962)
Hydromolysite (1962)
Hydroniojarosite (1962)
——, & Dusmatov, V. D.
Calcybeborosilite (1970)
Kanaekanite (1970)
——, & Rusakova, L. D.
Kafehydrocyanite (1973)
——, *v*. Gevorkyan, S. V.
Povondra, P., *v*. Johan, Z.
——, *v*. Novak, F.
Powell, A. R., *v*. Schoeller, W. R.
Prager, C.
Korteite (1923?)
Wathlingenite (1923)
Pratt, J. H.
Mitchellite (1889)
Pirssonite (1896)
——, & Foote, H. W.
Wellsite (1897)
——, *v*. Hidden, W. E.
Prchazka, K., *v*. Haranczyk, C.
Prewitt-Hopkins, J.
Torreyite (1949)
Prider, R. T.
Kali-magnesio-katophorite (1939)
Magnophorite (1939)
Noonkanbahite (1965)
——, *v*. Wade, A.
Prien, E. L., *v*. Frondel, C.
Prinz, M., *v*. Anderson, A. T.
——, *v*. Lovering, J. F.
Prior, G. T.
Calc-clinobronzite (1920)
Calc-clinoenstatite (1920)
Calc-clinohypersthene (1920)
Teallite (1904)
——, & Coomaraswamy, A. K.
Serendibite (1902)
——, & Smith, G. F. H.
Fermorite (1910)
——, & Spencer, L. J.
Iodembolite (1902)
——, *v*. Hussak, E.
Pritchard, G. B.
Mooraboolite (1901)
Proctor, P. D., *v*. Erd, R. C.
Protas, J.
Compreignacite (1964)
Wölsendorfite (1957)
——, *v*. Gaudefroy, C.
——, *v*. Guillemin, C.
——, *v*. Sierra Lopez, J.
Pryce, M. W.
Holtite (1971)
——, & Just, J.
Glaukosphaerite (1974)
——, *v*. Bridge, P. J.
Pu, Wang, & Shou-Tsuen, Juan
Fluor-antigorite (1965)
Puchelt, H., & Müller, G.
Coelestobaryt (1964)
Pudovkina, Z. G., *v*. Efimov, A. F.
Pudovkina, Z. V., *v*. Kapustin, Yu. L.
——, *v*. Khomyakov, A. P.
Pufahl, O.
Duftite (1920)
Germanite (1922)
Puffer, J. H., *v*. Haji-Vassiliou, A.
Pukhalsky, L. Ch., *v*. Melkov, V. G.
Pulou, R., *v*. Capdecomme, L.
Pyatenko, Yu. A., *v*. Borodin, L. S.

Schreyer, W.—*cont.*
Clinoeulite (1978)
Schröder, E., *v.* Muthmann, W.
Schubnel, H.-J., *v.* Johan, Z.
——, *v.* Pierrot, R.
Schüller, A., & Ottemann, J.
Castaingite (1963)
——, & Wohlmann, E.
Betekhtinite (1955)
Schulz, H., *v.* Menzel, H.
Schulze, E.
Pilite (1895)
Schulze, H. O.
Argentopercylite
Cuproiodargyrite (1892)
Schwarz, R., & Trageser, G.
Anhydrokaolin (1932)
Prokaolin (1932)
Sclar, C. B., & Drovenik, M.
Lazarevićite (1960)
Scott, A., *v.* Mellor, J. W.
Scott, E. R. D.
Haxonite (1971)
——, *v.* Buchwald, V. F.
Scott, T. R., *v.* Davey, P. T.
——, *v.* McAndrew, J.
Scrivenor, J. B., & Shenton, J. C.
Thorotungstite (1927)
Seaman, W. A., *v.* Kraus, E. H.
Searle, A. B.
Pelinite (1912)
Šebor, J.
Bilinite (1913)
Sederholm, J. J.
Maltesite (1896)
Sedletzky, I. D.
Goeschwitzite (1940)
Gepherite (1940)
Hydrogedroitzite (1940)
Hydrohalloysite (1940)
Hydromontmorillonite (1940)
Hydronontronite (1940)
Hydropyrophyllite (1940)
Magnymontmorillonite (1951)
Metaquartz (1940)
Prokaolin (1940)
Promontmorillonite (1937)
Sarospatakite (1940)
——, & Samodurov, P. S.
Magny-monothermite (1949)
——, & Yusupova, S. M.
Ablykite (1940)
——, *v.* Antipov-Karatajev, I. N.
Seeliger, E., & Mücke, A.
Donathite (1969)
Ernstite (1970)
Para-schachnerite (1972)
Schachnerite (1972)
——, *v.* Strunz, H.
Segnit, E. R.
Calciocelsian (1946)
Sekanina, J.
Koktaite (1948)
Letovicite (1932)
Rosickyite (1931)
Semenov, E. I.
Erikite (1959)
Gel-anatase (1957)
Gel-bertrandite (1957)
Gel-cassiterite (1960)
Gel-cristobalite (1960)
Gelgoethite (1960)
Gel-rutile (1957)
Gel-thorite (1960)
Gel-zircon (1960)
Gerasimovskite (1957)
Hydropolylithionite (1959)
Hydrorinkite (1969)
Kupletskite (1956)
Mangan-belyankinite (1957)
Metaberyllite (1972)
Mn-palygorskite (1969)
Mn-sepiolite (1969)
Monohydrocalcite (1964)
Niobanatase (1957)
Niobobelyankinite (1957)
Silicorhabdophane (1959)
Silicosmirnovskite (1959)
Spherobertrandite (1957)
Titanorhabdophane (1959)
Tundrite (1963)
——, Bonshtedt-Kupletskaya, E. M., Moleva, V. A., & Sludskaya, N. N.
Vinogradovite (1956)
——, & Burova, T. A.
Labuntsovite (1955)
——, & Bykova, A. V.
Beryllosodalite (1960)
——, & Chang, P'ei-shan
Huanghoite (1961)
——, Dusmatov, V. D., Khomyakov, A. P., Boronkov, A. A., & Kazakova, M. E.
Darapiozite (1975)
——, & Kazakova, M. E.
Hydrothorite (1961)
——, ——, & Bukin, V. I.
Ilimaussite (1968)
——, ——, & Simonov, V. I.
Seidozerite (1958)
——, Khomyakov, A. P., & Bykova, A. V.
Magbasite (1965)
——, Kulakov, M. P., Kostynina, L. P., Kazakova, M. E., & Dudykina, A. S.
Sc-perrierite (1966)
——, Maksimova, N. V., & Petersen, O. V.
Sørensenite (1965)
——, Organova, N. I., & Kukharchik, M. V.
Metalomonosovite (1961)
Metamurmanite (1961)
——, Sørensen, H., Bezsmertnaya, M. S., & Novorossova, L. E.
Chalcothallite (1967)
——, Spitzyn, A. N., & Burova, Z. N.
Hydropyrochlore (1963)
——, & Tikhonenkov, I. P.
Hydrocatapleiite (1962)

Shubnikova, I. M.—*cont.*
Uxporite (1934)
——, *v.* Fersman, A. E.
Shu-chang, Wang, & Xue-yen, Hsu
Magnodravite (1966)
Shu-Chien, Chen, *v.* Tzu-Chiang, Chien
Shu-Jen, Lin, *v.* Tsu-Hsiang, Yu
Shumkova, N. G., *v.* Gruzdev, V. S.
Shumyatskaya, N. G., *v.* Eskova, E. M.
Shuvalova, N. I., *v.* Nefedov, E. I.
Shvartsev, S. L., *v.* Oleinikov, B. V.
Shyusareva, M. N.
Uklonskovite (1964)
Sidelnikova, V. D., *v.* Chernikov, A. A.
Sidorenko, G. A.
Paraphane (1960)
——, *v.* Ginzburg, A. I.
——, *v.* Ginzburg, I. V.
——, *v.* Gorzhevskaya, S. A.
——, *v.* Kopchenova, E. V.
——, *v.* Kudrina, M. A.
——, *v.* Kupriyanova, I. I.
——, *v.* Novikova, M. I.
——, *v.* Rogova, V. P.
——, *v.* Shpanov, E. P.
——, *v.* Skorobogatova, H. V.
——, *v.* Skvortsova, K. V.
——, *v.* Yakontova, L. K.
——, *v.* Zdorik, T. B.
Sidorenko, M.
Hydrotroilite (1901)
Sidorenko, O. V., *v.* Eskova, E. M.
Sidorov, A. F., *v.* Yalovai, A. A.
Siegl, W.
Plumbodolomite (1936)
Sierra Lopez, J., Leal, G., Pierrot, R., Laurent, Y., Protas, J., & Dusausov, Y.
Rodalquilarite (1965)
Siftar, D., & Jurkovic, I.
Viterite (1961)
Sigmund, A.
Sphaeromagnesite (1909)
Siivola, J., *v.* Sahama, Th. G.
——, *v.* Vorma, Atso
Silanteva, N. I., *v.* Kopchenova, E. V.
——, *v.* Skvortsova, K. V.
Sillitoe, R. H., *v.* Clark, A. H.
Šimek, A., *v.* Jaeger, F. M.
Siminov, V. I., *v.* Portnov, A. M.
——, *v.* Semenov, E. I.
Simmons, W. B., *v.* Peacor, D. R.
Simonov, M. A., & Belov, N. V.
Zn,Cd-chkalovite (1965)
——, Yamnova, N. A., Kazanskaya, E. V., Egorov-Tismenko, Yu. K., & Belov, N. V.
Hexahydroborite (1976)
——, *v.* Belokoneva, E. L.
——, *v.* Shashkin, D. P.
Simons, F. S., & Straczek, J. A.
Delatorreite (1958)
——, *v.* Smith, R. L.
Simpson, E. S.
Beresofskite (1932)
Calciotantalite (1907)
Ferantigorite (1937)
Goongarrite (1924)
Hydroallanite (1938)
Hydrogadolinite (1938)
Hydrothorite (1928)
Leucophosphite (1932)
Maitlandite (1930)
Manganilmenite (1929)
Metasimpsonite (1938)
Nicolayite (1930)
Picrochromite (1920)
Picrocollite (1928)
Pilbarite (1910)
Tantalopolycrase (1938)
Tanteuxenite (1928)
——, & LeMesurier, C. R.
Minyulite (1933)
Sindeeva, I. D.
Tyrrellite (1959)
Sinkova, L. A., *v.* Aleksandrov, I. V.
Sinyugina, G. P., *v.* Portnov, A. M.
Sjögren, H.
Celsian (1895)
Mauzeliite (1895)
Natron-richterite (1892)
Potash-richterite (1895)
Soda-berzeliite (1895)
Soda-richterite (1892)
Tilasite (1895)
Skakovsky, N. K.
Ferrohumite (1929)
Skerl, A. C., & Bannister, F. A.
Lusakite (1934)
Skinner, B. J., Erd, R. C., & Grimaldi, F. S.
Greigite (1964)
——, Jambor, J. L., & Ross, M.
Mckinstryite (1966)
——, *v.* Chao, E. C. T.
——, *v.* Hak, J.
Skornyakov, P. I.
Dnieprovskite (1944)
Skorobogatova, H. V., Sidorenko, G. A., Dorofeeva, K. A., & Stolyarova, T. I.
Plumbopyrochlore (1966)
Skropyshev, A. V.
Evenkite (1953)
Skvortsova, K. V., & Siderenko, G. A.
Sedovite (1965)
——, ——, Dara, A. D., Silanteva, N. I., & Medoeva, M. M.
Femolite (1964)
——, ——, Nesterova, Yu. S., Arapova, G. A., Dara, A. D., & Rybakova, L. I.
Sodium betpakdalite (1971)
——, *v.* Kopchenova, E. V.
Slack, J. F., *v.* Radtke, A. S.
Slánsky, E., *v.* Johan, Z.
Slavík, F.
Ježekite (1914)

Slavík, F.—*cont.*
  Lacroixite (1914)
  Roscherite (1914)
Slawson, C. B., *v.* Kraus, E. H.
Slivko, M. M., *v.* Pavlishin, V. I.
Sludskaya, N. N., *v.* Semenov, E. I.
Slyusareva, M. N.
  Hydroglauberite (1963)
Smeeth, W. F.
  Bababudanite (1911)
Smelyanskaya, G. A., *v.* Razumnaya, E. G.
Smirnov, S. S.
  Cuproboulangerite (1933)
Smirnova, E. M., *v.* Genkin, A. D.
Smirnova, N. G., *v.* Yurgenson, G. A.
Smith, C. M., *v.* McDonnell, C. C.
Smith, G. F. H.
  Paralaurionite (1899)
  Paratacamite (1905)
——, *v.* Prior, G. T.
——, *v.* Solly, R. H.
Smith, J. V., & Mason, B.
  Majorite (1970)
——, & Tuttle, O. F.
  Tetra-kalsilite (1957)
——, *v.* Chao, E. C. T.
——, *v.* Sahama, Th. G.
——, *v.* Steele, I. M.
Smith, M. L.
  Metadelrioïte (1970)
——, & Frondel, C.
  Bannisterite (1968)
Smith, R. L., Simons, F. S., & Vlisidis, A. C.
  Hidalgoite (1953)
Smith, W. C., Bannister, F. A., & Hey, M. H.
  Banalsite (1944)
  Cymrite (1949)
  Pennantite (1946)
——, *v.* Bannister, F. A.
Smith, W. L., Stone, J., Riska, D. D., & Levine, H.
  Doverite (1955)
Smolyaninov, N. A., & Isakov, E. N.
  Paragearksutite (1946)
Smolyaninova, N. N., Moleva, V. A., & Organova, N. I.
  Zincsilite (1960)
Smorchkov, I. E., *v.* Gorzhevskaya, S. A.
Smulikowski, K.
  Pholidoide (1936)
  Skolite (1936)
Smuts, J., Steyn, J. G. D., & Boeyens, J. C. A.
  Iscorite (1969)
Smythe, J. A., & Dunham, K. C.
  Ferromangandolomite (1947)
Snetsinger, K. G.
  Barium-vanadium muscovite (1966)
  Erlichmanite (1971)
  Osarsite (1972)
——, *v.* Keil, K.
Snow, R. B.
  Manganese-cordierite (1943)
Sobolev, N. D.
  Asbophite (1930)
——, *v.* Soboleva, M. V.
Sobolev, N. V., *v.* Sobolev, V. S.
Sobolev, V. S.
  Grosspydite (1960)
——, & Sobolev, N. V.
  Chromdisthene (1967)
Soboleva, L. N., *v.* Bezsmertnaya, M. S.
Soboleva, M. V., & Sobolev, N. D.
  Rezhikite (1959)
Sochipanova, O. V., *v.* Karpenko, V. S.
Soellner, J.
  Deeckeite (1913)
  Rhönite (1907)
Söhnge, G., *v.* Strunz, H.
Soklakov, A. I., & Dorfman, M. D.
  Ferrichinglusuite (1964)
  Manganchinglusuite (1964)
  Natrohisingerite (1964)
Soknev, V. S., *v.* Indolev, L. N.
Sokolov, G. A.
  Sungulite (1925)
Sokolov, Yu. A., *v.* Vertushkov, G. N.
Sokolova, E. P., *v.* Rimskaya-Korsakova, O. M.
Sokolova, G. V., *v.* Yablokova, S. V.
Sokolova, M. N., Dobrovol'skaya, M. G., Organova, N. I., & Dmitrik, A. L.
  Rasvumite (1970)
Solly, R. H.
  Baumhauerite (1902)
  Bowmanite (1904)
  Hutchinsonite (1904)
  Legenbachite (1904)
  Liveingite (1901)
  Marrite (1904)
  Smithite (1905)
  Trechmannite (1904)
——, & Smith, G. F. H.
  Hatchite (1912)
Solntseva, L. S., *v.* Portnov, A. M.
Sommerfeldt, E.
  Metanhydrite (1907)
Sørensen, C., Semenov, E. I., Bezsmertnaya, M. S., & Khalezova, E. B.
  Cuprostibite (1969)
Sørensen, H.
  Tugtupite (1962)
——, *v.* Dano, M.
——, *v.* Semenov, E. I.
Sorita, E., *v.* Sakurai, K.
Sorrell, C. A.
  Lead feldspar (1962)
Sosman, R. B.
  Coesite (1954)
  Keatite (1954)
Souza-Brandão, V.
  Cryptoclase (1909)
  Orthomimic felspars (1909)
  Para-orthose (1909)
Spangenberg, K.
  Aluminatchromit (1943)
  Ferritchromit (1943)

Strunz, H.—*cont.*
Arsen-stibiconite (1957)
Arsenuranocircite (1957)
Arsenvanadinite (1941)
Bacillarite (1941)
Barium-aragonite (1941)
β-Bassanite (1966)
Beryllium-Sodalith (1966)
Bismostibnit (1966)
Bleiantimonspießglanze (1957)
Bleiarsenspießglanze (1957)
Bleikupferarsen (1966)
Bleikupferspießglanze (1957)
Bleisilberspießglanze (1957)
Bleispießglanze (1957)
Bleiwismutspießglanze (1957)
Bromchlorargyrite (1941)
Cadmiumspat (1941)
Ca-Illite (1957)
Calciobaryt (1941)
Calcio-jarosite (1957)
Calciorinkite (1970)
Calcium-barium-mimetite (1941)
Calciumkatapleit (1966)
Calcium-pyromorphite (1941)
Calcium-strontianite (1941)
Carobbiite (1956)
Cerium-ankerite (1941)
Ce-Vesuvian (1970)
Chalcokyanite (1961)
Chrombiotit (1957)
Chrome-antigorite (1941)
Chromitspinelle (1957)
Chrom-Lanarkit (1966)
Chrommuscovit (1957)
Chudobaite (1960)
Clino-antigorite (1957)
Clinoberthierine (1966)
Cordobaite (1957)
Creniadite (1957)
Cs-Beryll (1966)
Cuprobismuthit (1957)
Cyclowollastonit (1965)
Eisen(III)-Spinelle (1970)
Eisen-Berlinit (1966)
Fabulit (1966)
Fauserite (1957)
Ferrialunogen (1941)
Ferri-berthierine (1957)
Ferripalygorskite (1957)
Ferrisaponite (1957)
Ferrisepiolite (1956)
Ferri-sericite (1957)
Ferro-berthierine (1957)
Ferro-friedelite (1957)
Ferromagnesite (1941)
Ferrospinel (1957)
Ferrozincrhodochrosite (1957)
Freboldite (1957)
Groutellit (1966)
Hagendorfite (1954)
Hexabolite (1949)
Hoch-Bassanit (1970)
Hoch-Schapbachit (1944)
Hydrocassiterite (1957)
Hydroparagonite (1957)
Hydroserpentine (1970)
Iridiumplatin (1941)
Kalium-Richterit (1966)
Kaliumstruvit (1957)
Kaolin-Chamosit (1957)
Katophorit (1957)
Klinochrysotil (1957)
Klinotscheffkinit (1957)
Kupferspießglanze (1957)
Laspeyrit (1957)
Laueite (1954)
Leukophosphatit (1970)
Mafurit (1957)
Magnesiomargarit (1966)
Magnesioniobit (1966)
Magnesiospinel (1957)
Magnesiotriplit (1966)
Magnesium-hydromuscovite (1957)
Magnesium-phosphoruranite (1941)
Mangan-antigorite (1957)
Mangan-Arfvedsonit (1966)
Mangan-chrysotile (1957)
Mangan-Cummingtonit (1966)
Manganmelanterit (1941)
Mangan-niobite (1941)
Manganogel (1957)
Manganoniobite (1957)
Manganoxyapatite (1941)
Meta-bassetite (1957)
Metalaumontite (1957)
Metamontmorillonite (1957)
Meta-Na-Uranospinit (1966)
Meta-uranospinite (1957)
Metavermiculite (1957)
Metazeolith (1957)
Mn-Beljankinit (1966)
Molybdoscheelite (1941)
Na-Autunit (1966)
Na-Meta-Autunit (1966)
Natrium-Hewettit (1957)
Natronbiotite (1941)
Natron-Carnotit (1966)
Natron-Heulandit (1966)
Natron-mesomicrocline (1957)
Nichtspießglanze (1957)
Nickel-antigorite (1957)
Nickelchlorite (1957)
Nickel-chrysotile (1957)
Nickel-pyrite (1941)
Nickelspinel (1957)
Nitrokalite (1941)
Octobolite (1957)
Orthoberthierine (1966)
Orthotorbernite (1961)
Para-autunite (1957)
Para-Kupferglanz (1941)
Para-Silberglanz (1941)
Paratenorite (1941)
Para-uranite (1957)
Phosphate-schultenite (1957)

Tainter, M. L.—*cont.*
Elkonite (1940)
Takano, Y., *v.* Minato, H.
Takeda, H., & Burnham, C. W.
Fluor-polylithionite (1969)
Takubo, J., Ukai, Y., & Minato, T.
Kobeite (1950)
——, *et al.*
Yttrianite (1953)
Tanaka, N., *v.* Kinoshita, K.
Tanaka, T., *v.* Kami, K.
Tani, B., *v.* Chao, E. C. T.
Tanida, K., *v.* Nambu, M.
Tanjeloff, J.
Tanjeloffite (1971)
Tanton, T. L.
Conchilites (1944)
Tarkian, M., *v.* Stumpfl, E. F.
Tarney, J., Nicol, A. W., & Marriner, G. F.
Boron-melilite (1973)
Taşman, C. E.
Harbolite (1946)
Tatsuka, K., & Morimoto, N.
Pseudotetrahedrite (1973)
Taylor, C. M., *v.* Radtke, A. S.
Taylor, H. F. W., *v.* Buckle, E. R.
——, *v.* Carpenter, A. B.
——, *v.* Gard, J. A.
Taylor, R. E., *v.* Hurlbut, C. S.
Taylor, T. G., *v.* David, T. W. E.
Teeple, J. E.
Burkeite (1921)
Temple, A. K., *v.* Teufer, G.
Tenne, C. A.
Leonite (1896)
Tennyson, C. I., *v.* Strunz, H.
Terasaki, Y., *v.* Urashima, Y.
Terekhova, G. A., *v.* Gruzdev, V. S.
Termier, P.
Neotantalite (1902)
Tertian, R., *v.* Papée, D.
Terziev, G.
Hemusite (1965)
Kostovite (1966)
Teufer, G., & Temple, A. K.
Pseudorutile (1966)
Thilo, E.
Metatalc (1937)
——, & Rogge, G.
Mesoenstatite (1939)
——, *v.* Ramdohr, P.
Thomas, C. A.
Manganonatrolite (1947)
Thompson, C. S., *v.* Mumpton, F. A.
Thompson, M. E., Ingram, B., & Gross, E. B.
Abernathyite (1956)
——, Roach, C. H., & Meyrowitz, R.
Duttonite (1956)
Sherwoodite (1958)
Simplotite (1956)
——, & Sherwood, A. M.
Delrioite (1959)
——, Weeks, A. D., & Sherwood, A. M.
Rabbittite (1954)
——, *v.* Lindberg, M. L.
——, *v.* Weeks, A. D.
Thompson, R. B., *v.* Weeks, A. D.
Thompson, R. M.
Frohbergite (1946)
——, *v.* Berry, L. G.
——, *v.* Drummond, A. D.
——, *v.* Montgomery, J. H.
——, *v.* Peacock, M. A.
Thomson, T.
Junkerite (1836)
Levyine (1836)
Levyite (1836)
Thomssen, R. W., *v.* McLean, W. J.
——, *v.* White, J. S.
Thoreau, J.
Kobokobite (1957)
Uranolepidite (1933)
——, & Vaes, J. F.
Saleite (1932)
Thorpe, R. I., & Harris, D. C.
Mattagamite (1973)
Tellurantimony (1973)
——, *v.* Harris, D. C.
Threadgold, I. M., *v.* Cameron, E. N.
——, *v.* Hudson, D. R.
Thronber, M. R., *v.* Barker, W. W.
Thugutt, S. J.
Allophanoids (1911)
Cleïte (1945)
Courzite (1945)
Curzite (1949)
Epinatrolite (1911)
Episcolecite (1949)
Epithomsonite (1949)
Fojasite (1949)
Fożasyt (1949)
Janite (1933)
Kurtzite (1945)
Marburgite (1949)
Natronanorthite (1895)
——, *v.* Rosický, V.
Tien, Pei-Lin, Leavens, P. B., & Nelen, J. A.
Swinefordite (1975)
Tikhonenkov, I. P., & Kazakova, M. E.
Nioboloparite (1957)
——, *v.* Semenov, E. I.
Tikhonenkova, R. P., & Kazakova, M. E.
Vlasovite (1961)
Tikonowitch, M. N., *v.* Duparc, L.
Tilley, C. E.
Ferrogedrite (1939)
Ferropericlase (1951)
Harkerite (1948)
Hydrocalumite (1934)
Iron-wollastonite (1937)
Larnite (1929)
Magnesio-cummingtonite (1939)
Magnesium-wollastonite (1948)
Portlandite (1933)
Pseudo-sarcolite (1929)

Tilley, C. E.—*cont.*
Rankinite (1942)
Scawtite (1929)
Shannonite (1927)
——, & Henry, N. F. M.
Latiumite (1952)
——, & Vincent, H. C. G.
Bredigite (1948)
Tilmanns, E., *v.* Hofmeister, W.
Timofeevskii, D. A.
Teremkovite (1967)
Timoshenkov, I. M., Menshikov, Yu. P., Gannibal, L. F., & Bussen, I. V.
Natrosilite (1975)
Tobelko, K. I., *v.* Pavlenko, A. S.
Tokody, L.
Zinc-fauserite (1949)
——, Mándy, T., & Nemes-Varga, S.
Mauritzite (1957)
Tolun, R.
Bursaite (1955)
Tomás, L., & Folch, J.
Almeraite (1914)
Tomita, T.
Gokaite (1936)
Titanpigeonite (1933)
Torii, T., & Ossaka, J.
Antarcticite (1965)
Tóth, G., *v.* Zechmeister, L.
Toubes, R. O., *v.* Gordillo, C. E.
Toussaint, J., *v.* Pierrot, R.
Trageser, G., *v.* Schwarz, R.
Traill, R. J., & Boyle, R. W.
Hawleyite (1955)
——, *v.* Chao, E. C. T.
Treibs, A., & Steinmetz, H.
Graebeite (1933)
Trelles, R. A., *v.* Angelelli, V.
Tresham, A. E., *v.* Jobbins, E. A.
Trobeva, N. V., *v.* Evstigneeva, T. L.
Tröger, W. E.
Ferrolazulite (1952)
Hydroxyl-topaz (1952)
Magnesioarfvedsonite (1952)
Pigeonite-augite (1951)
Skorzalith (1952)
Syntagmatite (of Tröger) (1952)
Trojer, F.
Güggenite (1958)
Trömel, G., & Eitel, W.
Silikatapatit (1957)
Troneva, N. V., *v.* Evstigneeva, T. L.
——, *v.* Genkin, A. D.
Trotter, J., *v.* Drummond, A. D.
Tsan-Fu, Chuang, *v.* Van-Kang, Huang
Tsao, Rung-Lung, *v.* Peng, Chi-Jui
Tschernich, R. W., *v.* Wise, W. S.
Tschirch, A., & Aweng, E.
Allingite (1894)
——, & Kato, T.
Plaffeiite (1926)
Tschirwinsky, *v.* Čirvinsky
Tsepin, A. I., *v.* Evstigneeva, T. L.
——, *v.* Ivanitskii, R. V.
Tsertsvadze, Z. Ya., *v.* Gruzdev, V. S.
Tsu-Hsiang, Yu; Shu-Jen, Lin; Pao, Chao; Ching-Sung, Fang; & Chi-Shun, Huang
Daomanite (1974)
Dayingite (1974)
Fengluanite (1974)
Guanglinite (1974)
Hongquiite (1974)
Hongshiite (1974)
Malanite (1974)
Xingzhongite (1974)
Yanzhongite (1974)
Yixunite (1974)
Tsypkina, P. E., *v.* Nikolaeva, E. P.
Tu, Shao-Hua, *v.* Huang, Wen-Hui
Tućan, F.
Diasporogelite (1913)
Gajite (1911)
Haematogelite (1913)
Kljakite (1912)
Tu-Chih, K; Li, H-L; Hsieh, H-T; & Yin, S-S.
Zincobotryogen (1964)
Zincocopiapite (1964)
Tunell, G., Fahey, J. J., Daugherty, F. W., & Gibbs, G. V.
Gianellaite (1977)
Turco, G., *v.* Baumer, A.
Turner, F. J.
Seminephrite (1935)
Turner, W. H.
Terlinguaite (1900)
Tuttle, O. F., *v.* Smith, J. V.
Tvalchrelidze, A. A.
Gumbrine (1929)
Twelvetrees, W. H.
Petterdite (1901)
Tyler, S. A., *v.* Bailey, S. W.
Tyrrell, J. B., & Graham, R. P. D.
Yukonite (1913)
Tysseland, M., *v.* Raade, G.
Tzimbalenko, N. N., *v.* Ovchinnikov, L. N.
Tzu-Chiang, Chien; & Shu-Chien, Chen
Hydrochlorborite (1965)

Uhlemann, A.
Tolypite (1909)
Uhlig, J.
Fasernephrit (1910)
Ukai, Y., *v.* Takubo, J.
Ukhina, T. A., *v.* Ginzburg, A. I.
Ullmann, A. T.
Chillagite (1913)
Ulrich, F., & Munk, R.
Oligonsiderite (1936)
——, *v.* Jirkovský, R.
Umpleby, J. B., Schaller, W. T., & Larsen, E. S.
Custerite (1913)
Ungemach, H.
Amarillite (1933)
Lapparentite (1933)
Leucoglaucite (1933)

Yarilova, E. A., & Parfenova, E. I.
Polynite (1957)
——, *v.* Parfenova, E. I.
Yaroshch, I. A., *v.* Ivanov, A. A.
Yarzhemski, Ya. Ya.
Balavinskite (1966)
Kurgantaite (1952)
Preobrazhenskite (1956)
——, *v.* Blazko, L. N.
——, *v.* Kondrateva, V. V.
Ying-Chen, Jen, & Wan-Kang, Huang
Yenshanite (1973)
Yingnian, Xu, *v.* Xiongfei, Gu
Yin, Shu-Shen, *v.* Tu-Chih, Kuang
Yoder, H. S.
Aluminous-serpentine (1952)
Magnesian chamosite (1952)
——, & Keith, M. L.
Yttroalumite (1951)
Yttrogarnet (1951)
Yoder, H. S., Jr., *v.* Huckenholz, H. G.
Yorks, K. P., *v.* Fahey, J. J.
Yoshi, M., Aoki, Y., & Maeda, K.
Nambulite (1972)
Yoshida, M., *v.* Hayashi, H.
Yoshii, M., Maeda, K., Kato, T., Watanabe, T., Yui, S., Kato, A., & Nagashima, N.
Kinoshitalite (1973)
Yoshiki, B., & Matsumoto, K.
Hexacelsian (1951)
Yoshimura, J., *v.* Iimori, S.
Yoshimura, T.
Barium-albite (1939)
Dosulite (1967)
Iron-hornblende (1939)
Iron-knebelite (1939)
Iron-tephroite (1939)
Kasoite (1936)
Mangan-actinolite (1939)
Mangan-knebelite (1939)
Manganophyllite (1939)
Mangan-phlogopite (1939)
Mangan-tremolite (1939)
Picrogalaxite (1936)
Picroknebelite (1939)
Picrourbanite (1937)
Riebeckrichterite (1937)
Teineite (1939)
Todorokite (1934)
Tosalite (1967)
——, & Momoi, H.
Yamatoite (1964)
——, Shirozu, H., & Kimura, M.
Barium-adularia (1954)
Yoshinaga, M.
Sonolite (1963)
Yoshinaga, N., & Aomine, S.
Imogolite (1962)
Youell, R. F.
Pseudochlorite (1960)
——, *v.* Brindley, G. W.
Young, B. R., *v.* Hemi, C.
——, *v.* Jobbins, E. A.
Young, E. J., & Munson, E. L.
Fluor-chlor-oxy-apatite (1966)
——, Weeks, A. D., & Meyrowitz, R.
Coconinoite (1966)
Young, F., *v.* Kerr, P. F.
Yu, Cheng-Chih, *v.* Huang, Wen-Hui
Yuan, S.-F., *v.* Sun, W.-C.
Yudin, R. H., *v.* Gruzdev, V. S.
Yuen-Ming, Chang, *v.* Van-Kang, Huang
Yuferov, D. V., *v.* Shubnikova, O. M.
Yui, S., *v.* Yoshii, M.
Yun-Hue, Huang
Solanite (1965)
Yunosova, Z. R., *v.* Avrova, N. P.
Yur'ev, L. D., *v.* Goroshnikov, B. I.
Yurgenson, G. A.
Ferrochalcanthite (1971)
——, Smirnova, N. G., & Karenina, L. A.
Mirupolskite (1968)
Udokanite (1968)
Yurk, Yu. Yu.
Priazovite (1956) (1941?)
Yurkina, K. V., *v.* Khomyakov, A. P.
——, *v.* Malinko, S. V.
——, *v.* Razin, L. V.
Yushko-Zakharova, O. E.
Imgreïte (1964)
——, & Cheryaev, L. A.
Palladium bismuthide (1966)
——, *v.* Kulagov, E. A.
Yuskin, N. P., *v.* Goldin, B. A.
Yusupova, S. M., *v.* Sedletzky, I. D.

Zabavnikova, N. I., *v.* Bussen, I. V.
——, *v.* Menshikov, Yu. P.
Zaiteseva, R. I., *v.* Ankinovich, E. A.
Zak, C. S., *v.* Poplavko, E. M.
Žak, L., & Syneček, V.
Kettnerite (1956)
——, ——, & Hybler, J.
Krupkaite (1974)
Zalashkova, N. E.
Bismuthmicrolite (1957)
——, *v.* Beus, A. A.
Zalessky, M. D.
Kuckersite (1916)
Sapromyxite (1915)
Tomite (1915)
Zambonini, F.
Acmite-augite (1910)
Aluminioepidote (1920)
Arsenioardennite (1922)
Avogadrite (1926)
Bassanite (1910)
Calciobiotite (1919)
Calciocancrinite (1910)
Chromepidote (1920)
Delorenzite (1908)
Ferriepidote (1920)
Grothine (1913)
Hydroclinohumite (1919)